Managing Treatment-Resistant Depression

Managing Treatment-Resistant Depression

Road to Novel Therapeutics

Edited by

Joao Quevedo
Professor; Vice-Chair, Faculty Development and Outreach; Director, Translational Psychiatry Program; Director, Treatment-Resistant Mood Disorders Program; Faillace Department of Psychiatry and Behavioral Sciences, McGovern Medical School, The University of Texas Health Science Center at Houston, Houston, TX, United States

Patricio Riva-Posse
Emory University, Atlanta, GA, United States

William V. Bobo
Mayo Clinic, Jacksonville, FL, United States

ELSEVIER

ACADEMIC PRESS
An imprint of Elsevier

Academic Press is an imprint of Elsevier
125 London Wall, London EC2Y 5AS, United Kingdom
525 B Street, Suite 1650, San Diego, CA 92101, United States
50 Hampshire Street, 5th Floor, Cambridge, MA 02139, United States
The Boulevard, Langford Lane, Kidlington, Oxford OX5 1GB, United Kingdom

Notices
Knowledge and best practice in this field are constantly changing. As new research and experience broaden our understanding, changes in research methods, professional practices, or medical treatment may become necessary.

Practitioners and researchers must always rely on their own experience and knowledge in evaluating and using any information, methods, compounds, or experiments described herein. In using such information or methods they should be mindful of their own safety and the safety of others, including parties for whom they have a professional responsibility.

To the fullest extent of the law, neither the Publisher nor the authors, contributors, or editors, assume any liability for any injury and/or damage to persons or property as a matter of products liability, negligence or otherwise, or from any use or operation of any methods, products, instructions, or ideas contained in the material herein.

Library of Congress Cataloging-in-Publication Data
A catalog record for this book is available from the Library of Congress

British Library Cataloguing-in-Publication Data
A catalogue record for this book is available from the British Library

ISBN 978-0-12-824067-0

For information on all Academic Press publications
visit our website at https://www.elsevier.com/books-and-journals

Publisher: Nikki P. Levy
Acquisitions Editor: Joslyn Chaiprasert-Paguio
Editorial Project Manager: Susan Ikeda
Production Project Manager: Kiruthika Govindaraju
Cover Designer: Miles Hitchen

Typeset by STRAIVE, India

Contents

Contributors

Numbers in paraentheses indicate the pages on which the authors' contributions begin.

Scott T. Aaronson (319), Institute for Advanced Diagnostics and Therapeutics, Sheppard Pratt; Department of Psychiatry, University of Maryland School of Medicine, Baltimore, MD, United States

Giulia Adavastro (253), Department of Neuroscience, Rehabilitation, Ophthalmology, Genetics, Maternal and Child Health, Section of Psychiatry, University of Genoa, Genoa, Italy; Monitoring Center on Addiction, Mental Health DSM-DP, Bologna, Italy

Andrea Aguglia (253), Department of Neuroscience, Rehabilitation, Ophthalmology, Genetics, Maternal and Child Health, Section of Psychiatry, University of Genoa, Genoa, Italy; IRCCS Ospedale Policlinico San Martino, Genoa, Italy

Olusola Ajilore (407), Department of Psychiatry, University of Illinois College of Medicine, Chicago, IL, United States

Jonathan E. Alpert (559), Department of Psychiatry and Behavioral Sciences, Albert Einstein College of Medicine, Montefiore Medical Center, Bronx, NY, United States

Andrea Amerio (253), Department of Neuroscience, Rehabilitation, Ophthalmology, Genetics, Maternal and Child Health, Section of Psychiatry, University of Genoa, Genoa, Italy; IRCCS Ospedale Policlinico San Martino, Genoa, Italy

Mario Amore (253), Department of Neuroscience, Rehabilitation, Ophthalmology, Genetics, Maternal and Child Health, Section of Psychiatry, University of Genoa, Genoa, Italy; IRCCS Ospedale Policlinico San Martino, Genoa, Italy

Gregory Samuel Barber (319), Department of Psychiatry, University of Maryland School of Medicine, Baltimore, MD, United States

Jean-Marie Batail (269), Brain Stimulation Laboratory, Department of Psychiatry and Behavioral Sciences, Stanford University, Palo Alto, CA, United States

Mandakh Bekhbat (237), Department of Psychiatry and Behavioral Sciences, Emory University School of Medicine, Atlanta, GA, United States

Francesco Benedetti (505), Division of Neuroscience, Psychiatry and Psychobiology Unit, IRCCS San Raffaele Scientific Institute; Vita-Salute San Raffaele University, Milan, Italy

Jack Van Bezooyen (461), Department of Psychiatry and Behavioral Sciences, Emory University, Atlanta, GA, United States

Daniel Blumberger (M.) (309), General Adult Psychiatry and Health Systems Division; Temerty Centre for Therapeutic Brain Intervention, CAMH, Toronto; Department of Psychiatry, University of Toronto, Toronto, ON, Canada

William V. Bobo (23,107,137), Department of Psychiatry and Psychology, Mayo Clinic, Jacksonville, FL, United States

Olivia E. Bogucki (511), Division of Integrated Behavioral Health, Mayo Clinic, Rochester, MN, United States

Lucas Borrione (283), Service of Interdisciplinary Neuromodulation; Laboratory of Neuroscience and National Institute of Biomarkers in Psychiatry, Department and Institute of Psychiatry, University of São Paulo Medical School, São Paulo, Brazil

Chad Bousman (99), Department of Medical Genetics, University of Calgary, Calgary, AB, Canada

Joshua Brown (269), Department of Psychiatry and Human Behavior, Brown University School of Medicine, Providence, RI, United States; Department of Neurology, Brown University School of Medicine, Providence, RI, United States

Lisa C. Brown (99), Great Scott! Consulting, Brooklyn, NY, United States

Andre R. Brunoni (283), Service of Interdisciplinary Neuromodulation; Laboratory of Neuroscience and National Institute of Biomarkers in Psychiatry, Department and Institute of Psychiatry, University of São Paulo Medical School; Center for Clinical and Epidemiological Research & Interdisciplinary Center for Applied Neuromodulation, University Hospital, University of São Paulo, São Paulo, Brazil

Giovanna Canepa (253), Department of Neuroscience, Rehabilitation, Ophthalmology, Genetics, Maternal and Child Health, Section of Psychiatry, University of Genoa, Genoa, Italy

Sara V. Carlini (519), The Zucker Hillside Hospital, Women's Behavioral Health Research, Northwell Health, Glen Oaks, NY, United States

Mohit Chauhan (511), Division of Integrated Behavioral Health, Mayo Clinic, Jacksonville, FL, United States

Xiangchuan Chen (237), Department of Psychiatry and Behavioral Sciences, Emory University School of Medicine, Atlanta, GA, United States

John Coetzee (269), Brain Stimulation Laboratory, Department of Psychiatry and Behavioral Sciences, Stanford University; Adamson Lab, Headache Center for Excellence, Rehabilitation Service, VA Palo Alto Health Care System, Palo Alto, CA, United States

Eleanor Cole (269), Department of Psychiatry and Behavioral Sciences, Stanford University, Palo Alto, CA, United States

Charles R. Conway (481), Department of Psychiatry, Washington University School of Medicine, St. Louis, MO, United States

Ana Paula Costa (493), Faillace Department of Psychiatry and Behavioral Sciences at McGovern Medical School, The University of Texas Health Science Center at Houston (UTHealth), Houston, TX, United States

Cristina Cusin (13), Department of Psychiatry, Depression Clinical and Research Program (DCRP), Massachusetts General Hospital; Department of Psychiatry, Harvard Medical School, Boston, MA, United States

Sara Dallaspezia (505), Division of Neuroscience, Psychiatry and Psychobiology Unit, IRCCS San Raffaele Scientific Institute, Milan, Italy

Zafiris Daskalakis (J.) (309), Department of Psychiatry, School of Medicine UC San Diego Health, University of California San Diego, San Diego, CA, United States

Melissa DeFilippis (393), Department of Psychiatry and Behavioral Sciences, University of Texas Medical Branch, Galveston, TX, United States

Kristina M. Deligiannidis (519), The Zucker Hillside Hospital, Women's Behavioral Health Research, Northwell Health, Glen Oaks; Departments of Psychiatry and Obstetrics & Gynecology at the Zucker School of Medicine at Hofstra/Northwell, Hempstead; Institute of Behavioral Science, The Feinstein Institutes for Medical Research, Manhasset, NY, United States

Alexandre Paim Diaz (379), Center of Excellence on Mood Disorders; Translational Psychiatry Program, Faillace Department of Psychiatry and Behavioral Sciences, McGovern Medical School, The University of Texas Health Science Center at Houston (UTHealth), Houston, TX, United States

Karina Drake (293), Department of Psychiatry, University of Colorado School of Medicine, University of Colorado, Boulder, CO, United States

Alexandra Dyer Merg (407), Department of Psychiatry, University of Illinois College of Medicine, Chicago, IL, United States

Tracy D. Eells (369), Department of Psychiatry and Behavioral Sciences, University of Louisville, Louisville, Kentucky

Rebecca M. Eichenbaum (519), The Zucker Hillside Hospital, Women's Behavioral Health Research, Northwell Health, Glen Oaks; Montefiore Medical Center, Bronx, NY, United States

Harris A. Eyre (99), Neuroscience-inspired Policy Initiative, Organisation for Economic Co-Operation and Development (OECD) and PRODEO Institute, Paris, France

Julie Farrington (193), Department of Psychiatry and Behavioral Sciences, Ascension Seton Shoal Creek Hospital, University of Texas at Austin Dell Medical School, Austin, TX, United States

Jennifer C. Felger (237), Department of Psychiatry and Behavioral Sciences, Emory University School of Medicine, Atlanta, GA, United States

Brisa S. Fernandes (379), Center of Excellence on Mood Disorders; Translational Psychiatry Program, Faillace Department of Psychiatry and Behavioral Sciences, McGovern Medical School, The University of Texas Health Science Center at Houston (UTHealth), Houston, TX, United States

Peter J. Franz (559), Department of Psychology, Harvard University, Cambridge, MA, United States

John F. Greden (99), University of Michigan Eisenberg Family Depression Center and National Network of Depression Centers, Ann Arbor, MI, United States

Roland R. Griffiths (221), Departments of Psychiatry and Neuroscience, Johns Hopkins University School of Medicine, Baltimore, MD, United States

Natalie Gukasyan (221), Department of Psychiatry and Behavioral Sciences, Center for Psychedelic and Consciousness Research, Johns Hopkins University School of Medicine, Baltimore, MD, United States

Ebrahim Haroon (237), Department of Psychiatry and Behavioral Sciences, Emory University School of Medicine, Atlanta, GA, United States

Rachel Hershenberg (415), Department of Psychiatry & Behavioral Sciences, Emory University, Atlanta, GA, United States

Steven K. Huprich, PhD (441), Division of Psychiatry and Behavioral Medicine, Michigan State University College of Human Medicine, Grand Rapids; Department of Psychology, University of Detroit Mercy, Detroit, MI, United States

Manuella P. Kaster (493), Department of Biochemistry, Federal University of Santa Catarina, Florianopolis, SC, Brazil

David J. Katzelnick (511), Division of Integrated Behavioral Health, Mayo Clinic, Rochester, MN, United States

Yardana Kaufman (519), The Zucker Hillside Hospital, Women's Behavioral Health Research, Northwell Health, Glen Oaks, NY, United States

Sidney H. Kennedy (81), Department of Psychiatry, University of Toronto, Toronto, Canada

Julia A. Koretski (355), Department of Psychiatry, University of Massachusetts Chan Medical School, Worcester, MA, United States

Sussann Kotara (193), Department of Psychiatry and Behavioral Sciences, U.T. Dell Medical Center Hospital, Austin, TX, United States

William B. Leasure (511), Division of Integrated Behavioral Health, Mayo Clinic, Rochester, MN, United States

Michael J. Lucido (237), Department of Psychiatry and Behavioral Sciences, Emory University School of Medicine, Atlanta, GA, United States

Rodrigo Machado-Vieira (493), Faillace Department of Psychiatry and Behavioral Sciences at McGovern Medical School, The University of Texas Health Science Center at Houston (UTHealth), Houston, TX, United States

Daniel F. Maixner (293), Department of Psychiatry, University of Michigan Medical School, University of Michigan, Ann Arbor, MI, United States

John C. Markowitz (369), Columbia University; New York State Psychiatric Institute, New York City, NY, United States

R. Hamish McAllister-Williams (1), Northern Centre for Mood Disorders, Translational and Clinical Research Institute, Newcastle University; Cumbria, Northumberland, Tyne and Wear NHS Foundation Trust, Newcastle upon Tyne, United Kingdom

Neeti D. Mehta (237), Department of Psychiatry and Behavioral Sciences, Emory University School of Medicine, Atlanta, GA, United States

Ashley K. Meyer (13), The Division of Digital Psychiatry, Beth Israel Deaconess Medical Center, Harvard Medical School, Boston, MA, United States

Brian J. Mickey (481), Department of Psychiatry; Department of Anesthesiology; Department of Bioengineering, The University of Utah, Salt Lake City, UT, United States

Andrew H. Miller (237), Department of Psychiatry and Behavioral Sciences, Emory University School of Medicine, Atlanta, GA, United States

Adriano H. Moffa (283), Laboratory of Neuroscience and National Institute of Biomarkers in Psychiatry, Department and Institute of Psychiatry, University of São Paulo Medical School, São Paulo, Brazil; School of Psychiatry & Black Dog Institute, University of New South Wales, Sydney, NSW, Australia

Ahmad Subhi Abu Mohammad (473), Department of Psychiatry, University of Michigan, Ann Arbor, MI, United States

Daniel J. Mueller (99), Campbell Family Mental Health Research Institute, Centre for Addiction and Mental Health, Institute of Medical Science, University of Toronto, Toronto, ON, Canada

Peter Nagele (481), Department of Anesthesia and Critical Care, The University of Chicago, Chicago, IL, United States

Ziad Nahas (345), Department of Psychiatry and Behavioral Sciences, University of Minnesota, Medical School, Minneapolis, MN, United States

Zui Narita (269), Department of Behavioral Medicine, National Institute of Mental Health, National Center of Neurology and Psychiatry, Tokyo, Japan

Romina Nejad (269), Brain Stimulation Laboratory, Department of Psychiatry and Behavioral Sciences, Stanford University, Palo Alto, CA, United States

Charles B. Nemeroff (193), Department of Psychiatry and Behavioral Sciences, Institute for Early Life Adversity Research, University of Texas at Austin Dell Medical School, Austin, TX, United States

Shehzad K. Niazi (23), Department of Psychiatry and Psychology, Mayo Clinic, Jacksonville, FL, United States

Sina Nikayin (205), Yale Depression Research Program; Interventional Psychiatry Service, Department of Psychiatry, Yale University, New Haven, CT, United States

Ben J.A. Palanca (481), Department of Psychiatry; Department of Anesthesiology, Washington University School of Medicine, St. Louis, MO, United States

Sagar V. Parikh (473), Department of Psychiatry, University of Michigan, Ann Arbor, MI, United States

Angela Phillips (269), Brain Stimulation Laboratory, Department of Psychiatry and Behavioral Sciences, Stanford University; Adamson Lab, Headache Center for Excellence, Rehabilitation Service, VA Palo Alto Health Care System, Palo Alto, CA, United States

William Pitchot (167), Department of Psychiatry, Centre Hospitalier Université de Liège, Liege, Belgium

Danielle Postorivo (61), Queensland Brain Institute, The University of Queensland, St Lucia, QLD, Australia

Joao Quevedo (379), Center of Excellence on Mood Disorders; Translational Psychiatry Program, Faillace Department of Psychiatry and Behavioral Sciences, McGovern Medical School, The University of Texas Health Science Center at Houston (UTHealth); Neuroscience Graduate Program, The University of Texas MD Anderson Cancer Center UTHealth Graduate School of Biomedical Sciences, Houston, TX, United States; Translational Psychiatry Laboratory, Graduate Program in Health Sciences, University of Southern Santa Catarina (UNESC), Criciúma, SC, Brazil

Mark Rapaport (415), Department of Psychiatry, University of Utah, Salt Lake City, UT, United States

Lais B. Razza (283), Service of Interdisciplinary Neuromodulation; Laboratory of Neuroscience and National Institute of Biomarkers in Psychiatry, Department and Institute of Psychiatry, University of São Paulo Medical School, São Paulo, Brazil

Jessica Reis (369), Department of Psychiatry and Behavioral Sciences, University of Louisville, Louisville, Kentucky

Elliott Richelson (107,137), Department of Psychiatry and Psychology, Mayo Clinic, Jacksonville, FL, United States

Patricio Riva-Posse (331), Department of Psychiatry and Behavioral Sciences, Emory University School of Medicine, Atlanta, GA, United States

Anthony J. Rothschild (355), Department of Psychiatry, University of Massachusetts Chan Medical School and Umass Memorial Healthcare, Worcester, MA, United States

Gregory Sahlem (269), Brain Stimulation Laboratory, Department of Psychiatry and Behavioral Sciences, Stanford University, Palo Alto, CA, United States

Gerard Sanacora (205), Yale Depression Research Program; Interventional Psychiatry Service, Department of Psychiatry, Yale University, New Haven, CT, United States

Marsal Sanches (379), Center of Excellence on Mood Disorders, Faillace Department of Psychiatry and Behavioral Sciences, McGovern Medical School, The University of Texas Health Science Center at Houston (UTHealth), Houston, TX, United States

Craig N. Sawchuk (511), Division of Integrated Behavioral Health, Mayo Clinic, Rochester, MN, United States

Gianluca Serafini (253), Department of Neuroscience, Rehabilitation, Ophthalmology, Genetics, Maternal and Child Health, Section of Psychiatry, University of Genoa; IRCCS Ospedale Policlinico San Martino, Genoa, Italy

Richard C. Shelton (47), Department of Psychiatry and Behavioral Neurobiology, School of Medicine, University of Alabama at Birmingham, Birmingham, AL, United States

Jair C. Soares (379), Center of Excellence on Mood Disorders; Translational Psychiatry Program, Faillace Department of Psychiatry and Behavioral Sciences, McGovern Medical School, The University of Texas Health Science Center at Houston (UTHealth), Houston, TX, United States

Daniel Souery (167), PSYPLURIEL at EPSYLON – Psychiatric Care Network; Laboratory of Psychological Medicine and Addictology, Faculty of Medicine, Université Libre de Bruxelles, Brussels, Belgium

Katy Stimpson (269), Brain Stimulation Laboratory, Department of Psychiatry and Behavioral Sciences, Stanford University, Palo Alto, CA, United States

Kimberly R. Stubbs (461), Department of Psychiatry, Wright State University, Dayton, OH, United States

Scott C. Tadler (481), Department of Psychiatry; Department of Anesthesiology, The University of Utah, Salt Lake City, UT, United States

Yi-lang Tang (461), Department of Psychiatry and Behavioral Sciences, Emory University, Atlanta; Substance Abuse Treatment Program, Atlanta VA Medical Center, Decatur, GA, United States

Michael E. Thase (175), Department of Psychiatry, Perelman School of Medicine, University of Pennsylvania, Corporal Michael J. Crescenz Veterans Affairs Medical Center, Philadelphia, PA, United States

Marisa Toups (551), Singularity Psychiatry and Mental Health, Austin, TX, United States

Michael T. Treadway (237), Department of Psychiatry and Behavioral Sciences, Emory University School of Medicine, Atlanta, GA, United States

Susannah Tye (61), Queensland Brain Institute, The University of Queensland, St Lucia, QLD, Australia

Robert J. Ursano (575), Department of Psychiatry, Uniformed Services University of the Health Sciences, Bethesda, MD, United States

Sophie R. Vaccarino (81), Institute of Medical Science, University of Toronto; Department of Psychiatry, University Health Network; Centre for Depression and Suicide Studies, Unity Health Toronto, Toronto, Canada

Victor Alejandro Valencia (407), Department of Psychiatry, University of Illinois College of Medicine, Chicago, IL, United States

Mary C. Vance (575), Department of Psychiatry, Uniformed Services University of the Health Sciences, Bethesda, MD, United States

Vitaliy L. Voytenko, PsyD (441), Division of Psychiatry and Behavioral Medicine, Michigan State University College of Human Medicine, Grand Rapids, MI; Alliance Clinical Associates, Wheaton, IL, United States

Karen Dineen Wagner (393), Department of Psychiatry and Behavioral Sciences, University of Texas Medical Branch, Galveston, TX, United States

Kamile Weischedel (415), Department of Psychiatry, University of Utah, Salt Lake City, UT, United States

Cory Weissman (R.) (309), Department of Psychiatry, School of Medicine UC San Diego Health, University of California San Diego, San Diego, CA, United States

James C. West (575), Department of Psychiatry, Uniformed Services University of the Health Sciences, Bethesda, MD, United States

Mark D. Williams (511), Division of Integrated Behavioral Health, Mayo Clinic, Rochester, MN, United States

Nolan Williams (269), Brain Stimulation Laboratory, Department of Psychiatry and Behavioral Sciences, Stanford University, Palo Alto, CA, United States

Jennifer Wood (369), Department of Psychiatry and Behavioral Sciences, University of Louisville, Louisville, Kentucky

Jesse H. Wright (369), Department of Psychiatry and Behavioral Sciences, University of Louisville, Louisville, Kentucky

Clarissa G. Yates (61), Queensland Brain Institute, The University of Queensland, St Lucia, QLD, Australia

Chapter 1

Defining treatment-resistant depression, difficult-to-treat depression, and staging treatment intensity

R. Hamish McAllister-Williams

Northern Centre for Mood Disorders, Translational and Clinical Research Institute, Newcastle University, Newcastle upon Tyne, United Kingdom; Cumbria, Northumberland, Tyne and Wear NHS Foundation Trust, Newcastle upon Tyne, United Kingdom

Introduction

Depression has enormous impact on individuals, their families, and society. Fortunately, there are many effective treatments options (Malhi and Mann, 2018). However, unfortunately, a significant proportion of patients do not achieve sustained remission, despite serial treatments (Rush et al., 2006). "Treatment-resistant depression (TRD)" is a well-established term and concept to define such patients. While the definition of TRD is arbitrary and there is a lack of complete consensus, it is relatively easy to assess in individual patients, and it has utility in defining populations of patients for research. "Difficult-to-treat depression (DTD)" is a recently proposed term that is argued to have semantic and conceptual advantages over TRD for many patients in the clinic. There is significant overlap between populations of depressed patients who would attract these "labels"; however, they are not identical. The major difference, however, between DTD and TRD relates to the implied models of care that arises from their respective concepts. TRD relates to the treatment of acute episodes of illness using serial acute treatment trials until one is successful in achieving symptomatic remission. DTD draws on the models of care for chronic physical health problems with waxing and waning symptoms such as arthritis, diabetes, and hypertension. It highlights the need for multidimensional treatment and a focus on improvements in psychosocial function and quality of life.

Treatment-resistant depression

Definition of TRD

TRD can be thought of as describing patients beyond a certain threshold in a course of serial treatment trials. There is, however, no universally accepted criterion for this threshold. A systematic review identified 150 randomized controlled trials, observational studies, and case series of patients defined as having "TRD" published between January 2016 and December 2017 (Brown et al., 2019). The most common element of the definition used in these studies was a failure to respond to two different treatments, though this was the case in only just over half of the definitions (50.3%: (Brown et al., 2019), see Fig. 1.1).

However, there is a complete lack of consensus regarding how different classes of medication are considered, the length and dose of treatment trial required, how nonpharmacological interventions are considered, whether the trials needed to be in the current episode or not, what the target outcomes of trials is, how this should be assessed, and whether trials in past episodes count or not (Brown et al., 2019; Gaynes et al., 2020). A systematic review has also highlighted the observation that only 17% of interventional studies recruited populations of patients meeting the most frequently specified criteria of two failed trails (Gaynes et al., 2020). This leads to a major issue with regard to the evidence base for the management of TRD. A systematic review of all psychological and pharmacological augmentation strategies was only able to evaluate 28 randomized controlled trials (RCTs) including just 18 different treatments in participants meeting the TRD criteria of prior failure of two previous treatments, despite a wide range being widely used clinically (Strawbridge et al., 2019).

So how should TRD be defined? One possible approach is to examine whether there is evidence of factors that distinguish patients with TRD compared with other patients with depression. Considering the data from the largest randomized

Managing Treatment-Resistant Depression. https://doi.org/10.1016/B978-0-12-824067-0.00001-3

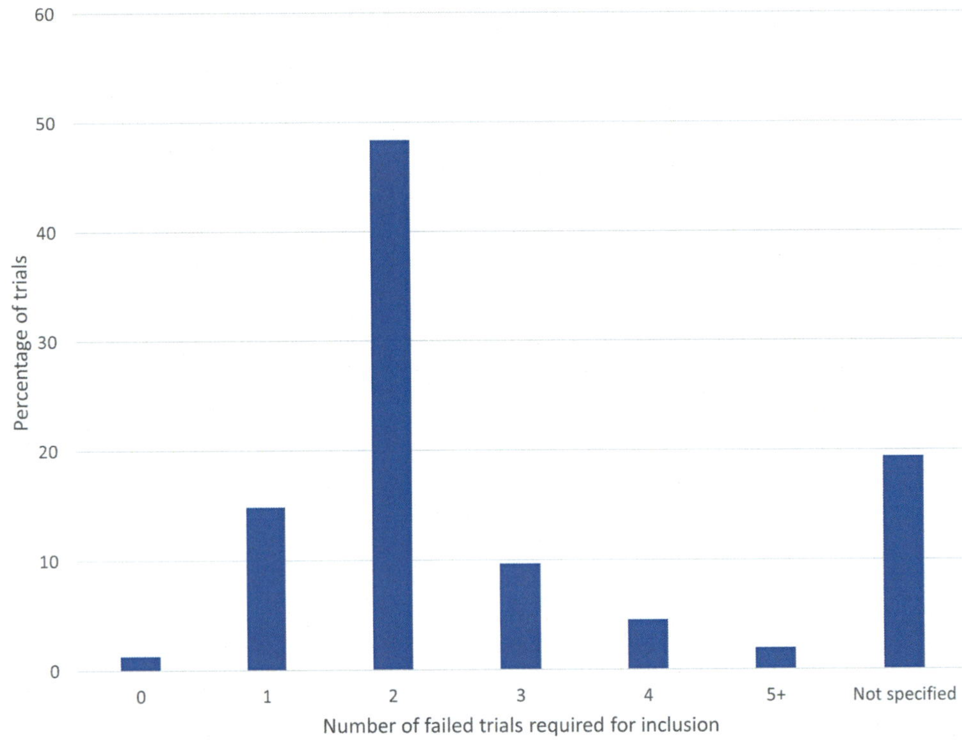

FIG. 1.1 Number of failed trials required for inclusion in studies of patients with "treatment-resistant depression." *(Data taken from Brown, S., Rittenbach, K., Cheung, S., McKean, G., MacMaster, F.P., Clement, F., 2019. Current and common definitions of treatment-resistant depression: findings from a systematic review and qualitative interviews. Can. J. Psychiatry 64(6), 380–387. https://doi.org/10.1177/0706743719828965.)*

trial in depression, the STAR*D study, it has been suggested that there is a marked change in remission rates between second and third step treatments supporting a view that TRD should be defined on the basis of two failed treatments (Conway et al., 2017) (see Fig. 1.2).

However, in a meta-analysis of over 4500 patients, a dichotomous or even a highly variable response to antidepressants does not seem to be the case, challenging the notion that there are distinct subgroups of treatment-responsive and treatment-resistant patients with depression (Guo et al., 2020). Big data sets can be used to identify factors associated with proxy markers of TRD, for example, the need for using ECT, vagus nerve stimulation (VNS), or deep brain stimulation (DBS). One such study identified that such patients could be defined by having received three or more antidepressants, or one or more antipsychotics, in the preceding year (Cepeda et al., 2018). However, the argument is somewhat circular in defining TRD on the basis of use of certain (neurostimulatory) treatments, and the derived criteria will vary between different health care environments whose treatment algorithms differ. The reality is that there are likely to be a multitude of reasons why some patients respond to medication and others do not, including pharmacokinetic and pharmacodynamic differences, and there are no clearly defined symptomatology-based, phenotypic, or biological boundaries between TRD and depression that is not treatment resistant (Dodd et al., 2020), precluding this as an avenue through which to define TRD—at least at the present time.

What about from a regulatory perspective? While regulators do set out some outline criteria, what companies do is determined at least in part by business considerations. Intranasal esketamine has received a license in many parts of the world for augmentation of antidepressants in patients with TRD. This was on the basis of pivotal RCTs in which TRD was defined as a failure to respond to at least one antidepressant in the current episode assessed retrospectively, plus failure of a different antidepressant taken at an adequate dose for a total duration of at least 6 weeks observed prospectively (Fedgchin et al., 2019). No response was defined as less than a 25% improvement in symptoms and adequacy of dose of drug assessed using the Massachusetts General Hospital antidepressant treatment response questionnaire (Desseilles et al., 2011). Currently, there are no other drugs licensed in Europe for TRD. An alternative approach was taken by the manufacturers of a long-acting formulation of quetiapine who obtained a license for its use in patients with "sub-optimal response" to antidepressants based on pivotal studies that recruited participants whose depression had failed to respond to just one or more antidepressants (Bauer et al., 2009). The decision between these two different positions is largely governed by marketing

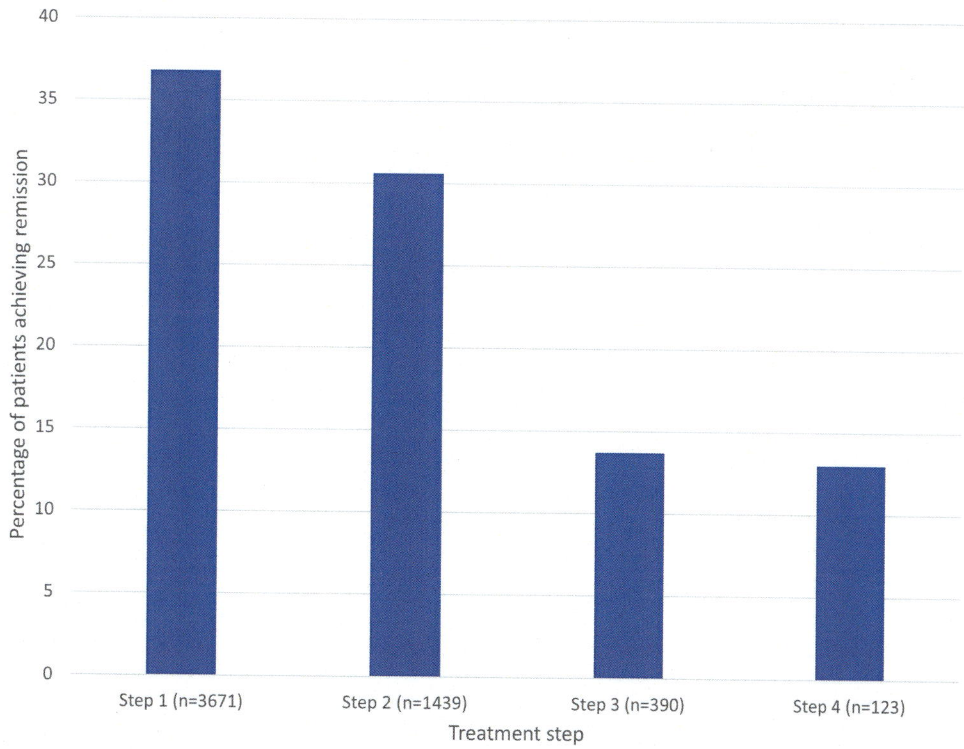

FIG. 1.2 Remission rates to first-, second-, third-, or fourth-line pharmacological treatments. *(Data from Rush, A.J., Trivedi, M.H., Wisniewski, S.R., Nierenberg, A.A., Stewart, J.W., Warden, D., Niederehe, G., Thase, M.E., Lavori, P.W., Lebowitz, B.D., McGrath, P.J., Rosenbaum, J.F., Sackeim, H.A., Kupfer, D.J., Luther, J., Fava, M., 2006. Acute and longer-term outcomes in depressed outpatients requiring one or several treatment steps: a STAR*D report. Am. J. Psychiatry 163(11), 1905–1917.)*

factors as well as a belief that the more previous treatments have been ineffective, the less likely a subsequent one is to work, based on the data from the STAR*D study (Rush et al., 2006).

Ultimately, the definition of TRD is arbitrary. In an attempt to address this, a European-funded partnership with industry has drawn together a number of international experts in the management of depression from across the globe and, using a Delphi technique, attempted to arrive at a consensus definition of TRD particularly, but not exclusively, for regulatory studies (Sforzini et al., 2021). There was strong consensus that the definition should be failure of at least two prior antidepressant treatments but only weak consensus that this should be based on treatments in the current episode or the last 2 years, whichever is longer. There was moderate consensus that an adequate trial of a medication was a minimum of 4 weeks at a minimally effective dose and that the minimum of two failed treatments needed to include two established treatments with different mechanisms of action, though how the latter is defined was not detailed. The consensus was that trials in TRD could include patients who had received psychotherapy or neurostimulation treatments, such as transcranial direct current stimulation (tDCS), repetitive transcranial magnetic stimulation (rTMS), and ECT, but not VNS or DBS. There was also consensus that trials should exclude comorbid personality disorders, other mental disorders, and active and severe substance misuse (Sforzini et al., 2021). If this consensus is adopted, then it will lead to more consistency in studies allowing for comparisons of treatments, but the downside is that the data will be less generalizable given high rates of comorbidity of all types in TRD populations (Dodd et al., 2020; McAllister-Williams et al., 2020).

Most definitions of TRD focus exclusively on unipolar depression, though some studies in TRD do include patients with a bipolar diagnosis (Brown et al., 2019). There is much less consistency regarding the definition of treatment-resistant bipolar depression (TRBD). The most common definition in the recent literature is failure of one treatment known to be effective (Gaynes et al., 2020). However, a group of experts in the management of bipolar disorder, using a Delphi technique, arrived at a consensus definition of TRBD as failure to reach sustained symptomatic remission for eight consecutive weeks after two different treatment trials, at adequate therapeutic doses, with at least two recommended monotherapy treatments or at least one monotherapy treatment and another combination treatment (Hidalgo-Mazzei et al., 2019). This has been operationalized in an ongoing trial as failure to respond to (or tolerate) two out of four treatments included in current guidelines for the treatment of bipolar depression (Goodwin et al., 2016): quetiapine, olanzapine, lamotrigine, and lurasidone (Azim et al., 2021).

Staging treatment-resistant depression

"Staging" TRD refers to operationalizing the differentiation of TRD from treatment-responsive depression and/or quantifying or categorizing the degree of treatment resistance. Several staging tools exist that have various pros and cons and are well reviewed by Ruhé et al. (2012) and Sackeim et al. (2019). The original purpose of these tools varies, as some being more orientated toward use in research studies and others being more intended for clinical practice. See Table 1.1.

The oldest staging tool is the Antidepressant Treatment History Form (ATHF) first published in 1990 (Ruhé et al., 2012; Sackeim et al., 2019), with several updates, most recently in 2019, in a short-form version (ATHF-SF) (Sackeim et al., 2019). The tool was originally introduced to assess degree of resistance in patients receiving ECT and is oriented toward use in research studies. In its original form, the tool assesses each treatment trial on a scale of 1–5, with scores of 1 and 2 indicating an inadequate trial and 3 and above being adequate, with higher scores indicating higher doses and/or augmentation of a base antidepressant. The tool is most typically used to describe the total number of "failed" (irrespective of adequacy) and "failed adequate" trials a patient has received. It is possible to sum the scores across treatment trials and describe the highest score for any trial administered, though these are rarely reported (Sackeim et al., 2019). The new ATHF-SR simplifies matters by defining a trial simply as adequate or not. For a trial of pharmacotherapy to be counted as adequate it needs to have been taken for at least 4 weeks at a dose specified in the ATHF-SR. This new version also collects information regarding neurostimulatory and psychotherapeutic treatment trials and defines what counts as adequate trials for these. As a result, the number of failed and number of failed adequate trials can be totaled or separated out by drug class and treatment modality. It is the only commonly used staging tool that has separate and specific criteria for psychotic and bipolar depression. For example, for a trial of an antidepressant to count as adequate in a patient with psychotic depression, it must have been combined with an adequate dose of an antipsychotic. The 2019 paper includes data collection tools, checklist, and instructions in supplementary material that is freely available for use.

The "Massachusetts General Hospital Staging model (MGH-s)" or "Antidepressant Treatment Response Questionnaire (ATRQ)" was first published by Fava (2003) and updated in 2011 (Desseilles et al., 2011). It is similar but somewhat simpler to use than the ATHF. Apart from ECT, it does not include any treatment other than pharmacotherapy. An adequate trial of an antidepressant is deemed to be at least 6 weeks (rather than 4 as in the ATHF) with the minimum dose for trials defined in the tool. Each failed trial in the current episode scores 1 point. As with the ATHF, there is no implicit hierarchy of efficacy of antidepressants nor preference for between- versus within-class switches. However, unlike the ATHF-SF, an account is taken of optimization of treatment trials with an extra 0.5 points added if the trial extended to at least 10 weeks and if the dose is increased to or above a specified "maximum" dose. Augmentation is dealt with by adding a further 0.5 points for any antidepressant that was augmented with at least one other drug. Only examples of possible augmentation agents are given, meaning that when used in research studies, these need to be operationally defined (Morriss et al., 2020), and it is unclear why augmentation with a medication supported by positive RCT evidence scores the same as the antidepressant trial being extended from 6 to 10 weeks. A failed trial of ECT adds a further 3 points, though it is unclear as to how this score has been arrived at. Nevertheless, despite these shortcomings, the ATRQ relatively straight forwardly produces a continuous score which might help differentiate degrees of treatment resistance.

The Thase and Rush staging model (TRSM) was developed as a guide for clinical practice (Thase and Rush, 1997). Rather than attempting to provide a continuous score for degree of treatment resistance, patients are categorized into one of five "stages." Stage I includes individuals whose depression has not responded to at least one adequate trial of an antidepressant; Stage II if nonresponse to at least two antidepressant from two distinct pharmacological classes; Stage III as Stage II plus failure of a trial of a tricyclic antidepressant (TCA); Stage IV as Stage III plus a failure of a trial of a monoamine oxidase inhibitor (MAOI); and Stage V as Stage IV plus failure of a course of bilateral ECT. The TRSM is both simple to use and implies a treatment algorithm for patients who don't respond to first line treatment. However, what constitutes an adequate trials is not defined and the TRSM has also been criticized on a number of grounds (Ruhé et al., 2012), including the implication that failure to respond to two different classes of antidepressants is harder to treat than failure to respond to two drugs of the same class, despite the evidence suggesting little difference in response rates when comparing between versus within class switches (Papakostas et al., 2008). It also implies a hierarchy of efficacy of different classes of antidepressants for which there is little evidence, and does not take into account augmentation strategies for which there is an evidence base (Ruhé et al., 2012).

Like the TRSM, the European Staging Model (ESM) does not produce a continuous score, but rather categorizes patients into three groups: "nonresponders," defined as having had a nonresponse to one adequate trial of an antidepressant or ECT; those with "TRD," defined as failure of response to two or more antidepressant trials from different classes; and those with "chronic resistant depression (CRD)" who have been resistance to several antidepressants, including augmentation strategies, over a period of at least 12 months (Souery et al., 1999). The TRD subgroup is subdivided by total duration of

TABLE 1.1 Features of staging tools for treatment-resistant depression.

Staging tool	Explicit operational criteria	Inclusion of nonpharmacological interventions	Separate criteria for depression subgroups	Criteria for combination and/or augmentation trials	Additional information used for TRD score
Antidepressant Treatment History Form (ATHF) (Sackeim et al., 1990)	Yes	ECT only	Bipolar vs unipolar MDE; psychotic vs nonpsychotic MDE	Yes	No
Antidepressant Treatment History Form: Short Form (ATHF-SF) (Sackeim et al., 2019)	Yes	Multiple brain stimulation and psychotherapy interventions	Bipolar vs unipolar MDE; psychotic vs nonpsychotic MDE	Yes	No
Massachusetts General Hospital Staging Model (MGH-s)/Antidepressant Treatment Response Questionnaire (ATRQ) (Fava, 2003; Desseilles et al., 2011)	Yes	ECT only	No	Yes	No
Thase and Rush Staging Model (TRSM) (Thase and Rush, 1997)	No	ECT only	No	No	No
European Staging Model (ESM) (Souery et al., 1999)	No	None	No	Yes	Duration of treatment
Conway et al. Staging Model (Conway et al., 2017)	No	Unspecified brain stimulation and psychotherapy	No	No	No
Maudsley Staging Model (MSM) (Fekadu et al., 2009a)	Yes	ECT only	No	Yes	Duration of episode, baseline symptom severity
Dutch extension of the MSM (Peeters et al., 2016)	Yes	ECT and psychotherapy	No	Yes	Duration of episode, baseline symptom severity, functional impairment, comorbid anxiety, comorbid personality disorder, psychosocial stress, location of treatment

ECT, electroconvulsive therapy; *MDE*, major depressive episode.
Adapted from Sackeim, H.A, Aaronson, S.T., Bunker, M.T., Conway, C.R., Demitrack, M.A., George, M.S., Prudic, J., Thase, M.E., Rush, A.J., 2019. The assessment of resistance to antidepressant treatment: rationale for the Antidepressant Treatment History Form: Short Form (ATHF-SF). J. Psychiatric Res. 113, 125–136. https://doi.org/10.1016/j.jpsychires.2019.03.021.

failed trails from 12 to 16 weeks for TRD1 to 36 weeks to 1 year for TRD5. As with the TRSM, the ESM has a number of weaknesses (Ruhé et al., 2012). It is far from clear why total duration of treatment trials takes on such an important element in categorizing degree of resistance. While it is true that duration of untreated episode of depression is inversely related to the probability of remission (de Diego-Adeliño et al., 2010), it is also clear that probability of remission decreases with each treatment step (Rush et al., 2006) though this is not taken into account. Similarly, the use of augmentation agents is only included in the definition of CRD and not the various degrees/stages of TRD.

The simplest staging tool developed for clinical usage is that published by Conway et al. (2017). This proposes two stages based on the findings from the STAR*D study: Stage I and Stage II TRD. Stage I is defined as "failure of 2 adequate dose-duration antidepressants or psychotherapy from different classes (either in combination or succession) in the current episode," while Stage II is defined as "failure of 3 or more adequate antidepressant or psychotherapy trials from different classes … ." This proposal was born out of an attempt to consider where in the treatment algorithm newer, more expensive, invasive and/or less safe treatments might be considered (see later).

The Maudsley Staging Model (MSM) takes a slightly different approach to the other tools (Fekadu et al., 2009a). It produces a continuous score using a simple scoring system, making it usable in routine clinical practice. What makes the tool different is that it incorporates duration and baseline severity of the current episode into the score. The total score is made up of a score of 1–3 for duration, 1–5 for baseline severity, 1–5 for number of antidepressant failures (a score of 1 for 1–2 failures and a score of 5 for more than 10 failures), with an additional 1 point if any augmentation has been used and 1 for a trial of ECT, giving a total of between 3 and 15. While there are criticisms of the MSM, including that the importance of augmentation is minimized, other neurostimulation and psychotherapy is not considered and that duration of episode and number of trialed medication are not independent variables (Ruhé et al., 2012), there is evidence that the MSM has utility in being able to predict patient outcomes. This was originally shown in a sample of 62 patients followed up naturalistically over an average of 2.5 years (Fekadu et al., 2009a) and then replicated in a sample of 643 patients assessed in the Netherlands Study of Depression and Anxiety at baseline and after 12 months (van Belkum et al., 2018). A Dutch extension of the MSM has been proposed (Peeters et al., 2016) with additional items of degree of functional impairment (rated on the Global Assessment of Function scale and scored 0–3), comorbid anxiety (scored 0, 0.5, or 1), comorbid personality disorder (scored 0, 0.5, or 1), psychosocial stressors (scoring 0 or 1), use of pharmacological augmentation (scored 0–3), use of psychotherapy (scored 0–2), and intensity of treatment (i.e., outpatient, day patient or inpatient, scored 0–2). This extension of the MSM is reported as outperforming the original MSM in prediction of future depressive symptomatology (Peeters et al., 2016).

In terms of thresholds, there is some evidence of differences in the performance of the tools. In a Swedish registry study of 127,108 patients on antidepressants, 19% were identified as having TRD using the ATRQ, 15% by the MSM and 9.5% by the ESM (Hägg et al., 2020). There was little in the way of differences in the characteristics of the patients identified by each tool.

There is no consensus as to which staging tool should be used in different circumstances. While tools such as the ATHF and ATRQ can aid systematic assessment of past treatments, staging tools were used to define TRD in only just over a third of studies (38.1%), the most commonly used being the TRSM, in a systematic review (Brown et al., 2019). The ATHF continues to be used in large studies of patients with significant degrees of treatment resistance, such as the RECOVER study of VNS (Conway et al., 2020) and, as previously mentioned, the ATRQ is also used in a number of past and current large RCTs of various treatments in patients with TRD (McAllister-Williams et al., 2016; Morriss et al., 2020). The European project to build consensus around the definition of TRD previously described (Sforzini et al., 2021) recommended use of the MSM on the basis of its predictive ability, though it should be noted that this European project is led by clinicians from the Maudsley and the other staging tools split opinion.

Utility and issues of the treatment-resistant depression categorization

An agreed consistent definition of TRD and its assessment/staging would be of great value in terms of research, allowing for comparison and meta-analyses of treatment trials, and for regulatory purposes (Conway et al., 2017; Malhi and Byrow, 2016). Given the very significant impact of number of previous failed treatments on likelihood of response (Rush et al., 2006) it is potentially inappropriate, for example, to compare different treatments in a network meta-analysis when they have been examined in populations defined in different ways. In addition, inherent in all definitions of TRD is the concept that the management of depression involves serial acute treatment trials (Rush et al., 2006). In some health care settings, it is clear that many patients receive suboptimal treatment, for example being continued on antidepressants to which they are not responding for long periods, without consideration of alternative or additional pharmacological options (Wiles et al., 2018). In such situations, considering further serial options may be extremely valuable (Kraus et al., 2019). Staging tools such as TRSM can help clinicians to sequence such treatment. Similarly, a staging tool such as the MSM which has predictive utility (Fekadu et al., 2009b; van Belkum et al., 2018) may be of value in clinical prognostication.

However, there are a number of issues with the definition and concept of TRD. There is a lack of consensus regarding how adequate treatment trials are defined, how outcomes to trials are assessed and how to consider trials of treatments of different modalities. Conceptually, TRD implies that it is possible to define patients with depression as either having a treatment-responsive or treatment nonresponsive illness. This dichotomization of patients is not supported by the evidence base (Dodd et al., 2020). Rather, the situation appears to be that both groups of patients are extremely heterogeneous.

Several novel psychotherapeutic, pharmacological and neurostimulatory options have recently joined the treatment armamentarium, with more in development. This is a boon, but forces the question of when and for whom to choose these newer, often more expensive, and/or invasive treatments (Conway et al., 2017; McAllister-Williams et al., 2018; Rush et al., 2019)? The situation varies between different health care settings, but the TRD threshold is not clearly the point at which management of depression radically changes. For example, in the United Kingdom, primary care physicians find it hard on occasion to access specialist mental health services (Telford et al., 2002) meaning that patients are potentially treated with three or four antidepressants as monotherapy before augmentation and novel options are considered. To help guide clinicians as to when to consider novel treatments, Conway et al. (2017) coupled their staging of TRD with possible next step treatment options. For Stage I TRD they suggested rTMS, Esketamine, nitrous oxide, buprenorphine or ECT in some case. For Stage II TRD they suggested ECT, VNS or DBS. Given that Stage II TRD in their model is reached after a failure of just three failed trials, this does not sit well with health care for example in the United Kingdom. As a result, a group of UK experts from primary, secondary and tertiary care proposed the concept of "multi-therapy resistant major depressive disorder (MTR-MDD)" (McAllister-Williams et al., 2018). The aim of this group was to propose a threshold around which it might be appropriate to consider more expensive, less evidence supported and/or more invasive treatments. The idea was to find a balance between identifying as early as possible the point when such "nonstandard" treatments should be considered against ensuring that more standard, safe, well evidence-based treatments have been at least considered. The MTR-MDD criteria were set as "a failure to respond, achieve remission, maintain a response/remission, tolerate, refused or contra-indication of … ." at least: two trials of a structured evidence-supported psychotherapy; four adequate trials of antidepressants; two adequate trials of augmentation (specifically including lithium, quetiapine, or aripiprazole); and a course of eight ECT treatments. Such a threshold is way above that of conventional TRD but it may be more appropriate for considering treatments such as VNS (McAllister-Williams et al., 2021). However, other treatments, for example augmentation with modafinil, might be considered well before the full MTR-MDD criteria are met (McAllister-Williams et al., 2018). This illustrates that rather than TRD being a single important point for clinical decision making, the reality is that different decisions will be made at various points along the treatment pathway, often well beyond the TRD threshold.

Difficult-to-treat depression

Why is an alternative concept needed?

While clinicians generally view TRD as being a failure to respond to two antidepressants they generally do not use or apply this definition in practice (Brown et al., 2019; Han et al., 2020). Clinicians tend to use terms and phrases such as "recurrent or relapsing depression," "complex depression," "difficult-to-treat depression," "multiple episodes of depression," "chronic depression," and "struggled with depression throughout their lives" (Brown et al., 2019). Clinicians also identify challenges in defining TRD (Brown et al., 2019). Firstly, the number of failed treatments is confounded by time given observations that increasing duration of depressive episode is associated with decreasing chances of remission (de Diego-Adeliño et al., 2010). Secondly, it is not clear what the target level of improvement is for each treatment trial. The issue of patients with "partially responsive depression (PRD)" has also been identified as an issue that needs consideration by the pharmaceutical regulators (Sforzini et al., 2021). Thirdly, patients defined as having TRD often have comorbidities which can influence the chance of improvement with treatment, something not taken into account in most definitions of TRD.

Experts in the treatment of mood disorders also recognize problems with the TRD concept in a publication written by representatives of the World Federation of Societies for Biological Psychiatry, Australasian Society for Bipolar and Depressive Disorders, International Society for Affective Disorders, Collegium Internationale Neuro-Psychopharmacologium, and the Canadian Network for Mood and Anxiety Treatments (Dodd et al., 2020). A key issue is that there are multiple reasons for failure to respond to antidepressants including "biological resistance, diagnostic error, limitations of current therapies, psychosocial variables, a past history of exposure to childhood maltreatment or abuse, job satisfaction, personality disorders, co-morbid mental and physical disorders, substance use or non-adherence to treatment," while only a subset of individuals do not respond to pharmacokinetic or pharmacodynamic issues related to a medication which may be resolved by substitution or addition of another (Dodd et al., 2020). The authors therefore recommend that in practice it is necessary to "deconstruct" an individual's depression to understand all the various individual predisposing, precipitating and perpetuating factors related to their depression and why it is nonresponsive to treatment. Only by doing so may it be possible to construct an individualized care plan.

Encapsulating these ideas is the proposal from Rush et al. (2019) that instead of considering whether a patient has "TRD" or not, a better term and concept is of "difficult-to-treat depression (DTD)," drawing on models of care for chronic waxing and waning physical health conditions such as rheumatoid arthritis (Luqmani et al., 2009) or the recovery model as applied, for example, to schizophrenia (Warner, 2009). Semantically, the term "TRD" raises the question as to who or what is "resistant" and runs the risk of leading to blame and/or nihilism. The term "difficult to treat" is proposed since it recognizes that there is a problem that needs addressing and implies that to address this a collaboration between patient, family and health care team is required to address it (McAllister-Williams et al., 2020; Rush et al., 2019).

Conceptually, the DTD model takes a chronic, rather than acute, illness approach with two fundamental differences with the TRD model of care. Firstly, the TRD model assumes that a patient remains unwell following a series of acute treatment trials because of nonresponse to each in turn. However, DTD recognizes multiple routes for there to be a poor treatment outcome. In addition to a lack of response or remission, there can be a situation of an inability to sustain this following initial improvement through relapse or recurrence of depression. Another clinically challenging situation is when a series of treatment trials are ended not through nonresponse but intolerance. STAR*D data demonstrates that with each increasing treatment step, there is not only a reduction in the rate of remission, but also an increase in rates of relapse and intolerance (see Fig. 1.2). It is critical that all three routes to a poor outcome to treatment are considered, not just one. Secondly, an implicit assumption of the TRD model is that depression is an acute illness and that the goal of treatment is remission. This is not an unreasonable position given that lack of full remission of symptoms is associated with both ongoing psychosocial functional impairment (Kennedy and Paykel, 2004; Romera et al., 2010) and a very significantly increased risk of relapse (Judd et al., 1998). Whenever possible, clinicians should strive for full remission. However, the reality is that this is not possible for all people suffering from depression as a result of the multitude of their predisposing, precipitating and perpetuating etiological factors (Dodd et al., 2020). The DTD model recognizes this reality and addresses some of the issues of applying a TRD model in clinical practice, such as a "try and try again approach" without consideration of the individual patient (Dodd et al., 2020) and the risk of missing diagnostic and therapeutic opportunities (Rush et al., 2019).

Definition

A group of international experts has published a consensus statement on the definition, assessment and management of DTD (McAllister-Williams et al., 2020). This group proposed a definition of DTD, or "suspected DTD" as *depression that continues to cause significant burden despite usual treatment efforts.* This definition is intentionally broad in scope due to match the heterogeneity of clinical scenarios. There are two main elements to the definition: what constitutes "significant burden" and "usual treatment efforts." In terms of burden, DTD recognizes that this may arise from (a) a failure in achieving response or remission acutely; (b) failure in sustaining response or remission; (c) a lack of functional restoration despite good symptomatic control; (d) unacceptable tolerability or nonadherence or rejection of the treatment option. Critically burden is defined by the patient. What constitutes "usual treatment efforts" depends on the professional treating the patient, the health care environment, local treatment guidelines etc. What is perceived as "difficult" in one setting may not be in another. It may appear inappropriate that a patient is deemed to have DTD in one setting but not another. However, the purpose of the term is to bring into play a different way of thinking about the illness management, including review of whether a referral to another setting, where it may not be perceived as so difficult, is in order. Also note that "treatment" relates to all and any forms of treatment for depression. It is acknowledged that the DTD "label" will encompass a very heterogeneous group of individuals (McAllister-Williams et al., 2020).

DTD model of care

A model of care for DTD is described in detail in McAllister-Williams et al. (2020). Essentially, if depression continues to cause significant burden despite standard first- and second-line treatment, then it may be considered "suspected DTD." Full remission may be achieved through increasing the dose, or switching, an antidepressant, or augmentation with another medication, psychotherapy, neurostimulatory treatment, psychosocial intervention or self-help strategy. The depression has proved to no longer be "difficult to treat." However, the further down the treatment algorithm the clinician and patient get without full remission and restoration of function, the more a DTD approach may be warranted.

The first step in managing suspected DTD is an assessment encompassing a review of the differential diagnosis, an identification and active treatment of all comorbidities (physical, psychiatric, substance misuse and iatrogenic), and a thorough assessment of the patient's individual etiological factors and reasons for nonresponse. The overall notion is to address any tractable reasons for the depression being difficult to treat in the hope that this turns the illness into a more easily treated depression. For example identification that the underlying illness is bipolar disorder and hence a drug like quetiapine may be more appropriate than an antidepressant (Goodwin et al., 2016), or identifying that a patient is perimenopausal management of which can make depression more treatable (Maki et al., 2019). Similarly, if a poor outcome results from ongoing interpersonal conflicts, then a course of interpersonal psychotherapy may be in order (Cuijpers et al., 2016).

A model of care based on the TRD concept has a single goal: remission. While optimization of symptom control and achieving remission if possible remains critical, the DTD includes two additional treatment goals: reducing risk and impact of relapse; and optimizing psychosocial function and quality of life, with this last goal in particular being defined by the patient themselves (McAllister-Williams et al., 2020). The DTD also describes several principles of care including enhancing engagement and shared decision making within an integrated service pathway while also encouraging self-management techniques. Measurement-based care, rating not only depressive symptoms but also psychosocial functioning and quality of life is advocated (McAllister-Williams et al., 2020). While the patient's treatment history is an important prognostic indicator, other factors are also critical in the understanding and management of DTD (Dodd et al., 2020). With the partial exception of the Dutch modification of the MSM, these are not included in any of the current TRD staging tools. Further research is required to explore the optimal way of objectively assessing degree of DTD. Critical the model recommends reassessments and consideration of treatment direction at frequent intervals at which the differential diagnosis and etiological factors are reviewed, rescreening for comorbidities is conducted and the use of all treatment modalities considered: pharmacological, neurostimulatory, psychotherapeutic, psychosocial, self-help etc. The aim is to leave no stone unturned in considering options to advance the three goals of treatment, seeking specialist input as necessary to achieve this.

Utility

Since some patients do not response acutely, and others do not sustain their responses despite multiple treatments (Judd et al., 1998; Rush et al., 2006), how long should clinicians go on recommending further acute treatment trials? If all available options and combinations are tried this would take more than a lifetime. Clearly, we should neither undertreat nor overtreat. Patients with clear acute episodic depression and those earlier in the treatment pathway may be appropriately managed using a TRD approach of repeated acute treatment trials. However, this approach may be less appropriate for patients with more persistent depression, those who are not entirely symptom free between episodes, or any for whom finding a tolerable treatment is proving challenging. When is it parsimonious to reconsider the differential diagnosis and case formulation, and to contemplate a different approach—perhaps to address psychosocial stressors or to develop ways to cope with or adjust to some of their symptoms? The DTD model is proposed as a way of trying to address these questions. The model also has advantages in emphasizing the "shared care" between the patient, their carers and family, and health care professionals, and the need to take a broad holistic approach to management. Recognizing that some patients have depression that is difficult to treat is acknowledging a clinical reality but also emphasizing that things may be difficult but that this does not mean impossible.

Conclusions

DTD and TRD are related, overlapping, concepts both with utility but also with key differences. Firstly, TRD is defined by a failure to respond to treatment, while DTD is defined by lack of acute phase response/remission or not sustaining the acute response/remission. Secondly, TRD is unidimensional resting solely on depressive symptom outcomes. In addition to symptoms, DTD considers psychosocial functioning and quality of life from a patient perspective. Thirdly, the action implied by the TRD concept is further acute treatment trials, while the DTD concept calls for reevaluation of the depression, a search for treatable biomedical and psychosocial causes of poor outcome, and consideration of a shift in treatment goals from remission to optimal management. Fundamentally, TRD is of most relevance for treatment trials, drug approval and commissioning of services where a nomothetic approach is necessarily required, while DTD is intended for clinical practice where an idiographic approach is most appropriate.

References

Azim, L., Hindmarch, P., Browne, G., Chadwick, T., Clare, E., Courtney, P., Dixon, L., Duffelen, N., Fouweather, T., Geddes, J.R., Goudie, N., Harvey, S.H., Helter, T., Holstein, E.M., Martin, G., Mawson, P., McCaffery, J., Morriss, R., Simon, J., et al., 2021. Study protocol for a randomised placebo-controlled trial of pramipexole in addition to mood stabilisers for patients with treatment resistant bipolar depression (the PAX-BD study). BMC Psychiatry 21 (1), 334. https://doi.org/10.1186/s12888-021-03322-y.

Bauer, M., Pretorius, H.W., Constant, E.L., Earley, W.R., Szamosi, J., Brecher, M., 2009. Extended-release quetiapine as adjunct to an antidepressant in patients with major depressive disorder: results of a randomized, placebo-controlled, double-blind study. J. Clin. Psychiatry 70 (4), 540–549.

Brown, S., Rittenbach, K., Cheung, S., McKean, G., MacMaster, F.P., Clement, F., 2019. Current and common definitions of treatment-resistant depression: findings from a systematic review and qualitative interviews. Can. J. Psychiatry 64 (6), 380–387. https://doi.org/10.1177/0706743719828965.

Cepeda, M.S., Reps, J., Fife, D., Blacketer, C., Stang, P., Ryan, P., 2018. Finding treatment-resistant depression in real-world data: how a data-driven approach compares with expert-based heuristics. Depress. Anxiety 35 (3), 220–228. https://doi.org/10.1002/da.22705.

Conway, C.R., George, M.S., Sackeim, H.A., 2017. Toward an evidence-based, operational definition of treatment-resistant depression: when enough is enough. JAMA Psychiatry 74 (1), 9–10. https://doi.org/10.1001/jamapsychiatry.2016.2586.

Conway, C.R., Olin, B., Aaronson, S.T., Sackeim, H.A., Bunker, M., Kriedt, C., Greco, T., Broglio, K., Vestrucci, M., Rush, A.J., 2020. A prospective, multi-center randomized, controlled, blinded trial of vagus nerve stimulation for difficult to treat depression: a novel design for a novel treatment. Contemp. Clin. Trials 95. https://doi.org/10.1016/j.cct.2020.106066, 106066.

Cuijpers, P., Donker, T., Weissman, M.M., Ravitz, P., Cristea, I.A., 2016. Interpersonal psychotherapy for mental health problems: a comprehensive meta-analysis. Am. J. Psychiatry 173 (7), 680–687. https://doi.org/10.1176/appi.ajp.2015.15091141.

de Diego-Adeliño, J., Portella, M.J., Puigdemont, D., Pérez-Egea, R., Alvarez, E., Pérez, V., 2010. A short duration of untreated illness (DUI) improves response outcomes in first-depressive episodes. J. Affect. Disord. 120 (1–3), 221–225. https://doi.org/10.1016/j.jad.2009.03.012.

Desseilles, M., Witte, J., Chang, T.E., Iovieno, N., Dording, C.M., Ashih, H., Nyer, M., Freeman, M.P., Fava, M., Mischoulon, D., 2011. Assessing the adequacy of past antidepressant trials: a clinician's guide to the antidepressant treatment response questionnaire. J. Clin. Psychiatry 72 (8), 1152–1154. https://doi.org/10.4088/JCP.11ac07225.

Dodd, S., Bauer, M., Carvalho, A.F., Eyre, H., Fava, M., Kasper, S., Kennedy, S.H., Khoo, J.-P., Lopez Jaramillo, C., Malhi, G.S., McIntyre, R.S., Mitchell, P.B., Castro, A.M.P., Ratheesh, A., Severus, E., Suppes, T., Trivedi, M.H., Thase, M.E., Yatham, L.N., et al., 2020. A clinical approach to treatment resistance in depressed patients: what to do when the usual treatments don't work well enough? World J. Biol. Psychiatry, 1–20. https://doi.org/10.1080/15622975.2020.1851052.

Fava, M., 2003. Diagnosis and definition of treatment-resistant depression. Biol. Psychiatry 53 (8), 649–659.

Fedgchin, M., Trivedi, M., Daly, E.J., Melkote, R., Lane, R., Lim, P., Vitagliano, D., Blier, P., Fava, M., Liebowitz, M., Ravindran, A., Gaillard, R., Ameele, H.V.D., Preskorn, S., Manji, H., Hough, D., Drevets, W.C., Singh, J.B., 2019. Efficacy and safety of fixed-dose esketamine nasal spray combined with a new oral antidepressant in treatment-resistant depression: results of a randomized, double-blind, active-controlled study (TRANSFORM-1). Int. J. Neuropsychopharmacol. 22 (10), 616–630. https://doi.org/10.1093/ijnp/pyz039.

Fekadu, A., Wooderson, S., Donaldson, C., Markopoulou, K., Masterson, B., Poon, L., Cleare, A.J., 2009a. A multidimensional tool to quantify treatment resistance in depression: the Maudsley staging method. J. Clin. Psychiatry 70 (2), 177–184.

Fekadu, A., Wooderson, S.C., Markopoulou, K., Cleare, A.J., 2009b. The Maudsley staging method for treatment-resistant depression: prediction of longer-term outcome and persistence of symptoms. J. Clin. Psychiatry 70 (7), 952–957. https://doi.org/10.4088/JCP.08m04728.

Gaynes, B.N., Lux, L., Gartlehner, G., Asher, G., Forman-Hoffman, V., Green, J., Boland, E., Weber, R.P., Randolph, C., Bann, C., Coker-Schwimmer, E., Viswanathan, M., Lohr, K.N., 2020. Defining treatment-resistant depression. Depress. Anxiety 37 (2), 134–145. https://doi.org/10.1002/da.22968.

Goodwin, G.M., Haddad, P.M., Ferrier, I.N., Aronson, J.K., Barnes, T., Cipriani, A., Coghill, D.R., Fazel, S., Geddes, J.R., Grunze, H., Holmes, E.A., Howes, O., Hudson, S., Hunt, N., Jones, I., Macmillan, I.C., McAllister-Williams, H., Miklowitz, D.R., Morriss, R., et al., 2016. Evidence-based guidelines for treating bipolar disorder: revised third edition recommendations from the British Association for Psychopharmacology. J. Psychopharmacol. 30 (6), 495–553. https://doi.org/10.1177/0269881116636545.

Guo, X., McCutcheon, R.A., Pillinger, T., Mizuno, Y., Natesan, S., Brown, K., Howes, O., 2020. The magnitude and heterogeneity of antidepressant response in depression: a meta-analysis of over 45,000 patients. J. Affect. Disord. 276, 991–1000. https://doi.org/10.1016/j.jad.2020.07.102.

Hägg, D., Brenner, P., Reutfors, J., Li, G., DiBernardo, A., Bodén, R., Brandt, L., 2020. A register-based approach to identifying treatment-resistant depression-comparison with clinical definitions. PLoS One 15 (7). https://doi.org/10.1371/journal.pone.0236434, e0236434.

Han, C., Wang, G., Chan, S., Kato, T., Ng, C.H., Tan, W., Zhang, L., Feng, Y., Liu, C.-Y., 2020. Definition and identification of patients with treatment-resistant depression in real-world clinical practice settings across Asia. Neuropsychiatr. Dis. Treat. 16, 2929–2941. https://doi.org/10.2147/NDT.S264799.

Hidalgo-Mazzei, D., Berk, M., Cipriani, A., Cleare, A.J., Florio, A.D., Dietch, D., Geddes, J.R., Goodwin, G.M., Grunze, H., Hayes, J.F., Jones, I., Kasper, S., Macritchie, K., McAllister-Williams, R.H., Morriss, R., Nayrouz, S., Pappa, S., Soares, J.C., Smith, D.J., et al., 2019. Treatment-resistant and multi-therapy-resistant criteria for bipolar depression: consensus definition. Br. J. Psychiatry 214 (1), 27–35. https://doi.org/10.1192/bjp.2018.257.

Judd, L.L., Akiskal, H.S., Maser, J.D., Zeller, P.J., Endicott, J., Coryell, W., Paulus, M.P., Kunovac, J.L., Leon, A.C., Mueller, T.I., Rice, J.A., Keller, M.B., 1998. Major depressive disorder: a prospective study of residual subthreshold depressive symptoms as predictor of rapid relapse. J. Affect. Disord. 50 (2–3), 97–108.

Kennedy, N., Paykel, E.S., 2004. Residual symptoms at remission from depression: impact on long-term outcome. J. Affect. Disord. 80 (2–3), 135–144.

Kraus, C., Kadriu, B., Lanzenberger, R., Zarate, C.A., Kasper, S., 2019. Prognosis and improved outcomes in major depression: a review. Transl. Psychiatry 9 (1), 127. https://doi.org/10.1038/s41398-019-0460-3.

Luqmani, R., Hennell, S., Estrach, C., Basher, D., Birrell, F., Bosworth, A., Burke, F., Callaghan, C., Candal-Couto, J., Fokke, C., Goodson, N., Homer, D., Jackman, J., Jeffreson, P., Oliver, S., Reed, M., Sanz, L., Stableford, Z., Taylor, P., et al., 2009. British Society for Rheumatology and British Health Professionals in Rheumatology guideline for the management of rheumatoid arthritis (after the first 2 years). Rheumatology (Oxford) 48 (4), 436–439. https://doi.org/10.1093/rheumatology/ken450a.

Maki, P.M., Kornstein, S.G., Joffe, H., Bromberger, J.T., Freeman, E.W., Athappilly, G., Bobo, W.V., Rubin, L.H., Koleva, H.K., Cohen, L.S., Soares, C.N., 2019. Guidelines for the evaluation and treatment of perimenopausal depression: summary and recommendations. J. Women's Health 28 (2), 117–134. https://doi.org/10.1089/jwh.2018.27099.mensocrec.

Malhi, G.S., Byrow, Y., 2016. Is treatment-resistant depression a useful concept? Evid. Based Ment. Health 19 (1), 1–3. https://doi.org/10.1136/eb-2015-102299.

Malhi, G.S., Mann, J.J., 2018. Depression. Lancet 392 (10161), 2299–2312. https://doi.org/10.1016/S0140-6736(18)31948-2.

McAllister-Williams, R.H., Christmas, D.M.B., Cleare, A.J., Currie, A., Gledhill, J., Insole, L., Malizia, A.L., McGeever, M., Morriss, R., Robinson, L.J., Scott, M., Stokes, P.R.A., Talbot, P.S., Young, A.H., 2018. Multiple-therapy-resistant major depressive disorder: a clinically important concept. Br. J. Psychiatry 212 (5), 274–278. https://doi.org/10.1192/bjp.2017.33.

McAllister-Williams, R.H., Anderson, I.M., Finkelmeyer, A., Gallagher, P., Grunze, H.C.R., Haddad, P.M., Hughes, T., Lloyd, A.J., Mamasoula, C., McColl, E., Pearce, S., Siddiqi, N., Sinha, B.N.P., Steen, N., Wainwright, J., Winter, F.H., Ferrier, I.N., Watson, S., 2016. Antidepressant augmentation with metyrapone for treatment-resistant depression (the ADD study): a double-blind, randomised, placebo-controlled trial. Lancet Psychiatry 3 (2), 117–127. https://doi.org/10.1016/S2215-0366(15)00436-8.

McAllister-Williams, R.H., Arango, C., Blier, P., Demyttenaere, K., Falkai, P., Gorwood, P., Hopwood, M., Javed, A., Kasper, S., Malhi, G.S., Soares, J.C., Vieta, E., Young, A.H., Papadopoulos, A., Rush, A.J., 2020. The identification, assessment and management of difficult-to-treat depression: an international consensus statement. J. Affect. Disord. 267, 264–282. https://doi.org/10.1016/j.jad.2020.02.023.

McAllister-Williams, R.H., Bulmer, S., Newton, K., Heath, K., Cousins, D.A., Currie, A., 2021. Assessment for vagus nerve stimulation in patients with difficult-to-treat depression: a model from the Newcastle Regional Affective Disorders Service (RADS). J. Affect. Disord. 280 (Pt A), 315–318. https://doi.org/10.1016/j.jad.2020.11.020.

Morriss, R., Webster, L., Abdelghani, M., Auer, D.P., Barber, S., Bates, P., Blamire, A., Briley, P.M., Brookes, C., Iwabuchi, S., James, M., Kaylor-Hughes, C., Lankappa, S., Liddle, P., McAllister-Williams, H., O'Neill-Kerr, A., Pszczolkowski Parraguez, S., Suazo Di Paola, A., Thomson, L., Walters, Y., 2020. Connectivity guided theta burst transcranial magnetic stimulation versus repetitive transcranial magnetic stimulation for treatment-resistant moderate to severe depression: study protocol for a randomised double-blind controlled trial (BRIGhTMIND). BMJ Open 10 (7). https://doi.org/10.1136/bmjopen-2020-038430, e038430.

Papakostas, G.I., Fava, M., Thase, M.E., 2008. Treatment of SSRI-resistant depression: a meta-analysis comparing within- versus across-class switches. Biol. Psychiatry 63 (7), 699–704.

Peeters, F.P.M.L., Ruhe, H.G., Wichers, M., Abidi, L., Kaub, K., van der Lande, H.J., Spijker, J., Huibers, M.J.H., Schene, A.H., 2016. The Dutch measure for quantification of treatment resistance in depression (DM-TRD): an extension of the Maudsley staging method. J. Affect. Disord. 205, 365–371. https://doi.org/10.1016/j.jad.2016.08.019.

Romera, I., Perez, V., Menchón, J.M., Delgado-Cohen, H., Polavieja, P., Gilaberte, I., 2010. Social and occupational functioning impairment in patients in partial versus complete remission of a major depressive disorder episode. A six-month prospective epidemiological study. Eur. Psychiatry 25 (1), 58–65. https://doi.org/10.1016/j.eurpsy.2009.02.007.

Ruhé, H.G., van Rooijen, G., Spijker, J., Peeters, F.P.M.L., Schene, A.H., 2012. Staging methods for treatment resistant depression. A systematic review. J. Affect. Disord. 137 (1–3), 35–45. https://doi.org/10.1016/j.jad.2011.02.020.

Rush, A.J., Trivedi, M.H., Wisniewski, S.R., Nierenberg, A.A., Stewart, J.W., Warden, D., Niederehe, G., Thase, M.E., Lavori, P.W., Lebowitz, B.D., McGrath, P.J., Rosenbaum, J.F., Sackeim, H.A., Kupfer, D.J., Luther, J., Fava, M., 2006. Acute and longer-term outcomes in depressed outpatients requiring one or several treatment steps: a STAR*D report. Am. J. Psychiatry 163 (11), 1905–1917.

Rush, A.J., Aaronson, S.T., Demyttenaere, K., 2019. Difficult-to-treat depression: a clinical and research roadmap for when remission is elusive. Aust. N. Z. J. Psychiatry 53 (2), 109–118. https://doi.org/10.1177/0004867418808585.

Sackeim, H.A., Prudic, J., Devanand, D.P., Decina, P., Kerr, B., Malitz, S., 1990. The impact of medication resistance and continuation pharmacotherapy on relapse following response to electroconvulsive therapy in major depression. J. Clin. Psychopharmacol. 10 (2), 96–104.

Sackeim, H.A., Aaronson, S.T., Bunker, M.T., Conway, C.R., Demitrack, M.A., George, M.S., Prudic, J., Thase, M.E., Rush, A.J., 2019. The assessment of resistance to antidepressant treatment: rationale for the Antidepressant Treatment History Form: Short Form (ATHF-SF). J. Psychiatr. Res. 113, 125–136. https://doi.org/10.1016/j.jpsychires.2019.03.021.

Sforzini, L., Worrell, C., Kose, M., Anderson, I.M., Aouizerate, B., Arolt, V., Bauer, M., Baune, B.T., Blier, P., Cleare, A.J., Cowen, P.J., Dinan, T.G., Fagiolini, A., FerrierHegerl, I.N.U., Krystal, A.D., Leboyer, M., McAllister-Williams, R.H., McIntyre, R.S., et al., 2021. A Delphi-method-based consensus guideline for definition of Treatment-Resistant Depression for clinical trials. Mol. Psychiatry. In press.

Souery, D., Amsterdam, J., de Montigny, C., Lecrubier, Y., Montgomery, S., Lipp, O., Racagni, G., Zohar, J., Mendlewicz, J., 1999. Treatment resistant depression: methodological overview and operational criteria. Eur. Neuropsychopharmacol. 9 (1–2), 83–91.

Strawbridge, R., Carter, B., Marwood, L., Bandelow, B., Tsapekos, D., Nikolova, V.L., Taylor, R., Mantingh, T., de Angel, V., Patrick, F., Cleare, A.J., Young, A.H., 2019. Augmentation therapies for treatment-resistant depression: systematic review and meta-analysis. Br. J. Psychiatry 214 (1), 42–51. https://doi.org/10.1192/bjp.2018.233.

Telford, R., Hutchinson, A., Jones, R., Rix, S., Howe, A., 2002. Obstacles to effective treatment of depression: a general practice perspective. Fam. Pract. 19 (1), 45–52.

Thase, M.E., Rush, A.J., 1997. When at first you don't succeed: sequential strategies for antidepressant nonresponders. J. Clin. Psychiatry 58 (Suppl. 13), 23–29.

van Belkum, S.M., Geugies, H., Lysen, T.S., Cleare, A.J., Peeters, F.P.M.L., Penninx, B.W.J.H., Schoevers, R.A., Ruhe, H.G., 2018. Validity of the Maudsley staging method in predicting treatment-resistant depression outcome using the Netherlands study of depression and anxiety. J. Clin. Psychiatry 79 (1). https://doi.org/10.4088/JCP.17m11475.

Warner, R., 2009. Recovery from schizophrenia and the recovery model. Curr. Opin. Psychiatry 22 (4), 374–380. https://doi.org/10.1097/YCO.0b013e32832c920b.

Wiles, N., Taylor, A., Turner, N., Barnes, M., Campbell, J., Lewis, G., Morrison, J., Peters, T.J., Thomas, L., Turner, K., Kessler, D., 2018. Management of treatment-resistant depression in primary care: a mixed-methods study. Br. J. Gen. Pract. 68 (675), e673–e681. https://doi.org/10.3399/bjgp18X699053.

Chapter 2

Clinical and epidemiological predictors of treatment-resistant depression

Cristina Cusin[a,b] and Ashley K. Meyer[c]

[a]*Department of Psychiatry, Depression Clinical and Research Program (DCRP), Massachusetts General Hospital, Boston, MA, United States,* [b]*Department of Psychiatry, Harvard Medical School, Boston, MA, United States,* [c]*The Division of Digital Psychiatry, Beth Israel Deaconess Medical Center, Harvard Medical School, Boston, MA, United States*

Introduction

Treatment-resistant depression (TRD) is common among individuals with major depressive disorder (MDD), with over one-third of patients being unable to reach an adequate level of symptomatic improvement after trying an antidepressant (AD) treatment (Fava, 2003). However, it is unclear which factors may lead to poor response to treatment, categorically defined as "treatment resistance." It is also unclear whether the lack of response to a medication or to a class of medications has clinical implications for long-term prognosis, or whether it has implications for the choice of the next AD treatment. Possible predictors of TRD have been investigated with different methodologies, from prospective clinical trials to large epidemiological studies; however, most of the available data addressing predictors of TRD is derived from post hoc analysis of data from randomized trials. The predictors derived from post hoc studies have significant limitations, including difficulties with replicating positive results across studies, and very few predictors have been tested prospectively. One of the major causes of conflicting results is the use of different definitions of TRD across studies, as the number of failed AD trials required to meet the definition of TRD varies substantially from "nonresponse to one prior treatment" to "nonresponse to three or more prior ADs." Additional factors such as type of treatment considered (i.e., medications, psychotherapy, ECT), duration, and adequacy of doses, may also differ across studies, further complicating the definition of TRD (Berlim and Turecki, 2007).

The purpose of this chapter is to review and summarize the literature reporting clinical and epidemiological factors that have been associated with treatment resistance in patients with MDD. Each of those factors has been investigated for possible correlation with the presence of the categorical phenotypes of "non-response" or "non-remission."

Clinical predictors of TRD that have been commonly explored include those related to the timeline of the illness, such as age of onset, duration and number of episodes, and chronicity (i.e., symptoms of depression persisting unabated for more than 2 years). Some of the other factors considered are related to markers of depression severity including, for example, psychiatric hospitalizations, suicide attempts, the presence of psychotic symptoms, or patterns of general medical or psychiatric comorbidity. Other factors investigated are related to specific features of MDD (i.e., melancholic, atypical, or anxious depression), family history, and personality features. Finally, epidemiological predictors of TRD include gender, ethnicity, psychosocial and environmental factors (i.e., employment, marital status, educational status, adverse life events), and global level of functioning.

Clinical predictors

Clinical subtypes

Although some studies have suggested that individuals with different depressive subtypes, including anxious, melancholic, atypical, mixed, and psychotic depression, respond differentially to ADs, there are inconsistencies in the literature (Kraus et al., 2019). For example, regarding the "atypical" subtype of depression, in the 1970s researchers reported superior efficacy of MAOI compared to tricyclics for patients with this subtype, and additional trials were conducted in the 1980s (Davidson, 2007); however, this association has not been investigated with more recently developed ADs and it is of limited utility for clinicians due to the infrequent use of MAOIs, as discussed in greater detail elsewhere in this

Managing Treatment-Resistant Depression. https://doi.org/10.1016/B978-0-12-824067-0.00002-5

volume. Regarding "melancholic" depression, some studies showed that patients with this subtype and higher depression severity may not respond as well to SSRIs compared to venlafaxine or tricyclic ADs (Bauer et al., 2013) and that melancholic features are associated with TRD (defined as failure to respond to at least two consecutive AD trials) (Souery et al., 2007), but these observations were not supported by other studies (Uher et al., 2011). In an uncontrolled parallel design study using sertraline, citalopram, and venlafaxine, there was no clear association between depressive subtype and clinical outcome (Arnow et al., 2015). It should be noted that, often, the percentage of patients classified as having a pure-form subtype (i.e., melancholic) or more than one subtype (mixed) is fairly similar (approximately one-third of the sample), further complicating the classification (Kraus et al., 2019). The presence of anxiety symptoms in the context of a major depressive episode, or so-called "anxious depression," seems to confer increased risk for nonresponse to ADs in a manner that is similar to having a comorbid anxiety disorder, and several groups have found that anxiety symptoms at baseline predicted poorer or delayed response to treatment (Russell et al., 2001; Andreescu et al., 2007; Cohen et al., 2009; Howland et al., 2009).

Depression severity

Among individuals treated with antidepressants, the severity of illness itself is commonly considered a factor contributing to resistance (Howland et al., 2009; Fava et al., 2002a; Perlis et al., 2003, 2004). In a prospective study that included multiple next-step stages of treatment for depressed antidepressant-treated patients, Agid and Lerer (2003) found that higher depression severity at baseline was associated with poorer response to SSRIs (fluoxetine and paroxetine). Similar results were reported by Souery et al. (2007) who found that higher levels of depression severity at the beginning of follow-up was associated with TRD in a European multicenter study, in addition to other factors including comorbid anxiety disorders, comorbid personality disorders, presence of recurrent episodes, number of psychiatric hospitalizations, early age at onset of index depressive episode, and a lifetime history of poor response to the first antidepressant prescribed. However, the association between depression severity and TRD has not been consistently replicated. For example, in a post hoc analysis of pooled data from three randomized trials of pharmacotherapy, psychotherapy, or combination therapy for depression, Van et al. (2008) reported that lower baseline depression severity, and not higher severity, was associated with non-response to treatment. Other researchers explored the relationship between symptom severity at baseline and response to escitalopram, compared with response to several other SSRIs and SNRIs combined (citalopram, duloxetine, fluoxetine, paroxetine, sertraline, and venlafaxine) (Kilts et al., 2009). In that study, higher depression severity was associated with lower response rates to all the other antidepressants combined, while there was no association between baseline symptom severity and response to escitalopram.

Chronicity

One of the more commonly reported factors associated with non-response to ADs is chronicity, or duration of the MDD episode of more than 2 years (Howland et al., 2009; Petersen et al., 2002; Sagud et al., 2013; Balestri et al., 2016). These findings of duration of episode overlap with those of Van et al. (2008), reporting nonresponse to psychotherapy associated with a longer duration of depression. Interestingly, this study explored psychotherapy, pharmacotherapy, and the combination of the two, and showed that patients with chronic depression exhibited a lower response to psychotherapy alone. However, the predictive effect of duration of episode was no longer significant when considering pharmacotherapy only. In a more recent, large, single-blind, 7-month prospective randomized trial, chronicity of depression did not appear to differentially impact acute or longer-term outcome with SSRI monotherapy or with combination of antidepressants (Sung et al., 2012). In other studies the number of previous hospitalizations, a possible marker of both severity and chronicity, was significantly associated with non-response, however, this may lead to circular definition, as non-response itself may lead to an increase in the rate of hospitalizations (Souery et al., 2007; Agid and Lerer, 2003).

Treatment adherence

It is intuitive that patients who prematurely self-discontinue treatment, or have poor compliance, may be at higher risk for depressive relapse or for developing a more chronic symptomatic course. A repeated pattern of interruptions of pharmacologic treatment may be due to several factors including poor access to care or financial barriers, and these may overall lead to worse prognosis or be correlated with other adverse prognostic factors in people with depression. In an interesting study that explored the relationship between adherence to depression treatment guidelines (4–9 months of AD therapy following acute symptom remission) and recurrence of depression in a sample of Medicaid beneficiaries with diagnosed depression

(Melfi et al., 1998), patients who stopped their treatment prematurely were at higher risk of experiencing a relapse, while those who continued with their AD treatment were the least likely to relapse. Similar findings were reported in a more recent study that suggested a fivefold increase in nonresponse associated with nonadherence to psychotherapeutic treatment (Van et al., 2008). Another variable to consider that is related to care not consistent with prescription guidelines and may be considered parallel to the general principle of adherence is the prescribing of inadequate doses of antidepressant treatment, which has been consistently associated with non-response (Berlim and Turecki, 2007; Nasso et al., 2011; Bock et al., 2009). However, in most cases, treatment non-adherence or under-prescribing (i.e., inadequate doses or inadequate duration of trials) are generally considered to be causes of "pseudo-resistance," apparent non-response due to inadequately conducted therapeutic trials, as opposed to depression that has responded poorly to vigorous, evidence-based, guideline-consistent treatment.

Psychiatric comorbidities

Anxiety disorders

The association between co-occurring anxiety disorders and higher likelihood of TRD has been replicated in several studies (Souery et al., 2007; Walker et al., 2000; Riedel et al., 2011; Schosser et al., 2012; Mogi and Yoshino, 2017) and, specifically, the presence of one or more comorbid anxiety disorders has been repeatedly associated with lower likelihood of response to ADs (Souery et al., 2007; Berlanga and Ortega-Soto, 1995; Petersen et al., 2001; Kiejna et al., 2010). As a specific example, the large multicenter study conducted by Souery et al. (2007) referenced earlier showed that poor response to antidepressant treatment was significantly associated with comorbid anxiety in general (OR 2.6) and panic disorder (OR 3.2) and social phobia (OR 2.1), specifically. These general observations were replicated in a large, multicenter, multistage clinical trial that was designed to study clinical and genetic factors associated with TRD, wherein comorbid anxiety disorders were only one of two significant clinical predictors (the other being lifetime history of non-response to an initial antidepressant trial) (Schosser et al., 2012).

Personality disorders

A comorbid diagnosis of personality disorder has been previously found to be associated with AD non-response (Souery et al., 2007; Papakostas et al., 2003); however, a straightforward clinical interpretation of these findings has been made more challenging by the results of a study by our group wherein nearly half of subjects classified as having maladaptive personality traits or having a diagnosis of personality disorder during an acute depressive episode will no longer meet criteria for personality disorder after successful treatment with antidepressants (Fava et al., 2002b). These results raise the question of whether maladaptive behaviors and cognitive patterns that are observed in at least a subset of acutely depressed patients are actually state-dependent depressive symptoms that respond well to AD treatment, rather than being trait-like characteristics that are more resistant to pharmacotherapy. Managing TRD in patients with comorbid personality disorders is discussed in greater detail elsewhere in this volume.

Childhood trauma

Childhood trauma has also been repeatedly associated with poor outcome with AD treatment and higher levels of depression severity in people with TRD. For example, in a sample of depressed adults, persons with TRD reported significantly greater levels of emotional abuse during childhood than individuals classified as having treatment-responsive depression (Kaplan and Klinetob, 2000). More recently, Tunnard et al. (2014) reported very high rates of childhood abuse (64% of the sample) in a cohort of patients with TRD who were previously psychiatrically hospitalized in an inpatient affective disorders unit. In that study, childhood adversity was associated with having an earlier age of depression onset, more recurrent depressive episodes, and a lifetime history of suicide attempts, findings that were also consistent with the results of a cohort study that documented a significant association between childhood adversity and depression severity in adults with TRD (Yrondi et al., 2020). Among participants in the International Study to Predict Optimized Treatment in Depression (i-SPOT) (Williams et al., 2016), a larger comparative effectiveness study, a history of trauma was not associated with the clinical outcome; however, in a subsample of this study, a history of abuse in early childhood was associated with lower likelihood of response. Interestingly, the neurobiological mechanisms implicated in treatment non-response may be influenced by trauma. In fact, neuroimaging research studies have suggested that early life stress may impact brain circuit development that correlates with depression later in life and can potentially correlate with an individual's capacity to respond to antidepressants (Williams et al., 2016).

Substance use

Major depressive disorder often co-occurs with alcohol (AUD) or substance use disorders (SUD), and the longitudinal course of these diseases is intertwined, as discussed in greater detail in Chapter 31. As appears to be the case with most other risk factors for TRD, associations with SUDs appear to vary between studies. For example, in a large, population-based registry study of people with diagnosed depression (13% classified as treatment-resistant) from Sweden, having any diagnosed SUD within 180 days preceding the initiation of AD treatment was associated with nearly double the risk for TRD (adjusted OR 1.86), with the greatest risk of treatment-resistance observed in those with diagnosed sedative, opioid, and mixed substance use disorders (Brenner et al., 2020). Similar results were reported in another claims-based retrospective cohort study from the United States (Cepeda et al., 2018). However, prospective studies have yielded different findings. In STAR*D, no significant differences were observed between patients with or without SUD in regard to time to achieve response, or rate of response to citalopram; however, those patients who endorsed both alcohol and drug use had significantly reduced rates of remission and significantly increased times to reach remission compared to the MDD group without SUD (Davis et al., 2010). In the Combining Medications to Enhance Depression Outcomes study (COMED) there were no significant differences between patients with MDD with or without SUD in terms of dose, time in treatment, response or remission (Davis et al., 2012). Overall, no study has specifically and prospectively investigated the comparative effectiveness of different AD treatments in patients with comorbid MDD and SUD.

Suicide risk

A group using the data pool from the European "Group for the Study of Resistant Depression" (GSRD), a multi-center study, identified suicidality as a key variable in treatment outcome (Kautzky et al., 2017). Cross-sectional data from the GSDR reported "suicide risk" as a risk factor linked to TRD (Schosser et al., 2012), with suicide risk defined according to the MINI diagnostic interview as the presence of one of the following items in the past month: believing it would be better being dead or wishing one's death, wishing to inflict harm on oneself, having thoughts of suicide, having a plan for suicide, making a suicide attempt, and ever having at least one-lifetime suicide attempt. Souery et al. (2007) also found suicidal risk to be significantly associated with TRD. Also for this possible predictor, it is unclear whether it represents a marker of severity of depression or an independent risk factor.

Medical comorbidities

The presence of medical comorbidities has been associated with worse overall prognosis during acute treatment of depression with ADs, and this result has been replicated independently in multiple studies (Keitner, 1991; Oslin, 2002; Swindle et al., 1998). In the STAR*D study, two-thirds of patients with MDD had at least one general medical condition (Rush, 2007) and these patients showed poorer response to treatment. Physical disabilities and medical comorbidities are likely to impact MDD outcomes, particularly in elderly individuals (Davis et al., 2010). In another study, higher burden of comorbid medical illness, as measured by the Cumulative Illness Rating Scale (CIRS) has been associated with poor treatment outcomes for major depression (Iosifescu et al., 2003). Another group found that patients with the melancholic subtype of MDD and with comorbid cardiovascular and endocrinological disorders experienced significantly worse response to antidepressant treatment compared to melancholic depression patients without these medical comorbidities (Pohle et al., 2009).

The presence of thyroid disease, even subclinical hypothyroidism (an abnormally elevated serum thyroid-stimulating hormone [TSH] but normal serum free thyroxine [fT4] level), has been repeatedly associated with poorer AD response and treatment-resistance in MDD (Hickie et al., 1996; Pae et al., 2009).

A recently published article from Bialek et al. (2019) discussed the potential role of autoimmune diseases in particular, suggesting that diminished therapeutic effects of ADs may be related to immune system dysfunction or to neuroinflammation.

Neurological illnesses are also often associated with depression. For example, depressive symptoms or MDD are common in patients with Parkinson's Disease (PD); however, very few studies investigated Parkinson's Disease (PD) as a specific predictor of TRD. A study conducted by Weintraub et al. (2003) found that among PD patients more than one-third of those who were being treated with an AD continued to meet criteria for MDD; however, it is not clear whether the response rate is significantly different from depressed patients without PD. Also, elevated levels of physical pain at baseline predicted a longer time to remit from depression, which may indicate more difficult-to-treat depression and a need for more aggressive treatment (Karp et al., 2005). Other conditions, such as sleep apnea, have been reported to be associated with nonresponse to AD treatment (Waterman et al., 2016).

AD response rates have been shown to be lower in patients affected by cardiovascular illness and, in turn, the lack of adequate response to AD treatment for post-myocardial infarction seems to be related to increased risk of subsequent cardiac events (de Jonge et al., 2007). As an additional complicating factor, the presence of comorbid anxiety can, in turn, exacerbate cardiac outcomes (Roest et al., 2010) and ultimately lead to inadequate response to AD treatment (Howland et al., 2008; Rush et al., 2008). A number of studies investigated the link between TRD and metabolic syndrome; however, the association between the two is still not clear. While the prevalence of metabolic syndrome appears consistently higher than normal in MDD patients (Vancampfort et al., 2014), in a cross-sectional study Sagud et al. (2013) did not detect an association between TRD and metabolic syndrome.

Similar to what has been observed in other medical conditions, although depression in patients with cancer is a common occurrence, in particular in some forms of cancer like pancreatic adenocarcinoma, no studies have investigated prospectively whether this may have prognostic implications, or to affect the decision-making process when choosing an antidepressant treatment.

Other factors

While family history of mood disorders is often associated with earlier age of onset, chronicity, suicidality and having multiple psychiatric comorbidities, it is unclear whether a positive family history of depression influences the effectiveness of treatment for MDD. In STAR*D, over half (55.6%) of the subjects reported a positive family history of depression; however, they did not differ in response or remission rates from those without family history (Husain et al., 2009). Similarly, in a Chinese study, family history of affective disorders was not associated with the effectiveness of duloxetine in MDD (Wang et al., 2015).

Epidemiological predictors

Socioeconomic status (SES)

Socioeconomic status (SES) has been shown to be associated with response to depression treatment in a study in older adults, or "late-life" depression (Cohen et al., 2006). Specifically, participants who lived in middle- or high-income neighborhoods were significantly more likely to respond to treatment than participants who lived in low-income neighborhoods (Reynolds, 1999, 2006). Cohen et al. (2009) found that treatment response rates were higher for residents of middle- and high-income, when aggregated, compared to those of low-income. Among patients treated with ADs, other social predictors of poor response to ADs or non-remission include lower level of functioning in both work and family life and being unmarried (Balestri et al., 2016). Other psychosocial and environmental factors associated with non-response were unemployment, lower educational status and adverse life events, while cognitive and global functioning were not found to be associated (De Carlo et al., 2016).

Age

Age was consistently found not to be associated with outcome when evaluating non-response to different ADs, or rate of remission (Howland et al., 2009; Fava et al., 2002a; Perlis et al., 2003, 2004; Nasso et al., 2011; Maron et al., 2009; Dudek et al., 2010; Fang, 2010; Rosso, 2012). Only a few studies found a correlation between advanced age and poorer outcome (Petersen et al., 2002; Sagud et al., 2013; Bock et al., 2009; Uher et al., 2009; Bergman et al., 2011) and for this reason, some authors still consider age as a potential negative predictor (De Carlo et al., 2016; Bares et al., 2012).

Gender

Epidemiological studies have repeatedly reported notable gender differences in AD response in MDD. It is well known that the prevalence of MDD in women is nearly twice that in men (Kessler et al., 1993; Scheibe et al., 2003), and depressive symptoms and illness trajectory appear to be different between men and women (Kessler et al., 1993). Studies dating back to the 1970s have reported gender differences in AD response, and this was supported by some more recent research (Keers and Aitchison, 2010). Studies have shown women receive more benefit from some AD treatments including sertraline (Kornstein et al., 2000; Morishita and Kinoshita, 2008), fluoxetine (Martényi, 2001), and citalopram (Berlanga and Flores-Ramos, 2006) while men respond preferentially to other drugs like imipramine (Kornstein et al., 2000) and maprotiline (Martényi, 2001). Similar findings have been consistent with STAR*D results, which reported that females were more likely to respond to citalopram compared to males (Trivedi et al., 2006).

These findings are complicated by the lack of reporting of subgroups, such as menopausal status, which is at times included as a covariate or moderator of the effects of gender on AD response (Keers and Aitchison, 2010). Within a study assessing post-menopausal women, high levels of luteinizing hormone (LH) and follicular stimulating hormone (FSH) were reported to be associated with poorer SSRI response (Zanardi et al., 2007). In other studies, AD response rates have been shown to improve with hormone replacement therapy (HRT), estrogen replacement therapy (ERT), or estrogen replacement (Morgan, 2005; Nagata et al., 2005; Schmidt, 2000; Schneider et al., 2001).

Nevertheless, the data on association between gender and AD response is still controversial. When nine AD studies were reevaluated by Quitkin et al. (2002), there were no reported sex differences in treatment efficacy. No gender differences were found in later AD efficacy studies (Hildebrandt et al., 2003), including a prospective and retrospective study investigating SSRI and TCA response (Parker et al., 2003). A review article published in 2016 found that several additional studies occurring between 2002 and 2012 found no predictive role of gender for response or remission (De Carlo et al., 2016).

Race and ethnicity

Race and ethnicity have been shown to play a role in AD response rates; however, it is very difficult to draw any conclusion as the majority of the clinical trials enroll participants who are not representative of different races and ethnicities. The STAR*D study demonstrated black participants experienced poorer AD responses than white participants (Murphy et al., 2013). In a secondary analysis of the STAR*D study (Lesser et al., 2017), differences in response or remission rates were not found to be significantly different. Overall data on potential associations between ethnicity and non-remission are inconclusive due to a low number of studies and lack of diversity within samples.

Discussion

The research for predictors in TRD has been complicated by the use of inconsistent definitions of TRD and retrospective analyses, while the prospective studies, or studies in which patients are randomized to a certain treatment based on the presence or absence of a sociodemographic variable are difficult to conduct. It is also likely that multiple predictors interact with each other, increasing or decreasing the likelihood of response to treatment. There are no predictors clearly associated with choosing a specific type of treatment, such as augmentation with atypical antipsychotic, or switching to TMS or ECT. Inconsistent findings have made the translation of findings from research studies difficult to translate to clinical care or programs, policies, and other public health interventions. Some of the predictors of poor prognosis have been independently replicated and may be clinically useful, such as the presence of comorbid anxiety disorders, medical comorbidities, poor adherence to treatment, and the presence of suicidality. Unfortunately, most of the other variables have been only inconsistently associated with TRD. Moreover, some of the "predictors" (i.e., chronicity) are detected in patients who already have failed several treatments, who may have more severe illness or are dealing with adverse circumstances, but it is not clear whether the illness was already severe since the onset of it was modified by several attempts at treating it. Finally, while many of the predictors of TRD reviewed in this chapter may characterize subtypes of depression that is more difficult to treat, they are usually not helpful, in-and-of themselves, for choosing the next-step of treatment. Further research is needed in the form of well-designed, large-scale, prospective comparative effectiveness studies in well-characterized and broadly-phenotyped samples of depressed patients.

References

Agid, O., Lerer, B., 2003. Algorithm-based treatment of major depression in an outpatient clinic: clinical correlates of response to a specific serotonin reuptake inhibitor and to triiodothyronine augmentation. Int. J. Neuropsychopharmacol. 6 (1), 41–49.

Andreescu, C., et al., 2007. Effect of comorbid anxiety on treatment response and relapse risk in late-life depression: controlled study. Br. J. Psychiatry 190, 344–349.

Arnow, B.A., et al., 2015. Depression subtypes in predicting antidepressant response: a report from the iSPOT-D trial. Am. J. Psychiatry 172 (8), 743–750.

Balestri, M., et al., 2016. Socio-demographic and clinical predictors of treatment resistant depression: a prospective European multicenter study. J. Affect. Disord. 189, 224–232.

Bares, M., et al., 2012. The early improvement of depressive symptoms as a potential predictor of response to antidepressants in depressive patients who failed to respond to previous antidepressant treatments. Analysis of naturalistic data. Eur. Psychiatry 27 (7), 522–527.

Bauer, M., et al., 2013. World Federation of Societies of Biological Psychiatry (WFSBP) guidelines for biological treatment of unipolar depressive disorders, part 1: update 2013 on the acute and continuation treatment of unipolar depressive disorders. World J. Biol. Psychiatry 14 (5), 334–385.

Bergman, J., et al., 2011. Efficacy of bupropion XR in treatment-resistant elderly patients: a case series study. Clin. Neuropharmacol. 34 (1), 17–20.

Berlanga, C., Flores-Ramos, M., 2006. Different gender response to serotonergic and noradrenergic antidepressants. A comparative study of the efficacy of citalopram and reboxetine. J. Affect. Disord. 95 (1–3), 119–123.

Berlanga, C., Ortega-Soto, H.A., 1995. A 3-year follow-up of a group of treatment-resistant depressed patients with a MAOI/tricyclic combination. J. Affect. Disord. 34 (3), 187–192.

Berlim, M.T., Turecki, G., 2007. Definition, assessment, and staging of treatment-resistant refractory major depression: a review of current concepts and methods. Can. J. Psychiatr. 52 (1), 46–54.

Bialek, K., et al., 2019. Major depressive disorders accompanying autoimmune diseases—response to treatment. Prog. Neuro-Psychopharmacol. Biol. Psychiatry 95, 109678.

Bock, C., et al., 2009. Do stressful life events predict medical treatment outcome in first episode of depression? Soc. Psychiatry Psychiatr. Epidemiol. 44 (9), 752–760.

Brenner, P., et al., 2020. Substance use disorders and risk for treatment resistant depression: a population-based, nested case-control study. Addiction 115 (4), 768–777.

Cepeda, M.S., Reps, J., Ryan, P., 2018. Finding factors that predict treatment-resistant depression: results of a cohort study. Depress. Anxiety 35 (7), 668–673.

Cohen, A., et al., 2006. Social inequalities in response to antidepressant treatment in older adults. Arch. Gen. Psychiatry 63 (1), 50–56.

Cohen, A., et al., 2009. Socioeconomic status and anxiety as predictors of antidepressant treatment response and suicidal ideation in older adults. Soc. Psychiatry Psychiatr. Epidemiol. 44 (4), 272–277.

Davidson, J.R., 2007. A history of the concept of atypical depression. J. Clin. Psychiatry 68 (Suppl. 3), 10–15.

Davis, L.L., et al., 2010. Does comorbid substance use disorder impair recovery from major depression with SSRI treatment? An analysis of the STAR*D level one treatment outcomes. Drug Alcohol Depend. 107 (2–3), 161–170.

Davis, L.L., et al., 2012. Effect of concurrent substance use disorder on the effectiveness of single and combination antidepressant medications for the treatment of major depression: an exploratory analysis of a single-blind randomized trial. Depress. Anxiety 29 (2), 111–122.

De Carlo, V., Calati, R., Serretti, A., 2016. Socio-demographic and clinical predictors of non-response/non-remission in treatment resistant depressed patients: a systematic review. Psychiatry Res. 240, 421–430.

de Jonge, P., et al., 2007. Nonresponse to treatment for depression following myocardial infarction: association with subsequent cardiac events. Am. J. Psychiatry 164 (9), 1371–1378.

Dudek, D., et al., 2010. Risk factors of treatment resistance in major depression: association with bipolarity. J. Affect. Disord. 126 (1–2), 268–271.

Fang, Y., et al., 2010. Comparisons of the efficacy and tolerability of extended-release venlafaxine, mirtazapine, and paroxetine in treatment-resistant depression: a double-blind, randomized pilot study in a Chinese population. J. Clin. Psychopharmacol. 30 (4), 357–364.

Fava, M., 2003. Diagnosis and definition of treatment-resistant depression. Biol. Psychiatry 53 (8), 649–659.

Fava, M., et al., 2002a. Double-blind study of high-dose fluoxetine versus lithium or desipramine augmentation of fluoxetine in partial responders and nonresponders to fluoxetine. J. Clin. Psychopharmacol. 22 (4), 379–387.

Fava, M., et al., 2002b. Personality disorders and depression. Psychol. Med. 32 (6), 1049–1057.

Hickie, I., et al., 1996. Clinical and subclinical hypothyroidism in patients with chronic and treatment-resistant depression. Aust. N. Z. J. Psychiatry 30 (2), 246–252.

Hildebrandt, M.G., et al., 2003. Are gender differences important for the clinical effects of antidepressants? Am. J. Psychiatry 160 (9), 1643–1650.

Howland, R.H., et al., 2008. Factors predicting reduced antidepressant response: experience with the SNRI duloxetine in patients with major depression. Ann. Clin. Psychiatry 20 (4), 209–218.

Howland, R.H., et al., 2009. Concurrent anxiety and substance use disorders among outpatients with major depression: clinical features and effect on treatment outcome. Drug Alcohol Depend. 99 (1–3), 248–260.

Husain, M.M., et al., 2009. Family history of depression and therapeutic outcome: findings from STAR*D. J. Clin. Psychiatry 70 (2), 185–195.

Iosifescu, D.V., et al., 2003. The impact of medical comorbidity on acute treatment in major depressive disorder. Am. J. Psychiatry 160 (12), 2122–2127.

Kaplan, M.J., Klinetob, N.A., 2000. Childhood emotional trauma and chronic posttraumatic stress disorder in adult outpatients with treatment-resistant depression. J. Nerv. Ment. Dis. 188 (9), 596–601.

Karp, J.F., et al., 2005. Pain predicts longer time to remission during treatment of recurrent depression. J. Clin. Psychiatry 66 (5), 591–597.

Kautzky, A., et al., 2017. A new prediction model for evaluating treatment-resistant depression. J. Clin. Psychiatry 78 (2), 215–222.

Keers, R., Aitchison, K.J., 2010. Gender differences in antidepressant drug response. Int. Rev. Psychiatry 22 (5), 485–500.

Keitner, G.I., et al., 1991. 12-month outcome of patients with major depression and comorbid psychiatric or medical illness (compound depression). Am. J. Psychiatry 148 (3), 345–350.

Kessler, R.C., et al., 1993. Sex and depression in the national comorbidity survey. I: lifetime prevalence, chronicity and recurrence. J. Affect. Disord. 29 (2–3), 85–96.

Kiejna, A., et al., 2010. The utility of mood disorder questionnaire for the detection of bipolar diathesis in treatment-resistant depression. J. Affect. Disord. 124 (3), 270–274.

Kilts, C.D., et al., 2009. Baseline severity of depression predicts antidepressant drug response relative to escitalopram. Expert. Opin. Pharmacother. 10 (6), 927–936.

Kornstein, S.G., et al., 2000. Gender differences in treatment response to sertraline versus imipramine in chronic depression. Am. J. Psychiatry 157 (9), 1445–1452.

Kraus, C., et al., 2019. Prognosis and improved outcomes in major depression: a review. Transl. Psychiatry 9 (1), 127.

Lesser, I.M., et al., 2007. Ethnicity/race and outcome in the treatment of depression: results from STAR*D. Med. Care 45 (11), 1043–1051.

Maron, E., et al., 2009. Effects of bupropion augmentation in escitalopram-resistant patients with major depressive disorder: an open-label, naturalistic study. J. Clin. Psychiatry 70 (7), 1054–1056.

Martényi, F., et al., 2001. Gender differences in the efficacy of fluoxetine and maprotiline in depressed patients: a double-blind trial of antidepressants with serotonergic or norepinephrinergic reuptake inhibition profile. Eur. Neuropsychopharmacol. 11 (3), 227–232.

Melfi, C.A., et al., 1998. The effects of adherence to antidepressant treatment guidelines on relapse and recurrence of depression. Arch. Gen. Psychiatry 55 (12), 1128–1132.

Mogi, T., Yoshino, A., 2017. The multiple diagnoses of comorbid anxiety disorders and higher interpersonal sensitivity predict treatment-resistant depression. Asian J. Psychiatr. 26, 131–135.

Morgan, M.L., et al., 2005. Estrogen augmentation of antidepressants in perimenopausal depression: a pilot study. J. Clin. Psychiatry 66 (6), 774–780.

Morishita, S., Kinoshita, T., 2008. Predictors of response to sertraline in patients with major depression. Hum. Psychopharmacol. 23 (8), 647–651.

Murphy, E., et al., 2013. Race, genetic ancestry and response to antidepressant treatment for major depression. Neuropsychopharmacology 38 (13), 2598–2606.

Nagata, H., Nozaki, M., Nakano, H., 2005. Short-term combinational therapy of low-dose estrogen with selective serotonin re-uptake inhibitor (fluvoxamine) for oophorectomized women with hot flashes and depressive tendencies. J. Obstet. Gynaecol. Res. 31 (2), 107–114.

Nasso, E.D., et al., 2011. Clinical and demographic predictors of improvement during duloxetine treatment in patients with major depression: an open-label study. Clin. Drug Investig. 31 (6), 385–405.

Oslin, D.W., et al., 2002. Association between medical comorbidity and treatment outcomes in late-life depression. J. Am. Geriatr. Soc. 50 (5), 823–828.

Pae, C.U., et al., 2009. Thyroid hormones affect recovery from depression during antidepressant treatment. Psychiatry Clin. Neurosci. 63 (3), 305–313.

Papakostas, G.I., et al., 2003. Somatic symptoms in treatment-resistant depression. Psychiatry Res. 118 (1), 39–45.

Parker, G., et al., 2003. Gender differences in response to differing antidepressant drug classes: two negative studies. Psychol. Med. 33 (8), 1473–1477.

Perlis, R.H., et al., 2003. Clinical and sociodemographic predictors of response to augmentation, or dose increase among depressed outpatients resistant to fluoxetine 20 mg/day. Acta Psychiatr. Scand. 108 (6), 432–438.

Perlis, R.H., et al., 2004. Effect of medical comorbidity on response to fluoxetine augmentation or dose increase in outpatients with treatment-resistant depression. Psychosomatics 45 (3), 224–229.

Petersen, T., et al., 2001. Treatment resistant depression and axis I co-morbidity. Psychol. Med. 31 (7), 1223–1229.

Petersen, T., et al., 2002. Treatment-resistant depression and axis II comorbidity. Psychother. Psychosom. 71 (5), 269–274.

Pohle, K., et al., 2009. Medical comorbidity affects antidepressant treatment response in patients with melancholic depression. Psychother. Psychosom. 78 (6), 359–363.

Quitkin, F.M., et al., 2002. Are there differences between women's and men's antidepressant responses? Am. J. Psychiatry 159 (11), 1848–1854.

Reynolds 3rd, C.F., et al., 1999. Nortriptyline and interpersonal psychotherapy as maintenance therapies for recurrent major depression: a randomized controlled trial in patients older than 59 years. JAMA 281 (1), 39–45.

Reynolds 3rd, C.F., et al., 2006. Maintenance treatment of major depression in old age. N. Engl. J. Med. 354 (11), 1130–1138.

Riedel, M., et al., 2011. Clinical predictors of response and remission in inpatients with depressive syndromes. J. Affect. Disord. 133 (1–2), 137–149.

Roest, A.M., et al., 2010. Prognostic association of anxiety post myocardial infarction with mortality and new cardiac events: a meta-analysis. Psychosom. Med. 72 (6), 563–569.

Rosso, G., et al., 2012. A randomized, single-blind, comparison of duloxetine with bupropion in the treatment of SSRI-resistant major depression. J. Affect. Disord. 136 (1–2), 172–176.

Rush, A.J., 2007. STAR*D: what have we learned? Am. J. Psychiatry 164 (2), 201–204. PMID: 17267779.

Rush, A.J., et al., 2008. Selecting among second-step antidepressant medication monotherapies: predictive value of clinical, demographic, or first-step treatment features. Arch. Gen. Psychiatry 65 (8), 870–880.

Russell, J.M., et al., 2001. Effect of concurrent anxiety on response to sertraline and imipramine in patients with chronic depression. Depress. Anxiety 13 (1), 18–27.

Sagud, M., et al., 2013. The lack of association between components of metabolic syndrome and treatment resistance in depression. Psychopharmacology 230 (1), 15–21.

Scheibe, S., et al., 2003. Are there gender differences in major depression and its response to antidepressants? J. Affect. Disord. 75 (3), 223–235.

Schmidt, P.J., et al., 2000. Estrogen replacement in perimenopause-related depression: a preliminary report. Am. J. Obstet. Gynecol. 183 (2), 414–420.

Schneider, L.S., Small, G.W., Clary, C.M., 2001. Estrogen replacement therapy and antidepressant response to sertraline in older depressed women. Am. J. Geriatr. Psychiatry 9 (4), 393–399.

Schosser, A., et al., 2012. European group for the study of resistant depression (GSRD)—where have we gone so far: review of clinical and genetic findings. Eur. Neuropsychopharmacol. 22 (7), 453–468.

Souery, D., et al., 2007. Clinical factors associated with treatment resistance in major depressive disorder: results from a European multicenter study. J. Clin. Psychiatry 68 (7), 1062–1070.

Sung, S.C., et al., 2012. The impact of chronic depression on acute and long-term outcomes in a randomized trial comparing selective serotonin reuptake inhibitor monotherapy versus each of 2 different antidepressant medication combinations. J. Clin. Psychiatry 73 (7), 967–976.

Swindle Jr., R.W., Cronkite, R.C., Moos, R.H., 1998. Risk factors for sustained nonremission of depressive symptoms: a 4-year follow-up. J. Nerv. Ment. Dis. 186 (8), 462–469.

Trivedi, M.H., et al., 2006. Evaluation of outcomes with citalopram for depression using measurement-based care in STAR*D: implications for clinical practice. Am. J. Psychiatry 163 (1), 28–40.

Tunnard, C., et al., 2014. The impact of childhood adversity on suicidality and clinical course in treatment-resistant depression. J. Affect. Disord. 152–154, 122–130.

Uher, R., et al., 2009. Differential efficacy of escitalopram and nortriptyline on dimensional measures of depression. Br. J. Psychiatry 194 (3), 252–259.

Uher, R., et al., 2011. Melancholic, atypical and anxious depression subtypes and outcome of treatment with escitalopram and nortriptyline. J. Affect. Disord. 132 (1–2), 112–120.

Van, H.L., et al., 2008. Identifying patients at risk of complete nonresponse in the outpatient treatment of depression. Psychother. Psychosom. 77 (6), 358–364.

Vancampfort, D., et al., 2014. Metabolic syndrome and metabolic abnormalities in patients with major depressive disorder: a meta-analysis of prevalences and moderating variables. Psychol. Med. 44 (10), 2017–2028.

Walker, E.A., et al., 2000. Predictors of outcome in a primary care depression trial. J. Gen. Intern. Med. 15 (12), 859–867.

Wang, S., et al., 2015. Comparison of the effectiveness of duloxetine in depressed patients with and without a family history of affective disorders in first-degree relatives. Shanghai Arch. Psychiatry 27 (4), 237–245.

Waterman, L., Stahl, S.T., Buysse, D.J., Lenze, E.J., Blumberger, D., Mulsant, B., Butters, M., Gebara, M.A., Reynolds 3rd, C.F., Karp, J.F., 2016. Self-reported obstructive sleep apnea is associated with nonresponse to antidepressant pharmacotherapy in late-life depression. Depress Anxiety 33 (12), 1107–1113.

Weintraub, D., et al., 2003. Recognition and treatment of depression in Parkinson's disease. J. Geriatr. Psychiatry Neurol. 16 (3), 178–183.

Williams, L.M., et al., 2016. Childhood trauma predicts antidepressant response in adults with major depression: data from the randomized international study to predict optimized treatment for depression. Transl. Psychiatry 6 (5), e799.

Yrondi, A., et al., 2020. Childhood maltreatment and clinical severity of treatment-resistant depression in a French cohort of outpatients (FACE-DR): one-year follow-up. Depress. Anxiety 37 (4), 365–374.

Zanardi, R., et al., 2007. Response to SSRIs and role of the hormonal therapy in post-menopausal depression. Eur. Neuropsychopharmacol. 17 (6–7), 400–405.

Chapter 3

The economic burden of treatment-resistant depression: Cost-of-illness perspective

Shehzad K. Niazi and William V. Bobo

Department of Psychiatry and Psychology, Mayo Clinic, Jacksonville, FL, United States

Introduction

Depression affects more than 264 million people worldwide (GBD 2017 Disease and Injury Incidence and Prevalence Collaborators, 2018), making it one of the most prevalent of all medical conditions (Bromet et al., 2011). The sheer prevalence of depression is troubling, in and of itself. It is a chronic, relapsing illness that causes considerable burden from persisting residual deficits in energy level, motivation, cognitive capacity, and other emotional functions necessary for optimum functioning. It thus comes as no surprise that depression is a major cause of psychosocial dysfunction and impairment (Murray and Lopez, 1997; World Health Organization (WHO), 2004). This dysfunction increases in a step-wise manner according to depression severity (Judd et al., 2000). The World Health Organization (WHO) considers depression, broadly-defined, to be the single largest factor contributing to global disability in disability-adjusted life-years (DALYs)—the sum of life-years lost due to premature death and years lived with disability (Smith, 2014).

The high prevalence of depression and its adverse effects on day-to-day functioning cause considerable societal burden. Burden-of-disease studies have consistently documented profoundly high economic costs of depressive disorders (Berto et al., 2000; Broadhead et al., 1990; Greenberg and Birnbaum, 2005; Kessler et al., 2006; Luppa et al., 2007; Olesen et al., 2012). For instance, in 2010, national health care spending in the United States reached nearly $2.6 trillion (or $8383 per person) (Centers for Medicare and Medicaid Services (CMS), 2020); and direct and indirect costs associated with diagnosed major depressive disorder (MDD) accounted for a significant proportion of this expenditure. Over a 5-year span, the total economic burden of MDD among US adults (aged 18–64 years) increased by 21.5%—from $173.2 billion per year in 2005 to $210.5 billion per year in 2010 (Greenberg et al., 2015). The mere presence of clinically significant depressive symptoms, short of formally diagnosed MDD (defined as a Center for Epidemiological Studies-Depression (CES-D) scale total score ≥ 16 (Radloff, 1977)), has been associated with significantly decreased annual salaries and increased unemployment, as compared with persons lacking clinically significant depressive symptoms (Whooley et al., 2002). This finding remained significant, even after adjustment for the effects of sociodemographic variables, initial income, type of employment, and smoking status.

The high societal burden and economic impact of diagnosed depression and subthreshold depressive symptoms raises the possibility of even greater incremental costs associated with treatment-resistant illness. That expectation is somewhat intuitive, given the cumulative costs of pharmaceuticals and other failed interventions in treatment-resistant patients (Crown et al., 2002). However, as reviewed elsewhere in this volume, treatment resistance in depressed patients is consistently associated with greater symptom burden and proneness to relapses (Fekadu et al., 2009), higher rates of hospitalization and utilization of outpatient services (Crown et al., 2002), higher rates of premature mortality (Bergfeld et al., 2018; Fekadu et al., 2009), higher burden from comorbid mental health and general medical conditions (Kornstein and Schneider, 2001), and greater functional impairment and disproportionately higher rates of early workforce exit (Demyttenaere and Van Duppen, 2019; Madsen et al., 2020), compared to those with treatment-responsive depression. Each of these factors brings additional costs to patients, their families, and society. By some estimates, treatment resistance accounts for up to half of

Managing Treatment-Resistant Depression. https://doi.org/10.1016/B978-0-12-824067-0.00003-7

the total cost of depression treatment (Rost et al., 1998)—a disproportionately high percentage of the total cost burden from depression given a 20%–35% prevalence of treatment resistance in depressed patients (Fava and Davidson, 1996; Fekadu et al., 2009; Nemeroff, 2007; Rizvi et al., 2014).

In this chapter, we provide a broad overview of published studies of the cost of illness (COI) associated with treatment-resistant depression, beginning with a general overview of COI study methods. Limitations of existing studies and gaps in the current literature will be highlighted, and future directions in COI research as they pertain to treatment-resistant depression will be discussed.

Economic burden from a cost-of-illness perspective

Overview of the cost-of-illness perspective

As the title implies, this chapter addresses the burden of treatment-resistant depression from a COI perspective. When discussing the economic burden of virtually any illness state, we are concerned with, at minimum, the value (costs and outcomes) of healthcare resources that are produced and consumed. At its simplest level, the goal of COI studies is to identify and measure all such costs that are associated with a particular illness or condition. The endpoint of COI studies is a monetary value—the sum total of the estimated values of all the components of cost burden associated with the given condition within a specified time period. Depending on the size and representativeness of the underlying population studied, COI studies are often thought of as providing estimates of the burden of a given disease to society (Byford et al., 2000). In that sense, having accurate knowledge about COI is necessary for understanding the extent of a given health problem in economic terms—information that is needed to gain the attention of policymakers in order to appropriately direct limited healthcare resources (Knapp and Wong, 2020). Below, we provide a succinct summary of basic methodological considerations for the COI studies. We will review these in greater detail in the context of treatment-resistant depression. Interested readers are directed to several excellent reviews and commentaries for a more comprehensive coverage of COI methodology (Byford et al., 2000; Clabaugh and Ward, 2008; Greenberg and Birnbaum, 2005; Jo, 2014; Larg and Moss, 2011; Luppa et al., 2007; Céilleachair et al., 2013; Onukwugha et al., 2016; Tarricone, 2007).

Types of costs

To derive relatively unbiased estimates of societal burden using a COI approach, all of the relevant costs of a particular condition or disease must be ascertained. In reality, this can seldom be accomplished. Nevertheless, for purposes of our discussion, the relevant costs that may be considered in COI studies can be broken down into three general groups: direct costs, indirect costs, and intangible costs (Jo, 2014). Each is briefly defined below.

Direct costs

Direct costs are incurred when providers diagnose and treat illnesses. These include the costs of hospitalization, outpatient visits, diagnostic tests, pharmaceuticals, psychosocial treatments, preventive care, ancillary services, and other healthcare interventions or services (Clabaugh and Ward, 2008). Direct costs also include those associated with commonly-used nonhealthcare resources such as transportation, informal cares, cost of relocation due to illness, childcare expenses, and related expenditures (Jo, 2014; Oliva-Moreno et al., 2017). Direct costs are usually the easiest of the three types of costs to estimate, given the availability of structured billing information or standardized cost data for most healthcare services, tests, and treatments. Therefore, direct costs are the most common types of costs ascertained in COI studies.

Indirect costs

Indirect costs are those related primarily to work productivity losses due to absenteeism (missed days of work due to illness or injury, disability, or discretionary time), presenteeism (reduced productivity when working), and early mortality. There is far less structured or standardized information available on indirect costs for most illnesses. Therefore, researchers infer or impute indirect costs. For example, Greenberg et al. (2015) counted time off due to disability as a missed full day of work in their depression COI study. An outpatient office visit on a working day was assigned the equivalent of a half-day missed (even if the cohort member's office visit occurred outside of working hours or on breaks), and hospital stays and emergency room visits were counted as a full day of missed work. They multiplied each whole- or half-day missed by the individual's daily wage to estimate indirect costs due to absenteeism. In that same study (Greenberg et al., 2015), researchers assigned a value of 6.1 times the cost of injury/illness-related absenteeism to calculate the presenteeism costs (Stewart et al., 2003). As highlighted by this example, indirect cost estimation is often more of an approximation of the likely "true" value than is the estimation of direct costs.

Intangible costs

Intangible costs involve disease-related pain and suffering or loss of functioning, typically assessed by measuring their adverse impacts on quality of life or related metric. Researchers and other stakeholders typically measure the adverse impacts of illnesses on the quality of life or related metrics to calculate the intangible dimensions illness. Estimating the monetary value of intangible costs constitutes a considerable methodological challenge, given the absence of a well-defined market. Measuring effects on quality of life or conducting extensive surveys that assess respondents' willingness or desire to avoid certain morbidities and what they are willing to pay to do so are often required to derive intangible cost estimates. As such, methods for estimating intangible costs in monetary terms are often more difficult to employ in COI studies than those used for determining direct and indirect costs.

Methodologic considerations in cost-of-illness studies

Data sources

Numerous types of data exist for conducting COI studies. These include patient or caregiver surveys, prospective studies (e.g., prospective cohort studies, clinical trials, etc.), chart review studies, and retrospective cohort studies that utilize large administrative claims databases. Each type of data source has certain advantages and disadvantages. For instance, surveys and prospective studies have the advantage of yielding data that were collected specifically for research purposes, thus providing some assurance of high data quality and completeness. However, statistical power may be limited for some questions due to insufficient enrollment, poor participation, or high attrition; and survey respondents or prospective study participants may not be sufficiently representative of the underlying population of interest.

Retrospective chart review and administrative database studies have the advantages of convenience (due to the lack of need to recruit study cohorts and follow them over time when longitudinal data is required) and high statistical power due to the large sizes of study cohorts that can be created using data that are often already available in a structured form. However, the information in health records or administrative databases was collected for purposes other than for conducting research and are thus subject to misclassification and to limitations related to incomplete or missing information on key study variables (Wang and Kattan, 2020).

Exposure groups: People with the disease of interest and those without

To contextualize COI data and increase its interpretability, disease-associated costs are often compared between at least two groups of individuals—those with the disease of interest, and those without the disease of interest—over a defined time period. In the case of treatment-resistant depression, COI comparisons are often made between people with depression that meet a prespecified definition of treatment resistance and those with depression who do not meet the prespecified definition of treatment resistance. Some studies include a second referent group composed of persons with no evidence of depression during the observation period. Researchers can use such an approach to compare raw (unadjusted) costs. However, persons with treatment-resistant depression are much different than persons with treatment-responsive depression, analogous to how people with a disease of interest and those without the disease may differ on a number of baseline characteristics. This difference in baseline characteristics may systematically influence cost differences between groups. As such, investigators will often attempt to balance exposure groups (or adjust cost estimates) on these other factors using matched designs, multivariable modeling approaches, other types of modeling approaches, or a combination of these.

Endpoints

As mentioned earlier, the "inputs" for COI studies are the types of costs associated with specific exposure groups (e.g., persons with treatment-resistant depression vs persons with treatment-responsive depression, etc.) and the "outputs" (endpoints) for COI studies are the comparative costs associated with each, in monetary terms. The comparative costs of illness are calculated for a defined time period, often expressed as a monetary value per patient per year (PPPY) or per patient per month (PPPM). For each exposure group, costs are typically presented in aggregate (e.g., total direct healthcare costs for patients with treatment-resistant depression vs those same costs for people with treatment-responsive depression). However, cost outputs may be stratified according to the type of cost (direct costs, indirect costs, etc.) or costs in specific subgroups (e.g., behavioral health costs, depression-related costs, or intervention-specific costs such as those related to hospitalization, etc.).

Heterogeneity across cost-of-illness studies

As briefly reviewed above and elsewhere, there are several approaches to estimating the economic burden from MDD and other depressive syndromes (Greenberg and Birnbaum, 2005). COI studies are often descriptive in nature, unlike other types of studies that estimate illness burden using more standardized measures such as years of life lost, years lived with disability, disability-adjusted life-years, and health-related quality of life (GBD 2016 and HALE Collaborators, 2017; National Academy of Sciences, 1998). As of this writing, there is no current consensus on which specific elements or measures should be used to derive COI estimates for depression. As a result, individual COI studies are subject to wide variation (heterogeneity) with regard to the types of data used, the perspective of the study (e.g., patients, families/caregivers, employers, insurance providers, etc.), the types of costs that are calculated (e.g., direct costs, indirect costs, etc.), how calculated costs are stratified (e.g., behavioral health-specific costs, depression-specific costs, costs related to suicide, etc.), the duration of follow-up, the types of depression included (e.g., major depression only, multiple unipolar depressive syndromes, unipolar and bipolar depressive syndromes, etc.), the type(s) of payor(s) from which data on costs of treatments and cares are derived (commercial insurance databases, public insurance databases, prospective datasets, etc.), and other factors (Onukwugha et al., 2016).

Not surprisingly, cost estimates can vary widely for a given condition, given the methodological heterogeneity in COI studies. The heterogeneity in cost estimates for even a single condition raises questions about the reliability and utility of COI studies for guiding practice and policy decisions (Rice, 1994; Tarricone, 2007). This may be particularly true of COI studies for depression—a condition that in-and-of-itself is subject to high levels of interindividual variation in clinical presentation and treatment response (Goldberg, 2011; Lynch et al., 2020). In our opinion, this is perhaps the biggest limitation of COI studies of depression. Despite this limitation, it remains the case that COI studies provide important estimates of the burden of depression and treatment-resistant depression from a societal perspective—both in terms of the magnitude of costs in economic terms and how those costs are distributed.

The cost of treatment-resistant depression

As mentioned earlier, people with treatment-resistant depression have greater morbidity burden, higher levels of psychosocial dysfunction, and higher rates of early mortality than people with treatment-responsive depression. Such disparities raise the possibility of substantially greater incremental COI burden with treatment-resistant depression. Below, we provide a review of the literature describing the comparative costs-of-illness between cohorts of patients with diagnosed (or "likely") treatment-resistant depression and those classified as having treatment-responsive depression.

Systematic reviews

Two published reviews focused on COI studies in people with treatment-resistant depression in adults (Johnston et al., 2019; Mrazek et al., 2014). The first report, written by Mrazek et al. (2014), presented a summary of incremental COI associated with treatment-resistant depression. A total of 62 studies of various designs published between January 1996 and May 2011 were reviewed, including clinical trials (30% of reviewed studies), other prospective studies (30% of reviewed studies), retrospective cohort studies (20% of reviewed studies), and others. Persons classified as having treatment-resistant depression (TRD) had an average of 4.7 unsuccessful courses of antidepressive treatments involving an average of 2.1 separate medication classes. Treatment resistance was defined as failing to respond to one or more therapeutic antidepressant trials of ≥ 6 weeks duration at an appropriate dose. Direct costs included those related to medications, general medical and psychiatric hospitalizations, emergency room visits, and physician visits. Costs were reported in 2012 US dollars. Yearly direct healthcare per-patient costs were $5481 higher for individuals with TRD ($13,196), compared to those who were classified as having treatment-responsive depression ($7715). TRD was also associated with higher indirect costs (TRD, $4048 vs treatment-responsive depression, $2876—a difference of $4048) due to lost work productivity than treatment-responsive depression. The estimated annual added societal cost burden associated with TRD was $29–$48 billion, assuming a 12%–20% prevalence of treatment resistance among depressed patients.

The second report by Johnston et al. (2019) reviewed 12 studies that investigated the incremental direct and indirect COI associated with treatment-resistant depression. Most of the reviewed studies (all published between 2004 and 2014) were retrospective claims data analyses, while two studies each were cohort studies or retrospective chart reviews. One report included cost data that was collected concurrently as part of a clinical trial. Because of inconsistencies in the way that treatment-resistant depression was defined in the individual studies as well as the methodological heterogeneity summarized above, a synthesized estimate of incremental cost burden associated with treatment-resistant depression was not presented. However, across COI studies, increasing levels of treatment resistance or nonresponse to treatment was associated with higher direct and indirect costs.

Individual studies of costs-of-illness related to treatment-resistant depression

In this section, we provide an updated descriptive review of studies that investigated the incremental COI associated with treatment-resistant depression. We searched PubMed/Medline databases (January 1969 through February 2021) using the following search terms: depression, major depressive disorder, MDD, major depression, treatment-resistant, treatment-resistant depression, costs, cost of illness, direct costs, and indirect costs. Bibliographies of individual reports were also searched in order to identify additional references of potential relevance. We included individual studies that focused on comparing estimated COI between: (a) individuals with treatment-resistant depression and those with nontreatment-resistant depression; (b) depressed individuals with varying levels of treatment resistance; (c) various clinical states (e.g., actively ill vs recovered) in people with treatment-resistant depression; and (d) caregivers of individuals with treatment-resistant depression and caregivers of individuals without treatment-resistant depression.

A total of 32 individual reports are included in this review. As was the case with the report by Johnston et al. (2019), heterogeneity between studies with respect to data sources, populations of interest, definitions of treatment-resistant depression, analytic approaches, types of included costs, and other methodological features prevented us from conducting a quantitative synthesis. Therefore, we summarize below the key methodological features of included studies (Table 3.1).

Study characteristics

Sources of data

The characteristics of all 32 COI studies included in this chapter are summarized in Table 3.1. The majority of studies (n = 26) were conducted in North America (Amos et al., 2018; Benson et al., 2020; Cai et al., 2020; Crown et al., 2002; Feldman et al., 2013; Gibson et al., 2010; Greenberg et al., 2004; Ivanova et al., 2010; Khandker et al., 2008; Kubitz et al., 2013; Li et al., 2020; Line et al., 2019; McIntyre et al., 2020; Olchanski et al., 2013; Olfson et al., 2018; Pilon et al., 2019a, b, c, 2020; Russell et al., 2004; Schultz and Joish, 2009; Shrestha et al., 2020; Sussman et al., 2019; Zhdanava et al., 2020, 2021), nearly all of which were retrospective cohort studies that analyzed data from large, automated administrative claims databases. One exception was a cross-sectional survey-based study by Lerner and colleagues (2020) that analyzed costs to employed caregivers of persons with treatment-resistant depression. North American-based administrative claims database studies focused mainly on commercially insured populations (Amos et al., 2018; Cai et al., 2020; Crown et al., 2002; Gibson et al., 2010; Greenberg et al., 2004; Ivanova et al., 2010; Khandker et al., 2008; Kubitz et al., 2013; Li et al., 2020; Olchanski et al., 2018; Pilon et al., 2019b, 2020; Russell et al., 2004; Schultz and Joish, 2009; Shrestha et al., 2020; Zhdanava et al., 2020, 2021). Far fewer studies focused on publicly-insured persons (Benson et al., 2020; Feldman et al., 2013; McIntyre et al., 2020; Pilon et al., 2019a, c) or included a mix of commercially- and publicly-insured individuals (Cai et al., 2020; Lin et al., 2019; Sussman et al., 2019). Studies that were conducted outside of North America included retrospective chart review studies from the United Kingdom (Denee et al., 2020), Brazil (Lepine et al., 2012), and Spain (Sicras-Mainar et al., 2014); two retrospective administrative claims-based cohort studies from Japan (Mahlich et al., 2018) and South Korea (Shin et al., 2020); and one analysis of cost data from a clinical trial (Von Knorring et al., 2006).

TABLE 3.1 Characteristics of cost-of-illness (COI) studies of treatment-resistant depression (TRD).

References	Setting	Study design (data source)	Population	TRD definition	Time frame (study dates)	COI reported	Cost components	Confounding management
Amos et al. (2018)	North America, commercially-insured	Retrospective cohort[a]	MDD, ICD-9 TRD, *n*=6411 Non-TRD, *n*=6411 Non-MDD, *n*=6411	Third AD treatment course[b]	2 years of follow-up (2009–15)	PPPY (2015 $US)	Direct[c,d] and indirect	PS-matching; adjustment (comorbidity, baseline HCC)
Benson et al. (2020)	North America, publicly -insured, aged 65 + years	Retrospective cohort[a]	MDD, ICD-9 TRD, *n*=178 Non-TRD, *n*=178 Non-MDD, *n*=178	Third AD treatment course[b]	1 year of follow-up (2012–15)	PPPY ($US)	Direct[c,e]	Matching (demographics, year); adjustment (demographics, CGI score, comorbidity, baseline HCC)
Cai et al. (2020)	North America, commercially- and publicly-insured	Retrospective cohort[a]	MDD, ICD-9 & ICD-10 TRD, *n*=3317 Non-TRD, *n*=45,123	Third AD treatment course[b]	Variable follow-up duration (2010–15)	Per patient costs during newly observed pharmacologically treated MDD episode (2016 $US)	Direct[c]	Unadjusted
Crown et al. (2002)	North America, commercially-insured	Retrospective cohort[a]	Depression, ICD-9[f,g] Hosp. TRD, *n*=483 Outpt. TRD, *n*=2887 Non-TRD, *n*=7335	Hosp. (switch or augment initial AD)[h] Outpt. (third AD treatment course)[b]	Minimum 9 months of follow-up (1995–2000)	PPPY (2000 $US)	Direct[c,e]	Unadjusted
Denee et al. (2020)	Europe, GP and psychiatry practices	Retrospective chart review (physicians contributed up to 10 patients each)	MDD diagnosed by physician TRD, *n*=295[i]	≥2 AD regimens	2 years of follow-up (2016–18)	PPPM, Health state-specific[i] (£UK)	Direct[c,i]	Unadjusted
Feldman et al. (2013)	North America, publicly-insured	Retrospective cohort[a]	Depression, ICD-9[f] TRD, *n*=4639 Non-TRD, *n*=7524 Gen. pop., *n*>36 million	Algorithm based on number of medication management visits, psychiatric hospitalization, and ECT treatment	2 years of follow-up (2001–09)	PPPY (2010 $US)	Direct[c]	Unadjusted

Gibson et al. (2010)	North America, commercially-insured	Retrospective cohort[a]	Depression, ICD-9[f] TRD, *n*=22,593[j] Non-TRD, *n*=22,593	Algorithm based on MGH scale criteria[j]	1 year of follow-up (2000–07)	PPPY (2006 $US)	Direct[c,d,j]	PS-matching; adjustment (demographics, type of health plan, number of Psychiatric Diagnosis Groups, comorbidity, year of index date)
Greenberg et al. (2004)	North America, commercially-insured, single employer	Retrospective cohort[a]	Depression, ICD-9[f] TRD, *n*=180 Non-TRD, *n*=1512 Non-MDD, 10% random sample of employee population	Algorithm based on medication changes, dose increases, and use of specific treatments (ECT, MAOIs)	1 year of follow-up (1996–98)	PPPY (1998 $US)	Direct[c,e] and indirect	Unadjusted
Ivanova et al. (2010)	N. America, commercially-insured	Retrospective cohort[a]	MDD, ICD-9 TRD, *n*=2312 Non-TRD, *n*=2312	Third AD treatment course[b] and separate algorithm based on AD switching, dose titration, and use of specific treatments (ECT, MAOIs)	2 years of follow-up (2004–07)	PPPY (2007 $US)	Direct[c,e] and indirect	Adjustment (demographics, comorbidities, baseline medication and service use)
Khandker et al. (2008)	N. America, commercially-insured	Retrospective cohort[a]	Depression, ICD-9[k] Total, *n*=48,950[l] Switch, *n*=2378 No switch, *n*=46,572	Switch from an index AD	1 year of follow-up (2002)	PPPY (2002–03 $US)	Direct[c,e]	Adjustment (demographics, payer, provider specialty)
Kubitz et al. (2013)	North America, commercially-insured	Retrospective cohort[a]	Depression, ICD-9[f] TRD, *n*=3314 Non-TRD, *n*=44,520	Third AD treatment course[b]	Per episode of treatment for depression	Cost per episode of treatment for depression ($US)	Direct[c], depression only	Unadjusted
Lepine et al. (2012)	South America, commercially- and publicly-insured, single hospital	Retrospective chart review	MDD, ICD-10 TRD, *n*=90 Non-TRD, *n*=122	Algorithm based on AD switching, dose titration, and use of specific treatments (ECT, MAOIs)	5 years of follow-up (1997–2002)	PPPY (2010 $R and $US)	Direct[c,d]	Unadjusted

Continued

TABLE 3.1 Characteristics of cost-of-illness (COI) studies of treatment-resistant depression (TRD)—cont'd

References	Setting	Study design (data source)	Population	TRD definition	Time frame (study dates)	COI reported	Cost components	Confounding management
Lerner et al. (2020)	North America, employed caregivers	Cross-sectional survey	Caregivers of patients with: TRD, n=169 Health conditions other than TRD, n=1070	Third AD treatment course or one AD plus an augmentation course	–	Cost to caregivers (annualized based on US Bureau of Labor Statistics estimated annual average hourly wage in 2018)	Direct costs of caregiving	Unadjusted
Li et al. (2020)	North America, commercially-insured	Retrospective cohort[a]	Depression, ICD-9[f] TRD, n=2370 Non-TRD, n=9289	Third AD treatment course[b]	1 year of follow-up (2013–14)	PPPY (2017 $US)	Direct[c,d], including direct costs to patients	PS-matching; adjustment (baseline cost)
Lin et al. (2019)	North America, all payers	Retrospective cohort[a]	MDD, ICD-9 & ICD-10 TRD, n=45,127 Non-TRD, n=159,448	Receiving ECT, TMS, VNS, or any AD plus one of four approved adjuncts for TRD (aripiprazole, brexpiprazole, olanzapine, quetiapine)	Minimum 6 months of follow-up (2012–15)	PPPY ($US)	Direct medical[c,e] and suicidal ideation/ attempt-related	PS-matching
Mahlich et al. (2018)	East Asia, commercially-insured	Retrospective cohort[a]	Depression, ICD-10[k] TRD, n=137 Non-TRD, n=1006	Third AD treatment course[b]	Minimum 1 year of follow-up	Total cost per patient and PPPY (¥JPN)	Direct	Unadjusted
McIntyre et al. (2020)	North America, commercially-insured	Retrospective cohort[a]	Depression, ICD-10[f] TRD, n=277 Non-TRD, n=68,861 (1108 matched) Non-MDD, n=1108 (all matched)	Third AD treatment course[b]	2 years of follow-up (2005–17)	PPPY (2016–17 $CAN)	Direct[c,d]	PS-matching, and matching on fiscal year
Olchanski et al. (2013)	North America, commercially-insured	Retrospective cohort[a]	Chronic MDD[l], ICD-9 TRD, n=24,415 Non-TRD, n=58,697	Four or more courses of AD treatment	Per episode of treatment for MDD (2001–09)	Per patient cost, normalized to duration of MDD episode	Direct[c,e]	Unadjusted
Olfson et al. (2018)	North America, publicly-insured	Retrospective cohort[a]	MDD, ICD-9 TRD, n=1503 Non-TRD, n=4298	Third AD treatment course[b]	1 year of follow-up (2008–14)	Per patient cost ($US)	Direct[c,e]	Adjustment (demographics, year, health plan type)

Pilon et al. (2019a)	North America, publicly-insured	Retrospective cohort[a]	Depression, ICD-9 and ICD-10[f] TRD, *n*=3224 Non-TRD, *n*=3224	Third AD treatment course[b]	2 years of follow-up (2010–16)	PPPY (2017 $US)	Direct[c,d]	Adjustment (baseline healthcare costs and CCI score)
Pilon et al. (2019b)	North America, commercially-insured	Retrospective cohort[a]	Depression, ICD-9[f] Mild TRD, *n*=455 Moderate TRD, *n*=2153 Severe TRD, *n*=1455	Third AD treatment course[b,m]	2 years of follow-up (2009–15)	PPPY (2015 $US)	Direct[c,d]	Adjustment (demographics, year, type of healthcare plan, relationship to healthcare plan holder, and comorbidity)
Pilon et al. (2019c)	North America, publicly-insured	Retrospective cohort[a]	Depression, ICD-9 and ICD-10[f] TRD, *n*=14,170 Non-TRD, *n*=40,235 Non-MDD, *n*=223,799	Third AD treatment course[b]	2 years of follow-up (2009–17)	PPPY (2016 $US)	Direct[c,e]	PS-matching; adjustment (for baseline total healthcare costs and comorbidity)
Pilon et al. (2020)	North America, commercially-insured	Retrospective cohort[a]	Depression, ICD-9[f] TRD, *n*=1582 Non-TRD, *n*=1582 Non-MDD, *n*=1582	Third AD treatment course[b]	2 years of follow-up (2011–17)	PPPY (2017 $US)	Direct[c,e]	PS-matching; adjustment (baseline healthcare costs, comorbidity)
Russell et al. (2004)	North America, private pay	Retrospective cohort[a]	Depression, ICD-9[f] TRD, *n*=7377	Third AD treatment course[b] or at least one AD treatment regimen change plus a depression related hospitalization, ECT, or suicide attempt	Costs compared for time periods between the index date and the time of each subsequent depression medication regimen change	Total healthcare expenditures per month (2000 $US)	Direct[c,e]	Unadjusted
Schultz and Joish (2009)	North America, commercially-insured	Retrospective cohort[a]	MDD, ICD-9 Switch, *n*=2931 Augment, *n*=109 Maintain, *n*=4233	Treatment-naïve patients who switched from, augmented, or maintained first-line AD	1 year of follow-up (2002–06)	PPPY costs ($US)	Direct[c,e] and indirect	Adjustment (demographics, payer type, type of insurance plan, comorbidity, baseline psychotropic medication use, baseline healthcare costs)

Continued

TABLE 3.1 Characteristics of cost-of-illness (COI) studies of treatment-resistant depression (TRD)—cont'd

References	Setting	Study design (data source)	Population	TRD definition	Time frame (study dates)	COI reported	Cost components	Confounding management
Shin et al. (2020)	East Asia, publicly-insured	Retrospective cohort[a]	MDD or persistent depressive disorder, ICD-10 TRD, *n*=34,812 Non-TRD, *n*=799,882	Third AD treatment course[b]	1 year of follow-up (2012)	PPPY costs (2012 KRW)	Direct medical[c,e] and suicide-related	Unadjusted
Shrestha et al. (2020)	North America, commercially-insured	Retrospective cohort[a]	MDD, ICD-9 and ICD-10 TRD, *n*=27,595 Non-TRD, *n*=83,949	Third AD treatment course[b]	1 year of follow-up (2006–17)	PPPY (2018 $US)	Direct[c,e]	Adjustment (demographics, type of insurance plan, comorbidity, baseline healthcare costs)
Sicras-Mainar et al. (2014)	Europe, single hospital system	Retrospective chart review	MDD, all treated with venlafaxine, *n*=1115 (cohort members not formally defined as TRD or non-TRD)	Cohort divided into remission, response, and nonresponse groups	1 year of follow-up (2008–10)	PPPY (Euros)	Direct and indirect	Adjusted (demographics, resource utilization, comorbidity)
Sussman et al. (2019)	North America, commercially- and publicly-insured	Retrospective cohort[a]	Depression, ICD-9 & ICD-10[f] TRD, *n*=1112 Non-TRD, *n*=10,734	Third AD treatment course[b]	1 year of follow-up (2008–16)	PPPY (2016 $US)	Direct[c,e]	PS-matching
von Knorring et al. (2006)	Europe, data from randomized trial	Secondary data analysis	MDD, DSM-IV criteria Nonresponder, *n*=344 Responder, *n*=687	Nonresponder to trial of sertraline	2 years of follow-up (1999–2003)	PPPY (20002/2003 SEK)	Direct and indirect	Adjustment (demographics)
Zhdanava et al. (2020)	North America, commercially-insured	Retrospective cohort[a]	Depression, ICD-9 and ICD-10[f] TRD, *n*=2317 Non-TRD, *n*=2317 Non-MDD, *n*=2317	Third AD treatment course[b]	2 years of follow-up (2009–17)	PPPY (2017 $US)	Direct[c,d] and indirect	Exact matching (availability of work loss data); PS-matching
Zhdanava et al. (2021)	North America, commercially-insured	Retrospective cohort[a]	Depression, ICD-9 and ICD-10[f] TRD, *n*=3166 Non-TRD, *n*=3166 Non-MDD, *n*=3166	Third AD treatment course[b]	2 years of follow-up (2009–17)	PPPY (2017 $US)	Direct[c,n] and indirect	Exact matching (availability of work loss data); PS-matching

AD, antidepressant; *$CAN*, Canadian dollars; *CGI*, Clinical Global Impression scale; *DSM-IV*, Diagnostic and Statistical Manual of Mental Disorders, 4th edition; *ECT*, electroconvulsive therapy; *GP*, general (typically primary care) practitioner; *HCC*, healthcare costs; *Hosp. TRD*, hospitalized patients with treatment-resistant depression; *ICD-9*, International Classification of Diseases, 9th Edition; *ICD-10*, International Classification of Diseases, 10th Edition; *¥JPN*, Japanese yen; *KRW*, South Korean won; *MAOI*, monoamine oxidase inhibitor antidepressant; *MDD*, major depressive disorder; *Medicare Suppl.*, Medicare Supplemental database; *Outpt. TRD*, ambulatory (nonhospitalized) patients with treatment-resistant depression; *PPPM*, per patient per month costs; *PPPY*, per patient per year costs; *PS*, propensity score; *TMS*, transcranial magnetic stimulation; *TRD*, treatment-resistant depression; *$R*, Brazilian Reais; *SEK*, Swedish kronor; *£UK*, pounds; *$US*, US dollars; *VNS*, vagal nerve stimulation.

[a] *Data source was a linked administrative claims database.*

[b] *Indicates evidence of inadequate response to two or more prior AD treatment regimens. A course of antidepressant treatment is typically defined as the use (or evidence of use based on claims data) of an antidepressant medication at an approved dose for a predefined minimum period of time (e.g., 6 weeks, or a prescription filled with at least 6 weeks of supply, etc.) before switching to a new antidepressant or using augmentation therapy.*

[c] *Includes total direct medical costs and direct medical costs by category of cost type (e.g., inpatient, outpatient, emergency room, pharmacy, etc.).*

[d] *Includes direct medical costs stratified into mental-health-related and nonmental-health-related categories.*

[e] *Includes direct medical costs stratified into mental-health-related and depression-specific categories.*

[f] *Cohort included persons with (or persons with diagnosed) MDD, persistent depressive disorder, unspecified depression, and adjustment disorders.*

[g] *Cohort include persons with (or persons diagnosed as having) both unipolar and bipolar depression.*

[h] *TRD defined in hospitalized patients as having evidence of switching or augmenting an initial AD course at least once, in addition to having evidence of being hospitalized for depression or a suicide attempt.*

[i] *Costs were calculated for—and were then compared between—three health states that were defined by study physicians: active major depressive episode (moderate to severe symptoms of major depressive disorder after two or more regimens of different ADs), remission (free of depressive symptoms, or only minimal symptomatology, occurring no more than 6 months), and recovery (asymptomatic beyond 6 months). Here, TRD will be defined as being in an active major depressive episode.*

[j] *Definition of TRD was defined using an algorithm based on the Massachusetts General Hospital (MGH) staging method. The algorithm assigned 1 point for each AD trial (defined as two or more fills for the same AD) and a half-point for each optimization strategy, the latter of which included extending treatment with the same AD (at least three fills), increasing the dose, or adding an adjunctive atypical antipsychotic drug, mood stabilizer, or psychostimulant. A total score > 3 was used to define TRD. MGH scores were used to define five groups of depressed persons: 0–3 (referent group), 3.5–4, 4.5–5, 5.5–6, and ≥ 6.5 (higher values approximating more severe treatment resistance).*

[k] *Includes persons (or persons with diagnosed) MDD, persistent depressive disorder, and unspecified depression.*

[l] *Indicated evidence of major depressive episode lasting 2 years of longer, based on claims data.*

[m] *TRD cases were stratified into mild, moderate, and severe based on ICD-9 fifth digit code (excluded if severity status was unspecified).*

[n] *Includes direct medical costs related to co-occurring anxiety and/or substance use disorders.*

Study populations

All administrative claims-based retrospective cohort studies utilized International Classification of Diseases 9th edition (ICD-9) or 10th edition (ICD-10) diagnosis codes to identify individuals with diagnosed MDD (Amos et al., 2018; Benson et al., 2020; Cai et al., 2020; Ivanova et al., 2010; Lin et al., 2019; Olfson et al., 2018; Schultz and Joish, 2009; Shin et al., 2020; Shrestha et al., 2020) or a mixture of unipolar depressive syndromes—typically MDD, persistent depressive disorder (formerly known as dysthymic disorder), unspecified depression (or depressive disorder, not otherwise specified) and, in some cases, adjustment disorders with depressed mood (Crown et al., 2002; Feldman et al., 2013; Gibson et al., 2010; Greenberg et al., 2004; Khandker et al., 2008; Kubitz et al., 2013; Li et al., 2020; Mahlich et al., 2018; McIntyre et al., 2020; Pilon et al., 2019a,b,c, 2020; Russell et al., 2004; Sussman et al., 2019; Zhdanava et al., 2020, 2021). One study included both diagnosed unipolar and bipolar depressive syndromes (Crown et al., 2002). One administrative claims-based study focused on persons with evidence of chronic depression, defined as having evidence of a major depressive episode lasting 2 years or longer (Olchanski et al., 2013). The chart review, clinical trial, and survey studies identified depressed patients based on physician diagnoses (Denee et al., 2020; Lepine et al., 2012; Sicras-Mainar et al., 2014; von Knorring et al., 2006) or self-report from caregivers (Lerner et al., 2020).

Definitions of treatment-resistant depression

Various computerized algorithms were applied to identify individuals with treatment-resistant depression in administrative claims-based retrospective cohort studies. Broadly, most such studies utilized automated prescription claims data to identify courses of antidepressant treatment that included evidence of antidepressant monotherapy at a given dose, adjustments in antidepressant dose, or the use of combination pharmacotherapy. For example, in most cases, when the claims data indicated an alteration in antidepressant therapy (e.g., a change in antidepressants or the addition of a second psychotropic medication commonly used to augment antidepressants) after a prespecified period of time, such occurrences were considered to represent an inadequate response to the original antidepressant. Therefore, changes in antidepressant treatment marked the beginning of a subsequent antidepressant treatment course. Using this approach, the majority of retrospective cohort studies defined treatment-resistant depression as having evidence of a third antidepressant treatment course (i.e., failure of two prior treatment courses) (Amos et al., 2018; Benson et al., 2020; Cai et al., 2020; Crown et al., 2002; Kubitz et al., 2013; Li et al., 2020; Mahlich et al., 2018; McIntyre et al., 2020; Olfson et al., 2018; Pilon et al., 2019a,b,c, 2020; Shin et al., 2020; Shrestha et al., 2020; Sussman et al., 2019; Zhdanava et al., 2020, 2021). Two administrative claims-based studies required only evidence of switching from an index antidepressant (Khandker et al., 2008; Schultz and Joish, 2009), while one such study required evidence of four or more courses of antidepressant treatment to define treatment resistance (Olchanski et al., 2013). Five administrative claims-based retrospective cohort studies used other types of computerized algorithms that were based on the number of antidepressant treatment courses, hospitalizations for mental health reasons, and/or specific interventions usually reserved for treatment-resistant depressed patients (most commonly, electroconvulsive therapy, other types of neuromodulatory treatment, and monoamine oxidase inhibitors) (Feldman et al., 2013; Greenberg et al., 2004; Ivanova et al., 2010; Lin et al., 2019; Russell et al., 2004). One retrospective cohort study used a prescription claims-based algorithm modeled after the Massachusetts General Hospital (MGH) staging method for treatment-resistant depression (Fava, 2003), wherein a certain number of points was assigned for each course of antidepressant treatment or antidepressant optimization strategy (such as raising the daily dose or using an augmenting agent) (Gibson et al., 2010).

Treatment resistance was defined in chart review studies as having evidence of at least two prior courses of antidepressant treatment failure (Denee et al., 2020), or evidence of multiple alterations in antidepressant treatment or use of electroconvulsive therapy or monoamine oxidase inhibitors (Lepine et al., 2012). In a cross-sectional caregiver survey, treatment resistance was defined on the basis of a self-reported third antidepressant treatment course or a single antidepressant treatment course plus augmentation therapy (Lerner et al., 2020). In one prospective study (von Knorring et al., 2006), TRD was defined as failure to respond adequately to sertraline. One retrospective chart review study divided a cohort of venlafaxine-treated persons with MDD into remission, response, and nonresponse groups without explicitly defining cohort members as being treatment-resistant or nontreatment-resistant (Sicras-Mainar et al., 2014).

Costs and cost components

Nearly all papers reported per-patient-per-year (PPPY) or per-patient-per month (PPPM) direct medical costs. Direct costs included total medical costs and costs within categories defined by service type (e.g., hospitalization, outpatient visits, emergency room care, drug costs, etc.). Several papers also reported costs stratified into behavioral health-related costs (Amos et al., 2018; Gibson et al., 2010; Lepine et al., 2012; Li et al., 2020; McIntyre et al., 2020; Zhdanava et al., 2020, 2021) and depression-specific costs (Benson et al., 2020; Crown et al., 2002; Greenberg et al., 2004; Ivanova et al., 2010;

Khandker et al., 2008; Lin et al., 2019; Olchanski et al., 2013; Olfson et al., 2018; Pilon et al., 2019a,b,c, 2020; Russell et al., 2004; Schultz and Joish, 2009; Shin et al., 2020; Shrestha et al., 2020; Sussman et al., 2019). Far fewer studies reported both direct and indirect costs within a single cohort of individuals (Amos et al., 2018; Greenberg et al., 2004; Ivanova et al., 2010; Schultz and Joish, 2009; Sicras-Mainar et al., 2014; von Knorring et al., 2006; Zhdanava et al., 2020, 2021). One study focused on direct costs of caregiving (Lerner et al., 2020).

Analytic approach

As shown in Table 3.1, 13 of the reviewed papers presented unadjusted results. The remainder of studies deployed a variety of methods for the management of confounding factors. In general, these consisted of direct matching or propensity score (PS)-matching (Lin et al., 2019; McIntyre et al., 2020; Susssman et al., 2019), statistical adjustment (Ivanova et al., 2010; Khandker et al., 2008; Olfson et al., 2018; Pilon et al., 2019a, b; Schultz and Joish, 2009; Shrestha et al., 2020; Sicras-Mainar et al., 2014; Von Knorring et al., 2006), or a combination of approaches to address residual confounding (Amos et al., 2018; Benson et al., 2020; Gibson et al., 2010; Li et al., 2020; Pilon et al., 2019c, 2020; Zhdanava et al., 2020, 2021). For most PS-matched analyses, the PS was calculated as the conditional probability of being classified as a treatment-resistant case, given a variety of covariates including sociodemographic characteristics, calendar or fiscal year, healthcare plan type, various measures of comorbidity from co-occurring illnesses, and other factors. Similar variables were used for adjusting statistical models, many of which also included adjustment for baseline levels of service use or estimated health-care costs before the beginning of follow-up in longitudinal studies.

In administrative claims-based retrospective cohort studies, an index date was assigned that defined the beginning of follow-up. In most cases, the index date was the date of a first qualifying antidepressant prescription (Amos et al., 2018; Benson et al., 2020; Crown et al., 2002; Ivanova et al., 2010; Khandker et al., 2008; Li et al., 2020; Mahlich et al., 2018; Olfson et al., 2018; Pilon et al., 2019a,b,c, 2020; Schultz and Joish, 2009; Shrestha et al., 2020; Sussman et al., 2019; Zhdanava et al., 2020, 2021). For other studies, the index date or time period was defined on the basis of a qualifying depression diagnosis (Cai et al., 2020; Kubitz et al., 2013; Mahlich et al., 2018; McIntyre et al., 2020; Olchanski et al., 2013; Russell et al., 2004; Shin et al., 2020), a prespecified time period following an initial antidepressant prescription fill (Gibson et al., 2010), or a specific event, such as a hospital admission, hospital discharge, or surgical procedure (Crown et al., 2002; Feldman et al., 2013; Lin et al., 2019). For most administrative claims-based studies, the postindex period constituted the time window from which information on health care services and other outcome data was collected—information that was subsequently used to estimate the associated costs. A specified time preceding the index date was used to provide information on the "baseline" characteristics of cohort members. In many cases, information on "baseline" characteristics was used to create matched cohorts, calculate propensity scores, and/or adjust cost estimates.

Treatment-resistant vs nontreatment-resistant depression

Direct healthcare costs

The results of studies assessing the direct medical costs of treatment-resistant depression, relative to treatment-responsive depression or no depression, are summarized in Table 3.2. In all studies, direct health care costs were numerically higher in treatment-resistant depression cohorts than in nontreatment-resistant cohorts—a trend that applied to both total costs and costs that were stratified into behavioral-health-related and depression-specific direct costs. Total and stratified direct COI were higher for both treatment-resistant and treatment-responsive depressed groups than nondepressed persons for the limited number of studies that included nondepressed controls.

The estimated direct costs of illness varied substantially between studies. Therefore, relative COI comparisons between treatment-resistant and nontreatment-resistant cohorts are provided in Table 3.2, which may be regarded as an estimate of the relative incremental cost of treatment resistance in treated depressed patients. The total direct costs in the treatment-resistant depression cohorts were about 1.1 times (11%) to 2.6 times (164%) higher than the total direct costs in the nontreatment-resistant cohorts, with the exception of one prospective study that documented total direct costs in patients with treated-resistant depression that were 5.2 times that of nontreatment-resistant patients (Von Knorring et al., 2006). Across most studies, behavioral-health-related (nine studies) and depression-specific (nine studies) direct costs were about 1.4 times (40%) to 3.3 times (230%) higher in treatment-resistant depression cohorts than in nontreatment-resistant cohorts.

Indirect costs

A total of six studies compared indirect costs between treatment-resistant depression and nontreatment-resistant (or treatment-responsive) depression cohorts (Table 3.3). As was the case with direct costs, indirect costs were numerically higher in treatment-resistant depression cohorts than in nontreatment-resistant cohorts. Indirect costs were 1.7 times (70%)

TABLE 3.2 Summary of results from cost-of-illness (COI) studies assessing the direct medical costs of treatment-resistant depression (TRD), including total costs and behavioral-health (BH)-related and depression (Dep.)-specific categories.

References	Direct cost						COI comparison (TRD vs non-TRD)			
	TRD		Non-TRD		Nondepressed		Ratio		% Difference	
	Group	Cost	Group	Cost	Group	Cost	Group	Ratio	Group	Difference (%)
Amos et al. (2018)	Total BH-related	$17,261 $5325	Total BH-related	$9790 $2019	Total BH-related	$4782 $196	Total BH-related	1.76 2.64	Total BH-related	76.3 163.7
Benson et al. (2020)	Total BH-related Dep.-specific	$24,543 $8553 $3578	Total BH-related Dep.-specific	$16,059 $2762 $1073	Total BH-related	$8058 $124	Total BH-related Dep.-specific	1.53 3.10 3.33	Total BH-related Dep.-specific	52.8 209.7 233.5
Cai et al. (2020)	Total	$19,626[a]	Total	$7440[a]	–	–	Total	2.64	Total	163.8
Feldman et al. (2013)	Total	$9116[b]	Total	$8185[b]	–	–	Total	1.11	Total	11.4
Greenberg et al. (2004)	Total	$8368	Total	$3571	Total	$2359	Total	2.34	Total	134.3
Ivanova et al. (2010)	Total BH-related	$20,754 $13,349	Total BH-related	$4663 $1876	–	–	Total BH-related	1.55 2.49	Total BH-related	55.4 148.6
Khandker et al. (2008)	Total Dep.-specific	$6836[c] $4968[c]	Total Dep.-specific	$1225[c] $562[c]	–	–	Total Dep.-specific	1.38 2.18	Total Dep.-specific	37.6 118.0
Lepine et al. (2012)	Total	$5520	Total	$3042	–	–	Total	1.81	Total	81.5
Li et al. (2020)	Total	$12,726[d]	Total	$8881[d]	–	–	Total	1.43	Total	43.3
Lin et al. (2019)	Total Dep.-specific	$8681[e] $6632[e]	Total Dep.-specific	$2314[e] $1608[e]	–	–	Total Dep.-specific	1.31 1.44	Total Dep.-specific	30.9 43.9
Mahlich et al. (2018)	Total	¥1.01M	Total	¥0.64M	–	–	Total	1.57	Total	57.1
McIntyre et al. (2020)	Total	$20,998	Total	$17,951	–	–	Total	1.17	Total	17.0
Olchanski et al. (2013)	Total	$85,937[f]	Total	$49,920[f]	–	–	Total	1.72	Total	72.1
Olfson et al. (2018)	Total BH-related Dep.-specific	$18,982 $5703 $3112	Total BH-related Dep.-specific	$11,642 $2999 $1910	–	–	Total BH-related Dep.-specific	1.63 1.90 1.63	Total BH-related Dep.-specific	63.0 90.1 62.9

Pilon et al. (2019a)	Total BH-related Dep.-specific	$25,059 $7698 $5106	Total BH-related Dep.-specific	$19,945 $5437 $3690	Total BH-related Dep.-specific	$14,410 $1991 $424	Total BH-related Dep.-specific	1.26 1.42 1.38	Total BH-related Dep.-specific	25.6 41.6 38.4
Pilon et al. (2019c)	Total BH-related Dep.-specific	$16,654 $2861 $1224	Total BH-related Dep.-specific	$12,133 $1644 $696	Total BH-related Dep.-specific	$5812 $495 $59	Total BH-related Dep.-specific	1.37 1.74 1.76	Total BH-related Dep.-specific	37.3 74.0 75.9
Pilon et al. (2020)	Total BH-related Dep.-specific	$25,807 $3673 $1810	Total BH-related Dep.-specific	$13,701 $1496 $830	Total BH-related Dep.-specific	$8500 $1251 $72	Total BH-related Dep.-specific	1.88 2.46 2.18	Total BH-related Dep.-specific	88.4 145.5 118.1
Shin et al. (2020)	Total Dep.-specific	6610.5 KRW 3461.6 KRW	–	–	Total Dep.-specific	1273.0 KRW 697.7 KRW	Total Dep.-specific	5.19 2.15	Total Dep.-specific	419.3 115.0
Sicras-Mainar et al. (2014)	Total	$2276[g]	Total	$1738[g]	–	–	Total	1.31	Total	31.0
Sussman et al. (2019)	Total Dep.-specific	$9890 $2740	–	–	Total Dep.-specific	$6848 $1322	Total Dep.-specific	1.44 2.07	Total Dep.-specific	44.4 107.3
von Knorring et al. (2006)	Total Dep.-specific	491 kr 437 kr	–	–	Total Dep.-specific	299 kr 254 kr	Total Dep.-specific	1.64 1.72	Total Dep.-specific	64.2 72.0
Zhdanava et al. (2020)	Total BH-related	$22,541 $4270	Total BH-related	$17,450 $1458	Total BH-related	$10,047 $69	Total BH-related	1.29 2.93	Total BH-related	29.2 192.9
Zhdanava et al. (2021)	Total BH-related	$16,674 $6493[h]	Total BH-related	$10,945 $1834[h]	Total BH-related	$6493 $522[h]	Total BH-related	1.52 1.46	Total BH-related	52.3 45.6

All cost values are per patient per year (PPPY) costs, unless otherwise specified.

[a] *Per patient costs during a newly observed pharmacologically-treated MDD episode are presented. The mean(SD) duration of a newly-observed MDD episode was 571±285 days in the TRD group and 200±198 days in the non-TRD group. Mean per patient per month (PPPM) costs were comparable between TRD and non-TRD groups ($1031 vs $1115).*

[b] *Data shown only for persons with no negative events during follow-up, defined as no psychiatric hospitalization, emergency room use, other hospitalization, electroconvulsive therapy, or diagnoses for poisoning or self-injury). Total costs for individuals in the subgroup of patients with negative events were higher for those with TRD than non-TRD ($38,807 vs $28,948, ratio 1.34, % difference 34.1%).*

[c] *Data shown only persons for whom the index antidepressant was a selective serotonin reuptake inhibitor.*

[d] *Data shown only for total healthcare costs in the first year of follow-up, including the sum of costs to payers and to patients.*

[e] *Data are hospital costs for matched samples of cohort members meeting operational definition of TRD and those classified as having non-TRD.*

[f] *Per patient cost, normalized to the duration of the depressive episode.*

[g] *This study included persons who received treatment with venlafaxine. Cohort members were divided into three groups: those who did not receive concomitant treatment with medications that were substrates or inhibitors of CYP2D6, those who received concomitant CYP2D6 substrates (venlafaxine/substrate group), and those who received concomitant CYP2D6 inhibitors (venlafaxine/inhibitor group). Data shown in the table are only for the subgroup of cohort members without CYP2D6 substrates or inhibitors. The same general trend in total costs also held for the venlafaxine/inhibitor and venlafaxine/substrate groups.*

h Behavioral health-related costs are for cohort members with a diagnosed anxiety or substance use disorder.

TABLE 3.3 Summary of results from cost-of-illness (COI) studies assessing the indirect medical costs of treatment-resistant depression (TRD).

	Indirect cost			COI comparison (TRD vs non-TRD)		COI comparison in people with TRD (direct vs indirect)[a]	
References	TRD	Non-TRD	Nondepressed	Ratio	% Difference	Ratio	% Difference
Amos et al. (2018)	$4576	$2686	$906	1.70	70.4	3.77	277.2
Greenberg et al. (2004)	$6622	$3094	$1684	2.14	114.0	1.26	26.4
Ivanova et al. (2010)	$11,793	$7610	–	1.55	55.0	1.76	76.0
Sicras-Mainar et al. (2014)	$978	$454	–	2.15	115.4	2.33	132.7
von Knorring et al. (2006)	437 kr	254 kr	–	1.72	72.0	1.12	12.4
Zhdanava et al. (2020)	$15,337	$6661	$5014	2.30	130.3	1.47	47.0

All cost values are per patient per year (PPPY) costs, unless otherwise specified.

[a] *Data on direct costs for comparisons between direct and indirect costs in people with TRD may be found in Table 3.2.*

to 2.3 times (130%) higher in treatment-resistant depression cohorts than in nontreatment-resistant cohorts. Each of these studies included both indirect and direct cost estimates. For treatment-resistant depression across all six studies, direct costs were about 1.1 times (12%) to 3.8 times (277%) higher than indirect costs.

Costs associated with varying levels of treatment resistance or depression severity

Four administrative claims-based retrospective cohort studies investigated the costs associated with varying levels of treatment resistance or estimated depression severity within treatment-resistant cohorts.

Increasing levels of treatment resistance

Gibson et al. (2010) applied a claims-based algorithm modeled after the Massachusetts General Hospital (MGH) clinical staging approach to stratify cohort members into increasing levels of treatment resistance (Fava, 2003). The adapted MGH staging approach assigned each cohort member a certain number of points for each adequate therapeutic trial of antidepressants (1 point each), each augmentation or dose-optimization strategy (a half-point each), and use of ECT (3 points). A score of 3 points or more defined treatment resistance, broadly. Defined this way, treatment-resistant depression was associated with a 40% higher cost of medical care. Cohort members were further classified into treatment resistance subgroups at five levels, with higher score ranges indicating greater levels of treatment resistance: 0–3, 3.5–4.0, 4.5–5.0, 5.5–6.0, and 6.5+. There was a linear relationship between MGH score group and annual direct medical costs: Compared to cohort members with MGH-scores below 3.5, persons with MGH scores of 3.5–4.0, 4.5–5.0, 5.5–6.0, and >6 had medical care costs that were 21.4%, 27.0%, 39.7%, and 48.9% higher, respectively (P<.001 for all comparisons). When MGH scores were analyzed as a continuous variable, each 1-unit increase in score was associated with a 7.9% increase in direct medical costs.

A similar study by Russell et al. (2004) used an administrative claims-based algorithm to define treatment-resistant depressed patients (two or more courses of antidepressant treatment) and substratify them on the total number of changes in their depression medication regimens. Each member of the study cohort had a minimum of two and a maximum of eight depression medication regimen changes. Mean monthly total direct healthcare expenditures were calculated and compared for time windows beginning at the date on which cohort members met criteria for treatment-resistant depression and ending at the date of each ensuing depression regimen change. Total monthly healthcare costs increased by 104% from the second to eighth change in antidepressant regimen—from $571/month to $1165/month. The mean total monthly healthcare costs increased in a linear fashion with increasing number of antidepressant medication regimen changes.

Higher levels of depression severity

Two studies indirectly investigated the effects of higher levels of depression severity on healthcare costs among persons with evidence of treatment-resistant depression. Crown et al. (2002) investigated the impact of treatment-resistant depression on healthcare utilization and costs in a cohort of 3370 individuals with treatment-resistant depression and 7335 individuals who are selected at random with diagnosed depression but who did not meet criteria for treatment resistance. Depression-related hospitalization of hospitalization with evidence of attempted suicide served as a proxy for high depression severity. For the study, a claims-based algorithm was used to define hospitalized (n=483) and outpatient (n=2887) treatment-resistant cases. Hospitalized treatment-resistant patients had evidence of switching from or augmenting an initial antidepressant medication at least once with a depression-related hospitalization or an ICD-9–coded suicide attempt. Outpatient treatment-resistant patients had evidence of switching from or augmenting an initial antidepressant medication at least twice but without being hospitalized. Not surprisingly, total direct annualized healthcare costs were significantly higher in the hospitalized treatment-resistant cohort ($42,344) than the outpatient treatment-resistant cohort ($10,241) or the nontreatment-resistant control group ($6512). Similar trends were observed for total depression-related costs and general medical costs. Total pharmaceutical costs were nearly identical for the hospitalized treatment-resistant and outpatient treatment-resistant groups ($3251 vs $3513); and were higher for both treatment-resistant groups than the nontreatment-resistant group ($1713).

Pilon et al. (2019b) identified a cohort of 6411 individuals with evidence of treatment-resistant depression and divided them into subgroups on based on ICD-9 fifth digit codes corresponding to mild (n=455), moderate (n=2153), and severe depression (n=1455). Treatment resistance was defined as having evidence of failing to respond to two or more antidepressant treatment courses, including augmentation. Mean PPPY total healthcare costs were $18,911, $16,885, and $12,123, respectively, for severe, moderate, and mild depressed treatment-resistant cohorts.

State-specific costs in patients with treatment-resistant depression

One retrospective chart review study investigated healthcare costs associated with active depression, as compared with remitted or recovered states, in people with diagnosed treatment-resistant depression (Denee et al., 2020). Physicians were recruited and were asked to identify up to 10 eligible patients with treatment-resistant depression. A total of 295 patients with treatment-resistant depression were identified and study data were abstracted from their medical records. Treatment resistance was defined on the basis of having failed to respond to at least two separate therapeutic trials of antidepressants, adequately dosed and of sufficient duration, during a current depressive episode. Study physicians were asked to provide the dates for changes in healthcare state, using prespecified definitions, for each subject. An actively symptomatic state was defined as having moderate or severe symptoms. Remission was defined as having no symptoms of depression, or only minimal symptoms, for up to 6 months. Recovery was defined as being depression symptom-free for greater than 6 months. Information on healthcare utilization included primary and secondary care visits, use of crisis services, emergency room visits, hospitalizations, medications, and nonpharmacological interventions such as electroconvulsive therapy, transcranial magnetic stimulation, and psychotherapy. Associated costs were calculated for each of the three specific healthcare states. Per-patient-per-month (PPPM) direct costs were highest in the actively symptomatic state (£992), followed by remission (£254) and recovery (£99).

Caregiver cost burden

One study focused on workplace costs for caregivers of people with treatment-resistant depression, as assessed by a cross-sectional survey (Lerner et al., 2020). A total of 169 respondents who were employed caregivers of persons with treatment-resistant depression were identified. Treatment resistance was defined on the basis of a care recipient having taken two or more prescribed antidepressants at an adequate dose and for an adequate duration of time, or having taken one antidepressant with an augmentation medicine. Survey items that accounted for direct costs of caregiving included the amount of time spent in a caregiving role and out-of-pocket caregiving expenses. The latter included costs of home health care, housekeeping, baby-sitting, food or meal delivery, grounds keeping, transportation, employment or vocational training, financial management, legal services, and other family or household services. The comparator group consisted of 1070 survey respondents who were caregivers for persons without treatment-resistant depression. The total unadjusted direct caregiving cost for treatment-resistant depression was higher than for no treatment-resistant depression for service payments ($2759 vs $1282) and for caregiving hours ($30,308 vs $18,953). The proportion of treatment-resistant depression caregivers with out-of-pocket expenses for caregiving services was twice that of the no treatment-resistant depression caregivers. Unadjusted indirect costs of caregiving were also higher in the treatment-resistant depression caregiver group than the nontreatment-resistant depression caregiver group ($10,098 vs $7959).

Discussion

The COI studies reviewed in this chapter converge on a number of important themes. All reviewed studies documented higher incremental costs in treatment-resistant depressed persons than individuals classified as having treatment-responsive depression. This trend was maintained for both direct and indirect costs, in specific cost subgroups (such as behavioral health-related and depression-specific costs), and across a variety of patient populations, payer types, and definitions of depression and treatment resistance. In this chapter, we highlighted relative cost differences between treatment-resistant and treatment-responsive depression, not absolute differences in cost, given the wide variation in methodological approaches and the consequent variation in cost estimates between studies. Regardless, the cost differences between treatment-resistant and treatment-responsive cohorts in nearly all of the reviewed studies were striking. The importance of these observed differences in direct and indirect costs between treatment-resistant and treatment-responsive depression rests in the high prevalence of depressed patients that may be expected meet the most commonly-deployed definition of treatment resistance in the reviewed studies—having failed to respond adequately to two or more courses of antidepressant treatment.

Taken together, the findings from the reviewed studies in this chapter make a compelling case for a significantly greater investment in developing new and more effective approaches to the management of depressed patients who respond poorly to first- and second-line antidepressants. As reviewed elsewhere in this volume, decades of preclinical and clinical research have provided a number of therapeutic options for treatment-resistant patients, and laudable efforts are presently underway to develop new and more rapidly-acting treatments and to individualize or make best use of the therapeutic options that already exist. Much hope rests in these ongoing research efforts to not only relieve suffering at the individual level, but also alleviate the substantial cost burden of treatment-resistant depression. Novel treatments, enhanced therapeutic approaches, and more sophisticated care delivery models will incur their own costs. Given the staggering costs of treatment-resistant depression and its chronic and unrelenting course for many patients, treatments that are very costly (such as novel pharmacological interventions or noninvasive or invasive neuromodulation strategies) may be proven to be cost-beneficial if they can provide sustained benefit and functional recovery. Still, in view of the limited resources available to combat treatment resistance in depressed patients, detailed cost-outcome analyses for these advances will be needed.

The brief review of COI studies presented in this chapter is subject to a number of limitations. This chapter was written from the COI perspective, which considers only the explicit costs that are associated with the utilization or consumption of resources or the opportunity costs associated with work absences or lower-quality work due to treatment-resistant depression. The COI approach does not generally consider the aspects of treatment-resistant depression that are more difficult to monetize, such as burdens related to poor quality of life, adverse social outcomes, other morbidities, and early mortality. There is a paucity of studies investigating the costs to families and to caregivers of people with treatment-resistant depression. As such, only a relatively narrow picture of depression's "true" cost can be captured. Additional perspectives (beyond the economic burden to society) and approaches (such as measuring quality of life, years lived with disability, disability-adjusted life-years, and others) must be integrated with COI data in order to appropriately broaden our understanding of the added burden of treatment resistance in depressed patients (Knapp and Wong, 2020; Onukwugha et al., 2016).

The COI studies reviewed in this chapter were diverse in terms of methodological approaches and analytic methods, as reviewed above, and had a reasonably broad representation across different payer types—both commercial and public. But our review focused on studies in adult populations, and nearly all of the published studies were from North America. There were strikingly few studies from Europe, Asia, and Latin America, and none from other parts of the world. In addition, important subgroups of patients could not be "pulled out" from population-level datasets, which prevented COI comparisons across strata defined by depressive subtypes, levels of depression severity, and differing demographic characteristics, including socioeconomic status. These are important limitations because certain types of depressive syndromes (e.g., anxious depression, etc.), particular comorbidities (personality disorders, substance use disorders, etc.), and low socioeconomic status are predictors of poor response to treatment with antidepressants (Bagby et al., 2002; Fava et al., 2008; Jain et al., 2013; Kennedy, 2008; Reich, 2003; Wu et al., 2013). Although four studies at least attempted to use proxy measures of severity—the number of antidepressant treatment courses (Gibson et al., 2010; Russell et al., 2004), psychiatric hospitalization (Crown et al., 2002), and specific diagnosis codes corresponding to mild, moderate, and severe illness (Pilon et al., 2019b)—these served as only indirect measures at best, without validation against a clinical gold standard.

An additional limitation is the lack of a standardized approach for conducting COI studies and a lack of standardization regarding the definition of treatment-resistant depression (Jo, 2014; Trevino et al., 2014). On the one hand, lack of standardization is a clear disadvantage inasmuch as drawing meaningful comparisons between individual studies is made more difficult to accomplish, and the ability to translate the findings from individual studies to other populations or clinical contexts is profoundly limited. These factors negatively impact the utility of COI studies for guiding resource allocation and call for the development of common data elements and definitions, particularly within electronic health records, administrative claims databases, and other large-scale data sources (Kilbourne et al., 2018). Despite the lack of standardized study methodology and definitions of treatment resistance that may have resulted in variable absolute cost estimates and differing absolute cost differences, TRD was still consistently associated with higher total, direct, and indirect costs than treatment-responsive depression, and the relative cost differences between the two were also remarkably consistent.

The COI approaches in the reviewed studies did not consider the effects of beneficial interventions on direct or indirect costs of TRD or estimates of cost savings that might be expected if successful treatment is initiated. Indeed, COI studies of treatment-resistant depression did not typically include estimates of the utility of resources that are used for treatment or disease management. Specifically, none of the reviewed studies provided comparative costs or cost-utility estimates between alternative treatment approaches. These limitations further detract from the ability to represent the true cost of treatment-resistant depression and prevent the examination of how limited resources may be best utilized from an economic perspective.

And finally, the majority of reviewed studies relied on retrospective data sources such as health records and information contained in large administrative claims databases. These retrospective data sources enable the canvassing of large groups or populations of patients and contain longitudinal records (health records or claims) that permit the creation of cohorts of patients with evidence of diagnosed depression and treatment-resistant illness. Diagnoses and presumed indications for treatment are based on records of medical care and prescribed treatments. However, such records can only ascertain costs related to patients who come to clinical care. This is less of a limitation when studying treatment-resistant depression since the presence of treatment resistance, however defined, is conditional on having received medical care. Moreover, it would make little sense from an economic or policy perspective to compare the costs of treatment-resistant depression with the costs of depression that is undiagnosed or untreated. However, an additional limitation that is more difficult to overcome is that retrospective data sources contain information that was collected for purposes outside of conducting research and are thus subject to error (due to inaccurate diagnostic or coding practices or noncompliance with prescribed treatments). Changes in antidepressant treatment course are presumed in such studies to reflect poor treatment response, which is a significant limitation given that clinicians alter antidepressant treatment for a number of additional reasons, such as addressing adverse effects, poor adherence, tachyphylaxis, or potential drug-drug interactions. Indeed, as many as 60% of depressed patients initially classified as being treatment-resistant may be "pseudo-resistant," i.e., only apparently poorly-responsive to antidepressants due to misdiagnosis, inadequate treatment, or other factors (Trevino et al., 2014). Absent the ability to verify the reason(s) for altering a course of antidepressant treatment and, in most cases, to also ascertain failures to respond to psychosocial treatments (such as evidence-supported psychotherapy), it is very difficult to estimate the cost differences between treatment-resistant and treatment-responsive depressed cohorts that are truly attributable to poor treatment response.

Conclusion

Despite the limitations above, COI studies of treatment-resistant depression consistently provide evidence of an exorbitant cost burden associated with treatment-resistant depression. Based on our review, the total direct and indirect costs of treatment-resistant depression are likely to be 1.5–3 times higher than for treatment-responsive depression. The cost burden in depressed patients appears to increase with increasing levels of treatment resistance and is higher for more severely-ill than less-symptomatic treatment-resistant depressed patients. These studies provide an increasingly clear picture of the societal cost of treatment resistance in depressed patients that come to clinical attention despite wide variation in study methodologies, data sources, payer types, and definitions of treatment resistance. Limited evidence suggests that costs to caregivers for people with treatment-resistant depression may also be conspicuously high. No single COI study can provide a comprehensive picture of the "true" cost of treatment-resistant depression. Therefore, COI studies as a whole must be interpreted in the light of findings from different populations, and from studies that used alternative approaches to the investigation of illness burden. In particular, more investigations of COI associated with treatment-resistant depression are needed from patient populations outside of North America. As researchers develop and test new approaches for managing TRD, comparative cost-outcome studies will become increasingly necessary given the high cost of TRD and the limited resources available for combating the problem.

References

Amos, T.B., Tandon, N., Lefebvre, P., Pilon, D., Kamstra, R.L., Pivneva, I., Greenberg, P.E., 2018. Direct and indirect cost burden and change of employment status in treatment-resistant depression: a matched-cohort study using a US commercial claims database. J. Clin. Psychiatry 79 (2). https://doi.org/10.4088/JCP.17m11725.

Bagby, R.M., Ryder, A.G., Cristi, C., 2002. Psychosocial and clinical predictors of response to pharmacotherapy for depression. J. Psychiatry Neurosci. 27 (4), 250–257.

Benson, C., Szukis, H., Sheehan, J.J., Alphs, L., Yuce, H., 2020. An evaluation of the clinical and economic burden among older adult medicare-covered beneficiaries with treatment-resistant depression. Am. J. Geriatr. Psychiatry 28 (3), 350–362. https://doi.org/10.1016/j.jagp.2019.10.012.

Bergfeld, I.O., Mantione, M., Figee, M., Schuurman, P.R., Lok, A., Denys, D., 2018. Treatment-resistant depression and suicidality. J. Affect. Disord. 235, 362–367. https://doi.org/10.1016/j.jad.2018.04.016.

Berto, P., D'Ilario, D., Ruffo, P., Di Virgilio, R., Rizzo, F., 2000. Depression: cost-of-illness studies in the international literature, a review. J. Ment. Health Policy Econ., 3–10. https://doi.org/10.1002/1099-176x(200003)3:1<3::aid-mhp68>3.0.co;2-h.

Broadhead, W.E., Blazer, D.G., George, L.K., Tse, C.K., 1990. Depression, disability days, and days lost from work in a prospective epidemiologic survey. JAMA 264 (19), 2524–2528. https://doi.org/10.1001/jama.1990.03450190056028.

Bromet, E., Andrade, L.H., Hwang, I., Sampson, N.A., Alonso, J., de Girolamo, G., de Graaf, R., Demyttenaere, K., Hu, C., Iwata, N., Karam, A.N., Kaur, J., Kostyuchenko, S., Lépine, J.P., Levinson, D., Matschinger, H., Mora, M.E.M., Browne, M.O., Posada-Villa, J., et al., 2011. Cross-national epidemiology of DSM-IV major depressive episode. BMC Med. 9. https://doi.org/10.1186/1741-7015-9-90.

Byford, S., Torgerson, D.J., Raftery, J., 2000. Cost of illness studies. BMJ 320 (7245), 1335. https://doi.org/10.1136/bmj.320.7245.1335.

Cai, Q., Sheehan, J.J., Wu, B., Alphs, L., Connolly, N., Benson, C., 2020. Descriptive analysis of the economic burden of treatment resistance in a major depressive episode. Curr. Med. Res. Opin. 36 (2), 329–335. https://doi.org/10.1080/03007995.2019.1671087.

Céilleachair, A.J.Ó., Hanly, P., Skally, M., O'Neill, C., Fitzpatrick, P., Kapur, K., Staines, A., Sharp, L., 2013. Cost comparisons and methodological heterogeneity in cost-of-illness studies: the example of colorectal cancer. Med. Care 51 (4), 339–350. https://doi.org/10.1097/MLR.0b013e3182726c13.

Centers for Medicare & Medicaid Services (CMS), 2020. National Health Expenditure Fact Sheet. https://www.cms.gov/Research-Statistics-Data-and-Systems/Statistics-Trends-and-Reports/NationalHealthExpendData/NHE-Fact-Sheet.

Clabaugh, G., Ward, M.M., 2008. Cost-of-illness studies in the United States: a systematic review of methodologies used for direct cost. Value Health 11 (1), 13–21. https://doi.org/10.1111/j.1524-4733.2007.00210.x.

Crown, W.H., Finkelstein, S., Berndt, E.R., Ling, D., Poret, A.W., Rush, A.J., Russell, J.M., 2002. The impact of treatment-resistant depression on health care utilization and costs. J. Clin. Psychiatry 63 (11), 963–971. https://doi.org/10.4088/JCP.v63n1102.

Demyttenaere, K., Van Duppen, Z., 2019. The impact of (the concept of) treatment-resistant depression: an opinion review. Int. J. Neuropsychopharmacol. 22 (2), 85–92. https://doi.org/10.1093/ijnp/pyy052.

Denee, T., Ming, T., Waller, J., Bailey, T., Rajkovic-Hooley, O., Middleton-Dalby, C., Le, H.H., Zhang, Q., McCrone, P., Taylor, D., 2020. A retrospective chart review study to quantify the monthly medical resource use and costs of treating patients with treatment resistant depression in the United Kingdom. Curr. Med. Res. Opin. https://doi.org/10.1080/03007995.2020.1857580.

Fava, M., 2003. Diagnosis and definition of treatment-resistant depression. Biol. Psychiatry 53 (8), 649–659. https://doi.org/10.1016/S0006-3223(03)00231-2.

Fava, M., Davidson, K.G., 1996. Definition and epidemiology of treatment-resistant depression. Psychiatr. Clin. North Am. 19 (2), 179–200. https://doi.org/10.1016/S0193-953X(05)70283-5.

Fava, M., Rush, A.J., Alpert, J.E., Balasubramani, G.K., Wisniewski, S.R., Carmin, C.N., Biggs, M.M., Zisook, S., Leuchter, A., Howland, R., Warden, D., Trivedi, M.H., 2008. Difference in treatment outcome in outpatients with anxious versus nonanxious depression: a STAR*D report. Am. J. Psychiatry 165 (3), 342–351. https://doi.org/10.1176/appi.ajp.2007.06111868.

Fekadu, A., Wooderson, S.C., Markopoulo, K., Donaldson, C., Papadopoulos, A., Cleare, A.J., 2009. What happens to patients with treatment-resistant depression? A systematic review of medium to long term outcome studies. J. Affect. Disord. 116 (1–2), 4–11. https://doi.org/10.1016/j.jad.2008.10.014.

Feldman, R.L., Dunner, D.L., Muller, J.S., Stone, D.A., 2013. Medicare patient experience with vagus nerve stimulation for treatment-resistant depression. J. Med. Econ. 16 (1), 62–74. https://doi.org/10.3111/13696998.2012.724745.

GBD 2016 and HALE Collaborators, 2017. Global, regional, and national disability-adjusted life-years (DALYs) for 333 diseases and injuries and healthy life expectancy (HALE) for 195 countries and territories, 1990–2016: a systematic analysis for the Global Burden of Disease Study 2016. Lancet 390, 1260–1344.

GBD 2017 Disease and Injury Incidence and Prevalence Collaborators, 2018. Global, regional, and national incidence, prevalence, and years lived with disability for 354 diseases and injuries for 195 countries and territories, 1990–2017: a systematic analysis for the Global Burden of Disease Study 2017. Lancet 392, 1789–1858.

Gibson, T.B., Jing, Y., Carls, G.S., Kim, E., Bagalman, J.E., Burton, W.N., Tran, Q.V., Pikalov, A., Goetzel, R.Z., 2010. Cost burden of treatment resistance in patients with depression. Am. J. Manag. Care 16 (5), 370–377.

Goldberg, D., 2011. The heterogeneity of "major depression". World Psychiatry 10 (3), 226–228. https://doi.org/10.1002/j.2051-5545.2011.tb00061.x.

Greenberg, P.E., Birnbaum, H.G., 2005. The economic burden of depression in the US: societal and patient perspectives. Expert Opin. Pharmacother. 6 (3), 369–376. https://doi.org/10.1517/14656566.6.3.369.

Greenberg, P., Corey-Lisle, P.K., Birnbaum, H., Marynchenko, M., Claxton, A., 2004. Economic implications of treatment-resistant depression among employees. Pharmacoeconomics 22 (6), 363–373. https://doi.org/10.2165/00019053-200422060-00003.

Greenberg, P.E., Fournier, A.-A., Sisitsky, T., 2015. The economic burden of adults with major depressive disorder in the United States (2005 and 2010). J. Clin. Psychiatry 76, 155–162.

Ivanova, J.I., Birnbaum, H.G., Kidolezi, Y., Subramanian, G., Khan, S.A., Stensland, M.D., 2010. Direct and indirect costs of employees with treatment-resistant and non-treatment-resistant major depressive disorder. Curr. Med. Res. Opin. 26 (10), 2475–2484. https://doi.org/10.1185/03007995.2010.517716.

Jain, F.A., Hunter, A.M., Brooks, J.O., Leuchter, A.F., 2013. Predictive socioeconomic and clinical profiles of antidepressant response and remission. Depress. Anxiety 30 (7), 624–630. https://doi.org/10.1002/da.22045.

Jo, C., 2014. Cost-of-illness studies: concepts, scopes, and methods. Clin. Mol. Hepatol. 20 (4), 327–337. https://doi.org/10.3350/cmh.2014.20.4.327.

Johnston, K.M., Powell, L.C., Anderson, I.M., Szabo, S., Cline, S., 2019. The burden of treatment-resistant depression: a systematic review of the economic and quality of life literature. J. Affect. Disord. 242, 195–210. https://doi.org/10.1016/j.jad.2018.06.045.

Judd, L.L., Akiskal, H.S., Zeller, P.J., Paulus, M., Leon, A.C., Maser, J.D., Endicott, J., Coryell, W., Kunovac, J.L., Mueller, T.I., Rice, J.P., Keller, M.B., 2000. Psychosocial disability during the long-term course of unipolar major depressive disorder. Arch. Gen. Psychiatry 57 (4), 375–380. https://doi.org/10.1001/archpsyc.57.4.375.

Kennedy, S.H., 2008. Treating each and every depressed patient. J. Psychopharmacol. (Oxford, England) 22 (7), 19–23. https://doi.org/10.1177/0269881108093270.

Kessler, R.C., Akiskal, H.S., Ames, M., Birnbaum, H., Greenberg, P., Hirschfeld, R.M.A., Jin, R., Merikangas, K.R., Simon, G.E., Wang, P.S., 2006. Prevalence and effects of mood disorders on work performance in a nationally representative sample of U.S. workers. Am. J. Psychiatry 163 (9), 1561–1568. https://doi.org/10.1176/ajp.2006.163.9.1561.

Khandker, R.K., Kruzikas, D.T., McLaughlin, T.P., 2008. Pharmacy and medical costs associated with switching between venlafaxine and SSRI antidepressant therapy for the treatment of major depressive disorder. J. Manag. Care Pharm. 14 (5), 426–441. https://doi.org/10.18553/jmcp.2008.14.5.426.

Kilbourne, A.M., Beck, K., Spaeth-Rublee, B., Ramanuj, P., O'Brien, R.W., Tomoyasu, N., Pincus, H.A., 2018. Measuring and improving the quality of mental health care: a global perspective. World Psychiatry 17 (1), 30–38. https://doi.org/10.1002/wps.20482.

Knapp, M., Wong, G., 2020. Economics and mental health: the current scenario. World Psychiatry 19 (1), 3–14. https://doi.org/10.1002/wps.20692.

Kornstein, S.G., Schneider, R.K., 2001. Clinical features of treatment-resistant depression. J. Clin. Psychiatry 62 (16), 18–35.

Kubitz, N., Mehra, M., Potluri, R.C., Garg, N., Cossrow, N., 2013. Characterization of treatment resistant depression episodes in a cohort of patients from a US commercial claims database. PLoS One 8 (10). https://doi.org/10.1371/journal.pone.0076882.

Larg, A., Moss, J., 2011. Cost-of-illness studies. Pharmacoeconomics, 653–671. https://doi.org/10.2165/11588380-000000000-00000.

Lepine, B.A., Moreno, R.A., Campos, R.N., Couttolenc, B.F., 2012. Depressão resistente ao tratamento aumenta os custos e utilização de recursos. Rev. Bras. Psiquiatr. 34 (4), 379–388. https://doi.org/10.1016/j.rbp.2012.05.009.

Lerner, D., Lavelle, T.A., Adler, D., Chow, W., Chang, H., Godar, S.C., Rogers, W.H., 2020. A population-based survey of the workplace costs for caregivers of persons with treatment-resistant depression compared with other health conditions. J. Occup. Environ. Med. 62 (9), 746–756. https://doi.org/10.1097/JOM.0000000000001957.

Li, G., Zhang, L., DiBernardo, A., Wang, G., Sheehan, J.J., Lee, K., Reutfors, J., Zhang, Q., 2020. A retrospective analysis to estimate the healthcare resource utilization and cost associated with treatment-resistant depression in commercially insured US patients. PLoS One 15 (9). https://doi.org/10.1371/journal.pone.0238843.

Lin, J., Szukis, H., Sheehan, J., Alphs, L., Menges, B., Lingohr-Smith, M., Benson, C., 2019. Economic burden of treatment-resistant depression among patients hospitalized for major depressive disorder in the United States. Psychiatr. Res. Clin. Pract., 68–76. https://doi.org/10.1176/appi.prcp.20190001.

Luppa, M., Heinrich, S., Angermeyer, M.C., König, H.H., Riedel-Heller, S.G., 2007. Cost-of-illness studies of depression. A systematic review. J. Affect. Disord. 98 (1–2), 29–43. https://doi.org/10.1016/j.jad.2006.07.017.

Lynch, C.J., Gunning, F.M., Liston, C., 2020. Causes and consequences of diagnostic heterogeneity in depression: paths to discovering novel biological depression subtypes. Biol. Psychiatry 88 (1), 83–94. https://doi.org/10.1016/j.biopsych.2020.01.012.

Madsen, K., Petersen, L., Plana-Ripoll, O., 2020. Early labor force exits in patients with treatment-resistant depression: an assessment of work years lost in a Danish nationwide register-based cohort study. Ther. Adv. Psychopharmacol. 10, 1–13.

Mahlich, J., Tsukazawa, S., Wiegand, F., 2018. Estimating prevalence and healthcare utilization for treatment-resistant depression in Japan: a retrospective claims database study. Drugs Real World Outcomes 5 (1), 35–43. https://doi.org/10.1007/s40801-017-0126-5.

McIntyre, R.S., Millson, B., Power, G.S., 2020. Burden of Treatment Resistant Depression (TRD) in patients with major depressive disorder in Ontario using Institute for Clinical Evaluative Sciences (ICES) databases: economic burden and healthcare resource utilization. J. Affect. Disord. 277, 30–38. https://doi.org/10.1016/j.jad.2020.07.045.

Mrazek, D.A., Hornberger, J.C., Altar, C.A., Degtiar, I., 2014. A review of the clinical, economic, and societal burden of treatment-resistant depression: 1996–2013. Psychiatr. Serv. 65 (8), 977–987. https://doi.org/10.1176/appi.ps.201300059.

Murray, C.J.L., Lopez, A.D., 1997. Global mortality, disability, and the contribution of risk factors: global burden of disease study. Lancet 349 (9063), 1436–1442. https://doi.org/10.1016/S0140-6736(96)07495-8.

National Academy of Sciences, 1998. Summarizing Population Health: Directions for the Development and Application of Population Metrics. National Academies Press.

Nemeroff, C.B., 2007. Prevalence and management of treatment-resistant depression. J. Clin. Psychiatry 68 (8), 17–25.

Olchanski, N., McInnis Myers, M., Halseth, M., Cyr, P.L., Bockstedt, L., Goss, T.F., Howland, R.H., 2013. The economic burden of treatment-resistant depression. Clin. Ther. 35 (4), 512–522. https://doi.org/10.1016/j.clinthera.2012.09.001.

Olesen, J., Gustavsson, A., Svensson, M., Wittchen, H.U., Jönsson, B., 2012. The economic cost of brain disorders in Europe. Eur. J. Neurol. 19 (1), 155–162. https://doi.org/10.1111/j.1468-1331.2011.03590.x.

Olfson, M., Amos, T.B., Benson, C., McRae, J., Marcus, S.C., 2018. Prospective service use and health care costs of Medicaid beneficiaries with treatment-resistant depression. J. Manag. Care Spec. Pharm. 24 (3), 226–236. https://doi.org/10.18553/jmcp.2018.24.3.226.

Oliva-Moreno, J., Trapero-Bertran, M., Peña-Longobardo, L.M., del Pozo-Rubio, R., 2017. The valuation of informal care in cost-of-illness studies: a systematic review. Pharmacoeconomics 35 (3), 331–345. https://doi.org/10.1007/s40273-016-0468-y.

Onukwugha, E., McRae, J., Kravetz, A., Varga, S., Khairnar, R., Mullins, C.D., 2016. Cost-of-illness studies: an updated review of current methods. Pharmacoeconomics 34 (1), 43–58. https://doi.org/10.1007/s40273-015-0325-4.

Pilon, D., Joshi, K., Sheehan, J.J., Zichlin, M.L., Zuckerman, P., Lefebvre, P., Greenberg, P.E., 2019a. Burden of treatment-resistant depression in medicare: a retrospective claims database analysis. PLoS One 14 (10). https://doi.org/10.1371/journal.pone.0223255.

Pilon, D., Sheehan, J.J., Szukis, H., Morrison, L., Zhdanava, M., Lefebvre, P., Joshi, K., 2019b. Is clinician impression of depression symptom severity associated with incremental economic burden in privately insured US patients with treatment resistant depression? J. Affect. Disord. 255, 50–59. https://doi.org/10.1016/j.jad.2019.04.100.

Pilon, D., Sheehan, J.J., Szukis, H., Singer, D., Jacques, P., Lejeune, D., Lefebvre, P., Greenberg, P.E., 2019c. Medicaid spending burden among beneficiaries with treatment-resistant depression. J. Comp. Eff. Res. 8 (6), 381–391. https://doi.org/10.2217/cer-2018-0140.

Pilon, D., Szukis, H., Joshi, K., Singer, D., Sheehan, J.J., Wu, J.W., Lefebvre, P., Greenberg, P., 2020. US integrated delivery networks perspective on economic burden of patients with treatment-resistant depression: a retrospective matched-cohort study. Pharmacoecon. Open 4 (1), 119–131. https://doi.org/10.1007/s41669-019-0154-z.

Radloff, L., 1977. The CES-D scale: a self- report depression scale for research in the general population. Appl. Psychol. Measur. 1, 385–401.

Reich, J., 2003. The effect of axis II disorders on the outcome of treatment of anxiety and unipolar depressive disorders: a review. J. Pers. Disord. 17 (5), 387–405. https://doi.org/10.1521/pedi.17.5.387.22972.

Rice, D.P., 1994. Cost-of-illness studies: fact or fiction? Lancet 344 (8936), 1519–1520. https://doi.org/10.1016/S0140-6736(94)90342-5.

Rizvi, S.J., Grima, E., Tan, M., Rotzinger, S., Lin, P., McIntyre, R.S., Kennedy, S.H., 2014. Treatment-resistant depression in primary care across Canada. Can. J. Psychiatry 59 (7), 349–357. https://doi.org/10.1177/070674371405900702.

Rost, K., Zhang, M., Fortney, J., Smith, J., Smith, G.R., 1998. Expenditures for the treatment of major depression. Am. J. Psychiatry 155 (7), 883–888. https://doi.org/10.1176/ajp.155.7.883.

Russell, J.M., Hawkins, K., Ozminkowski, R.J., Orsini, L., Crown, W.H., Kennedy, S., Finkelstein, S., Berndt, E., Rush, A.J., 2004. The cost consequences of treatment-resistant depression. J. Clin. Psychiatry 65 (3), 341–347. https://doi.org/10.4088/JCP.v65n0309.

Schultz, J., Joish, V., 2009. Costs associated with changes in antidepressant treatment in a managed care population with major depressive disorder. Psychiatr. Serv. 60 (12), 1604–1611. https://doi.org/10.1176/ps.2009.60.12.1604.

Shin, D., Kim, N.W., Kim, M.J., Rhee, S.J., Park, C.H.K., Kim, H., Yang, B.R., Kim, M.S., Choi, G.J., Koh, M., Ahn, Y.M., 2020. Cost analysis of depression using the national insurance system in South Korea: a comparison of depression and treatment-resistant depression. BMC Health Serv. Res. 20 (1). https://doi.org/10.1186/s12913-020-05153-1.

Shrestha, A., Roach, M., Joshi, K., Sheehan, J.J., Goutam, P., Everson, K., Heerlein, K., Jena, A.B., 2020. Incremental health care burden of treatment-resistant depression among commercial, Medicaid, and Medicare payers. Psychiatr. Serv. 71 (6), 593–601. https://doi.org/10.1176/appi.ps.201900398.

Sicras-Mainar, A., Guijarro, P., Armada, B., Blanca-Tamayo, M., Navarro-Artieda, R., 2014. Influence of the CYP2D6 isoenzyme in patients treated with venlafaxine for major depressive disorder: clinical and economic consequences. PLoS One 9 (11). https://doi.org/10.1371/journal.pone.0090453.

Smith, K., 2014. Mental health: a world of depression. Nature 515 (7526), 181. https://doi.org/10.1038/515180a.

Stewart, W.F., Ricci, J.A., Chee, E., Hahn, S.R., Morganstein, D., 2003. Cost of lost productive work time among US workers with depression. JAMA 289 (23), 3135–3144. https://doi.org/10.1001/jama.289.23.3135.

Sussman, M., Menzin, J., O'Sullivan, A.K., Shah, A., Olfson, M., 2019. Economic burden of treatment-resistant depression on the U.S. Health care system. J. Manag. Care Spec. Pharm. *25* (7), 823–835. https://doi.org/10.18553/jmcp.2019.25.7.823.

Tarricone, R., 2007. Cost-of-illness analysis: what room in health economics? Health Policy 77, 61–63.

Trevino, K., McClintock, S.M., Fischer, N.M.D., Vora, A., Husain, M.M., 2014. Defining treatment-resistant depression: a comprehensive review of the literature. Ann. Clin. Psychiatry 26 (3), 222–232. https://www.aacp.com/pdf%2F0814%2F0814ACP_Trevino.pdf.

von Knorring, L., Akerblad, A.-C., Bengtsson, F., Carlsson, A., Ekselius, L., 2006. Cost of depression: effect of adherence and treatment response. Eur. Psychiatry, 349–354. https://doi.org/10.1016/j.eurpsy.2006.04.005.

Wang, X., Kattan, M.W., 2020. Cohort studies: design, analysis, and reporting. Chest 158 (1), S72–S78. https://doi.org/10.1016/j.chest.2020.03.014.

Whooley, M.A., Kiefe, C.I., Chesney, M.A., Markovitz, J.H., Matthews, K., Hulley, S.B., 2002. Depressive symptoms, unemployment, and loss of income: the CARDIA study. Arch. Intern. Med. 162 (22), 2614–2620. https://doi.org/10.1001/archinte.162.22.2614.

World Health Organization (WHO), 2004. The Global Burden of Disease: 2004 Update. https://www.who.int/healthinfo/global_burden_disease/GBD_report_2004update_full.pdf?ua=1.

Wu, Z., Chen, J., Yuan, C., Hong, W., Peng, D., Zhang, C., Cao, L., Fang, Y., 2013. Difference in remission in a Chinese population with anxious versus nonanxious treatment-resistant depression: a report of OPERATION study. J. Affect. Disord. 150 (3), 834–839. https://doi.org/10.1016/j.jad.2013.03.012.

Zhdanava, M., Kuvadia, H., Joshi, K., Daly, E., Pilon, D., Rossi, C., Morrison, L., Lefebvre, P., Nelson, C., 2020. Economic burden of treatment-resistant depression in privately insured U.S. patients with physical conditions. J. Manag. Care Spec. Pharm. *26* (8), 996–1007. https://doi.org/10.18553/jmcp.2020.20017.

Zhdanava, M., Kuvadia, H., Joshi, K., Daly, E., Pilon, D., Rossi, C., Morrison, L., Lefebvre, P., Nelson, C., 2021. Economic burden of treatment-resistant depression in privately insured US patients with co-occurring anxiety disorder and/or substance use disorder. Curr. Med. Res. Opin. 37 (1), 123–133. https://doi.org/10.1080/03007995.2020.1844645.

Chapter 4

The neurobiology of treatment-resistant depression

Richard C. Shelton
Department of Psychiatry and Behavioral Neurobiology, School of Medicine, University of Alabama at Birmingham, Birmingham, AL, United States

Introduction

Multiple treatment modalities are effective for major depressive disorder (MDD), including medications, certain psychotherapies, and brain stimulation approaches such as electroconvulsive therapy (ECT) or repetitive transcranial magnetic stimulation (rTMS). However, treatment resistance, that is lack of full symptom remission after one or more rounds of antidepressant treatments of adequate dose and duration, is the rule, not the exception with MDD (Harris et al., 2020). For example, the NIMH-supported Sequenced Treatment Alternatives to Relieve Depression (STAR*D) study was a large trial that treated depressed patients across four treatment steps, beginning with citalopram monotherapy (Rush et al., 2006). Only about one-third of patients remitted with initial treatment with citalopram, and another third to second-level treatments with either monotherapies, combinations of medications, or combined citalopram and cognitive behavioral therapy. By the third and fourth steps, only 13.7% and 13% of patients remitted, respectively. The cumulative remission rate was only 67%, and over half of remitted patients relapsed during a 12-month naturalistic follow-up phase (Rush et al., 2006). Clearly, something is preventing response in many people and causing others to relapse. The purpose of this chapter will be to address the neurobiological impediments to antidepressant treatment response.

Why do people get depressed?

Allostasis is the normal physiological adaptation to maintain homeostasis in face of stressors (McEwen and Stellar, 1993; McEwen, 1998). The cumulative effect of stress on the body is referred to as allostatic load (McEwen, 2006). The responses to threats of homeostasis, such as activation of the autonomic nervous system and the hypothalamic–pituitary–adrenal axis, and cytokine responses are adaptive in the short-term (McEwen and Wingfield, 2003). In fact, short-term stressors that are perceived to be controllable have a positive effect on brain health, enhancing cognitive and physical performance (Dhabhar, 2018). However, chronic allostatic loading is harmful and is associated with a range of conditions, including abdominal obesity, hypertension, diabetes, cardiovascular disease, arthritis, and depressive symptoms (Mattei et al., 2010; Kobrosly and van Wijngaarden, 2010; Kobrosly et al., 2014). Persistent activation of endocrine, cardiovascular, metabolic, and emotional responses increases risk for and progression of disease (Peters et al., 2017). Brain allostatic loading is a known factor in depression (Kobrosly et al., 2014). For example, Kobrosly et al. (2014) found that several aspects of allostatic responses, including HPA activity, immune response, anabolic activity, and cardiovascular disease, were associated with depressive symptoms, after adjusting for demographic, socioeconomic, and other health-related factors in older adults. How, then, does that lead to depression?

Much work over the last two decades has focused on the dynamic regulation of neuronal dendrites, spines, and synapses in brain, collectively referred to as neuroplasticity. Spines are protrusions from dendrites that form synapses with other cells (Fig. 4.1). A single neuron can have thousands of spines and synapses. These change very dynamically in response to environmental cues. The formation, maintenance, and regression of spines and synapses is regulated by a wide range of factors; chief among these are neurotrophins such as brain derived neurotrophic factor (BDNF) and related trophic factors such as insulin-like growth factor-1 (IGF-1) and vascular endothelial growth factor (VEGF) (Deyama and Duman, 2020). Interestingly, acute stress that is short-term and controllable actually increases BDNF along with spines and synapses (Schmitt et al., 2016). However, chronic, unpredictable, and uncontrollable stress reduces BDNF synthesis and release and regression of dendrites and spines, a process that is associated with depression-like behavior in animal models (Licznerski and Duman, 2013).

Managing Treatment-Resistant Depression. https://doi.org/10.1016/B978-0-12-824067-0.00004-9

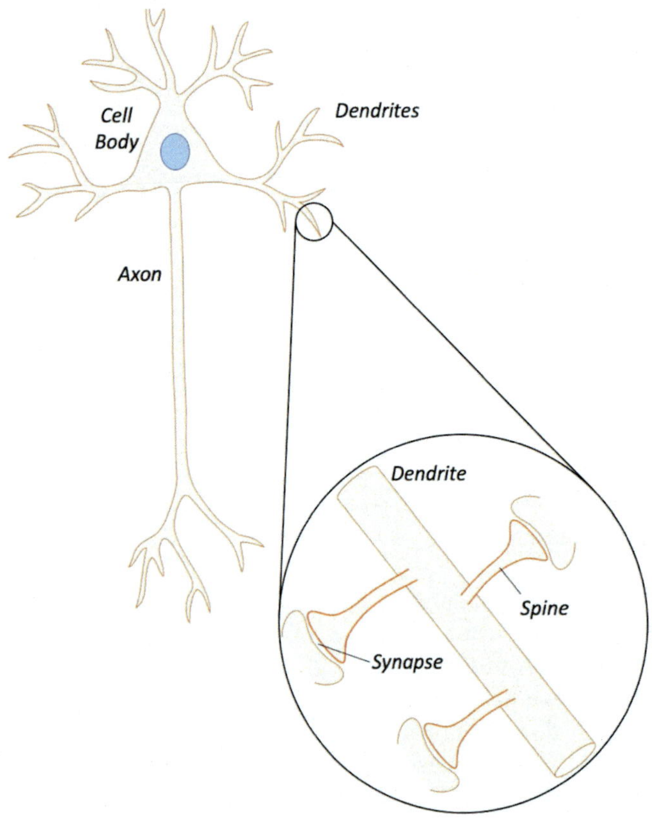

FIG. 4.1 Dendrites, spines, and synapses. Brain-derived neurotrophic factor binds to its receptor, TrkB, and initiates a series of steps that result in the products of proteins that are assembled into dendritic spines and synapses. Dendrites are innervated by axon terminals from principal neurons and interneurons.

The effects of stress on the brain are complex and vary by region. For example, stress and glucocorticoids result in regression of dendrites and loss of dendritic spines and synapses in hippocampus and frontal cortex (McEwen et al., 2015). Conversely, acute stress increases spines, and chronic stress leads to enlargement of dendrites in the basolateral amygdala (McEwen et al., 2015). However, the net loss of synapses in frontal cortex and hippocampus, particularly those connecting pyramidal and other neurons with interneurons, appears to underlie depression.

The process of removal of synapses, spines, and dendrites is a type of macroautophagy, usually just referred to as autophagy, which is an adaptive process by which cells remove components. A structure called a phagophore engulfs cellular material forming an autophagosome, which merges with a lysosome, resulting in the degrading of proteins. Various types of cell stress, including toxins, microorganisms, metabolic stress, and ischemia, activate autophagy, resulting in the clearing of damaged proteins or cell remodeling (Peker and Gozuacik, 2020). This process is tonically suppressed by the mammalian (or mechanistic) target of rapamycin complex 1 (mTORC1), discussed in greater detail below. A variety of metabolic factors can suppress mTORC1 activity, releasing the autophagic process and leading to loss of spines and synapses (Peker and Gozuacik, 2020). Therefore, maintaining mTORC1 activity is critical in preserving spines and synapses.

Interneurons serve as functional regulators of pyramidal and related neurons (referred to as principal neurons), which mediate communication both within and between brain regions. Some types of interneurons release gamma-aminobutyric acid (GABA), the principal inhibitory transmitter in the nervous system, which functions as a key regulator of brain activity by "gating" both input and output of principal neurons (Duman et al., 2019). Glutamate is the main excitatory output transmitter of principal neurons, which is balanced by GABA. Chronic stress and the depressed state that results from stress involve loss of GABA synapses on principal neurons and an imbalance in input and output. This excitatory (glutamate) to inhibitory (GABA) imbalance occurs because of a reduction in principal neuron dendrites, spines, and synapses (Duman et al., 2019; McKlveen et al., 2019). A result of stress is a loss of inhibitory control by GABA interneurons.

BDNF and other trophic factors increase dendritic volume and density, and lead to the formation of spines and synapses. It also induces cell replication of neurons in the hippocampus (neurogenesis) (Schmidt and Duman, 2007). However, continued input by BDNF is also required for the maintenance of dendrites, spines, and synapses. Chronic stress and the

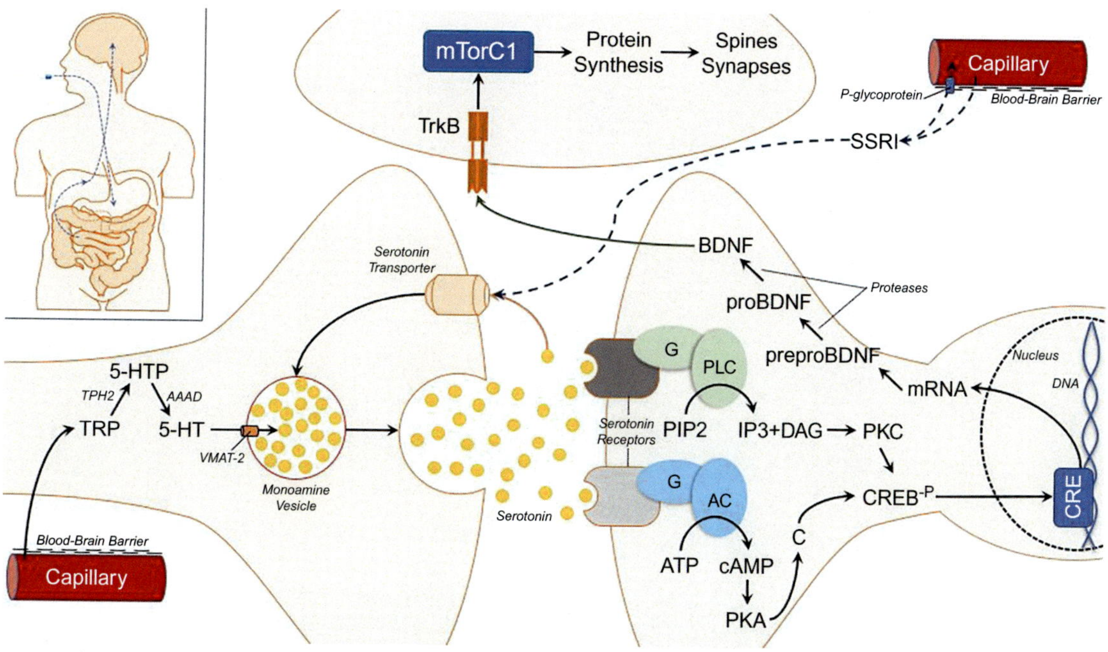

FIG. 4.2 The complex mechanisms of action of standard antidepressant medications. As described in detail in the text, most antidepressant mechanisms act via a very complicated series of steps. In the case of SSRIs, the precursor of 5-HT is actively transported across the blood brain into neurons and converted in two steps to 5-HT. It is stored in presynaptic vesicles prior to being released into the synapse and binding to postsynaptic receptors. This initiates a series of intracellular steps that lead to increased transcription of preproBDNF, which is metabolized by proteases to proBDNF, and eventually BDNF itself. Upon release, BDNF binds to its receptor TrkB, which is linked to the mTORC1 complex and subsequent protein synthesis. Abbreviations: 5-HT, serotonin (5-hydroxytryptamine); 5-HTP, 5-hydroxytryptophan; AAAD, aromatic amino acid decarboxylase; AC, adenylate cyclase; BDNF, brain-derived neurotrophic factor; C, PKA catalytic subunit; CRE, cyclic AMP response element; CREB, cyclic AMP response element-binding protein; G, guanine nucleotide-binding protein; mRNA, messenger RNA; mTORC1, mammalian target of rapamycin 1; PKA, protein kinase A; TPH2, tryptophan hydroxylase 2; TrkB, tyrosine receptor kinase B; TRP, tryptophan; VMAT2, vesicular monoamine transporter 2.

resulting depression-like behaviors in rodents are associated with reduced synthesis and release of BDNF, which results in progressive loss of spines and synapses (Duman and Monteggia, 2006). Antidepressant treatments appear to reverse this process and increase BDNF activity. The binding of BDNF to TrkB initiates a cascade of intracellular signaling molecules that ultimately activate mTORC1 (Figs. 4.1–4.3) (Hay and Sonenberg, 2004). The complex consists of the mTOR protein itself, along with regulatory-associated protein of mTOR (Raptor), and mammalian lethal with SEC13 protein 8 (nLST8). Proline-rich AKT1 substrate 40 kDa (PRAS40) interacts and activates mTORC1 following phosphorylation by RAC-alpha serine/threonine-protein kinase (AKT). AKT also inhibits the TSC1/2 protein complex and inhibitor of mTORC1 (Fig. 4.3). mTORC1 regulates the synthesis of proteins, including those that form spines and synapses (Duman and Voleti, 2012). Stress inhibits this process in part by increasing cortisol which binds to the glucocorticoid receptor (GR) and inhibits mTORC1 (Polman et al., 2012). Under normal circumstances, mTORC1 counterbalances this by inhibition of GR gene transcription. However, high or chronic stress unbalances this system resulting in reduced mTORC1 activity. Activation of the BDNF–TrkB–mTORC1 pathway leads to the genesis and maintenance of dendrites, spines, and synapses. Stress counters this by inhibition of both BDNF and mTORC1. As we will see below, antidepressant treatments help to restore BDNF and mTORC1 activity, leading to normalization of dendrites, spines, and synapses (Fig. 4.2).

How do antidepressant treatments work?

Antidepressant treatments work by widely divergent proximal mechanisms—contrast, for example, the effects of selective serotonin (5-HT) reuptake inhibitors (SSRIs) with electroconvulsive therapy (ECT) or repetitive transcranial magnetic stimulation (rTMS). However, the net effect of most or possibly all effective antidepressant treatments is to increase the activity of trophic factors, particularly BDNF and its receptor tyrosine receptor kinase B (TrkB) (also known as tropomyosin receptor kinase B) (Nibuya et al., 1995; Wang et al., 2011). Direct injections of BDNF and other trophins like neurotrophin-3 have antidepressant-like effects in animal models of depression when injected directly into the brain (Shirayama et al., 2002). Inhibition of BDNF (Duman and Monteggia, 2006), TrkB (Saarelainen et al., 2003), or mTORC1 (Xu et al., 2018) as well as other trophins like IGF-1 (Hoshaw et al., 2008) and VEGF (Deyama et al., 2019) block the effects of antidepressant

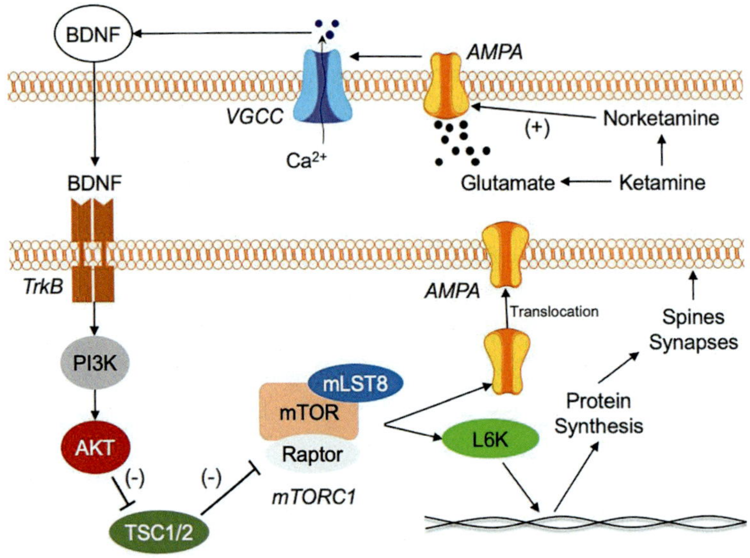

FIG. 4.3 Mechanisms regulating the production of dendritic spines and proteins. AMPA receptors are linked to voltage-gated calcium channels. Activation of AMPA results in an influx of calcium, which stimulates the rapid release of BDNF. BDNF binding to its receptor, TrkB initiates a cascade of events, activating mTORC1, leading to synthesis of spine and synapse proteins, and translocation of AMPA receptors to the cell surface. (Simplified diagram.) Abbreviations: AKT, RAC-alpha serine/threonine-protein kinase; AMPA, α-amino-3-hydroxy-5-methyl-4-isoxazolepropionic acid; BDNF, brain-derived neurotrophic factor; Ca^{2+}, calcium ions; mLST8, mammalian lethal with SEC13 protein 8; mTORC1, mammalian target of rapamycin 1; PI3K, phosphatidylinositol 3-kinase; PRAS40, proline-rich AKT1 substrate 40 kDa; Raptor, regulatory-associated protein of mTOR; TrkB, tyrosine receptor kinase B; TSC1/2, tuberous sclerosis 1/2 complex; VGCC, voltage-gated calcium channels.

treatments. Moreover, a series of studies indicated that depressed patients have lower BDNF levels prior to treatments, which increase after treatment (Sen et al., 2008). However, antidepressant medications such as SSRIs do not directly affect these factors. How, then, do they produce effects on trophins and related molecules?

Fig. 4.2 illustrates the complex mechanisms of action of standard antidepressant treatments, focusing on the actions of SSRIs. Tryptophan (TRP) is taken up into the presynaptic neuron and is converted to 5-hydroxytryptophan (5-HTP) by tryptophan hydroxylase (TPH) (primarily TPH2 in brain). 5-HTP is converted to 5-HT by aromatic amino acid decarboxylase (AAAD). Free 5-HT is taken up into presynaptic vesicles by vesicular monoamine transporter 2 (VMAT2). The vesicles merge with the plasma membrane, releasing 5-HT into the synapse. There is a similar process with norepinephrine (NE), which is required by many antidepressant drugs. The NE precursor amino acid tyrosine is taken up by the presynaptic neuron and converted initially to L-DOPA by tyrosine hydroxylase (TH), then to dopamine (DA) by AAAD, and finally to NE by dopamine beta hydroxylase (DBH) (Fig. 4.2). The vesicles merge with the plasma membrane, releasing 5-HT into the synapse. There is a similar process with norepinephrine (NE), which is required by many antidepressant drugs. The NE precursor amino acid tyrosine is taken up by the presynaptic neuron and converted initially to L-DOPA by tyrosine hydroxylase (TH), then to dopamine (DA) by AAAD, and finally to NE by dopamine beta hydroxylase (DBH). DA and NE are also taken up into vesicles by VMAT2 and released into the synapse.

5-HT binds to receptors that activate adenylate cyclase, which converts adenosine triphosphate (ATP) to cyclic adenosine monophosphate (cyclic AMP) through several 5-HT receptors (5-HT_4, 5-HT_6, and 5-HT_7). SSRIs block 5-HT reuptake, which increases 5-HT interactions with postsynaptic receptors (Fig. 4.2). Activation of the 5-HT receptor leads to coupling with a guanine nucleotide-binding protein (g-protein) which, in turn activates adenylate cyclase, producing cyclic AMP. The latter then interacts with protein kinase A (PKA), which releases its catalytic subunit (C), which phosphorylates the transcriptional factor cyclic AMP response element binding protein (CREB). Similar downstream effects can occur with 5-HT binding to receptors that activate phospholipase-C (PLC) (Jiang et al., 2016) (e.g., the 5-HT_2 family of receptors), which catalyzes the conversion of phosphatidylinositol 4,5-bisphosphate (PIP2) to inositol triphosphate (IP3) and diacylglycerol (DAG). The latter activates protein kinase C, which phosphorylates CREB. Drugs that increase synaptic NE have similar actions via NE beta-2 receptors that are also positively coupled to AC (Shelton, 2000). Phosphorylated CREB then translocates to the nucleus where it binds to a DNA enhancer region called a cyclic AMP response element (CRE) in the promoters of specific genes. This increases the transcription of a wide range of genes, including those coding for BDNF, TrkB (Shelton, 2000; Condorelli et al., 1994; Duman et al., 1995), and VEGF (Wu et al., 2007). This system is also affected by other trophins such as NT-3 (Saini et al., 2004) and IGF-1, which increase CREB phosphorylation (Zuloaga et al.,

2013). The downstream effects of antidepressants on BDNF and other trophic factors appear to be required to normalize the reduced dendrites, spines, and synapses in the depressed brain and restore normal mood (Duman and Duman, 2015). Other effective antidepressant treatments such as ECT (Duman and Vaidya, 1998) and rTMS (Chen et al., 2019) also affect BDNF, TrkB, and other trophic elements.

Cells maintain homeostasis through the balance of activating and inhibiting processes. While the effects described above are occurring, there are counteracting elements that dampen cell signaling. For example, when NE or 5-HT transporters are blocked, the transmitters not only bind to postsynaptic receptors, but also to presynaptic autoreceptors: NE alpha-1, and 5-HT 1A (5-HT$_{1A}$) and 1B (5-HT$_{1B}$) receptors (Blier, 2001). These inhibit the release of NE and 5-HT respectively. Although these receptors do eventually downregulate with time, they slow the actions of antidepressants. Blier and colleagues (Blier, 2001) conducted a series of experiments, first in rodents and then in human with the 5-HT$_{1A}$ antagonist pindolol showing that it prevented the acute reduction in 5-HT release seen after short-term administration of SSRIs and accelerated the antidepressant response. The postsynaptic receptors coupled to the effector mechanisms described above (e.g., adenylate cyclase) downregulate after repeated activation by neurotransmitters by internalization (called receptor trafficking), making them vulnerable to degrading enzymes (Clague and Urbé, 2001), and by uncoupling from g-proteins. In addition, there are a large number of enzymes within neurons that metabolize or inhibit the elements described earlier. For example, there are enzymes that degrade drugs, proteins, and transmitters. Others inhibit protein action through the process of either phosphorylation (i.e., adding a phosphoryl group [$P^+O_3^{2-}$]) at serine, threonine, or tyrosine amino acids in proteins, or dephosphorylation. A well-known example of the latter is the inactivation of cyclic AMP by phosphodiesterases, but there are many other instances. These compensatory mechanisms maintain homeostasis in the cell but work against the mechanisms of action of standard antidepressants.

There are mechanisms of action of standard antidepressants that extend beyond uptake blockade. These effects include acting as antagonists of 5-HT$_{2A}$ (mirtazapine, trazodone, nefazodone, and the atypical antipsychotics), 5-HT$_3$ (mirtazapine, vortioxetine, and the atypical antipsychotics clozapine and olanzapine), 5-HT$_7$ (vortioxetine) receptors, and acting as partial agonists (nefazodone, buspirone, vilazodone, and the atypical antipsychotics aripiprazole, asenapine, brexpiprazole, cariprazine, clozapine, and ziprasidone) or full agonists (vortioxetine) of 5-HT$_{1A}$. These effects confer anxiolytic- and antidepressant-like effects in animal models. Some of these effects appear to be direct, perhaps not involving neurotrophic actions. For example, acute activation of 5-HT$_{1A}$, or blockade of 5-HT$_{2A}$ in the amygdala produces robust anxiolytic effects. In other instances, the effects are similar to the ones described for uptake inhibitors earlier. For example, activation of 5-HT$_{2A}$ receptors inhibits BDNF mRNA synthesis, which is reversed with 5-HT$_{2A}$ selective antagonists (Vaidya et al., 1997). Therefore, some of the antidepressant effects of these mechanisms may depend on a downstream neurotrophic response.

Why do antidepressants fail?

Psychological and social factors

Fig. 4.2 illustrates some of the factors involved in determining if someone will respond to a medication or not. Beginning with the inset at the upper left—there are patient-level factors that affect response that have nothing to do with physiology. Could the patient afford the medication? Did they actually take it consistently? Did they continue to take it long enough for it to work? Could they tolerate any side effects that occurred? A third or more of patients will discontinue an antidepressant within the first 10 weeks of treatment (Linden et al., 2000). This increases to two-thirds or higher by 6 months (Sheehan et al., 2008). Early termination of medication occurs often in patients without consulting their health care professional. Collectively, these patient factors are the most common reason for antidepressant treatment failure.

There are also clinician factors that influence outcome. The relationship between prescriber and patient, referred to as the therapeutic alliance, is an important part of the outcome because it affects the patient's therapeutic engagement and adherence to treatment. As many as one-third of prescriptions are never filled, and half of medications taken chronically are not taken as prescribed (Viswanathan et al., 2012). Clinicians are sometimes not thoughtful in prescribing antidepressants. Some use only a single, starting dose of a medication and do not titrate based on response; this is unlike treatments for other diseases like hypertension or diabetes in which regular measurement of outcomes is done and treatment adjusted accordingly. This is related in part to the fact that for antidepressant treatment, outcome is not usually measured (Trivedi et al., 2020). Prescribers also often do not take mechanism of action into consideration when selecting a medication. For example, is a third SSRI likely to be helpful when the first two weren't? Finally, psychotherapy for depression is seldom evidence-based (Bruijniks et al., 2018; Harris et al., 2015). The clinician impacts the outcome of treatment almost as much as the patient.

Non-CNS physiological causes

There are physiological reasons outside of the brain for why a medication does not work. For example, is the medication adequately absorbed? The absorption of an orally-administered drug depends on a range of factors, including residence time in the intestine and gut pH (Viswanathan et al., 2017). Some drugs are absorbed better when taken with a meal containing fat, but the absorption rate is slowed for most drugs taken with food. Once absorbed, all antidepressant drugs undergo first-pass (and subsequent) metabolism. Many antidepressants are metabolized by cytochrome enzymes, which vary widely between individuals, ranging from poor to ultrarapid metabolism (Nassan et al., 2016). This is particularly true for two enzymes that metabolize many psychotropics: cytochrome P450 (CYP) 2C19 and CYP2D6 (Tornio and Backman, 2018). This is why doses range so widely, and why recommended doses only represent guidelines, not absolutes. Venlafaxine, for example, is a CYP2D6 substrate. The enzyme activity can range from poor to ultrarapid metabolizer status. This means that some people will not be able to tolerate even the lowest available dose, 37.5 mg, while others never achieve an adequate blood level at 375 mg. Further, most drugs are metabolized by several CYP enzymes, which expand the possibility of CYP enzyme involvement in nonresponse. Prescribers should be aware of the various factors influencing drug absorption, metabolism, distribution, and elimination.

Neuronal and related cellular causes

Returning to Fig. 4.2: There is an enormous number of factors that must be working adequately for any antidepressant treatment to work. We could even say that it is surprising that antidepressants work at all, given the number of factors that influence the outcome. This is why it is hard to implicate any single factor in influencing the outcome of antidepressant treatments. For example, some, but not all studies have found positive but very modest associations of polymorphic variants of the serotonin transporter gene (which goes by the designation *SLC6A4*) and outcome of serotonin reuptake inhibitor treatment (Porcelli et al., 2012). The most common polymorphic variant involves a region in the gene promoter referred to as the serotonin-transporter-linked polymorphic region (5-HTTLPR) (Heils et al., 1995; Prasad et al., 2005). This is a variable number tandem repeat region of the promoter with two common forms, short and long, referring to the number of repeated segments. The long variant is more active and is associated with a higher level of gene transcription and, therefore, more of the transporter. This is also a common G/A variant of the long allele, commonly expressed as L_G and L_A; the G variant of the long allele is also associated with reduced transcription. It has been hypothesized in the past that the variants associated with reduced availability of the serotonin transporter may reduce the effectiveness of serotonin reuptake inhibitors (SRIs). In fact, while this is true of most studies, the association of 5-HTTLPR variants and clinical outcome has been very weak. However, this is not surprising because antidepressants could fail for many other reasons in people who do not carry the loss of function 5-HTTLPR or other *SLC6A4* variants, weakening the associations.

Beginning from the left in Fig. 4.2, tryptophan (TRP) is the precursor of serotonin, which is of course required for the actions of the SSRIs. TRP is an essential amino acid in humans and is derived exclusively from dietary sources. TRP is transported across the blood–brain barrier (BBB) by a carrier protein that also transports isoleucine, leucine, phenylalanine, tyrosine, and valine (Mehedint and Gulledge, 2014). Competition for this carrier limits the amount of TRP that enters the brain, and, therefore, the amount of 5-HT that can be made. The actions of SSRIs are sensitive to TRP levels in the body. Delgado et al. showed that the antidepressant benefits in depressed patients who had responded to SSRIs could be rapidly reversed by depletion of TRP (Delgado et al., 1990, 1999). This procedure consisted of administering a TRP-free amino acid mixture, which was hypothesized to be lead rapidly to protein synthesis peripherally, which, then, results in a 70%–80% depletion of plasma TRP (Delgado et al., 1999). This temporarily reversed the actions of SSRIs within hours, which recovered after giving TRP. These findings suggest that SSRI antidepressant response is sensitive to TRP levels.

Why would the action of SSRIs be so sensitive to TRP levels? In a typical physiological state, most of the 5-HT that is released into the synapse is taken by up by serotonin transporters and stored for future use in presynaptic vesicles. There is, of course, some loss in this process requiring new synthesis of 5-HT. However, the system is relatively efficient in conserving 5-HT. When a 5-HT uptake inhibitor is given, this process is disrupted, and it appears that most of the 5-HT is not taken up. Kinetic studies have shown that a 5-HT transporter deficiency (which mimics the SSRI effect) significantly reduces reuptake of 5-HT. Mice lacking one copy of the 5-HT transporter gene (i.e., heterozygous, +/), a kind of physiological imitation of the effects of SSRIs, show substantial reductions in the rate and total amount of 5-HT reuptake (Perez et al., 2007). In addition, when an SSRI is given chronically to TPH2-deficient mice, in which 5-HT synthesis is reduced by 80%, there is a dramatic reduction in neuronal serotonin of about 99%! (Siesser et al., 2013) By contrast, the control mice without the TPH2 reduction showed no depletion after SSRI administration. Therefore, 5-HT has to be continuously synthesized for SSRIs to work. Therefore, TRP availability and conversion to 5-HT are important determinants of SRI response. In addition, the speed at which relapse occurs after TRP depletion shows that the effect of an SSRI is very "fragile"—that is, it can be reversed by a short-term reduction in 5-HT availability. A similar effect is seen with the depletion of the NE precursor tyrosine on the response to NE reuptake inhibitors in some studies (Delgado et al., 1993; Miller et al., 1996). However, the effect may be somewhat less robust than with TRP depletion (Delgado et al., 1993; Miller et al., 1996) and study results have been inconsistent (Porter et al., 2005; McTavish et al., 2005; Zhu et al., 2005, 2006, 2010).

Are there natural physiological equivalents to TRP depletion? One possible set of mechanisms is systemic inflammation that occurs in inflammatory diseases such as rheumatoid arthritis or lupus erythematosus but is also associated with obesity and metabolic diseases like type 2 diabetes (Shelton and Miller, 2010). Systemic inflammatory diseases are known risk factors for depression and both inflammation and obesity reduce the response to SSRIs (Shelton and Miller, 2010). Inflammation depletes TRP and 5-HT and reduces synaptic 5-HT by several processes (Fig. 4.4). The upper panel of Fig. 4.4 illustrates two different mechanisms for reduced 5-HT synaptic signaling. A variety of inflammatory factors

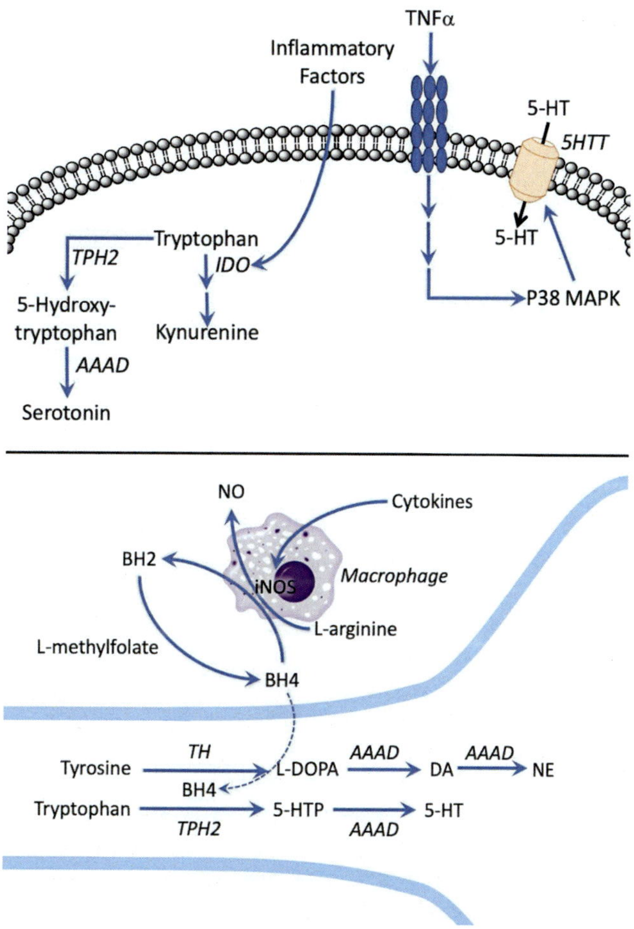

FIG. 4.4 Theoretical mechanisms linking systemic inflammation to treatment resistance. Top panel: Inflammatory factors such as cytokines (e.g., tumor necrosis factor alpha [TNFa], interferons, and bacterial endotoxins [lipopolysaccharide]) reduce synaptic 5-HT by at least two mechanisms. Inflammation activates the enzyme indoleamine-2,3-dioxygenase (IDO), or a related liver enzyme tryptophan deoxygenase, which convert available tryptophan (TRP) to kynurenine and, ultimately, other metabolites like quinolinic acid. IDO activation depletes TRP resulting in reduced synthesis of 5-HT. TNFa also activates P38 MAP kinase (P38 MAPK) via a series of proteins. P38 MAPK phosphorylates the 5-HT transporter, which increases the rate at which the transporter takes up 5-HT. Bottom panel: Tetrahydrobiopterin (BH4) is a requisite cofactor for the enzymes tryptophan hydroxylase (TPH), which is the rate-limiting step in 5-HT synthesis, and tyrosine hydroxylase (TH), the rate-limiting enzyme in the synthesis of dopamine (DA) and norepinephrine (NE). It is also a cofactor for inducible nitric oxide synthase (iNOS) that is activated by cytokines to convert L-arginine to nitric oxide which is involved in host defense. BH4 can be depleted, reducing the synthesis of catecholamines and serotonin, which are required for the actions of most antidepressants. Abbreviations: Top panel: 5-HT, serotonin; 5-HTT, serotonin transporter; AAAD, aromatic amino acid decarboxylase; IDO, indoleamine-2,3-dioxygenase; P38 MAPK, P38 mitogen-activated protein kinase; TNF-r, TNF receptor; TNFa, tumor necrosis factor alpha; TPH2, tryptophan hydroxylase 2. Bottom panel: 5-HT, serotonin; 5-HTP, 5-hydroxytryptophan; AAAD, aromatic amino acid decarboxylase; BH2, dihydrobiopterin; BH4, tetrahydrobiopterin; DA, dopamine; DBH, dopamine beta hydroxylase; iNOS, inducible nitric oxide synthase; L-DOPA, levodopa; NE, norepinephrine; NO, nitric oxide; TH, tyrosine hydroxylase; TPH2, tryptophan hydroxylase.

activate the enzyme indoleamine-2,3-dioxygenase (IDO), which converts TRP to kynurenine (Capuron and Miller, 2011). Persistent IDO activation leads to TRP and 5-HT depletion (Wichers and Maes, 2004), theoretically reducing response to antidepressants. Interestingly, functional single-nucleotide polymorphisms (SNPs) of the *IDO1* gene are associated with antidepressant response (Cutler et al., 2012). A second pathway specifically affects synaptic 5-HT levels. Tumor necrosis factor alpha (TNF) and other inflammatory mediators activate the enzyme P38 MAP kinase (P38 MAPK) (Fig. 4.4). This in turn phosphorylates the serotonin transporter (5-HTT), which increases the rate of 5-HT reuptake into the presynaptic neuron (Zhu et al., 2005, 2006, 2010). These mechanisms and others like chronic systemic inflammation can lead to reduced 5-HT signaling and response to serotonergic antidepressants.

Inflammatory cytokines also activate inducible nitric oxide synthase (iNOS) in macrophages and other tissues that catalyzes the production of nitric oxide (NO) from L-arginine (Fig. 4.4, bottom panel) (Green et al., 1994). NO has a range of effects in cells, but cytokine induction of iNOS generates high levels of NO which reacts with superoxides, forming cytotoxic peroxynitrite, which is involved in antimicrobial response (Lee et al., 2017). Tetrahydrobiopterin (BH4) is a cofactor for all NOS species including iNOS, which converts BH2 to dihydrobiopterin (BH2) (McNeill and Channon, 2012). iNOS activation in response to relative BH4 depletion. That is significant because BH4 is also a cofactor for both tryptophan and tyrosine hydroxylases, the rate-limiting enzymes in the synthesis of 5-HT and norepinephrine (NE) and dopamine (DA), respectively. Inflammation-induced depletion of BH4 is thought to be a mechanism that reduces responsiveness of antidepressants that require either 5-HT or NE, which include the SSRIs, serotonin and norepinephrine reuptake inhibitors (SNRIs), and tricyclics (Haroon et al., 2012; Miller et al., 2013). L-5-methyltetrahydrofolate (LMF) is required for the regeneration of BH4 from BH2. These relationships were the basis of two studies testing the effects of LMF 15 mg per day combined with an antidepressant in SSRI nonresponders (Papakostas et al., 2012). Overall, the effects of LMF 15 mg were greater than placebo, but relatively modest. However, secondary analyses showed that both obesity and the elevated levels of the inflammatory factors C-reactive protein and TNF were all associated with greater LMF effects (Shelton et al., 2015). The differences in outcome comparing LMF to placebo was -4.66 points in patients with a BMI of ≥ 30, while those <30 actually had a marginally worse outcome with LMF ($+0.63$ points). The results are consistent with the inflammation–iNOS–BH4 hypothesis in SSRI nonresponse and suggest that normalization of BH4 may restore SRI response (Vancassel et al., 2018).

A large number of genetic studies have investigated various components of the mechanistic pathways for antidepressants acting via 5-HT and/or NE. Positive associations have been found between specific gene SNPs and antidepressant response, including 5-HTTLPR (Porcelli et al., 2012; Kato and Serretti, 2010), another variant of the 5-HT transporter referred to as STin2, *TPH1* (Kato and Serretti, 2010), *TPH2* (Gao et al., 2012), the genes for the 5-HT_{1A} (*HTR1A*) (Kato and Serretti, 2010), and 5-HT_{2A} (*HTR2A*) receptors (Kato and Serretti, 2010; Laje et al., 2009), and the *BDNF* gene (Kato and Serretti, 2010). Genes for proteins not shown in the illustration but reportedly associated with antidepressant response include catechol-*O*-methyltransferase (*COMT*), an enzyme that metabolizes NE (Szegedi et al., 2005), a chaperon protein for the glucocorticoid receptor, *FKBP5* (Lekman et al., 2008), glutamate receptor ionotropic kainate 4 (*GRIK4*) (Laje et al., 2009), and potassium channel subfamily K member 2 (*KCNK2*) (Laje et al., 2009), among others (El-Hage et al., 2013). The association of specific gene SNPs or haplotypes with antidepressant response has generally been weak and variable with frequent failures to replicate, for the reasons described earlier. It should be noted that the effects of a single SNP within a population of depressed patients may be weak because of the "diluting" effect of other SNPs, the effect in an individual may be much greater. The cellular mechanisms of all standard antidepressants are so complicated, the likelihood of failure of *some* mechanism is relatively high. Even when all the steps are working correctly there are many cellular factors that counteract the signal described earlier. At best, the effects of standard antidepressants are weak, indirect, and slow (Duman and Voleti, 2012). Contrast that, for example, with a beta blocker for hypertension. The drug acts directly on the target site of action that usually has a rapid blood pressure lowering effect. Standard antidepressants require repeated interactions of transmitters (5-HT or NE) with receptors over time with slowly accumulating effects. This is probably why antidepressant medications take so long to work even when they are effective—4–6 weeks or more to achieve maximum effect.

Why is ketamine different?

Ketamine is an anesthetic drug first introduced in 1970 and still commonly used today (Li and Vlisides, 2016). Ketamine acts by blocking the *N*-methyl-D-aspartate (NMDA) receptor for glutamate, the most abundant transmitter in the CNS. High dose ketamine causes sedation, but a relatively low dose of ketamine (0.5 mg/kg) infused slowly by IV over 40 min produces a rapid antidepressant action, maximizing in many people with depression within 1 day (Zarate Jr. et al., 2006). This occurs even in many patients with treatment-resistant depression (Zarate Jr. et al., 2006). The benefit slowly dissipates, but it can be prolonged by repeated administration (Singh et al., 2016). Ketamine has other pharmacological effects; for example, it is an

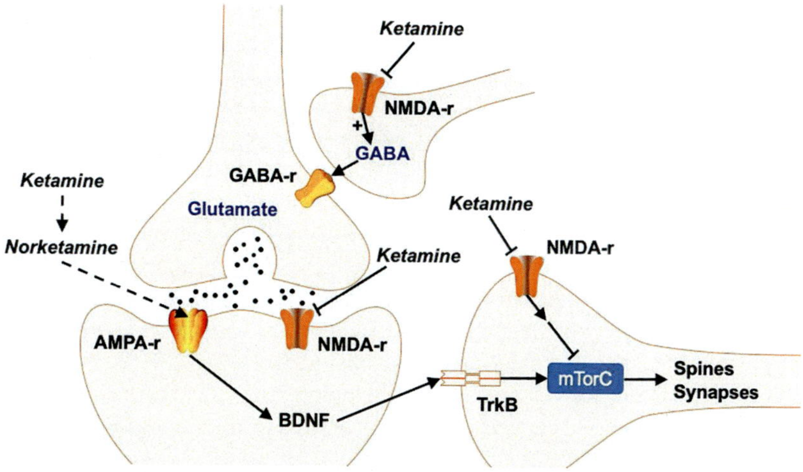

FIG. 4.5 Putative mechanisms of action of ketamine for TRD. Ketamine is an effective treatment for the majority of patients with TRD. GABA releasing interneurons innervate and tonically inhibit glutamate neurons. NMDA receptors on GABA neurons stimulate the release of GABA. Blockade of NMDA receptors reduce the firing rate of GABA neurons and results in a large release of glutamate. The transmitter binds to AMPA receptors, stimulating the release of BDNF, which binds to TrkB receptors, activating mTORC1 leading to the synthesis of spine and synaptic proteins. Abbreviations: AMPA-r, α-amino-3-hydroxy-5-methyl-4-isoxazolepropionic acid receptor; BDNF, brain-derived neurotrophic factor; GABA, gamma-aminobutyric acid; NMDA-R, *N*-methyl-D-aspartate receptor; TrkB, tyrosine receptor kinase B.

agonist for mu opioid receptors which may contribute to its antidepressant effect (Williams et al., 2018), although binding affinity for the mu receptor is about 1/40th of the potency for NMDA (Mathew and Zarate, 2016). The *S*(+)-enantiomer of ketamine called esketamine has been approved in the US by the FDA for major depression following two failed antidepressants and one trial of psychotherap (Daly et al., 2018; Deyama and Duman, 2020; Popova et al., 2019).

Under normal physiological conditions, glutamate-containing neurons are tonically inhibited by GABA interneurons (Fig. 4.5). These interneurons are activated by glutamate NMDA receptors, a feedback inhibitory mechanism for glutamate neurons (Gerhard et al., 2020). Ketamine blocks NMDA receptors on the GABA interneurons. This results in a large and rapid release of glutamate, which binds to the α-amino-3-hydroxy-5-methyl-4-isoxazolepropionic acid (AMPA) receptor. AMPA is positively coupled with voltage-gated calcium channels, and activation results in an influx of calcium ions, which triggers a rapid release of BDNF (Fig. 4.3). Ketamine is also metabolized to norketamine, which is agonist for glutamate AMPA receptors (Fig. 4.5) (Zanos et al., 2018). NMDA receptors also inhibit mTORC1 via activation of striatal-enriched tyrosine phosphatase (STEP) leading to dephosphorylation and inactivation of extracellular signal regulated kinase 1/2 (ERK1/2), an activator of mTORC1 (Burket et al., 2015). Activation of mTORC1 increases expression of spine and synapse proteins, and also causes the trafficking of AMPA receptors to the cell surface, which enhances AMPA effects (Wang et al., 2006). So, rather than the weak and slow downstream effect of standard antidepressants, ketamine is able to bypass the regular antidepressant mechanisms and directly cause the release of BDNF. This rapidly induces growth of spines and synapses via the mTOR mechanism, which is thought to be its main mechanism of action (Li et al., 2010).

Conclusions

Most depressed patients will respond to standard antidepressant treatments, but the process is slow and difficult for the majority. Antidepressant treatments, probably including traditional antidepressant medications, ECT, rTMS, and psychotherapy, act via a complex set of mechanisms involving precursor amino acids, neurotransmitters, postsynaptic receptor-medicated effectors, and gene expression with a slow accumulation of neurotrophins. These factors are released from neurons and bind to neurotrophic receptors, activating mTORC1, which leads to protein synthesis and spine and synapse formation. The process typically is a slow one, and there are many opposing forces. As described earlier, these include enzymes that degrade drugs, transmitters, and proteins; phosphorylation or dephosphorylation of proteins leading to inactivation; internalization of receptors; and other processes that slow the downstream effects.

Depressive episodes are often precipitated and maintained by external and internal stressors which, among other mechanisms, can inhibit mTORC1 activity via GRs. In fact, many and possibly most patients will have better long-term outcomes if they learn to mitigate life stresses via psychotherapy. The evidence-based psychotherapies (e.g., cognitive-behavioral therapy and interpersonal psychotherapy) have common elements of actively engaging and reducing stressors and changing self-defeating thought patterns. Psychotherapy requires learning and memory, and increasing synaptic plasticity via the mechanisms described above may facilitate response. This seems to be particularly true among more severely ill patients (Hollon et al., 2014). Increased cognitive load, learning, and memory all stimulate and are dependent on plasticity. Therefore, medications plus psychotherapy essentially are targeting the same mechanisms through different means: using "bottom-up" mechanisms—medication directly stimulating the development of spines and synapses at a cellular level—and "top-down" mechanisms (psychotherapy)—an activity-dependent means of increasing synaptic plasticity (Takeuchi et al., 2013).

The recent US FDA approval of intranasal esketamine as a treatment for TRD has ushered in a more direct path to treatment response—directly stimulating the release of BDNF followed in short order by spine and synapse formation (Duman and Duman, 2015). Other NMDA receptor antagonists or negative allosteric modulators have been tested and, thus far, have all failed in pivotal trials (Sanacora et al., 2017). It is not clear whether the lack of success is because of failed (but not negative) studies due to a high placebo effect, a relative lack of potency at NMDA receptor inhibition, or if other mechanisms of action than NMDA may add to the effect of ketamine. These include mu-opioid receptor agonism (Williams et al., 2018) or metabolism of ketamine to norketamine, a potent AMPA agonist (Zanos et al., 2018). The introduction of ketamine for TRD has at least ushered in a period of renewed interest in developing drugs that don't simply use the old mechanisms of monoamine uptake or receptor blockade. This may usher in a generation of truly new, more effective treatments for TRD.

References

Blier, P., 2001. Pharmacology of rapid-onset antidepressant treatment strategies. J. Clin. Psychiatry 62 (Suppl 15), 12–17.

Bruijniks, S.J.E., Franx, G., Huibers, M.J.H., 2018. The implementation and adherence to evidence-based protocols for psychotherapy for depression: the perspective of therapists in Dutch specialized mental healthcare. BMC Psychiatry 18 (1), 190.

Burket, J.A., Benson, A.D., Tang, A.H., Deutsch, S.I., 2015. NMDA receptor activation regulates sociability by its effect on mTOR signaling activity. Prog. Neuro-Psychopharmacol. Biol. Psychiatry 60, 60–65.

Capuron, L., Miller, A.H., 2011. Immune system to brain signaling: neuropsychopharmacological implications. Pharmacol. Ther. 130, 226–238.

Chen, X., Chen, S., Liang, W., Ba, F., 2019. Administration of repetitive transcranial magnetic stimulation attenuates Aβ (1-42)-induced Alzheimer's disease in mice by activating β-catenin signaling. Biomed. Res. Int. 2019, 1431760.

Clague, M.J., Urbé, S., 2001. The interface of receptor trafficking and signalling. J. Cell Sci. 114 (17), 3075.

Condorelli, D.F., Dell'Albani, P., Mudò, G., Timmusk, T., Belluardo, N., 1994. Expression of neurotrophins and their receptors in primary astroglial cultures: induction by cyclic AMP-elevating agents. J. Neurochem. 63 (2), 509–516.

Cutler, J.A., Rush, A.J., McMahon, F.J., Laje, G., 2012. Common genetic variation in the indoleamine-2,3-dioxygenase genes and antidepressant treatment outcome in major depressive disorder. J. Psychopharmacol. 26 (3), 360–367.

Daly, E.J., Singh, J.B., Fedgchin, M., et al., 2018. Efficacy and safety of intranasal esketamine adjunctive to oral antidepressant therapy in treatment-resistant depression: a randomized clinical trial. JAMA Psychiatry 75 (2), 139–148.

Delgado, P.L., Charney, D.S., Price, L.H., Aghajanian, G.K., Landis, H., Heninger, G.R., 1990. Serotonin function and the mechanism of antidepressant action. Reversal of antidepressant-induced remission by rapid depletion of plasma tryptophan. Arch. Gen. Psychiatry 47 (5), 411–418.

Delgado, P.L., Miller, H.L., Salomon, R.M., et al., 1993. Monoamines and the mechanism of antidepressant action: effects of catecholamine depletion on mood of patients treated with antidepressants. Psychopharmacol. Bull. 29 (3), 389–396.

Delgado, P.L., Miller, H.L., Salomon, R.M., et al., 1999. Tryptophan-depletion challenge in depressed patients treated with desipramine or fluoxetine: implications for the role of serotonin in the mechanism of antidepressant action. Biol. Psychiatry 46 (2), 212–220.

Deyama, S., Bang, E., Kato, T., Li, X.Y., Duman, R.S., 2019. Neurotrophic and antidepressant actions of brain-derived neurotrophic factor require vascular endothelial growth factor. Biol. Psychiatry 86 (2), 143–152.

Deyama, S., Duman, R.S., 2020. Neurotrophic mechanisms underlying the rapid and sustained antidepressant actions of ketamine. Pharmacol. Biochem. Behav. 188, 172837.

Dhabhar, F.S., 2018. The short-term stress response – mother nature's mechanism for enhancing protection and performance under conditions of threat, challenge, and opportunity. Front. Neuroendocrinol. 49, 175–192.

Duman, C.H., Duman, R.S., 2015. Spine synapse remodeling in the pathophysiology and treatment of depression. Neurosci. Lett. 601, 20–29.

Duman, R.S., Monteggia, L.M., 2006. A neurotrophic model for stress-related mood disorders. Biol. Psychiatry 59 (12), 1116–1127.

Duman, R.S., Vaidya, V.A., 1998. Molecular and cellular actions of chronic electroconvulsive seizures. J. ECT 14 (3), 181–193.

Duman, R.S., Voleti, B., 2012. Signaling pathways underlying the pathophysiology and treatment of depression: novel mechanisms for rapid-acting agents. Trends Neurosci. 35 (1), 47–56.

Duman, R.S., Vaidya, V.A., Nibuya, M., Morinobu, S., Fitzgerald, L.R., 1995. Review: stress, antidepressant treatments, and neurotrophic factors: molecular and cellular mechanisms. Neuroscientist 1 (6), 351–360.

Duman, R.S., Sanacora, G., Krystal, J.H., 2019. Altered connectivity in depression: GABA and glutamate neurotransmitter deficits and reversal by novel treatments. Neuron 102 (1), 75–90.

El-Hage, W., Leman, S., Camus, V., Belzung, C., 2013. Mechanisms of antidepressant resistance. Front. Pharmacol. 4, 146.

Gao, J., Pan, Z., Jiao, Z., et al., 2012. TPH2 gene polymorphisms and major depression—a meta-analysis. PLoS One 7 (5), e36721.

Gerhard, D.M., Pothula, S., Liu, R.-J., et al., 2020. GABA interneurons are the cellular trigger for ketamine's rapid antidepressant actions. J. Clin. Invest. 130 (3), 1336–1349.

Green, S.J., Scheller, L.F., Marletta, M.A., et al., 1994. Nitric oxide: cytokine-regulation of nitric oxide in host resistance to intracellular pathogens. Immunol. Lett. 43 (1–2), 87–94.

Haroon, E., Raison, C.L., Miller, A.H., 2012. Psychoneuroimmunology meets neuropsychopharmacology: translational implications of the impact of inflammation on behavior. Neuropsychopharmacology 37, 137–162.

Harris, M.G., Hobbs, M.J., Burgess, P.M., et al., 2015. Frequency and quality of mental health treatment for affective and anxiety disorders among Australian adults. Med. J. Aust. 202 (4), 185–189.

Harris, M.G., Kazdin, A.E., Chiu, W.T., et al., 2020. Findings from world mental health surveys of the perceived helpfulness of treatment for patients with major depressive disorder. JAMA Psychiatry 77, 830–841.

Hay, N., Sonenberg, N., 2004. Upstream and downstream of mTOR. Genes Dev. 18 (16), 1926–1945.

Heils, A., Teufel, A., Petri, S., et al., 1995. Functional promoter and polyadenylation site mapping of the human serotonin (5-HT) transporter gene. J. Neural Transm. Gen. Sect. 102 (3), 247–254.

Hollon, S.D., DeRubeis, R.J., Fawcett, J., et al., 2014. Effect of cognitive therapy with antidepressant medications vs antidepressants alone on the rate of recovery in major depressive disorder: a randomized clinical trial. JAMA Psychiatry 71, 1151–1164.

Hoshaw, B.A., Hill, T.I., Crowley, J.J., et al., 2008. Antidepressant-like behavioral effects of IGF-I produced by enhanced serotonin transmission. Eur. J. Pharmacol. 594 (1–3), 109–116.

Jiang, D.-G., Jin, S.-L., Li, G.-Y., et al., 2016. Serotonin regulates brain-derived neurotrophic factor expression in select brain regions during acute psychological stress. Neural Regen. Res. 11 (9), 1471–1479.

Kato, M., Serretti, A., 2010. Review and meta-analysis of antidepressant pharmacogenetic findings in major depressive disorder. Mol. Psychiatry 15 (5), 473–500.

Kobrosly, R., van Wijngaarden, E., 2010. Associations between immunologic, inflammatory, and oxidative stress markers with severity of depressive symptoms: an analysis of the 2005-2006 National Health and nutrition examination survey. Neurotoxicology 31 (1), 126–133.

Kobrosly, R.W., van Wijngaarden, E., Seplaki, C.L., Cory-Slechta, D.A., Moynihan, J., 2014. Depressive symptoms are associated with allostatic load among community-dwelling older adults. Physiol. Behav. 123, 223–230.

Laje, G., Perlis, R.H., Rush, A.J., McMahon, F.J., 2009. Pharmacogenetics studies in STAR*D: strengths, limitations, and results. Psychiatr. Serv. 60, 1446–1457.

Lee, M., Rey, K., Besler, K., Wang, C., Choy, J., 2017. Immunobiology of nitric oxide and regulation of inducible nitric oxide synthase. Results Probl. Cell Differ. 62, 181–207.

Lekman, M., Laje, G., Charney, D., et al., 2008. The FKBP5-gene in depression and treatment response—an association study in the Sequenced Treatment Alternatives to Relieve Depression (STAR*D) cohort. Biol. Psychiatry 63 (12), 1103–1110.

Li, N., Lee, B., Liu, R.J., et al., 2010. mTOR-dependent synapse formation underlies the rapid antidepressant effects of NMDA antagonists. Science 329 (5994), 959–964.

Li, L., Vlisides, P.E., 2016. Ketamine: 50 years of modulating the mind. Front. Hum. Neurosci. 10, 612.

Licznerski, P., Duman, R.S., 2013. Remodeling of axo-spinous synapses in the pathophysiology and treatment of depression. Neuroscience 251, 33–50.

Linden, M., Gothe, H., Dittmann, R.W., Schaaf, B., 2000. Early termination of antidepressant drug treatment. J. Clin. Psychopharmacol. 20 (5), 523–530.

Mathew, S.J., Zarate, C.A., 2016. Ketamine for Treatment-Resistant Depression: The First Decade of Progress. Springer Publishing, New York, NY, USA.

Mattei, J., Demissie, S., Falcon, L.M., Ordovas, J.M., Tucker, K., 2010. Allostatic load is associated with chronic conditions in the Boston Puerto Rican health study. Soc. Sci. Med. 70 (12), 1988–1996.

McEwen, B.S., 1998. Stress, adaptation, and disease. Allostasis and allostatic load. Ann. N. Y. Acad. Sci. 840, 33–44.

McEwen, B.S., 2006. Sleep deprivation as a neurobiologic and physiologic stressor: allostasis and allostatic load. Metab. Clin. Exp. 55 (10 Suppl 2), S20–S23.

McEwen, B.S., Stellar, E., 1993. Stress and the individual. Mechanisms leading to disease. Arch. Intern. Med. 153 (18), 2093–2101.

McEwen, B.S., Wingfield, J.C., 2003. The concept of allostasis in biology and biomedicine. Horm. Behav. 43 (1), 2–15.

McEwen, B.S., Bowles, N.P., Gray, J.D., et al., 2015. Mechanisms of stress in the brain. Nat. Neurosci. 18 (10), 1353–1363.

McKlveen, J.M., Moloney, R.D., Scheimann, J.R., Myers, B., Herman, J.P., 2019. "Braking" the prefrontal cortex: the role of glucocorticoids and interneurons in stress adaptation and pathology. Biol. Psychiatry 86 (9), 669–681.

McNeill, E., Channon, K.M., 2012. The role of tetrahydrobiopterin in inflammation and cardiovascular disease. Thromb. Haemost. 108 (5), 832–839.

McTavish, S.F.B., Mannie, Z.N., Harmer, C.J., Cowen, P.J., 2005. Lack of effect of tyrosine depletion on mood in recovered depressed women. Neuropsychopharmacology 30 (4), 786–791.

Mehedint, M.G., Gulledge, A., 2014. Nutritional impact on cognitive development. In: Reference Module in Biomedical Sciences, 2014 ed. Elsevier, Inc., Amsterdam, Netherlands.

Miller, H.L., Delgado, P.L., Salomon, R.M., et al., 1996. Clinical and biochemical effects of catecholamine depletion on antidepressant-induced remission of depression. Arch. Gen. Psychiatry 53 (2), 117–128.

Miller, A.H., Haroon, E., Raison, C.L., Felger, J.C., 2013. Cytokine targets in the brain: impact on neurotransmitters and neurocircuits. Depress. Anxiety 30 (4), 297–306.

Nassan, M., Nicholson, W.T., Elliott, M.A., Rohrer Vitek, C.R., Black, J.L., Frye, M.A., 2016. Pharmacokinetic prescribing guidelines for antidepressants: a template for psychiatric precision medicine. Mayo Clin. Proc. 91 (7), 897–907.

Nibuya, M., Morinobu, S., Duman, R.S., 1995. Regulation of BDNF and trkB mRNA in rat brain by chronic electroconvulsive seizure and antidepressant drug treatments. J. Neurosci. 15, 7539–7547.

Papakostas, G.I., Shelton, R.C., Zajecka, J.M., et al., 2012. L-Methylfolate as adjunctive therapy for SSRI-resistant major depression: results of two randomized, double-blind, parallel-sequential trials. Am. J. Psychiatry 169 (12), 1267–1274.

Peker, N., Gozuacik, D., 2020. Autophagy as a cellular stress response mechanism in the nervous system. J. Mol. Biol. 432 (8), 2560–2588.

Perez, X.A., Bressler, A.J., Andrews, A.M., 2007. Determining serotonin and dopamine uptake rates in synaptosomes using high-speed chronoamperometry. In: Michael, A.C., Borland, L.M. (Eds.), Electrochemical Methods for Neuroscience. CRC Press/Taylor & Francis.

Peters, A., McEwen, B.S., Friston, K., 2017. Uncertainty and stress: why it causes diseases and how it is mastered by the brain. Prog. Neurobiol. 156, 164–188.

Polman, J.A., Hunter, R.G., Speksnijder, N., et al., 2012. Glucocorticoids modulate the mTOR pathway in the hippocampus: differential effects depending on stress history. Endocrinology 153 (9), 4317–4327.

Popova, V., Daly, E.J., Trivedi, M., et al., 2019. Efficacy and safety of flexibly dosed esketamine nasal spray combined with a newly initiated oral antidepressant in treatment-resistant depression: a randomized double-blind active-controlled study. Am. J. Psychiatry 176 (6), 428–438.

Porcelli, S., Fabbri, C., Serretti, A., 2012. Meta-analysis of serotonin transporter gene promoter polymorphism (5-HTTLPR) association with antidepressant efficacy. Eur. Neuropsychopharmacol. 22 (4), 239–258.

Porter, R.J., Mulder, R.T., Joyce, P.R., Luty, S.E., 2005. Tryptophan and tyrosine availability and response to antidepressant treatment in major depression. J. Affect. Disord. 86 (2–3), 129–134.

Prasad, H.C., Zhu, C.B., McCauley, J.L., Samuvel, D.J., Ramamoorthy, S., Shelton, R.C., Blakely, R.D., 2005. Human serotonin transporter variants display altered sensitivity to protein kinase G and p38 mitogen-activated protein kinase. Proc. Natl. Acad. Sci. USA 102, 11545–11550.

Rush, A.J., Trivedi, M.H., Wisniewski, S.R., et al., 2006. Acute and longer-term outcomes in depressed outpatients requiring one or several treatment steps: a STAR*D report. Am. J. Psychiatry 163 (11), 1905–1917.

Saarelainen, T., Hendolin, P., Lucas, G., et al., 2003. Activation of the TrkB neurotrophin receptor is induced by antidepressant drugs and is required for antidepressant-induced behavioral effects. J. Neurosci. 23 (1), 349.

Saini, H.S., Gorse, K.M., Boxer, L.M., Sato-Bigbee, C., 2004. Neurotrophin-3 and a CREB-mediated signaling pathway regulate Bcl-2 expression in oligodendrocyte progenitor cells. J. Neurochem. 89 (4), 951–961.

Sanacora, G., Johnson, M.R., Khan, A., et al., 2017. Adjunctive lanicemine (AZD6765) in patients with major depressive disorder and history of inadequate response to antidepressants: a randomized, placebo-controlled study. Neuropsychopharmacology 42 (4), 844–853.

Schmidt, H.D., Duman, R.S., 2007. The role of neurotrophic factors in adult hippocampal neurogenesis, antidepressant treatments and animal models of depressive-like behavior. Behav. Pharmacol. 18 (5–6), 391–418.

Schmitt, K., Holsboer-Trachsler, E., Eckert, A., 2016. BDNF in sleep, insomnia, and sleep deprivation. Ann. Med. 48 (1–2), 42–51.

Sen, S., Duman, R., Sanacora, G., 2008. Serum brain-derived neurotrophic factor, depression, and antidepressant medications: meta-analyses and implications. Biol. Psychiatry 64 (6), 527–532.

Sheehan, D.V., Keene, M.S., Eaddy, M., Krulewicz, S., Kraus, J.E., Carpenter, D.J., 2008. Differences in medication adherence and healthcare resource utilization patterns: older versus newer antidepressant agents in patients with depression and/or anxiety disorders. CNS Drugs 22 (11), 963–973.

Shelton, R.C., 2000. Intracellular mechanisms of antidepressant drug action. Harv. Rev. Psychiatry 8 (4), 161–174.

Shelton, R.C., Miller, A.H., 2010. Eating ourselves to death (and despair): the contribution of adiposity and inflammation to depression. Prog. Neurobiol. 91 (4), 275–299.

Shelton, R.C., Falola, M., Li, L., Zajecka, J., Fava, M., Papakostas, G.I., 2015. The pro-inflammatory profile of depressed patients is (partly) related to obesity. J. Psychiatr. Res. 70, 91–97.

Shirayama, Y., Chen, A.C., Nakagawa, S., Russell, D.S., Duman, R.S., 2002. Brain-derived neurotrophic factor produces antidepressant effects in behavioral models of depression. J. Neurosci. 22 (8), 3251–3261.

Siesser, W.B., Sachs, B.D., Ramsey, A.J., et al., 2013. Chronic SSRI treatment exacerbates serotonin deficiency in humanized Tph2 mutant mice. ACS Chem. Neurosci. 4 (1), 84–88.

Singh, J.B., Fedgchin, M., Daly, E.J., et al., 2016. A double-blind, randomized, placebo-controlled, dose-frequency study of intravenous ketamine in patients with treatment-resistant depression. Am. J. Psychiatry 173 (8), 816–826.

Szegedi, A., Rujescu, D., Tadic, A., et al., 2005. The catechol-O-methyltransferase Val108/158Met polymorphism affects short-term treatment response to mirtazapine, but not to paroxetine in major depression. Pharmacogenomics J. 5 (1), 49–53.

Takeuchi, T., Duszkiewicz, A.J., Morris, R.G.M., 2013. The synaptic plasticity and memory hypothesis: encoding, storage and persistence. Philos. Trans. R. Soc. Lond. Ser. B Biol. Sci. 369 (1633), 20130288.

Tornio, A., Backman, J.T., 2018. Cytochrome P450 in pharmacogenetics: an update. Adv. Pharmacol. 83, 3–32.

Trivedi, M.H., Papakostas, G.I., Jackson, W.C., Rafeyan, R., 2020. Implementing measurement-based care to determine and treat inadequate response. J. Clin. Psychiatry 81 (3).

Vaidya, V.A., Marek, G.J., Aghajanian, G.K., Duman, R.S., 1997. 5-HT2A receptor-mediated regulation of brain-derived neurotrophic factor mRNA in the hippocampus and the neocortex. J. Neurosci. 17 (8), 2785–2795.

Vancassel, S., Capuron, L., Castanon, N., 2018. Brain kynurenine and BH4 pathways: relevance to the pathophysiology and treatment of inflammation-driven depressive symptoms. Front. Neurosci. 12, 499.

Viswanathan, M., Golin, C.E., Jones, C.D., et al., 2012. Interventions to improve adherence to self-administered medications for chronic diseases in the United States: a systematic review. Ann. Intern. Med. 157 (11), 785–795.

Viswanathan, P., Muralidaran, Y., Ragavan, G., 2017. Challenges in oral drug delivery: a nano-based strategy to overcome. In: Andronescu, E., Grumezescu, A.M. (Eds.), Nanostructures for Oral Medicine. Elsevier, pp. 173–201 (Chapter 7).

Wang, Y., Barbaro, M.F., Baraban, S.C., 2006. A role for the mTOR pathway in surface expression of AMPA receptors. Neurosci. Lett. 401 (1), 35–39.

Wang, H.Y., Crupi, D., Liu, J., et al., 2011. Repetitive transcranial magnetic stimulation enhances BDNF-TrkB signaling in both brain and lymphocyte. J. Neurosci. 31 (30), 11044–11054.

Wichers, M.C., Maes, M., 2004. The role of indoleamine 2,3-dioxygenase (IDO) in the pathophysiology of interferon-alpha-induced depression. J. Psychiatry Neurosci. 29 (1), 11–17.

Williams, N.R., Heifets, B.D., Blasey, C., et al., 2018. Attenuation of antidepressant effects of ketamine by opioid receptor antagonism. Am. J. Psychiatry 175 (12), 1205–1215.

Wu, D., Zhau, H.E., Huang, W.C., et al., 2007. cAMP-responsive element-binding protein regulates vascular endothelial growth factor expression: implication in human prostate cancer bone metastasis. Oncogene 26 (35), 5070–5077.

Xu, D., Sun, Y., Wang, C., et al., 2018. Hippocampal mTOR signaling is required for the antidepressant effects of paroxetine. Neuropharmacology 128, 181–195.

Zanos, P., Moaddel, R., Morris, P.J., et al., 2018. Ketamine and ketamine metabolite pharmacology: insights into therapeutic mechanisms. Pharmacol. Rev. 70 (3), 621–660.

Zarate Jr., C.A., Singh, J.B., Carlson, P.J., et al., 2006. A randomized trial of an N-methyl-D-aspartate antagonist in treatment-resistant major depression. Arch. Gen. Psychiatry 63 (8), 856–864.

Zhu, C.B., Blakely, R.D., Hewlett, W.A., 2006. The proinflammatory cytokines interleukin-1beta and tumor necrosis factor-alpha activate serotonin transporters. Neuropsychopharmacology 31 (10), 2121–2131.

Zhu, C.B., Carneiro, A.M., Dostmann, W.R., Hewlett, W.A., Blakely, R.D., 2005. p38 MAPK activation elevates serotonin transport activity via a trafficking-independent, protein phosphatase 2A-dependent process. J. Biol. Chem. 280 (16), 15649–15658.

Zhu, C.B., Lindler, K.M., Owens, A.W., Daws, L.C., Blakely, R.D., Hewlett, W.A., 2010. Interleukin-1 receptor activation by systemic lipopolysaccharide induces behavioral despair linked to MAPK regulation of CNS serotonin transporters. Neuropsychopharmacology 35, 2510–2520.

Zuloaga, R., Fuentes, E.N., Molina, A., Valdés, J.A., 2013. The cAMP response element binding protein (CREB) is activated by insulin-like growth factor-1 (IGF-1) and regulates myostatin gene expression in skeletal myoblast. Biochem. Biophys. Res. Commun. 440 (2), 258–264.

Chapter 5

Modeling treatment-resistant depression in the preclinical setting

Susannah Tye, Clarissa G. Yates, and Danielle Postorivo
Queensland Brain Institute, The University of Queensland, St Lucia, QLD, Australia

Introduction

This is an exciting time in the field of psychiatry. The advent of rapid-acting antidepressants (ADs) like ketamine has sparked a paradigm shift in how we think about the biological basis of treatment-resistant depression (TRD) and has inspired creative solutions in preclinical model development. The purpose of this chapter is to review current and emerging preclinical research trends capable of characterizing the mechanisms mediating and moderating AD response. As novel approaches continue to emerge, preclinical models will collectively help untangle the heterogeneity inherent within current disease classification systems. In turn, these advances will create new opportunities to more effectively target specific treatments to underlying pathophysiologies using precision medicine approaches currently applied in other fields. Ultimately, the integration of preclinical modeling with translational clinical research is set to transform the way we conceptualize TRD. However, bridging this gap will require refinement of current preclinical research strategies and the establishment of interdisciplinary collaborations across psychiatry, neuroscience, bioinformatics, molecular pharmacology, and other fields.

Common mechanisms of antidepressant response in depression

First-line ADs work primarily by blocking monoamine reuptake mechanisms, which almost instantly elevate monoamine neurotransmitter levels at the synapse. Yet, for a long time, there remained an apparent disconnect between the presumed neurobiological mechanisms of these ADs (overcoming a deficit in monoamine neurotransmitter levels) and time to clinical AD response (several weeks). Mark Hyman and Eric Nestler proposed a model of initiation and adaptation in 1996 (Hyman and Nestler, 1996) to help explain this gap, suggesting that AD responses emerged via indirect pathways dependent on the initiation of molecular signaling cascades that, in turn, coordinate cellular neurotrophic adaptations over weeks to months (Fig. 5.1).

Synaptic dysfunction in depression

Preclinical models have demonstrated that AD mechanisms work on a backdrop of synaptic dysfunction induced through various exposures, including chronic stress and/or inflammation, and accelerated by underlying genetic vulnerabilities (Price et al., 2018). Neurons and glia make up the majority of cells in the brain and form complex neural circuits. Effective communication within these circuits requires highly integrated electrochemical signaling, which depends on efficient and effective synaptic communication. At the synapse, stored neurotransmitters are released and bind to receptors on pre- and postsynaptic membranes to trigger intracellular changes and ongoing propagation of the electrochemical signal throughout the circuit. The brain's highly adaptive nature enables active synapses to become strengthened, thereby increasing the efficiency of signal transduction along this same pathway in the future, while inactive synapses are pruned, effectively whittling away unused connections. In depression, this process is maladaptive, with excessive synaptic pruning occurring in specific brain regions such as the prefrontal cortex (PFC) and hippocampus (Duman and Aghajanian, 2012; Liu et al., 2017). This disrupts activity within the mesocorticolimbic mood regulatory circuit to impact a range of essential functions, including motivation, perception, cognition, and emotion regulation (Duman, 2002; Vose and Stanton, 2017). Recent work suggests that both long-term activity-dependent changes in synaptic strength (synaptic plasticity) and shifting set points for these that result in relatively "hard-wired" changes that persist over the longer term (metaplasticity) and are essential for establishing or reversing neurobiological adaptations contributing to depression (Fig. 5.1).

Managing Treatment-Resistant Depression. https://doi.org/10.1016/B978-0-12-824067-0.00005-0

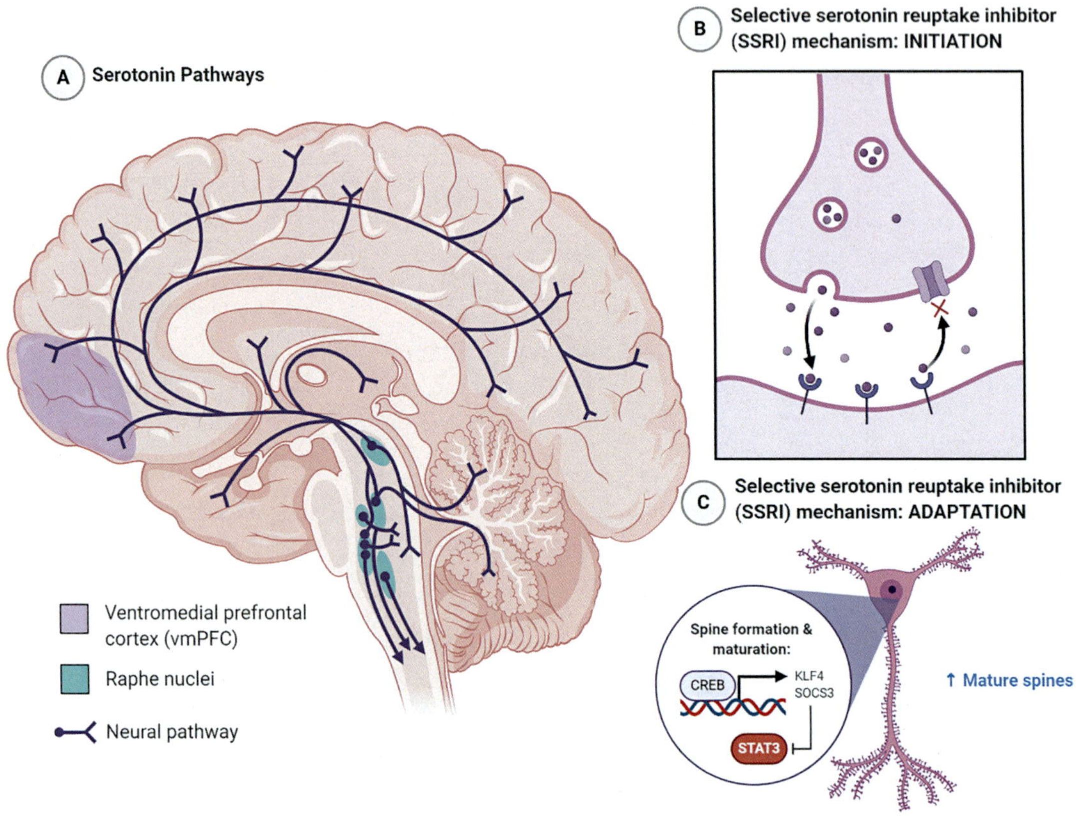

FIG. 5.1 Proposed initiation and adaptation mechanisms of slower-acting antidepressants, such as selective serotonin reuptake inhibitors (SSRIs). (A) Over time, SSRIs induce changes in connectivity and information processing within the mesocorticolimbic neural pathway. (B) SSRIs act at the serotonergic synapse to block reuptake mechanisms and increase neurotransmitter availability at the cleft. These changes initiate cellular adaptations (C) upregulating gene transcription and promoting dendritic spine grown and the development and maturation of new synaptic connections.

Synapses are situated on dendritic spines, which are dynamic structures that functionally serve as the locus of long-term synaptic plasticity (Borczyk et al., 2019). At the tip of most dendritic spines, there is a dense region that harbors glutamate receptors and other protein complexes, namely a postsynaptic density (PSD), or postsynaptic synapse (Okabe, 2007). The size of a dendritic spine is proportional to the size of its PSD, the number of glutamate receptors, and, in turn, the synaptic strength or capacity for synaptic communication. For example, the volume of the spine-head is directly proportional to the number of postsynaptic receptors (Nusser et al., 1998), and the presynaptic number of docked vesicles (Schikorski and Stevens, 1999). Synaptic plasticity modifies information flow throughout the circuit and the consequent cognitive and behavioral outputs. Prolonged exposure to stress and inflammation dynamically regulate dendritic complexity, particularly within the PFC and hippocampus, regions that play an important role in the depression circuit (Duman, 2002; Vose and Stanton, 2017; Mayberg, 1997). The loss of synaptic connectivity in these critical nodes of the mood circuit profoundly impacts mood and cognitive function. This inability to maintain synaptic integrity and plasticity in depression represents a failure of one of the most fundamental and critical functions of the brain, namely the ability to communicate and store information and make appropriate, adaptive, and efficient responses to subsequent inputs (Vose and Stanton, 2017). While this critical brain function has been best studied for its role in learning and memory, it is now clear that it also plays a crucial role in the pathophysiology of depression and serves as the underlying mechanism of AD action (Deyama and Duman, 2020; Price et al., 2018; Vose and Stanton, 2017; Castren and Hen, 2013). Failure to effectively engage this mechanism is now thought to underlie the poor clinical response profiles observed in TRD (Price et al., 2018). The reasons for this target engagement failure are likely many and varied. Preclinical models will be critical for both the discovery of these rate-limiting mechanisms and their translation into innovative biomarker and treatment strategies for the clinic.

Antidepressant regulation of synaptic plasticity

Antidepressant treatments elicit region-specific structural changes via the induction of dendritic arborization, spinogenesis, improved cell survival, enhanced hippocampal neurogenesis, angiogenesis, and gliogenesis, a dynamic process known as synaptic plasticity (Liu et al., 2017; Castren and Hen, 2013). Synaptic plasticity is an essential neurobiological mechanism

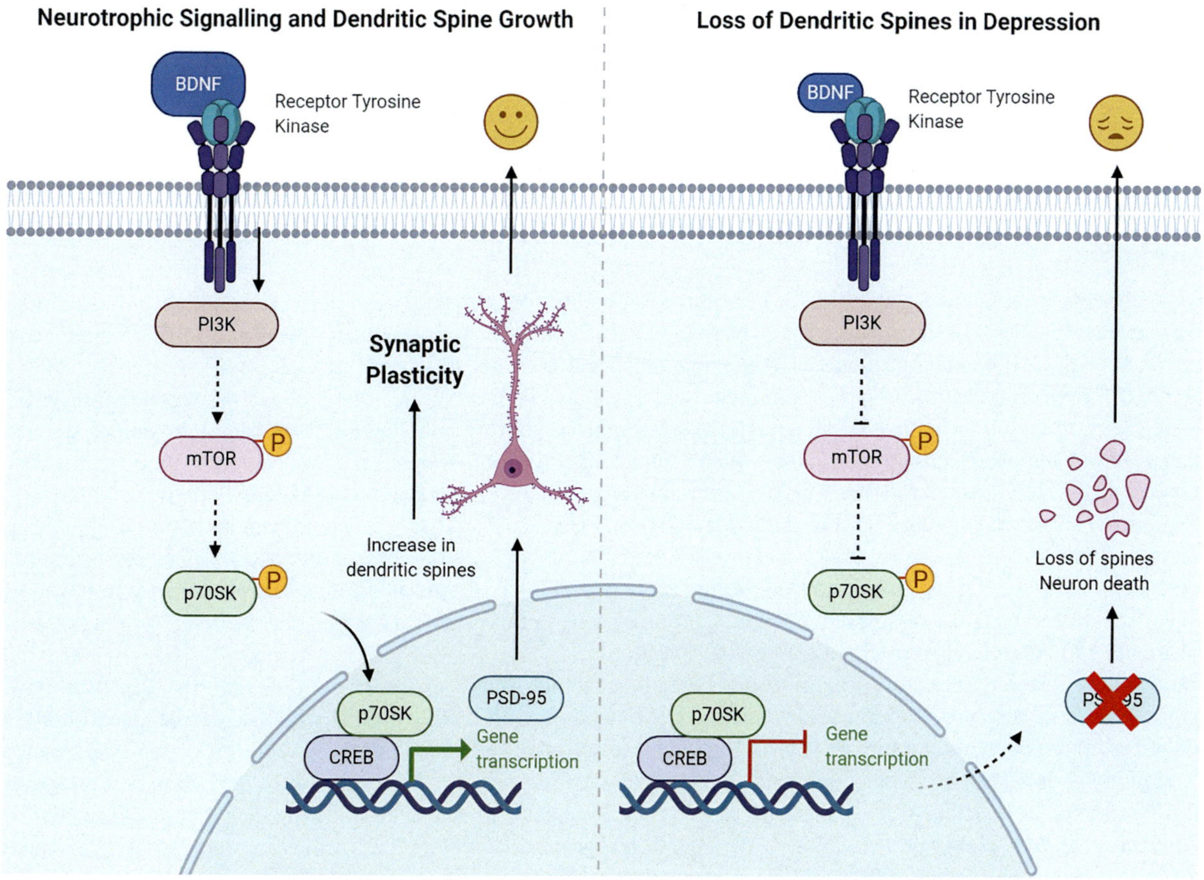

FIG. 5.2 Synaptic dysfunction in depression. Synapses form at the end of dendritic spines and become stabilized through activity (left). Neurotrophic signaling plays a critical role in upregulating dendritic spine growth. In depression, this system is attenuated, resulting in the downregulation of dendritic spines and synaptic connection (right). This dysfunction is reversed with antidepressants.

that plays a vital role in brain development, maturation, and ongoing life-long learning processes and is implicated in several major neuropsychiatric and neurodegenerative disorders (Forrest et al., 2018; Martella et al., 2018; Dorostkar et al., 2015). Synaptic plasticity occurs in part through activity-dependent, long-lasting modifications of synaptic strength, more commonly referred to as long-term potentiation (LTP) and long-term depression (LTD). The strengthening of synapses through LTP is triggered by *N*-methyl-D-aspartate (NMDA) receptors (NMDARs). A large NMDAR-dependent increase in dendritic spine calcium concentrations leads to activation of intracellular signaling cascades involving several protein kinases, most importantly Ca^{2+}/calmodulin-dependent protein kinase (CaMK). This leads to increased single-channel conductance of synaptic α-amino-3-hydroxy-5-methyl-4-isoxazole propionic acid receptor (AMPARs) and synaptic proteins such as PSD-95. In parallel, structural changes within the synapse occur, such that the size of the PSD and dendritic spine are increased, leading to an increase in synaptic strength, maintenance and ultimately impacting synaptic plasticity (Fig. 5.2).

Classical ADs elicit plasticity via three main routes: receptor blockade, reuptake inhibition, and inhibition of enzymes that degrade monoamines, particularly monoamine oxidase. The resulting excess of monoamine neurotransmitter availability at the synapse results in the regulation of postsynaptic G protein-coupled receptors (metabotropic receptors), which couple to a variety of second messenger systems, including the cAMP-protein kinase A (PKA)-cAMP response element-binding (CREB) pathway (Bourin et al., 2002). These actions result in slow adaptive processes that impact gene transcription and brain-derived neurotrophic growth factor (BDNF) signaling to ultimately alter synaptic regulatory elements that work to restore monoamine balance at the synapse (Hyman and Nestler, 1996; Castren and Hen, 2013; Dias et al., 2003; Schmidt and Duman, 2007; Shirayama et al., 2002).

In contrast, rapid-acting ADs such as ketamine, which functions as an NMDAR antagonist, enable synaptic plasticity via more direct activation of neurotrophic mechanisms (Deyama and Duman, 2020). A distinct difference between ketamine and slower-acting classical ADs is that ketamine triggers a rapid (within hours) upregulation of growth factor signaling and dendritic spine development in the PFC (Murrough et al., 2013; Zarate et al., 2006). Studies have provided evidence of ketamine activation of AMPAR, an ionotropic transmembrane glutamate receptor that mediates fast synaptic transmission and is essential for the generation of LTP. The activation of AMPAR leads to the release of BDNF and downstream activation of growth and metabolic signaling cascades, such as the phosphoinositide 3-kinase (PI3K/Akt/mTOR) signaling pathway and the MAPK/ERK signaling pathway in the limbic system to mediate AD effects (Zanos and Gould, 2018). The downstream actions of AMPA receptor activation lead to an increase in neurotrophins such as brain-derived neurotrophic factor (BDNF) levels and activation of its receptor tropomyosin receptor kinase B (TrkB) (Fig. 5.2).

Neurogenesis

Antidepressants stimulate the generation of newborn neurons from neural stem cells in the adult hippocampus, a process known as neurogenesis (Boldrini et al., 2009; Malberg et al., 2000), which takes about 4 weeks to form synaptic connections (Toni et al., 2007) and contributes to antidepressant-like effects in rodents (Denny et al., 2012; Kee et al., 2007; Hill et al., 2015). In mammals, this process is limited to select regions, the best-studied of which is the subgranular layer of the dentate gyrus of the hippocampus. Herein, adult newborn neurons are formed from quiescent neural progenitor stem cells, which differentiate into mature granule cells, which become integrated into the existing hippocampal circuitry. It is becoming increasingly clear that neurogenesis in the hippocampus and other regions plays a key role in both the pathophysiology and treatment of mood disorders (Hill et al., 2015; Sahay and Hen, 2007). There is a significant reduction in hippocampal volume in depressed relative to healthy subjects (Sheline, 1996; Sheline et al., 1996), with contrasting increases in volume in other regions, including the amygdala (MacMillan et al., 2003). Such structural differences are confirmed in postmortem studies, demonstrating, for example, more significant reductions in hippocampal histological sections from depressed patients relative to control subjects (Stockmeier et al., 2004).

Hippocampal overexpression of neurotrophic signaling factors BDNF or CREB mimics both the structural and behavioral effects of sustained antidepressant treatment (Shirayama et al., 2002). In line with this, studies have shown that antidepressant drugs, including tricyclic antidepressants and selective serotonin reuptake inhibitors, increase neuronal proliferation in the dentate gyrus when administered chronically (Malberg et al., 2000). However, the rate of cell proliferation and programmed cell death in the dentate gyrus of mice chronically treated with antidepressants does not appear to be dependent on BDNF. However, activation of the BDNF receptor TrkB is essential for the long-term survival of newborn neurons (Sairanen et al., 2005). These effects are dependent on chronic administration of antidepressant therapy, consistent with the time course of antidepressant treatments (Russo-Neustadt and Chen, 2005). This suggests that the regulations of BDNF and TrkB are involved in antidepressant treatment response and that this is directly reversing a critical pathophysiological feature of the illness. Whether this is a common mechanism of AD response remains to be determined.

Studying antidepressant response in preclinical models

Significant advancements have been made toward our understanding of the fundamental neurobiological mechanisms mediating AD response over the last decade, on the back of discovering the rapid-acting AD potential of ketamine. While this has made clear the critical role that neurotrophic mechanisms play in this process, the factors that interfere with this to effectively limit AD response are not well understood. Suitable preclinical scientific models are needed to address this gap. Development and use of depression models is challenging due to the complex nature of the disorder, and much has been written on their limitations (Nestler and Hyman, 2010; Iqbal et al., 2020; Becker et al., 2021). No one model can represent the full complexity of the human condition, particularly TRD. Arguably, to hold such aspirations is somewhat misplaced, and preclinical approaches that focus instead on the biological mechanisms of AD response and nonresponse could more effectively inform on the mechanisms contributing to treatment resistance. In this way, each model must be within the context of its underlying strength as part of a greater whole, referring back to the human condition for verification and optimization (Becker et al., 2021). The factors that can interfere with AD response are many and varied, and include biological factors such as genetic (Cuellar-Barboza et al., 2020; Athreya et al., 2019), inflammation (Miller and Raison, 2016; Arteaga-Henriquez et al., 2019; Mocking et al., 2017; Wong et al., 2008), neurotrophins (Brand et al., 2015; Lieb et al., 2018; Allen et al., 2015; Lopez et al., 2013), gut-brain interactions (Butler et al., 2019; Cryan and Dinan, 2012), neuroendocrine dysregulation (i.e., glucocorticoid resistance) (Oglodek et al., 2014), and bioenergetic processes (Price et al., 2018; Mehta et al., 2013). Consequently, the development of useful preclinical tools to investigate AD mechanisms must take these into account, collectively contributing to the development of a multi-scale, systems biology approach to addressing treatment resistance in depression (Akil et al., 2018). However, as for the human condition, there will be heterogeneity in the presence and impact of these. Thus a more direct focus on how these factors impact neurotrophic responses to ADs may enable us to more effectively screen for these treatment response interrupters in a manner that can enable personalized treatment options, potentially in combination with other clinical tools or recommendations that remove or reduce biological barriers to plasticity, and in turn AD response.

Animal models

Our understanding of the specific neurobiological mechanisms contributing to AD response within the context of animal models of depression positions us optimally to determine which factors functionally limit this molecular machinery's effective engagement. While never effectively recapitulating the human condition of depression, animal models offer the unique capacity to dissect out factors capable of altering the biological efficacy of ADs (Reus et al., 2019; Willner and Belzung, 2015). When used in partnership with clinical paradigms, they add significant value to the scientific endeavor to understand and more effectively treat what we currently, and somewhat pessimistically, define as TRD.

Advantages of research with animal model systems include:

1. high degree of experimental control over the environment;
2. ability to monitor biological effects across central and peripheral compartments in a highly selective, cell-specific manner;
3. focus on pathways of communication between central and peripheral compartments (including nervous and immune systems); and
4. assessment of integrated neurobiological and biobehavioral processes.

Without a doubt, experimental animal models are an indispensable preclinical research tool. To effectively model human depression, they are required to have face (phenotypic similarities), construct (neurobiological similarities), and predictive (treatment response similarities) validity (Nestler and Hyman, 2010). The latter requirement has particularly limited our capacity to develop valid models of TRD. Rodent models have long used different pathogenic factors to elicit depression-like behaviors (Table 5.1), including stress (Krishnan et al., 2007; Walker et al., 2013), genetic mechanisms (Aleksandrova et al., 2019; Vollmayr et al., 2004; Muller and Holsboer, 2006; Tarantino et al., 2011), inflammation (Couch et al., 2016; Zhao et al., 2019; Dantzer, 2017), aging (Kelley et al., 2013; Malatynska et al., 2012), and targeted network disruption (Muir et al., 2019). These models have also been used to study the biological basis of AD responses, including AD treatment resistance (Willner and Belzung, 2015; Walker et al., 2013, 2019; Kim et al., 2016a). These rodent models are used to determine how pathogenic mechanisms (e.g., chronic stress exposure) affect the brain to impact behavioral responses. In contrast, behavioral tests, or screening models, are used to access specific behavioral features that can be changed due to treatments. There is increasing attention being placed on improving our current psychiatric and translational paradigms to model human conditions more effectively (Willner and Belzung, 2015).

Rodent animal models

Rodent models are powerful tools for investigating the mechanistic role of biological factors associated with AD resistance in clinical and epidemiological studies. Early animal research in depression aimed to elucidate psychological processes influencing vulnerability to depression-like behavior. Building on this, more recently, animal models have been used to address neurobiological mechanisms regulating behavioral states of relevance to depression. These rodent models of "depression" have historically been validated via antidepressant reversal of "depression-like" behaviors, limiting their potential for use in the study of TRD. Alternative approaches are now needed to explore the mechanisms governing antidepressant nonresponse (Arteaga-Henriquez et al., 2019). This critical next step builds logically on the now well-established mechanisms of antidepressant response, allowing for the discovery of the pathophysiological processes contributing to poor antidepressant outcomes.

Rodent models of TRD, as for depression more broadly, involve a specific manipulation (a procedure used to elicit abnormal behavior) and an experimental measure (a behavioral output, often associated with biomarkers, used to measure AD response; Fig. 5.3).

Rodent behavioral tests of AD response

Progress toward modeling AD resistance and studying potential novel therapies therein requires a paradigm shift away from stress-based models of depression-like behaviors (Willner et al., 2013), with a new focus on screens of AD response and nonresponse (Willner and Belzung, 2015). Indeed, it makes intuitive sense that the ideal models to investigate the potential induction and/or reversal of factors contributing to TRD will be precisely those models in which conventional

TABLE 5.1 Rodent models of antidepressant resistance.

Models	Depressive-like behavior	Biological alterations	Antidepressant response
Wistar Kyoto rats (Will et al., 2003; Willner and Belzung, 2015; Willner et al., 2019)	Increased learned helplessness; increased immobility in the forced swim test; decreased activity in the open field test	Dysregulation of the hypothalamic-pituitary-adrenal (HPA) and hypothalamic-pituitary-thyroid (HPT) axes	Resistant to fluoxetine. Responsive to ketamine
Congenital learned helplessness (cLH) rats (Vollmayr et al., 2004; Willner and Belzung, 2015)	Deficient reward perception, depressive-like cognitive bias, increased LH and anhedonic; respond less to sucrose and show hyperactivity during open field test	Increased activity in the ACC and habenula; higher glutamate/GABA ratios in the hippocampus and prefrontal cortex	Responsive to deep brain stimulation
High anxiety behavior (HAB) mice (Willner and Belzung, 2015)	Increased immobility in the forced swim test	Reduced hippocampal neurogenesis	Response to chronic noradrenergic drugs (reboxetine or desipramine), but resistant to chronic selective serotonin reuptake inhibitors (SSRIs) including fluoxetine, paroxetine and citalopram. Good response to deep brain stimulation
5-HT1a mice (Pattij et al., 2002; Santarelli et al., 2003)	Increased anxiety and response to stress	Physiological responses to stress	No response to chronic SSRIs
5-HT1b mice (Vinkers et al., 2011)	Increased locomotor response to stress	Stress-induced locomotor responses	Response to SSRIs
Unpredictable chronic mild stress (UCMS) mice (Willner, 2016; Willner et al., 1987)	Impairment of rewarded behavior	Hypothalamic pituitary adrenal dysregulation, brain derived neurotrophic factor (BDNF) alterations, reduced monoamine levels	Response to antidepressants except when on a high-fat diet when they display resistance to fluoxetine. Good response to ketamine
Cannabinoid receptor type 1 knockout (CB1 KO) mice (Rácz et al., 2015; Robbe et al., 2002; Steiner et al., 2008)	Anhedonia, increased response to chronic mild stress	Impaired cannabinoid receptor type 1 affecting mesoaccumbens dopamine signaling and plasticity	Reduced response to desipramine or paroxetine subchronic administration
Organic Cation Transporter 2 (OCT2) null mutant mice (Bacq et al., 2012)	Increased anxiety, increased response to chronic mild stress and immobility in the forced swim test	Decreased levels of serotonin and noradrenaline	Response to noradrenaline and serotonin ADs, but resistance to venlafaxine
Dopamine β-hydroxylase (DBH) deficient mice (Cryan et al., 2001, 2004; Swoap et al., 2004; Thomas and Palmiter, 1997, 1998)	Increased response to stress	Lack of norepinephrine and epinephrine; low heart rate, hypotensive, attenuated circadian blood pressure rhythm	Resistance to acute administration of desipramine, reboxetine pargyline, bupropion and the SSRIs fluoxetine, sertraline and paroxetine. Response to citalopram
Hypothalamic pituitary adrenal (HPA) axis disruption (Kale et al., 2021; Kim et al., 2016a,b, Kitamura et al., 2002, 2010, Walker et al., 2013, 2015)	Anhedonia	Increased levels of serotonin, norepinephrine and epinephrine following stress; low dopamine	Resistance to tricyclic antidepressants, reversed with lithium augmentation. Bupropion response and partial ketamine response observed. Responsive to deep brain stimulation

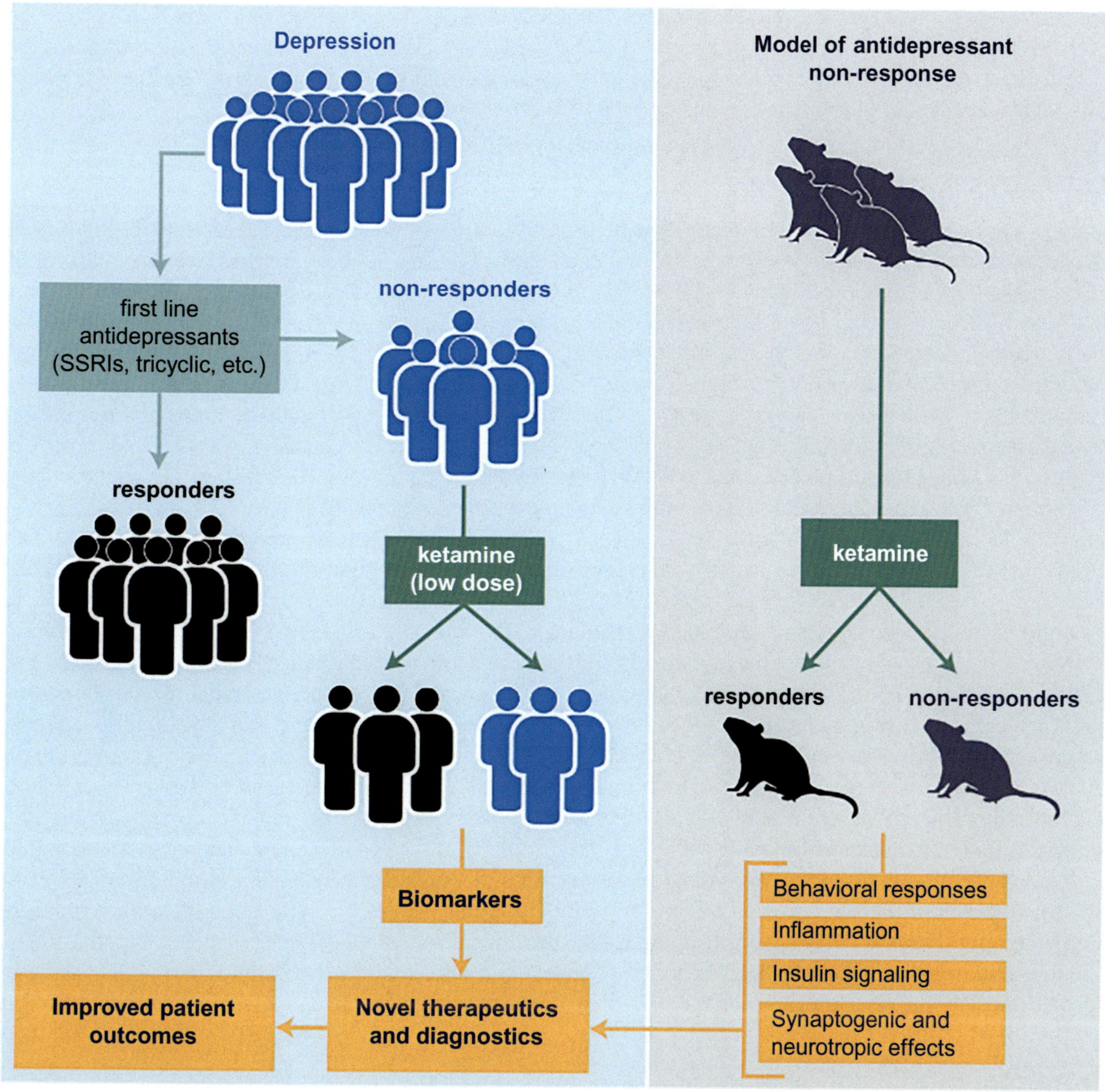

FIG. 5.3 Translational preclinical models for discovery of moderators of antidepressant response. Quantifying factors contributing to antidepressant resistance in rodent models, across central and peripheral measures, can inform biomarker approaches for use in clinical populations. Collectively this translational approach will lead to novel therapeutics and diagnostics, to improve patient outcomes. Rapid-acting antidepressants are ideal for this approach given their acute therapeutic actions.

antidepressants are ineffective. For decades behavioral screening tests for ADs have been used primarily for the development of new treatments. Some of the procedures that are most frequently used in AD research, such as the forced swim test (FST) (Porsolt et al., 1977a,b) and the tail suspension test (TST) (Steru et al., 1985), involve acute or chronic treatment administration and provide effective first pass screens of AD effects on active versus passive behavioral coping responses to inescapable stress. An AD-induced shift toward increased active stress coping (climbing or swimming in the FST or climbing in the TST) serves as a proxy behavioral readout for AD effects on key regions of the neural network, including the infralimbic and prelimbic PFC, as well as the mesoaccumbens dopamine systems, the stimulation of which alone is sufficient to enhance active coping behaviors.

Even when administered acutely, drug effects in this paradigm are considered predictive of longer-term AD efficacy in human patients. When used to test novel compounds, these behavioral screens can be undertaken in nondepressed animals. Thus, the FST behavioral assay is considered a useful, predictive screening tool for antidepressant-like activity, rather than an animal model of depression per se (Hyman and Nestler, 1996). When misrepresented as the latter, these tests rightly draw intense criticism. However, their utility for rapid AD screens should not be hastily dismissed. Indeed, the ability to demonstrate efficacy to specific AD treatments at baseline and then block this through induction of AD resistance enables direct ascertainment of the neurobiological mechanisms implicated in AD nonresponse, which may have physiologic relevance for TRD. Such models have demonstrated utility in the search for mechanisms blocking AD efficacy and uncovering related biomarkers of response. Within this context, rapid behavioral screens of AD response can help discover mechanisms moderating AD nonresponse of relevance to clinical TRD.

Translational rodent behavioral tests

Moving from predictive AD screens to translational behavioral models of TRD is essential for understanding the core pathophysiological underpinnings of distinct behavioral phenotypic features. Such approaches have been informed by the United States National Institute of Mental Health Research Domain Criteria (RDoC) framework, which aims, in part, to develop research paradigms with broad trans-species applications (Insel et al., 2010). In so doing, this approach aims to integrate biological, behavioral, and clinical data to explain transdiagnostic dimensions of human functioning that span the full behavioral spectrum. Importantly, through the development of behavioral tests that can be used across several species, from rodents to humans, this approach allows for discovery of neurobiological underpinnings of specific behavioral features (e.g., distinct reward processes dependent on specific neurobiological mechanisms (Der-Avakian et al., 2016). In the context of preclinical models of TRD, this will help establish a more complete picture of how discrete clinical presentations can impact AD treatment response, potentially allowing for the development of a set of diagnostic screening tools for treatment selection. In this way, by simulating essentially the same process across species, the behavioral models share necessary homology with specific aspects of psychiatric disorders and provide significant translational value (Table 5.2).

This newer generation of translational tasks has emerged, theoretically and procedurally analogous across species, to enable parallel testing between human and rodent behaviors (Der-Avakian et al., 2016). This enhanced conformity between cross-species tasks will facilitate the investigation of the neurobiological mechanisms underlying discrete behaviors impacted by ADs. Such approaches are directly improving our understanding of the biological mechanisms of specific phenotypic features, such as inflammation-induced disruption of motivational systems, as one example. Understanding which neurobiological processes are affected can then directly inform treatment selection. However, in contrast to the high-throughput rapid AD screens described above, high levels of sophistication are necessary for cross-species translational behavioral paradigms to probe increasingly discrete aspects of the behavioral task that are not confounded by other factors. For example, in studying motivated behavior, enhancing the specificity of the reward construct being assessed will improve the likelihood of identifying discrete neurobiological mechanisms that underlie the specific reward process (Der-Avakian et al., 2016). However, this often comes at the cost of weeks to months of behavioral training necessary to perform the behavior effectively. Ultimately, this approach allows for the preclinical animal model to test specific hypotheses regarding novel treatment strategies that have a high likelihood of successfully translating to the clinic. It is anticipated that these cross-species translational research approaches will lead to the development of safe and effective treatments for TRD.

Other animal models of depression

While depression-like behavioral responses and antidepressant-induced changes in stress coping have long been studied in rodents, other model species can be used to investigate evolutionarily conserved neurobiological underpinnings of these responses. Nonhuman primates, for example, have been used to bridge the gap between rodents and humans, enabling higher-order behavioral assessments and greater neurobiological homology with the human brain (Barr et al., 2003). However, experimental numbers are limited, and associated costs are high, inherently limiting their use for broader AD screening and biomarker-related research for TRD. Maternal (Harro, 2013) or social separation (Worlein, 2014) paradigms, as well as inflammatory challenge, (Felger et al., 2007) have been used in nonhuman primate models to reflect specific features of human depression (Vellucci, 1990), including anhedonia and lethargy (Alexander et al., 2019). As for rodent models, nonhuman primate models of depression have been pharmacologically validated by reversing these depression-like behavioral traits using common antidepressants, which work on a similar time course to that observed clinically (Rasmussen and Reite, 1982). This, by definition, limits their utility for use in the study of TRD. However, other approaches, including the observational studies of spontaneous depression-like behaviors in macaques (Kalidindi et al., 2017) or specific models of AD-disrupting elements, such as cytokine-induced depression (Felger et al., 2007), could prove useful in bridging this gap.

Alternative cost-effective, high-throughput TRD models capable of screening promising lead compounds for drug discovery are needed. Zebrafish represent a novel preclinical model organism for the study of neural mechanism involved in behaviors of relevance to psychiatric disorders such as anxiety, addiction, autism and depression (Demin et al., 2018; Khan et al., 2017). This model affords important advantages, lending itself not just to low-cost, high-throughput approaches, but enabling development of sophisticated reporter lines that provide real-time insight into the biological mechanisms

TABLE 5.2 Behavioral tests of antidepressant response.

Behavioral test	Measurement	Validity	Value
– Forced swim test	– Exert effort to escape or passive coping	– Predictive validity of antidepressant response	– Useful as a first-pass, high-throughput drug screen
– Tail suspension test	– Exert effort to escape or passive coping	– Sensitive to acute and chronic antidepressant treatment	– Behavioral measures of active and passive stress coping are regulated by specific neural circuits involved in stress coping
		– Engages neural circuitry involved in stress coping and mood regulation in humans	– Can be used to quantify biological mechanisms and markers associated with antidepressant-induced shifts in stress coping response
		– Detects acute responses to antidepressants that predict long-term therapeutic adaptations	– Predictive antidepressant-response validty only. Lack of face validity prevents use to model depression-like behavior
– Learned helplessness	Exert effort to escape or passive coping	– Predictive and face validity of antidepressant response	– Useful for quantifying biological mechanisms and markers associated with passive stress coping (i.e., perceived helplessness)
		– Sensitive to acute and chronic antidepressant treatment effects	– Valid cross-species behavioral paradigms available, with relevance to clinical feature of helplessness/hopelessness
		– Engages neural circuitry involved in stress coping and mood regulation in humans	– Can serve as an animal model of depression-like behavior
Effort-related choice task	Choice to exert effort for more salient reward	– Face validity of antidepressant response of relevance to specific features (i.e., anhedonia)	– Useful for quantifying biological mechanisms and markers associated with specific deficits in motivated behavior relevant to a subset of depressed patients
		– Sensitive to acute and chronic antidepressant treatment	– Detects deficits in dopaminergic signaling, which may predict better outcomes to treatments targeted to the dopamine, rather than noradrenergic and serotonergic systems
		– Engages neural circuitry involved in motivation and mood regulation in humans	– Can be used as a cross-species behavioral task, with relevance to clinical feature of anhedonia
		– Sensitive to deficits in ventral striatum activity	

Continued

TABLE 5.2 Behavioral tests of antidepressant response—cont'd

Behavioral test	Measurement	Validity	Value
Novelty suppressed feeding	Motivation to seek reward relative to the anxiety induced by the challenge	– Predictive validity of antidepressant and anxiolytic response	– Useful as a first-pass, high-throughput drug screen
		– Sensitive to acute and chronic antidepressant treatment effects	– Contrasting effects of anxiety can suppress motivated drive to approach and consume food
		– Engages neural circuitry involved in regulating motivation and anxiety in humans	– Provides direct measure of antidepressant effects on anxiety-like and motivated behavior
			– Can be used to quantify biological mechanisms and markers associated with antidepressant-induced shifts in motivation and anxiety-related behavior
			– Cannot be used to model depression-like behavior
Elevated plus maze (or similar)	Anxiety-related avoidance behavior	– Predictive validity of anxiolytic response	– Useful as a first-pass drug screen for effects on anxiety-related behaviors
Light/dark box		– Engages neural circuitry involved in regulating motivation and anxiety in humans	– Provides direct measure of anxiolytic effects
			– Can be used to quantify biological mechanisms and markers associated with antidepressant-induced shifts in anxiety-related behavior
			– Cannot be used to model depression-like behavior

impacting behavioral outputs. Such models can complement work done in rodent modes, and indeed, many of the behavioral paradigms used in zebra fish research are based on rodent protocols described above (Meshalkina et al., 2017). The novel tank test (NTT), in which anxiety-like behavior is represented by diving to avoid the unprotected upper area, and the light-dark test (LDT), in which the animals prefer the darkened area of the apparatus (Khan et al., 2017; Meshalkina et al., 2017). Similar tests for depression-like behaviors, such as the zebrafish learned helplessness paradigm are not as well-established (de Abreu et al., 2018). New zebra fish models are needed for development of high-throughput TRD drug screens. As for rodent models, such behavioral screens would first need to demonstrate, and then perturb, behavioral responses to available treatments; prior to screening for new AD compounds.

Alternative cost-effective, high-throughput TRD models capable of screening promising lead compounds for drug discovery are needed. Zebrafish represent a novel preclinical model organism to study the neural mechanism involved in behaviors of relevance to psychiatric disorders such as anxiety, addiction, autism, and depression (Demin et al., 2018; Khan et al., 2017). This model affords important advantages, lending itself not just to low-cost, high-throughput approaches, but enabling the development of sophisticated reporter lines that provide real-time insight into the biological mechanisms impacting behavioral outputs. Such models can complement work done in rodent modes, and indeed, many of the behavioral paradigms used in zebrafish research are based on rodent protocols described above (Meshalkina et al., 2017). The novel tank test (NTT), in which anxiety-like behavior is represented by diving to avoid the unprotected upper area, and the light-dark test (LDT), in which the animals prefer the darkened area of the apparatus (Khan et al., 2017; Meshalkina et al., 2017). Similar tests for depression-like behaviors, such as the zebrafish learned helplessness paradigm, are not as well-established (de Abreu et al., 2018). New zebrafish models are needed for the development of high-throughput TRD drug screens. As for rodent models, such behavioral screens would first need to demonstrate, and then perturb, behavioral responses to available treatments; before screening for new AD compounds.

Stress, genetic and pharmacological approaches are possible manipulations for TRD model development (de Abreu et al., 2018), with other work already demonstrating the potential to elicit related pathophysiological states (i.e., increased anxiety-like behavior, cortisol and inflammatory mediators, and decreased and monoamine levels) (de Abreu et al., 2018; Manuel et al., 2014; Chakravarty et al., 2013; Song et al., 2018). Perhaps most promisingly, ADs can reverse these physiological and behavioral effects as for rodent models (Song et al., 2018). However, an important distinction between zebrafish and mammalian model responses to stress is that they demonstrate opposite neurotrophic responses to stress, with neurotrophic factor levels elevated in zebrafish but downregulated in rodents, nonhuman primates, and human subjects (Song et al., 2018). Given the critical role that neurotrophic signaling plays in AD response, this represents an important distinction. However, this may reflect critical differences in neuroanatomical structures across the models, with relatively great real-estate dedicated to limbic brain functions in zebrafish. Indeed, while downregulation of neurotrophic factors are observed in the PFC and hippocampus of mammals during depression-like states, contrasting upregulation of neurotrophic signaling is also observed in limbic areas such as the amygdala, which is highly sensitive to stress and plays a critical role in promoting anxiety-like behaviors. Indeed, fish possess a structure homologous to the mammalian amygdala (Perathoner et al., 2016). Thus, zebrafish may offer a more focused, complementary model organism, lending itself to quantify stress-related behavioral changes more effectively. Still, their utility in biological psychiatry requires further development for more in-depth biobehavioral characterization (Demin et al., 2019).

Cellular models

Cellular models provide an opportunity to examine AD-induced molecular signaling cascades, providing direct access to patient-specific biological responses and a direct translational link between animal and human studies. Early work in laboratory cell lines has helped to define the molecular actions of ADs. For example, much work on AD-induced effects on cellular vulnerability and neurotrophic signaling has been undertaken in cell lines, such as PC12 cells. This cell line is derived from cells of embryonic origin from the rat adrenal medulla and was initially cultured by Greene and Tischler (1976). These neuroblastic cells can be differentiated into neuron-like cells, sharing many fundamental properties with neurons, including the ability to produce and release neurotransmitter vesicles and activate neurotrophic signaling to promote dendritic spine growth. The PC12 cell line has proven to be highly versatile and easy to use to study pro- or antidepressant-induced drug actions on neuroprotection and neuroplasticity associated markers, together with corresponding impacts on cellular viability, apoptosis, and plasticity (Zeng et al., 2016; Gong et al., 2018). Indeed, these and related studies have helped define the exquisite insight we have into the molecular and cellular mechanisms through which ADs act on neuronal cells. However, a significant challenge for the field is to translate these preclinical models to the clinic to identify patient-specific factors that may impact cellular responses to ADs. For this reason, patient-specific cell-lines have been developed to investigate these processes and advance precision medicine approaches for TRD.

Induced pluripotent stem cells

Until relatively recently, the developmental path of cellular differentiation and maturation was thought to be irreversible such that a differentiated cell could not regress to a state of pluripotency and then redifferentiate into a different cell type. However, this biological dogma was disrupted in 2006, when Shinya Yamanaka demonstrated that "cellular reprogramming" was indeed possible, allowing specialized cells, such as skin cells, to be transformed into a pluripotent state through the transient expression of four genes (Oct3/4, Sox2, c-Myc, and Klf4) (Shi et al., 2008; Takahashi and Yamanaka, 2006; Takahashi et al., 2007). These cells, aptly named "induced pluripotent stem cells" (iPSCs), afford the pluripotency necessary for developing specialized human neuronal cell lines, without the ethical implications associated with the use of embryonic tissue and with the valuable option of deriving the original cells directly from patients. Since 2006, cellular reprogramming capabilities have increasingly improved, increasing the efficiency of reprogramming outcomes and resulting in the development of pluripotent cells free from genome alterations and suitably reliable to be considered "clinical grade" (Li and Izpisua Belmonte, 2018). The development of iPSCs has revolutionized opportunities for preclinical modeling of TRD, affording new opportunities to use patient-derived cells that can be used to quantify individualized drug actions on

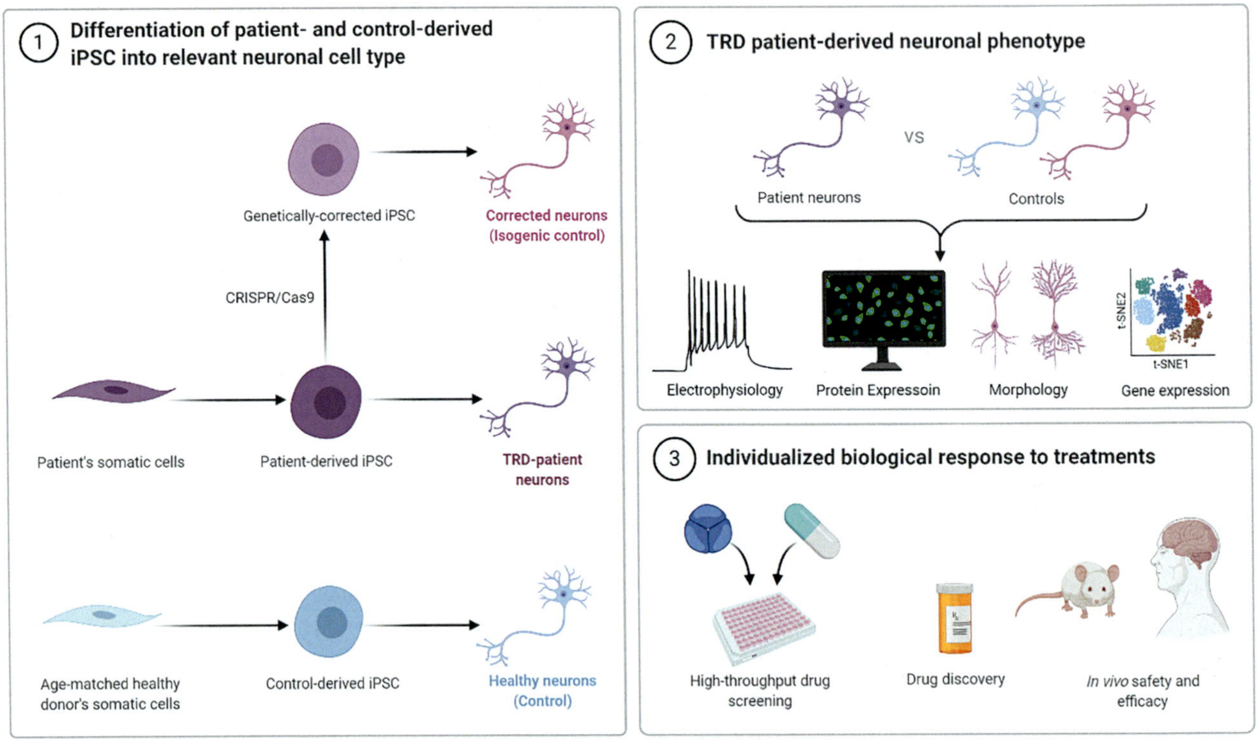

FIG. 5.4 iPSC technology for investigating patient specific factors impacting antidepressant response. (1) iPSCs can be derived from patient cells and transformed into neuronal cells. (2) Cellular phenotyping and testing of cellular response to antidepressants can inform on genetic mechanisms of TRD, and (3) aid in the identification of drug discovery and individualized treatments.

molecular pathways targeted by ADs to confer neuroprotection and plasticity. The advantage of this approach is that cells carry the same genetic material as the patients requiring treatment (Hamazaki et al., 2017). Thus, iPSCs provide valuable tools for developing patient-derived brain cell models capable of producing either neuronal and/or glial cell lines and are particularly valuable for pharmacogenomics studies (Fig. 5.4).

Several iPSC lines have been derived from fibroblasts and peripheral blood mononuclear cells (PBMCs) from patients with schizophrenia, autism spectrum disorders, major depressive disorder, and bipolar spectrum disorders (Corti et al., 2015; Rowe and Daley, 2019; Marcatili et al., 2016). iPSCs provide a unique opportunity to study how single-nucleotide polymorphisms (SNPs) impact critical neuronal functions, including electrophysiological properties, neurotransmitter release and reuptake dynamics, as well as subcellular (i.e., mitochondria) and molecular (i.e., neurotrophic signaling) responses to treatment on a previously inaccessible level. Indeed, patient-derived neurons, developed using iPSC technology, have been successfully utilized in studies from patients with schizophrenia, bipolar disorder, and autism spectrum disorder, each of which has a high heritability component (Brennand et al., 2011; Mertens et al., 2015; Marchetto et al., 2017; Yoon et al., 2014; Madison et al., 2015). However, the use of iPSCs for studies in depression is more challenging, likely in part due to the clinical and genetic heterogeneity (Wray et al., 2018). However, the impact of environmental risk factors, such as childhood abuse, and any epigenetic imprinting or broader physiological conditions such as disrupted glucocorticoid feedback that can impact cellular functions underpinning AD response are eliminated as part of the iPSC protocol. The two available iPSC-based studies in depression are from the same research group and focused on TRD. In these studies, researchers stratified patients based on SSRI response and studied serotonergic neurotransmission (Vadodaria et al., 2019a,b). They observed serotonin-induced hyperactivity in SSRI-nonresponsive cells, relative to neurons derived from iPSCs generated from healthy or SSRI-responsive patient cells. The investigators were then able to demonstrate that pharmacological blockade of serotonin receptors with Lurasidone partially revered the serotonin-induced neuronal hyperactivity. Based on these biological characteristics, a mechanistic mismatch may have occurred with the original SSRI treatment plan, suggesting instead that alternate treatment options, such as the serotonergic antagonist, may have improved clinical efficacy in these individuals (Vadodaria et al., 2019b). The subsequent study also demonstrated critical distinctions in neurite growth and morphology in serotonergic neurons derived from iPSCs from SSRI nonresponsive individuals relative to healthy controls and SSRI responsive individuals. Via the direct regulation of specific genes, the team demonstrated fundamental mechanisms governing these inherent structural differences (Vadodaria et al., 2019a). The observed differences in neurite length and morphology in cells from SSRI-responsive and nonresponsive patients further imply that intrinsic structural and functional differences are present in the circuitry of TRD patients (Marcatili et al., 2020). This reiterates the need to link preclinical screening models with translational behavioral measures that provide quantifiable measures of specific neural circuits' functional integrity.

Cocultured cellular models

In the brain, interdependent cell subtypes work together to coregulate synaptic responses and plasticity. Preclinical cell culture modeling of TRD can take this complexity into account by coculturing multiple cell types. In combination with iPSC reprogramming techniques, it is now possible to produce patient-derived synaptic models, such as the glutamatergic synapse, by incorporating glutamatergic cells, GABAergic interneurons, and astrocytes (Marcatili et al., 2020). Such patient-derived "brain in a dish" models can be used to determine patient-specific effects of ADs across several levels, including:

1. Morphological: area of the soma, level of branching, dendritic spines density, etc.;
2. Functional: functional electrophysiological or neurochemical responses to stimulus;
3. Molecular imaging: calcium signaling, protein trafficking, etc.;
4. Metabolic flux: shifts in glucose metabolism, etc.;
5. Molecular signaling: gene and protein expression;
6. Antigenic: expression of cellular markers for neuronal subtype, etc.

This framework enhances the biological sophistication of patient-derived cellular models and has the potential to enable the identification of specific mechanisms of AD resistance that lie at the intersection of synaptic communication between cell types. To appropriately understand the relevance of the multitude of factors at play here, multiple neuroscience methods need to be applied in combination with sophisticated bioinformatics approaches for multilevel analyses. Collectively, these approaches will help to identify the interdependent mechanism moderating AD response/nonresponse (Lin et al., 2016).

High-throughput peripheral blood mononuclear cell assays

While the use of patient-derived iPSCs models creates exciting new opportunities for studying the mechanisms of human TRD, this approach remains relatively low throughput, limiting its potential to characterize the complex neuronal mechanisms involved in TRD broadly. Moreover, the reprogramming necessary for producing iPSCs, and then differentiating them into neuronal cells requires altering the epigenetic code. Thus, while these cells are useful for looking at the impact of individual genetic factors on neural functioning, they do not account for environmental influences that can serve as potent regulators of cellular responses. Peripheral blood mononuclear cells (PBMCs) provide an alternative to iPSCs in providing patient-derived cells for phenotypic analysis, with the advantage of carrying both the genetic and epigenetic code. When analyzed directly from the patient, they maintain a level of physiologic memory that can inform on cellular response to specific treatment exposures (Mehta et al., 2013). PBMCs can be used to characterize gene expression at baseline (Bekhbat et al., 2020) or following drug treatment (Mehta et al., 2013), or can be cultured and exposed to specific stimuli to assess molecular and cellular responses of relevance to TRD (Walker et al., 2019; Roy et al., 2021). While early criticism for this approach indicates that this cell type was too far removed from the brain to inform neurotrophic responses of relevance to antidepressant response, recent studies have shown that PBMCs serve as valid proxies for AD-induced effects on the brain associated with treatment response (Walker et al., 2019; Roy et al., 2021). Such approaches have been used effectively for drug discovery in complex disorders such as diabetes (Konstantopoulos et al., 2011; Sithara et al., 2017) and mood disorders (Kidnapillai et al., 2020). Such approaches will no doubt prove beneficial in screening for novel treatments for TRD.

Multiscale reverse translation modeling of TRD

A reverse translation approach based upon the recognized complexity of human TRD has the opportunity to address the converging yet heterogeneous factors contributing to poor antidepressant outcomes. Such an approach must necessarily employ a range of animal models employing different aspects of the human condition. While progress has been made, human TRD has not been optimally modeled in the preclinical setting to date. As noted earlier, this is partly because the field has relied heavily on AD response as a tool to validate the depression-like phenotypes induced by particular exposures, such as stress. This, in turn, has left a significant gap in our understanding of those factors contributing to AD nonresponse. There remains a significant opportunity to leverage clinical insights to improve on currently available models of TRD. For example, genetic variation moderates the biological and behavioral responses to chronic stress, but existing animal studies have not effectively modeled such gene × environment interactions. Such studies would complement the iPSC work described above to provide further insight into the biological systems impact and consequences for AD drug effects on the brain and behavior. Further, the impact of multiple hits and/or early life stress on reactions to a novel stressor, hormonal or inflammatory challenge has not been sufficiently studied (Gray et al., 2014; Datson et al., 2013), given these interactions are similarly implicated in poor AD outcomes (Price et al., 2018).

The biological and behavioral impacts of multiple stress exposures *and* multiple courses of antidepressant treatment, both typical of clinical cases of TRD, are even less well explored. Additionally, the concurrent impacts of poor diet choices (high sugar and/or high fat), circadian disruption, and/or inflammation as multisystem insults have not been adequately explored. Such factors may be missed opportunities for modeling the pathophysiological underpinnings of poor AD outcomes, given both childhood adversity, circadian disruption, impaired metabolic capacity, inflammation, and stress in adulthood predict increased risk for TRD following consecutive antidepressant treatments (Thase, 2011; Nanni et al., 2012; Williams et al., 2016; Chen et al., 2018; Kruse et al., 2018; Woo et al., 2016; Toups et al., 2013). Systematic exploration of the impact of such multisystem challenges will provide the basis for novel and translationally relevant preclinical models examining the mechanisms moderating AD responses. Such models provide a unique opportunity to study convergence and divergence of biological changes associated with AD nonresponse to inform multiple levels of analyses, from behavior to neural circuitry to cellular, molecular and genetic mechanisms, consistent with RDoC approaches.

Conclusions

Preclinical models of TRD are urgently needed, given approximately 30% AD nonresponse rates in the clinical population across treatment modalities (Nierenberg and Amsterdam, 1990; Fava, 2003; Rush et al., 2006). Existing models have made reasonable progress in recent years; however, they have not been used to their full potential for the discovery of central and peripheral biomarkers of treatment resistance, nor for novel drug development. Preclinical models of TRD, particularly those screening AD drug responses, offer real potential for developing rational biomarker strategies that can inform treatment selection (Price et al., 2018; Kalueff et al., 2015). The focus must move away from using available ADs (e.g., SSRIs, tricyclics, etc.) to validate models. Instead, models where AD efficacy is either low, to begin with, or wanes with repeated stressors are more translationally relevant. As molecular mechanisms moderating treatment responsivity are determined, alternate pharmacological approaches can be pursued, potentially repurposing treatments already approved for other indications (Price et al., 2018).

Building on this understanding of the common mechanisms of AD actions across distinct pharmacological drug classes, we have a new opportunity to discover those mechanisms blocking antidepressant response. We can utilize recent innovations in animal and cellular models to understand what rate-limiting factors may functionally block these responses to treatment with an eye toward neurotrophic signaling. This, too, represents a paradigm shift, as we shift our focus from modeling depression-like behavior in rodents to studying antidepressant response. It is essential to recognize the two may be interdependent processes with overlapping behavioral consequences. As such, we can explore mechanisms of antidepressant response in new ways, which removes some of the constraints necessary for modeling depression. When modeling antidepressant response, predictive validity becomes a key metric of interest as we work toward elucidating fundamental mechanisms underpinning these responses at the molecular, cellular, and systems level. Having face validity at the newer biological level becomes more critical than having face validity at the biological level. Of course, moving toward an era of genuinely translational behavioral paradigms will be increasingly beneficial when translating from their preclinical environment to the clinic. However, this also necessitates innovations in clinical neuroscience and clinical trial where we mirror the behavioral paradigms and the neurobiological deficits underlying them. Using these approaches will move us into the new era of psychiatry in which we can more effectively target each patient's unique pathophysiology with precision medicine approaches. Such approaches have the potential to change our definition of treatment-resistant depression. No longer will it be viewed simply as an algorithmic trial and error process where two failed therapeutic attempts will render someone with a pessimistic prognosis of treatment-resistant depression. Instead, we will be looking to the individual and attempting to understand what mechanisms unique to their physiology limit their capacity to mount neurotrophic responses to treatments. Moreover, with this individualized molecular, biochemical road map, we will be positioned as a field to target these roadblocks and overcome them or find new pathways around them to illicit adaptive newer traffic capacity at the cellular level.

Our vision, arguably, should be focused on shifting the curve of treatment response for each patient from nonresponse to full therapeutic capacity. This is not to negate the other contributing social and psychological factors that limit therapeutic recovery capacity from depression. Without a doubt, there are critical there is a critical hierarchy of needs that every individual requires. Antidepressants cannot solve significant life stresses such as loss of job or homelessness or the death of a loved one. These require critical psychological and social interventions to help an individual in the recovery process. However, hand in hand with these, we must appreciate the power of biology to shift one's ability to adapt and cope with these sorts of stresses. With all else held equal, shifting the biological baseline provides individuals with increased capacity to navigate stress and repair the brain from the cumulative wear and tear that occurs with chronic stress and associated inflammation and metabolic dysfunction. Preclinical models of TRD will be essential for developing a clear and effective road map individualized treatment approaches that target the biological factors currently limiting AD response to enable recovery from depression.

References

Akil, H., Gordon, J., Hen, R., Javitch, J., Mayberg, H., McEwen, B., Meaney, M.J., Nestler, E.J., 2018. Treatment resistant depression: a multi-scale, systems biology approach. Neurosci. Biobehav. Rev. 84, 272–288.

Aleksandrova, L.R., Wang, Y.T., Phillips, A.G., 2019. Evaluation of the Wistar-Kyoto rat model of depression and the role of synaptic plasticity in depression and antidepressant response. Neurosci. Biobehav. Rev. 105, 1–23.

Alexander, L., Gaskin, P.L.R., Sawiak, S.J., Fryer, T.D., Hong, Y.T., Cockcroft, G.J., Clarke, H.F., Roberts, A.C., 2019. Fractionating blunted reward processing characteristic of anhedonia by over-activating primate subgenual anterior cingulate cortex. Neuron 101 (2), 307–320. e6.

Allen, A.P., Naughton, M., Dowling, J., Walsh, A., Ismail, F., Shorten, G., Scott, L., McLoughlin, D.M., Cryan, J.F., Dinan, T.G., Clarke, G., 2015. Serum BDNF as a peripheral biomarker of treatment-resistant depression and the rapid antidepressant response: a comparison of ketamine and ECT. J. Affect. Disord. 186, 306–311.

Arteaga-Henriquez, G., Simon, M.S., Burger, B., Weidinger, E., Wijkhuijs, A., Arolt, V., Birkenhager, T.K., Musil, R., Muller, N., Drexhage, H.A., 2019. Low-grade inflammation as a predictor of antidepressant and anti-inflammatory therapy response in MDD patients: a systematic review of the literature in combination with an analysis of experimental data collected in the EU-MOODINFLAME consortium. Front. Psychiatry 10, 458.

Athreya, A.P., Neavin, D., Carrillo-Roa, T., Skime, M., Biernacka, J., Frye, M.A., Rush, A.J., Wang, L., Binder, E.B., Iyer, R.K., Weinshilboum, R.M., Bobo, W.V., 2019. Pharmacogenomics-driven prediction of antidepressant treatment outcomes: a machine-learning approach with multi-trial replication. Clin. Pharmacol. Ther. 106 (4), 855–865.

Bacq, A., Balasse, L., Biala, G., Guiard, B., Gardier, A.M., Schinkel, A., Louis, F., Vialou, V., Martres, M.P., Chevarin, C., Hamon, M, Giros, B., Gautron, S., 2012. Organic cation transporter 2 controls brain norepinephrine and serotonin clearance and antidepressant response. Mol. Psychiatry 17 (9), 926–939.

Barr, C.S., Newman, T.K., Becker, M.L., Parker, C.C., Champoux, M., Lesch, K.P., Goldman, D., Suomi, S.J., Higley, J.D., 2003. The utility of the non-human primate; model for studying gene by environment interactions in behavioral research. Genes Brain Behav. 2 (6), 336–340.

Becker, M., Pinhasov, A., Ornoy, A., 2021. Animal models of depression: what can they teach us about the human disease? Diagnostics (Basel) 11 (1), 123.

Bekhbat, M., Treadway, M.T., Goldsmith, D.R., Woolwine, B.J., Haroon, E., Miller, A.H., Felger, J.C., 2020. Gene signatures in peripheral blood immune cells related to insulin resistance and low tyrosine metabolism define a sub-type of depression with high CRP and anhedonia. Brain Behav. Immun. 88, 161–165.

Boldrini, M., Underwood, M.D., Hen, R., Rosoklija, G.B., Dwork, A.J., John Mann, J., Arango, V., 2009. Antidepressants increase neural progenitor cells in the human hippocampus. Neuropsychopharmacology 34 (11), 2376–2389.

Borczyk, M., Sliwinska, M.A., Caly, A., Bernas, T., Radwanska, K., 2019. Neuronal plasticity affects correlation between the size of dendritic spine and its postsynaptic density. Sci. Rep. 9 (1), 1693.

Bourin, M., David, D.J., Jolliet, P., Gardier, A., 2002. Mechanism of action of antidepressants and therapeutic perspectives. Therapie 57 (4), 385–396.

Brand, S.J., Moller, M., Harvey, B.H., 2015. A review of biomarkers in mood and psychotic disorders: a dissection of clinical vs. preclinical correlates. Curr. Neuropharmacol. 13 (3), 324–368.

Brennand, K.J., Simone, A., Jou, J., Gelboin-Burkhart, C., Tran, N., Sangar, S., Li, Y., Mu, Y., Chen, G., Yu, D., McCarthy, S., Sebat, J., Gage, F.H., 2011. Modelling schizophrenia using human induced pluripotent stem cells. Nature 473 (7346), 221–225.

Butler, M.I., Sandhu, K., Cryan, J.F., Dinan, T.G., 2019. From isoniazid to psychobiotics: the gut microbiome as a new antidepressant target. Br. J. Hosp. Med. (Lond.) 80 (3), 139–145.

Castren, E., Hen, R., 2013. Neuronal plasticity and antidepressant actions. Trends Neurosci. 36 (5), 259–267.

Chakravarty, S., Reddy, B.R., Sudhakar, S.R., Saxena, S., Das, T., Meghah, V., Brahmendra Swamy, C.V., Kumar, A., Idris, M.M., 2013. Chronic unpredictable stress (CUS)-induced anxiety and related mood disorders in a zebrafish model: altered brain proteome profile implicates mitochondrial dysfunction. PLoS One 8 (5), e63302.

Chen, M.H., Li, C.T., Lin, W.C., Hong, C.J., Tu, P.C., Bai, Y.M., Cheng, C.M., Su, T.P., 2018. Rapid inflammation modulation and antidepressant efficacy of a low-dose ketamine infusion in treatment-resistant depression: a randomized, double-blind control study. Psychiatry Res. 269, 207–211.

Corti, S., Faravelli, I., Cardano, M., Conti, L., 2015. Human pluripotent stem cells as tools for neurodegenerative and neurodevelopmental disease modeling and drug discovery. Expert Opin. Drug Discovery 10 (6), 615–629.

Couch, Y., Trofimov, A., Markova, N., Nikolenko, V., Steinbusch, H.W., Chekhonin, V., Schroeter, C., Lesch, K.P., Anthony, D.C., Strekalova, T., 2016. Low-dose lipopolysaccharide (LPS) inhibits aggressive and augments depressive behaviours in a chronic mild stress model in mice. J. Neuroinflammation 13 (1), 108.

Cryan, J.F., Dinan, T.G., 2012. Mind-altering microorganisms: the impact of the gut microbiota on brain and behaviour. Nat. Rev. Neurosci. 13 (10), 701–712.

Cryan, J.F., Dalvi, A., Jin, S.H., Hirsch, B.R., Lucki, I., Thomas, S.A., 2001. Use of dopamine-beta-hydroxylase-deficient mice to determine the role of norepinephrine in the mechanism of action of antidepressant drugs. J. Pharmacol. Exp. Ther. 298 (2), 651–657.

Cryan, J.F., O'Leary, O.F., Jin, S.H., Friedland, J.C., Ouyang, M., Hirsch, B.R., Page, M.E., Dalvi, A., Thomas, S.A., Lucki, I., 2004. Norepinephrine-deficient mice lack responses to antidepressant drugs, including selective serotonin reuptake inhibitors. Proc. Natl. Acad. Sci. USA 101 (21), 8186–8191.

Cuellar-Barboza, A.B., McElroy, S.L., Veldic, M., Singh, B., Kung, S., Romo-Nava, F., Nunez, N.A., Cabello-Arreola, A., Coombes, B.J., Prieto, M., Betcher, H.K., Moore, K.M., Winham, S.J., Biernacka, J.M., Frye, M.A., 2020. Potential pharmacogenomic targets in bipolar disorder: considerations for current testing and the development of decision support tools to individualize treatment selection. Int. J. Bipolar Disord. 8 (1), 23.

Dantzer, R., 2017. Role of the kynurenine metabolism pathway in inflammation-induced depression: preclinical approaches. Curr. Top. Behav. Neurosci. 31, 117–138.

Datson, N.A., van den Oever, J.M., Korobko, O.B., Magarinos, A.M., de Kloet, E.R., McEwen, B.S., 2013. Previous history of chronic stress changes the transcriptional response to glucocorticoid challenge in the dentate gyrus region of the male rat hippocampus. Endocrinology 154 (9), 3261–3272.

de Abreu, M.S., Friend, A.J., Demin, K.A., Amstislavskaya, T.G., Bao, W., Kalueff, A.V., 2018. Zebrafish models: do we have valid paradigms for depression? J. Pharmacol. Toxicol. Methods 94 (Pt 2), 16–22.

Demin, K.A., Meshalkina, D.A., Kysil, E.V., Antonova, K.A., Volgin, A.D., Yakovlev, O.A., Alekseeva, P.A., Firuleva, M.M., Lakstygal, A.M., de Abreu, M.S., Barcellos, L.J.G., Bao, W., Friend, A.J., Amstislavskaya, T.G., Rosemberg, D.B., Musienko, P.E., Song, C., Kalueff, A.V., 2018. Zebrafish models relevant to studying central opioid and endocannabinoid systems. Prog. Neuro-Psychopharmacol. Biol. Psychiatry 86, 301–312.

Demin, K.A., Sysoev, M., Chernysh, M.V., Savva, A.K., Koshiba, M., Wappler-Guzzetta, E.A., Song, C., De Abreu, M.S., Leonard, B., Parker, M.O., Harvey, B.H., Tian, L., Vasar, E., Strekalova, T., Amstislavskaya, T.G., Volgin, A.D., Alpyshov, E.T., Wang, D., Kalueff, A.V., 2019. Animal models of major depressive disorder and the implications for drug discovery and development. Expert Opin. Drug Discovery 14 (4), 365–378.

Denny, C.A., Burghardt, N.S., Schachter, D.M., Hen, R., Drew, M.R., 2012. 4- to 6-week-old adult-born hippocampal neurons influence novelty-evoked exploration and contextual fear conditioning. Hippocampus 22 (5), 1188–1201.

Der-Avakian, A., Barnes, S.A., Markou, A., Pizzagalli, D.A., 2016. Translational assessment of reward and motivational deficits in psychiatric disorders. Curr. Top. Behav. Neurosci. 28, 231–262.

Deyama, S., Duman, R.S., 2020. Neurotrophic mechanisms underlying the rapid and sustained antidepressant actions of ketamine. Pharmacol. Biochem. Behav. 188, 172837.

Dias, B.G., Banerjee, S.B., Duman, R.S., Vaidya, V.A., 2003. Differential regulation of brain derived neurotrophic factor transcripts by antidepressant treatments in the adult rat brain. Neuropharmacology 45 (4), 553–563.

Dorostkar, M.M., Zou, C., Blazquez-Llorca, L., Herms, J., 2015. Analyzing dendritic spine pathology in Alzheimer's disease: problems and opportunities. Acta Neuropathol. 130 (1), 1–19.

Duman, R.S., 2002. Pathophysiology of depression: the concept of synaptic plasticity. Eur. Psychiatry 17 (Suppl. 3), 306–310.

Duman, R.S., Aghajanian, G.K., 2012. Synaptic dysfunction in depression: potential therapeutic targets. Science 338 (6103), 68–72.

Fava, M., 2003. Diagnosis and definition of treatment-resistant depression. Biol. Psychiatry 53 (8), 649–659.

Felger, J.C., Alagbe, O., Hu, F., Mook, D., Freeman, A.A., Sanchez, M.M., Kalin, N.H., Ratti, E., Nemeroff, C.B., Miller, A.H., 2007. Effects of interferon-alpha on rhesus monkeys: a nonhuman primate model of cytokine-induced depression. Biol. Psychiatry 62 (11), 1324–1333.

Forrest, M.P., Parnell, E., Penzes, P., 2018. Dendritic structural plasticity and neuropsychiatric disease. Nat. Rev. Neurosci. 19 (4), 215–234.

Gong, S., Zhang, J., Guo, Z., Fu, W., 2018. Senkyunolide A protects neural cells against corticosterone-induced apoptosis by modulating protein phosphatase 2A and alpha-synuclein signaling. Drug Des. Devel. Ther. 12, 1865–1879.

Gray, J.D., Rubin, T.G., Hunter, R.G., McEwen, B.S., 2014. Hippocampal gene expression changes underlying stress sensitization and recovery. Mol. Psychiatry 19 (11), 1171–1178.

Greene, L.A., Tischler, A.S., 1976. Establishment of a noradrenergic clonal line of rat adrenal pheochromocytoma cells which respond to nerve growth factor. Proc. Natl. Acad. Sci. U. S. A. 73 (7), 2424–2428.

Hamazaki, T., El Rouby, N., Fredette, N.C., Santostefano, K.E., Terada, N., 2017. Concise review: induced pluripotent stem cell research in the era of precision medicine. Stem Cells 35 (3), 545–550.

Harro, J., 2013. Animal models of depression vulnerability. Curr. Top. Behav. Neurosci. 14, 29–54.

Hill, A.S., Sahay, A., Hen, R., 2015. Increasing adult hippocampal neurogenesis is sufficient to reduce anxiety and depression-like behaviors. Neuropsychopharmacology 40 (10), 2368–2378.

Hyman, S.E., Nestler, E.J., 1996. Initiation and adaptation: a paradigm for understanding psychotropic drug action. Am. J. Psychiatry 153 (2), 151–162.

Insel, T., Cuthbert, B., Garvey, M., Heinssen, R., Pine, D.S., Quinn, K., Sanislow, C., Wang, P., 2010. Research domain criteria (RDoC): toward a new classification framework for research on mental disorders. Am. J. Psychiatry 167 (7), 748–751.

Iqbal, J., Adu-Nti, F., Wang, X., Qiao, H., Ma, X.M., 2020. Sex difference in depression: which animal models mimic it. Behav. Neurosci. 134 (3), 248–266.

Kale, R.P., Nguyen, T.T.L., Price, J.B., Yates, N.J., Walder, K., Berk, M., Sillitoe, R.V., Kouzani, A.Z., Tye, S.J., 2021. Mood regulatory actions of active and sham nucleus accumbens deep brain stimulation in antidepressant resistant rats. Front. Hum. Neurosci. 15, 644921.

Kalidindi, A., Kelly, S.D., Singleton, K.S., Guzman, D., Merrill, L., Willard, S.L., Shively, C.A., Neigh, G.N., 2017. Reduced marker of vascularization in the anterior hippocampus in a female monkey model of depression. Physiol. Behav. 172, 12–15.

Kalueff, A.V., Stewart, A.M., Nguyen, M., Song, C., Gottesman, I.I., 2015. Targeting drug sensitivity predictors: new potential strategies to improve pharmacotherapy of human brain disorders. Prog. Neuro-Psychopharmacol. Biol. Psychiatry 63, 76–82.

Kee, N., Teixeira, C.M., Wang, A.H., Frankland, P.W., 2007. Preferential incorporation of adult-generated granule cells into spatial memory networks in the dentate gyrus. Nat. Neurosci. 10 (3), 355–362.

Kelley, K.W., O'Connor, J.C., Lawson, M.A., Dantzer, R., Rodriguez-Zas, S.L., McCusker, R.H., 2013. Aging leads to prolonged duration of inflammation-induced depression-like behavior caused by Bacillus Calmette-Guerin. Brain Behav. Immun. 32, 63–69.

Khan, K.M., Collier, A.D., Meshalkina, D.A., Kysil, E.V., Khatsko, S.L., Kolesnikova, T., Morzherin, Y.Y., Warnick, J.E., Kalueff, A.V., Echevarria, D.J., 2017. Zebrafish models in neuropsychopharmacology and CNS drug discovery. Br. J. Pharmacol. 174 (13), 1925–1944.

Kidnapillai, S., Bortolasci, C.C., Udawela, M., Panizzutti, B., Spolding, B., Connor, T., Sanigorski, A., Dean, O.M., Crowley, T., Jamain, S., Gray, L., Scarr, E., Leboyer, M., Dean, B., Berk, M., Walder, K., 2020. The use of a gene expression signature and connectivity map to repurpose drugs for bipolar disorder. World J. Biol. Psychiatry 21 (10), 775–783.

Kim, Y., McGee, S., Czeczor, J.K., Walker, A.J., Kale, R.P., Kouzani, A.Z., Walder, K., Berk, M., Tye, S.J., 2016a. Nucleus accumbens deep-brain stimulation efficacy in ACTH-pretreated rats: alterations in mitochondrial function relate to antidepressant-like effects. Transl. Psychiatry 6 (6), e842.

Kim, Y., Morath, B., Hu, C., Byrne, L.K., Sutor, S.L., Frye, M.A., Tye, S.J., 2016b. Antidepressant actions of lateral habenula deep brain stimulation differentially correlate with CaMKII/GSK3/AMPK signaling locally and in the infralimbic cortex. Behav. Brain Res. 306, 170–177.

Kitamura, Y., Araki, H., Gomita, Y., 2002. Influence of ACTH on the effects of imipramine, desipramine and lithium on duration of immobility of rats in the forced swim test. Pharmacol. Biochem. Behav. 71 (1–2), 63–69.

Kitamura, Y., Yagi, T., Kitagawa, K., Shinomiya, K., Kawasaki, H., Asanuma, M., Gomita, Y., 2010. Effects of bupropion on the forced swim test and release of dopamine in the nucleus accumbens in ACTH-treated rats. Naunyn Schmiedebergs Arch. Pharmacol. 382 (2), 151–158.

Konstantopoulos, N., Foletta, V.C., Segal, D.H., Shields, K.A., Sanigorski, A., Windmill, K., Swinton, C., Connor, T., Wanyonyi, S., Dyer, T.D., Fahey, R.P., Watt, R.A., Curran, J.E., Molero, J.C., Krippner, G., Collier, G.R., James, D.E., Blangero, J., Jowett, J.B., Walder, K.R., 2011. A gene expression signature for insulin resistance. Physiol. Genomics 43 (3), 110–120.

Krishnan, V., Han, M.H., Graham, D.L., Berton, O., Renthal, W., Russo, S.J., Laplant, Q., Graham, A., Lutter, M., Lagace, D.C., Ghose, S., Reister, R., Tannous, P., Green, T.A., Neve, R.L., Chakravarty, S., Kumar, A., Eisch, A.J., Self, D.W., Lee, F.S., Tamminga, C.A., Cooper, D.C., Gershenfeld, H.K., Nestler, E.J., 2007. Molecular adaptations underlying susceptibility and resistance to social defeat in brain reward regions. Cell 131 (2), 391–404.

Kruse, J.L., Congdon, E., Olmstead, R., Njau, S., Breen, E.C., Narr, K.L., Espinoza, R., Irwin, M.R., 2018. Inflammation and improvement of depression following electroconvulsive therapy in treatment-resistant depression. J. Clin. Psychiatry 79 (2), 17m11597.

Li, M., Izpisua Belmonte, J.C., 2018. Deconstructing the pluripotency gene regulatory network. Nat. Cell Biol. 20 (4), 382–392.

Lieb, K., Dreimuller, N., Wagner, S., Schlicht, K., Falter, T., Neyazi, A., Muller-Engling, L., Bleich, S., Tadic, A., Frieling, H., 2018. BDNF plasma levels and BDNF exon IV promoter methylation as predictors for antidepressant treatment response. Front. Psychiatry 9, 511.

Lin, M., Lachman, H.M., Zheng, D., 2016. Transcriptomics analysis of iPSC-derived neurons and modeling of neuropsychiatric disorders. Mol. Cell. Neurosci. 73, 32–42.

Liu, W., Ge, T., Leng, Y., Pan, Z., Fan, J., Yang, W., Cui, R., 2017. The role of neural plasticity in depression: from hippocampus to prefrontal cortex. Neural Plast. 2017, 6871089.

Lopez, J.P., Mamdani, F., Labonte, B., Beaulieu, M.M., Yang, J.P., Berlim, M.T., Ernst, C., Turecki, G., 2013. Epigenetic regulation of BDNF expression according to antidepressant response. Mol. Psychiatry 18 (4), 398–399.

MacMillan, S., Szeszko, P.R., Moore, G.J., Madden, R., Lorch, E., Ivey, J., Banerjee, S.P., Rosenberg, D.R., 2003. Increased amygdala: hippocampal volume ratios associated with severity of anxiety in pediatric major depression. J. Child Adolesc. Psychopharmacol. 13 (1), 65–73.

Madison, J.M., Zhou, F., Nigam, A., Hussain, A., Barker, D.D., Nehme, R., van der Ven, K., Hsu, J., Wolf, P., Fleishman, M., O'Dushlaine, C., Rose, S., Chambert, K., Lau, F.H., Ahfeldt, T., Rueckert, E.H., Sheridan, S.D., Fass, D.M., Nemesh, J., Mullen, T.E., Daheron, L., McCarroll, S., Sklar, P., Perlis, R.H., Haggarty, S.J., 2015. Characterization of bipolar disorder patient-specific induced pluripotent stem cells from a family reveals neurodevelopmental and mRNA expression abnormalities. Mol. Psychiatry 20 (6), 703–717.

Malatynska, E., Steinbusch, H.W., Redkozubova, O., Bolkunov, A., Kubatiev, A., Yeritsyan, N.B., Vignisse, J., Bachurin, S., Strekalova, T., 2012. Anhedonic-like traits and lack of affective deficits in 18-month-old C57BL/6 mice: implications for modeling elderly depression. Exp. Gerontol. 47 (8), 552–564.

Malberg, J.E., Eisch, A.J., Nestler, E.J., Duman, R.S., 2000. Chronic antidepressant treatment increases neurogenesis in adult rat hippocampus. J. Neurosci. 20 (24), 9104–9110.

Manuel, R., Gorissen, M., Zethof, J., Ebbesson, L.O., van de Vis, H., Flik, G., van den Bos, R., 2014. Unpredictable chronic stress decreases inhibitory avoidance learning in Tuebingen long-fin zebrafish: stronger effects in the resting phase than in the active phase. J. Exp. Biol. 217 (Pt 21), 3919–3928.

Marcatili, M., Marsoner, F., D'Agostino, A., Karnavas, T., Bottai, D., Scarone, S., Conti, L., 2016. Establishment of an induced pluripotent stem cell (iPSC) line from a patient with Clozapine-responder Schizophrenia. Stem Cell Res. 17 (3), 630–633.

Marcatili, M., Sala, C., Dakanalis, A., Colmegna, F., D'Agostino, A., Gambini, O., Dell'Osso, B., Benatti, B., Conti, L., Clerici, M., 2020. Human induced pluripotent stem cells technology in treatment resistant depression: novel strategies and opportunities to unravel ketamine's fast-acting antidepressant mechanisms. Ther. Adv. Psychopharmacol. 10, 2045125320968331.

Marchetto, M.C., Belinson, H., Tian, Y., Freitas, B.C., Fu, C., Vadodaria, K., Beltrao-Braga, P., Trujillo, C.A., Mendes, A.P.D., Padmanabhan, K., Nunez, Y., Ou, J., Ghosh, H., Wright, R., Brennand, K., Pierce, K., Eichenfield, L., Pramparo, T., Eyler, L., Barnes, C.C., Courchesne, E., Geschwind, D.H., Gage, F.H., Wynshaw-Boris, A., Muotri, A.R., 2017. Altered proliferation and networks in neural cells derived from idiopathic autistic individuals. Mol. Psychiatry 22 (6), 820–835.

Martella, G., Bonsi, P., Johnson, S.W., Quartarone, A., 2018. Synaptic plasticity changes: hallmark for neurological and psychiatric disorders. Neural Plast. 2018, 9230704.

Mayberg, H.S., 1997. Limbic-cortical dysregulation: a proposed model of depression. J. Neuropsychiatr. Clin. Neurosci. 9 (3), 471–481.

Mehta, D., Raison, C.L., Woolwine, B.J., Haroon, E., Binder, E.B., Miller, A.H., Felger, J.C., 2013. Transcriptional signatures related to glucose and lipid metabolism predict treatment response to the tumor necrosis factor antagonist infliximab in patients with treatment-resistant depression. Brain Behav. Immun. 31, 205–215.

Mertens, J., Wang, Q.W., Kim, Y., Yu, D.X., Pham, S., Yang, B., Zheng, Y., Diffenderfer, K.E., Zhang, J., Soltani, S., Eames, T., Schafer, S.T., Boyer, L., Marchetto, M.C., Nurnberger, J.I., Calabrese, J.R., Odegaard, K.J., McCarthy, M.J., Zandi, P.P., Alda, M., Nievergelt, C.M., Pharmacogenomics of Bipolar Disorder S, Mi, S., Brennand, K.J., Kelsoe, J.R., Gage, F.H., Yao, J., 2015. Differential responses to lithium in hyperexcitable neurons from patients with bipolar disorder. Nature 527 (7576), 95–99.

Meshalkina, D.A., Kysil, E.V., Warnick, J.E., Demin, K.A., Kalueff, A.V., 2017. Adult zebrafish in CNS disease modeling: a tank that's half-full, not half-empty, and still filling. Lab Anim. (NY) 46 (10), 378–387.

Miller, A.H., Raison, C.L., 2016. The role of inflammation in depression: from evolutionary imperative to modern treatment target. Nat. Rev. Immunol. 16 (1), 22–34.

Mocking, R.J.T., Nap, T.S., Westerink, A.M., Assies, J., Vaz, F.M., Koeter, M.W.J., Ruhe, H.G., Schene, A.H., 2017. Biological profiling of prospective antidepressant response in major depressive disorder: associations with (neuro)inflammation, fatty acid metabolism, and amygdala-reactivity. Psychoneuroendocrinology 79, 84–92.

Muir, J., Lopez, J., Bagot, R.C., 2019. Wiring the depressed brain: optogenetic and chemogenetic circuit interrogation in animal models of depression. Neuropsychopharmacology 44 (6), 1013–1026.

Muller, M.B., Holsboer, F., 2006. Mice with mutations in the HPA-system as models for symptoms of depression. Biol. Psychiatry 59 (12), 1104–1115.

Murrough, J.W., Perez, A.M., Pillemer, S., Stern, J., Parides, M.K., aan het Rot, M., Collins, K.A., Mathew, S.J., Charney, D.S., Iosifescu, D.V., 2013. Rapid and longer-term antidepressant effects of repeated ketamine infusions in treatment-resistant major depression. Biol. Psychiatry 74 (4), 250–256.

Nanni, V., Uher, R., Danese, A., 2012. Childhood maltreatment predicts unfavorable course of illness and treatment outcome in depression: a meta-analysis. Am. J. Psychiatry 169 (2), 141–151.

Nestler, E.J., Hyman, S.E., 2010. Animal models of neuropsychiatric disorders. Nat. Neurosci. 13 (10), 1161–1169.

Nierenberg, A.A., Amsterdam, J.D., 1990. Treatment-resistant depression: definition and treatment approaches. J. Clin. Psychiatry 51 (Suppl), 39–47. discussion 48–50.

Nusser, Z., Lujan, R., Laube, G., Roberts, J.D., Molnar, E., Somogyi, P., 1998. Cell type and pathway dependence of synaptic AMPA receptor number and variability in the hippocampus. Neuron 21 (3), 545–559.

Oglodek, E., Szota, A., Just, M., Mos, D., Araszkiewicz, A., 2014. The role of the neuroendocrine and immune systems in the pathogenesis of depression. Pharmacol. Rep. 66 (5), 776–781.

Okabe, S., 2007. Molecular anatomy of the postsynaptic density. Mol. Cell. Neurosci. 34 (4), 503–518.

Pattij, T., Groenink, L., Hijzen, T.H., Oosting, R.S., Mae, R.A., van der Gugten, J., Olivier, B., 2002. Autonomic changes associated with enhanced anxiety in 5-HT(1A) receptor knockout mice. Neuropsychopharmacology 27 (3), 380–390.

Perathoner, S., Cordero-Maldonado, M.L., Crawford, A.D., 2016. Potential of zebrafish as a model for exploring the role of the amygdala in emotional memory and motivational behavior. J. Neurosci. Res. 94 (6), 445–462.

Porsolt, R.D., Bertin, A., Jalfre, M., 1977a. Behavioral despair in mice: a primary screening test for antidepressants. Arch. Int. Pharmacodyn. Ther. 229 (2), 327–336.

Porsolt, R.D., Le Pichon, M., Jalfre, M., 1977b. Depression: a new animal model sensitive to antidepressant treatments. Nature 266 (5604), 730–732.

Price, J.B., Bronars, C., Erhardt, S., Cullen, K.R., Schwieler, L., Berk, M., Walder, K., McGee, S.L., Frye, M.A., Tye, S.J., 2018. Bioenergetics and synaptic plasticity as potential targets for individualizing treatment for depression. Neurosci. Biobehav. Rev. 90, 212–220.

Rácz, I., Nent, E., Erxlebe, E, Zimmer, A., 2015. CB1 receptors modulate affective behaviour induced by neuropathic pain. Brain Res. Bull. 114, 42–48.

Rasmussen, K.L., Reite, M., 1982. Loss-induced depression in an adult macaque monkey. Am. J. Psychiatry 139 (5), 679–681.

Reus, G.Z., de Moura, A.B., Borba, L.A., Abelaira, H.M., Quevedo, J., 2019. Strategies for treatment-resistant depression: lessons learned from animal models. Mol. Neuropsychiatry 5 (4), 178–189.

Robbe, D., Kopf, M., Remaury, A., Bockaert, J., Manzoni, O.J., 2002. Endogenous cannabinoids mediate long-term synaptic depression in the nucleus accumbens. Proc. Natl. Acad. Sci. USA 99 (12), 8384–8388.

Rowe, R.G., Daley, G.Q., 2019. Induced pluripotent stem cells in disease modelling and drug discovery. Nat. Rev. Genet. 20 (7), 377–388.

Roy, A.V., Thai, M., Klimes-Dougan, B., Westlund Schreiner, M., Mueller, B.A., Albott, C.S., Lim, K.O., Fiecas, M., Tye, S.J., Cullen, K.R., 2021. Brain entropy and neurotrophic molecular markers accompanying clinical improvement after ketamine: preliminary evidence in adolescents with treatment-resistant depression. J. Psychopharmacol. 35 (2), 168–177.

Rush, A.J., Trivedi, M.H., Wisniewski, S.R., Nierenberg, A.A., Stewart, J.W., Warden, D., Niederehe, G., Thase, M.E., Lavori, P.W., Lebowitz, B.D., McGrath, P.J., Rosenbaum, J.F., Sackeim, H.A., Kupfer, D.J., Luther, J., Fava, M., 2006. Acute and longer-term outcomes in depressed outpatients requiring one or several treatment steps: a STAR*D report. Am. J. Psychiatry 163 (11), 1905–1917.

Russo-Neustadt, A.A., Chen, M.J., 2005. Brain-derived neurotrophic factor and antidepressant activity. Curr. Pharm. Des. 11 (12), 1495–1510.

Sahay, A., Hen, R., 2007. Adult hippocampal neurogenesis in depression. Nat. Neurosci. 10 (9), 1110–1115.

Sairanen, M., Lucas, G., Ernfors, P., Castren, M., Castren, E., 2005. Brain-derived neurotrophic factor and antidepressant drugs have different but coordinated effects on neuronal turnover, proliferation, and survival in the adult dentate gyrus. J. Neurosci. 25 (5), 1089–1094.

Santarelli, L., Saxe, M., Gross, C., Surget, A., Battaglia, F., Dulawa, S., Weisstaub, N., Lee, J., Duman, R., Arancio, O., Belzung, C., Hen, R., 2003. Requirement of hippocampal neurogenesis for the behavioral effects of antidepressants. Science 301 (5634), 805–809.

Schikorski, T., Stevens, C.F., 1999. Quantitative fine-structural analysis of olfactory cortical synapses. Proc. Natl. Acad. Sci. U. S. A. 96 (7), 4107–4112.

Schmidt, H.D., Duman, R.S., 2007. The role of neurotrophic factors in adult hippocampal neurogenesis, antidepressant treatments and animal models of depressive-like behavior. Behav. Pharmacol. 18 (5–6), 391–418.

Sheline, Y.I., 1996. Hippocampal atrophy in major depression: a result of depression-induced neurotoxicity? Mol. Psychiatry 1 (4), 298–299.

Sheline, Y.I., Wang, P.W., Gado, M.H., Csernansky, J.G., Vannier, M.W., 1996. Hippocampal atrophy in recurrent major depression. Proc. Natl. Acad. Sci. U. S. A. 93 (9), 3908–3913.

Shi, Y., Desponts, C., Do, J.T., Hahm, H.S., Scholer, H.R., Ding, S., 2008. Induction of pluripotent stem cells from mouse embryonic fibroblasts by Oct4 and Klf4 with small-molecule compounds. Cell Stem Cell 3 (5), 568–574.

Shirayama, Y., Chen, A.C., Nakagawa, S., Russell, D.S., Duman, R.S., 2002. Brain-derived neurotrophic factor produces antidepressant effects in behavioral models of depression. J. Neurosci. 22 (8), 3251–3261.

Sithara, S., Crowley, T.M., Walder, K., Aston-Mourney, K., 2017. Gene expression signature: a powerful approach for drug discovery in diabetes. J. Endocrinol. 232 (2), R131–R139.

Song, C., Liu, B.P., Zhang, Y.P., Peng, Z., Wang, J., Collier, A.D., Echevarria, D.J., Savelieva, K.V., Lawrence, R.F., Rex, C.S., Meshalkina, D.A., Kalueff, A.V., 2018. Modeling consequences of prolonged strong unpredictable stress in zebrafish: complex effects on behavior and physiology. Prog. Neuro-Psychopharmacol. Biol. Psychiatry 81, 384–394.

Steiner, M.A., Marsicano, G., Nestler, E.J., Holsboer, F., Lutz, B., Wotjak, C.T., 2008. Antidepressant-like behavioral effects of impaired cannabinoid receptor type 1 signaling coincide with exaggerated corticosterone secretion in mice. Psychoneuroendocrinology 33 (1), 54–67.

Steru, L., Chermat, R., Thierry, B., Simon, P., 1985. The tail suspension test: a new method for screening antidepressants in mice. Psychopharmacology 85 (3), 367–370.

Stockmeier, C.A., Mahajan, G.J., Konick, L.C., Overholser, J.C., Jurjus, G.J., Meltzer, H.Y., Uylings, H.B., Friedman, L., Rajkowska, G., 2004. Cellular changes in the postmortem hippocampus in major depression. Biol. Psychiatry 56 (9), 640–650.

Takahashi, K., Yamanaka, S., 2006. Induction of pluripotent stem cells from mouse embryonic and adult fibroblast cultures by defined factors. Cell 126 (4), 663–676.

Swoap, S.J., Weinshenker, D., Palmiter, R.D., Garber, G., 2004. Dbh(-/-) mice are hypotensive, have altered circadian rhythms, and have abnormal responses to dieting and stress. Am. J. Physiol. Regul. Integr. Comp. Physiol. 286 (1), R108–R113.

Takahashi, K., Tanabe, K., Ohnuki, M., Narita, M., Ichisaka, T., Tomoda, K., Yamanaka, S., 2007. Induction of pluripotent stem cells from adult human fibroblasts by defined factors. Cell 131 (5), 861–872.

Tarantino, L.M., Sullivan, P.F., Meltzer-Brody, S., 2011. Using animal models to disentangle the role of genetic, epigenetic, and environmental influences on behavioral outcomes associated with maternal anxiety and depression. Front. Psychiatry 2, 44.

Thase, M.E., 2011. Treatment-resistant depression: prevalence, risk factors, and treatment strategies. J. Clin. Psychiatry 72 (5), e18.

Thomas, S.A., Palmiter, R.D., 1997. Disruption of the dopamine beta-hydroxylase gene in mice suggests roles for norepinephrine in motor function, learning, and memory. Behav. Neurosci. 1113 (3), 579–589.

Thomas, S.A., Palmiter, R.D., 1998. Examining adrenergic roles in development, physiology, and behavior through targeted disruption of the mouse dopamine beta-hydroxylase gene. Adv. Pharmacol. 42, 57–60.

Toni, N., Teng, E.M., Bushong, E.A., Aimone, J.B., Zhao, C., Consiglio, A., van Praag, H., Martone, M.E., Ellisman, M.H., Gage, F.H., 2007. Synapse formation on neurons born in the adult hippocampus. Nat. Neurosci. 10 (6), 727–734.

Toups, M.S., Myers, A.K., Wisniewski, S.R., Kurian, B., Morris, D.W., Rush, A.J., Fava, M., Trivedi, M.H., 2013. Relationship between obesity and depression: characteristics and treatment outcomes with antidepressant medication. Psychosom. Med. 75 (9), 863–872.

Vadodaria, K.C., Ji, Y., Skime, M., Paquola, A.C., Nelson, T., Hall-Flavin, D., Heard, K.J., Fredlender, C., Deng, Y., Elkins, J., Dani, K., Le, A.T., Marchetto, M.C., Weinshilboum, R., Gage, F.H., 2019a. Altered serotonergic circuitry in SSRI-resistant major depressive disorder patient-derived neurons. Mol. Psychiatry 24 (6), 808–818.

Vadodaria, K.C., Ji, Y., Skime, M., Paquola, A., Nelson, T., Hall-Flavin, D., Fredlender, C., Heard, K.J., Deng, Y., Le, A.T., Dave, S., Fung, L., Li, X., Marchetto, M.C., Weinshilboum, R., Gage, F.H., 2019b. Serotonin-induced hyperactivity in SSRI-resistant major depressive disorder patient-derived neurons. Mol. Psychiatry 24 (6), 795–807.

Vellucci, S.V., 1990. Primate social behavior—anxiety or depression? Pharmacol. Ther. 47 (2), 167–180.

Vinkers, C.H., Groenink, L., Pattij, T., Olivier, B., Bouwknecht, J.A., 2011. 5-HT(1A) receptor sensitivity in 5-HT(1B) receptor KO mice is unaffected by chronic fluvoxamine treatment. Eur. J. Pharmacol. 667 (1–3), 250–257.

Vollmayr, B., Bachteler, D., Vengeliene, V., Gass, P., Spanagel, R., Henn, F., 2004. Rats with congenital learned helplessness respond less to sucrose but show no deficits in activity or learning. Behav. Brain Res. 150 (1–2), 217–221.

Vose, L.R., Stanton, P.K., 2017. Synaptic plasticity, metaplasticity and depression. Curr. Neuropharmacol. 15 (1), 71–86.

Walker, A.J., Burnett, S.A., Hasebe, K., McGillivray, J.A., Gray, L.J., McGee, S.L., Walder, K., Berk, M., Tye, S.J., 2013. Chronic adrenocorticotrophic hormone treatment alters tricyclic antidepressant efficacy and prefrontal monoamine tissue levels. Behav. Brain Res. 242, 76–83.

Walker, A.J., Foley, B.M., Sutor, S.L., McGillivray, J.A., Frye, M.A., Tye, S.J., 2015. Peripheral proinflammatory markers associated with ketamine response in a preclinical model of antidepressant-resistance. Behav. Brain. Res. 293, 198–202.

Walker, A.J., Price, J.B., Borreggine, K., Sutor, S.L., Gogos, A., McGillivray, J.A., Frye, M.A., Tye, S.J., 2019. Insulin-stimulated mTOR activation in peripheral blood mononuclear cells associated with early treatment response to lithium augmentation in rodent model of antidepressant-resistance. Transl. Psychiatry 9 (1), 113.

Will, C.C., Aird, F., Redei, E.E., 2003. Selectively bred Wistar-Kyoto rats: an animal model of depression and hyper-responsiveness to antidepressants. Mol. Psychiatry. 8 (11), 925–932.

Williams, L.M., Debattista, C., Duchemin, A.M., Schatzberg, A.F., Nemeroff, C.B., 2016. Childhood trauma predicts antidepressant response in adults with major depression: data from the randomized international study to predict optimized treatment for depression. Transl. Psychiatry 6, e799.

Willner, P., 2016. The chronic mild stress (CMS) model of depression: history, evaluation and usage. Neurobiol. Stress 6, 78–93.

Willner, P., Belzung, C., 2015. Treatment-resistant depression: are animal models of depression fit for purpose? Psychopharmacology 232 (19), 3473–3495.

Willner, P., Towell, A., Sampson, D., Sophokleous, S., Muscat, R., 1987. Reduction of sucrose preference by chronic unpredictable mild stress, and its restoration by a tricyclic antidepressant. Psychopharmacology (Berl) 93 (3), 358–364.

Willner, P., Scheel-Kruger, J., Belzung, C., 2013. The neurobiology of depression and antidepressant action. Neurosci. Biobehav. Rev. 37 (10 Pt 1), 2331–2371.

Willner, P., Gruca, P., Lason, M., Tota-Glowczyk, K., Litwa, E., Niemczyk, M., Papp, M., 2019. Validation of chronic mild stress in the Wistar-Kyoto rat as an animal model of treatment-resistant depression. Behav. Pharmacol. 30 (2 and 3-Spec Issue), 239–250.

Wong, M.L., Dong, C., Maestre-Mesa, J., Licinio, J., 2008. Polymorphisms in inflammation-related genes are associated with susceptibility to major depression and antidepressant response. Mol. Psychiatry 13 (8), 800–812.

Woo, Y.S., Seo, H.J., McIntyre, R.S., Bahk, W.M., 2016. Obesity and its potential effects on antidepressant treatment outcomes in patients with depressive disorders: a literature review. Int. J. Mol. Sci. 17 (1), 80. https://pubmed.ncbi.nlm.nih.gov/26771598/.

Worlein, J.M., 2014. Nonhuman primate models of depression: effects of early experience and stress. ILAR J. 55 (2), 259–273.

Wray, N.R., Ripke, S., Mattheisen, M., Trzaskowski, M., Byrne, E.M., Abdellaoui, A., Adams, M.J., Agerbo, E., Air, T.M., Andlauer, T.M.F., Bacanu, S.A., Baekvad-Hansen, M., Beekman, A.F.T., Bigdeli, T.B., Binder, E.B., Blackwood, D.R.H., Bryois, J., Buttenschon, H.N., Bybjerg-Grauholm, J., Cai, N., Castelao, E., Christensen, J.H., Clarke, T.K., Coleman, J.I.R., Colodro-Conde, L., Couvy-Duchesne, B., Craddock, N., Crawford, G.E., Crowley, C.A., Dashti, H.S., Davies, G., Deary, I.J., Degenhardt, F., Derks, E.M., Direk, N., Dolan, C.V., Dunn, E.C., Eley, T.C., Eriksson, N., Escott-Price, V., Kiadeh, F.H.F., Finucane, H.K., Forstner, A.J., Frank, J., Gaspar, H.A., Gill, M., Giusti-Rodriguez, P., Goes, F.S., Gordon, S.D., Grove, J., Hall, L.S., Hannon, E., Hansen, C.S., Hansen, T.F., Herms, S., Hickie, I.B., Hoffmann, P., Homuth, G., Horn, C., Hottenga, J.J., Hougaard, D.M., Hu, M., Hyde, C.L., Ising, M., Jansen, R., Jin, F., Jorgenson, E., Knowles, J.A., Kohane, I.S., Kraft, J., Kretzschmar, W.W., Krogh, J., Kutalik, Z., Lane, J.M., Li, Y., Li, Y., Lind, P.A., Liu, X., Lu, L., MacIntyre, D.J., MacKinnon, D.F., Maier, R.M., Maier, W., Marchini, J., Mbarek, H., McGrath, P., McGuffin, P., Medland, S.E., Mehta, D., Middeldorp, C.M., Mihailov, E., Milaneschi, Y., Milani, L., Mill, J., Mondimore, F.M., Montgomery, G.W., Mostafavi, S., Mullins, N., Nauck, M., Ng, B., Nivard, M.G., Nyholt, D.R., O'Reilly, P.F., Oskarsson, H., Owen, M.J., Painter, J.N., Pedersen, C.B., Pedersen, M.G., Peterson, R.E., Pettersson, E., Peyrot, W.J., Pistis, G., Posthuma, D., Purcell, S.M., Quiroz, J.A., Qvist, P., Rice, J.P., Riley, B.P., Rivera, M., Saeed Mirza, S., Saxena, R., Schoevers, R., Schulte, E.C., Shen, L., Shi, J., Shyn, S.I., Sigurdsson, E., Sinnamon, G.B.C., Smit, J.H., Smith, D.J., Stefansson, H., Steinberg, S., Stockmeier, C.A., Streit, F., Strohmaier, J., Tansey, K.E., Teismann, H., Teumer, A., Thompson, W., Thomson, P.A., Thorgeirsson, T.E., Tian, C., Traylor, M., Treutlein, J., Trubetskoy, V., Uitterlinden, A.G., Umbricht, D., Van der Auwera, S., van Hemert, A.M., Viktorin, A., Visscher, P.M., Wang, Y., Webb, B.T., Weinsheimer, S.M., Wellmann, J., Willemsen, G., Witt, S.H., Wu, Y., Xi, H.S., Yang, J., Zhang, F., eQtlgen, andMe, Arolt, V., Baune, B.T., Berger, K., Boomsma, D.I., Cichon, S., Dannlowski, U., de Geus, E.C.J., DePaulo, J.R., Domenici, E., Domschke, K., Esko, T., Grabe, H.J., Hamilton, S.P., Hayward, C., Heath, A.C., Hinds, D.A., Kendler, K.S., Kloiber, S., Lewis, G., Li, Q.S., Lucae, S., PFA, M., Magnusson, P.K., Martin, N.G., McIntosh, A.M., Metspalu, A., Mors, O., Mortensen, P.B., Muller-Myhsok, B., Nordentoft, M., Nothen, M.M., O'Donovan, M.C., Paciga, S.A., Pedersen, N.L., Penninx, B., Perlis, R.H., Porteous, D.J., Potash, J.B., Preisig, M., Rietschel, M., Schaefer, C., Schulze, T.G., Smoller, J.W., Stefansson, K., Tiemeier, H., Uher, R., Volzke, H., Weissman, M.M., Werge, T., Winslow, A.R., Lewis, C.M., Levinson, D.F., Breen, G., Borglum, A.D., Sullivan, P.F., Major Depressive disorder working Group of the Psychiatric Genomics C, 2018. Genome-wide association analyses identify 44 risk variants and refine the genetic architecture of major depression. Nat. Genet. 50 (5), 668–681.

Yoon, K.J., Nguyen, H.N., Ursini, G., Zhang, F., Kim, N.S., Wen, Z., Makri, G., Nauen, D., Shin, J.H., Park, Y., Chung, R., Pekle, E., Zhang, C., Towe, M., Hussaini, S.M., Lee, Y., Rujescu, D., St Clair, D., Kleinman, J.E., Hyde, T.M., Krauss, G., Christian, K.M., Rapoport, J.L., Weinberger, D.R., Song, H., Ming, G.L., 2014. Modeling a genetic risk for schizophrenia in iPSCs and mice reveals neural stem cell deficits associated with adherens junctions and polarity. Cell Stem Cell 15 (1), 79–91.

Zanos, P., Gould, T.D., 2018. Intracellular signaling pathways involved in (S)- and (R)-ketamine antidepressant actions. Biol. Psychiatry 83 (1), 2–4.

Zarate Jr., C.A., Singh, J.B., Carlson, P.J., Brutsche, N.E., Ameli, R., Luckenbaugh, D.A., Charney, D.S., Manji, H.K., 2006. A randomized trial of an N-methyl-D-aspartate antagonist in treatment-resistant major depression. Arch. Gen. Psychiatry 63 (8), 856–864.

Zeng, B., Li, Y., Niu, B., Wang, X., Cheng, Y., Zhou, Z., You, T., Liu, Y., Wang, H., Xu, J., 2016. Involvement of PI3K/Akt/FoxO3a and PKA/CREB signaling pathways in the protective effect of fluoxetine against corticosterone-induced cytotoxicity in PC12 cells. J. Mol. Neurosci. 59 (4), 567–578.

Zhao, X., Cao, F., Liu, Q., Li, X., Xu, G., Liu, G., Zhang, Y., Yang, X., Yi, S., Xu, F., Fan, K., Ma, J., 2019. Behavioral, inflammatory and neurochemical disturbances in LPS and UCMS-induced mouse models of depression. Behav. Brain Res. 364, 494–502.

Chapter 6

Treatment algorithms for treatment-resistant depression

Sophie R. Vaccarino[a,b,c] and Sidney H. Kennedy[d,*]

[a]*Institute of Medical Science, University of Toronto, Toronto, Canada,* [b]*Department of Psychiatry, University Health Network, Toronto, Canada,* [c]*Centre for Depression and Suicide Studies, Unity Health Toronto, Toronto, Canada,* [d]*Department of Psychiatry, University of Toronto, Toronto, Canada*

Introduction

An algorithm is a stepwise solution used for problem solving. In medicine, the clinical algorithm, most commonly represented as a flow chart, is used in clinical decision-making processes to address diagnosis and treatment (Margolis, 1983). Since the 1990s, many clinical algorithms have emerged as a consequence of evidence-based medicine (EBM), which is defined as the integration of "individual clinical expertise with the best available external clinical evidence from systematic research" (Sackett, 1997). The "EBM movement" took off in the early 1990s after the term was coined by Gordon Guyatt in 1991 at McMaster University, Canada (Guyatt, 1991), and several highly-influential texts were subsequently published on the topic by David Sackett and colleagues (Sackett, 1997; Sackett and Rosenberg, 1995; Sackett et al., 1996, 1997). Over the past few decades, there has been a push toward the use of EBM in mainstream clinical practice. While efficacy and cost-effectiveness of practicing EBM is well-established, there is still a challenge to ensure adherence in clinical settings (Kilbourne et al., 2004). It is especially difficult for researchers to replicate the results of systematically-designed randomized-controlled trials in real-world clinical practice (Pincus et al., 2001). The simplicity and ease-of-use of treatment algorithms may present a solution.

Algorithms differ from clinical practice guidelines (CPGs) in that CPGs provide users with a synthesis of up-to-date data and clinical recommendations to direct their decision making for a specific disease or disorder, whereas algorithms apply this information to create a comprehensive step-by-step formula for the user to follow. While CPGs provide general recommendations and information about the safety and efficacy of potential treatment modalities, algorithms are highly specialized and standardized to provide specific strategies and steps for the treatment of a clinical condition based on information patterns (Adli et al., 2003). Often, CPGs will include algorithms to synthesize data for ease of use (Kennedy et al., 2016; Malhi et al., 2015; Taylor et al., 2018). Algorithms in psychiatry aim to minimize inconsistencies in prescribing patterns among clinicians, increase likelihood of remission, minimize side effects, and decrease overall costs for the healthcare system.

In certain medical fields, such as oncology, the integration of genetic and other biological markers to identify diagnostic subtypes has enhanced sensitivity and specificity of treatment selection (Chang et al., 2016; Galiè et al., 2004; Reck and Rabe, 2017). However, the heterogeneity of mood disorder etiology and symptomatology makes it significantly more difficult to apply these techniques in the treatment of depression. Failure to identify valid and reliable biomarkers of MDD, that may be used in conjunction with symptom-based features, has hindered the progress of precision psychiatry and therefore the development of reliable treatment algorithms with high sensitivity and specificity (Fernandes et al., 2017; Kapur et al., 2012; Kennedy and Ceniti, 2018). In the case of treatment-resistant depression (TRD), failure to develop a gold-standard, universal definition of TRD has further hindered this progress (Vaccarino and Kennedy, 2020). Nevertheless, in the past two decades, several TRD treatment algorithms have been developed, many of which address treatment strategies across varying stages of treatment resistance.

In this chapter, we summarize the major treatment algorithms of the past three decades, including the seminal algorithms of the 1990s and early 2000s, as well as more modern approaches. The potential benefits of using treatment algorithms, both on an individual patient and societal/economic level, are then examined. Further, we will discuss important considerations when developing and implementing treatment guidelines to treat depression in clinical practice. We end the chapter by considering the future of treatment algorithms for depression, particularly the potential role of pharmacogenetics and computational analysis to enhance algorithm sensitivity and specificity in the world of precision psychiatry.

* Senior author.

Managing Treatment-Resistant Depression. https://doi.org/10.1016/B978-0-12-824067-0.00006-2

Early treatment algorithms: GAP, TMAP, and STAR*D

The German Algorithm Project

Established in 1990, the German Algorithm Project (GAP) was the first attempt by a clinical research group to systematically create a treatment algorithm for depression (Adli et al., 2002; Linden et al., 1994). The algorithm was developed by a consensus group of psychiatrists in the Department of Psychiatry at the Free University of Berlin (*Freie Universität Berlin*), combining available literature at the time with clinical expertise, and was specifically designed for inpatient treatment. This was achieved through the development of a "standardized stepwise drug treatment regimen" (SSTR; Adli et al., 2002; Linden et al., 1994) (Fig. 6.1).

Among 119 patients enrolled in an observational study to test the efficacy, feasibility, and acceptance of this SSTR, 34% achieved remission (Bech-Rafaelsen Melancholia Scale (BMRS) score ≤ 5), and an additional 34% met response criteria (change in BMRS score $\geq 50\%$) (Adli et al., 2002). However, there was no control group to compare the SSTR to treatment-as-usual (TAU), limiting the generalizability of these results. Further, this study was limited by low enrolment and high dropout rates: more than half of patients meeting initial inclusion criteria were not enrolled because their individual treatment needs could not be met by the SSTR, suggesting a reluctance on the part of their physicians to accept algorithm-guided treatment; and one-third of participants dropped-out of the study, most commonly due to intolerable drug side effects (Adli et al., 2002). The SSTR was developed and tested at a single university site which may have led to bias due to lack of diversity among the algorithm developers. Finally, the algorithm only provided a single treatment option at each stage, not allowing for adjustments or differences in specific drug choices based on individual differences (Adli et al., 2002).

Addressing several of these issues, two additional phases of the GAP study were completed. The second phase finished in 2000 and compared SSTR to TAU (Bauer et al., 2009). This study integrated knowledge and insights developed during GAP Phase 1 into an updated SSTR, and also included novel therapeutic agents that had been approved since GAP1, specifically paroxetine and venlafaxine. Participants in the SSTR group achieved remission significantly earlier than TAU (7 vs 12 weeks, on average). However, there were no significant differences in depressive severity scores between the two groups at any time point, neither was there a significant difference between remission rates. A higher drop-out rate in the SSTR versus TAU group (45% vs 16%) was primarily due to physician noncompliance in the SSTR group; withdrawals for physician noncompliance were only applicable to the SSTR group, leading to a biased drop-out rate (Bauer et al., 2009).

The third phase of GAP was a large-scale, naturalistic study that included participants from 10 psychiatric departments across Germany and tested three different SSTRs against each other and TAU (Adli et al., 2017). If the first antidepressant treatment failed, one of the three SSTRs were randomly prescribed: lithium augmentation, dose escalation, or antidepressant switch. Although patients treated according to any SSTR achieved remission more quickly than TAU, the lithium-augmentation group performed the worst of the three SSTRs. However, there was no difference in remission or response rates at any time point between SSTR and TAU. Similar to the previous two phases, drop-out rates were high in the algorithm-treated groups (41%–43%) compared to TAU (19%) (Adli et al., 2017).

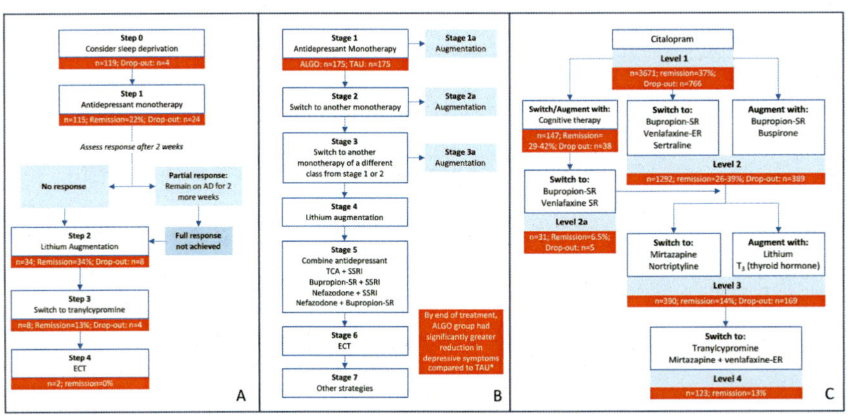

FIG. 6.1 Early treatment algorithms: Adapted from (A) German Algorithm Project, (B) Texas Medication Algorithm Project, (C) Sequenced Treatment Alternatives to Relieve Depression. *ECT*, electroconvulsive therapy; *TCA*, tricyclic antidepressant; *SSRI*, selective serotonin reuptake inhibitor; *SR*, sustained release; *ER*, extended release; **for TMAP*, "n" at each phase and specific remission/response rates were not provided.

Texas Medication Algorithm Project

Addressing many of the limitations of the German Algorithm Project, the Texas Medication Algorithm Project (TMAP) began in 1995 and was the first controlled trial to evaluate the effectiveness of algorithm-based treatment in clinical practice (Gilbert et al., 1998; Rush et al., 1999; Trivedi et al., 2004b). The developers hypothesized that algorithm-guided treatment would lead to faster and more significant symptom improvement, better functioning, and a lower side effect burden over 1 year, compared to TAU (Trivedi et al., 2004b). The TMAP algorithm was developed using a formal consensus, where academic psychiatrists, psychopharmacology specialists, other physicians, mental health consumers, and family members met two discuss the sequence of treatment over a two-and-a-half day conference (Crismon et al., 1999). Importantly, the TMAP algorithm suggests clinicians take into account individual patient factors, allowing for more flexibility and clinical expertise when determining the next-step for treatment (Crismon et al., 1999; Gilbert et al., 1998; Rush et al., 1999; Trivedi et al., 2004b).

A total of 547 participants diagnosed with MDD were enrolled in the original TMAP study (Trivedi et al., 2004b). While both algorithm-treated (ALGO) and TAU groups improved after 3 months of treatment, there was a significantly greater reduction in depression severity in ALGO compared to TAU, and this difference remained significant at every follow-up point, up to 1 year (the last timepoint measured). Algorithm-driven treatment was especially effective in comparison to TAU among participants with more severe depressive symptoms and greater functional impairment at baseline.

However, there were limitations to this study design (Trivedi et al., 2004b). There was no measurement of physician adherence to the algorithm. The TMAP guidelines required extra and ongoing training of physicians, making algorithm-guided treatment more time consuming for physicians to use compared to TAU (e.g., because of extra paperwork), which may have been detrimental to physician adherence. Further, no randomization techniques were used to assign participants, clinics, or physicians to ALGO or TAU (Trivedi et al., 2004b).

The Sequenced Treatment Alternatives to Relieve Depression

Sequenced Treatment Alternatives to Relieve Depression (STAR*D) is an influential study in the field of psychiatry. The goal of this trial was to determine the most effective treatment for individuals with MDD who had an unsatisfactory outcome following initial and subsequent treatment trials (Fava et al., 2003; Rush et al., 2004, 2006). An important aspect of this study was the decision to use "remission" (defined as a score of ≤ 5 on the Quick Inventory of Depressive Symptomatology—Clinician Rating 16-item [QIDS-C]) instead of "response" as the primary outcome measure (Fava et al., 2003; Rush et al., 2004, 2006). Participants who responded to a treatment trial (defined as a $\geq 50\%$ reduction in QIDS score), but did not achieve remitter status, were combined with the "nonresponders" and moved on to the next treatment level. In other studies where response is used as the primary outcome, responders with clinically significant residual symptoms would be combined with those who achieved remission and not further treated for residual symptoms. STAR*D represents an important shift in psychiatry, particularly in the treatment of mood disorders, where the goal of treatment transitioned from antidepressant response to remission of symptoms and a subsequent return to premorbid levels of functioning (Fava et al., 2003; Rush et al., 2004, 2006; Trivedi et al., 2005).

Citalopram was selected as the first treatment (level 1) because of its low risk of discontinuation-related side effects and good representation of the SSRI class (Gilbert et al., 1998; Rush et al., 1999; Trivedi et al., 2004b, 2006). Individuals who did not achieve remission with citalopram were randomized to receive one of a variety of switching or augmentation strategies during level 2 (Fava et al., 2003; Rush et al., 2004, 2006). This pattern was repeated twice more in nonremitters. While the Berlin Algorithm Project and TMAP evaluated the clinical benefit of a specific, predeveloped treatment algorithm, STAR*D set out to determine comparative efficacy of specific agents or treatment strategies, using an algorithm-like approach (Fava et al., 2003; Rush et al., 2004, 2006).

Nonremitters during level 1 who tolerated citalopram received augmentation with bupropion, buspirone, or cognitive therapy (CT) (Fava et al., 2003; Rush et al., 2004, 2006). The use of bupropion as an adjunct was supported by a survey of 400 psychiatrists who endorsed bupropion as their first-choice adjunctive strategy with an SSRI (Fredman et al., 2000). At the time of STAR*D's publication, the evidence for this combination strategy was based on anecdotal reports, case series, and small open-label trials (Fava et al., 2003; Rush et al., 2004, 2006). STAR*D presented the opportunity to empirically test the effectiveness of bupropion using a large study cohort. Today (2020), while still widely prescribed as an adjunct antidepressant to SSRIs, the evidence supporting bupropion as an adjunct remains weak (Patel et al., 2016). At the time of STAR*D, preliminary evidence supported the efficacy of buspirone as an adjunctive agent in those resistant to SSRIs (Appelberg et al., 2001; Fava et al., 2003; Rush et al., 2004, 2006) however, in a subsequent metaanalysis, there was no evidence of efficacy for buspirone augmentation (Davies et al., 2019). At the time STAR*D was launched, the evidence appeared insufficient to justify adding an atypical antipsychotic agent.

Nonremitters during level 1 who did not tolerate citalopram were switched to bupropion, venlafaxine, sertraline, or CT. These medications were selected to test both intraclass (SSRI-to-SSRI [sertraline]) and interclass (SSRI to non-SSRI [bupropion or venlafaxine]) switching strategies (Fava et al., 2003; Rush et al., 2004, 2006). A switch to CT is supported in the literature as a reasonable next-step option following antidepressant nonresponse (Fava et al., 2003; Keller, 2001; Rush et al., 2004, 2006). Overall, these stage 2 strategies were selected to test prevalent clinical beliefs and common clinical practices for initial SSRI nonresponse (Fava et al., 2003; Rush et al., 2004, 2006).

In treatment level 3, those who failed to remit in the previous treatment stage received either adjunctive lithium or triiodothyronine/T_3 augmentation (among the earliest reported antidepressant augmentation strategies) (Dé Montigny et al., 1981; Joffe et al., 1993), or were switched to mirtazapine or nortriptyline, in an attempt to study how these treatment options compare (Fava et al., 2003; Rush et al., 2004, 2006). Finally, level 4 compared treatment strategies that are typically reserved for more treatment-resistant cases, namely the monoamine oxidase inhibitor (MAOI) tranylcypromine or mirtazapine plus venlafaxine. This combination was expected to have a lower side effect burden and greater safety profile compared to the MAOI (Fava et al., 2003; Fiedorowicz and Swartz, 2004; Rush et al., 2004, 2006; Stahl, 2000). An important limitation of this study is that participants were allowed to select their own treatment options for steps 2–4, which may have biased the outcome. Further, the attrition rate was quite high, at 37% overall (from Level 1 to Level 4) (Rush et al., 2006).

While STAR*D did not test a specific algorithm per se, it did provide valuable data on the effectiveness of antidepressants and augmentation agents that could be used for future algorithm development. With no clear "winner," a definitive "best treatment algorithm" could not be developed. The cumulative remission rate in this study was 67% (Gaynes et al., 2009), with likelihood of remission decreasing significantly with number of failed treatments: 37% and 31% of participants remitted after one or two treatment trials, respectively, but the remission rate dropped to 14% and 13% after three and four antidepressant trials, respectively (Rush et al., 2006).

The results from STAR*D have informed many subsequent guidelines, algorithms, and other publications on the treatment of mood disorders (Kennedy et al., 2016,b; Malhi et al., 2015; Taylor et al., 2018).

Subsequent treatment algorithms: 2015–20

Since the seminal TMAP and STAR*D studies, interest in treatment algorithms for depression seems to have waned, with attention shifting to CPGs. Guidelines incorporate more detailed descriptions of clinical evidence and grade recommendations according to efficacy and safety considerations, while allowing clinicians to use these data to create their own treatment plans. However, these are often biased toward depression in the primary-care setting and do not provide progressive steps to deal with treatment-resistance.

Among treatment guidelines published since 2015, several have included algorithms to synthesize data presented in the CPG. The most influential algorithms include the Royal Australian and New Zealand College of Physicians (RANZCP), Canadian Network for Mood and Anxiety Treatments (CANMAT), and the Maudsley Prescribing Guidelines (MPG) algorithms (Kennedy et al., 2016,b; Malhi et al., 2015; Taylor et al., 2018). Other treatment algorithms published since 2015 are summarized (Table 6.1), as well as CPGs without algorithms (Table 6.2) (Bauer et al., 2013, 2015; Bennabi et al., 2019; Cleare et al., 2015; Feng et al., 2019; Gelenberg et al., 2010; Giakoumatos and Osser, 2019; McIntyre, 2018; NICE, 2018; Qaseem et al., 2016; Seo et al., 2018; The Management of Major Depressive Disorder Working Group, 2016).

The Royal Australian and New Zealand College of Physicians clinical practice guidelines for mood disorders (RANZCP)

The RANZCP guidelines are based on both literature review and expert consensus (Malhi et al., 2015). Their treatment algorithm recommends psychosocial and psychological interventions as first and second treatment steps, primarily addressing community primary-care strategies for MDD. Further, these guidelines reflect regional differences in practice patterns in treating more resistant depression. For example, the authors include an MAOI as a third-line monotherapy before considering augmentation strategies, in contrast to other recent guidelines, where adjunctive atypical antipsychotics are recommended before an MAOI, due to the tolerability and safety issues, as well as the general lack of experience among prescribers in using MAOIs (Bayes and Parker, 2018; Kennedy et al., 2016,b; Malhi et al., 2015; Taylor et al., 2018).

Canadian Network for Mood and Anxiety Treatments

While known primarily as a comprehensive CPG, the CANMAT guidelines also include an algorithm for pharmacotherapies (Kennedy et al., 2016; Malhi et al., 2015; Taylor et al., 2018). These guidelines were developed for use by psychiatrists

TABLE 6.1 Other treatment algorithms for depression (since 2015).

World Federation of Societies of Biological Psychiatry (**Bauer et al., 2013, 2015**)
Choose initial AD based on individual patient factors: (a) intolerance: switch to another AD with evidence of better tolerability; (b) inadequate response after 2–4 weeks: increase dose (where appropriate); (c) inadequate response: augmentation strategy* if initial antidepressant is an SSRI, first try combining with a presynaptic autoreceptor inhibitor (e.g., mirtazapine) before trying an augmentation strategy. Inadequate response: AD switch or consider ECT* *1st choice: lithium, quetiapine, or aripiprazole; 2nd choice: T3, T4, or olanzapine + fluoxetine. Note: Consider adding psychotherapy at any time during treatment*
Department of Veterans Affairs and Department of Defence (**The Management of Major Depressive Disorder Working Group, 2016**)
(a) Mild/moderate MDD: monotherapy or pharmacotherapy + psychotherapy; (b) severe/complicated MDD: refer to speciality care or pharmacotherapy + psychotherapy; (c) remission: continuation, maintenance treatment, and relapse prevention; (d) no remission: provide referral to higher level of care/speciality care
Florida Medicaid Drug Therapy Management Program (**McIntyre, 2018**)
Psychotherapy*, SSRI, SNRI, vortioxetine, bupropion, or mirtazapine for 4 weeks: (a) partial response: continue for another 2–4 weeks or treat as no response; (B) no response: dose optimization, evaluate adherence, AD switch, monotherapy + psychotherapy, or combine current AD with aripiprazole, brexpiprazole, or another AD. No response and/or poor tolerability: evaluate adherence, seek psychiatric consultation, and/or try TCA, MAOI, ECT, TMS, or (SSRI or SNRI) + (quetiapine or lithium or T_3 or L-methylfolate or SAMe). No response and/or poor tolerability: reevaluate diagnosis, switch to MAOI, L-methylfolate augmentation, other neuromodulatory approach, intravenous ketamine, or one of the following triple-drug combinations: (SSRI or SNRI) + mirtazapine + bupropion (SSRI or SNRI) + mirtazapine + lithium (SSRI or SNRI) + bupropion + atypical antipsychotic **Evidence-based*, i.e., *CBT, IPT, or behavioral activation*
Korean Society for Affective Disorders (**Seo et al., 2018**)
(a) Mild/moderate MDD: AD monotherapy; (b) severe MDD: AD monotherapy or AD + atypical antipsychotic. No response: AD switch, AD combination, or augment with atypical antipsychotic. No response: AD combination, augment with or switch to atypical antipsychotic, or try another augmentation strategy
Psychopharmacology Algorithm Project at Harvard South Shore Program (**Giakoumatos and Osser, 2019**)
Nonmelancholic depression sertraline, escitalopram, or bupropion (if not previously tried). No/partial response: switch to another of the above ADs, a dual-action agent (venlafaxine, mirtazapine), TMS, SAMe, or St. John's wort; or augment with omega-3 fatty acid, L-methylfolate, SAMe, light therapy, quetiapine, risperidone, aripiprazole, bupropion, mirtazapine, lithium or T_3. No/partial response: switch to TCA or venlafaxine + mirtazapine, augment with ECT, or try another of the above options. If the patient has atypical features, first try an MAOI (selegiline or phenelzine) or SSRI + aripiprazole. Severe melancholic depression. Determine if there is an urgent indication for ECT (a) urgent indication: try ECT, or ketamine if ECT is refused or unsuccessful; (b) no urgent indication: try venlafaxine, mirtazapine, or TCA. No/partial response: try another of venlafaxine, mirtazapine, or TCA; or augment with lithium or T_3

AD, antidepressant; *SSRI*, selective serotonin reuptake inhibitor; *ECT*, electroconvulsive therapy; *T3/T4*, thyroid hormone; *MDD*, major depressive disorder; *SNRI*, serotonin-norepinephrine reuptake inhibitor; *TCA*, tricyclic antidepressant; *TCA*, tricyclic antidepressant; *MAOI*, monoamine oxidase inhibitor; *TMS*, transcranial magnetic stimulation; *SAMe*, *S*-adenosyl-L-methionine; *CBT*, cognitive behavioral therapy; *IPT*, interpersonal therapy.

and other mental health professionals to span the spectrum from primary-care physicians to general psychiatrists and mood disorder specialists dealing with more complex cases and treatment methods. Recommendations are based on published data, specifically metaanalyses and systematic reviews, where available and include expert opinion and consensus when study data are insufficient (Kennedy et al., 2016). CANMAT addresses the use of adjunct treatment strategies and suggests circumstances when it is more appropriate to switch or apply adjunctive therapy (Kennedy et al., 2016). For example, if a patient experiences side effects with an initial antidepressant, switching is recommended over adding an adjunct agent. Further, if a patient has already had two or more antidepressant trials, adjunct medication is recommended over switching. Finally, CANMAT details how long to wait for response and remission before trying an optimization, switching, or adjunctive strategy: 2–4 weeks for response and a further 6–8 weeks for remission (Kennedy et al., 2016). While CANMAT includes comprehensive, evidence-based guidelines for nonpharmacological treatments, there are no strategies to integrate or sequence nonpharmacotherapies into a comprehensive algorithm (Kennedy et al., 2016).

Maudsley Prescribing Guidelines

The MPG was introduced in 1994, and by 2018 it is in its 13th edition (Kennedy et al., 2016; Malhi et al., 2015; Taylor et al., 2018). Not only are these guidelines used as a reference in the United Kingdom, but they are also influential in other countries, including China, Japan, and the United States (Taylor et al., 2015). The MPG was developed by combining data

TABLE 6.2 Clinical guidelines without algorithms (since 2010).

American Psychiatric Association (Gelenberg et al., 2010)
Based on literature review, developed by psychiatrists, and reviewed by 15 organizations and an Independent Review Panel Provides guidance for acute, continuation, and maintenance treatment, as well as treatment discontinuation Addresses pharmacotherapy, somatic therapies, psychotherapy, and complementary and alternative medicine Guidelines for optimization, switching, augmentation, and combination strategies Guidelines for special populations
British Association for Psychopharmacology (Cleare et al., 2015)
Based on literature review and consensus meeting with experts in depressive disorders Provides guidance on acute treatment with antidepressants, psychological and behavioral treatments, physical treatments, and complementary and alternative medicine Guidelines for treatment nonresponse and relapse prevention/treatment Guidelines for special populations
American College of Physicians (Qaseem et al., 2016)
Based on randomized-controlled trial data Reviews psychotherapies, complementary and alternative medicine, exercise, and second-generation antidepressants Discusses briefly switching and augmentation strategies
The National Institute for Health and Care Excellence (NICE, 2018)
Recommends tailoring care and treatment based on depressive severity Discusses general principles of care and first-line treatments Guidelines for addressing poor response to treatment, electroconvulsive therapy, and relapse prevention
Chinese Society of Psychiatry (Feng et al., 2019)
Based on recent literature and other, international guidelines Discusses diagnosis and acute, continuation, and maintenance treatment
French Association for Biological Psychiatry and Neuropsychopharmacology (Bennabi et al., 2019)
Specifically developed for treatment-resistant depression Based on scientific evidence and expert clinicians' opinions Discusses identification of treatment resistance, indications for hospitalization, and relapse prevention Guidelines for antidepressant medications, switching strategies, adjuvant treatments, combination strategies, psychotherapy, and brain stimulation

from previously published guidelines with expert opinion and clinical experience, and is heavily influenced by recommendations from the National Institute for Health and Care Excellence's (NICE) clinical guideline for depression in adults, which are integrated into the MPG text (National Institute for Health and Care Excellence, 2009). NICE works closely with the United Kingdom's National Health Service to provide UK physicians with treatment recommendations that are both evidence based and influenced by economic factors.

The MPG is presented as a comprehensive textbook of over 800 pages addressing treatments for MDD, bipolar disorder, and schizophrenia (Taylor et al., 2015). Drug treatments for TRD are summarized in a single algorithm. The algorithm first addresses strategies for initial treatment nonresponse, which would likely be useful for primary-care physicians. This is followed by treatment options for refractory depression, which contain more complex and investigational treatments, and therefore, we recommend them for use by a psychiatrist or other physician specialized in mood disorder treatment (Fig. 6.2).

Similar to the RANZP, the MPG is limited by the fact that it does not suggest adjunctive strategies until later stages, instead suggesting changing the antidepressant three times before trying an adjunct medication. Further, the MPG algorithm does not outline treatment steps for partial response; in cases of partial response, an adjunct agent is typically recommended over a full switch, whereas a switch may be recommended in cases of complete nonresponse or poor tolerability (Kennedy et al., 2016). However, unlike the RANZP and similar to CANMAT, the MPG algorithm provides an extensive list of potential treatment options for refractory depression, including less studied or more controversial treatment modalities, such as pramipexole and estrogen, that may be tried when other options have been exhausted. Interestingly, the MPG suggests intravenous ketamine as a second-line option for treatment-refractory depression. This drug has shown robust antidepressant

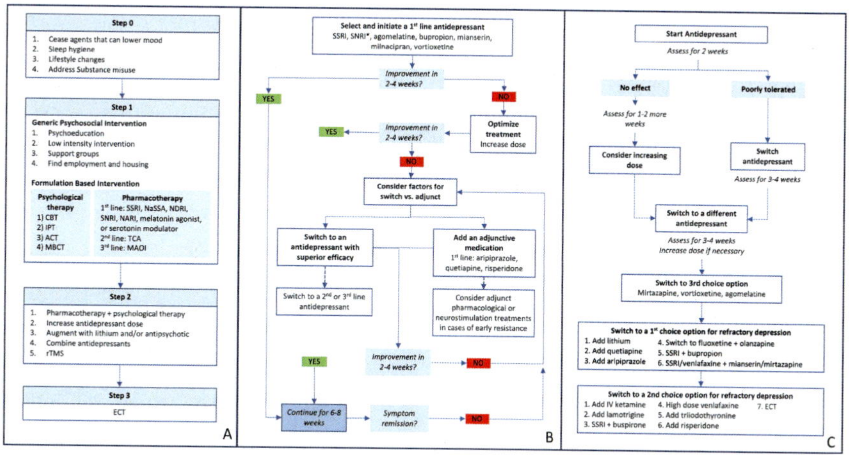

FIG. 6.2 Modern treatment algorithms: Adapted from (A) The Royal Australian and New Zealand College of Physicians clinical practice guidelines for mood disorders, (B) Canadian Network for Mood and Anxiety Treatment, (C) Maudsley Prescribing Guideline. *CBT*, cognitive behavioral therapy; *IPT*, interpersonal therapy; *ACT*, acceptance and commitment therapy; *MBCT*, mindfulness-based cognitive therapy; *SSRI*, selective serotonin reuptake inhibitor; *NaSSA*, noradrenergic and specific serotonergic antidepressant; *NARI*, norepinephrine reuptake inhibitor; *TCA*, tricyclic antidepressant; *MAOI*, monoamine oxidase inhibitor; *rTMS*, repetitive transcranial magnetic stimulation; *ECT*, electroconvulsive therapy; *SNRI**, serotonin-norepinephrine reuptake inhibitor (excluding levomilnacipran); *IV*, intravenous.

effects in pilot studies (Mathew et al., 2010; Murrough et al., 2013; Zarate et al., 2006). However, the use of intravenous ketamine for the treatment of depression remains "off-label." At the time of publication of the MPG 13th edition, intranasal esketamine had not been approved for use; the guideline therefore states that "intranasal ketamine may become available and supplant the intravenous form" (Taylor et al., 2015). Since publication, intranasal esketamine has been approved in Europe, the United States (2019) and Canada (2020) for treatment-resistant depression.

The benefits of treatment algorithms

The emergence of CPGs, followed by treatment algorithms, allows busy practitioners, who realistically cannot integrate all the emerging data, to implement best practices (Fava et al., 2003; Rush et al., 2004, 2006). Over the past four decades, there has been a significant increase in the number of treatments available for MDD, as well as potential treatment combination and augmentation strategies (Adli et al., 2003; Bauer et al., 2019). It is a challenge for physicians to make sense of constantly emerging data and understand which of the numerous treatment strategies is best for their patients. Algorithms are designed to aid physicians with decision making by synthesizing available data and presenting them in a practical format, thus enabling physicians to consistently practice evidence-based medicine.

It is assumed that the implementation of treatment algorithms will decrease prescribing variance among physicians, and increase the appropriateness of prescribed medications, thereby improving the quality of patient outcomes and reducing the prevalence of treatment resistance (Adli et al., 2006; Gilbert et al., 1998; Gottlieb et al., 1990; Margolis, 1983). In turn, by increasing response/remission rates and decreasing likelihood of treatment resistance, algorithms are expected to improve the cost efficiency of clinical practice and manage burgeoning health care costs (Gilbert et al., 1998; National Institute for Health and Clinical Excellence, 2011; Rush et al., 1999).

Further, algorithms can minimize the frequency of "pseudo-resistance," where nonresponse is a result of poor quality treatment or poor treatment adherence (Adli et al., 2006). Common causes of "pseudo-resistance" are inadequate dosing, early discontinuation of treatment, or insufficient use of the wide variety of available treatment options for depression (Bauer et al., 2019). Treatment algorithms not only provide guidance on which antidepressant to prescribe, but also at what dose and for how long. Further, algorithms provide guidance on when a physician should consider, in cases of apparent nonresponse, either increasing the dose of the current antidepressant; adding an adjunctive medication; or switching to a new antidepressant altogether (Kennedy et al., 2016).

Increased visit frequency and positive outcomes

While several studies have found algorithms to be superior to TAU in improving patient outcomes (Guo et al., 2015), explanations for these findings remain unclear—are algorithms effective because of the specific treatment recommendations, or is it because of the increased frequency of, and structured approach to, clinical visits? (Bauer et al., 2019). In a review of electronic health records and administrative claims data of 250 million people treated for depression, diabetes, or hypertension conducted by "Observational Health Data Sciences and Informatics," 11% of individuals with depression followed a treatment pathway that was not shared by any other patient in the database (Hripcsak et al., 2016). The authors of this report suggest that this heterogeneity may represent ineffective differences in clinical practice, trial-and-error prescribing patterns, and/or failure among clinicians to agree on a most effective treatment pathway (Hripcsak et al., 2016). Therefore, the specific treatment recommendations provided in algorithms may reduce this heterogeneity in prescribing patters, and, in turn, improve patient outcomes.

Data regarding the association between visit frequency and response to antidepressant medication are scarce. In psychotherapy studies, more frequent visits with a health care practitioner are associated with greater improvement in symptom severity and higher rates of recovery. Further, an adequate number of follow-up visits after initiation of antidepressant therapy is associated with sufficient duration of antidepressant treatment (Jones et al., 2006). In studies of late-life depression, there is a positive relationship between visit frequency and response rate among patients receiving placebo treatment, but not active antidepressant therapy (Kevin Delucchi, 2015). This may suggest that more frequent visits enhance the positive placebo effect with antidepressant therapy (Kirsch, 2014, 2019; Rief et al., 2009). Nonetheless, data show that individuals receiving antidepressant treatment, on average, receive much less follow-up monitoring than is recommended (Morrato et al., 2008; Simon et al., 2001; Stettin et al., 2006). For example, in a study of nearly 85,000 patients receiving antidepressant therapy, only 15% of patients received the recommended level of follow-up, according to the U.S. Food and Drug Administration, during their first month of treatment (Stettin et al., 2006).

While it is most likely that both of these factors explain the superiority of algorithms over TAU, this presents an interesting avenue for future research. Perhaps, the efficacy of algorithms does not lie in the practice of evidence-based medicine per se, but the improved patient-clinician relationship that results from this practice via the increased frequency and quality of clinical visits.

Cost efficacy of algorithms

In the United States, the annual economic burden of MDD exceeds $200 billion, with 55% attributable to indirect costs, specifically lost work productivity (Greenberg et al., 2015). Algorithms are expected to decrease both direct and indirect costs of depression on society by increasing rates of remission and time to recovery.

In Phase 2 and 3 of the German Algorithm Project, algorithm-guided treatment was associated with significantly lower costs than TAU (Ricken et al., 2011, 2018). Using a cost-effectiveness analysis, which compares direct treatment costs with clinical outcome (i.e., average direct treatment cost divided by remission rate), cost per remission was twice as much in the TAU group compared to the ALGO group (Ricken et al., 2011, 2018). This was mainly attributable to less treatment days and lower treatment costs in the ALGO group than in the TAU group (Ricken et al., 2011, 2018).

In contrast, algorithm guided treatment was actually *more* expensive than TAU in TMAP. Over 1 year, the average cost per patient receiving algorithm-guided treatment was over $10,000 (USD) more than TAU (Kashner et al., 2006). Unlike GAP, TMAP combined medication algorithms with clinical care coordinator visits, patient education sessions, enhanced clinical documentation, and expert consensus (Crismon et al., 1999; Gilbert et al., 1998). These additional aspects of treatment incur many more charges than standard of care, which likely explains the stark difference between costs associated with TMAP versus GAP algorithms (Kashner et al., 2006). Further, GAP was completed in Germany and TMAP in the United States, where healthcare systems are administered quite differently. Therefore, differences in costs between studies may be explained by regional differences (Ridic et al., 2012).

Unfortunately, the indirect costs of algorithm-guided treatment versus standard of care have not been compared, likely due to feasibility issues.

Algorithm development

An effective algorithm should provide users with guidance to decide which treatments should be used and at what dosages and frequencies, and in what order different treatments should be prescribed in cases of treatment-resistance (Adli et al., 2006). In order to achieve this, researchers should have a predefined goal and scope for their algorithm (Adli et al., 2006; Shekelle et al., 1999). First and foremost, based on the standards of 2020, the desired outcome of any algorithm should be achieving symptomatic remission and functional recovery (Trivedi and Kleiber, 2001). Exceptions to this include patients with persistent or chronic depression; in these cases, a chronic disease management approach may be more realistic and effective, where improved functioning and quality of life are the goals of treatment (Keitner and Mansfield, 2012; Kennedy et al., 2016).

Treatment algorithms should adhere to the principle of greatest efficacy and lowest side effect burden—i.e., the more effective and safer a treatment, the earlier in the algorithm it should be tried (Trivedi and Kleiber, 2001). One must also consider what stage(s) of disease progression the algorithm will address (e.g., initial diagnosis, early treatment, treatment resistance) (Adli et al., 2006; Shekelle et al., 1999). Finally, prior to algorithm development, researchers must determine how outcome(s) will be measured in order to test the efficacy of an algorithm: both clinician-rated depression scales and patient-reported outcomes should be included (Adli et al., 2006).

While there are many different methods to determine which treatments are the most favorable, most consistently through the use of consensus groups (Adli et al., 2002; Crismon et al., 1999; Kennedy et al., 2016), it is also essential that treatment algorithms projects rely on explicit scientific evidence from high-quality, randomized controlled clinical trials (Cipriani et al., 2018). However, high-quality evidence does not always exist. In these situations, expert consensus and data from smaller, open-label or observational studies may be used to guide recommendations; it is important to inform users that these recommendations are based on weaker evidence. In situations where recommendations are based solely on the opinions of a development group, bias can be reduced by forming a diverse algorithm development group that involves different mental health and medical professionals (e.g., primary-care practitioners, psychiatrists, and allied health care practitioners), community stakeholders, patients, and family members, all from a wide variety of geographic and socioeconomic regions (Shekelle et al., 1999). Further, clear guidelines on how expert opinions will be assessed and incorporated into algorithm development should be outlined before starting the process, if the group decides to use expert opinion at all (Shekelle et al., 1999).

Unfortunately, even if extensive clinical data existed for every treatment option, it would still be impossible to create a completely objective treatment algorithm free of all bias and opinion, since members of the algorithm development group must interpret and synthesize the data. Therefore, transparency regarding the data used and how they are interpreted is crucial. Not only does this increase the credibility of the algorithm, but can also prevent misinterpretation on the part of the clinician.

There are several ways to ensure transparency and decrease bias when interpreting and synthesizing data in algorithm development. Firstly, there should be preestablished rules to rate quality of evidence in support of a particular treatment (i.e., level of evidence). Based on the level of evidence combined with other factors, the strength of recommendation for each treatment can also be established (Shekelle et al., 1999). Typically, the higher the quality of evidence, the stronger the recommendation; however, certain medications with high quality evidence for efficacy may not be highly recommended for various reasons, such as side effect burden or risk of drug-drug interactions.

Secondly, the development group must consider how benefits and drawbacks/harms of a treatment will be weighed (Adli et al., 2003). As previously mentioned, algorithms should first recommend treatments with the highest efficacy and lowest side effect burden—but what if a treatment demonstrates a very high efficacy and a significant side effect burden? As an example, this could apply to some "atypical antipsychotic" drugs with good efficacy but high side effect burden. Further, where treatment algorithms also aim to enhance cost efficiency, this could result in a treatment with high efficacy and high cost being downgraded in favor of a less expensive therapy with a greater side effect burden. Group members must establish beforehand how they will deal with these important issues as they will inevitably arise.

Typically, it is better to provide more than one treatment option at each level of an algorithm. This allows clinicians to follow a personalized approach to treatment, while working within the confines of an algorithm. A specific treatment may be chosen based on factors such as patient preference, medication availability, and tolerability issues. Since algorithms are meant to assist with clinical decision making, rather than provide rigid instructions, it is important that clinicians can adapt algorithm-guided treatment to fit the specific needs of their patients.

Another dimension that developers may consider including in their treatment algorithm is an evidence-based approach to medication-related side effects or patients with specific concerns (Trivedi and Kleiber, 2001). For example, an algorithm may suggest that patients who experience sexual dysfunction during treatment with an SSRI are switched to, or augmented with, bupropion (Bennabi et al., 2019; Cleare et al., 2015; Feng et al., 2019; Gelenberg et al., 2010; MacKinnon, 2019; NICE, 2018; Qaseem et al., 2016), or that patients with MDD and comorbid sleep disturbances receive mirtazapine (Kennedy et al., 2016).

It is also important to consider the specific obstacles and constraints that clinicians face and that may affect their ability to follow the treatment algorithm, such as socioeconomic constraints, time restrictions, and availability of required equipment and/or medications (Adli et al., 2006; Green et al., 2007; Shekelle et al., 1999). Developers may provide recommendations to the clinician regarding organizational restructuring and/or improvements to clinical practice that would allow for algorithm implementation to be successful; this may be achieved by standardizing processes and procedures used in clinic, or developing protocols for algorithm implementation (Fischer et al., 2016; Lugtenberg et al., 2009).

Finally, in order to keep up to date with emerging literature, algorithms should be updated periodically, typically every few years (Kennedy et al., 2016; Shekelle et al., 1999; Taylor et al., 2018) (Table 6.3).

TABLE 6.3 An algorithm for developing algorithms.

1. Form a diverse algorithm development group
2. Before starting development consider:
What is the goal and scope? Which stage(s) of disease progression will be addressed? How will outcome(s) be measured? How will expert opinion be assessed and incorporated? How will benefits and harms of treatments be weighed?
3. Develop the algorithm
Follow the rule of greatest efficacy and lowest side effect burden. Use data from high-quality randomized-controlled trials. If unavailable, expert consensus, and/or open-label or observational trials may be used, provide more than one treatment option at each step where possible. Carefully determine critical decision points. Consider obstacles and constraints that clinicians might face upon implementation. Include recommendations for implementing the algorithm. Be transparent about methodology. Preestablish rules to rate quality of evidence and strength of recommendations. Inform users when randomized-controlled trial data were not used to guide recommendations. Explain how benefits and harms were weighed
4. Provide clinicians with comprehensive training
5. Update the algorithm every few years and/or as new evidence becomes available

Critical decision points

Critical decision points (CDPs) are predefined timepoints where patient outcomes are measured, and the next treatment decision is activated (Adli et al., 2006; Bauer et al., 2019). Results of these outcome measures can help the clinician determine whether there should be a dose increase, decrease, or no change; or if a switch or augmentation strategy may be applied (Adli et al., 2006). Side effects should also be measured at CDPs, as this may guide next treatment steps (Adli et al., 2006).

The frequency of CDPs should be carefully determined, as they should allow adequate time to elapse for previous adjustments to take effect before making the next adjustment (Adli et al., 2006). If CDPs occur too often, this may not allow enough time for treatments to show efficacy, as could be the case with slower acting antidepressants (Adli et al., 2006). Further, visits occurring too often would not be cost-efficient. Conversely, if CDPs occur too infrequently, patients may remain on an ineffective regimen for too long (Adli et al., 2006). Importantly, CDPs should be feasible within a clinic and should be adaptable to the needs of each individual clinical setting (Adli et al., 2006). For example, a physician may be too busy to see each patient every few weeks. A certain amount of flexibility is required in these CDPs so that clinicians will adhere to them. Finally, CDPs will vary based on the specific treatment prescribed (e.g., faster acting treatments will require sooner visits than those that are slower acting). In general, in the early stages of algorithm-guided treatment, CDPs occur every 2–4 weeks (Kennedy et al., 2016; Trivedi and Kleiber, 2001). Appropriately timed evaluation of treatment response may increase the likelihood that a patient will achieve remission during their first treatment (Crismon et al., 1999; Trivedi and Kleiber, 2001).

Implementing algorithms in clinical practice

Physicians should understand that an algorithm is not a rigid set of rules designed to replace clinical decision making, but is a support tool to aid in the clinical decision-making process as one part of a multifaceted disease management program created by both physician and patient (Adli et al., 2006; Bauer et al., 2019). Such a program may include personalized treatment approaches, patient and family education programs, and clinical expertise (Adli et al., 2006). Clinicians must understand when a treatment recommended in the algorithm would be appropriate and when it would not be, based on individual patient and situational factors.

Before launching a treatment algorithm in clinical practice, developers should ensure that physicians who will be using it have received comprehensive training. Simply providing a clinician with a treatment algorithm (e.g., by mail or published in a research journal) is unlikely to enhance patient care (Azocar et al., 2003). Algorithm developers should train clinicians on the algorithm, in person, and provide initial supervision, to ensure clinicians are adhering to the steps and applying them correctly (Adli et al., 2006). Further, ongoing education and retraining on the algorithm will improve outcomes and adherence (Fischer et al., 2016).

Clinicians must also be willing and able to adjust their treatment approach in order to follow the algorithm; this may involve not only a change in their prescribing patterns, but also increased visit frequency, longer visits with patients, and the use of standardized outcome measures or other assessments/surveys (Adli et al., 2006). Structural changes to clinical practice may be required; for example, a chronic disease management program may need to be implemented in order to enhance patient-physician engagement (e.g., through more frequent visits), increase treatment adherence, reduce attrition, and promote symptomatic and functional recovery (Bauer et al., 2019).

Implementation at the organizational level

Effective algorithm implementation does not just require adjustments on the part of the physician, but the "healthcare organization" as a whole. Prior to guideline implementation across a healthcare network, perceived barriers to implementation, such as economic constraints or unavailability of treatment, should be identified and addressed (Feder et al., 1999). Specific algorithms may be more suitable for one organization than another, depending on the organization's unique needs. For example, a primary-care organization might prefer an algorithm, such as the RANZP, that focuses on easier-to-use, first-line interventions like psychotherapy and antidepressant monotherapy (Malhi et al., 2015). On the other hand, a psychiatrist's office may benefit more from CANMAT, which outlines more complex treatment options for greater degrees of treatment-resistance (Kennedy et al., 2016). Organizations looking to implement algorithms in clinical practice should compare effectiveness, efficiency, and feasibility across different algorithms and identify which will best address the specific needs of their organization and be most conducive to their patient population (Feder et al., 1999).

Engaging the patient

Shared decision making between the clinician and patient leads to lower rates of treatment discontinuation and greater treatment satisfaction (Hopwood, 2020; Loh et al., 2007). It is especially important to engage the patient in decision making when there is more than one reasonable treatment option, when different treatments show a similar efficacy-to-side-effect ratio, and when the possible benefits and harms of different medications may be particularly favorable or aversive to specific patients (e.g., the decreased risk of antidepressant-related sexual dysfunction with bupropion may be of particular importance to one patient, while the sleep-promoting effects of mirtazapine may be more beneficial for another) (Hopwood, 2020; Loh et al., 2007). In algorithm-guided treatment, there is typically more than one treatment option at each stage, so involving the patient in decision making is highly recommended (Adli et al., 2006; Bauer et al., 2019).

Obstacles to implementation

Several obstacles to algorithm implementation, both at the systemic and clinician level, continue to impede their widespread use, particularly in psychiatry (Cabana et al., 1999; Fischer et al., 2016; Forsner et al., 2010; Yang et al., 2013).

Barriers to algorithm implementation include ...

Physicians' lack of knowledge of algorithms
- Frequent visits not feasible in clinical practice
- Specific treatments unavailable
- Economic constraints and/or lack of available resources
- No access to specialists in less conventional or complex treatments
- Physicians' hesitancy to try algorithms
- Physicians' beliefs that they lack certain skills or expertise
- Belief that algorithms cannot address heterogeneous patient population
- Lack of economic incentive

Some treatments recommended in algorithms for MDD, particularly in later stages of TRD, require close monitoring by an experienced psychiatrist due to the higher side effect burden and/or increased safety risk. Examples include MAOIs or neurostimulation techniques such as rTMS or ECT. In certain regions, particularly rural areas, access to experts in these specific treatment modalities may not be available (Dauenhauer et al., 2011; Health Quality Ontario, 2016). Further, specific medications, psychological treatments, or somatic therapies recommended in an algorithm may not be available (Adli et al., 2006). If certain recommendations are unavailable, the treating physician should be trained on alternative treatment options and how to identify the best alternative based on availability and individual patient factors.

Strict critical decision points mean that patients must come in frequently, especially in the acute phase of treatment, often on a biweekly basis (Adli et al., 2017; Kennedy et al., 2016; Taylor et al., 2018). This may not be economically or strategically manageable for many patients or clinics, especially those in densely populated areas with a high demand for their services. Further, it is to be expected that some practitioners will be hesitant to try treatment algorithms, guided by misbeliefs about the algorithm's role in the medical system. For example, physicians may believe that algorithms are unsafe or inappropriate for their patients (Adli et al., 2006). This may be based on the falsehood that algorithms undermine clinical expertise and do not allow room for changes based on individual patient characteristics. On the other hand, clinicians may think they lack the proper skills or have the resources to implement an algorithm (Fischer et al., 2016). This is why training of clinicians on an algorithm, and routine follow-up and retraining, is essential. This ensures that the clinician is implementing the algorithm as intended and is able to understand when the algorithm should be followed and when clinical expertise should take over.

Mental health professionals may also doubt the efficacy of treatment algorithms because of the heterogeneity of their patients' symptomatology—"one size fits all" treatment does not exist for MDD, so why would a single treatment algorithm be effective for such a diverse population? While many psychiatrists may believe that developing an individualized treatment plan for each unique patient would lead to better outcomes, researchers actually hypothesize the opposite: it is suggested that an unsystematic approach to depression treatment may be a major contributor to the high rates of antidepressant nonresponse (Bauer et al., 2019). However, it is important that algorithm users understand that all recommendations are based on group trends—i.e., which treatment options work for the majority of patients. Therefore, particular patients, even entire subsets of patients (e.g., those with symptoms of atypical depression), may not be adequately addressed by the algorithm. Since algorithms are so specific compared to other decision-support tools, there is an even greater need for knowledgeable and experienced algorithm users who know when to follow, adjust, or ignore altogether algorithm recommendations for the benefit of the patient (Bauer et al., 2009, 2019; Rush, 2001).

The implementation of algorithms into clinical practice may be further limited by a lack of economic incentives to do so: increased length and frequency of visits is not cost-efficient (Dijkstra et al., 2000). Increasing financial incentives for physicians, such as implementing fee-for-service payments (Gosden et al., 2000), and for clinics/hospitals, such as financial rewards for sending physicians to algorithm training, may increase algorithm adherence (Berthiaume et al., 2004).

The future of algorithms

Over the past few decades, there has been a growing interest in precision psychiatry and the identification of reliable biomarkers of antidepressant treatment response (Kennedy et al., 2016; Lam et al., 2016b). To date, no reliable biomarkers of antidepressant response have been identified (Gadad et al., 2018). This is likely attributable to the heterogeneity of MDD (Hasler, 2010; Hasler et al., 2004; Kapur et al., 2012; Kennedy and Ceniti, 2018; Lam et al., 2016a; Mayberg, 2003). However, with advances in technology, including pharmacogenetics and neuroimaging, there are grounds for optimism that precision psychiatry is advancing (Drysdale et al., 2017). By identifying reliable biomarkers of antidepressant treatment response, clinicians would be able to determine which treatment has the highest probability of efficacy for an individual patient based on specific factors, such as their unique genetic profile, neurobiology, or clinical presentation.

Given the heterogeneity in MDD etiology, pathology, symptomatology, and treatment options, it is very unlikely that just a few key biomarkers alone will be able to guide treatment decision making. On the contrary, many different clinical and biological biomarkers will likely have to be considered when choosing the appropriate treatment. Algorithms may be an effective and efficient way for physicians to make sense of all of this information, by providing step-by-step instructions on how to consider different clinical, biological, and genetic factors when deciding on the best treatment for their patient.

Pharmacogenomic-guided algorithms

Pharmacogenomics (PGx) in the treatment of TRD are discussed in detail in the next chapter of this book. Here, we briefly discuss how pharmacogenetic testing has emerged as a component of algorithm-guided treatments. To date, the most compelling findings for PGx in guiding clinical decision making for TRD are for allelic variations in genes encoding for two cytochrome P450 enzymes: CYP2D6 and CYP2C19, which are responsible for metabolism of most antidepressants and antipsychotics (International Society of Psychiatric Genetics (ISPG), 2019). Current literature suggests that PGx testing in the treatment of depression is most beneficial for patients who experienced a high degree of adverse reactions based on poor metabolism or patients who failed to achieve response based on rapid metabolism (International Society of Psychiatric Genetics (ISPG), 2019).

Several clinical trials have tested the efficacy of PGx-guided treatment for depression (Brennan et al., 2015; Hall-Flavin et al., 2012, 2013; Han et al., 2018; Winner et al., 2013). In these studies, DNA tests are run, and, based on results of these tests and variations present in key genes associated with antidepressant response, certain medications are either promoted as associated with a higher likelihood of response and/or tolerability; or cautioned against due to increased risk of nonresponse and/or adverse events. In several clinical trials and metaanalyses, PGx-guided treatment was associated with greater antidepressant response compared to TAU (Bousman et al., 2019; Hall-Flavin et al., 2012, 2013; Han et al., 2018; Rosenblat et al., 2018). However, a more recent study found no benefit of PGx-guided treatment (Shan et al., 2019). Currently, the evidence regarding the benefits of PGx-guided treatment is inconclusive (Li and Loshak, 2020).

Computerized algorithms

Computerization of treatment algorithms would likely be conducive to the practice of precision psychiatry. As research regarding biomarkers of antidepressant response increases exponentially and is eventually implemented in clinical practice, it will become increasingly difficult for clinicians to absorb all the available literature and to practice evidence-based medicine. Advanced computer systems have the potential to integrate all of this constantly emerging information into a comprehensive and easy-to-use format (Trivedi et al., 2002).

Data show that physician use and adherence to guidelines increase when they are delivered in a computerized format (Gilbert et al., 1998; Lobach and Hammond, 1997; Rush et al., 1999; Trivedi et al., 2004a). Computer algorithms are typically easier to use and less time-consuming than paper formats (Trivedi et al., 2004a). In an initial study using a computerized version of TMAP, the authors found that patients treated according to this algorithm experienced significantly greater reduction in depressive symptoms compared to TAU (Kurian et al., 2009). Computerized algorithm programs allow information to be constantly updated and can provide physicians with ongoing education and support regarding algorithm use (Kurian et al., 2009). Eventually, artificial intelligence may be able to continuously adapt to integrate emerging data and individual patient characteristics.

A good example: The antidepressant advisor

The antidepressant advisor (ADeSS) is a computerized decision-support tool and algorithm for use by general practitioners treating patients with early treatment resistance in the United Kingdom (Harrison et al., 2020). While efficacy/feasibility trials are still ongoing at the time of writing this chapter, the ADeSS aims to improve adherence to algorithm-guided treatment by creating an easy-to-use computerized system that can be easily implemented within existing frameworks of UK primary-care facilities. The ADeSS will be the first "scientifically evaluated and pragmatic" antidepressant decision-support tool for use in primary care across the United Kingdom (Harrison et al., 2020). Importantly, the ADeSS developers strive for widespread feasibility in their algorithm, so that it may be used by primary-care physicians as seamlessly as possible. For example, the ADeSS can be integrated into an already existing computer system used by general practitioners across the United Kingdom. Further, the algorithm may be followed in just 10 min, the typical appointment time for general practitioners in the United Kingdom. A common concern regarding algorithm implementation is that busy practitioners simply do not have enough time to treat each patient according to cumbersome algorithms, which typically require longer and more frequent visits (Adli et al., 2006). Specifically designing algorithms to be used easily and quickly may increase algorithm compliance among reluctant physicians.

The ADeSS makes strides toward implementing precision psychiatry within their guidelines by providing prescribing instructions based on an individual's biological characteristics. For example, if body mass index is >25, hemoglobin A1c levels are abnormal, or the patient has a history of diabetes, the guideline recommends vortioxetine be prescribed; if none of these conditions is met, mirtazapine is recommended (Harrison et al., 2020). By providing specific instructions, primary-care physicians with limited experience in treating depression are still able to provide evidence-based recommendations to their patients. This would mean that a larger percentage of patients can be managed in a primary-care setting, instead of being referred to a mental health specialist, which reduces burden on the patient and the healthcare system (Harrison et al., 2020).

It is important to note that this prototype of the ADeSS only provides recommendations for monotherapy; physicians are instructed to try switching to several different antidepressant monotherapies in cases on treatment resistance before referring to a psychiatrist who may try augmentation and/or combination strategies. This strategy allows all monotherapy options (besides more complex options, such as MAOIs) to be exhausted before consulting a psychiatrist. NICE guidelines state combination/augmentation strategies should not be initiated by a general practitioner without consulting a psychiatrist. However, there are certain situations where combination/augmentation strategies may be recommended before switching to a new antidepressant, such as in cases of partial response, which the ADeSS cannot address (National Collaborating Centre for Mental Health (UK), 2010).

Conclusion

Algorithms for the treatment of resistant depression aim to decrease idiosyncratic prescribing patterns among primary-care physicians and psychiatrists, promote the practice of evidence-based medicine, and reduce treatment costs. Following the widespread interest in treatment algorithms for depression in the late 1990s to early 2000s (i.e., GAP, TMAP, and STAR*D), there has been a shift from interest in algorithms toward more general clinical practice guidelines. Therefore, more data on algorithm efficacy, particularly from modern algorithms which implement newer treatment modalities, are required before a definitive conclusion can be made regarding their superiority over "treatment as usual."

Several factors must be considered when developing an algorithm, such as who will be the primary user, how will outcomes be measured, and how will the algorithm be implemented in clinical practice? Algorithm developers and implementors should consider situational factors that may influence algorithm adherence, such as medication costs, time constraints of physicians, and socioeconomic factors. Finally, algorithms are not meant to replace clinical expertise, but instead act as an aid to clinical decision making, integrated within a multifaceted treatment approach that involves active patient engagement, personalized treatment approaches, patient and family education programs, and clinical expertise.

References

Adli, M., Berghofer, A., Linden, M., Helmchen, H., Muller-Oerlinghausen, B., Mackert, A., Stamm, T., Bauer, M., 2002. Effectiveness and feasibility of a standardized stepwise drug treatment regimen algorithm for inpatients with depressive disorders. J. Clin. Psychiatry 63 (9), 782–790. https://doi.org/10.4088/JCP.v63n0906.

Adli, M., Rush, A.J., Moller, H.-J., Bauer, M., 2003. Algorithms for optimizing the treatment of depression: making the right decision at the right time. Pharmacopsychiatry 36, S222–S229. https://doi.org/10.1055/s-2003-45134.

Adli, M., Bauer, M., Rush, A.J., 2006. Algorithms and collaborative-care systems for depression: are they effective and why? Biol. Psychiatry 59 (11), 1029–1038. https://doi.org/10.1016/j.biopsych.2006.05.010.

Adli, M., Wiethoff, K., Baghai, T.C., Fisher, R., Seemüller, F., Laakmann, G., Brieger, P., Cordes, J., Malevani, J., Laux, G., Hauth, I., Möller, H.-J., Kronmüller, K.-T., Smolka, M.N., Schlattmann, P., Berger, M., Ricken, R., Stamm, T.J., Heinz, A., Bauer, M., 2017. How effective is algorithm-guided treatment for depressed inpatients? Results from the randomized controlled multicenter German Algorithm Project 3 trial. Int. J. Neuropsychopharmacol. 20 (9), 721–730. https://doi.org/10.1093/ijnp/pyx043.

Appelberg, B.G., Syvalahti, E.K., Koskinen, T.E., Mehtonen, O.-P., Muhonen, T.T., Naukkarinen, H.H., 2001. Patients with severe depression may benefit from buspirone augmentation of selective serotonin reuptake inhibitors. J. Clin. Psychiatry 62 (6), 448–452. https://doi.org/10.4088/JCP.v62n0608.

Azocar, F., Cuffel, B., Goldman, W., McCarter, L., 2003. The impact of evidence-based guideline dissemination for the assessment and treatment of major depression in a managed behavioral health care organization. J. Behav. Health Serv. Res. 30 (1), 109–118. https://doi.org/10.1007/BF02287816.

Bauer, M., Pfennig, A., Linden, M., Smolka, M.N., Neu, P., Adli, M., 2009. Efficacy of an algorithm-guided treatment compared with treatment as usual. J. Clin. Psychopharmacol. 29 (4), 327–333. https://doi.org/10.1097/JCP.0b013e3181ac4839.

Bauer, M., Pfennig, A., Severus, E., Whybrow, P.C., Angst, J., Möller, H.-J., 2013. World Federation of Societies of Biological Psychiatry (WFSBP) guidelines for biological treatment of unipolar depressive disorders, part 1: update 2013 on the acute and continuation treatment of unipolar depressive disorders. World J. Biol. Psychiatry 14 (5), 334–385. https://doi.org/10.3109/15622975.2013.804195.

Bauer, M., Severus, E., Köhler, S., Whybrow, P.C., Angst, J., Möller, H.-J., 2015. World Federation of Societies of Biological Psychiatry (WFSBP) guidelines for biological treatment of unipolar depressive disorders. Part 2: maintenance treatment of major depressive disorder—update 2015. World J. Biol. Psychiatry 16 (2), 76–95. https://doi.org/10.3109/15622975.2014.1001786.

Bauer, M., Rush, A., Ricken, R., Pilhatsch, M., Adli, M., 2019. Algorithms for treatment of major depressive disorder: efficacy and cost-effectiveness. Pharmacopsychiatry 52 (03), 117–125. https://doi.org/10.1055/a-0643-4830.

Bayes, A.J., Parker, G.B., 2018. Comparison of guidelines for the treatment of unipolar depression: a focus on pharmacotherapy and neurostimulation. Acta Psychiatr. Scand. 137 (6), 459–471. https://doi.org/10.1111/acps.12878.

Bennabi, D., Charpeaud, T., Yrondi, A., Genty, J.-B., Destouches, S., Lancrenon, S., Alaïli, N., Bellivier, F., Bougerol, T., Camus, V., Dorey, J.-M., Doumy, O., Haesebaert, F., Holtzmann, J., Lançon, C., Lefebvre, M., Moliere, F., Nieto, I., Rabu, C., et al., 2019. Clinical guidelines for the management of treatment-resistant depression: French recommendations from experts, the French Association for Biological Psychiatry and Neuropsychopharmacology and the fondation FondaMental. BMC Psychiatry 19 (1), 262. https://doi.org/10.1186/s12888-019-2237-x.

Berthiaume, J.T., Tyler, P.A., Ng-Osorio, J., LaBresh, K.A., 2004. Aligning financial incentives with "get with the guidelines" to improve cardiovascular care. Am. J. Manage. Care 10 (7 Pt 2), 501–504.

Bousman, C.A., Arandjelovic, K., Mancuso, S.G., Eyre, H.A., Dunlop, B.W., 2019. Pharmacogenetic tests and depressive symptom remission: a meta-analysis of randomized controlled trials. Pharmacogenomics 20 (1), 37–47. https://doi.org/10.2217/pgs-2018-0142.

Brennan, F.X., Gardner, K.R., Lombard, J., Perlis, R.H., Fava, M., Harris, H.W., Scott, R., 2015. A naturalistic study of the effectiveness of pharmacogenetic testing to guide treatment in psychiatric patients with mood and anxiety disorders. Prim. Care Companion CNS Disord. https://doi.org/10.4088/PCC.14m01717.

Cabana, M.D., Rand, C.S., Powe, N.R., Wu, A.W., Wilson, M.H., Abboud, P.-A.C., Rubin, H.R., 1999. Why don't physicians follow clinical practice guidelines? JAMA 282 (15), 1458. https://doi.org/10.1001/jama.282.15.1458.

Chang, S.S., Boorjian, S.A., Chou, R., Clark, P.E., Daneshmand, S., Konety, B.R., Pruthi, R., Quale, D.Z., Ritch, C.R., Seigne, J.D., Skinner, E.C., Smith, N.D., McKiernan, J.M., 2016. Diagnosis and treatment of non-muscle invasive bladder cancer: AUA/SUO guideline. J. Urol. 196 (4), 1021–1029. https://doi.org/10.1016/j.juro.2016.06.049.

Cipriani, A., Furukawa, T.A., Salanti, G., Chaimani, A., Atkinson, L.Z., Ogawa, Y., Leucht, S., Ruhe, H.G., Turner, E.H., Higgins, J.P.T., Egger, M., Takeshima, N., Hayasaka, Y., Imai, H., Shinohara, K., Tajika, A., Ioannidis, J.P.A., Geddes, J.R., 2018. Comparative efficacy and acceptability of 21 antidepressant drugs for the acute treatment of adults with major depressive disorder: a systematic review and network meta-analysis. Lancet 391 (10128), 1357–1366. https://doi.org/10.1016/S0140-6736(17)32802-7.

Cleare, A., Pariante, C., Young, A., Anderson, I., Christmas, D., Cowen, P., Dickens, C., Ferrier, I., Geddes, J., Gilbody, S., Haddad, P., Katona, C., Lewis, G., Malizia, A., McAllister-Williams, R., Ramchandani, P., Scott, J., Taylor, D., Uher, R., 2015. Evidence-based guidelines for treating depressive disorders with antidepressants: a revision of the 2008 British Association for Psychopharmacology guidelines. J. Psychopharmacol. 29 (5), 459–525. https://doi.org/10.1177/0269881115581093.

Crismon, M.L., Trivedi, M., Pigott, T.A., Rush, A.J., Hirschfeld, R.M., Kahn, D.A., DeBattista, C., Nelson, J.C., Nierenberg, A.A., Sackeim, H.A., Thase, M.E., 1999. The Texas Medication Algorithm Project: report of the Texas Consensus Conference Panel on Medication Treatment of Major Depressive Disorder. J. Clin. Psychiatry 60 (3), 142–156. http://www.ncbi.nlm.nih.gov/pubmed/10192589.

Dauenhauer, L.E., Chauhan, P., Cohen, B.J., 2011. Factors that influence electroconvulsive therapy referrals. J. ECT 27 (3), 232–235. https://doi.org/10.1097/YCT.0b013e3181f9789c.

Davies, P., Ijaz, S., Williams, C.J., Kessler, D., Lewis, G., Wiles, N., 2019. Pharmacological interventions for treatment-resistant depression in adults. Cochrane Database Syst. Rev. 12 (12), CD10557. https://doi.org/10.1002/14651858.CD010557.pub2.

Dé Montigny, C., Grunberg, F., Mayer, A., Deschenes, J.-P., 1981. Lithium induces rapid relief of depression in tricyclic antidepressant drug non-responders. Br. J. Psychiatry 138 (3), 252–256. https://doi.org/10.1192/bjp.138.3.252.

Dijkstra, R.F., Braspenning, J.C.C., Uiters, E., van Ballegooie, E., Grol, R.T.P.M., 2000. Perceived barriers to the implementation of diabetes guidelines in hospitals in The Netherlands. Neth. J. Med. 56 (3), 80–85. https://doi.org/10.1016/S0300-2977(99)00123-0.

Drysdale, A.T., Grosenick, L., Downar, J., Dunlop, K., Mansouri, F., Meng, Y., Fetcho, R.N., Zebley, B., Oathes, D.J., Etkin, A., Schatzberg, A.F., Sudheimer, K., Keller, J., Mayberg, H.S., Gunning, F.M., Alexopoulos, G.S., Fox, M.D., Pascual-Leone, A., Voss, H.U., et al., 2017. Resting-state connectivity biomarkers define neurophysiological subtypes of depression. Nat. Med. 23 (1), 28–38. https://doi.org/10.1038/nm.4246.

Fava, M., Rush, A.J., Trivedi, M.H., Nierenberg, A.A., Thase, M.E., Sackeim, H.A., Quitkin, F.M., Wisniewski, S., Lavori, P.W., Rosenbaum, J.F., Kupfer, D.J., for the STAR∗D Investigators Group, 2003. Background and rationale for the Sequenced Treatment Alternatives to Relieve Depression (STAR∗D) study. Psychiatr. Clin. North Am. 26 (2), 457–494. https://doi.org/10.1016/S0193-953X(02)00107-7.

Feder, G., Eccles, M., Grol, R., Griffiths, C., Grimshaw, J., 1999. Clinical guidelines: using clinical guidelines. BMJ 318 (7185), 728–730. https://doi.org/10.1136/bmj.318.7185.728.

Feng, Y., Xiao, L., Wang, W.-W., Ungvari, G.S., Ng, C.H., Wang, G., Xiang, Y.-T., 2019. Guidelines for the diagnosis and treatment of depressive disorders in China: the second edition. J. Affect. Disord. 253, 352–356. https://doi.org/10.1016/j.jad.2019.04.104.

Fernandes, B.S., Williams, L.M., Steiner, J., Leboyer, M., Carvalho, A.F., Berk, M., 2017. The new field of 'precision psychiatry'. BMC Med. 15 (1), 80. https://doi.org/10.1186/s12916-017-0849-x.

Fiedorowicz, J.G., Swartz, K.L., 2004. The role of monoamine oxidase inhibitors in current psychiatric practice. J. Psychiatr. Pract. 10 (4), 239–248. https://doi.org/10.1097/00131746-200407000-00005.

Fischer, F., Lange, K., Klose, K., Greiner, W., Kraemer, A., 2016. Barriers and strategies in guideline implementation—a scoping review. Healthcare 4 (3), 36. https://doi.org/10.3390/healthcare4030036.

Forsner, T., Hansson, J., Brommels, M., Wistedt, A.Å., Forsell, Y., 2010. Implementing clinical guidelines in psychiatry: a qualitative study of perceived facilitators and barriers. BMC Psychiatry 10 (1), 8. https://doi.org/10.1186/1471-244X-10-8.

Fredman, S.J., Fava, M., Kienke, A.S., White, C.N., Nierenberg, A.A., Rosenbaum, J.F., 2000. Partial response, nonresponse, and relapse with selective serotonin reuptake inhibitors in major depression. J. Clin. Psychiatry 61 (6), 403–408. https://doi.org/10.4088/JCP.v61n0602.

Gadad, B.S., Jha, M.K., Czysz, A., Furman, J.L., Mayes, T.L., Emslie, M.P., Trivedi, M.H., 2018. Peripheral biomarkers of major depression and antidepressant treatment response: current knowledge and future outlooks. J. Affect. Disord. 233, 3–14. https://doi.org/10.1016/j.jad.2017.07.001.

Galiè, N., Seeger, W., Naeije, R., Simonneau, G., Rubin, L.J., 2004. Comparative analysis of clinical trials and evidence-based treatment algorithm in pulmonary arterial hypertension. J. Am. Coll. Cardiol. 43 (12), S81–S88. https://doi.org/10.1016/j.jacc.2004.02.038.

Gaynes, B.N., Warden, D., Trivedi, M.H., Wisniewski, S.R., Fava, M., Rush, A.J., 2009. What did STAR*D teach us? Results from a large-scale, practical, clinical trial for patients with depression. Psychiatr. Serv. 60 (11), 1439–1445. https://doi.org/10.1176/ps.2009.60.11.1439.

Gelenberg, Alan J., Freeman, Marlene P., Markowitz, John C., Rosenbaum, Jerrold F., Thase, Michael E., Trivedi, Madhukar H., Van Rhoads, Richard S., 2010. American Psychiatric Association Practice Guideline for the Treatment of Patients With Major Depressive Disorder, Third. American Psychiatric Association, Arlington, VA.

Giakoumatos, C.I., Osser, D., 2019. The Psychopharmacology Algorithm Project at the Harvard South Shore Program. Harv. Rev. Psychiatry 27 (1), 33–52. https://doi.org/10.1097/HRP.0000000000000197.

Gilbert, D.A., Altshuler, K.Z., Rago, W.V., Shon, S.P., Crismon, M.L., Toprac, M.G., Rush, A.J., 1998. Texas Medication Algorithm Project: definitions, rationale, and methods to develop medication algorithms. J. Clin. Psychiatry 59 (7), 345–351. http://www.ncbi.nlm.nih.gov/pubmed/9714262.

Gosden, T., Forland, F., Kristiansen, I., Sutton, M., Leese, B., Giuffrida, A., Sergison, M., Pedersen, L., 2000. Capitation, salary, fee-for-service and mixed systems of payment: effects on the behaviour of primary care physicians. Cochrane Database Syst. Rev. 3, CD002215. https://doi.org/10.1002/14651858.CD002215.

Gottlieb, L.K., Margolis, C.Z., Schoenbaum, S.C., 1990. Clinical practice guidelines at an HMO: development and implementation in a quality improvement model. QRB Qual. Rev. Bull. 16 (2), 80–86. https://doi.org/10.1016/S0097-5990(16)30343-8.

Green, L.A., Wyszewianski, L., Lowery, J.C., Kowalski, C.P., Krein, S.L., 2007. An observational study of the effectiveness of practice guideline implementation strategies examined according to physicians' cognitive styles. Implement. Sci. 2 (1), 41. https://doi.org/10.1186/1748-5908-2-41.

Greenberg, P.E., Fournier, A.-A., Sisitsky, T., Pike, C.T., Kessler, R.C., 2015. The economic burden of adults with major depressive disorder in the United States (2005 and 2010). J. Clin. Psychiatry 76 (02), 155–162. https://doi.org/10.4088/JCP.14m09298.

Guo, T., Xiang, Y.-T., Xiao, L., Hu, C.-Q., Chiu, H.F.K., Ungvari, G.S., Correll, C.U., Lai, K.Y.C., Feng, L., Geng, Y., Feng, Y., Wang, G., 2015. Measurement-based care versus standard care for major depression: a randomized controlled trial with blind raters. Am. J. Psychiatry 172 (10), 1004–1013. https://doi.org/10.1176/appi.ajp.2015.14050652.

Guyatt, G.H., 1991. Evidence-based medicine. ACP J. Club 1 (2), A16. https://doi.org/10.7326/ACPJC-1991-114-2-A16.

Hall-Flavin, D.K., Winner, J.G., Allen, J.D., Jordan, J.J., Nesheim, R.S., Snyder, K.A., Drews, M.S., Eisterhold, L.L., Biernacka, J.M., Mrazek, D.A., 2012. Using a pharmacogenomic algorithm to guide the treatment of depression. Transl. Psychiatry 2 (10), e172. https://doi.org/10.1038/tp.2012.99.

Hall-Flavin, D.K., Winner, J.G., Allen, J.D., Carhart, J.M., Proctor, B., Snyder, K.A., Drews, M.S., Eisterhold, L.L., Geske, J., Mrazek, D.A., 2013. Utility of integrated pharmacogenomic testing to support the treatment of major depressive disorder in a psychiatric outpatient setting. Pharmacogenet. Genomics 23 (10), 535–548. https://doi.org/10.1097/FPC.0b013e3283649b9a.

Han, C., Wang, S.-M., Bahk, W.-M., Lee, S.-J., Patkar, A.A., Masand, P.S., Mandelli, L., Pae, C.-U., Serretti, A., 2018. A pharmacogenomic-based antidepressant treatment for patients with major depressive disorder: results from an 8-week, randomized, single-blinded clinical trial. Clin. Psychopharmacol. Neurosci. 16 (4), 469–480. https://doi.org/10.9758/cpn.2018.16.4.469.

Harrison, P., Carr, E., Goldsmith, K., Young, A.H., Ashworth, M., Fennema, D., Barrett, B., Zahn, R., 2020. Study protocol for the antidepressant advisor (ADeSS): a decision support system for antidepressant treatment for depression in UK primary care: a feasibility study. BMJ Open 10 (5), e035905. https://doi.org/10.1136/bmjopen-2019-035905.

Hasler, G., 2010. Pathophysiology of depression: do we have any solid evidence of interest to clinicians? World Psychiatry 9 (3), 155–161. https://doi.org/10.1002/j.2051-5545.2010.tb00298.x.

Hasler, G., Drevets, W.C., Manji, H.K., Charney, D.S., 2004. Discovering endophenotypes for major depression. Neuropsychopharmacology 29 (10), 1765–1781. https://doi.org/10.1038/sj.npp.1300506.

Health Quality Ontario, 2016. Repetitive transcranial magnetic stimulation for treatment-resistant depression: an economic analysis. Ont. Health Technol. Assess. Ser. 16 (6), 1–51. http://www.hqontario.ca/evidence/publications-and-ohtacrecommendations/ontario-health-technology-assessment-series/econ-rtms.

Hopwood, M., 2020. The shared decision-making process in the pharmacological management of depression. The Patient 13 (1), 23–30. https://doi.org/10.1007/s40271-019-00383-w.

Hripcsak, G., Ryan, P.B., Duke, J.D., Shah, N.H., Park, R.W., Huser, V., Suchard, M.A., Schuemie, M.J., DeFalco, F.J., Perotte, A., Banda, J.M., Reich, C.G., Schilling, L.M., Matheny, M.E., Meeker, D., Pratt, N., Madigan, D., 2016. Characterizing treatment pathways at scale using the OHDSI network. Proc. Natl. Acad. Sci. U. S. A. 113 (27), 7329–7336. https://doi.org/10.1073/pnas.1510502113.

International Society of Psychiatric Genetics (ISPG), 2019. Genetic Testing Statement: Genetic Testing and Psychiatric Disorders: A statement for the International Society of Psychiatric Genetics.

Joffe, R.T., Singer, W., Levitt, A.J., MacDonald, C., 1993. A placebo-controlled comparison of lithium and triiodothyronine augmentation of tricyclic antidepressants in unipolar refractory depression. Arch. Gen. Psychiatry 50 (5), 387–393. https://doi.org/10.1001/archpsyc.1993.01820170065008.

Jones, L.E., Turvey, C., Carney-Doebbeling, C., 2006. Inadequate follow-up care for depression and its impact on antidepressant treatment duration among veterans with and without diabetes mellitus in the Veterans Health Administration. Gen. Hosp. Psychiatry 28 (6), 465–474. https://doi.org/10.1016/j.genhosppsych.2006.08.002.

Kapur, S., Phillips, A.G., Insel, T.R., 2012. Why has it taken so long for biological psychiatry to develop clinical tests and what to do about it? Mol. Psychiatry 17 (12), 1174–1179. https://doi.org/10.1038/mp.2012.105.

Kashner, T.M., Rush, A.J., Crismon, M.L., Toprac, M., Carmody, T.J., Miller, A.L., Trivedi, M.H., Wicker, A., Suppes, T., 2006. An empirical analysis of cost outcomes of the Texas Medication Algorithm Project. Psychiatr. Serv. 57 (5), 648–659. https://doi.org/10.1176/ps.2006.57.5.648.

Keitner, G.I., Mansfield, A.K., 2012. Management of treatment-resistant depression. Psychiatr. Clin. N. Am. 35 (1), 249–265. https://doi.org/10.1016/j.psc.2011.11.004.

Keller, M.B., 2001. Pharmacotherapy and psychotherapy: new data on the long-term treatment of chronic depression. In: 154th Annual Meeting of the American Psychiatric Association.

Kennedy, S.H., Ceniti, A.K., 2018. Unpacking major depressive disorder: from classification to treatment selection. Can. J. Psychiatry 63 (5), 308–313. https://doi.org/10.1177/0706743717748883.

Kennedy, S.H., Lam, R.W., McIntyre, R.S., Tourjman, S.V., Bhat, V., Blier, P., Hasnain, M., Jollant, F., Levitt, A.J., MacQueen, G.M., McInerney, S.J., McIntosh, D., Milev, R.V., Müller, D.J., Parikh, S.V., Pearson, N.L., Ravindran, A.V., Uher, R., 2016. Canadian Network for Mood and Anxiety Treatments (CANMAT) 2016 clinical guidelines for the management of adults with major depressive disorder: Section 3. Pharmacological Treatments. Can. J. Psychiatry 61 (9), 540–560. https://doi.org/10.1177/0706743716659417.

Kevin Delucchi, J.C.N., 2015. Relationship of number of clinic visits to response during placebo-controlled antidepressant trials in late life depression. J. Depress. Anxiety s1. https://doi.org/10.4172/2167-1044.S1-008.

Kilbourne, A.M., Schulberg, H.C., Post, E.P., Rollman, B.L., Belnap, B.H., Pincus, H.A., 2004. Translating evidence-based depression management services to community-based primary care practices. Milbank Q. 82 (4), 631–659. https://doi.org/10.1111/j.0887-378X.2004.00326.x.

Kirsch, I., 2014. Antidepressants and the placebo effect. Z. Psychol. 222 (3), 128–134. https://doi.org/10.1027/2151-2604/a000176.

Kirsch, I., 2019. Placebo effect in the treatment of depression and anxiety. Front. Psych. 10. https://doi.org/10.3389/fpsyt.2019.00407.

Kurian, B.T., Trivedi, M.H., Grannemann, B.D., Claassen, C.A., Daly, E.J., Sunderajan, P., 2009. A computerized decision support system for depression in primary care. Primary Care Compan. J. Clin. Psychiatry 11 (4), 140–146. https://doi.org/10.4088/PCC.08m00687.

Lam, R.W., Milev, R., Rotzinger, S., Andreazza, A.C., Blier, P., Brenner, C., Daskalakis, Z.J., Dharsee, M., Downar, J., Evans, K.R., Farzan, F., Foster, J.A., Frey, B.N., Geraci, J., Giacobbe, P., Feilotter, H.E., Hall, G.B., Harkness, K.L., et al., 2016a. Discovering biomarkers for antidepressant response: protocol from the Canadian biomarker integration network in depression (CAN-BIND) and clinical characteristics of the first patient cohort. BMC Psychiatry 16 (1). https://doi.org/10.1186/s12888-016-0785-x.

Lam, R.W., Kennedy, S.H., Parikh, S.V., MacQueen, G.M., Milev, R.V., Ravindran, A.V., 2016b. Canadian Network for Mood and Anxiety Treatments (CANMAT) 2016 clinical guidelines for the management of adults with major depressive disorder. Can. J. Psychiatry 61 (9), 506–509. https://doi.org/10.1177/0706743716659061.

Li, K.X., Loshak, H., 2020. Pharmacogenomic Testing in Depression: A Review of Clinical Effectiveness, Cost-Effectiveness, and Guidelines. https://cadth.ca/sites/default/files/pdf/htis/2020/RC1232RxGxtestingdepressionFinal.pdf.

Linden, M., Helmchen, H., Mackert, A., Müller-Oerlinghausen, B., 1994. Structure and feasibility of a standardized stepwise drug treatment regimen (SSTR) for depressed inpatients. Pharmacopsychiatry 27 (S 1), 51–53. https://doi.org/10.1055/s-2007-1014329.

Lobach, D.F., Hammond, W.E., 1997. Computerized decision support based on a clinical practice guideline improves compliance with care standards. Am. J. Med. 102 (1), 89–98. https://doi.org/10.1016/S0002-9343(96)00382-8.

Loh, A., Simon, D., Wills, C.E., Kriston, L., Niebling, W., Härter, M., 2007. The effects of a shared decision-making intervention in primary care of depression: a cluster-randomized controlled trial. Patient Educ. Couns. 67 (3), 324–332. https://doi.org/10.1016/j.pec.2007.03.023.

Lugtenberg, M., Burgers, J.S., Westert, G.P., 2009. Effects of evidence-based clinical practice guidelines on quality of care: a systematic review. Qual. Saf. Health Care 18 (5), 385–392. https://doi.org/10.1136/qshc.2008.028043.

MacKinnon, D.F., 2019. Depression in adults. In: BMJ Best Practice. https://bestpractice.bmj.com/topics/en-us/55/pdf/55/Depressioninadults.pdf.

Malhi, G.S., Bassett, D., Boyce, P., Bryant, R., Fitzgerald, P.B., Fritz, K., Hopwood, M., Lyndon, B., Mulder, R., Murray, G., Porter, R., Singh, A.B., 2015. Royal Australian and New Zealand College of Psychiatrists clinical practice guidelines for mood disorders. Aust. N. Z. J. Psychiatry 49 (12), 1087–1206. https://doi.org/10.1177/0004867415617657.

Margolis, C.Z., 1983. Uses of clinical algorithms. JAMA 249 (5), 627. https://doi.org/10.1001/jama.1983.03330290049028.

Mathew, S.J., Murrough, J.W., Aan Het Rot, M., Collins, K.A., Reich, D.L., Charney, D.S., 2010. Riluzole for relapse prevention following intravenous ketamine in treatment-resistant depression: a pilot randomized, placebo-controlled continuation trial. Int. J. Neuropsychopharmacol. 13 (1), 71–82. https://doi.org/10.1017/S1461145709000169.

Mayberg, H.S., 2003. Modulating dysfunctional limbic-cortical circuits in depression: towards development of brain-based algorithms for diagnosis and optimised treatment. Br. Med. Bull. 65 (1), 193–207. https://doi.org/10.1093/bmb/65.1.193.

McIntyre, R.S., 2018. Pharmacological treatment of major depressive disorder: 2017-2018 pdate summary. 2017–2018 Florida Best Practice Psychotherapeutic Medication Guidelines for Adults. The University of South Florida, Florida Medicaid Drug Therapy Management Program sponsored by the Florida Agency for Health Care Administration.

Morrato, E.H., Libby, A.M., Orton, H.D., DeGruy, F.V., Brent, D.A., Allen, R., Valuck, R.J., 2008. Frequency of provider contact after FDA advisory on risk of pediatric suicidality with SSRIs. Am. J. Psychiatry 165 (1), 42–50. https://doi.org/10.1176/appi.ajp.2007.07010205.

Murrough, J.W., Perez, A.M., Pillemer, S., Stern, J., Parides, M.K., Aan Het Rot, M., Collins, K.A., Mathew, S.J., Charney, D.S., Iosifescu, D.V., 2013. Rapid and longer-term antidepressant effects of repeated ketamine infusions in treatment-resistant major depression. Biol. Psychiatry 74 (4), 250–256. https://doi.org/10.1016/j.biopsych.2012.06.022.

National Collaborating Centre for Mental Health (UK), 2010. Depression: The Treatment and Management of Depression in Adults (Updated Edition). British Psychological Society.

National Institute for Health and Care Excellence, 2009. Depression in Adults: Recognition and Management. Clinical Guideline [CG90].

National Institute for Health and Clinical Excellence, 2011. Assessing Cost Impact: Methods Guide. https://www.nice.org.uk/Media/Default/About/what-we-do/Into-practice/Costing_Manual_update_050811.pdf.

NICE, 2018. Depression in Adults: Treatment and Management, NICE Guideline: Short Version, Draft for Second Consultation, May 2018.

Patel, K., Allen, S., Haque, M.N., Angelescu, I., Baumeister, D., Tracy, D.K., 2016. Bupropion: a systematic review and meta-analysis of effectiveness as an antidepressant. Ther. Adv. Psychopharmacol. 6 (2), 99–144. https://doi.org/10.1177/2045125316629071.

Pincus, H.A., Pechura, C.M., Elinson, L., Pettit, A.R., 2001. Depression in primary care: linking clinical and systems strategies. Gen. Hosp. Psychiatry 23 (6), 311–318. https://doi.org/10.1016/S0163-8343(01)00165-7.

Qaseem, A., Barry, M.J., Kansagara, D., 2016. Nonpharmacologic versus pharmacologic treatment of adult patients with major depressive disorder: a clinical practice guideline from the American College of Physicians. Ann. Intern. Med. 164 (5), 350. https://doi.org/10.7326/M15-2570.

Reck, M., Rabe, K.F., 2017. Precision diagnosis and treatment for advanced non-small-cell lung cancer. N. Engl. J. Med. 377 (9), 849–861. https://doi.org/10.1056/NEJMra1703413.

Ricken, R., Wiethoff, K., Reinhold, T., Schietsch, K., Stamm, T., Kiermeir, J., Neu, P., Heinz, A., Bauer, M., Adli, M., 2011. Algorithm-guided treatment of depression reduces treatment costs—results from the randomized controlled German Algorithm Project (GAPII). J. Affect. Disord. 134 (1–3), 249–256. https://doi.org/10.1016/j.jad.2011.05.053.

Ricken, R., Wiethoff, K., Reinhold, T., Stamm, T.J., Baghai, T.C., Fisher, R., Seemüller, F., Brieger, P., Cordes, J., Laux, G., Hauth, I., Möller, H.-J., Heinz, A., Bauer, M., Adli, M., 2018. A standardized stepwise drug treatment algorithm for depression reduces direct treatment costs in depressed inpatients—Results from the German Algorithm Project (GAP3). J. Affect. Disord. 228, 173–177. https://doi.org/10.1016/j.jad.2017.11.051.

Ridic, G., Gleason, S., Ridic, O., 2012. Comparisons of health care systems in the United States, Germany and Canada. Mater. Sociomed. 24 (2), 112. https://doi.org/10.5455/msm.2012.24.112-120.

Rief, W., Nestoriuc, Y., Weiss, S., Welzel, E., Barsky, A.J., Hofmann, S.G., 2009. Meta-analysis of the placebo response in antidepressant trials. J. Affect. Disord. 118 (1–3), 1–8. https://doi.org/10.1016/j.jad.2009.01.029.

Rosenblat, J.D., Lee, Y., McIntyre, R.S., 2018. The effect of pharmacogenomic testing on response and remission rates in the acute treatment of major depressive disorder: a meta-analysis. J. Affect. Disord. 241, 484–491. https://doi.org/10.1016/j.jad.2018.08.056.

Rush, A.J., 2001. Practice guidelines and algorithms. In: Treatment of Depression. Bridging the 21st Century. American Psychiatric Press, pp. 213–240.

Rush, A.J., Rago, W.V., Crismon, M.L., Toprac, M.G., Shon, S.P., Suppes, T., Miller, A.L., Trivedi, M.H., Swann, A.C., Biggs, M.M., Shores-Wilson, K., Kashner, T.M., Pigott, T., Chiles, J.A., Gilbert, D.A., Altshuler, K.Z., 1999. Medication treatment for the severely and persistently mentally ill. J. Clin. Psychiatry 60 (5), 284–291. https://doi.org/10.4088/JCP.v60n0503.

Rush, A.J., Fava, M., Wisniewski, S.R., Lavori, P.W., Trivedi, M.H., Sackeim, H.A., Thase, M.E., Nierenberg, A.A., Quitkin, F.M., Kashner, T.M., Kupfer, D.J., Rosenbaum, J.F., Alpert, J., Stewart, J.W., McGrath, P.J., Biggs, M.M., Shores-Wilson, K., Lebowitz, B.D., Ritz, L., et al., 2004. Sequenced treatment alternatives to relieve depression (STAR*D): rationale and design. Control. Clin. Trials 25 (1), 119–142. https://doi.org/10.1016/S0197-2456(03)00112-0.

Rush, A.J., Trivedi, M.H., Wisniewski, S.R., Nierenberg, A.A., Stewart, J.W., Warden, D., Niederehe, G., Thase, M.E., Lavori, P.W., Lebowitz, B.D., McGrath, P.J., Rosenbaum, J.F., Sackeim, H.A., Kupfer, D.J., Luther, J., Fava, M., 2006. Acute and longer-term outcomes in depressed outpatients requiring one or several treatment steps: a STAR*D report. Am. J. Psychiatry 163 (11), 1905–1917. https://doi.org/10.1176/ajp.2006.163.11.1905.

Sackett, D.L., 1997. Evidence-based medicine. Semin. Perinatol. 21 (1), 3–5. https://doi.org/10.1016/S0146-0005(97)80013-4.

Sackett, D.L., Rosenberg, W.C., 1995. On the need for evidence-based medicine. J. Public Health 17 (3), 330–334. https://doi.org/10.1093/oxfordjournals.pubmed.a043127.

Sackett, D.L., Rosenberg, W.M.C., Gray, J.A.M., Haynes, R.B., Richardson, W.S., 1996. Evidence based medicine: what it is and what it isn't. BMJ 312 (7023), 71–72. https://doi.org/10.1136/bmj.312.7023.71.

Sackett, D.L., Richardson, S.W., Rosenberg, W., Haynes, R.B., 1997. Evidence-Based Medicine: How to Practice and Teach EBM, first ed. Churchill Livingstone, London.

Seo, J.S., Bahk, W.-M., Wang, H.R., Woo, Y.S., Park, Y.-M., Jeong, J.-H., Kim, W., Shim, S.-H., Lee, J.G., Jon, D.-I., Min, K.J., 2018. Korean medication algorithm for depressive disorders 2017: third revision. Clin. Psychopharmacol. Neurosci. 16 (1), 67–87. https://doi.org/10.9758/cpn.2018.16.1.67.

Shan, X., Zhao, W., Qiu, Y., Wu, H., Chen, J., Fang, Y., Guo, W., Li, L., 2019. Preliminary clinical investigation of combinatorial pharmacogenomic testing for the optimized treatment of depression: a randomized single-blind study. Front. Neurosci. 13. https://doi.org/10.3389/fnins.2019.00960.

Shekelle, P.G., Woolf, S.H., Eccles, M., Grimshaw, J., 1999. Clinical guidelines: developing guidelines. BMJ 318 (7183), 593–596. https://doi.org/10.1136/bmj.318.7183.593.

Simon, G.E., Von Korff, M., Rutter, C.M., Peterson, D.A., 2001. Treatment process and outcomes for managed care patients receiving new antidepressant prescriptions from psychiatrists and primary care physicians. Arch. Gen. Psychiatry 58 (4), 395. https://doi.org/10.1001/archpsyc.58.4.395.

Stahl, S.M., 2000. Essential Psychopharmacology: Neuroscientific Basis and Practical Applications, second ed. Cambridge University Press.

Stettin, G.D., Yao, J., Verbrugge, R.R., Aubert, R.E., 2006. Frequency of follow-up care for adult and pediatric patients during initiation of antidepressant therapy. Am. J. Manag. Care 12 (8), 453–461.

Taylor, D.M., Panton, C., Kapur, S., 2015. The Maudsley Prescribing Guidelines in Psychiatry, twelfth ed. Wiley Blackwell.

Taylor, D., Barnes, T.R.E., Young, A.H., 2018. The Maudsley Prescribing Guidelines in Psychiatry, thirteenth ed. Wiley Blackwell.

The Management of Major Depressive Disorder Working Group, 2016. VA/DoD Clinical Practice Guideline for the Management of Major Depressive Disorder. Department of Veterans Affairs and Department of Defense.

Trivedi, M.H., Kleiber, B.A., 2001. Using treatment algorithms for the effective management of treatment-resistant depression. J. Clin. Psychiatry 62 (Suppl. 1), 25–29.

Trivedi, M.H., Kern, J.K., Voegtle, T.M., Baker, S.M., Altshuler, K.Z., 2002. Computerized medication algorithms in behavioural health care. In: Behavioural Healthcare Informatics. Springer-Verlag, pp. 53–63, https://doi.org/10.1007/978-0-387-21586-0.

Trivedi, M.H., Kern, J.K., Grannemann, B.D., Altshuler, K.Z., Sunderajan, P., 2004a. A computerized clinical decision support system as a means of implementing depression guidelines. Psychiatr. Serv. 55 (8), 879–885. https://doi.org/10.1176/appi.ps.55.8.879.

Trivedi, M.H., Rush, A.J., Crismon, M.L., Kashner, T.M., Toprac, M.G., Carmody, T.J., Key, T., Biggs, M.M., Shores-Wilson, K., Witte, B., Suppes, T., Miller, A.L., Altshuler, K.Z., Shon, S.P., 2004b. Clinical results for patients with major depressive disorder in the Texas Medication Algorithm Project. Arch. Gen. Psychiatry 61 (7), 669. https://doi.org/10.1001/archpsyc.61.7.669.

Trivedi, M.H., Kleiber, B., Greer, T.L., 2005. Remission and recovery in depression treatment. Drug Dev. Res. 65 (4), 335–343. https://doi.org/10.1002/ddr.20034.

Trivedi, M.H., Rush, A.J., Wisniewski, S.R., Nierenberg, A.A., Warden, D., Ritz, L., Norquist, G., Howland, R.H., Lebowitz, B., McGrath, P.J., Shores-Wilson, K., Biggs, M.M., Balasubramani, G.K., Fava, M., 2006. Evaluation of outcomes with citalopram for depression using measurement-based care in STAR*D: implications for clinical practice. Am. J. Psychiatry 163 (1), 28–40. https://doi.org/10.1176/appi.ajp.163.1.28.

Vaccarino, S.R., Kennedy, S.H., 2020. Treatment resistant depression. Ketamine for Treatment Resistant Depression. Academic Press, pp. 33–84.

Winner, J.G., Carhart, J.M., Altar, C.A., Allen, J.D., Dechairo, B.M., 2013. A prospective, randomized, double-blind study assessing the clinical impact of integrated pharmacogenomic testing for major depressive disorder. Discov. Med. 16 (89), 219–227. http://www.ncbi.nlm.nih.gov/pubmed/24229738.

Yang, J., Han, C., Yoon, H.-K., Pae, C.-U., Kim, M.-J., Park, S.-Y., Ahn, J., 2013. Experiences and barriers to implementation of clinical practice guideline for depression in Korea. BMC Psychiatry 13 (1), 150. https://doi.org/10.1186/1471-244X-13-150.

Zarate Jr., C.A., Singh, J., Carlson, P., Brutsche, N., Ameli, R., Luckenbaugh, D., 2006. A randomized trial of an N-methyl-D-aspartate antagonist in treatment-resistant major depression. Arch. Gen. Psychiatry 63.

Chapter 7

Tools to aid precision treatments to prevent or manage treatment-resistant depression (TRD): Pharmacogenomics, machine learning, and artificial intelligence

Lisa C. Brown[a], Daniel J. Mueller[b], Harris A. Eyre[c], Chad Bousman[d], and John F. Greden[e]

[a]*Great Scott! Consulting, Brooklyn, NY, United States,* [b]*Campbell Family Mental Health Research Institute, Centre for Addiction and Mental Health, Institute of Medical Science, University of Toronto, Toronto, ON, Canada,* [c]*Neuroscience-inspired Policy Initiative, Organisation for Economic Co-Operation and Development (OECD) and PRODEO Institute, Paris, France,* [d]*Department of Medical Genetics, University of Calgary, Calgary, AB, Canada,* [e]*University of Michigan Eisenberg Family Depression Center and National Network of Depression Centers, Ann Arbor, MI, United States*

Introduction

One aspirational approach to improve treatment response in major depressive disorder (MDD) and treatment-resistant depression (TRD) is to utilize biomarkers to aid accuracy of diagnosis and selection of more precise treatments (Dodd et al., 2020). One must also appreciate the role of correct diagnosis in utilizing biomarkers to improve treatment response. Biomarkers for diagnosis, such as polygenic risk scores, may also improve treatment decisions and therefore outcomes for individuals with depression and comorbid illnesses. Areas such as biomarkers in pharmacogenomics, epigenomics, proteomics, metabolomics, and microbiomics as well as algorithms such as machine learning and artificial intelligence may hold insights into matching a more effective treatment to each individual patient (Baune, 2019). In this chapter, we will focus on the roles of pharmacogenomics (PGx) and algorithms to aid in treatment decisions to improve patient outcomes in MDD and TRD. Specifically, we will review the gradually improving performances of PGx tests and briefly describe progress in strategies to improve them further using artificial intelligence (AI) and machine learning (ML). We will also discuss relevant considerations for use of such tests such as optimal clinical times for administering PGx tests; recent trends in insurance coverage and federal programs for biomarker tests; and updates of FDA- drug labels for these new biomarkers. The "PGx" abbreviation in this chapter will be inclusive of both pharmacogenomic and pharmacogenetic approaches. The term "algorithm" will be inclusive of algorithmic approaches such as machine learning (ML) and artificial intelligence (AI).

Brief history of pharmacogenomic tests and algorithms for clinical use

The link between genes and mood disorders have been recognized since the Hippocratic era, but clinical approaches designed to leverage an individual's genetic information to personalize therapy have taken centuries to develop (Angst and Marneros, 2001; Müller and Rizhanovsky, 2020).

For most major medical disorders, biomarkers are demonstrated to be pivotal—even indispensable—for progress. Examples abound in oncology, cardiovascular medicine, pulmonary medicine, neurology, and other arenas (Olivier et al., 2019). The potential value of biomarker measures when treating cancers and other medical disorders gained credibility during the 1950s and helped fuel interest in PGx tests for brain-behavior illnesses. During cancer treatments, individual differences in genes and enzymes were shown to alter metabolism, clinical effectiveness and toxicity of antineoplastic medications (Miteva-Marcheva et al., 2020). Genetic outliers were found to have poorer clinical outcomes, sometimes because of ultrarapid metabolism requiring greater dosing, sometimes because of genetic patterns that led to poor metabolism, side effects, poor adherence, and the need for lower doses. The value of such information for treatment selections became evident, and clinical desires for biomarkers became more widespread.

Managing Treatment-Resistant Depression. https://doi.org/10.1016/B978-0-12-824067-0.00007-4

Despite a growing desire for biomarkers, progress has been particularly slow in developing precision-based biomarkers to aid antidepressant selection and adjustments for the treatment of MDD, and for resolving TRD should it occur. Consequently, physicians are forced to make important clinical treatment decisions based primarily upon a collection of subjective and often erroneous clinical determinations for antidepressants such as: (1) "MDD depressions generally are the 'same illness' so can be treated using the same treatments"; (2) "One-size treatment tends to fit all so I will use my favorite"; and (3) "that didn't work, so let's try this." These factors are common launching pads for the development of TRD. Fortunately, research advances are accelerating, and a variety of PGx tests and algorithms are beginning to show promise (Arnow et al., 2015; Bousman et al., 2019).

Facilitating this acceleration is a current estimate that suggests genomic variation explains approximately 80% of the variability in drug efficacy and adverse effects for most medical illnesses (Kim et al., 2021). Likewise, the past decade has been witness to a growing recognition that an individual's genes play an important role in psychotropic treatment response, in part attributed to Dr. David Mrazek's pivotal book on Pharmacogenomics (Mrazek, 2010, India slides, p. 3). He defined psychiatric pharmacogenomics as "the study of how gene variations influence the responses of a patient to treatment with psychotropic medications." Mrazek emphasized that gene actions stemming from variations in pharmaco*kinetic* (PK) genes helped determine how one's *body affects the drug* such as metabolizing rapidly or poorly, whereas variations in pharmacodynamic (PD) genes predominantly changed how the *drug affects the body,* potentially altering clinical response. Importantly, Mrazek and other investigators clarified that pharmacodynamic genes and pharmacokinetic genes *interactively* assert their composite actions, that medication effects are regulated by a *multitude* of genes and enzymes, and that most clinical decisions could be improved only by integrating and considering these *composite* considerations. However, most evidence in the field of pharmacogenomics relies on single gene-drug pairs and this composite or combinatorial approach is limited (PMID: 31004441) (Fig. 7.1).

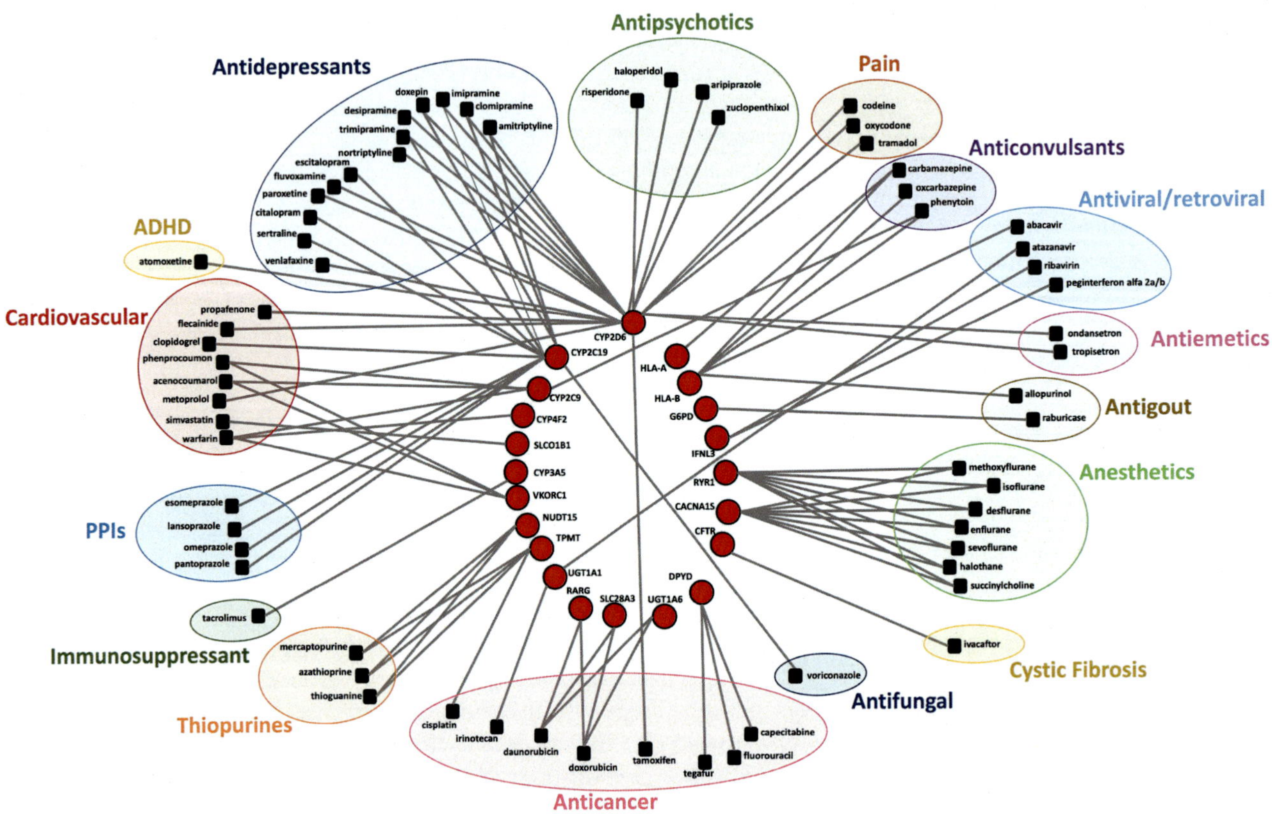

FIG. 7.1 Pharmacogenomic gene drug pairs with actionable clinical practice guidelines.

Evidence for pharmacogenomic testing in clinical practice

Methodological shortcomings used in nascent PGx tests and early publications were quite recognizable. Many were understandable because of absent or minimal sustained funding support, and studies were often plagued by small samples, lack of standardized clinical evaluations, limited focus upon single genes or enzymes, short durations of follow-up, no blinded ratings, and an array of other reasons. Commercial or parochial origins of emerging PGx tests for mood disorders also generated concerns (Zubenko et al., 2018; Bousman and Müller, 2018).

Efforts to minimize or overcome research shortcomings and accelerate PGx tests focus on both studies from academic/healthcare institutions and commercial PGx companies. The majority of evidence for prospective clinical utility of PGx testing has come from commercially sponsored clinical randomized controlled trials. Multiple meta-analyses have analyzed the overall effect of these studies and PGx testing efficacy in depression (PMID: 30520364).

Of note, multiple studies did show improved outcomes with PGx testing when decisions were congruent (or aligned) with the PGx test report. This indicates that individuals with gene-drug interactions may benefit the most from PGx testing (Greden et al., 2019; Pérez et al., 2017; Perlis et al., 2020; Thase et al., 2019). "Congruence" generally is used to mean "in agreement," "in harmony" or "compatible." "Incongruence" implies its absence. While few clinicians traditionally use the terms "congruency" or "incongruency" with regard to medications, many have known about the consequences of prescribing genomically incongruent medications, i.e., those that may not be compatible with the recipient's genomic composition. An example would be a medication for which FDA approval was based on average rates of metabolism, but when given to an individual with "incongruent" genetic composition, the results could be ineffective or undesirable. Incongruent medications for an individual commonly result in more adverse events and side effects, lower adherence, more medication discontinuations, less likelihood to achieve improved responses by failure to alter underlying, more achieving lower remission rates, and poor, disappointing outcomes. These all contribute to TRD and enable its devastating consequences.

Importantly, some experts have argued that RCTs should not be the gold standard in evaluating PGx testing because it does not mirror real clinical practice (Caudle et al., 2016; Huddart et al., 2019). While open-label studies risk bias without a "placebo" standard of care arm, implementation mirrors real clinical practice. One of the largest, prospective clinical trials, Individualized Medicine: Pharmacogenetic Assessment & Clinical Treatment (IMPACT), evaluated the implementation of PGx testing in both primary care provider (PCP) settings and specialist psychiatry settings in almost 2000 patients and found that patient outcomes were actually better in individuals under the care of a PCP compared to a psychiatrist, indicating that PGx testing can be used in both specialist and nonspecialist settings (Tanner et al., 2018).

For individuals with MDD, the gold standard algorithm was when MDD treatments were compared in the Sequenced Treatment Alternatives to Relieve Depression (STAR*D) study. This study showed that with each medication trial, response and remission decreases while side effects increase. Interestingly, Fan et al. performed a PGx analysis simulating the STAR*D algorithm to determine potential gene–drug interactions (Fan and Bousman, 2020). The authors found that in the first step of STAR*D, which includes the prescription of citalopram, ~30% of patients had genetic implications that would change how the medication was administered. In the second step, ~21% had actionable gene-drug interactions. These data beg the question, that if PGx had been used to guide treatment decisions, would outcomes in STAR*D have been improved?

PGx testing to help guide medication selection appears to meaningfully help eliminate or reduce exposure to incongruent medications and to guide clinicians into avoiding their use. When making determinations about congruity vs. incongruity, PGx tests appear to more objectively integrate contributing factors, whether those be pharmacokinetic or pharmacodynamic. Consequently, should combinatorial PGx testing identify medications with greater likelihood for incongruency, that medication ideally should be avoided if possible, even should it be a "favorite" treatment for the clinician. Incongruency matters. Switching to achieve congruency matters.

PGx economic studies

Expenses for biomarker tests are areas of concern for many. Multiple Health Economic Outcomes Research (HEOR) studies have been performed in order to determine the economic validity and utility of PGx testing. Prospective, retrospective, and modeling studies have been performed. The largest prospective study to date, evaluated medication cost savings in individuals with psychiatric disorders receiving psychotropic medication prescriptions with and without guidance from PGx testing (Winner et al., 2015). The study included over 13,000 patients and found that PGx testing resulted in savings of $1035.60 USD per year compared to standard of care. Interestingly, prescribing that was congruent with the test resulted in even greater savings of $2774.53 compared to standard of care. Equally interesting, a subanalysis of this study also found that individuals receiving psychotropic prescriptions from PCPs congruent with the PGx test saved $3988 per year compared to $1308 per year under the care of psychiatrists (Brown et al., 2017). These data support the results of the IMPACT study that PCPs can not only implement PGx testing effectively, but implementation is also cost saving. A large, retrospective study of health insurance claims found that in over 3500 individuals, those whose treatment was guided by PGx testing had significantly less healthcare utilization, including fewer emergency room visits and in-patient hospitalizations (Perlis et al., 2018).

Some systems such as Medicare and private health insurance companies have now begun to reimburse portions or total PGx testing expenses for specific patients. Most often, this reimbursement is for individuals with depression who have failed at least one medication and whose clinician is considering a medication with a known gene-drug interaction (CMS, 2020). Cost and insurance coverage variables may turn out to be among the most influential in determining future refinements, more widespread use, and accessibility of PGx testing.

Algorithms for medication guidance

One limitation to PGx testing is that it tends to be diagnostic and clinically agnostic, meaning it only takes into account specific gene-drug interactions alone, without considering other features of the individual such as diagnosis, symptoms, concomitant medications, etc. Other approaches to try to identify markers of disease and treatment response have also taken a singular approach. Algorithmic tools that take multiple features into account may improve the use of PGx testing or be valuable tools on their own. These tools can also incorporate patient reported features such as measurement-based symptom scales. Three algorithmic pathways with great promise for guiding medication treatment (and potentially even diagnosis) are machine learning (ML), artificial intelligence (AI), and polygenic risk scores (PRS). Here we discuss the current landscape for ML, AI, and PRS in guiding medication decisions. In 2021, the FDA released an action plan for AI/ML-based software as a medical device for regulating AI and ML algorithms in healthcare (FDA, 2021).

Machine learning for medication guidance

Machine Learning is the ability to create or train a model and then apply that model to a new set of data. For example, Chekroud et al. took patient reported data from the STAR*D study to identify features predictive of medication response (Chekroud et al., 2016). The authors trained the model on these data then applied the output to the results of the Combining Medications to Enhance Depression Outcomes (CO-MED) trial to determine if the model could also predict outcomes in this study. Overall, the model was able to predict outcomes with 59.6% accuracy. Another study sought to use ML to predict TRD (Nunez et al., 2021). The group trained the ML algorithm on STAR*D and applied the algorithm to the Canadian Biomarker Integration Network in Depression (CAN-BIND1) dataset. The ML algorithm predicted the risk for TRD with 70%–73% accuracy.

Even more valuable tools are those that integrate genomic data and ML. Athreya et al. trained a model on baseline depression data from Mayo Clinic Pharmacogenomics Research Network Antidepressant Medication Pharmacogenomic Study (PGRN-AMPS) and in conjunction with genome-wide association date, applied the model to STAR*D and the International SSRI Pharmacogenomics Consortium (ISPC) (Athreya et al., 2019). The study showed that when using mutations and baseline depression score, trained ML predicted remission and response in PGRN-AMPs with an AUC > 0.7 and in STAR*D/ISPC with an AUC > 69%. This study showed that ML can be used to combine clinical features and PGx.

AI for medication guidance

In order to understand if AI is acceptable and can be implemented in care, a recent group evaluated the acceptability and impact of an AI-based clinical decision support tool (Benrimoh et al., 2021). The investigators reported that the AI tool would increase patient understanding and was valuable in guiding treatment decisions.

Genome-wide association studies and polygenic risk scores (PRS) for diagnosis and medication guidance

With the surge in genomic medicine and technologies, polygenic risk scores (PRS) have become the next big biomarker in psychiatry for both diagnostics and medication response (Murray et al., 2021; Wray et al., 2021). PRS is the use of genome-wide sequencing to associate multiple variants in multiple genes with a specific trait. One of the first large-scale PRS studies was published in 2017 for schizophrenia and identified 108 loci associated with the disease (PGC, 2014). Since then, PRSs have been evaluated for other diseases such as postpartum depression, MDD, Bipolar Disorder, attention deficit hyperactivity disorder, and autism spectrum disorders (Byrne et al., 2014; Howard et al., 2019; Wray et al., 2018; Stahl et al., 2019; Demontis et al., 2019; Grove et al., 2019). Interestingly, investigators have also sought to apply disease PRS for medication response. One study showed that individuals with higher PRS for schizophrenia had more schizophrenia symptoms post-antipsychotic treatment, indicating a genetic load against treatment response (Zhang et al., 2019). Another study showed that individuals with MDD and a high PRS for schizophrenia were less likely to respond to treatment (Fanelli et al., 2021). Finally, another study showed that genetic liability for depression increased risk for suicide (Mullins et al., 2019). Impute. Me is an open-source program that allows individuals to upload their genomic information and retrieve PRS for various traits such as anxiety, athleticism, asthma, etc. (Folkersen et al., 2020).

A note on black box algorithms

Dr. David Mrazek was the first to utilize the fact that multiple genes affect multiple medications at once, and therefore many commercial companies have created proprietary PGx algorithms to implement this information into clinical practice (Altar et al., 2015). However, much controversy over the secrecy of these algorithms exists (Bousman and Eyre, 2020). While companies may publish clinical data supporting these algorithms, the lack of transparency means that others cannot replicate or validate the algorithms (Shelton et al., 2020). Clinicians and healthcare decison-makers can argue that they want to know what is in the "black box" if it's going to be used in patient care. Proponents, however, have argued that the black box approach is aligned with the idiosyncratic manner by which most clinicians treat patients, the unknown mechanisms of many medications, and the conventional commercial roles in developing new treatments or tests (Holm, 2019; Topol, 2019). Efforts must be made to increase the transparency, explainability, and value of black-box algorithms. We recently provided recommendations to encourage this (Bousman and Eyre, 2020).

Future directions for PGx and algorithm testing

While the evidence for PGx testing efficacy to guide medication decisions in individuals with depression who have failed one medication is strong, efficacy in other populations and varying clinical features is unknown. Many questions remain for the implementation of ML and AI algorithms. Future questions that need to be addressed with PGx testing and algorithms are outlined below:

Should PGx/algorithm testing be done proactively or retroactively, before first treatment with medications or only after failure to respond?

Will PGx/algorithm results help predict vulnerability for and/or aid in reducing suicides?

Will PGx/algorithm testing improve identification of individuals whose antidepressant or mood-stabilizing medications could be removed safely following resolution of depressive episode?

Will PGx/algorithms be useful in guiding treatment selections for individuals with other diagnoses than MDD such as Bipolar Spectrum Disorders, schizophrenia, anxiety, etc.?

Can PGx/algorithm testing help predict manic switches in MDD and bipolar disorders?

Will PGx/algorithm testing help identify individuals who may benefit from placebo?

Will PGx/algorithm testing results remain consistent over time among individuals who continue to suffer additional episodes? Might such a clinical pattern suggest a different etiology cause and the need for different precise treatments?

Will ethnicity differences alter interpretation of PGx/algorithm test results? Data from other pharmacogenomic assessments such as coagulation studies suggest this will be the case.

Will PGx/algorithm test results be significantly altered or made invalid by cooccurring medical illnesses and/or medications?

Can the PGx/algorithm fields adhere to responsible innovation in neurotechnology principles?

Disclosures

LCB is a stockholder and former employee of Myriad Genetics, former employee and current consultant to Tempus Labs, Principal Consultant for Great Scott! Consulting, founding member of Pharmacogenetics Research Network (PGRN), and member of the Clinical Pharmacogenetics Implementation Consortium (CPIC). CAB is the founder and shareholder of Sequence2Script Inc. and a member of the CPIC and Pharmacogene Variation Consortium (PharmVar). HAE is a former employee of CNSdose.

References

Altar, C.A., et al., 2015. Clinical utility of combinatorial pharmacogenomics-guided antidepressant therapy: evidence from three clinical studies. Mol. Neuropsychiatry 1 (3), 145–155.

Angst, J., Marneros, A., 2001. Bipolarity from ancient to modern times: conception, birth and rebirth. J. Affect. Disord. 67 (1–3), 3–19.

Arnow, B.A., et al., 2015. Depression subtypes in predicting antidepressant response: a report from the iSPOT-D trial. Am. J. Psychiatry 172 (8), 743–750.

Athreya, A.P., et al., 2019. Pharmacogenomics-driven prediction of antidepressant treatment outcomes: a machine-learning approach with multi-trial replication. Clin. Pharmacol. Ther. 106 (4), 855–865.

Baune, 2019. Personalized Psychiatry, first ed.

Benrimoh, D., et al., 2021. Using a simulation centre to evaluate preliminary acceptability and impact of an artificial intelligence-powered clinical decision support system for depression treatment on the physician-patient interaction. BJPsych Open 7 (1), e22.

Bousman, C.A., Eyre, H.A., 2020. "Black box" pharmacogenetic decision-support tools in psychiatry. Braz. J. Psychiatry 42 (2), 113–115.

Bousman, C.A., Müller, D.J., 2018. Pharmacogenetics in psychiatry: a companion, rather than competitor, to protocol-based care. JAMA Psychiat. 75 (10), 1090.

Bousman, C.A., et al., 2019. Pharmacogenetic tests and depressive symptom remission: a meta-analysis of randomized controlled trials. Pharmacogenomics 20 (1), 37–47.

Brown, L.C., et al., 2017. Economic utility: combinatorial pharmacogenomics and medication cost savings for mental health care in a primary care setting. Clin. Ther. 39 (3), 592–602.e1.

Byrne, E.M., et al., 2014. Applying polygenic risk scores to postpartum depression. Arch. Womens Ment. Health 17 (6), 519–528.

Caudle, K.E., et al., 2016. Evidence and resources to implement pharmacogenetic knowledge for precision medicine. Am. J. Health Syst. Pharm. 73 (23), 1977–1985.

Chekroud, A.M., et al., 2016. Cross-trial prediction of treatment outcome in depression: a machine learning approach. Lancet Psychiatry 3 (3), 243–250.

CMS, 2020. Local Coverage Determination (LCD): MolDX: Pharmacogenomics Testing (L38294). Available from https://www.cms.gov/medicare-coverage-database/details/lcd-details.aspx?LCDId=38294&ver=16.

Demontis, D., et al., 2019. Discovery of the first genome-wide significant risk loci for attention deficit/hyperactivity disorder. Nat. Genet. 51 (1), 63–75.

Dodd, S., et al., 2020. A clinical approach to treatment resistance in depressed patients: what to do when the usual treatments don't work well enough? World J. Biol. Psychiatry, 1–20.

Fan, M., Bousman, C.A., 2020. Estimating the potential impact of CYP2C19 and CYP2D6 genetic testing on protocol-based care for depression in Canada and the United States. Mol. Neuropsychiatry 5 (Suppl. 1), 27–33.

Fanelli, G., et al., 2021. Higher polygenic risk scores for schizophrenia may be suggestive of treatment non-response in major depressive disorder. Prog. Neuro-Psychopharmacol. Biol. Psychiatry 108, 110170.

FDA, 2021. Artificial Intelligence/Machine Learning (AI/ML)-Based.:Jf/<X Software as a Medical Device (SaMD) Action Plan.

Folkersen, L., et al., 2020. Impute.me: an open-source, non-profit tool for using data from direct-to-consumer genetic testing to calculate and interpret polygenic risk scores. Front. Genet. 11, 578.

Greden, J.F., et al., 2019. Impact of pharmacogenomics on clinical outcomes in major depressive disorder in the GUIDED trial: a large, patient- and rater-blinded, randomized, controlled study. J. Psychiatr. Res. 111, 59–67.

Grove, J., et al., 2019. Identification of common genetic risk variants for autism spectrum disorder. Nat. Genet. 51 (3), 431–444.

Holm, E.A., 2019. In defense of the black box. Science 364 (6435), 26–27.

Howard, D.M., et al., 2019. Genome-wide meta-analysis of depression identifies 102 independent variants and highlights the importance of the prefrontal brain regions. Nat. Neurosci. 22 (3), 343–352.

Huddart, R., et al., 2019. Are randomized controlled trials necessary to establish the value of implementing pharmacogenomics in the clinic? Clin. Pharmacol. Ther. 106 (2), 284–286.

Kim, J.A., Ceccarelli, R., Lu, C.Y., 2021. Pharmacogenomic biomarkers in US FDA-approved drug labels (2000-2020). J. Pers. Med. 11 (3), 179. https://doi.org/10.3390/jpm11030179.

Miteva-Marcheva, N.N., et al., 2020. Application of pharmacogenetics in oncology. Biomarker Res. 8, 32.

Mrazek, D., 2010. Psychiatric Pharmacogenomics. Oxford University Press.

Müller, D.J., Rizhanovsky, Z., 2020. From the origins of pharmacogenetics to first applications in psychiatry. Pharmacopsychiatry 53 (4), 155–161.

Mullins, N., et al., 2019. GWAS of suicide attempt in psychiatric disorders and association with major depression polygenic risk scores. Am. J. Psychiatry 176 (8), 651–660.

Murray, G.K., et al., 2021. Could polygenic risk scores be useful in psychiatry?: a review. JAMA Psychiatry 78 (2), 210–219.

Nunez, J.J., et al., 2021. Replication of machine learning methods to predict treatment outcome with antidepressant medications in patients with major depressive disorder from STAR*D and CAN-BIND-1. PLoS One 16 (6), e0253023.

Olivier, M., et al., 2019. The need for multi-omics biomarker signatures in precision medicine. Int. J. Mol. Sci. 20 (19).

Pérez, V., et al., 2017. Efficacy of prospective pharmacogenetic testing in the treatment of major depressive disorder: results of a randomized, double-blind clinical trial. BMC Psychiatry 17 (1), 250.

Perlis, R.H., et al., 2018. Pharmacogenetic testing among patients with mood and anxiety disorders is associated with decreased utilization and cost: a propensity-score matched study. Depress. Anxiety 35 (10), 946–952.

Perlis, R.H., et al., 2020. Randomized, controlled, participant- and rater-blind trial of pharmacogenomic test-guided treatment versus treatment as usual for major depressive disorder. Depress. Anxiety 37 (9), 834–841.

PGC, 2014. Biological insights from 108 schizophrenia-associated genetic loci. Nature 511 (7510), 421–427.

Shelton, R.C., et al., 2020. Combinatorial pharmacogenomic algorithm is predictive of citalopram and escitalopram metabolism in patients with major depressive disorder. Psychiatry Res. 290, 113017.

Stahl, E.A., et al., 2019. Genome-wide association study identifies 30 loci associated with bipolar disorder. Nat. Genet. 51 (5), 793–803.

Tanner, J.A., et al., 2018. Combinatorial pharmacogenomics and improved patient outcomes in depression: treatment by primary care physicians or psychiatrists. J. Psychiatr. Res. 104, 157–162.

Thase, M.E., et al., 2019. Impact of pharmacogenomics on clinical outcomes for patients taking medications with gene-drug interactions in a randomized controlled trial. J. Clin. Psychiatry 80 (6).

Topol, E.J., 2019. High-performance medicine: the convergence of human and artificial intelligence. Nat. Med. 25 (1), 44–56.

Winner, J.G., et al., 2015. Combinatorial pharmacogenomic guidance for psychiatric medications reduces overall pharmacy costs in a 1 year prospective evaluation. Curr. Med. Res. Opin. 31 (9), 1633–1643.

Wray, N.R., et al., 2018. Genome-wide association analyses identify 44 risk variants and refine the genetic architecture of major depression. Nat. Genet. 50 (5), 668–681.

Wray, N.R., et al., 2021. From basic science to clinical application of polygenic risk scores: a primer. JAMA Psychiat. 78 (1), 101–109.

Zhang, J.P., et al., 2019. Schizophrenia polygenic risk score as a predictor of antipsychotic efficacy in first-episode psychosis. Am. J. Psychiatry 176 (1), 21–28.

Zubenko, G.S., Sommer, B.R., Cohen, B.M., 2018. On the marketing and use of pharmacogenetic tests for psychiatric treatment. JAMA Psychiat. 75 (8), 769–770.

Further reading

Bousman, C.A., Dunlop, B.W., 2018. Genotype, phenotype, and medication recommendation agreement among commercial pharmacogenetic-based decision support tools. Pharmacogenomics J. 18 (5), 613–622.

Bradley, P., et al., 2018. Improved efficacy with targeted pharmacogenetic-guided treatment of patients with depression and anxiety: a randomized clinical trial demonstrating clinical utility. J. Psychiatr. Res. 96, 100–107.

Brown, L., et al., 2020. The clinical utility of combinatorial pharmacogenomic testing for patients with depression: a meta-analysis. Pharmacogenomics 21 (8), 559–569.

Forester, B.P., et al., 2020. Combinatorial pharmacogenomic testing improves outcomes for older adults with depression. Am. J. Geriatr. Psychiatry 28 (9), 933–945.

Rosenblat, J.D., Lee, Y., McIntyre, R.S., 2018. The effect of pharmacogenomic testing on response and remission rates in the acute treatment of major depressive disorder: a meta-analysis. J. Affect. Disord. 241, 484–491.

Singh, A.B., 2015. Improved antidepressant remission in major depression via a pharmacokinetic pathway polygene pharmacogenetic report. Clin. Psychopharmacol. Neurosci. 13 (2), 150–156.

Winner, J.G., et al., 2013. A prospective, randomized, double-blind study assessing the clinical impact of integrated pharmacogenomic testing for major depressive disorder. Discov. Med. 16 (89), 219–227.

Chapter 8

Tricyclic antidepressants for treatment-resistant depression

William V. Bobo and Elliott Richelson
Department of Psychiatry and Psychology, Mayo Clinic, Jacksonville, FL, United States

Introduction

As a group, the tricyclic antidepressants (TCAs) comprise the second-oldest class of antidepressants, beginning with the serendipitous discovery in the late 1950s that imipramine, investigated initially for its sedative properties, was effective for patients with severely depressed mood, marked psychomotor retardation, lethargy, hopelessness, and a diurnal fluctuation in symptom severity (Kuhn, 1958; Lehmann et al., 1958). A large number of TCAs were subsequently introduced, including the tertiary-amine agents (amitriptyline, dothiepin, doxepin, trimipramine, lofepramine, and clomipramine) and the secondary-amine drugs (nortriptyline, desipramine, and protriptyline). The TCAs are so named because of a common polycyclic structure consisting of three fused rings and attached side chains (Fig. 8.1). Other drugs, called tetracyclics (maprotiline and amoxapine), were introduced in the 1970s. They are structurally related to TCAs, but have four rings instead of three (Wells and Gelenberg, 1981).

The introduction and wide propagation of TCAs represented a safety advance over monoamine oxidase inhibitors (MAOIs), the clinical use of which was limited by potentially severe and life-threatening drug-drug and drug-nutrient interactions (reviewed elsewhere in this volume). As such, the TCAs replaced MAOIs as the drugs of choice for the pharmacological management of depression during the 1970s and 1980s (Fangmann et al., 2008). However, the TCAs were associated with a range of treatment-limiting side effects, clinically significant effects on cardiac conduction, and life-threatening complications in the setting of toxicity and overdose. Beginning in the early 1990s, TCAs were supplanted by selective serotonin reuptake inhibitors (SSRIs) and, subsequently, other newer-generation antidepressants with improved side-effect and safety profiles and comparable antidepressive efficacy (Boyce and Judd, 1999). Indeed, based on data from the National Disease and Therapeutic Index, nearly half of depressed patients visiting office-based practices in the United States were prescribed tricyclic antidepressants in 1987; however, by 2001, only 2% of depressed patients received TCAs (Stafford et al., 2001).

Despite the reduced utilization of TCAs for depression, TCAs are highly efficacious treatments for depression and several other indications—both within psychiatry and outside of psychiatry (Schneider et al., 2019). Pertinent to this chapter, TCAs are frequently overlooked as therapeutic options for depressed patients who respond poorly to newer-generation antidepressants. Here, we review the pharmacology and evidence supporting the use of TCAs for depression, with an emphasis on treatment-resistant illness. We will close the chapter by providing practical advice on the effective use of TCAs for patients with major depression who have responded poorly to other types of antidepressants. The tetracyclic antidepressants, maprotiline and amoxapine, will not be reviewed in detail in this chapter since they are rarely used in clinical practice and, in our view, do not have good evidence supporting their use for treatment-resistant depression.

Pharmacological properties

Mechanism of action

Individual TCAs bind with varying potencies to norepinephrine and serotonin transporters located in nerve terminals in the brain (Richelson, 1996; Tatsumi et al., 1997). The TCAs also bind at nanomolar concentrations to various types of receptors and interact with a number of ion channels located throughout the central nervous system (CNS) and peripherally (Cusack et al., 1994). However, the blockade of norepinephrine and serotonin reuptake is thought to be the primary mechanism by which TCAs exert their antidepressive effects. That said, serotonin and norepinephrine transporter blockade occur acutely

Managing Treatment-Resistant Depression. https://doi.org/10.1016/B978-0-12-824067-0.00008-6

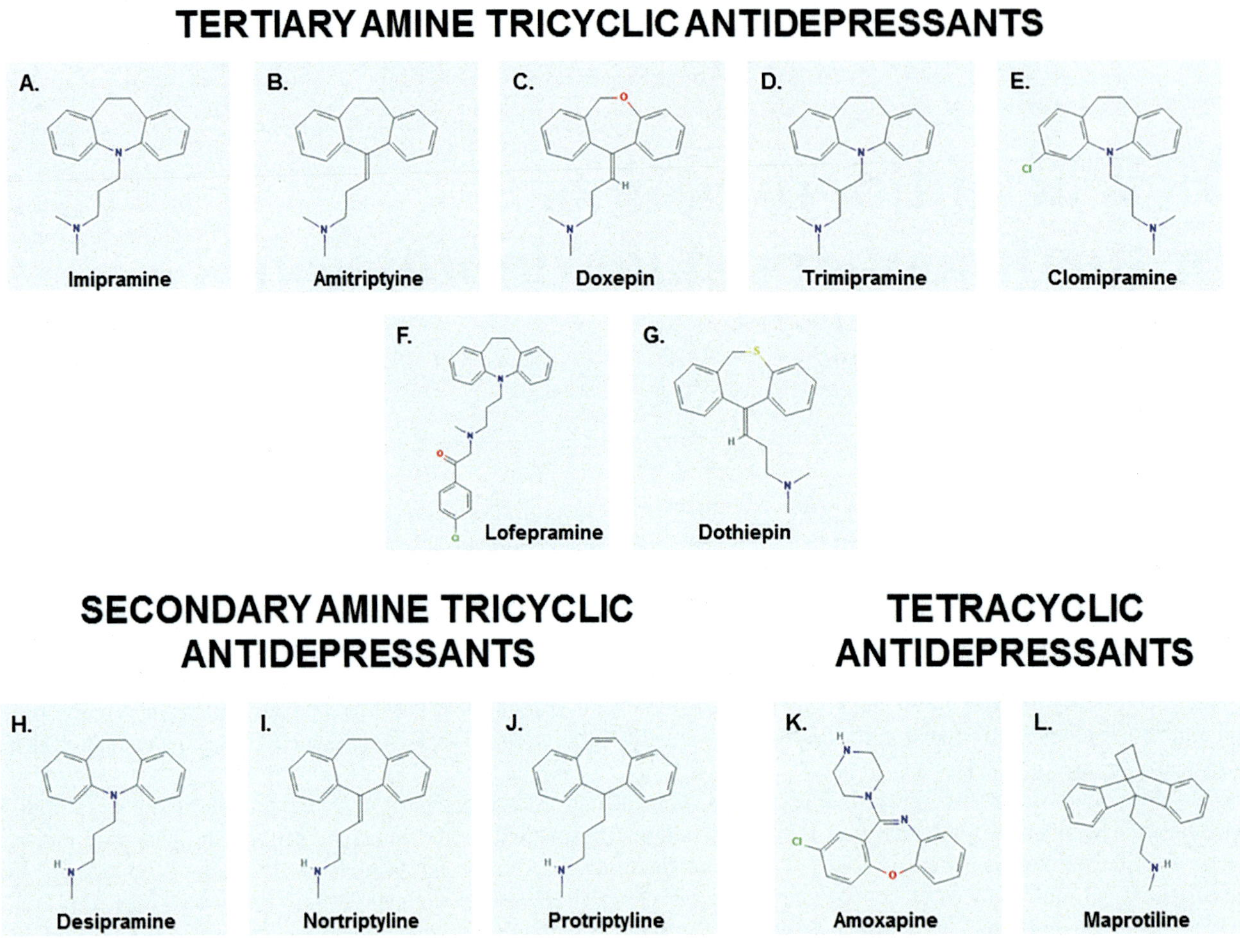

FIG. 8.1 Chemical structures of tertiary- and secondary-amine tricyclic antidepressants (TCAs) and tetracyclic antidepressants. This figure shows the chemical structures of tertiary-amine TCAs (A–G), secondary-amine TCAs (H–I), and tetracyclic antidepressants (K, L). For tertiary-amine TCAs, the side chain nitrogens are fully substituted. For secondary-amine TCAs, the side chain nitrogens have only one substituted hydrogen with a methyl group. *(Source for images: PubChem (https://pubchem.ncbi.nlm.nih.gov/).)*

following exposure to TCAs, resulting in a rapid increase in the synaptic availability of these neurotransmitters. Clearly, these effects do not correlate in time with the 3–6 weeks that are often required for depressed patients who are treated with TCAs to show clinical improvement. The reason for this time lag in therapeutic response with chronic administration of tricyclic and other antidepressants is thought to be due to desensitization and downregulation of selected serotonin and adrenergic receptors in a manner that better approximates the lag time to therapeutic effect of conventional antidepressants (Baker et al., 1992; Piñeyro and Blier, 1999). The link between these pharmacological activities and the delayed response to TCAs is theoretical and may not ultimately explain the therapeutic action for these drugs for treating depression.

Secondary- and tertiary-amine compounds

As mentioned earlier, TCAs are characterized structurally by their three, fused-ring structure and the amine on the attached side chains. The side chain nitrogen is either fully substituted (in this case, the hydrogen atoms have been replaced by two methyl groups) and called a tertiary amine; or has only one substituted hydrogen with a methyl group, making it a secondary amine (Fig. 8.1).

The difference in chemical structure between tertiary- and secondary-amine TCAs appears to predict pharmacological properties that are of clinical relevance. For example, as noted earlier, TCAs bind with varying affinities to serotonin and norepinephrine transporters. In general, tertiary-amine TCAs display more potent serotonin transporter blockade than secondary-amine TCAs (Cusack et al., 1994; Richelson, 2003; Richelson and Nelson, 1984), as illustrated in McTavish and Benfield (1990) (Table 8.1).

TABLE 8.1 **Relative binding affinities for tricyclic and selected other antidepressants for selected pharmacological targets.**

Drug name	Monoamine transporters			Receptors (selected)		
	5-HTT	NET	DAT	Histamine H1	Adrenergic α1	Muscarinic
Tertiary-amine tricyclic antidepressants						
Imipramine	●●●●●○	●●●○○○	○○○○○○	●●●●○○	●●○○○○	●●○○○○
Amitriptyline	●●●●○○	●●●○○○	○○○○○○	●●●●●○	●●●○○○	●●●○○○
Clomipramine	●●●●●●	●●●○○○	○○○○○○	●●●○○○	●●●○○○	●●●○○○
Doxepin	●●●○○○	●●●○○○	…	●●●●●●	●●○○○○	●●●○○○
Trimipramine	●●○○○○	●○○○○○	○○○○○○	●●●●●●	●●●○○○	●●●○○○
Secondary-amine tricyclic antidepressants						
Desipramine	●●●○○○	●●●●●○	○○○○○○	●●○○○○	●●○○○○	●●○○○○
Nortriptyline	●●●○○○	●●●●○○	○○○○○○	●●●○○○	●●●○○○	●●○○○○
Protriptyline	●●●○○○	●●●●●○	○○○○○○	●●●○○○	●●○○○○	●●●○○○
Other antidepressants (for purposes of comparison)						
Citalopram	●●●●●○	○○○○○○	…	●○○○○○	●○○○○○	○○○○○○
Fluoxetine	●●●●●○	●●○○○○	○○○○○○	○○○○○○	○○○○○○	○○○○○○
Sertraline	●●●●●●	●○○○○○	●●●○○○	…	●●○○○○	●○○○○○
Duloxetine	●●●●●●	●●●●○○	●○○○○○	○○○○○○	○○○○○○	○○○○○○
Venlafaxine	●●●●○○	●○○○○○	○○○○○○	…	…	…
Bupropion	…	○○○○○○	●○○○○○	○○○○○○	○○○○○○	…

5-HTT, serotonin transporter; *DAT*, dopamine transporter; *NET*, norepinephrine transporter. Symbols ○○○○○○ to ●●●●●● indicate increasing levels of potency based on target binding profiles (K_i[nmol/L]) reported or summarized in the references below. (…) indicates no effect. All drugs act as antagonists at these pharmacological targets.

TCAs have additional pharmacodynamic properties that broadly differ between tertiary- and secondary-amine agents that are clinically relevant (Table 8.2). For instance, TCAs bind with varying affinities to central and peripheral receptors that influence the propensity for various adverse effects (Richelson, 2003). This includes binding to histamine H1 receptors, adrenergic α-1 receptors, and muscarinic receptors, where they act as antagonists. Broadly, the H1 antagonist activity of TCAs is thought to contribute to their sedative properties and to their propensity to stimulate appetite, causing body weight gain (Richelson, 1979; Salvi et al., 2016). Antagonist activity at α-1 receptors contributes to the risk of orthostatic hypotension, reflex tachycardia, sedation, sexual dysfunction (in addition to sexual dysfunction related to serotonin reuptake blockade), dizziness, and fall risk in the elderly (Khawam et al., 2006). Muscarinic cholinergic receptor antagonist effects of TCAs include blurry vision, dry mouth, constipation, urinary retention, increased heart rate (due to vagolytic effects), and cognitive dysfunction or confusion in susceptible patients. The risks for each of these types of side effects with particular TCAs correlate with their receptor binding affinities at these targets (Richelson, 2001; Cusack et al., 1994; Richelson, 2003; Richelson and Nelson, 1984; Stanton et al., 1993).

Additional pharmacodynamic properties

TCAs have binding activities at various ion channels that are clinically important. In experimental models, TCAs demonstrate dose-dependent suppression of the delayed rectifier potassium current (IK_r). Amitriptyline, imipramine, and doxepin have also been shown to block the human ether-a-go-go-related gene (hERG) potassium channel (Duncan et al., 2007; Teschemacher et al., 1999). These mechanisms are associated with a slowing of cardiac conduction, and are thought to contribute to an increased risk of potentially fatal ventricular arrhythmias (including *torsades de pointes*) in the setting of

TABLE 8.2 Interactions of selected tricyclic antidepressants with various CYP450 isoenzymes.

	CYP1A2	CYP2D6	CYP2C9	CYP2C19	CYP3A4
Amitriptyline	Substrate	Substrate	Substrate	Substrate Inhibitor (potent)	Substrate
Clomipramine	Substrate	Substrate	...	Substrate Inhibitor (potent)	...
Desipramine	...	Substrate Inhibitor (weak)	...	... Inhibitor (weak)	...
Doxepin	Substrate	Substrate Inhibitor (relevance unclear)	Substrate	Substrate	Substrate
Imipramine	Substrate	Substrate	...	Substrate Inhibitor (potent)	...
Nortriptyline	...	Substrate Inhibitor (weak)	...	... Inhibitor (weak)	...

Inhibitor = medication is a competitive or noncompetitive inhibitor of the specified CYP450 isoenzyme. Substrate = medication is a substrate of the specified CYP450 isoenzyme. ... indicates that medication is not a substrate of the specified CYP450 isoenzyme.

TCA toxicity (Alvarez and Pahissa, 2010; Roose and Glassman, 1989). A number of pharmacological actions in addition to norepinephrine reuptake blockade may contribute to the beneficial effects of TCAs for neuropathic pain syndromes, including inhibition of sodium channels, calcium channels, and glutamate NMDA receptors (Obata, 2017).

Pharmacokinetics

In general, TCAs are rapidly absorbed from the gastrointestinal tract following oral administration, with peak concentrations in the blood reached within 2–8 h. In most cases, the presence of food does not affect absorption pharmacokinetics. Once absorbed, TCAs undergo first-pass metabolism and bind extensively to plasma proteins. Nevertheless, the impact of total and free concentrations of most TCAs on therapeutic response is similar enough that it is not necessary to measure the free fractions of TCAs (Brøsen and Gram, 1988; Nelson et al., 1983).

TCAs are metabolized in the liver by demethylation and hydroxylation (Gillman, 2007). Selected agents, such as imipramine and amitriptyline, are demethylated into active metabolites. Therefore, when interpreting plasma levels for imipramine, the total plasma concentration consists of the concentration of imipramine plus the concentration of desipramine, its monomethylated counterpart. Similarly, the total plasma concentration of amitriptyline consists of the combined concentrations of amitriptyline and nortriptyline. The tertiary-amine TCA, lofepramine (not available in many countries, including the United States and Canada), is structurally related to imipramine and, like imipramine, includes the secondary-amine TCA, desipramine, as a main metabolite (Strandgården and Gunnarsson, 1994).

Several factors govern the metabolism and elimination of TCAs. Therefore, the elimination half-lives of TCAs vary widely, from as little as 6 h with doxepin to as much as 56 h with nortriptyline. As mentioned earlier and as shown in Table 8.2, TCAs are metabolized by several cytochrome P450 enzymes, principally CYP2D6 and 2C19, with relatively minor contributions from other CYP450 isoenzymes including CYP1A2 and 3A4 (Gillman, 2007). Therefore, varying exposure to TCAs can be expected in people who are pharmacogenetic poor, intermediate, or ultrarapid metabolizers at these CYP450 isoenzymes. Similarly, wide variations in TCA exposure at a given daily dose can be expected in patients who take medications that inhibit or induce these hepatic isoenzymes (Taylor, 1995). Greater-than-expected plasma TCA concentrations also occur in people with varying degrees of hepatic compromise (Mullish et al., 2014). The importance of these pharmacokinetic factors rests in the possibility that, even at typical daily doses, elevated risk for severe side-effects and toxicity may occur when TCA metabolism is inhibited, or the risk of poor treatment response may increase when TCA metabolism is induced (Ereshefsky et al., 1995, Mauri et al., 2014; Preskorn et al., 1990).

Only a small fraction of TCAs are eliminated in their active form in the urine. Therefore, major adjustments in the dose of most TCAs are not necessary in people with less than advanced-stage renal insufficiency. Even so, the renal clearance of TCAs is reduced as a function of normal aging, which is thought to confer a much higher risk of toxicity from TCAs in older adults, as compared to young people (Rudorfer and Potter, 1999). None of the TCAs are considered dialyzable, likely due to their lipophilicity, with tight protein binding, and large volumes of distribution. Drug-specific guidance for dosing TCAs in people in various stages of chronic kidney disease is provided below.

Indications

The list of evidence-supported indications for TCAs is lengthy and includes the acute and long-term treatment of major depression in both specialty and general medical treatment settings (Anderson, 1998; Arroll et al., 2005; Bakker et al., 2002; Cipriani et al., 2018; Hirschfeld and Judge, 2000; Kerihuel and Dreyfus, 1991; MacGillivray et al., 2003). TCAs are also effective for the treatment of persistent depressive disorder (De Lima and Moncrieff, 2000), formerly known as dysthymic disorder, as well as other chronic depressive states, as reviewed in greater detail later. Beyond unipolar depressive syndromes, TCAs are effective for treating other psychiatric conditions including obsessive-compulsive disorder and body dysmorphic disorder (mainly clomipramine), posttraumatic stress disorder, generalized anxiety disorder (except for clomipramine), panic disorder, selected symptoms of bulimia nervosa, and treatment-resistant enuresis in children (Bighelli et al., 2018; Bobo et al., 2019; Caldwell et al., 2013; Fineberg and Gale, 2005; Hay and Claudino, 2010; Ipser et al., 2009; Ipser and Stein, 2011; Puetz et al., 2015; Schmitt et al., 2005). As shown in Table 8.4, TCAs also have a variety of nonpsychiatric indications that are supported by clinical evidence, including the management or prevention of chronic headache syndromes, various neuropathic pain syndromes, chronic low back pain, fibromyalgia, other diffuse pain syndromes, functional abdominal and gastrointestinal symptoms, urticaria/pruritus (doxepin), and insomnia (Argoff 2011; Bendtsen et al., 2000; Bobo et al., 2019; Chou and Huffman, 2007; Dharmshaktu et al., 2011; Drake et al., 1999; Finnerup et al., 2015; Ford et al., 2015; Griebeler et al., 2014; Jackson et al., 2010; Moore et al., 2016; Ruepert et al., 2011; Schneider et al., 2019). The remainder of this section focuses on a review of the evidence supporting the use of TCAs for major depression, other unipolar depressive syndromes, and treatment-resistant depression.

Depressive disorders, broadly defined

The efficacy of TCAs for depression in adults is well-established based on the results of numerous randomized trials, mainly involving people with major depressive disorder. At least three metaanalyses and one systematic review that supported the development of evidence-based guidelines for the treatment of depression in adults have shown clear advantages of TCAs over placebo in short-term trials and in studies lasting up to 12 months (Arroll et al., 2005; Furukawa et al., 2002; Joffe et al., 1996; National Institute for Health and Clinical Excellence, 2010). Pooled response rates from these reviews ranged from 45% to 60% for short-term studies ($\leq$12 weeks). In a comprehensive review of 65 randomized trials, TCAs and placebo had similar dropout rates for any cause; however, the risk of dropout owing to adverse effects is significantly higher with TCAs than with placebo (8173 patients; RR 4.02, 95% CI 3.46–4.67) (National Institute for Health and Clinical Excellence, 2010).

For adults with depressive disorders, TCAs are as efficacious as SSRIs and venlafaxine (Cipriani et al., 2011). Although it has been suggested that SSRIs are less-effective than TCAs for the acute treatment of major depression, a recent metaanalysis of 89 heterogeneous head-to-head trials (15,435 patients) lasting up to 12 weeks found no significant difference in the relative responder rate between SSRIs and TCAs (relative response rate (RRR) 1.03, 95% CI 0.97–1.09, number needed to treat (NNT) 72.6) (Undurraga and Baldessarini, 2017). The authors speculated that earlier findings of significantly greater efficacy of TCAs over SSRIs in acutely-depressed patients may have been due to lower placebo response rates from earlier trials, and not necessarily differential antidepressive efficacy. Dropout rates from lack of efficacy are similar between TCAs, SSRIs and SNRIs (<10% for all antidepressant types); however, rates of treatment discontinuation due to adverse effects were higher with TCAs (20%) than with SSRIs (8%) or SNRIs (6%) (Machado et al., 2006).

As reviewed elsewhere in this volume, the results of multiple metaanalyses suggest that irreversible monoamine oxidase inhibitors may have an efficacy advantage over TCAs for depression with atypical features (Cipriani et al., 2007; Henkel et al., 2006; Ricken et al., 2017; Thase et al., 1995). However, TCAs may be particularly effective for melancholic depression, as reviewed immediately below.

Severe and melancholic depression

Major depression with melancholic features, or melancholic depression, is a depressive subtype characterized by prominent anhedonia, lack of emotional reactivity to pleasant stimuli, and marked psychomotor slowing (Martino et al., 2019). A number of other clinical features may be present including a diurnal variation in symptom severity (worse in the morning), terminal insomnia, appetite suppression leading to weight loss, and excessive and ruminative guilt (Leventhal and Rehm, 2005; Rush and Weissenburger, 1994). Melancholic depression has been associated with greater symptom severity, higher risk for psychotic features, and more adverse cognitive effects than other depressive subtypes (Quinn et al., 2012; Zimmerman and Spitzer, 1989). As such, melancholic depression is considered to be an especially severe depressive subtype. The results of several studies suggest that melancholic and severely-depressed but nonmelancholic patients tend to respond more readily to antidepressants with dual effects on serotonergic and noradrenergic neurotransmission, including venlafaxine and selected TCAs (Bauer et al., 2013; Perry 1996; Roose et al., 1994; Schatzberg 1999), although not all studies support these conclusions (Bobo et al., 2011; Rush et al., 2008). Nevertheless, in an early meta-analysis that compared the antidepressive effects of TCAs and SSRIs in depressed patients with high and low depression severity based on a median split of baseline depression rating scale scores, TCAs were more efficacious than SSRIs in the high-severity group and in analyses restricted to inpatient studies (Anderson and Tomenson, 1994). In a more recent meta-analysis of 25 trials comparing antidepressant response rates in people with melancholic (2597 patients) and nonmelancholic depression (5016 patients) who were treated with TCAs or serotonin reuptake inhibitors (including SSRIs and SNRIs), overall antidepressive response rates did not differ significantly between melancholic and nonmelancholic patients (Undurraga et al., 2020). However, response rates were higher with TCAs than with serotonin reuptake inhibitors in both melancholic and nonmelancholic subjects.

Persistent depressive disorder and chronic depression

The efficacy of TCAs for persistent depressive disorder is supported by the results of four systematic reviews and metaanalyses. The first review supported the efficacy of antidepressants in general (TCAs included) for persistent depressive disorder (De Lima and Moncrieff, 2000), and the second review documented similar efficacy for TCAs, SSRIs, and MAOIs—but higher dropout rates with TCAs than SSRIs—after 4–12 weeks of treatment (De Lima and Hotopf, 2003). The third review included a network meta-analysis of data from 45 trials totaling 5806 patients with persistent depressive disorder (Kriston et al., 2014). The analysis was restricted to trials that were at least 8 weeks in duration and had a sample size of at least 200 patients. Compared with placebo, imipramine was associated with a significantly higher response rate (13 trials, OR 4.53, 95% CI 2.80–7.34) and a higher but nonstatistically significant increase in the rate of all-cause dropout, a proxy for treatment acceptability (nine trials, OR 1.26, 95% CI 0.81–1.96). The clinical effects of other TCAs were not evaluated. Finally, a fourth meta-analysis of 34 studies (4769 patients with persistent depressive disorder) focused on comparative rates of adverse effects across several classes of antidepressants (Meister et al., 2016). Both the odds of experiencing adverse events and treatment discontinuation due to adverse effects were significantly higher for TCAs than for placebo (for discontinuation, OR 3.98, 95% CI 2.54–6.21).

The clinical effects of TCAs for chronic major depression and major depression + persistent depressive disorder (double depression) have been investigated to a limited extent. Imipramine and sertraline were similarly effective (but sertraline was better-tolerated) in a 12-week randomized trial of 635 patients with chronic major depression or double depression (Keller et al., 1998). In the second phase of the study, patients who responded poorly to imipramine ($n=51$) or sertraline ($n=117$) were crossed over to the alternative medication (Thase et al., 2002). Response rates were higher for patients who switched from imipramine to sertraline than those who switched from sertraline to imipramine (60% vs 44%). Rates of early treatment discontinuation owing to adverse effects were lower with sertraline.

Long-term maintenance studies of TCAs for chronic depressive states are few. Desipramine was shown to be effective for maintaining symptomatic remission and preventing relapse in a cohort of patients with persistent depressive disorder or double depression for up to 20 weeks (Kocsis et al., 1995). Patients who remained well were then randomized to a 24-month placebo-controlled maintenance trial. Relapse rates were higher with placebo (52%) than desipramine (15%) (Kocsis et al., 1996). When the analysis was restricted to patients with persistent depressive disorder only, relapse rates were 46% with placebo and 0% with desipramine (Miller et al., 2001).

Clinical studies in people with treatment-resistant depression

In modern practice, TCAs are usually reserved for depressed patients who fail to respond to newer antidepressants. Our clinical experience suggests that patients who are being considered for their first trial of a TCA have failed to respond to multiple SSRIs, SNRIs, other antidepressants (including bupropion and mirtazapine), and newer or nonconventional antidepressants (including vortioxetine, vilazodone, and even ketamine). Yet, as reviewed below, there are surprisingly few controlled studies of TCAs for depressed patients who failed to respond to multiple therapeutic antidepressant trials.

Broad estimates of TCA efficacy

Based on a systematic review of randomized trials of reasonably good quality, clinicians can expect a positive treatment response in approximately 17%–49% of depressed patients after switching to TCAs following a poor response to at least one prior therapeutic antidepressant trial (Ruhé et al., 2006). However, the broader portfolio of studies that investigated the clinical effects of TCAs for patients with a history of poor response to other antidepressants indicates an even wider variation in clinical effectiveness (Table 8.3).

TCAs after failing to respond to a single antidepressant trial

Several studies have investigated the antidepressive efficacy of TCAs after failure to respond to a single antidepressant trial. To our knowledge, the first such study was a randomized cross-over trial that initially enrolled 122 patients with major depression (Peselow et al., 1989). Of these, 44 patients who responded poorly to initial double-blind treatment with imipramine, paroxetine, or placebo were crossed over to the alternative antidepressant (placebo failures were crossed over to paroxetine) and were followed for 6 weeks. Eleven of 15 patients (73%) who originally failed to respond to paroxetine eventually responded to imipramine, while 5 of the 10 patients (50%) who originally failed to respond to imipramine responded to paroxetine. Following cross-over, imipramine was associated with significantly greater reduction in depressive symptoms and clinical global state. Data on adverse effects were not reported.

In a larger 12-week randomized cross-over study, the clinical effects of imipramine and sertraline were compared in 168 patients with chronic depression who had previously failed to respond to a prospective trial of imipramine or sertraline (Thase et al., 2002). There were no statistically significant differences in remission rates between patients who were switched to imipramine compared with those who were switched to sertraline (23% vs 32%). Treatment discontinuation owing to adverse effects was higher for patients who were switched to imipramine than those who were switched to sertraline (9% vs 0%).

In a 10-week, open, randomized switch study that enrolled 112 depressed patients with prior nonresponse to venlafaxine (Navarro et al., 2019), subjects were randomized to imipramine (dose adjusted to achieve a plasma level of 175–300 ng/mL) or mirtazapine augmentation of venlafaxine. Remission rates were significantly higher after switching to imipramine than with adjunctive mirtazapine (71% vs 39%). Completion rates in the study were very high, as only five patients randomized to imipramine and two subjects in the combination therapy group were unable to complete all study visits. Only three patients, in total, dropped out due to adverse effects.

TCAs after failing to respond to multiple antidepressant trials

Surprisingly few published studies have investigated the effects of TCAs in patients with higher levels of treatment resistance than those reviewed above. In an open single-arm trial, 92 patients with treatment-resistant depression with an average of 2.3 failed prior antidepressant trials of adequate design and duration were switched to nortriptyline and were followed for 6 weeks (Nierenberg et al., 2003), Blood levels of nortriptyline were drawn at week 2, and the dose was raised if blood levels were < 100 ng/mL. Approximately 40% of patients responded to nortriptyline after 6 weeks, although 35% of subjects dropped out of the study. At the end of follow-up, the mean dose and drug level of nortriptyline was 121.2 mg/day and 101.0 ng/mL, respectively. There was no significant difference between responders and nonresponders with respect to nortriptyline dose or drug level.

In a randomized trial, 235 patients with treatment-resistant major depression were treated with nortriptyline (mean dose 97 mg/day) or mirtazapine (mean dose 42 mg/day) for 14 weeks (Fava et al., 2006). Individuals in this study had previously failed to remit following 12 weeks of citalopram monotherapy and a subsequent trial of either citalopram augmentation therapy (with bupropion or buspirone) or monotherapy with sertraline, venlafaxine extended-release, or bupropion sustained-release. Remission rates were low with both nortriptyline (19.8%) and mirtazapine (12.3%). There were no significant between-group differences in tolerability measures or adverse effects. Although dosing by nortriptyline drug level was allowed, it was also not required and was not done for the majority of nortriptyline-treated subjects.

A two-phase randomized, open trial compared the clinical effects of desipramine (minimum 150 mg/day) and citalopram (minimum 40 mg/day) in 189 patients who failed to respond to a previous therapeutic antidepressant trial (Souery

TABLE 8.3 Summary of clinical trials of tricyclic antidepressants for treatment-resistant depression (TRD).

Study	Sample	Design	Comparator groups	TRD definition	Key findings
Peselow et al. (1989)	25 Adults with TRD (DSM-III MDD)	Randomized cross-over (6 weeks)	Imipramine (mean daily dose/drug level not reported) vs paroxetine (mean daily dose not reported)	Cross-over to alternative treatment after failing to respond to imipramine or paroxetine	Higher response rate (CGI-I score 1–2 or ≥50% improvement in HAM-D) after crossing over to imipramine than to paroxetine (73% vs 50%)
Thase et al. (2002)	168 Adult outpatients with chronic TRD (DSM-IIIR chronic MDD ± comorbid PDD)	Randomized cross-over trial (12 weeks)	Imipramine 50–300 mg/day (mean 221 mg/day) vs sertraline 50–200 mg/day (mean dose 163 mg/day)	Cross-over to alternative treatment after failing to respond to imipramine or sertraline	Lower rate of response (CGI-I score 1–2, ≥50% improvement in 24-item HAM-D to final score ≤15, and CGI-S ≤3) after crossing over to imipramine than sertraline (44% vs 60%)
Nierenberg et al. (2003)	92 Adult outpatients with TRD (DSM-IIIR MDD)	Open single-arm trial (6 weeks)	Nortriptyline, dosing guided by drug level (mean 121 mg/day, mean blood level 101.0 ng/mL)	Failed to respond to 1–5 (mean 2.3) previous adequate antidepressant trials	Response rate (≥50% decrease in 17-item HAM-D) was 42% after 6 weeks of nortriptyline treatment
Fava et al. (2006)	235 Adult outpatients with TRD (DSM-IV MDD)	Randomized equipoise-stratified parallel-group trial (14 weeks)	Nortriptyline up to 200 mg/day (mean 97 mg/day)[a] vs mirtazapine up to 60 mg/day (mean 42 mg/day)	Failed to tolerate or remit following two prospective antidepressive intervention trials[b]	Comparably low response rates (≥50% improvement in 16-item QIDS-SR) with nortriptyline and mirtazapine (17% vs 13%)
Gervasoni et al. (2009)	20 Adult outpatients with TRD (DSM-IV MDD)	Open single-arm trial (4 weeks)	Clomipramine 150 mg/day	Failed to remit following prospective trials of paroxetine and either venlafaxine or paroxetine augmented by lithium	Response rate (≥50% decrease in MADRS) was 40% (five remitters and three responders) after 4 weeks of clomipramine treatment
Souery et al. (2011)	189 Adults with TRD (DSM-IV MDD)	RCT with secondary cross-over (4 weeks)	Desipramine (minimum dose 150 mg/day) vs citalopram (minimum dose 40 mg/day)	Failure to respond to antidepressant followed before prospective trial of desipramine or citalopram (phase 1), then randomized to continuation on same medication or cross over to alternative medication after failing to respond (phase 2)	No significant difference in response rate (≥50% decrease in 17-item HAM-D) between desipramine and citalopram (55% vs 54%) in phase 1; numerically higher response rate after switching to desipramine than after switching to citalopram (50% vs 38%) in phase 2
Navarro et al. (2019)	112 Hospitalized and nonhospitalized adults with TRD (DSM-IV MDD)	Randomized open trial (10 weeks)	Imipramine 20–100 mg/day vs venlafaxine + add-on mirtazapine 30 mg/day	Failure to respond to prospective trial of venlafaxine	Higher remission rates (17-item HAM-D <8) after crossing over to imipramine than combination therapy (71% vs 39%)

[a] *Clinicians were not required to use nortriptyline blood levels as a guide to dosing; as such, blood drug levels were obtained for only 34% of participants randomized to nortriptyline.*

[b] *The 235 patients in this study entered into the switching arm of Phase 3 of the Sequenced Treatment Alternatives to Relieve Depression (STAR*D) trial. All participants failed to tolerate or remit following an initial trial of citalopram (Phase 1), and failed to tolerate or remit following Phase 2 treatment that consisted of a trial of alternative monotherapy (sertraline, bupropion, venlafaxine, cognitive therapy) or augmentation of citalopram (with bupropion, buspirone, or cognitive therapy).*

CGI-I, Clinical Global Impressions improvement subscale; *DSM-III*, according to criteria specified in the Diagnostic and Statistical Manual for Mental Disorders, Third edition; *DSM-IIIR*, according to criteria specified in the Diagnostic and Statistical Manual for Mental Disorders, Third edition (revised); *DSM-IV*, according to criteria specified in the Diagnostic and Statistical Manual for Mental Disorders, Fourth edition; *HAM-D*, Hamilton Depression Rating Scale; *MADRS*, Montgomery-Asberg Depression Rating Scale; *MDD*, major depressive disorder; *PDD*, persistent depressive disorder (formerly referred to as dysthymic disorder); *QIDS-SR*, subject-rated version of the Quick Inventory of Depressive Symptomatology; *TRD*, treatment-resistant depression.

et al., 2011). After 4 weeks, patients who did not respond to their initially assigned treatment were randomized to the same antidepressant or the alternative drug for 4 additional weeks. In the second phase, continuation of the same antidepressant was associated with higher remission rates than switching. In the switching arms of the study, there were numerically higher rates of positive response after switching from citalopram to desipramine than switching from desipramine to citalopram (50.0% vs 37.5%). Completion rates across all four study arms in the second phase of the study were high (ranging from 78% to 94%); however, the sample sizes in each of the study arms were very small, ranging from just nine patients in the desipramine-to-citalopram switching arm to 23 subjects in the citalopram continuation arm, thus limiting conclusions that can be drawn from the reported differential response rates between groups.

A large retrospective chart review study compared the clinical outcomes of monotherapy with TCAs and monoamine oxidase inhibitors in 94 patients with treatment-resistant unipolar depression (Kim et al., 2019). A little over half of the study subjects had received two or more therapeutic trials of TCAs and/or monoamine oxidase inhibitors on at least two separate occasions prior to study entry. Forty-seven of the 147 treatment outcome observations in this study involved TCAs that included desipramine (57%), clomipramine (15%), doxepin (13%), imipramine (11%), and protriptyline (4%). The results of analyses that examined the interaction between antidepressant treatment and the number of prior therapeutic antidepressant trials suggested that monoamine oxidase inhibitors were more effective than TCAs for people who had fewer prior trials; however, the differences in outcome between treatments diminished as the number of prior therapeutic trials increased.

Adverse effects and toxicology

Adverse effects

Changes in body weight

Several TCAs are associated with appetite stimulation and clinically significant weight gain. Increases in body weight can occasionally occur in patients who report no changes in appetite, although a detailed history will often elicit changes in the types of food consumed and reduced physical activity, sometimes due to sedation, which can unfavorably tip the balance between caloric intake and expenditure. As noted earlier, both sedation and weight gain have been linked to pharmacologic antagonist effects at the histamine H1 receptor (Salvi et al., 2016). Therefore, broadly speaking, all TCAs can cause weight gain, but tertiary-amine tricyclic antidepressants have a greater propensity for causing significant body weight gain than do secondary-amine tricyclic antidepressants (Serretti and Mandelli, 2010). Even so, the effects of tricyclic and other antidepressants on body weight over longer-term treatment are highly variable in clinical practice. In a systematic review and meta-analysis of antidepressant effects on body weight, nortriptyline was associated with a mean gain of +1.24 kg in studies with 4 or more months of follow-up; however, the increase in body weight with nortriptyline lacked statistical significance owing to a wide 95% confidence interval (−0.51 to 2.99 kg; three nonheterogeneous studies, $n = 100$ patients) (Serretti and Mandelli, 2010). Amitriptyline, by contrast, was associated with a clear increase in body weight (2.24 kg) over the same time period that was statistically significant (95% CI 1.82–2.66 kg; four nonheterogeneous studies, $n = 170$ patients). We recommend assessing body weight at baseline and periodically during the course of treatment, regardless of which TCA is used.

Central nervous system effects

TCAs are associated with a variety of effects on the CNS. These include sedation, inattention, other adverse cognitive effects due to sedation and anticholinergic activity, headaches, delirium secondary to anticholinergic activity, and a fast-frequency hand tremor (Trindade et al., 1998). Because of sedation, TCAs are often dosed at nighttime, which can be effective for promoting sleep in depressed patients. This is especially so for tertiary-amine TCAs, which are typically more sedating than secondary-amine agents due to higher-potency antagonist effects at histamine H1 receptors (Mayers and Baldwin, 2005). TCAs have been associated with a range of adverse cognitive effects including impairments in attention, psychomotor speed, coordination, and memory. Anticholinergic delirium is dose-dependent and most often occurs in elderly patients who are prescribed multiple drugs with anticholinergic activity (Oxman, 1996). Hand tremors may improve over time with continuous treatment or with lowering the dose (Morgan and Sethi, 2005; Raethjen et al., 2001). However, spontaneous resolution of hand tremors does not always occur and lowering the dose may not be possible for many patients. Patients with problematic hand tremors who require continued treatment with a TCA may benefit from low-dose propranolol (Duma and Fung, 2019), as long as there are no relative contraindications such as existing hypotension, a history of sick sinus syndrome or second- or third-degree atrioventricular block, certain types of chronic obstructive pulmonary disease, and clinically significant bradycardia (Wasielewski et al., 1998).

Like many other antidepressants, TCAs can lower seizure threshold. As such, TCAs may be associated with an increase in the risk of breakthrough seizures in patients with epilepsy, and may precipitate seizures in overdose situations or in patients who are otherwise at an increased risk of having seizures. The latter includes patients with a history or brain injury or those who are taking other drugs that lower the seizure threshold. Of the tricyclic and tetracyclic antidepressants, maprotiline, amoxapine, and clomipramine have the highest risk of seizure occurrence and should not be used in people with epilepsy or those who are at high risk of seizures (Habibi et al., 2016). The risk of seizures with clomipramine is dose-related (0.5% in people taking <250 mg/day and 1.7% in people taking ≥300 mg/day) (McTavish and Benfield, 1990). For other TCAs, the seizure risk is thought to be much lower (Landmark et al., 2016), and if treatment with a TCA is needed, they can often be administered to patients with seizure disorders with cautious dosing and careful monitoring.

Other adverse effects involving the CNS include neuropsychiatric symptoms (confusion, dysphoria, restlessness, anxiety, agitation, vivid dreams/nightmares, and hypomania), numbness, tingling/paresthesias, weakness, problems with coordination and balance, and extrapyramidal symptoms (Godfrey, 1996).

Cardiovascular effects

The most common cardiovascular adverse effect from TCAs is tachycardia, which is usually mild and without symptoms. Tachycardia caused by TCAs may be related to antimuscarinic (vagolytic) effects, blockade of norepinephrine transporters, reflex tachycardia secondary to adrenergic α-1 receptor blockade, or a combination of these mechanisms.

Orthostatic hypotension occurs in approximately 5%–10% of patients who are treated with TCAs. Orthostatic hypotension is most problematic in elderly patients (where it may increase the risk of falls), patients who are dehydrated, or patients who are prescribed multiple concomitant antihypertensive medications (Sultana et al., 2015; Waring, 2012). The risk of clinically significant orthostatic hypotension from TCAs is also increased in individuals with cardiac conduction disturbances, patients taking concomitant medications that cause orthostasis or inhibit the metabolism of TCAs, and those with evidence of impaired left ventricular function (Glassman, 1984). Patients with orthostatic hypotension can continue to take TCAs if treatment is clearly beneficial, there are no contraindications, and alternatives to TCAs have been exhausted or are otherwise infeasible. Of all TCAs, nortriptyline may be associated with the lowest risk of clinically significant orthostatic hypotension (Roose et al., 1981; Thayssen et al., 1981), and may be worth considering in such cases following a careful assessment of risks, in consultation with a colleague in cardiovascular medicine if necessary.

TCAs can slow cardiac conduction in a manner similar to class Ia antiarrhythmic drugs (Glassman et al., 1993). Consequently, combining TCAs with antiarrhythmic agents or other drugs known to slow cardiac conduction can cause progressive lengthening of the QT interval, prolongation of the PR interval, widening of the QRS complex, and increasing degrees of atrioventricular block (Roose et al., 1987; Sala et al., 2009; Vohra et al., 1975). Overdoses of TCAs can cause fatal cardiac arrhythmias, even as sole ingestions (Henry et al., 1995; Kerr et al., 2001; Thanacoody and Thomas, 2005), as reviewed in greater detail later. No single TCA is thought to have less impact on cardiac conduction than others (Glassman and Preud'homme, 1993). Therefore, caution is warranted when using any TCA in patients with cardiac conduction abnormalities (Glassman et al., 1993). TCAs should generally be avoided in patients with coronary artery disease, and should not be given to patients with acute coronary syndrome (Cohen et al., 2000; Lichtman et al., 2008). TCAs also should not be prescribed to patients with a personal or family history of QT prolongation. All medications should be reviewed for potential impact on cardiac conduction before initiating treatment with a TCA.

Gastrointestinal effects

TCAs can cause a range of gastrointestinal adverse effects including nausea, vomiting, anorexia (as well as weight gain), epigastric pain, loose stools, and dysgeusia (Godfrey, 1996). Muscarinic receptor blockade can lead to constipation, especially in people taking multiple anticholinergic medications, individuals who are physically inactive, and individuals who are dehydrated (Remick, 1988). In more extreme cases, TCAs have been associated with rare treatment-emergent ileus (George, 1980), which requires urgent attention. Asymptomatic increases in liver function enzymes may occur with the use of TCAs, but drug-induced hepatitis is a rare complication of treatment (Voican et al., 2014). Clomipramine should generally be avoided in patients with existing hepatic disease since, among the TCAs, it has been shown to have the highest risk of drug-induced liver injury (Friedrich et al., 2016). Imipramine and amitriptyline should also be avoided in patients with known liver failure (Voican et al., 2014). It is recommended that treatment with TCAs be stopped when serum ALT levels increase to three times the upper limit of normal or higher, or when there are unexplained increases in bilirubin levels to two times the upper limit of normal or higher (Voican et al., 2014).

Genitourinary effects

TCAs can cause a variety of adverse genitourinary effects including urinary retention (and possibly overflow incontinence) due to blockade of muscarinic receptors and sexual dysfunction (Verhamme et al., 2008; Werneke et al., 2006). The latter includes reduced libido, arousal, erectile dysfunction, and delayed orgasms in men (Rothschild, 2000). In women, TCAs can cause reduced libido, problems with arousal, and delayed orgasms (Werneke et al., 2006). Since anticholinergic medications can weaken detrusor muscle contractions (Drake et al., 1998), TCAs with higher binding affinities at muscarinic receptors are best avoided in patients with benign prostatic hypertrophy and other causes of obstructed urine flow.

Other adverse effects

Additional adverse effects that are commonly encountered in people taking TCAs include blurring of vision (due to anticholinergic activity), diaphoresis (especially at nighttime), photosensitivity, hyperpigmentation, and other dermatological reactions (rash, petechiae, edema, and alopecia) (Bliss and Warnock, 2013; Godfrey, 1996; Marcy and Britton, 2005). TCAs can sometimes be a rare cause of secondary cataracts (Fu et al., 2018). TCAs have variable effects on sleep patterns and sleep architecture. Nearly all TCAs suppress REM sleep (except trimipramine) while preserving slow-wave sleep (Mayers and Baldwin, 2005). Some of the more sedating TCAs improve sleep latency, total sleep time, and sleep continuity (Roth et al., 2007), and are thus sometimes used as soporific agents at low doses, even in the absence of depression (Walsh, 2004). Oppositely, insomnia can also be observed with nearly all TCAs, with the possible exception of amitriptyline (Mayers and Baldwin, 2005).

Discontinuation syndrome

Missed doses or rapid discontinuation of TCAs can lead to a discontinuation syndrome characterized by flu-like symptoms, headache, lethargy, abdominal cramping and pain, appetite disturbances, loose stools, nausea/vomiting, insomnia, vivid dreams/nightmares, dizziness, balance problems, tremulousness, other movement problems, low mood, anxiety, and agitation (Warner et al., 2006). Particularly for tertiary-amine TCAs, some discontinuation symptoms may also be attributable to cholinergic rebound (Henssler et al., 2019). The symptoms of tricyclic discontinuation syndrome occur within a few days of stopping medication (or missing consecutive doses) and usually resolve within a few weeks if treatment is not resumed (Fava, 2006). Discontinuation symptoms resolve within a day or two of resuming treatment with the same TCA. It is not clear if using high doses of TCAs or a long duration of TCA treatment are risk factors for the occurrence or greater severity of the discontinuation syndrome. Our clinical experience suggests that patients taking higher doses of TCAs for at least 3–4 months may be at greatest risk. Regardless, patients should be warned about the possibility of discontinuation symptoms with all TCAs if they are stopped suddenly, discontinued too rapidly, or if consecutive doses are missed.

Overdose

Overdoses on TCAs are medical emergencies, regardless of whether they occur as single ingestions or mixed overdoses with other agents (Henry et al., 1995; Kerr et al., 2001). The case-fatality rates from TCA toxicities are significantly higher than toxicities from other antidepressants, such as SSRIs, venlafaxine, and mirtazapine (Hawton et al., 2010). Clinical features of TCA overdose including acute mental status changes, drowsiness, confusion, lethargy, hypotension (or even blood pressure increases with some agents), orthostatic dizziness, tachycardia, and urinary retention. The most important toxic effect of TCAs is the slowing of cardiac conduction, leading to increased risk of severe hypotension and potentially fatal arrhythmias. Although desipramine toxicity has been associated with higher case-fatality rates than toxicities from amitriptyline, nortriptyline, and imipramine (Amitai and Frischer, 2004), it should be emphasized that all TCAs affect cardiac conduction and are considered to be dangerous in overdose (Hawton et al., 2010). Other potentially serious consequences of toxic ingestions of TCAs include altered mental status (often with accompanying anticholinergic signs), respiratory depression, seizures, and coma (Judge and Rentmeester, 2013).

Drug interactions

There are two broad types of clinically significant drug-drug interactions for people taking TCAs. The first type of interaction is pharmacodynamic, which can have direct bearing on the risk of adverse effects. The most important pharmacodynamic drug interactions with TCAs relate to elevated risk of serotonin toxicity, clinically significant prolongation of the QT interval, excessive sedation, symptomatic orthostatic hypotension, and severe or burdensome anticholinergic effects.

The second type of interaction is pharmacokinetic, which pertains not only to the risk of adverse effects, but also to the dosing of particular agents. The most important pharmacokinetic drug interactions with TCAs are related to the inhibition or induction of key CYP450 isoenzymes. What follows is a broad overview of each type of drug-drug interaction, beginning with pharmacodynamic drug interactions.

Pharmacodynamic drug interactions

Serotonin toxicity

As discussed elsewhere in this volume, certain tertiary-amine TCAs (clomipramine and imipramine, in particular) can increase the risk of serotonin toxicity (also known as "serotonin syndrome") when combined with monoamine oxidase inhibitors (Gillman, 2006). That being said, selected TCAs (other than clomipramine and imipramine) and monoamine oxidase inhibitors have been safely combined and can thus be carefully considered in expert hands when alternative approaches cannot be exercised (Gillman, 2006; Thomas et al., 2015). Specific guidance may be found in Chapter 9. The risks of serotonin toxicity are less-clear when combining other serotonin-potentiating medications with TCAs that display potent serotonin reuptake blockade, so great care must be exercised when multiple serotonergic medications are prescribed. The antibiotic, linezolid, acts as a nonselective monoamine oxidase inhibitor (Quinn and Stern, 2009), and may precipitate serotonin toxicity when combined with TCAs that potently block serotonin transporters. Because severe serotonin toxicity is life-threatening, we recommend conducting a thorough discussion of such risk with patients and maintaining vigilance for associated signs and symptoms (see Chapter 9) when TCAs are combined with other serotonergic medications.

QT interval prolongation

TCAs should generally be avoided in patients who are taking other medications that slow cardiac conduction or prolong the QT interval in order to minimize the risk of potentially fatal ventricular arrhythmias, including *torsades de pointes*. The list of interacting drugs with respect to QT interval prolongation clearly includes antiarrhythmic drugs, but due caution is also extended to numerous medications, the list of which has become extensive over the past 20 years. Broadly, such drugs include selected antihistamines, antipsychotic drugs, antimicrobials, methadone, and others (Nachimuthu et al., 2012). A number of online references provide downloadable lists of specific medications that increase the risk of QT interval prolongation (see section "Additional resources").

Sedation and orthostatic hypotension

Increased risk of sedation may be expected when TCAs are combined with other sedating medications (e.g., benzodiazepines, barbiturates, antihistamines, and antipsychotic drugs, etc.) and even alcohol. Heightened risk of sedation can, in turn, increase the risk of falls, especially when increased sedation cooccurs with orthostatic hypotension (Sultana et al., 2015). The α-1 antagonist effects of TCAs can potentiate the blood pressure lowering effects of other antihypertensives, including prazosin, terazosin, doxazosin, and labetalol (Richelson, 2001), which can increase the risk of hypotensive episodes. Oppositely, TCAs can antagonize the hypotensive effects of the centrally acting α-2 agonist, clonidine (which is thought to be related to the blockade of α-2 receptors by TCAs) (Van Spanning and Van Zwieten, 1973). TCAs can also reverse the antihypertensive effects of guanethidine and guanadrel due to the inhibition of norepinephrine transporters in peripheral adrenergic neurons, thus preventing uptake of these agents into their principal site of action (Richelson, 2001).

Anticholinergic effects

TCAs that potently bind to muscarinic cholinergic receptors should be avoided in people taking other medications with anticholinergic side effects. This is because the risks of clinically significant anticholinergic symptoms, acute complications related to anticholinergic effects (e.g., precipitation of acute angle closure glaucoma, urinary retention, adynamic ileus, and delirium), and anticholinergic toxicity increase when multiple medications with anticholinergic side effects are taken concomitantly (Spina and Perucca, 1994).

Pharmacokinetic drug interactions

Inhibition of CYP450 isoenzymes by TCAs

The tertiary-amine TCAs, amitriptyline, clomipramine, and imipramine, potently inhibit CYP2C19 (Table 8.4) and have the potrential to increase the exposure to drugs that are CYP2C19 substrates (English et al., 2012; Gillman, 2007). In a study of 11 depressed patients, the tertiary-amine TCA, doxepin, was shown to inhibit CYP2D6 (Szewczuk-Bogusławska et al., 2004). The clinical relevance of doxepin inhibition of CYP2D6 is unclear, however, given that none of the 10 study enrollees who were classified as phenotypic CYP2D6 extensive (normal) metabolizers at baseline changed CYP2D6 metabolizer phenotype status after 2 weeks of exposure to doxepin. Secondary-amine TCAs are only weak inhibitors of cytochrome P450 isoenzymes (CYP2D6 and 2C19) and thus have only limited risk of causing pharmacokinetic drug-drug interactions (Gillman, 2007).

Effects of CYP450 inhibition and induction by other medications

As noted earlier, most TCAs are extensively metabolized in the liver by hydroxylation and demethylation, and all are substrates of various CYP450 isoenzymes (Table 8.2). Therefore, exposure to TCAs are significantly influenced by CYP450 enzyme inhibitors and inducers (Gillman, 2007). These pharmacokinetic interactions are often clinically significant when they result in the lack of efficacy of certain TCAs (when taken with other drugs that accelerate the activity of these enzymes) or an increased risk of adverse effects including toxic reactions, cardiac arrhythmias, seizures, and delirium (when taken with enzyme inhibitors) (Ereshefsky et al., 1995; Flockhart, 1995).

For desipramine and nortriptyline, drugs that inhibit or induce CYP2D6 are the most important (Ereshefsky et al., 1995). Desipramine drug levels, in particular, appear to be highly sensitive to the inhibition or induction of CYP2D6 because desipramine has no alternative metabolic pathways (Brøsen et al., 1993; Preskorn et al., 1994). For the metabolism of tertiary-amine TCAs, more CYP450 enzymes are involved, including CYP1A2, 3A4, and 2C19. The higher complexity of tertiary-amine TCA metabolism makes clinically significant interactions harder to predict compared with secondary-amine TCAs. That said, drugs that are known to potently inhibit or induce multiple CYP450 isoenzymes may be expected to cause clinically significant or even serious interactions with tertiary-amine and other TCAs. The SSRI, fluvoxamine, is probably best avoided in patients taking TCAs because it is a potent inhibitor of CYP1A2, 2C19, and 3A4. Fluvoxamine markedly increases the concentration of clomipramine and may significantly increase exposures to other TCAs (Gillman, 2007). More drug-specific information is available to prescribers by using the online references described in the section "Additional resources."

To reduce the chances of clinically significant drug interactions, we recommend that patients' medication lists be updated at each clinical visit. Prescribers should be especially alert for potent inhibitors or inducers of CYP2D6 in patients taking nortriptyline or desipramine and for potent inhibitors/inducers (or combinations of medications that, together, result in potent inhibition/induction) of multiple CYP450 isoenzymes—especially CYP1A2, 2D6, 3A4, and 2C19. Patients should be frequently reminded to alert all prescribers to the fact that they are taking a TCA, and should be encouraged to consult with their TCA-prescribing clinicians before taking any new medications, including over-the-counter drugs.

Clinical use

In modern practice, depressed patients are usually at a relatively advanced stage of treatment resistance by the time therapeutic trials of TCAs are considered. In our experience, the crowded market for antidepressants has made it such that fewer and fewer clinicians are now as familiar with the use of TCAs for treating depressed patients. As a result, TCAs are often overlooked due to a lack of prescriber familiarity. In this section, we provide practical advice on the use of TCAs for people with treatment-resistant depression, organized around the following subtopics: Transitioning onto TCA therapy, selecting a TCA, initiating treatment, monitoring therapeutic responses and side effects, optimizing the effectiveness of TCAs, and transitioning off of TCAs when necessary.

Transitioning onto TCA therapy

Overall place in therapy

TCAs are reasonable to consider, as discussed earlier, for the pharmacological treatment of several unipolar depressive syndromes, including major depressive disorder, persistent depressive disorder, and cooccurring major depression and persistent depressive disorder. As reviewed earlier, TCAs may be particularly effective for severe or melancholic depression. However, TCAs are second-line agents for depression based on higher adverse effect burden and poorer tolerability than first-line agents. Many patients with treatment-resistant depression present with comorbid conditions that can confound the evaluation of depressive symptoms and obscure or interfere with the effects of treatment (Keitner et al., 1991). In that sense, TCAs may be particularly attractive options for depressed patients who have cooccurring problems for which TCAs have also been shown to be effective (Zetin et al., 2006). A partial list of such conditions is provided in Table 8.4.

TABLE 8.4 Mental health and general medical considerations for the use of tricyclic antidepressants (TCAs) for patients with treatment-resistant major depression and comorbid conditions.

TCAs may be helpful	TCAs can be used with caution	TCAs are generally avoided
Comorbid mental health conditions/disorders		
Anxiety disorders Generalized anxiety disorder Except for clomipramine Panic disorder Posttraumatic stress disorder Body dysmorphic disorder Mainly clomipramine Bulimia nervosa, selected symptoms Dysthymic disorder (double depression) Insomnia Primary or secondary		Bipolar depression (especially as monotherapy) Patients at high risk of suicide Patients who frequently overdose Patients who are not reliable Problems taking medications as prescribed Anticipated difficulties complying with drug monitoring procedures
Comorbid general medical conditions		
Functional gastrointestinal syndromes Selected chronic pain syndromes Chronic low back pain Fibromyalgia syndrome Neuropathic pain syndromes, selected Migraine headache prophylaxis Premature ejaculation If SSRIs or SNRIs are ineffective Dermatoses, selected Urticaria Pruritus	Orthostatic hypotension, preexisting Once underlying cause is addressed Diabetes mellitus Hypomagnesemia or hypokalemia, mild Once underlying cause is addressed Avoid if hypokalemia is severe Balance/gait problems Once underlying causes are addressed Brain injury Avoid if cognitive impairments or seizure risk are severe Chronic kidney disease (CKD) Low doses advised for some agents (clomipramine, desipramine) in patients with more advanced CKD Obesity Secondary-amine TCAs are preferred Benign prostatic hypertrophy and other urinary obstruction Once underlying cause is addressed	Acute angle-closure glaucoma Arrhythmias Preexisting QT interval prolongation Right bundle branch block History of ventricular arrhythmias Long QT syndrome, family history of sudden cardiac death, family history of congestive heart failure, any cause coronary heart disease Contraindicated in patients with acute coronary syndrome Can be used with caution in some postmyocardial infarction patients, but should generally be avoided if possible Dementia Other clinically significant cognitive disorder or impairment Hepatic disease Avoid clomipramine, amitriptyline, and imipramine Epilepsy Avoid clomipramine, amoxapine, and maprotiline

Psychiatric factors

TCAs should generally be avoided in patients who are suicidal and those with a history of self-destructive impulsivity due to the potential for fatal overdoses. If it is necessary to treat such patients with TCAs, then the risks of lethal overdoses may be reduced by collaborating with family members or pharmacies to make available only a limited (nonlethal) amount of medication available to patients. The amount of medication that defines "nonlethal" depends on the specific TCA under consideration and a variety of other factors. For most adults, a supply totaling <2.5 mg/kg is typically nonlethal (Woolf et al., 2007). However, this serves as only a broad and conservative guideline, and does not consider additional variables that influence TCA lethality including patient age (higher risk in children and elderly adults), underlying medical illnesses, concomitant medications taken by the patient, and other factors.

TCAs are far less attractive as a treatment option for depressed patients who are suspected of having bipolar spectrum disorders. Such patients may have a prior history of subthreshold hypomanic episodes, evidence of mixed depressive episodes, a strong family history of validly diagnosed bipolar spectrum disorders, and a history of antidepressant-induced manic or hypomanic episodes (Bobo, 2017). A detailed history, including history obtained from collateral informants, may be enough to diagnose retrospectively past hypomanic and mixed depressive episodes. The importance of conducting such a detailed history rests in the observations that patients with bipolar spectrum disorders often respond poorly to antidepressants (Bobo, 2017). TCAs have been associated with higher risk of polar mood switching than other antidepressants in patients with bipolar spectrum disorders (Gijsman et al., 2004; Peet, 1994).

General medical factors

TCAs should be avoided in patients with preexisting heart disease, prolonged QT interval, history of ventricular arrhythmias, congenital long QT syndrome, right bundle branch block, severe electrolyte disturbances (including hypokalemia), and other cardiac conduction defects that can result in a prolonged QT interval. TCAs should also be generally avoided in patients who are at high risk for coronary heart disease, and those with a family history of sudden cardiac death or QT interval prolongation. Special care should be taken when evaluating the risks of TCA use in hospitalized or elderly patients (especially patients with symptomatic orthostasis or problems with balance and gait), and in patients with renal or hepatic impairment, epilepsy, and heightened seizure risk in the absence of known epilepsy. Due to anticholinergic effects, TCAs should be used very cautiously in people with benign prostatic hypertrophy, other causes of preexisting urinary retention, angle-closure glaucoma, dementia, and other forms of cognitive impairment.

Pharmacological factors

Clearly, a history of hypersensitivity reactions to a specific TCA contraindicates further use of that specific drug, although it may be possible to use a different TCA with cautious dosing and close observation if no other treatment alternative is exercisable. TCAs as a group are considered to be contraindicated with MAOIs, although the risk of serotonin toxicity is increased when certain TCAs that act as potent serotonin reuptake inhibitors (e.g., clomipramine and imipramine) are combined with MAOIs. Other TCAs can be safely combined with MAOIs in expert hands, with close monitoring (see Chapter 9). Medications that inhibit or induce CYP450 isoenzymes responsible for the metabolism of TCAs, drugs that cause QT interval prolongation, medications that can cause electrolyte disturbances, and agents that result in additive sedation, increased risk of orthostasis, or significant anticholinergic side effects should generally be avoided in patients who are taking TCAs (see the section "Drug interactions").

Selecting a TCA

TCAs are equally efficacious for depressed patients, so the selection of a particular TCA depends mainly on prior treatment history, the need for sedation (in order to promote sleep), and differences between agents in adverse effect and drug interaction profiles. If there is a history of good response to a particular TCA, then it would be reasonable to consider another trial of that same agent, especially if it was well-tolerated. This assumes, of course, that no new psychiatric, general medical, or pharmacological factors have arisen that would directly or relatively contraindicate such treatment. Moderate to severe insomnia is frequently reported by patients with treatment-resistant depression. In such cases, the sedating effects of tertiary-amine tricyclic antidepressants may be preferred. Secondary-amine TCAs, such as desipramine or protriptyline, can be prioritized when a less-sedating TCA is desired. Clomipramine is an especially attractive option for people with treatment-resistant depression and comorbid obsessive-compulsive disorder (OCD), especially when comorbid OCD also responds poorly to SSRIs or SNRIs. Secondary-amine TCAs should be prioritized over tertiary-amine agents for patients who are poorly tolerant of orthostasis, anticholinergic effects, weight gain, or sedation.

Initiating treatment with a TCA

Prior to initiating treatment with any TCA, it is advisable to conduct a thorough review the patient's medications, including over the counter drugs and those taken on an as-needed basis. In some cases, it may be necessary to communicate the clinical necessity of treatment with a TCA to the prescribers of potentially interacting medications so that alternatives can be considered. Patients should be asked specifically about a personal history of heart disease and about any current cardiac symptoms such as palpitations, syncope/presyncope, shortness of breath, dyspnea on exertion, and chest pain. A baseline EKG is recommended for any patient who is over the age of 40 years, for patients taking QT-prolonging drugs or drugs that cause electrolyte disturbances, and for patients with a personal or family history of cardiac disease, regardless of age. In most cases, baseline laboratory studies are not required before initiating TCAs. However, for patients with known or suspected renal or hepatic disease, baseline tests of electrolytes, renal function, and liver function are advised. Baseline electrolyte concentrations should also be considered for people with a history of fluid and/or electrolyte disturbances, people with known cardiac disease, and patients who are taking medications that cause electrolyte disturbances.

Initial dosing and titration

Starting doses and suggested titration schedules for depressed but otherwise reasonably healthy adults are provided in Table 8.5. Broadly speaking, the dosing schedules for the most commonly-used tertiary-amine TCAs, imipramine and amitriptyline, also apply to desipramine, trimpramine, and doxepin. The starting doses of nortriptyline and protriptyline are generally lower and the titration schedules for these drugs are more conservative. All TCAs are usually started as a single bedtime dose to take advantage of sedating properties. The exception is protriptyline, which is started as a single dose in the morning. Like protriptyline, desipramine can have activating rather than sedating effects for some patients, and may be better tolerated if started as a single morning dose if that occurs.

The choice of starting doses and the speed of titration for individual patients depend on a number of factors including the patient's age, concomitant medications, medical and psychiatric comorbidities, and treatment setting. Generally, lower doses and slower titration schedules than those shown in Table 8.5 are needed for elderly patients and for patients who are taking multiple medications or have more medical comorbidities. In psychiatric hospital settings, the initial dosing and titration of most TCAs can be more aggressive, although due caution is always warranted. The same cannot be said of

TABLE 8.5 Starting doses, titration schedules, therapeutic doses, and blood drug levels for selected tricyclic antidepressants.

Drug name	Usual starting dose[a]	Dose titration[b]	Usual therapeutic dose[c] (mg/day)	Blood drug level[d] (ng/mL)
Amitriptyline	25–75 mg at bedtime	25–50 mg every 3–5 days	100–300	80–200
Clomipramine	25 mg at bedtime	Slowly over 14 days to 100 mg/day	100–300	175–450
Desipramine	25 mg/day (in the morning or at bedtime)	25–50 mg every 3–5 days	100–300	>125
Doxepin	25 mg at bedtime	25–50 mg every 3–5 days	100–300	50–150
Imipramine	25 mg at bedtime	25–50 mg every 3–5 days	100–300	175–300
Nortriptyline	25 mg at bedtime	25–50 mg every 3–5 days	50–200	50–170
Protriptyline	10 mg in the morning	Increase to 15 mg after 5–7 days, and as clinically indicated afterwards	50–200	...
Trimipramine	25 mg at bedtime	25–50 mg every 3–5 days	100–300	150–350

[a]The starting dose of tricyclic antidepressants (TCAs) depends on several factors including but not limited to treatment setting (may be able to start at higher doses, with total daily doses divided BID or TID, in supervised psychiatric settings), age (lower starting doses, e.g., amitriptyline 10 mg at bedtime, etc. in elderly patients), comorbid medical conditions (may require lower starting doses), interacting drugs, and pharmacogenetic metabolizer status at selected CYP450 isoenzymes.
[b]The rate at which the dose of TCAs is increased is largely determined by treatment setting (faster titration schedules can be used in supervised settings) and tolerability. Some patients will require a slower titration (with more time in between dosage increases or smaller incremental dose increases (e.g., increase in 10 mg increments of amitriptyline instead of 25–50 mg, etc.), especially elderly and medically ill patients, patients who are poor metabolizers at selected CYP450 isoenzymes, and those who are taking multiple medications.
[c]Lower doses are often sufficient for patients who are elderly, who are poor metabolizers at selected CYP450 isoenzymes, and who are taking medications that inhibit selected CYP450 isoenzymes.
[d]Therapeutic levels have been established for imipramine (a sum of imipramine and desipramine levels) and nortriptyline. For the remaining TCAs, no clear or definitive association between blood drug levels and therapeutic response in people with major depressive disorder has been established. The suggested drug level for amitriptyline, clomipramine, doxepin, trimipramine, and desipramine are based on expert recommendations. Approximate therapeutic drug levels are unknown for protriptyline.

patients admitted to general medical hospital settings, where medically compromised status, the use of multiple drugs (often given parenterally), and other factors may limit the use of TCAs or, at the very least, call for the cautious administration of these agents. In the general hospital setting, consultation with appropriate medical specialists and specialty pharmacists may be necessary.

Patients with renal or hepatic disease

Close follow-up is required when TCAs are used for treating depression in patients with renal or hepatic disease. European Renal Best Practice (ERBP) guidelines for dosing antidepressants in patient with renal disease advise that no dosing adjustments are needed with amitriptyline, nortriptyline, or doxepin in patients with stage 3–5 (estimated glomerular filtration rate (eGFR) <60 mL/min) chronic kidney disease (Nagler et al., 2012). In the case of desipramine, ERBP guidelines recommend no dosing adjustments for patients with stage 3 and 4 chronic kidney disease and a starting dose of 25 mg/day and "careful" titration in patients with stage 5 chronic kidney disease (eGFR <15 mL/min). For clomipramine, no dosing adjustment is recommended for stage 3 chronic kidney disease; however, for patients in stages 4 and 5 (eGFR <30 mL/min), the recommended starting dose is 10 mg/day with "careful" titration.

There are no dosing guidelines available for patients with hepatic insufficiency. As reviewed earlier, clomipramine should generally be avoided in patients with existing hepatic disease and imipramine and amitriptyline should be avoided in patients with known liver failure (Friedrich et al., 2016; Voican et al., 2014). When treatment with a TCA is necessary for patients with liver disease, agents other than clomipramine, imipramine, and amitriptyline can be carefully considered, ideally in consultation with a gastroenterologist or hepatologist. Treatment is best initiated at the lowest available doses with slow titration guided by clinical response, tolerability, and plasma drug concentrations (given the likelihood of impaired drug metabolism). A reasonable target dose may be approximately half of a typical therapeutic dose for otherwise healthy patients, depending on the degree of hepatic impairment. However, this is only a rough guideline based on clinical experience.

Monitoring therapeutic response and adverse effects

Therapeutic response

Although 6–8 weeks is the minimum time interval needed to define an adequate therapeutic trial of antidepressants, both clinical experience and empirical evidence suggests that improvements in mood, affect, and neurovegetative functioning can occur within 2–3 weeks of treatment (Posternak and Zimmerman, 2005). Further improvements may unfold after several more weeks of treatment. In our clinical experience, signs of early improvement are often observed by others before the patient begins to report a positive change in mood. The use of validated depression rating scales for the periodic measurement of depressive symptoms can be used as a supplement to patient self-report and clinical judgment when evaluating the clinical effectiveness of TCAs.

Adverse effects

Patients should be routinely monitored for adverse effects during all phases of treatment with TCAs. During the dose-titration phase, sedation, orthostasis, and anticholinergic effects may become problematic. For some patients, these side effects are treatment-limiting even at very low doses. Some degree of tolerance to the anticholinergic effects of TCAs may develop after several weeks of treatment; however, tolerance to sedating effects of TCAs does not tend to develop, even after years of continuous use (Godfrey, 1996). This means that patients who encounter sedation that threatens compliance may benefit from switching to a less-sedating TCA such as protriptyline or desipramine. Dysgeusia that occurs during the titration phase can be managed by wrapping the TCA tablet in a small piece of bread and swallowing them whole (Godfrey, 1996).

All patients who are started on TCAs should be screened for new-onset cardiac symptoms such as palpitations, shortness of breath, syncope/presyncope, dizziness, and chest pain. If these occur, it is best to stop treatment with the TCA and obtain an EKG. In some cases, consultation with a cardiologist may be needed. Periodic EKG monitoring is also recommended for patients who are taking higher doses of a TCA (>150 mg/day of amitriptyline or equivalent doses of other agents in nongeriatric adults), patients with known cardiac disease, and patients who are taking other medications that affect cardiac conduction. Routine monitoring of electrolytes (especially potassium) is also recommended for patients who require treatment with a TCA and are taking diuretics or QT-prolonging drugs. Periodic assessment of body weight is also recommended, especially for patients with obesity, type II diabetes, lipidoses, and other cardiovascular risk factors.

Optimizing the effectiveness of TCAs

Optimizing the TCA dose

A meta-analysis of 41 heterogeneous trials of tricyclic antidepressants (mainly amitriptyline and imipramine) totaling 2013 patients compared the clinical effects of standard- (> 100 mg/day) and low-dose strategies (75–100 mg/day) with placebo (Furukawa et al., 2003). This study reported that standard-dose TCA treatment was not superior to low-dose treatment. Standard-dose TCA treatment was associated with a higher dropout rate owing to adverse effects. However, interindividual differences in steady-state drug concentrations can vary widely (5- to 10-fold, in many cases) between patients taking the same TCA at a given dose (Hicks et al., 2013; Hockings et al., 2020; Mattuzas et al., 1982). The optimal dosage of most TCAs for treating depressed adults is difficult to predict, so the dose of TCAs must be individualized (Lydiard, 1985).

We initially manage partial responses to TCAs by optimizing the dose and recommend that practitioners avoid discontinuing TCAs owing to an apparent lack of efficacy until at least 4 weeks have transpired at the maximum tolerated dose or at doses that yield therapeutic drug concentrations for selected agents (see Table 8.5). When the clinical response to nortriptyline is inadequate, it is often useful to measure a steady-state trough drug level and adjust the dose with the goal of achieving a therapeutic concentration (Hiemke et al., 2011; Perry et al., 1994). An imipramine plasma concentration >175 ng/mL has been correlated with clinical response and may thus serve as a useful guideline in cases of suboptimal response where imipramine concentrations are below that threshold. For other TCAs, the relationship between drug levels and clinical response are more conflicting. Some have advocated for dosing desipramine to achieve a minimum blood level of 125 ng/mL (Lydiard, 1985). Others have suggested that clomipramine blood concentration of >150 μg/L (and perhaps no greater than 450 μg/L) can be used to optimize therapeutic response (Balant-Gorgia et al., 1991). The measurement of plasma drug concentrations should occur 8–12 h after the last dose after achieving steady state—a process that typically requires at least 5–7 days, or 5 half-lives, at a given daily dose.

Combining medications

When optimizing the dose of TCAs is not possible (usually due to inefficacy or intolerable side effects), adding a second agent to ongoing treatment with a TCA may enhance therapeutic efficacy and is therefore worth considering before switching to another antidepressant. This is especially true if there is at least some evidence of a partial antidepressive response. Psychosocial therapies, reviewed elsewhere in this volume, are also very important adjuncts to pharmacotherapy for patients with treatment-resistant depression. These therapies should not be overlooked. Below, we provided brief review of pharmacological adjuncts to TCAs.

Adjunctive lithium

Lithium is one of the best-studied pharmacological adjuncts to TCAs, the efficacy of which is supported by at least two metaanalyses that included patients with treatment-resistant depression. The first meta-analysis (10 placebo-controlled trials of lithium augmentation in patients with treatment-resistant depression) documented significantly higher response rates with adjunctive lithium than placebo (40% vs 17%; OR 3.11, 95% CI 1.80–5.37; NNT 5) (Crossley and Bauer, 2007). Eight of the 10 trials involved the addition of lithium to a tricyclic or other type of antidepressant. A second meta-analysis of nine trials (237 patients) with low heterogeneity compared the effects of lithium augmentation in depressed patients who were treated with first-generation antidepressants (mainly TCAs) or second-generation antidepressants (primarily SSRIs) (Nelson et al., 2014). The odds of positive response were significantly higher with lithium than with placebo, both overall (OR 2.89, 95% CI 1.65–5.05) and in patients who were taking TCAs (OR 2.80, 95% CI 1.40–5.59) and second-generation antidepressants (OR 3.06, 95% CI 1.19–7.88). Sensitivity analyses could not be performed to explore the effects of the number of prior failed antidepressant trials.

At least two randomized trials focused on the clinical effects of adjunctive lithium in patients with relatively advanced stages of treatment-resistant depression. In the third phase of the STAR*D trial, 142 patients with MDD who failed to remit after prospective treatment with citalopram and a subsequent switch or augmentation trial were randomized to adjunctive lithium or thyroid hormone (T3) (Nierenberg et al., 2006). Remission rates were low with Li (13%–16%) and with T3 (25%); however, these pharmacological adjuncts were added to ongoing treatment with SSRIs (citalopram or sertraline), sustained-release bupropion, or extended-release venlafaxine—not TCAs. In a 6-week randomized trial, 35 patients with treatment-resistant depression who had failed to respond to at least one (and up to 5) therapeutic antidepressant trials and a 6-week prospective trial of nortriptyline were randomized to receive lithium or placebo as an adjunct to ongoing treatment with nortriptyline (Nierenberg et al., 2003). Rates of treatment response were comparably low in both groups (13% vs 20%). Protocol details describing the dosing of lithium and mean lithium doses and drug levels were not provided.

All-in-all, there is still a paucity of data from controlled studies that describe the clinical effects adjunctive lithium for TCA-treated depressed patients with relatively advanced stages of therapy resistance. Existing studies point to relatively lower effectiveness of lithium added to ongoing treatment with antidepressants, in general, for depressed patients who have failed multiple prior therapeutic intervention trials. The latter observation is in keeping with general observations of an inverse correlation between effectiveness rates with any given intervention for depressed patients and the number of prior failed therapeutic trials. Despite these limitations, we think that a therapeutic trial of adjunctive lithium is reasonable to consider for patients who have a suboptimal response to TCAs. When combined with TCAs, lithium is usually dosed at 600–900 mg/day or higher (or lithium level of $\geq$0.4 mEq/L), usually divided into twice-daily or thrice-daily doses (Fava, 2001). The same general precautions apply when lithium is used as an adjunct to antidepressants in unipolar depressed patients as for lithium use in people with bipolar disorders. These include assessment of renal and thyroid function and obtaining an EKG at baseline, and periodic monitoring of these indices over time. Patients treated with adjunctive lithium should be counseled about the importance of maintaining adequate hydration, avoidance of interacting drugs, and early signs of potential lithium toxicity.

Adjunctive thyroid hormone (T3)

There is evidence supporting the use of T3 for accelerating the onset of antidepressive effects and enhancing the antidepressive effects of ongoing treatment with antidepressants, including TCAs (Altshuler et al., 2001; Aronson et al., 1996; Joffe et al., 1996). With regard to the latter, a meta-analysis of eight heterogeneous studies (292 patients) of add-on therapy with T3 in depressed patients with suboptimal responses to TCAs, adjunctive T3 (25–50 μg/day) was associated with a significantly higher rate of positive treatment response as compared with control conditions that included placebo or adjunctive thyroxine (T4) (RRR 2.09, 95% CI 1.31–3.32, NNT 4.3) and a pooled absolute difference in response rate (T3 vs controls) of 23% (Aronson et al., 1996). The clinical translation of these pooled results is limited by the heterogeneity between individual studies.

Nevertheless, adjunctive T3 is worth considering for patients with suboptimal antidepressive response to TCAs who have no absolute or relative contraindications to the use of thyroid hormones. Such contraindications include recent myocardial infarction, coronary heart disease/angina, preexisting hyperthyroidism, thyrotoxicosis, diabetes, pituitary hormone deficiency, adrenal insufficiency, hypertension, and osteoporosis. Appropriate consultations can be obtained in order to appropriately weigh risks and benefits before starting adjunctive T3 where there is uncertainty. For most depressed patients, T3 is started at a dose of 25 μg/day and can be increased to 50 μg/day if no adverse effects are reported or observed.

Other strategies

A number of other therapeutic adjuncts are often tried, albeit with limited or at times an absence of empiric support, making it difficult to recommend any of them for routine consideration for patients who are taking TCAs. These include the use of adjunctive bupropion and buspirone—neither of which has been clearly shown to be effective for TCA-treated depressed patients with suboptimal antidepressive response.

Support for the use of various psychostimulants, such as methylphenidate or dextroamphetamine, to augment TCAs is based primarily on anecdotal reports and small case series (Carvalho et al., 2009; Wharton et al., 1971). To our knowledge, there are no controlled studies of modafinil or armodafinil added to TCAs for patients with treatment-resistant depression.

Augmentation of antidepressants with selected atypical antipsychotic drugs has become a popular management strategy although controlled evidence only supports their use with SSRI or SNRI antidepressants—not necessarily TCAs (Zhou et al., 2015). Only a very small number of subjects taking TCAs were included in the few available reports of adjunctive atypical antipsychotics (all involving risperidone) that were not restricted to primary treatment with SSRIs or SNRIs (Keitner et al., 2009; Mahmoud et al., 2007; Reeves et al., 2008). As such, it is not possible to derive reasonably valid inferences about the effectiveness of adjunctive atypical antipsychotics for treatment-resistant depressed patients who are taking TCAs. Furthermore, the risk of QT interval prolongation is likely increased by combining some atypical antipsychotic drugs (especially quetiapine and ziprasidone) with TCAs. The clinical effects of TCA-atypical antipsychotic drug combination therapy in people with treatment-resistant psychotic and nonpsychotic major depression is an important area in need of further study.

Combination therapy with noradrenergic TCAs (such as nortriptyline or desipramine) and SSRIs has been investigated. For example, a randomized trial of 41 patients with major depression and either a suboptimal response or no response to prospective treatment with 20 mg/day of fluoxetine compared the effects of higher-dose fluoxetine (40–60 mg/day), lithium augmentation, and desipramine augmentation (Fava et al., 1994). High-dose fluoxetine and lithium augmentation were associated with the best treatment outcomes for patients who had no prior response to fluoxetine. In a larger study

(101 depressed patients) of similar design conducted by the same group, there were no significant differences in rates of positive treatment response between high-dose fluoxetine (35%), lithium augmentation (13%), and desipramine augmentation (26%) (Fava et al., 2002). There was no significant effect of desipramine (or lithium) blood concentration on treatment response. A third randomized trial compared the effects of fluoxetine monotherapy (20 mg/day), desipramine monotherapy, and fluoxetine plus desipramine in 38 inpatients with major depression over 6 weeks (Nelson et al., 2004). Desipramine was dosed according to plasma drug levels. Remission rates were higher with combination therapy (54%) than monotherapy with fluoxetine (7%) or desipramine (0%). Across treatment groups, 39%–43% of the subjects were classified as treatment-resistant, defined as failure to respond to at least one prior pharmacotherapeutic trial of adequate dose and duration. However, the study results were impacted by neither treatment-resistance status nor drug plasma levels. A fourth randomized trial of 45 patients with major depression documented significantly higher rates of positive treatment response with citalopram (40 mg/day)-nortriptyline (50 mg/day) combination therapy than citalopram (40 mg/day) alone (65% vs 32%) after 8 weeks of treatment (Raisi et al., 2007). There were no significant between-group differences in most adverse effects, including orthostatic hypotension or occurrence of bradycardia or tachycardia. No information on changes in EKG parameters was provided. It should be noted that these studies generally describe adding TCAs to SSRIs rather than the other way around. Nevertheless, when combining SSRIs and secondary-amine TCAs, low starting doses and careful dose adjustments are recommended, mindful of potential pharmacokinetic drug-drug interactions. In particular, CYP450 enzyme inhibition, especially at CYP2D6, may increase the risk of TCA side effects or toxicity. Combinations of TCAs and citalopram are best avoided due to a pharmacodynamic increase in the risk of QT interval prolongation (Beach et al., 2014). In our clinical experience, combination therapy with SSRIs and tertiary-amine TCAs is rarely encountered, as it makes little sense to do so pharmacologically and only increases the risk of significant drug-drug interactions.

The combination of TCAs and monoamine oxidase inhibitors has also been studied in randomized trials. This topic is reviewed in detail in Chapter 9. The propensity for potentially serious adverse effects, including weight gain, significant orthostasis and serotonin toxicity, must always be kept in mind when exercising this treatment option. Imipramine and clomipramine should not be prescribed with monoamine oxidase inhibitors due to high risk of severe serotonin toxicity (Gillman, 2006). Even though amitriptyline binds with greater affinity to serotonin transporters than secondary-amine TCAs, combination therapy with amitriptyline and the monoamine oxidase inhibitors, tranylcypromine and phenelzine, has been described (Davidson et al., 1978; Ferreira-Garcia et al., 2018; Schmauss et al., 1988). It may be safer to add monoamine oxidase inhibitors (starting at low doses, with slow titration) to established TCA treatment (excluding clomipramine and imipramine) rather than proceeding in the reverse order (White et al., 1982).

We are aware of a single trial of *S*-adenosyl-L-methionine (SAMe) for accelerating the onset of antidepressive effects of imipramine (Berlanga et al., 1992); however, we are aware of no trials of SAMe for patients with well-defined treatment-resistant depression following a suboptimal antidepressive response to TCAs. Similarly, we are unaware of any studies of adjunctive L-methylfolate, omega-3-fatty acids, St. Johnswort, or other nutraceuticals for augmenting a partial response or nonresponse to TCAs. Extracts of St. Johnswort can inhibit serotonin reuptake and weakly inhibit monoamine oxidase (Henderson et al., 2002). Therefore, the same warnings about combining TCAs with monoamine oxidase inhibitors also apply to the use of St. Johnswort. We are unaware of any controlled investigations of other potential adjuncts to TCAs for patients with treatment-resistant depression, including dopamine agonists, melatonin agonists, or hormonally based therapeutics.

Switching to another TCA

At present, the effectiveness of switching to an alternative TCA following failure to respond to another TCA in people with treatment-resistant depression is unclear. In our experience, switching between tertiary-amine TCAs or between secondary-amine TCAs does not often lead to significant clinical improvement, especially if the therapeutic trial of the first TCA was rigorously conducted with good compliance over an adequate period of time—at least 4 weeks at maximum tolerated doses or at doses that yield adequate drug levels for selected agents. On the other hand, patients may improve clinically when switching between TCAs because of alleviation of intolerable side effects, rather than inefficacy. In such cases, the first TCA can be withdrawn while the second TCA is initiated at a low dose and titrated slowly based on clinical response and tolerability. In our experience, many patients tolerate secondary-amine TCAs better than tertiary-amine TCAs, although the opposite may occur on occasion. Therefore, it is reasonable to consider switching between these broad classes of TCAs in cases of failure to achieve a good therapeutic outcome after a well-executed trial of an agent from the other class. Partial responses to the new TCA can be managed by dose optimization or add-on therapies, as discussed earlier.

Transitioning off of TCAs

When there is a need to stop treatment with a TCA, a slow tapering over several weeks is recommended, if possible, to avoid the occurrence of discontinuation syndrome. In our experience, patients will often develop feelings of anxiety or distress related to transitioning from one medication to another. These symptoms are fairly easy to distinguish from discontinuation symptoms with time and reassurance. Patients will also frequently fear that they are experiencing relapses of their mood or anxiety disorder when they are really experiencing discontinuation symptoms. The relatively rapid onset and offset of discontinuation symptoms (if the TCA is restarted) makes it fairly easy to differentiate the presence of a discontinuation syndrome from relapses of mood or anxiety disorders, the latter of which require several weeks to unfold after stopping medication (Rosenbaum and Zajecka, 1997). For many patients, monoamine oxidase inhibitors are the next therapeutic step following the failure to respond to TCAs. In such cases, 14 or more days must transpire after the last dose of TCAs before monoamine oxidase inhibitors can be safely started.

Patients who are receiving electroconvulsive therapy

As reviewed elsewhere in this volume, electroconvulsive therapy (ECT) is a mainstay for the acute- and sometimes maintenance-phase management of treatment-resistant depression (UK ECT Review Group, 2003). However, relapse rates are high (51%) within the first year following successful acute treatment with ECT, many of whom (38%) become ill again within 6 months (Jelovac et al., 2013). The risk of post-ECT relapse highlights the importance of the prevention of such outcomes with vigorous pharmacotherapy for patients with treatment-resistant depression. To date, the largest body of evidence for the pharmacological prevention of post-ECT relapse exists for TCAs. A meta-analysis of 32 studies summarized the effect of antidepressant continuation on post-ECT relapse rates at 3 and 6 months, and documented a large and significant reduction in 6-month relapse with TCAs compared with placebo (RR 0.44, 95% CI 0.29–0.66, NNT 3.3) (Jelovac et al., 2013). Treatment with nortriptyline + lithium was shown to have a significantly lower rate or relapse (39%) than nortriptyline monotherapy (60%) or placebo (84%) in a 24-week randomized trial of 290 patients with major depression following ECT (Sackeim et al., 2001). Further work in this important area is required, however, given a lack of clarity as to whether a possible post-ECT relapse-preventive effect applies to all TCAs or only specific agents, and at least one report documenting high 6-month relapse rates (50%) despite the use of lithium with nortriptyline or venlafaxine following remission from ECT (Prudic et al., 2013).

Summary

TCAs are highly effective antidepressants with a long history of use for treating a variety of unipolar depressive syndromes. TCAs are perhaps best known for treating severe or melancholic depressive states; however, they are also efficacious for the treatment of chronic depressive states, and for ambulatory patients with major depression who are seen in specialty and primary care settings. For decades, the TCAs were the most commonly used medications for the treatment of major depression, but were later supplanted by the SSRIs and newer antidepressants based on a less-advantageous tolerability profile and potential fatality in overdose. The results of controlled studies show that as many as half of patients with major depression who fail to respond to at least one prior therapeutic antidepressant trial achieve a positive treatment response after switching to TCAs. However, response rates vary between individual studies and, not surprisingly, response rates following a switch to TCAs are much lower in studies of depressed patients with more advanced stages of treatment-resistance than patients who have failed only a single prior therapeutic antidepressant trial. Still, TCAs are viable switch options for patients who fail to respond to newer antidepressants. Their safe use depends on clinician knowledge of common adverse effects and an awareness of potentially significant and, in some cases, severe drug interactions and safety issues associated with TCAs. The efficacy profile of TCAs extends across several psychiatric and nonpsychiatric indications, making them potentially attractive options for treating a range of conditions that frequently cooccur with depressive syndromes.

Additional resources

Although this chapter provides a reasonably detailed review and discussion of the use of TCAs for patients with treatment-resistant depression, readers may be interested in additional sources of information. Below, we provide a selected list of references and other resources that are available free of charge. These resources and references we have found to be helpful to clinicians and patients.

- There are a number of drug interaction screening tools that are available to clinicians via subscription (e.g., Micromedex, PEPID, Facts & Comparisons, Lexicomp, etc.) or free or charge (e.g., Medscape, Drugs.com, Epocrates, WebMD, RxList, etc.) over the Internet. We advise checking with your institution, if applicable, to verify an existing prescription. It is important to note that the performance characteristics of both subscription-based and free internet tools can vary substantially (Monteith and Glenn, 2019; Schjott et al., 2019). Nevertheless, we think that these tools are suitable for screening potentially severe drug interactions as an initial step. Final prescribing decisions should be based on a thorough review of risks and benefits for all drug combinations under consideration.
- There are also a number of internet-based sources that provide vetted lists of medications that have the potential to cause QT interval prolongation and/or torsades de pointes (e.g., https://crediblemeds.org,www.QTFactors.org, https://www.cpsbc.ca/files/pdf/Methadone-Drugs-QT-Interval-Prolongation.pdf, and others).
- A downloadable CYP450 drug interaction table and searchable database for identifying potential drug-specific interactions at various CYP450 isoenzymes can be found at: https://drug-interactions.medicine.iu.edu/Home.aspx.

References

Altshuler, L.L., Bauer, M., Frye, M.A., Gitlin, M.J., Mintz, J., Szuba, M.P., Leight, K.L., Whybrow, P.C., 2001. Does thyroid supplementation accelerate tricyclic antidepressant response? A review and meta-analysis of the literature. Am. J. Psychiatr. 158 (10), 1617–1622. https://doi.org/10.1176/appi.ajp.158.10.1617.

Alvarez, P.A., Pahissa, J., 2010. QT alterations in psychopharmacology: proven candidates and suspects. Curr. Drug Saf. 5 (1), 97–104. https://doi.org/10.2174/157488610789869265.

Amitai, Y., Frischer, H., 2004. Excess fatality from desipramine and dosage recommendations. Ther. Drug Monit. 26 (5), 468–473. https://doi.org/10.1097/00007691-200410000-00002.

Anderson, I.M., 1998. SSRIS versus tricyclic antidepressants in depressed inpatients: a meta- analysis of efficacy and tolerability. Depress. Anxiety 7 (1), 11–17. https://doi.org/10.1002/(SICI)1520-6394(1998)7:1+<11::AID-DA4>3.0.CO;2-I.

Anderson, I.M., Tomenson, B.M., 1994. The efficacy of selective serotonin re-uptake inhibitors in depression: a meta-analysis of studies against tricyclic antidepressants. J. Psychopharmacol. 8 (4), 238–249. https://doi.org/10.1177/026988119400800407.

Argoff, C.E., 2011. Review of current guidelines on the care of postherpetic neuralgia. Postgrad. Med. 123 (5), 134–142. https://doi.org/10.3810/pgm.2011.09.2469.

Aronson, R., Offman, H.J., Joffe, R.T., David Naylor, C., 1996. Triiodothyronine augmentation in the treatment of refractory depression: a meta-analysis. Arch. Gen. Psychiatry 53 (9), 842–848. https://doi.org/10.1001/archpsyc.1996.01830090090013.

Arroll, B., Macgillivray, S., Ogston, S., Reid, I., Sullivan, F., Williams, B., Crombie, I., 2005. Efficacy and tolerability of tricyclic antidepressants and SSRIs compared with placebo for treatment of depression in primary care: a meta-analysis. Ann. Fam. Med. 3 (5), 449–456. https://doi.org/10.1370/afm.349.

Bakker, A., Van Balkom, A.J.L.M., Spinhoven, P., 2002. SSRIs vs. TCAs in the treatment of panic disorder: a meta-analysis. Acta Psychiatr. Scand. 106 (3), 163–167. https://doi.org/10.1034/j.1600-0447.2002.02255.x.

Balant-Gorgia, A.E., Gex-Fabry, M., Balant, L.P., 1991. Clinical pharmacokinetics of clomipramine. Clin. Pharmacokinet. 20 (6), 447–462. https://doi.org/10.2165/00003088-199120060-00002.

Bauer, M., Pfennig, A., Severus, E., Whybrow, P.C., Angst, J., Möller, H.J., Adli, M., Benkert, O., Bschor, T., Tadić, A., Holsboer-Trachsler, E., Anderson, I., Baldwin, D., Cookson, J.C., Grunze, H., Katona, C., Paykel, E.S., Tylee, A., Ayuso-Gutierrez, J.L., et al., 2013. World Federation of Societies of Biological Psychiatry (WFSBP) guidelines for biological treatment of unipolar depressive disorders, part 1: update 2013 on the acute and continuation treatment of unipolar depressive disorders. World J. Biol. Psychiatry 14 (5), 334–385. https://doi.org/10.3109/15622975.2013.804195.

Baumann, P., Ulrich, S., Eckermann, G., Gerlach, M., Kuss, H.J., Laux, G., Müller-Oerlinghausen, B., Rao, M.L., Riederer, P., Zernig, G., Hiemke, C., 2005. The AGNP-TDM Expert Group Consensus Guidelines: focus on therapeutic monitoring of antidepressants. Dialogues Clin. Neurosci. 7 (3), 231–247.

Beach, S.R., Kostis, W.J., Celano, C.M., Januzzi, J.L., Ruskin, J.N., Noseworthy, P.A., Huffman, J.C., 2014. Meta-analysis of selective serotonin reuptake inhibitor-associated QTc prolongation. J. Clin. Psychiatry 75 (5), e441–e449. https://doi.org/10.4088/JCP.13r08672.

Bendtsen, L., Jensen, R., 2000. Amitriptyline reduces myofascial tenderness in patients with chronic tension-type headache. Cephalalgia 20 (6), 603–610. https://doi.org/10.1046/j.1468-2982.2000.00087.x.

Berlanga, C., Ortega-Soto, H.A., Ontiveros, M., Senties, H., 1992. Efficacy of S-adenosyl-L-methionine in speeding the onset of action of imipramine. Psychiatry Res. 44 (3), 257–262. https://doi.org/10.1016/0165-1781(92)90029-3.

Bighelli, I., Castellazzi, M., Cipriani, A., Girlanda, F., Guaiana, G., Koesters, M., Turrini, G., Furukawa, T.A., Barbui, C., 2018. Antidepressants versus placebo for panic disorder in adults. Cochrane Database Syst. Rev. 2018 (4). https://doi.org/10.1002/14651858.CD010676.pub2.

Bliss, S.A., Warnock, J.K., 2013. Psychiatric medications: adverse cutaneous drug reactions. Clin. Dermatol. 31 (1), 101–109. https://doi.org/10.1016/j.clindermatol.2011.11.014.

Bobo, W.V., 2017. The diagnosis and management of bipolar I and II disorders: clinical practice update. Mayo Clin. Proc. 92 (10), 1532–1551. https://doi.org/10.1016/j.mayocp.2017.06.022.

Bobo, W.V., Chen, H., Trivedi, M.H., Stewart, J.W., Nierenberg, A.A., Fava, M., Kurian, B.T., Warden, D., Morris, D.W., Luther, J.F., Husain, M.M., Cook, I.A., Lesser, I.M., Kornstein, S.G., Wisniewski, S.R., Rush, A.J., Shelton, R.C., 2011. Randomized comparison of selective serotonin reuptake inhibitor (escitalopram) monotherapy and antidepressant combination pharmacotherapy for major depressive disorder with melancholic features: a CO-MED report. J. Affect. Disord. 133 (3), 467–476. https://doi.org/10.1016/j.jad.2011.04.032.

Bobo, W.V., Grossardt, B.R., Lapid, M.I., Leung, J.G., Stoppel, C., Takahashi, P.Y., Hoel, R.W., Chang, Z., Lachner, C., Chauhan, M., Flowers, L., Brue, S.M., Frye, M.A., St. Sauver, J., Rocca, W.A., Sutor, B., 2019. Frequency and predictors of the potential overprescribing of antidepressants in elderly residents of a geographically defined U.S. population. Pharmacol. Res. Perspect. 7 (1). https://doi.org/10.1002/prp2.461.

Boyce, P., Judd, F., 1999. The place for the tricyclic antidepressants in the treatment of depression. Aust. N. Z. J. Psychiatry 33 (3), 323–327. https://doi.org/10.1046/j.1440-1614.1999.00580.x.

Brøsen, K., Gram, L.F., 1988. First-pass metabolism of imipramine and desipramine: impact of the sparteine oxidation phenotype. Clin. Pharmacol. Ther. 43 (4), 400–406. https://doi.org/10.1038/clpt.1988.50.

Brøsen, K., Hansen, J.G., Nielsen, K.K., Sindrup, S.H., Gram, L.F., 1993. Inhibition by paroxetine of desipramine metabolism in extensive but not in poor metabolizers of sparteine. Eur. J. Clin. Pharmacol. 44 (4), 349–355. https://doi.org/10.1007/BF00316471.

Caldwell, P.H.Y., Deshpande, A.V., VonGontard, A., 2013. Management of nocturnal enuresis. BMJ 347. https://doi.org/10.1136/bmj.f6259 (Online).

Carvalho, A.F., Machado, J.R., Cavalcante, J.L., 2009. Augmentation strategies for treatment-resistant depression. Curr. Opin. Psychiatry 22 (1), 7–12. https://doi.org/10.1097/YCO.0b013e32831be9ef.

Chou, R., Huffman, L.H., 2007. Medications for acute and chronic low back pain: a review of the evidence for an American Pain Society/American college of physicians clinical practice guideline. Ann. Intern. Med. 147 (7), 505–514. https://doi.org/10.7326/0003-4819-147-7-200710020-00008.

Cipriani, A., Geddes, J.R., Furukawa, T.A., Barbui, C., 2007. Metareview on short-term effectiveness and safety of antidepressants for depression: an evidence-based approach to inform clinical practice. Can. J. Psychiatr. 52 (9), 553–562. https://doi.org/10.1177/070674370705200903.

Cipriani, A., Barbui, C., Butler, R., 2011. Depression in adults: drug and physical treatments. BMJ Clin. Evid. 5.

Cipriani, A., Furukawa, T.A., Salanti, G., Chaimani, A., Atkinson, L.Z., Ogawa, Y., Leucht, S., Ruhe, H.G., Turner, E.H., Higgins, J.P.T., Egger, M., Takeshima, N., Hayasaka, Y., Imai, H., Shinohara, K., Tajika, A., Ioannidis, J.P.A., Geddes, J.R., 2018. Comparative efficacy and acceptability of 21 antidepressant drugs for the acute treatment of adults with major depressive disorder: a systematic review and network meta-analysis. Lancet 391 (10128), 1357–1366. https://doi.org/10.1016/S0140-6736(17)32802-7.

Cohen, H.W., Gibson, G., Alderman, M.H., 2000. Excess risk of myocardial infarction in patients treated with antidepressant medications: association with use of tricyclic agents. Am. J. Med. 108 (1), 2–8. https://doi.org/10.1016/S0002-9343(99)00301-0.

Crossley, N.A., Bauer, M., 2007. Acceleration and augmentation of antidepressants with lithium for depressive disorders: two meta-analyses of randomized, placebo-controlled trials. J. Clin. Psychiatry 68 (6), 935–940. https://doi.org/10.4088/JCP.v68n0617.

Cusack, B., Nelson, A., Richelson, E., 1994. Binding of antidepressants to human brain receptors: focus on newer generation compounds. Psychopharmacology 114 (4), 559–565. https://doi.org/10.1007/BF02244985.

Davidson, J., McLeod, M., Law-Yone, B., Linnoila, M., 1978. A comparison of electroconvulsive therapy and combined phenelzine-amitriptyline in refractory depression. Arch. Gen. Psychiatry 35 (5), 639–642. https://doi.org/10.1001/archpsyc.1978.01770290121011.

De Lima, M., Hotopf, M., 2003. A comparison of active drugs for the treatment of dysthymia. Cochrane Database Syst. Rev. 2003 (3).

De Lima, M., Moncrieff, M., 2000. Drugs versus placebo for dysthymia. Cochrane Database Syst. Rev. (4).

Dharmshaktu, P., Tayal, V., Kalra, B.S., 2011. Efficacy of antidepressants as analgesics. J. Clin. Pharmacol. 52, 6–17.

Drake, M.J., Nixon, P.M., Crew, J.P., 1998. Drug-induced bladder and urinary disorders. Drug Saf. 19 (1), 45–55. https://doi.org/10.2165/00002018-199819010-00004.

Drake, L.A., Cohen, L., Gillies, R., Flood, J.G., Riordan, A.T., Phillips, S.B., Stiller, M.J., 1999. Pharmacokinetics of doxepin in subjects with pruritic atopic dermatitis. J. Am. Acad. Dermatol. 41 (2 I), 209–214. https://doi.org/10.1016/S0190-9622(99)70051-4.

Duma, S.R., Fung, V.S.C., 2019. Drug-induced movement disorders. Aust. Prescr. 42 (2), 56–61. https://doi.org/10.18773/austprescr.2019.014.

Duncan, R.S., McPate, M.J., Ridley, J.M., Gao, Z., James, A.F., Leishman, D.J., Leaney, J.L., Witchel, H.J., Hancox, J.C., 2007. Inhibition of the HERG potassium channel by the tricyclic antidepressant doxepin. Biochem. Pharmacol. 74 (3), 425–437. https://doi.org/10.1016/j.bcp.2007.04.024.

English, B.A., Dortch, M., Ereshefsky, L., Jhee, S., 2012. Clinically significant psychotropic drug-drug interactions in the primary care setting. Curr. Psychiatry Rep. 14 (4), 376–390. https://doi.org/10.1007/s11920-012-0284-9.

Ereshefsky, L., Riesenman, C., Lam, Y.W.F., 1995. Antidepressant drug interactions and the cytochrome P450 system: the role of cytochrome P450 2D6. Clin. Pharmacokinet. 29 (1), 10–19. https://doi.org/10.2165/00003088-199500291-00004.

Fangmann, P., Assion, H.J., Juckel, G., González, C.A., López-Muñoz, F., 2008. Half a century of antidepressant drugs: on the clinical introduction of monoamine oxidase inhibitors, tricyclics, and tetracyclics. Part II: tricyclics and tetracyclics. J. Clin. Psychopharmacol. 28 (1), 1–4. https://doi.org/10.1097/jcp.0b013e3181627b60.

Fava, M., 2001. Augmentation and combination strategies in treatment-resistant depression. J. Clin. Psychiatry 62 (18), 4–11.

Fava, M., 2006. Prospective studies of adverse events related to antidepressant discontinuation. J. Clin. Psychiatry 67 (4), 14–21.

Fava, M., Rosenbaum, J.F., McGrath, P.J., Stewart, J.W., Amsterdam, J.D., Quitkin, F.M., 1994. Lithium and tricyclic augmentation of fluoxetine treatment for resistant major depression: a double-blind, controlled study. Am. J. Psychiatry 151 (9), 1372–1374. https://doi.org/10.1176/ajp.151.9.1372.

Fava, M., Alpert, J., Nierenberg, A., Lagomasino, I., Sonawalla, S., Tedlow, J., Worthington, J., Baer, L., Rosenbaum, J.F., 2002. Double-blind study of high-dose fluoxetine versus lithium or desipramine augmentation of fluoxetine in partial responders and nonresponders to fluoxetine. J. Clin. Psychopharmacol. 22 (4), 379–387. https://doi.org/10.1097/00004714-200208000-00008.

Fava, M., Rush, A.J., Wisniewski, S.R., Nierenberg, A.A., Alpert, J.E., McGrath, P.J., Thase, M.E., Warden, D., Biggs, M., Luther, J.F., Niederehe, G., Ritz, L., Trivedi, M.H., 2006. A comparison of mirtazapine and nortriptyline following two consecutive failed medication treatments for depressed outpatients: a STAR*D report. Am. J. Psychiatry 163 (7), 1161–1172.

Ferreira-Garcia, R., Da Rocha Freire, R.C., Appolinário, J.C., Levitan, M.N., Halkjær-Lassen, R.D., Bueno, J.R., Nardi, A.E., 2018. Tranylcypromine plus amitriptyline for electroconvulsive therapy-resistant depression. J. Clin. Psychopharmacol. 38 (5), 502–504. https://doi.org/10.1097/JCP.0000000000000945.

Fineberg, N.A., Gale, T.M., 2005. Evidence-based pharmacotherapy of obsessive-compulsive disorder. Int. J. Neuropsychopharmacol. 8 (1), 107–129. https://doi.org/10.1017/S1461145704004675.

Finnerup, N.B., Attal, N., Haroutounian, S., McNicol, E., Baron, R., Dworkin, R.H., Gilron, I., Haanpää, M., Hansson, P., Jensen, T.S., Kamerman, P.R., Lund, K., Moore, A., Raja, S.N., Rice, A.S.C., Rowbotham, M., Sena, E., Siddall, P., Smith, B.H., Wallace, M., 2015. Pharmacotherapy for neuropathic pain in adults: a systematic review and meta-analysis. Lancet Neurol. 14 (2), 162–173. https://doi.org/10.1016/S1474-4422(14)70251-0.

Flockhart, D.A., 1995. Drug interactions and the cytochrome P450 system: the role of cytochrome P450 2C19. Clin. Pharmacokinet. 29 (1), 45–52. https://doi.org/10.2165/00003088-199500291-00008.

Ford, A., Luthra, P., Tack, J., Boeckxstaens, G., Moayyedi, P., Talley, N., 2017. Efficacy of psychotropic drugs in functional dyspepsia: systematic review and meta-analysis. Gut, 411–420. https://doi.org/10.1136/gutjnl-2015-310721.

Friedrich, M.E., Akimova, E., Huf, W., Konstantinidis, A., Papageorgiou, K., Winkler, D., Toto, S., Greil, W., Grohmann, R., Kasper, S., 2016. Drug-induced liver injury during antidepressant treatment: results of AMSP, a drug surveillance program. Int. J. Neuropsychopharmacol. 19 (4). https://doi.org/10.1093/ijnp/pyv126.

Fu, Y., Dai, Q., Zhu, L., Wu, S., 2018. Antidepressants use and risk of cataract development: a systematic review and meta-analysis. BMC Ophthalmol. 18 (1). https://doi.org/10.1186/s12886-018-0699-0.

Furukawa, T.A., McGuire, H., Barbui, C., 2002. Meta-analysis of the effects and side-effects of low dosage tricyclic antidepressants in depression: systematic review. BMJ 325, 991–999.

Furukawa, T., McGuire, H., Barbui, C., 2003. Low dosage tricyclic antidepressants for depression. Cochrane Database Syst. Rev. 3, CD003197 (Online).

George, C.F., 1980. Drugs causing intestinal obstruction: a review. J. R. Soc. Med. 73 (3), 200–204. https://doi.org/10.1177/014107688007300309.

Gervasoni, N., Aubry, J.-M., Gex-Fabry, M., et al., 2009. Is there a place for tricyclic antidepressants and subsequent augmentation strategies in obtaining remission for patients with treatment resistant depression? Pharmacol. Res. 59 (3), 202–206.

Gijsman, H.J., Geddes, J.R., Rendell, J.M., Nolen, W.A., Goodwin, G.M., 2004. Antidepressants for bipolar depression: a systematic review of randomized, controlled trials. Am. J. Psychiatr. 161 (9), 1537–1547. https://doi.org/10.1176/appi.ajp.161.9.1537.

Gillman, P.K., 2006. A review of serotonin toxicity data: implications for the mechanisms of antidepressant drug action. Biol. Psychiatry 59 (11), 1046–1051. https://doi.org/10.1016/j.biopsych.2005.11.016.

Gillman, P.K., 2007. Tricyclic antidepressant pharmacology and therapeutic drug interactions updated. Br. J. Pharmacol. 151 (6), 737–748. https://doi.org/10.1038/sj.bjp.0707253.

Glassman, A.H., 1984. Cardiovascular effects of tricyclic antidepressants. Annu. Rev. Med. 35, 503–511. https://doi.org/10.1146/annurev.me.35.020184.002443.

Glassman, A.H., Preud'homme, X.A., 1993. Review of the cardiovascular effects of heterocyclic antidepressants. J. Clin. Psychiatry 54 (2), 16–22.

Glassman, A.H., Roose, S.P., Bigger, J.T., 1993. The safety of tricyclic antidepressants in cardiac patients: risk-benefit reconsidered. JAMA 269 (20), 2673–2675. https://doi.org/10.1001/jama.1993.03500200087039.

Godfrey, R.G., 1996. A guide to the understanding and use of tricyclic antidepressants in the overall management of fibromyalgia and other chronic pain syndromes. Arch. Intern. Med. 156 (10), 1047–1052. https://doi.org/10.1001/archinte.156.10.1047.

Griebeler, M.L., Morey-Vargas, O.L., Brito, J.P., Tsapas, A., Wang, Z., Carranza Leon, B.G., Phung, O.J., Montori, V.M., Murad, M.H., 2014. Pharmacologic interventions for painful diabetic neuropathy: an umbrella systematic review and comparative effectiveness network meta-analysis. Ann. Intern. Med. 161 (9), 639–649. https://doi.org/10.7326/M14-0511.

Habibi, M., Hart, F., Bainbridge, J., 2016. The impact of psychoactive drugs on seizures and antiepileptic drugs. Curr. Neurol. Neurosci. Rep. 16, 71.

Hawton, K., Bergen, H., Simkin, S., Cooper, J., Waters, K., Gunnell, D., Kapur, N., 2010. Toxicity of antidepressants: rates of suicide relative to prescribing and non-fatal overdose. Br. J. Psychiatry 196 (5), 354–358. https://doi.org/10.1192/bjp.bp.109.070219.

Hay, P., Claudino, A., 2010. Bulimia nervosa. BMJ Clin. Evid. 7, 1009. https://www.ncbi.nlm.nih.gov/pmc/articles/PMC3275326/pdf/2010-1009.pdf.

Henderson, L., Yue, Q.Y., Bergquist, C., Gerden, B., Arlett, P., 2002. St John's wort (Hypericum perforatum): drug interactions and clinical outcomes. Br. J. Clin. Pharmacol. 54 (4), 349–356. https://doi.org/10.1046/j.1365-2125.2002.01683.x.

Henkel, V., Mergl, R., Allgaier, A.K., Kohnen, R., Möller, H.J., Hegerl, U., 2006. Treatment of depression with atypical features: a meta-analytic approach. Psychiatry Res. 141 (1), 89–101. https://doi.org/10.1016/j.psychres.2005.07.012.

Henry, J.A., Alexander, C.A., Sener, E.K., 1995. Relative mortality from overdose of antidepressants. BMJ 310 (6974), 221. https://doi.org/10.1136/bmj.310.6974.221.

Henssler, J., Heinz, A., Brandt, L., Bschor, T., 2019. Absetz- und Rebound-Phänomene bei Antidepressiva. Dtsch. Arztebl. Int. 116 (20), 355–361. https://doi.org/10.3238/arztebl.2019.0355.

Hicks, J.K., Swen, J.J., Thorn, C.F., Sangkuhl, K., Kharasch, E.D., Ellingrod, V.L., Skaar, T.C., Müller, D.J., Gaedigk, A., Stingl, J.C., 2013. Clinical pharmacogenetics implementation consortium guideline for CYP2D6 and CYP2C19 genotypes and dosing of tricyclic antidepressants. Clin. Pharmacol. Ther. 93 (5), 402–408. https://doi.org/10.1038/clpt.2013.2.

Hiemke, C., Baumann, P., Bergemann, N., Conca, A., Dietmaier, O., Egberts, K., Fric, M., Gerlach, M., Greiner, C., Gründer, G., Haen, E., Havemann-Reinecke, U., Jaquenoud Sirot, E., Kirchherr, H., Laux, G., Lutz, U.C., Messer, T., Mller, M.J., Pfuhlmann, B., et al., 2011. AGNP consensus guidelines for therapeutic drug monitoring in psychiatry: update 2011. Pharmacopsychiatry 44 (6), 195–235. https://doi.org/10.1055/s-0031-1286287.

Hirschfeld, R.M.A., Judge, R., 2000. Antidepressants in long-term therapy: a review of tricyclic antidepressants and selective serotonin reuptake inhibitors. Acta Psychiatr. Scand. Suppl. 101 (403), 35–38.

Hockings, J.K., Pasternak, A.L., Erwin, A.L., Mason, N.T., Eng, C., Hicks, J.K., 2020. Pharmacogenomics: an evolving clinical tool for precision medicine. Cleve. Clin. J. Med. 87 (2), 91–99. https://doi.org/10.3949/ccjm.87a.19073.

Ipser, J., Stein, D.J., 2011. Evidence-based pharmacotherapy of posttraumatic stress disorder (PTSD). Int. J. Neuropsychopharmacol. 15, 825–840.

Ipser, J.C., Sander, C., Stein, D.J., 2009. Pharmacotherapy and psychotherapy for body dysmorphic disorder. Cochrane Database Syst. Rev. 1. https://doi.org/10.1002/14651858.CD005332.pub2.

Jackson, J.L., Shimeall, W., Sessums, L., DeZee, K.J., Becher, D., Diemer, M., Berbano, E., O'Malley, P.G., 2010. Tricyclic antidepressants and headaches: systematic review and meta-analysis. BMJ 341 (7778), 869. https://doi.org/10.1136/bmj.c5222 (Online).

Jelovac, A., Kolshus, E., McLoughlin, D.M., 2013. Relapse following successful electroconvulsive therapy for major depression: a meta-analysis. Neuropsychopharmacology 38 (12), 2467–2474. https://doi.org/10.1038/npp.2013.149.

Joffe, R., Sokolov, S., Streiner, D., 1996. Antidepressant treatment of depression: a meta analysis. Can. J. Psychiatr. 41 (10), 613–616. https://doi.org/10.1177/070674379604101002.

Judge, B.S., Rentmeester, L.L., 2013. Antidepressant overdose-induced seizures. Psychiatr. Clin. N. Am. 36 (2), 245–260. https://doi.org/10.1016/j.psc.2013.02.004.

Keitner, G.I., Ryan, C.E., Miller, I.W., Kohn, R., Epstein, N.B., 1991. 12-Month outcome of patients with major depression and comorbid psychiatric or medical illness (compound depression). Am. J. Psychiatr. 148 (3), 345–350. https://doi.org/10.1176/ajp.148.3.345.

Keitner, G.I., Garlow, S.J., Ryan, C.E., Ninan, P.T., Solomon, D.A., Nemeroff, C.B., Keller, M.B., 2009. A randomized, placebo-controlled trial of risperidone augmentation for patients with difficult-to-treat unipolar, non-psychotic major depression. J. Psychiatr. Res. 43 (3), 205–214. https://doi.org/10.1016/j.jpsychires.2008.05.003.

Keller, M.B., Gelenberg, A.J., Hirschfeld, R.M.A., Rush, A.J., Thase, M.E., Kocsis, J.H., Markowitz, J.C., Fawcett, J.A., Koran, L.M., Klein, D.N., Russell, J.M., Kornstein, S.G., McCullough, J.P., Davis, S.M., Harrison, W.M., 1998. The treatment of chronic depression, part 2: a double-blind, randomized trial of sertraline and imipramine. J. Clin. Psychiatry 59 (11), 598–607. https://doi.org/10.4088/JCP.v59n1107.

Kerihuel, J.C., Dreyfus, J.F., 1991. Meta-analyses of the efficacy and tolerability of the tricyclic antidepressant lofepramine. J. Int. Med. Res. 19 (3), 183–201. https://doi.org/10.1177/030006059101900304.

Kerr, G.W., McGuffie, A.C., Wilkie, S., 2001. Tricyclic antidepressant overdose: a review. Emerg. Med. J. 18 (4), 236–241. https://doi.org/10.1136/emj.18.4.236.

Khawam, E.A., Laurencic, G., Malone, D.A., 2006. Side effects of antidepressants: an overview. Cleve. Clin. J. Med. 73 (4), 351–361. https://doi.org/10.3949/ccjm.73.4.351.

Kim, T., Xu, C., Amsterdam, J.D., 2019. Relative effectiveness of tricyclic antidepressant versus monoamine oxidase inhibitor monotherapy for treatment-resistant depression. J. Affect. Disord. 250, 199–203. https://doi.org/10.1016/j.jad.2019.03.028.

Kocsis, J.H., Friedman, R.A., Markowitz, J.C., Miller, N., Gniwesch, L., Bram, J., 1995. Stability of remission during tricyclic antidepressant continuation therapy for dysthymia. Psychopharmacol. Bull. 31 (2), 213–216.

Kocsis, J., Friedman, R., Markowitz, J., Leon, A., Miller, N., Gniwesch, L., Parides, M., 1996. Maintenance therapy for chronic depression. Arch. Gen. Psychiatry 769. https://doi.org/10.1001/archpsyc.1996.01830090013002.

Kriston, L., Von Wolff, A., Westphal, A., Hölzel, L.P., Härter, M., 2014. Efficacy and acceptability of acute treatments for persistent depressive disorder: a network meta-analysis. Depress. Anxiety 31 (8), 621–630. https://doi.org/10.1002/da.22236.

Kuhn, R., 1958. The treatment of depressive states with G 22355 (imipramine hydrochloride). Am. J. Psychiatry 115 (5), 459–464. https://doi.org/10.1176/ajp.115.5.459.

Landmark, C., Henning, O., Johannessen, S., 2016. Proconvulsant effects of antidepressants—what is the current evidence? Epilepsy Behav., 287–291. https://doi.org/10.1016/j.yebeh.2016.01.029.

Lehmann, H.E., Cahn, C.H., De Verteuil, R.L., 1958. The treatment of depressive conditions with imipramine (G 22355). Can. Psychiatr. Assoc. J. 3 (4), 155–164. https://doi.org/10.1177/070674375800300401.

Leventhal, A.M., Rehm, L.P., 2005. The empirical status of melancholia: implications for psychology. Clin. Psychol. Rev. 25 (1), 25–44. https://doi.org/10.1016/j.cpr.2004.09.001.

Lichtman, J., Bigger, J., Blumenthal, J., Frasure-Smith, N., Kaufmann, P., Lesperance, F., Mark, D., Sheps, D., Barr Taylor, C., Froelicher, E., 2008. Depression and coronary heart disease: recommendations for screening, referral, and treatment: a science advisory from the American Heart Association Prevention Committee of the Council on Cardiovascular Nursing, Council on Clinical Cardiology, Council on Epidemiology and Prevention, and Interdisciplinary Council on Quality of Care and Outcomes Research: endorsed by the American Psychiatric Association. Circulation 118 (17), 1768–1775.

Lydiard, R.B., 1985. Tricyclic-resistant depression: treatment resistance or inadequate treatment? J. Clin. Psychiatry 46 (10), 412–417.

MacGillivray, S., Arroll, B., Hatcher, S., Ogston, S., Reid, I., Sullivan, F., Williams, B., Crombie, I., 2003. Efficacy and tolerability of selective serotonin reuptake inhibitors compared with tricyclic antidepressants in depression treated in primary care: systematic review and meta-analysis. Br. Med. J. 326 (7397), 1014–1017.

Machado, M., Iskedjian, M., Ruiz, I., Einarson, T.R., 2006. Remission, dropouts, and adverse drug reaction rates in major depressive disorder: a meta-analysis of head-to-head trials. Curr. Med. Res. Opin. 22 (9), 1825–1837. https://doi.org/10.1185/030079906X132415.

Mahmoud, R.A., Pandina, G.J., Turkoz, I., Kosik-Gonzalez, C., Canuso, C.M., Kujawa, M.J., Gharabawi-Garibaldi, G.M., 2007. Risperidone for treatment-refractory major depressive disorder: a randomized trial. Ann. Intern. Med. 147 (9), 593–602. https://doi.org/10.7326/0003-4819-147-9-200711060-00003.

Marcy, T.R., Britton, M.L., 2005. Antidepressant-induced sweating. Ann. Pharmacother. 39 (4), 748–752. https://doi.org/10.1345/aph.1E564.

Martino, D.J., Szmulewicz, A.G., Valerio, M.P., Parker, G., 2019. Melancholia: an attempt at definition based on a review of empirical data. J. Nerv. Ment. Dis. 207 (9), 792–798. https://doi.org/10.1097/NMD.0000000000001090.

Mattuzas, W., Javaid, J.I., Glass, R., Davis, J.M., Ross, J.A., Uhlenhuth, E.H., 1982. Plasma concentrations of imipramine and clinical response among depressed outpatients. J. Clin. Psychopharmacol. 2 (2), 140–142.

Mauri, M.C., Fiorentini, A., Paletta, S., Altamura, A.C., 2014. Pharmacokinetics of antidepressants in patients with hepatic impairment. Clin. Pharmacokinet. 53 (12), 1069–1081. https://doi.org/10.1007/s40262-014-0187-5.

Mayers, A.G., Baldwin, D.S., 2005. Antidepressants and their effect on sleep. Human Psychopharmacol. 20 (8), 533–559. https://doi.org/10.1002/hup.726.

McTavish, D., Benfield, P., 1990. Clomipramine: an overview of its pharmacological properties and a review of its therapeutic use in obsessive compulsive disorder and panic disorder. Drugs 39 (1), 136–153. https://doi.org/10.2165/00003495-199039010-00010.

Meister, R., Von Wolff, A., Mohr, H., Härter, M., Nestoriuc, Y., Hölzel, L., Kriston, L., 2016. Comparative safety of pharmacologic treatments for persistent depressive disorder: a systematic review and network meta-analysis. PLoS One 11 (5). https://doi.org/10.1371/journal.pone.0153380.

Miller, N.L., Kocsis, J.H., Leon, A.C., Portera, L., Dauber, S., Markowitz, J.C., 2001. Maintenance desipramine for dysthymia: a placebo-controlled study. J. Affect. Disord. 64 (2–3), 231–237. https://doi.org/10.1016/S0165-0327(00)00256-1.

Monteith, S., Glenn, T., 2019. A comparison of potential psychiatric drug interactions from six drug interaction database programs. Psychiatry Res. 275, 366–372. https://doi.org/10.1016/j.psychres.2019.03.041.

Moore, R.A., Derry, S., Aldington, D., Cole, P., Wiffen, P.J., 2015. Amitriptyline for neuropathic pain in adults. Cochrane Database Syst. Rev. 2017 (10). https://doi.org/10.1002/14651858.CD008242.pub3.

Morgan, J.C., Sethi, K.D., 2005. Drug-induced tremors. Lancet Neurol. 4 (12), 866–876. https://doi.org/10.1016/S1474-4422(05)70250-7.

Mullish, B.H., Kabir, M.S., Thursz, M.R., Dhar, A., 2014. Review article: depression and the use of antidepressants in patients with chronic liver disease or liver transplantation. Aliment. Pharmacol. Ther. 40 (8), 880–892. https://doi.org/10.1111/apt.12925.

Nachimuthu, S., Assar, M.D., Schussler, J.M., 2012. Drug-induced QT interval prolongation: mechanisms and clinical management. Ther. Adv. Drug Saf. 3 (5), 241–253. https://doi.org/10.1177/2042098612454283.

Nagler, E.V., Webster, A.C., Vanholder, R., Zoccali, C., 2012. Antidepressants for depression in stage 3–5 chronic kidney disease: a systematic review of pharmacokinetics, efficacy and safety with recommendations by European Renal Best Practice (ERBP). Nephrol. Dial. Transplant. 27 (10), 3736–3745. https://doi.org/10.1093/ndt/gfs295.

National Institute for Health and Clinical Excellence, 2010. Depression: The Treatment and Management of Depression in Adults. (updated edition), 2010 https://www.ncbi.nlm.nih.gov/books/NBK63748/.

Navarro, V., Boulahfa, I., Obach, A., Jerez, D., Diaz-Ricart, M., Gast, C., Guarch, J., 2019. Switching to imipramine versus add-on mirtazapine in venlafaxine-resistant major depression: a 10-week randomized open study. J. Clin. Psychopharmacol. 39 (1), 63–66. https://doi.org/10.1097/JCP.0000000000000988.

Nelson, J., Bock, J., Jatlow, P.I., 1983. Clinical implications of 2-hydroxydespiramine plasma concentration. Clin. Pharmacol. Ther. 33, 183–189.

Nelson, J.C., Mazure, C.M., Jatlow, P.I., Bowers, M.B., Price, L.H., 2004. Combining norepinephrine and serotonin reuptake inhibition mechanisms for treatment of depression: a double-blind, randomized study. Biol. Psychiatry 55 (3), 296–300. https://doi.org/10.1016/j.biopsych.2003.08.007.

Nelson, J.C., Baumann, P., Delucchi, K., Joffe, R., Katona, C., 2014. A systematic review and meta-analysis of lithium augmentation of tricyclic and second generation antidepressants in major depression. J. Affect. Disord. 168, 269–275. https://doi.org/10.1016/j.jad.2014.05.053.

Nierenberg, A.A., Papakostas, G.I., Petersen, T., Kelly, K.E., Iacoviello, B.M., Worthington, J.J., Tedlow, J., Alpert, J.E., Fava, M., 2003. Nortriptyline for treatment-resistant depression. J. Clin. Psychiatry 64 (1), 35–39. https://doi.org/10.4088/JCP.v64n0108.

Nierenberg, A.A., Fava, M., Trivedi, M.H., Wisniewski, S.R., Thase, M.E., McGrath, P.J., Alpert, J.E., Warden, D., Luther, J.F., Niederehe, G., Lebowitz, B., Shores-Wilson, K., Rush, A.J., 2006. A comparison of lithium and T3 augmentation following two failed medication treatments for depression: a STAR*D report. Am. J. Psychiatr. 163 (9), 1519–1530. https://doi.org/10.1176/ajp.2006.163.9.1519.

Obata, H., 2017. Analgesic mechanisms of antidepressants for neuropathic pain. Int. J. Mol. Sci. 18 (11). https://doi.org/10.3390/ijms18112483.

Oxman, T.E., 1996. Antidepressants and cognitive impairment in the elderly. J. Clin. Psychiatry 57 (5), 38–44.

Peet, M., 1994. Induction of mania with selective serotonin re-uptake inhibitors and tricyclic antidepressants. Br. J. Psychiatry 164, 549–550. https://doi.org/10.1192/bjp.164.4.549.

Perry, P.J., 1996. Pharmacotherapy for major depression with melancholic features: relative efficacy of tricyclic versus selective serotonin reuptake inhibitor antidepressants. J. Affect. Disord. 39 (1), 1–6. https://doi.org/10.1016/0165-0327(96)00014-6.

Perry, P., Zeilmann, C., Arndt, S.V., 1994. Tricyclic antidepressant plasma concentrations: an estimate of their sensitivity and specificity as a predictor of response. J. Clin. Psychopharmacol. 14, 230–240.

Peselow, E.D., Filippi, A.M., Goodnick, P., Barouche, F., Fieve, R.R., 1989. The short- and long-term efficacy of paroxetine HCl: B. data from a double-blind crossover study and from a year-long term trial vs. imipramine and placebo. Psychopharmacol. Bull. 25 (2), 272–276.

Piñeyro, G., Blier, P., 1999. Autoregulation of serotonin neurons: role in antidepressant drug action. Pharmacol. Rev. 51 (3), 533–591.

Posternak, M.A., Zimmerman, M., 2005. Is there a delay in the antidepressant effect? A meta-analysis. J. Clin. Psychiatry 66 (2), 148–158. https://doi.org/10.4088/JCP.v66n0201.

Preskorn, S.H., Beber, J.H., Faul, J.C., Hirschfeld, R.M.A., 1990. Serious adverse effects of combining fluoxetine and tricyclic antidepressants. Am. J. Psychiatry 147 (4), 532.

Preskorn, S.H., Alderman, J., Chung, M., Harrison, W., Messig, M., Harris, S., 1994. Pharmacokinetics of desipramine coadministered with sertraline or fluoxetine. J. Clin. Psychopharmacol. 14 (2), 90–98. https://doi.org/10.1097/00004714-199404000-00002.

Prudic, J., Haskett, R.F., McCall, W.V., Isenberg, K., Cooper, T., Rosenquist, P.B., Mulsant, B.H., Sackeim, H.A., 2013. Pharmacological strategies in the prevention of relapse after electroconvulsive therapy. J. ECT 29 (1), 3–12. https://doi.org/10.1097/YCT.0b013e31826ea8c4.

Puetz, T.W., Youngstedt, S.D., Herring, M.P., 2015. Effects of pharmacotherapy on combat-related PTSD, anxiety, and depression: a systematic review and meta-regression analysis. PLoS One 10 (5). https://doi.org/10.1371/journal.pone.0126529.

Quinn, C., Harris, A., Kemp, A., 2012. The interdependence of subtype and severity: contributions of clinical and neuropsychological features to melancholia and non-melancholia in an outpatient sample. J. Int. Neuropsychol. Soc. 18 (2), 361–369. https://doi.org/10.1017/S1355617711001858.

Quinn, D.K., Stern, T.A., 2009. Linezolid and serotonin syndrome. Prim. Care Companion J. Clin. Psychiatry 11 (6), 353–356. https://doi.org/10.4088/PCC.09r00853.

Raethjen, J., Lemke, M.R., Lindemann, M., Wenzelburger, R., Krack, P., Deuschl, G., 2001. Amitriptyline enhances the central component of physiological tremor. J. Neurol. Neurosurg. Psychiatry 70 (1), 78–82. https://doi.org/10.1136/jnnp.70.1.78.

Raisi, F., Habibi, N., Ali Nasehi, A., Akhondzadeh, S., 2007. Combination of citalopram and nortriptyline in the treatment of severe major depression: a double-blind, placebo-controlled trial. Therapy 4, 187–192.

Reeves, H., Batra, S., May, R.S., Zhang, R., Dahl, D.C., Li, X., 2008. Efficacy of risperidone augmentation to antidepressants in the management of suicidality in major depressive disorder: a randomized, double-blind, placebo-controlled pilot study. J. Clin. Psychiatry 69 (8), 1228–1236. https://doi.org/10.4088/JCP.v69n0805.

Remick, R.A., 1988. Anticholinergic side effects of tricyclic antidepressants and their management. Prog. Neuropsychopharmacol. Biol. Psychiatry 12 (2–3), 225–231. https://doi.org/10.1016/0278-5846(88)90039-5.

Richelson, E., 1979. Tricyclic antidepressants and histamine H1 receptors. Mayo Clin. Proc. 54 (10), 669–674.

Richelson, E., 1996. Synaptic effects of antidepressants. J. Clin. Psychopharmacol. 16 (3). https://doi.org/10.1097/00004714-199606002-00001.

Richelson, E., 2001. Pharmacology of antidepressants. Mayo Clin. Proc. 76 (5), 511–527. https://doi.org/10.4065/76.5.511.

Richelson, E., 2003. Interaction of antidepressants with neurotransmitter transporters and receptors and their clinical relevance. J. Clin. Psychiatry 64 (13), 5–13.

Richelson, E., Nelson, A., 1984. Antagonism by antidepressants of neurotransmitter receptors of normal human brain in vitro. J. Pharmacol. Exp. Ther. 230 (1), 94–102.

Ricken, R., Ulrich, S., Schlattmann, P., Adli, M., 2017. Tranylcypromine in mind (part II): review of clinical pharmacology and meta-analysis of controlled studies in depression. Eur. Neuropsychopharmacol. 27 (8), 714–731. https://doi.org/10.1016/j.euroneuro.2017.04.003.

Roose, S., Glassman, A.H., 1989. Cardiovascular effects of tricyclic antidepressants in depressed patients with and without heart disease. J. Clin. Psychiatry 7, 1–18.

Roose, S.P., Glassman, A.H., Siris, S.G., Walsh, B.T., Brumisctext, R.L., Wright, L.B., 1981. Comparison of imipramine- and misctextrtriptyline-induced orthostatic hypotension: a meaningful difference. J. Clin. Psychopharmacol. 1 (5), 316–319. https://doi.org/10.1097/00004714-198109000-00008.

Roose, S.P., Glassman, A.H., Giardina, E.G.V., Walsh, B.T., Woodring, S., Bigger, J.T., 1987. Tricyclic antidepressants in depressed patients with cardiac conduction disease. Arch. Gen. Psychiatry 44 (3), 273–275. https://doi.org/10.1001/archpsyc.1987.01800150093011.

Roose, S.P., Glassman, A.H., Attia, E., Woodring, S., 1994. Comparative efficacy of selective serotonin reuptake inhibitors and tricyclics in the treatment of melancholia. Am. J. Psychiatr. 151 (12), 1735–1739. https://doi.org/10.1176/ajp.151.12.1735.

Rosenbaum, J.F., Zajecka, J., 1997. Clinical management of antidepressant discontinuation. J. Clin. Psychiatry 58 (7), 37–40.

Roth, T., Rogowski, R., Hull, S., Schwartz, H., Koshorek, G., Corser, B., Seiden, D., Lankford, A., 2007. Efficacy and safety of doxepin 1 mg, 3 mg, and 6 mg in adults with primary insomnia. Sleep 30 (11), 1555–1561. https://doi.org/10.1093/sleep/30.11.1555.

Rothschild, A.J., 2000. Sexual side effects of antidepressants. J. Clin. Psychiatry 61 (11), 28–36.

Rudorfer, M.V., Potter, W.Z., 1999. Metabolism of tricyclic antidepressants. Cell. Mol. Neurobiol. 19 (3), 373–409. https://doi.org/10.1023/A:1006949816036.

Ruepert, L., Quartero, A.O., de Wit, N.J., van der Heijden, G.J., Rubin, G., Muris, J.W., 2011. Bulking agents, antispasmodics and antidepressants for the treatment of irritable bowel syndrome. Cochrane Database Syst. Rev. 2011 (8). https://doi.org/10.1002/14651858.CD003460.pub3.

Ruhé, H.G., Huyser, J., Swinkels, J.A., Schene, A.H., 2006. Switching antidepressants after a first selective serotonin reuptake inhibitor in major depressive disorder: a systematic review. J. Clin. Psychiatry 67 (12), 1836–1855. https://doi.org/10.4088/JCP.v67n1203.

Rush, A.J., Weissenburger, J.E., 1994. Melancholic symptom features and DSM-IV. Am. J. Psychiatr. 151 (4), 489–498. https://doi.org/10.1176/ajp.151.4.489.

Rush, A.J., Wisniewski, S.R., Warden, D., Luther, J.F., Davis, L.L., Fava, M., Nierenberg, A.A., Trivedi, M.H., 2008. Selecting among second-step antidepressant medication monotherapies: predictive value of clinical, demographic, or first-step treatment features. Arch. Gen. Psychiatry 65 (8), 870–881. https://doi.org/10.1001/archpsyc.65.8.870.

Sachs, G.S., Nierenberg, A.A., Calabrese, J.R., Marangell, L.B., Wisniewski, S.R., Gyulai, L., Friedman, E.S., Bowden, C.L., Fossey, M.D., Ostacher, M.J., Ketter, T.A., Patel, J., Hauser, P., Rapport, D., Martinez, J.M., Allen, M.H., Miklowitz, D.J., Otto, M.W., Dennehy, E.B., Thase, M.E., 2007. Effectiveness of adjunctive antidepressant treatment for bipolar depression. N. Engl. J. Med. 356 (17), 1711–1722. https://doi.org/10.1056/NEJMoa064135.

Sackeim, H.A., Haskett, R.F., Mulsant, B.H., Thase, M.E., Mann, J.J., Pettinati, H.M., Greenberg, R.M., Crowe, R.R., Cooper, T.B., Prudic, J., 2001. Continuation pharmacotherapy in the prevention of relapse following electroconvulsive therapy: a randomized controlled trial. J. Am. Med. Assoc. 285 (10), 1299–1307. https://doi.org/10.1001/jama.285.10.1299.

Sala, M., Lazzaretti, M., Vidovich, G., Caverzasi, E., Barale, F., d'Allio, G., Brambilla, P., 2009. Electrophysiological changes of cardiac function during antidepressant treatment. Ther. Adv. Cardiovasc. Dis. 3 (1), 29–43. https://doi.org/10.1177/1753944708096282.

Salvi, V., Mencacci, C., Barone-Adesi, F., 2016. H1-histamine receptor affinity predicts weight gain with antidepressants. Eur. Neuropsychopharmacol. 26 (10), 1673–1677. https://doi.org/10.1016/j.euroneuro.2016.08.012.

Schatzberg, A.F., 1999. Antidepressant effectiveness in severe depression and melancholia. J. Clin. Psychiatry 60 (4), 14–22.

Schjott, J., Schjott, P., Assmus., 2019. Analysis of consensus among drug interaction databases with regard to combinations of psychotropics. Basic Clin. Pharmacol. Toxicol. 126, 126–132.

Schmauss, M., Kapfhammer, H.P., Meyr, P., Hoff, P., 1988. Combined MAO-inhibitor and tri- (tetra) cyclic antidepressant treatment in therapy resistant depression. Prog. Neuropsychopharmacol. Biol. Psychiatry 12 (4), 523–532. https://doi.org/10.1016/0278-5846(88)90111-X.

Schmitt, R., Kratz Gazalle, F.K., Silva De Lima, M.S., Cunha, A., Souza, J., Kapczinski, F., 2005. The efficacy of antidepressants for generalized anxiety disorder: a systematic review and meta-analysis. Rev. Bras. Psiquiatr. 27 (1), 18–24. https://doi.org/10.1590/S1516-44462005000100007.

Schneider, J., Patterson, M., Jimenez, X.F., 2019. Beyond depression: other uses for tricyclic antidepressants. Cleve. Clin. J. Med. 86 (12), 807–814. https://doi.org/10.3949/ccjm.86a.19005.

Serretti, A., Mandelli, L., 2010. Antidepressants and body weight: a comprehensive review and meta-analysis. J. Clin. Psychiatry 71 (10), 1259–1272. https://doi.org/10.4088/JCP.09r05346blu.

Souery, D., Serretti, A., Calati, R., Oswald, P., Massat, I., Konstantinidis, A., Linotte, S., Kasper, S., Montgomery, S., Zohar, J., Mendlewicz, J., 2011. Citalopram versus desipramine in treatment resistant depression: effect of continuation or switching strategies. A randomized open study. World J. Biol. Psychiatry 12 (5), 364–375. https://doi.org/10.3109/15622975.2011.590225.

Spina, E., Perucca, E., 1994. Newer and older antidepressants: a comparative review of drug interactions. CNS Drugs 2 (6), 479–497. https://doi.org/10.2165/00023210-199402060-00008.

Stafford, R.S., MacDonald, E.A., Finkelstein, S.N., 2001. National patterns of medication treatment for depression, 1987 to 2001. Prim. Care Companion J. Clin. Psychiatry 3 (6), 232–235. https://doi.org/10.4088/PCC.v03n0611.

Stanton, T., Bolden-Watson, C., Cusack, B., Richelson, E., 1993. Antagonism of the five cloned human muscarinic cholinergic receptors expressed in CHO-K1 cells by antidepressants and antihistaminics. Biochem. Pharmacol. 45 (11), 2352–2354. https://doi.org/10.1016/0006-2952(93)90211-E.

Strandgården, K., Gunnarsson, P.O., 1994. Metabolism of lofepramine and imipramine in liver microsomes from rat and man. Xenobiotica 24 (8), 703–711. https://doi.org/10.3109/00498259409043271.

Sultana, J., Spina, E., Trifirò, G., 2015. Antidepressant use in the elderly: the role of pharmacodynamics and pharmacokinetics in drug safety. Expert Opin. Drug Metab. Toxicol. 11 (6), 883–892. https://doi.org/10.1517/17425255.2015.1021684.

Szewczuk-Bogusławska, M., Kiejna, A., Beszłej, J.A., Orzechowska-Juzwenko, K., Milejski, P., 2004. Doxepin inhibits CYP2D6 activity in vivo. Pol. J. Pharmacol. 56 (4), 491–494.

Task Force on the Use of Laboratory Tests in Psychiatry, 1985. Tricyclic antidepressants—blood level measurements and clinical outcome: an APA task force report. Am. J. Psychiatr. 142, 155–162.

Tatsumi, M., Groshan, K., Blakely, R.D., Richelson, E., 1997. Pharmacological profile of antidepressants and related compounds at human monoamine transporters. Eur. J. Pharmacol. 340 (2–3), 249–258. https://doi.org/10.1016/S0014-2999(97)01393-9.

Taylor, D., 1995. Selective serotonin reuptake inhibitors and tricyclic antidepressants in combination. Interactions and therapeutic uses. Br. J. Psychiatry 167, 575–580. https://doi.org/10.1192/bjp.167.5.575.

Teschemacher, A.G., Seward, E.P., Hancox, J.C., Witchel, H.J., 1999. Inhibition of the current of heterologously expressed HERG potassium channels by imipramine and amitriptyline. Br. J. Pharmacol. 128 (2), 479–485. https://doi.org/10.1038/sj.bjp.0702800.

Thanacoody, H.K.R., Thomas, S.H.L., 2005. Tricyclic antidepressant poisoning: cardiovascular toxicity. Toxicol. Rev. 24 (3), 205–214. https://doi.org/10.2165/00139709-200524030-00013.

Thase, M.E., Trivedi, M.H., Rush, A.J., 1995. MAOIs in the contemporary treatment of depression. Neuropsychopharmacology 12 (3), 185–219. https://doi.org/10.1016/0893-133X(94)00058-8.

Thase, M.E., John Rush, A., Howland, R.H., Kornstein, S.G., Kocsis, J.H., Gelenberg, A.J., Schatzberg, A.F., Koran, L.M., Keller, M.B., Russell, J.M., Hirschfeld, R.M.A., LaVange, L.M., Klein, D.N., Fawcett, J., Harrison, W., 2002. Double-blind switch study of imipramine or sertraline treatment of antidepressant-resistant chronic depression. Arch. Gen. Psychiatry 59 (3), 233–239. https://doi.org/10.1001/archpsyc.59.3.233.

Thayssen, P., Bjerre, M., Kragh-Sorensen, P., Moller, M., Petersen, O., Kristensen, C., Gram, L., 1981. Cardiovascular effects of imipramine and nortriptyline in elderly patients. Psychopharmacology, 360–364. https://doi.org/10.1007/bf00432748.

Thomas, S.J., Shin, M., McInnis, M.G., Bostwick, J.R., 2015. Combination therapy with monoamine oxidase inhibitors and other antidepressants or stimulants: strategies for the management of treatment-resistant depression. Pharmacotherapy 35 (4), 433–449. https://doi.org/10.1002/phar.1576.

Trindade, E., Menon, D., Topfer, L.A., Coloma, C., 1998. Adverse effects associated with selective serotonin reuptake inhibitors and tricyclic antidepressants: a meta-analysis. CMAJ 159 (10), 1245–1252.

UK ECT Review Group, 2003. Efficacy and safety of electroconvulsive therapy in depressive disorders: a systematic review and meta-analysis. Lancet 361, 799–808.

Undurraga, J., Baldessarini, R.J., 2017. Direct comparison of tricyclic and serotonin-reuptake inhibitor antidepressants in randomized head-to-head trials in acute major depression: systematic review and meta-analysis. J. Psychopharmacol. 31 (9), 1184–1189. https://doi.org/10.1177/0269881117711709.

Undurraga, J., Vázquez, G.H., Tondo, L., Baldessarini, R.J., 2020. Antidepressant responses in direct comparisons of melancholic and non-melancholic depression. J. Psychopharmacol. 34 (12), 1335–1341. https://doi.org/10.1177/0269881120953983.

Van Spanning, H., Van Zwieten, P., 1973. The interference of tricyclic antidepressants with the central hypotensive effect of clonidine. Eur. J. Pharmacol. 24, 402.

Verhamme, K.M.C., Sturkenboom, M.C.J.M., Stricker, B.H.C., Bosch, R., 2008. Drug-induced urinary retention: incidence, management and prevention. Drug Saf. 31 (5), 373–388. https://doi.org/10.2165/00002018-200831050-00002.

Vohra, J., Burrows, G., Hunt, D., Sloman, G., 1975. The effect of toxic and therapeutic doses of tricyclic antidepressant drugs on intracardiac conduction. Eur. J. Cardiol. 3 (3), 219–227.

Voican, C.S., Corruble, E., Naveau, S., Perlemuter, G., 2014. Antidepressant-induced liver injury: a review for clinicians. Am. J. Psychiatr. 171 (4), 404–415. https://doi.org/10.1176/appi.ajp.2013.13050709.

Walsh, J.K., 2004. Drugs used to treat insomnia in 2002: regulatory-based rather than evidence-based medicine. Sleep 27 (8), 1441–1442. https://doi.org/10.1093/sleep/27.8.1441.

Waring, S., 2012. Clinical use of antidepressant therapy and associated cardiovascular risk. Drug Healthc. Patient Saf. 4 (1), 93–101. https://doi.org/10.2147/DHPS.S28804.

Warner, C.H., Bobo, W., Warner, C., Reid, S., Rachal, J., 2006. Antidepressant discontinuation syndrome. Am. Fam. Physician 74 (3), 449–456. http://www.aafp.org/afp/20060801/449.pdf.

Wasielewski, P.G., Burns, J.M., Koller, W.C., 1998. Pharmacologic treatment of tremor. Mov. Disord. 13 (3), 90–100. Lippincott Williams and Wilkins https://doi.org/10.1002/mds.870131316.

Wells, B.G., Gelenberg, A.J., 1981. Chemistry, pharmacology, pharmacokinetics, adverse effects, and efficacy of the antidepressant maprotiline hydrochloride. Pharmacotherapy 1 (2), 121–138. https://doi.org/10.1002/j.1875-9114.1981.tb03559.x.

Werneke, U., Northey, S., Bhugra, D., 2006. Antidepressants and sexual dysfunction. Acta Psychiatr. Scand. 114 (6), 384–397. https://doi.org/10.1111/j.1600-0447.2006.00890.x.

Wharton, R.N., Perel, J.M., Dayton, P.G., Malitz, S., 1971. A potential clinical use for methylphenidate with tricyclic antidepressants. Am. J. Psychiatry 127 (12), 1619–1625. https://doi.org/10.1176/ajp.127.12.1619.

White, K., Razani, J., Simpson, G., 1982. Combined MAOI-tricyclic antidepressant treatment: a controlled trial. Psychopharmacol. Bull. 18 (4), 180–181.

Woolf, A.D., Erdman, A.R., Nelson, L.S., Caravati, E.M., Cobaugh, D.J., Booze, L.L., Wax, P.M., Manoguerra, A.S., Scharman, E.J., Olson, K.R., Chyka, P.A., Christianson, G., Troutman, W.G., 2007. Tricyclic antidepressant poisoning: an evidence-based consensus guideline for out-of-hospital management. Clin. Toxicol. 45 (3), 203–233. https://doi.org/10.1080/15563650701226192.

Zetin, M., Hoepner, C.T., Bjornson, L., 2006. Rational antidepressant selection: applying evidence-based medicine to complex real-world patients. Psychopharmacol. Bull. 39 (1), 38–104.

Zhou, X., Keitner, G.I., Qin, B., Ravindran, A.V., Bauer, M., Del Giovane, C., Zhao, J., Liu, Y., Fang, Y., Zhang, Y., Xie, P., 2015. Atypical antipsychotic augmentation for treatment-resistant depression: a systematic review and network meta-analysis. Int. J. Neuropsychopharmacol. 18 (11). https://doi.org/10.1093/ijnp/pyv060.

Zimmerman, M., Spitzer, R.L., 1989. Melancholia: from DSM-III to DSM-III-R. Am. J. Psychiatr. 146 (1), 20–28. https://doi.org/10.1176/ajp.146.1.20.

Chapter 9

Monoamine oxidase inhibitors for treatment-resistant depression

William V. Bobo and Elliott Richelson
Department of Psychiatry and Psychology, Mayo Clinic, Jacksonville, FL, United States

Introduction

Oral irreversible inhibitors of the intracellular enzyme, monoamine oxidase, have been in clinical use for over 60 years. As a group, they constitute the oldest class of antidepressants. The monoamine oxidase inhibitors (MAOIs) were developed decades ago out of the search for novel antituberculous drugs. One such drug, iproniazid, was serendipitously discovered to have mood-elevating properties in patients with tuberculosis (López-Muñoz and Alamo, 2009). These observations, first described in the 1950s, were confirmed in subsequent trials (López-Muñoz et al., 2007). From there, a series of pivotal studies showed that iproniazid was a potent inhibitor of monoamine oxidase; that inhibition of monoamine oxidase increased the levels of serotonin, norepinephrine, and dopamine in the brain; and that iproniazid was capable of reversing depressive states induced by reserpine, a drug that depletes catecholamines and serotonin from central and peripheral nerve terminals (López-Muñoz and Alamo, 2009; Pletscher, 1991).

Following the discovery of iproniazid's antidepressive effects, other irreversible MAOIs were developed, including the hydrazine compounds, phenelzine and isocarboxazid, and the nonhydrazine drug, tranylcypromine. Iproniazid soon fell out of favor as a pharmaceutic due to reports of hepatotoxicity and nephrotoxicity (Rosenblum et al., 1960). The remaining MAOIs became the predominant form of treatment for a variety of depressive states until they, too, were temporarily withdrawn from the market due to hepatotoxicity (in the case of the hydrazine compounds) and deaths due to hypertensive crises and intracranial hemorrhages (all MAOIs) when combined with sympathomimetic drugs or after the consumption of foods containing high concentrations of tyramine (Hendley and Snyder, 1968; Pletscher, 1991). As knowledge accumulated regarding the mechanisms of hypertensive reactions to selected drugs and tyramine-containing foodstuff in people taking MAOIs, they were subsequently reintroduced and have since remained important therapeutic options for patients suffering from a wide variety of mood and anxiety disorders.

Unlike other antidepressants that function primarily as inhibitors of monoamine transporters (mainly inhibition of the norepinephrine transporter, the serotonin transporter, or both), MAOIs are inhibitors of the enzyme that degrades the monoamine neurotransmitters—a distinctive and important mechanism that allows MAOIs to increase synaptic availability of all three neurotransmitters (serotonin, norepinephrine, and dopamine) implicated in the pathophysiology of depressive states and the hypothesized mechanisms of action of antidepressants (Stahl and Felker, 2008). The MAOI class of antidepressants is diverse (Table 9.1) and includes three orally administered irreversible inhibitors of monoamine oxidase (phenelzine, isocarboxazid, and tranylcypromine), a transdermally administered irreversible MAOI (transdermal selegiline, an older drug that is marketed in pill form to treat Parkinson's disease), and an orally administered reversible inhibitor of monoamine oxidase (moclobemide, which is not available for clinical use in the United States).

Despite being highly effective, the clinical use of MAOIs has been limited by waning clinical experience and a poor understanding of dietary restrictions and drug interactions—to the point that they are now rarely prescribed by most psychiatrists (Balon et al., 1999; Gillman, 2017; Shulman et al., 2009). Because of infrequent use, drug manufacturers have reduced production and thus have limited the availability of some MAOIs, phenelzine in particular, throughout the world (Gillman, 2020). MAOIs have been relegated to third-line or lower status in nearly all practice guidelines for the pharmacological treatment of major depression (American Psychiatric Association, 2010; Bauer et al., 2013; Cleare et al., 2015; Kennedy et al., 2016; Malhi et al., 2015; Spijker and Nolen, 2010). Nevertheless, there has been renewed interest in the clinical use of MAOIs given the high rates of treatment resistance in people with major depression, the clinical effectiveness of selected irreversible MAOIs for treatment-resistant depressive states (reviewed below), and changes in food preparation techniques and

Managing Treatment-Resistant Depression. https://doi.org/10.1016/B978-0-12-824067-0.00009-8

TABLE 9.1 Drugs classified as monoamine oxidase inhibitors (MAOIs).

Drug name	Brand names (United States)	Reversibility[a]	MAO isoenzyme selectivity[b]	
			MAO-A	MAO-B
Phenelzine	Nardil	Irreversible	Yes	Yes
Tranylcypromine	Parnate	Irreversible	Yes	Yes
Isocarboxazid	Marplan	Irreversible	Yes	Yes
Selegiline, transdermal	EMSAM[c]	Irreversible[d]	Yes, but only at higher doses	Yes
Moclobemide	...[e]	Reversible	Yes	No
Brofaromine	...[f]	Reversible	Yes	No
Clorgyline	...[f]	Reversible	Yes	No

[a] *Irreversible inhibition of MAO indicates that the drug binds to the enzyme permanently, rendering it inactive for its entire lifespan. Reversible inhibition of MAO indicates that the drug can be displaced from its binding site by tyramine.*
[b] *"Yes" in both columns indicates that the drug is a nonselective inhibitor of monoamine oxidase-A (MAO-A) and monoamine oxidase-B (MAO-B). "Yes" in only one box indicates that the drug is a selective inhibitor of that MAO isoenzyme.*
[c] *Selegiline is available in both oral and transdermal forms. EMSAM is the transdermal form of selegiline that has regulatory approval as an antidepressant drug.*
[d] *Selegiline selectively inhibits MAO-B at low doses (30–60 mg/day of oral selegiline, or ≤ 6 mg/day of transdermal selegiline [EMSAM]). At higher doses, selegiline inhibits both MAO-A and MAO-B.*
[e] *Not marketed in the United States.*
[f] *Developed but not brought to market.*

hygiene regulations that have greatly increased the safety of MAOIs in people who consume modern diets (Gillman, 2018). This chapter reviews the clinical pharmacology and efficacy of MAOIs for treatment-resistant depression and provides practical advice on the safe and effective use of this often-misunderstood, yet highly effective class of antidepressant medications.

Pharmacological properties

Mechanism of action

MAOIs inhibit the enzyme, monoamine oxidase, leading to a rapid increase in the synaptic transmission of serotonin, norepinephrine, and dopamine (Stahl and Felker, 2008). This pharmacological effect occurs acutely (Millan, 2006), and does not correlate in time with the 3–6-week delay in the onset of antidepressive effects after starting MAOI treatment. The chronic administration of MAOIs, on the other hand, may lead to secondary downregulation of monoamine receptors, an effect that appears to parallel the delayed onset of antidepressive effects with MAOIs (Baker et al., 1992; Piñeyro and Blier, 1999). Whether these chronic effects of MAOI administration better explain the therapeutic action of these drugs for depression is unclear at this time.

Specificity and irreversibility of monoamine oxidase inhibition

The more clinically relevant aspects of MAOI pharmacology involve specificity for various monoamine oxidase isoenzymes and irreversibility of enzyme inhibition. There are two monoamine oxidase isoenzymes: Monoamine oxidase-A (MAO-A) and monoamine oxidase-B (MAO-B). MAO-A is found in the brain, gut, liver, pancreas, heart, placenta, and skin; MAO-B is found in the brain, platelets, and lymphocytes (Billett, 2004; Shih et al., 1999). Levels of serotonin, dopamine, and norepinephrine are increased by inhibition of MAOI-A, whereas inhibition of MAO-B increases levels of dopamine and phenylethylamine (Billett, 2004; Fowler et al., 2005; Shih et al., 1999), as summarized in Table 9.2.

Inhibition of MAO-A is necessary for the antidepressive effect of MAOIs due to its specificity for serotonin and norepinephrine (Ulrich et al., 2017). Even though both MAO-A and MAO-B metabolize tyramine, the irreversible inhibition of MAO-A in the small intestine is believed to be the key mechanism by which potentially severe hypertensive reactions due to tyramine overaccumulation occur (Youdim and Weinstock, 2004). Ingested tyramine is "detoxified" while passing through the gut wall from its metabolism by MAO-A and never reaches the bloodstream. When MAO-A is inhibited, tyramine does reach the bloodstream, where it is thought to push out NE from peripheral neurons that regulate blood vessel constriction, thereby elevating blood pressure.

TABLE 9.2 Selected substrates for monoamine oxidase isoforms (MAO-A, MAO-B).

	MAO-A	MAO-B
Serotonin	Yes	No
Norepinephrine	Yes	No
Dopamine	Yes	Yes
Tyramine	Yes	No
Melatonin	Yes	Yes
Phenylethylamine	No	Yes

The MAOIs can be subgrouped based on their selectivity for specific monoamine oxidase isoenzymes and whether they display irreversible or reversible binding to monoamine oxidase. Phenelzine, isocarboxazid, and tranylcypromine irreversibly inhibit MAO-A and MAO-B, meaning that these drugs render MAO-A and MAO-B permanently inactive (Rudorfer, 1992). Following irreversible inhibition, normal monoamine oxidase activity can be restored only after new enzyme is synthesized, a process that requires approximately 14–28 days (Lotufo-Neto et al., 1999). The relative selectivity of selegiline for monoamine oxidase isoenzymes is dose-dependent. At low doses, selegiline is highly selective for MAO-B, where it acts as an irreversible inhibitor (Magyar, 2011). At higher doses (i.e., > 6 mg/24 h of transdermal selegiline), selectivity for MAO-B is lost and both MAO-A and MAO-B are irreversibly inhibited (van den Berg, 2012). Moclobemide is a selective reversible inhibitor of MAO-A (RIMA). The "reversibility" of MAO inhibition with moclobemide refers to the fact that it can be displaced by high concentrations of dietary tyramine (Lavian et al., 1993). Therefore, MAO-A activity is recovered more quickly and the pressor effect of concomitantly ingested tyramine appears to be less pronounced with the use of moclobemide than with the use of irreversible MAOIs (da Prada et al., 1988). Combining moclobemide with serotonergic agents can, however, lead to severe serotonin toxicity (Isbister et al., 2003).

Other nonselective irreversible MAOIs (nialamide), reversible (brofaromine, toloxatone, comoxatone, befloxatone) and irreversible (clorgyline) MAO-A selective drugs, and irreversible MAO-B selective drugs (pargyline) have been developed. However, these drugs are unavailable, untested, or not routinely available for clinical use in depressed patients and will not be discussed further.

Additional pharmacodynamic properties

The irreversible nonselective MAOIs, phenelzine and tranylcypromine, have additional properties that may be clinically relevant. In addition to inhibiting monoamine oxidase, phenelzine is thought to potentiate gamma-aminobutyric acid (GABA)-ergic neurotransmission by inhibiting GABA transaminases (Todd and Baker, 2008), the principle group of enzymes responsible for the degradation of GABA. This mechanism is believed to contribute to phenelzine's sedating and anxiolytic effects (Paslawski et al., 1996). Tranylcypromine, on the other hand, has a chemical structure that is similar to amphetamine and has been shown to produce a stimulant-like effect (Gentil et al., 1978). There is debate in the literature as to whether tranylcypromine is in fact metabolized into amphetamine, with the consensus being that it is not (Baker et al., 1999).

Pharmacokinetics

Phenelzine, isocarboxazid, and tranylcypromine

There is surprisingly little information about the pharmacokinetic properties of most orally administered irreversible MAOIs. All are readily absorbed from the gastrointestinal tract and generally reach peak concentrations within a few hours of administration (Knud and Lene, 2016; Mallinger and Smith, 1991; Ulrich et al., 2017). In general, oral MAOIs are extensively metabolized and have relatively short elimination half-lives—generally in the range of 1.5–4 h. Phenelzine and isocarboxazid are acetylated into largely inactive metabolites (Gillman, 2011; Knud and Lene, 2016; Ulrich et al., 2017). However, drug metabolizing characteristics have no impact on the duration of the antidepressive effects of MAOIs. Instead, the duration of antidepressive effects of MAOIs is primarily dependent upon the inhibition of monoamine oxidase and the rate of monoamine oxidase synthesis. There are no known sex-specific effects on drug absorption, distribution, metabolism, or elimination.

Transdermal selegiline

The form of selegiline that is approved for use in people with depression is delivered transdermally. About 25%–30% of the selegiline content of the patch is delivered systemically over 24 h, reaching steady-state concentrations within 5 days (van den Berg, 2012). The release of less than one-third of the active drug contained in a patch over 24 h raises the possibility that a single patch could be kept on for up to 3 days—a clinical hypothesis that, to our knowledge, has not been empirically investigated. Because additional selegiline may be delivered beyond 24 h when using the transdermal form, it is important to warn patients to avoid wearing more than one patch at any given time. Once absorbed transdermally, selegiline is distributed extensively without accumulating in the skin (van den Berg, 2012). Selegiline undergoes hepatic metabolism by several cytochrome CYP450 isoenzymes, including CYP2B6, CYP2C9, CYP2A4/5, and others (Azzaro et al., 2007a,b). The mean elimination half-life of transdermal selegiline ranges from 15 to 25 h (Azzaro et al., 2007a,b). As is the case for oral MAOIs, the degradation characteristics of transdermal selegiline do not influence the duration of its antidepressive effects, and sex does not appear to influence its pharmacokinetic properties.

Moclobemide

Moclobemide is rapidly absorbed following oral dosing and reaches peak plasma concentrations within 1 h (Nair et al., 1993). Moclobemide is extensively metabolized in the liver by CYP2C19 and appears to inhibit CYP2C19, CYP2D6, and CYP1A2 isoenzymes (Gram et al., 1995). The mean elimination half-life of moclobemide is 1.8 h for CYP2C19 extensive (normal) metabolizers and is 4 h for CYP2C19 poor metabolizers (Isbister et al., 2003). The pharmacokinetic properties of moclobemide do not appear to vary according to sex.

Indications

MAOIs are effective for a wide range of psychiatric disorders (Robinson et al., 1973). They are most commonly used for treating major depressive disorder (MDD), particularly MDD with atypical features and treatment-resistant cases that fail to respond to first-line and tricyclic antidepressants. They are also effective treatments for persistent depressive disorder (formerly referred to as dysthymic disorder) and mixed anxiety and depressive states (Paykel et al., 1982; Ravaris et al., 1976), as well as social anxiety disorder (Ravindran and Stein, 2010), panic disorder (Jefferson, 1997; Norman and Burrows, 1989; Ravindran and Stein, 2010), posttraumatic stress disorder (Albucher and Liberzon, 2002), and selected symptoms of bulimia nervosa (Flament et al., 2012). Below, we discuss the clinical evidence supporting the use of MAOIs for treating MDD and other unipolar depressive syndromes, with an emphasis on treatment-resistant cases. The use of MAOIs for treatment-resistant bipolar depression is discussed in a separate chapter.

Depressive disorders, broadly defined

Numerous studies support the effectiveness of MAOIs for a variety of unipolar depressive syndromes, including MDD. Rather than reviewing each of the individual studies separately, we summarize below the main results of four key metaanalyses of MAOIs for depressive disorders, defined broadly.

Phenelzine, isocarboxazid, and tranylcypromine were all shown to be about equally effective in a metaanalysis of heterogeneous randomized trials that enrolled patients with a range of depressive disorders (Thase et al., 1995). Response rates were 57.9% with phenelzine (14 outpatient studies and 7 inpatient studies, totaling 981 treated subjects), 60.1% with isocarboxazid (five outpatient studies and four inpatient studies, totaling 434 treated subjects), and 52.6% for tranylcypromine (five outpatient studies and four inpatient studies, totaling 293 treated subjects). All of the MAOIs were more effective than placebo in outpatient trials. In subgroup analyses, the MAOIs as a whole were found to be superior to tricyclic antidepressants for atypical depression, and were found to be effective for patients who were poorly responsive to tricyclic antidepressants. An updated metaanalysis focused on the clinical effectiveness of tranylcypromine for broadly defined depressive states (Ricken et al., 2017). Treatment response was defined as achieving > 50% improvement in total scores on depression rating scales or a predetermined cut-off score on measures of clinical global state. Tranylcypromine was superior to placebo for achieving a positive treatment response (pooled logOR 0.51, 95% CI 0.03 to 0.99, four studies) and was equal in efficacy to other antidepressants—mainly, tricyclic antidepressants (pooled logOR 0.21, 95% CI – 0.13 to 0.54, 10 studies).

The efficacy of transdermal selegiline for MDD is supported by a metaanalysis of five short-term randomized trials, totaling 1289 subjects (Robinson et al., 2007). Four studies used a fixed dose of 6 mg/24 h, and one trial used flexible dosing in the 6 mg/24 h to 12 mg/24 h range. After adjusting for baseline depression severity, transdermal selegiline was found to be superior to placebo for improving core depressive symptoms (depressed mood, guilt, impairment in work and activities, psychomotor retardation, psychic anxiety, and general somatic symptoms), as well as reverse vegetative symptoms (hypersomnia and overeating). In two pivotal randomized trials, response rates (defined as ≥ 50% reduction in depression scale scores) were significantly higher with transdermal selegiline than placebo at 6 weeks (37.5% vs 22.7%) and at 8 weeks (33.1% vs 20.8%) (Amsterdam, 2003; Bodkin and Amsterdam, 2002).

Finally, in a metaanalysis of over 700 depressed patients, moclobemide was found to be superior to placebo and comparable in efficacy to tricyclic antidepressants (Angst and Stabl, 1992). The results of pooled analyses suggested that moclobemide may be more efficacious for stress-induced depression than for depression without an apparent external cause (referred to as "endogenous" depression). Additional evidence suggests that moclobemide may be as effective as selective serotonin reuptake inhibitor (SSRI) antidepressants for depressed outpatients (Williams et al., 1993). However, moclobemide was found to be less-effective than clomipramine, a tricyclic antidepressant with very high affinity for the human serotonin transporter (Tatsumi et al., 1997), in two large multicenter trials (Danish University Antidepressant Group, 1993; Larsen et al., 1991).

Depression with atypical features

An estimated 15%–30% of people with MDD have atypical features (Asnis et al., 1995; Quitkin, 2002; Thase, 2007). This subtype of depression is characterized by preserved reactivity of mood—that is, the ability for mood to improve in response to positive life events—and at least two of the following symptoms: weight gain or appetite stimulation, hypersomnia, severe fatigue causing the sensation of intense heaviness in the upper and lower extremities, and rejection sensitivity (Davidson, 2007; Gaynes, 2009). Rejection sensitivity in this case is not confined to the depressive episode. Instead, it is a stable and enduring character trait marked by feelings of unease around others and proneness to feelings of inadequacy or being negatively judged (Parker, 2007).

Substantial evidence supports the effectiveness of irreversible oral MAOIs for depression with atypical features, as compared with tricyclic antidepressants, suggesting a preferential response profile (Liebowitz et al., 1984, 1988; McGrath et al., 1993; Stewart et al., 1989). Subgroup analyses from two of the metaanalyses reviewed earlier showed superior antidepressive responses to irreversible oral MAOIs compared with tricyclic antidepressants for participants with atypical depression (Ricken et al., 2017; Thase et al., 1995). In a separate metaanalysis of eight randomized trials (six phenelzine studies, two moclobemide studies) that focused specifically on the clinical effects of MAOIs for depression with atypical features, MAOIs were found to be superior to both placebo (ES 0.45, 95% CI 0.35–0.60) and tricyclic antidepressants (ES 0.27, 95% CI 0.16–0.42) (Henkel et al., 2006). These findings were similar to those of an additional pooled analysis of data from clinical trials showing superiority of MAOIs over tricyclic antidepressants, but not SSRIs, for depression with atypical features (Cipriani et al., 2007).

At present, the neurobiological basis for the apparent efficacy advantage of MAOIs over tricyclic antidepressants in people with atypical depression is unknown. An interesting positron emission tomography (PET) study of 42 people with MDD showed that so-called reversed neurovegetative symptoms (which are features of atypical depression) were associated with elevated MAO-A *VT*, a proxy measure for MAO-A density (Chiuccariello et al., 2014). Such findings may provide some insights into a putative mechanism underlying the preferential response to MAOIs over tricyclic antidepressants in people with MDD with atypical features, although it does not explain the apparent equivalence in efficacy between MAOIs and SSRIs.

Randomized trials of transdermal selegiline for depression with atypical features are sparse. In a post hoc analysis of data from five acute-phase trials of transdermal selegiline for MDD (Pae et al., 2014), the presence of atypical features was defined on the basis of specific individual item scores on the 28-item version of the Hamilton Depression Rating Scale (Hamilton, 1967). A total of 352 subjects satisfied the criteria for atypical depression. Transdermal selegiline was nearly equally efficacious for subjects with atypical and nonatypical depression. Randomized trials assessing the efficacy of transdermal selegiline for people with clinically defined atypical depression are still needed.

Individual studies of moclobemide for depression with atypical features have not yet pointed to a clear efficacy advantage compared with other antidepressant classes. For example, in a 12-week randomized trial comparing the antidepressive effects of moclobemide (300–450 mg/day) and sertraline (50–100 mg/day) in 197 outpatients with atypical depression, both groups improved from baseline but no significant differences in response rates (67.5% with moclobemide vs 77.5% with sertraline) were observed (Søgaard et al., 1999). In a 6-week randomized trial of 167 depressed outpatients (17 of whom had atypical features), moclobemide was no more effective than clomipramine (Larsen et al., 1991). There were no significant between-group differences in the distribution of people with atypical depression.

Persistent depressive disorder and chronic depression

There is evidence supporting the use of oral irreversible MAOIs for the treatment of persistent depressive disorder and other forms of chronic depression. For instance, a metaanalysis of 15 placebo-controlled trials showed similar efficacy for

MAOIs, tricyclic antidepressants, SSRIs, and other antidepressants for people with persistent depressive disorder (Lima and Moncrieff, 2000). The number needed to treat (NNT) for positive treatment response (vs placebo) was 2.9 for MAOIs, as compared with 4.3 for tricyclic antidepressants and 4.7 for SSRIs, suggesting a large clinical effect size for MAOIs and a possible small efficacy advantage when compared with the remaining classes of antidepressants.

Individual studies of phenelzine for persistent depressive disorder showed greater effectiveness than imipramine and placebo (Harrison et al., 1986; Stewart et al., 1993; Vallejo et al., 1987); however, imipramine was superior to phenelzine in subgroup analyses focused on patients with melancholic rather than atypical depressive features (Vallejo et al., 1987). In a pooled analysis of data from clinical studies of patients with chronic depression or comorbid MDD and dysthymic disorder, phenelzine was found to be significantly more effective than imipramine and placebo (Stewart et al., 1989). For the small number of subjects with persistent depressive disorder, there were no significant differences in efficacy between phenelzine and imipramine, although only imipramine was superior to placebo.

Regarding the long-term treatment of chronic depressive states, the efficacy of phenelzine is supported by the results of a randomized maintenance trial of 60 people with chronic atypical depression (Stewart et al., 1997). Enrollees had previously responded to either phenelzine or imipramine, and were randomly assigned to continued treatment with the same medication or switch to placebo (Stewart et al., 1993). During the maintenance phase of the trial, recurrence rates were lower with phenelzine than imipramine (23% vs 41%), and only phenelzine resulted in significantly greater efficacy than placebo for preventing depressive recurrences.

There are fewer randomized trials of isocarboxazid for chronic depression or persistent depressive disorder. In a pooled analysis of data from a three-site randomized trial, the enrolled participants with "minor" depression (a term that encompassed chronic forms of depression including persistent depressive disorder) as well as MDD, treatment with isocarboxazid resulted in a numerically higher response rate than placebo (58% vs 44%) (Davidson et al., 1988). Drug-placebo differences favoring isocarboxazid were larger and statistically significant, however, for the subgroup of 95 patients with MDD.

We were not able to locate randomized trials of tranylcypromine or transdermal selegiline for persistent depressive disorder. However, there is evidence supporting the clinical effectiveness of moclobemide. In a large network metaanalysis of 45 randomized trials (> 5300 patients) that examined the clinical effectiveness of 28 drugs for persistent depressive disorder, moclobemide was found to be superior to placebo (clinical response, OR 6.98, 95% CI 3.68–13.74), with no between-group differences in premature discontinuation for any cause as a proxy for treatment acceptability (Kriston et al., 2014). One randomized trial focused specifically on patients with comorbid MDD and dysthymic disorder and showed a higher response rate with moclobemide than fluoxetine (71% vs 38%) after 6 weeks of treatment, although both drugs were comparable on other depression treatment outcomes (Duarte et al., 1996).

Clinical studies in people with treatment-resistant depression

There is evidence that some MAOIs are effective treatments for depressed patients that have failed to respond to prior therapeutic trials of other antidepressants, although empirical data from randomized trials of subjects with well-defined treatment-resistant depression are relatively sparse (McAllister-Williams et al., 2020). The available evidence supporting the use of MAOIs for treatment-resistant unipolar depression is summarized in Table 9.3.

Metaanalyses of MAOIs for treatment-resistant depression

Two reviews document the effectiveness of phenelzine and tranylcypromine for patients with depression and a history of poor response to tricyclic antidepressants. In the report by Thase et al. (1995) presented earlier, the authors conducted a review of controlled and uncontrolled studies of MAOIs for tricyclic-resistant depression, and concluded that an estimated 50% of such patients respond favorably to MAOIs. When only tricyclic-resistant patients with atypical depression were considered, the estimated positive response rate with MAOIs increased to 70%. These estimated response rates may be inflated, however, by the inclusion of uncontrolled studies and by uncertainty about the number of therapeutic trials of antidepressants prior to tricyclic antidepressant failure—information that would be important for determining whether included subjects were also resistant to other types of antidepressants.

The more recent metaanalysis by Ricken et al. (2017), also reviewed earlier, confirmed the efficacy of tranylcypromine in patients who failed to respond to tricyclic antidepressants and SSRIs. Within that subgroup of depressed patients, tranylcypromine was associated with significantly greater likelihood of positive treatment response than placebo (logOR 2.83, 95% CI 1.49–4.16, one study) and nonestablished antidepressants (pooled logOR 1.98, 95% CI 0.91–3.05, four studies). Tranylcypromine was as effective as treatment with other MAOIs or combination therapy with venlafaxine and mirtazapine (pooled logOR 0.37, 95% CI − 0.87 to 0.14, four studies).

TABLE 9.3 Summary of clinical trials of monoamine oxidase inhibitors (MAOIs) for treatment-resistant depression.

Study	Sample	Design	Comparator groups	TRD definition	Key findings
Georgotas et al. (1983)	30 Elderly adults (ages 60–84 years) with TRD	Open, uncontrolled trial (7 weeks)	Phenelzine 30–75 mg/day (mean 49.5 mg/day)	Failed to respond to TCA (37% also resistant to ECT)	65% responded to phenelzine (HAM-D $\leq$ 10 at study termination)
Kok et al. (2007)	29 Elderly hospitalized adults with TRD	Open randomized trial (6 weeks)	Phenelzine 30–60 mg/day (mean 46.0 mg/day) vs venlafaxine or TCA plus lithium	Failed to respond to TCA or venlafaxine monotherapy	Response rates ($\geq$ 50% decrease in 17-item HAM-D) higher with phenelzine (46.7%) than comparator (7.1%)
McGrath et al. (1993)	66 Adults with TRD (nonmelancholic, mood-reactive, DSM-III)	RCT (6 weeks) with secondary cross-over	Phenelzine (up to 90 g/day) vs imipramine (up to 300 mg/day)	Failed to respond to initial trial of phenelzine, imipramine, or placebo[a]	Higher response rates (much improved or very much improved according to CGI-I) with phenelzine (51.2%) than imipramine (32.0%)
McGrath et al. (1987)	26 Adults with TRD	Randomized cross-over trial (6 weeks)	Phenelzine 60–90 mg/day (mean 72.0 mg/day) vs imipramine 200–300 mg/day (mean 275.0 mg/day)	Failed to respond to phenelzine, imipramine, or placebo prior to cross-over	Higher response rate (much improved or very much improved according to CGI-I) with phenelzine (65.4%) than imipramine (28.6%)
Birkenhäger et al. (2004)	67 Hospitalized adults with TRD	RCT (5 weeks)	Phenelzine 20–100 mg/day vs tranylcypromine 20–100 mg/day	Failure to respond to TCA or fluvoxamine	Comparable response rates ($\geq$ 50% decrease in 17-item HAM-D) between phenelzine (47.4%) and tranylcypromine (43.6%)
McGrath et al. (2006)	109 Adults with TRD	RCT (12 weeks)	Tranylcypromine 10–60 mg/day (mean 36.9 mg/day) vs mirtazapine (mean 35.7 mg/day)/venlafaxine (mean 275.0 mg/day) combination therapy	Failure to remit (or tolerate) three prospective antidepressant treatment trials	Low response rates ($\geq$ 50% decrease in QIDS-SR) for both tranylcypromine (12.1%) and combination (23.5%)
Nolen et al. (1993)	39 Hospitalized adults with TRD (59% with melancholic depression)	RCT (4 weeks)	Tranylcypromine 20–100 mg/day (mean 81.2 mg/day) vs brofaromine 50–250 mg/day (mean 215.9 mg/day)	Failure to respond to nortriptyline or maprotiline	Lower response rate ($\geq$ 50% decrease in 17-item HAM-D) with tranylcypromine (29.4%) than brofaromine (45.5%), but similar HDSS response rates (52.9% vs 45.5%) and proportion rated as "very much improved" on CGI-I (58.9% vs 59.1%)

Continued

TABLE 9.3 Summary of clinical trials of monoamine oxidase inhibitors (MAOIs) for treatment-resistant depression—cont'd

Study	Sample	Design	Comparator groups	TRD definition	Key findings
Nolen et al. (1988)	47 Hospitalized patients with TRD	RCT (4 weeks) with secondary cross-over	Tranylcypromine 40–100 mg/day (mean 78.0 mg/day) vs L-5HTP up to 200 mg/day (mean 191.7 mg/day) in phase 1; nomifensine 50–250 mg/day (235.0 mg/day) in phase 2	Failure to respond to fluoxetine or oxaprotiline (phase 1); nonresponders in phase 1 crossed over into phase 2[b]	Higher response rate (≥ 50% decrease in 17-item HAM-D) with tranylcypromine (57.1%) than L-5HTP (0.0%) in phase 1; higher response rate with tranylcypromine (62.5%) than nomifensine (0.0%) in phase 2
Nolen et al. (1985)	26 Hospitalized patients with TRD	Randomized cross-over trial (4 weeks)	Tranylcypromine 20–100 mg/day (mean 82.0 mg/day) vs L-5HTP 20–100 mg/day (mean 182.0 mg/day)	Failure to respond to at least one antidepressant trial of adequate design, followed by failure to respond to fluvoxamine or oxaprotiline and sleep deprivation	Higher response rate (≥ 50% decrease in 17-item HAM-D) with tranylcypromine (57.7%) than L-5HTP (0.0%)
Thase et al. (1992)	40 Patients with TRD (67.5% with atypical features)	Open, uncontrolled trial (6 weeks) with two historical control groups	Tranylcypromine 20–60 mg/day (mean 38.5 mg/day) vs phenelzine 30–60 mg/day (mean 60.0 mg/day)	Failure to respond to 16 weeks of imipramine + interpersonal psychotherapy	Comparable response rates (≥ 50% decrease in 17-item HAM-D) with tranylcypromine (55.0%) and phenelzine (57.5%)
Volz et al. (1994)	90 Hospitalized patients with TRD (mixture of unipolar [82.2% with MDD] and bipolar patients [4.4%]), 10.0% with atypical features	RCT (6 weeks)	Tranylcypromine 20–30 mg/day (mean 26.2 mg/day) vs brofaromine 100–150 mg/day (mean 133.0 mg/day)	Failure to respond to two or more classes of non-MAOI antidepressants	Comparable response rates (≥ 50% decrease in 21-item HAM-D) between tranylcypromine (72.3%) and brofaromine (73.0%)

CGI-I, Clinical Global Impressions improvement subscale; *DSM-III*, according to criteria specified in the Diagnostic and Statistical Manual for Mental Disorders, Third edition; *HAM-D*, Hamilton Depression Rating Scale; *L-5HTP*, 5-hydroxytryptamine, a serotonin precursor; *MDD*, major depressive disorder; *TCA*, tricyclic antidepressant; *TRD*, treatment-resistant depression.

[a] *A total of 66 patients crossed over to the opposite active treatment following initial failure to respond to phenelzine or imipramine.*

[b] *A total of 18 nonresponders during phase 1 were eligible for phase 2 treatment. Eight of the ten patients who were eligible for tranylcypromine received treatment and five of the remaining patients who were eligible for nomifensine received treatment during phase 2.*

Individual studies of MAOIs for treatment-resistant depression

Phenelzine, isocarboxazid, and tranylcypromine

In individual studies of MAOIs for treatment-resistant depression (Table 9.3), rates of positive treatment response ranged from approximately 13% to 67%. Such wide variation in response rates is the likely result of high degrees of heterogeneity between trials, making it difficult to provide a broad interpretation of these results. For instance, while MAOIs have been tested in patients who have responded poorly to prior antidepressant trials, particularly tricyclic antidepressants, the definition of treatment-resistance varies substantially across the individual studies. For most studies, treatment resistance was defined as either failing to respond to tricyclic antidepressants (Georgotas et al., 1983; Nolen et al., 1993; Thase et al., 1992), at least one prior therapeutic trial involving tricyclic or nontricyclic antidepressants (Birkenhager et al., 2004; Kok et al., 2007; McGrath et al., 1993; Nolen et al., 1985, 1988), or two or more classes of antidepressants (McGrath et al., 2006; Volz et al., 1994). Two cross-over studies involved depressed patients who failed to respond to an MAOI or tricyclic antidepressant (McGrath et al., 1987, 1993). All studies were controlled using either cross-over or traditional parallel-group designs, with the exceptions of one open uncontrolled study (Georgotas et al., 1983) and one study that used two historical control groups (Thase et al., 1992). A very wide range of doses (10–100 mg/day) was investigated in flexible-dose studies of phenelzine and tranylcypromine. Definitions of treatment response also varied from study to study. Importantly, none of the studies in Table 9.3 focused on the effects of isocarboxazid, transdermal selegiline, or moclobemide in patients with treatment-resistant depression.

Methodological variations aside, a number of the existing studies of MAOIs for treatment-resistant depression included patients with a relatively small number of prior treatment failures. This fact may limit the translational impact of study results on modern practice where patients who are being considered for MAOI treatment will have failed to respond to multiple classes of antidepressants; and, very likely, multiple agents within each class. Therefore, rather than conducting an exhaustive review of each of the trials summarized in Table 9.3, the remaining discussion will focus on the few clinical studies of MAOIs for patients with well-characterized treatment-resistant depression who are in the latter stages of treatment resistance and may better-resemble depressed patients for whom treatment with MAOIs would warrant serious consideration in current clinical practice.

In the fourth phase of the Sequenced Treatment Alternatives to Relieve Depression (STAR*D) study, 109 patients with MDD were randomized to next-step treatment with either tranylcypromine or combination therapy with venlafaxine and mirtazapine (McGrath et al., 2006). Study participants previously failed to respond to three carefully conducted prospective antidepressant trials beginning with citalopram (phase 1); augmentation of citalopram or cognitive behavioral therapy (phase 2); and switching to bupropion, sertraline, venlafaxine, or CBT without citalopram (phase 3). The response rates for tranylcypromine and combination therapy with venlafaxine and mirtazapine were low (12.1% vs 23.5%). Remission rates were even lower (6.9% vs 13.7%). After restricting the analyses to the 21 patients with atypical depression, there were no significant between-group differences in rates of positive treatment response. These findings have been criticized due to the low doses of tranylcypromine used in this trial (mean dose 36.9 mg/day) (Nolen et al., 2007), which may have at least partially accounted for the low response and remission rates with tranylcypromine in contrast with what appeared to be more aggressively dosed combination therapy with venlafaxine and mirtazapine. Despite the low mean tranylcypromine dose, dropout rates due to adverse effects were proportionally higher with tranylcypromine than combination therapy (41.4% vs 21.6%). Additionally, a higher proportion of tranylcypromine-treated patients entered into the study due to previous medication intolerance (rather than treatment failure) than those randomized to combination therapy (41% vs 22%) (Wingo and Ghaemi, 2007). This may have introduced some bias toward higher risk of premature discontinuation from the study due to adverse effects and poorer rates of treatment response in the tranylcypromine group. The number of prior trials (a proxy for the stage of treatment resistance) was not statistically correlated with the likelihood of a positive treatment outcome with tranylcypromine. It should be noted, however, that several patients who did achieve remission with tranylcypromine had a history of previous failure to respond to bitemporal electroconvulsive therapy (ECT).

In an interesting retrospective chart review study, information was abstracted from 400 records to test the hypothesis that MAOI therapy would be similarly effective for patients with treatment-resistant depression who had failed to respond to at least four prior therapeutic trials of antidepressants, as compared to those who had failed to respond to three or fewer trials (Amsterdam and Shults, 2005). Over half (59%) of the 59 unique patients in the chart review sample had MDD, while the rest had treatment-resistant bipolar I (2%), bipolar II (34%), or unspecified bipolar depression (5%). These 59 patients had a total of 75 MAOI treatment trials with tranylcypromine (n = 47, mean peak dose 57 mg/day), phenelzine (n = 12, mean peak dose 58 mg/day), isocarboxazid (n = 13 mean peak dose 47 mg/day), and oral selegiline (n = 3, mean peak dose 53 mg/day). After an average of 39 weeks of follow-up, MAOI treatment yielded a favorable response, defined as a CGI-change score of 1 (very much improved) or 2 (much improved), in 56% of patients who responded poorly to fewer than four prior antidepressant trials and 12% of those who failed to respond to four or more previous trials. This study was limited by its retrospective design, small sample size, lack of a non-MAOI comparator group, and inability to verify the adequacy of prior antidepressant trials.

In a longitudinal study, 118 people with confirmed treatment-resistant depression (77 with unipolar depressive disorders) who were initially managed in a psychiatric hospital setting were started on various antidepressive treatments and were followed for an average of 39 months (Fekadu et al., 2012). Most of the study participants eventually achieved remission (60.2%), 48.3% of whom had a sustained recovery defined as continued remission for at least 6 months. After controlling for other treatments, there was a positive association between treatment with a MAOI and remission—both at hospital discharge and at the end of long-term follow-up. The relationship between MAOIs and remission was especially strong for people with MDD at discharge (OR 6.49, 95% CI 1.62–25.91) and at the end of follow-up (OR 4.78, 95% CI 1.15–19.85). The mean daily doses of phenelzine, tranylcypromine, isocarboxazid, and moclobemide were 70.0, 32.5, 40.0, and 650.0 mg, respectively.

We were unable to locate studies that focused exclusively on the clinical effects of MAOIs in people with treatment-resistant MDD who failed to respond to ECT, still broadly considered a gold standard therapeutic for drug-resistant depression (Kellner et al., 2012). Limited information is available about MAOI effectiveness in ECT-resistant depression from a handful of the trials that included depressed patients who had previously failed to respond to ECT during the current episode. For example, in a 7-week uncontrolled trial of elderly patients with treatment-resistant MDD who had failed to respond to treatment with a tricyclic antidepressant, over one-third of enrollees had also failed to respond to ECT (Georgotas et al., 1983). The overall rate of positive response (defined as a HAMD total score $\leq$ 10 at termination) was 66.7%. However, response rates in the ECT-refractory subgroup were not reported.

Transdermal selegiline

Randomized trials of transdermal selegiline for patients with well-defined treatment-resistant depression are needed. One small randomized trial of oral selegiline (60 mg/day) enrolled 16 elderly subjects who had failed to respond after at least two therapeutic antidepressant trials (Sunderland et al., 1994). Study participants had failed to respond to an average of 4.2 antidepressants prior to treatment with selegiline. A total of 10 participants had failed to respond to oral MAOIs and six participants had failed to respond to ECT. During the 3-week randomized treatment phase, there were significant improvements in depressive symptoms, as well as global measures of depression and anxiety, with oral selegiline relative to placebo. No serious adverse effects were reported. The response rate with oral selegiline was 50%, based on a 17-item HAM-D total score of 11 or lower, which is reasonably high given the 3-week duration of the study. However, the durability of treatment response beyond 3 weeks was not reported, and the generalizability of findings from this study to nongeriatric adults with treatment-resistant depression or to the use of transdermal selegiline for treatment-resistant depression is unclear.

Moclobemide

To our knowledge, there are no randomized trials of moclobemide for well-defined treatment-resistant depression. Moclobemide may be expected to have tolerability advantages over irreversible MAOIs (Fulton et al., 1996); however, moclobemide and other RIMAs may be less-effective antidepressants than orally administered irreversible MAOIs such as phenelzine and tranylcypromine (Lotufo-Neto et al., 1999). Randomized trials are needed to determine whether or not moclobemide is worth considering as a treatment option for even early stage treatment-resistant depression.

Adverse effects and toxicology

Common adverse effects

Phenelzine, isocarboxazid, and tranylcypromine

Common adverse effects for most oral irreversible MAOIs occur at the early stages of treatment and include insomnia, sedation, orthostatic hypotension, dizziness, headache, dry mouth, loose stools, blurry vision, appetite changes, and nausea (Knud and Lene, 2016; Thase, 2012). In our experience, many patients will adapt to these adverse effects over time, although several weeks may be required. Orthostatic hypotension is the most common dose-limiting adverse effect, especially in elderly persons. Tranylcypromine can sometimes cause a transient increase in blood pressure that typically resolves after a few hours. Longer-term adverse effects can include muscular pain, sexual dysfunction, edema, and weight (Thase, 2012). The risk of weight gain is generally lower with tranylcypromine than with phenelzine (Cantu and Korek, 1988). Oral MAOIs do not directly block muscarinic acetylcholine receptors and, therefore, do not cause anticholinergic side-effects, per se. In this regard, there are no known adverse effects on cognitive functioning with the long-term use of oral irreversible MAOIs.

Transdermal selegiline

Common adverse effects of transdermal selegiline include headache, insomnia, and local skin irritation (itching, redness, and swelling) at the patch application site (Amsterdam and Bodkin, 2006). Transdermal selegiline is associated with a very low risk of sexual dysfunction and weight gain, even with longer-term treatment (Clayton et al., 2007).

Uncommon or rare adverse effects

Some MAOIs have been associated with rare but potentially severe adverse effects. Up to 3% of patients treated with MAOIs (including moclobemide) experience asymptomatic mild elevations in liver function enzymes (Voican et al., 2014). However, oral irreversible MAOIs are associated with a higher risk of hepatotoxicity than newer antidepressants, such as SSRIs (LiverTox, 2020a; Lucena et al., 2003). The risk of MAOI-associated hepatotoxicity is small overall, but appears to be higher with the hydrazine class than with tranylcypromine (LiverTox, 2020a, b, c). Although acute liver injury due to MAOIs is usually self-limited, cases of progressive, severe, and fatal acute liver injury have been reported, particularly for phenelzine. Selegiline has been reported to cause hepatic enzyme elevations in up to 40% of patients who receive long-term treatment, but these were usually mild and self-limited, without leading to clinically apparent acute liver injury. Therefore, selegiline is currently classified as an unlikely cause of clinically apparent liver injury (LiverTox, 2020d).

There are rare reports of drug-induced lupus erythematosus associated with phenelzine (Rubin, 2005). Symptoms of phenelzine-associated lupus are similar to those observed in cases of systemic lupus erythematosus. These include flu-like symptoms, myalgias, arthralgias, skin rash, and fever. Antinuclear antibody and antihistamine antibody testing may also be positive. The symptoms of MAOI-associated lupus typically resolve within 2 weeks following medication discontinuation, either spontaneously or with the use of nonsteroidal antiinflammatory medications or oral corticosteroids. Transdermal selegiline has been associated with reports of flu-like symptoms although, to our knowledge, there are no published reports of well-defined cases of drug-induced lupus.

Other rare or uncommon adverse effects associated with mainly oral irreversible MAOIs include severe daytime somnolence, mental status changes, weakness, involuntary muscle jerks, urinary hesitancy, and nystagmus. Hydrazine MAOIs (phenelzine and isocarboxazid) may cause inactivation of vitamin B6 that can lead to paresthesias. Oral irreversible MAOIs may cause generally asymptomatic electrocardiographic changes including bradycardia and shortening of the PR and corrected QT intervals (McGrath et al., 1987).

Discontinuation syndrome

Discontinuation syndromes occur rarely with immediate or rapid discontinuation of MAOIs, especially when they are used at usual effective doses for depression. However, the risk may be increased on immediate or rapid discontinuation of MAOIs in people who have taken higher doses and those who have been on long-term maintenance treatment. Common symptoms include nausea, vomiting, anxiety, diaphoresis, chills, and nightmares. More severe symptoms described in case reports include hallucinosis, agitation, paranoia, movement disorders, convulsions, and delirium (Warner et al., 2006). Therefore, whenever feasible, tapering over 4 weeks or more is recommended when discontinuing treatment with MAOIs. Longer tapering periods may be needed if discontinuation syndrome symptoms occur while reducing the dose. In cases of strongly suspected hypertensive crisis or serotonin toxicity, immediate discontinuation of MAOIs (and interacting drugs) is needed regardless of the risk of precipitating a discontinuation syndrome.

Overdose

Phenelzine, isocarboxazid, and tranylcypromine

Overdoses on oral irreversible MAOIs are potentially fatal, and thus require emergency assessment and management. Common symptoms observed with MAOI overdose include altered mental status, hyperthermia, increases in blood pressure and heart rate, and increased respiratory rate. Hyperglycemia and leukocytosis may also occur. Severe toxicity associated with large overdoses may be characterized by hyperpyrexia, seizures, respiratory depression, cardiovascular collapse, and death. Although deaths due to MAOI overdoses are rare, ingestions of 2–3 mg/kg or more are generally considered life-threatening (Linden et al., 1984).

Transdermal selegiline

There is a paucity of overdose toxicity data for transdermal selegiline. A review of postmarketing data from 29,141 unique patients who were treated with transdermal selegiline yielded a total of four cases of completed suicide and four additional cases of attempted suicide (Pae et al., 2012). All completed suicides involved the ingestion of multiple drugs including contraindicated substances like crystal methamphetamine. It is unclear whether transdermal selegiline is safer in overdose than orally administered selegiline or other orally administered MAOIs. The loss of pharmacological selectivity for MAOI-B at high doses of selegiline make it likely that the signs, symptoms and risks associated with overdoses of transdermal selegiline may be similar to overdoses of oral nonselective MAOIs. Therefore, transdermal selegiline overdose is as much a medical emergency as is an overdose on oral nonselective MAOIs.

Moclobemide

Signs and symptoms of moclobemide toxicity include mental status changes, dizziness, headaches, muscle rigidity and seizures (Isbister et al., 2003). The effects of moclobemide overdose as a sole ingestion are thought to be relatively minor; however, severe serotonin toxicity can occur in the setting of moclobemide overdose with the coingestion of serotonin-potentiating drugs such as SSRIs (Isbister et al., 2003), as reviewed further below.

Interactions

Pharmacokinetic interactions

There are few clinically relevant pharmacokinetic drug-drug or drug-nutrient interactions for most MAOIs. Phenelzine displays weak but irreversible inhibition of CYP2C9 and 3A4 in vitro; however, there are no clinically relevant drug reactions related to such activity (Gillman, 2011). Tranylcypromine is a competitive inhibitor of CYP2C19 although, at therapeutically relevant doses, the significance of this effect is also unknown (Ulrich et al., 2017). At antidepressive doses, tranylcypromine is a more potent inhibitor of CYP2A6, which can result in increased plasma nicotine levels and possibly affect the metabolism of a few commonly used medications such as tamoxifen and propofol (Ulrich et al., 2017).

Pharmacodynamic interactions

The inhibition of MAO-A, a pharmacological activity shared by nearly all MAOIs used to treat depression, in combination with certain medications or tyramine-containing foodstuffs, can lead to pharmacodynamic interactions that are highly significant and, in some cases, potentially life-threatening. These include serotonin toxicity when MAOIs are combined with certain serotonergic medications, hypertensive crises when they are combined with some sympathomimetic agents, and hypertensive responses following the ingestion of a sufficient amount of dietary tyramine. Frequent assessment for potential drug interactions and adherence to dietary guidelines is needed to avoid these outcomes. Fortunately, such interactions are predictable and relatively easy to avoid.

Serotonin toxicity

Serotonin toxicity (sometimes referred to as serotonin syndrome) is a potentially life-threatening toxicity syndrome that is precipitated by the concomitant use of MAOIs and selected serotonin-potentiating drugs, leading to excessive central and peripheral serotonergic activity (Boyer and Shannon, 2005). Clinical signs and symptoms of serotonin toxicity are grouped into three categories: altered mental status (confusion, agitation, excitement); autonomic hyperactivity (fever, diaphoresis, mydriasis, tachypnea, tachycardia); and neuromuscular hyperactivity (tremor, clonus, myoclonus, hyperreflexia, hypertonicity) (Gillman, 2006). The successful management of serotonin toxicity depends on the rapid recognition of the signs of serotonin toxicity, immediate cessation of all serotonergic agents, and the availability of supportive care—the latter of which includes intravenous hydration, cooling blankets or ice packs for hyperthermia, benzodiazepines for myoclonus, and airway protection or respiratory support (including mechanical ventilation) when required (Boyer and Shannon, 2005). Stabilization in an intensive care unit may be needed for severe cases.

Serotonin toxicity is precipitated, for purposes of our discussion, by any combination of MAOIs and drugs that display serotonin reuptake inhibition (Gillman, 2006). As shown in Table 9.4, this includes not only SSRI or serotonin and norepinephrine reuptake inhibitor (SNRI) antidepressants, but also selected tricyclic antidepressants (TCAs, e.g., clomipramine and imipramine), newer antidepressants (e.g., vortioxetine and vilazodone), and selected nonpsychotropics (Flockhart,

TABLE 9.4 Selected list of drugs that are contraindicated when taken with monoamine oxidase inhibitors (MAOIs) and potential substitutes.

Toxicity syndrome	Contraindicated drugs	Potential substitutes to consider
Serotonin toxicity	Antidepressants, SSRIs and SNRIs (including milnacipran and levomilnacipran)	None within the SSRI or SNRI class
	Antidepressants, selected TCAs: clomipramine, imipramine	With caution: amitriptyline, nortriptyline, desipramine, doxepin, trimipramine
	Antidepressants, other: vilazodone, vortioxetine	With caution: bupropion, mirtazapine, trazodone
	L-Tryptophan, 5-hydroxytryptophan	
	Analgesics, selected: dextropropoxyphene, fentanyl, meperidine, methadone, pentazocine, tapentadol, tramadol	As indicated: acetaminophen, aspirin, buprenorphine, butorphanol, codeine, hydrocodone, morphine, nalbuphine, nonsteroidal antiinflammatory drugs, oxycodone
	Cough medications: dextromethorphan-containing	Cough medicines that contain codeine or expectorants that do not contain dextromethorphan
Hypertensive reaction	Antihistamines, selected: chlorpheniramine, brompheniramine	Diphenhydramine, loratadine cetirizine
	Nasal decongestants, selected: pseudoephedrine, phenylpropanolamine, phenylephrine	
	Local anesthetics, selected: epinephrine-containing	Nonepinephrine-containing local anesthetics
	Bronchodilators, selected: ephedrine-containing, epinephrine-containing	
	Hemorrhoid products, selected: phenylephrine-containing	

This table summarizes the main types of medications that are contraindicated or may otherwise lead to potentially dangerous interactions with MAOIs. Prescribers should check all drug interactions before prescribing MAOIs or when new medications are being considered for patients who are already taking MAOIs.

2012; Gillman, 2006; Grady and Stahl, 2012; Pray, 2007; Sun-Edelstein et al., 2008; Ulrich et al., 2017). These drugs should be avoided in people who are being treated with any MAOIs (Isbister et al., 2003; Richard et al., 1997). In our experience, most atypical antipsychotic drugs that are used to augment minimal or partial therapeutic responses to antidepressants can be safely combined with irreversible MAOIs. Ziprasidone is probably best avoided, however, because it blocks serotonin reuptake pumps (Davis and Markham, 1997; Tatsumi et al., 1999). Case literature describes the safe coadministration of irreversible MAOIs and other commonly used adjuncts, such as trazodone (Thomas et al., 2015). Lithium is thought to have serotonergic properties, but there are several reports of the safe use of lithium for improving the antidepressive effects of irreversible MAOIs without precipitating serotonin toxicity (Fein et al., 1988; Thomas et al., 2015; Ulrich et al., 2017). It is often recommended that people taking MAOIs avoid augmentation with buspirone, a serotonin 5-HT1A partial agonist, although the association between buspirone and the risk of serotonin toxicity in people taking MAOIs is not clear (Gillman, 2010). A more detailed discussion of augmentation of irreversible MAOIs is provided below.

It is important to recommend that patients taking irreversible MAOIs avoid opioids like meperidine, tramadol, and methadone that can elevate synaptic levels of serotonin. However, analgesics such as morphine, codeine, and oxycodone can be safely taken with MAOIs. Triptans, which cause vasoconstriction by stimulating serotonin receptors, are generally avoided in people taking MAOIs; however, despite regulatory warnings, the association between triptans and serotonin toxicity in people taking antidepressants, including MAOIs, is controversial (Gillman, 2010). Antitussive preparations that contain dextromethorphan should be avoided.

In the case of moclobemide treatment, it is critically important to avoid the concomitant use of serotonin reuptake inhibitor drugs due to heightened risk of severe serotonin toxicity (Isbister et al., 2003). A number of cases of severe and life-threatening toxic responses to moclobemide at clinically relevant doses when combined with serotonin reuptake inhibitors have been reported (Gillman, 2004; Hawley et al., 1996).

Hypertensive crisis due to interactions with sympathomimetics

When combined with MAOIs, certain sympathomimetic drugs can cause excessive adrenergic stimulation, resulting in dangerous elevations in blood pressure (Grady and Stahl, 2012). This interaction is important for both direct-acting sympathomimetics that activate catecholaminergic receptors (agonists) and indirect sympathomimetics that may inhibit degradation, block reuptake, or stimulate release of catecholamines. These properties are found in a variety of drugs that range from stimulants to anorectics to cough and cold medications (Flockhart, 2012), as shown in Table 9.4. Indirect sympathomimetics are contraindicated in people taking MAOIs, while low doses of direct sympathomimetics used in nasal sprays/drops and local anesthetics may be relatively safe (Ulrich et al., 2017). However, the nasal decongestants containing pseudoephedrine or phenylpropanolamine, and local anesthetics that contain epinephrine are contraindicated in patients taking MAOIs. In our experience, it is important to recommend that patients avoid taking weight loss products, even nutraceuticals, until these can be reviewed by their prescribing clinician for safe use with irreversible MAOIs. Some stimulants (e.g., methylphenidate, modafinil, armodafinil) can be safety combined with MAOIs, as described in case reports, although due caution, rigorously applied informed consent, close monitoring, and expertise in the use of MAOIs is warranted (Israel, 2015; Thomas et al., 2015). It is recommended that the antihistamines, chlorpheniramine and brompheniramine, be avoided in people taking MAOIs. Data for chlorpheniramine show it to be relatively potent at blocking the human serotonin transporter, while weak at the norepinephrine transporter (Tatsumi et al., 1997). There are no analogous data for brompheniramine. On the other hand, the antihistamine, diphenhydramine, shows only weak blockade at both transporters (Tatsumi et al., 1999).

Hypertensive reactions due to excessive tyramine consumption

Tyramine is a biogenic amine that is almost absent in fresh food, but begins to accumulate after spoilage or with fermentation. Excessive amounts of tyramine can displace endogenous stores of norepinephrine in the periphery, causing contraction of blood vessels, with concomitant increase in blood pressure and in cardiac output. The average person can consume about 400 mg of tyramine before observing a meaningful increase in blood pressure. The irreversible inhibition of MAO-A increases the body's sensitivity to the hypertensive effects of tyramine, thus greatly lowering the amount of tyramine that can be ingested without blood pressure increases (Bieck and Antonin, 1988). For people taking irreversible MAOIs, tyramine ingestions as low as 8–10 mg per serving can cause an increase in systolic blood pressure by 30 mmHg or greater (Azzaro et al., 2006), while the ingestion of ≤ 6 mg of tyramine per serving is generally considered safe (Fiedorowicz and Swartz, 2004).

Fortunately, improvements in food production techniques and hygiene regulations have made it such that high levels of tyramine are seldom encountered (Gillman, 2018). Still, there is a high enough degree of interindividual variability in sensitivity to tyramine-induced pressor responses that clinicians must still maintain due vigilance (Gillman, 2018). As a matter of general principle, the main foods to avoid during treatment with irreversible MAOIs are improperly stored or aged meat, cured meat products, fava bean pods, banana peels, marmite, sauerkraut, certain soy sauces, tofu, soy bean condiments, and tap beer (Fiedorowicz and Swartz, 2004; Flockhart, 2012; Gardner et al., 1996; Gelenberg, 1998; Gillman, 2018; Grady and Stahl, 2012; Stahl and Felker, 2008; Wimbiscus et al., 2010). Bottled beer and most red or white wines are considered safe when consumed moderately. Some foods on the previously "banned" list, such as chocolate and avocados, are now considered to be relatively safe to eat in moderation (Table 9.5).

As noted earlier, dietary restrictions are unnecessary for patients taking moclobemide or transdermal selegiline at a dose of 6 mg/24 h (Fulton et al., 1996). However, in our experience, many patients taking transdermal selegiline will eventually require doses higher than 6 mg/24 h. At doses above 6 mg/24 h, adherence to dietary guidelines is recommended, although data suggest that, even at a dose of 12 mg/24 h, tyramine with a meal does not cause clinically significant increases in blood pressure (Azzaro et al., 2006).

Clinical use

There are several options for managing patients with treatment-resistant MDD. For patients with a partial positive response to a given antidepressant, it makes most sense in our opinion to increase the antidepressant dose (as tolerated) or augment with a second drug if dose optimization is poorly tolerated or minimally effective. For people with minimal or no response,

TABLE 9.5 Relative levels of caution for selected types of food or beverages in people taking monoamine oxidase inhibitors.

	Relative level of caution based on tyramine content		
	Low (allowed)[a,b]	Medium (caution)[a,c]	High (usually avoided)[a,d]
Food group			
Cheese/yogurt	Fresh cheese Processed cheese slices Cottage cheese Ricotta cheese Cream cheese Yogurt	Mozzarella (lower risk if fresh) Cured milk products (yogurt, kefir, buttermilk—up to 4 oz. per day)	Aged and fermented cheeses (bleu, cheddar, gouda, muenster, detabrick, brie, camembert, emmentaler, gruyere, parmesan, provolone, Roquefort, stilton, and liderkranz Foods containing these cheeses (some types of pizzas, casseroles, salad dressings, and quiche) Cheese soup Cheese sauces
Meat/alternative protein sources	Fresh or processed meat or poultry Eggs	Fresh sausage and pepperoni (1 oz. or less per day) Meat with tenderizer added Bacon, sausage, hot dogs, corned beef, bologna, ham, lunch meats with nitrates or nitrites added	Aged/fermented meat or poultry (such as chorizo, salchichon, other fermented sausages, corned beef) Smoked meat or poultry Aged beef or chicken liver, pate Air-dried sausages (pepperoni, summer sausage, salami, pastrami, mortadella) Meat extracts Meat products at or beyond expiration date
Fish/seafood	Fresh or processed fish		Caviar Smoked or pickled fish Cured fish Dried fish Fish paste (including anything made with) Shrimp paste (including anything made with)
Fruit/vegetables/beans	Fresh, frozen, or canned fruit/vegetables Legumes Nuts Peanut butter Bananas (without peels)	Raspberries Avocados (one large fruit or less)	Fermented vegetables (kimchi, sauerkraut) Spoiled, decomposed, fermented, or overripe fruit (including raisins) Broad (fava) bean pods Chinese pea pods Banana peels (bananas themselves are OK) Pickles Olives Fermented soy products (soy sauce, teriyaki, fermented tofu, miso, tamari, natto, shoyu, tempeh, soybean paste)
Breads/cereals/spreads	Whole wheat breads Enriched white bread Rolls, crackers, quick breads Cooked cereal Dry cereal Pasta Rice Grits Brewer's yeast Baker's yeast	Homemade yeast bread Sourdough bread	Marmite Vegemite
Oils/salad dressing	Salad dressings without aged cheese Vegetable oil		Products made with high-risk ingredients (including aged cheese)
Other	Fresh gravy	Pizza (small quantities if made with part-skim mozzarella or similar fresh cheese, no air-dried sausage or aged pepperoni)	Monosodium glutamate Bouillon and bouillon-based recipes/soups Yeast extracts Chocolate Tyramine-containing dietary supplements

Continued

TABLE 9.5 Relative levels of caution for selected types of food or beverages in people taking monoamine oxidase inhibitors—cont'd

	Relative level of caution based on tyramine content		
	Low (allowed)[a,b]	Medium (caution)[a,c]	High (usually avoided)[a,d]
Beverage type			
Alcohol-containing		Canned or bottled beer Most wine (in moderation) Clear spirits Soup made with beer or wine	Tap beer Alcohol-free beer Ale Chianti Vermouth wines Sherry Champagne Mixed drinks
Milk	Plain milk Soy milk		Kefir
Other	Juice Carbonated drinks Decaffeinated coffee and tea	Coffee Tea Chocolate drinks	Kombucha

[a] *Based on tyramine content, for the risk of precipitating hypertensive reactions. This table serves as a general guideline. Foods that contain only small amounts of tyramine and are considered medium or even low risk (in some cases) can become problematic if large quantities are consumed by people taking monoamine oxidase inhibitors (MAOIs).*
[b] *Foods labeled as low in risk are generally considered safe to consume by people taking MAOIs.*
[c] *Foods labeled as medium in risk can be consumed with caution, in moderation or in small amounts as specified, by people taking MAOIs.*
[d] *Foods labeled as high in risk should generally be avoided by people taking MAOIs.*

switching to a new antidepressant may make more sense. MAOIs are among the switch options that are usually reserved for people who have minimal or no response to multiple trials of antidepressants, psychotropic drug combinations, and combined pharmacotherapy-psychotherapy. Generally, it can be recommended that MAOIs be routinely considered after a third failed medication trial (Birkenhäger et al., 2004; Sen et al., 2011); however, in our experience, patients who initiate MAOI therapy have often failed five or more therapeutic trials of conventional antidepressants. In many cases, such patients have also responded poorly to ECT, transcranial magnetic stimulation, and even intravenous or intranasal ketamine.

In our opinion, the advanced stages of treatment resistance that we have observed in patients who have often never been offered a therapeutic trial of MAOIs indicates a high degree of reluctance among many practitioners, including psychiatrists, to provide MAOIs to their patients. We believe that such reluctance stems mainly from a lack of experience with these drugs and a relative lack of simple, pragmatic guidance on their use. Therefore, this section provides practical advice on the use of MAOIs for people with treatment-resistant depression, and is organized around the following subtopics: Transitioning onto MAOI therapy, selecting a MAOI, initiating treatment, monitoring therapeutic responses and side-effects, optimizing the effectiveness of MAOIs, and transitioning off of MAOIs when necessary. Some special circumstances are also discussed.

Transitioning onto MAOI therapy

Tapering and washout from prior medication

Depressed patients who consent to treatment with MAOIs must taper off of their current antidepressant and all other medications that interact with MAOIs. Once the tapering is complete, a minimum of five half-lives of these medications must transpire before the first dose of MAOIs can be given (referred to as a "washout"). On average, about 2–3 weeks is required to washout from most antidepressants and other interacting drugs after they are completely discontinued (Wimbiscus et al., 2010). In the case of fluoxetine, a washout period of 5–6 weeks is required before initiating MAOIs due to the long elimination half-lives of both the parent compound and its pharmacologically active metabolite (Keks et al., 2016).

Interventions to consider during the taper and washout

The tapering and washout periods, combined together, will last several weeks and may not be tolerated by patients with more severe depressive symptoms. Therapeutic options to consider during the tapering and washout period include nonserotonergic antidepressants (including secondary amine tricyclic antidepressants that act mainly as norepinephrine reuptake inhibitors), electroconvulsive therapy, or transcranial magnetic stimulation. The use of general anesthesia for electroconvulsive therapy is not contraindicated in people who are transitioning onto MAOIs (Naguib and Koorn, 2002). However, as pointed out above, many patients with treatment-resistant depression for whom MAOIs are being considered will have already responded poorly to these treatments. The use of ketamine (intravenous) or esketamine (intranasal) as a therapeutic bridge to MAOIs has not been systematically investigated, so we do not generally recommend this practice unless the risks of undertreated depression outweigh the potential risks of MAOI initiation in severely ill ketamine-treated patients. There is a theoretical risk of sympathetic overactivation during the transition to MAOIs during the course of antidepressive treatment with ketamine (Bobo et al., 2020). This is due to the possible exaggeration of ketamine-induced transient (up to 4 h) increases in systolic blood pressure (Riva-Posse et al., 2018). Therefore, if this strategy is being considered, very close monitoring is warranted, and consultation with an anesthesiologist prior to the initiation of MAOIs in patients receiving ketamine for depression is recommended.

On the other hand, the tapering and washout prior to initiating MAOI therapy may be an ideal time to educate patients about the low tyramine diet and the avoidance of interacting drugs, and to rehearse adhering to dietary guidelines, alert all prescribers (including dentists) to the fact that they are about to start treatment with a MAOI, and report all changes in prescribed and over-the-counter medications to the clinician who will be prescribing MAOIs. We and others recommend that patients purchase a medical alert bracelet indicating MAOI use in case of medical emergencies during which the patient may not be conscious or otherwise able to provide their own medical histories (Fiedorowicz and Swartz, 2004).

Selecting a MAOI

There is no good evidence that one irreversible MAOI is clearly superior to another. So treatment selection may depend primarily on the availability of specific agents and subtle differences in adverse effect profile between drugs. Among the irreversible MAOIs, tranylcypromine appears to have a more stimulating effect (i.e., a higher risk of insomnia and activation) and lower risk of weight gain, edema, and sedation than does phenelzine. In addition, tranylcypromine is not associated with the inactivation of vitamin B6, and may not carry the risk of treatment-emergent rare hepatotoxicity—characteristics that may contrast it with the hydrazine derivatives, phenelzine and isocarboxazid. On the other hand, sedation and the potentiation of GABA neurotransmission with phenelzine may be more desirable for some patients. There is very little evidence supporting the use of transdermal selegiline for treatment-resistant depression. Nevertheless, for patients who experience gastrointestinal distress with oral MAOIs, transdermal selegiline may be better-tolerated (Sen et al., 2011). As there is no good evidence supporting moclobemide for treatment-resistant depression, we do not generally recommend it as a next-step consideration.

Initiating treatment with a MAOI

Baseline laboratory and other assessments

For the majority of patients, no baseline laboratory studies or screening procedures are needed before starting MAOIs. As noted earlier, pyridoxine (vitamin B6) deficiency can be an adverse effect of hydrazine MAOIs. In some cases, it may be reasonable to check pyridoxine levels at baseline and then after one, 6 and 12 months, if the patient is taking phenelzine or isocarboxazid. Periodically checking pyridoxine levels is not needed with tranylcypromine. For some phenelzine- or isocarboxazid-treated patients, the use of preemptive B6 supplementation (50–200 mg/day) can be considered. Given the association between phenelzine and rare hepatotoxicity, obtaining baseline liver function tests and repeating these tests periodically during treatment has been recommended (Voican et al., 2014).

Simple dosing schemes

The initial dosing of MAOIs is relatively straightforward. Phenelzine can be started at a dose of 15 mg three times daily in healthy patients, or at lower doses if clinically necessary. From there, the dose can be slowly increased based on clinical response and tolerability to the usual therapeutic range of 45–90 mg, split three times daily. As a clinical "rule of thumb," the phenelzine target dosage can be estimated as 1 mg/kg, based on early studies showing the relationship between the dose of phenelzine and inhibition of platelet MAO-B (Robinson et al., 1978). Tranylcypromine and isocarboxazid can be started at a dose of 10 mg twice daily in healthy patients, or as a single dose of 10 mg daily in the morning if required. The second dose of the day is typically taken 2–4 h after the first morning dose. The dose of tranylcypromine and isocarboxazid can then be increased by 10–20 mg per week to a usual target of 30–60 mg, split twice daily, often with most of the daily dose taken first thing in the morning.

Treatment with transdermal selegiline is usually initiated using the 6 mg/24 h patch. Based on clinical response, a dose increase to the 9 mg/24 h patch and, if needed, the 12 mg/24 h patch can be considered. Proper use of the medication includes maintaining good general habits for the use of transdermal medications. These include washing one's hands before and after applying the patches; applying the patches to flat, dry, smooth and clean areas of the body; avoiding freshly shaved skin; and rotating the application site daily.

Initial monitoring of blood pressure

The dose titration of oral MAOIs is usually based on clinical response and tolerability, with orthostatic hypotension being the most common dose-limiting adverse effect. Therefore, starting at low doses and conducting slow and careful dose titration is especially important for elderly patients, those who are taking multiple medications that could perpetuate postural decreases in blood pressure, and in situations where dehydration risk is increased. Taking orthostatic vital signs at home twice daily (and keeping a diary of blood pressure readings to be reviewed at clinic visits) may be useful for preventing injuries and falls. If orthostasis threatens tolerability, a slower dose titration, the use of divided doses, increasing oral fluid intake, lowering the dose of antihypertensive drugs, and taking time to rise slowly against gravity (moving from lying to sitting or sitting to standing) are simple initial maneuvers that can be recommended (Fiedorowicz and Swartz, 2004). This is important because most patients develop tolerance to the orthostatic effects of MAOIs after a few weeks. Only rarely is consultation with a cardiovascular medicine specialist or specific treatment (e.g., pressure stockings, concomitant mineralocorticoids, etc.) required to manage orthostatic hypotension due to MAOIs.

Orthostatic blood pressure readings as a guide to dosing

Mild orthostatic blood pressure changes can, in many cases, be used to identify a minimal effective dose of oral MAOIs (Gillman, 2018). A recommended procedure is to take a seated blood pressure reading after 5 min of rest, and then take blood pressure measurements immediately upon standing and again (still standing) after a 1–2-min delay—and to do this twice daily after starting MAOI treatment, observing for a standing systolic blood pressure drop of about 10–15 mmHg (Gillman, 2018). For example, tranylcypromine can be started at a dose of 10–20 mg in divided doses. If tolerated, the dose can be increased to 30 mg in divided doses, and orthostatic vital signs can be assessed. If a drop in blood pressure is observed, then it is reasonable to continue the 30 mg dose for at least 4 weeks to determine its antidepressive effects. Otherwise, if there were no drop in blood pressure and there are no limiting side-effects after 7 days of treatment at the 30 mg dose, then the dose of tranylcypromine can be increased to 40 mg in divided doses. If a blood pressure drop is observed at 40 mg, then it is reasonable to continue that dose for at least 4 weeks in order to assess its antidepressive effects. If no blood pressure drop occurs after 7 days of treatment on the 40 mg dose, if there are no treatment-limiting side-effects, and if rapid improvement is essential, then it may be reasonable to increase the dose to 50 mg in divided doses.

Patients with renal and hepatic disease

The manufacturers of phenelzine, isocarboxazid, and tranylcypromine recommend avoiding these medications in patients with clinically significant renal disease. However, European Renal Best Practice (ERBP) guidelines advise that no dosing adjustments are necessary with phenelzine or isocarboxazid in patients with stage 3–5 (estimated glomerular filtration rate (eGFR) < 60 mL/min) chronic kidney disease (Nagler et al., 2012). In people with stage 4–5 (eGFR < 30 mL/min) chronic kidney disease, ERBP guidelines recommend initiating tranylcypromine at doses of 30 mg/day or less with a "slow and careful titration" (Nagler et al., 2012). Neither transdermal selegiline nor moclobemide appear to require dose modifications in people with renal insufficiency (Culpepper, 2013; Nagler et al., 2012; Stoeckel et al., 1990). Still, we advise a cautious approach to dose adjustment with all MAOIs in order to avoid potential problems with tolerability, including orthostasis, that threaten adherence.

There is very little information available regarding dose adjustments in patients with chronic liver disease who are taking MAOIs. When using MAOIs in patients with existing chronic liver disease, hepatic enzymes should be regularly monitored, although formal guidance regarding the frequency of testing is lacking (Voican et al., 2014). Given the association between selected MAOIs and hepatotoxicity (mainly hydrazines), some have recommended that oral irreversible MAOIs be avoided in patients with chronic liver disease (Livingston and Livingston, 1996). MAOI treatment may indeed be best avoided in patients with chronic liver disease and ALT levels that are > 3 times the upper limit of normal, and they should be stopped in any patient who develops signs of liver injury, new-onset clinically significant transaminase elevations (ALT > 3 times the upper limit of normal), or unexplained increases in bilirubin concentration (> 2 times the upper limit of normal) (Voican et al., 2014). Although there is a lack of data regarding the dosing of transdermal selegiline in people with severe hepatic impairment, no dose adjustment is recommended for patients with moderate (Child-Pugh class A or B) hepatic impairment (Lee and Chen, 2007). The dose of moclobemide should be reduced by one-half or one-third for patients with hepatic cirrhosis to prevent drug accumulation (Stoeckel et al., 1990).

Monitoring therapeutic response and adverse effects

General guidance

The therapeutic effects of MAOIs typically take 3–6 weeks to unfold, a timeframe which is similar to other antidepressants. In our experience, some patients require a longer period of time to manifest a positive treatment response, an observation that is also commonly found with other antidepressants. Scores on validated depressive symptom measures that are available in the public domain can be obtained at baseline and approximately every 4 weeks as a supplement to clinical judgment as to when dose adjustments or other alterations in the treatment plan may be needed.

Spontaneously reported adverse effects should be elicited at every visit, and it is advisable to review home blood pressure records kept by the patient and to measure a fresh set of orthostatic vital signs in the clinic. These procedures are especially important for elderly patients and people who are taking antihypertensive medications or medications that can increase the risk for orthostasis. As noted earlier, phenelzine and isocarboxazid can inactivate pyridoxine, which may cause treatment-emergent paresthesias (Rabkin et al., 1984). Fortunately, pyridoxine deficiencies can be detected in the clinical laboratory, and can be corrected by supplementing with as little as 50 mg per day.

MAOIs can cause varying adverse effects on sleep patterns. For treatment-emergent insomnia, soporific medications such as trazodone, doxepin (only at low doses because, at higher doses, doxepin blocks human serotonin and norepinephrine transporters (Tatsumi et al., 1997)), quetiapine, melatonin, and zolpidem (and other so-called "z-drugs") can be safely used by people taking MAOIs (Gillman, 2006; Thomas et al., 2015). There are also case reports of excessive daytime sleepiness with some MAOIs (Keck et al., 1992), which can be reduced with the careful use of methylphenidate or modafinil (which lacks pressor effects) if these adverse effects threaten adherence in patients who are otherwise benefitting from treatment (Israel, 2015; Thomas et al., 2015).

MAOIs can cause adverse changes in body weight, although selegiline is thought to be weight-neutral, and the risk of clinically significant weight gain appears to be higher with phenelzine than with tranylcypromine. The mechanisms underlying body weight gain with MAOIs can include increases in adiposity or weight gain due to edema. As noted earlier, it is important for MAOI-treated patients to avoid the use of weight loss agents, including over-the-counter products and nutraceuticals, without first consulting with their prescribing clinician.

Specific guidance: Serotonin toxicity and hypertensive crises due to drug interactions

The clinical presentation and management of serotonin toxicity and hypertensive crisis due to interactions with sympathomimetic drugs were reviewed earlier. The list of interacting drugs (Table 9.4) that can precipitate either type of reaction in people taking MAOIs is rather finite and dangerous interactions are relatively easy to avoid, especially with the availability of computer programs and apps for checking drug interactions that are widely available over the internet to clinicians and patients alike. But the prevention of serotonin toxicity and hypertensive crises relies upon maintaining accurate records of all prescription and nonprescription medications taken by the patient. Patients who are taking MAOIs should be repeatedly reminded to consult with their healthcare providers before starting any new medications, including over-the-counter drugs.

Specific guidance: Dietary tyramine ingestion

As reviewed earlier, treatment with MAOIs can increase the body's sensitivity to dietary tyramine for increasing blood pressure. It is therefore important to assess MAOI-treated patients for adherence to dietary guidelines, ask patients to keep dietary records (especially at the beginning of treatment), and ask patients to monitor their blood pressure at home twice-daily. Clinically, the tyramine reaction in patients taking MAOIs presents as a gradual rise in blood pressure starting 30–60 min after ingestion. The rise in blood pressure may be accompanied by chest discomfort/tightness, palpitations, nausea, vomiting, headache (that does not improve with the use of over-the-counter analgesics), tinnitus, neck stiffness, pallor, and pupillary dilation. Most such reactions are mild and resolve after 2–4 h, but patients will need to monitor their blood pressure at home until it normalizes. If symptoms potentially related to tyramine ingestions occur, it is important to ask patients about the specific signs and symptoms they are experiencing (and when they started), what they have eaten (and when), and what medicines they are taking (including over-the-counter drugs). The availability of detailed blood pressure readings, dietary logs, and updated medication lists makes the task of reviewing what foods (and medicines) were taken or consumed, relatively simple.

If symptoms arise that are potentially attributable to tyramine ingestions, patients should monitor their blood pressure at home, observing for a gradual normalization of blood pressure. Even elevations in blood pressure that reach the level of hypertensive urgency (SBP > 180 mmHg) can resolve after 1–2 h without intervention (Gillman, 2018). Emergency room care is recommended if there is evidence of end-organ damage due to large blood pressure elevations, if a large tyramine ingestion is suspected (100 mg or more), or if blood pressure exceeds 220 mmHg (Gillman, 2018). The aim of treatment in such cases is a gradual lowering of blood pressure. Rapid reductions in blood pressure can lead to potentially catastrophic adverse outcomes. Nifedipine should not be used due to the potential for severe hypotension and increased risk of myocardial infarction, cardiac conduction disturbances, and stroke (Grossman et al., 1996). Low-dose benzodiazepines can have blood pressure-lowering effects due to reduction in anxiety symptoms (Gillman, 2018).

Prevention of adverse tyramine reactions can be achieved by adherence to basic dietary guidelines, such as those provided in Table 9.5. General guidelines are typically sufficient given the limited amounts of tyramine in foods ingested by the majority of MAOI-treated patients. Moreover, many tyramine-rich foods can be tolerated if they are ingested in small amounts (Gillman, 2018). A number of other good practices are worth highlighting. It is important for patients to understand the importance of avoiding food spoilage and knowing correct portion sizes (Gillman, 2018). Cooking or heating foods or drinks that are high in tyramine content will not reduce tyramine levels. Therefore, cooked foods should be consumed right away and leftovers should be avoided if possible, even when refrigerated. Foods should not be purchased if expiration dates have passed and should be frozen or stored at the correct refrigeration temperature. Fresh meats, poultry, and fish should be consumed on the same day without tenderizing, or should be frozen immediately. Fresh cheeses should always be stored at the proper temperature and should be eaten within a few weeks.

Optimizing the effectiveness of MAOIs

In this section, we provide a discussion of the options that may be considered for addressing an inadequate or partial antidepressive response to MAOIs. Broadly, these measures are the same for MAOIs as they are for other antidepressants, and include dose optimization (usually a dose increase), combining MAOIs with other medications, and switching medications. However, there are a number of issues related specifically to MAOIs that may not necessarily apply to other types of antidepressants.

In our experience, the most common strategy for addressing a partial antidepressive response to MAOIs is increasing the dose, as long as it is well-tolerated. There is evidence supporting an increase in the dose of MAOIs when the overall level of benefit is insufficient. As reviewed by Thase et al. (1995), three dose-response studies of phenelzine (Ravaris et al., 1976; Tyrer et al., 1980) and isocarboxazid (Davidson et al., 1984) showed that higher doses of these medications were associated with more favorable response rates. The recommended minimum daily dose thresholds from that review were 45 mg for phenelzine, 30 mg for isocarboxazid, and 30 mg for tranylcypromine, with optimal responses occurring at 75–90 mg/day for phenelzine (again, as discussed above, 1 mg/kg body weight as an estimated target), 40–60 mg/day for isocarboxazid, and 40–60 mg/day for tranylcypromine (Thase et al., 1995). As reviewed earlier, it is important to avoid increasing the dose of MAOIs too rapidly to avoid overwhelming side-effects such as orthostatic hypotension. Unfortunately, there is no established relationship between blood drug levels and clinical efficacy for treating depressed patients with MAOIs. Although the degree of MAO inhibition can be estimated by measuring platelet MAO-B levels (Quitkin et al., 1979), this is seldom done in clinical practice. Moreover, depressive relapses can occur while taking MAOIs despite consistently high levels of platelet MAO inhibition (Thase et al., 1995). Therefore, we do not recommend the routine measurement of platelet MAO inhibition as a guide for therapeutic monitoring.

High-dose tranylcypromine

An extension of dose optimization with MAOIs is the escalation of doses beyond the upper end of the usual therapeutic dose range, the rationale for which is the possibility that very high doses may be required to optimize MAO inhibition in some patients (Guze et al., 1987). This strategy has been best-studied with tranylcypromine, and is supported by the results of two open-label clinical trials that documented better efficacy with tranylcypromine at 70–120 mg/day (Amsterdam and Berwish, 1989; Nolen, 1989). Although more definitive evidence is lacking, the careful execution of high-dose treatment with tranylcypromine can be considered for patients with highly refractory depression and no better alternatives. Before doing so, we recommend a reassessment of the case for potentially correctable factors that may be contributing to treatment resistance (including misdiagnosis, incomplete diagnoses or inadequate case formulation, failure to optimize other treatment modalities, substance use, poor adherence to treatment, poor tolerability of the current dose of tranylcypromine, etc.), a thorough review of past treatments and outcomes with each, and obtaining a second opinion from a colleague, ideally with experience in the use of MAOIs. The informed consent process should include a discussion about the "off-label" dosing of tranylcypromine and careful documentation of that discussion in the patient's chart. Very close follow-up and adverse effect monitoring is necessary when using high-dose tranylcypromine. Due to a lack of clinical evidence, we are hesitant to recommend high-dose strategies for other MAOIs.

Combining medications

When increasing the dose is ineffective or infeasible, it may be practical or even necessary to consider adding a second drug to ongoing but only minimally or partially effective treatment with a MAOI. We and others believe that this strategy should generally be used by practitioners with expertise in the use of MAOIs, and should be reserved for treatment-resistant patients who have failed to respond to (or cannot tolerate) less-potentially hazardous treatment alternatives, including electroconvulsive therapy (Thomas et al., 2015). With due vigilance, a number of medications that are often combined with other antidepressants may be combined with MAOIs. These medications include selected antidepressants (bupropion, mirtazapine, and tricyclic antidepressants other than clomipramine or imipramine), soporific medications (trazodone, short-acting benzodiazepines, z-drugs), longer-acting benzodiazepines, atypical antipsychotic drugs (except for ziprasidone), mood stabilizers (lithium, divalproex, lamotrigine), selected stimulants (such as methylphenidate, modafinil, and armodafinil), pramipexole, alpha-2-delta ligands (gabapentin, pregabalin), thyroid hormones, and selected nutraceuticals (L-methylfolate, *S*-adenosyl-L-methionine [SAMe], *N*-acetylcysteine, and omega-3-fatty acids) (Gillman, 2006; Hullett and Bidder, 1983; Israel, 2015; Joffe, 1988; Meyer et al., 2017; Quante and Zeugmann, 2012; Thomas et al., 2015). Unfortunately, there is little to no good clinical evidence supporting the effectiveness of the majority of these drugs in combination with MAOIs. Therefore, with any pharmacological augmentation strategy, expert consultation should be considered and slow titration with careful monitoring of clinical response, vital signs, and adverse effects is strongly encouraged.

Efficacy rates of 60%–90% have been reported for combined treatment with MAOIs and selected tricyclic antidepressants in depressed patients who responded poorly to monotherapy with either type of medication (Amsterdam and Chopra, 2001). In addition, a retrospective chart review study of 94 treatment-resistant depressed inpatients (78 with unipolar depression, 94 with bipolar depression) showed that, after treatment with tranylcypromine plus a tricyclic or tetracyclic antidepressant, 68% of patients were judged as having good or very good improvement in depressive symptoms over a mean follow-up duration of 22 days (Schmauss et al., 1988). In that study, treatment resistance was defined as failing to respond to at least two therapeutic trials of tricyclic or tetracyclic antidepressants lasting 3–4 weeks each. The adverse effect profile of combination therapy was reported as being similar to tricyclic or heterocyclic antidepressant monotherapy, which was somewhat surprising given that the study drugs included imipramine and clomipramine.

Although it is clear which tricyclic antidepressants must be avoided in patients taking MAOIs (imipramine, clomipramine) in order to avoid precipitating serotonin toxicity (Gillman, 2006), it is less clear which tricyclic antidepressant is the most effective augmenting agent. In the retrospective study by Schmauss et al. (1988), amitriptyline plus tranylcypromine was the most effective combination. In a long-term retrospective cohort study, 31 patients with ECT-resistant MDD were given open-label tranylcypromine plus amitriptyline and were followed prospectively (Ferreira-Garcia et al., 2018). Both medications were started at the same time at low doses (10 mg of tranylcypromine and 25 mg of amitriptyline) and were increased simultaneously to 80 mg/day of tranylcypromine and 150 mg/day of amitriptyline over 4 weeks. After 12 weeks, 25 (81%) treated patients achieved a positive antidepressive response. Of the 25 responders, 19 were maintained on long-term drug treatment with the same regimen at the maximum tolerated dose. None of these individuals experience a depressive recurrence during an average 9.4 years of follow-up. We cannot conclude that amitriptyline has special advantages over other tricyclic antidepressants in combination with MAOIs given the lack of well-designed comparator trials. Interestingly, amitriptyline has been shown to inhibit tyramine uptake—an effect that may, in turn, decrease the risk of developing hypertensive responses to tyramine in people taking MAOIs (Pare et al., 1982). This effect is not unique to amitriptyline, as other tricyclic antidepressants that act as norepinephrine reuptake inhibitors, including nortriptyline, which is a metabolite of amitriptyline, and desipramine, have also been shown to block tyramine uptake (Gillman, 2011). Regardless, the combination of MAOIs and tricyclic antidepressants is ultimately limited by low-quality evidence and the high potential for adverse effects, especially those stemming from tricyclic-associated antiadrenergic and anticholinergic effects (Thomas et al., 2015), necessitating judicious patient selection, very careful dose titration, and close follow-up.

Lithium is an accepted augmenting agent for depressed patients who have failed to respond adequately to a therapeutic trial of MAOIs, similar to its role for augmenting other antidepressants (Nelson et al., 2014). The evidence supporting the use of lithium for augmenting MAOIs, however, is only based on small studies and case series. For instance, rapid improvement in depressive symptoms with the use of adjunctive lithium was reported in a small case series of three depressed patients who responded poorly to phenelzine monotherapy (Nelson and Byck, 1982). In a small open trial, 11 of 12 hospitalized patients with treatment-resistant MDD improved on a combination of tranylcypromine (30–60 mg/day) and lithium (at plasma drug levels of 0.6–1.3 mEq/L), eight of whom were rated as being much or very much improved (Price et al., 1985). In a short-term randomized trial, 29 elderly depressed inpatients who failed to respond to a therapeutic trial of at least one tricyclic antidepressant or venlafaxine were randomized to 6 weeks of treatment with either lithium augmentation of venlafaxine or tricyclic antidepressant or phenelzine monotherapy (Kok et al., 2007). After completing the trial, patients could enter a 2-year treatment phase, at which time they could be switched to treatment with phenelzine (after unsuccessful lithium augmentation), an alternative tricyclic antidepressant, lithium augmentation of phenelzine (following unsuccessful phenelzine treatment), or electroconvulsive therapy (for patients who failed to respond to any of these pharmacological switch options). During the 2-year follow up period, five patients who responded poorly to phenelzine monotherapy during the randomized treatment phase achieved remission of depressive symptoms after adding lithium to phenelzine. One patient had to stop phenelzine due to marked elevation in hepatic transaminases and one patient had to start thyroid supplementation due to lithium-induced hypothyroidism. Serotonin toxicity is a theoretical risk with lithium augmentation of MAOIs because the efficacy of adjunctive lithium is thought to be related to enhanced serotonergic neurotransmission (Bschor et al., 2003). As such, close monitoring for the signs and symptoms of serotonin toxicity is warranted.

Switching to another MAOI

As mentioned earlier, there is no clear evidence of differential efficacy between MAOIs for depressed patients (Birkenhäger et al., 2004; Tulen et al., 2006). That said, hydrazine and nonhydrazine MAOIs may be different enough from one another pharmacologically and with respect to certain adverse effects that a therapeutic trial of the alternative may be worth considering when one agent is poorly tolerated or ineffective. There is little guidance in the literature regarding best practices for switching between MAOIs. When switching, it is generally recommended that the first MAOI be tapered gradually over at least 4 weeks (or longer if discontinuation symptoms such occur). Once the initial MAOI is stopped, treatment with the substituted MAOI can then be initiated after 14 days starting at a low initial dose, followed by careful titration. We do not generally recommend using moclobemide as a next-step treatment option following a poor clinical response to irreversible MAOIs. This is primarily due to a lack of clear evidence of efficacy for treatment-resistant depressed patients. However, when switching from moclobemide to an irreversible MAOI, it is recommended that moclobemide be tapered gradually over at least 4 weeks and stopped. The irreversible MAOI can be started a day later at a low initial dose followed by careful titration.

Transitioning off of MAOIs

When transitioning off of treatment with MAOIs, a gradual tapering rather than sudden discontinuation is advised as a matter of routine in order to avoid precipitating a discontinuation syndrome. The MAOI dose should be gradually tapered over at least 4 weeks (and longer if symptoms of discontinuation syndrome occur). The exception is, of course, when MAOIs have to be stopped rapidly due to suspected serotonin toxicity or hypertensive crisis. At least 14 days should pass after stopping MAOIs before initiating treatment with a new antidepressant or other medications that are known to interact with MAOIs (including over-the-counter medicines).

Elective surgery and invasive dental procedures

It is commonly advised to stop MAOI treatment 10–14 days before undergoing elective surgery with general anesthesia in order avoid drug interactions that could lead to hemodynamic instability or serotonin toxicity (Huyse et al., 2006). However, such interactions are uncommon and occur with only a small number of medications (Pass and Simpson, 2004). Also, stopping MAOIs before elective surgery can increase the possibility of depressive relapse (Abdi et al., 1996), making such a recommendation infeasible for some patients with a history of severe treatment-resistant depressive illness. Moreover, there is no absolute contraindication to general anesthesia for most surgical patients who take MAOIs (El-Ganzouri et al., 1985; Smith et al., 1996; Stack et al., 1988; Wells and Bjorksten, 1989). In a case-control study of 51 people treated with tranylcypromine or moclobemide and 149 MAOI nonusers, there were no significant increases in risk of hypertension, bradycardia, or tachycardia in comparisons between tranylcypromine and moclobemide as well as MAOI nonusers (van Haelst et al., 2012). The frequency of hypotensive episodes was lower with tranylcypromine than nonuser controls, and there were no occurrences of serotonin toxicity. These results confirmed the findings from previous studies, cited by the study authors, which suggested that irreversible MAOIs can be continued during surgery without a significant increase in the risk of hemodynamic instability (Ebrahim et al., 1993; El-Ganzouri et al., 1985; Wong, 1986).

Nevertheless, members of the surgical team, especially anesthesiologists, need to be aware of any current or recent use of MAOIs due to the risk of hypotension and the need to avoid the intraoperative use of contraindicated drugs. The latter includes the avoidance of sympathomimetic vasoconstrictors such as epinephrine in order to prevent potentially life-threatening pressor effects and the avoidance of meperidine, which could lead to serotonin toxicity. For invasive dental procedures, local epinephrine-containing oral anesthetics are often used and, in such cases, MAOI treatment will need to be stopped at least 2 weeks prior. Alternative oral anesthetics that do not contain epinephrine are generally considered acceptable, and can be considered when MAOI treatment cannot be interrupted. It is reasonable to stop moclobemide on the day of surgery (Peck et al., 2010). Prescribers of MAOIs should be involved in discussions related to risks of continuing versus temporarily discontinuing MAOIs in anticipation of elective surgery or invasive dental procedures, in direct collaboration with anesthesiologists and dentists, whenever possible.

Patients with substance use disorders

MAOIs should be avoided in patients who are using cocaine, methamphetamines, and other stimulants to avoid precipitating hypertensive crises (Grady and Stahl, 2012). We apply the same warning to people with active substance use disorders, in general. This recommendation is based on the likelihood that comorbid substance use disorders are strongly associated with compliance problems in people with mental health conditions (Herbeck et al., 2005), and may therefore be a barrier regarding adherence to important aspects of care while taking MAOIs, including keeping clinical visits, monitoring diet and blood pressure at home, and following dietary and nutritional guidelines. Beyond cocaine and amphetamines, any consumption of contraindicated beverages (sherry, tap beer, etc.), the overconsumption of other types of alcohol deemed low-risk but only in moderation, and the use of other types of contraindicated intoxicating substances (such dextromethorphan and 3,4-methylenedioxymethamphetamine (MDMA)) may be anticipated to present significant health risks in patients taking MAOIs.

There is little guidance in the literature regarding the use of MAOIs in patients with a history of substance use disorders who are in recovery or have a history of durable abstinence. For dually diagnosed patients with therapy-resistant depression who are in need of MAOI treatment, we will often require at least 6 months of full recovery, ideally in an environment that supports the patient's efforts at maintaining durable abstinence, often with at least one additional source of support such as mutual aid groups (e.g., Alcoholics Anonymous, Cocaine- or Narcotics Anonymous, Pills Anonymous, SMART Recovery, Rational Recovery, etc.) or a dual diagnosis or alcohol/drug therapist.

For alcohol-dependent persons on antidipsotropic medications, neither naltrexone nor acamprosate are contraindicated for patients taking MAOIs. Therefore, for opioid-dependent patients taking naltrexone maintenance therapy, MAOIs are also not contraindicated. The manufacturers of buprenorphine and methadone, however, recommend against coadministering these medications with MAOIs due to concerns about precipitating serotonin toxicity. For emergency situations, there does not appear to be a contraindication to administering naloxone in MAOI-treated patients.

Summary

MAOIs are effective antidepressants with a broad-spectrum efficacy profile that encompasses several forms of depression, including chronic and treatment-resistant depression, mixed depressive and anxious states, bipolar major depression (reviewed separately in this volume), and a range of anxiety disorders. Pooled analyses of data from controlled trials suggest that depressed patients who failed to respond to other antidepressants, tricyclic antidepressants in particular, may respond to oral irreversible MAOIs. However, definitions of treatment resistance vary widely across individual studies, and the relatively high estimated efficacy rates with tranylcypromine and phenelzine for treatment-resistant depression reported in early reviews may be exaggerated due to the inclusion of open trials and a number of other factors. Older trials of MAOIs are limited by relatively few treatment failures prior to initiating MAOIs and a paucity of studies comparing MAOIs against antidepressants available for clinical use after the mid-1980s, thus limiting the translational impact of older studies on modern practice. Clinical evidence is lacking with respect to the efficacy of isocarboxazid, transdermal selegiline, and moclobemide for people with well-defined treatment-resistant depression. Nevertheless, phenelzine, tranylcypromine, and possibly also isocarboxazid (because of its pharmacological similarity to phenelzine) can be considered for patients with treatment-resistant depression who fail to respond to at least three therapeutic trials of antidepressive treatments, including patients with atypical and nonatypical depression, and those who respond poorly to electroconvulsive therapy. The safe use of MAOIs requires that clinicians be mindful of common adverse effects and their management, and that they ably assist patients in avoiding potentially serious drug-drug and drug-nutrient interactions. Fortunately, the list of contraindicated drugs is finite and alternative medications are available in most cases. Advancements in food preparation and storage methods have limited the propensity for tyramine overexposure for MAOI-treated patients who consume modern diets. Greater experience in the use of MAOIs, and expert consultation and guidance as needed for difficult cases (i.e., where high MAOI doses or combination therapy with other psychotropic medications is required), will hopefully lead to more clinicians offering MAOIs to patients who are suffering from treatment-resistant depression.

Additional resources

Although this chapter provides a reasonably detailed review and discussion of the use of MAOIs for patients with treatment-resistant depression, readers may be interested in additional sources of information. Below, we provide a selected list of references and other resources that are available free of charge that we have found to be helpful to clinicians and patients.

The PsychoTropical Research website (https://psychotropical.com) is a comprehensive and searchable web-based resource for detailed information on serotonin toxicity and the clinical use of MAOIs. This resource is frequently updated and includes numerous downloadable monographs and podcasts detailing the safe use of MAOIs. One such monograph provides an exhaustive review of the tyramine content of cheeses, aged meats, sauces, and other foods.

There are a number of drug interaction screening tools that are available to clinicians via subscription (e.g., Micromedex, PEPID, Facts & Comparisons, Lexicomp, etc.) or free or charge (e.g., Medscape, Drugs.com, Epocrates, WebMD, RxList, etc.) over the Internet. We advise checking with your institution, if applicable, to verify an existing subscription. It is important to note that the performance characteristics of both subscription-based and free internet tools can vary substantially (Monteith and Glenn, 2019; Schjott et al., 2019). Nevertheless, we think that these tools are suitable for screening potentially severe drug interactions as an initial step. Final prescribing decisions should be based on a thorough review of risks and benefits for all drug combinations under consideration.

Patient and clinician-oriented infographics on serotonin syndrome are available via web-based links provided by the College of Family Physicians of Canada, both for patients (https://www.cfp.ca/content/cfp/suppl/2018/10/11/64.10.720.DC1/SS_Inforgraphic_Patient.pdf) and for clinicians (https://www.cfp.ca/content/cfp/suppl/2018/10/11/64.10.720.DC1/SS_Infographic_Clinician.pdf).

Useful, patient-friendly guidelines for adhering to a low tyramine diet may be found on the internet that are available for downloading, including:

- https://www.health.qld.gov.au/__data/assets/pdf_file/0020/145802/oncol_maoi.pdf
- http://www.sussexpartnership.nhs.uk/sites/default/files/documents/microsoft_word_-_maoi_leaflet_-_may_2013.pdf
- https://edu.cdhb.health.nz/Patients-Visitors/patient-information-pamphlets/Documents/MAOI-Diet-Pil-747.pdf
- https://prd-medweb-cdn.s3.amazonaws.com/documents/neurology/files/Tyramine%20Menu%20Book%2006227101.pdf

There are several other web sites that feature recipes that are labeled as being "low tyramine" or even "MAOI safe." We advise patients to avoid using such recipes until they can be reviewed by their prescribers. Portion size as well as the estimated relative tyramine content of the ingredients should be taken into account.

References

Abdi, S., Fishman, S.M., Messner, E., 1996. Acute exacerbation of depression after discontinuation of monoamine oxidase inhibitor prior to cardiac surgery. Anesth. Analg. 83 (3), 656–657. https://doi.org/10.1213/00000539-199609000-00041.

Albucher, R.C., Liberzon, I., 2002. Psychopharmacological treatment in PTSD: a critical review. J. Psychiatr. Res. 36 (6), 355–367. https://doi.org/10.1016/S0022-3956(02)00058-4.

American Psychiatric Association, 2010. Practice guidelines for the treatment of patients with major depressive disorder (third edition). Am. J. Psychiatry 167, 1–118.

Amsterdam, J.D., 2003. A double-blind, placebo-controlled trial of the safety and efficacy of selegiline transdermal system without dietary restrictions in patients with major depressive disorder. J. Clin. Psychiatry 64 (2), 208–214. https://doi.org/10.4088/JCP.v64n0216.

Amsterdam, J.D., Berwish, N.J., 1989. High dose tranylcypromine therapy for refractory depression. Pharmacopsychiatry 22 (1), 21–25. https://doi.org/10.1055/s-2007-1014572.

Amsterdam, J.D., Bodkin, J.A., 2006. Selegiline transdermal system in the prevention of relapse of major depressive disorder: a 52-week, double-blind, placebo-substitution, parallel-group clinical trial. J. Clin. Psychopharmacol. 26 (6), 579–586. https://doi.org/10.1097/01.jcp.0000239794.37073.70.

Amsterdam, J.D., Chopra, M., 2001. Monoamine oxidase inhibitors revisited. Psychiatr. Ann. 31 (6), 361–370. https://doi.org/10.3928/0048-5713-20010601-05.

Amsterdam, J.D., Shults, J., 2005. MAOI efficacy and safety in advanced stage treatment-resistant depression—a retrospective study. J. Affect. Disord. 89 (1–3), 183–188. https://doi.org/10.1016/j.jad.2005.06.011.

Angst, J., Stabl, M., 1992. Efficacy of moclobemide in different patient groups: a meta-analysis of studies. Psychopharmacology (Berl) 106 (1), S109–S113. https://doi.org/10.1007/BF02246250.

Asnis, G.M., McGinn, L.K., Sanderson, W.C., 1995. Atypical depression: clinical aspects and noradrenergic function. Am. J. Psychiatry 152 (1), 31–36. https://doi.org/10.1176/ajp.152.1.31.

Azzaro, A.J., Van DenBerg, C.M., Blob, L.F., Kemper, E.M., Sharoky, M., Oren, D.A., Campbell, B.J., 2006. Tyramine pressor sensitivity during treatment with the selegiline transdermal system 6 mg/24 h in healthy subjects. J. Clin. Pharmacol. 46 (8), 933–944. https://doi.org/10.1177/0091270006289852.

Azzaro, A.J., Ziemniak, J., Kemper, E., Campbell, B.J., Van DenBerg, C., 2007a. Pharmacokinetics and absolute bioavailability of selegiline following treatment of healthy subjects with the selegiline transdermal system (6 mg/24 h): a comparison with oral selegiline capsules. J. Clin. Pharmacol. 47 (10), 1256–1267. https://doi.org/10.1177/0091270007304779.

Azzaro, A.J., Ziemniak, J., Kemper, E., Campbell, B.J., Vandenberg, C., 2007b. Selegiline transdermal system: an examination of the potential for CYP450-dependent pharmacokinetic interactions with 3 psychotropic medications. J. Clin. Pharmacol. 47 (2), 146–158. https://doi.org/10.1177/0091270006296151.

Baker, G.B., Coutts, R.T., McKenna, K.F., Sherry-McKenna, R.L., 1992. Insights into the mechanisms of action of the MAO inhibitors phenelzine and tranylcypromine: a review. J. Psychiatry Neurosci. 17 (5), 206–214.

Baker, G.B., Urichuk, L.J., McKenna, K.F., Kennedy, S.H., 1999. Metabolism of monoamine oxidase inhibitors. Cell. Mol. Neurobiol. 19 (3), 411–426. https://doi.org/10.1023/A:1006982732681.

Balon, R., Mufti, R., Arfken, C.L., 1999. A survey of prescribing practices for monoamine oxidase inhibitors. Psychiatr. Serv. 50 (7), 945–947. https://doi.org/10.1176/ps.50.7.945.

Bauer, M., Pfennig, A., Severus, E., Whybrow, P.C., Angst, J., Möller, H.J., Adli, M., Benkert, O., Bschor, T., Tadić, A., Holsboer-Trachsler, E., Anderson, I., Baldwin, D., Cookson, J.C., Grunze, H., Katona, C., Paykel, E.S., Tylee, A., Ayuso-Gutierrez, J.L., et al., 2013. World Federation of Societies of Biological Psychiatry (WFSBP) guidelines for biological treatment of unipolar depressive disorders, part 1: update 2013 on the acute and continuation treatment of unipolar depressive disorders. World J. Biol. Psychiatry 14 (5), 334–385. https://doi.org/10.3109/15622975.2013.804195.

Bieck, P.R., Antonin, K.H., 1988. Oral tyramine pressor test and the safety of monoamine oxidase inhibitor drugs: comparison of brofaromine and tranylcypromine in healthy subjects. J. Clin. Psychopharmacol. 8 (4), 237–245.

Billett, E.E., 2004. Monoamine oxidase (MAO) in human peripheral tissues. NeuroToxicology 25 (1–2), 139–148. https://doi.org/10.1016/S0161-813X(03)00094-9.

Birkenhäger, T.K., Van Den Broek, W.W., Mulder, P.G., Bruijn, J.A., Moleman, P., 2004. Efficacy and tolerability of tranylcypromine versus phenelzine: a double-blind study in antidepressant-refractory depressed inpatients. J. Clin. Psychiatry 65 (11), 1505–1510. https://doi.org/10.4088/JCP.v65n1110.

Bobo, W., Riva-Posse, P., Goes, F., Parkih, S.V., 2020. Next-step treatment considerations for patients with treatment-resistant depression that responds to low-dose ketamine. Focus 18, 181–192.

Bodkin, J.A., Amsterdam, J.D., 2002. Transdermal selegiline in major depression: a double-blind, placebo-controlled, parallel-group study in outpatients. Am. J. Psychiatry 159 (11), 1869–1875. https://doi.org/10.1176/appi.ajp.159.11.1869.

Boyer, E.W., Shannon, M., 2005. Current concepts: the serotonin syndrome. N. Engl. J. Med. 352 (11), 1112–1120. https://doi.org/10.1056/NEJMra041867.

Bschor, T., Lewitzka, U., Sasse, J., Adli, M., Köberle, U., Bauer, M., 2003. Lithium augmentation in treatment-resistant depression: clinical evidence, serotonergic and endocrine mechanisms. Pharmacopsychiatry 36 (3), S230–S234.

Cantu, T.G., Korek, J.S., 1988. Monoamine oxidase inhibitors and weight gain. Drug Intell. Clin. Pharm. 22 (10), 755–759. https://doi.org/10.1177/106002808802201002.

Chiuccariello, L., Houle, S., Miler, L., Cooke, R.G., Rusjan, P.M., Rajkowska, G., Levitan, R.D., Kish, S.J., Kolla, N.J., Ou, X., Wilson, A.A., Meyer, J.H., 2014. Elevated monoamine oxidase A binding during major depressive episodes is associated with greater severity and reversed neurovegetative symptoms. Neuropsychopharmacology 39 (4), 973–980. https://doi.org/10.1038/npp.2013.297.

Cipriani, A., Geddes, J.R., Furukawa, T.A., Barbui, C., 2007. Metareview on short-term effectiveness and safety of antidepressants for depression: an evidence-based approach to inform clinical practice. Can. J. Psychiatry 52 (9), 553–562. https://doi.org/10.1177/070674370705200903.

Clayton, A.H., Campbell, B.J., Favit, A., Yang, Y., Moonsammy, G., Piontek, C.M., Amsterdam, J.D., 2007. Symptoms of sexual dysfunction in patients treated for major depressive disorder: a meta-analysis comparing selegiline transdermal system and placebo using a patient-rated scale. J. Clin. Psychiatry 68 (12), 1860–1866. https://doi.org/10.4088/JCP.v68n1205.

Cleare, A., Pariante, C.M., Young, A.H., Anderson, I.M., Christmas, D., Cowen, P.J., Dickens, C., Ferrier, I.N., Geddes, J., Gilbody, S., Haddad, P.M., Katona, C., Lewis, G., Malizia, A., McAllister-Williams, R.H., Ramchandani, P., Scott, J., Taylor, D., Uher, R., 2015. Evidence-based guidelines for treating depressive disorders with antidepressants: a revision of the 2008 British Association for Psychopharmacology guidelines. J. Psychopharmacol. 29 (5), 459–525. https://doi.org/10.1177/0269881115581093.

Culpepper, L., 2013. Reducing the burden of difficult-to-treat major depressive disorder: revisiting monoamine oxidase inhibitor therapy. Prim. Care Companion CNS Disord. 15 (5). https://doi.org/10.4088/PCC.13r01515.

da Prada, M., Zurcher, G., Wuthrich, I., Haefely, W.E., 1988. On tyramine, food, beverages and the reversible MAO inhibitor moclobemide. J. Neural Transm. 26, 31–56.

Danish University Antidepressant Group, 1993. Moclobemide: a reversible MAOI-A-inhibitor showing weaker antidepressant effect than clomipramine in a controlled multicenter study. J. Affect. Disord. 28, 105–116.

Davidson, J., Miller, R., Turnbull, C.D., Belyea, M., Strickland, R., 1984. An evaluation of two doses of isocarboxazid in depression. J. Affect. Disord. 6 (2), 201–207. https://doi.org/10.1016/0165-0327(84)90025-9.

Davidson, J.R., 2007. A history of the concept of atypical depression. J. Clin. Psychiatry 68, 10–15.

Davidson, J.R.T., Giller, E.L., Zisook, S., Overall, J.E., 1988. An efficacy study of isocarboxazid and placebo in depression, and its relationship to depressive nosology. Arch. Gen. Psychiatry 45 (2), 120–127. https://doi.org/10.1001/archpsyc.1988.01800260024003.

Davis, R., Markham, A., 1997. Ziprasidone. CNS Drugs 8 (2), 153–159. https://doi.org/10.2165/00023210-199708020-00006.

Duarte, A., Mikkelsen, H., Delini-Stula, A., 1996. Moclobemide versus fluoxetine for double depression: a randomized double-blind study. J. Psychiatr. Res. 30 (6), 453–458. https://doi.org/10.1016/S0022-3956(96)00030-1.

Ebrahim, Z.Y., O'Hara, J., Borden, L., Tetzlaff, J., 1993. Monoamine oxidase inhibitors and elective surgery. Cleve. Clin. J. Med. 60 (2), 129–130. https://doi.org/10.3949/ccjm.60.2.129.

El-Ganzouri, A.R., Ivankovich, A.D., Braverman, B., McCarthy, R., 1985. Monoamine oxidase inhibitors: should they be discontinued preoperatively? Anesth. Analg. 64 (6), 592–596.

Fein, S., Paz, V., Rao, N., LaGrassa, J., 1988. The combination of lithium carbonate and MAOI in refractory depressions. Am. J. Psychiatry 145 (2), 249–250. https://doi.org/10.1176/ajp.145.2.249.

Fekadu, A., Rane, L.J., Wooderson, S.C., Markopoulou, K., Poon, L., Cleare, A.J., 2012. Prediction of longer-term outcome of treatment-resistant depression in tertiary care. Br. J. Psychiatry 201 (5), 369–375. https://doi.org/10.1192/bjp.bp.111.102665.

Ferreira-Garcia, R., Da Rocha Freire, R.C., Appolinário, J.C., Levitan, M.N., Halkjær-Lassen, R.D., Bueno, J.R., Nardi, A.E., 2018. Tranylcypromine plus amitriptyline for electroconvulsive therapy-resistant depression. J. Clin. Psychopharmacol. 38 (5), 502–504. https://doi.org/10.1097/JCP.0000000000000945.

Fiedorowicz, J.G., Swartz, K.L., 2004. The role of monoamine oxidase inhibitors in current psychiatric practice. J. Psychiatr. Pract. 10 (4), 239–248. https://doi.org/10.1097/00131746-200407000-00005.

Flament, M.F., Bissada, H., Spettigue, W., 2012. Evidence-based pharmacotherapy of eating disorders. Int. J. Neuropsychopharmacol. 15 (2), 189–207. https://doi.org/10.1017/S1461145711000381.

Flockhart, D.A., 2012. Dietary restrictions and drug interactions with monoamine oxidase inhibitors: an update. J. Clin. Psychiatry 73 (1), 17–24. https://doi.org/10.4088/JCP.11096su1c.03.

Fowler, J.S., Logan, J., Volkow, N.D., Wang, G.J., 2005. Translational neuroimaging: positron emission tomography studies of monoamine oxidase. Mol. Imaging Biol. 7 (6), 377–387. https://doi.org/10.1007/s11307-005-0016-1.

Fulton, B., Benfield, P., Chouinard, G., Evans, L., Kragh-Sorensen, P., Laux, G., Maquire, K., Nair, N.P.V., Priest, R.G., Reynaert, C., Silverstone, T., 1996. Moclobemide. An update of its pharmacological properties and therapeutic use. Drugs 52 (3), 450–474. https://doi.org/10.2165/00003495-199652030-00013.

Gardner, D.M., Shulman, K.I., Walker, S.E., Tailor, S.A.N., 1996. The making of a user friendly MAOI diet. J. Clin. Psychiatry 57 (3), 99–104.

Gaynes, B.N., 2009. Identifying difficult-to-treat depression: differential diagnosis, subtypes, and comorbidities. J. Clin. Psychiatry 70 (6), 10–15. https://doi.org/10.4088/JCP.8133su1c.02.

Gelenberg, A.J., 1998. The MAOI diet. Biol. Ther. Psychiatry News. 21 (2).

Gentil, V., Alevizos, B., Felix-Gentil, M., Lader, M., 1978. Single-dose effects of tranylcypromine on psychophysiological measures in normals. Br. J. Clin. Pharmacol. 5 (6), 536–538. https://doi.org/10.1111/j.1365-2125.1978.tb01672.x.

Georgotas, A., Friedman, E., McCarthy, M., Mann, J., Krakowski, M., Siegel, R., Ferris, S., 1983. Resistant geriatric depressions and therapeutic response to monoamine oxidase inhibitors. Biol. Psychiatry 18 (2), 195–205.

Gillman, P., 2004. Moclobemide and the risk of serotonin toxicity (or serotonin syndrome). CNS Drug Rev., 83–85. https://doi.org/10.1111/j.1527-3458.2004.tb00005.x.

Gillman, P.K., 2006. A review of serotonin toxicity data: implications for the mechanisms of antidepressant drug action. Biol. Psychiatry 59 (11), 1046–1051. https://doi.org/10.1016/j.biopsych.2005.11.016.

Gillman, P.K., 2010. Triptans, serotonin agonists, and serotonin syndrome (serotonin toxicity): a review: review article. Headache 50 (2), 264–272. https://doi.org/10.1111/j.1526-4610.2009.01575.x.

Gillman, P.K., 2011. Advances pertaining to the pharmacology and interactions of irreversible nonselective monoamine oxidase inhibitors. J. Clin. Psychopharmacol. 31 (1), 66–74. https://doi.org/10.1097/JCP.0b013e31820469ea.

Gillman, K., 2017. Much ado about nothing: monoamine oxidase inhibitors, drug interactions, and dietary tyramine. CNS Spectr. 22 (5), 385–387. https://doi.org/10.1017/S1092852916000651.

Gillman, P.K., 2018. A reassessment of the safety profile of monoamine oxidase inhibitors: elucidating tired old tyramine myths. J. Neural Transm. 125 (11), 1707–1717. https://doi.org/10.1007/s00702-018-1932-y.

Gillman, K., 2020. MAOIs-International Shortage. Available online at: https://psychotropical.com/maois-international-shortage/. Accessed 19 February 2022.

Grady, M.M., Stahl, S.M., 2012. Practical guide for prescribing MAOIs: debunking myths and removing barriers. CNS Spectr. 17 (1), 2–10. https://doi.org/10.1017/S109285291200003X.

Gram, L.F., Guentert, T.W., Grange, S., Vistisen, K., Brøsen, K., 1995. Moclobemide, a substrate of CYP2C19 and an inhibitor of CYP2C19, CYP2D6, and CYP1A2: a panel study. Clin. Pharmacol. Ther. 57 (6), 670–677. https://doi.org/10.1016/0009-9236(95)90230-9.

Grossman, E., Messerli, F.H., Grodzicki, T., Kowey, P., 1996. Should a moratorium be placed on sublingual nifedipine capsules given for hypertensive emergencies and pseudoemergencies? JAMA 276 (16), 1328–1331. https://doi.org/10.1001/jama.276.16.1328.

Guze, B.H., Baxter, L.R., Rego, J., 1987. Refractory depression treated with high doses of a monoamine oxidase inhibitor. J. Clin. Psychiatry 48 (1), 31–32.

Hamilton, M., 1967. Development of a rating scale for primary depressive illness. Br. J. Soc. Clin. Psychol. 6 (4), 278–296. https://doi.org/10.1111/j.2044-8260.1967.tb00530.x.

Harrison, W., Rabkin, J., Stewart, J.W., McGrath, P.J., Tricamo, E., Quitkin, F., 1986. Phenelzine for chronic depression: a study of continuation treatment. J. Clin. Psychiatry 47 (7), 346–349.

Hawley, C.J., Quick, S.J., Ratnam, S., Pattinson, H.A., McPhee, S., 1996. Safety and tolerability of combined treatment with moclobemide and SSRIs: a systematic study of 50 patients. Int. Clin. Psychopharmacol. 11 (3), 187–191. https://doi.org/10.1097/00004850-199609000-00005.

Hendley, E.D., Snyder, S.H., 1968. Relationship between the action of monoamine oxidase inhibitors on the noradrenaline uptake system and their antidepressant efficacy [16]. Nature 220 (5174), 1330–1331. https://doi.org/10.1038/2201330a0.

Henkel, V., Mergl, R., Allgaier, A.K., Kohnen, R., Möller, H.J., Hegerl, U., 2006. Treatment of depression with atypical features: a meta-analytic approach. Psychiatry Res. 141 (1), 89–101. https://doi.org/10.1016/j.psychres.2005.07.012.

Herbeck, D.M., Fitek, D.J., Svikis, D.S., Montoya, I.D., Marcus, S.C., West, J.C., 2005. Treatment compliance in patients with comorbid psychiatric and substance use disorders. Am. J. Addict. 14 (3), 195–207. https://doi.org/10.1080/10550490590949488.

Hullett, F., Bidder, T., 1983. Phenelzine plus triiodothyronine combination in a case of refractory depression. J. Nerv. Ment. Dis. 171 (5), 318–320. https://doi.org/10.1097/00005053-198305000-00011.

Huyse, F.J., Touw, D.J., Van Schijndel, R.S., De Lange, J.J., Slaets, J.P.J., 2006. Psychotropic drugs and the perioperative period: a proposal for a guideline in elective surgery. Psychosomatics 47 (1), 8–22. https://doi.org/10.1176/appi.psy.47.1.8.

Isbister, G.K., Hackett, L.P., Dawson, A.H., Whyte, I.M., Smith, A.J., 2003. Moclobemide poisoning: toxicokinetics and occurrence of serotonin toxicity. Br. J. Clin. Pharmacol. 56 (4), 441–450. https://doi.org/10.1046/j.1365-2125.2003.01895.x.

Israel, J.A., 2015. Combining stimulants and monoamine oxidase inhibitors: a reexamination of the literature and a report of a new treatment combination. Prim. Care Companion CNS Disord. 17 (6), 423–426. https://doi.org/10.4088/PCC.15br01836.

Jefferson, J.W., 1997. Antidepressants in panic disorder. J. Clin. Psychiatry 58 (2), 20–24.

Joffe, R.T., 1988. Triiodothyronine potentiation of the antidepressant effect of phenelzine. J. Clin. Psychiatry 49 (10), 409–410.

Keck, P., Hudson, J., Cunningham, S., Harlyn, R., Aizley, G., McElroy, S., Vuckovic, A., 1992. Polysomnography features of excessive daytime somnolence associated with monoamine oxidase inhibitor treatment. Ann. Clin. Psychiatry, 49–53. https://doi.org/10.3109/10401239209148025.

Keks, N., Hope, J., Keogh, S., 2016. Switching and stopping antidepressants. Aust. Prescr. 39 (3), 76–83. https://doi.org/10.18773/austprescr.2016.039.

Kellner, C.H., Greenberg, R.M., Murrough, J.W., Bryson, E.O., Briggs, M.C., Pasculli, R.M., 2012. ECT in treatment-resistant depression. Am. J. Psychiatry 169 (12), 1238–1244. https://doi.org/10.1176/appi.ajp.2012.12050648.

Kennedy, S.H., Lam, R.W., McIntyre, R.S., Tourjman, S.V., Bhat, V., Blier, P., Hasnain, M., Jollant, F., Levitt, A.J., MacQueen, G.M., McInerney, S.J., McIntosh, D., Milev, R.V., Müller, D.J., Parikh, S.V., Pearson, N.L., Ravindran, A.V., Uher, R., 2016. Canadian Network for Mood and Anxiety Treatments (CANMAT) 2016 clinical guidelines for the management of adults with major depressive disorder: section 3. Pharmacological Treatments. Can. J. Psychiatry 61 (9), 540–560. https://doi.org/10.1177/0706743716659417.

Knud, L.J., Lene, K.N., 2016. The monoamine oxidase inhibitor isocarboxazid is a relevant treatment option in treatment-resistant depression-experience-based strategies in Danish psychiatry. Health Care Curr. Rev. https://doi.org/10.4172/2375-4273.1000168.

Kok, R.M., Vink, D., Heeren, T.J., Nolen, W.A., 2007. Lithium augmentation compared with phenelzine in treatment-resistant depression in the elderly: an open, randomized, controlled trial. J. Clin. Psychiatry 68 (8), 1177–1185. https://doi.org/10.4088/jcp.v68n0803.

Kriston, L., Von Wolff, A., Westphal, A., Hölzel, L.P., Härter, M., 2014. Efficacy and acceptability of acute treatments for persistent depressive disorder: a network meta-analysis. Depress. Anxiety 31 (8), 621–630. https://doi.org/10.1002/da.22236.

Larsen, J.K., Gjerris, A., Holm, P., Andersen, J., Bille, A., Christensen, E.M., Høyer, E., Jensen, H., Mejlhede, A., Langagergaard, A., Laursen, A.L., Nilakantan, B., Olafsson, K., Severin, B., Rafaelsen, O.J., 1991. Moclobemide in depression: a randomized, multicentre trial against isocarboxazide and clomipramine emphasizing atypical depression. Acta Psychiatr. Scand. 84 (6), 564–570. https://doi.org/10.1111/j.1600-0447.1991.tb03196.x.

Lavian, G., Finberg, J., Youdim, M.B., 1993. The advent of a new generation of monoamine oxidase inhibitor antidepressants: pharmacologic studies with moclobemide and brofaromine. Clin. Neuropharmacol. 16 (Suppl 2), S1–S7.

Lee, K.C., Chen, J.J., 2007. Transdermal selegiline for the treatment of major depressive disorder. Neuropsychiatr. Dis. Treat. 3 (5), 527–537.

Liebowitz, M.R., Quitkin, F.M., Stewart, J.W., Mcgrath, P.J., Harrison, W., Rabkin, J., Tricamo, E., Markowitz, J.S., Klein, D.F., 1984. Phenelzine v imipramine in atypical depression: a preliminary report. Arch. Gen. Psychiatry 41 (7), 669–677. https://doi.org/10.1001/archpsyc.1984.01790180039005.

Liebowitz, M.R., Quitkin, F.M., Stewart, J.W., McGrath, P.J., Harrison, W.M., Markowitz, J.S., Rabkin, J.G., Tricamo, E., Goetz, D.M., Klein, D.F., 1988. Antidepressant specificity in atypical depression. Arch. Gen. Psychiatry 45 (2), 129–137. https://doi.org/10.1001/archpsyc.1988.01800260037004.

Lima, M., Moncrieff, J., 2000. A comparison of drugs versus placebo for the treatment of dysthymia. Cochrane Database Syst. Rev. (2), CD001130.

Linden, C.H., Rumack, B.H., Strehlke, C., 1984. Monoamine oxidase inhibitor overdose. Ann. Emerg. Med. 13 (12), 1137–1144. https://doi.org/10.1016/S0196-0644(84)80339-X.

LiverTox, 2012. LiverTox: Clinical and Research Information on Drug-Induced Liver Injury. National Institute of Diabetes and Digestive and Kidney Diseases, Bethesda, MD. https://www.ncbi.nlm.nih.gov/books/NBK548891/.

LiverTox, 2020a. LiverTox: Clinical and Research Information on Drug-Induced Liver Injury [Internet]. Phenelzine (updated April 8, 2020). Available online at: https://www.ncbi.nlm.nih.gov/books/NBK548402/. Accessed 21 August 2020.

LiverTox, 2020b. LiverTox: Clinical and Research Information on Drug-Induced Liver Injury [Internet]. Tranylcypromine (updated April 8, 2020). Available online at: https://www.ncbi.nlm.nih.gov/books/NBK548572/. Accessed 21 August 2020.

LiverTox, 2020c. LiverTox: Clinical and Research Information on Drug-Induced Liver Injury [Internet]. Isocarboxazid (updated April 8, 2020). Available online at: https://www.ncbi.nlm.nih.gov/books/NBK548518/. Accessed 21 August 2020.

LiverTox, 2020d. LiverTox: Clinical and Research Information on Drug-Induced Liver Injury [Internet]. Selegiline (updated April 8, 2020). Available online at: https://www.ncbi.nlm.nih.gov/books/NBK548891/. Accessed 21 August 2020.

Livingston, M.G., Livingston, H.M., 1996. Monoamine oxidase inhibitors: an update on drug interactions. Drug Saf. 14 (4), 219–227. https://doi.org/10.2165/00002018-199614040-00002.

López-Muñoz, F., Alamo, C., 2009. Monoaminergic neurotransmission: the history of the discovery of antidepressants from 1950s until today. Curr. Pharm. Des. 15 (14), 1563–1586. https://doi.org/10.2174/138161209788168001.

López-Muñoz, F., Álamo, C., Juckel, G., Assion, H.J., 2007. Half a century of antidepressant drugs—on the clinical introduction of monoamine oxidase inhibitors, tricyclics, and tetracyclics. Part I: monoamine oxidase inhibitors. J. Clin. Psychopharmacol. 27 (6), 555–559. https://doi.org/10.1097/jcp.0b013e3181bb617.

Lotufo-Neto, F., Trivedi, M., Thase, M.E., 1999. Meta-analysis of the reversible inhibitors of monoamine oxidase type A moclobemide and brofaromine for the treatment of depression. Neuropsychopharmacology 20 (3), 226–247. https://doi.org/10.1016/S0893-133X(98)00075-X.

Lucena, M., Carvajal, A., Andrade, R.J., Velasco, A., 2003. Antidepressant-induced hepatotoxicity. Expert Opin. Drug Saf. 2 (3), 249–262. https://doi.org/10.1517/14740338.2.3.249.

Magyar, K., 2011. The pharmacology of selegiline. In: International Review of Neurobiology. Vol. 100. Academic Press Inc, pp. 65–84, https://doi.org/10.1016/B978-0-12-386467-3.00004-2.

Malhi, G.S., Bassett, D., Boyce, P., Bryant, R., Fitzgerald, P.B., Fritz, K., Hopwood, M., Lyndon, B., Mulder, R., Murray, G., Porter, R., Singh, A.B., 2015. Royal Australian and New Zealand college of psychiatrists clinical practice guidelines for mood disorders. Aust. N. Z. J. Psychiatry 49 (12), 1087–1206. https://doi.org/10.1177/0004867415617657.

Mallinger, A.G., Smith, E., 1991. Pharmacokinetics of monoamine oxidase inhibitors. Psychopharmacol. Bull. 27 (4), 493–502.

McAllister-Williams, R.H., Arango, C., Blier, P., Demyttenaere, K., Falkai, P., Gorwood, P., Hopwood, M., Javed, A., Kasper, S., Malhi, G.S., Soares, J.C., Vieta, E., Young, A.H., Papadopoulos, A., Rush, A.J., 2020. The identification, assessment and management of difficult-to-treat depression: an international consensus statement. J. Affect. Disord. 267, 264–282. https://doi.org/10.1016/j.jad.2020.02.023.

McGrath, P.J., Blood, D.K., Stewart, J.W., Harrison, W., Quitkin, F.M., Tricamo, E., Markowitz, J., 1987. A comparative study of the electrocardiographic effects of phenelzine, tricyclic antidepressants, mianserin, and placebo. J. Clin. Psychopharmacol. 7 (5), 335–339. https://doi.org/10.1097/00004714-198710000-00009.

McGrath, P.J., Stewart, J.W., Nunes, E.V., Ocepek-Welikson, K., Rabkin, J.G., Quitkin, F.M., Klein, D.F., 1993. A double-blind crossover trial of imipramine and phenelzine for outpatients with treatment-refractory depression. Am. J. Psychiatry 150 (1), 118–123. https://doi.org/10.1176/ajp.150.1.118.

McGrath, P.J., Stewart, J.W., Fava, M., Trivedi, M.H., Wisniewski, S.R., Nierenberg, A.A., Thase, M.E., Davis, L., Biggs, M.M., Shores-Wilson, K., Luther, J.F., Niederehe, G., Warden, D., Rush, A.J., 2006. Tranylcypromine versus venlafaxine plus mirtazapine following three failed antidepressant medication trials for depression: a STAR*D report. Am. J. Psychiatry 163 (9), 1531–1541. https://doi.org/10.1176/ajp.2006.163.9.1531.

Meyer, J.M., Cummings, M.A., Proctor, G., 2017. Augmentation of phenelzine with aripiprazole and quetiapine in a treatment-resistant patient with psychotic unipolar depression: case report and literature review. CNS Spectr. 22 (5), 391–396. https://doi.org/10.1017/S1092852916000821.

Millan, M.J., 2006. Multi-target strategies for the improved treatment of depressive states: conceptual foundations and neuronal substrates, drug discovery and therapeutic application. Pharmacol. Ther. 110 (2), 135–370. https://doi.org/10.1016/j.pharmthera.2005.11.006.

Monteith, S., Glenn, T., 2019. A comparison of potential psychiatric drug interactions from six drug interaction database programs. Psychiatry Res. 275, 366–372. https://doi.org/10.1016/j.psychres.2019.03.041.

Nagler, E.V., Webster, A.C., Vanholder, R., Zoccali, C., 2012. Antidepressants for depression in stage 3–5 chronic kidney disease: a systematic review of pharmacokinetics, efficacy and safety with recommendations by European Renal Best Practice (ERBP). Nephrol. Dial. Transplant. 27 (10), 3736–3745. https://doi.org/10.1093/ndt/gfs295.

Naguib, M., Koorn, R., 2002. Interactions between psychotropics, anaesthetics and electroconvulsive therapy: implications for drug choice and patient management. CNS Drugs 16 (4), 229–247. https://doi.org/10.2165/00023210-200216040-00003.

Nair, N.P.V., Ahmed, S.K., Ng Ying Kin, N.M.K., 1993. Biochemistry and pharmacology of reversible inhibitors of MAO-A agents: focus on moclobemide. J. Psychiatry Neurosci. 18 (5), 214–225.

Nelson, J.C., Byck, R., 1982. Rapid response to lithium in phenelzine non-responders. Br. J. Psychiatry 141 (1), 85–86. https://doi.org/10.1192/bjp.141.1.85.

Nelson, J.C., Baumann, P., Delucchi, K., Joffe, R., Katona, C., 2014. A systematic review and meta-analysis of lithium augmentation of tricyclic and second generation antidepressants in major depression. J. Affect. Disord. 168, 269–275. https://doi.org/10.1016/j.jad.2014.05.053.

Nolen, W.A., 1989. Tranylcypromine in depression resistant to cyclic antidepressions. Prog. Neuropsychopharmacol. Biol. Psychiatry 13 (1–2), 155–158. https://doi.org/10.1016/0278-5846(89)90012-2.

Nolen, W.A., Van de Putte, J.J., Dijken, W.A., Kamp, J.S., 1985. L-5HTP in depression resistant to re-uptake inhibitors. An open comparative study with tranylcypromine. Br. J. Psychiatry 147 (JULY), 16–22. https://doi.org/10.1192/bjp.147.1.16.

Nolen, W.A., van de Putte, J.J., Dijken, W.A., Kamp, J.S., Blansjaar, B.A., Kramer, H.J., Haffmans, J., 1988. Treatment strategy in depression: II. MAO inhibitors in depression resistant to cyclic antidepressants: two controlled crossover studies with tranylcypromine versus L-5-hydroxytryptophan and nomifensine. Acta Psychiatr. Scand. 78 (6), 676–683. https://doi.org/10.1111/j.1600-0447.1988.tb06403.x.

Nolen, W., Judith Haffmans, P., Bouvy, P., Duivenvoorden, H., 1993. Monoamine oxidase inhibitors in resistant major depression. J. Affect. Disord., 189–197. https://doi.org/10.1016/0165-0327(93)90104-r.

Nolen, W.A., van den Broek, W.W., Birkenhäger, T.K., 2007. Treatment with low doses of tranylcypromine resulted in a disappointing remission rate [1]. Am. J. Psychiatry 164 (3), 524. https://doi.org/10.1176/ajp.2007.164.3.524.

Norman, T.R., Burrows, G.D., 1989. Monoamine oxidase, monoamine oxidase inhibitors, and panic disorder. J. Neural Transm. Suppl. 28, 53–63.

Pae, C.U., Bodkin, J.A., Portland, K.B., Thase, M.E., Patkar, A.A., 2012. Safety of selegiline transdermal system in clinical practice: analysis of adverse events from postmarketing exposures. J. Clin. Psychiatry 73 (5), 661–668. https://doi.org/10.4088/JCP.12m07648.

Pae, C.U., Patkar, A.A., Jang, S., Portland, K.B., Jung, S., Nelson, J.C., 2014. Efficacy and safety of selegiline transdermal system (STS) for the atypical subtype of major depressive disorder: pooled analysis of 5 short-term, placebo-controlled trials. CNS Spectr. 19 (4), 324–329. https://doi.org/10.1017/S1092852913000655.

Pare, C.M.B., Hallstrom, C., Kline, N., Cooper, T.B., 1982. Will amitriptyline prevent the "cheese" reaction of monoamine-oxidase inhibitors? Lancet 320 (8291), 183–186. https://doi.org/10.1016/S0140-6736(82)91030-3.

Parker, G.B., 2007. Atypical depression: a valid subtype? J. Clin. Psychiatry 68 (3), 18–22. https://doi.org/10.4088/jcp.0307e08.

Paslawski, T., Treit, D., Baker, G.B., George, M., Coutts, R.T., 1996. The antidepressant drug phenelzine produces antianxiety effects in the plus-maze and increases in rat brain GABA. Psychopharmacology (Berl) 127 (1), 19–24. https://doi.org/10.1007/BF02805970.

Pass, S.E., Simpson, R.W., 2004. Discontinuation and reinstitution of medications during the perioperative period. Am. J. Health Syst. Pharm. 61 (9), 899–914. https://doi.org/10.1093/ajhp/61.9.899.

Paykel, E.S., Rowan, P.R., Parker, R.R., Bhat, A.V., 1982. Response to phenelzine and amitriptyline in subtypes of outpatient depression. Arch. Gen. Psychiatry 39 (9), 1041–1049. https://doi.org/10.1001/archpsyc.1982.04290090035008.

Peck, T., Wong, A., Norman, E., 2010. Anaesthetic implications of psychoactive drugs. Contin. Educ. Anaesth. Crit. Care Pain 10 (6), 177–181. https://doi.org/10.1093/bjaceaccp/mkq037.

Piñeyro, G., Blier, P., 1999. Autoregulation of serotonin neurons: role in antidepressant drug action. Pharmacol. Rev. 51 (3), 533–591.

Pletscher, A., 1991. The discovery of antidepressants: a winding path. Experientia 47 (1), 4–8. https://doi.org/10.1007/BF02041242.

Pray, W.S., 2007. Interactions between nonprescription products and psychotropic medications. U.S. Pharmacist 32 (11), 12–15. http://www.uspharmacist.com/index.asp?show=article&page=8_2168.htm.

Price, L.H., Charney, D.S., Heninger, G.R., 1985. Efficacy of lithium-tranylcypromine treatment in refractory depression. Am. J. Psychiatry 142 (5), 619–623. https://doi.org/10.1176/ajp.142.5.619.

Quante, A., Zeugmann, S., 2012. Tranylcypromine and bupropion combination therapy in treatment-resistant major depression: a report of 2 cases. J. Clin. Psychopharmacol. 32 (4), 572–574. https://doi.org/10.1097/JCP.0b013e31825de0a7.

Quitkin, F.M., 2002. Depression with atypical features: diagnostic validity, prevalence, and treatment. Prim. Care Companion J. Clin. Psychiatry 4 (3), 94–99. https://doi.org/10.4088/pcc.v04n0302.

Quitkin, F., Rifkin, A., Klein, D.F., 1979. Monoamine oxidase inhibitors: a review of antidepressant effectiveness. Arch. Gen. Psychiatry 36 (7), 749–760. https://doi.org/10.1001/archpsyc.1979.01780070027003.

Rabkin, J., Quitkin, F., Harrison, W., Tricamo, E., McGrath, P., 1984. Adverse reactions to momisctextamine oxidase inhibitors. Part I. A comparative study. J. Clin. Psychopharmacol. 4 (5), 270–278.

Ravaris, C.L., Nies, A., Robinson, D.S., Ives, J.O., Lamborn, K.R., Korson, L., 1976. A multiple-dose, controlled study of phenelzine in depression-anxiety states. Arch. Gen. Psychiatry 33 (3), 347–350. https://doi.org/10.1001/archpsyc.1976.01770030057008.

Ravindran, L.N., Stein, M.B., 2010. The pharmacologic treatment of anxiety disorders: a review of progress. J. Clin. Psychiatry 71 (7), 839–854. https://doi.org/10.4088/JCP.10r06218blu.

Richard, I.H., Kurlan, R., Tanner, C., Factor, S., Hubble, J., Suchowersky, O., Waters, C., 1997. Serotonin syndrome and the combined use of deprenyl and an antidepressant in Parkinson's disease. Neurology 48 (4), 1070–1077. https://doi.org/10.1212/wnl.48.4.1070.

Ricken, R., Ulrich, S., Schlattmann, P., Adli, M., 2017. Tranylcypromine in mind (part II): review of clinical pharmacology and meta-analysis of controlled studies in depression. Eur. Neuropsychopharmacol. 27 (8), 714–731. https://doi.org/10.1016/j.euroneuro.2017.04.003.

Riva-Posse, P., Reiff, C.M., Edwards, J.A., Job, G.P., Galendez, G.C., Garlow, S.J., Saah, T.C., Dunlop, B.W., McDonald, W.M., 2018. Blood pressure safety of subanesthetic ketamine for depression: a report on 684 infusions. J. Affect. Disord. 236, 291–297. https://doi.org/10.1016/j.jad.2018.02.025.

Robinson, D.S., Nies, A., Ravaris, C.L., Lamborn, K.R., 1973. The monoamine oxidase inhibitor, phenelzine, in the treatment of depressive-anxiety states: a controlled clinical trial. Arch. Gen. Psychiatry 29 (3), 407–413. https://doi.org/10.1001/archpsyc.1973.04200030093015.

Robinson, D.S., Nies, A., Ravaris, C.L., Ives, J.O., Bartlett, D., 1978. Clinical pharmacology of Phenelzine. Arch. Gen. Psychiatry 35 (5), 629–635. https://doi.org/10.1001/archpsyc.1978.01770290111010.

Robinson, D.S., Gilmor, M.L., Yang, Y., Moonsammy, G., Azzaro, A.J., Oren, D.A., Campbell, B.J., 2007. Treatment effects of selegiline transdermal system on symptoms of major depressive disorder: a meta-analysis of short-term, placebo-controlled, efficacy trials. Psychopharmacol. Bull. 40 (3), 15–28.

Rosenblum, L.E., Korn, R.J., Zimmerman, H.J., 1960. Hepatocellular jaundice as a complication of iproniazid therapy. A.M.A. Arch. Intern. Med. 105 (4), 583–593. https://doi.org/10.1001/archinte.1960.00270160081010.

Rubin, R.L., 2005. Drug-induced lupus. Toxicology 209 (2), 135–147. Elsevier Ireland Ltd https://doi.org/10.1016/j.tox.2004.12.025.

Rudorfer, M.V., 1992. Monoamine oxidase inhibitors: reversible and irreversible. Psychopharmacol. Bull. 28 (1), 45–57.

Schjott, J., Schjott, P., Assmus, 2019. Analysis of consensus among drug interaction databases with regard to combinations of psychotropics. Basic Clin. Pharmacol. Toxicol. 126, 126–132.

Schmauss, M., Kapfhammer, H.P., Meyr, P., Hoff, P., 1988. Combined MAO-inhibitor and tri- (tetra) cyclic antidepressant treatment in therapy resistant depression. Prog. Neuropsychopharmacol. Biol. Psychiatry 12 (4), 523–532. https://doi.org/10.1016/0278-5846(88)90111-X.

Sen, S., Mickey, B., Greden, J., 2011. Psychopharmacological road map for treatment resistant depression. In: Greden, J., Riba, M., McInnis, M. (Eds.), Treatment Resistant Depression: A Roadmap for Effective Care. American Psychiatric Publishing, Inc, pp. 33–49.

Shih, J.C., Chen, K., Ridd, M.J., 1999. Monoamine oxidase: from genes to behavior. Annu. Rev. Neurosci. 22, 197–217. https://doi.org/10.1146/annurev.neuro.22.1.197.

Shulman, K.I., Fischer, H.D., Herrmann, N., Huo, C.Y., Anderson, G.M., Rochon, P.A., 2009. Current prescription patterns and safety profile of irreversible monoamine oxidase inhibitors: a population-based cohort study of older adults. J. Clin. Psychiatry 70 (12), 1681–1686. https://doi.org/10.4088/JCP.08m05041blu.

Smith, M.S., Muir, H., Hall, R., 1996. Perioperative management of drug therapy: clinical considerations. Drugs 51 (2), 238–259. https://doi.org/10.2165/00003495-199651020-00005.

Søgaard, J., Lane, R.M., Latimer, P., Behnke, K., Christiansen, P.E., Nielsen, B., Ravindran, A.V., Reesal, R.T., Goodwin, D.P., 1999. A 12-week study comparing moclobemide and sertraline in the treatment of outpatients with atypical depression. J. Psychopharmacol. 13 (4), 406–414. https://doi.org/10.1177/026988119901300412.

Spijker, J., Nolen, W.A., 2010. An algorithm for the pharmacological treatment of depression. Acta Psychiatr. Scand., 180–189. https://doi.org/10.1111/j.1600-0447.2009.01492.x.

Stack, C.G., Rogers, P., Linter, S.P.K., 1988. Monoamine oxidase inhibitors and anaesthesia: a review. Br. J. Anaesth. 60 (2), 222–227. https://doi.org/10.1093/bja/60.2.222.

Stahl, S.M., Felker, A., 2008. Monoamine oxidase inhibitors: a modern guide to an unrequited class of antidepressants. CNS Spectr. 13 (10), 855–870. https://doi.org/10.1017/s1092852900016965.

Stewart, J.W., Mcgrath, P.J., Quitkin, F.M., Harrison, W., Markowitz, J., Wager, S., Leibowitz, M.R., 1989. Relevance of DMS-III depressive subtype and chronicity of antidepressant efficacy in atypical depression: differential response to phenelzine, imipramine, and placebo. Arch. Gen. Psychiatry 46 (12), 1080–1087. https://doi.org/10.1001/archpsyc.1989.01810120022005.

Stewart, J.W., McGrath, P.J., Quitkin, F.M., Rabkin, J.G., Harrison, W., Wager, S., Nunes, E., Ocepek-Welikson, K., Tricamo, E., 1993. Chronic depression: response to placebo, imipramine, and phenelzine. J. Clin. Psychopharmacol. 13 (6), 391–396.

Stewart, J.W., Tricamo, E., McGrath, P.J., Quitkin, F.M., 1997. Prophylactic efficacy of phenelzine and imipramine in chronic atypical depression: likelihood of recurrence on discontinuation after 6 months' remission. Am. J. Psychiatry 154 (1), 31–36. https://doi.org/10.1176/ajp.154.1.31.

Stoeckel, K., Pfefen, J.P., Mayersohn, M., Schoerlin, M.P., Andressen, C., Ohnhaus, E.E., Frey, F., Guentert, T.W., 1990. Absorption and disposition of moclobemide in patients with advanced age or reduced liver or kidney function. Acta Psychiatr. Scand. *82* (360 S), 94–97. https://doi.org/10.1111/j.1600-0447.1990.tb05346.x.

Sunderland, T., Cohen, R.M., Molchan, S., Lawlor, B.A., Mellow, A.M., Newhouse, P.A., Tariot, P.N., Mueller, E.A., Murphy, D.L., 1994. High-dose selegiline in treatment-resistant older depressive patients. Arch. Gen. Psychiatry 51 (8), 607–615. https://doi.org/10.1001/archpsyc.1994.03950080019003.

Sun-Edelstein, C., Tepper, S.J., Shapiro, R.E., 2008. Drug-induced serotonin syndrome: a review. Expert Opin. Drug Saf. 7 (5), 587–596. https://doi.org/10.1517/14740338.7.5.587.

Tatsumi, M., Groshan, K., Blakely, R.D., Richelson, E., 1997. Pharmacological profile of antidepressants and related compounds at human monoamine transporters. Eur. J. Pharmacol. 340 (2–3), 249–258. https://doi.org/10.1016/S0014-2999(97)01393-9.

Tatsumi, M., Jansen, K., Blakely, R.D., Richelson, E., 1999. Pharmacological profile of neuroleptics at human monoamine transporters. Eur. J. Pharmacol. 368 (2–3), 277–283. https://doi.org/10.1016/S0014-2999(99)00005-9.

Thase, M.E., 2007. Recognition and diagnosis of atypical depression. J. Clin. Psychiatry 68 (8), 11–16.

Thase, M.E., 2012. The role of monoamine oxidase inhibitors in depression treatment guidelines. J. Clin. Psychiatry 73 (1), 10–16. https://doi.org/10.4088/JCP.11096su1c.02.

Thase, M.E., Frank, E., Mallinger, A.G., Hamer, T., Kupfer, D.J., 1992. Treatment of imipramine-resistant recurrent depression, III: efficacy of monoamine oxidase inhibitors. J. Clin. Psychiatry 53 (1), 5–11.

Thase, M.E., Trivedi, M.H., Rush, A.J., 1995. MAOIs in the contemporary treatment of depression. Neuropsychopharmacology 12 (3), 185–219. https://doi.org/10.1016/0893-133X(94)00058-8.

Thomas, S.J., Shin, M., McInnis, M.G., Bostwick, J.R., 2015. Combination therapy with monoamine oxidase inhibitors and other antidepressants or stimulants: strategies for the management of treatment-resistant depression. Pharmacotherapy 35 (4), 433–449. https://doi.org/10.1002/phar.1576.

Todd, K.G., Baker, G.B., 2008. Neurochemical effects of the monoamine oxidase inhibitor phenelzine on brain GABA and alanine: a comparison with vigabatrin. J. Pharm. Pharm. Sci. 11 (2). https://doi.org/10.18433/j34s38.

Tulen, J.H.M., Volkers, A.C., Van Den Broek, W.W., Bruijn, J.A., 2006. Sustained effects of phenelzine and tranylcypromine on orthostatic challenge in antidepressant-refractory depression [8]. J. Clin. Psychopharmacol. 26 (5), 542–544. https://doi.org/10.1097/01.jcp.0000236657.08663.ae.

Tyrer, P., Gardner, M., Lambourn, J., Whitford, M., 1980. Clinical and pharmacokinetic factors affecting response to phenelzine. Br. J. Psychiatry 136 (4), 359–365. https://doi.org/10.1192/bjp.136.4.359.

Ulrich, S., Ricken, R., Adli, M., 2017. Tranylcypromine in mind (part I): review of pharmacology. Eur. Neuropsychopharmacol. 27 (8), 697–713. https://doi.org/10.1016/j.euroneuro.2017.05.007.

Vallejo, J., Gasto, C., Catalan, R., Salamero, M., 1987. Double-blind study of imipramine versus phenelzine in melancholias and dysthymic disorders. Br. J. Psychiatry 151, 639–642. https://doi.org/10.1192/bjp.151.5.639.

van den Berg, C., 2012. The transdermal delivery system of monoamine oxidase inhibitors. J. Clin. Psychiatry 73 (1), 25–30.

van Haelst, I., van Klei, W., Doodeman, J., 2012. Antidepressive treatment with monoamine oxidase inhibitors and the occurrence of intraoperative hemodynamic events: a restrospective observational cohort study. J. Clin. Psychiatry 73, 1193–1199.

Voican, C.S., Corruble, E., Naveau, S., Perlemuter, G., 2014. Antidepressant-induced liver injury: a review for clinicians. Am. J. Psychiatry 171 (4), 404–415. https://doi.org/10.1176/appi.ajp.2013.13050709.

Volz, H.P., Faltus, F., Magyar, I., Möller, H.J., 1994. Brofaromine in treatment-resistant depressed patients—a comparative trial versus tranylcypromine. J. Affect. Disord. 30 (3), 209–217. https://doi.org/10.1016/0165-0327(94)90081-7.

Warner, C.H., Bobo, W., Warner, C., Reid, S., Rachal, J., 2006. Antidepressant discontinuation syndrome. Am. Fam. Physician 74 (3), 449–456. http://www.aafp.org/afp/20060801/449.pdf.

Wells, D.G., Bjorksten, A.R., 1989. Monoamine oxidase inhibitors revisited. Can. J. Anaesth. 36 (1), 64–74. https://doi.org/10.1007/BF03010890.

Williams, R., Edwards, R.A., Newburn, G.M., Mullen, R., Menkes, D.B., Segkar, C., 1993. A double-blind comparison of moclobemide and fluoxetine in the treatment of depressive disorders. Int. Clin. Psychopharmacol. 7 (3), 155–158. https://doi.org/10.1097/00004850-199300730-00006.

Wimbiscus, M., Kostenko, O., Malone, D., 2010. MAO inhibitors: risks, benefits, and lore. Cleve. Clin. J. Med. 77 (12), 859–882. https://doi.org/10.3949/ccjm.77a.09103.

Wingo, A.P., Ghaemi, S.N., 2007. STAR*D level IV methodology [6]. Am. J. Psychiatry 164 (4), 681. https://doi.org/10.1176/ajp.2007.164.4.681.

Wong, K.C., 1986. Preoperative discontinuation of monoamine oxidase inhibitor therapy: an old wives' tale? Semin. Anesth. 2, 145.

Youdim, M.B.H., Weinstock, M., 2004. Therapeutic applications of selective and non-selective inhibitors of monoamine oxidase A and B that do not cause significant tyramine potentiation. Neurotoxicology 25 (1–2), 243–250. https://doi.org/10.1016/S0161-813X(03)00103-7.

Chapter 10

Switching antidepressants in patients with treatment-resistant depression

Daniel Souery[a,b] and William Pitchot[c]

[a]*PSYPLURIEL at EPSYLON – Psychiatric Care Network, Brussels, Belgium,* [b]*Laboratory of Psychological Medicine and Addictology, Faculty of Medicine, Université Libre de Bruxelles, Brussels, Belgium,* [c]*Department of Psychiatry, Centre Hospitalier Université de Liège, Liege, Belgium*

Introduction

Despite the availability of many therapeutic strategies, the optimal treatment of depression remains a major challenge. The overall effectiveness of antidepressants remains limited, and in almost 30%–50%, the response to treatment will be insufficient or absent (Al-Harbi, 2012; Bschor et al., 2014). The concept of treatment-resistant depression (TRD) is often used to characterize these complex situations where more than one antidepressant trial is needed to achieve response or remission (Souery et al., 1999). This implies that in many cases it will be necessary to change the antidepressant, even if the choice of the first treatment was adequate based on the clinical picture. Given the high prevalence of major depression and the significant risk of poor or insufficient response to the first or second treatment step, any clinician involved in the treatment of depression, whether psychiatrist or general practitioner, will very often be confronted with patients with TRD. This reality is even more frequent for psychiatrists who practice in hospital care units dedicated to difficult-to-treat cases.

This issue raises several clinically important questions. When it is necessary to consider a change of treatment, how to achieve this change of treatment and especially what therapeutic choice after a first failure? The second-step strategy after nonresponse to an antidepressant remains a real clinical challenge. Although different strategies are recommended, it is difficult to predict their efficacy. The current lack of biological predictors of treatment response leaves the clinician to rely on some clinical factors that have been associated to TRD. Severity of the depressive episode, anxiety disorders comorbidity or high suicidal risk are among key clinical factors associated to lack of response to antidepressant therapy (Bennabi et al., 2019; Kornstein and Schneider, 2001; Souery et al., 2007).

At present, the choice of second step strategy is essentially based on the clinic and the prescriber's experience. Recommended strategies include optimization (high-dose antidepressant therapy) (Adli et al., 2005), augmentation (especially with lithium or second-generation antipsychotics) (Bschor, 2014; Nelson and Papakostas, 2009), combining two antidepressants (Henssler et al., 2016), electroconvulsive therapy (UK ECT Review Group, 2003), various psychotherapeutic approaches (Huhn et al., 2014), novel or experimental treatments (Ionescu et al., 2015), and switching to a different antidepressant (Bschor and Baethge, 2010). Ionescu et al. (2015) suggested the "SACO" mnemonic (Switching therapies, Augmentation, Combination and Optimization) to aid providers in choosing the next option in TRD. There are many reviews of the literature on these different strategies, but it remains difficult to have clear recommendations on which attitude to adopt as a second step. To our knowledge, there are no completely reliable clinical indications to guide the physician in the attitude to adopt.

Switching to a different antidepressant is a common strategy, mainly in the situation of complete nonresponse to the first antidepressant and in case of tolerability issues. The issue of switching antidepressant in TRD is fundamental and needs to be addressed from different perspectives. At first, this question is a basic element of the definition of the concept of TRD. It refers to the number of antidepressant trials required before considering resistance to treatment. In parallel, the second step antidepressant choice is also qualitatively important and raises the question of considering a change in antidepressant class. For example, it may make sense to switch to an SNRI after failure to an SSRI. For many clinicians, there will be a preconception that they need to switch to an antidepressant with a different mechanism of action in the hope of achieving a better result. This issue has been the subject of much debate. A metaanalysis concluded that there is no significant benefit to switch across class (Ruhé et al., 2006).

Managing Treatment-Resistant Depression. https://doi.org/10.1016/B978-0-12-824067-0.00010-4

In this chapter, we propose to address these issues from a clinical perspective, focusing on key elements of the switching strategy in TRD: implication in the definition of TRD, how to implement the treatment switch and choice of second step treatment.

Switching antidepressant and TRD definition

Is switching part of the TRD concept?

How is the question of switching between antidepressants crucial in defining the concept of TRD? For decades, since the 1970s, the scientific literature on the subject has abounded with proposals to define TRD. In the early days, extraordinarily complex definitions were proposed, all very elaborated and clever but impractical and often impossible to apply in the clinic. This first wave of definitions of TRD was also of little use for possible clinical trials aimed at evaluating the efficacy of a treatment or therapeutic strategy in TRD. It was a time, moreover, when there were no studies of this type and no recognition of a treatment in this indication. The basic principle of this first generation of definitions was to consider a depression as resistant to treatment only if the patient had been treated by multiple drugs or biological therapies, with the imperative to have received what was considered at the time as the most powerful treatments. At that time tricyclic antidepressants (TCA), monoamine oxidase inhibitors (MAOI), lithium, and electro convulsive therapy (ECT) were among the treatments to be incorporated in any TRD definition but all of them in different sequences and with various durations based on empirical assumptions. The main thing was therefore to consider TRD only if the depressive episode had benefited from these treatments.

Switching in staging models of TRD

Subsequently, more pragmatic proposals appeared, it was the era of staging models to better characterize TRD. These models are clinically extremely useful and have significantly contributed to a better understanding of the different stages of resistance to treatment in depression (Kasper et al., 2013, 2020). They give more space to the notion of "switching" regardless of the bias on relative effectiveness among the options used. Although particularly useful in the clinic, these models are however little or not used in clinical trials. They have therefore never contributed to demonstrating the efficacy of a treatment or a therapeutic strategy in TRD. Moreover, they are not considered by the regulatory authorities in charge of recognizing a specific indication in TRD, neither in Europe nor in the United States.

Switching antidepressant and clinical trials in TRD

The concept of switching antidepressant finally gained in importance when it came to a definition of TRD that could be used in clinical trials to demonstrate efficacy in TRD as a specific recognized indication. Lately, in response to the need to validate treatment strategies or specific medication in TRD, regulatory authorities both in Europe and in the United States elaborated their own recommendations to be used in clinical trials. Things were then significantly simplified, and for practical reasons TRD was defined as nonresponse to two consecutive trials of two different antidepressants at an adequate dose and duration. In this vision of TRD, what matters most is the switch in treatment. When, how and with which other antidepressant to make this switch?

At first, it was required to change antidepressant class. That is, using an antidepressant with a different mechanism of action than the one that did not work on the first attempt. This requirement was mainly based on clinical practice. It is more common to switch antidepressant classes in the hope of having a better chance of success. This practice, although theoretically meaningful, has never really been demonstrated through robust clinical data. The European Medical Agency (EMA) Guideline on clinical investigation of medicinal products in the treatment of depression consider monotherapy in patients with TRD as a separate claim and proposes clinical trial design and definition of TRD. TRD is considered, *when treatment with at least two different antidepressant agents prescribed in adequate dosages for adequate duration and with adequate affirmation of treatment adherence showed lack of clinically meaningful improvement* (EMA, 2013). This pragmatic definition differs from the complex available staging models but is mainly intended to be used within clinical trials as a reference to characterize patients included based on the number and type of previous treatments. This definition differs from previous versions where two products of different pharmacological classes were requested. This important revision was in line with the data showing no advantage in favor of switching to a different class of antidepressant (Bschor and Baethge, 2010; Papakostas et al., 2008; Rush et al., 2006a; Souery et al., 2011a,b).

The efficacy of Esketamine nasal spray, the newly approved medication in TRD was investigated using this definition of TRD (Popova et al., 2019). As the study definition of TRD did not require the second antidepressant to be from a different class, a patient who had received two consecutive SSRIs or SNRIs was considered treatment resistant.

Switching within the same class or switching across classes?

Beyond theoretical considerations and the impact on the definition of TRD, the question of switching antidepressant class is fundamental from a practical standpoint. The history of this debate is supported by a limited number of studies, but most of them point in the same direction. The evidence in favor of using antidepressants from two different classes is weak. Papakostas et al. (2008) published the first metaanalysis on data comparing switching strategies for depressed patients who failed to respond to a SSRI, a second SSRI or a different class of antidepressant. Results suggested a marginal benefit of switching from one class of antidepressant to another on remission rates. It is worth mentioning that not all antidepressant classes have been considered but only Venlafaxine, Mirtazapine, and Bupropion have been included. Thereafter, several reports showed no advantage in favor of switching to a different class of antidepressant (Bschor and Baethge, 2010; Ruhé et al., 2006; Rush et al., 2006a; Souery et al., 2011b). Evidences concerning the switch between SSRI and TCA classes are relatively limited: a between-class switch from SSRI to TCA response rates of approximately 16.5%–48.5% has been reported, with lower response rates in studies including treatment-resistant patients; for the between-class switch from TCA to SSRI response rates were of 4%–75%; finally, response rates of within-class TCA switch were 9% and 30% (Ruhé et al., 2006). Concerning within-class SSRI, switch response rates varied between 26.7% and 72%. In a prospective randomized open study in TRD the effect of continuation with the same treatment or switching to a different class of antidepressant using the SSRI citalopram and desipramine, a tricyclic antidepressant which has predominant noradrenaline reuptake properties, a statistically significant inferiority of switching across class was observed (Souery et al., 2011b). Results from the Sequenced Treatment Alternatives to Relieve Depression (STAR*D) trial that included depressive patients who were switched to a second step antidepressant after a SSRI failure demonstrated that all treatments in the second step were comparable (Rush et al., 2020). These included bupropion sustained release, sertraline and venlafaxine. The STAR*D trial is essential in understanding the different sequences of the trajectory of antidepressant treatments in TRD. The data from this study shed light on the prognosis of response to various antidepressant treatments following treatment failure (Rush et al., 2020).

Is switching better than continuation of the same antidepressant?

Although a switch in treatment is a recognized option in resistant depression, it is useful to compare it to simply continuing the first antidepressant. Due to the latency of action of antidepressants, and especially in cases of partial response to treatment, the remission to the first antidepressant may appear late. In this case, it would be desirable to prolong the antidepressant pending a late response. In clinical practice, antidepressants are often switched (Lamy et al., 2015). Most studies looking at the outcome of different second step antidepressants found comparable response rates (Fava et al., 2006; McGrath et al., 1987; Rush et al., 2006b). These studies, while informative in revealing comparable efficacy among diverse antidepressants do not contribute to answer the question of whether switching is better than continuation or placebo. Bschor et al. have published a very comprehensive metaanalysis on this issue (Al-Harbi, 2012; Bschor et al., 2018). Their first observation is that there are very few studies that have been carried out with rigorous methodology. They report the results of 12 studies, of which only four are randomized studies using strict criteria comparing the outcome of the continuation of the initial antidepressant and a new antidepressant. Due to the limited number of studies, it is evident that the comparisons are made on a small number of antidepressants which are not representative of all switching possibilities between antidepressants. Taking this limitation into account, this metaanalysis concludes that there is no high-level evidence that switching the antidepressant is effective when compared to simply continuing the initial antidepressant. Two studies demonstrated a statistically significant inferiority of switching (Bose et al., 2012; Souery et al., 2011b).

Practical and clinical issues in switching antidepressants

Diagnostic and clinical reassessment

The issue of diagnosis is fundamental and could be the basis of the vast majority of difficult-to-treat clinical situations. At this stage of our knowledge in the diagnosis of depression, it is likely that conventional antidepressant treatments are not sufficient to target the full heterogeneity of clinical manifestations of major depression. Most antidepressants have limited action on certain symptomatic dimensions of depression without affecting the overall clinical manifestations. Furthermore, it is possible that our diagnostic reading grid, based mostly on subjective assessments, does not identify the symptomatic dimensions that would be the real target of antidepressants. Every clinician must remain humble and consider the limits of antidepressant treatments in a double constraint. The part of resistance to treatment related to treatment choice and the part of "resistance" associated with the inability to target certain dimensions of psychopathology. This clinical posture allows a more comprehensive analysis of therapeutic choices that will not focus on treatment alone with an antidepressant molecule. It is undeniable that the combination of an antidepressant with, for example, psychotherapy or an anxiolytic drug treatment will be more likely to be effective. Acting on all dimensions of symptomatology, but also on personal and psychosocial elements is essential before evoking a real resistance to treatment.

Some practical and clinical considerations should be checked before switching antidepressant. These prerequisites are essential to minimize the risk of aggravation in case of inadequate discontinuation of the first antidepressant. It is also important to consider clinical dimensions that will increase the chances of success of the second antidepressant. In case of insufficient response to a treatment, it is advisable to seek for the possible causes of ineffectiveness, such as a possible noncompliance, the occurrence of undesirable side effects which reduce the chance of response, drug interactions or a possible psychiatric or somatic comorbidity. Because of the wide variety of forms of depression, care should be taken to ensure that the patients have received the antidepressant that best matches their clinical condition. After treatment failure, it is necessary to reevaluate in a detailed and rigorous manner the clinical parameters which could better guide the treatment. Clinicians should carefully consider factors associated with or predictive of TRD. Some clinical variables have been associated with poorer treatment outcomes and resistance to treatment. Identifying these variables must be a prerequisite for any therapeutic choice. A therapeutic choice based on the meticulous analysis of the clinical dimensions associated with resistance depression avoids long wanderings before realizing the inadequacy of a treatment. Personality disorders, melancholic features, severity, anxious depression, and/or comorbidity with anxiety disorders are among the most frequent clinical factors associated with TRD. Evaluating TRD must include clarification of the diagnosis and potential risk factors for TRD in addition to evaluation of treatment adherence, presence of comorbidities, medication dose and duration (Ionescu et al., 2015).

Optimize dose and duration before switching

In the absence of a demonstrated therapeutic response, it is essential to optimize the ongoing treatment before considering switching. A simple option is to increase the dose of the antidepressant to reach an effective or the maximal tolerated therapeutic dose. Patients should be informed of potential side effects and should be informed of potential change of antidepressant. The clinical effect of an antidepressant usually appears within the first 2 weeks of treatment, but in many cases significant response may appear later, and it is recommended to wait 4–6 weeks before judging the effectiveness of a treatment (McIntyre, 2010; Papakostas et al., 2006).

Initial improvement in the first weeks of antidepressant treatment may be early predictor of remission. It is difficult to find consensus in the literature about the optimal timing of a switch. Although the adequate duration of the acute phase of treatment is 4–8 weeks, some evidence suggests that one should not wait that long to switch to another treatment if the patient's clinical condition has not significantly improved. Many clinicians wait for 6–12 weeks before considering a switch. This attitude of inertia despite the lack of response can have serious consequences on treatment compliance and leaves the patient with a significant degree of symptoms over a long period of time, thus increasing the risk of complications (e.g., suicidal behavior) and chronicity. A literature search reviewing the evidence (randomized controlled trials (RCTs) that examine early switching strategies) for the predictive value of early improvement at 2–4 weeks to predict final antidepressant response concluded that lack of early improvement (e.g., $<20\%$ reduction in a depression scale score) at 2–4 weeks can be an accurate predictor to identify eventual nonresponders (Kudlow et al., 2014). Only about one in five patients with lack of improvement at 4 weeks will have a response by 8 weeks. Based on these data, the authors recommend earlier assessment for improvement. If there is no indication of early improvement at 2–4 weeks after starting an antidepressant, a change in management can be considered. A metaanalysis of antidepressant naturalistic studies (nine studies; 6185 patients) investigating whether a partial decrease in depressive symptoms by week 4 was associated with response and remission by weeks 6–14 showed that such an early improvement is not a reliable outcome predictor in TRD (Olgiati et al., 2018).

Optimal switching conditions and switching rules

If the decision is made to change the treatment, the switch of antidepressant must be carried out under optimal conditions and there are some important rules to follow in how to make this change. Switching from one antidepressant to another must be a carefully considered decision and the patient must be informed on potential consequences. The most frequent problems are related to possible withdrawal symptoms due to the discontinuation of the first molecule and undesirable effects in connection with possible drug interactions with the new antidepressant. It is common to wrongly attribute to the introduction of a new treatment side effects which are in fact linked to the discontinuation of the previous treatment. The switching phase is also more at risk to a resurgence of anxiety, depressive symptoms, or relapse.

There are rules for safely switching from one molecule to another, these depend on the pharmacokinetic and pharmacodynamic characteristics of the first and second molecule. The half-life of molecules, their metabolism by cytochromes, their specific side effects are all criteria to be considered. The characteristics related to the patient should also be noted: history of adverse effects or discontinuation syndrome with a given molecule, age, pregnancy, renal or hepatic failure, severity of depression, combined effects of two treatments.

Switching methods

Several strategies are available for switching between antidepressants (Gauthier et al., 2018; Jefferson, 2000; Keks et al., 2016). The conservative or progressive switch allows for a gradual reduction of the first antidepressant. The new antidepressant is started after a drug-free period of two to five half-lives of the first molecule. Variants of this approach can be applied by reducing the length or missing the wash-out period. The new antidepressant is started according to its dose recommendation or gradually starting at a low dose. In the direct switch method, the first antidepressant is simply stopped and the second one is directly initiated at the usual therapeutic dose. The likelihood of discontinuation symptoms and drug interactions is higher and will mainly depend on the half-life of the first antidepressant. The cross-taper switch technique is frequently used and consists in gradually reducing the dose of the first antidepressant while introducing the second one at low dose. In this method, the patient is taking two antidepressants simultaneously, significantly increasing the risk of adverse effects. In a switch scenario where two antidepressants with the same or partially similar mechanism of action, a direct switch may be considered with minimal risks. For example, switching from an SSRI to another, switching from an SSRI to an SNRI, or from and SNRI to another SNRI.

Each method will have its advantages and disadvantages depending on the drug context, the individual parameters of the patient and the severity of the symptoms. Beyond the proposed method, the experience of the physician will be essential as well as the quality of the follow-up during this sensitive period. It is often necessary to reassess the original plan and adapt the method based on the outcome. This requires in-depth knowledge of the antidepressant pharmacopeia including the mechanism of action, metabolism, pharmacokinetics, interactions, and side effect profile of each molecule.

Pharmacokinetics and switching strategies in TRD

Genetic factors may interfere with treatment response to antidepressants. Genetic variation in drug-metabolizing genes may influence drug response and may predispose patients to TRD. These are for the most part highly penetrant and monogenic. More than 90 genetic variants in the CYP2D6 isoenzyme have been identified (Nebert and Dieter, 2000). CYP2D6 is one of the cytochromes P450 isoenzyme responsible for metabolizing most of the antidepressants (Ingelman-Sundberg, 2005). CYP1A2, CYP2B6, CYP2C9, CYP2C19, and CYP3A4 are also among cytochrome P450 isoenzymes involved in antidepressants metabolism. Variants within these genes define different group of individuals based on enzyme activity: poor metabolizers (PMs), intermediate metabolizers (IMs), extensive metabolizers (EMs), and ultrarapid metabolizers (UMs). Clinically, UMs are less likely to show positive response to treatment.

Choice of second step antidepressant

General consideration

Due to the high number of possible antidepressants change scenarios from an ineffective first choice, it is almost impossible to find controlled comparative trials in the literature for each possibility. Although switching is the most preferred strategy, little evidence exists to guide the clinician's decision to switch at the level of class or mechanism of action but also at the level of which molecule within a class. Because of this reality, the effectiveness of different switching possibilities is considered comparable, whether from one class to a different class or from one molecule to another. The choice is essentially based on the clinician's experience.

Contribution of the STAR*D study

A systematic review of studies evaluating the efficacy of changing antidepressant medication after failure with an SSRI showed that 50% of patients who failed to respond to one SSRI responded to a second (Ruhé et al., 2006). This relatively optimistic result is in contradiction with other studies, notably the STAR*D (Sequenced Treatment Alternatives to Relieve Depression) study, a large study funded by the NIMH (National Institute of Mental Health) evaluating the treatment of depression, showing a significantly lower response rate (around 27%) in patients treated with sertraline after failure with citalopram (Rush et al., 2006a,b). The main objective of the STAR*D program is to determine the most effective treatment if the first-line antidepressant does not induce a significant response. Following ineffective treatment with a SSRI, there is little evidence on the effectiveness of a medication change strategy. In the STAR*D study, 2876 severely depressed patients undergoing psychiatric and general medical care were treated with citalopram as first-line therapy. At the end of the study, approximately 30% of patients were considered to be in remission after 6–7 weeks. In a second stage, 727 patients who did not respond to citalopram were treated with either bupropion 400 mg, sertraline 200 mg or venlafaxine 375 mg (Fava et al., 2006). This change in treatment has enabled one in four patients to achieve complete remission. The interest of these three medications seemed quite equivalent.

In level 3 of STAR*D, switching to mirtazapine monotherapy ($n=114$) was compared to nortriptyline monotherapy ($n=121$) in a group of 235 patients who did not respond to citalopram and a switch to another antidepressant or an augmentation with another pharmacological option (Fava et al., 2006). There was no significant difference in response and in remission rates. In another study, 150 patients who failed to respond to two antidepressants were switched to mirtazapine, venlafaxine, or paroxetine. There was no significant difference in terms of remission rates (Fang et al., 2010).

Efficacy of switching: Other data

Clinically, there is a clear trend to replace the SSRI with an antidepressant with a different mechanism of action such as tricyclics, antidepressants with a mixed serotonergic and noradrenergic action (venlafaxine, duloxetine, mirtazapine), products with an effect on the dopaminergic and/or noradrenergic system (bupropion, reboxetine), and MAOIs. In a meta-analysis of five randomized clinical studies, switching from one SSRI to another class of antidepressant (venlafaxine, mirtazapine, and bupropion) was associated with a higher rate of remission than switching to another SSRI (Papakostas et al., 2008). Response rates obtained by replacing an SSRI with a combination product ranged from 28% to 50%. Switching to bupropion resulted in a response in 26%–35% of cases. Depending on the studies, switching from an SSRI to a tricyclic gave response rates between 16.5% and 48.5%. In two open studies, nonresponders to SSRI were treated with Venlafaxine with response rates of 58% and 87% (De Montigny et al., 1999; Kaplan, 2002). In another study, one third of 84 patients suffering from a highly resistant major depression were considered to be full or partial responders after 12 weeks of venlafaxine treatment (Nierenberg et al., 1994). In a double-blind study, response (52% vs 33%) and remission (42% vs 20%) rates were higher with venlafaxine than paroxetine in patients resistant to two antidepressant treatments (Poirier and Boyer, 1999). The switch to mirtazapine did not appear to be an interesting option as it has been demonstrated in two studies (Baldomero et al., 2005).

The use of MAOIs in cases of resistant depression has been widespread in clinical practice in relation to a perception of this class of antidepressant as particularly effective in difficult cases. Some studies have shown a high response rate with phenelzine or tranylcypromine in patients with resistance to tricyclics (McGrath et al., 1987; Thase et al., 1992). However, the risks associated with their prescription (hypertensive crises), their scarcity and the need for a period of 1–2 weeks of SSRI withdrawal before their introduction explain the lack of interest in this type of strategy.

Conclusions

The change of one antidepressant by another in resistant depression remains a commonly used option in clinical practice. However, there is little evidence to guide the choice of antidepressant after a therapeutic failure, and the data are not sufficient to produce reliable recommendations. The most contributory lessons from the available data still reveal important guidelines of clinical utility. The main elements that can determine the success of a treatment change lie more in the manner and timing of making this change more than in the choice of antidepressant. The same is true for the identification of predictive variables of resistance to treatment that allow from the early stages of a treatment to make the best choices in selecting the best option. Finally, the most important thing is to be able to integrate all the options into the therapeutic plan, including other possibilities such as combination and augmentation strategies. And at every step never give up and avoid therapeutic inertia.

References

Adli, M., Baethge, C., Heinz, A., Langlitz, N., Bauer, M., 2005. Is dose escalation of antidepressants a rational strategy after a medium-dose treatment has failed? A systematic review. Eur. Arch. Psychiatry Clin. Neurosci. 255 (6), 387–400.

Al-Harbi, K.S., 2012. Treatment-resistant depression: therapeutic trends, challenges, and future directions. Patient Prefer. Adherence 6, 369–388. https://doi.org/10.2147/PPA.S29716.

Baldomero, E.B., Ubago, J.G., Cercós, C.L., Ruiloba, J.V., Calvo, C.G., López, R.P., 2005. Venlafaxine extended release versus conventional antidepressants in the remission of depressive disorders after previous antidepressant failure: ARGOS study. Depress. Anxiety 22 (2), 68–76.

Bennabi, D., Yrondi, A., Charpeaud, T., Genty, J.B., Destouches, S., Lancrenon, S., Allaili, N., Bellivier, F., Bougerol, T., Camus, V., Doumy, O., Dorey, J.M., Haesebaert, F., Holtzmann, J., Lançon, C., Lefebvre, M., Moliere, F., Nieto, I., Rabu, C., et al., 2019. Clinical guidelines for the management of depression with specific comorbid psychiatric conditions French recommendations from experts (the French association for biological psychiatry and neuropsychopharmacology and the fondation Fonda Mental). BMC Psychiatry 19 (1). https://doi.org/10.1186/s12888-019-2025-7.

Bose, A., Tsai, J., Li, D., 2012. Early non-response in patients with severe depression: escitalopram up-titration versus switch to duloxetine. Clin. Drug Investig. 32 (6), 373–385. https://doi.org/10.2165/11631890-000000000-00000.

Bschor, T., 2014. Lithium in the treatment of major depressive disorder. Drugs 74 (8), 855–862. https://doi.org/10.1007/s40265-014-0220-x.

Bschor, T., Baethge, C., 2010. No evidence for switching the antidepressant: systematic review and meta-analysis of RCTs of a common therapeutic strategy: review. Acta Psychiatr. Scand. 121 (3), 174–179. https://doi.org/10.1111/j.1600-0447.2009.01458.x.

Bschor, T., Bauer, M., Adli, M., 2014. Chronische und therapieresistente depression—Diagnostik und Stufentherapie. Dtsch. Arztebl. Int. 111 (45), 766–776. https://doi.org/10.3238/arztebl.2014.0766.

Bschor, T., Kern, H., Henssler, J., Baethge, C., 2018. Switching the antidepressant after nonresponse in adults with major depression: a systematic literature search and meta-analysis. J. Clin. Psychiatry 79 (1). https://doi.org/10.4088/JCP.16r10749.

De Montigny, C., Silverstone, P.H., Debonnel, G., Blier, P., Bakish, D., 1999. Venlafaxine in treatment-resistant major depression: a Canadian multicenter, open-label trial. J. Clin. Psychopharmacol. 19 (5), 401–406. https://doi.org/10.1097/00004714-199910000-00003.

EMA, 2013. Guideline on Clinical Investigation of Medicinal Products in the Treatment of Depression. https://www.ema.europa.eu/en/documents/scientific-guideline/guideline-clinical-investigation-medicinal-products-treatment-depression_en.pdf.

Fang, Y., Yuan, C., Xu, Y., Chen, J., Wu, Z., Cao, L., Yi, Z., Hong, W., Wang, Y., Jiang, K., Gao, K., Cui, X., Nierenberg, A.A., 2010. Comparisons of the efficacy and tolerability of extended-release venlafaxine, mirtazapine, and paroxetine in treatment-resistant depression: a double-blind, randomized pilot study in a Chinese population. J. Clin. Psychopharmacol. 30 (4), 357–364. https://doi.org/10.1097/JCP.0b013e3181e7784f.

Fava, M., Rush, A.J., Wisniewski, S.R., Nierenberg, A.A., Alpert, J.E., McGrath, P.J., Thase, M.E., Warden, D., Biggs, M., Luther, J.F., Niederehe, G., Ritz, L., Trivedi, M.H., 2006. A comparison of mirtazapine and nortriptyline following two consecutive failed medication treatments for depressed outpatients: a STAR*D report. Am. J. Psychiatry 163 (7), 1161–1172. https://doi.org/10.1176/ajp.2006.163.7.1161.

Gauthier, C., Abdel-Ahad, P., Gaillard, R., 2018. Recommandations pour switcher et arrêter les antidépresseurs. Encéphale 44 (4), 379–386. https://doi.org/10.1016/j.encep.2018.08.001.

Henssler, J., Bschor, T., Baethge, C., 2016. Combining antidepressants in acute treatment of depression: a meta-analysis of 38 studies including 4511 patients. Can. J. Psychiatry 61 (1), 29–43. https://doi.org/10.1177/0706743715620411.

Huhn, M., Tardy, M., Spineli, L.M., Kissling, W., Förstl, H., Pitschel-Walz, G., Leucht, C., Samara, M., Dold, M., Davis, J.M., Leucht, S., 2014. Efficacy of pharmacotherapy and psychotherapy for adult psychiatric disorders: a systematic overview of meta-analyses. JAMA Psychiatry 71 (6), 706–715. https://doi.org/10.1001/jamapsychiatry.2014.112.

Ingelman-Sundberg, M., 2005. Genetic polymorphisms of cytochrome P450 2D6 (CYP2D6): clinical consequences, evolutionary aspects and functional diversity. Pharmacogenomics J. 5 (1), 6–13. https://doi.org/10.1038/sj.tpj.6500285.

Ionescu, D.F., Rosenbaum, J.F., Alpert, J.E., 2015. Pharmacological approaches to the challenge of treatment-resistant depression. Dialogues Clin. Neurosci. 17 (2), 111–126.

Jefferson, J.W., 2000. Strategies for switching antidepressants to achieve maximum efficacy adolescents. J. Clin. Psychiatry 69 (1), 124–135.

Kaplan, E.M., 2002. Efficacy of venlafaxine in patients with major depressive disorder who have unsustained or no response to selective serotonin reuptake inhibitors: an open-label, uncontrolled study. Clin. Ther. 24 (7), 1194–1200. https://doi.org/10.1016/S0149-2918(02)80029-7.

Kasper, S., Montgomery, S., Souery, D., Pitchot, W., 2013. Definitions and predictors of treatment-resistant depression. In: Kasper, S., Montgomery, S. (Eds.), Treatment-resistant Depression. John Wiley & Sons, Hoboken, NJ.

Kasper, S., Cubała, W.J., Fagiolini, A., Ramos-Quiroga, J.A., Souery, D., Young, A.H., 2020. Practical recommendations for the management of treatment-resistant depression with esketamine nasal spray therapy: basic science, evidence-based knowledge and expert guidance. World J. Biol. Psychiatry, 1–15. https://doi.org/10.1080/15622975.2020.1836399.

Keks, N., Hope, J., Keogh, S., 2016. Switching and stopping antidepressants. Aust. Prescr. 39 (3), 76–83. https://doi.org/10.18773/austprescr.2016.039.

Kornstein, S.G., Schneider, R.K., 2001. Clinical features of treatment-resistant depression. J. Clin. Psychiatry 62 (16), 18–35.

Kudlow, P.A., McIntyre, R.S., Lam, R.W., 2014. Early switching strategies in antidepressant non-responders: current evidence and future research directions. CNS Drugs 28 (7), 601–609. https://doi.org/10.1007/s40263-014-0171-5.

Lamy, F.X., Chollet, J., Clay, E., Brignone, M., Rive, B., Saragoussi, D., 2015. Pharmacotherapeutic strategies for patients treated for depression in UK primary care: a database analysis. Curr. Med. Res. Opin. 31 (4), 795–807. https://doi.org/10.1185/03007995.2015.1020362.

McGrath, P.J., Stewart, J.W., Harrison, W., Quitkin, F.M., 1987. Treatment of tricyclic refractory depression with a monoamine oxidase inhibitor antidepressant. Psychopharmacol. Bull. 23 (1), 169–172.

McIntyre, R.S., 2010. When should you move beyond first-line therapy for depression? J. Clin. Psychiatry 71 (Suppl. 1), 16–20. https://doi.org/10.4088/JCP.9104su1c.03.

Nebert, D.W., Dieter, M.Z., 2000. The evolution of drug metabolism. Pharmacology 61 (3), 124–135.

Nelson, J.C., Papakostas, G.I., 2009. Atypical antipsychotic augmentation in major depressive disorder: a meta-analysis of placebo-controlled randomized trials. Am. J. Psychiatry 166 (9), 980–991. https://doi.org/10.1176/appi.ajp.2009.09030312.

Nierenberg, A.A., Feighner, J.P., Rudolph, R., Cole, J.O., Sullivan, J., 1994. Venlafaxine for treatment-resistant unipolar depression. J. Clin. Psychopharmacol. 14 (6), 419–423. https://doi.org/10.1097/00004714-199412000-00008.

Olgiati, P., Serretti, A., Souery, D., Dold, M., Kasper, S., Montgomery, S., Zohar, J., Mendlewicz, J., 2018. Early improvement and response to antidepressant medications in adults with major depressive disorder. Meta-analysis and study of a sample with treatment-resistant depression. J. Affect. Disord. 227, 777–786. https://doi.org/10.1016/j.jad.2017.11.004.

Papakostas, G.I., Perlis, R.H., Scalia, M.J., Petersen, T.J., Fava, M., 2006. A meta-analysis of early sustained response rates between antidepressants and placebo for the treatment of major depressive disorder. J. Clin. Psychopharmacol. 26 (1), 56–60.

Papakostas, G.I., Fava, M., Thase, M.E., 2008. Treatment of SSRI-resistant depression: a meta-analysis comparing within- versus across-class switches. Biol. Psychiatry 63 (7), 699–704. https://doi.org/10.1016/j.biopsych.2007.08.010.

Poirier, M.F., Boyer, P., 1999. Venlafaxine and paroxetine in treatment-resistant depression. Double-blind, randomised comparison. Br. J. Psychiatry 175, 12–16. https://doi.org/10.1192/bjp.175.1.12.

Popova, V., Daly, E.J., Trivedi, M., Cooper, K., Lane, R., Lim, P., Mazzucco, C., Hough, D., Thase, M.E., Shelton, R.C., Molero, P., Vieta, E., Bajbouj, M., Manji, H., Drevets, W.C., Singh, J.B., 2019. Efficacy and safety of flexibly dosed esketamine nasal spray combined with a newly initiated oral antidepressant in treatment-resistant depression: a randomized double-blind active-controlled study. Am. J. Psychiatry 176 (6), 428–438. https://doi.org/10.1176/appi.ajp.2019.19020172.

Ruhé, H.G., Huyser, J., Swinkels, J.A., Schene, A.H., 2006. Switching antidepressants after a first selective serotonin reuptake inhibitor in major depressive disorder: a systematic review. J. Clin. Psychiatry 67 (12), 1836–1855. https://doi.org/10.4088/JCP.v67n1203.

Rush, A.J., Trivedi, M.H., Wisniewski, S.R., Nierenberg, A.A., Stewart, J.W., Warden, D., Niederehe, G., Thase, M.E., Lavori, P.W., Lebowitz, B.D., McGrath, P.J., Rosenbaum, J.F., Sackeim, H.A., Kupfer, D.J., Luther, J., Fava, M., 2006a. Acute and longer-term outcomes in depressed outpatients requiring one or several treatment steps: a STAR*D report. Am. J. Psychiatry 163 (11), 1905–1917. https://doi.org/10.1176/ajp.2006.163.11.1905.

Rush, A.J., Trivedi, M.H., Wisniewski, S.R., Stewart, J.W., Nierenberg, A.A., Thase, M.E., Ritz, L., Biggs, M.M., Warden, D., Luther, J.F., Shores-Wilson, K., Niederehe, G., Fava, M., 2006b. Bupropion-SR, sertraline, or venlafaxine-XR after failure of SSRIs for depression. N. Engl. J. Med. 354 (12), 1231–1242.

Rush, A.J., South, C., Jha, M.K., Jain, S.B., Trivedi, M.H., 2020. What to expect when switching to a second antidepressant medication following an ineffective initial SSRI: a report from the randomized clinical STAR*D study. J. Clin. Psychiatry 81 (5). https://doi.org/10.4088/JCP.19m12949.

Souery, D., Amsterdam, J., De Montigny, C., Lecrubier, Y., Montgomery, S., Lipp, O., Racagni, G., Zohar, J., Mendlewicz, J., 1999. Treatment resistant depression: methodological overview and operational criteria. Eur. Neuropsychopharmacol. 9 (1–2), 83–91. https://doi.org/10.1016/S0924-977X(98)00004-2.

Souery, D., Oswald, P., Massat, I., Bailer, U., Bollen, J., Demyttenaere, K., Kasper, S., Lecrubier, Y., Montgomery, S., Serretti, A., Zohar, J., Mendlewicz, J., 2007. Clinical factors associated with treatment resistance in major depressive disorder: results from a European multicenter study. J. Clin. Psychiatry 68 (7), 1062–1070. https://doi.org/10.4088/jcp.v68n0713.

Souery, D., Serretti, A., Calati, R., Oswald, P., Massat, I., Konstantinidis, A., Linotte, S., Bollen, J., Demyttenaere, K., Kasper, S., Lecrubier, Y., Montgomery, S., Zohar, J., Mendlewicz, J., 2011a. Switching antidepressant class does not improve response or remission in treatment-resistant depression. J. Clin. Psychopharmacol. 31 (4), 512–516. https://doi.org/10.1097/JCP.0b013e3182228619.

Souery, D., Serretti, A., Calati, R., Oswald, P., Massat, I., Konstantinidis, A., Linotte, S., Kasper, S., Montgomery, S., Zohar, J., Mendlewicz, J., 2011b. Citalopram versus desipramine in treatment resistant depression: effect of continuation or switching strategies: a randomized open study. World J. Biol. Psychiatry 12 (5), 364–375. https://doi.org/10.3109/15622975.2011.590225.

Thase, M.E., Frank, E., Mallinger, A.G., Hamer, T., Kupfer, D.J., 1992. Treatment of imipramine-resistant recurrent depression, III: efficacy of monoamine oxidase inhibitors. J. Clin. Psychiatry 53 (1), 5–11.

UK ECT Review Group, 2003. Efficacy and safety of electroconvulsive therapy in depressive disorders: a systematic review and meta-analysis. Lancet 361 (9360), 799–808.

Chapter 11

Adjunctive strategies for treatment-resistant depression

Michael E. Thase
Department of Psychiatry, Perelman School of Medicine, University of Pennsylvania, Corporal Michael J. Crescenz Veterans Affairs Medical Center, Philadelphia, PA, United States

Overview

The problem of nonresponse to antidepressant medications is essentially as old as this area of therapeutics: Even during the 1960s and 1970s, it was known that the likelihood of response to a tricyclic antidepressant (TCA) was far from certain, and perhaps 40% to 50% of depressed patients would not obtain significant symptom relief despite an adequate course of treatment (Thase and Rush, 1995). Among the other treatments then available, two decades of clinical experience established that a substantial proportion of depressed patients who did not benefit from TCAs would respond when the antidepressant was switched to a monoamine oxidase inhibitor (MAOI), which was—at the time—the principal alternate class of medication (Thase and Rush, 1995; Thase et al., 1995). Yet others were found to benefit from a carefully orchestrated combination of a TCA and MAOI (Thase and Rush, 1995; White and Simpson, 1981). And, of course, electroconvulsive therapy (ECT)—the first treatment for depression with proven efficacy—continued to provide a strong nonpharmacological option for antidepressant nonresponders with more severe or psychotic depressive syndromes (Thase and Rush, 1995; Devanand et al., 1991). The options for TCA nonresponders increased substantially in the late 1980s and early 1990s as a newer generation of antidepressants was introduced. Within a few years of their introduction, the selective serotonin reuptake inhibitors (SSRIs) and serotonin norepinephrine reuptake inhibitors (SNRIs) and several other newer generation medications, including bupropion and mirtazapine, supplanted the TCAs and MAOIs as first- and second-line therapies for most depressed outpatients. The therapeutic optimism that characterized those golden days of the psychopharmacology is reflected by the appraisal of Thase and Rush (1997), who estimated that up to 90% of treatment-seeking depressed patients could be expected to benefit from a four-step sequence of treatments consisting of sequential trials of newer and older generation antidepressants and ECT.

Despite a sound clinical basis and pragmatic rationale for such a four-step regimen, there continues to be a great unmet need for effective therapies for depressed patients who do not obtain a clear-cut response to monotherapy with several trials of antidepressants. As even newer options have been introduced, progressively fewer psychiatrists are willing to initiate therapy with MAOIs and even comfort with prescribing TCAs is waning among the rank and file. Similarly, ECT is now seldom used (Wilkinson et al., 2018), particularly for those patients who are still able to work and/or not markedly impaired in their social roles. Moreover, many depressed patients obtain some degree of symptomatic relief from an adequate course of therapy with a first- or second-line antidepressant, even if they have not experienced a marked response or full symptomatic remission. For such patients, there exists another broad array of strategies known as augmentation or adjunctive therapies.

By convention, an adjunct is a medication that is not indicated as a monotherapy for treatment of depression but has been shown to enhance or augment response when added to ongoing antidepressant therapy (Thase and Rush, 1995, 1997). A number of the medications that were first used to augment response to TCAs and MAOIs in the 1960s and 1970s are still in use 40+ years later, including lithium salts, thyroid hormones and psychostimulants.Options also have broadened with introduction of a second generation of antipsychotic medications (SGAs). A number of other candidate strategies also have been evaluated. This chapter will review the major adjunctive options now in use. In keeping with current regulatory conventions, the more descriptive terms "adjunct" and "adjunctive therapy" will be used preferentially because, for most of the options, the mechanism of action of the second agent has not actually been shown to "augment" the mechanism of action (MoA) of the primary antidepressant. However, it should also be recognized that many clinicians consider these terms to be generally synonymous.

Managing Treatment-Resistant Depression. https://doi.org/10.1016/B978-0-12-824067-0.00011-6

General principles of adjunctive therapy

In terms of strategic decisions, clinicians usually must choose among three options for their patients who have had an inadequate response to an adequate course of antidepressant therapy: (1) switch to another antidepressant, (2) add a second antidepressant to craft a combination therapy, or (3) add another medication known to augment or enhance the antidepressant (which, by convention, is a drug that is not an approved antidepressant). The rationale for and evidence base supporting use of switching strategies and for combining antidepressants are covered in detail elsewhere in this volume. Although switching antidepressants was once the preferred option for most patients who had not responded to one or two courses of antidepressant therapy (Thase and Rush, 1995), the results of the massive Sequenced Treatment Alternatives to Relieve Depression (STAR*D) study (Rush et al., 2006) suggested that adjunctive and combination strategies may have advantages in terms of speed of effect and overall probability of benefit even after a single antidepressant trial. These findings, coupled with growing evidence of the efficacy of a number of SGAs for adjunctive therapy following inadequate response to antidepressant medication, fostered growing enthusiasm for use of combination and adjunctive strategies in contemporary practice guidelines and treatment algorithms (Cleare et al., 2015; Bauer et al., 2013a; Kennedy et al., 2016; Malhi et al., 2018).

In the present context, a course of adjunctive therapy should begin as soon as it is determined that an adequate trial of a standard antidepressant has not produced a meaningful level of symptom relief. In this context, there is reasonable agreement that an inadequate response can be defined by less than a 50% reduction in symptom intensity. In randomized controlled trials (RCTs) this criterion can be defined fairly precisely using standardized scales such as the Hamilton Rating Scale for Depression (HAM-D) or Montgomery Asberg Depression Rating Scale (MADRS). Although briefer self-report scales are gaining some use in every day practice, an adequate response is often based on the subjective appraisal that the patient is "minimally improved" or "unchanged" as a result of the treatment in question. There is also fair agreement that the treated individual must have adhered to a "therapeutic dose" of the antidepressant medication (i.e., a dose with established efficacy in placebo controlled trials) for an "adequate duration." There is less consensus about how many weeks constitute an adequate duration and whether or not the dose needs to be increased to the top of the therapeutic range (if tolerability permits). Nevertheless, defining adequacy as a 4–6 weeks course of antidepressant therapy at the maximally tolerated dose within the therapeutic range for a particular antidepressant is consistent with both older suggestions (Thase and Rush, 1995, 1997; Cleare et al., 2015; Bauer et al., 2013a; Kennedy et al., 2016; Malhi et al., 2018).

As the prescribing behavior of the STAR*D clinicians demonstrated (Rush et al., 2006), the major indication for pursuing an adjunctive strategy instead of switching to a different antidepressant is when the index antidepressant is both well-tolerated and partially effective (e.g., a reduction in symptom intensity between 25% and 50%). In such cases the STAR*D clinicians were about four times more likely to opt for an adjunctive strategy than a switching option (Rush et al., 2006). When tolerability is not problematic, our patients' experiences during washouts and cross-titrations provide a continual reminder us just how uncomfortable or even arduous a fortnight (or longer) of treatment transition can be.

Taking into account both older and newer strategies, clinicians have a broad range of options for adjunctive therapy. Sometimes the adjunct is selected to treat a particularly troublesome persistent symptom, such as insomnia or fatigue. Given the high prevalence of anxiety in MDD, whether as a symptom of the depressive episode or and as a cooccurring disorder, the association with antidepressant nonresponse (Fava et al., 2008) and development of treatment resistance (Souery et al., 2007) has substantial importance. Introduced at the same time as the TCAs and MAOIs, the benzodiazepines have for 60 years provided clinicians a rapidly effective option for the anxious depressed patients and they are often prescribed from the outset of treatment. As the benzodiazepines and the newer more selective hypnotic drugs such as zolpidem are indicated for treatment of anxiety and insomnia, whether or not associated with MDD, this approach is probably better thought of as coprescription rather than adjunctive therapy. Other examples of coprescription include psychostimulant therapy for the patient with attention deficit disorder, narcolepsy or idiopathic hypersomnolence. When the cooccurring condition is subsyndromal or recognized after the fact, the line between coprescription and adjunctive therapy blurs.

Ideally, the choice among particular adjunctive strategies is based on a careful assessment of the patient's past treatment history, phenomenology or other considerations. For example, prescribing adjunctive thyroid hormone is arguably the correct next step when it is known that patient has an elevated level of thyroid stimulating hormone (TSH). However, if the TSH value had not been obtained, one could argue that the adding a psychostimulant or dopamine partial agonist to the somewhat helpful SSRI was a solid choice for a patient with significant fatigue, psychomotor slowing and a tendency to oversleep. Consider the case of a patient with persistent anxiety and insomnia despite a partial response to an SNRI. Although prescription of a sedative-hypnotic medication or combination antidepressant therapy might appear to be clinically indicated, what if the clinician had obtained the history that the patient was raised by adoptive parents and that both her biological father and one older sibling had bipolar I disorder? Or, what if the pretreatment assessment had supplemented by a self-report screening inventory for bipolar disorder, which revealed a score at the 95th percentile for risk for bipolarity?

In those scenarios, a much stronger could have been made for adjunctive treatment with lithium or a SGA, even though the patient had never experienced a diagnosable episode of hypomania.

Preferred options for adjunctive therapy

Second-generation antipsychotics

The value of antipsychotic medications for a select subgroup of depressed patients was recognized shortly following the introduction of medications such as chlorpromazine and thioridazine; the potential for additive benefit in combination with antidepressants was explicitly recognized by the marketing of several proprietary formulations, including amitriptyline plus perphenazine and tranylcypromine plus trifluoperazine. However, by the late 1980s, use of these "first-generation" antipsychotics was largely delimited to treatment of more severely depressed patients, particularly those experiencing psychotic features. This practice changed within a few years of the introduction of risperidone and olanzapine, the first members of the SGA class. The SGAs are now the best proven option for adjunctive therapy following inadequate response to antidepressant therapy of MDD (Nelson and Papakostas, 2009; Zhou et al., 2015a; Strawbridge et al., 2019). In the United States, three SGAs (aripiprazole, quetiapine, and brexpiprazole) are approved by the Food and Drug Administration (FDA) for the broader indication of inadequate response to antidepressants, and a fourth—olanzapine, specifically in combination with fluoxetine—has FDA approval for the stricter classification of treatment-resistant depression (TRD). Yet another SGA, risperidone, has established adjunctive efficacy in controlled trials (Nelson and Papakostas, 2009) even though the manufacturer opted not to seek FDA approval for this indication. As there are positive placebo controlled studies of adjunctive therapy with a number of other SGAs, including ziprasidone (Papakostas et al., 2015), cariprazine (Durgam et al., 2016) and pimavanserin (Fava et al., 2019), it may well be that most if not all of the SGAs has efficacy as adjuncts to antidepressants.

Looking across studies, there are several key clinical observations. First, significant drug placebo differences are routinely observed within the first 2 weeks of adjunctive therapy with SGAs. The SGAs thus have a faster onset of effect than conventional antidepressants. Second, the dose needed to exert such a rapid effect tends to be lower than the minimal dose needed for antipsychotic efficacy in patients with schizophrenia or mania. For aripiprazole, for example, the modal dose for antidepressant responders is 5 mg/day, whereas 15 mg/day is generally considered to be the minimal effective dose for treatment of schizophrenia or mania (Jing et al., 2013). Third, unlike conventional antidepressants, the SGAs appear to be comparably effective for patients with or without prominent anxiety symptoms. This observation does not mean that adjunctive SGA therapy is specifically indicated for patients with high levels of anxiety, but does reflect the clinical observation that they are useful for a broad range of patients who do not benefit from antidepressant (see, for example, Thase et al. (2018)). And fourth, although there are marked differences between the various members of the SGA class in terms of side effects and risks of weight gain and other metabolic complications (Zhou et al., 2015a; Thase, 2016), their efficacy—as determined by drug versus placebo differences in RCTs—appears to be roughly comparable.

It may also be that some—if not most—of the SGAs could be antidepressants if prescribed as monotherapy. In this respect, it is noteworthy that the efficacy of quetiapine monotherapy in MDD has been established by multiple RCTs in doses ranging from 50 to 300 mg/day (Weisler and McIntyre, 2013). Although approved for this indication in some countries, the US FDA did not recommend approval of quetiapine monotherapy of MDD because of concerns about side effects and the potential for longer term complications, including tardive dyskinesia. The potential role of SGA monotherapy for a carefully selected subset of depressed patients is underscored by the fact that four members of the class—olanzapine, quetiapine, lurasidone, and cariprazine—have established efficacy for treatment of bipolar depression (Bahji et al., 2020). In the case of lurasidone, monotherapy efficacy also has been shown in one placebo controlled study of patients with MDD with mixed features (Suppes et al., 2016).

Although inspection of effect sizes strongly supports the notion that these medications have similar efficacy as adjunctive therapies, it must also be noted that there is a dearth of data from comparative RCTs. Indeed, there is only one placebo controlled RCT in the published literature comparing two SGAs as adjunctive therapies for antidepressant nonresponders (Hobart et al., 2018). In this trial, 503 MDD outpatients with a history of nonresponse to antidepressant therapy in the current depressive episode as well as nonresponse to a prospective, 8–10 week trial of antidepressant monotherapy were randomized to 6 weeks of double blind adjunctive therapy with brexpiprazole, quetiapine or placebo. At study endpoint, they found that brexpiprazole (titrated to 1–3 mg/day) was efficacious (versus an adjunctive placebo; mean MADRS difference 1.5 points, $P<.008$), whereas quetiapine (titrated to 150–300 mg/day) was not (mean MADRS difference 0.3 points). One important limitation in the interpretability of this study is that there were no planned comparisons between the two SGAs. In fact, the quetiapine group was included in the design to determine "assay sensitivity" and was only one half the size of the brexpiprazole and placebo groups. Thus, it cannot be said that adjunctive brexpiprazole therapy was significantly more effective than adjunctive quetiapine therapy in this study.

There is also a dearth of studies of the comparative efficacy of SGAs and other clinically relevant options. Only two large scale studies exist, one of quetiapine and the other of aripiprazole, contrasting adjunctive SGA therapy against proven strategies. In the first, Bauer et al. (2013b) randomized 681 outpatients with MDD with a history of antidepressant nonresponse to 6 weeks of open label therapy; about one-third of the sample was switched quetiapine (300 mg/day), one-third received quetiapine as an adjunct (300 mg/day), and one-third received adjunctive therapy with lithium salts (target blood level: 0.6–1.0 mEq/L). The latter intervention was chosen as the existing standard of adjunctive therapy. They found that adjunctive therapy with quetiapine was significantly more effective than the other two options after 4 days of treatment ($P<.01$), which is consistent with the notion that this strategy has a more rapid onset of benefit. After 6 weeks of therapy, the adjunctive quetiapine group continued to show a larger effect than the adjunctive lithium group (difference in MADRS scores of 2.32; nominal [unadjusted] $P<.05$). However, as the primary hypothesis of this study was a strict test of noninferiority, the conclusion was that adjunctive therapy with quetiapine was *not inferior* to adjunctive lithium therapy. There also did not appear to be a meaningful difference in tolerability between the adjunctive interventions.

In an even larger study (Mohamed et al., 2017), 1522 Veterans with MDD who had not responded to at least one adequate trial of antidepressant monotherapy in the current episode were randomized to up to 36 weeks of treatment with one of three strategies: switch to bupropion, combined therapy with bupropion and the current therapy (typically an SSRI or SNRI) or adjunctive therapy with aripiprazole (dosage titrated from 2 to 15 mg/day). They found adjunctive aripiprazole to be significantly more effective than switching to antidepressant monotherapy with bupropion (week 12 response rates: 74% vs 62%, respectively); comparisons between adjunctive aripiprazole arm and combination antidepressant therapy yielded intermediate differences that generally did not reach statistical significance.

Longer term considerations. Although we are now have more than two decades of experience with some of the older SGAs, there are still some important questions about their use as adjunctive therapies for patients with MDD and it seems likely that these questions will not be answered by clinical trial data (Thase, 2016). Perhaps most important is the uncertainty about just how long an effective adjunctive therapy be continued. Clinical experience informs us that many patients who respond to adjunctive SGA therapy will relapse if the adjunct is withdrawn during the first few months after response, so a standard of practice has emerged for patients to routinely receive at least 4 to 6 months of adjunctive therapy. The need for a continuation phase of treatment forces clinicians to be vigilant to minimize the hazards of weight gain and other metabolic side effects, which are likely to be cumulative across at least the first 6 months of therapy. At some point of continuation therapy, clinicians also must begin to monitor for signs of treatment-emergent dyskinetic movements. In fact, in one study in which patients who responded to adjunctive aripiprazole (mean dose 10 mg/day) could receive up to 1 year of continuation therapy, 4 of 987 patients (about 0.4%) developed dyskinesias (Berman et al., 2011). Importantly, in all four cases the dyskinetic movements first diminished and then ceased after withdrawal of aripiprazole, which suggests that clinician vigilance and ongoing monitoring can have rewarding outcomes. The risk of weight gain in this study (4.4 kg) also was not trivial, with about 37% among the 303 patients who completed a year of adjunctive therapy gaining at least 7% body weight.

There are simply not enough data on longer term adjunctive treatment with SGAs in MDD patients. The study of Brunner and colleagues (Brunner et al., 2014), a double-blind relapse prevention study that enrolled 444 outpatients with TRD who responded to an 8-week open-label course of olanzapine-fluoxetine combination and who did not relapse across 12 weeks of continuation therapy, is the largest of its kind. They found that, after 20 weeks of treatment, the risk of relapse over the next 6 months of double blind therapy was about 32% for the patients who were withdrawn from olanzapine (double blind) and maintained on fluoxetine monotherapy, compared to 16% among those who continued to take the combination of olanzapine and fluoxetine. One patient in the combined treatment arm developed dyskinetic movements during the 6 months of continued therapy, which suggests a risk of treatment-emergent dyskinesia during longer term olanzapine therapy that is roughly comparable to that reported for aripiprazole. Weight gain was a much larger problem in this trial, with about a 4 kg difference in mean weight at the end of the 6 month trial, accompanied by and subtler changes in selected metabolic parameters, including fasting triglycerides. Moreover, looking across treatment phases, 56% of the patients who received 10+ months of therapy with OFC experienced a weight gain of at least 7% of their pretreatment weight. Although open label data collected during ongoing adjunctive therapy with other SGAs suggests somewhat lower risks (Thase, 2016), it is clear that careful monitoring of weight and metabolic parameters is indicated during longer therapy.

Lithium salts

The therapeutic effects of lithium salts, principally lithium carbonate and lithium citrate, were recognized for patients with recurrent mood disorders from that outset of the first generation of psychopharmacology. Moreover, lithium—typically in doses of 600–1200 mg/day—was the first adjunctive medication found to be efficacious for antidepressant nonresponders on the basis of multiple placebo controlled studies (Thase and Rush, 1995; Strawbridge et al., 2019; Zhou et al., 2015b). Originally described to have a remarkably rapid onset of benefit (i.e., within 48 h (Dé Montigny et al., 1981)) in patients who had not responded to 3–4 weeks of TCA therapy, subsequent reports suggested that many patients needed a more conventional 6–8 weeks of treatment with serum levels (e.g., 0.5–0.8 mEq/L; see, for example, Thase et al. (1989)). In the first metaanalysis of placebo controlled trials, Crossley and Bauer (2007) identified 10 eligible studies and confirmed a meaningful advantage for adjunctive lithium salts as compared to the double blind placebo. With an average benefit estimated as an odds ratio of about 3, Crossley and Bauer's findings suggested that one might expect up to a 20% advantage over and above placebo-expectancy effects, which would correspond to a number needed to treat of about 5. It is noteworthy that only two of the RCTs included in Crossley and Bauer's metaanalysis studied patients who were taking SSRIs. Thus, their findings and conclusions may be more relevant to adjunctive therapy of TCAs.

Several subsequent metaanalyses have permitted estimations of benefit to be updated and refined, in part by including several more recent studies, and, in part, by use of newer analytic methods to estimate relative efficacy (Strawbridge et al., 2019; Zhou et al., 2015b). Zhou and colleagues (Zhou et al., 2015b), for example, found that the average benefit of adjunctive lithium therapy was smaller than originally estimated by Crossley and Bauer (Odds Ratio = 1.56). In terms of the average benefit over and above placebo, an effect of size magnitude would suggest an average benefit of only about 10% absolute advantage, which corresponds to a number needed to treat of about 10. In their network metaanalysis, Zhou et al. (2015b) estimated that the benefit was generally comparable to the estimated effect sizes of the SGAs. Consistent with the findings of the RCT conducted by Bauer and colleagues (Weisler and McIntyre, 2013) discussed earlier, Zhou et al. (2015b) concluded that the tolerability of adjunctive lithium therapy also was comparable to adjunctive SGA therapy across 4–8 weeks of treatment (Fig. 11.1).

Before the introduction of the SGAs, adjunctive lithium therapy was the best-proven adjunctive therapy for patients with difficult to treat depression and was generally considered to be more effective than the major alternative of the era, adjunctive therapy with thyroid hormones (Thase and Rush, 1995, 1997). With this question in mind, the STAR*D research project included a head-to-head trial that directly compared adjunctive therapy with lithium carbonate (up to 900 mg/day) and thyroid hormone (triiodothyronine [T_3], up to 50 μg/day)(Nierenberg et al., 2006). Like all of the STAR*D comparisons, a randomized, open label and rated-blinded design was used to compare the adjunctive therapies across 12 weeks of open label treatment. The study enrolled 142 MDD patients who had not responded to about 6 months of treatment, including a 12 week course of citalopram monotherapy, followed by either a second course of antidepressant treatment (with sertraline, venlafaxine or bupropion) or "augmentation" of citalopram by either buspirone or bupropion. Neither adjunctive therapy was particularly effective, with intent to treat remission rates of 16% and 25%, for lithium and thyroid respectively, across up to 12 weeks of therapy. By contemporary standards, a 9% nominal difference in remission rates in patients with difficult

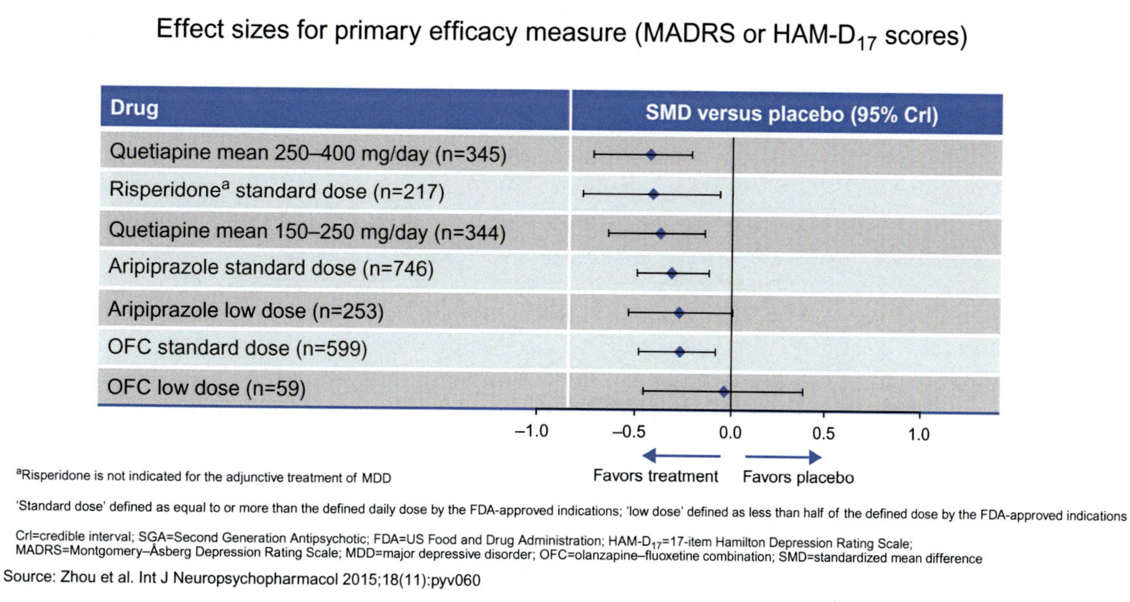

FIG. 11.1 Network metaanalysis of efficacy of SGAs for adjunctive treatment of MDD.

to treat depression would be strongly suggestive of an advantage for thyroid hormone, the STAR*D study suffered from a higher than expected attrition rate and statistical power was not as high as planned. Nevertheless, tests of several secondary outcomes pertaining to acceptability and side effects did reveal significant advantages for the patients who received adjunctive thyroid hormone. The unexpectedly poor showing of adjunctive lithium therapy in STAR*D was not limited to tolerability: prescribers also had problems implementing lithium therapy, with infrequent monitoring of serum levels and a tendency to prescribe lower than recommended lithium doses.

Perhaps heralded by the STAR*D findings but more likely a consequence of prescribers' growing confidence in the value of combined antidepressant therapy and adjunctive SGA therapy, the role of lithium salts in the management of patients with difficult to treat forms of MDD has diminished over the past 15 years. Nevertheless, there important reasons to continue to consider this time-tested strategy for selected patients, particularly for patients at more advanced levels of treatment resistance. Beyond efficacy in MDD that is comparable to SGAs (Strawbridge et al., 2019; Zhou et al., 2015b), other potential reasons to consider lithium salts include its utility across phases of treatment for bipolar affective disorder (Grunze et al., 2010) and its established efficacy for relapse/recurrence prevention for patients with recurrent (unipolar) MDD (Souza and Goodwin, 1991; Cipriani et al., 2006). Of particular relevance to patients with more advanced levels of TRD is prophylactic efficacy of the combination of lithium and the TCA nortriptyline following successful treatment with ECT (Sackeim et al., 2001). Finally, although evidence of lithium's "antisuicide" effect is largely derived from studies of bipolar disorder (Benard et al., 2016), a review conducted under the auspices of the Cochrane Collaboration concluded that this benefit was relevant to all patients with recurrent mood disorders (Cipriani et al., 2013).

Longer-Term Considerations. If effective, adjunctive lithium may be continued for months for relapse prevention or years for maintenance therapy. Later emerging side effects include skin changes (including acne and psoriasis) and weight gain, which may suggest development of hypothyroidism. For this reason, thyroid stimulating hormone (TSH) levels are monitored periodically and, when clinically indicated, on the basis of newly emergent symptoms or side effects. Onset of symptoms such as excessive thirst or polyuria may reflect an impairment of renal concentrating ability, which in turn is a consequence of lithium-induced interstitial fibrosis of the kidneys. This "not uncommon" consequence of years of lithium therapy is arguably the most serious complication of maintenance treatment and, if not addressed, may eventually cause chronic renal failure. Following standards of care developed for maintenance lithium therapy of bipolar disorder (Grunze et al., 2010; Yatham et al., 2018), periodic monitoring of blood levels and measures of renal function, including electrolytes, plasma BUN and creatinine levels and 24 h urinary creatinine clearance levels are recommended to identify patients at risk for renal complications of adjunctive lithium therapy.

Thyroid hormones

The importance of the thyroid axis in the maintenance of well-being and of disturbances of thyroid hormone—both hypothyroidism and hyperthyroidism—in the pathogenesis of mood disorders were known early in the first generation of antidepressant therapy (Lipton et al., 1972; Whybrow and Mendels, 1969). On the basis of small studies by Prange and colleagues (Prange Jr et al., 1969, 1972; Wilson et al., 1970), it was hypothesized that adjunctive therapy with thyroid hormone could accelerate response to TCAs such as imipramine or even be used as an adjunct to convert nonresponders into remitters. Several smaller RCTs tested this hypothesis and the results of a metaanalysis (Aronson et al., 1996) suggested that adjunctive therapy with T_3 (25–50 μg/day) significantly lowered depressive symptoms. However, the overall effect on response rates in the four placebo controlled trials (OR = 1.53) did not reach conventional standards of statistical significance (see Fig. 11.2).

That the efficacy of adjunctive therapy with thyroid hormone was comparable to that of adjunctive therapy with lithium carbonate and SGAs. As noted earlier, results of the STAR*D comparison tended to favor T_3 over adjunctive lithium therapy, particularly with respect to ease of implementation and better tolerability (Fig. 11.3). However, the most recent systematic review of adjunctive thyroid therapy concluded there were too few studies in well-documented TRD to establish efficacy (Lorentzen et al., 2020).

It is noteworthy that most of evidence pertaining to adjunctive therapy with thyroid hormone in MDD is derived from studies utilizing triiodothyronine (T_3)instead of the more commonly prescribed thyroxine (T_4); this is because it was known early on that the brain preferentially uses T_3 (Joffe et al., 1995). Although a small body of work can be marshaled to support this practice (Joffe et al., 1995), most people have the enzymatic capacity to readily convert T_4 to T_3; this underpins the practice of internists to prefer T_4 for treatment of hypothyroidism (Khandelwal and Tandon, 2012) and weakens the rationale for psychiatrists' preferential use of T_3. With this issue in mind, the results of a reanalysis of an RCT (Cooper-Kazaz et al., 2007) investigating the use of adjunctive T_3 to accelerate the antidepressant effects of the SSRI sertraline are particularly interesting, albeit preliminary. The reanalysis focused on whether the benefit of adjunctive T_3 was primarily experienced by patients who had the less functional polymorphism of the type 1 form of the enzyme deiodinase enzyme (DIO1-C785T); this subgroup might be expected to have low brain levels of T_3 (Cooper-Kazaz et al., 2009). As hypothesized, they found a

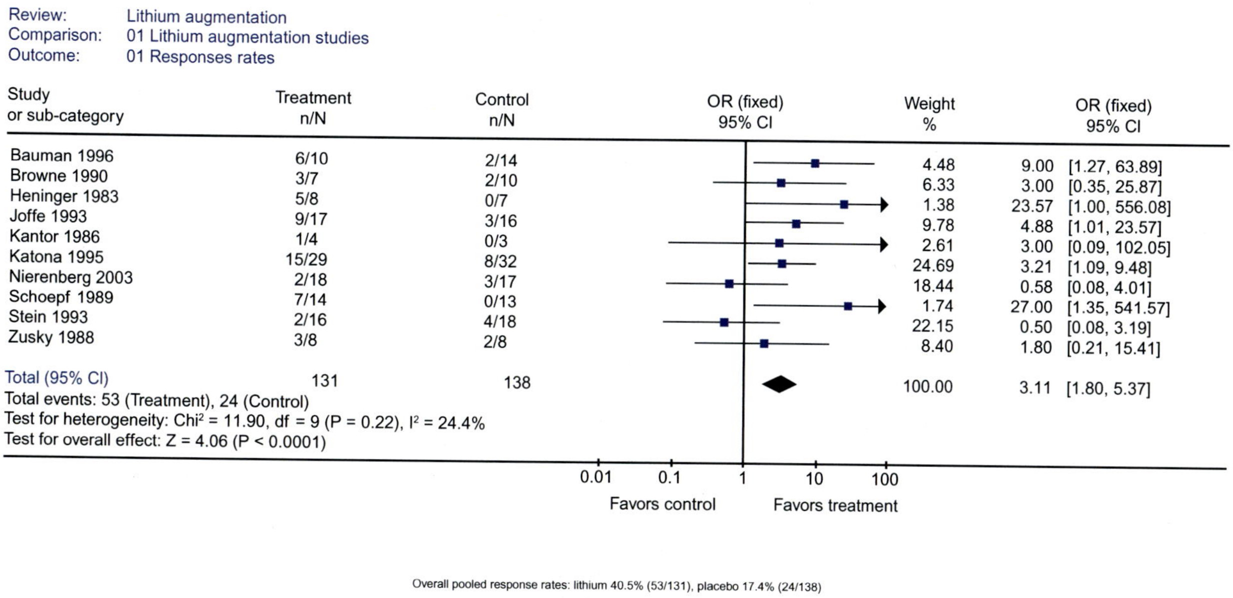

FIG. 11.2 Metaanalysis of placebo-controlled studies of adjunctive therapy with lithium salts.

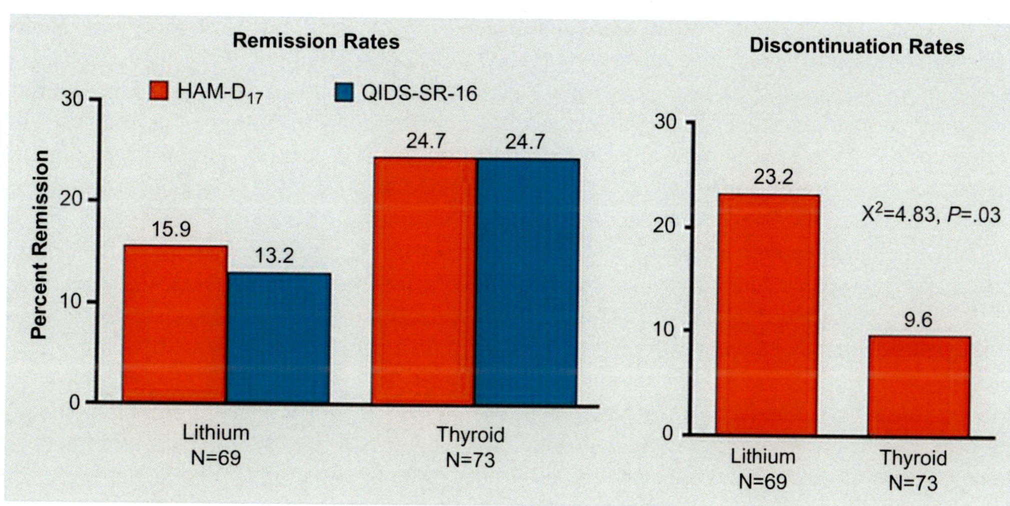

Source: Nierenberg AA, et al. Am J Psychiatry. 2006;163(9):1519-1530.

FIG. 11.3 Adjunctive therapy with thyroid hormone in STAR*D: results of level 3 comparison with lithium.

significant treatment x genotype interaction such that the entire advantage favoring adjunctive T_3 over placebo appeared to be explained by subgroup of patients who had at least one copy of DIO1-C785T. If replicated, this finding could lead to a more personalized and efficient use of adjunctive T_3 therapy for antidepressant nonresponders. In the meantime, evidence pertaining to safety and ease of implementation are reassuring and it is at the least clinically reasonable consider this strategy for patients have a history of thyroid dysfunction and/or who have borderline high or elevated TSH levels (Thase and Rush, 1995; Dwyer et al., 2020).

Longer-term considerations. As there have been no longer term studies of adjunctive thyroid hormone in patients with difficult to treat MDD, decisions about whether or not continue treatment must be made on an individual basis, taking into account the magnitude of the benefit and side effects. For patients who have had a good response and have experienced minimal or no side effects, one consideration during continued therapy is to ensure that the thyroid axis is not chronically suppressed, which is a risk factor for development of osteoporosis. In practice, periodic monitoring of TSH values is sufficient to implement this recommendation, with that are below 0.5 iu (Khandelwal and Tandon, 2012; Dwyer et al., 2020; Bekkering et al., 2019).

Other options for adjunctive therapy

Psychostimulants

The psychoactive effects of amphetamine salts were first recognized in the 1930s, about 25 years before the discovery of the first antidepressants (Rasmussen, 2015). However, for most depressed patients the favorable effects that racemic amphetamine, its more potent stereoisomer and the several related drugs that followed had on mood, energy or motivation were often short-lived; even those who obtained more dramatic early responses sometimes developed tolerance within weeks of starting treatment (Rasmussen, 2015). By the early 1970s, it was clear that, whatever the benefits that psychostimulants might have for the treatment of depression, their effects were less reliable than those of the TCAs or MAOIs. Growing awareness of the potential of psychostimulants for misuse and abuse further marginalized their role in most clinicians' treatment hierarchies (Thase and Rush, 1995; Rasmussen, 2015). As such, the psychostimulants were largely reserved for third- or fourth-line use, often as adjuncts to ongoing antidepressant therapy. Nevertheless, it has been consistently documented over the last 50 years that a small proportion of highly treatment-resistant depressed patients obtain palliative benefits from psychostimulants, typically as adjuncts to conventional antidepressants (Thase and Rush, 1995; Parker et al., 2013).

In current practice, clinicians may pick among several psychostimulants, including D-amphetamine, immediate or sustained release formulations of mixed amphetamine salts, and lisdexamfetamine, and methylphenidate, which also is available in immediate and sustained release formulations. These drugs are all considered Schedule II controlled substances by the US Drug Enforcement Agency (DEA), which necessitates a fair amount of oversight by both the prescriber and the pharmacy. Two more recently introduced drugs, modafinil and its stereoisomer armodafinil, are typically grouped with the classical psychostimulants, although these drugs have a more benign side effect profile and less potential for misuse and abuse (Murillo-Rodríguez et al., 2018); both of these alertness promoting drugs are classified as Schedule IV drugs by the DEA.

Metaanalyses have confirmed clinical impressions that psychostimulants have at least short-term antidepressant effects for patients with both MDD and bipolar depression (Goss et al., 2013; McIntyre et al., 2017; Tsapakis et al., 2020). However, evidence from well-controlled studies has not established adjunctive efficacy in patients with a documented history of nonresponse to standard antidepressants. This point is underscored by the failure of several well-funded clinical development programs for newer formulations of psychostimulants to obtain regulatory approval for adjunctive therapy in MDD. The discrepancy between metaanalytic evidence of an antidepressant effect and the failure of a number of potentially pivotal phase III trials could result from two possibilities: (1) psychostimulants have a beneficial effect on several depressive symptoms, but not a broad enough effect to show overall efficacy or (2) stimulants may have a meaningful effect for a small proportion of depressed patients, but have little or no benefit for most. The case series of Parker and colleagues (Parker et al., 2013), which described a highly resistant group of patients treated at a tertiary care specialty program, suggests that the first explanation may be accurate—only about 20% of their patients were judged to obtain great benefit from open label therapy.

Longer-term considerations. The lack of regulatory approval for use in MDD, coupled and the potential for misuse, abuse and drug dependence, provide a solid argument to minimize the use of psychostimulants as adjunctive therapy for depressed patients. Of course, cotherapy is indicated for a subset of patients with difficult to treat depressive disorders who suffer from comorbid conditions such as attention deficit hyperactivity disorder, sleep apnea, narcolepsy, or binge-eating disorder. And, as illustrated by Parker and colleagues (Parker et al., 2013), there is another subgroup of patients with more severe, treatment-resistant depressive disorders who do obtain sustained benefits with these medications. The principal longer term consideration in such cases is to ensure that there is not upward drift in dosing over time, which would reflect tolerance, and that the medication is being responsibly managed. In many parts of the United States, these goals are facilitated by state-sponsored data bases that register prescriptions for all controlled substances.

Nutraceuticals

The so-called nutraceuticals, which are standardized and purified pharmaceutical versions of common nutrients, have broad popularity in many societies, both as alternatives to conventional medications or as adjuncts to standard therapies. Systematic reviews and metaanalyses of treatment research on various nutraceuticals generally provide cautiously optimistic appraisals about the antidepressant effects of a select group of nutraceutical compounds, including *S*-adenosylmethionine (SAMe), folate and methylfolate, omega-3 fatty acids (primarily EPA or ethyl-EPA), vitamin D and tryptophan (including 5-HTP) (Ravindran et al., 2016; Sarris et al., 2016; Haller et al., 2019). Whereas evidence of antidepressant efficacy is generally modest, these reviews are largely reassuring about the safety and tolerability of nutraceuticals for depressed patients (Ravindran et al., 2016; Sarris et al., 2016; Haller et al., 2019).

With specific reference to adjunctive therapy of MDD following inadequate response to antidepressants, Sarris and colleagues (Sarris et al., 2016) identified 40 RCTs of adjunctive therapy with various nutraceuticals that met minimal standards for inclusion; the best studied strategies in their systematic review were folic acid (nine studies, including two studies of L-methylfolate), omega-3 fatty acids (eight studies), tryptophan (eight studies, including one study of 5-HTP) and SAMe (four studies). Individual metaanalyses of the adjunctive studies of tryptophan and SAMe were not possible: in the former case, the literature was older and did not conform to current standards of reporting and, in the latter case, the studies were too heterogeneous to permit a valid pooling of the data. Among the nutraceuticals that could be assessed by metaanalysis, the studies of adjunctive omega-3 revealed a significant, moderate effect as compared to placebo, although the validity of this finding was weakened by significant heterogeneity across studies. The metaanalysis of RCTs of folic acid likewise revealed significant heterogeneity, although the difference between drug and placebo across studies was not significant (Sarris et al., 2016).

Given the heterogeneity of antidepressant effects for adjunctive folic acid, it is noteworthy that L-methylfolate, not folate, is used in the so-called one carbon cycle for synthesis of monoamines. Conversion of dietary folic acid to L-methylfolate requires the enzyme methylenetetrahydrofolate reductase (MTHFR), which is under genomic control; about 40% of the population has at least one copy of the less functional polymorphism of this gene (Wan et al., 2018). Thus, it is possible that an adequate intake of dietary folic acid or even the impact of adjunctive folic acid might not ensure adequate amounts of L-methylfolate in the brain. Two RCTs—reported in one paper (Papakostas et al., 2012)—evaluated adjunctive therapy with L-methylfolate (7.5–15 mg/day) in antidepressant nonresponders with MDD; both studies used a sequential multistage design. In the first trial, 148 outpatients with SSRI-resistant major depressive disorder were enrolled in a 60-day study divided into two 30-day periods. Patients were randomly assigned, in a 2:3:3 ratio, to receive L-methylfolate for 60 days (7.5 mg/day for 30 days followed by 15 mg/day for 30 days), placebo for 30 days followed by L-methylfolate (7.5 mg/day) for 30 days, or placebo for 60 days. SSRI dosages were kept constant throughout the study. In the second trial, with 75 patients, the design was identical to the first, except that the L-methylfolate dosage was 15 mg/day during both 30-day periods. In the first trial, no significant difference was observed in outcomes between the treatment groups. In the second trial, adjunctive L-methylfolate at 15 mg/day showed significantly greater efficacy compared with continued SSRI therapy plus placebo on both primary outcome measures (response rate and degree of change in depression symptom score) and two secondary outcome measures of symptom severity. The number needed to treat for response was approximately six in favor of adjunctive L-methylfolate at 15 mg/day. L-Methylfolate was well tolerated, with rates of adverse events no different from those reported with placebo. Results of secondary analyses of this dataset indicated that the benefit of adjunctive L-methylfolate therapy was largely delimited to those who were carriers of the less functional MTHFR genotype (Papakostas et al., 2014); this effect was further moderated by obesity and a subset of biomarkers of inflammation (Shelton et al., 2015).

It is not clear what role nutraceuticals should play in the management of treatment-resistant depression. On the one hand, omega-3 fatty acids, SAMe, and methylfolate are safe and well-tolerated compounds; they also are generally not particularly expensive, even if purchased over the counter. On the other hand, none of the nutraceuticals have the kind of replicated efficacy in well-controlled studies of patients with carefully documented treatment-resistant depression that leads to high rankings in practice guidelines or systematic reviews (see, for example, Strawbridge et al. (2019) or Zhou et al. (2015b)). One potential area for study is a potential link between the putative mechanisms of action of several of the nutraceuticals and the role of inflammation in both chronicity of depression and nonresponse to antidepressants (Hoepner et al., 2021).

Antiinflammatory strategies

There has been growing evidence over the past decade that the notorious heterogeneity of the depressive disorders may be partly parsed by the impact of inflammation or, more likely, the pathogenic impact of the mediators of inflammation (Hoepner et al., 2021; Raison and Miller, 2013). Up to one half of depressed patients may show one or more of the so-called inflammatory biomarkers, and both cytokines and nonspecific biomarkers such as C-Reactive Protein (C-RP) have been associated with both chronicity and nonresponse to antidepressant medications (Haroon et al., 2018; Kopschina Feltes et al., 2017). Moreover, although the literature is far from conclusive, several recent metaanalyses of effects of a broad range of interventions, patient populations and durations of treatment suggest that medications that are known to suppress or attenuate inflammation have a modest effect on improving depressive symptoms (Köhler-Forsberg et al., 2019; Bai et al., 2020). For example, the metaanalysis of Köhler-Forsberg and colleagues (Köhler-Forsberg et al., 2019) identified eight placebo controlled RCTs of cytokine inhibitors in patients ($n = 3345$) with depressive symptoms (Köhler-Forsberg et al., 2019). The participants in these studies generally suffered from illnesses such as psoriasis or rheumatoid arthritis and the drugs studied, which included adalimumab, ixekizumab, or methotrexate, were typically prescribed without concomitant antidepressant medication. They found that therapy with cytokine inhibitors resulted in a modest improvement of depressive symptoms (as compared to placebo). However, confidence in the overall effect was diminished by significant heterogeneity across studies. With respect to tolerability, these drugs are certainly not as innocuous as an Omega-3 fatty acid or L-methylfolate, but are considered to have a favorable benefit/risk ratio for treatment of significant medical illnesses.

The handfuls of RCTs that have examined the impact of antiinflammatory medications in depressed patients with documented nonresponse to antidepressant therapy have yielded mixed results (Raison et al., 2013; Nettis et al., 2021; McIntyre et al., 2019; Husain et al., 2020). For example, the single-site study of Raison and colleagues (Raison et al., 2013), which evaluated the antidepressant effect of three infusions with the tumor necrosis factor (TNF) antagonist infliximab (5 mg/kg) across 12 weeks, found no effect overall, but did observe a significant improvement in depressive symptoms in the subset of patients with C-RP levels of 5 mg/L or higher. In this subgroup, which comprised slightly more than one-third of the sample, almost twice as many of the infliximab-treated patients responded as compared the placebo treated patients (62% [8/13] vs 33% [3/9]). Of note, baseline concentrations of TNF and its soluble receptors were significantly higher in infliximab-treated responders, who also showed significantly greater decreases in CRP levels at week 12 compared with placebo-treated responders.

A similar result was observed of placebo controlled RCT of the minocycline (Nettis et al., 2021), an antibiotic descendent of tetracycline thought to have immune suppressing and potentially neuroprotective effects. An earlier metaanalysis of three small placebo controlled RCTs ($n = 158$) suggested that minocycline (100–400 mg/day) may have moderate to large antidepressant effects (Rosenblat and McIntyre, 2018). The study of Nettis and colleagues (Nettis et al., 2021) randomized 39 MDD patients who had not responded to at least one adequate course of antidepressant therapy in the current episode to 4 weeks of treatment with either minocycline (200 mg/day) or placebo; all patients had C-RP levels of at least 1 mg/L. At study endpoint, the groups did not differ significantly with respect to symptom scores or response rates. A post hoc stratification on the basis of CRP levels (<3 mg/L versus ≥ 3 mg/L), revealed that minocycline was effective for patients with higher CRP levels (mean change in HAM-D-17 scores 12.00 points [sd = 6.45] compared with 3.50 points [sd = 4.34] among those treated with placebo. By contrast, adjunctive minocycline was of no benefit compared to placebo among those with lower CRP levels; both groups showed minimal improvement on the HAMD.

Antiinflammatory strategies also failed to show clear-cut efficacy in two well-designed placebo controlled RCTs of antidepressant-resistant bipolar depression (McIntyre et al., 2019; Husain et al., 2020); these trials infliximab (McIntyre et al., 2019) and celecoxib and minocycline, singly and in combination (Husain et al., 2020). Although the primary test of efficacy failed, the study of infliximab infusion therapy (three infusions of 5 mg/kg across 12 weeks) did observe a significant treatment × time × subgroup interaction among the patients with a past history of abuse or trauma during childhood.

Antiinflammatory therapies cannot be recommended at this time for adjunctive treatment of MDD. There is enough evidence to suggest that this approach may be helpful for a subset of patients with laboratory indicators such as elevated CRP. It may be that further research on this general approach will identify important, pathophysiologically relevant treatment options for small subsets of antidepressant nonresponders.

Notable failures, disappointments, and near misses

The figurative road to Mecca for adjunctive therapies of MDD is littered by a fairly large number of notable failures, several remarkable disappointments and one near miss. What follows is a brief summary of the stories of four drugs that were thought to hold promise as adjunctive therapies for antidepressant nonresponders, but did not show sufficient efficacy for regulatory approval.

Buspirone

Given the association between high levels of anxiety and antidepressant nonresponse and the relevance of the pre- and postsynaptic serotonin receptor sensitivity to response to SSRIs, the possibility that adjunctive therapy with buspirone may have beneficial effects for patients with difficult to treat depression received careful investigation. Buspirone—a high affinity agonist of 5-HT_{1A} receptors and a partial agonist of presynaptic 5-HT_{1A} receptors, which are inhibitory autoreceptors (Loane and Politis, 2012)—was approved by the FDA in 1986 for treatment of anxiety (more specifically in current context, generalized anxiety disorder). Unlike the benzodiazepines, buspirone has no abuse liability and is minimally sedating, though it also is not useful on an "as needed" basis and has little impact on sleep. Like first-generation antidepressants, buspirone therapy typically must be up-titrated and requires 4–6 weeks to exert therapeutic effects.

Following a number of clinical reports and case series, three RCTs of adjunctive buspirone therapy were completed in patients with difficult to treat depression (Landén et al., 1998; Appelberg et al., 2001; Trivedi et al., 2006). In the first (Landén et al., 1998), 117 MDD patients who had not responded to treatment with citalopram (40 mg/day) or paroxetine (30 mg/day) were randomized to 4 weeks of adjunctive therapy with either buspirone (titrated up to 60 mg/day) or placebo. At endpoint, the groups had comparable outcomes; there were no significant differences on measures of depressive symptoms.

The second study (Appelberg et al., 2001) randomized 102 MDD patients who had not responded to a least 6 weeks of SSRI treatment ($N=102$) to an additional 6 weeks of treatment with either buspirone (titrated up to 30 mg/day) or placebo. The group that receive active buspirone showed significantly greater improvement at week 1, but by week 6 the treatment groups had roughly comparable outcomes and did not show statistically significant differences on any outcome measure. Results of a secondary analysis suggested that adjunctive buspirone therapy had a significant effect among a subset of patients with high pretreatment symptom severity scores.

The third trial (Trivedi et al., 2006) was conducted as a component of the second stage of the large STAR*D project. A total of 565 MDD patients participated; all had not responded to an average of 11.9 weeks of citalopram therapy (mean final dose: 55 mg/day). The trial was open patients who were agreed to continue to take citalopram and accept randomization for an additional 12 weeks of treatment with either adjunctive buspirone (titrated up to 60 mg/day) or the addition of bupropion (titrated to up to 400 mg/day). At endpoint, the two groups did not meaningfully differ on the primary outcome: approximately 30% of the patients in each group had a final HAMD score of 7 or less. The group that received treatment with adjunctive buspirone did significantly worse on several secondary outcomes, including higher levels of self-reported depressive symptoms at study endpoint and a greater dropout rate due to side effects (21% vs 13%). A detailed post hoc analysis of these data by Bech and colleagues (Bech et al., 2012) focused on a novel concept of benefit of adjunctive therapy, namely a "triangle" of desirable effects defined by reduction of depressive symptoms, fewer side effects and improved social functioning. According to Bech et al.'s reanalysis, adjunctive buspirone was significantly inferior to the combination of bupropion and citalopram.

In summary, the results of three RCTs indicate that buspirone is not a particularly effective adjunct to pharmacotherapy with SSRIs. Interestingly, another drug that modulates 5-HT$_{1A}$ receptor sensitivity—pindolol—also failed to show adjunctive efficacy despite a clear rationale supported by a large body of preclinical research (see Zhou et al. (2015b) for meta-analytic review). Because buspirone is approved for treatment of anxiety and may have some value for reducing sexual side effects, clinicians continue to prescribe it. Nevertheless, it is not recommended for adjunctive therapy of antidepressant nonresponders.

Dexmecamylamine

Given decades of research and anecdotal evidence suggesting that modulation of nicotinic acetylcholine receptors may have antidepressant effects—perhaps through enhancement of dopaminergic neurotransmission (Philip et al., 2010)—the therapeutic potential of dexmecamylamine, a nicotinic channel modulator, was evaluated in preclinical studies and showed promising effects (Lippiello et al., 2008). Subsequently, the results of a placebo controlled Phase II study of adjunctive therapy of MDD patients who were citalopram nonresponders were announced in a press release to be unequivocally positive (https://www.nomura.com/resources/npv/pdfs/15-07-09-targacept.pdf). In retrospect, it is noteworthy that the actual results of this Phase II study were never published and, even more ominously, it appears that they were never even posted on the US regulatory website (https://www.clinicaltrials.gov/ct2/show/record/NCT00692445?term=TC-5214&cond=depression&draw=2&rank=1). However, the field only had to wait a couple of years for the disappointment engendered by the subsequent series of negative Phase III trials (Vieta et al., 2014; Möller et al., 2015; Tummala et al., 2015). The findings of the metaanalysis of Wang et al. (2016) succinctly summarize this story: the odds ratio for the likelihood of benefit for adjunctive dexmecamylamine was 1.01, which is about as close to being identical to placebo as mathematically possible.

Selective norepinephrine reuptake inhibitors

By the end of the first decade of the 21st century it was widely believed that response to SSRIs could be enhanced by norepinephrine reuptake inhibitors (NRIs). This belief was directly supported by the work of Nelson and colleagues (Nelson et al., 1991, 2004) on combination therapy with desipramine and fluoxetine and indirectly supported by evidence that tertiary amine TCAs and SNRIs might have greater efficacy than SSRIs (Anderson, 1998; Thase et al., 2001), perhaps more specifically for patients with more severe depressive symptoms (Anderson, 1998; Thase et al., 2007). Given the potential cardiovascular side effects of desipramine and other secondary amine TCAs and the growing preference of clinicians for adjunctive strategies (as opposed to switching to an SNRI or tertiary amine TCA), there was interest in determining if one of the novel, more selective NRIs might prove useful for adjunctive therapy of SSRI nonresponders.

Among potential candidates, edivoxetine received a thorough evaluation for adjunctive efficacy of MDD across Phase II (Ball et al., 2014) and Phase III (Ball et al., 2016; Martinez et al., 2015; Oakes et al., 2015; Stauffer et al., 2017) programs. However, despite glimpses of benefit on particular depressive symptoms or aspects of social functioning (Ball et al., 2014; Stauffer et al., 2017), efficacy as an adjunctive therapy on primary outcome measures could not be established. Perhaps most discouraging was the consistent documentation of side effects that are characteristic of NRIs, which indicates that the drug failed to exert antidepressant effects even though patients received adequate drug exposure (Ball et al., 2014, 2016; Martinez et al., 2015; Oakes et al., 2015). This would suggest that there also is little reason to try the other FDA approved selective NRI, atomoxetine, for adjunctive therapy of SSRI nonresponders, unless of course the patient had comorbid ADHD. For patients who appear to warrant treatment with the combination of an SSRI and a norepinephrine inhibitor, at this moment, it would appear that switching to venlafaxine, duloxetine, or a tertiary amine TCA or combining therapy with nortriptyline or desipramine would be better justified by the literature.

Buprenorphine + samidorphan combination

Although some people experience as an emotionally painful state and opioids have known antidepressant activity, their use in major depressive disorder (MDD) has been severely limited by risks of misuse, abuse, and addiction (Ehrich et al., 2015). Clinical experience with the μ and κ opioid partial agonist buprenorphine (BUP) suggested that beneficial effects of opioids on mood might be retained with a compound that had a lower potential for abuse liability (Karp et al., 2014; Ahmadi et al., 2018; Serafini et al., 2018). The availability of samidorphan (SAM), a potent μ and κ opioid antagonist, created the opportunity to evaluate whether the putative antidepressant effects of BUP could be harnessed without any risks for misuse or abuse (Ehrich et al., 2015).

The first RCT of adjunctive therapy with BUP + SAM (also known as ALKS 5461) was conducted in 141 MDD patients with a history of an inadequate response to one or two trials of antidepressant therapy in the current depressive episode (Fava et al., 2016). This study used a two-stage sequential parallel comparison design; each phase consisted of 4 weeks of double blind treatment (Fava et al., 2016). Participants were randomly assigned to receive adjunctive treatment with 2 mg/2 mg of BUP + SAM, 8 mg/8 mg of BUP + SAM or placebo/placebo. In this design, nonresponders to placebo in the first phase were rerandomized to the three options in the second stage. Compared with the placebo group, the 2 mg/2 mg dosage group showed a significant reduction in depressive symptoms; the magnitude of improvement in the 8 mg/8 mg dosage group was smaller and did not differ significantly from the placebo group. Both doses of BUP + SAM were judged to be well tolerated and, at the end of up to 8 weeks of therapy with BUP + SAM there was no evidence of opioid withdrawal.

Subsequent work focused on the 2 mg/2 mg dosage of ALKS 5461 and provided further evidence that this formulation had low abuse liability in both healthy volunteers (Pathak et al., 2019) and MDD (Thase et al., 2019; Fava et al., 2020); no withdrawal symptoms were noted even after months of continuation phase therapy (Thase et al., 2019). However, the adjunctive efficacy of ALKS 5461 (2 mg/2 mg) was not consistently established by a series of placebo controlled Phase III trials (Fava et al., 2020; Zajecka et al., 2019). Two of these studies, known as FORWARD-4 and FORWARD-5 (Fava et al., 2020), employed the same sequential parallel-comparison design as used in Phase II. The FORWARD-5 trial found a significant effect on the primary endpoint favoring the adjunctive BUP + SAM group, although the drug versus placebo difference was relatively small (1.9 points [P = .026]). The results of the FORWARD-4 study were very similar, although in this case a relatively modest drug versus placebo difference did not meet conventional criteria for statistical significance (1.8 points [P = .109]). In both studies, secondary analyses of other depression endpoints were supportive of efficacy. A pooled analysis of the two studies demonstrated an overall effect size (Hedges' $g = 0.22$) at the end of treatment that also indicated that this intervention was efficacious, albeit with a modest effect. Common side effects of ALKS 5461 included nausea, constipation, dizziness, vomiting, somnolence, fatigue, and sedation.

Despite evidence of efficacy, the FDA reviewers of the New Drug Application for ALKS 5461 for adjunctive therapy of MDD did not recommend approval, citing the inconsistency of results in the Phase III program and, with respect to the positive studies, the relative small signal for benefit (FDA Briefing Document, 2018). The FDA reviewer's concerns were heightened by a number of unknowns about the safety of this medication related to its, mechanism of action, including the lack of experience with patients with cooccurring alcoholism or substance use disorders. The reviewers strongly encouraged additional research to better establish efficacy and provide additional information about safety in potentially at risk populations.

The FDA's appraisal that ALKS 5461 had insufficient efficacy was supported by a recent metaanalysis of controlled studies of BUP, which included five of the RCTs from the ALKS 5461 program (Dinoff et al., 2020). In the aftermath of the FDA decision, the manufacturer suspended further research on ALKS 5461 and turned its attention to other indications for samidorphan. Although the combination of BUP and SAM may never be commercially available, there remains an unmet need and, at this time, off label prescription of buprenorphine probably is the best option (see, for example, Madison and Eitan, 2020). Until options with established efficacy are available, BUP at low doses is clearly the best studied opioid for use in the treatment of depressed patients and has a reasonable safety profile.

Summary

The adjunctive therapies for MDD represent a diverse group of medications that, although not approved as monotherapies, are intended to enhance the effectiveness of standard first and second line antidepressants. Three adjunctive therapies have been sufficiently researched to establish both efficacy and safety when used as adjuncts to SSRIs: newer generation antipsychotics (particularly aripiprazole, brexpiprazole, and quetiapine), lithium salts and the thyroid hormone triiodothyronine. A fourth antipsychotic, olanzapine, is specifically approved for treatment in combination with fluoxetine and the use of others is supported by placebo controlled studies. The benefits of the SGAs must be balanced against their risks, including weight gain, adverse metabolic changes and extrapyramidal symptoms. Adjunctive therapy with lithium and T_3 provide time-tested alternatives for patients with difficult to treat depression and may be particularly helpful for subsets of depressed patients. A large number of other strategies have been introduced and tested, with varying degrees of evidence of objective benefit and a number of medications that were thought to have considerable promise have failed to show reliable benefit. Research focusing on one potential target for adjunctive intervention, inflammation (both measured peripherally and inferred centrally), is gaining momentum and brings together several of the nutraceutical options and various drugs already indicated for treatment of autoimmune or inflammatory disorders. As we get deeper into the third decade of the 21st century, it may be that strategies that permit a more personalized selection of therapeutics will prove to be even more helpful for selecting among adjunctive treatment strategies than first line therapies.

References

Ahmadi, J., Jahromi, M.S., Ehsaei, Z., 2018. The effectiveness of different singly administered high doses of buprenorphine in reducing suicidal ideation in acutely depressed people with co-morbid opiate dependence: a randomized, double-blind, clinical trial. Trials 19 (1), 462. https://doi.org/10.1186/s13063-018-2843-9. PMID: 30157924; PMCID: PMC6114789.

Anderson, I.M., 1998. SSRIs versus tricyclic antidepressants in depressed inpatients: a meta-analysis of efficacy and tolerability. Depress. Anxiety 7 (Suppl. 1), 11–17.

Appelberg, B.G., Syvälahti, E.K., Koskinen, T.E., et al., 2001. Patients with severe depression may benefit from buspirone augmentation of selective serotonin reuptake inhibitors: results from a placebo-controlled, randomized, double-blind, placebo wash-in study. J. Clin. Psychiatry 62 (6), 448–452.

Aronson, R., Offman, H.J., Joffe, R.T., Naylor, C.D., 1996. Triiodothyronine augmentation in the treatment of refractory depression. A meta-analysis. Arch. Gen. Psychiatry 53 (9), 842–848.

Bahji, A., Ermacora, D., Stephenson, C., Hawken, E.R., Vazquez, G., 2020. Comparative efficacy and tolerability of pharmacological treatments for the treatment of acute bipolar depression: a systematic review and network meta-analysis. J. Affect. Disord. 15 (269), 154–184.

Bai, S., Guo, W., Feng, Y., Deng, H., Li, G., Nie, H., Guo, G., Yu, H., Ma, Y., Wang, J., Chen, S., Jing, J., Yang, J., Tang, Y., Tang, Z., 2020. Efficacy and safety of anti-inflammatory agents for the treatment of major depressive disorder: a systematic review and meta-analysis of randomised controlled trials. J. Neurol. Neurosurg. Psychiatry 91 (1), 21–32.

Ball, S., Dellva, M.A., D'Souza, D.N., Marangell, L.B., Russell, J.M., Goldberger, C., 2014. A double-blind, placebo-controlled study of edivoxetine as an adjunctive treatment for patients with major depressive disorder who are partial responders to selective serotonin reuptake inhibitor treatment. J. Affect. Disord. 167, 215–223.

Ball, S.G., Ferguson, M.B., Martinez, J.M., Pangallo, B.A., Nery, E.S., Dellva, M.A., Sparks, J., Zhang, Q., Liu, P., Bangs, M., Goldberger, C., 2016. Efficacy outcomes from 3 clinical trials of edivoxetine as adjunctive treatment for patients with major depressive disorder who are partial responders to selective serotonin reuptake inhibitor treatment. J. Clin. Psychiatry 77 (5), 635–642.

Bauer, M., et al., 2013a. World Federation of Societies of Biological Psychiatry (WFSBP) guidelines for biological treatment of unipolar depressive disorders, part 1: update 2013 on the acute and continuation treatment of unipolar depressive disorders. World J. Biol. Psychiatry 14 (5), 334–385.

Bauer, M., Dell'osso, L., Kasper, S., Pitchot, W., Dencker Vansvik, E., Köhler, J., Jørgensen, L., Montgomery, S.A., 2013b. Extended-release quetiapine fumarate (quetiapine XR) monotherapy and quetiapine XR or lithium as add-on to antidepressants in patients with treatment-resistant major depressive disorder. J. Affect. Disord. 151 (1), 209–219.

Bech, P., Fava, M., Trivedi, M.H., Wisniewski, S.R., Rush, A.J., 2012. Outcomes on the pharmacopsychometric triangle in bupropion-SR vs. buspirone augmentation of citalopram in the STAR*D trial. Acta Psychiatr. Scand. 125 (4), 342–348.

Bekkering, G.E., Agoritsas, T., Lytvyn, L., Heen, A.F., Feller, M., Moutzouri, E., Abdulazeem, H., Aertgeerts, B., Beecher, D., Brito, J.P., Farhoumand, P.D., Singh Ospina, N., Rodondi, N., van Driel, M., Wallace, E., Snel, M., Okwen, P.M., Siemieniuk, R., Vandvik, P.O., Kuijpers, T., Vermandere, M.,

2019. Thyroid hormones treatment for subclinical hypothyroidism: a clinical practice guideline. BMJ 14 (365). https://doi.org/10.1136/bmj.l2006, l2006 (PMID: 31088853).

Benard, V., Vaiva, G., Masson, M., Geoffroy, P.A., 2016. Lithium and suicide prevention in bipolar disorder. Encéphale 42 (3), 234–241.

Berman, R.M., Thase, M.E., Trivedi, M.H., Hazel, J.A., Marler, S.V., McQuade, R.D., Carson, W., Baker, R.A., Marcus, R.N., 2011. Long-term safety and tolerability of open-label aripiprazole augmentation of antidepressant therapy in major depressive disorder. Neuropsychiatr. Dis. Treat. 7, 303–312.

Brunner, E., Tohen, M., Osuntokun, O., Landry, J., Thase, M.E., 2014. Efficacy and safety of olanzapine/fluoxetine combination vs fluoxetine monotherapy following successful combination therapy of treatment-resistant major depressive disorder. Neuropsychopharmacology 39 (11), 2549–2559.

Cipriani, A., Smith, K., Burgess, S., Carney, S., Goodwin, G., Geddes, J., 2006. Lithium versus antidepressants in the long-term treatment of unipolar affective disorder. Cochrane Database Syst. Rev. 2006 (4). https://doi.org/10.1002/14651858.CD003492.pub2, CD003492. PMID: 17054174; PMCID: PMC7003995.

Cipriani, A., Hawton, K., Stockton, S., Geddes, J.R., 2013. Lithium in the prevention of suicide in mood disorders: updated systematic review and meta-analysis. BMJ 27 (346). https://doi.org/10.1136/bmj.f3646, f3646.

Cleare, A., et al., 2015. Evidence-based guidelines for treating depressive disorders with antidepressants: a revision of the 2008 British Association for Psychopharmacology guidelines. J. Psychopharmacol. 29 (5), 459–525.

Cooper-Kazaz, R., Apter, J.T., Cohen, R., Karagichev, L., Muhammed-Moussa, S., Grupper, D., Drori, T., Newman, M.E., Sackeim, H.A., Glaser, B., Lerer, B., 2007. Combined treatment with sertraline and liothyronine in major depression: a randomized, double-blind, placebo-controlled trial. Arch. Gen. Psychiatry 64 (6), 679–688.

Cooper-Kazaz, R., van der Deure, W.M., Medici, M., Visser, T.J., Alkelai, A., Glaser, B., Peeters, R.P., Lerer, B., 2009. Preliminary evidence that a functional polymorphism in type 1 deiodinase is associated with enhanced potentiation of the antidepressant effect of sertraline by triiodothyronine. J. Affect. Disord. 116 (1–2), 113–116.

Crossley, N.A., Bauer, M., 2007. Acceleration and augmentation of antidepressants with lithium for depressive disorders: two meta-analyses of randomized, placebo-controlled trials. J. Clin. Psychiatry 68 (6), 935–940.

Dé Montigny, C., Grunberg, F., Mayer, A., Deschenes, J.P., 1981. Lithium induces rapid relief of depression in tricyclic antidepressant drug non-responders. Br. J. Psychiatry 138, 252–256.

Devanand, D.P., Sackeim, H.A., Prudic, J., 1991. Electroconvulsive therapy in the treatment-resistant patient. Psychiatr. Clin. North Am. 14 (4), 905–923.

Dinoff, A., Lynch, S.T., Sekhri, N., Klepacz, L., 2020. A meta-analysis of the potential antidepressant effects of buprenorphine versus placebo as an adjunctive pharmacotherapy for treatment-resistant depression. J. Affect. Disord. 271, 91.

Durgam, S., Earley, W., Guo, H., Li, D., Németh, G., Laszlovszky, I., Fava, M., Montgomery, S.A., 2016. Efficacy and safety of adjunctive cariprazine in inadequate responders to antidepressants: a randomized, double-blind, placebo-controlled study in adult patients with major depressive disorder. J. Clin. Psychiatry 77 (3), 371–378.

Dwyer, J.B., Aftab, A., Radhakrishnan, R., Widge, A., Rodriguez, C.I., Carpenter, L.L., Nemeroff, C.B., McDonald, W.M., Kalin, N.H., APA Council of Research Task Force on Novel Biomarkers and Treatments, 2020. Hormonal treatments for major depressive disorder: state of the art. Am. J. Psychiatry 177 (8), 686–705.

Ehrich, E., Turncliff, R., Du, Y., Leigh-Pemberton, R., Fernandez, E., Jones, R., Fava, M., 2015. Evaluation of opioid modulation in major depressive disorder. Neuropsychopharmacology 40 (6), 1448–1455.

Fava, M., Rush, A.J., Alpert, J.E., Balasubramani, G.K., Wisniewski, S.R., Carmin, C.N., Biggs, M.M., Zisook, S., Leuchter, A., Howland, R., Warden, D., Trivedi, M.H., 2008. Difference in treatment outcome in outpatients with anxious versus nonanxious depression: a STAR*D report. Am. J. Psychiatry 165 (3), 342–351.

Fava, M., Memisoglu, A., Thase, M.E., Bodkin, J.A., Trivedi, M.H., de Somer, M., Du, Y., Leigh-Pemberton, R., DiPetrillo, L., Silverman, B., Ehrich, E., 2016. Opioid modulation with buprenorphine/samidorphan as adjunctive treatment for inadequate response to antidepressants: a randomized double-blind placebo-controlled trial. Am. J. Psychiatry 173 (5), 499–508.

Fava, M., Dirks, B., Freeman, M.P., Papakostas, G.I., Shelton, R.C., Thase, M.E., Trivedi, M.H., Liu, K., Stankovic, S., 2019. A phase 2, randomized, double-blind, placebo-controlled study of adjunctive pimavanserin in patients with major depressive disorder and an inadequate response to therapy (CLARITY). J. Clin. Psychiatry 80 (6). https://doi.org/10.4088/JCP.19m12928, 19m12928. PMID: 31556975.

Fava, M., Thase, M.E., Trivedi, M.H., Ehrich, E., Martin, W.F., Memisoglu, A., Nangia, N., Stanford, A.D., Yu, M., Pathak, S., 2020. Opioid system modulation with buprenorphine/samidorphan combination for major depressive disorder: two randomized controlled studies. Mol. Psychiatry 25 (7), 1580–1591.

FDA Briefing Document, November 1, 2018. Psychopharmacologic Drugs Advisory Committee (PDAC) and Drug Safety and Risk Management (DSaRM) Advisory Committee Meeting. https://www.fda.gov/media/121359/download.

Goss, A.J., Kaser, M., Costafreda, S.G., Sahakian, B.J., Fu, C.H., 2013. Modafinil augmentation therapy in unipolar and bipolar depression: a systematic review and meta-analysis of randomized controlled trials. J. Clin. Psychiatry 74 (11), 1101–1107.

Grunze, H., Vieta, E., Goodwin, G.M., Bowden, C., Licht, R.W., Möller, H.J., Kasper, S., WFSBP Task Force on Treatment Guidelines for Bipolar Disorders, 2010. The world federation of societies of biological psychiatry (WFSBP) guidelines for the biological treatment of bipolar disorders: update 2010 on the treatment of acute bipolar depression. World J. Biol. Psychiatry 11 (2), 81–109.

Haller, H., Anheyer, D., Cramer, H., Dobos, G., 2019. Complementary therapies for clinical depression: an overview of systematic reviews. BMJ Open 9 (8). https://doi.org/10.1136/bmjopen-2018-028527, e028527. PMID: 31383703; PMCID: PMC6686993.

Haroon, E., Daguanno, A.W., Woolwine, B.J., Goldsmith, D.R., Baer, W.M., Wommack, E.C., Felger, J.C., Miller, A.H., 2018. Antidepressant treatment resistance is associated with increased inflammatory markers in patients with major depressive disorder. Psychoneuroendocrinology 95, 43–49.

Hobart, M., Skuban, A., Zhang, P., Josiassen, M.K., Hefting, N., Augustine, C., Brewer, C., Sanchez, R., McQuade, R.D., 2018. Efficacy and safety of flexibly dosed brexpiprazole for the adjunctive treatment of major depressive disorder: a randomized, active-referenced, placebo-controlled study. Curr. Med. Res. Opin. 34 (4), 633–642.

Hoepner, C.T., McIntyre, R.S., Papakostas, G.I., 2021. Impact of supplementation and nutritional interventions on pathogenic processes of mood disorders: a review of the evidence. Nutrients 13 (3), 767. https://doi.org/10.3390/nu13030767. PMID: 33652997; PMCID: PMC7996954.

Husain, M.I., Chaudhry, I.B., Khoso, A.B., Husain, M.O., Hodsoll, J., Ansari, M.A., Naqvi, H.A., Minhas, F.A., Carvalho, A.F., Meyer, J.H., Deakin, B., Mulsant, B.H., Husain, N., Young, A.H., 2020. Minocycline and celecoxib as adjunctive treatments for bipolar depression: a multicentre, factorial design randomised controlled trial. Lancet Psychiatry 7 (6), 515–527.

Jing, Y., Guo, Z., Kalsekar, I., Forbes, R.A., Hebden, T., Thase, M.E., 2013. Dosing patterns of aripiprazole and quetiapine for adjunctive treatment of major depressive disorder (2006-2010). Int. Clin. Psychopharmacol. 28 (2), 87–90.

Joffe, R.T., Sokolov, S.T., Singer, W., 1995. Thyroid hormone treatment of depression. Thyroid 5 (3), 235–239.

Karp, J.F., Butters, M.A., Begley, A.E., Miller, M.D., Lenze, E.J., Blumberger, D.M., Mulsant, B.H., Reynolds 3rd., C.F., 2014. Safety, tolerability, and clinical effect of low-dose buprenorphine for treatment-resistant depression in midlife and older adults. J. Clin. Psychiatry 75 (8), e785–e793. https://doi.org/10.4088/JCP.13m08725. PMID: 25191915; PMCID: PMC4157317.

Kennedy, S.H., et al., 2016. Canadian Network for Mood and Anxiety Treatments (CANMAT) 2016 clinical guidelines for the management of adults with major depressive disorder: section 3. Pharmacological treatments. Can. J. Psychiatry 61 (9), 540–560.

Khandelwal, D., Tandon, N., 2012. Overt and subclinical hypothyroidism: who to treat and how. Drugs 72 (1), 17–33.

Köhler-Forsberg, O., Lydholm, C.N., Hjorthøj, C., Nordentoft, M., Mors, O., Benros, M.E., 2019. Efficacy of anti-inflammatory treatment on major depressive disorder or depressive symptoms: meta-analysis of clinical trials. Acta Psychiatr. Scand. 139 (5), 404–419.

Kopschina Feltes, P., Doorduin, J., Klein, H.C., Juárez-Orozco, L.E., Dierckx, R.A., Moriguchi-Jeckel, C.M., de Vries, E.F., 2017. Anti-inflammatory treatment for major depressive disorder: implications for patients with an elevated immune profile and non-responders to standard antidepressant therapy. J. Psychopharmacol. 31 (9), 1149–1165.

Landén, M., Björling, G., Agren, H., et al., 1998. A randomized, double-blind, placebo-controlled trial of buspirone in combination with an SSRI in patients with treatment-refractory depression. J. Clin. Psychiatry 59 (12), 664–668.

Lippiello, P.M., Beaver, J.S., Gatto, G.J., James, J.W., Jordan, K.G., Traina, V.M., Xie, J., Bencherif, M., 2008. TC-5214 (S-(+)-mecamylamine): a neuronal nicotinic receptor modulator with antidepressant activity. CNS Neurosci. Ther. 14 (4), 266–277.

Lipton, M.A., Prange Jr., A.J., Wilson, I.C., 1972. Sources of contribution to the understanding of the etiology of mood disorders. Proc. Annu. Meet. Am. Psychopathol. Assoc. 60, 117–131.

Loane, C., Politis, M., 2012. Buspirone: what is it all about? Brain Res. 21 (1461), 111–118.

Lorentzen, R., Kjær, J.N., Østergaard, S.D., Madsen, M.M., 2020. Thyroid hormone treatment in the management of treatment-resistant unipolar depression: a systematic review and meta-analysis. Acta Psychiatr. Scand. 141 (4), 316–326.

Madison, C.A., Eitan, S., 2020. Buprenorphine: prospective novel therapy for depression and PTSD. Psychol. Med. 50 (6), 881–893. https://doi.org/10.1017/S0033291720000525 [Epub 2020 Mar 24. PMID: 32204739].

Malhi, G.S., Outhred, T., Hamilton, A., Boyce, P.M., Bryant, R., Fitzgerald, P.B., Lyndon, B., Mulder, R., Murray, G., Porter, R.J., Singh, A.B., Fritz, K., 2018. Royal Australian and New Zealand College of Psychiatrists clinical practice guidelines for mood disorders: major depression summary. Med. J. Aust. 208 (4), 175–180.

Martinez, J.M., Ferguson, M.B., Pangallo, B.A., Oakes, T.M., Sparks, J., Dellva, M.A., Zhang, Q., Liu, P., Bangs, M., Ahl, J., Goldberger, C., 2015. Safety and tolerability of edivoxetine as adjunctive treatment to selective serotonin reuptake inhibitor antidepressants for patients with major depressive disorder. Drugs Context 13 (4). https://doi.org/10.7573/dic.212279, 212279. PMID: 26005493; PMCID: PMC4434861.

McIntyre, R.S., Lee, Y., Zhou, A.J., Rosenblat, J.D., Peters, E.M., Lam, R.W., Kennedy, S.H., Rong, C., Jerrell, J.M., 2017. The efficacy of psychostimulants in major depressive episodes: a systematic review and Meta-analysis. J. Clin. Psychopharmacol. 37 (4), 412–418.

McIntyre, R.S., Subramaniapillai, M., Lee, Y., Pan, Z., Carmona, N.E., Shekotikhina, M., Rosenblat, J.D., Brietzke, E., Soczynska, J.K., Cosgrove, V.E., Miller, S., Fischer, E.G., Kramer, N.E., Dunlap, K., Suppes, T., Mansur, R.B., 2019. Efficacy of adjunctive infliximab vs placebo in the treatment of adults with bipolar I/II depression: a randomized clinical trial. JAMA Psychiat. 76 (8), 783–790.

Mohamed, S., Johnson, G.R., Chen, P., Hicks, P.B., Davis, L.L., Yoon, J., Gleason, T.C., Vertrees, J.E., Weingart, K., Tal, I., Scrymgeour, A., Lawrence, D.D., Planeta, B., Thase, M.E., Huang, G.D., Zisook, S., The VAST-D Investigators, Rao, S.D., Pilkinton, P.D., Wilcox, J.A., Iranmanesh, A., Sapra, M., Jurjus, G., Michalets, J.P., Aslam, M., Beresford, T., Anderson, K.D., Fernando, R., Ramaswamy, S., Kasckow, J., Westermeyer, J., Yoon, G., D'Souza, D.C., Larson, G., Anderson, W.G., Klatt, M., Fareed, A., Thompson, S.I., Carrera, C.J., Williams, S.S., Juergens, T.M., Albers, L.J., Nasdahl, C.S., Villarreal, G., Winston, J.L., Nogues, C.A., Connolly, K.R., Tapp, A., Jones, K.A., Khatkhate, G., Marri, S., Suppes, T., LaMotte, J., Hurley, R., Mayeda, A.R., Niculescu 3rd, A.B., Fischer, B.A., Loreck, D.J., Rosenlicht, N., Lieske, S., Finkel, M.S., Little, J.T., 2017. Effect of antidepressant switching vs augmentation on remission among patients with major depressive disorder unresponsive to antidepressant treatment: the VAST-D randomized clinical trial. JAMA 318 (2), 132–145.

Möller, H.J., Demyttenaere, K., Olausson, B., Szamosi, J., Wilson, E., Hosford, D., Dunbar, G., Tummala, R., Eriksson, H., 2015. Two phase III randomised double-blind studies of fixed-dose TC-5214 (dexmecamylamine) adjunct to ongoing antidepressant therapy in patients with major depressive disorder and an inadequate response to prior antidepressant therapy. World J. Biol. Psychiatry 16 (7), 483–501.

Murillo-Rodríguez, E., Barciela Veras, A., Barbosa Rocha, N., Budde, H., Machado, S., 2018. An overview of the clinical uses, pharmacology, and safety of modafinil. ACS Chem. Nerosci. 9 (2), 151–158.

Nelson, J.C., Papakostas, G.I., 2009. Atypical antipsychotic augmentation in major depressive disorder: a meta-analysis of placebo-controlled randomized trials. Am. J. Psychiatry 166 (9), 980–991.

Nelson, J.C., Mazure, C.M., Bowers Jr., M.B., Jatlow, P.I., 1991. A preliminary, open study of the combination of fluoxetine and desipramine for rapid treatment of major depression. Arch. Gen. Psychiatry 48 (4), 303–307.

Nelson, J.C., Mazure, C.M., Jatlow, P.I., Bowers Jr., M.B., Price, L.H., 2004. Combining norepinephrine and serotonin reuptake inhibition mechanisms for treatment of depression: a double-blind, randomized study. Biol. Psychiatry 55 (3), 296–300.

Nettis, M.A., Lombardo, G., Hastings, C., Zajkowska, Z., Mariani, N., Nikkheslat, N., Worrell, C., Enache, D., McLaughlin, A., Kose, M., Sforzini, L., Bogdanova, A., Cleare, A., Young, A.H., Pariante, C.M., Mondelli, V., 2021. Augmentation therapy with minocycline in treatment-resistant depression patients with low-grade peripheral inflammation: results from a double-blind randomised clinical trial. Neuropsychopharmacology 46 (5), 939–948.

Nierenberg, A.A., Fava, M., Trivedi, M.H., Wisniewski, S.R., Thase, M.E., McGrath, P.J., Alpert, J.E., Warden, D., Luther, J.F., Niederehe, G., Lebowitz, B., Shores-Wilson, K., Rush, A.J., 2006. A comparison of lithium and T(3) augmentation following two failed medication treatments for depression: a STAR*D report. Am. J. Psychiatry 163 (9), 1519–1530.

Oakes, T.M., Dellva, M.A., Waterman, K., Greenbaum, M., Poppe, C., Goldberger, C., Ahl, J., Perahia, D.G., 2015. Edivoxetine compared to placebo as adjunctive therapy to selective serotonin reuptake inhibitors in the prevention of symptom reemergence in major depressive disorder. Curr. Med. Res. Opin. 31 (6), 1179–1189.

Papakostas, G.I., Shelton, R.C., Zajecka, J.M., Etemad, B., Rickels, K., Clain, A., Baer, L., Dalton, E.D., Sacco, G.R., Schoenfeld, D., Pencina, M., Meisner, A., Bottiglieri, T., Nelson, E., Mischoulon, D., Alpert, J.E., Barbee, J.G., Zisook, S., Fava, M., 2012. L-methylfolate as adjunctive therapy for SSRI-resistant major depression: results of two randomized, double-blind, parallel-sequential trials. Am. J. Psychiatry 169 (12), 1267–1274.

Papakostas, G.I., Shelton, R.C., Zajecka, J.M., Bottiglieri, T., Roffman, J., Cassiello, C., Stahl, S.M., Fava, M., 2014. Effect of adjunctive L-methylfolate 15 mg among inadequate responders to SSRIs in depressed patients who were stratified by biomarker levels and genotype: results from a randomized clinical trial. J. Clin. Psychiatry 75 (8), 855–863.

Papakostas, G.I., Fava, M., Baer, L., Swee, M.B., Jaeger, A., Bobo, W.V., Shelton, R.C., 2015. Ziprasidone augmentation of escitalopram for major depressive disorder: efficacy results from a randomized, double-blind, placebo-controlled study. Am. J. Psychiatry 172 (12), 1251–1258.

Parker, G., Brotchie, H., McClure, G., Fletcher, K., 2013. Psychostimulants for managing unipolar and bipolar treatment-resistant melancholic depression: a medium-term evaluation of cost benefits. J. Affect. Disord. 151 (1), 360–364.

Pathak, S., Vince, B., Kelsh, D., Shram, M.J., Setnik, B., Lu, H., Nangia, N., Stanford, A.D., Ehrich, E., 2019. Abuse potential of buprenorphine/samidorphan combination compared to buprenorphine and placebo: a phase 1 randomized controlled trial. J. Clin. Pharmacol. 59 (2), 206–217.

Philip, N.S., Carpenter, L.L., Tyrka, A.R., Price, L.H., 2010. Nicotinic acetylcholine receptors and depression: a review of the preclinical and clinical literature. Psychopharmacology (Berl) 212, 1–12.

Prange Jr., A.J., Wilson, I.C., Rabon, A.M., Lipton, M.A., 1969. Enhancement of imipramine antidepressant activity by thyroid hormone. Am. J. Psychiatry 126 (4), 457–469.

Prange Jr., A.J., Wilson, I.C., Knox, A.E., McClane, T.K., Breese, G.R., Martin, B.R., Alltop, L.B., Lipton, M.A., 1972. Thyroid-imipramine clinical and chemical interaction: evidence for a receptor deficit in depression. J. Psychiatr. Res. 9 (3), 187–205.

Raison, C.L., Miller, A.H., 2013. Role of inflammation in depression: implications for phenomenology, pathophysiology and treatment. Mod. Trends Pharmacopsychiatry 28, 33–48.

Raison, C.L., Rutherford, R.E., Woolwine, B.J., Shuo, C., Schettler, P., Drake, D.F., Haroon, E., Miller, A.H., 2013. A randomized controlled trial of the tumor necrosis factor antagonist infliximab for treatment-resistant depression: the role of baseline inflammatory biomarkers. JAMA Psychiat. 70 (1), 31–41.

Rasmussen, N., 2015. Amphetamine-type stimulants: the early history of their medical and non-medical uses. Int. Rev. Neurobiol. 120, 9–25.

Ravindran, A.V., Balneaves, L.G., Faulkner, G., Ortiz, A., McIntosh, D., Morehouse, R.L., Ravindran, L., Yatham, L.N., Kennedy, S.H., Lam, R.W., MacQueen, G.M., Milev, R.V., Parikh, S.V., CANMAT Depression Work Group, 2016. Canadian network for mood and anxiety treatments (CANMAT) 2016 clinical guidelines for the management of Adults with major depressive disorder: section 5. Complementary and alternative medicine treatments. Can. J. Psychiatry 61 (9), 576–587.

Rosenblat, J.D., McIntyre, R.S., 2018. Efficacy and tolerability of minocycline for depression: a systematic review and meta-analysis of clinical trials. J. Affect. Disord. 227, 219–225.

Rush, A.J., Trivedi, M.H., Wisniewski, S.R., Nierenberg, A.A., Stewart, J.W., Warden, D., Niederehe, G., Thase, M.E., Lavori, P.W., Lebowitz, B.D., McGrath, P.J., Rosenbaum, J.F., Sackeim, H.A., Kupfer, D.J., Luther, J., Fava, M., 2006. Acute and longer-term outcomes in depressed outpatients requiring one or several treatment steps: a STAR*D report. Am. J. Psychiatry 163 (11), 1905–1917.

Sackeim, H.A., Haskett, R.F., Mulsant, B.H., Thase, M.E., Mann, J.J., Pettinati, H.M., Greenberg, R.M., Crowe, R.R., Cooper, T.B., Prudic, J., 2001. Continuation pharmacotherapy in the prevention of relapse following electroconvulsive therapy: a randomized controlled trial. JAMA 285 (10), 1299–1307.

Sarris, J., Murphy, J., Mischoulon, D., Papakostas, G.I., Fava, M., Berk, M., Ng, C.H., 2016. Adjunctive nutraceuticals for depression: a systematic review and meta-analyses. Am. J. Psychiatry 173 (6), 575–587.

Serafini, G., Adavastro, G., Canepa, G., De Berardis, D., Valchera, A., Pompili, M., Nasrallah, H., Amore, M., 2018. The efficacy of buprenorphine in major depression, treatment-resistant depression and suicidal behavior: a systematic review. Int. J. Mol. Sci. 19 (8), 2410. https://doi.org/10.3390/ijms19082410. PMID: 30111745; PMCID: PMC6121503.

Shelton, R.C., Pencina, M.J., Barrentine, L.W., Ruiz, J.A., Fava, M., Zajecka, J.M., Papakostas, G.I., 2015. Association of obesity and inflammatory marker levels on treatment outcome: results from a double-blind, randomized study of adjunctive L-methylfolate calcium in patients with MDD who are inadequate responders to SSRIs. J. Clin. Psychiatry 76 (12), 1635–1641.

Souery, D., Oswald, P., Massat, I., Bailer, U., Bollen, J., Demyttenaere, K., Kasper, S., Lecrubier, Y., Montgomery, S., Serretti, A., Zohar, J., Mendlewicz, J., Group for the Study of Resistant Depression, 2007. Clinical factors associated with treatment resistance in major depressive disorder: results from a European multicenter study. J. Clin. Psychiatry 68 (7), 1062–1070.

Souza, F.G., Goodwin, G.M., 1991. Lithium treatment and prophylaxis in unipolar depression: a meta-analysis. Br. J. Psychiatry 158, 666–675.

Stauffer, V.L., Liu, P., Goldberger, C., Marangell, L.B., Nelson, C., Gorwood, P., Fava, M., 2017. Is the noradrenergic symptom cluster a valid construct in adjunctive treatment of major depressive disorder? J. Clin. Psychiatry 78 (3), 317–323.

Strawbridge, R., Carter, B., Marwood, L., Bandelow, B., Tsapekos, D., Nikolova, V.L., Taylor, R., Mantingh, T., de Angel, V., Patrick, F., Cleare, A.J., Young, A.H., 2019. Augmentation therapies for treatment-resistant depression: systematic review and meta-analysis. Br. J. Psychiatry 214 (1), 42–51. https://doi.org/10.1192/bjp.2018.233. Epub 2018 Nov 20. PMID: 30457075.

Suppes, T., Silva, R., Cucchiaro, J., Mao, Y., Targum, S., Streicher, C., Pikalov, A., Loebel, A., 2016. Lurasidone for the treatment of major depressive disorder with mixed features: a randomized, double-blind, placebo-controlled study. Am. J. Psychiatry 173 (4), 400–407.

Thase, M.E., 2016. Adverse effects of second-generation antipsychotics as adjuncts to antidepressants: are the risks worth the benefits? Psychiatr. Clin. North Am. 39 (3), 477–486.

Thase, M.E., Rush, A.J., 1995. Treatment resistant depression. In: Bloom, F.E., Kupfer, D.J. (Eds.), Psychopharmacology: The Fourth Generation of Progress. Raven Press, New York, NY, pp. 1081–1097.

Thase, M.E., Rush, A.J., 1997. When at first you don't succeed...sequential strategies for antidepressant nonresponders. J. Clin. Psychiatry 58 (Suppl. 13), 23–29.

Thase, M.E., Kupfer, D.J., Frank, E., Jarrett, D.B., 1989. Treatment of imipramine-resistant recurrent depression: II. An open clinical trial of lithium augmentation. J. Clin. Psychiatry 50 (11), 413–417.

Thase, M.E., Trivedi, M.H., Rush, A.J., 1995. MAOIs in the contemporary treatment of depression. Neuropsychopharmacology 12, 185–219.

Thase, M.E., Entsuah, A.R., Rudolph, R.L., 2001. Remission rates during treatment with venlafaxine or selective serotonin reuptake inhibitors. Br. J. Psychiatry 178, 234–241.

Thase, M.E., Pritchett, Y.L., Ossanna, M.J., Swindle, R.W., Xu, J., Detke, M.J., 2007. Efficacy of duloxetine and selective serotonin reuptake inhibitors: comparisons as assessed by remission rates in patients with major depressive disorder. J. Clin. Psychopharmacol. 27 (6), 672–676.

Thase, M.E., Weiller, E., Zhang, P., Weiss, C., McIntyre, R.S., 2018. Adjunctive brexpiprazole in patients with major depressive disorder and anxiety symptoms: post hoc analyses of three placebo-controlled studies. Neuropsychiatr. Dis. Treat. 20 (15), 37–45.

Thase, M.E., Stanford, A.D., Memisoglu, A., Martin, W., Claxton, A., Bodkin, J.A., Trivedi, M.H., Fava, M., Yu, M., Pathak, S., 2019. Results from a long-term open-label extension study of adjunctive buprenorphine/samidorphan combination in patients with major depressive disorder. Neuropsychopharmacology 44 (13), 2268–2276.

Trivedi, M.H., Fava, M., Wisniewski, S.R., Thase, M.E., Quitkin, F., Warden, D., Ritz, L., Nierenberg, A.A., Lebowitz, B.D., Biggs, M.M., Luther, J.F., Shores-Wilson, K., Rush, A.J., STAR*D Study Team, 2006. Medication augmentation after the failure of SSRIs for depression. N. Engl. J. Med. 354 (12), 1243–1252.

Tsapakis, E.M., Preti, A., Mintzas, M.D., Fountoulakis, K.N., 2020. Adjunctive treatment with psychostimulants and stimulant-like drugs for resistant bipolar depression: a systematic review and meta-analysis. CNS Spectr. 9, 1–12.

Tummala, R., Desai, D., Szamosi, J., Wilson, E., Hosford, D., Dunbar, G., Eriksson, H., 2015. Safety and tolerability of dexmecamylamine (TC-5214) adjunct to ongoing antidepressant therapy in patients with major depressive disorder and an inadequate response to antidepressant therapy: results of a long-term study. J. Clin. Psychopharmacol. 35 (1), 77–81.

Vieta, E., Thase, M.E., Naber, D., D'Souza, B., Rancans, E., Lepola, U., Olausson, B., Szamosi, J., Wilson, E., Hosford, D., Dunbar, G., Tummala, R., Eriksson, H., 2014. Efficacy and tolerability of flexibly-dosed adjunct TC-5214 (dexmecamylamine) in patients with major depressive disorder and inadequate response to prior antidepressant. Eur. Neuropsychopharmacol. 24 (4), 564–574. https://doi.org/10.1016/j.euroneuro.2013.12.008. Epub 2013 Dec 21. PMID: 24507016.

Wan, L., Li, Y., Zhang, Z., Sun, Z., He, Y., Li, R., 2018. Methylenetetrahydrofolate reductase and psychiatric diseases. Transl. Psychiatry 8 (1), 242. https://doi.org/10.1038/s41398-018-0276-6. PMID: 30397195; PMCID: PMC6218441.

Wang, H.R., Woo, Y.S., Bahk, W.M., 2016. Ineffectiveness of nicotinic acetylcholine receptor antagonists for treatment-resistant depression: a meta-analysis. Int. Clin. Psychopharmacol. 31 (5), 241–248.

Weisler, R., McIntyre, R.S., 2013. The role of extended-release quetiapine fumarate monotherapy in the treatment of patients with major depressive disorder. Expert Rev. Neurother. 13 (11), 1161–1182.

White, K., Simpson, G., 1981. Combined MAOI-tricyclic antidepressant treatment: a reevaluation. J. Clin. Psychopharmacol. 1 (5), 264–282.

Whybrow, P.C., Mendels, J., 1969. Toward a biology of depression: some suggestions from neurophysiology. Am. J. Psychiatry 125 (11), 1491–1500.

Wilkinson, S.T., Agbese, E., Leslie, D.L., Rosenheck, R.A., 2018. Identifying recipients of electroconvulsive therapy: data from privately insured Americans. Psychiatr. Serv. 69 (5), 542–548.

Wilson, I.C., Prange Jr., A.J., McClane, T.K., Rabon, A.M., Lipton, M.A., 1970. Thyroid-hormone enhancement of imipramine in nonretarded depressions. N. Engl. J. Med. 282 (19), 1063–1067.

Yatham, L.N., Kennedy, S.H., Parikh, S.V., Schaffer, A., Bond, D.J., Frey, B.N., Sharma, V., Goldstein, B.I., Rej, S., Beaulieu, S., Alda, M., MacQueen, G., Milev, R.V., Ravindran, A., O'Donovan, C., McIntosh, D., Lam, R.W., Vazquez, G., Kapczinski, F., McIntyre, R.S., Kozicky, J., Kanba, S., Lafer, B., Suppes, T., Calabrese, J.R., Vieta, E., Malhi, G., Post, R.M., Berk, M., 2018. Canadian network for mood and anxiety treatments (CANMAT) and International Society for Bipolar Disorders (ISBD) 2018 guidelines for the management of patients with bipolar disorder. Bipolar Disord. 20 (2), 97–170.

Zajecka, J.M., Stanford, A.D., Memisoglu, A., Martin, W.F., Pathak, S., 2019. Buprenorphine/samidorphan combination for the adjunctive treatment of major depressive disorder: results of a phase III clinical trial (FORWARD-3). Neuropsychiatr. Dis. Treat. 4 (15), 795–808.

Zhou, X., Keitner, G.I., Qin, B., Ravindran, A.V., Bauer, M., Del Giovane, C., Zhao, J., Liu, Y., Fang, Y., Zhang, Y., Xie, P., 2015a. Atypical antipsychotic augmentation for treatment-resistant depression: a systematic review and network meta-analysis. Int. J. Neuropsychopharmacol. 18 (11), pyv060. https://doi.org/10.1093/ijnp/pyv060. PMID: 26012350; PMCID: PMC4756722.

Zhou, X., Ravindran, A.V., Qin, B., Del Giovane, C., Li, Q., Bauer, M., Liu, Y., Fang, Y., da Silva, T., Zhang, Y., Fang, L., Wang, X., Xie, P., 2015b. Comparative efficacy, acceptability, and tolerability of augmentation agents in treatment-resistant depression: systematic review and network meta-analysis. J. Clin. Psychiatry 76 (4), e487–e498. https://doi.org/10.4088/JCP.14r09204. PMID: 25919841.

Chapter 12

Combining antidepressants to address treatment-resistant depression

Julie Farrington[a], **Sussann Kotara**[b], **and Charles B. Nemeroff**[c]

[a]*Department of Psychiatry and Behavioral Sciences, Ascension Seton Shoal Creek Hospital, University of Texas at Austin Dell Medical School, Austin, TX, United States,* [b]*Department of Psychiatry and Behavioral Sciences, U.T. Dell Medical Center Hospital, Austin, TX, United States,* [c]*Department of Psychiatry and Behavioral Sciences, Institute for Early Life Adversity Research, University of Texas at Austin Dell Medical School, Austin, TX, United States*

As discussed in Chapter 1, it is helpful to define treatment-resistant depression (TRD). Although one defines treatment resistance, it is clear that this is a very significant clinical problem. We now know from multiple randomized controlled clinical trials and years of clinical experience that only a minority of patients are able to attain remission after treatment with an adequate dose of an antidepressant (Fife et al., 2019).

Definitions of treatment resistance range from inadequate response to at least one antidepressant trial of adequate doses and duration, to failure to respond to a minimum of four different antidepressant treatments, including medications, evidence-based psychotherapy, or electroconvulsive therapy, administered at adequate doses and duration during the episode. Response has traditionally been defined as a 50% improvement on a dimensional rating scale of depressive symptoms such as the Hamilton Rating Scale for Depression (HAM-D) or the Montgomery-Asberg Depression Rating Scale (MADRS). However, the holy grail of depression treatment is remission, defined as a complete return to premorbid functioning arbitrarily defined as a HAM-D of 7 or less or an MADRS score of 10 or less. One might argue that even these are too liberal definitions of remission as those cutoffs would include patients with considerable morbidity (Schoenbaum et al., 2002; Judd et al., 2000).

Reasons for why the remission rates in major depressive disorder (MDD) range from 28% in STAR*D (Gaynes et al., 2008) to 50% in the PReDICT study of never-treated depressed patients remain obscure except the most obvious explanation, as recently reviewed by Nemeroff, is our very poor understanding of the pathophysiology of this serious and common psychiatric disorder (Nemeroff, 2020). Be that as it may, clinicians have a variety of evidence-based treatments to choose from when presented with a clear antidepressant monotherapy failure. These evidence-based strategies comprise a substantial portion of this volume. There are also exciting new developments in terms of novel treatments including theta burst accelerated TMS and novel antidepressants (Nemeroff, 2020).

Conceptualization of treatment resistance and modeling the approach, as addressed in Chapters 2 and 6 respectively, are first steps to aid the clinician in achieving the goal of remission. There are multiple facets to creating a person-centered approach and our subspecialty of medicine allows us a very open-ended journey to knowing and helping patients as individuals. An optimal treatment plan combines the complex matrices of the person and the treatments. Other chapters provide in depth discussions of subtype of depression (unipolar, bipolar, psychotic); the macro- and microtime dimensions (child and adolescent, geriatric, pregnancy and postpartum, chronotherapeutics), the more common comorbidities (anxiety disorders, personality disorders, substance abuse, and chronic pain), and the potential role of pharmacogenomics, the latter clearly not realized. Faced with the many evidence-based treatments for patients who have failed antidepressant monotherapy, the clinician has little to guide him/her as there are no validated predictors of treatment response. Thus lithium, T3, pramipexole, a variety of atypical antipsychotics and several neuromodulation modalities are all available, many FDA approved.

In this chapter, we focus on the extant data concerning the use of combination antidepressant therapy.

Managing Treatment-Resistant Depression. https://doi.org/10.1016/B978-0-12-824067-0.00012-8

Brief overview of antidepressants

Tricyclic antidepressants and monoamine oxidase inhibitors were developed in the 1950s and were the first FDA-approved antidepressants. They have continued utility in the treatment of MDD (see Chapters 9 and 10 respectively) and in combination therapy. Antidepressants are typically categorized by their purported mechanism of action, namely their affinity for neurotransmitter receptors and transporters.. The main systems with which the existing antidepressant armamentarium interacts are serotonin, norepinephrine, dopamine, glutamate, gamma-aminobutyric acid (GABA), acetylcholine, and histamine neural circuits. The first three systems are believed to represent the predominant sites that mediate their therapeutic effects, while the latter two are viewed as mediating side effects.

Within this classification system are selective serotonin reuptake inhibitors (SSRIs), serotonin agonist/antagonist reuptake inhibitors (SARIs), serotonin and norepinephrine reuptake inhibitors (SNRIs), selective norepinephrine reuptake inhibitors (NRIs) and other miscellaneous agents such as bupropion and trazodone. Not surprisingly the strategies for combining antidepressants is largely based on the notion that combining antidepressants with distinctly different mechanisms of action will result in a more powerful antidepressant effect. To that end, understanding the pharmacological difference among antidepressants in their affinity for receptors and transporters may be helpful in developing a rational combination therapy strategy.

Limits of the monoamine hypothesis are the comparable efficacies of antidepressants despite widely varying potencies for monoaminergic effects and disparate mechanisms of action. Multiple monoamines may influence final common pathways relevant to depression. Additionally, the precise role of glutamate and GABA neural circuits as well as others, such as the corticotropin-releasing hormone and brain derived neurotropic factor, remain obscure.

Below are tables with comparable receptor affinities and monoamine oxidase inhibitor comparisons (Table 12.1).

The monoamine oxidase inhibitors are broad-acting via enzyme inhibition, blocking breakdown of norepinephrine, serotonin and dopamine. Their use requires dietary restrictions and drug interaction monitoring. The members of this class all have relatively short half-lives, affecting dosing frequency and side effects. However, irreversible inhibition of enzymes lasts 2 weeks (Table 12.2).

Receptor overview

The complexity of receptors may hold future promise in guiding the decision process of choosing antidepressant therapy. Research areas include allosteric and orthosteric binding site modulation. Below, we review monoamine receptor structures and activity. Serotonin receptors are found in both the central and peripheral nervous systems as well as in platelets and the gastrointestinal tract. They are divided into 7 families of G protein-coupled receptors except for the 5-HT3 receptor, a ligand-gated ion channel, which activates an intracellular second messenger cascade to produce an excitatory or inhibitory response. As illustrated in Tables 12.3 and 12.4, serotonergic circuits have broad physiological and behavioral effects, including several (mood, appetite, sleep) that are part and parcel of the MDD syndrome.

Norepinephrine receptors fall into two broad categories, alpha and beta adrenergic receptors: $\alpha1$, $\alpha2$, $\beta1$, $\beta2$, and $\beta3$. All of these function as G protein-coupled receptors. Alpha-2 receptors usually have inhibitory effects (decrease in the amount of norepinephrine released). Alpha-1 receptors and beta receptors usually have excitatory effects. Norepinephrine may be thought of as broad acting, regulating many aspects of mood via other neurotransmitter systems, as well as impacting energy levels (Table 12.5).

What is the evidence for the efficacy of combining antidepressants?

Reviewing the research looking at efficacy of various antidepressants, along with the spectrum of variables that influence the diagnostic criteria for major depression, reveals inconsistent results. One metaanalysis of 234 studies showed no clinically relevant differences in efficacy of antidepressants (Gartlehner et al., 2011) and others have found some antidepressants such as amitriptyline, mirtazapine and the SNRIs to have higher treatment efficacy effective than others (Cipriani et al., 2018). STAR*D (5), a National Institute of Mental Health-funded study addressed treating depression to remission that was completed in 2006 (see Chapter 7 for in depth review). The four sequential treatment levels covered monotherapy, combination therapy, switching therapy and augmentation therapy and represented SSRI, SNRI, MAOI, tricyclic and tetracyclic antidepressant classes. A total of 4041 individuals from a combination of psychiatry sites (Henssler et al., 2018) and primary care sites (Thomas et al., 2015) participated. Remission rates after open label citalopram monotherapy up to 40 mg per day were 28% using the HAM-D. Within level two, remission rates for combination antidepressant treatment of adding bupropion to citalopram were approximately 27%. The combination treatment of mirtazapine and venlafaxine in level 4 resulted in a remission rate of 13%. The good news from this study is that with each successive strategy, more and more

TABLE 12.1 Antidepressants and relative affinities (Ki) for monoamine transporters (Vaishnavi et al., 2004; Owens et al., 1997; Foley and De Santy, 2006; Zhou, 2004; Sanchez et al., 2014).

Drug	Serotonin transporter (nM)	Norepinephrine transporter (nM)	Dopamine transporter (nM)	Notes
Fluoxetine	1	660	4180	Active metabolite, long half-lives; P450 inhibition
Sertraline	0.29	420	25	Some dopaminergic activity
Paroxetine	0.07–0.2	40–85	490	1st trimester risks; discontinuation syndrome; K_i M_1: 72 nM
Citalopram	1.6	6190	–	QTc prolongation risk
Escitalopram	80	6190	–	QTc prolongation risk
Fluvoxamine	2.5	1427	–	Short half-life; P450 inhibition; not FDA approved for MDD
Venlafaxine	7.8	1920	–	Discontinuation syndrome; serotonin: norepinephrine 30:1
Desvenlafaxine	115	3152	–	Serotonin: norepinephrine 14:1
Duloxetine	0.8	7.5	240	Serotonin: norepinephrine 10:1
Milnacipran	8.44	22	–	Serotonin: norepinephrine 2:1
Levomilnacipran	11.2	99.2	–	NMDA K_i: 1.7 μM serotonin: norepinephrine 1:3
Trazodone	69	30,751	–	$5HT_2$ antagonist; mCPP metabolite
Nefazodone	549	713	–	$5HT_2$ antagonist; no histamine activity; mCPP metabolite; cases of liver failure
Vilazodone	0.1	56	37	Food
Vortioxetine	1.6	113	–	$5HT_{1a}$ agonism; $5HT_{3,7}$ antagonism
Atomoxetine	77	5	1451	Cardiovascular effects; not FDA approved for MDD
Bupropion	28.69 μM	1.37 μM	526	Active metabolites, possible VMAT activity
Imipramine	1.4	37	8500	D_2 antagonism; shortens delta wave stage sleep
Desipramine	17.6	3.5	3190	Relatively selective for norepinephrine transporter; mACh K_i: 66–198
Amitriptyline	2.8–4.3	19–35	320	mACh K_i: 9.6
Nortriptyline	15–18	1.8–4.4	1140	Relatively selective for norepinephrine transporter; mACh K_i: 37
Mirtazapine	> 10K	4600	> 10K	a2 K_i:18 nM

patients attained remission. The bad news is that the side effect burden rose at each level. Another concerning outcome was that the harder it was for patients to attain remission, the higher the likelihood of subsequent relapse. Subsequent studies of remission rates with monotherapy average 50% or less, even in treatment naive patients. As referenced earlier, Dunlop et al. compared antidepressant and CBT monotherapy in a randomized controlled trial with 344 patients with major depression with no treatment history and found remission rates of 46.7% with escitalopram and 54.7% with duloxetine (Dunlop et al., 2017).

TABLE 12.2 MAOIs.

Phenelzine	May need divided daily dosing, pyridoxine supplementation
Tranylcypromine	Amphetamine structure
Isocarboxazid	Twice a day dosing
Selegiline	Transdermal patch avoids MAO-B inhibition in GI tract; loses MAO-B selectivity at higher doses

TABLE 12.3 Serotonin receptors.

Family	Type	Mechanism	Potential
5-HT1	Gi/Go-protein coupled	Decreasing levels of cAMP	Inhibitory
5-HT2	Gq/G1-protein coupled	Increasing levels of IP3 & DAG	Excitatory
5-HT3	Ligand-gated Na^+ and K^+ cation channel	Depolarizing plasma membrane	Excitatory
5-HT4	Gs-protein coupled	Increasing levels of cAMP	Excitatory
5-HT5	Gi/Go-protein coupled	Decreasing levels of cAMP	Inhibitory
5-HT6	Gs-protein coupled	Increasing levels of cAMP	Excitatory
5-HT7	Gs-protein coupled	Increasing levels of cAMP	Excitatory

cAMP, cyclic adenosine monophosphate; *IP3*, inositol triphosphate; *DAG*, diacylglycerol.

TABLE 12.4 Serotonin receptor subtypes.

Receptor	Function	CNS	Blood vessels	Platelets	PNS	Smooth muscle	GI
5HT1A	Mood	Y	Y	N	N	N	N
	Anxiety						
	Appetite						
	Nausea						
	Sleep						
	Memory						
	Penile erection						
	Sexual behavior						
	Sociability						
	Impulsivity						
	Addiction						
	Aggression						
	Cardiovascular function						
	Pupil dilation						
	Nociception						
	Thermoregulation						

TABLE 12.4 Serotonin receptor subtypes—cont'd

Receptor	Function	CNS	Blood vessels	Platelets	PNS	Smooth muscle	GI
5HT1B	Anxiety	Y	Y	N	N	N	N
	Learning/memory						
	Penile erection						
	Addiction						
	Aggression						
	Vasoconstriction						
5HT1D	Anxiety	Y	Y	N	N	N	N
	Locomotion						
	Vasoconstriction						
5HT1E	Unknown	Y	Y	N	N	N	N
5HT1F	Migraine	Y	N	N	N	N	N
5HT2A	Mood	Y	Y	Y	Y	Y	Y
	Anxiety						
	Appetite						
	Sleep						
	Cognition						
	Learning/memory						
	Perception						
	Imagination						
	Sexual behavior						
	Addiction						
	Thermoregulation						
	Vasoconstriction						
5HT2B	Anxiety	Y	Y	Y	Y	Y	Y
	Appetite						
	GI motility						
	Sleep						
	Cardiovascular function						
	Vasoconstriction						
5HT2C	Mood	Y	Y	Y	Y	Y	Y
	Anxiety						
	Appetite						
	GI motility						
	Sleep						
	Penile erection						
	Sexual behavior						
	Addiction						
	Thermoregulation						
	Vasoconstriction						

Continued

TABLE 12.4 Serotonin receptor subtypes—cont'd

Receptor	Function	CNS	Blood vessels	Platelets	PNS	Smooth muscle	GI
5HT3 (A-E)	Anxiety	Y	N	N	Y	N	Y
	GI motility						
	Emesis						
	Nausea						
	Learning/memory						
	Addiction						
5HT4	Mood	Y	N	N	Y	N	Y
	Anxiety						
	Appetite						
	GI motility						
	Learning/memory						
	Respiration						
5HT5A	Sleep	Y	N	N	N	N	N
5HT6	Mood	Y	N	N	N	N	N
	Anxiety						
	Cognition						
	Learning/memory						
5HT7	Mood	Y	Y	N	N	N	Y
	Anxiety						
	Sleep						
	Memory						
	Thermoregulation						
	vasoconstriction						
	Respiration						

TABLE 12.5 Norepinephrine receptors.

Family	Receptor	Type	Mechanism
Alpha	α1	Gq-coupled	Increase IP3 and calcium by activating phospholipase C
	α2	Gi/Go-coupled	Decrease cAMP by inhibiting adenylate cyclase
Beta	β1, β2, β3	Gs-coupled	Increase cAMP by activating adenylate cyclase

Research studies on the effectiveness of combining antidepressants are actually remarkably meager. The end points include abatement of symptoms (defined by rating scales such as PHQ-9, HAM-D), time to remission, time to relapse/sustaining remission and adverse drug reactions.

One of the earliest combination antidepressant strategies was investigated by Amsterdam and colleagues, namely tricyclic antidepressants (TCAs) and monoamine oxidase inhibitors (MAOIs), and recently described in a retrospective review of 2500 patient records in a treatment-resistant depression clinic. They identified 54 subjects who received either a combination of 2 TCAs, TCA added to an MAOI or MAOI added to a TCA. A significant number of patients benefited from the TCA-MAOI combination and noteworthy was the absence of any hypertensive crisis or serotonin syndrome (Amsterdam and Kim, 2019). Schipper et al. described a case report documenting the efficacy of the TCA-MAOI combination and moreover reviewed the literature concluding that there was clearly some evidence that patients who had failed either TCA or MAOI monotherapy did respond to the combination (Schipper et al., 2015). Finally, Thomas et al. comprehensively reviewed this literature and included those findings with those of 29 adults treated with an MAOI in combination with a TCA or psychostimulant. In this case series, 21% of the patients improved significantly and there were no serious side effects (Thomas et al., 2015).

Nelson, et al. conducted an early study in 1991 on the combination of fluoxetine and desipramine, hypothesizing that the combinations would elicit a more rapid down-regulation of beta-adrenergic receptors that would drive a faster rate of response (Nelson et al., 1991). This open study was comprised of 14 patients who received the combination of antidepressants and 20 who received desipramine alone in a 4 week trial. The median dose of desipramine in the desipramine-fluoxetine group was 125 mg and the median dose of desipramine in the desipramine-only group was 175 mg, but the dose ranges were very large. Desipramine plasma levels were considerably higher in the combination group than in the desipramine alone group, undoubtedly due to fluoxetine inhibition of the cytochrome P4502D6 isoenzyme resulting in inhibition of desipramine metabolism. The fluoxetine-desipramine group responded more quickly and completely than the desipramine alone group. The study was followed by a double-blind randomized controlled trial in 2004 of 38 patients which again revealed markedly higher rates of remission for the combination therapy, an effect not attributed to higher desipramine levels in the combination group or to a history of prior treatment resistance (Nelson et al., 2004).

A number of investigators have attempted to review the more recent literature on combinations of antidepressants in the management of depression. Palaniyappan et al. identified three randomized controlled trials of SSRI-TCA combination therapy that included 181 patients, four open trials of 85 patients and three case series of 40 patients. The concatenation of results suggested that additional data would be necessary to determine whether this strategy was superior to other evidence-based treatments for treatment-resistant depression. They also reviewed the data on the combination of SSRIs with the reversible MAOI, moclobemide, the combination of SSRIs and trazodone and the combination of SSRIs with reboxetine, an NE reuptake inhibitor. All of these studies were conducted with relatively small numbers of patients with the conclusion, not surprisingly, that some patients benefited from this combination strategy (Palaniyappan et al., 2018). Feetham reviews studies from 1988 through 2012 addressing both outcome and treatment steps. Her synopsis appropriately concluded that the data, taken together, was not convincing (Feetam, 2012). In a metaanalysis review from 2016, Henssler et al. reviewed 8688 international articles and identified 38 studies between 1977 and 2013 with a total of 4511 patients diagnosed with depressive order according to standard operationalized criteria. Within the 38 studies, 32 were randomized and 20 were double-blinded. Seventeen of the trials explicitly recruited nonresponders to antidepressant monotherapy. The review favored combining reuptake inhibitors with an antidepressant with alpha-2 antagonism action, such as mirtazapine or mianserin (Henssler et al., 2018).

Blier and colleagues, arguably the leading investigators in this field, have conducted multiple studies and have recently comprehensively reviewed their findings (Farah and Blier, 2021). These include a comparison of monotherapy and the combination of mirtazapine and paroxetine. One of this study's strengths is highlighting the importance of adequate dosing levels. They found a remission rate of 43% with combination therapy compared to ~20% for the two monotherapy groups (Blier et al., 2009). In another study, they compared fluoxetine monotherapy with the combination of mirtazapine with either fluoxetine, venlafaxine or bupropion. At 6 weeks, remission rates of HAM-D score below 8 were 25% for fluoxetine monotherapy and 52%, 58%, and 46% for the respective mirtazapine combinations (Blier et al., 2010). In contrast, Rush et al. found no difference in monotherapy with escitalopram and combination therapies with escitalopram and bupropion or venlafaxine and mirtazapine at either 12 weeks or 7 month treatment periods (Rush et al., 2011). Carpenter et al. completed a small placebo-controlled study of mirtazapine augmentation after 4+ weeks of monotherapy. A total of 26 patients with persistent major depression on monotherapy with bupropion 450 mg, citalopram 30 to 60 mg., fluoxetine 40 to 50 mg, fluvoxamine 300 mg, paroxetine 30 to 40 mg, sertraline 100 to 200 mg or venlafaxine 200 to 300 mg. were randomized to receive placebo or mirtazapine 15 to 30 mg. Patients receiving mirtazapine augmentation had a remission rate of 45.4% after 4 weeks versus 13.3% remission rate for patients receiving placebo (Carpenter et al., 2002).

Table 12.6 summarizes the major combination antidepressant randomized controlled studies conducted.

Along the time dimension of treatment, one can infer that the diminishing response to subsequent interventions supports the hypothesis that the longer an episode persists, the more difficult it is to achieve remission. Combining antidepressants as the initial treatment approach has been suggested as a potentially effective strategy to improve remission rates. Stewart et al. initiated treatment with bupropion and escitalopram in the hope that starting with two antidepressants would

TABLE 12.6 Randomized control studies—Summary.

Study	Design	Medications	Timeframe	Participants (N, dx, setting)	Outcome (p) rating scale	Other (side effect burden, adverse events, quality of life, functioning, attrition)
Combining Medications to Enhance Depression Outcomes (CO-MED): Acute and Long-Term Outcomes of a Single-Blind Randomized Study (Rush et al., 2011)	Single blind prospective	12 tetracyclic: venlafaxine-ER to 300 mg + mirtazapine to 45 mg	12 weeks 7 months	665 moderate to severe MDD, nonpsychotic; chronic and/or recurrent out-patients (primary care—6; psychiatric—9)	12 weeks remission rate: a. 38.8%, b. 38.9% (0.13), c. 37.7% (0.88). 7 month remission rate: a. 46.0%, b. 46.6% (0.38), c. 41.8% (0.77) QIDS-SR	Mean no. worsening adverse events (p): a. 4.7, b. 4.8 (0.47), c. 5.4 (<0.05)
Combination of antidepressant medications from treatment initiation for major depressive disorder: a double-blind randomized study (Blier et al., 2010)	Double blind	a. fluoxetine 20 mg, b. fluoxetine 20 mg + mirtazapine 30 mg, c. venlafaxine 225 mg + mirtazapine 30 mg, d. bupropion 150 mg + mirtazapine 30 mg	6 weeks	105 moderate to severe major depression (DSM-IV; HAM-D score)	Remission: a. 25% (0.011), b. 52% (0.011), c. 58% (0.011), d. 46% (0.011) HAM-D	Relapse from remission with switch back to monotherapy after 6 weeks: a. 27%, b. 56%, c. 50%, d. 24%
Mirtazapine and paroxetine in major depression: a comparison of monotherapy versus their combination from treatment initiation (Blier et al., 2009)	Double blind	a. mirtazapine 30–45 mg, b. paroxetine 20–30 mg, c. mirtazapine 30–45 mg + paroxetine 20–30 mg	6 weeks	61 unipolar depression newspaper ad. recruitment; PCP consultation requests; in-patient unit	Remission: a. 19% (>0.05), b. 26% (>0.05), c. 43% (>0.05) MADRS, HAM-D17	Adverse events in $>50\%$ pts. a. none, b. none, c. dry mouth; somnolence
Combining norepinephrine and serotonin reuptake inhibition mechanisms for treatment of depression: a double-blind, randomized study (Nelson et al., 2004)	Double blind	a. fluoxetine 20 mg, b. desipramine—dose f (serum level), c. fluoxetine 20 mg + desipramine	6 weeks	38 nonpsychotic, unipolar major depression in-patient	Remission: a. 7.1% (0.001), b. 0% (0.001), c. 53.8% (0.001) MADRS, HAM-D17	Combined treatment did not result in more rapid response
A double-blind, placebo-controlled study of antidepressant augmentation with mirtazapine (Carpenter et al., 2002)	Double blind	a. mirtazapine 15–30 mg, b. placebo	4 weeks	26 major depression (DSM-IV); out-patient	Remission: a. 45.4% (0.068), b. 13.3% (0.068) HAM-D17	% reporting moderate or severe side effect (no statistically significant group differences) weight gain. a. 27%, b. 20%. Sedation a. 13%, b. 9%

increase the rate of remission. They used a rapid dose titration schedule to achieve doses of escitalopram of 40 mg per day and bupropion of 400–450 mg per day, by day 15 of the study and enrolled 55 MDD patients. While limited by the absence of a control group and nonblinded raters, 16 (33%) of the patients remitted by week 2, and 31 (63%) by week 8 (Stewart et al., 2009).

Using the receptor pathway paradigm, one can merge the individual characteristics of the antidepressants to have a framework for making a well-informed choice of combining antidepressants for each patient.

There are some broad parameters to first consider. An absolute contraindication is combining SSRIs with a nonselective, irreversible MAOIs. Other combinations, such as a tricyclic with an MAOI, while not absolutely contraindicated, do require careful management and patient education. Dose adjustment may be required if one antidepressant inhibits or accelerates the metabolism of the other.

In conclusion, there is clearly a role for combination antidepressant therapy in the management of treatment-resistant depression. Unlike the large data base available for the use of atypical antipsychotic drug augmentation which is robust because of the requisite requirements for FDA approval, combination antidepressant clinical studies have not been funded by pharmaceutical companies. Although NIMH has funded a small portion of the extant data base, it remains quite meager compared to other strategies to manage this common clinical conundrum. It is unlikely that this situation will be remedied in the near future and as such we must make clinical decisions based on the extant literature and in the end, in part, our own clinical experience. A minority of patients achieve remission with antidepressant monotherapy, and the studies that do exist show that a meaningful percentage of patients respond to combined antidepressant therapy after failing to remit with monotherapy. A more expedient trial of combination antidepressant therapy could decrease the time to achieve remission and reduce the morbidity from prolonged depression. Though widely utilized, antipsychotic agents in combination therapy for depression have significant side effect risks, particularly the well-known risks of metabolic derangements and the more rare potential for movement disorders. The need for additional lab testing when using antipsychotics creates its own burden to patients and clinicians. The side effect profile of some antidepressants can be used in a way that is advantageous to targeting specific symptoms that are particularly problematic for patients, such as utilizing mirtazapine to aid sleep or stimulate appetite, or bupropion for its activating properties. There is certainly sufficient evidence to support the use of combination antidepressant treatment in treatment-resistant patients with MDD especially in view of its remarkably benign side effect profile, compared, for example, with certain atypical antipsychotic agents.

Financial disclosures

Charles B. Nemeroff, M.D., Ph.D.

Research/Grants: National Institutes of Health (NIH).
Consulting: ANeuroTech (division of Anima BV), Signant Health, Magstim, Inc., Navitor Pharmaceuticals, Inc., Intra-Cellular Therapies, Inc., EMA Wellness, Acadia Pharmaceuticals, Sage, BioXcel Therapeutics, Silo Pharma, XW Pharma, Neuritek, Engrail Therapeutics, Corcept Therapeutics Pharmaceuticals Company, SK Life Science, Alfasigma, Pasithea Therapeutic Corp., EcoR1.
Stockholder: Xhale, Seattle Genetics, Antares, BI Gen Holdings, Inc., Corcept Therapeutics Pharmaceuticals Company, EMA Wellness, TRUUST Neuroimaging.
Scientific Advisory Boards: ANeuroTech (division of Anima BV), Brain and Behavior Research Foundation (BBRF), Anxiety and Depression Association of America (ADAA), Skyland Trail, Signant Health, Laureate Institute for Brain Research (LIBR), Inc., Magnolia CNS, Heading Health, TRUUST Neuroimaging, Pasithea Therapeutic Corp.
Board of Directors: Gratitude America, ADAA, Xhale Smart, Inc., Lucy Scientific Discovery, Inc.
Patents: Method and devices for transdermal delivery of lithium (US 6,375,990B1).
Method of assessing antidepressant drug therapy via transport inhibition of monoamine neurotransmitters by ex vivo assay (US 7,148,027B2).
Speakers Bureau: None.

Julie Farrington, M.D.

Research/Grants: Cancer Research Institute of Texas (CPRIT)—3/1/2022.

Sussann Kotara, M.D.

None.

References

Amsterdam, J.D., Kim, T.T., 2019. Relative effectiveness of monoamine oxidase inhibitor and tricyclic antidepressant combination therapy for treatment-resistant depression. J. Clin. Psychopharmacol. 39 (6), 649–652. https://doi.org/10.1097/jcp.0000000000001130. 31688401.

Blier, P., Gobbi, G., Turcotte, J.E., de Montigny, C., Boucher, N., Hebert, C., Debonnel, G., 2009. Mirtazapine and paroxetine in major depression: a comparison of monotherapy versus their combination from treatment initiation. Eur. Neuropsychopharmacol. 19, 457–465. Crossref, Medline, Google Scholar Paroxetine + mirtazapine > paroxetine or mirtazapine monotherapy.

Blier, P., Ward, H.E., Tremblay, P., Laberge, L., Hébert, C., Bergeron, R., 2010. Combination of antidepressant medications from treatment initiation for major depressive disorder: a double-blind randomized study. Am. J. Psychiatry 167, 281–288.

Carpenter, L.L., Yasmin, S., Price, L.H., 2002. A double-blind, placebo-controlled study of antidepressant augmentation with mirtazapine. Biol. Psychiatry 51 (2), 183–188. https://doi.org/10.1016/s0006-3223(01)01262-8.

Cipriani, A., Furukawa, T.A., Salanti, G., Chaimani, A., Atkinson, L.Z., Ogawa, Y., Leucht, S., Ruhe, H.G., Turner, E.H., Higgins, J.P.T., Egger, M., Takeshima, N., Hayasaka, Y., Imai, H., Shinohara, K., Tajika, A., Ioannidis, J.P.A., Geddes, J.R., 2018. Comparative efficacy and acceptability of 21 antidepressant drugs for the acute treatment of adults with major depressive disorder: a systematic review and network meta-analysis. Lancet 391, 1357–1366.

Dunlop, B.W., Kelley, M.E., Aponte-Rivera, V., Mletzko-Crowe, T., Kinkead, B., Ritchie, J.C., Nemeroff, C.B., Craighead, W.E., Mayberg, H.S., PReDICT Team, 2017. Effects of patient preferences on outcomes in the predictors of remission in depression to individual and combined treatments (PReDICT) Study. Am. J. Psychiatry 174 (6), 546–556. https://doi.org/10.1176/appi.ajp.2016.16050517 (Epub 2017 Mar 24).

Farah and Blier, 2021. Textbook of Mood Disorders. APA Press.

Feetam, C., 2012. Are two antidepressants always better than one? Prog. Neurol. Psychiatry. https://www.progressnp.com/.

Fife, D., Rep, J., Cepeda, M.S., Stang, P., Blacketer, M., Singh, J., 2019. Corrigendum to "Treatment resistant depression incidence estimates from studies of health insurance databases depend strongly on the details of the operating definition". Heliyon 5 (3), e01198. Published online 2019 Mar 7 https://doi.org/10.1016/j.heliyon.2019.e01198. PMC6406969. PMID: 30899820 https://www.ncbi.nlm.nih.gov/pmc/articles/PMC6406969/.

Foley, K.F., De Santy, K.P., 2006. Bupropion: Pharmacology and therapeutic applications. Expert. Rev. Neurother. 6 (9), 1249–1265. https://doi.org/10.1586/14737175.6.9.1249. Source PubMed.

Gartlehner, G., et al., 2011. Comparative benefits and harms of second-generation antidepressants for treating major depressive disorder: an updated meta-analysis. Ann. Intern. Med. 155 (11), 772–785.

Gaynes, B., Rush, A.J., Trivedi, M.H., Wisniewski, S.R., 2008. The STAR*D study: Treating depression in the real world. Cleve. Clin. J. Med. 75 (1), 57–66. https://doi.org/10.3949/ccjm.75.1.57.

Henssler, T.J., Kurschus, M., Franklin, J., Bschor, T., Baethge, C., 2018. Trajectories of acute antidepressant efficacy: how long to wait for response? A systematic review and meta-analysis of long-term, placebo-controlled acute treatment. J. Clin. Psychiatry 79 (3). https://doi.org/10.4088/JCP.17r11470, 17r11470.

Judd, L.L., Paulus, M.J., Schettler, P.J., Akiskal, H.S., Endicott, J., Leon, A.C., Maser, J.D., Mueller, T., Solomon, D.A., Keller, M.B., 2000. Does incomplete recovery from first lifetime major depressive episode herald a chronic course of illness? Am. J. Psychiatry. https://doi.org/10.1176/appi.ajp.157.9.1501. Published Online:1 Sep.

Nelson, J.C., Mazure, C.M., Bowers Jr., M.B., Jatlow, P.I., 1991. A preliminary, open study of the combination of fluoxetine and desipramine for rapid treatment of major depression. Arch. Gen. Psychiatry 48 (4), 303–307. https://doi.org/10.1001/archpsyc.1991.01810280019002. 2009031. DOI: 10.1001/archpsyc.1991.01810280019002.

Nelson, J.C., Mazure, C.M., Jatlow, P.I., Bowers Jr., M.B., Price, L.H., 2004. Combining norepinephrine and serotonin reuptake inhibition mechanisms for treatment of depression: a double-blind, randomized study. Biol. Psychiatry 55 (3), 296–300. https://doi.org/10.1016/j.biopsych.2003.08.007. 14744472. DOI: 10.1016/j.biopsych.2003.08.007.

Nemeroff, C.B., 2020. The state of our understanding of the pathophysiology and optimal treatment of depression: glass half full or half empty? Am. J. Psychiatry 177, 671–685.

Owens, M.J., Morgan, W.N., Plott, S.J., Nemeroff, C.B., 1997. Neurotransmitter receptor and transporter binding profile of antidepressants and their metabolites. J. Pharmacol. Exp. Ther. 283 (3), 1305–1322.

Palaniyappan, L., Insole, L., Ferrier, N., 02 January 2018. Combining Antidepressants: A Review of Evidence. Published Online by Cambridge University Press. https://www.cambridge.org/core/journals/advances-in-psychiatric-treatment/article/combining-antidepressants-a-review-of-evidence/10A101DFD409DCB1FDCBB951F391F7AE.

Rush, S.J., Trivedi, M.H., Stewart, J.W., Nierenberg, A.A., Fava, M., Kurian, B.T., Warden, D., Morris, D.W., Luther, J.F., Husain, M.M., Cook, I.A., Shelton, R.C., Lesser, I.M., Kornstein, S.G., Wisniewski, S.R., 2011. Combining medications to enhance depression outcomes (CO-MED): acute and long-term outcomes of a single-blind randomized. Am. J. Psychiatry 168 (7), 689–701.

Sanchez, C., Reines, E.H., Montgomery, S.A., 2014. A comparative review of escitalopram, paroxetine, and sertraline: are they all alike? Int. Clin. Psychopharmacol. 29 (4), 185–196.

Schipper, P., Vanmolkot, L., Peeters, F.P., 2015. Combining a classic monoamine oxidase inhibitor with a tricyclic antidepressant in therapy resistant depression: a case report and literature review. Tijdschr. Psychiatr. 58 (12), 886–890. Language:dut 27976787.

Schoenbaum, M., Unützer, J., McCaffrey, D., Duan, N., Sherbourne, C., Wells, K.B., 2002. The effects of primary care depression treatment on Patients' clinical status and employment. Health Serv. Res. 37 (5), 1145–1158. https://doi.org/10.1111/1475-6773.01086. PMC1464020 12479490.

Stewart, J.W., McGrath, P.J., Deliyannides, R.A., Quitkin, F.M., 2009. Does dual antidepressant therapy as initial treatment hasten and increase remission from depression? J. Psychiatr. Pract. 15, 337–345.

Thomas, S.J., Shin, M., McInnis, M.G., Bostwick, J.R., 2015. Combination therapy with monoamine oxidase inhibitors and other antidepressants or stimulants: strategies for the management of treatment-resistant depression pharmacotherapy. J. Human Pharmacol. Drug Therapy 35 (4). https://doi.org/10.1002/phar.1576. First published: 17 April.

Vaishnavi, S.N., Nemeroff, C.B., Plott, S.J., Rao, S.G., et al., 2004. Milnacipran: a comparative analysis of human monoamine uptake and transporter binding affinity. Biol. Psychiatry 55, 320–322.

Zhou, J., 2004. Norepinephrine transporter inhibitors and their therapeutic potential. Drugs Future 29 (12), 1235–1244. https://www.ncbi.nlm.nih.gov/pmc/articles/PMC1518795/.

Chapter 13

Ketamine/esketamine for treatment-resistant depression

Sina Nikayin[a,b] and Gerard Sanacora[a,b]

[a]*Yale Depression Research Program, Department of Psychiatry, Yale University, New Haven, CT, United States,* [b]*Interventional Psychiatry Service, Department of Psychiatry, Yale University, New Haven, CT, United States*

Introduction

In recent years, ketamine, its individual enantiomers, and related compounds have been a major focus of research and clinical development for the treatment of depression. With the United States Food and Drug Administration (USFDA) approval of esketamine for the indication of treatment-resistant depression in 2019, use of the treatment has rapidly increased in clinical settings. As such, it is important for clinicians and researchers to be familiar with the current state of knowledge surrounding efficacy, safety, and proposed mechanisms of action of this treatment modality. It is equally important to be aware of the remaining unanswered questions related to the treatment in order to provide the best possible care to patients and promote meaningful translational research. In this chapter, we briefly review the history of ketamine, explore proposed mechanisms of action, review data pertaining to efficacy and safety, and consider "real-world" usage and the challenges to implementation. Future directions related to the optimization of ketamine and the development of related compounds will also be discussed.

History

Ketamine discovery

Like most meaningful discoveries, the road leading to the development of ketamine as a treatment of mood disorders was long and winding, punctuated by dead ends and fueled by serendipity. In 1956, V. Harold Maddox, a chemist at Parke-Davis, created a new chemical process which resulted in the creation of phencyclidine (Maddox et al., 1965). Initially used as an anesthetic in animals (Chen et al., 1959), phencyclidine was later tried in humans (Greifenstein et al., 1958). Although it was generally considered a safe anesthetic, it produced a severe emergence delirium in some patients (Domino, 2010), and later studies raised concerns that the drug could cause psychotomimetic symptoms similar to schizophrenia (Rosenbaum et al., 1959). As such, phencyclidine was deemed inappropriate for human use, motivating work to identify related compounds that might be more appropriate for clinical applications. One of these compounds, CI-581, a short-acting agent with favorable anesthetic properties, was synthesized by Calvin Stevens in 1962. Considering the compound contained both a ketone and an amine in its structure, it was named ketamine (Mion, 2017). Domino and Corssen first reported on the use of ketamine in humans in a study conducted on prisoners at the Jackson Prison in Michigan in 1964 (Domino et al., 1965), coining the term "dissociative anesthetic," that remains in use today.

For over 50 years, ketamine has been used as an anesthetic agent. Originally patented for veterinary use in 1963 and receiving a USFDA indication under the name Ketalar for human use as an anesthetic treatment in 1970 (Drugs@FDA, n.d.), it rapidly became a favored field anesthetic given its sympathomimetic properties and analgesic effects that outlast its anesthetic effect (White et al., 1982). Ketamine was first placed on the World Health Organizations Essential Medicines List in 1985 related to its ease of use and safety profile that make it especially appealing in underresourced health systems. In recent years, illicit and recreational use of ketamine in some countries has prompted renewed evaluations of ketamine's role in the worldwide medical system and the possible need for increased levels of control. However, four reviews of the drug over the past 15 years by the World Health Organization (WHO) consistently recommended that ketamine should not be placed under increased international control that could limit access to the drug (World Health Organization, 2015).

Managing Treatment-Resistant Depression. https://doi.org/10.1016/B978-0-12-824067-0.00013-X

Rationale for use in depression

A series of evolving hypotheses and discoveries inspired the studies that ultimately demonstrated ketamine's antidepressant properties. To our knowledge, the first reported study of ketamine's potential use in the treatment of psychiatric illnesses was done in 1973 in Shiraz, Iran. Khorramzadeh and Lotfy treated 100 patients carrying various psychiatric diagnoses with increasing doses of ketamine. Astonishingly, they reported improvement in psychiatric symptoms in 95% of patients treated with 0.4–0.6 mg/kg or slightly higher (0.7–1.0 mg/kg) doses of ketamine (Khorramzadeh and Lotfy, 1973). The rationale behind this trial was based primarily on the psychological effects of ketamine and the potential for it to improve the psychodynamic healing process, as the underlying neurobiological effects of ketamine had not even been well understood at this time.

In the early 1990s, Skolnick and colleagues at the NIH explored the effects of *N*-methyl-D-aspartate (NMDA) glutamate receptor antagonists on the depression-like behaviors caused by inescapable stress in mice (Trullas and Skolnick, 1990). Based on the knowledge that NMDA receptors are required for long-term potentiation (LTP), which is impaired by inescapable stress, and also by the fact that traditional antidepressants are effective in reversing the depressive-like behaviors caused by inescapable stress, they hypothesized that NMDA receptor antagonists could produce similar antidepressant effects in mice. Using standard rodent behavioral assays they were able to successfully demonstrate antidepressant-like effects with a competitive antagonist, a noncompetitive antagonist, and a partial agonist of the NMDA receptor.

Around the time the findings of the NIH lab were being published, evidence demonstrating notable stress-related effects on glutamatergic neurotransmission was mounting. Studies showing acute stress-related increases the extracellular glutamate levels, especially in hippocampus, amygdala and PFC and dramatic alterations in the membrane trafficking of NMDA and α-amino-3-hydroxy-5-methyl-4-isoxazolepropionic acid (AMPA) receptors in the prefrontal cortex were increasing interest in the potential role of glutamate in the pathogenesis and pathophysiology of mood disorders (Popoli et al., 2011). Meanwhile, in the chronic stress setting there was evidence of changes in the regulation of glutamate release in PFC following exposure to chronic stressors. In sum, the emerging series of preclinical studies provided increasing supportive evidence to suggest glutamate's contribution to pathogenic pathways leading to depression and the potential to target the system in the development of novel treatments for mood disorders. This background paired with the recognition that altered function of cortical circuits, that are predominantly glutamatergic in nature, were critical to the pathogenesis and pathophysiology of mood disorders were specifically cited as the driving impetus (Krystal et al., 2019) for the seminal study by Berman et al. (2000) that first demonstrated the antidepressant effects of ketamine in a small controlled clinical trial.

Since the publication of the initial findings suggesting ketamine possessed rapid and robust antidepressant effects there has been a tremendous amount of reverse "bedside to bench" translation providing more information on its mechanisms of antidepressant action. Multiple new mechanisms of action have been proposed in addition to those hypothesized previously, and it is possible, in fact likely, that several mechanisms uniquely contribute to the overall antidepressant effect. At present, evidence of ketamine's proximal effect as an NMDA receptor antagonist via binding to its phencyclidine site (Ravikrishnan et al., 2018; Lumsden et al., 2019) serving as the initial step in antidepressant process remains best supported by the current evidence. The antagonism of NMDA receptors was originally postulated to protect against the excitotoxic effects of elevated levels of extracellular glutamate levels and the associated downstream consequences. However, it was also noted that selective NMDA receptor antagonism can decrease the firing of the inhibitory GABAergic neurons, resulting in transiently increased levels of glutamate release, at times referred to as a glutamate surge. The increased levels of glutamate released into the synaptic space in turn engage other non-NMDA glutamate receptors such as the α-amino-3-hydroxy-5-methyl-4-isoxazolepropionic acid (AMPA) receptor. Activation of the AMPA receptor is believed to activate several intracellular signaling cascades and eventually facilitate the process of synaptogenesis (Duman and Aghajanian, 2012; Duman et al., 2016). Other mechanisms such as effects on opioid receptors (Zanos et al., 2016), immune function (Cui et al., 2019; Li et al., 2020), and monoaminergic pathways (Iro et al., 2021), among others (Wei et al., 2021), have also been reported, but have not been as extensively studied. Beyond the more proximal effects of ketamine's pharmacology there is also great interest in understanding how the effects are transduced through changes in brain circuitry (Abdallah et al., 2021; Shinohara et al., 2020). The range of possible critical mechanisms of antidepressant action and the unique properties of the various forms of ketamine should be considered in review of the efficacy data presented below.

Efficacy

The relation of molecular structure and pharmacology to function and response

When considering the data on efficacy presented below it is important to remember ketamine is a chiral molecule comprised of two enantiomers, S-ketamine (esketamine) and R-ketamine (arketamine). The racemic mixture was initially used for clinical purposes and continues to be the most common form used for anesthetic purposes in North America. However, the

S-ketamine enantiomer is also commonly used in Europe for anesthesia based on reports suggesting improved tolerance (Trimmel et al., 2018). R-ketamine is not currently approved for clinical use but remains under development for potential use in neuropsychiatric disorders. The specific properties of the individual enantiomers in relation to the clinical benefits and untoward effects associated with the treatment of mood disorders are currently an area of great interest.

Racemic

The use of racemic ketamine as a therapeutic agent in mood disorders has now been studied for more than two decades and has repeatedly been demonstrated to have efficacy with a generally acceptable safety profile in the treatment of major depressive episodes in numerous small to moderate sized clinical trials. Recent meta-analytic studies (Kryst et al., 2020; McIntyre et al., 2020, 2021; Sanacora et al., 2017) confirmed the early findings of a ketamine's significant efficacy advantage over control from 24h to 7 days post administration of a single dose by both continuous severity measures and binary response and remission rates. The peak response was most commonly reported to be seen over the first 24h, but while attenuated, the antidepressant effect was still present with small to medium effect sizes 1-week following the administration of a single dose. Typical response rates ranged between studies but were most commonly in the range of 50% for treatment-resistant depressed (TRD) patients and greater for non-TRD patients with odds ratios of response typically approximating five overall compared to placebo for the first week after treatment. There are fewer randomized controlled studies examining repeated dosing with racemic ketamine. However, the limited studies that are available show relatively large advantages of ketamine over placebo for periods of 2–3 weeks with repeated dosing.

S-Ketamine

The use of individual ketamine enantiomers was initially reported by White et al. (1985) who compared them to the effects of racemic ketamine. They found that S-ketamine produced anesthetic effects with only about half the dose required for racemic ketamine. The quality of anesthesia was similar to the racemic mixture, but the recovery with individual isomers was reported to be persistently shorter. After these findings, S-ketamine was considered as a more potent anesthetic agent, and has been approved and regularly used as an anesthetic in Europe (European Medicines Agency, 2018). The fact that S-ketamine has three to four times the affinity compared to R-ketamine for the PCP binding site of the NMDA receptor is believed to explain the increased potency of the enantiomer (Vollenweider et al., 1997).

While the use of ketamine in TRD predates esketamine, the most vigorously studied formulation to date for the treatment of TRD is intranasally administered esketamine. Several large multicenter double-blind randomized controlled trials of intranasal esketamine in conjunction with another oral antidepressant were conducted as part of the submission package for USFDA approval in the treatment of TRD (Papakostas et al., 2020). To address a major limitation of ketamine treatment access, the need for intravenous (IN) drug delivery, IN delivery was considered as a possible alternative method of delivery. Nasal insufflation is considered simpler and more comfortable for patients but requires a significant volume of medication to be administered due to issues with delivery and bioavailability. Esketamine, with its higher affinity for the NMDA receptors and consequent greater expected potency (Vollenweider et al., 1997; Oye et al., 1992; Muller et al., 2016), was reasoned to require lower volumes that could more practically be administered intranasally. In the proof-of-concept phase of development, esketamine's antidepressant effects were initially studied using IV delivery. This smaller study supported the idea that esketamine could have similar antidepressant responses with reduced doses and plasma levels compared to racemic ketamine (Singh et al., 2016a).

With the initial data suggest esketamine's efficacy in MDD, a series of clinical trials were launched to examine the efficacy and safety of intranasal esketamine in the treatment of TRD. These studies consisted of two phase three randomized controlled trials in adults younger than 65 years of age (Fedgchin et al., 2019) with another in adults older than 65 (Ochs-Ross et al., 2020). Additionally, a randomized withdrawal study was conducted (Daly et al., 2019), along with a long-term 1 year open-label study (Wajs et al., 2020). Collectively, results from these studies which included over 1000 participants provided adequate efficacy and safety data to lead the USFDA to approve esketamine IN for treatment of TRD alongside an oral antidepressant in March 2019 (FDA, 2019a). It is important to note that as part of the approval, the USFDA required a Risk Evaluation and Mitigation Strategy (REMS) program to monitor the treatment and to ensure it was only conducted in appropriate settings and with adequate monitoring and oversight (FDA, 2019b).

R-Ketamine

As mentioned earlier, in comparison to S-ketamine, R-ketamine has a lower affinity for the PCP site of the NMDA receptor. However, it also weakly interacts with the sigma sites. In addition, R-ketamine produces unique alterations in cerebral metabolic rates of glucose across brain regions, especially in left insula and temporomedial cortex. It also seems not to produce the same quality of psychotomimetic symptoms as racemic or S-ketamine, but rather has been reported to produce a state of relaxation (Vollenweider et al., 1997). Interestingly, there have been reports of R-ketamine having superior antidepressant-like responses in rodent models of antidepressant action (Yang et al., 2015). Other rodent studies have also suggested that a metabolite of R-ketamine, (2*R*,6*R*)-hydroxynorketamine (RR-HNK), is both necessary and sufficient to produce the antidepressant-like response, in part suggesting NMDA receptor activity was not central to the antidepressant response as the metabolite has very little affinity for the receptor (Zanos et al., 2016). However, contrasting findings have also been reported regarding the efficacy of this metabolite in other preclinical studies (Yamaguchi et al., 2018). In sum, the combined findings of R-ketamine possibly having reduced negative acute effects on cognition and emotion as well as possible superior antidepressant efficacy in rodent models have been the foundation for the recent interest in R-ketamine as a possible treatment for mood disorders. To date, the limited data available seem promising (Leal et al., 2020); however, further investigations are clearly required to assess the true safety and efficacy of this treatment in humans.

In summary, there is now clear evidence demonstrating the efficacy of both racemic and esketamine in the short-term treatment of symptoms in patients with major depressive disorder, especially those considered resistant to standard forms of oral antidepressant medications. Although there are some attempts to compare the relative efficacy of the two formulations (racemic ketamine delivered IV and esketamine delivered IN) (Bahji et al., 2021) great caution should be used in interpreting these reports as comparisons across trials are notoriously complicated and potentially misleading (Drevets et al., 2021). At present, only intranasal esketamine spray has been approved for use in the treatment of adults with treatment-resistant depression and for the treatment of depressive symptoms in adults diagnosed with major depressive disorder experiencing suicidal ideation.

Dose

General

Many attempts have been made to assess the effects of ketamine dose on various physiological states in both humans and animals. The findings suggest a clear dose-dependent response to ketamine with an excitatory effect on regional brain function at lower doses and a diffuse inhibitory response with higher, anesthetic doses of ketamine. Preclinical studies done in rats using in vivo microdialysis and magnetic resonance spectroscopy show that ketamine at lower doses increases glutamate efflux and cycling with corresponding cellular and behavioral changes suggestive of antidepressant-like effects (Chowdhury et al., 2017; Moghaddam et al., 1997). However, at higher doses, the opposite is observed, with a decrease in neurotransmitter cycling and a corresponding disappearance of the cellular and behavioral changes suggestive of antidepressant effect (Chowdhury et al., 2017; Li et al., 2010; Kim and Monteggia, 2020). This preclinical data suggest that ketamine may have a relatively narrow therapeutic window of peak efficacy as an antidepressant.

Another important factor affecting dosage and possibly antidepressant efficacy is the bioavailability of ketamine. Depending on the route of administration, the bioavailability can be significantly affected; the highest bioavailability of is achieved with IV administration (100%), whereas IN administration can have 30%–50% bioavailability (Zanos et al., 2018). Considering these effects on ketamine bioavailability it is extremely important to not only consider the dose being used but also the route of administration.

Racemic

Remarkably, there is limited high quality clinical data available in humans to assess the dose-response relationship with racemic ketamine doses delivered intravenously. A dose ranging trial of 99 participants across six US academic sites compared the effects of a single 0.1 mg, 0.2 mg, 0.5 mg, and 1 mg/kg dose of ketamine in the treatment of depression. It showed the 0.5 and 1.0 mg doses to be effective compared to midazolam even after appropriate adjustments, but only the 0.5 mg dose was significantly superior to placebo across all secondary outcomes, the 1.0 mg dose was only significant with one of the five secondary outcomes (Fava et al., 2020). While this study was not sufficiently powered to detect even moderately sized differences in the antidepressant effect associated with the various doses, and only investigated the response after a single infusion for a relatively short duration of time, it is not inconsistent with the rodent data suggesting the existence of a nonlinear dose-response curve in which the ratio of clinical benefit to adverse events may not favor higher dosing.

S-Ketamine

With regards to esketamine, a double-blind dose-finding study sponsored by Janssen was conducted to assess the optimum dosage of IV esketamine. The results of this multicenter study which compared 0.2 mg/kg versus 0.4 mg/kg IV showed that the lower dosage maintained efficacy while reducing the dose-dependent adverse effects (Singh et al., 2016a). In the later phase studies with nasal insufflation, there was no clear evidence of superior response between the 56 and 84 mg dosing regimens, but there appeared to be a diminished response to the 28 mg dose. The current recommendation for esketamine dosing in the treatment of TRD consists of starting with the 56 mg dose at the first treatment, which may be adjusted to 84 mg based on clinical response at subsequent sessions. For major depression associated with suicidal ideation (MDSI), which is considered to demand a faster onset of effect, the recommended starting dose is 84 mg, which can be reduced to 56 mg in patients who have difficulty tolerating the higher dose.

Frequency

There are also limited data available to guide the choice of dosing frequency for treatment with ketamine or esketamine at this time. An early study explored the effect of initiating ketamine 0.5 mg/kg IV with different dosing frequencies, comparing treatment twice a week to thrice a week dosing (Singh et al., 2016b). The study found both regimens to produce a rapid and sustained response to the ketamine treatment compared to placebo over the 2-week treatment period but were not able to detect any meaningful difference between the two dosing frequencies. Although the study was of short duration and not sufficiently powered to determine smaller differences between the two groups the failure to provide evidence suggesting the more frequent dosing was any more beneficial than the twice a week dosing supported the use of lower frequency dosing as the preferable treatment plan. The twice weekly dosing regimen was adopted in the Phase 3 clinical trials for IN esketamine for TRD and for MDSI and continues to serve as the standard recommendation for clinical dosing over the first month of treatment (twice weekly for 4 weeks for both indications) (SPRAVATO, 2020). Interestingly, a survey of providers administering ketamine IV in clinical settings throughout North America performed prior to the approval of esketamine treatment and widespread knowledge of the findings from the dosing frequency study showed that most providers were offering the initial treatments either twice or thrice weekly (Wilkinson et al., 2017a). This affords some level of experiential evidence that clinicians independently arrived at this dosing frequency. Although the optimal dosing frequency has yet to be established for ketamine or esketamine treatments, it is important to note that there is no high-quality evidence currently available to suggest more frequent dosing is either safe or effective for the treatment of mood disorders. Studies examining alternative dosing frequency strategies are ongoing and will likely provide more information in the near future.

The other major issue regarding the frequency of ketamine and/or esketamine dosing surrounds the continuation/maintenance course treatment following the acute phase. The previously noted survey of ketamine providers showed that the majority of patients treated prior to the esketamine approval were receiving maintenance treatments either on an every 3-week or monthly interval schedule (Wilkinson et al., 2017a). The long-term continuation (Daly et al., 2019; Wajs et al., 2020) and randomized withdrawal studies (Daly et al., 2019) completed as part of the esketamine Phase 3 studies used a regimen whereby patients, after completing the 4-week course of twice a week dosing, moved to weekly dosing for 1 month and then dosing at either weekly or every other week dosing based on symptom severity. The currently recommended frequency of esketamine for acute treatment of both indications is twice weekly for 4 weeks. For TRD, the recommendation is for once weekly treatment during weeks 5–8 and every 1–2 weeks beyond that time, individualized to each patient to maintain response or remission (SPRAVATO, 2020). For MDSI, continuation beyond 4 weeks is based on the clinician's determination, and there are currently no systematic data to guide continuation (Table 13.1).

Setting

Given concerns surrounding acute adverse events, especially hemodynamic changes and dissociative symptoms, monitoring of ketamine/esketamine treatment is an important factor in providing the safest treatment possible. In a survey of providers administering IV ketamine in the community, most providers reported monitoring heart rate, pulse oximetry and blood pressure with various frequencies during the treatment period.

For esketamine, the recommendation for monitoring during the treatment includes monitoring the patient by a healthcare provider for at least 2 h, followed by an assessment to assure patient is safe to leave. This includes assessment of blood pressure throughout the treatment, especially 40 min after treatment and afterwards as warranted to assure return of blood pressure to baseline.

Considering the above, the treatment setting should be able to support adequate monitoring of vital signs, with a healthcare provider present at all times during treatment. In addition, while data regarding the effects of the environment on the treatment are limited, it is advisable to provide a comfortable environment with minimal distractions considering the dissociative effects of the medication.

TABLE 13.1 Recommended dosing and frequency of intranasal esketamine and IV ketamine.

Treatment	Indication	Phase	Frequency	Dose	Duration	Goal	Next step
Esketamine intranasal	TRD	Induction/ acute	Twice weekly	Start at 56 mg, may increase to 84 mg	Eight treatments	Response[a]	Responder: Go to continuation
							Nonresponder: Stop
		Continuation[b]	Weekly	56 mg or 84 mg	Four treatments	Maintain response/ achieve remission	Consider maintenance
		Maintenance	Every 1–2 weeks, consider decreasing frequency as tolerated	56 mg or 84 mg	Varies	Prevent relapse	Consider stopping (individualized)
	MDSI	Induction/ acute	Twice weekly	Start at 84 mg, may decrease to 56 mg	Eight treatments	Response	Responder: consider continuation[c]
							Nonresponder: Stop
		Continuation	No specific data available for MDSI. May consider similar approach as TRD	56 mg or 84 mg	Varies	Maintain response/ achieve remission	Consider maintenance
		Maintenance		56 mg or 84 mg		Prevent relapse	Consider stopping (individualized)
Ketamine IV[d]	TRD[e]	Induction/ acute	Twice weekly	0.5 mg/kg[f]	—Six to eight treatments	Response	Responder: Go to continuation
							Nonresponder: Stop
		Continuation	Weekly	0.5 mg/kg	Four treatments	Maintain response/ achieve remission	Consider maintenance
		Maintenance	Every 2 weeks, then gradually increase up to every 4–6 weeks	0.5 mg/kg	Varies	Prevent relapse	Consider stopping (individualized)

[a] *Usually considered to be a 50% reduction in symptoms.*

[b] *Spravato prescribing information does not differentiate between continuation and maintenance phases.*

[c] *Spravato prescribing information does not provide any specific recommendation with regards to continuation or maintenance in MDSI; given lack of systematic data, it is advisable for clinicians to consider appropriateness of a similar approach to TRD on a case by case basis.*

[d] *Not FDA-indicated for depression. Limited data available for doses other than 0.5 mg/kg, and regarding optimum dosing frequency.*

[e] *Also consider a similar approach for other diagnoses on an individualized basis.*

[f] *May use ideal body weight instead of weight for patients with a BMI > 30.*

Consideration of diagnostic and comorbidity factors

The strongest and most abundant evidence for the use of ketamine and esketamine in psychiatric illnesses relates to its use in the treatment of adults with treatment-resistant major depressive disorder (Popova et al., 2019; Short et al., 2018). However, limited data are available on the use of ketamine on other disorders such as bipolar depression (McCloud et al., 2015), mood disorders with psychotic features (da Frota Ribeiro et al., 2016), PTSD (Feder et al., 2021), and OCD (Rodriguez et al., 2013; Murrough et al., 2015a).

The presence of certain comorbidities has become of specific concern when treating patients with TRD. Extremely limited data available to date suggest that ketamine or esketamine may be used in patients with primary psychotic disorders (Bartova et al., 2018) or depression with psychotic features (da Frota Ribeiro et al., 2016; Ajub and Lacerda, 2018). However, these data are extremely scant and is imperative to assess risks and benefits when considering treatment options for these patients as no large-scale studies have yet been completed to examine the safety and efficacy of the treatment for these specific depressive subtypes. There is also justifiable concern about using ketamine or esketamine in patients with a history or current substance use disorders. Although a few studies suggest ketamine may actually be used beneficially to reduce the severity of substance use disorders (Jones et al., 2018; Ivan Ezquerra-Romano et al., 2018; Dakwar et al., 2019), these reports should be considered preliminary and caution should be exercised in providing the treatments to patients with histories of significant substance use disorders until more data are available.

Perhaps the most clinically important diagnostic and/or comorbid issue to consider with the use of the treatments is the presence of suicidal ideations and the risk of suicidal behavior. Since ketamine was shown to be effective in rapidly addressing depressive symptoms, and as suicidal ideation and behavior is one of the most severe and urgent presentations of MDD, there was obvious motivation to look into the specific effects of ketamine on addressing these symptoms. Initial data coming from studies of ketamine in depression showed promising potential effects in suicidal ideation (Price et al., 2009; DiazGranados et al., 2010; Murrough et al., 2015b; Price et al., 2014; Ballard et al., 2014). A metaanalysis of patient level data from 11 studies of single-dose ketamine use in depression (Wilkinson et al., 2018a) showed a significant reduction in suicidal ideation with possible independence from the improvement in depressive symptoms (Ballard et al., 2014; Wilkinson et al., 2018a). Some small studies tried to directly examine the efficacy of ketamine in the treatment of suicidal ideation (Murrough et al., 2015b; Grunebaum et al., 2018; Price and Mathew, 2015; Witt et al., 2020). The data from these trials suggested that ketamine may be able to rapidly reduce suicidal ideation, though the results were not conclusive.

Given the promising data with racemic ketamine related to suicidal ideation and the evidence of esketamine's efficacy in TRD patients, a series of trials were initiated to formally assess the effects of intranasal esketamine in the treatment of patients with major depressive disorder experiencing clinically meaningful suicidal ideation or behavior. These consisted of an initial phase 2 randomized controlled trial (Canuso et al., 2018), which showed greater rapid reduction in depressive symptoms in these patients with esketamine, along with a significant reduction of suicidal ideation at some of the endpoints. Afterwards, two phase 3 trials (Fu et al., 2020; Ionescu et al., 2020) were conducted to assess these findings at a larger scale. Both studies met their respective primary endpoints, showing statistically significant reduction in depressive symptoms with esketamine plus standard of care treatment over placebo plus standard of care treatment at 4 and 24h after dosing. However, while a notable and rapid decrease of suicidal ideation was noted in both studies, the difference between the two treatment groups did not reach statistical significance. Though at first glance the failure to find a significant difference in suicidal ideation measures between the two groups may seem concerning, it is important to consider a few major factors that complicate the interpretation of these results. Firstly, every subject in the trial, regardless of study drug assignment, received high-level standard of care treatment that included an inpatient hospital stay. Thus, the trials were unique in nature for antidepressant trials by comparing the study drugs against the background of a newly initiated active treatment rather than placebo; secondly, both groups in these trials achieved rapid reduction in suicidal ideation making it difficult to show one treatment to be superior to another; these facts along with the limited tools available to assess short-term *change* in suicidal ideation and risk should be considered when interpreting the results.

The USFDA, after review of these findings, concluded that these results in combination demonstrated the benefits of this treatment in patients with major depression and suicidal ideation or behavior and the risks were adequately understood, and approved the use of esketamine in this population in 2020 (FDA, 2020c).

Concomitant treatments

The data on the effects of other medications and treatments on ketamine treatment (whether augmenting or inhibiting) remain limited. However, those most commonly studied in the literature are briefly outlined.

Potential augmenting treatments

Lithium

Both lithium and ketamine are believed to have an inhibitory effect on Glycogen Synthase Kinase 3 (Jope, 2003; Beurel et al., 2011). Reports from rodent studies suggest that the combination of ketamine with lithium may result in synergistic and enduring antidepressant effects (Ghasemi et al., 2010). However, the use of lithium after ketamine treatment in 34 patients with at least partial response to ketamine did not significantly increase the duration of ketamine's antidepressant effect in the treatment of MDD (Costi et al., 2019). A second study conducted 36 patients with bipolar depression (Xu et al., 2015) also showed no meaningful difference in effects of ketamine between the valproate and lithium groups. Further studies are needed to better understand the relationship between lithium and ketamine in treatment of depression but at present there is no convincingly supportive clinical evidence.

Riluzole

Based on riluzole's effects as a glutamatergic modulator it has also been considered to potentially prolong the effects of ketamine treatment. However, a study of 42 subjects comparing riluzole to placebo treatment after a single IV ketamine treatment failed to show any significant improvement in the riluzole group in comparison to placebo (Ibrahim et al., 2012). Another similar trial was stopped for futility after enrolling 14 patients (Mathew et al., 2010).

Transcranial magnetic stimulation

Ketamine has been shown to reduce the resting motor threshold and active motor thresholds in transcranial magnetic stimulation (TMS) and may lead to an increase in the amplitude of EMG response after magnetic stimulation (Di Lazzaro et al., 2003). A few case reports in TRD and bipolar depression have suggested a synergistic effect for ketamine and TMS (Best, 2014; Best et al., 2015; Best and Griffin, 2015). However, the data are insufficient to provide any meaningful guidance on the safety or efficacy of this treatment combination at this time.

CBT psychotherapy

As mentioned previously, some of the early studies exploring ketamine's use in psychiatric disorders focused on the drug's potential to facilitate psychotherapeutic interventions. Given ketamine's effect on neuroplasticity, it is conceivable that it may provide a mechanism to enhance the effectiveness of psychotherapeutic treatments. An open label trial of 16 patients receiving a 4-session course of ketamine treatment suggests that CBT may be able to sustain the improvements achieved with ketamine treatment (Wilkinson et al., 2017b). Case reports have also shown potential benefits of CBT and ketamine in other disorders such as obsessive compulsive disorder (Adams et al., 2017). Several studies examining the effectiveness and possible mechanisms underlying this treatment combination are currently underway.

Inhibiting

Lamotrigine

In preclinical studies, lamotrigine has been shown to decrease the craving for ketamine abuse (Lee et al., 2019). In humans, it has been suggested to decrease perceptual changes in healthy subjects (Anand et al., 2000), and in patients undergoing surgery (Maheshwari et al., 2021), though this findings was not present in another study involving patients with TRD (Mathew et al., 2010). Given these limited findings, adjustment of lamotrigine dosing in patients receiving ketamine should be considered only after careful assessment of risks and benefits of such modification.

Benzodiazepines

Animal studies have suggested that inhibition of NMDA receptors on GABAergic interneurons may be a critical component in ketamine's mechanism of antidepressant action. Additionally, glucose utilization increases observed by ketamine in the limbic system may be attenuated or prevented by administration of benzodiazepines such as diazepam (Eintrei et al., 1999), suggesting that concomitant use of benzodiazepines may attenuate or inhibit the antidepressant response to ketamine. In humans, a secondary analysis of the results of two studies involving 47 patients, benzodiazepines were associated with nonresponse and significantly worse outcomes on days 3 and 7 (Andrashko et al., 2020). Another post-hoc analysis of 10 patients with TRD showed a difference in the dose of benzodiazepines in responder versus nonresponder groups (Frye et al., 2015), although the same study did not find a significant difference when comparing GABAergic medication users versus nonusers. The larger phase 3 studies with esketamine did not report any specific attenuating effects of benzodiazepines although use in these studies was constrained. Given the theoretical concerns and preliminary findings, it is advisable to use benzodiazepines with caution in patients undergoing ketamine treatment.

Naltrexone

A double-blind cross over study of patients with TRD treated with ketamine and naltrexone/placebo found a significant reduction of antidepressant effects in an interim analysis of 14 participants with no associated change in dissociative effects (Williams et al., 2018). Notwithstanding the potential importance of this early finding, other reports have shown no interference in the antidepressant effects of ketamine by naltrexone, buprenorphine or methadone (Yoon et al., 2019; Marton et al., 2019). At this point it appears prudent to closely evaluate the potential benefits and risks, including the possibility of reduced response rates, if planning on starting ketamine or esketamine treatments in patients using mu opioid receptor antagonists.

Drug-drug and drug-food pharmacokinetic effects

Ketamine is a substrate of cytochromes CYP2B6 and CYP3A4, and as such can be affected by most inhibitors of these cytochromes such as grapefruit juice and clarithromycin (CYP3A4) and ticlopidine (CYP2B6), and inducers such as rifampin or St. John's Wart. Of note, due to the first-pass effect, the levels are more significantly affected if ketamine is administered orally (Peltoniemi et al., 2016).

Given these factors, it is important to consider the effects of inhibitors or inducers of these cytochrome (e.g., thioridazine (CYP2B6 inhibitor), carbamazepine (CYP3A4 and CYP2B6 inducer), and nefazodone (CYP3A4 inhibitor)) when administering ketamine.

Overall, given the limited data currently available in this area, it is best to make decisions on treatment augmentation or discontinuation of ongoing medications with care and by considering all factors relevant for each individual patient.

Safety

Personal

As noted, concerns over ketamine use in several countries had previously prompted several reviews of the drug by the WHO. Although the organization consistently recommended that ketamine should not be placed under increased international control which could limit access to the drug (World Health Organization, 2015), several issues related to the broader use of ketamine remain of concern. These concerns can be considered at the more basic physiological level or at the psychological level. Issues surrounding the broader use of the treatment could also be deliberated at the individual and societal levels.

Physiological

One of the more easily tractable and manageable side effects of ketamine and esketamine are effects on hemodynamic measures. Ketamine can cause an increase in pulse rate and blood pressure, which in the latter case seems to be dose-dependent (Szarmach et al., 2019; Riva-Posse et al., 2018). A recent report of 66 patients receiving 684 total ketamine infusions at 0.5 mg/kg over 40 min at Emory University found the greatest increases in blood pressure were measured at 30 min, averaging 3.28 mmHg systolic and 3.17 mmHg diastolic. Blood pressures returned to baseline during postinfusion monitoring. Of note, hypertensive patients at baseline exhibited higher blood pressure peaks during the infusions (Riva-Posse et al., 2018). A second study of 135 patients with unipolar and bipolar depression who received a total of 741 infusions under similar dosing conditions also reported maximum blood pressure and pulse values 30–40 min after initiating infusions. In this case, the largest mean systolic/diastolic blood pressure increases were 7.4/6.0 mmHg, and the largest mean pulse increase was 1.9 beats per min (Zhou et al., 2021). In pooled data from 97 patients receiving 205 doses of ketamine IV from three different studies, a substantial increase in blood pressure 19.6 systolic and 13.4 mm/Hg diastolic was observed (Wan et al., 2015). These findings, while quite variable, are generally in line with those reported in the dose testing study noted previously where blood pressure values were higher in the ketamine 0.5 and 1.0 mg/kg groups compared to the active placebo midazolam (Fava et al., 2020).

Much more comprehensive hemodynamic data have been collected with esketamine IN spray, including data from 1708 esketamine-treated adults with TRD in six trials initiating the treatment along with a new oral antidepressant medication (Doherty et al., 2020). In the two randomized studies of adults aged 18–65, each 4 weeks in duration with two treatments per week, the largest mean maximum systolic/diastolic postdose BP increases were 13.3 systolic and 6.1 diastolic with esketamine compared to 8.7 and 4.9 mmHg for placebo, and in a short-term study of adults $\geq$65 years of age the increase was 16.0 systolic and 11.1 diastolic for esketamine compared to 9.5 systolic and 6.8 mmHg diastolic for placebo. Most importantly, cardiovascular adverse events were relatively rare and only three cardiovascular adverse events related to BP increase were reported as serious and severe. Similar to racemic ketamine infusions hemodynamic increases reached the maximum postdose value within ~40 min of esketamine dosing and returned to the predose range by ~1.5 h of dosing. Also similar to the ketamine studies, patients with underlying hypertension appeared to be at greatest risk of having significantly elevated blood pressures with treatment. No clinically relevant effect on ECG parameters was observed.

Based on these findings, vital sign monitoring before, during, and for a short period of time after dosing is highly recommended to assure a safe treatment and a return to baseline values before discharge. It is important to note that the large majority of the studies contained in the literature carefully screened and excluded patients from participation if they had underlying medical conditions. In light of this, it is especially important to complete a thorough evaluation process that includes screening for cardiovascular risk factors prior to initiating treatment with either ketamine or esketamine.

Another physiological side effect reported mainly with long-term use or misuse of ketamine is ulcerative or interstitial cystitis. In the longer-term clinical studies with esketamine, no cases of interstitial cystitis were reported, including in patients followed up for a year after initiating treatment (SPRAVATO, 2020). There are not enough prospectively collected data related to interstitial cystitis with racemic ketamine to comment on its safety at this point; however, there is no clearly defined reason to believe that treatment at similar frequencies and resulting in similar plasma levels as reported in the esketamine studies should produce dramatically different effects.

The brain is the other major organ believed to be at risk with ongoing ketamine exposure (Morgan and Curran, 2006), and concerns of persistent cognitive impairment are warranted with repeated ketamine use. The limited data currently available with short- and extended-term racemic ketamine treatment of mood disorders suggest either no change or possibly even marginal improvement, but has not provided any evidence meaningful declines in cognitive processes (Wilkinson et al., 2018b; Shiroma et al., 2020; Diamond et al., 2014). Again, much more extensive and comprehensive data have been collected with regards to esketamine IN spray's effects on cognition. Similar to the existing data with racemic IV administration, the data suggest either no change or possibly even marginal improvement in most measures without any meaningful concerns, although there was note of a decline in reaction time in patients over the age of 65 (Wajs et al., 2020). In short, dosing with either racemic ketamine 0.5 mg/kg IV delivered over 40 min or esketamine IN at either 56 mg or 84 mg according to the published dosing regimens of up to thrice weekly for initiation with no more than weekly for extended time periods does not appear to be associated with any meaningful decline in cognitive function at this time. However, there are little or no data to suggest safety for more frequent dosing or higher doses, or for longer-term treatments. It is important to continue to maintain some level of surveillance to ensure safety.

Psychological

The acute effects of ketamine on cognition and perception are well documented (Krystal et al., 2005). Dissociation is one of the most common effects of ketamine provided at doses used for the treatment of depression (Short et al., 2018). Some level of dissociation is reported by the majority of patients receiving either ketamine or esketamine for the first time. However, the effect is usually mild to moderate and not distressing if the patient is appropriately prepared. The presence and severity of dissociation is usually assessed with a psychometric scale, most commonly the Clinician-Administered Dissociative States Scale (CADSS) and typically decreases with repeated treatments. Given the frequency of this side effect, specifically monitoring for resolution of any dissociative symptoms after treatment is advisable. The question of whether the dissociative effects are associated with the antidepressant properties of ketamine and esketamine treatment remains in debate (Ballard and Zarate, 2020). However, the dissociation does not appear to be highly correlated with the clinical improvement and especially not beyond the first treatment (Wajs et al., 2020).

The acute anxiogenic or dysphoric effects of ketamine also remain an issue of concern. Ketamine use as an anesthetic has been somewhat limited by patients reporting anxiety and dysphoria on awakening, sometimes considered part of an emergence delirium. It is not uncommon that patients can experience elevated levels of anxiety with the initial treatments with either ketamine or esketamine. However, experience suggests that education prior to the first treatment session, and having trained staff readily available to assure patients during the treatment markedly lowers the number of patients who terminate treatment due to this adverse event.

Yet another important concern surrounding ketamine and esketamine treatment relates to its effects on cognition and alertness is the impact it can have on patients' ability to drive or operate machinery. Two studies assessed the ability of patients to drive after receiving nasal esketamine. In one study, driving performance of subjects after single-dose and multidose treatment was similar to placebo after 18 and 6 h, respectively. In another study, the driving performance of subjects after a single 84 mg dose of esketamine was similar to placebo at 8 h, but two subjects discontinued the driving test after receiving nasal esketamine. Given these findings and until more information is available patients are recommended to avoid driving or operating heavy machinery until after a restful night of sleep (SPRAVATO, 2020).

Addiction

Raising concerns at both the individual and societal levels is the fact that ketamine is a known drug of abuse. Ketamine has been used as a recreational drug, mostly in "rave" settings, for several decades (Liu et al., 2016). It has also been shown to reinforce self-administration in preclinical studies (De Luca and Badiani, 2011). It is considered a controlled substance in many countries, and is classified as schedule III in the United States (Code of Federal Regulations, 1973) and as class B in United Kingdom (Legislation.gov.uk, 2014). Given these facts, the risks of abuse should always be considered when contemplating the use of ketamine for the treatment of TRD. At the same time, the clinical data to date including an open label 1-year safety study of nasal esketamine have not indicated any increase in substance use to be associated with the treatment under controlled conditions (Freedman et al., 2018).

Societal

At the societal level, drug diversion becomes a potential concern for the expansion of ketamine treatment. Due to its use as a recreational substance, ketamine has an underground market with a street value of between \$20 and \$25 per dose (STREETDRUGS, 2021). While esketamine treatment is closely monitored by a REMS program, and the majority of ketamine providers only administer the treatment in clinic, thus it is necessary to consider the risk of diversion in the use of ketamine or esketamine, and to monitor as appropriate for this phenomenon.

Implementation

Considering the various safety and societal concerns surrounding the use of the drug the USFDA approval for esketamine included a Risk Evaluation and Mitigation Strategy (REMS) program (FDA, 2019b). This program has specific requirements for healthcare settings and pharmacies providing nasal esketamine. Healthcare settings need to be certified by the program to ensure that esketamine is only dispensed and administered in the healthcare setting, that patients are enrolled in the REMS program, and that treatment is done under direct observation of a healthcare provider and patients are monitored for at least 2 h after treatment. Pharmacies require certification and must ensure that nasal esketamine is only made available to healthcare settings certified to provide this treatment (SPRAVATO, 2020). By extension, it would appear similar mitigation strategies should be used for racemic ketamine treatment until further data are presented.

The complexity and costs associated with the unique demands of ketamine/esketamine delivery and compliance with the risk mitigation strategies have dramatically limited access to the treatment. At present, third-party payer coverage for the treatments remains somewhat limited and can vary regionally. This has led to a situation whereby specialty clinics offering ketamine/esketamine and related services have appeared in a variety of unique settings, offering treatment predominantly to self-paying individuals (Wilkinson et al., 2017a). The specific issues related to this development have stirred controversy and led to some division in the field. However, along with other developments such as the use of Brexanolone treatment for postpartum depression and a growing number of neurostimulation therapeutics, the increasing use of ketamine and esketamine treatments in psychiatry signals a possible evolution within the field and the need for specialized training in these developing areas within psychiatry.

Interventional psychiatry is an emerging subspecialty of psychiatry that focuses on advanced diagnostic and treatment modalities to help patients with psychiatric disorders that have been resistant to "traditional" treatments such as oral medications and psychotherapy. With advancements in various technological and more medically based psychiatric treatment modalities in the past two decades, the need for clinicians with training and experience in implementing these treatments is rapidly becoming a necessity in the field of psychiatry. The modalities under this umbrella are generally treatments that require higher levels of care for diagnosis, administration or monitoring of treatment. Neuromodulation modalities such as electroconvulsive therapy (ECT), transcranial magnetic stimulation (TMS), deep brain stimulation (DBS) and vagal nerve stimulation (VNS) among others are the treatments most commonly considered to be in this category. However, other modalities such as ketamine, esketamine, and psychedelic treatments also seem appropriate, not only because they require a specialized setting for their administration, but also because they require further training and experience to be appropriately implemented. Interventional psychiatrists are not only able to offer these treatments, but perhaps more importantly are able to provide consultation and advice to patients and their primary psychiatric providers about the most appropriate modality for their needs.

Conclusion

As seen in the paragraphs above, it is difficult to draw a direct line neatly connecting the discovery process leading to the development of ketamine as an antidepressant, and ultimately the USFDA indication of esketamine for treatment-resistant depression and major depressive disorder associated with suicidal ideation. The hypotheses related to the treatment's mechanisms of action have evolved over time and continue to be refined. The initial focus on the use of ketamine as an antidepressant had a heavier focus on its dissociative and acute psychological effects, suggesting a capacity by ketamine to render the patient more open to psychological change. The later efforts, which are most closely associated with the final stages of clinical development, focused more on the underlying neurobiological mechanisms including ketamine's effects on glutamate neurotransmission, neuroplasticity, and neurocircuitry. While the exact mechanisms of ketamine's antidepressant action remain unknown and are probably multifactorial, there is a growing consensus related to its primary effects on glutamate neurotransmission contributing to these effects.

There is now overwhelming evidence of the short-term clinical benefits of treatment with both ketamine and esketamine. Large, longer-termed controlled studies examining the safety and efficacy of the treatment remains limited to esketamine IN treatment. Balancing the optimism generated by this new class of antidepressants offering rapid onset of action to previously treatment-resistant patients, and to patients experiencing suicidal ideation, are the unique complexities associated with the delivery of the treatment. Results of several ongoing studies will soon provide information regarding the efficacy and safety of other formulations and variants of ketamine and related compounds.

References

Abdallah, C.G., et al., 2021. A robust and reproducible connectome fingerprint of ketamine is highly associated with the connectomic signature of antidepressants. Neuropsychopharmacology 46 (2), 478–485.

Adams, T.G., Bloch, M.H., Pittenger, C., 2017. Intranasal ketamine and cognitive-behavioral therapy for treatment-refractory obsessive-compulsive disorder. J. Clin. Psychopharmacol. 37 (2), 269–271.

Ajub, E., Lacerda, A.L.T., 2018. Efficacy of esketamine in the treatment of depression with psychotic features: a case series. Biol. Psychiatry 83 (1), e15–e16.

Anand, A., et al., 2000. Attenuation of the neuropsychiatric effects of ketamine with lamotrigine: support for hyperglutamatergic effects of N-methyl-D-aspartate receptor antagonists. Arch. Gen. Psychiatry 57 (3), 270–276.

Andrashko, V., et al., 2020. The antidepressant effect of ketamine is dampened by concomitant benzodiazepine medication. Front. Psych. 11, 844.

Bahji, A., Vazquez, G.H., Zarate Jr., C.A., 2021. Comparative efficacy of racemic ketamine and esketamine for depression: a systematic review and meta-analysis. J. Affect. Disord. 278, 542–555.

Ballard, E.D., Zarate Jr., C.A., 2020. The role of dissociation in ketamine's antidepressant effects. Nat. Commun. 11 (1), 6431.

Ballard, E.D., et al., 2014. Improvement in suicidal ideation after ketamine infusion: relationship to reductions in depression and anxiety. J. Psychiatr. Res. 58, 161–166.

Bartova, L., et al., 2018. Rapid antidepressant effect of S-ketamine in schizophrenia. Eur. Neuropsychopharmacol. 28 (8), 980–982.

Berman, R.M., et al., 2000. Antidepressant effects of ketamine in depressed patients. Biol. Psychiatry 47 (4), 351–354.

Best, S.R., 2014. Combined ketamine/transcranial magnetic stimulation treatment of severe depression in bipolar I disorder. J. ECT 30 (4), e50–e51.

Best, S.R., Griffin, B., 2015. Combination therapy utilizing ketamine and transcranial magnetic stimulation for treatment-resistant depression: a case report. Int. J. Neurosci. 125 (3), 232–234.

Best, S.R., Griffin, B.P., Pavel, D.G., 2015. Ketamine and transcranial magnetic stimulation treatment for bipolar II disorder: a case report. J .Med. Case Rep. 9, 73.

Beurel, E., Song, L., Jope, R.S., 2011. Inhibition of glycogen synthase kinase-3 is necessary for the rapid antidepressant effect of ketamine in mice. Mol. Psychiatry 16 (11), 1068–1070.

Canuso, C.M., et al., 2018. Efficacy and safety of intranasal esketamine for the rapid reduction of symptoms of depression and suicidality in patients at imminent risk for suicide: results of a double-blind, randomized, placebo-controlled study. Am. J. Psychiatry 175 (7), 620–630.

Chen, G., et al., 1959. The pharmacology of 1-(1-phenylcyclohexyl) piperidine-HCl. J. Pharmacol. Exp. Ther. 127, 241–250.

Chowdhury, G.M., et al., 2017. Transiently increased glutamate cycling in rat PFC is associated with rapid onset of antidepressant-like effects. Mol. Psychiatry 22 (1), 120–126.

Code of Federal Regulations, 1973. Available from: https://ecfr.federalregister.gov/current/title-21/chapter-II/part-1308. (date of access is 4/21/2021).

Costi, S., et al., 2019. Lithium continuation therapy following ketamine in patients with treatment resistant unipolar depression: a randomized controlled trial. Neuropsychopharmacology 44 (10), 1812–1819.

Cui, W., et al., 2019. Crosstalk between inflammation and glutamate system in depression: signaling pathway and molecular biomarkers for ketamine's antidepressant effect. Mol. Neurobiol. 56 (5), 3484–3500.

da Frota Ribeiro, C.M., et al., 2016. The use of ketamine for the treatment of depression in the context of psychotic symptoms: to the editor. Biol. Psychiatry 79 (9), e65–e66.

Dakwar, E., et al., 2019. A single ketamine infusion combined with mindfulness-based behavioral modification to treat cocaine dependence: a randomized clinical trial. Am. J. Psychiatry 176 (11), 923–930.

Daly, E.J., et al., 2019. Efficacy of esketamine nasal spray plus oral antidepressant treatment for relapse prevention in patients with treatment-resistant depression: a randomized clinical trial. JAMA Psychiat. 76 (9), 893–903.

De Luca, M.T., Badiani, A., 2011. Ketamine self-administration in the rat: evidence for a critical role of setting. Psychopharmacology (Berl) 214 (2), 549–556.

Di Lazzaro, V., et al., 2003. Ketamine increases human motor cortex excitability to transcranial magnetic stimulation. J. Physiol. 547 (Pt 2), 485–496.

Diamond, P.R., et al., 2014. Ketamine infusions for treatment resistant depression: a series of 28 patients treated weekly or twice weekly in an ECT clinic. J. Psychopharmacol. 28 (6), 536–544.

DiazGranados, N., et al., 2010. Rapid resolution of suicidal ideation after a single infusion of an N-methyl-D-aspartate antagonist in patients with treatment-resistant major depressive disorder. J. Clin. Psychiatry 71 (12), 1605–1611.

Doherty, T., et al., 2020. Cardiac safety of esketamine nasal spray in treatment-resistant depression: results from the clinical development program. CNS Drugs 34 (3), 299–310.

Domino, E.F., 2010. Taming the ketamine tiger. 1965. Anesthesiology 113 (3), 678–684.

Domino, E.F., Chodoff, P., Corssen, G., 1965. Pharmacologic effects of Ci-581, a new dissociative anesthetic, in man. Clin. Pharmacol. Ther. 6, 279–291.

Drevets, W.C., et al., 2021. Comments to Drs. Bahji, Vazquez, and Zarate. J. Affect. Disord. 283, 262–264.

Drugs@FDA, FDA Approved Drugs—Ketamine. Available from: https://www.accessdata.fda.gov/scripts/cder/daf/index.cfm?event=overview.process&ApplNo=016812.

Duman, R.S., Aghajanian, G.K., 2012. Synaptic dysfunction in depression: potential therapeutic targets. Science 338 (6103), 68–72.

Duman, R.S., et al., 2016. Synaptic plasticity and depression: new insights from stress and rapid-acting antidepressants. Nat. Med. 22 (3), 238–249.

Eintrei, C., Sokoloff, L., Smith, C.B., 1999. Effects of diazepam and ketamine administered individually or in combination on regional rates of glucose utilization in rat brain. Br. J. Anaesth. 82 (4), 596–602.

European Medicines Agency, 2018. Article 57 Database. Available from: https://www.ema.europa.eu/en/human-regulatory/post-authorisation/data-medicines-iso-idmp-standards/public-data-article-57-database. (accessed date 05/01/2021).

Fava, M., et al., 2020. Double-blind, placebo-controlled, dose-ranging trial of intravenous ketamine as adjunctive therapy in treatment-resistant depression (TRD). Mol. Psychiatry 25 (7), 1592–1603.

FDA, 2019a. Approval Letter for NDA 211243 for treatment of treatment-resistant depression. Available from: https://www.accessdata.fda.gov/drugsatfda_docs/appletter/2019/211243Orig1s000ltr.pdf. (accessed date 12/19/2020).

FDA, 2019b. Approved Risk Evaluation and Mitigation Strategies (REMS)—Spravato (Esketamine). Available from: https://www.accessdata.fda.gov/scripts/cder/rems/index.cfm?event=IndvRemsDetails.page&REMS=386. (accessed date 11/11/2020).

FDA, 2020. Approval Letter for NDA 211243/S-004 for Treatment of Depressive Symptoms in Adults With MDD With Acute Suicidal Ideation or Behavior. Available from: https://www.accessdata.fda.gov/drugsatfda_docs/appletter/2020/211243Orig1s004ltr.pdf. (accessed date 12/19/2020).

Feder, A., et al., 2021. A randomized controlled trial of repeated ketamine administration for chronic posttraumatic stress disorder. Am. J. Psychiatry 178 (2), 193–202.

Fedgchin, M., et al., 2019. Efficacy and safety of fixed-dose esketamine nasal spray combined with a new oral antidepressant in treatment-resistant depression: results of a randomized, double-blind, active-controlled study (TRANSFORM-1). Int. J. Neuropsychopharmacol. 22 (10), 616–630.

Freedman, R., et al., 2018. Can a framework be established for the safe use of ketamine? Am. J. Psychiatry 175 (7), 587–589.

Frye, M.A., Blier, P., Tye, S.J., 2015. Concomitant benzodiazepine use attenuates ketamine response: implications for large scale study design and clinical development. J. Clin. Psychopharmacol. 35 (3), 334–336.

Fu, D.J., et al., 2020. Esketamine nasal spray for rapid reduction of major depressive disorder symptoms in patients who have active suicidal ideation with intent: double-blind, randomized study (ASPIRE I). J. Clin. Psychiatry 81 (3), 19m13191.

Ghasemi, M., Raza, M., Dehpour, A.R., 2010. NMDA receptor antagonists augment antidepressant-like effects of lithium in the mouse forced swimming test. J. Psychopharmacol. 24 (4), 585–594.

Greifenstein, F.E., et al., 1958. A study of a 1-aryl cyclo hexyl amine for anesthesia. Anesth. Analg. 37 (5), 283–294.

Grunebaum, M.F., et al., 2018. Ketamine for rapid reduction of suicidal thoughts in major depression: a midazolam-controlled randomized clinical trial. Am. J. Psychiatry 175 (4), 327–335.

Ibrahim, L., et al., 2012. Course of improvement in depressive symptoms to a single intravenous infusion of ketamine vs add-on riluzole: results from a 4-week, double-blind, placebo-controlled study. Neuropsychopharmacology 37 (6), 1526–1533.

Ionescu, D.F., et al., 2020. Esketamine nasal spray for rapid reduction of depressive symptoms in patients with major depressive disorder who have active suicide ideation with intent: results of a phase 3, double-blind, randomized study (ASPIRE II). Int. J. Neuropsychopharmacol. 24 (1), 22–31.

Iro, C.M., et al., 2021. Repeated but not single administration of ketamine prolongs increases of the firing activity of norepinephrine and dopamine neurons. Int. J. Neuropsychopharmacol. 24 (7), 570–579.

Ivan Ezquerra-Romano, I., et al., 2018. Ketamine for the treatment of addiction: evidence and potential mechanisms. Neuropharmacology 142, 72–82.

Jones, J.L., et al., 2018. Efficacy of ketamine in the treatment of substance use disorders: a systematic review. Front. Psych. 9, 277.

Jope, R.S., 2003. Lithium and GSK-3: one inhibitor, two inhibitory actions, multiple outcomes. Trends Pharmacol. Sci. 24 (9), 441–443.

Khorramzadeh, E., Lotfy, A.O., 1973. The use of ketamine in psychiatry. Psychosomatics 14 (6), 344–346.

Kim, J.W., Monteggia, L.M., 2020. Increasing doses of ketamine curtail antidepressant responses and suppress associated synaptic signaling pathways. Behav. Brain Res. 380, 112378.

Kryst, J., et al., 2020. Efficacy of single and repeated administration of ketamine in unipolar and bipolar depression: a meta-analysis of randomized clinical trials. Pharmacol. Rep. 72 (3), 543–562.

Krystal, J.H., et al., 2005. Preliminary evidence of attenuation of the disruptive effects of the NMDA glutamate receptor antagonist, ketamine, on working memory by pretreatment with the group II metabotropic glutamate receptor agonist, LY354740, in healthy human subjects. Psychopharmacology (Berl) 179 (1), 303–309.

Krystal, J.H., et al., 2019. Ketamine: a paradigm shift for depression research and treatment. Neuron 101 (5), 774–778.

Leal, G.C., et al., 2020. Intravenous arketamine for treatment-resistant depression: open-label pilot study. Eur. Arch. Psychiatry Clin. Neurosci. 271 (3), 577–582.

Lee, M.Y., et al., 2019. Lamotrigine attenuates the motivation to self-administer ketamine and prevents cue- and prime-induced reinstatement of ketamine-seeking behavior in rats. Drug Alcohol Depend. 194, 257–263.

Legislation.gov.uk, 2014. Available from: https://www.legislation.gov.uk/uksi/2014/1106/made. (accessed date 04/21/2021).

Li, N., et al., 2010. mTOR-dependent synapse formation underlies the rapid antidepressant effects of NMDA antagonists. Science 329 (5994), 959–964.

Li, Q.S., et al., 2020. Genome-wide association study and polygenic risk score analysis of esketamine treatment response. Sci. Rep. 10 (1), 12649.

Liu, Y., et al., 2016. Ketamine abuse potential and use disorder. Brain Res. Bull. 126 (Pt 1), 68–73.

Lumsden, E.W., et al., 2019. Antidepressant-relevant concentrations of the ketamine metabolite (2R,6R)-hydroxynorketamine do not block NMDA receptor function. Proc. Natl. Acad. Sci. U. S. A. 116 (11), 5160–5169.

Maddox, V.H., Godefroi, E.F., Parcell, R.F., 1965. The synthesis of phencyclidine and other 1-arylcyclohexylamines. J. Med. Chem. 8, 230–235.

Maheshwari, K., et al., 2021. Lamotrigine for reducing ketamine-induced psychologic disturbances: a pilot randomized and blinded trial. J. Clin. Anesth. 68, 110074.

Marton, T., et al., 2019. Concurrent use of buprenorphine, methadone, or naltrexone does not inhibit ketamine's antidepressant activity. Biol. Psychiatry 85 (12), e75–e76.

Mathew, S.J., et al., 2010. Riluzole for relapse prevention following intravenous ketamine in treatment-resistant depression: a pilot randomized, placebo-controlled continuation trial. Int. J. Neuropsychopharmacol. 13 (1), 71–82.

McCloud, T.L., et al., 2015. Ketamine and other glutamate receptor modulators for depression in bipolar disorder in adults. Cochrane Database Syst. Rev. 9, CD011611.

McIntyre, R.S., et al., 2020. The effect of intravenous, intranasal, and oral ketamine in mood disorders: a meta-analysis. J. Affect. Disord. 276, 576–584.

McIntyre, R.S., et al., 2021. Synthesizing the evidence for ketamine and esketamine in treatment-resistant depression: an international expert opinion on the available evidence and implementation. Am. J. Psychiatry. p. appiajp202020081251.

Mion, G., 2017. History of anaesthesia: the ketamine story—past, present and future. Eur. J. Anaesthesiol. 34 (9), 571–575.

Moghaddam, B., et al., 1997. Activation of glutamatergic neurotransmission by ketamine: a novel step in the pathway from NMDA receptor blockade to dopaminergic and cognitive disruptions associated with the prefrontal cortex. J. Neurosci. 17 (8), 2921–2927.

Morgan, C.J., Curran, H.V., 2006. Acute and chronic effects of ketamine upon human memory: a review. Psychopharmacology (Berl) 188 (4), 408–424.

Muller, J., et al., 2016. Ketamine enantiomers in the rapid and sustained antidepressant effects. Ther. Adv. Psychopharm. 6 (3), 185–192.

Murrough, J.W., et al., 2015a. Neurocognitive effects of ketamine and association with antidepressant response in individuals with treatment-resistant depression: a randomized controlled trial. Neuropsychopharmacology 40 (5), 1084–1090.

Murrough, J.W., et al., 2015b. Ketamine for rapid reduction of suicidal ideation: a randomized controlled trial. Psychol. Med. 45 (16), 3571–3580.

Ochs-Ross, R., et al., 2020. Efficacy and safety of esketamine nasal spray plus an oral antidepressant in elderly patients with treatment-resistant depression-TRANSFORM-3. Am. J. Geriatr. Psychiatry 28 (2), 121–141.

Oye, I., Paulsen, O., Maurset, A., 1992. Effects of ketamine on sensory perception: evidence for a role of N-methyl-D-aspartate receptors. J. Pharmacol. Exp. Ther. 260 (3), 1209–1213.

Papakostas, G.I., et al., 2020. Efficacy of esketamine augmentation in major depressive disorder: a meta-analysis. J. Clin. Psychiatry 81 (4), 19r12889.

Peltoniemi, M.A., et al., 2016. Ketamine: a review of clinical pharmacokinetics and pharmacodynamics in anesthesia and pain therapy. Clin. Pharmacokinet. 55 (9), 1059–1077.

Popoli, M., et al., 2011. The stressed synapse: the impact of stress and glucocorticoids on glutamate transmission. Nat. Rev. Neurosci. 13 (1), 22–37.

Popova, V., et al., 2019. Efficacy and safety of flexibly dosed esketamine nasal spray combined with a newly initiated oral antidepressant in treatment-resistant depression: a randomized double-blind active-controlled study. Am. J. Psychiatry 176 (6), 428–438.

Price, R.B., Mathew, S.J., 2015. Does ketamine have anti-suicidal properties? Current status and future directions. CNS Drugs 29 (3), 181–188.

Price, R.B., et al., 2009. Effects of intravenous ketamine on explicit and implicit measures of suicidality in treatment-resistant depression. Biol. Psychiatry 66 (5), 522–526.

Price, R.B., et al., 2014. Effects of ketamine on explicit and implicit suicidal cognition: a randomized controlled trial in treatment-resistant depression. Depress. Anxiety 31 (4), 335–343.

Ravikrishnan, A., et al., 2018. Region-specific expression of NMDA receptor GluN2C subunit in parvalbumin-positive neurons and astrocytes: analysis of GluN2C expression using a novel reporter model. Neuroscience 380, 49–62.

Riva-Posse, P., et al., 2018. Blood pressure safety of subanesthetic ketamine for depression: a report on 684 infusions. J. Affect. Disord. 236, 291–297.

Rodriguez, C.I., et al., 2013. Randomized controlled crossover trial of ketamine in obsessive-compulsive disorder: proof-of-concept. Neuropsychopharmacology 38 (12), 2475–2483.

Rosenbaum, G., et al., 1959. Comparison of sernyl with other drugs: simulation of schizophrenic performance with sernyl, LSD-25, and amobarbital (amytal) sodium; I. Attention, motor function, and proprioception. AMA Arch. Gen. Psychiatry 1, 651–656.

Sanacora, G., et al., 2017. A consensus statement on the use of ketamine in the treatment of mood disorders. JAMA Psychiat. 74 (4), 399–405.

Shinohara, R., Aghajanian, G.K., Abdallah, C.G., 2020. Neurobiology of the rapid-acting antidepressant effects of ketamine: impact and opportunities. Biol. Psychiatry 90 (2), 85–95.

Shiroma, P.R., et al., 2020. Neurocognitive performance of repeated versus single intravenous subanesthetic ketamine in treatment resistant depression. J. Affect. Disord. 277, 470–477.

Short, B., et al., 2018. Side-effects associated with ketamine use in depression: a systematic review. Lancet Psychiatry 5 (1), 65–78.

Singh, J.B., et al., 2016a. Intravenous esketamine in adult treatment-resistant depression: a double-blind, double-randomization, placebo-controlled study. Biol. Psychiatry 80 (6), 424–431.

Singh, J.B., et al., 2016b. A double-blind, randomized, placebo-controlled, dose-frequency study of intravenous ketamine in patients with treatment-resistant depression. Am. J. Psychiatry 173 (8), 816–826.

SPRAVATO, 2020. Prescribing Information. Available from: https://www.janssenlabels.com/package-insert/product-monograph/prescribing-information/SPRAVATO-pi.pdf. (accessed date 04/21/2021).

STREETDRUGS, 2021. Ketamine. 04/21/2021; Available from: https://streetdrugs.org/ketamine/.

Szarmach, J., et al., 2019. Short-term ketamine administration in treatment-resistant depression: focus on cardiovascular safety. Psychiatr. Danub. 31 (Suppl. 3), 585–590.

Trimmel, H., et al., 2018. S(+)-ketamine: current trends in emergency and intensive care medicine. Wien. Klin. Wochenschr. 130 (9–10), 356–366.

Trullas, R., Skolnick, P., 1990. Functional antagonists at the NMDA receptor complex exhibit antidepressant actions. Eur. J. Pharmacol. 185 (1), 1–10.

Vollenweider, F.X., et al., 1997. Differential psychopathology and patterns of cerebral glucose utilisation produced by (S)- and (R)-ketamine in healthy volunteers using positron emission tomography (PET). Eur. Neuropsychopharmacol. 7 (1), 25–38.

Wajs, E., et al., 2020. Esketamine nasal spray plus oral antidepressant in patients with treatment-resistant depression: assessment of long-term safety in a phase 3, open-label study (SUSTAIN-2). J. Clin. Psychiatry 81 (3), 19m12891.

Wan, L.B., et al., 2015. Ketamine safety and tolerability in clinical trials for treatment-resistant depression. J. Clin. Psychiatry 76 (3), 247–252.

Wei, Y., Chang, L., Hashimoto, K., 2021. Molecular mechanisms underlying the antidepressant actions of arketamine: beyond the NMDA receptor. Mol. Psychiatry. https://doi.org/10.1038/s41380-021-01121-1.

White, P.F., Way, W.L., Trevor, A.J., 1982. Ketamine—its pharmacology and therapeutic uses. Anesthesiology 56 (2), 119–136.

White, P.F., et al., 1985. Comparative pharmacology of the ketamine isomers. Studies in volunteers. Br. J. Anaesth. 57 (2), 197–203.

Wilkinson, S.T., et al., 2017a. A survey of the clinical, off-label use of ketamine as a treatment for psychiatric disorders. Am. J. Psychiatry 174 (7), 695–696.

Wilkinson, S.T., et al., 2017b. Cognitive behavior therapy may Sustain antidepressant effects of intravenous ketamine in treatment-resistant depression. Psychother. Psychosom. 86 (3), 162–167.

Wilkinson, S.T., et al., 2018a. The effect of a single dose of intravenous ketamine on suicidal ideation: a systematic review and individual participant data meta-analysis. Am. J. Psychiatry 175 (2), 150–158.

Wilkinson, S.T., et al., 2018b. Acute and longer-term outcomes using ketamine as a clinical treatment at the yale psychiatric hospital. J. Clin. Psychiatry 79 (4).

Williams, N.R., et al., 2018. Attenuation of antidepressant effects of ketamine by opioid receptor antagonism. Am. J. Psychiatry 175 (12), 1205–1215.

Witt, K., et al., 2020. Ketamine for suicidal ideation in adults with psychiatric disorders: a systematic review and meta-analysis of treatment trials. Aust. N. Z. J. Psychiatry 54 (1), 29–45.

World Health Organization, 2015. Essential Medicines and Health Products. Available from: https://www.who.int/medicines/access/controlled-substances/recommends_against_ick/en/. (accessed date 04/21/2021).

Xu, A.J., et al., 2015. Lithium and valproate levels do not correlate with ketamine's antidepressant efficacy in treatment-resistant bipolar depression. Neural Plast. 2015, 858251.

Yamaguchi, J.I., et al., 2018. (2R,6R)-Hydroxynorketamine is not essential for the antidepressant actions of (R)-ketamine in mice. Neuropsychopharmacology 43 (9), 1900–1907.

Yang, C., et al., 2015. R-Ketamine: a rapid-onset and sustained antidepressant without psychotomimetic side effects. Transl. Psychiatry 5, e632.

Yoon, G., Petrakis, I.L., Krystal, J.H., 2019. Association of combined naltrexone and ketamine with depressive symptoms in a case series of patients with depression and alcohol use disorder. JAMA Psychiat. 76 (3), 337–338.

Zanos, P., et al., 2016. NMDAR inhibition-independent antidepressant actions of ketamine metabolites. Nature 533 (7604), 481–486.

Zanos, P., et al., 2018. Ketamine and ketamine metabolite pharmacology: insights into therapeutic mechanisms. Pharmacol. Rev. 70 (3), 621–660.

Zhou, Y.L., et al., 2021. Cardiovascular effects of repeated subanaesthetic ketamine infusion in depression. J. Psychopharmacol. 35 (2), 159–167.

Chapter 14

Psychedelic agents for treatment-resistant depression

Natalie Gukasyan[a] and Roland R. Griffiths[b]

[a]*Department of Psychiatry and Behavioral Sciences, Center for Psychedelic and Consciousness Research, Johns Hopkins University School of Medicine, Baltimore, MD, United States,* [b]*Departments of Psychiatry and Neuroscience, Johns Hopkins University School of Medicine, Baltimore, MD, United States*

Introduction

Classic psychedelics are serotonergic agonists that can produce profound changes in perception, mood, and cognition. The past two decades have been marked by a renewal of interest in these compounds for their potential to produce rapid and long-lasting improvements in a variety of mental health conditions including treatment-resistant depression (TRD). Psychedelics have been used sacramentally in non-Western cultures for thousands of years but did not receive widespread attention or recognition in the West until the 1940s when the psychoactive properties of lysergic acid diethylamide (LSD) were recognized by Swiss chemist Albert Hofmann. A first wave of clinical research in the 1940s–1970s suggested that these compounds may have had considerable therapeutic potential, but studies were halted in response to political and cultural pressures resulting from widespread use of psychedelics outside of clinics and laboratories (Hofmann, 1983). With the passage of the Controlled Substances Act in the United States, psychedelics were relegated to Schedule I, a category reserved for substances with no recognized clinical benefit and substantial associated risk. Thus, for several decades following this shutdown, psychedelics were known primarily as illicit drugs of abuse with little attention paid to previously documented clinical benefits.

After a decades-long hiatus in clinical experimentation involving these substances, a second wave of research began in the late 1990s with increasing numbers of high-quality human and laboratory studies being published. Phase I studies demonstrated that psychedelics had a favorable safety profile, and growing numbers of phase II studies are demonstrating therapeutic potential for various mental health conditions including cancer-related psychosocial distress, alcohol and nicotine use disorders, major depressive disorder (MDD), and TRD (Bogenschutz et al., 2015; Carhart-Harris et al., 2018; Davis et al., 2020b; Griffiths et al., 2016; Johnson et al., 2014).

This chapter will focus on the more recent literature with classic psychedelic compounds, whose effects are primarily mediated by serotonin 2A (5-HT_{2A}) receptor agonism or partial agonism. A variety of terms have been used to refer to these substances since their first description in medical literature. Initially called "psychotomimetic" agents, this term fell out of favor when it became apparent that their effects could not accurately be described as psychotic phenomena. "Hallucinogen" later gained popularity and is still widely used today including in the Diagnostic and Statistical Manual (DSM). However, this term suggests that the primary effect of these drugs is to produce true hallucinations, when such phenomena are in fact uncommon with most drugs in this class. An additional weakness of this label is that it is often used to refer to a broader category of psychoactive drugs that work by other mechanisms, such as 3,4-Methylenedioxymethamphetamine (MDMA), and dissociatives like ketamine and phencyclidine. The term "psychedelic" was coined by English psychiatrist Humphrey Osmond in the 1950s and is currently preferred among researchers working with these compounds (Osmond, 1957). The Greek root *delos*, meaning "to manifest," is thought to more adequately capture these drugs' potential "mind-manifesting" or therapeutic effects.

A number of key differences set psychedelic-assisted treatment apart from most currently available psychiatric treatments. Unlike first-line pharmacotherapies for MDD, which can take weeks of continuous use to produce clinically meaningful improvement, psychedelics appear to be capable of improving symptoms rapidly, as soon as one day after a single drug administration. Large effect sizes are apparent after just one or two doses and many patients appear to experience sustained improvement for months without the need for repeated drug administration. Additionally, treatment with psychedelics is typically delivered under the supervision of two therapists or "facilitators" with substantial adjunctive psychotherapy

Managing Treatment-Resistant Depression. https://doi.org/10.1016/B978-0-12-824067-0.00014-1

before and after drug administration (Johnson et al., 2008). This model is thought to be necessary for safety and efficacy, especially as clinical research with these drugs has shifted from healthy volunteers to clinical populations. This hybrid model contrasts with the typical delivery of care in the United States, where pharmacotherapy and psychotherapy have been gradually siloed over the last several decades. It has also posed challenges to conducting "gold-standard" placebo-controlled clinical trials with these drugs.

Though the use of these substances remains prohibited outside of laboratory settings in most of North America and Europe, multiple groups are currently conducting phase II and III studies to determine whether psychedelics can be a safe and effective treatment for MDD. Psychedelics are of particular interest in the possible treatment of TRD given that their mechanisms of action are distinct from existing treatments. In this chapter, we will review the chemistry, mechanisms of action, and recent clinical trial data on these compounds with special focus on their applications in individuals with TRD.

Chemistry

Most psychedelics relevant to our discussion of TRD are classified as tryptamines, which can be identified by their indole ring and structural similarity to serotonin and its precursor tryptophan. Naturally occurring tryptamines include *N,N*-dimethyltryptamine (*N,N*-DMT), 5-methoxy-*N,N*-dimethyltryptamine (5-MeO-DMT), psilocybin, and psilocybin's active metabolite psilocin. The synthetic ergoline LSD is also a tryptamine psychedelic. The second category of classic psychedelics is that of phenethylamines, of which mescaline is the naturally occurring exemplar.

Early clinical research focused principally on LSD, whose psychoactive properties were accidentally discovered in 1943 by Swiss chemist Albert Hoffman. In the process of attempting to isolate clinically useful compounds from the ergot fungus, lysergic acid was discovered (Hofmann, 1983). Hofmann developed a synthetic process that allowed him to create novel ergolines by combining lysergic acid with a variety of other compounds. His 25th synthetic compound, LSD-25, was made with diethylamine. LSD-25 did not immediately appear to have any useful effects in animal models, and sat untouched for several years until Hoffman unintentionally ingested it and discovered its striking psychoactive effects. It was hypothesized that LSD's effect profile may have had phenomenological similarity to psychosis and thus could be of interest to psychiatric researchers and clinicians. Sandoz Laboratories (now Novartis), subsequently distributed samples of LSD to clinicians and researchers worldwide under the trade name Delysid.

Most contemporary clinical research with psychedelics has used tryptamines other than LSD. Psilocybin has been most commonly used, and though it occurs naturally in the *Psilocybe* genus of mushrooms (Schultes and Hofmann, 1992), most research with psilocybin is conducted using synthetically prepared drug. *N,N*-DMT also occurs naturally in small concentrations in many plants and animals. It is the principal constituent of ayahuasca, a plant decoction that is used ceremonially in cultures in the Amazon Basin (Luna, 2011). Ayahuasca is most commonly prepared using a combination of *Psychotria viridis*, which contains *N,N*-DMT, and the *Banisteriopsis caapi* vine, containing harmala alkaloids. When taken alone, orally ingested *N,N*-DMT has no effect due to rapid metabolism by peripheral monoamine oxidase (MAO). Harmala alkaloids have an MAO-inhibiting effect and when taken together with *N,N*-DMT allow it to pass through peripheral circulation and exert effects centrally. Though ayahuasca has been studied for its therapeutic potential in MDD and TRD (Osório et al., 2015; Palhano-Fontes et al., 2019), it is an unlikely candidate for clinical use in the West due to the challenges of standardizing the dosage of a heterogeneous plant brew.

In some contexts, compounds like MDMA and glutamatergic dissociatives like ketamine and dextromethorphan have been discussed for their "psychedelic" effects, but due to their differing mechanisms of action, these will be excluded from this discussion. It is worthwhile noting, however, that a phase III trial of MDMA for the treatment of PTSD is currently ongoing and uses an integrated psychotherapy approach similar to that which is commonly used with classic psychedelics (Clinicaltrials.gov identifier NCT04077437).

Pharmacokinetics

Psychedelic compounds vary in their pharmacokinetic profiles. Here we will briefly review the pharmacokinetics of the three psychedelic compounds most commonly encountered in clinical psychedelic research.

Psilocybin

Following oral administration, psilocybin is absorbed in the stomach and the intestinal tract, where it is rapidly dephosphorylated by alkaline phosphatases to its active metabolite, psilocin. Onset of effects occurs by 30 min following moderate oral doses (0.28–0.43 mg/kg), with peak effects occurring around 2–3 h, and total duration of effects lasting somewhat more than 6 h (Griffiths et al., 2011). Noticeable subjective effects occur at plasma concentrations of about 4–6 μg/mL (Passie et al., 2002). Laboratory studies have demonstrated that doses of up to 0.6 mg/kg have been well tolerated in healthy adults (Brown et al., 2017). The LD_{50} of psilocybin is estimated to be on the order of grams per kilogram in humans, indicating a very large therapeutic index (Passie et al., 2002). Psilocin is renally excreted with a terminal elimination half-life of roughly 3 h (Hasler et al., 2002).

LSD

LSD has good oral bioavailability. Following a typical dose (100–250 μg), onset of subjective effects occurs within 30–45 min and peak effects are reached within 1–2.5 h. Total duration of effects is about 9–12 h (Holze et al., 2020). One study with a small number of subjects found oral absorption to be somewhat decreased when taken with a full meal (Upshall and Wailling, 1972). The longer duration of action makes LSD somewhat less practical for clinical use when compared to psilocybin. LSD undergoes hepatic metabolism and renal excretion of metabolites. LSD has a half-life of roughly 3.6 h (Dolder et al., 2017). LSD binds 5-HT_{2A} and 5-HT_{2B} receptors for an extended period of time, which explains its protracted period of effects despite its short half-life (Wacker et al., 2017). Similar to psilocybin, the LD_{50} of LSD is orders of magnitude larger than the effective dose. A case series of 8 individuals who accidentally took massive overdoses of LSD after mistaking the substance for cocaine found that the patients suffered from coma, hyperthermia, and gastric bleeding, though all 8 survived without long-term consequences despite having reached serum LSD concentrations of 1000–7000 μg/100 mL (Klock et al., 1974).

N,N-DMT

As noted elsewhere, *N,N*-DMT has limited oral bioavailability due to rapid breakdown by gastrointestinal MAO. However, parenteral administration, usually achieved by smoking or intravenous injection, readily produces strong but short-lived psychedelic effects. Intravenous administration at 0.2–0.4 mg/kg produces almost immediate onset of intense perceptual effects that resolve within 30 min (Strassman et al., 1994). In addition to MAO metabolism, parenterally administered *N,N*-DMT undergoes hepatic CYP-based metabolism, which is less rapid (Riba et al., 2015). *N,N*-DMT is the chief psychedelic constituent of the plant brew ayahuasca. Riba et al. (2003) have described the pharmacology of ayahuasca, which contains harmaline, harmine, and tetrahydroharmine—compounds that inhibit MAO and thus allow *N,N*-DMT to be absorbed and to exert effects centrally. Because different ayahuasca preparations vary in their composition, pharmacokinetics may be variable between batches. Typical doses described in research have ranged from 0.3 to 1 mg/kg of *N,N*-DMT with varying amounts of MAO inhibiting compounds (Palhano-Fontes et al., 2019; Riba et al., 2001). Initial effects are noticeable at 30–45 min after oral administration of ayahuasca, with peak effects occurring 1.5–2 h after ingestion (Riba et al., 2003). Effects gradually taper until about 6 h after drug ingestion.

Mechanisms of action

Research from a variety of disciplines has informed the current understanding of the mechanisms of action of psychedelics. While understanding of the biological mechanisms of acute psychedelic effects has improved substantially in the past two decades, the mechanisms of long-term therapeutic effects remain less clear. For example, multiple lines of evidence have demonstrated that 5-HT_{2A}–mediated signaling is crucial for the acute subjective effects of psilocybin. However, we do not yet know whether psychedelics directly induce longer-term changes in serotonergic functioning or whether such changes are associated with clinical improvement weeks or months after intervention. We also do not yet know whether subjective experience of the acute drug state is necessary for therapeutic efficacy, or if subjective effects are merely an epiphenomenon of some neurobiological mechanism that is actually driving clinical improvement. A second key point is that psychedelics are typically delivered together with a psychotherapeutic intervention. Contextual factors or psychological "set and setting" have long been believed to be important for treatment efficacy, though empirical study of this idea remains challenging. Both psychological and neurobiological factors are likely at play in the immediate and sustained treatment effects, and it is likely that these are synergistic.

Biological mechanisms of action

5-HT_{2A} receptor agonism is central to the effects of psychedelics. These receptors are present throughout the brain but exist in high concentrations in the claustrum and layers III and V of many cortical areas, especially the frontal and temporal lobes (Pazos et al., 1987; Wong et al., 1987). Low concentrations are found in lower structures like the basal ganglia and amygdala. Blocking 5-HT_{2A} receptors with a selective antagonist such as ketanserin abolishes subjective effects of classic psychedelics in humans (Vollenweider et al., 1998). Furthermore, binding affinity for 5-HT_{2A} receptors is correlated with the intensity of observed drug effects (Glennon et al., 1984). Preclinical evidence suggests that downstream glutamatergic signaling at mGluR2/3 receptors secondary to 5-HT_{2A} agonism is also necessary for the action of psychedelics (Moreno et al., 2011, 2013). Blocking or knocking out these receptors effectively negates observed psychedelic effects. It is notable that 5-HT_{2A} receptors are also present in large quantities elsewhere in the body including the peripheral nervous system, the cardiovascular system, and in platelets (Nagatomo et al., 2004; Tokunaga et al., 1998). The effect of psychedelics in other tissues is an ongoing area of study.

Individual psychedelics vary in their binding affinity to other receptors, though these are generally thought to be less critical in the production of drug effects than 5-HT_{2A} receptors. For example, psilocin binds to 5-HT_{2C}, 5-HT_{1A}, and 5-HT_{1B} receptors (Nichols, 2004). LSD on the other hand has partial agonism at 5-HT_{2A} receptor, and also binds 5-HT_{1A}, 5-HT_{2C}, 5-HT_{2B}, and dopamine D1 and D2 receptors (Watts et al., 1995). *N,N*-DMT, which is found in minute quantities in the human body, acts as an endogenous sigma-1 agonist in addition to its 5-HT agonism (Fontanilla et al., 2009). The exact role of these receptors in the action of psychedelics is still being determined. 5-HT_{1A} and 5-HT_{2C} in particular may be of particular importance (Nichols, 2004).

Changes in serotonin and glutamate signaling may explain the clinical efficacy of psychedelics for major depressive disorder. Similar to the proposed mechanism of action of typical serotonergic antidepressants, improvements in mood may be mediated by direct effects of psychedelics at 5-HT receptors. 5-HT_{2A} agonism leads to receptor desensitization and internalization (Gray and Roth, 2001; Hanley and Hensler, 2002). Downstream glutamate signaling may also produce antidepressant effects in a manner similar to glutamatergic antidepressants such as ketamine (Vollenweider and Kometer, 2010). Furthermore, $5HT_{2A}$ and mGlu2/3 sites have interactive and potentially synergistic effects, suggesting that psychedelics may result in more powerful antidepressant effects than a drug that works by other mechanism alone (Celada et al., 2013).

Functional magnetic resonance imaging (fMRI) studies have revealed that psilocybin acutely decreases resting state functional connectivity in the default mode network (DMN), as well as areas of the brain involved in emotion processing including the amygdala and the anterior cingulate cortex (Carhart-Harris et al., 2012, 2014; Kraehenmann et al., 2015). Decreases in DMN connectivity appear to persist at least 2 days following psilocybin administration in healthy subjects (Smigielski et al., 2019). Conversely, Carhart-Harris et al. (2017) found that DMN connectivity was increased one week after psilocybin administration. Barrett and colleagues (Barrett et al., 2020) found that resting state functional connectivity increased 1 week after psilocybin administration in healthy subjects, and that this pattern persisted at the one-month follow-up timepoint. Whether classic psychedelics are unique in their ability to produce such changes in functional connectivity has been challenged by recent findings that hallucinogens that work by other mechanisms also seem to be able to produce such changes (Doss et al., 2020; Müller et al., 2020). The DMN has been implicated in the pathophysiology of MDD in that individuals with depression may exhibit abnormally high levels of connectivity in the DMN (Leibenluft and Pine, 2013). DMN hyperconnectivity, and low levels of connectivity between the DMN and the cognitive control network have been associated with higher response rates to MDD treatment (Dichter et al., 2015). DMN connectivity has been shown to normalize following transcranial magnetic stimulation treatment for MDD (Anderson et al., 2016).

Psychedelics have also been shown to affect amygdala function. Increased amygdala response to negative affective stimuli has been associated with depression. A recent fMRI study in 12 healthy volunteers demonstrated that amygdala function in response to affective stimuli varied over the course of 1 month following psilocybin administration (Barrett et al., 2020). At 1 week, amygdala response to facial affect stimuli were reduced compared to baseline. These findings were associated with increased positive affect as measured by the Profile of Mood States (POMS). Though amygdala response returned to baseline at 1 month, mean POMS remained elevated compared to baseline.

Psychedelic compounds have also been found to produce potent antiinflammatory effects, which may be a key mechanism of action for the improvement of TRD and other conditions involving inflammation. Although serotonin is known to have pro-inflammatory effects throughout the body, 5-HT_{2A} agonism by classic psychedelics such as LSD and (*R*)-2,5-dimethoxy-4-iodoamphetamine (DOI) has been shown to suppress TNFα expression and nuclear translocation of NF-κB in vitro, with similar findings in vivo (Nau Jr et al., 2013; Yu et al., 2008). Flanagan and Nichols (2018) hypothesize that these effects can be explained by functional selectivity, meaning that different agonists at the same receptor can produce different downstream effects. In the case of psychedelics, binding at 5-HT_{2A} may result in recruiting antiinflammatory pathways rather than pro-inflammatory pathways typically activated by serotonin. Inflammation is hypothesized as one pathophysiologic pathway contributing to TRD. Patients with TRD have been found to have elevated levels of inflammatory markers including interleukin-6 (IL-6), which may have promise as a marker of treatment response (and notably is also reduced by suppression of TNFα) (Yang et al., 2019).

Subjective experiences and psychological mechanisms of action

Other explanations for the immediate and long-term effects of psychedelics are related to the complex subjective experiences produced by these drugs, which some people interpret as deeply meaningful, psychologically insightful, emotionally rich, or spiritual in nature (Griffiths et al., 2006, 2008). This holds significance for TRD, which for some patients may be related to a sense of chronic adversity or demoralization, personality vulnerabilities, and other factors that would understandably be unresponsive to traditional forms of pharmacotherapy (Thase et al., 2001). A variety of instruments have been developed to systematically assess the complex phenomenology of psychedelic experience, including the Mystical Experience Questionnaire (MEQ-30), a validated measure reflecting several qualities of the mystical experience—unity, sacredness or preciousness, authenticity, deeply felt positive mood, transcendence of space or time, and ineffability (Barrett et al., 2015). Though some researchers regard the description of these experiences as mystical to be unscientific (Roseman et al., 2018) this description does not imply a supernatural explanation for these phenomena. Several studies have demonstrated dose-dependent relationships between psychedelics and measures of mystical qualities of the experience, with greater mystical qualities being correlated with better long-term outcomes (Yaden and Griffiths, 2020). Davis et al., (2020b) found that in a sample of 24 adults undergoing psilocybin-assisted psychotherapy for MDD, there was a moderate correlation ($r=0.41$, $p<0.5$) between changes in GRID-Hamilton Depression Rating Scale (GRID-HAMD) scores at 4 weeks and peak MEQ-30 score.

MacLean et al. (2011) demonstrated changes in the personality domain of openness that persisted at 14 months following psilocybin administration. Such effects have also been documented as a result of prolonged SSRI treatment, intensive psychotherapy, and contemplative practice, but had not previously been demonstrated secondary to an acute drug intervention (Costa Jr et al., 2005; Piedmont, 2001; Sahdra et al., 2011). The authors examined the relationship between mystical experiences and changes in openness, which revealed that individuals who had had a "complete" mystical experience (>60% of maximum score on each subscale of the States of Consciousness Questionnaire) were significantly more likely to show increases in trait openness assessed with a standardized personality inventory. Those who had an "incomplete" or absent mystical experience were not significantly different from baseline. Low trait openness has been identified as a risk factor for TRD (Takahashi et al., 2013). Erritzoe and colleagues Erritzoe and colleagues (2018) examined changes in personality structure following psilocybin administration in 20 patients with TRD and found that at 3-month follow-up, participants had decreases in neuroticism, and increases in extraversion and openness.

Other psychological changes may also mediate outcomes or otherwise be clinically relevant. Increases in cognitive flexibility have been demonstrated following psilocybin administration and had a mediating effect on long term benefits for mood and anxiety (Davis et al., 2020a). Increases in capacity for mindfulness secondary to ayahuasca have also been reported (Soler et al., 2016).

Clinical studies

Early evidence of efficacy of psychedelics in depressive disorders

The first wave of psychedelic research from the 1940s through 1970s generated numerous published accounts of treatment outcomes for depressive illness. Unfortunately, much of this literature was in the form of case reports or case series, and most clinical trials suffered from lack of methodological rigor (Savage et al., 1966). Further, generalizing those earlier results to the clinical entity we now know as MDD is challenging due to differing diagnostic procedures and terminology used by clinicians of that era. The first report of a psychedelic used to treat a depressive illness was published by Condrau in 1949 and described the use of gradually escalating doses of LSD (Condrau, 1949). Despite some improvement, Condrau concluded that the treatment did not demonstrate substantial efficacy. Subsequent reports of psychedelics for the treatment of so called "depressive," "neurotic," and "psychoneurotic" disorders were subsequently published by other researchers.

Rucker and colleagues completed a systematic review of the extensive literature from this period and summarized findings from 20 studies on psychedelic treatment effects in patients with depressive disorders (Rucker et al., 2016). The majority of studies used LSD, with a minority using mescaline and none reporting on findings with psilocybin. A variety of treatment approaches, dosing and dosing regimens were used. A notable distinction was made between "psycholytic" therapies, which employed low doses of psychedelic to facilitate regular psychotherapy sessions, vs. "psychedelic therapy," which resembled the high dose treatment used in contemporary studies. 19 of 21 studies reported on the number of patients who seemed to improve—of a combined 423 patients, 335 (79.2%) were reported to have improved. Unfortunately, calculating effect size from these studies is not possible because only one research group used a continuous measure of depressive severity. Whether improvement was sustained was also not apparent. Ascertaining the risk of adverse events from this body of literature was challenging due to inconsistencies in classifying and reporting such findings.

Fortunately, the second wave of psychedelic research has been characterized by much greater attention to methodological rigor. Initial studies in healthy subjects convincingly established that psychedelics could be administered safely in a laboratory setting (Carter et al., 2005; Hasler et al., 2004; Strassman et al., 1994). In 2006, Griffiths et al. demonstrated that a single dose of psilocybin in healthy volunteers produced mystical type experiences that were associated with increases in positive mood at the 2 month follow-up assessment (Griffiths et al., 2006). Three subsequent studies examined the effects of psilocybin in individuals with cancer-related psychological distress and demonstrated rapid and sustained decreases in depressive symptoms (Griffiths et al., 2016; Grob et al., 2011; Ross et al., 2016). Later studies demonstrated similar findings in patients with MDD and TRD (Carhart-Harris et al., 2016; Davis et al., 2020b).

The standard model of treatment

The majority of contemporary clinical studies using psychedelics have adhered to a similar format and set of safety guidelines (Johnson et al., 2008). Following screening, participants are paired with two clinicians, also called "facilitators," who are ideally both present for all clinical visits. This begins with one or more preparatory visits lasting a total of 4–8 h during which the participant and facilitators review the participant's biographical history, discuss potential effects of psychedelics, and build rapport. Drug administration is provided in a room decorated in a comfortable home-like manner (Fig. 14.1).

Contemporary clinical studies with psychedelics

Among the second wave of psychedelic research, the first evidence that psychedelics could have efficacy for depressive disorders came from studies in individuals suffering psychological distress secondary to advanced stage cancer. Grob et al. (2011) completed a small ($N=12$) randomized, placebo-controlled crossover study of psilocybin for the treatment of anxiety in patients with advanced-stage cancer. Beck Depression Inventory (BDI) differences following placebo and experimental treatment were not significantly different, but BDI scores dropped by nearly 30% at 1 month and reached significance at 6 months following intervention for the whole group. These results were later expanded upon in larger studies at Johns Hopkins and New York University (Griffiths et al., 2016; Ross et al., 2016). Griffiths et al. (2016) employed a randomized double-blind crossover design comparing the effects of very low dose (1 or 3 mg/70 kg) versus moderate- to high-dose (22 or 30 mg/70 kg) psilocybin in individuals with depression and anxiety symptoms in the setting of terminal cancer ($N=51$). Eighty-four percent of participants met criteria for a treatment response (≥ 50 reduction on GRID-HAMD compared to baseline) following the high dose session. Sustained response was present in 78% of participants at 6 months,

FIG. 14.1 Typical set-up for a psychedelic-assisted treatment room. A standard session room used for drug administration. A patient lies on the couch with eyeshades and headphones. Two facilitators are at bedside. *(From Johnson, M., Richards, W., Griffiths, R. (2008). Human hallucinogen research: guidelines for safety. J. Psychopharmacol. 22(6), 603–620. https://journals.sagepub.com/doi/abs/10.1177/0269881108093587.)*

with 65% being in remission. Ross et al. (2016) completed a double-blind, placebo-controlled crossover study ($N=29$) comparing the effects of a 0.3 mg/kg (equivalent to 21 mg/70 kg) dose of psilocybin to an active placebo condition (niacin). Similar decreases in depression symptoms were reported with large effect sizes as measured by the BDI and HADS were reported.

Efficacy for MDD and TRD

Extending the results from prior populations with cancer-related psychological distress, Carhart-Harris et al. (2016, 2018) completed an open-label pilot study of two doses of psilocybin in 20 participants with TRD. Treatment resistance was defined as having failed medication trials lasting 6 weeks or longer from at least two classes of antidepressants. Participants had a mean estimated illness duration of 17.7 years (SD 8.4). The preliminary report (Carhart-Harris et al., 2016) on the first 12 patients indicated a mean of 4.7 prior medication trials, though the later report on the complete group did not include this information. The majority had attempted at least one form of psychotherapy. A 10 mg "safety dose" and a moderately high dose of 25 mg were given about 1 week apart in combination with supportive psychotherapy before and after drug administration. Mean BDI score at baseline was 34.5 (SD 7.3), which corresponded with severe depressive symptoms. One week following the high dose psilocybin administration, mean BDI decreased to 11.8 (SD 11.1; Cohen's $d=2.5$). At three-month follow up, mean BDI was 15.2 (SD 11.0; Hedge's $g=2.0$). Mean BDI was 19.2 (SD 13.9; Cohen's $d=1.4$) at 3 months, and 19.5 (SD 13.9; Cohen's $d=1.4$) at 6 months.

Psilocybin was generally well tolerated in this study, with only mild side effects such as transient anxiety, nausea, and headache. Three participants experienced transient paranoia during the acute drug administration period. One participant experienced a return of significant depressive symptoms during the 3-month follow-up period and was referred to a general practitioner for treatment. Notably, six participants resumed psychiatric medications between the 3- and 6-month timepoint, five resumed psychotherapy, and five participants sought and received additional psilocybin-containing products from sources outside of the study. Removing the five participants who received additional psilocybin from analyses did not significantly alter the main results. Despite the limitations of the open-label design, this study served as a proof of concept that psilocybin can be safely administered to individuals with TRD, and suggested that it may be of clinical utility. However, the resumption of other forms of treatment and the use of psilocybin outside of the study limit the conclusions.

Davis et al., (2020b) conducted a randomized, waitlist-controlled trial of two moderate to high doses of psilocybin for MDD. Though this study did not explicitly test the intervention in a treatment-resistant group, the mean number of years with depression was 21.5 (SD 12.2), and the mean number of months in the current depressive episode was 24.4 (SD 22.0). Eligible participants were randomized to an immediate treatment group or an 8-week delay period, during which they were monitored by in-person or phone visits. All participants received two moderate to high weight-based doses of psilocybin about 2 weeks apart (20 mg/70 kg and 30 mg/70 kg). Staff provided approximately 13 h of supportive psychotherapy during the preparatory and follow-up period. There were large significant differences in GRID-Hamilton Depression Rating Scale (GRID-HAMD) scores between the immediate and delayed treatment groups at study weeks 5 and 8. For the full sample, GRID-HAMD scores (mean and SD) decreased from 22.8 (3.9) to 8.9 (7.4) at 4 weeks after the second psilocybin administration. At 4 weeks after the second psilocybin administration, 71% of participants had a clinically significant response to the intervention (≥50% reduction in GRID-HAMD score compared to baseline), and 58% were in remission (≤7 GRID-HAMD score). A different measure of depression (QIDS-SR) documented large, rapid, and significant decrease in depression score from baseline to day 1 after session the first psilocybin session, which remained statistically significantly reduced through the week 4 follow-up. Psilocybin was well tolerated with no serious adverse events. Almost all of participants reported transient challenging emotional experiences during the drug administration session; however, other studies have shown that such experiences after psilocybin are not uncommon. About one third of participants reported mild to moderate headache after the subjective psilocybin effects had subsided on session days.

Carhart-Harris et al. (2021) conducted a randomized controlled trial using a double dummy design to compare the antidepressant effects of escitalopram to those of psychedelic-assisted treatment. Patients with moderate to severe MDD were recruited ($N=59$) and randomized to receive either (1) two 25 mg doses of psilocybin in a therapeutic setting, plus 6 weeks of daily placebo, or (2) two doses of placebo in a therapeutic setting, plus 6 weeks of daily escitalopram. Though treatment-resistance was not a criterion for study enrollment, participants had a history of longstanding depressive illness (mean of 22 years in the psilocybin group and 15 years in the escitalopram group). Mean number of prior antidepressant trials was 2.2 (SD 1.6) in the psilocybin group and 1.8 (SD 1.5) in the escitalopram group. Baseline depressive symptoms as measured by the Quick Inventory of Depressive Symptoms Self-Report (QIDS-SR) was 14.5 (SD 3.9) in the psilocybin group and 16.4 (SD 4.1) in the escitalopram group. There were no significant differences in QIDS-SR score between the two treatment groups at 6 weeks. Though secondary outcome measures including other measures of depression such as BDI and HAMD generally favored the psilocybin group, these analyses were not adjusted for multiple comparisons.

The antidepressant effects of ayahuasca have also been studied. Osório et al. (2015) reported promising findings in a small ($N=6$) open label pilot of ayahuasca for recurrent depression. In that study, ayahuasca significantly reduced depressive symptoms from baseline to 1, 7, and 21 days after drug administration using the Hamilton Rating Scale for Depression (HAM-D) and several other measures of depression. Palhano-Fontes et al. (2019) subsequently reported a randomized parallel group placebo-controlled study of ayahuasca in individuals with TRD ($N=29$). Inclusion criteria required that participants had failed trials of antidepressants from at least two distinct pharmacologic classes, and participants had a mean of almost 4 medication trials. Participants in the ayahuasca arm received a weight-based dose containing 0.36 mg/kg of *N,N*-DMT. At days 1, 2, and 7 after the session, participants in the ayahuasca group showed significant decreases in depression compared to those in the placebo group, as assessed with the MADRS. A similar difference between the groups was shown on the HAMD at day 7 after the session. Furthermore on the HAMD at day 7, there was a significant difference between the response rate in the ayahuasca vs. placebo groups (57% vs. 20%, respectively, $p=0.04$) and a trend toward significance when comparing remission rates (43% vs. 13%, respectively, $p=0.07$). There were no serious adverse events recorded. The authors noted that, though participants generally felt safe during treatment, the experience was often physically and psychologically uncomfortable, with 57% experiencing vomiting during the session.

Adverse effects and toxicology

When used in supervised settings with substances of known purity, psychedelics have a low risk of serious adverse reactions (Johnson et al., 2018). Contrary to long-held popular beliefs about the dangers of psychedelics, these drugs have a very low abuse liability. Preclinical model of drug self-administration in animals have been largely negative (Johnson et al., 2018). However, others have highlighted the need to better characterize abuse liability of psychedelics in vulnerable populations (Reiff et al., 2020).

Physiological adverse effects

Somatic effects and physiological adverse events are generally mild. Transient increases in blood pressure are common via 5-HT_{2A} and 5-HT_{1B}–mediated vasoconstriction. For this reason, individuals with uncontrolled hypertension have been excluded from most clinical trials using psychedelics. Nausea, loss of appetite, dizziness, and tremor are also common (Nichols, 2004). For psychedelics other than ayahuasca, vomiting is rare. Transient, dose-dependent delayed onset headache is a common side effect of treatment with moderate-high dose psilocybin (Johnson et al., 2012). Seizures have been reported secondary to ingestion of both LSD and psilocybin-containing mushrooms (Fisher and Ungerleider, 1967; McCawley et al., 1962). For this reason, individuals with epilepsy have historically been excluded from clinical trials with psychedelics.

Cases of serotonin syndrome have also been reported secondary to use of psilocybin-containing mushrooms (Suzuki, 2016). However, no episodes have been described following psychedelic use in clinical research settings to date. Risk of serotonin syndrome can be reduced by discontinuing other serotonergic medications prior to drug administration. Other case reports of serious but rare adverse events secondary to psychedelics have included rhabdomyolysis, cortical blindness, and vasospasm leading to lower extremity ischemia (Berrens et al., 2010; Bickel et al., 2005; Raval et al., 2008).

There has been concern that psychedelics may cause cardiac valvular dysfunction via agonism at 5-HT_{2B} receptors, which are densely distributed in cardiac valves and regulate growth in these tissues. Other drugs with 5-HT_{2B} agonism have been associated with valvular heart disease including fenfluramine and MDMA, though it seems that chronic use is required to produce clinically significant changes in valvular function (Roth, 2007). No cases of valvular heart disease have been attributed directly to use of classic psychedelics to date. The typical course of treatment with psychedelics has involved just a few administrations of the drug and would ostensibly impart a low level of risk, if any. However, other patterns of use may place individuals at greater risk. Growing numbers of recreational users are now taking "microdoses" or very low doses of psychedelics multiple times per week on a chronic bases for supposed performance-enhancing purposes (Rosenbaum et al., 2020). Even at low doses, it is possible that repeated exposure over time could increase risk for cardiac complications.

Psychological adverse effects

Dysphoric or anxious reactions, also called "challenging experiences" are not uncommon in the setting of moderate-high dose psychedelic use and may be more common among those with depression. Griffiths et al. (2016) found that of 51 participants receiving high dose psilocybin for low mood in the setting of life-threatening cancer, 32% and 26% of their volunteers experienced a transient but significant episode of psychological discomfort or anxiety respectively. One participant reported transient paranoid ideation. In Carhart-Harris et al. (2018) study of psilocybin for TRD, 12 of 19 (63%) of participants reported transient anxiety lasting minutes during acute drug effects, and three participants experienced transient paranoia. In the Davis et al., (2020b) study of psilocybin for MDD, most participants endorsed experiencing a variety of potentially unpleasant or challenging emotions and physical sensations during psilocybin session (Davis et al., 2020b). Such effects do not typically last beyond the period of acute drug action and respond well to reassurance. High levels of neuroticism as measured by the Big Five Inventory have been associated with higher likelihood of having challenging experiences with psychedelics among a large group of naturalistic users, which may be relevant for sufferers of TRD (Barrett et al., 2017). Interestingly, individuals who have challenging experiences under the influence of psychedelics often report that the experience was meaningful and resulted in an overall benefit (Carbonaro et al., 2016).

Challenging psychological experiences appear to be more problematic when occurring in an unsupervised natural setting. In a survey of 1993 individuals who self-identified as having had challenging experiences under the influence of psilocybin in naturalistic settings, 2.6% reported that they put themselves or others at risk of physical harm (Carbonaro et al., 2016), 2.7% reported seeking urgent medical attention as a result of the experience, and 7.6% sought help for ongoing psychological distress after the experience.

Precipitation of mania and psychosis is perhaps the most serious psychiatric concern associated with psychedelics (Abraham et al., 1996). Cohen (1960) surveyed investigators who had administered LSD or mescaline and found the rate of prolonged adverse reactions to be low: only 1 case among approximately 1200 participants had developed psychotic symptoms lasting longer than 48 h. This single adverse event occurred with a participant whose identical twin had schizophrenia and thus may have been genetically predisposed to psychosis. The reaction resolved in 5 days. Though the risk of psychosis is quite low, it cannot be excluded completely, and for this reason most modern psychedelic clinical trials have excluded individuals with personal or family history of bipolar or schizophrenia spectrum disorders. Though transient paranoia can occur during acute drug effects, no cases of protracted psychotic or manic syndromes have been reported in human subjects' research with psychedelics since the 1990s to date.

Hallucinogen persisting perceptual disorder (HPPD) is defined in the DSM-V as a reexperiencing of drug-like perceptual effects following cessation of hallucinogen use (American Psychiatric Association and American Psychiatric Association, 2013). To meet criteria for this disorder, the symptoms must also be distressing or cause impairment in functioning, and must not be accounted for by another disorder. Qualitative research has identified two possible subtypes of HPPD: 1) a syndrome characterized by brief "flashbacks" of perceptual changes, with long symptom-free intervals between episodes, and 2) a chronic syndrome in which symptoms are present over months or years in varying intensity (Halpern et al., 2016). The true prevalence of Type I HPPD is difficult to estimate. Halpern et al. (2016) have highlighted the fact that though millions of doses of psychedelics have been consumed since the 1960s, few large case series have been published. However, early case series have estimated that prevalence may be as high as 1:20 in regular users in the 1960s and 1970s (Cohen, 1960; Horowitz, 1969). The prevalence of Type 2 HPPD is thought to be very rare, and is estimated to be approximately 1:50,000 (Grinspoon and Bakalar, 1979). No clear risk factors have been defined, but qualitative interviews with HPPD sufferers indicate that personal or family history of anxiety, preexisting tinnitus, eye floaters, and concentration difficulties may be associated with development of the disorder (Halpern et al., 2016). Baggott et al. (2011) conducted a survey of hallucinogen users with unusual visual disturbances that occurred outside of the context of acute drug effects. They found that such experiences were positively associated with the number of prior episodes of hallucinogen use. Compared to other psychedelics, past LSD use was more commonly associated with unusual visual experiences. Notably, there have been no reports of cases of HPPD reported in association with administration of a psychedelic in a clinical trial between the 1990s and the time of this writing.

Drug interactions

Common psychotropic medications can produce clinically significant interactions with psychedelics. Given that psychedelic effects are mediated primarily by 5-HT_{2A} agonism, most known interactions involve drugs that have an effect on serotonin signaling (Stenbæk et al., 2020). However, it is important to note that due to variability in receptor binding profiles, interactions may vary somewhat between different psychedelics, e.g., LSD is known to bind the dopamine D2 receptor while psilocin has no affinity at this site (Creese et al., 1975). Survey and case report data suggest that chronic use of selective serotonin reuptake inhibitors (SSRIs) may dampen subjective effects from both psilocybin and LSD, which is of particular relevance given the high rates of use of these medications in patients with TRD (Ruck J. Strassman et al., 1994). This interaction is not completely understood, but can likely be attributed to the fact that chronic SSRI use can result in desensitization and downregulation of 5-HT_{2A} receptor binding sites (Carr and Lucki, 2011; Yamauchi et al., 2006).

Serotonin syndrome is a risk when multiple serotonergic drugs are used together, though has rarely been reported in the context of psychedelic use. One case report documented the experience of an individual using ayahuasca during treatment with 20 mg of fluoxetine daily, which resulted in symptoms suggestive of serotonin syndrome including confusion, diaphoresis, tremors, and nausea and vomiting (Callaway and Grob, 1998). Among psychedelics, ayahuasca in particular may carry higher risk for serotonin syndrome or other interactions due to the MAO inhibiting properties of some of its constituents. Animal studies have shown that chronicity of SSRI use may have differential impact on psychedelic effects—while *chronic* SSRI treatment reduced signs of intoxication from 5-MeO-DMT, *acute* SSRI use potentiated signs of LSD intoxication in rats (Fiorella et al., 1996; Stolz et al., 1983). Thus it is possible that risk of potentiated psychedelic effects or serotonin syndrome may be higher in the period immediately following initiation of an SSRI.

Other psychotropic medications have demonstrated variable effects when used in combination with psychedelics. Similar to SSRIs, MAO inhibitors have been observed to reduce subjective effects of LSD (Bonson and Murphy, 1996; Grof and Dytrych, 1965; Resnick et al., 1964). Conversely, tricyclic antidepressants (TCAs) appear to potentiate the effects of LSD, which may be related to TCA-mediated increases in dopamine sensitivity in combination with LSD's action at dopamine receptors (Bonson and Murphy, 1996). Among mood stabilizing medications, Bonson et al. (1996) found that individuals taking lithium reported stronger than expected effects from LSD. Curiously, Vollenweider et al. (1998) reported that compared to subjects receiving psilocybin alone, those who were given the first-generation antipsychotic haloperidol in combination with psilocybin reported higher rates of psychotic-type symptoms including paranoia, loss of control over body and thought, and thought disorder. Conversely, they observed that risperidone effectively blocked the development of psychotic symptoms associated with psilocybin.

Among other drugs, one case report has suggested that allopurinol may dampen the subjective effects of psychedelics (Strassman, 1992). The mechanism here is less clear but may be related to allopurinol's inhibition of tryptophan pyrrolase, which may increase central levels of serotonin. Other drugs that are known to inhibit tryptophan pyrrolase include yohimbine, which is known mostly for its α_2-adrenergic antagonism (Papeschi et al., 1971). Yohimbine has also been found to increase central serotonin levels and to antagonize effects of LSD in rats (Dwoskin et al., 1988).

Due to the known interactions with other psychotropics and especially antidepressants, contemporary studies on the effects of psychedelics have typically required subjects to discontinue all psychiatric drugs for a washout period of at least five half-lives. However, changes in 5-HT signaling and receptor density following cessation of antidepressant treatment are not well understood and may persist beyond the acute washout period. This is evidenced by the fact that antidepressant discontinuation symptoms can persist well beyond the washout period following cessation of treatment (Stockmann et al., 2018). Changes in serotonin signaling impact the efficacy of psychedelics in individuals with chronic use, and this warrants further research.

Summary and future directions

Psychedelic-assisted psychotherapy has shown promise for the treatment of MDD and TRD, but larger placebo-controlled studies are needed to determine whether psilocybin has a safety and efficacy profile that would support its widespread clinical use. Two entities, Usona Institute and Compass Pathways are currently conducting phase II studies of their formulations of psilocybin for MDD (ClinicalTrials.gov identifiers 03866174 & 03775200). The latter will compare the effects of low, medium, and high doses of psilocybin on depressive symptoms in people with TRD, and is scheduled to be completed in December 2021. Both Usona and Compass have received "breakthrough therapy" designation by the United States FDA, which is intended to expedite the FDA's role in development and review of promising drugs and devices. Should these studies yield favorable results, legislative measures would be required in order to allow the clinical use of these substances. In the United States, psilocybin would need to be reclassified from its current Schedule I designation under the Controlled Substances Act. A recent review and analysis of psilocybin's abuse potential suggested that a reasonable reclassification would be Schedule IV, which currently includes hypnotic agents including benzodiazepines (Johnson et al., 2018).

There are many avenues for future research. Clinical studies to date have provided some understanding of the effects of one or two doses of psilocybin in combination with a fixed schedule of psychotherapy, but the understanding of the risks or benefits of subsequent dosing (i.e., repeated dosing weeks or months after the initial intervention in the event of relapse) is limited. Improvements to the current model of treatment are also needed. Current phase II studies require direct involvement of two clinicians (at least one of whom has doctoral level clinical training) during all phases of preparation, drug administration, and follow up care. This represents the costliest component of psychedelic-assisted treatment and would present a significant barrier to accessibility. Even if psychedelic-assisted treatment may offer significant cost savings over the long term compared the current standard of care, high upfront costs are a potential problem. It is not clear at this time whether the FDA will impose specific requirements regarding the type and nature of staff involvement required to administer psychedelics. Other models for psilocybin-assisted therapy have been proposed including psilocybin-assisted group therapy and 1:1 monitoring with as-needed back-up by an additional clinician, but further feasibility studies are needed (Anderson et al., 2020; Rucker et al., 2019).

The effects of psychedelics on bipolar depression have not yet been evaluated. Despite concerns outlined earlier with respect to the risk of precipitating mania or psychosis in at-risk populations, bipolar depression has recently gained attention as a potential clinical target for treatment with psychedelics. Psilocybin and other psychedelics may pose too great of a risk of precipitating mania in individuals with bipolar 1 disorder, but risks may be lower in individuals with bipolar 2 disorder. A current exploratory mixed methods study is underway at University of California San Francisco, and at least one clinical trial for this indication is planned at the time of this writing (ClinicalTrials.gov identifier 04433845).

A practice that has not yet been studied extensively for its effects on mood is that of microdosing psychedelics. Microdosing, which has becoming increasingly popular in the last decade, involves repeated administration of very low doses of psychedelics on an intermittent schedule. Survey data of individuals who have engaged in microdosing found that 71.8% of respondents indicated benefit for depression, and 56.6% reported improvements in anxiety (Cameron et al., 2020). A recent animal study found that chronic microdoses of *N,N*-DMT in rats resulted in an "antidepressant-like phenotype" characterized by improvements in fear extinction and weight gain (Cameron et al., 2019). Human laboratory studies of microdosing LSD have demonstrated safety, but thus far have not demonstrated a substantial effect on mood, perhaps due to a limitations in duration of treatment (Bershad et al., 2019, 2020).

Finally, although psilocybin is furthest on the path to possible approval for clinical use, the safety and efficacy of many other psychedelic drugs have yet to be studied. For example, 5-MeO-DMT has a duration of action between 30 and 90 min, which could make it more conducive for administration in existing outpatient settings. A comparative study examining differences between LSD, psilocybin, and mescaline in healthy subjects is currently underway (Clinicaltrials.gov identifier 04227756). The coming decades will no doubt see an ongoing increase in clinical and research interest in these substances, and they may hold unique promise for those suffering from depression that has failed to respond to conventional treatment.

Acknowledgments

Effort for authors was provided by The Center for Psychedelic and Consciousness Research which is funded by the Steven and Alexandra Cohen Foundation, Tim Ferriss, Matt Mullenweg, Craig Nerenberg, and Blake Mycoskie, and by NIH grant T32DA07209.

References

Abraham, H.D., Aldridge, A.M., Gogia, P., 1996. The psychopharmacology of hallucinogens. Neuropsychopharmacology 14 (4), 285–298.

American Psychiatric Association, American Psychiatric Association, 2013. Diagnostic and Statistical Manual of Mental Disorders: DSM-5, fifth ed. American Psychiatric Association. http://swbplus.bsz-bw.de/bsz381494357inh.htm. http://www.PsychiatryOnline.org. http://dsm.psychiatryonline.org/doi/book/10.1176/appi.books.9780890425596.

Anderson, B.T., Danforth, A., Daroff, R, Stauffer, C., Ekman, E., Agin-Liebes, G., Trope, A., Boden, M.T., Dilley, J., Mitchell, J., Woolley, J., 2020. Psilocybin-assisted group therapy for demoralized older long-term AIDS survivor men: An open-label safety and feasibility pilot study. EClinicalMedicine 27, 100538. https://doi.org/10.1016/j.eclinm.2020.100538. 33150319.

Anderson, R.J., Hoy, K.E., Daskalakis, Z.J., Fitzgerald, P.B., 2016. Repetitive transcranial magnetic stimulation for treatment resistant depression: re-establishing connections. Clin. Neurophysiol. 127 (11), 3394–3405. https://doi.org/10.1016/j.clinph.2016.08.015.

Baggott, M.J., Coyle, J.R., Erowid, E., Erowid, F., Robertson, L.C., 2011. Abnormal visual experiences in individuals with histories of hallucinogen use: a web-based questionnaire. Drug Alcohol Depend. 114 (1), 61–67. https://doi.org/10.1016/j.drugalcdep.2010.09.006.

Barrett, F.S., Johnson, M.W., Griffiths, R.R., 2015. Validation of the revised mystical experience questionnaire in experimental sessions with psilocybin. J. Psychopharmacol 29 (11), 1182–1190. https://doi.org/10.1177/0269881115609019. 26442957.

Barrett, F.S., Johnson, M.W., Griffiths, R.R., 2017. Neuroticism is associated with challenging experiences with psilocybin mushrooms. Personal. Individ. Differ. 117, 155–160.

Barrett, F.S., Doss, M.K., Sepeda, N.D., Pekar, J.J., Griffiths, R.R., 2020. Emotions and brain function are altered up to one month after a single high dose of psilocybin. Sci. Rep. 10 (1), 1–14.

Berrens, Z., Lammers, J., White, C., 2010. Rhabdomyolysis after LSD ingestion. Psychosomatics 51 (4), 356.e3.

Bershad, A.K., Schepers, S.T., Bremmer, M.P., Lee, R., de Wit, H., 2019. Acute subjective and behavioral effects of microdoses of lysergic acid diethylamide in healthy human volunteers. Biol. Psychiatry 86 (10), 792–800.

Bershad, A.K., Preller, K.H., Lee, R., Keedy, S., Wren-Jarvis, J., Bremmer, M.P., de Wit, H., 2020. Preliminary report on the effects of a low dose of LSD on resting-state amygdala functional connectivity. Biol. Psychiatry 5 (4), 461–467.

Bickel, M., Ditting, T., Watz, H., Roesler, A., Weidauer, S., Jacobi, V., Gueller, S., Betz, C., Fichtlscherer, S., Stein, J., 2005. Severe rhabdomyolysis, acute renal failure and posterior encephalopathy after 'magic mushroom'abuse. Eur J Emerg Med 12 (6), 306–308.

Bogenschutz, M.P., Forcehimes, A.A., Pommy, J.A., Wilcox, C.E., Barbosa, P.C., Strassman, R.J., 2015. Psilocybin-assisted treatment for alcohol dependence: a proof-of-concept study. J. Psychopharmacol. 29 (3), 289. https://doi.org/10.1177/0269881114565144.

Bonson, K.R., Murphy, D.L., 1996. Alterations in responses to LSD in humans associated with chronic administration of tricyclic antidepressants, monoamine oxidase inhibitors or lithium. Behav. Brain Res. 73 (1–2), 229–233. https://doi.org/10.1016/0166-4328(96)00102-7. 8788508.

Bonson, K.R., Buckholtz, J.W., Murphy, D.L., 1996. Chronic administration of serotonergic antidepressants attenuates the subjective effects of LSD in humans. Neuropsychopharmacology 14 (6), 425. http://search.ebscohost.com/login.aspx?direct=true&db=cmedm&AN=8726753&site=ehost-live&scope=site.

Brown, R.T., Nicholas, C.R., Cozzi, N.V., Gassman, M.C., Cooper, K.M., Muller, D., Thomas, C.D., Hetzel, S.J., Henriquez, K.M., Ribaudo, A.S., 2017. Pharmacokinetics of escalating doses of oral psilocybin in healthy adults. Clin. Pharmacokinet. 56 (12), 1543–1554.

Callaway, J.C., Grob, C.S., 1998. Ayahuasca preparations and serotonin reuptake inhibitors: a potential combination for severe adverse interactions. J. Psychoactive Drugs 30 (4), 367–369.

Cameron, L.P., Benson, C.J., DeFelice, B.C., Fiehn, O., Olson, D.E., 2019. Chronic, intermittent microdoses of the psychedelic N, N-dimethyltryptamine (DMT) produce positive effects on mood and anxiety in rodents. ACS Chem. Neurosci. 10 (7), 3261–3270.

Cameron, L.P., Nazarian, A., Olson, D.E., 2020. Psychedelic microdosing: prevalence and subjective effects. J. Psychoactive Drugs 52 (2), 113–122.

Carbonaro, T.M., Bradstreet, M.P., Barrett, F.S., MacLean, K.A., Jesse, R., Johnson, M.W., Griffiths, R.R., 2016. Survey study of challenging experiences after ingesting psilocybin mushrooms: acute and enduring positive and negative consequences. J. Psychopharmacol. 30 (12), 1268. https://doi.org/10.1177/0269881116662634.

Carhart-Harris, R.L., Erritzoe, D., Williams, T., Stone, J.M., Reed, L.J., Colasanti, A., Tyacke, R.J., Leech, R., Malizia, A.L., Murphy, K., Hobden, P., Evans, J., Feilding, A., Wise, R.G., Nutt, D.J., 2012. Neural correlates of the psychedelic state as determined by fMRI studies with psilocybin. Proc. Natl. Acad. Sci. U. S. A. 109 (6), 2138–2143. https://doi.org/10.1073/pnas.1119598109. 22308440.

Carhart-Harris, R.L., Leech, R., Hellyer, P.J., Shanahan, M., Feilding, A., Tagliazucchi, E., Chialvo, D.R., Nutt, D., 2014. The entropic brain: a theory of conscious states informed by neuroimaging research with psychedelic drugs. Front. Hum. Neurosci. 8, 20. https://doi.org/10.3389/fnhum.2014.00020. 24550805.

Carhart-Harris, R.L., Bolstridge, M., Rucker, J., Day, C.M.J., Erritzoe, D., Kaelen, M., Bloomfield, M., Rickard, J.A., Forbes, B., Feilding, A., Taylor, D., Pilling, S., Curran, V.H., Nutt, D.J., 2016. Psilocybin with psychological support for treatment-resistant depression: an open-label feasibility study. Lancet Psychiatry 3 (7), 619–627. https://doi.org/10.1016/S2215-0366(16)30065-7.

Carhart-Harris, R.L., Roseman, L., Bolstridge, M., Demetriou, L., Pannekoek, J.N., Wall, M.B., Tanner, M., Kaelen, M., McGonigle, J., Murphy, K., Leech, R., Curran, H.V., Nutt, D.J., 2017. Psilocybin for treatment-resistant depression: FMRI-measured brain mechanisms. Sci. Rep. 7 (1). https://doi.org/10.1038/s41598-017-13282-7.

Carhart-Harris, R.L., Bolstridge, M., Day, C., Rucker, J., Watts, R., Erritzoe, D.E., Kaelen, M., Giribaldi, B., Bloomfield, M., Pilling, S., 2018. Psilocybin with psychological support for treatment-resistant depression: six-month follow-up. Psychopharmacology 235 (2), 399–408.

Carhart-Harris, R., Giribaldi, B., Watts, R., Baker-Jones, M., Murphy-Beiner, A., Murphy, R., Martell, J., Blemings, A., Erritzoe, D., Nutt, D.J., 2021. Trial of psilocybin versus escitalopram for depression. N. Engl. J. Med. 384 (15), 1402–1411. https://doi.org/10.1056/NEJMoa2032994.

Carr, G.V., Lucki, I., 2011. The role of serotonin receptor subtypes in treating depression: a review of animal studies. Psychopharmacology 213 (2–3), 265–287.

Carter, O.L., Burr, D.C., Pettigrew, J.D., Wallis, G.M., Hasler, F., Vollenweider, F.X., 2005. Using psilocybin to investigate the relationship between attention, working memory, and the serotonin 1A and 2A receptors. J. Cogn. Neurosci. 17 (10), 1497–1508.

Celada, P., Puig, M., Artigas, F., 2013. Serotonin modulation of cortical neurons and networks. Front. Integr. Neurosci. 7, 25.

Cohen, S., 1960. Lysergic acid diethylamide: side effects and complications. J. Nerv. Ment. Dis. 130 (Jan), 30–40.

Condrau, G., 1949. Clinical experiences in mental patients with lysergic acid diethylamide. Acta Psychiatr. Neurol. 24, 9.

Costa Jr., P.T., Bagby, R.M., Herbst, J.H., McCrae, R.R., 2005. Personality self-reports are concurrently reliable and valid during acute depressive episodes. J. Affect. Disord. 89 (1–3), 45–55.

Creese, I., Burt, D.R., Snyder, S.H., 1975. The dopamine receptor: differential binding of d-LSD and related agents to agonist and antagonist states. Life Sci. 17 (11), 1715–1719.

Davis, A.K., Barrett, F.S., Griffiths, R.R., 2020a. Psychological flexibility mediates the relations between acute psychedelic effects and subjective decreases in depression and anxiety. J. Contextual Behav. Sci. 15, 39–45.

Davis, A.K., Barrett, F.S., May, D.G., Cosimano, M.P., Sepeda, N.D., Johnson, M.W., Finan, P.H., Griffiths, R.R., 2020b. Effects of psilocybin-assisted therapy on major depressive disorder: a randomized clinical trial. JAMA Psychiatry. https://doi.org/10.1001/jamapsychiatry.2020.3285.

Dichter, G.S., Gibbs, D., Smoski, M.J., 2015. A systematic review of relations between resting-state functional-MRI and treatment response in major depressive disorder. J. Affect. Disord. 172, 8–17.

Dolder, P.C., Schmid, Y., Steuer, A.E., Kraemer, T., Rentsch, K.M., Hammann, F., Liechti, M.E., 2017. Pharmacokinetics and pharmacodynamics of lysergic acid diethylamide in healthy subjects. Clin. Pharmacokinet. 56 (10), 1219–1230.

Doss, M.K., May, D.G., Johnson, M.W., Clifton, J.M., Hedrick, S.L., Prisinzano, T.E., Barrett, & F., 2020. The acute effects of the atypical dissociative hallucinogen Salvinorin A on functional connectivity in the human. Brain 10, 1–12.

Dwoskin, L.P., Neal, B.S., Sparber, S.B., 1988. Evidence for antiserotonergic properties of yohimbine. Pharmacol. Biochem. Behav. 31 (2), 321–326.

Erritzoe, D., Roseman, L., Nour, M.M., MacLean, K., Kaelen, M., Nutt, D.J., Carhart-Harris, R.L., 2018. Effects of psilocybin therapy on personality structure. Acta Psychiatr. Scand. 138 (5), 368–378. https://doi.org/10.1111/acps.12904.

Fiorella, D., Helsley, S., Rabin, R.A., Winter, J.C., 1996. Potentiation of LSD-induced stimulus control by fluoxetine in the rat. Life Sci. 59 (18), PL283–PL287.

Fisher, D.D., Ungerleider, J.T., 1967. Grand mal seizures following ingestion of LSD. Calif. Med. 106 (3), 210–211.

Flanagan, T.W., Nichols, C.D., 2018. Psychedelics as anti-inflammatory agents. Int. Rev. Psychiatry 30 (4), 363–375.

Fontanilla, D., Johannessen, M., Hajipour, A.R., Cozzi, N.V., Jackson, M.B., Ruoho, A.E., 2009. The hallucinogen N,N-dimethyltryptamine (DMT) is an endogenous sigma-1 receptor regulator. Science 323 (5916), 934–937. https://doi.org/10.1126/science.1166127.

Glennon, R.A., Titeler, M., McKenney, J.D., 1984. Evidence for 5-HT2 involvement in the mechanism of action of hallucinogenic agents. Life Sci. 35 (25), 2505–2511.

Gray, J.A., Roth, B.L., 2001. Paradoxical trafficking and regulation of 5-HT(2A) receptors by agonists and antagonists. Brain Res. Bull. 56 (5), 441.

Griffiths, R.R., Richards, W.A., McCann, U., Jesse, R., 2006. Psilocybin can occasion mystical-type experiences having substantial and sustained personal meaning and spiritual significance. Psychopharmacology 187 (3), 268–283. https://doi.org/10.1007/s00213-006-0457-5.

Griffiths, R.R, Richards, W.A., Johnson, M.W., McCann, U.D., Jesse, R., 2008. Mystical-type experiences occasioned by psilocybin mediate the attribution of personal meaning and spiritual significance 14 months later. J. Psychopharmacol. 22 (6), 621–632. https://doi.org/10.1177/0269881108094300. 18593735.

Griffiths, R.R., Johnson, M., Richards, W., Richards, B., McCann, U., Jesse, R., 2011. Psilocybin occasioned mystical-type experiences: immediate and persisting dose-related effects. Psychopharmacology 218 (4), 649.

Griffiths, R.R., Johnson, M.W., Carducci, M.A., Umbricht, A., Richards, W.A., Richards, B.D., Cosimano, M.P., Klinedinst, M.A., 2016. Psilocybin produces substantial and sustained decreases in depression and anxiety in patients with life-threatening cancer: a randomized double-blind trial. J. Psychopharmacol. 30 (12), 1181–1197.

Grinspoon, L., Bakalar, J.B., 1979. Psychedelic Drugs Reconsidered. Basic Books, New York.

Grob, C.S., Danforth, A.L., Chopra, G.S., Hagerty, M., McKay, C.R., Halberstad, A.L., Greer, G.R., 2011. Pilot study of psilocybin treatment for anxiety in patients with advanced-stage cancer. Arch. Gen. Psychiatry 68 (1), 71–78. https://doi.org/10.1001/archgenpsychiatry.2010.116.

Grof, S, Dytrych, Z, 1965. Blocking of LSD reaction by premedication with Niamid. Act. Nerv. Super. (Praha) 7 (3), 306. 5882136.

Halpern, J.H., Lerner, A.G., Passie, T., 2016. A review of hallucinogen persisting perception disorder (HPPD) and an exploratory study of subjects claiming symptoms of HPPD. In: Behavioral Neurobiology of Psychedelic Drugs. Springer, pp. 333–360.

Hanley, N.R., Hensler, J.G., 2002. Mechanisms of ligand-induced desensitization of the 5-hydroxytryptamine(2A) receptor. J. Pharmacol. Exp. Ther. 300 (2), 468.

Hasler, F., Bourquin, D., Brenneisen, R., Vollenweider, F.X., 2002. Renal excretion profiles of psilocin following oral administration of psilocybin: a controlled study in man. J. Pharm. Biomed. Anal. 30 (2), 331–339.

Hasler, F., Grimberg, U., Benz, M.A., Huber, T., Vollenweider, F.X., 2004. Acute psychological and physiological effects of psilocybin in healthy humans: a double-blind, placebo-controlled doseeffect study. Psychopharmacology 172 (2), 145.

Hofmann, A., 1983. LSD, my Problem Child: Reflections on Sacred Drugs, Mysticism, and Science. J.P. Tarcher, Boston, p. 209.

Holze, F., Vizeli, P., Müller, F., Ley, L., Duerig, R., Varghese, N., Eckert, A., Borgwardt, S., Liechti, M.E., 2020. Distinct acute effects of LSD, MDMA, and d-amphetamine in healthy subjects. Neuropsychopharmacology 45 (3), 462–471.

Horowitz, M.J., 1969. Flashbacks: recurrent intrusive images after the use of LSD. Am. J. Psychiatr. 126 (4), 565–569.

Johnson, M.W., Richards, W.A., Griffiths, R.R., 2008. Human hallucinogen research: guidelines for safety. J. Psychopharmacol. 22 (6), 603–620.

Johnson, M.W., Sewell, R.A., Griffiths, R.R., 2012. Psilocybin dose-dependently causes delayed, transient headaches in healthy volunteers. Drug Alcohol Depend. 123 (1), 132–140.

Johnson, M.W., Garcia-Romeu, A., Cosimano, M.P., Griffiths, R.R., 2014. Pilot study of the 5-HT 2A R agonist psilocybin in the treatment of tobacco addiction. J. Psychopharmacol. 28 (11), 983–992. https://doi.org/10.1177/0269881114548296.

Johnson, M.W., Griffiths, R.R., Hendricks, P.S., Henningfield, J.E., 2018. The abuse potential of medical psilocybin according to the 8 factors of the controlled substances act. Neuropharmacology 142, 143–166.

Klock, J.C., Boerner, U., Becker, C.E., 1974. Coma, hyperthermia and bleeding associated with massive LSD overdose: a report of eight cases. West. J. Med. 120 (3), 183.

Kraehenmann, R., Preller, K.H., Scheidegger, M., Pokorny, T., Bosch, O.G., Seifritz, E., Vollenweider, F.X., 2015. Psilocybin-induced decrease in amygdala reactivity correlates with enhanced positive mood in healthy volunteers. Biol. Psychiatry 78 (8), 572–581. https://doi.org/10.1016/j.biopsych.2014.04.010. 24882567.

Leibenluft, E., Pine, D.S., 2013. Resting state functional connectivity and depression: in search of a bottom line. Biol. Psychiatry 74 (12), 868–869.

Luna, L.E., 2011. Indigenous and mestizo use of ayahuasca: an overview. In: The Ethnopharmacology of Ayahuasca. 2. Transworld Research Network, p. 1.

MacLean, K.A., Johnson, M.W., Griffiths, R.R., 2011. Mystical experiences occasioned by the hallucinogen psilocybin lead to increases in the personality domain of openness. J. Psychopharmacol. 25 (11), 1453–1461. https://doi.org/10.1177/0269881111420188.

McCawley, E.L., Brummett, R.E., Dana, G.W., 1962. Convulsions from psilocybe mushroom poisoning. Proc. West. Pharmacol. Soc. 5, 27.

Moreno, J.L., Holloway, T., Albizu, L., Sealfon, S.C., González-Maeso, J., 2011. Metabotropic glutamate mGlu2 receptor is necessary for the pharmacological and behavioral effects induced by hallucinogenic 5-HT2A receptor agonists. Neurosci. Lett. 493 (3), 76–79.

Moreno, J.L., Holloway, T., Rayannavar, V., Sealfon, S.C., González-Maeso, J., 2013. Chronic treatment with LY341495 decreases 5-HT2A receptor binding and hallucinogenic effects of LSD in mice. Neurosci. Lett. 536, 69–73.

Müller, F., Holze, F., Dolder, P., Ley, L., Vizeli, P., Soltermann, A., Liechti, M.E., Borgwardt, S., 2020. MDMA-induced changes in within-network connectivity contradict the specificity of these alterations for the effects of serotonergic hallucinogens. Neuropsychopharmacology. https://doi.org/10.1038/s41386-020-00906-2.

Nagatomo, T., Rashid, M., Muntasir, H.A., Komiyama, T., 2004. Functions of 5-HT2A receptor and its antagonists in the cardiovascular system. Pharmacol. Ther. 104 (1), 59–81.

Nau Jr., F., Yu, B., Martin, D., Nichols, C.D., 2013. Serotonin 5-HT 2A receptor activation blocks TNF-α mediated inflammation in vivo. PLoS One 8 (10), e75426.

Nichols, D.E., 2004. Hallucinogens. Pharmacol. Ther. 101 (2), 131–181.

Osmond, H., 1957. A review of the clinical effects of psychotomimetic agents. Ann. N. Y. Acad. Sci. 66 (3), 418–434.

Osório, F.d.L., Sanches, R.F., Macedo, L.R., Dos Santos, R.G., Maia-de-Oliveira, J.P., Wichert-Ana, L., De Araujo, D.B., Riba, J., Crippa, J.A., Hallak, J.E., 2015. Antidepressant effects of a single dose of ayahuasca in patients with recurrent depression: a preliminary report. Braz. J. Psychiatry 37 (1), 13–20.

Palhano-Fontes, F., Barreto, D., Onias, H., Andrade, K.C., Novaes, M.M., Pessoa, J.A., Mota-Rolim, S.A., Osório, F.L., Sanches, R., Dos Santos, R.G., 2019. Rapid antidepressant effects of the psychedelic ayahuasca in treatment-resistant depression: a randomized placebo-controlled trial. Psychol. Med. 49 (4), 655–663.

Papeschi, R., Sourkes, T.L., Youdim, M., 1971. The effect of yohimbine on brain serotonin metabolism, motor behavior and body temperature of the rat. Eur. J. Pharmacol. 15 (3), 318–326.

Passie, T., Seifert, J., Schneider, U., Emrich, H.M., 2002. The pharmacology of psilocybin. Addict. Biol. 7 (4), 357. https://doi.org/10.1080/1355621021000005937.

Pazos, A., Probst, A., Palacios, J.M., 1987. Serotonin receptors in the human brain—IV. Autoradiographic mapping of serotonin-2 receptors. Neuroscience 21 (1), 123–139.

Piedmont, R.L., 2001. Cracking the plaster cast: big five personality change during intensive outpatient counseling. J. Res. Pers. 35 (4), 500–520.

Raval, M.V., Gaba, R.C., Brown, K., Sato, K.T., Eskandari, M.K., 2008. Percutaneous transluminal angioplasty in the treatment of extensive LSD-induced lower extremity vasospasm refractory to pharmacologic therapy. J. Vasc. Interv. Radiol. 19 (8), 1227–1230.

Reiff, C.M., Richman, E.E., Nemeroff, C.B., Carpenter, L.L., Widge, A.S., Rodriguez, C.I., Kalin, N.H., McDonald, W.M., Work Group on Biomarkers and Novel Treatments, a D. of the A. P. A. C. of R, 2020. Psychedelics and psychedelic-assisted psychotherapy. Am. J. Psychiatr. 177 (5), 391–410.

Resnick, O., Krus, D.M., Raskin, M., 1964. LSD-25 action in normal subjects treated with a monoamine oxidase inhibitor. Life Sci. (1962) 3, 1207–1214. https://doi.org/10.1016/0024-3205(64)90037-2. 14231769.

Riba, J., Rodríguez-Fornells, A., Urbano, G., Morte, A., Antonijoan, R., Montero, M., Callaway, J.C., Barbanoj, M.J., 2001. Subjective effects and tolerability of the South American psychoactive beverage Ayahuasca in healthy volunteers. Psychopharmacology 154 (1), 85–95.

Riba, J., Valle, M., Urbano, G., Yritia, M., Morte, A., Barbanoj, M.J., 2003. Human pharmacology of ayahuasca: subjective and cardiovascular effects, monoamine metabolite excretion, and pharmacokinetics. J. Pharmacol. Exp. Ther. 306 (1), 73–83.

Riba, J., McIlhenny, E.H., Bouso, J.C., Barker, S.A., 2015. Metabolism and urinary disposition of N, N-dimethyltryptamine after oral and smoked administration: a comparative study. Drug Test. Anal. 7 (5), 401–406.

Roseman, L., Nutt, D.J., Carhart-Harris, R.L., 2018. Quality of acute psychedelic experience predicts therapeutic efficacy of psilocybin for treatment-resistant depression. Front. Pharmacol. 8. https://doi.org/10.3389/fphar.2017.00974.

Rosenbaum, D., Weissman, C., Anderson, T., Petranker, R., Dinh-Williams, L.-A., Hui, K., Hapke, E., 2020. Microdosing psychedelics: demographics, practices, and psychiatric comorbidities. J. Psychopharmacol. 34 (6), 612–622.

Ross, S., Bossis, A., Guss, J., Agin-Liebes, G., Malone, T., Cohen, B., Mennenga, S.E., Belser, A., Kalliontzi, K., Babb, J., Su, Z., Corby, P., Schmidt, B.L., 2016. Rapid and sustained symptom reduction following psilocybin treatment for anxiety and depression in patients with life-threatening cancer: a randomized controlled trial. J. Psychopharmacol. 30 (12), 1165. https://doi.org/10.1177/0269881116675512.

Roth, B.L., 2007. Drugs and valvular heart disease. N. Engl. J. Med. 356 (1), 6–9.

Rucker, J.J., Jelen, L.A., Flynn, S., Frowde, K.D., Young, A.H., 2016. Psychedelics in the treatment of unipolar mood disorders: a systematic review. J. Psychopharmacol. 30 (12), 1220–1229.

Rucker, J., Young, A., Bird, C., Daniel, A., Lennard-Jones, M., Marwood, L., Posadas, K.G., Reinholdt, F., Saldarini, F., Stansfield, S., 2019. Psilocybin administration to healthy participants: safety and feasibility in a placebo-controlled study. Reflections 4 (3), 2.

Sahdra, B.K., MacLean, K.A., Ferrer, E., Shaver, P.R., Rosenberg, E.L., Jacobs, T.L., Zanesco, A.P., King, B.G., Aichele, S.R., Bridwell, D.A., 2011. Enhanced response inhibition during intensive meditation training predicts improvements in self-reported adaptive socioemotional functioning. Emotion 11 (2), 299.

Savage, C., Fadiman, J., Mogar, R., Allen, M.H., 1966. The effects of psychedelic (LSD) therapy on values, personality, and behavior. Int. J. Neuropsychiatry 2 (3), 241–254.

Schultes, R.E., Hofmann, A., 1992. Plants of the Gods: Their Sacred, Healing, and Hallucinogenic Powers Pdf.

Smigielski, L., Scheidegger, M., Kometer, M., Vollenweider, F.X., 2019. Psilocybin-assisted mindfulness training modulates self-consciousness and brain default mode network connectivity with lasting effects. NeuroImage 196, 207–215.

Soler, J., Elices, M., Franquesa, A., Barker, S., Friedlander, P., Feilding, A., Pascual, J.C., Riba, J., 2016. Exploring the therapeutic potential of Ayahuasca: acute intake increases mindfulness-related capacities. Psychopharmacology 233 (5), 823–829.

Stenbæk, D.S., Madsen, M.K., Ozenne, B., Kristiansen, S., Burmester, D., Erritzoe, D., Knudsen, G.M., Fisher, P.M., 2020. Brain serotonin 2A receptor binding predicts subjective temporal and mystical effects of psilocybin in healthy humans. J. Psychopharmacol. 35 (4), 459–468. 0269881120959609.

Stockmann, T., Odegbaro, D., Timimi, S., Moncrieff, J., 2018. SSRI and SNRI withdrawal symptoms reported on an internet forum. Int. J. Risk Saf. Med. 29 (3–4), 175–180. https://doi.org/10.3233/JRS-180018.

Stolz, J.F., Marsden, C.A., Middlemiss, D.N., 1983. Effect of chronic antidepressant treatment and subsequent withdrawal on [3*H*]-5-hydroxytryptamine and [3 H]-spiperone binding in rat frontal cortex and serotonin receptor mediated behaviour. Psychopharmacology 80 (2), 150–155.

Strassman, R.J., 1992. Human hallucinogen interactions with drugs affecting serotonergic neurotransmission. Neuropsychopharmacology 7 (3), 241.

Strassman, R.J., Qualls, C.R., Uhlenhuth, E.H., Kellner, R., 1994. Dose-response study of N, N-dimethyltryptamine in humans: II. Subjective effects and preliminary results of a new rating scale. Arch. Gen. Psychiatry 51 (2), 98–108.

Suzuki, K., 2016. Three cases of acute serotonin syndrome due to psilocybin mushroom poisoning. Jpn. J. Toxicol. 29 (1), 33.

Takahashi, M., Shirayama, Y., Muneoka, K., Suzuki, M., Sato, K., Hashimoto, K., 2013. Low openness on the revised NEO personality inventory as a risk factor for treatment-resistant depression. PLoS One 8 (9), e71964.

Thase, M.E., Friedman, E.S., Howland, R.H., 2001. Management of treatment-resistant depression: psychotherapeutic perspectives. J. Clin. Psychiatry 62, 18–24.

Tokunaga, A., Saika, M., Senba, E., 1998. 5-HT2A receptor subtype is involved in the thermal hyperalgesic mechanism of serotonin in the periphery. Pain 76 (3), 349–355.

Upshall, D.G., Wailling, D.G., 1972. The determination of LSD in human plasma following oral administration. Clin. Chim. Acta 36 (1), 67–73.

Vollenweider, F.X., Kometer, M., 2010. The neurobiology of psychedelic drugs: implications for the treatment of mood disorders. Nat. Rev. Neurosci. 11 (9), 642.

Vollenweider, F.X., Vollenweider-Scherpenhuyzen, M.F., Bäbler, A., Vogel, H., Hell, D., 1998. Psilocybin induces schizophrenia-like psychosis in humans via a serotonin-2 agonist action. Neuroreport 9 (17), 3897–3902.

Wacker, D., Wang, S., McCorvy, J.D., Betz, R.M., Venkatakrishnan, A.J., Levit, A., Lansu, K., Schools, Z.L., Che, T., Nichols, D.E., 2017. Crystal structure of an LSD-bound human serotonin receptor. Cell 168 (3), 377–389.e12.

Watts, V.J., Mailman, R.B., Lawler, C.P., Neve, K.A., Nichols, D.E., 1995. LSD and structural analogs: pharmacological evaluation at D 1 dopamine receptors. Psychopharmacology 118 (4), 401–409.

Wong, D.F., Lever, J.R., Hartig, P.R., Dannals, R.F., Villemagne, V., Hoffman, B.J., Wilson, A.A., Ravert, H.T., Links, J.M., Scheffel, U., 1987. Localization of serotonin 5-HT2 receptors in living human brain by positron emission tomography using N1-([11C]-methyl)-2-Br-LSD. Synapse 1 (5), 393–398.

Yaden, D.B., Griffiths, R.R., 2020. The subjective effects of psychedelics are necessary for their enduring therapeutic effects. ACS Pharmacol. Transl. Sci. 4, 568–572.

Yamauchi, M., Miyara, T., Matsushima, T., Imanishi, T., 2006. Desensitization of 5-HT2A receptor function by chronic administration of selective serotonin reuptake inhibitors. Brain Res. 1067 (1), 164–169.

Yang, C., Wardenaar, K.J., Bosker, F.J., Li, J., Schoevers, R.A., 2019. Inflammatory markers and treatment outcome in treatment resistant depression: a systematic review. J. Affect. Disord. 257, 640–649. https://doi.org/10.1016/j.jad.2019.07.045.

Yu, B., Becnel, J., Zerfaoui, M., Rohatgi, R., Boulares, A.H., Nichols, C.D., 2008. Serotonin 5-hydroxytryptamine2A receptor activation suppresses tumor necrosis factor-α-induced inflammation with extraordinary potency. J. Pharmacol. Exp. Ther. 327 (2), 316–323.

Chapter 15

Inflammation and treatment resistance: Mechanisms and treatment implications

Andrew H. Miller, Michael J. Lucido, Mandakh Bekhbat, Neeti D. Mehta, Xiangchuan Chen, Michael T. Treadway, Jennifer C. Felger, and Ebrahim Haroon

Department of Psychiatry and Behavioral Sciences, Emory University School of Medicine, Atlanta, GA, United States

Introduction

Increasing data indicate that the immune system in general and inflammation in particular may play a role in the pathophysiology of psychiatric disorders including depression and anxiety (Miller and Raison, 2016). These data include evidence of low-grade inflammation in subgroups of patients with a variety of psychiatric diseases as well as effects of inflammatory mediators on neurotransmitter systems and neurocircuits that regulate behavior (Miller and Raison, 2016). Interestingly, there appears to be a special relationship between increased inflammation and treatment response, best documented in depression (Strawbridge et al., 2015, 2018; Haroon et al., 2018a; Lucido et al., 2021; Haroon et al., 2012). Data indicate that increased inflammatory markers prior to initiation of therapy predict poor treatment response to conventional antidepressants (Strawbridge et al., 2015; Cattaneo et al., 2016), and treatment-resistant depressed patients exhibit increased markers of inflammation (Strawbridge et al., 2018; Haroon et al., 2018a; Chamberlain et al., 2019). Moreover, a rich literature describes the multiple pathways by which inflammatory factors can disrupt neurotransmitter metabolism and growth factors and ultimately neurocircuits that are the foundations of treatment response to conventional antidepressant medications (Miller and Raison, 2016; Lucido et al., 2021). Increased inflammation is also associated with a better response to ketamine and electroconvulsive therapy (ECT) (Yang et al., 2015; Kruse et al., 2018), suggesting that a primary effect of inflammation is to undermine the efficacy of conventional antidepressant treatments, while increasing the responsiveness to alternative treatment strategies.

Given the relationship between inflammation and pathophysiologic mechanisms related to depression and other psychiatric disorders, studies using drugs with antiinflammatory properties have been investigated (Kohler-Forsberg et al., 2019). In general, there is modest evidence that antiinflammatory therapies may have potential to treat depression including treatment-resistant depression (TRD). However, due to multiple confounds involving both the designs and the drugs used in these studies, the results are almost impossible to interpret, and therefore, the field is bereft of meaningful clinical trial data to serve as the foundation for future research (Miller and Raison, 2016; Miller, 2020; Miller and Raison, 2015).

The goal of this chapter is to first provide the basis for the hypothesis that inflammation plays a role in depression, including a review of the neurobiological and immunological mechanisms involved. Next, we will address the special relationship between inflammation and treatment resistance. Finally, we will review relevant clinical trials that have targeted inflammation in depression including TRD and make recommendations regarding the drugs and clinical trials' designs of future studies in this area. It is the position of the authors that there is great promise in further understanding the translational relevance of the rich database that has been developed regarding the impact of inflammation on the brain and behavior. Indeed, as will be described, patients with increased inflammation can be readily identified, and given our knowledge regarding inflammatory pathways to pathology, we believe that targeting inflammation and/or its downstream effects on the brain will provide an important platform for precision psychiatry. By the same token, we also recognize that failure to integrate what is known about inflammation and its role in depression and other psychiatric disorders will result in a missed opportunity based on outdated clinical trial methodologies, drugs with multiple off-target effects, and outcomes that are not well-aligned with our understanding of the impact of inflammation on the brain.

Managing Treatment-Resistant Depression. https://doi.org/10.1016/B978-0-12-824067-0.00015-3

Foundation for the hypothesis of inflammation's role in depression

Patients with depression exhibit all the cardinal features of a chronic inflammatory response

The earliest studies examining the relationship between the immune system and depression were derived from human and laboratory animal studies demonstrating that a variety of stressors could inhibit lymphocyte proliferation in vitro (Stein et al., 1991). Indeed, numerous studies demonstrated that peripheral blood immune cells from depressed patients exhibited decreased in vitro B- and T-cell responses as well as decreased natural killer (NK) cell activity (Stein et al., 1991). These data suggested that depression was associated with suppression of acquired immune responses. Interestingly, however, data began to surface that in contrast to acquired immune responses, there was evidence of activation of the innate immune response as revealed by increased concentrations of acute phase proteins (Maes et al., 1992). This early work served as the foundation for subsequent studies demonstrating increased markers of an activated innate immune inflammatory response including increased inflammatory cytokines, increased chemokines, and increased cellular adhesion molecules as well as increased acute phase proteins. Hundreds of papers have been published in this area, and meta-analyses of the literature have established reliable elevations of the inflammatory cytokines interleukin (IL)-6 and tumor necrosis factor (TNF) and the acute phase protein C-reactive protein (CRP) in the peripheral blood of depressed patients (Dowlati et al., 2010; Howren et al., 2009). Postmortem studies have also revealed evidence of an inflammatory response in the brain of depressed suicide victims including increased expression of toll-like receptors (TLRs) and other inflammatory signaling molecules in brain parenchyma (Pandey et al., 2014), activation of microglia and astrocytes (Steiner et al., 2008; Rao et al., 2010; Torres-Platas et al., 2011, 2014a, 2016), and the presence of perivascular monocytes (Torres-Platas et al., 2014b; Enache et al., 2019). In addition, positron emission tomography (PET) using ligands that bind to the translocator protein (TSPO), which is expressed in microglia as well as other cell types in the brain, has provided evidence of microglial activation in depression (Setiawan et al., 2015), although the specificity of these findings for neuroinflammation has yet to be established (Lee et al., 2020a). Finally, several large, longitudinal, epidemiological studies have shown that CRP and other inflammatory mediators can predict the development of depression, with less compelling data suggesting that depression leads to increased inflammation (Mac Giollabhui et al., 2021).

Relationship with treatment nonresponse

Relevant to the response to antidepressants, a meta-analysis of the literature suggested that increased inflammatory biomarkers prior to treatment with conventional antidepressants (e.g., selective serotonin reuptake inhibitors (SSRIs) and serotonin norepinephrine reuptake inhibitors (SNRIs)) is predictive of a reduced therapeutic response in ambulatory depressed patients (Strawbridge et al., 2015). Moreover, studies have indicated that treatment-resistant depressed patients exhibit increased inflammatory markers including IL-6, TNF, and CRP (Haroon et al., 2018a; Chamberlain et al., 2019). Molecular markers of inflammation including mRNA for the purinergic receptor P2RX7, IL-1beta, TNF, the CXC chemokine CXCL12, and migration inhibitory factor have also been shown to predict antidepressant nonresponse to SSRIs and SNRIs (Cattaneo et al., 2016, 2020). In addition, two studies have suggested that CRP may predict a differential response to antidepressants, with both studies finding a poor response to escitalopram in individuals with a CRP > 1 mg/L (Uher et al., 2014; Jha et al., 2017). One of these studies found bupropion and the other found nortriptyline to lead to better responsiveness in the subgroup with CRP > 1 mg/L. Interestingly, increased inflammatory markers also have been found to predict a positive response to ketamine and ECT (Yang et al., 2015; Kruse et al., 2018). It should be noted, however, that most studies identifying inflammatory markers as predictors of treatment response have used post hoc analyses. Few studies have a priori determined that inflammatory markers can predict response to antidepressant or antiinflammatory treatment. Taken together, these data suggest that inflammatory markers have the potential to help guide treatment decision-making, especially as it relates to treatment response. Moreover, as discussed below, increased inflammatory markers appear to have special relevance to well-known predictors of treatment nonresponse including obesity, early life stress, medical illnesses, and anxiety (Miller and Raison, 2016; Haroon et al., 2012).

Administration of inflammatory cytokines/stimuli causes depressive symptoms

A second major body of data that supports the relationship between inflammation and depression, while also addressing the cause-and-effect nature of this relationship, is work demonstrating that administration of inflammatory cytokines (e.g., interferon (IFN)-alpha) or inflammatory stimuli (e.g., endotoxin or typhoid or influenza vaccination) leads to behavioral changes that are consistent with those seen in patients with depression (Miller and Raison, 2016). For example, chronic treatment with IFN-alpha leads to symptoms that meet criteria for major depression in 30%–50% of subjects (e.g., patients with cancer or infectious diseases) depending on the dose (Raison et al., 2006; Capuron et al., 2002). Moreover, comparison of symptoms induced by IFN-alpha with symptoms from otherwise healthy depressed individuals yield few meaningful differences between the two, although IFN-alpha-treated patients exhibit an overrepresentation of neurovegetative symptoms (Capuron et al., 2009). Endotoxin and typhoid and influenza vaccination, typically administered acutely to healthy volunteers in laboratory settings, also lead to symptoms of depression, albeit short-lived (Eisenberger et al., 2010; Brydon et al., 2008; Harrison et al., 2009). The capacity of these inflammatory stimuli to induce a behavioral syndrome like depression provides compelling data that inflammation can cause depression. Based on these data, much has been learned about the central nervous system (CNS) mechanisms by which inflammation affects neurotransmitter systems and neurocircuits in the brain to change behavior (Miller et al., 2013).

Inhibition of inflammation reduces depressive symptoms

Further supporting the cause-and-effect relationship between inflammation and depression are studies indicating that inhibition of inflammation can reduce depressive symptoms. The most convincing data come from patients with autoimmune and inflammatory disorders who have been treated with anticytokine therapies that both reduce inflammation and reduce depressive symptoms (Kappelmann et al., 2018; Wittenberg et al., 2020). In many of the studies, the decrease in depressive symptoms is independent of the improvement in disease activity, although the possibility that behavioral effects are tied to the response of the underlying disease process remains a consideration. Studies also suggest that drugs with antiinflammatory properties including celecoxib, aspirin, minocycline, statins, pioglitazone, and omega-3 fatty acids have antidepressant effects in otherwise healthy individuals (Kohler-Forsberg et al., 2019). Nevertheless, as indicated earlier, there are many complications in the clinical trial designs and the drugs used that make interpretation of these data challenging (Miller and Raison, 2016; Miller, 2020; Miller and Raison, 2015).

Clinical predictors of treatment resistance and inflammation

Several clinical factors have been shown to be predictors of antidepressant treatment nonresponse and are related to inflammation. These include obesity, early life stress, medical illness, and anxiety (Miller and Raison, 2016; Raison et al., 2013a). There is a dose-response relationship between body mass index and inflammatory markers. In addition, early life stress has been reliably associated with increased inflammation and an increased inflammatory response to stress in exposed individuals (Danese and Lewis, 2017). A rich literature also describes the strong relationship between markers of inflammation and medical illnesses including diabetes, cardiovascular disease, and cancer (Furman et al., 2019). Indeed, elevated CRP has been shown to be a reliable predictor of the development of these disorders, and the American Heart Association has defined levels of inflammatory risk based on CRP with a CRP < 1 mg/L representing low risk, a CRP between 1 and 3 mg/L representing moderate risk, and a CRP > 3 mg/L representing high risk (Ridker, 2003). Finally, data also strongly implicate increased inflammatory markers in patients with anxiety disorders (Michopoulos et al., 2016).

Mechanisms by which inflammation affects antidepressant treatment response

There are many pathways by which inflammation can undermine the fundamental mechanisms by which conventional antidepressants act including effects on monoamine metabolism, glutamate metabolism and neurogenesis and neural plasticity (Fig. 15.1). These effects of inflammation in turn influence neural circuits in the brain, especially those that regulate motivation and motor activity as well as arousal, anxiety, and alarm (Miller and Raison, 2016; Miller et al., 2013).

Monoamine metabolism

SSRIs and SNRIs act in large part by increasing the synaptic availability of the monoamine neurotransmitters serotonin and norepinephrine. Inflammation has multiple effects on monoamine metabolism including decreasing monoamine synthesis and release and increasing monoamine reuptake (Fig. 15.1A) (Haroon et al., 2012; Lucido et al., 2021; Miller et al., 2013; Miller and Raison, 2016). Through effects on tetrahydrobiopterin (BH4), inflammation decreases monoamine synthesis (Haroon et al., 2012; Lucido et al., 2021; Miller et al., 2017). BH4 is an essential enzyme cofactor for the enzymes responsible for synthesis of the monoamines. These enzymes include tryptophan hydroxylase, which converts tryptophan to 5-hydroxytryptophan, a metabolic intermediate in the synthesis of serotonin; phenylalanine hydroxylase, which converts phenylalanine to tyrosine; and tyrosine hydroxylase which converts tyrosine to L-dihydroxyphenylalanine (L-DOPA) a metabolic intermediate in the synthesis of dopamine (DA) and norepinephrine. Relevant to inflammation, BH4 also serves

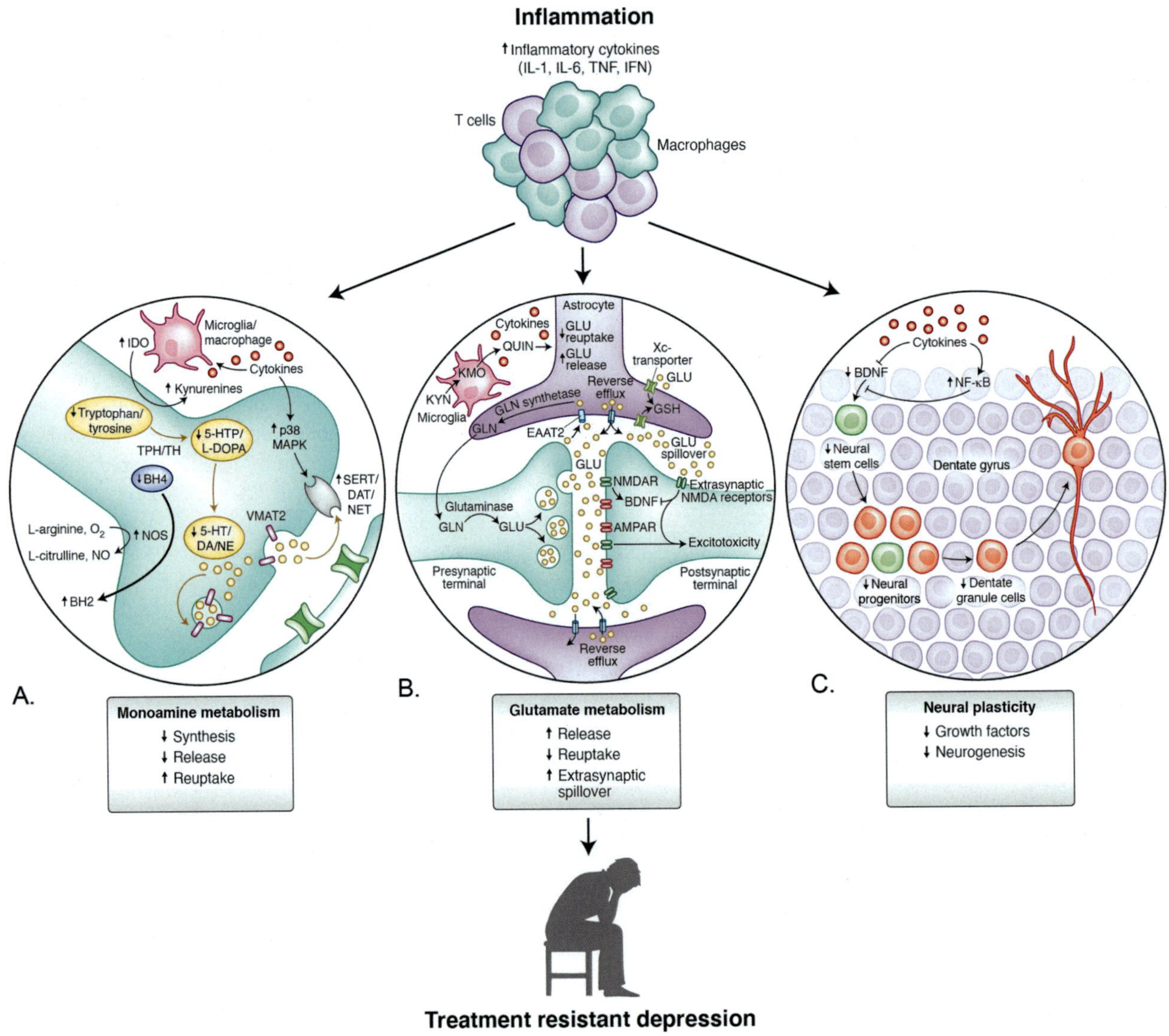

FIG. 15.1 Inflammation undermines the mechanism of action of conventional antidepressants leading to treatment-resistant depression. Inflammation and the release of inflammatory cytokines from myeloid and/or lymphoid cell populations including tumor necrosis factor (TNF), interleukin (IL)-1, IL-6 and interferon (IFN) (alpha and gamma) lead to activation of pathways that can sabotage and circumvent the mechanism of action of conventional antidepressants including serotonin and serotonin/norepinephrine reuptake inhibitors. (A) Inflammation can reduce monoamine availability by decreasing the synthesis and release and increasing the reuptake of monoamines. Inflammation decreases the availability of tetrahydrobiopterin (BH4) through oxidation as well as generation of dihydrobiopterin (BH2) in the conversion of arginine to nitric oxide (NO) by nitric oxide synthase (NOS). BH4 is an essential enzyme cofactor for enzymes responsible for the synthesis of serotonin, dopamine and norepinephrine including tryptophan hydroxylase (TPH) and tyrosine hydroxylase (TH). Inflammation can also decrease the vesicular monoamine transporter 2 (VMAT2) and thus decrease the packaging and release of monoamines from their vesicles, while also activating mitogen activated protein kinases (MAPK) including p38 MAPK that can increase the expression and function of monoamine reuptake pumps including the serotonin transporter (SERT), dopamine (DA) transporter (DAT), and the norepinephrine (NE) transporter (NET). (B) Through effects on astrocytes and the induction of quinolinic acid (QUIN) from kynurenine (KYN) via kynurenine monooxygenase (KMO), inflammation can also cause decreased reuptake and increased release of glutamate (GLU) from astrocytes as well as lead to reverse efflux of GLU from GLU transporters (excitatory amino acid transporter 2—EAAT2). In addition, inflammation-induced oxidative stress leads to the production of the antioxidant glutathione (GSH), which is formed by the extrusion of one molecule of GLU for one molecule of cysteine by the exchange (Xc) transporter. GLU is recycled as glutamine (GLN) by GLN synthetase and glutaminase to GLU in glutaminergic neurons. Build-up of GLU in the synapse leads to extrasynaptic spillover and binding of GLU to extrasynaptic *N*-methyl-D-aspartate receptors (NMDAR) that can decrease growth factors such as brain derived neurotrophic factor (BDNF) and ultimately lead to excitotoxicity. Of relevance to treatment resistance, GLU is not a target of conventional antidepressants. (C) Finally, direct effects of inflammatory cytokines on BDNF through the inflammatory signaling molecule nuclear factor kappa B (NF-kB) can decrease neurogenesis and neural plasticity, critical components of the response to conventional antidepressants. *5-HTP*, serotonin; *AMPAR*, alpha-amino-3-hydroxy-5-methyl-4-isoxazolepropionic acid (AMPA) receptors.

as an enzyme cofactor for nitric oxide synthase (NOS), which leads to the production of nitric oxide, a key mediator of pathogen resolution (Haroon et al., 2012; Miller and Raison, 2016). Thus, BH4 can be usurped by excessive activation of NOS in the context of an immune challenge and in the process is converted to its inactive form BH2. BH4 is also highly sensitive to oxidative stress that is prevalent in the context of an inflammatory response, and as a result is irreversibly degraded into the inert dihydroxyanthopterin (Haroon et al., 2012). Of note, increased IL-6 in the CSF of patients receiving IFN-alpha was associated with decreased CSF BH4 (Felger et al., 2013a). Moreover, the phenylalanine to tyrosine ratio (a potential index of BH4 activity) was associated with increased scores on the Multidimensional Fatigue Inventory and decreased CSF DA (Felger et al., 2013a).

Inflammation is also associated with decreased monoamine release. Studies have shown that administration of inflammatory stimuli to humans and laboratory animals leads to reduced release of DA (Felger and Treadway, 2017). For example, nonhuman primates administered the inflammatory cytokine IFN alpha for 4 weeks exhibit decreased release of DA in response to amphetamine and K+ in the striatum as determined by in vivo microdialysis (Felger et al., 2013b). In vivo microdialysis has also demonstrated decreased release of DA following intraperitoneal administration of IL-6 in the nucleus accumbens (part of the ventral striatum) in rodents (Yohn et al., 2016). In human studies, neuroimaging using F18-labeled DOPA demonstrated decreased release of radiolabeled ligand from DAergic neurons (Capuron et al., 2012). Interestingly, consistent with inflammation's effects on DA synthesis, administration of the immediate DA precursor L-DOPA via reverse microdialysis was found to restore the DA response to amphetamine after chronic (4 weeks) IFN-alpha administration to non-human primates (Felger et al., 2015). Of note, both IL-1 and TNF have been found to reduce expression of the vesicular monoamine transporter 2, which is responsible for packaging DA into vesicles prior to release (Kazumori et al., 2004).

Finally, inflammation has been shown to increase the expression and activity of the serotonin and norepinephrine transporters (Zhu et al., 2006, 2010). These effects are mediated by p38 mitogen activated protein kinase (MAPK), and blockade of p38 MAPK has been shown to reverse the effects of endotoxin on depressive-like behavior in rodents, while also reversing endotoxin's reduction of extracellular serotonin (Zhu et al., 2006, 2010). Similar effects of MAPK on the DA transporter have been described (Moron et al., 2003).

Taken together, the effects of inflammation on monoamine metabolism serve to reduce monoamine availability and thus undermine a primary effect of conventional antidepressants, which is to increase synaptic monoamines.

Glutamate metabolism

Another pathway by which inflammation can undermine the effects of conventional antidepressants is to affect pathways that are not the targets of these drugs (Fig. 15.1B). One such pathway is related to the metabolism of glutamate, an excitatory amino acid, whose concentration is carefully controlled within the synapse (Lucido et al., 2021; Haroon et al., 2017). Astrocytes are a major target of inflammation with well-established effects on excitatory amino acid transporters (EAAT) especially EAAT2, which is the primary glutamate transporter on astrocytes that removes glutamate from the tripartite synapse, consisting of pre- and postsynaptic glutamatergic neurons and astrocytes (Haroon et al., 2017). Several inflammatory cytokines including TNF and IL-1beta have been shown to decrease the expression of EAAT2 on astrocytes and can lead to reverse efflux of glutamate from these transporters (Ida et al., 2008; Matute et al., 2006; Tilleux and Hermans, 2007). In addition, inflammation can activate indoleamine 2,3 dioxygenase (IDO) which converts tryptophan to kynurenine (KYN) (Haroon et al., 2012). KYN is in turn transported to the brain through the large neutral amino acid transporter (LAT2) in the blood-brain-barrier (BBB) where it is converted by microglia into quinolinic acid (QUIN). QUIN can cause astrocytes to decrease the reuptake of glutamate and increase its release (Haroon et al., 2017). Inflammation also leads to increased oxidative stress that induces astrocytes to produce the antioxidant glutathione through a mechanism called the xC transporter that extrudes one molecule of glutamate in exchange for one molecule of cysteine (Haroon et al., 2017). Taken together, these and other effects of inflammation on glutamate metabolism can lead to increased synaptic glutamate and ultimately spillover of glutamate into the extrasynaptic space. Once in the extrasynaptic space, glutamate can bind to extrasynaptic *N*-methyl-D-aspartate (NMDA) receptors which in some reports can lead to decreased production of growth factors including brain derived neurotrophic factor (BDNF) and ultimately excitotoxicity (Haroon et al., 2017). Extrasynaptic spillover of glutamate can also influence local firing patterns of adjacent neurons leading to reduced synaptic precision and ultimately chaotic and incoherent local and interregional neuronal signaling and connectivity (Haroon et al., 2017). For example, patients with major depression and increased inflammation as indexed by CRP in conjunction with elevated basal ganglia glutamate as measured by magnetic resonance spectroscopy exhibit decreased regional homogeneity (ReHo), a measure of local coherence in firing of neighboring voxels in the brain (Haroon et al., 2018b). Moreover, in depressed patients with both increased CRP and basal ganglia glutamate, multiple regions in the brain exhibit decreased ReHo including networks that are involved in motivation in association with symptoms of anhedonia (Haroon et al., 2018b). Depressed patients with increased CRP plus decreased ReHo also exhibit treatment resistance to antidepressant therapy (Haroon et al., 2018b). Thus, alterations in glutamate as a function of the impact of inflammation on astrocytes and glutamate metabolism is another mechanism by which inflammation can undermine the effects of conventional antidepressants, which have limited effects on glutamate neurotransmission.

Neural plasticity

Several studies have indicated that neurogenesis plays an important role in the response to conventional antidepressant therapy (Miller et al., 2009). Indeed, blockade of neurogenesis in the brain using radiation for example can abrogate the antidepressant effects of conventional antidepressants seen in laboratory animals exposed to chronic stress (Santarelli et al., 2003). Relevant in this regard is that inflammation also has been shown to inhibit neurogenesis both in vitro and in vivo and in the context of chronic stress (Fig. 15.1C) (Borsini et al., 2015). For example, laboratory animals exposed to chronic mild stress exhibit increased IL-1 and decreased neurogenesis in conjunction with the development of depressive-like behavior (Goshen et al., 2008; Koo and Duman, 2008). Blockade of IL-1 reverses the effects of chronic stress on depressive-like behavior in these animals, an effect ultimately mediated by inhibiting the inflammatory transcription factor nuclear factor kappa B (NF-kB), which decreases BDNF, a primary driver of neurogenesis (Goshen et al., 2008; Koo et al., 2010). The inhibitory effects of inflammation on neurogenesis represent another possible pathway by which inflammation can sabotage the effects of conventional antidepressants which are in part dependent on stimulating the growth and development of new neurons.

Impact of inflammation on neurocircuits

Given the impact of inflammation on neurotransmitter systems and neurogenesis/neural plasticity, much attention has been paid to the neural circuits that are affected by inflammation to lead to behavioral change. Based on a variety of neuroimaging strategies during inflammatory states, the circuits in the brain most reliably affected by inflammation involve those that regulate motivation and motor activity and those that regulate arousal, anxiety, and alarm (Miller and Raison, 2016; Davies et al., 2021; Raison and Miller, 2013). For example, administration of the inflammatory cytokine IFN-alpha to patients with hepatitis C for 4 weeks was shown to significantly decrease activation of the ventral striatum (VS) using functional magnetic resonance imaging (fMRI) and a hedonic reward (gambling) task (Capuron et al., 2012). Decreases in VS activation in turn were correlated with increased anhedonia. Similar findings have been observed in healthy volunteers acutely exposed to endotoxin or typhoid vaccination, demonstrating the reproducibility of the impact of inflammation on these brain circuits (Eisenberger et al., 2010; Harrison et al., 2016). In patients with major depression (MD), increased inflammation as indexed by CRP was also found to be associated with decreased functional connectivity between the VS and ventromedial prefrontal cortex (vmPFC) in conjunction with symptoms of anhedonia (Felger et al., 2016). Increased CRP was also found in MD patients to be associated with decreased connectivity between dorsal striatum and motor regions of the PFC in relation to symptoms of psychomotor slowing (Felger et al., 2016). These effects of inflammation on circuits regulating motivation and motor activity complement data demonstrating that inflammatory stimuli also activate circuits involving the dorsal anterior cingulate cortex (dACC), insula, hippocampus, and amygdala in association with symptoms of anxiety (Davies et al., 2021; Capuron et al., 2005). Indeed, increased inflammation as measured by CRP in patients with MD was associated with decreased connectivity between PFC and amygdala in conjunction with anxiety symptoms (Mehta et al., 2018). Increased inflammatory responses to stress have also been associated with increased activation of the dACC in association with task-based emotional distress during fMRI (Slavich et al., 2010). Interestingly, the effects of inflammation on these circuits relate in part to data demonstrating the impact of stress and inflammation on blood-brain barrier integrity in the region of the nucleus accumbens/VS and the trafficking of immune cells to brain vasculature and meninges including areas surrounding the amygdala (Menard et al., 2017; Weber et al., 2017; McKim et al., 2018; Dudek et al., 2020). Taken together, these data indicate that the effects of inflammation on specific neurocircuitry provide an excellent platform to establish relevant outcome variables for therapeutic interventions designed to reverse inflammation's effects on the brain.

Therapeutic targets addressing TRD and inflammation

Given the many mechanisms by which inflammation can affect neurotransmitter systems, neural plasticity and neurocircuits in the brain to lead to treatment nonresponse, there is a target rich environment supporting opportunities for development of new treatments for TRD (Lucido et al., 2021). The most obvious target is inflammation itself followed by the metabolic pathways within immune cells that drive the production and release of inflammatory mediators. Downstream effects of inflammation on the brain are also viable treatment options including most notably the impact of inflammation on DA and glutamate. Moreover, recent data suggest that novel technological developments in neuromodulation may be relevant regarding inflammation and treatment resistance. Finally, although not explicitly discussed in this chapter, lifestyle interventions also have a role in reducing inflammation including diet and exercise as well as stress reduction techniques. Aggressive management of diseases associated with inflammation is additionally relevant in this regard.

Inflammation

A multitude of studies have examined the efficacy of a variety of drugs with antiinflammatory properties in the treatment of major depression (Kohler-Forsberg et al., 2019). Although some studies have specifically addressed the role of these agents in the context of TRD, most of the literature has focused on patients with MD who are presumably treatment responsive or are only partially treatment responsive. A vast array of agents has been tested including COX 1 and 2 inhibitors (e.g., aspirin, nonsteroidal antiinflammatory drugs (NSAIDs), celecoxib), omega-3 fatty acids, minocycline, statins, pioglitazone, and cytokine antagonists (Kohler-Forsberg et al., 2019; Kappelmann et al., 2018; Cai et al., 2020). Metaanalyses of the data generated from these trials suggests that there may be modest efficacy of these agents in heterogeneous populations of patients with depression (Kohler-Forsberg et al., 2019). In general, however, these studies have been small, and given that only a subpopulation of depressed patients have increased inflammation, it is surprising that any effect has been observed at all. Indeed, in the largest trial to date, no effect was found for either celecoxib or minocycline compared to placebo (Husain et al., 2020). In addition, longitudinal, epidemiological studies with large populations examining whether chronic exposure to antiinflammatory agents (commonly aspirin, NSAIDS, and statins) reduces the development of depression have been mixed and largely negative (Berk et al., 2020; Glaus et al., 2015; Molero et al., 2019; Veronese et al., 2018). As discussed in detail below, there are many pitfalls in the studies that have been conducted to date, and therefore, the field is bereft of interpretable data to move the field forward. Nevertheless, several trials have yielded promising data using inflammatory biomarkers in post hoc fashion to identify patients more likely to respond, and a few trials have used inflammatory makers prospectively to identify likely responders. For example, two of the anticytokine trials (one of which was on TRD subjects) found that patients with increased inflammation at baseline were more likely to respond (Raison et al., 2013b; Salvadore et al., 2018), whereas a third study found early life stress (often associated with inflammation) was a predictor of response (Lee et al., 2020b; McIntyre et al., 2019). In addition, a recent trial using minocycline found that TRD patients with a CRP ≥ 3 mg/L were more likely to respond to minocycline augmentation than patients with a CRP < 3 mg/L (Nettis et al., 2021). Moreover, increased brain TSPO binding at baseline, possibly reflective of microglial activation, was associated with a greater response to celecoxib in an open label study (Attwells et al., 2020).

There are many antiinflammatory drugs available to be used in future trials including those targeting novel pathways such as Janus kinase (JAK)1 and JAK2, p38 MAPK, the inflammasome (via inhibition of P2X7 receptor), and anticytokine drugs targeting T-cell cytokines such as anti-IL-17 and anti-IL-12/23 as well as the anticytokine drugs that block TNF, IL-1, and IL-6 (Lucido et al., 2021). However, the most pressing issue at this juncture is to conduct trials that leverage the knowledge of the effects of inflammation on the brain and behavior to conduct trials that will yield interpretable data (see later).

Immunometabolism

Upon activation, immune cells undergo shifts in metabolism that involve a relative increase in glycolysis and a reduction in oxidative phosphorylation (O'Neill et al., 2016). This metabolic reprogramming, which occurs in the presence of oxygen (aerobic glycolysis), is referred to as the Warburg effect and has also been described in cancer cells (O'Neill et al., 2016). Of note, glycolysis is highly energy inefficient generating 2 ATP per molecule of glucose compared to 34 ATP per molecule of glucose during oxidative phosphorylation. The shift to glycolysis allows rapid cellular activation and proliferation without the need to generate more mitochondria where oxidative phosphorylation occurs. The role of metabolic reprogramming in the response to antiinflammatory medications was first observed in the context of TRD patients being treated with the TNF antagonist infliximab (Mehta et al., 2013; Bekhbat et al., 2018). Gene expression analysis of peripheral blood immune cells revealed that infliximab responders versus nonresponders exhibited an overrepresentation of not only pathways related to TNF and NF-kB but also pathways related to glycolysis and gluconeogenesis as well as lipid metabolism, which is also associated with immunometabolic reprogramming (Mehta et al., 2013). More recent studies have revealed that molecular pathways in peripheral blood immune cells including hypoxia-inducible factor 1 alpha, protein kinase B (also known as AKT), and phosphoinositide 3-kinase/AKT/mammalian target of rapamycin (mTOR) as well as insulin signaling are associated with anhedonia in patients with high inflammation (CRP > 3 mg/L) and correlate with functional connectivity and reward-related pathways that are associated with motivational deficits (Bekhbat et al., 2020; Goldsmith et al., 2020). These data are consistent with the impact of inflammation on symptoms of anhedonia and expand consideration of the immunologic pathways involved in inflammation to include cellular immunometabolism.

Of relevance to treatment targets for TRD, drug development in autoimmune and inflammatory disorders is targeting glycolysis and related molecular pathways as novel treatments, in contradistinction to targeting the molecular pathways and products of the inflammatory response itself (Lucido et al., 2021). Drugs of interest in this regard include dimethyl fumarate (used in multiple sclerosis), which inhibits the enzyme glyceraldehyde 3-phosphate dehydrogenase that occupies a central position in the glycolytic pathway (Kornberg et al., 2018); imatinib (used for cancer) which inhibits tyrosine kinases that in turn inhibit hexokinase (HK) and glucose-6-phosphate dehydrogenase, enzymes that promote glycolysis (Gottschalk et al., 2004; AlAsfoor et al., 2018), and lonidamine, which inhibits mitochondrially bound HK (de Lartigue, 2017; Song et al., 2019). In addition, recent data suggest that rapamycin (an mTOR complex 1 inhibitor) prolonged the antidepressant effects of ketamine in patients with MD, consistent with the prediction of response to ketamine by inflammatory mediators including IL-6 (Abdallah et al., 2020). Finally, given the interaction between insulin resistance and inflammation in association with anhedonia, insulin sensitizing agents have been considered and, in some cases, tested in depression including thiazolidinediones (e.g., pioglitazone), which stimulates peroxisome proliferator-activated receptor gamma, and metformin, which promotes 5′ adenosine monophosphate-activated protein kinase (Moulton et al., 2018).

Neurotransmitter targets

Dopamine

As noted earlier, the effects of inflammation on DA synthesis, release and reuptake appear to converge on alterations in ventral striatal responses to rewarding stimuli as well as objective and clinical measures of motivational deficits (Lucido et al., 2021; Felger and Treadway, 2017; Felger and Miller, 2012). Therefore, increasing the availability of DA is a reasonable strategy to reverse inflammation's effects on behavior, while also providing a target in the brain as reflected by reward circuitry involving the VS and vmPFC (Felger et al., 2016). Indeed, as noted earlier, two studies in depression have shown that increased CRP (> 1 mg/L) was associated with a poor response to escitalopram, but a much better response to nortriptyline or bupropion (Uher et al., 2014; Jha et al., 2017). These latter drugs both impact DA metabolism, possibly providing the rationale for their relative success in the face of increased inflammation.

In terms of DA synthesis, supplementation of BH4 with various precursor molecules can facilitate the conversion of phenylalanine to tyrosine and tyrosine to L-DOPA (Haroon et al., 2012). Such molecules include folic acid, L-methylfolate and *S*-adenosyl methionine (Miller et al., 2017). Relevant in this regard, treatment of inadequate responders to SSRIs with L-methylfolate led to improvement in depressive symptoms in those individuals with increased baseline levels of inflammatory markers including CRP, TNF, and IL-6 in conjunction with markers of metabolic dysfunction including BMI and leptin (Shelton et al., 2015). These data not only support the notion that inflammation and metabolism may interact to disrupt treatment responsiveness but also suggest that depressed patients with this immunometabolic profile may be most responsive to drugs targeted to DA in general and DA synthesis in particular. Of note, L-DOPA, the immediate precursor of DA, has also been shown to improve psychomotor speed and depressive symptoms in elderly depressed individuals who are often treatment resistant (Rutherford et al., 2019). Additionally, L-DOPA is being tested for its capacity to reverse decreased functional connectivity between VS and vmPFC in patients with MD and increased inflammation (NCT02513485).

Other medications that affect DA availability are those that increase DA release and inhibit synaptic reuptake. Drugs that increase DA release include amphetamine, methamphetamine, and lisdexamfetamine. Although there is limited data supporting the efficacy of these agents in TRD, and no data on patients with TRD and increased inflammation, studies have found efficacy of lisdexamfetamine as an augmentation strategy to treat persistent symptoms of depression in MD patients (Trivedi et al., 2013; Madhoo et al., 2014). In addition, several drugs have the capacity to block DA reuptake including bupropion (discussed earlier) as well as methylphenidate and modafinil. Methylphenidate has been shown to reverse the effects of IL-6 on effort-based motivation in laboratory animals (Yohn et al., 2016), while improving fatigue in cancer patients who often exhibit increased inflammatory markers (Rojí and Centeno, 2017; Xiao et al., 2017). Modafinil has also been found to reduce fatigue in patients with depression or TRD (Fava et al., 2005), although the efficacy of both methylphenidate and modafinil as antidepressants is not well established. Of note, bupropion is currently being tested (vs escitalopram) for superior efficacy for anhedonia in MD patients with CRP ≥ 2 mg/L (NCT04352101).

Finally, there has been interest in the use of DA agonists to treat patients with TRD. Indeed, pramipexole has been shown to improve depressive symptoms in TRD patients at high doses. Pramipexole has also shown antidepressant efficacy in patients with Parkinson's disease as well as patients with MD with special benefit for symptoms of anhedonia and psychomotor retardation (Lemke et al., 2006; Fawcett et al., 2016; Escalona and Fawcett, 2017). Finally, aripiprazole and brexpiprazole both exhibit partial D2/3 agonism and both are FDA approved as augmentation strategies for patients with

TRD (Fava et al., 2012; Al Shirawi et al., 2017). Whether TRD patients with increased inflammation may be especially responsive to these medications remains untested.

Glutamate

Given the profound antidepressant effects of ketamine, a NMDA antagonist, and the prediction of ketamine response by inflammatory markers including IL-6 and CRP in TRD patients as well as a laboratory animal model of TRD (Yang et al., 2015; Walker et al., 2015), there is reason to believe that drugs that target glutamate and its receptors may be especially relevant for TRD patients with increased inflammation (Wilkinson and Sanacora, 2019). Indeed, studies in laboratory animals have shown that ketamine can block the development of endotoxin-induced depressive symptoms (Walker et al., 2013). However, despite the success of ketamine and its recently FDA-approved (*S*)-enantiomer esketamine, drugs targeting glutamate and its NMDA, AMPA, and metabotropic glutamate receptors have yet to show efficacy in antidepressant trials to date. For example, memantine, an NMDA antagonist, has failed to show efficacy in multiple trials in MD (Kishi et al., 2017), and lanicemine and GLYX-13 (rapastinel) did not separate from placebo, despite initial promise in disorders including depression (Wilkinson and Sanacora, 2019; Sanacora et al., 2017). There are numerous other glutamate receptor drugs in development, however, review of these agents is outside the scope of this chapter (Wilkinson and Sanacora, 2019). Nevertheless, the possibility that these glutamatergic drugs may be uniquely effective in patients with increased inflammation has yet to be examined. Indeed, the compound AXS-05, a combination of bupropion and the NMDA receptor antagonist dextromethorphan, may be of special interest, given its potential capacity to address both DA and glutamate pathologies in patients with TRD and increased inflammation. AXS-05 has shown initial promise and has been given FDA Breakthrough Therapy designation (Anderson et al., 2019). Finally, riluzole is an interesting compound in that it has been shown to decrease presynaptic glutamate release and increase the activity of EAAT2 (Banasr et al., 2010). Riluzole has also been shown in laboratory animal models of chronic unpredictable stress to mitigate anhedonic behavior (Banasr et al., 2010). Riluzole has exhibited evidence of benefit in at least two small studies in TRD (Zarate et al., 2004; Sanacora et al., 2007), in addition to augmentation of the response to ketamine, delaying the time to relapse (Ibrahim et al., 2012). Current studies are underway to examine the efficacy of riluzole in treating depressive symptoms in the context of increased inflammation in cancer patients (NCT02796755).

Another target relevant to glutamate is the IDO pathway and the generation of QUIN. As noted earlier, QUIN is largely derived from peripherally-produced KYN that is transported to the brain through LAT-2 in the BBB (Walker et al., 2019). Leucine is an amino acid that competes with KYN for transport into the brain and was shown to block the development of depressive-like behavior following the administration of endotoxin to rodents (Walker et al., 2019). Leucine is currently being tested in patients with MD (NCT03079297). Inhibition of IDO is another relevant strategy to reduce QUIN. Several IDO inhibitors are currently in clinical testing (Prendergast et al., 2017). However, the majority of current trials of IDO inhibitors are in cancer where the blockade of IDO activity spares tryptophan from being converted to KYN (Prendergast et al., 2017; Vacchelli et al., 2014). Tryptophan is an essential amino acid for T-cell activation, which is critical in the immunologic response to tumors.

Neuromodulation

The impact of inflammation on neural circuitry invokes the potential use of neuromodulation strategies to restore the integrity of these circuits by stimulating or inhibiting relevant brain regions using a variety of techniques including ECT, repetitive transcranial magnetic stimulation, deep brain stimulation, and MRI-guided ultrasound. As noted earlier, increased inflammatory markers including IL-6 has been shown to predict response to ECT (Kruse et al., 2018). Response to these other neuromodulation strategies has yet to be evaluated in humans.

Although these strategies focus on the downstream effects of inflammation on the brain, new neuromodulation technologies have been developed that can influence the immune system itself. Indeed, stimulation of the afferent vagus nerve can inhibit inflammatory responses through the release of acetylcholine, which binds to the alpha 7 subunit of the nicotinic receptor on immune cells that in turn inhibits NF-kB (Pavlov et al., 2020). Such stimulation of vagal afferents has been shown to reduce mortality following endotoxin administration in laboratory animals, and more recently has received FDA approval for use in rheumatoid arthritis (Pavlov et al., 2020). This use of vagal nerve stimulation is the first example of a neuromodulation technique being employed to target the immune system. Of note, implanted vagal nerve stimulation devices used for depression generate efferent signals to the brain and are therefore not designed to target the immune system. Nevertheless, recent work has shown that transcutaneous vagal nerve stimulation, which stimulates both efferent and afferent vagal nerve fibers, blocks stress-induced production of peripheral blood IL-6 and IFN-gamma in patients with PTSD and may have therapeutic potential (Bremner et al., 2020). Finally, recent data suggest that exposure to lights flickering at specific frequencies can drive relative frequencies in the brain, which in turn can influence the activation state of microglia (Garza et al., 2020). This technology has only been applied to laboratory animals but holds intriguing promise for regulating CNS immune responses in neuropsychiatric disorders.

Future directions

Based on the information provided above, there are many opportunities to advance strategies targeting inflammation and its downstream effects on the brain to address depressed patients who have experienced nonresponse to conventional antidepressant therapies. Although the lion's share of research on drugs in this area has not focused on patients with treatment resistance, there is sufficient data to suggest that drugs affecting inflammation's impact on behavior may be especially relevant to patients with TRD, given that these patients are more likely to exhibit increased inflammatory markers and increased inflammatory markers predict treatment nonresponse to conventional antidepressants. Moreover, as noted earlier, increased inflammatory markers predict response to treatments including ECT and ketamine that are typically reserved for TRD patients.

Unfortunately, despite the many trials that have been conducted using antiinflammatory medications to treat patients with depression and TRD, there are significant problems in study design, and the drugs employed as well as the outcome variables make interpretation of the extant literature virtually impossible. The lack of attention to the unique aspects of a pathophysiologic pathway with defined effects on the brain has led to a literature poorly aligned with the subgroup of patients that are most likely to respond and if anything jeopardizes the field. Indeed, if what is known about inflammation in depression is not incorporated into clinical trial design, the field will not advance, and an extraordinary opportunity to develop new treatments for depression will be lost.

Clinical trial design

The opportunities afforded clinical trial designs by virtue of our understanding of the role of inflammation in depression are manifold (Table 15.1).

Immunophenotyping inflammation in depression

Finally, one area that has drawn increasing interest includes the relevant components of the immune response that drive inflammation in depression and other psychiatric disorders. Considerable data exists that myeloid cells including monocytes/macrophages, dendritic cells and neutrophils exhibit activation of inflammatory signaling molecules including NF-kB that sustain inflammatory responses (Miller and Raison, 2016; Felger and Miller, 2020). Nevertheless, a growing literature

TABLE 15.1 Design considerations for clinical trials of drugs to mitigate the impact of inflammation on the brain.

Sample
Enrich sample for patients with high inflammation
Stratify sample by high and low inflammation[a]
Use blood, CSF, or brain markers of inflammation to enrich or stratify subjects
Outcome variables
Use clinical and objective assessments of symptom domains aligned with inflammation's effects on the brain including Positive and Negative Valence Systems (e.g., anhedonia, psychomotor slowing, anxiety)
Target engagement
Monitor decrease in inflammatory markers in blood, CFS, or brain (e.g., translocator protein binding, free water diffusion in white matter) and associate with treatment response
Monitor interaction of treatment with downstream effects of inflammation on neurotransmitters/neurocircuits affected by inflammation (e.g., reward circuitry, threat circuitry) using neuroimaging

[a]*Match/mismatch design enhances assessment of specificity (to inflammation).*
CSF, cerebrospinal fluid.

supports the role of T cells including increased representation of highly proinflammatory TH17 cells in depression (Felger and Miller, 2020; Beurel and Lowell, 2018). Interestingly in this regard, a recent study using a variety of clustering strategies of depressed patients with and without inflammation indexed by CRP found that there were two clusters of depressed patients with increased inflammation, one group primarily represented by myeloid cells and the other by lymphoid cells (Lynall et al., 2020). These data suggest that there may be distinct immunophenotypes that drive inflammation in depression, or alternatively, that there is a longitudinal trajectory from innate immune (myeloid) cell responses to inflammatory responses driven more by lymphoid components as the disease progresses. In either case, obtaining more clarity on the immunology of inflammation in depression is necessary if uniquely immune-targeted treatments are to be successfully developed. Moreover, it may be that CRP is not specific enough to guide patient and treatment selection, especially as it relates to specific immune targets. Just as fever only identifies the presence of an infection and not the type, so does CRP only identify inflammation but not the immunologic source.

References

Abdallah, C.G., Averill, L.A., Gueorguieva, R., et al., 2020. Modulation of the antidepressant effects of ketamine by the mTORC1 inhibitor rapamycin. Neuropsychopharmacology 45 (6), 990–997.

Al Shirawi, M.I., Edgar, N.E., Kennedy, S.H., 2017. Brexpiprazole in the treatment of major depressive disorder. Clin. Med. Insights Ther. 9, 1179559X17731801.

AlAsfoor, S., Rohm, T.V., Bosch, A.J.T., et al., 2018. Imatinib reduces non-alcoholic fatty liver disease in obese mice by targeting inflammatory and lipogenic pathways in macrophages and liver. Sci. Rep. 8 (1), 15331.

Anderson, A., Iosifescu, D.V., Jacobson, M., et al., 2019. Efficacy and safety of AXS-05, an oral NMDA receptor antagonist with multimodal activity, in major depressive disorder: results of a phase 2, double-blind, active-controlled trial. W43. In: Paper Presented at ASCP Annual Meeting.

Attwells, S., Setiawan, E., Rusjan, P.M., et al., 2020. Translocator protein distribution volume predicts reduction of symptoms during open-label trial of celecoxib in major depressive disorder. Biol. Psychiatry 88 (8), 649–656.

Banasr, M., Chowdhury, G.M., Terwilliger, R., et al., 2010. Glial pathology in an animal model of depression: reversal of stress-induced cellular, metabolic and behavioral deficits by the glutamate-modulating drug riluzole. Mol. Psychiatry 15 (5), 501–511.

Bekhbat, M., Chu, K., Le, N.A., et al., 2018. Glucose and lipid-related biomarkers and the antidepressant response to infliximab in patients with treatment-resistant depression. Psychoneuroendocrinology 98, 222–229.

Bekhbat, M., Treadway, M.T., Goldsmith, D.R., et al., 2020. Gene signatures in peripheral blood immune cells related to insulin resistance and low tyrosine metabolism define a sub-type of depression with high CRP and anhedonia. Brain Behav. Immun. 88, 161–165.

Berk, M., Woods, R.L., Nelson, M.R., et al., 2020. Effect of aspirin vs placebo on the prevention of depression in older people: a randomized clinical trial. JAMA Psychiatry 77 (10), 1012–1020.

Beurel, E., Lowell, J.A., 2018. Th17 cells in depression. Brain Behav. Immun. 69, 28–34.

Borsini, A., Zunszain, P.A., Thuret, S., Pariante, C.M., 2015. The role of inflammatory cytokines as key modulators of neurogenesis. Trends Neurosci. 38 (3), 145–157.

Bremner, J.D., Gurel, N.Z., Jiao, Y., et al., 2020. Transcutaneous vagal nerve stimulation blocks stress-induced activation of interleukin-6 and interferon-γ in posttraumatic stress disorder: a double-blind, randomized, sham-controlled trial. Brain Behav. Immun. Health 9, 100138.

Brydon, L., Harrison, N.A., Walker, C., Steptoe, A., Critchley, H.D., 2008. Peripheral inflammation is associated with altered substantia nigra activity and psychomotor slowing in humans. Biol. Psychiatry 63 (11), 1022–1029.

Cai, D.B., Zheng, W., Zhang, Q.E., et al., 2020. Minocycline for depressive symptoms: a meta-analysis of randomized, double-blinded, placebo-controlled trials. Psychiatry Q. 91, 451–461.

Capuron, L., Gumnick, J.F., Musselman, D.L., et al., 2002. Neurobehavioral effects of interferon-alpha in cancer patients: phenomenology and paroxetine responsiveness of symptom dimensions. Neuropsychopharmacology 26 (5), 643–652.

Capuron, L., Pagnoni, G., Demetrashvili, M., et al., 2005. Anterior cingulate activation and error processing during interferon-alpha treatment. Biol. Psychiatry 58 (3), 190–196.

Capuron, L., Fornwalt, F.B., Knight, B.T., Harvey, P.D., Ninan, P.T., Miller, A.H., 2009. Does cytokine-induced depression differ from idiopathic major depression in medically healthy individuals? J. Affect. Disord. 119 (1–3), 181–185.

Capuron, L., Pagnoni, G., Drake, D.F., et al., 2012. Dopaminergic mechanisms of reduced basal ganglia responses to hedonic reward during interferon alfa administration. Arch. Gen. Psychiatry 69 (10), 1044–1053.

Cattaneo, A., Ferrari, C., Uher, R., et al., 2016. Absolute measurements of macrophage migration inhibitory factor and interleukin-1-beta mRNA levels accurately predict treatment response in depressed patients. Int. J. Neuropsychopharmacol. 19 (10), pyw045.

Cattaneo, A., Ferrari, C., Turner, L., et al., 2020. Whole-blood expression of inflammasome- and glucocorticoid-related mRNAs correctly separates treatment-resistant depressed patients from drug-free and responsive patients in the BIODEP study. Transl. Psychiatry 10 (1), 232.

Chamberlain, S.R., Cavanagh, J., de Boer, P., et al., 2019. Treatment-resistant depression and peripheral C-reactive protein. Br. J. Psychiatry 214, 11–19.

Danese, A., Lewis, S., 2017. Psychoneuroimmunology of early life stress: the hidden wounds of childhood trauma? Neuropsychopharmacology 42, 99–114.

Davies, K.A., Cooper, E., Voon, V., Tibble, J., Cercignani, M., Harrison, N.A., 2021. Interferon and anti-TNF therapies differentially modulate amygdala reactivity which predicts associated bidirectional changes in depressive symptoms. Mol. Psychiatry 26, 5150–5160.

de Lartigue, J., 2017. Hallmark tumor metabolism becomes a validated therapeutic target. J. Community Support. Oncol. 15 (6), e47–e52.

Dowlati, Y., Herrmann, N., Swardfager, W., et al., 2010. A meta-analysis of cytokines in major depression. Biol. Psychiatry 67 (5), 446–457.

Dudek, K.A., Dion-Albert, L., Lebel, M., et al., 2020. Molecular adaptations of the blood-brain barrier promote stress resilience vs. depression. Proc. Natl. Acad. Sci. U. S. A. 117 (6), 3326–3336.

Eisenberger, N.I., Berkman, E.T., Inagaki, T.K., Rameson, L.T., Mashal, N.M., Irwin, M.R., 2010. Inflammation-induced anhedonia: endotoxin reduces ventral striatum responses to reward. Biol. Psychiatry 68 (8), 748–754.

Enache, D., Pariante, C.M., Mondelli, V., 2019. Markers of central inflammation in major depressive disorder: a systematic review and meta-analysis of studies examining cerebrospinal fluid, positron emission tomography and post-mortem brain tissue. Brain Behav. Immun. 81, 24–40.

Escalona, R., Fawcett, J., 2017. Pramipexole in treatment resistant-depression, possible role of inflammatory cytokines. Neuropsychopharmacology 42 (1), 363.

Fava, M., Thase, M.E., DeBattista, C., 2005. A multicenter, placebo-controlled study of modafinil augmentation in partial responders to selective serotonin reuptake inhibitors with persistent fatigue and sleepiness. J. Clin. Psychiatry 66, 85–93.

Fava, M., Mischoulon, D., Iosifescu, D., et al., 2012. A double-blind, placebo-controlled study of aripiprazole adjunctive to antidepressant therapy among depressed outpatients with inadequate response to prior antidepressant therapy (ADAPT-A study). Psychother. Psychosom. 81 (2), 87–97.

Fawcett, J., Rush, A.J., Vukelich, J., et al., 2016. Clinical experience with high-dosage pramipexole in patients with treatment-resistant depressive episodes in unipolar and bipolar depression. Am. J. Psychiatry 173 (2), 107–111.

Felger, J., Miller, A., 2012. Cytokine effects on the basal ganglia and dopamine function: the subcortical source of inflammatory malaise. Front. Neuroendocrinol. 33 (3), 315–327.

Felger, J.C., Miller, A.H., 2020. Identifying immunophenotypes of inflammation in depression: dismantling the monolith. Biol. Psychiatry 88 (2), 136–138.

Felger, J.C., Treadway, M.T., 2017. Inflammation effects on motivation and motor activity: role of dopamine. Neuropsychopharmacology 42 (1), 216–241.

Felger, J.C., Li, L., Marvar, P.J., et al., 2013a. Tyrosine metabolism during interferon-alpha administration: association with fatigue and CSF dopamine concentrations. Brain Behav. Immun. 31, 153–160.

Felger, J.C., Mun, J., Kimmel, H.L., et al., 2013b. Chronic interferon-α decreases dopamine 2 receptor binding and striatal dopamine release in association with anhedonia-like behavior in nonhuman primates. Neuropsychopharmacology 38 (11), 2179–2187.

Felger, J.C., Hernandez, C.R., Miller, A.H., 2015. Levodopa reverses cytokine-induced reductions in striatal dopamine release. Int. J. Neuropsychopharmacol. 18 (4), pyu084.

Felger, J.C., Li, Z., Haroon, E., et al., 2016. Inflammation is associated with decreased functional connectivity within corticostriatal reward circuitry in depression. Mol. Psychiatry 21 (10), 1358–1365.

Furman, D., Campisi, J., Verdin, E., et al., 2019. Chronic inflammation in the etiology of disease across the life span. Nat. Med. 25 (12), 1822–1832.

Garza, K.M., Zhang, L., Borron, B., Wood, L.B., Singer, A.C., 2020. Gamma visual stimulation induces a neuroimmune signaling profile distinct from acute neuroinflammation. J. Neurosci. 40 (6), 1211–1225.

Glaus, J., Vandeleur, C.L., Lasserre, A.M., et al., 2015. Aspirin and statin use and the subsequent development of depression in men and women: results from a longitudinal population-based study. J. Affect. Disord. 182, 126–131.

Goldsmith, D.R., Bekhbat, M., Le, N.A., et al., 2020. Protein and gene markers of metabolic dysfunction and inflammation together associate with functional connectivity in reward and motor circuits in depression. Brain Behav. Immun. 88, 193–202.

Goshen, I., Kreisel, T., Ben-Menachem-Zidon, O., et al., 2008. Brain interleukin-1 mediates chronic stress-induced depression in mice via adrenocortical activation and hippocampal neurogenesis suppression. Mol. Psychiatry 13 (7), 717–728.

Gottschalk, S., Anderson, N., Hainz, C., Eckhardt, S.G., Serkova, N.J., 2004. Imatinib (STI571)-mediated changes in glucose metabolism in human leukemia BCR-ABL-positive cells. Clin. Cancer Res. 10 (19), 6661–6668.

Haroon, E., Raison, C.L., Miller, A.H., 2012. Psychoneuroimmunology meets neuropsychopharmacology: translational implications of the impact of inflammation on behavior. Neuropsychopharmacology 37 (1), 137–162.

Haroon, E., Miller, A.H., Sanacora, G., 2017. Inflammation, glutamate, and glia: a trio of trouble in mood disorders. Neuropsychopharmacology 42 (1), 193–215.

Haroon, E., Daguanno, A.W., Woolwine, B.J., et al., 2018a. Antidepressant treatment resistance is associated with increased inflammatory markers in patients with major depressive disorder. Psychoneuroendocrinology 95, 43–49.

Haroon, E., Chen, X., Li, Z., et al., 2018b. Increased inflammation and brain glutamate define a subtype of depression with decreased regional homogeneity, impaired network integrity, and anhedonia. Transl. Psychiatry 8 (1), 189.

Harrison, N.A., Brydon, L., Walker, C., Gray, M.A., Steptoe, A., Critchley, H.D., 2009. Inflammation causes mood changes through alterations in subgenual cingulate activity and mesolimbic connectivity. Biol. Psychiatry 66 (5), 407–414.

Harrison, N.A., Voon, V., Cercignani, M., Cooper, E.A., Pessiglione, M., Critchley, H.D., 2016. A neurocomputational account of how inflammation enhances sensitivity to punishments versus rewards. Biol. Psychiatry 80 (1), 73–81.

Howren, M.B., Lamkin, D.M., Suls, J., 2009. Associations of depression with C-reactive protein, IL-1, and IL-6: a meta-analysis. Psychosom. Med. 71 (2), 171–186.

Husain, M.I., Chaudhry, I.B., Khoso, A.B., et al., 2020. Minocycline and celecoxib as adjunctive treatments for bipolar depression: a multicentre, factorial design randomised controlled trial. Lancet Psychiatry 7 (6), 515–527.

Ibrahim, L., Diazgranados, N., Franco-Chaves, J., et al., 2012. Course of improvement in depressive symptoms to a single intravenous infusion of ketamine vs add-on riluzole: results from a 4-week, double-blind, placebo-controlled study. Neuropsychopharmacology 37 (6), 1526–1533.

Ida, T., Hara, M., Nakamura, Y., Kozaki, S., Tsunoda, S., Ihara, H., 2008. Cytokine-induced enhancement of calcium-dependent glutamate release from astrocytes mediated by nitric oxide. Neurosci. Lett. 432 (3), 232–236.

Jha, M.K., Minhajuddin, A., Gadad, B.S., et al., 2017. Can C-reactive protein inform antidepressant medication selection in depressed outpatients? Findings from the CO-MED trial. Psychoneuroendocrinology 78, 105–113.

Kappelmann, N., Lewis, G., Dantzer, R., Jones, P.B., Khandaker, G.M., 2018. Antidepressant activity of anti-cytokine treatment: a systematic review and meta-analysis of clinical trials of chronic inflammatory conditions. Mol. Psychiatry 23 (2), 335–343.

Kazumori, H., Ishihara, S., Rumi, M.A., Ortega-Cava, C.F., Kadowaki, Y., Kinoshita, Y., 2004. Transforming growth factor-alpha directly augments histidine decarboxylase and vesicular monoamine transporter 2 production in rat enterochromaffin-like cells. Am. J. Physiol. Gastrointest. Liver Physiol. 286 (3), G508–G514.

Kishi, T., Matsunaga, S., Iwata, N., 2017. A meta-analysis of memantine for depression. J. Alzheimers Dis. 57 (1), 113–121.

Kohler-Forsberg, O., Nicolaisen Lydholm, C., Hjorthoj, C., Nordentoft, M., Mors, O., Benros, M.E., 2019. Efficacy of anti-inflammatory treatment on major depressive disorder or depressive symptoms: meta-analysis of clinical trials. Acta Psychiatr. Scand. 139, 404–419.

Koo, J.W., Duman, R.S., 2008. IL-1beta is an essential mediator of the antineurogenic and anhedonic effects of stress. Proc. Natl. Acad. Sci. U. S. A. 105 (2), 751–756.

Koo, J.W., Russo, S.J., Ferguson, D., Nestler, N.J., Duman, R.S., 2010. Nuclear factor-kb is a critical mediator of stress-impaired neurogenesis and depressive behavior. Proc. Natl. Acad. Sci. U. S. A. 107, 2669–2674.

Kornberg, M.D., Bhargava, P., Kim, P.M., et al., 2018. Dimethyl fumarate targets GAPDH and aerobic glycolysis to modulate immunity. Science 360 (6387), 449–453.

Kruse, J.L., Congdon, E., Olmstead, R., et al., 2018. Inflammation and improvement of depression following electroconvulsive therapy in treatment-resistant depression. J. Clin. Psychiatry 79 (2), 17m11597.

Lee, Y., Park, Y., Nam, H., Lee, J.W., Yu, S.W., 2020a. Translocator protein (TSPO): the new story of the old protein in neuroinflammation. BMB Rep. 53 (1), 20–27.

Lee, Y., Mansur, R.B., Brietzke, E., et al., 2020b. Efficacy of adjunctive infliximab vs. placebo in the treatment of anhedonia in bipolar I/II depression. Brain Behav. Immun. 88, 631–639.

Lemke, M.R., Brecht, H.M., Koester, J., Reichmann, H., 2006. Effects of the dopamine agonist pramipexole on depression, anhedonia and motor functioning in Parkinson's disease. J. Neurol. Sci. 248 (1), 266–270.

Lucido, M.J., Bekhbat, M., Goldsmith, D.R., Treadway, M.T., Haroon, E., Felger, J.C., Miller, A.H., 2021. Aiding and abetting anhedonia: impact of inflammation on the brain and pharmacological implications. Pharmacol. Rev. 73 (3), 1084–1117.

Lynall, M.E., Turner, L., Bhatti, J., et al., 2020. Peripheral blood cell-stratified subgroups of inflamed depression. Biol. Psychiatry 88 (2), 185–196.

Mac Giollabhui, N., Ng, T.H., Ellman, L.M., Alloy, L.B., 2021. The longitudinal associations of inflammatory biomarkers and depression revisited: systematic review, meta-analysis, and meta-regression. Mol. Psychiatry 26, 3302–3314.

Madhoo, M., Keefe, R.S., Roth, R.M., et al., 2014. Lisdexamfetamine dimesylate augmentation in adults with persistent executive dysfunction after partial or full remission of major depressive disorder. Neuropsychopharmacology 39 (6), 1388–1398.

Maes, M., Scharpe, S., Van Grootel, L., et al., 1992. Higher alpha 1-antitrypsin, haptoglobin, ceruloplasmin and lower retinol binding protein plasma levels during depression: further evidence for the existence of an inflammatory response during that illness. J. Affect. Disord. 24 (3), 183–192.

Matute, C., Domercq, M., Sanchez-Gomez, M.V., 2006. Glutamate-mediated glial injury: mechanisms and clinical importance. Glia 53 (2), 212–224.

McIntyre, R.S., Subramaniapillai, M., Lee, Y., et al., 2019. Efficacy of adjunctive infliximab vs placebo in the treatment of adults with bipolar I/II depression: a randomized clinical trial. JAMA Psychiatry 76, 783–790.

McKim, D.B., Weber, M.D., Niraula, A., et al., 2018. Microglial recruitment of IL-1beta-producing monocytes to brain endothelium causes stress-induced anxiety. Mol. Psychiatry 23 (6), 1421–1431.

Mehta, D., Raison, C.L., Woolwine, B.J., et al., 2013. Transcriptional signatures related to glucose and lipid metabolism predict treatment response to the tumor necrosis factor antagonist infliximab in patients with treatment-resistant depression. Brain Behav. Immun. 31, 205–215.

Mehta, N.D., Haroon, E., Xu, X., Woolwine, B.J., Li, Z., Felger, J.C., 2018. Inflammation negatively correlates with amygdala-ventromedial prefrontal functional connectivity in association with anxiety in patients with depression: preliminary results. Brain Behav. Immun. 73, 725–730.

Menard, C., Pfau, M.L., Hodes, G.E., et al., 2017. Social stress induces neurovascular pathology promoting depression. Nat. Neurosci. 20 (12), 1752–1760.

Michopoulos, V., Powers, A., Gillespie, C.F., Ressler, K.J., Jovanovic, T., 2016. Inflammation in fear- and anxiety-based disorders: PTSD, GAD, and beyond. Neuropsychopharmacology 42, 254–270.

Miller, A.H., 2020. Beyond depression: the expanding role of inflammation in psychiatric disorders. World Psychiatry 19 (1), 108–109.

Miller, A.H., Raison, C.L., 2015. Are anti-inflammatory therapies viable treatments for psychiatric disorders?: where the rubber meets the road. JAMA Psychiatry 72 (6), 527–528.

Miller, A.H., Raison, C.L., 2016. The role of inflammation in depression: from evolutionary imperative to modern treatment target. Nat. Rev. Immunol. 16 (1), 22–34.

Miller, A.H., Maletic, V., Raison, C.L., 2009. Inflammation and its discontents: the role of cytokines in the pathophysiology of major depression. Biol. Psychiatry 65 (9), 732–741.

Miller, A.H., Haroon, E., Raison, C.L., Felger, J.C., 2013. Cytokine targets in the brain: impact on neurotransmitters and neurocircuits. Depress. Anxiety 30 (4), 297–306.

Miller, A.H., Haroon, E., Felger, J.C., 2017. Therapeutic implications of brain-immune interactions: treatment in translation. Neuropsychopharmacology 42 (1), 334–359.

Molero, P., Ruiz-Estigarribia, L., Lahortiga-Ramos, F., et al., 2019. Use of non-steroidal anti-inflammatory drugs, aspirin and the risk of depression: the "Seguimiento Universidad de Navarra (SUN)" cohort. J. Affect. Disord. 247, 161–167.

Moron, J.A., Zakharova, I., Ferrer, J.V., et al., 2003. Mitogen-activated protein kinase regulates dopamine transporter surface expression and dopamine transport capacity. J. Neurosci. 23 (24), 8480–8488.

Moulton, C.D., Hopkins, C.W.P., Ismail, K., Stahl, D., 2018. Repositioning of diabetes treatments for depressive symptoms: a systematic review and meta-analysis of clinical trials. Psychoneuroendocrinology 94, 91–103.

Nettis, M.A., Lombardo, G., Hastings, C., et al., 2021. Augmentation therapy with minocycline in treatment-resistant depression patients with low-grade peripheral inflammation: results from a double-blind randomised clinical trial. Neuropsychopharmacology 46, 939–948.

O'Neill, L.A., Kishton, R.J., Rathmell, J., 2016. A guide to immunometabolism for immunologists. Nat. Rev. Immunol. 16 (9), 553–565.

Pandey, G.N., Rizavi, H.S., Ren, X., Bhaumik, R., Dwivedi, Y., 2014. Toll-like receptors in the depressed and suicide brain. J. Psychiatr. Res. 53, 62–68.

Pavlov, V.A., Chavan, S.S., Tracey, K.J., 2020. Bioelectronic medicine: from preclinical studies on the inflammatory reflex to new approaches in disease diagnosis and treatment. Cold Spring Harb. Perspect. Med. 10 (3), a034140.

Prendergast, G.C., Malachowski, W.P., DuHadaway, J.B., Muller, A.J., 2017. Discovery of IDO1 inhibitors: from bench to bedside. Cancer Res. 77 (24), 6795–6811.

Raison, C.L., Miller, A.H., 2013. The evolutionary significance of depression in Pathogen Host Defense (PATHOS-D). Mol. Psychiatry 18 (1), 15–37.

Raison, C.L., Capuron, L., Miller, A.H., 2006. Cytokines sing the blues: inflammation and the pathogenesis of depression. Trends Immunol. 27 (1), 24–31.

Raison, C.L., Felger, J.C., Miller, A.H., 2013a. Inflammation and treatment resistance in major depression: a perfect storm. Psychiatr. Times 30 (September 12). https://www.psychiatrictimes.com/view/inflammation-and-treatment-resistance-major-depression-perfect-storm.

Raison, C.L., Rutherford, R.E., Woolwine, B.J., et al., 2013b. A randomized controlled trial of the tumor necrosis factor antagonist infliximab for treatment-resistant depression: the role of baseline inflammatory biomarkers. JAMA Psychiatry 70 (1), 31–41.

Rao, J.S., Harry, G.J., Rapoport, S.I., Kim, H.W., 2010. Increased excitotoxicity and neuroinflammatory markers in postmortem frontal cortex from bipolar disorder patients. Mol. Psychiatry 15 (4), 384–392.

Ridker, P.M., 2003. Clinical application of C-reactive protein for cardiovascular disease detection and prevention. Circulation 107 (3), 363–369.

Rojí, R., Centeno, C., 2017. The use of methylphenidate to relieve fatigue. Curr. Opin. Support. Palliat. Care 11 (4), 299–305.

Rutherford, B.R., Slifstein, M., Chen, C., et al., 2019. Effects of L-DOPA monotherapy on psychomotor speed and [(11)C]Raclopride binding in high-risk older adults with depression. Biol. Psychiatry 86 (3), 221–229.

Salvadore, G., Nash, A., Bleys, C., Hsu, B., Saad, Z., Gause, A., Moyer, J., Xi, L., Manji, H., Van Nueten, L., Drevets, W., 2018. A Double-Blind, Placebo-Controlled, Multicenter Study of Sirukumab as Adjunctive Treatment to a Monoaminergic Antidepressant in Adults With Major Depressive Disorder. American College of Neuropsychopharmacology, Hollywood, FL.

Sanacora, G., Kendell, S.F., Levin, Y., et al., 2007. Preliminary evidence of riluzole efficacy in antidepressant-treated patients with residual depressive symptoms. Biol. Psychiatry 61 (6), 822–825.

Sanacora, G., Johnson, M.R., Khan, A., et al., 2017. Adjunctive lanicemine (AZD6765) in patients with major depressive disorder and history of inadequate response to antidepressants: a randomized, placebo-controlled study. Neuropsychopharmacology 42 (4), 844–853.

Santarelli, L., Saxe, M., Gross, C., et al., 2003. Requirement of hippocampal neurogenesis for the behavioral effects of antidepressants. Science 301 (5634), 805–809.

Setiawan, E., Wilson, A.A., Mizrahi, R., et al., 2015. Role of translocator protein density, a marker of neuroinflammation, in the brain during major depressive episodes. JAMA Psychiatry 72 (3), 268–275.

Shelton, R.C., Pencina, M.J., Barrentine, L.W., et al., 2015. Association of obesity and inflammatory marker levels on treatment outcome: results from a double-blind, randomized study of adjunctive L-methylfolate calcium in patients with MDD who are inadequate responders to SSRIs. J. Clin. Psychiatry 76 (12), 1635–1641.

Slavich, G.M., Way, B.M., Eisenberger, N.I., Taylor, S.E., 2010. Neural sensitivity to social rejection is associated with inflammatory responses to social stress. Proc. Natl. Acad. Sci. U. S. A. 107 (33), 14817–14822.

Song, G., Lu, Q., Fan, H., et al., 2019. Inhibition of hexokinases holds potential as treatment strategy for rheumatoid arthritis. Arthritis Res. Ther. 21 (1), 87.

Stein, M., Miller, A.H., Trestman, R.L., 1991. Depression, the immune system, and health and illness. Findings in search of meaning. Arch. Gen. Psychiatry 48 (2), 171–177.

Steiner, J., Bielau, H., Brisch, R., et al., 2008. Immunological aspects in the neurobiology of suicide: elevated microglial density in schizophrenia and depression is associated with suicide. J. Psychiatr. Res. 42 (2), 151–157.

Strawbridge, R., Arnone, D., Danese, A., Papadopoulos, A., Herane Vives, A., Cleare, A.J., 2015. Inflammation and clinical response to treatment in depression: a meta-analysis. Eur. Neuropsychopharmacol. 25 (10), 1532–1543.

Strawbridge, R., Hodsoll, J., Powell, T.R., et al., 2018. Inflammatory profiles of severe treatment-resistant depression. J. Affect. Disord. 246, 42–51.

Tilleux, S., Hermans, E., 2007. Neuroinflammation and regulation of glial glutamate uptake in neurological disorders. J. Neurosci. Res. 85 (10), 2059–2070.

Torres-Platas, S.G., Hercher, C., Davoli, M.A., et al., 2011. Astrocytic hypertrophy in anterior cingulate white matter of depressed suicides. Neuropsychopharmacology 36 (13), 2650–2658.

Torres-Platas, S.G., Comeau, S., Rachalski, A., et al., 2014a. Morphometric characterization of microglial phenotypes in human cerebral cortex. J. Neuroinflammation 11, 12.

Torres-Platas, S.G., Cruceanu, C., Chen, G.G., Turecki, G., Mechawar, N., 2014b. Evidence for increased microglial priming and macrophage recruitment in the dorsal anterior cingulate white matter of depressed suicides. Brain Behav. Immun. 42, 50–59.

Torres-Platas, S.G., Nagy, C., Wakid, M., Turecki, G., Mechawar, N., 2016. Glial fibrillary acidic protein is differentially expressed across cortical and subcortical regions in healthy brains and downregulated in the thalamus and caudate nucleus of depressed suicides. Mol. Psychiatry 21, 509–515.

Trivedi, M.H., Cutler, A.J., Richards, C., et al., 2013. A randomized controlled trial of the efficacy and safety of lisdexamfetamine dimesylate as augmentation therapy in adults with residual symptoms of major depressive disorder after treatment with escitalopram. J. Clin. Psychiatry 74 (8), 802–809.

Uher, R., Tansey, K.E., Dew, T., et al., 2014. An inflammatory biomarker as a differential predictor of outcome of depression treatment with escitalopram and nortriptyline. Am. J. Psychiatry 171 (12), 1278–1286.

Vacchelli, E., Aranda, F., Eggermont, A., et al., 2014. Trial watch: IDO inhibitors in cancer therapy. Oncoimmunology 3 (10), e957994.

Veronese, N., Koyanagi, A., Stubbs, B., et al., 2018. Aspirin and incident depressive symptoms: a longitudinal cohort study over 8 years. Int. J. Geriatr. Psychiatry 33, e193–e198.

Walker, A.K., Budac, D.P., Bisulco, S., et al., 2013. NMDA receptor blockade by ketamine abrogates lipopolysaccharide-induced depressive-like behavior in C57BL/6J mice. Neuropsychopharmacology 38 (9), 1609–1616.

Walker, A.J., Foley, B.M., Sutor, S.L., McGillivray, J.A., Frye, M.A., Tye, S.J., 2015. Peripheral proinflammatory markers associated with ketamine response in a preclinical model of antidepressant-resistance. Behav. Brain Res. 293, 198–202.

Walker, A.K., Wing, E.E., Banks, W.A., Dantzer, R., 2019. Leucine competes with kynurenine for blood-to-brain transport and prevents lipopolysaccharide-induced depression-like behavior in mice. Mol. Psychiatry 24 (10), 1523–1532.

Weber, M.D., Godbout, J.P., Sheridan, J.F., 2017. Repeated social defeat, neuroinflammation, and behavior: monocytes carry the signal. Neuropsychopharmacology 42 (1), 46–61.

Wilkinson, S.T., Sanacora, G., 2019. A new generation of antidepressants: an update on the pharmaceutical pipeline for novel and rapid-acting therapeutics in mood disorders based on glutamate/GABA neurotransmitter systems. Drug Discov. Today 24 (2), 606–615.

Wittenberg, G.M., Stylianou, A., Zhang, Y., et al., 2020. Effects of immunomodulatory drugs on depressive symptoms: a mega-analysis of randomized, placebo-controlled clinical trials in inflammatory disorders. Mol. Psychiatry 25 (6), 1275–1285.

Xiao, C., Miller, A.H., Felger, J., Mister, D., Liu, T., Torres, M.A., 2017. Depressive symptoms and inflammation are independent risk factors of fatigue in breast cancer survivors. Psychol. Med. 47 (10), 1733–1743.

Yang, J.J., Wang, N., Yang, C., Shi, J.Y., Yu, H.Y., Hashimoto, K., 2015. Serum interleukin-6 is a predictive biomarker for ketamine's antidepressant effect in treatment-resistant patients with major depression. Biol. Psychiatry 77 (3), e19–e20.

Yohn, S.E., Arif, Y., Haley, A., et al., 2016. Effort-related motivational effects of the pro-inflammatory cytokine interleukin-6: pharmacological and neurochemical characterization. Psychopharmacology 233 (19–20), 3575–3586.

Zarate Jr., C.A., Payne, J.L., Quiroz, J., et al., 2004. An open-label trial of riluzole in patients with treatment-resistant major depression. Am. J. Psychiatry 161 (1), 171–174.

Zhu, C.B., Blakely, R.D., Hewlett, W.A., 2006. The proinflammatory cytokines interleukin-1beta and tumor necrosis factor-alpha activate serotonin transporters. Neuropsychopharmacology 31 (10), 2121–2131.

Zhu, C.B., Lindler, K.M., Owens, A.W., Daws, L.C., Blakely, R.D., Hewlett, W.A., 2010. Interleukin-1 receptor activation by systemic lipopolysaccharide induces behavioral despair linked to MAPK regulation of CNS serotonin transporters. Neuropsychopharmacology 35 (13), 2510–2520.

Chapter 16

Opioid agents for treatment-resistant depression

Gianluca Serafini[a,b], Giulia Adavastro[a,c], Giovanna Canepa[a], Andrea Amerio[a,b], Andrea Aguglia[a,b], and Mario Amore[a,b]

[a]Department of Neuroscience, Rehabilitation, Ophthalmology, Genetics, Maternal and Child Health, Section of Psychiatry, University of Genoa, Genoa, Italy, [b]IRCCS Ospedale Policlinico San Martino, Genoa, Italy, [c]Monitoring Center on Addiction, Mental Health DSM-DP, Bologna, Italy

Introduction

Almost 70% of depressed individuals do not adequately respond to common antidepressants and need multiple therapeutic trials in order to achieve a full remission. Furthermore, according to the STAR*D study, patients who need more trials had the greatest risk of failure (Rush, 2007). This subgroup of subjects experiencing TRD is more likely to suffer from recurrent episodes and shows a higher risk to manifest substance abuse and/or suicidal behavior as well as more severe psychosocial impairment and physical disability when compared to individuals with major depressive disorder (MDD) (Holtzheimer and Mayberg, 2011; Serafini et al., 2018).

Treatment resistance may occur in patients previously responsive to treatments or gradually as a neuroprogressive condition (Thase and Schwartz, 2015), characterized by more severe recurrent episodes with a worse response to treatments and residual symptoms recognized as an index of poor outcome (Serafini, 2018).

Treatment strategies commonly involve antidepressant combinations, augmentation, or switching, but the use of non-pharmacological options such as vagus nerve stimulation, transcranial magnetic stimulation (TMS), and electroconvulsive therapy (ECT) has been also described although linked to complications such as the increased risk of toxicity due to pharmacological combination regimens or higher drug dosages, memory impairments with ECT, and others (Akil et al., 2018).

The difficulty in adequately treating TRD presumably reflects the multiple neurobiological mechanisms underlying its clinical presentation. As a matter of fact, monoamine depletion theories are able to explain TRD only partially; therefore, in recent years, studies focused on different mechanisms including cortisol levels, Brain-Derived Neurotrophic Factor (BDNF), pro-inflammatory cytokines, and new medications have been stressed. Importantly, existing research highlighted the importance of endogenous opioid system and its abnormalities in regulating mood, and thus playing a crucial role in affective disorders. These findings suggested that it may be possible to use drugs targeting this system for people with TRD (Kennedy et al., 2006).

It would be more correct to say that we should return to investigate the potential of drugs capable of acting on the opioid system, given the ancient use of the plant *Papaver somniferum* to improve mood dates back to the Sumerian age (3000 BCE). In addition, in the 19th century, opium, first introduced by a German psychiatrist, was known as one of the psychoactive medications used to treat melancholic states with agitation. Later, even Emil Kraepelin recommended opium tincture to treat dysphoria and agitated depression (Saxena and Bodkin, 2019). However, the use of these remedies commonly resulted in abuse and addiction; therefore, with the introduction of the monoamine oxidase inhibitors (MAOIs) in the 1950s, tricyclics (TCAs) in the 1960s, and SSRIs in the 1980s, the therapeutic potential of opioid derivatives for mood disorders has been long neglected (Tenore, 2008). Many compounds are now being studied with renewed interest for their potential benefit for depressed patients, including those with TRD.

Opioid agents and their potential in treatment-resistant depression

Various studies reported that the opioid system is involved in the modulation of pain, response to stress, reward (and therefore addiction), respiratory functioning, control of gastrointestinal motility, and immune and endocrine systems regulation—in particular, at the level of the hypothalamic-pituitary-adrenal (HPA) axis (Bodnar, 2010). The opioid system is partially regulated by three different receptors, μ-, δ-, and κ- (these receptors are coupled to G proteins that commonly

Managing Treatment-Resistant Depression. https://doi.org/10.1016/B978-0-12-824067-0.00016-5

bind a family of three opioid peptides, i.e., β-endorphin, enkephalins, and dynorphins), which are disseminated in the central and peripheral nervous systems (Serafini, 2018). M-opioid receptors (MORs), when activated, convey an analgesic effect but they are also involved in the modulation of behaviors which are crucial in the conservation of the human species, such as eating, reproduction and response to social stimuli (Peciña et al., 2019). Δ-opioid receptors (DORs) and κ-opioid receptors (KORs) modulate pain, but they also seem to play a role in depression: encephalin/DORs have a mood-enhancing activity while dynorphin/KORs—acting on dopaminergic neurons in nucleus accumbens—appear to be related to depressive-like behaviors (Knoll and Carlezon, 2010; Pradhan et al., 2011). Moreover, stress may lead to increased levels of dynorphins that bind to KORs, thus inducing symptoms like dysphoria and anhedonia (Knoll and Carlezon, 2010). Opioid receptors are highly expressed in cortical regions involved in the response to stressors and in the regulation of emotionally salient stimuli as well as subcortical limbic and paralimbic circuits that process stressful stimuli, reward, hedonic response, and mood control (Berrocoso et al., 2009; Ehrich et al., 2015).

As mentioned above, opioid receptors modulate important neurotransmitter systems that are related to major depression, such as the dopaminergic (DA), serotonergic (5-HT), and noradrenergic (NA) systems at multiple sites; synaptic plasticity and neurogenesis in the hippocampus; functioning of the hypothalamic-pituitary-adrenal gland (HPA) axis, a major neuroendocrine stress system, and regulation of BDNF activity (Lutz and Kieffer, 2013). Importantly, 50% of MDD patients experience cognitive dysfunctions in one or more cognitive domains (such as memory, attention, executive functions and processing speed), and cognitive impairment is a significant contributor (mediator or moderator) of functional impairment in MDD; thus, medications with the potential to target cognitive impairment may be useful in TRD. Cognitive deficits tend to be more frequent and pronounced in patients with TRD, and often do not improve with the remission of affective symptoms (Gregory et al., 2020). Although conventional antidepressants attenuate depression severity, they are linked to mixed outcomes in terms of long-term cognitive dysfunctions. In the context of mood, opioid antagonists may be useful for improving psychomotor responses, decision making, and attention; in addition, they may effectively target cognitive dysfunctions presumably by modulating DA neurotransmission within the mesolimbic circuit and specific cortical regions (Jacobson et al., 2018).

It is generally accepted that the opioid system may be dysregulated in depression (Peciña et al., 2019). Further evidence to this regard are provided by postmortem studies, highlighting the existence of endorphin deficiencies coupled with an increased density of μ-receptors in the brains of depressed patients who committed suicide (Gabilondo et al., 1995; Scarone et al., 1990). Moreover, from a clinical point of view, both traditional antidepressants and ECT indirectly modulate opioid neurotransmission (Berrocoso et al., 2009), and drugs that act as MOR agonists demonstrated a rapid mood improvement in patients with MDD or TRD (Ehrich et al., 2015).

Opioid agents: Mechanisms of action, pharmacokinetics and clinical studies

Among the opioidergic drugs that may be involved in the treatment of affective disorders, the following medications are currently used as pain-killers or in the setting of opioid addiction: *methadone*, *buprenorphine*, *tramadol* and *naltrexone*. Tianeptine, an atypical agonist of the μ-opioid receptors with clinically negligible effects on the δ- and κ-opioid receptors, has antidepressant and anxiolytic effects with a relative lack of sedative, anticholinergic, and cardiovascular adverse effects, and it has potential uses for treating anxiety, asthma, and irritable bowel syndrome. However, there are also new treatments under investigation which have been developed to treat major depression such as *ALKS-5461* (buprenorphine plus the μ-antagonist, samidorphan) and m-trifluoromethyl-diphenyl diselenide and Salvindolin—the latter having been tested only in preclinical models.

Below, the most relevant opioid agents used in TRD are summarized:

Methadone

Mechanism of action: Methadone is a diphenylpropylamine derivative. It is an agonist of MORs, and to a lesser extent KORs, but it may also bind to the *N*-methyl-D-aspartate receptor (NMDAR) where it acts as an antagonist (Ferrari et al., 2004). Methadone also acts as a serotonergic drug, being able to inhibit the serotonin transporter SERT (Tenore, 2008); this latter mechanism of action should be taken into account particularly when coprescribing other serotonergic drugs, such as MAOIs due to the risk of serotonin toxicity. Another mechanism of action that seems to be involved in the antidepressant effect of methadone is the saturation of upregulated endorphin receptors (due to lower endorphin levels in depressed subjects). This may stimulate dopamine neurons to release dopamine in order to restore hedonic tone, thus improving depressive symptoms (Tenore, 2008). It should also be noted that methadone has been shown to reduce cortisol levels immediately after its administration: in a study which was conducted in 1982 by Judd and colleagues, intramuscular administration of

10 mg of methadone resulted in a rapid reduction of cortisol levels and significant reduction of manic symptoms (Judd et al., 1982). As such, methadone may be potentially used either in unipolar and bipolar depression (in the latter case, the previously mentioned use of opiates by Emil Kraepelin for agitated depression is of particular interest).

Pharmacokinetics: Methadone is generally orally administered, has a high fat solubility, is metabolized by cytochrome P450 (CYP) 3A4, CYP2D6, and CYP2B6. The metabolites resulting from *N*-demethylation (2-ethylidene-1,5-dimethyl-3,3-diphenylpyrrolidine (EDDP) and 2-ethyl-5-methyl-3,3-diphenylpyrroline (EMDP)) are inactive and may be measured in urine (this may be useful for monitoring purposes). The excretion of methadone after biotransformation commonly occurs via renal and fecal elimination; the terminal half-life ranges between 8 and 59 h, with a significant degree of individual variability due to genetic differences regarding cytochromes related to methadone metabolism (Ferrari et al., 2004).

Clinical studies: Methadone's antidepressant action has been investigated mostly in opioid-dependent patients with comorbid MDD. For instance, Dyer and colleagues found that the antidepressant effect of methadone appears to be directly related to its circulating levels: the complete resolution of clinical symptoms is commonly achieved within 3 h while the effect diminishes as methadone levels decline (Dyer et al., 2001). In another study, Dean et al. compared a cohort of patients who were treated with methadone + placebo with another cohort who was prescribed methadone + fluoxetine and found a significant improvement of depressive symptoms in both groups; in addition, the antidepressant effect did not increase when adding a serotonergic drug (Dean et al., 2002). In another study, the same authors demonstrated a rapid improvement in depressive symptoms over several weeks in a group of 30 patients who were treated with methadone as well as in a control group (24 patients) who were treated with buprenorphine (Dean et al., 2004). Importantly, patients in methadone maintenance treatment who present with comorbid MDD generally require a 20%–50% higher dosage of this medication (Tenore, 2008). Regarding methadone daily dosage, patients with comorbid substance abuse/dependence disorder need higher dosages (70–150 mg per day vs. 30–40 mg per day in patients without comorbidity) in order to obtain a better treatment retention and stabilize their clinical symptoms (Tenore, 2008).

Efficacy for treatment-resistant depression: Unfortunately, studies on the antidepressive activity of methadone in TRD are very limited due to the occurrence of adverse effects related to the possibility of overdose. This factor may limit the potential utility of methadone for TRD patients. Further studies are needed to this specific regard.

Adverse effects and toxicology: Methadone has, unfortunately, a potential lethality risk related to overdose, which may result from respiratory depression/pulmonary edema or *ab-ingestis* pneumonia. Respiratory depression is unlikely with methadone mono-therapy, especially if the patient has developed tolerance; however, the risk of respiratory depression may be increased when methadone is coadministered with drugs inhibiting CYP3A4 (e.g., fluoxetine, fluvoxamine, ketoconazole, erythromycin) (Corkery et al., 2004). Another interaction causing toxicity is with MAOIs or other antidepressant medications which, when administered together with methadone, may cause an increase in the risk of serotonin toxicity (Tenore, 2008). Methadone may be also responsible for cardiac arrhythmias and cardiovascular collapse (Corkery et al., 2004). Common but less-dangerous adverse effects include sedation, gastrointestinal discomfort, constipation, and nausea (Gudin et al., 2015). The risk of withdrawal from methadone can be increased when it is prescribed concomitantly with opioid antagonists (e.g., naltrexone) or inducers of CYP3A4 (e.g., carbamazepine, rifampicin, fluconazole) (Corkery et al., 2004). Finally, an uncommon long-term adverse effect of methadone is an endocrinopathy known as the opioid-induced androgen deficiency (OPIAD), which may cause hypogonadism and hypotestosteronism due to an alteration of the hypothalamic-pituitary-gonadal/adrenal axis (Gudin et al., 2015).

Buprenorphine

Mechanism of action: Buprenorphine is a partial μ-agonist and δ –/κ-antagonist. KOR antagonism seems to be implicated in the antidepressant effect of this drug, especially in reducing anhedonia (Peciña et al., 2019). Others have suggested that the importance of MOR agonism action of buprenorphine is important in the setting of TRD, as demonstrated in animal models of depressive behavior (e.g., the learned helplessness model), where morphine or other μ(mu)-opioid peptide (MOP) agonists showed an antidepressant-like activity which can be antagonized by naloxone (Berrocoso et al., 2009). Moreover, based on its partial agonist properties, buprenorphine presents a lower risk of respiratory depression in overdose (when not associated with other Central Nervous System (CNS) depressants such as alcohol and benzodiazepines). Buprenorphine commonly acts as an exogenous endorphin but, similar to methadone, it also acts as an NMDA antagonist (Tenore, 2008).

Pharmacokinetics: Buprenorphine is a semisynthetic derivative of thebaine, which has been isolated from opium poppy (Kyzer and Wenthur, 2020). Buprenorphine is usually administered sublingually. It's a lipophilic molecule and, once in circulation, it is 96% bound to plasma proteins (Stefanowski et al., 2020). The elimination half-life of buprenorphine is about 20 h. Buprenorphine undergoes hepatic metabolism and is transformed into norbuprenorphine through N-dealkylation by CYP3A4 and CYP3A5. Since buprenorphine is metabolized by CYP3A4, caution should be used when it is prescribed with CYP3A4 inducers or inhibitors. Norbuprenorphine, an active metabolite that possesses a MOR agonist and partial KOR agonism action, is then conjugated with glucuronic acid and eliminated mostly through feces and, to a lesser extent, by urine (Kyzer and Wenthur, 2020).

Clinical studies: Due in part to potentially advantageous safety profiles, buprenorphine (together with ALKS-5461) are among the most investigated opioids in the treatment of depression. There are at least 15 studies both in patients with addiction (Ahmadi et al., 2018, 2019; Ahmadi and Abtahi, 2017; Dean et al., 2004; Gerra et al., 2006; Kosten, 1990; Striebel and Kalapatapu, 2014) as well as in drug-naive patients with MDD or TRD (Bodkin et al., 1995; Emrich et al., 1982; Karp et al., 2014; Nyhuis et al., 2008), as well as borderline personality disorder (Resnick and Falk, 1987) and posttraumatic stress disorder (PTSD) (Mongan and Callaway, 1990). Other studies investigated the efficacy of buprenorphine specifically for reducing suicidal ideation or nonsuicidal self-injury (Ahmadi et al., 2018; Ahmadi and Abtahi, 2017; Norelli et al., 2013; Striebel and Kalapatapu, 2014; Yovell et al., 2016). For further details, please see Table 16.1.

Efficacy for treatment-resistant depression: Available reports of the clinical effects of buprenorphine for TRD mostly used lower dosages of buprenorphine (or buprenorphine/naloxone or buprenorphine/ naltrexone in order to reduce the risk of addiction), ranging from 0.1 to 8 mg/day, with one study using a dose of 16 mg/day of this medication. The main results of these studies indicated that depressive symptoms and suicidal ideation improved rapidly after buprenorphine administration, similarly to the use of ketamine (Serafini, 2018). Relevantly, the antidepressant effects of buprenorphine occurred independently of its euphoric effects (Peciña et al., 2019).

Adverse effects and toxicology: Compared to methadone treatment, buprenorphine is safer and has been associated with lower rates of all-cause mortality; however, the combination of buprenorphine with CNS depressants (e.g., benzodiazepines or alcohol) may be dangerous. Other adverse effects of buprenorphine include constipation and sedation. With regard to the cardiovascular system, buprenorphine may slightly raise blood pressure and, at higher dosages, it may elongate the corrected QT interval (QTc) (Kyzer and Wenthur, 2020). Importantly, buprenorphine (similarly to methadone) is not exempt from the risk of diversion and addiction. For this reason, it is also administered in combination with MOR antagonists (samidorphan or naloxone) (Heo and Scott, 2018).

Alkermes or ALKS-5461

Mechanisms of action: ALKS-5461 is a 1:1 combination of buprenorphine and samidorphan designed for depression, unlike the other opioidergic agents that were developed for pain or addiction treatment. The three available dosages range from 0.5 mg /0.5 mg to 8 mg/8 mg per day. Samidorphan, a naltrexone derivative, binds with a high affinity to MOR, exerting an antagonism action. In addition, it has a weak partial agonism at KORs and DORs although, overall, ALKS-5461 if often considered a KOR antagonist (Ragguett et al., 2018).

Pharmacokinetics: As with buprenorphine, ALKS-5461 is administered sublingually in order to avoid first-pass metabolism. Samidorphan has an 8-h termination half-life, and 20% is excreted unchanged in urine. The rest undergoes N-dealkylation to RDC-9986, which has no known clinical activity (Kumar et al., 2019).

Clinical studies: One of the first reports of buprenorphine/samidorphan for depression was conducted by Ehrich et al. (2015). Although limited by the short duration of the study design (1 week), ALKS-5461 demonstrated a rapid antidepressive effect compared with placebo, particularly at a 1:1 buprenorphine/samidorphan ratio (Ehrich et al., 2015). Subsequent trials, FORWARD-1 to FORWARD-5, found that patients treated with buprenorphine/samidorphan at a dose of 2 mg/2 mg had a significant improvement in depressive symptoms, as measured by the Montgomery–Asberg Depression Rating Scale (MADRS) compared to placebo (Fava et al., 2016, 2020; Thase et al., 2019; Zajecka et al., 2019). A lower risk of abuse/dependence related to buprenorphine/samidorphan was associated with the 1:1 ratio (Peckham et al., 2018) and the antidepressive effects of the drug seem to persist for a long period—the improvement in MADRS scores has been observed for 56 weeks (Thase et al., 2019) (for further additional details, see Table 16.2).

Efficacy for treatment-resistant depression: Buprenorphine and ALKS-5461 seem to be basically equipotent (Saxena and Bodkin, 2019), although ALKS-5461 may have a lower risk of addiction (Pathak et al., 2019). However, clinical studies provided ambiguous results to this specific regard: a recent metaanalysis conducted by Dinoff et al. (2020) compared six randomized trials of buprenorphine (five of the six reports involved the use of ALKS-5461) in addition to an SSRI or serotonin and norepinephrine reuptake inhibitors (SNRI) for TRD patients without chronic pain or opioid addiction. The results did not show efficacy defined as superiority to placebo (Dinoff et al., 2020). ALKS-5461 has not yet been approved by the U.S. FDA, which in November 2018 voted against its approval, deeming it necessary to carry out a greater number of studies to evaluate its effectiveness (Alkermes, 2019; Ragguett et al., 2019). The manufacturer is still carrying out studies on the drug and, recently, in a phase III trial (the aforementioned FORWARD-5) ALKS-5461—at a dosage of 2 mg/2 mg—was superior to placebo in reducing depressive symptoms as assessed by MADRS-6 and MADRS-10 total scores (Fava et al., 2020).

Adverse effects and toxicology: The adverse effects of ALKS-5461 are similar to other opiates drugs; however, tolerability studies are very limited. Common adverse effects that emerged in the FORWARD 1 to 5 trials included headache, nausea, vomiting, diarrhea, fatigue, dizziness, and abuse potential (Ragguett et al., 2018). This drug is metabolized by CYP3A4. Therefore, it is necessary to take into account possible drug–drug interactions and pharmacogenetic profile at CYP3A4 (as discussed above). In order to be able to administer buprenorphine or ALKS-5461 to patients on chronic therapy with other opioids, it is necessary to provide the drug only after at least a partial wean because the high affinity of the drug for MORs may induce a displacement of the previous opiate, leading to a frank withdrawal syndrome (Ragguett et al., 2018).

Tramadol

Mechanisms of action: Tramadol is a synthetic opioid with a central analgesic action, but it may also exert anxiolytic and antidepressant effects (it has a chemical structure similar to that of venlafaxine) (Barber, 2011; Berrocoso et al., 2009; Saxena and Bodkin, 2019; Tenore, 2008). Tramadol is considered an atypical opiate since it has only a modest MOR agonism and a weak affinity to DORs and KORs, while it has a potent SNRI action. Caution should therefore be taken when adding this drug to other antidepressants (especially MAOi's) due to the risk of serotonin toxicity (Berrocoso et al., 2009; Saxena and Bodkin, 2019; Tenore, 2008). It should be carefully considered that tramadol is an NMDA-antagonist as well (Tenore, 2008).

Pharmacokinetics: Tramadol may be administered both orally or parentally. The oral bioavailability of tramadol is 75%; this drug is rapidly absorbed, reaching the peak blood concentration 2h after its assumption. Tramadol is 20% protein-bound. The half-life of this drug is about 6h: tramadol partially undergoes an extensive hepatic metabolism trough CYP2D6, which demethylates this drug into the active metabolite M1, while the remaining 30% of the medication is eliminated unchanged through urine. Relevantly, tramadol's elimination half-life tends to increase in patients with hepatic problems (Loredana et al., 2004).

Clinical studies: Most of the existing studies of tramadol for depression have been performed on animal models (Faron-Gorecka, 2004; Hopwood et al., 2001; Jesse et al., 2010), while those in humans were derived according to observations on patients who used tramadol as a pain reliever. Spencer in 2000 reported the case of a patient with depression and suicidal ideation who was treated with tramadol for back pain: a rapid improvement of mood was reported after its administration that unfortunately rapidly disappeared after its interruption (the patient attempted suicide) (Spencer, 2000). In another case study, a patient with TRD refractory to ECT responded well to tramadol monotherapy (Andrew and Verduin, 2001). More recently, Rougemont-Bucking and colleagues (2017) reported the successful antidepressant use of tramadol in a patient treated for pain, previously treated with other medications for major depression and PTSD; this treatment was successfully continued at a dosage between 15 and 70mg per day in accordance to mood, administered no more than 3days in a row for no greater than 5days per week. The same authors also reported the case of a female with relapsing depressive episodes and anxiety comorbid with alcohol addiction. When treated with a dosage of 50 to 100mg per day of tramadol, she was able to stop alcohol consumption and reported improvement in mood (Rougemont-Bucking et al., 2017). In addition, the antidepressive efficacy of tramadol in patients with TRD has been demonstrated in phase II trials that showed therapeutic effects similar to amitriptyline, but with better tolerability (E-therapeutics, 2009, 2016).

Efficacy in treatment-resistant depression: Although tramadol appears to be a promising and potentially rapidly effective compound (which might also be useful in augmentation, especially in the first treatment phase of typical antidepressants) for depression, there are still no available studies of sufficient size and quality to recommend it as a therapeutic option for patients with TRD. Further studies in patients with TRD are therefore needed.

Adverse effects and toxicology: Although tramadol may be used as an analgesic medication at 400mg daily (Pierre et al., 1994), the therapeutic doses needed to attenuate depressive symptoms may be lower. Above 400mg daily, there is an higher risk of seizure (particularly in patients with a history of epilepsy or treated concomitantly with other antidepressants, opiates or drugs known to lower seizure threshold) (Pierre et al., 1994; Rougemont-Bucking et al., 2017). Based on its lower degrees of μ-agonism, tramadol seems to induce less tolerance than other opioids; therefore, there may be no need to increase the dosage, even with longer-term treatments. Indeed, the risk of inducing dependence is lower when compared to other opioid drugs (Berrocoso et al., 2009; Pierre et al., 1994). Furthermore, after interruption, only 1 to 3 patients per 100,000 undergo withdrawal symptoms (Barber, 2011). Tramadol does not appear to induce significant respiratory depression and cardiovascular effects (Loredana et al., 2004). Other reported adverse effects are constipation and dysphoria (Dyer et al., 2001). Other long-term adverse effects may include weight gain and sexual dysfunction (Barber, 2011). The risk of inducing mania in patients with a bipolar diathesis is unknown but should not be ignored (Rougemont-Bucking et al., 2017).

TABLE 16.1 Summary of the most relevant studies concerning buprenorphine and depression or suicidal behavior.

Reference	Study design	Psychiatric condition	Sample	Treatment
Emrich et al. (1982)	Double-blind, placebo-controlled clinical trial	MDD	10 adults	0.2 mg sublingual BUP
Resnick and Falk (1987)	Single-blind, placebo controlled clinical trial	Nine subjects had borderline personality disorder, six had not specified disorders	15 adults	15 patients: sublingual BUP starting at 0.2 mg till 12.3 mg
Kosten (1990)	Observational prospective study	Opioid addiction	40 adults (mean age 30.7 years; 31 M and 38 F)	Sublingual BUP mean dose 3.2 mg (range from 2 to 8 mg)
Mongan and Callaway (1990)	Placebo controlled clinical trial	Veterans with previous substance abuse and multiple diagnoses (PTSD, MDD, Panic Disorder, Bipolar Depression, Borderline, Schizophrenia, Adjustment Disorder), plus a normal subject	12 adults (age from 37 to 52 years; all M)	Sublingual BUP 0.3 mg or placebo. Four subjects in open tests and eight subjects in double blind test
Bodkin et al. (1995)	Open label clinical study	TRD	Ten adults (3 completed 4 weeks and 4 completed 6 weeks, 3 dropped-out; of 7 who completed at least 4 weeks the mean age was 44 years), four were M and three F	Intranasal or sublingual BUP starting at 0.15 mg and titrated until a maximum of 1.8 mg (final average dose was 1.26 mg)
Dean et al. (2004)	Randomized, double-blind, placebo controlled clinical trial	Opioid addiction	Overall, 147 adults (mean age 29.5 years, 91 M and 56 F)	68 patients: 4 mg sublingual BUP and placebo methadone syrup. 79 patients: 30 mg methadone syrup and placebo BUP tablet. The therapies were started with the dosage indicated above and then individually titrated
Gerra et al. (2006)	Controlled observational prospective study	Opioid addiction	Overall, 60 adults (mean age 31.5 years; 45 M and 15 F)	30 patients: NAL (50 mg). 30 patients: sublingual BUP 4 mg + NAL 50 mg
Nyhuis et al. (2008)	Observational prospective study	TRD	Six adults (mean age 45.8 years; 2 M and 4 F)	Sublingual BUP ranging from 0.8 to 2.0 mg
Norelli et al. (2013)	Case-report	Treatment-resistant NSSI	Six adults (mean age 30.3 years; 3 M and 3 F)	Personalized dose of BUP (from 0.5 mg to 6 mg) and BUP/NAL (2 mg/0.5 mg to 1 mg/0.25 mg)
Striebel and Kalapatapu (2014)	Case-report	Chronic suicidal ideation; TRD; chronic back pain; opioid dependence	One adult (61 years, F)	16 mg BUP + 4 mg naloxone

Study duration	Psychometric instruments	Main conclusions	Limitations/shortcomings
5–8 days	HAM-D; IMPS; VBS	Four patients showed more than 50% reduction in depression, two a moderate response, and four a slight reduction	Short duration. Lack of a control group. The small sample size does not allow the generalization of the main findings.
1 month	HAM-D	Nine out of the fifteen patients showed 30%–50% reduction on HAM-D. These nine patients met the diagnostic criteria for Borderline personality disorder while the others had different diagnosis	Short duration. The small sample size does not allow the generalization of the main findings. The diagnostic heterogeneity with a lack of a precise description of all diagnoses and group characteristics.
1 month	BDI; SDS	Significant mood improvement since the end of the first week, that continues over the second week. Depressive symptoms steadily declined during the month	Short duration. Lack of control group. The small sample size of the study. Diagnostic heterogeneity. Depression was not included as a primary outcome. Self-report measures.
6 h	POMS	Nine patients responded with a relief in anxiety, depression and hostility	Short duration. Groups receiving placebo or drug were not described in detail. The small sample size does not allow the generalization of the main results. Diagnostic heterogeneity. Lack of exclusion criteria. A part from POMS, lack of standardized measures.
6 weeks	HAM-D; ADDS; POMS	Six of seven patients achieved marked clinical improvement by the end of 1 week. This improvement in all the indicated parameters remained stable until the end of the trial	Short duration. Lack of control group and placebo group. The small sample size.
3 months	BDI	Depressive symptoms improved in all subjects, with no difference between methadone and buprenorphine groups	Diagnostic heterogeneity. Depression was not a primary outcome. Self-report measures. Lack of a baseline psychiatric evaluation. Lack of specified exclusion criteria.
3 months	SCL-90; VAS for craving	Despite both groups showed a significant decrease in the scores of irritability, depression, tiredness and psychosomatic symptoms, the BUP/NAL group presented a major improvement	Lack of randomization and placebo control group. Diagnostic heterogeneity. Lack of baseline psychiatric evaluation. Depression was not a primary outcome. Lack of exclusion criteria. Self-report measures.
7 days	HAM-D; BDI	All patients improved over the week and five of six achieved a complete remission	Short duration. Lack of a control group. The small sample size. Lack of exclusion criteria.
Different duration for each patients (from 14 days to 11 months)	–	Five out of six patients improved their mood and NSSI behavior	Short duration. Lack of a control group. The small sample size. Lack of exclusion criteria.
3 months	–	The patients manifested cessation of suicidal thoughts, a decrease of pain and a reduction in depression	The small sample size. Lack of a control group. Lack of exclusion criteria. Lack of standardized measures.

Continued

TABLE 16.1 Summary of the most relevant studies concerning buprenorphine and depression or suicidal behavior—cont'd

Reference	Study design	Psychiatric condition	Sample	Treatment
Karp et al. (2014)	Open label clinical study	TRD	Fifteen adults (mean age 60.7 years; 7 M and 8 F)	BUP titrated from 0.2 to 1.6 mg (mean dosage 0.4 mg). 13 patients used BUP as an augmentation therapy while 2 patients used BUP as a mono therapy
Yovell et al. (2016)	Randomized, double-blind, placebo controlled clinical trial	Suicidal ideation	Overall, 88 adults (mean age 37.3 years, 25 M and 63 F)	57 patients: BUP (0.1 or 0.2 mg/day. Once a week, the daily dose could be raised of 0.1–0.2 mg increments until a maximal dose of 0.8 mg); 31 patients: placebo
Ahmadi and Abtahi (2017)	Case-report	Chronic suicidal ideation in a substance-induced depressive disorder	One adult (25 years, M)	8-mg single dose of sublingual BUP
Ahmadi et al. (2018)	Randomized, double-blind, clinical trial	Suicidal patients with opioid dependence and MDD	Overall, 51 adults (mean age 32, 8 years, all M)	Overall, 16 patients: 32 mg sublingual BUP. 17 patients: 64 mg sublingual BUP. 14 patients: 96 mg sublingual BUP. BUP was administered as a one-time dose. Four patients refused medication
Ahmadi et al. (2020)	Randomized, double-blind, placebo-controlled clinical trial	Opioid dependence	Overall, 61 adults (mean age 36.5 years, all M)	21 patients: placebo. 20 patients: 16 mg sublingual BUP. 20 patients: 32 mg sublingual BUP. BUP was administered as a one-time dose

Abbreviations: *ADDS*, Atypical Depression Diagnostic Scale; *BDI*, Beck Depression Inventory; *BSIS*, Beck Suicide Intent Scale; *BSSI*, Beck Scale for Suicidal Ideation; *BUP*, buprenorphine; *HAM-D*, Hamilton Rating Scale for Depression; *IMPS*, Inpatient Multidimensional Psychiatric Scale; *MADRS*, Montgomery–Åsberg Depression Rating Scale; *MDD*, major depressive disorder; *NAL*, naltrexone; *NSSI*, nonsuicidal self-injury; *POMS*, profile of mood states; *SCL-90*, symptom checklist-90; *SDS*, short depression scale; *SSI*, scale for suicide ideation; *SPS*, Suicide Probability Scale; *TRD*, treatment-resistant depression; *VAS*, Visual Analog Scale; *VBS*, Verlaufs-Beurteilungs-Skala.

Study duration	Psychometric instruments	Main conclusions	Limitations/shortcomings
2 months	MADRS; SSI	Improvement in depression started within the first week of treatment and was sustained during the exposure to BUP. Once BUP was discontinued, MADRS score was found increased at reassessment (16th week)	Short duration. The small sample size. Lack of a control group and placebo group.
1 month	BSSI; BDI; SPS	Patients in the buprenorphine group had a greater reduction in suicidal ideation, suicide likelihood and depression scores than patients in placebo group	Short duration. Diagnostic heterogeneity. Self-report measures.
10 days	BSIS; BDI	The patient had a rapid reduction and cessation of suicidal thoughts and depressive symptoms	Short duration. Lack of a control group. The small sample size. Lack of exclusion criteria.
3 days	BSSI	In all 3 groups a significant reduction in suicidal ideations was observed	Short duration. Lack of a placebo control group. Lack of female gender in the study sample. Depression was not a primary outcome. Self-report measures.
4 days	BSSI	Suicidal ideation decreased significantly in the 32 mg BUP group.	Short duration. Lack of female gender in the study sample. Depression was not a primary outcome. Self-report measures.

TABLE 16.2 Summary of the most relevant studies concerning ALKS-5461 use in TRD patients.

Reference	Study design	Psychiatric condition	Sample	Treatment
Ehrich et al. (2015)	Randomized, double-blind, placebo controlled clinical trial	TRD	Thirty-two adults (mean age of 51.5 years, 17 M and 15 F)	Fourteen patients: BUP/SAM 8:1 dose-ratio (starting dose 2 mg/0.25 mg to final dose 4 mg/0.5 mg). 14 patients: BUP/SAM 1:1 dose-ratio (starting dose 4 mg / 4 mg to final dose 8 mg/8 mg). 4 patients: placebo. All patients remained on SSRI or SNRI background therapy throughout the treatment period
Fava et al. (2016)	Randomized, double-blind, placebo controlled clinical trial	TRD	Overall, 142 adults (mean age 45.8 years, 96 F and 46 M)	In stage 1: 24 patients BUP/SAM 2 mg/2 mg; 19 patients BUP/SAM 8 mg/8 mg and 98 patients placebo. All patients continued on their current antidepressant therapy and on the same dosage throughout the course of the study.
Fava et al. (2020)	2 randomized, double-blind, placebo controlled multicenter clinical trials (FORWARD-4 and 5)	TRD	FORWARD-4: 385 patients. FORWARD-5: 407 patients	FORWARD-4: adjunctive BUP/SAM 2 mg/2 mg (60 patients, mean age 46.2 years, 20M and 40 F) or 0.5 mg/0.5 mg (59 patients, mean age 45 years, 21 M and 38 F) or placebo (265 patients, mean age 45.8 years, 83 M and 182 F). FORWARD-5: adjunctive BUP/SAM 2 mg/2 mg (63 patients, mean age 42.9 years, 21 M and 42 F) or 1 mg/1 mg (63 patients, mean age 45.1 years, 21 M and 42 F) or placebo (280 patients, mean age 45.7 years, 87 M and 193 F)
Thase et al. (2019)	Open-label study to evaluate long-term safety and tolerability (FORWARD-2)	TRD	Overall, 1486 adult patients (mean age 46.5 years, 522 M and 964 F)	All patients continued their current drug for depression throughout the study plus BUP/SAM 2 mg/2 mg
Zajecka et al. (2019)	Multicenter, randomized, double-blind, placebo-controlled study (FORWARD-3)	TRD	Overall, 429 patients: group 1 (HAM-D > 20) 399 patients + group 2 (HAM-D = 18 or 19) 30 patients	All patients continued their current drug for depression throughout the study plus BUP/SAM 2 mg/2 mg

Abbreviations: *BUP*, buprenorphine; *CGI-S*, clinical global impressions severity scale; *HAM-D*, Hamilton Rating Scale for Depression; *MADRS*, Montgomery–Åsberg Depression Rating Scale; *NAL*, naltrexone; *SAM*, samidorphan; *TRD*, treatment-resistant depression.

Trial duration	Psychometric Instruments	Main conclusions	Limitations/ shortcomings
1 week	HAM-D; MADRS; VAS	Patients in the 1:1 ratio group exhibited significant improvement in depressive symptoms after 7 days. The antidepressant effect was greatest in this group	Short duration. The small sample size. Self-report measures.
1 month	HAM-D; MADRS; CGI-S	Patients in the 2:2 ratio group, compared with patients in the placebo group, showed significantly greater improvements. There was also evidence of improvement in the 8:8 ratio group, although this improvement did not achieve statistical significance	Short duration.
Stage 1: 5 weeks. Stage 2: 6 weeks	MADRS-6; MADRS-10	FORWARD-4: the BUP/Sam 2 mg/2 mg group had the greater reduction in MADRS-10 from baseline to week 5, but without statistical significance. FORWARD-5: the BUP/SAM 2 mg/2 mg group had the greater reduction in MADRS-6 from week 3 to the end of the trial, but without statistical significance. The 1 mg/1 mg group showed a nonstatistical significantly improvement	Short duration.
52 weeks	MADRS; CGI-S	The drug is well tolerated and associated with a low risk of abuse. Some evidences of withdrawal symptoms with abrupt discontinuation were reported. The antidepressant effect continued until the end of study	–
10 weeks	HAM-D; CGI-S; MADRS-10	Patients treated with adjunctive BUP/SAM 2 mg/2 mg showed improvements in depressive symptom score vs. baseline on the MADRS-10, although with no statistical significance compared to placebo	Short duration.

Naltrexone

Mechanisms of action: Naltrexone is a synthetic congener of oxymorphone with μ-competitive antagonist (and to a lesser extent a κ−/δ-antagonist) action but, at very low doses, it seems to be able to reverse desensitization at D2/D3 receptors, thus exerting a dopaminergic effect (Davis, 2016; Saxena and Bodkin, 2019).

Pharmacokinetics: Naltrexone is rapidly and completely absorbed after oral administration. The average elimination half-life is 6h. It is metabolized mainly to 6-beta-naltrexol (first-pass effect in liver) and is then eliminated through renal excretion (Crabtree, 1984; Davis, 2016).

Clinical studies: Limited studies are available on the use of naltrexone in subjects with depressive symptoms, although this drug has been approved for alcohol abuse/dependence. An open-label study in patients with comorbid alcohol abuse/dependence and bipolar disorder reported a significant improvement in depressive/manic symptoms and alcohol craving (Brown et al., 2006). Later, Murphy et al. (2014) carried out a double-blind, placebo-controlled study on 32 depressed bipolar patients. In this trial, naltrexone was used as augmentation therapy to lithium, other anticonvulsants and antipsychotics at a dose of 50 mg per day for 12 weeks, but it did not exert mood stabilizing properties when compared to placebo (Murphy et al., 2014). More recently, Mischoulon and colleagues showed that, in 12 patients with recurrent MDD, augmentation of common antidepressants with low-dose (1 mg/day) naltrexone for 3 weeks showed some improvement in mood that was not statistically significant when compared to placebo (Mischoulon et al., 2017). Another study investigated naltrexone extended release + bupropion in a cohort of 25 subjects with binge eating disorder and comorbid depressive symptoms, highlighting similar improvements for both eating and depressive symptoms (Guerdjikova et al., 2017).

Efficacy for treatment-resistant depression: There are no studies of naltrexone as a primary or adjunctive treatment in patients with well-defined TRD.

Adverse effects and toxicology: Naltrexone is considered a relatively safe medication, associated with minor adverse effects like gastrointestinal problems and small increases in blood pressure. Naltrexone is associated with a low risk of hepatotoxicity. Importantly, patients treated with opioid agents should be detoxified for at least 5 days before taking naltrexone to avoid withdrawal syndrome (Crabtree, 1984).

Summary and future directions

In summary, there are still important unmet needs in the treatment of depression as a high percentage of treated patients do not achieve complete remission, and many such patients are recognized to exert TRD. The monoamine theory alone cannot explain the pathophysiology of MDD, and abnormalities in other systems such as the endogenous opioid system which have been found both in preclinical and clinical models of depression. Therefore, research has been directed toward the use of opioidergic drugs, several of which appear to exert antidepressant properties. Among the various opioidergic medications described in this overview, the most investigated and promising compound is buprenorphine, alone or as a combination treatment (i.e., ALKS-5461, a combination of buprenorphine and samidorphan). However, further studies on larger samples of depressed subjects without a history of substance abuse/dependence are needed in order to establish its short- and longer-term efficacy for patients with TRD.

References

Ahmadi, J., Abtahi, S., 2017. Eight-milligram single dose of buprenorphine as an expeditious treatment for severe suicidal ideation: a case report. Iran Red Crescent Med J 19 (9). https://doi.org/10.5812/ircmj.59999.

Ahmadi, J., Jahromi, M.S., Ghahremani, D., London, E.D., 2018. Single high-dose buprenorphine for opioid craving during withdrawal. Trials 19 (1). https://doi.org/10.1186/s13063-018-3055-z.

Ahmadi, M.A., Moradi, G., Khoshravesh, S., Roshani, D., Qaderi, N., Amjadian, M., 2020. The determinants of methadone maintenance treatment in opioids dependence users: a survival analysis study. Med. J. Islam. Repub. Iran 34, 149.

Ahmadi, J., Sarani, E., Jahromi, M., 2019. Rapid effect of a single-dose buprenorphine on reduction of opioid craving and suicidal ideation: a randomized, double blind, placebo-controlled study. Tzu Chi Med. J. 32, 58–64.

Akil, H., Gordon, J., Hen, R., Javitch, J., Mayberg, H., McEwen, B., Meaney, M.J., Nestler, E.J., 2018. Treatment resistant depression: a multi-scale, systems biology approach. Neurosci. Biobehav. Rev. 84, 272–288. https://doi.org/10.1016/j.neubiorev.2017.08.019.

Alkermes, 2019. Alkermes receives complete response letter from U.S. Food and Drug Administration for ALKS 5461 new drug application.

Andrew, S.N., Verduin, M.L., DeGraw, J.D., 2001. Treatment of refractory major depression with tramadol monotherapy. J. Clin. Psychiatry, 205–206. https://doi.org/10.4088/JCP.v62n0312b.

Barber, J., 2011. Examining the use of tramadol hydrochloride as an antidepressant. Exp. Clin. Psychopharmacol. 19 (2), 123–130. https://doi.org/10.1037/a0022721.

Berrocoso, E., Sánchez-Blázquez, P., Garzón, J., Mico, J.A., 2009. Opiates as antidepressants. Curr. Pharm. Des. 15 (14), 1612–1622. https://doi.org/10.2174/138161209788168100.

Bodkin, J.A., Zornberg, G.L., Lukas, S.E., Cole, J.O., 1995. Buprenorphine treatment of refractory depression. J. Clin. Psychopharmacol. 15 (1), 49–57. https://doi.org/10.1097/00004714-199502000-00008.

Bodnar, R.J., 2010. Endogenous opiates and behavior: 2009. Peptides 31 (12), 2325–2359. https://doi.org/10.1016/j.peptides.2010.09.016.

Brown, E.S., Beard, L., Dobbs, L., Rush, A.J., 2006. Naltrexone in patients with bipolar disorder and alcohol dependence. Depress. Anxiety 23 (8), 492–495. https://doi.org/10.1002/da.20213.

Corkery, J.M., Schifano, F., Ghodse, A.H., Oyefeso, A., 2004. The effects of methadone and its role in fatalities. Hum. Psychopharmacol. 19 (8), 565–576. https://doi.org/10.1002/hup.630.

Crabtree, B.L., 1984. Review of naltrexone, a long-acting opiate antagonist. Clin. Pharm. 3 (3), 273–280.

Davis, M.P., 2016. Pharmacokinetic and pharmacodynamic evaluation of oxycodone and naltrexone for the treatment of chronic lower back pain. Expert Opin. Drug Metab. Toxicol. 12 (7), 823–831. https://doi.org/10.1080/17425255.2016.1191469.

Dean, A.J., Bell, J., Mascord, D.J., Parker, G., Christie, M.J., 2002. A randomised, controlled trial of fluoxetine in methadone maintenance patients with depressive symptoms. J. Affect. Disord. 72 (1), 85–90. https://doi.org/10.1016/S0165-0327(01)00419-0.

Dean, A.J., Bell, J., Christie, M.J., Mattick, R.P., 2004. Depressive symptoms during buprenorphine vs. methadone maintenance: findings from a randomised, controlled trial in opioid dependence. Eur. Psychiatry 19 (8), 510–513. https://doi.org/10.1016/j.eurpsy.2004.09.002.

Dinoff, A., Lynch, S.T., Sekhri, N., Klepacz, L., 2020. A meta-analysis of the potential antidepressant effects of buprenorphine versus placebo as an adjunctive pharmacotherapy for treatment-resistant depression. J. Affect. Disord. 271, 91–99. https://doi.org/10.1016/j.jad.2020.03.089.

Dyer, K.R., White, J.M., Foster, D.J.R., Bochner, F., Menelaou, A., Somogyi, A.A., 2001. The relationship between mood state and plasma methadone concentration in maintenance patients. J. Clin. Psychopharmacol. 21 (1), 78–84. https://doi.org/10.1097/00004714-200102000-00014.

Ehrich, E., Turncliff, R., Du, Y., Leigh-Pemberton, R., Fernandez, E., Jones, R., Fava, M., 2015. Evaluation of opioid modulation in major depressive disorder. Neuropsychopharmacology 40 (6), 1448–1455. https://doi.org/10.1038/npp.2014.330.

Emrich, H.M., Vogt, P., Herz, A., 1982. Possible antidepressive effects of opioids: action of buprenorphine. Ann. N. Y. Acad. Sci. 398 (1), 108–112. https://doi.org/10.1111/j.1749-6632.1982.tb39483.x.

E-therapeutics, 2009. http://www.firstwordpharma.com/node/39762#axzz54JR5qryu.

E-therapeutics, 2016. http://www.fiercebiotech.com/biotech/e-therapeutics-reports-top-line-phase-iib-results-major-depressive-disorder.

Faron-Gorecka, A., 2004. Effects of tramadol on alpha2-adrenergic receptors in the rat brain. Brain Res. 1016 (2), 263–267.

Fava, M., Memisoglu, A., Thase, M.E., Bodkin, J.A., Trivedi, M.H., De Somer, M., Du, Y., Leigh-Pemberton, R., DiPetrillo, L., Silverman, B., Ehrich, E., 2016. Opioid modulation with buprenorphine/samidorphan as adjunctive treatment for inadequate response to antidepressants: a randomized double-blind placebo-controlled trial. Am. J. Psychiatr. 173 (5), 499–508. https://doi.org/10.1176/appi.ajp.2015.15070921.

Fava, M., Thase, M.E., Trivedi, M.H., Ehrich, E., Martin, W.F., Memisoglu, A., Nangia, N., Stanford, A.D., Yu, M., Pathak, S., 2020. Opioid system modulation with buprenorphine/samidorphan combination for major depressive disorder: two randomized controlled studies. Mol. Psychiatry 25 (7), 1580–1591. https://doi.org/10.1038/s41380-018-0284-1.

Ferrari, A., Coccia, C.P.R., Bertolini, A., Sternieri, E., 2004. Methadone – metabolism, pharmacokinetics and interactions. Pharmacol. Res. 50 (6), 551–559. https://doi.org/10.1016/j.phrs.2004.05.002.

Gabilondo, A.M., Javier Meana, J., García-Sevilla, J.A., 1995. Increased density of μ-opioid receptors in the postmortem brain of suicide victims. Brain Res. 682 (1–2), 245–250. https://doi.org/10.1016/0006-8993(95)00333-L.

Gerra, G., Fantoma, A., Zaimovic, A., 2006. Naltrexone and buprenorphine combination in the treatment of opioid dependence. J. Psychopharmacol. 20 (6), 806–814. https://doi.org/10.1177/0269881106060835.

Gregory, E., Torres, I.J., Ge, R., Blumberger, D.M., Downar, J.H., Daskalakis, Z.J., Lam, R.W., Vila-Rodriguez, F., 2020. Predictors of cognitive impairment in treatment-resistant depression. J. Affect. Disord. 274, 593–601. https://doi.org/10.1016/j.jad.2020.05.101.

Gudin, J.A., Laitman, A., Nalamachu, S., 2015. Opioid related endocrinopathy. Pain Med. 16, S9–S15. https://doi.org/10.1111/pme.12926.

Guerdjikova, A.I., Walsh, B., Shan, K., Halseth, A.E., Dunayevich, E., McElroy, S.L., 2017. Concurrent improvement in both binge eating and depressive symptoms with naltrexone/bupropion therapy in overweight or obese subjects with major depressive disorder in an open-label, uncontrolled study. Adv. Ther. 34 (10), 2307–2315. https://doi.org/10.1007/s12325-017-0613-9.

Heo, Y.A., Scott, L.J., 2018. Buprenorphine/naloxone (Zubsolv ®): a review in opioid dependence. CNS Drugs 32 (9), 875–882. https://doi.org/10.1007/s40263-018-0560-2.

Holtzheimer, P.E., Mayberg, H.S., 2011. Stuck in a rut: rethinking depression and its treatment. Trends Neurosci. 34 (1), 1–9. https://doi.org/10.1016/j.tins.2010.10.004.

Hopwood, S.E., Owesson, C.A., Callado, L.F., McLaughlin, D.P., Stamford, J.A., 2001. Effects of chronic tramadol on pre- and post-synaptic measures of monoamine function. J. Psychopharmacol. 15 (3), 147–153. https://doi.org/10.1177/026988110101500301.

Jacobson, M.L., Wulf, H.A., Browne, C.A., Lucki, I., 2018. Opioid modulation of cognitive impairment in depression. In: Progress in Brain Research. vol. 239. Elsevier B.V, pp. 1–48, https://doi.org/10.1016/bs.pbr.2018.07.007.

Jesse, C.R., Wilhelm, E.A., Bortolatto, C.F., Nogueira, C.W., 2010. Evidence for the involvement of the noradrenergic system, dopaminergic and imidazoline receptors in the antidepressant-like effect of tramadol in mice. Pharmacol. Biochem. Behav. 95 (3), 344–350. https://doi.org/10.1016/j.pbb.2010.02.011.

Judd, L.L., Parker, D.C., Janowsky, D.S., Segal, D.S., Risch, S.C., Huey, L.Y., 1982. The effect of methadone on the behavioral and neuroendocrine responses of manic patients. Psychiatry Res. 7 (2), 163–170. https://doi.org/10.1016/0165-1781(92)90089-L.

Karp, J.F., Butters, M., Begley, A., Miller, M., Lenze, E.J., Blumberger, D., Mulsant, B., Reynolds III, C.F., 2014. Safety, tolerability, and clinical effect of low-dose buprenorphine for treatment-resistant depression in mid-life and older adults. J. Clin. Psychiatry 75, e785-93.

Kennedy, S.E., Koeppe, R.A., Young, E.A., Zubieta, J.K., 2006. Dysregulation of endogenous opioid emotion regulation circuitry in major depression in women. Arch. Gen. Psychiatry 63 (11), 1199–1208. https://doi.org/10.1001/archpsyc.63.11.1199.

Knoll, A.T., Carlezon, W.A., 2010. Dynorphin, stress, and depression. Brain Res. 1314, 56–73. https://doi.org/10.1016/j.brainres.2009.09.074.

Kosten, T.R., 1990. Neurobiology of abused drugs opioids and stimulants. J. Nerv. Ment. Dis. 178 (4), 217–227. https://doi.org/10.1097/00005053-199004000-00001.

Kumar, V., Lu, H., Hard, M., von Moltke, L., 2019. Characterization of the pharmacokinetics of Samidorphan in healthy volunteers: absolute bioavailability and the effect of food and age. Drugs R&D 19 (3), 277–287. https://doi.org/10.1007/s40268-019-00280-5.

Kyzer, J.L., Wenthur, C.J., 2020. Classics in chemical neuroscience: buprenorphine. ACS Chem. Neurosci. 11 (10), 1385–1399. https://doi.org/10.1021/acschemneuro.0c00100.

Loredana, S., Lorenzo, P., Mario, E., 2004. Tramadolo: profilo farmacologico, terapeutico e farmacoeconomico. Farmecon. Health Econ. Ther. Pat., 151–162. https://doi.org/10.7175/fe.v5i3.794.

Lutz, P.E., Kieffer, B.L., 2013. Opioid receptors: distinct roles in mood disorders. Trends Neurosci. 36 (3), 195–206. https://doi.org/10.1016/j.tins.2012.11.002.

Mischoulon, D., Hylek, L., Yeung, A.S., Clain, A.J., Baer, L., Cusin, C., Ionescu, D.F., Alpert, J.E., Soskin, D.P., Fava, M., 2017. Randomized, proof-of-concept trial of low dose naltrexone for patients with breakthrough symptoms of major depressive disorder on antidepressants. J. Affect. Disord. 208, 6–14. https://doi.org/10.1016/j.jad.2016.08.029.

Mongan, L., Callaway, E., 1990. Buprenorphine responders. Biol. Psychiatry 28 (12), 1078–1080. https://doi.org/10.1016/0006-3223(90)90619-D.

Murphy, B.L., Ravichandran, C., Babb, S.M., Cohen, B.M., 2014. Naltrexone in bipolar disorder with depression a double-blind, placebo-controlled study. J. Clin. Psychopharmacol. 34 (6), 749–751. https://doi.org/10.1097/JCP.0000000000000222.

Norelli, L.J., Smith, H.S., Sher, L., Blackwood, T.A., 2013. Buprenorphine in the treatment of non-suicidal self-injury: a case series and discussion of the literature. Int. J. Adolesc. Med. Health 25 (3), 323–330. https://doi.org/10.1515/ijamh-2013-0069.

Nyhuis, P.W., Gastpar, M., Scherbaum, N., 2008. Opiate treatment in depression refractory to antidepressants and electroconvulsive therapy. J. Clin. Psychopharmacol. 28 (5), 593–595. https://doi.org/10.1097/JCP.0b013e31818638a4.

Pathak, S., Vince, B., Kelsh, D., Shram, M.J., Setnik, B., Lu, H., Nangia, N., Stanford, A.D., Ehrich, E., 2019. Abuse potential of buprenorphine/Samidorphan combination compared to buprenorphine and placebo: a phase 1 randomized controlled trial. J. Clin. Pharmacol. 59 (2), 206–217. https://doi.org/10.1002/jcph.1280.

Peciña, M., Karp, J.F., Mathew, S., Todtenkopf, M.S., Ehrich, E.W., Zubieta, J.K., 2019. Endogenous opioid system dysregulation in depression: implications for new therapeutic approaches. Mol. Psychiatry 24 (4), 576–587. https://doi.org/10.1038/s41380-018-0117-2.

Peckham, A.M., Cruz, A.D.L., Dufresne, R.L., 2018. Kappa opioid receptor antagonism: are opioids the answer for treatment resistant depression? Ment. Health Clin., 175–183. https://doi.org/10.9740/mhc.2018.07.175.

Pierre, D., Laurence, C., Jules, D., 1994. The pharmacology of tramadol. Drugs, 3–7. https://doi.org/10.2165/00003495-199400471-00003.

Pradhan, A.A., Befort, K., Nozaki, C., Gavériaux-Ruff, C., Kieffer, B.L., 2011. The delta opioid receptor: an evolving target for the treatment of brain disorders. Trends Pharmacol. Sci. 32 (10), 581–590. https://doi.org/10.1016/j.tips.2011.06.008.

Ragguett, R.M., Rong, C., Rosenblat, J.D., Ho, R.C., McIntyre, R.S., 2018. Pharmacodynamic and pharmacokinetic evaluation of buprenorphine + samidorphan for the treatment of major depressive disorder. Expert Opin. Drug Metab. Toxicol. 14 (4), 475–482. https://doi.org/10.1080/17425255.2018.1459564.

Ragguett, R.M., Tamura, J.K., McIntyre, R.S., 2019. Keeping up with the clinical advances: depression. CNS Spectrums 24 (1), 28–36. https://doi.org/10.1017/S1092852919001159.

Resnick, R., Falk, F., 1987. Buprenorphine: pilot trials in border- line patients and opiate dependence: treatment of a common disorder? In: Harris, L.S. (Ed.), NIDA Research Monograph: Problems of Drug Dependence. NIDA.

Rougemont-Bucking, A., Gamma, F., Panksepp, J., 2017. Use of tramadol in psychiatric care: a comprehensive review and report of two cases. Swiss Med. Wkly. 147.

Rush, A.J., 2007. STAR*D: what have we learned? Am. J. Psychiatry 164 (2), 201–204. https://doi.org/10.1176/appi.ajp.164.2.201.

Saxena, P.P., Bodkin, J.A., 2019. Opioidergic agents as antidepressants: rationale and promise. CNS Drugs 33 (1), 9–16. https://doi.org/10.1007/s40263-018-0584-7.

Scarone, S., Gambini, O., Calabrese, G., Sacerdote, P., Bruni, M., Carucci, M., Panerai, A.E., 1990. Asymmetrical distribution of beta-endorphin in cerebral hemispheres of suicides: preliminary data. Psychiatry Res. 32 (2), 159–166. https://doi.org/10.1016/0165-1781(90)90082-G.

Serafini, G., 2018. The efficacy of buprenorphine in major depression, treatment-resistant depression and suicidal behavior: a systematic review. Int. J. Mol. Sci. 19, 2410.

Serafini, G., Nebbia, J., Cipriani, N., Conigliaro, C., Erbuto, D., Pompili, M., Amore, M., 2018. Number of illness episodes as predictor of residual symptoms in major depressive disorder. Psychiatry Res. 262, 469–476. https://doi.org/10.1016/j.psychres.2017.09.026.

Spencer, C., 2000. The efficacy of intramuscular tramadol as a rapid-onset antidepressant [4]. Aust. N. Z. J. Psychiatry 34 (6), 1032–1033. https://doi.org/10.1080/000486700284.

Stefanowski, B., Antosik-Wójcińska, A., Święcicki, Ł., 2020. The use of buprenorphine in the treatment of drug-resistant depression - an overview of the studies. Psychiatr. Pol. 54 (2), 199–207. https://doi.org/10.12740/PP/102658.

Striebel, J.M., Kalapatapu, R.K., 2014. The anti-suicidal potential of buprenorphine: a case report. Int. J. Psychiatry Med. 47 (2), 169–174. https://doi.org/10.2190/PM.47.2.g.

Tenore, P.L., 2008. Psychotherapeutic benefits of opioid agonist therapy. J. Addict. Dis. 27 (3), 49–65. https://doi.org/10.1080/10550880802122646.

Thase, M.E., Schwartz, T.L., 2015. Choosing medications for treatment-resistant depression based on mechanism of action. J. Clin. Psychiatry 76 (6), 720–727. https://doi.org/10.4088/JCP.14052ah2c.

Thase, M.E., Stanford, A.D., Memisoglu, A., Martin, W., Claxton, A., Bodkin, J.A., Trivedi, M.H., Fava, M., Yu, M., Pathak, S., 2019. Results from a long-term open-label extension study of adjunctive buprenorphine/samidorphan combination in patients with major depressive disorder. Neuropsychopharmacology 44 (13), 2268–2276. https://doi.org/10.1038/s41386-019-0451-3.

Yovell, Y., Bar, G., Mashiah, M., Baruch, Y., Briskman, I., Asherov, J., Lotan, A., Rigbi, A., Panksepp, J., 2016. Ultra-low-dose buprenorphine as a time-limited treatment for severe suicidal ideation: a randomized controlled trial. Am. J. Psychiatr. 173 (5), 491–498. https://doi.org/10.1176/appi.ajp.2015.15040535.

Zajecka, J.M., Stanford, A.D., Memisoglu, A., Martin, W.F., Pathak, S., 2019. Buprenorphine/samidorphan combination for the adjunctive treatment of major depressive disorder: results of a phase III clinical trial (FORWARD-3). Neuropsychiatr. Dis. Treat. 15, 795–808. https://doi.org/10.2147/NDT.S199245.

Chapter 17

Transcranial magnetic stimulation for treatment-resistant depression

Jean-Marie Batail[a,*], Zui Narita[b,*], John Coetzee[a,c], Joshua Brown[d,e], Romina Nejad[a], Katy Stimpson[a], Angela Phillips[a,c], Eleanor Cole[f], Gregory Sahlem[a,†], and Nolan Williams[a,†]

[a]*Brain Stimulation Laboratory, Department of Psychiatry and Behavioral Sciences, Stanford University, Palo Alto, CA, United States,* [b]*Department of Behavioral Medicine, National Institute of Mental Health, National Center of Neurology and Psychiatry, Tokyo, Japan,* [c]*Adamson Lab, Headache Center for Excellence, Rehabilitation Service, VA Palo Alto Health Care System, Palo Alto, CA, United States,* [d]*Department of Psychiatry and Human Behavior, Brown University School of Medicine, Providence, RI, United States,* [e]*Department of Neurology, Brown University School of Medicine, Providence, RI, United States,* [f]*Department of Psychiatry and Behavioral Sciences, Stanford University, Palo Alto, CA, United States*

Introduction

Although diverse treatments are currently available for depression, one class of treatments that holds significant promise, especially for treatment-resistant depression (TRD) in the near future, is neuromodulation. This growing category of treatment techniques includes interventions, either electrical, chemical, or mechanical in nature, that directly modify the function of the nervous system (Krames et al., 2009). The oldest form of neuromodulation, electroconvulsive therapy (ECT), is still the most effective treatment available for depression (Kellner et al., 2006; Milev et al., 2016), and has been joined by a variety of newer techniques. The possibility of directly modifying activity in discrete brain areas or systems has a number of advantages over traditional first-line treatments for depression, not least among these being the focal nature of neuromodulatory treatments, their speed of action, and their potentially lower cost (compared to the cost of untreated or undertreated depression in treatment-resistant individuals) (Glass, 2003; Luppa et al., 2007; Nguyen and Gordon, 2015).

Neuromodulatory treatments for depression can be divided into two broad categories: invasive and noninvasive. Invasive treatments include procedures such as vagus nerve stimulation (VNS) and deep brain stimulation (DBS), and are typically reserved for the most severe cases, given the risks involved with surgery. The noninvasive techniques, such as ECT, repetitive transcranial magnetic stimulation (rTMS), and transcranial electrical stimulation (TES), have a potentially broader clinical application than the invasive techniques given the significantly lower risks and costs associated with them. This chapter will focus in particular on the application of rTMS for TRD.

Compared to ECT, rTMS has several advantages. It does not require general anesthesia, which is perhaps the most dangerous aspect of administering ECT (Lagasse, 2002). It also does not require seizure induction, meaning that it is substantially less disruptive to the lives of patients undergoing this treatment (e.g., does not impair daily activities, does not prevent driving during the recovery period). Because of the more focal nature of rTMS, it is free of the cognitive and memory impairments that are associated with ECT (Martin et al., 2015). Finally, rTMS does not come with the cultural stigma that has long been attached to ECT (Payne and Prudic, 2009) and has likely prevented more widespread use of the technique (Wilkinson et al., 2018).

Several large randomized controlled studies have investigated the efficacy of rTMS for depression and have repeatedly found active stimulation to be more effective than sham stimulation (Levkovitz et al., 2015; Loo and Mitchell, 2005; O'Reardon et al., 2007). Although it is not as effective as ECT in its current conventional form (Cusin and Dougherty, 2012), rTMS is significantly more tolerable than ECT (Cusin and Dougherty, 2012), and for many, this is an acceptable tradeoff. In this chapter, we first introduce the theoretical basis of rTMS. Then, we present a state of the art of rTMS practice in the indication of TRD. Finally, we open the discussion on the perspectives of development of this technique through the use of biomarkers and the optimization of stimulation parameters.

* Co-first authors.

† Co-senior authors.

Managing Treatment-Resistant Depression. https://doi.org/10.1016/B978-0-12-824067-0.00017-7

Historical overview

Since the development of magnetic stimulators as a clinical tool (Barker et al., 1985), it was known that electromagnetic stimulation could stimulate neurons (Devinsky, 1993). The first formal work in the area combined TMS with electromyography, to measure neural conduction time (Hallett and Cohen, 1989), and to compare the brain electrophysiological response in different psychiatric disorders (Abarbanel et al., 1996). These early works focused on the motor system, and TMS, used here as a probe, utilized a single-pulse format and was able to induce a motor response (Hallett and Cohen, 1989). Contemporary to these studies, further work has confirmed that "patterned, or repetitive" simulation of the motor cortex is able to predictably alter motor function (Avanzino et al., 2008; Pascual-Leone et al., 1998). Utilizing this knowledge, this work initiated on the motor cortex was extended to the prefrontal cortex in healthy volunteers and was found to have an effect on mood and emotions (Avanzino et al., 2008; Pascual-Leone et al., 1998).

One of the first reports of rTMS being used to treat depression was a study in which daily rTMS was applied to the left prefrontal cortex of six patients with TRD (George et al., 1995), resulting in significant reduction in depressive symptoms and an increase in brain metabolic demand, as measured by positron emission tomography. Large-scale controlled studies (George et al., 1996; Pascual-Leone et al., 1996) later led to FDA approval of high frequency (HF) rTMS for treatment-resistant forms of depression. Additionally, through functional magnetic resonance imaging (fMRI) and electroencephalography (EEG), researchers have observed secondary effects from rTMS on brain regions that are connected to but distant from the site of stimulation (Hallett et al., 2017; Li et al., 2004). This secondary activation of connected regions is hypothesized to be essential for the antidepressant effects of rTMS (Fox et al., 2012; Siddiqi et al., 2020).

Proposed mechanisms: How does TMS work?

In TMS, an electric charge is stored in a capacitor and discharged through a wire that is wound around a coil (Rossi et al., 2009). The current is discharged in an intermittent manner, causing a pulsed magnetic field in the local area around the coil. This magnetic field is time varying, resulting in the induction of an electrical field with a magnitude proportional to the rate of change of the magnetic field (in accordance with Faraday's law) (Rossi et al., 2009). When applied transcranially, the coil is placed near the scalp, and the electromagnetic field it generates passes through the intervening skin and bone, inducing an electrical current in the brain, which when delivered with a sufficient intensity leads to action potentials in cortical neurons.

If TMS is applied over the motor cortex, the induced action potentials can produce movements in muscles. By targeting specific muscles, such as those of the fingers or hand, and systematically varying the intensity of the coil output, a motor threshold (MT) can be established. In this manner, the "dose" of stimulation can be individualized to each person. The most important variables in determining how TMS affects the brain are the shape of the coil, placement of the coil, frequency of the pulsation, and the intensity of the magnetic field (Deng et al., 2013; Rossi et al., 2009). The nomenclature describes several different ways to deliver TMS: one stimulus at a time (single-pulse), pairs of pulses (paired-pulse), and repeated pulses called trains (repetitive TMS) (Rossi et al., 2009). The former two are usually used in experimental neurophysiology (cortical mapping, conduction time, cortico-cortical interactions). rTMS can be divided into conventional rTMS, characterized by *the application of regularly repetitive single TMS pulses in a fixed frequency* (i.e., 10 Hz, 1 Hz), and burst patterns of rTMS. This latter form of rTMS is composed of high frequency trains (bursts) interleaved with short pauses. Theta burst is the most widely known patterned form of rTMS and refers to bursts of 50 Hz rTMS repeated at a rate of 5 Hz, either in a continuous (cTBS) or intermittent (iTBS) trains. The effects of rTMS on brain excitability can also be measured in the human motor system via electromyography (EMG). A substantial literature supports the generalization that HF ($\geq$5 Hz)/iTBS rTMS tends to increase brain excitability, while low-frequency (LF) (~1 Hz)/cTBS rTMS tends to inhibit it (Hoogendam et al., 2010).

Despite the clear evidence of behavioral, system and network plasticity induced by rTMS, one of the major limitations in the field is that we do not yet have a clear understanding of how rTMS alters neuronal plasticity. Existing research supports the view that HF-rTMS increases the efficiency of communication between neurons, a type of synaptic plasticity known as long-term potentiation (LTP). The most specific evidence for this hypothesis comes from animal studies. Using the same experimental approach as with classical LTP experiments (which use electrical stimulation), Vlachos and colleagues found that 10 Hz repetitive magnetic stimulation (rMS) over mouse hippocampal slices produced LTP-like potentiation (Vlachos et al., 2012). More specifically, hallmark characteristics of LTP, including prolonged (over several hours) potentiation mediated by GluA1-containing AMPA receptors, were accompanied by dendritic spine growth, and required activation of NMDA receptors, sodium, and calcium channels. The same group also reported that 10 Hz rMS "cuts" the metaphorical "brakes"; that is, it reduces GABAergic tone by reducing expression of $GABA_A$ receptor alpha-2 subunits through destabilization of its major scaffolding protein, gephyrin (Lenz et al., 2016).

How these animal studies translate to humans is less clear; however, several studies have tested the effects of blocking various key steps in the synaptic plasticity pathway in humans (most often using pharmacologic approaches at the NMDA receptor). The NMDA receptor is well-known to be important for LTP (Collingridge et al., 1983; Tang et al., 1999). Using nonclinical protocols designed to enhance potentiation (i.e., pairing an rTMS pulse with a pulse on the median nerve, and pairing rTMS with an ischemic nerve block), several studies have found that NMDA receptor antagonists blocked LTP-like enhancement (Stefan et al., 2002; Suppa et al., 2013; Ziemann et al., 1998). Of greater clinical relevance, TBS protocols which produced robust bidirectional plasticity changes were blocked by the NMDA receptor antagonist, memantine (Huang et al., 2005, 2007). Whether the dependence of rTMS-induced plasticity on NMDA receptor activity is specific to LTP is less clear. The NMDA receptor partial agonist, d-cycloserine (DCS) was not by itself *sufficient* to enhance TBS-induced potentiation (Selby et al., 2019; Teo et al., 2007). On the other hand, NMDA receptor activity increased by DCS was sufficient to enhance potentiation with 10 Hz rTMS, hinting that these two clinical forms of TMS may work through different mechanisms (Brown et al., 2020).

In summary, TMS causes bidirectional local and distant circuit-level changes in the brain, but the cellular mechanisms remain unknown.

Parameters, response and remission rates, and durability

The US Food and Drug Administration (FDA) first approved the use of rTMS for treatment-resistant depression in 2008. The original approval included the use of 3000 pulses at 10 Hz stimulation sessions, in 4-s pulse trains with 26-s intertrain intervals for figure of 8 coils and the use of 1980 pulses at 18 Hz stimulation sessions, in 2-s pulse trains with 20-s intertrain intervals for the H-coil (Levkovitz et al., 2009; McClintock et al., 2018; O'Reardon et al., 2007). Treatment utilizing both coils is applied to the left dorsolateral prefrontal cortex (DLPFC) at 120% of a person's resting motor threshold for a total of 20–30 sessions spanning 4–6 weeks (O'Reardon et al., 2007).

There have been three multisite randomized controlled trials assessing the efficacy of rTMS for treatment-resistant depression (McClintock et al., 2018). In these trials, remission was defined as no longer meeting criteria for a clinical diagnosis of depression and response was defined as decreasing depressive symptoms by at least 50% from baseline. The reported response rates ranged from 18.1% to 38.4% and remission rates ranged from 7.1% to 32.6% for active rTMS compared to response rates ranging from 11% to 21.4% and remission rates ranging from 5.1% to 14.6% for sham rTMS (George et al., 2010; Levkovitz et al., 2015; O'Reardon et al., 2007). Both the treatment resistance level of the patients enrolled in these studies as well as the response and remission rates are similar to those treated in the level 2 with next step psychopharmacology in the Sequenced Treatment Alternatives to Relieve Depression (STAR*D) trial (Howland, 2008; Rush et al., 2006; Warden et al., 2007). However, it should be noted that these findings were reported after 15–30 rTMS treatment sessions, while it has been suggested to be more effective with more treatment sessions (Chen et al., 2020; Fitzgerald et al., 2018; Loo et al., 2007; Sonmez et al., 2019). Follow-up clinical effectiveness research has further suggested higher rates of response and remission in real world settings (Carpenter et al., 2012; Dunner et al., 2014; Sackeim et al., 2020).

Also, some works have addressed the question of whether it is possible to increase the efficacy of rTMS by increasing the number of pulses per session or by increasing the number of sessions delivered each day. A recent randomized controlled trial compared standard and high dose rTMS (including both high and low frequency protocols) in 300 patients and concluded that increasing the number of pulses in individual sessions does not appear to result in increased antidepressant efficacy (Fitzgerald et al., 2020). Studies looking at number of sessions delivered each day, however, have suggested that increasing the number of sessions in a day may result in a greater and faster antidepressant effect (Chen et al., 2020; Fitzgerald et al., 2018; Loo et al., 2007; Sonmez et al., 2019). A recent study compared standard and accelerated HF-rTMS protocols. Whereas there was no difference of efficacy between both protocols, accelerated form of HF-rTMS has been suggested to achieve a faster antidepressant action (Fitzgerald et al., 2018). In addition, a recent metaanalysis reported that accelerated HF-TMS protocols were not superior to traditional rTMS but more studies may be needed to draw definitive conclusions (Sonmez et al., 2019).

About a decade of research has led to the development of TBS in the treatment of depression (Suppa et al., 2016). The seminal randomized, multicenter study by Blumberger and colleagues demonstrated that just 3 min of TBS is noninferior to 37.5 min of traditional 10 Hz rTMS (Blumberger et al., 2018). This trial compared the effectiveness of the traditional rTMS protocol to the newer iTBS protocol in a noninferiority study. In this study, traditional rTMS achieved a response rate of 49% and remission rate of 32%, whereas patients that received iTBS treatment achieved a response rate of 47% and a remission rate of 27%, showing that this accelerated form of rTMS treatment was statistically noninferior to traditional treatment (Blumberger et al., 2018). The evidence from this study led to FDA clearance to use the iTBS protocol for the treatment of depression in 2018. Since then, there are a number of groups working to optimize different TBS parameters in hopes of improving remission and response rates in depressed individuals.

Considering the high relapse rate for major depressive disorder, it is necessary to understand the durability of the antidepressant response following a successful course of rTMS. Despite the fact that current FDA-approved rTMS treatment protocols do not recommend it yet, maintenance protocols for other antidepressant treatments have been shown to significantly extend the duration of remission and reduce the rate of relapse for ECT (Bailine et al., 2019; Kellner et al., 2006; Petrides et al., 2011; Sackeim et al., 2001), psychotherapy (Frank, 1991; Klein et al., 2004; Paykel, 2001), and antidepressant medication (Kaymaz et al., 2008; Kupfer, 1992). Maintenance rTMS therapy has been shown beneficial for prolonging the antidepressant effects of acute rTMS treatment as well (Perera et al., 2016; Senova et al., 2019; Wang et al., 2017). Initial studies investigating the impact of maintenance rTMS treatment have used various types of maintenance protocols ranging from 3 weeks of tapering after the acute treatment phase (Janicak et al., 2010) to 6 years of 1–2 sessions per week (O'Reardon et al., 2005). As well as these differences in maintenance course duration, the maintenance protocols have differed in the frequency at which maintenance sessions are delivered (Benadhira et al., 2005; Haesebaert et al., 2018; O'Reardon et al., 2005; Rachid, 2018). Maintenance rTMS can be provided in a prospective, scheduled regimen or on an as needed basis. Pharmacotherapy, psychotherapy, exercise, or a combination of these treatments have also been used to maintain the antidepressant effects of an acute course of rTMS treatment (Kedzior et al., 2015). To date, although promising, only a few studies have reported the potential benefit of the combination of antidepressants and rTMS and have resulted in mixed results (Fitzgerald, 2019; Haesebaert et al., 2018). Nevertheless, consensus recommendations converge toward the use of these medications after a successful course of rTMS (McClintock et al., 2018). Also, the combined use of mindfulness-based cognitive behavioral therapy could be beneficial (Fitzgerald, 2019; Sathappan et al., 2019) as well as the combination of aerobic exercise and rTMS (Hendrikse et al., 2017).

In conclusion, maintenance rTMS trials show great promise in prolonging the antidepressant response in some individuals with major depression, albeit more trials are needed to determine the optimal dose and duration of maintenance rTMS treatments to prolong durability.

Effect on treatment-resistant bipolar depression

Due to the significant challenges in the treatment and management of bipolar disorder (BD), novel treatments are needed (Geddes and Miklowitz, 2013; Reinares et al., 2014). As HF-rTMS has shown efficacy in the treatment of depressive episodes among individuals with MDD, clinical trials and "real world," naturalistic reports have found a similar antidepressant response during episodes of depression in BD (Dell'Osso et al., 2009; Gold et al., 2019; Phillips et al., 2020). Similar to the potential side-effects of antidepressant medications in individuals with BD, some have reported hypomanic or manic "affective switching" during or after a course of rTMS (Dell'Osso et al., 2015; Dolberg et al., 2002; Tamas et al., 2007; Tavares et al., 2017). In a recent review compiling 10 cases of affective switching, no clear risk factors were identified (Hett and Marwaha, 2020). Others have explored the use of HF- and LF-rTMS in the treatment of mania and mixed states, with varying results (Grisaru et al., 1998; Pallanti et al., 2014; Pathak et al., 2015). In addition to mood-related outcomes, HF-rTMS has also shown to improve neurocognitive function in BD (Yang et al., 2019). Gold et al., (2019) recently reviewed original trials highlighting the use of rTMS, including iTBS and bilateral simulation paradigms, across all phases of BD. A recent RCT, testing the superiority of active versus sham daily L-DLPFC iTBS over 4 weeks in bipolar depression, was stopped due to futility. This study randomized 37 participants (18 active/19 sham). No difference has been shown between groups on the primary outcome (MADRS total score) and two manic episodes has been reported, one in the active group and one during the open label phase (McGirr et al., 2021). This negative result raised few questions specifically the need to better understand the pathophysiology of this particular depressive phenotype (i.e., bipolar depression) that could inform a disorder-specific targeting (Camprodon, 2021). Further trials including larger samples, sham controlled trials, and targeted stimulation parameters are needed.

Potential biomarkers

Given the response and remission rates, and durability, rTMS is a validated treatment for TRD. Nevertheless, no established biomarker is yet available on this condition.

Three main large scale networks appear to be involved in the pathophysiology of depression, including the: (1) central executive network (CEN); (2) default mode network (DMN); and (3) salience network (SN) (Greicius et al., 2007; Hamilton et al., 2015; Sheline et al., 2010; Wise et al., 2017). Previous studies showed the hyperconnectivity of DMN in depression, which may be normalized with effective treatments (Liston et al., 2014; Philip et al., 2018). Also, an increased connectivity between DMN and sgACC has been suggested in depression (Berman et al., 2011; Chai et al., 2016; Greicius et al., 2007; Hamilton et al., 2015). Reduced activity of the dorsolateral prefrontal cortex (DLPFC), which is included in the CEN, has been reported in depression (Baxter et al., 1989; Martinot et al., 1990). This abnormal activity may potentially be changed by effective treatments (Brody et al., 2001; Koenigs and Grafman, 2009; Mayberg et al., 2000). Another study demonstrated the increased connectivity of DLPFC-amygdala (Han et al., 2015). Furthermore, the activity of sgACC has reportedly been increased in depression (Drevets et al., 2008), which is correlated with the reduced activity of the DLPFC (Baeken et al., 2014; Fox et al., 2012). Improvements of depression are associated with the reduction of sgACC activity (Dougherty et al., 2003; Drevets et al., 2008; Mayberg et al., 2000; Nobler et al., 2001).

HF-rTMS increases cortical excitability over a cortical target (Chen et al., 2013; Chung et al., 2015; Fitzgerald et al., 2018). HF-left DLPFC (L-DLPFC) rTMS has been hypothesized to improve depression by upregulating the CEN (Brody et al., 2001; Koenigs and Grafman, 2009; Mayberg et al., 2000). Also, the upregulation of CEN may be related to the down-regulation of DMN (Sridharan et al., 2008). These effects may be correlated with the accuracy of targeting (Fischer et al., 2016; Luber et al., 2017). The hypoactivity in LDPFC and hyperactivity in sgACC in depression are normalized by rTMS, which are indeed consistent with clinical effects.

A recent study employed functional MRI to individually target the L-DLPFC most anticorrelated with sgACC and induced remission in 90% of patients with severe TRD (Cole et al., 2020), which may confirm the importance of connectivity between these regions. These interventions also decreased the connectivity of sgACC-DMN after intervention (Xiao et al., 2019). Another group used a novel approach classifying patients with depression based on functional connectivity (Drysdale et al., 2017). In this study, hierarchical clustering was used, i.e., patients were classified into biotypes of similar functional connectivity. A specific biotype was demonstrated to predict the effect of dorsomedial prefrontal cortex (DMPFC) rTMS on depression. Also, responders and nonresponders were distinguished by baseline functional connectivity that involved DMPFC, left amygdala, L-DLPFC, bilateral orbitofrontal cortex and posterior cingulate cortex. The combination of biotype and functional connectivity best predicted the treatment effect of rTMS on depression.

Optimal stimulation parameters

As discussed earlier in this chapter, stimulation parameters are suggested to be important for the effect of rTMS. Several different parameters can be set to optimize the neural and clinical response, such as the number of pulses per session, number of pulses per course, the way these pulses are delivered (traditional-fixed frequency rTMS, or patterned burst TBS), the number of sessions per day, the spacing between sessions, and the intensity of the stimulation. First, studies have hypothesized that delivering a higher dose (i.e., number of pulses) may result in improved clinical outcome (Kaster et al., 2018; Yip et al., 2017). Despite that a recent trial has failed to show any benefit from increasing the number of pulses per session with traditional rTMS (Fitzgerald et al., 2020), it seems that some evidence supports this assumption for iTBS. Indeed, a study reported a dose-dependent response of iTBS with a greater enhancement of motor cortical excitability with 1800 pulses protocol compared with 600 and 1200 pulses (Nettekoven et al., 2014). In addition, basic science works have shown that the administration of 1800 iTBS pulses resulted in modulating brain activity in the intended direction with either increased motor cortex excitability (Peinemann et al., 2004) or decreased DMN-ACC functional connectivity (Singh et al., 2019). Note that this optimal 1800 pulses iTBS dose has been successfully tested in blinded trials in depression (Hallett et al., 2017; Li et al., 2014). Second, from works in human neurophysiology (Huang et al., 2005; Larson and Lynch, 1986), the pattern of the stimulation session (iTBS) has been indirectly demonstrated to be more effective than HF-rTMS in modulating motor cortex excitability (Huang et al., 2005; Maeda et al., 2000). The above-mentioned depression trial THREE-D concluded that patterned iTBS stimulation protocol, using less number of pulses for a shorter duration of session, was as effective as a traditional 10 Hz rTMS course (Blumberger et al., 2018). Third, the pattern of the stimulation course (intersession interval—ISI) had also shown benefits on both clinical and neurophysiological response to stimulation. Inspired by neurophysiology theory of learning, spaced training has more robust effects at the cellular and molecular level (Smolen et al., 2016). Basic science works have reported that a 50–90 min ISI resulted in a cumulative effect on dendritic spine enlargement whereas ISI $\leq$ 40 min didn't (Smolen et al., 2016). These properties have been reproduced in the human motor cortex with an increase of cortical excitability induced by iTBS with ISI > 30 min (Nettekoven et al., 2014, 2015; Tse et al., 2018). Moreover, the number of daily TBS sessions may also be of importance. Compared to a single one, two spaced stimulation sessions have greater and long-lasting effects (Goldsworthy et al., 2012, 2015) with beneficial clinical effects in depressed patients (Mielacher et al., 2019). Nevertheless, this last finding remains controversial as reported by a metaanalytic work on accelerated traditional rTMS protocols (Sonmez et al., 2019). Interestingly, contrary to traditional rTMS, high-dose and accelerated forms of iTBS, when combined, seem to provide a strong and fast antidepressant effect (Cole et al., 2022; Duprat et al., 2016). Fourth, the stimulation intensity is another important parameter. The distance between the targeted area and the coil may influence the brain response of the motor cortex and justify adjusting the intensity to the depth (Stokes et al., 2005). Using TBS stimulation, the optimal intensity for prefrontal regions seems to respond to an inverse U-shape relationship with best effect using subthreshold level (75%–90% of resting motor threshold) of stimulation on neurophysiological response (Alkhasli et al., 2019; Chung et al., 2018), while it doesn't seem to be the case with traditional rTMS (Pascual-Leone et al., 1994).

Recently, one study used these optimized parameters combining a high-dose (90,000 pulses in total) patterned iTBS and functional MRI targeted rTMS course in 22 treatment-resistant depressed participants (Cole et al., 2020). In this open-label study, participants achieved a remission rate of up to 90% with a significant and rapid reduction in depression scores and suicidality. These preliminary results are very promising given the evidence of a robust and rapid effect of network-based and optimized rTMS technique. These open label data were confirmed by a randomized controlled trial enrolling 29 TRD participants (14 in active group/15 in sham group) with an overall remission rate of 78.6% in the active group across the 4 weeks follow-up. This study reported strong effect sizes on the main outcome (MADRS score) were 1.7, 1.4, 1.8, 1.5, and 1.4, at immediate, week 1, week 2, week 3, and week 4 posttreatment, respectively (Cole et al., 2022). These results suggest that the future of TMS leads into a personalized, biomarker-based, and optimized stimulation approach.

Future directions and perspectives

Thirty-five years of rTMS research have led to a standardized and effective rTMS treatment course for TRD but further personalization, higher doses, and the development of a standardized maintenance treatment schedule may increase efficacy and durability further. Improved efficacy, alongside the minimal side effects (Loo et al., 2008) and data showing the cost-effectiveness of rTMS compared to other treatments for TRD (Nguyen and Gordon, 2015; Voigt et al., 2017; Zhao et al., 2018) will likely result in rTMS becoming more widely used; perhaps becoming a first-line treatment for depression (Baeken et al., 2019).

The future of rTMS, as many areas of medicine, is likely to be come increasingly personalized. The stimulation intensity of rTMS is already personalized in the current FDA-approved rTMS treatment course for TRD by measuring an individual's MT. Other stimulation parameters such as the frequency of stimulation, location of stimulation, number of pulses per session, intersession and interpulse interval are all kept constant across patients. Motor physiology studies have consistently shown high interindividual variability in response to rTMS protocols (Cheeran et al., 2008; Corp et al., 2020; Hamada et al., 2013; Hordacre et al., 2017; López-Alonso et al., 2014), suggesting optimal stimulation parameters differ across individuals. Individualizing stimulation frequency based on peak neural oscillatory activity in each individual has been shown to be more effective than standard stimulation protocols in schizophrenia (Jin et al., 2006) but initial studies in TRD show mixed results (Jin and Phillips, 2014; Leuchter et al., 2015). Retrospective studies suggest that the functional connectivity of the stimulated location predicts clinical outcome (Fox et al., 2012; Weigand et al., 2018) implying that individualizing stimulation location using functional MRI-guided targeting may increase efficacy. Studies using individualized functional MRI-guided targeting have shown high efficacy (Cole et al., 2022; Williams et al., 2018) but trials directly comparing different targeting methods need to be conducted. Motor physiology studies have provided evidence to support personalization of other stimulation parameters such as number of pulses per session (Daskalakis, 2014; de Jesus et al., 2014; Nettekoven et al., 2015) and intersession interval (Freitas et al., 2011; Oberman et al., 2014), but no clinical trials to date have investigated the clinical benefit of individualizing these parameters.

Development of a standardized tapering and maintenance rTMS treatment schedules may improve durability and cost effectiveness of rTMS treatment for TRD. Despite the high interindividual variability in clinical response to rTMS, predictors of response have not been widely studied and therefore no reliable predictors of response duration have been found (Miljevic et al., 2019; Senova et al., 2019). Once reliable clinical, genetic or neuroimaging biomarkers of response duration have been identified, it may be possible to add maintenance protocols for those individuals who are likely to have a shorter duration of response, e.g., patients with poor durability.

Eligibility criteria and accessibility of rTMS treatment for TRD are currently variable based on insurance coverage in the United States. Currently, nonresponse to three or more antidepressant medication trials is typically required for insurance coverage. Some insurance companies require higher numbers of failed medication trials and even an ECT trial before approving rTMS. Unlike antidepressant medications and ECT, rTMS has minimal side effects, the most common being discomfort at the stimulation site and headache, with incidence rates ranging from 39% to 65% (Blumberger et al., 2018; George et al., 2010; Levkovitz et al., 2015; Loo et al., 2008; O'Reardon et al., 2007). Longitudinal rTMS studies have demonstrated lower relapse rates for TRD patients 12 months after treatment without maintenance treatment compared to relapse rates reported for those who received ECT (Dunner et al., 2014; Janicak et al., 2010; Jelovac et al., 2013; Sackeim et al., 2001; Somani and Kar, 2019) which is currently the gold-standard treatment for TRD. Analyses into the cost-effectiveness of rTMS have shown this treatment is more cost-effective than antidepressant medications (Nguyen and Gordon, 2015; Voigt et al., 2017) due to the lower rates of retreatment and hospitalization. rTMS has particular utility for cases in which medication cannot be effectively metabolized or absorbed, when a patient is pregnant or when side effects of medications or ECT can be particularly damaging, e.g., heart disease. The utility of rTMS for these conditions, the durability of response, minimal side effects and superior cost-effectiveness, collectively suggest that rTMS should be accessible to patients earlier in their course of depression.

In conclusion, increased personalization of stimulation parameters and the addition of maintenance rTMS treatment in conjunction with the higher-pulse doses and individualized targeting already discussed in this book may increase the efficacy of rTMS treatment for TRD. The benefits of minimal side effects, superior durability, and cost-effectiveness compared to other available treatments for TRD suggest rTMS should be more readily available for TRD patients.

References

Abarbanel, J.M., Lemberg, T., Yaroslavski, U., Grisam, N., Belmaker, R.H., 1996. Electrophysiological responses to transcranial magnetic stimulation in depression and schizophrenia. Biol. Psychiatry 40 (2), 148–150. https://doi.org/10.1016/0006-3223(95)00664-8.

Alkhasli, I., Sakreida, K., Mottaghy, F.M., Binkofski, F., 2019. Modulation of Fronto-striatal functional connectivity using transcranial magnetic stimulation. Front. Hum. Neurosci. 13, 190. https://doi.org/10.3389/fnhum.2019.00190.

Avanzino, L., Bove, M., Trompetto, C., Tacchino, A., Ogliastro, C., Abbruzzese, G., 2008. 1-Hz repetitive TMS over ipsilateral motor cortex influences the performance of sequential finger movements of different complexity. Eur. J. Neurosci. 27 (5), 1285–1291. https://doi.org/10.1111/j.1460-9568.2008.06086.x.

Baeken, C., Brem, A.-K., Arns, M., Brunoni, A.R., Filipčić, I., Ganho-Ávila, A., Langguth, B., Padberg, F., Poulet, E., Rachid, F., Sack, A.T., Vanderhasselt, M.-A., Bennabi, D., 2019. Repetitive transcranial magnetic stimulation treatment for depressive disorders: current knowledge and future directions. Curr. Opin. Psychiatry 32 (5), 409–415. https://doi.org/10.1097/YCO.0000000000000533.

Baeken, C., Marinazzo, D., Wu, G.-R., Van Schuerbeek, P., De Mey, J., Marchetti, I., Vanderhasselt, M.-A., Remue, J., Luypaert, R., De Raedt, R., 2014. Accelerated HF-rTMS in treatment-resistant unipolar depression: insights from subgenual anterior cingulate functional connectivity. World J. Biol. Psychiatry 15 (4), 286–297. https://doi.org/10.3109/15622975.2013.872295.

Bailine, S.H., Sanghani, S.N., Petrides, G., 2019. Maintenance electroconvulsive therapy is not acute electroconvulsive therapy. J. ECT 35 (1), 1–2. https://doi.org/10.1097/YCT.0000000000000524.

Barker, A.T., Jalinous, R., Freeston, I.L., 1985. Non-invasive magnetic stimulation of human motor cortex. Lancet (London, England) 1 (8437), 1106–1107. https://doi.org/10.1016/s0140-6736(85)92413-4.

Baxter, L.R., Schwartz, J.M., Phelps, M.E., Mazziotta, J.C., Guze, B.H., Selin, C.E., Gerner, R.H., Sumida, R.M., 1989. Reduction of prefrontal cortex glucose metabolism common to three types of depression. Arch. Gen. Psychiatry 46 (3), 243–250. http://eutils. ncbi.nlm.nih.gov/entrez/eutils/elink.fcgi?dbfrom=pubmed&id=2784046&retmode=ref&cmd=prlinks.

Benadhira, R., Saba, G., Samaan, A., Dumortier, G., Lipski, H., Gastal, D., Kalalou, K., Verdon, C.-M., Januel, D., 2005. Transcranial magnetic stimulation for refractory depression. Am. J. Psychiatry 162 (1), 193. https://doi.org/10.1176/appi.ajp.162.1.193.

Berman, M.G., Peltier, S., Nee, D.E., Kross, E., Deldin, P.J., Jonides, J., 2011. Depression, rumination and the default network. Soc. Cogn. Affect. Neurosci. 6 (5), 548–555. https://doi.org/10.1093/scan/nsq080.

Blumberger, D.M., Vila-Rodriguez, F., Thorpe, K.E., Feffer, K., Noda, Y., Giacobbe, P., Knyahnytska, Y., Kennedy, S.H., Lam, R.W., Daskalakis, Z.J., Downar, J., 2018. Effectiveness of theta burst versus high-frequency repetitive transcranial magnetic stimulation in patients with depression (THREE-D): a randomised non-inferiority trial. Lancet 391 (10131), 1683–1692. https://doi.org/10.1016/S0140-6736(18)30295-2.

Brody, A.L., Saxena, S., Mandelkern, M.A., Fairbanks, L.A., Ho, M.L., Baxter, L.R., 2001. Brain metabolic changes associated with symptom factor improvement in major depressive disorder. Biol. Psychiatry 50 (3), 171–178.

Brown, J.C., DeVries, W.H., Korte, J.E., Sahlem, G.L., Bonilha, L., Short, E.B., George, M.S., 2020. NMDA receptor partial agonist, d-cycloserine, enhances 10 Hz rTMS-induced motor plasticity, suggesting long-term potentiation (LTP) as underlying mechanism. Brain Stimul. 13 (3), 530–532. https://doi.org/10.1016/j.brs.2020.01.005.

Camprodon, J.A., 2021. Therapeutic neuromodulation for bipolar disorder—the case for biomarker-driven treatment development. JAMA Netw. Open 4, e211055.

Carpenter, L.L., Janicak, P.G., Aaronson, S.T., Boyadjis, T., Brock, D.G., Cook, I.A., Dunner, D.L., Lanocha, K., Solvason, H.B., Demitrack, M.A., 2012. Transcranial magnetic stimulation (TMS) for major depression: a multisite, naturalistic, observational study of acute treatment outcomes in clinical practice. Depress. Anxiety 29 (7), 587–596. https://doi.org/10.1002/da.21969.

Chai, X.J., Hirshfeld-Becker, D., Biederman, J., Uchida, M., Doehrmann, O., Leonard, J.A., Salvatore, J., Kenworthy, T., Brown, A., Kagan, E., de los Angeles, C., Gabrieli, J.D.E., Whitfield-Gabrieli, S., 2016. Altered intrinsic functional brain architecture in children at familial risk of major depression. Biol. Psychiatry 80 (11), 849–858. https://doi.org/10.1016/j.biopsych.2015.12.003.

Cheeran, B., Talelli, P., Mori, F., Koch, G., Suppa, A., Edwards, M., Houlden, H., Bhatia, K., Greenwood, R., Rothwell, J.C., 2008. A common polymorphism in the brain-derived neurotrophic factor gene (*BDNF*) modulates human cortical plasticity and the response to rTMS: *BNDF* polymorphism modulates response to rTMS. J. Physiol. 586 (23), 5717–5725. https://doi.org/10.1113/jphysiol.2008.159905.

Chen, L., Hudaib, A.-R., Hoy, K.E., Fitzgerald, P.B., 2020. Efficacy, efficiency and safety of high-frequency repetitive transcranial magnetic stimulation applied more than once a day in depression: a systematic review. J. Affect. Disord. 277, 986–996. https://doi.org/10.1016/j.jad.2020.09.035.

Chen, Ashley C, Oathes, Desmond J, Chang, Catie, Bradley, Travis, Zhou, Zheng-Wei, Williams, Leanne M, Glover, Gary H, Deisseroth, Karl, Etkin, Amit, 2013. Causal interactions between fronto-parietal central executive and default-mode networks in humans. Proc Natl Acad Sci U S A 1091-6490. 110 (49), 19944–19949. https://doi.org/10.1073/pnas.1311772110.

Chung, S.W., Rogasch, N.C., Hoy, K.E., Fitzgerald, P.B., 2015. Measuring brain stimulation induced changes in cortical properties using TMS-EEG. Brain Stimul 8 (6), 1010–1020. https://doi.org/10.1016/j.brs.2015.07.029.

Chung, S.W., Rogasch, N.C., Hoy, K.E., Fitzgerald, P.B., 2018. The effect of single and repeated prefrontal intermittent theta burst stimulation on cortical reactivity and working memory. Brain Stimul. 11 (3), 566–574. https://doi.org/10.1016/j.brs.2018.01.002.

Cole, E.J., Stimpson, K.H., Bentzley, B.S., Gulser, M., Cherian, K., Tischler, C., Nejad, R., Pankow, H., Choi, E., Aaron, H., Espil, F.M., Pannu, J., Xiao, X., Duvio, D., Solvason, H.B., Hawkins, J., Guerra, A., Jo, B., Raj, K.S., Williams, N.R., 2020. Stanford accelerated intelligent neuromodulation therapy for treatment-resistant depression. Am. J. Psychiatry 177 (8), 716–726. https://doi.org/10.1176/appi.ajp.2019.19070720.

Cole, E.J., et al., 2022. Stanford neuromodulation therapy (SNT): a double-blind randomized controlled trial. Am. J. Psychiatry 179, 132–141.

Collingridge, G.L., Kehl, S.J., McLennan, H., 1983. Excitatory amino acids in synaptic transmission in the Schaffer collateral-commissural pathway of the rat hippocampus. J. Physiol. 334, 33–46. https://doi.org/10.1113/jphysiol.1983.sp014478.

Corp, D.T., Bereznicki, H.G.K., Clark, G.M., Youssef, G.J., Fried, P.J., Jannati, A., Davies, C.B., Gomes-Osman, J., Stamm, J., Chung, S.W., Bowe, S.J., Rogasch, N.C., Fitzgerald, P.B., Koch, G., Lazzaro, V.D., Pascual-Leone, A., Enticott, P.G., 2020. Large-scale analysis of interindividual variability in theta-burst stimulation data: Results from the 'Big TMS Data Collaboration. Brain Stimul. https://doi.org/10.1016/j.brs.2020.07.018.

Cusin, C., Dougherty, D.D., 2012. Somatic therapies for treatment-resistant depression: ECT, TMS, VNS, DBS. Biol. Mood Anxiety Disord. 2 (1), 14.

Daskalakis, Z.J., 2014. Theta-burst transcranial magnetic stimulation in depression: when less may be more. Brain 137 (7), 1860–1862. https://doi.org/10.1093/brain/awu123.

de Jesus, D.R., de Favalli, G.P.S., Hoppenbrouwers, S.S., Barr, M.S., Chen, R., Fitzgerald, P.B., Daskalakis, Z.J., 2014. Determining optimal rTMS parameters through changes in cortical inhibition. Clin. Neurophysiol. 125 (4), 755–762. https://doi.org/10.1016/j.clinph.2013.09.011.

Dell'Osso, B., Mundo, E., D'Urso, N., Pozzoli, S., Buoli, M., Ciabatti, M., Rosanova, M., Massimini, M., Bellina, V., Mariotti, M., Altamura, A.C., 2009. Augmentative repetitive navigated transcranial magnetic stimulation (rTMS) in drug-resistant bipolar depression. Bipolar Disord. 11 (1), 76–81. https://doi.org/10.1111/j.1399-5618.2008.00651.x.

Dell'Osso, B., Oldani, L., Camuri, G., Dobrea, C., Cremaschi, L., Benatti, B., Arici, C., Grancini, B., Altamura, A.C., 2015. Augmentative repetitive transcranial magnetic stimulation (rTMS) in the acute treatment of poor responder depressed patients: a comparison study between high and low frequency stimulation. Eur. Psychiatry 30 (2), 271–276. https://doi.org/10.1016/j.eurpsy.2014.12.001.

Deng, Z.-D., Lisanby, S.H., Peterchev, A.V., 2013. Electric field depth-focality tradeoff in transcranial magnetic stimulation: simulation comparison of 50 coil designs. Brain Stimul. 6 (1), 1–13. https://doi.org/10.1016/j.brs.2012.02.005.

Devinsky, O., 1993. Electrical and magnetic stimultion of the central nervous system: historical overview. Adv. Neurol. 63, 1–16.

Dolberg, O.T., Dannon, P.N., Schreiber, S., Grunhaus, L., 2002. Transcranial magnetic stimulation in patients with bipolar depression: a double blind, controlled study. Bipolar Disord 4 (Suppl 1), 94–95. https://doi.org/10.1034/j.1399-5618.4.s1.41.x.

Dougherty, D.D., Weiss, A.P., Cosgrove, G.R., Alpert, N.M., Cassem, E.H., Nierenberg, A.A., Price, B.H., Mayberg, H.S., Fischman, A.J., Rauch, S.L., 2003. Cerebral metabolic correlates as potential predictors of response to anterior cingulotomy for treatment of major depression. J. Neurosurg. 99 (6), 1010–1017. https://doi.org/10.3171/jns.2003.99.6.1010.

Drevets, W.C., Savitz, J., Trimble, M., 2008. The subgenual anterior cingulate cortex in mood disorders. CNS Spectr. 13 (8), 663–681. https://www.ncbi.nlm.nih.gov/pmc/articles/PMC2729429/.

Drysdale, A.T., Grosenick, L., Downar, J., Dunlop, K., Mansouri, F., Meng, Y., Fetcho, R.N., Zebley, B., Oathes, D.J., Etkin, A., Schatzberg, A.F., Sudheimer, K., Keller, J., Mayberg, H.S., Gunning, F.M., Alexopoulos, G.S., Fox, M.D., Pascual-Leone, A., Voss, H.U., Liston, C., 2017. Resting-state connectivity biomarkers define neurophysiological subtypes of depression. Nat. Med. 23 (1), 28–38. https://doi.org/10.1038/nm.4246.

Dunner, D.L., Aaronson, S.T., Sackeim, H.A., Janicak, P.G., Carpenter, L.L., Boyadjis, T., Brock, D.G., Bonneh-Barkay, D., Cook, I.A., Lanocha, K., Solvason, H.B., Demitrack, M.A., 2014. A multisite, naturalistic, observational study of transcranial magnetic stimulation for patients with pharmacoresistant major depressive disorder: durability of benefit over a 1-year follow-up period. J. Clin. Psychiatry 75 (12), 1394–1401. https://doi.org/10.4088/JCP.13m08977.

Duprat, R., Desmyter, S., Rudi, D.R., van Heeringen, K., Van den Abbeele, D., Tandt, H., Bakic, J., Pourtois, G., Dedoncker, J., Vervaet, M., Van Autreve, S., Lemmens, G.M.D., Baeken, C., 2016. Accelerated intermittent theta burst stimulation treatment in medication-resistant major depression: a fast road to remission? J. Affect. Disord. 200, 6–14. https://doi.org/10.1016/j.jad.2016.04.015.

Fischer, A.S., Keller, C.J., Etkin, A., 2016. The clinical applicability of functional connectivity in depression: pathways toward more targeted intervention. Biol. Psychiatry Cogn. Neurosci. Neuroimaging 1 (3), 262–270. https://doi.org/10.1016/j.bpsc.2016.02.004.

Fitzgerald, P.B., 2019. Is maintenance repetitive transcranial magnetic stimulation for patients with depression a valid therapeutic strategy? Clin. Pharmacol. Ther. 106 (4), 723–725. https://doi.org/10.1002/cpt.1566.

Fitzgerald, P.B., Hoy, K.E., Elliot, D., Susan McQueen, R.N., Wambeek, L.E., Daskalakis, Z.J., 2018. Accelerated repetitive transcranial magnetic stimulation in the treatment of depression. Neuropsychopharmacology 43 (7), 1565–1572. https://doi.org/10.1038/s41386-018-0009-9.

Fitzgerald, P.B., Hoy, K.E., Reynolds, J., Singh, A., Gunewardene, R., Slack, C., Ibrahim, S., Daskalakis, Z.J., 2020. A pragmatic randomized controlled trial exploring the relationship between pulse number and response to repetitive transcranial magnetic stimulation treatment in depression. Brain Stimul. 13 (1), 145–152. https://doi.org/10.1016/j.brs.2019.09.001.

Fox, M.D., Buckner, R.L., White, M.P., Greicius, M.D., Pascual-Leone, A., 2012. Efficacy of transcranial magnetic stimulation targets for depression is related to intrinsic functional connectivity with the subgenual cingulate. Biol. Psychiatry 72 (7), 595–603. https://doi.org/10.1016/j.biopsych.2012.04.028.

Frank, E., 1991. Efficacy of interpersonal psychotherapy as a maintenance treatment of recurrent depression: contributing factors. Arch. Gen. Psychiatry 48 (12), 1053. https://doi.org/10.1001/archpsyc.1991.01810360017002.

Freitas, C., Perez, J., Knobel, M., Tormos, J.M., Oberman, L., Eldaief, M., Bashir, S., Vernet, M., Peña-Gómez, C., Pascual-Leone, A., 2011. Changes in cortical plasticity across the lifespan. Front. Aging Neurosci. 3. https://doi.org/10.3389/fnagi.2011.00005.

George, M.S., Wassermann, E.M., Williams, W.A., Callahan, A., Ketter, T.A., Basser, P., Hallett, M., Post, R.M., 1995. Daily repetitive transcranial magnetic stimulation (rTMS) improves mood in depression. Neuroreport 6 (14), 1853–1856.

George, M.S., Wassermann, E.M., Williams, W.A., Steppel, J., Pascual-Leone, A., Basser, P., Hallett, M., Post, R.M., 1996. Changes in mood and hormone levels after rapid-rate transcranial magnetic stimulation (rTMS) of the prefrontal cortex. J. Neuropsychiatry Clin. Neurosci. 8 (2), 172–180. https://doi.org/10.1176/jnp.8.2.172.

Geddes, J.R., Miklowitz, D.J., 2013. Treatment of bipolar disorder. Lancet 381 (9878), 1672–1682. https://doi.org/10.1016/S0140-6736(13)60857-0. 23663953.

George, M.S., Lisanby, S.H., Avery, D., McDonald, W.M., Durkalski, V., Pavlicova, M., Anderson, B., Nahas, Z., Bulow, P., Zarkowski, P., Holtzheimer III, P.E., Schwartz, T., Sackeim, H.A., 2010. Daily left prefrontal transcranial magnetic stimulation therapy for major depressive disorder: a sham-controlled randomized trial. Arch. Gen. Psychiatry 67 (5), 507–516. https://doi.org/10.1001/archgenpsychiatry.2010.46.

Glass, R.M., 2003. Awareness about depression: important for all physicians. JAMA 289 (23), 3169–3170.

Gold, A.K., Ornelas, A.C., Cirillo, P., Caldieraro, M.A., Nardi, A.E., Nierenberg, A.A., Kinrys, G., 2019. Clinical applications of transcranial magnetic stimulation in bipolar disorder. Brain Behav. 9 (10). https://doi.org/10.1002/brb3.1419, e01419.

Goldsworthy, M.R., Pitcher, J.B., Ridding, M.C., 2015. Spaced noninvasive brain stimulation: prospects for inducing long-lasting human cortical plasticity. Neurorehabil. Neural Repair 29 (8), 714–721.

Goldsworthy, M.R., Pitcher, J.B., Ridding, M.C., 2012. The application of spaced theta burst protocols induces long-lasting neuroplastic changes in the human motor cortex. Eur. J. Neurosci. 35 (1), 125–134.

Greicius, M.D., Flores, B.H., Menon, V., Glover, G.H., Solvason, H.B., Kenna, H., Reiss, A.L., Schatzberg, A.F., 2007. Resting-state functional connectivity in major depression: abnormally increased contributions from Subgenual cingulate cortex and thalamus. Biol. Psychiatry 62 (5), 429–437. https://doi.org/10.1016/j.biopsych.2006.09.020.

Grisaru, N., Chudakov, B., Yaroslavsky, Y., Belmaker, R.H., 1998. Transcranial magnetic stimulation in mania: a controlled study. Am. J. Psychiatry 155 (11), 1608–1610. https://doi.org/10.1176/ajp.155.11.1608.

Haesebaert, F., Moirand, R., Schott-Pethelaz, A.-M., Brunelin, J., Poulet, E., 2018. Usefulness of repetitive transcranial magnetic stimulation as a maintenance treatment in patients with major depression. World J. Biol. Psychiatry 19 (1), 74–78. https://doi.org/10.1080/15622975.2016.1255353.

Hallett, M., Cohen, L.G., 1989. Magnetism: a new method for stimulation of nerve and brain. JAMA 262 (4), 538–541.

Hallett, M., Di Iorio, R., Rossini, P.M., Park, J.E., Chen, R., Celnik, P., Strafella, A.P., Matsumoto, H., Ugawa, Y., 2017. Contribution of transcranial magnetic stimulation to assessment of brain connectivity and networks. Clin. Neurophysiol. 128 (11), 2125–2139. https://doi.org/10.1016/j.clinph.2017.08.007.

Hamada, M., Murase, N., Hasan, A., Balaratnam, M., Rothwell, J.C., 2013. The role of interneuron networks in driving human motor cortical plasticity. Cereb. Cortex 23 (7), 1593–1605. https://doi.org/10.1093/cercor/bhs147.

Hamilton, J.P., Farmer, M., Fogelman, P., Gotlib, I.H., 2015. Depressive rumination, the default-mode network, and the dark matter of clinical neuroscience. Biol. Psychiatry 78 (4), 224–230. https://doi.org/10.1016/j.biopsych.2015.02.020.

Han, K., Chapman, S.B., Krawczyk, D.C., 2015. Altered amygdala connectivity in individuals with chronic traumatic brain injury and comorbid depressive symptoms. Front. Neurol. 6. https://doi.org/10.3389/fneur.2015.00231.

Hendrikse, J., Kandola, A., Coxon, J., Rogasch, N., Yücel, M., 2017. Combining aerobic exercise and repetitive transcranial magnetic stimulation to improve brain function in health and disease. Neurosci. Biobehav. Rev. 83, 11–20. https://doi.org/10.1016/j.neubiorev.2017.09.023.

Hett, D., Marwaha, S., 2020. Repetitive Transcranial Magnetic Stimulation in the Treatment of Bipolar Disorder. 2045125320973790, Ther. Adv. Psychopharmacol. 10. https://doi.org/10.1177/2045125320973790. 33282175.

Hoogendam, J.M., Ramakers, G.M.J., Di Lazzaro, V., 2010. Physiology of repetitive transcranial magnetic stimulation of the human brain. Brain Stimul. 3 (2), 95–118. https://doi.org/10.1016/j.brs.2009.10.005.

Hordacre, B., Goldsworthy, M.R., Vallence, A.-M., Darvishi, S., Moezzi, B., Hamada, M., Rothwell, J.C., Ridding, M.C., 2017. Variability in neural excitability and plasticity induction in the human cortex: a brain stimulation study. Brain Stimul. 10 (3), 588–595. https://doi.org/10.1016/j.brs.2016.12.001.

Howland, R.H., 2008. Sequenced treatment alternatives to relieve depression (STAR*D). Part 2: study outcomes. J. Psychosoc. Nurs. Ment. Health Serv. 46 (10), 21–24. https://doi.org/10.3928/02793695-20081001-05.

Huang, Y.-Z., Chen, R.-S., Rothwell, J.C., Wen, H.-Y., 2007. The after-effect of human theta burst stimulation is NMDA receptor dependent. Clin. Neurophysiol. 118 (5), 1028–1032. https://doi.org/10.1016/j.clinph.2007.01.021.

Huang, Y.-Z., Edwards, M.J., Rounis, E., Bhatia, K.P., Rothwell, J.C., 2005. Theta burst stimulation of the human motor cortex. Neuron 45 (2), 201–206. https://doi.org/10.1016/j.neuron.2004.12.033.

Janicak, P.G., Nahas, Z., Lisanby, S.H., Solvason, H.B., Sampson, S.M., McDonald, W.M., Marangell, L.B., Rosenquist, P., McCall, W.V., Kimball, J., O'Reardon, J.P., Loo, C., Husain, M.H., Krystal, A., Gilmer, W., Dowd, S.M., Demitrack, M.A., Schatzberg, A.F., 2010. Durability of clinical benefit with transcranial magnetic stimulation (TMS) in the treatment of pharmacoresistant major depression: assessment of relapse during a 6-month, multisite, open-label study. Brain Stimul. 3 (4), 187–199. https://doi.org/10.1016/j.brs.2010.07.003.

Jelovac, A., Kolshus, E., McLoughlin, D.M., 2013. Relapse following successful electroconvulsive therapy for major depression: A Meta-analysis. Neuropsychopharmacology 38 (12), 2467–2474. https://doi.org/10.1038/npp.2013.149.

Jin, Y., Phillips, B., 2014. A pilot study of the use of EEG-based synchronized transcranial magnetic stimulation (sTMS) for treatment of major depression. BMC Psychiatry 14, 13. https://doi.org/10.1186/1471-244X-14-13.

Jin, Y., Potkin, S.G., Kemp, A.S., Huerta, S.T., Alva, G., Thai, T.M., Carreon, D., Bunney, W.E., 2006. Therapeutic effects of individualized alpha frequency transcranial magnetic stimulation (alphaTMS) on the negative symptoms of schizophrenia. Schizophr. Bull. 32 (3), 556–561. https://doi.org/10.1093/schbul/sbj020.

Kaster, T.S., Daskalakis, Z.J., Noda, Y., Knyahnytska, Y., Downar, J., Rajji, T.K., Levkovitz, Y., Zangen, A., Butters, M.A., Mulsant, B.H., Blumberger, D.M., 2018. Efficacy, tolerability, and cognitive effects of deep transcranial magnetic stimulation for late-life depression: a prospective randomized controlled trial. Neuropsychopharmacology 43 (11), 2231–2238. https://doi.org/10.1038/s41386-018-0121-x.

Kaymaz, N., van Os, J., Loonen, A.J., Nolen, W.A., 2008. Evidence that patients with single versus recurrent depressive episodes are differentially sensitive to treatment discontinuation: A meta-analysis of placebo-controlled randomized trials. In: Database of Abstracts of Reviews of Effects (DARE): Quality-assessed Reviews [Internet]. Centre for Reviews and Dissemination (UK).

Kedzior, K.K., Reitz, S.K., Azorina, V., Loo, C., 2015. Durability of the antidepressant effect of the high-frequency repetitive transcranial magnetic stimulation (rTMS) in the absence of maintenance treatment in major depression: a systematic review and meta-analysis of 16 double-blind, randomized, sham-controlled trials. Depress. Anxiety 32 (3), 193–203. https://doi.org/10.1002/da.22339.

Kellner, C.H., Knapp, R.G., Petrides, G., Rummans, T.A., Husain, M.M., Rasmussen, K., Mueller, M., Bernstein, H.J., O'Connor, K., Smith, G., Biggs, M., Bailine, S.H., Malur, C., Yim, E., McClintock, S., Sampson, S., Fink, M., 2006. Continuation electroconvulsive therapy vs pharmacotherapy for relapse prevention in major depression: a multisite study from the consortium for research in electroconvulsive therapy (CORE). Arch. Gen. Psychiatry 63 (12). https://doi.org/10.1001/archpsyc.63.12.1337.

Klein, D.N., Santiago, N.J., Vivian, D., Blalock, J.A., Kocsis, J.H., Markowitz, J.C., McCullough, J.P., Rush, A.J., Trivedi, M.H., Arnow, B.A., Dunner, D.L., Manber, R., Rothbaum, B., Thase, M.E., Keitner, G.I., Miller, I.W., Keller, M.B., 2004. Cognitive-behavioral analysis system of psychotherapy as a maintenance treatment for chronic depression. J. Consult. Clin. Psychol. 72 (4), 681–688. https://doi.org/10.1037/0022-006X.72.4.681.

Koenigs, M., Grafman, J., 2009. The functional neuroanatomy of depression: distinct roles for ventromedial and dorsolateral prefrontal cortex. Behav. Brain Res. 201 (2), 239–243. https://doi.org/10.1016/j.bbr.2009.03.004.

Krames, E.S., Hunter Peckham, P., Rezai, A., Aboelsaad, F., 2009. What is neuromodulation? In: Neuromodulation. Elsevier, pp. 3–8. https://linkinghub.elsevier.com/retrieve/pii/B9780123742483000021.

Kupfer, D.J., 1992. Five-year outcome for maintenance therapies in recurrent depression. Arch. Gen. Psychiatry 49 (10), 769. https://doi.org/10.1001/archpsyc.1992.01820100013002.

Lagasse, R.S., 2002. Anesthesia safety: model or myth? A review of the published literature and analysis of current original data. Anesthesiology 97 (6), 1609–1617.

Larson, J., Lynch, G., 1986. Induction of synaptic potentiation in hippocampus by patterned stimulation involves two events. Science 232 (4753), 985–988.

Lenz, M., Galanis, C., Müller-Dahlhaus, F., Opitz, A., Wierenga, C.J., Szabó, G., Ziemann, U., Deller, T., Funke, K., Vlachos, A., 2016. Repetitive magnetic stimulation induces plasticity of inhibitory synapses. Nat. Commun. 7, 10020. https://doi.org/10.1038/ncomms10020.

Leuchter, A.F., Cook, I.A., Feifel, D., Goethe, J.W., Husain, M., Carpenter, L.L., Thase, M.E., Krystal, A.D., Philip, N.S., Bhati, M.T., Burke, W.J., Howland, R.H., Sheline, Y.I., Aaronson, S.T., Iosifescu, D.V., O'Reardon, J.P., Gilmer, W.S., Jain, R., Burgoyne, K.S., George, M.S., 2015. Efficacy and safety of low-field synchronized transcranial magnetic stimulation (sTMS) for treatment of major depression. Brain Stimul. 8 (4), 787–794. https://doi.org/10.1016/j.brs.2015.05.005.

Levkovitz, Y., Harel, E.V., Roth, Y., Braw, Y., Most, D., Katz, L.N., Sheer, A., Gersner, R., Zangen, A., 2009. Deep transcranial magnetic stimulation over the prefrontal cortex: evaluation of antidepressant and cognitive effects in depressive patients. Brain Stimul. 2 (4), 188–200. https://doi.org/10.1016/j.brs.2009.08.002.

Levkovitz, Y., Isserles, M., Padberg, F., Lisanby, S.H., Bystritsky, A., Xia, G., Tendler, A., Daskalakis, Z.J., Winston, J.L., Dannon, P., Hafez, H.M., Reti, I.M., Morales, O.G., Schlaepfer, T.E., Hollander, E., Berman, J.A., Husain, M.M., Sofer, U., Stein, A., Zangen, A., 2015. Efficacy and safety of deep transcranial magnetic stimulation for major depression: a prospective multicenter randomized controlled trial. World Psychiatry 14 (1), 64–73. https://doi.org/10.1002/wps.20199.

Li, X., Nahas, Z., Kozel, F.A., Anderson, B., Bohning, D.E., George, M.S., 2004. Acute left prefrontal transcranial magnetic stimulation in depressed patients is associated with immediately increased activity in prefrontal cortical as well as subcortical regions. Biol. Psychiatry 55 (9), 882–890. https://doi.org/10.1016/j.biopsych.2004.01.017.

Li, C.-T., Chen, M.-H., Juan, C.-H., Huang, H.-H., Chen, L.-F., Hsieh, J.-C., Tu, P.-C., Bai, Y.-M., Tsai, S.-J., Lee, Y.-C., 2014. Efficacy of prefrontal theta-burst stimulation in refractory depression: a randomized sham-controlled study. Brain 137 (7), 2088–2098.

Liston, C., Chen, A.C., Zebley, B.D., Drysdale, A.T., Gordon, R., Leuchter, B., Voss, H.U., Casey, B.J., Etkin, A., Dubin, M.J., 2014. Default mode network mechanisms of transcranial magnetic stimulation in depression. Biol. Psychiatry 76 (7), 517–526. https://doi.org/10.1016/j.biopsych.2014.01.023.

Loo, C.K., McFarquhar, T.F., Mitchell, P.B., 2008. A review of the safety of repetitive transcranial magnetic stimulation as a clinical treatment for depression. Int. J. Neuropsychopharmacol. 11 (1), 131–147. https://doi.org/10.1017/S1461145707007717.

Loo, C.K., Mitchell, P.B., 2005. A review of the efficacy of transcranial magnetic stimulation (TMS) treatment for depression, and current and future strategies to optimize efficacy. J. Affect. Disord. 88 (3), 255–267.

Loo, C.K., Mitchell, P.B., McFarquhar, T.F., Malhi, G.S., Sachdev, P.S., 2007. A sham-controlled trial of the efficacy and safety of twice-daily rTMS in major depression. Psychol. Med. 37 (3), 341–349. https://doi.org/10.1017/S0033291706009597.

López-Alonso, V., Cheeran, B., Río-Rodríguez, D., Fernández-del-Olmo, M., 2014. Inter-individual variability in response to non-invasive brain stimulation paradigms. Brain Stimul. 7 (3), 372–380. https://doi.org/10.1016/j.brs.2014.02.004.

Luber, B.M., Davis, S., Bernhardt, E., Neacsiu, A., Kwapil, L., Lisanby, S.H., Strauman, T.J., 2017. Using neuroimaging to individualize TMS treatment for depression: toward a new paradigm for imaging-guided intervention. Neuro Image 148, 1–7. https://doi.org/10.1016/j.neuroimage.2016.12.083.

Luppa, M., Heinrich, S., Angermeyer, M.C., König, H.-H., Riedel-Heller, S.G., 2007. Cost-of-illness studies of depression: a systematic review. J. Affect. Disord. 98 (1–2), 29–43.

Maeda, F., Keenan, J.P., Tormos, J.M., Topka, H., Pascual-Leone, A., 2000. Interindividual variability of the modulatory effects of repetitive transcranial magnetic stimulation on cortical excitability. Exp. Brain Res. 133 (4), 425–430. https://doi.org/10.1007/s002210000432. 10985677.

Martin, D.M., Gálvez, V., Loo, C.K., 2015. Predicting retrograde autobiographical memory changes following electroconvulsive therapy: relationships between individual, treatment, and early clinical factors. Int. J. Neuropsychopharmacol. 18 (12). https://doi.org/10.1093/ijnp/pyv067.

Martinot, J., Hardy, P., Feline, A., Huret, J., Mazoyer, B., Attar-Levy, D., Pappata, S., Syrota, A., 1990. Left prefrontal glucose hypometabolism in the depressed state: a confirmation. Am. J. Psychiatry 147 (10), 1313–1317. https://doi.org/10.1176/ajp.147.10.1313.

Mayberg, H.S., Brannan, S.K., Tekell, J.L., Silva, J.A., Mahurin, R.K., McGinnis, S., Jerabek, P.A., 2000. Regional metabolic effects of fluoxetine in major depression: serial changes and relationship to clinical response. Biol. Psychiatry 48 (8), 830–843.

McClintock, S.M., Reti, I.M., Carpenter, L.L., McDonald, W.M., Dubin, M., Taylor, S.F., Cook, I.A., O'Reardon, J., Husain, M.M., Wall, C., Krystal, A.D., Sampson, S.M., Morales, O., Nelson, B.G., Latoussakis, V., George, M.S., Lisanby, S.H., 2018. Consensus recommendations for the clinical application of repetitive transcranial magnetic stimulation (rTMS) in the treatment of depression. J. Clin. Psychiatry 79 (1). https://doi.org/10.4088/JCP.16cs10905.

McGirr, A., et al., 2021. Efficacy of active vs sham intermittent theta burst transcranial magnetic stimulation for patients with bipolar depression: a randomized clinical trial. JAMA Netw. Open 4, e210963.

Mielacher, C., Kiebs, M., Dellert, T., Metzner, A., Högenauer, H., Schultz, J., Lamm, C., Hurlemann, R., 2019. Once daily versus twice daily theta-burst stimulation in the treatment of major depression disorder. Brain Stimul. 12 (2), 397.

Milev, R.V., Giacobbe, P., Kennedy, S.H., Blumberger, D.M., Daskalakis, Z.J., Downar, J., Modirrousta, M., Patry, S., Vila-Rodriguez, F., Lam, R.W., MacQueen, G.M., Parikh, S.V., Ravindran, A.V., the CANMAT Depression Work Group, 2016. Canadian network for mood and anxiety treatments (CANMAT) 2016 clinical guidelines for the Management of Adults with major depressive disorder: section 4. Neurostimulation treatments. Can. J. Psychiatry 61 (9), 561–575. https://doi.org/10.1177/0706743716660033.

Miljevic, A., Bailey, N.W., Herring, S.E., Fitzgerald, P.B., 2019. Potential predictors of depressive relapse following repetitive transcranial magnetic stimulation: a systematic review. J. Affect. Disord. 256, 317–323. https://doi.org/10.1016/j.jad.2019.06.006.

Nettekoven, C., Volz, L.J., Kutscha, M., Pool, E.-M., Rehme, A.K., Eickhoff, S.B., Fink, G.R., Grefkes, C., 2014. Dose-dependent effects of theta burst rTMS on cortical excitability and resting-state connectivity of the human motor system. J. Neurosci. 34 (20), 6849–6859.

Nettekoven, C., Volz, L.J., Leimbach, M., Pool, E.-M., Rehme, A.K., Eickhoff, S.B., Fink, G.R., Grefkes, C., 2015. Inter-individual variability in cortical excitability and motor network connectivity following multiple blocks of rTMS. Neuroimage 118, 209–218. https://doi.org/10.1016/j.neuroimage.2015.06.004.

Nguyen, K.-H., Gordon, L.G., 2015. Cost-effectiveness of repetitive transcranial magnetic stimulation versus antidepressant therapy for treatment-resistant depression. Value Health 18 (5), 597–604. https://doi.org/10.1016/j.jval.2015.04.004.

Nobler, M.S., Oquendo, M.A., Kegeles, L.S., Malone, K.M., Campbell, C., Sackeim, H.A., Mann, J.J., 2001. Decreased regional brain metabolism after ECT. Am. J. Psychiatry 158 (2), 305–308. https://doi.org/10.1176/appi.ajp.158.2.305.

Oberman, L.M., Pascual-Leone, A., Rotenberg, A., 2014. Modulation of corticospinal excitability by transcranial magnetic stimulation in children and adolescents with autism spectrum disorder. Front. Hum. Neurosci. 8. https://doi.org/10.3389/fnhum.2014.00627.

O'Reardon, J.P., Blumner, K.H., Peshek, A.D., Pradilla, R.R., Pimiento, P.C., 2005. Long-term maintenance therapy for major depressive disorder with rTMS. J. Clin. Psychiatry 66 (12), 1524–1528. https://doi.org/10.4088/jcp.v66n1205.

O'Reardon, J.P., Solvason, H.B., Janicak, P.G., Sampson, S., Isenberg, K.E., Nahas, Z., McDonald, W.M., Avery, D., Fitzgerald, P.B., Loo, C., Demitrack, M.A., George, M.S., Sackeim, H.A., 2007. Efficacy and safety of transcranial magnetic stimulation in the acute treatment of major depression: a multisite randomized controlled trial. Biol. Psychiatry 62 (11), 1208–1216. https://doi.org/10.1016/j.biopsych.2007.01.018.

Pallanti, S., Grassi, G., Antonini, S., Quercioli, L., Salvadori, E., Hollander, E., 2014. rTMS in resistant mixed states: an exploratory study. J. Affect. Disord. 157, 66–71. https://doi.org/10.1016/j.jad.2013.12.024.

Pascual-Leone, A., Tormos, J.M., Keenan, J., Tarazona, F., Cañete, C., Catalá, M.D., 1998. Study and modulation of human cortical excitability with transcranial magnetic stimulation. J. Clin. Neurophysiol. 15 (4), 333–343. https://doi.org/10.1097/00004691-199807000-00005.

Pascual-Leone, A., Valls-Solé, J., Wassermann, E.M., Hallett, M., 1994. Responses to rapid-rate transcranial magnetic stimulation of the human motor cortex. Brain: A J. Neurol. 117 (Pt 4), 847–858. https://doi.org/10.1093/brain/117.4.847.

Pascual-Leone, A., Catala, M.D., Pascual, A.P.-L., 1996. Lateralized effect of rapid-rate transcranial magnetic stimulation of the prefrontal cortex on mood. Neurology 46 (2), 499–502.

Pathak, V., Sinha, V.K., Praharaj, S.K., 2015. Efficacy of adjunctive high frequency repetitive transcranial magnetic stimulation of right prefrontal cortex in adolescent mania: a randomized sham-controlled study. Clin. Psychopharmacol. Neurosci. 13 (3), 245–249. https://doi.org/10.9758/cpn.2015.13.3.245.

Paykel, E.S., 2001. Continuation and maintenance therapy in depression. Br. Med. Bull. 57 (1), 145–159. https://doi.org/10.1093/bmb/57.1.145.

Payne, N.A., Prudic, J., 2009. Electroconvulsive therapy: part I. A perspective on the evolution and current practice of ECT. J. Psychiatr. Pract. 15 (5), 346–368. https://doi.org/10.1097/01.pra.0000361277.65468.ef.

Peinemann, A., Reimer, B., Löer, C., Quartarone, A., Münchau, A., Conrad, B., Siebner, H.R., 2004. Long-lasting increase in corticospinal excitability after 1800 pulses of subthreshold 5 Hz repetitive TMS to the primary motor cortex. Clin. Neurophysiol. 115 (7), 1519–1526.

Perera, T., George, M.S., Grammer, G., Janicak, P.G., Pascual-Leone, A., Wirecki, T.S., 2016. The clinical TMS Society consensus review and treatment recommendations for TMS therapy for major depressive disorder. Brain Stimul. 9 (3), 336–346. https://doi.org/10.1016/j.brs.2016.03.010.

Petrides, G., Tobias, K.G., Kellner, C.H., Rudorfer, M.V., 2011. Continuation and maintenance electroconvulsive therapy for mood disorders: review of the literature. Neuropsychobiology 64 (3), 129–140. https://doi.org/10.1159/000328943.

Philip, N.S., Barredo, J., Van't Wout-Frank, M., Tyrka, A.R., Price, L.H., Carpenter, L.L., 2018. Network mechanisms of clinical response to transcranial magnetic stimulation in posttraumatic stress disorder and major depressive disorder. Biol. Psychiatry 83 (3), 263–272. https://doi.org/10.1016/j.biopsych.2017.07.021.

Phillips, A.L., Burr, R.L., Dunner, D.L., 2020. Repetitive transcranial magnetic stimulation in the treatment of bipolar depression: experience from a clinical setting. J. Psychiatr. Pract. 26 (1), 37–45. https://doi.org/10.1097/PRA.0000000000000447.

Rachid, F., 2018. Maintenance repetitive transcranial magnetic stimulation (rTMS) for relapse prevention in with depression: a review. Psychiatry Res. 262, 363–372. https://doi.org/10.1016/j.psychres.2017.09.009.

Reinares, M., Sánchez-Moreno, J., Fountoulakis, K.N., 2014. Psychosocial interventions in bipolar disorder: what, for whom, and when. J. Affect. Disord. 156, 46–55. https://doi.org/10.1016/j.jad.2013.12.017. 24439829.

Rossi, S., Hallett, M., Rossini, P.M., Pascual-Leone, A., 2009. Safety, ethical considerations, and application guidelines for the use of transcranial magnetic stimulation in clinical practice and research. Clin. Neurophysiol. 120 (12), 2008–2039. https://doi.org/10.1016/j.clinph.2009.08.016.

Rush, A.J., Trivedi, M.H., Wisniewski, S.R., Stewart, J.W., Nierenberg, A.A., Thase, M.E., Ritz, L., Biggs, M.M., Warden, D., Luther, J.F., Shores-Wilson, K., Niederehe, G., Fava, M., STAR*D Study Team, 2006. Bupropion-SR, sertraline, or venlafaxine-XR after failure of SSRIs for depression. N. Engl. J. Med. 354 (12), 1231–1242. https://doi.org/10.1056/NEJMoa052963.

Sackeim, H.A., Aaronson, S.T., Carpenter, L.L., Hutton, T.M., Mina, M., Pages, K., Verdoliva, S., West, W.S., 2020. Clinical outcomes in a large registry of patients with major depressive disorder treated with transcranial magnetic stimulation. J. Affect. Disord. 277, 65–74. https://doi.org/10.1016/j.jad.2020.08.005.

Sackeim, H.A., Haskett, R.F., Mulsant, B.H., Thase, M.E., Mann, J.J., Pettinati, H.M., Greenberg, R.M., Crowe, R.R., Cooper, T.B., Prudic, J., 2001. Continuation pharmacotherapy in the prevention of relapse following electroconvulsive therapy: A randomized controlled trial. JAMA 285 (10), 1299. https://doi.org/10.1001/jama.285.10.1299.

Sathappan, A.V., Luber, B.M., Lisanby, S.H., 2019. The dynamic duo: combining noninvasive brain stimulation with cognitive interventions. Prog. Neuropsychopharmacol. Biol. Psychiatry 89, 347–360. https://doi.org/10.1016/j.pnpbp.2018.10.006.

Selby, B., MacMaster, F.P., Kirton, A., McGirr, A., 2019. D-cycloserine blunts motor cortex facilitation after intermittent theta burst transcranial magnetic stimulation: a double-blind randomized placebo-controlled crossover study. Brain Stimul. 12 (4), 1063–1065. https://doi.org/10.1016/j.brs.2019.03.026.

Senova, S., Cotovio, G., Pascual-Leone, A., Oliveira-Maia, A.J., 2019. Durability of antidepressant response to repetitive transcranial magnetic stimulation: systematic review and meta-analysis. Brain Stimul. 12 (1), 119–128. https://doi.org/10.1016/j.brs.2018.10.001.

Sheline, Y.I., Price, J.L., Yan, Z., Mintun, M.A., 2010. Resting-state functional MRI in depression unmasks increased connectivity between networks via the dorsal nexus. Proc. Natl. Acad. Sci. U. S. A. 107 (24), 11020–11025. https://doi.org/10.1073/pnas.1000446107.

Siddiqi, S.H., Taylor, S.F., Cooke, D., Pascual-Leone, A., George, M.S., Fox, M.D., 2020. Distinct symptom-specific treatment targets for circuit-based neuromodulation. Am. J. Psychiatry 177 (5), 435–446. https://doi.org/10.1176/appi.ajp.2019.19090915.

Singh, A., Erwin-Grabner, T., Sutcliffe, G., Antal, A., Paulus, W., Goya-Maldonado, R., 2019. Personalized repetitive transcranial magnetic stimulation temporarily alters default mode network in healthy subjects. Sci. Rep. 9 (1), 5631. https://doi.org/10.1038/s41598-019-42067-3.

Smolen, P., Zhang, Y., Byrne, J.H., 2016. The right time to learn: mechanisms and optimization of spaced learning. Nat. Rev. Neurosci. 17 (2), 77.

Somani, A., Kar, S.K., 2019. Efficacy of repetitive transcranial magnetic stimulation in treatment-resistant depression: the evidence thus far. Gen. Psychiatry 32 (4). https://doi.org/10.1136/gpsych-2019-100074, e100074.

Sonmez, A.I., Camsari, D.D., Nandakumar, A.L., Voort, J.L.V., Kung, S., Lewis, C.P., Croarkin, P.E., 2019. Accelerated TMS for depression: a systematic review and meta-analysis. Psychiatry Res. 273, 770–781. https://doi.org/10.1016/j.psychres.2018.12.041.

Sridharan, D., Levitin, D.J., Menon, V., 2008. A critical role for the right fronto-insular cortex in switching between central-executive and default-mode networks. Proc. Natl. Acad. Sci. U. S. A. 105 (34), 12569–12574. https://doi.org/10.1073/pnas.0800005105.

Stefan, K., Kunesch, E., Benecke, R., Cohen, L.G., Classen, J., 2002. Mechanisms of enhancement of human motor cortex excitability induced by interventional paired associative stimulation. J. Physiol. 543 (Pt 2), 699–708. https://doi.org/10.1113/jphysiol.2002.023317.

Stokes, M.G., Chambers, C.D., Gould, I.C., Henderson, T.R., Janko, N.E., Allen, N.B., Mattingley, J.B., 2005. Simple metric for scaling motor threshold based on scalp-cortex distance: application to studies using transcranial magnetic stimulation. J. Neurophysiol. 94 (6), 4520–4527. https://doi.org/10.1152/jn.00067.2005.

Suppa, A., Biasiotta, A., Belvisi, D., Marsili, L., La Cesa, S., Truini, A., Cruccu, G., Berardelli, A., 2013. Heat-evoked experimental pain induces long-term potentiation-like plasticity in human primary motor cortex. Cerebral Cortex (New York, N.Y.: 1991) 23 (8), 1942–1951. https://doi.org/10.1093/cercor/bhs182.

Suppa, A., Huang, Y.-Z., Funke, K., Ridding, M.C., Cheeran, B., Di Lazzaro, V., Ziemann, U., Rothwell, J.C., 2016. Ten years of Theta burst stimulation in humans: established knowledge, unknowns and prospects. Brain Stimul. 9 (3), 323–335. https://doi.org/10.1016/j.brs.2016.01.006.

Tamas, R.L., Menkes, D., El-Mallakh, R.S., 2007. Stimulating research: a prospective, randomized, double-blind, sham-controlled study of slow transcranial magnetic stimulation in depressed bipolar patients. J. Neuropsychiatry Clin. Neurosci. 19 (2), 198–199. https://doi.org/10.1176/jnp.2007.19.2.198.

Tang, Y.P., Shimizu, E., Dube, G.R., Rampon, C., Kerchner, G.A., Zhuo, M., Liu, G., Tsien, J.Z., 1999. Genetic enhancement of learning and memory in mice. Nature 401 (6748), 63–69. https://doi.org/10.1038/43432.

Tavares, D.F., Myczkowski, M.L., Alberto, R.L., Valiengo, L., Rios, R.M., Gordon, P., de Sampaio-Junior, B., Klein, I., Mansur, C.G., Marcolin, M.A., Lafer, B., Moreno, R.A., Gattaz, W., Daskalakis, Z.J., Brunoni, A.R., 2017. Treatment of bipolar depression with deep TMS: results from a double-blind, randomized, parallel group, sham-controlled clinical trial. Neuropsychopharmacology 42 (13), 2593–2601. https://doi.org/10.1038/npp.2017.26.

Teo, J.T.H., Swayne, O.B., Rothwell, J.C., 2007. Further evidence for NMDA-dependence of the after-effects of human theta burst stimulation. Clin. Neurophysiol. 118 (7), 1649–1651. https://doi.org/10.1016/j.clinph.2007.04.010.

Tse, N.Y., Goldsworthy, M.R., Ridding, M.C., Coxon, J.P., Fitzgerald, P.B., Fornito, A., Rogasch, N.C., 2018. The effect of stimulation interval on plasticity following repeated blocks of intermittent theta burst stimulation. Sci. Rep. 8 (1), 1–10.

Vlachos, A., Müller-Dahlhaus, F., Rosskopp, J., Lenz, M., Ziemann, U., Deller, T., 2012. Repetitive magnetic stimulation induces functional and structural plasticity of excitatory postsynapses in mouse organotypic hippocampal slice cultures. J. Neurosci. Off. J. Soc. Neurosci. 32 (48), 17514–17523. https://doi.org/10.1523/JNEUROSCI.0409-12.2012.

Voigt, J., Carpenter, L., Leuchter, A., 2017. Cost effectiveness analysis comparing repetitive transcranial magnetic stimulation to antidepressant medications after a first treatment failure for major depressive disorder in newly diagnosed patients—a lifetime analysis. PLoS One 12 (10). https://doi.org/10.1371/journal.pone.0186950, e0186950.

Wang, H.-N., Wang, X.-X., Zhang, R.-G., Wang, Y., Cai, M., Zhang, Y.-H., Sun, R.-Z., Guo, L., Qiao, Y.-T., Liu, J.-C., He, H., Wang, Z.-H., Wan, Y.-C., Tan, Q.-R., Zhang, Z.-J., 2017. Clustered repetitive transcranial magnetic stimulation for the prevention of depressive relapse/recurrence: a randomized controlled trial. Transl. Psychiatry 7 (12), 1292. https://doi.org/10.1038/s41398-017-0001-x.

Warden, D., Rush, A.J., Trivedi, M.H., Fava, M., Wisniewski, S.R., 2007. The STAR*D project results: a comprehensive review of findings. Curr. Psychiatry Rep. 9 (6), 449–459. https://doi.org/10.1007/s11920-007-0061-3.

Weigand, A., Horn, A., Caballero, R., Cooke, D., Stern, A.P., Taylor, S.F., Press, D., Pascual-Leone, A., Fox, M.D., 2018. Prospective validation that Subgenual connectivity predicts antidepressant efficacy of transcranial magnetic stimulation sites. Biol. Psychiatry 84 (1), 28–37. https://doi.org/10.1016/j.biopsych.2017.10.028.

Wilkinson, S.T., Agbese, E., Leslie, D.L., Rosenheck, R.A., 2018. Identifying recipients of electroconvulsive therapy: data from privately insured Americans. Psychiatr. Serv. (Washington, D.C.) 69 (5), 542–548. https://doi.org/10.1176/appi.ps.201700364.

Williams, N.R., Sudheimer, K.D., Bentzley, B.S., Pannu, J., Stimpson, K.H., Duvio, D., Cherian, K., Hawkins, J., Scherrer, K.H., Vyssoki, B., DeSouza, D., Raj, K.S., Keller, J., Schatzberg, A.F., 2018. High-dose spaced theta-burst TMS as a rapid-acting antidepressant in highly refractory depression. Brain 141 (3), e18. https://doi.org/10.1093/brain/awx379.

Wise, T., Marwood, L., Perkins, A.M., Herane-Vives, A., Joules, R., Lythgoe, D.J., Luh, W.-M., Williams, S.C.R., Young, A.H., Cleare, A.J., Arnone, D., 2017. Instability of default mode network connectivity in major depression: a two-sample confirmation study. Transl. Psychiatry 7. https://doi.org/10.1038/tp.2017.40, e1105.

Xiao, X., Bentzley, B.S., Cole, E.J., Tischler, C., Stimpson, K.H., Duvio, D., Bishop, J.H., DeSouza, D.D., Schatzberg, A., Keller, C., Sudheimer, K.D., Williams, N.R., 2019. Functional connectivity changes with rapid remission from moderate-to-severe major depressive disorder. Clin. Trials. http://biorxiv.org/lookup/doi/10.1101/672154.

Yang, L.-L., Zhao, D., Kong, L.-L., Sun, Y.-Q., Wang, Z.-Y., Gao, Y.-Y., Li, N., Lu, L., Shi, L., Wang, X.-Y., Wang, Y.-M., 2019. High-frequency repetitive transcranial magnetic stimulation (rTMS) improves neurocognitive function in bipolar disorder. J. Affect. Disord. 246, 851–856. https://doi.org/10.1016/j.jad.2018.12.102.

Yip, A.G., George, M.S., Tendler, A., Roth, Y., Zangen, A., Carpenter, L.L., 2017. 61% of unmedicated treatment resistant depression patients who did not respond to acute TMS treatment responded after four weeks of twice weekly deep TMS in the Brainsway pivotal trial. Brain Stimul. 10 (4), 847–849. https://doi.org/10.1016/j.brs.2017.02.013.

Zhao, Y.J., Tor, P.C., Khoo, A.L., Teng, M., Lim, B.P., Mok, Y.M., 2018. Cost-effectiveness modeling of repetitive transcranial magnetic stimulation compared to electroconvulsive therapy for treatment-resistant depression in Singapore. Neuromodulation 21 (4), 376–382. https://doi.org/10.1111/ner.12723.

Ziemann, U., Hallett, M., Cohen, L.G., 1998. Mechanisms of deafferentation-induced plasticity in human motor cortex. J. Neurosci. Off. J. Soc. Neurosci. 18 (17), 7000–7007.

Chapter 18

tDCS for treatment-resistant depression

Lais B. Razza[a,b], Lucas Borrione[a,b], Adriano H. Moffa[b,d], and Andre R. Brunoni[a,b,c]

[a]*Service of Interdisciplinary Neuromodulation, Department and Institute of Psychiatry, University of São Paulo Medical School, São Paulo, Brazil,* [b]*Laboratory of Neuroscience and National Institute of Biomarkers in Psychiatry, Department and Institute of Psychiatry, University of São Paulo Medical School, São Paulo, Brazil,* [c]*Center for Clinical and Epidemiological Research & Interdisciplinary Center for Applied Neuromodulation, University Hospital, University of São Paulo, São Paulo, Brazil,* [d]*School of Psychiatry & Black Dog Institute, University of New South Wales, Sydney, NSW, Australia*

Introduction

Transcranial direct current stimulation (tDCS) is a noninvasive brain stimulation (NIBS) technique, whereby a subthreshold, continuous electrical current (typically 1–2 mA) is applied to the brain via two or more electrodes, placed over the scalp (Brunoni et al., 2019). The conventional tDCS device consists of a portable, battery-operated machine, connected through conductive cables to two electrodes (an anode and a cathode), that are cushioned in saline-humidified sponge pads (Fig. 18.1).

The device generates a current that passes through the skin, subcutaneous tissue, skull, meninges, cerebrospinal fluid, and finally, brain gray and white matter, with the direction of the electrical current flowing from the anode toward the cathode (Brunoni et al., 2019). In general terms, the neurons under the anode tend to be partially depolarized, increasing their excitability and facilitating synaptic transmission (Lefaucheur and Wendling, 2019). On the other hand, in the area under the cathode, the neurons tend to hyperpolarize, decreasing their excitability and inhibiting synaptic transmission (Lefaucheur and Wendling, 2019).

These processes take place, however, with tDCS not directly eliciting action potentials, but rather facilitating or inhibiting (modulating) the already ongoing neuronal activity taking place in the stimulated regions (Moffa et al., 2018). This biophysical change in the target region also tends to influence other brain areas that are anatomically or functionally connected to it (Fregni et al., 2021). It is with this underlying rationale that different electrode montages are considered when treating a disorder in which a specific brain region or network is involved (for a detailed description of the biophysical and network mechanisms of tDCS, please see "Mechanism(s) of action" section).

tDCS has been intensively researched in diverse neuropsychiatric disorders, with variable degrees of efficacy, but with consistent safety and tolerability profiles (Lefaucheur et al., 2017; Moffa et al., 2018). In the treatment of major depressive disorder (MDD) in particular, tDCS has been found to be superior to sham, with moderate effect sizes (Moffa et al., 2020; Mutz et al., 2019; Razza et al., 2020). Furthermore, pivotal trials have suggested tDCS' synergistic effects with sertraline (Brunoni et al., 2013b), albeit being inferior, as monotherapy, to escitalopram (Brunoni et al., 2017).

Currently, tDCS is not yet placed on the same level of clinical efficacy as other forms of NIBS for treatment-resistant depression (TRD), such as transcranial magnetic stimulation (TMS), which is an approved alternative after treatment failure with at least one antidepressant, or electroconvulsive therapy (ECT), for severe cases of depression, especially with psychotic symptoms (Milev et al., 2016).

However, some advantages associated with tDCS, in comparison with other NIBS techniques (i.e., TMS and ECT), are its relatively low cost, portability, easiness of use, and the possibility of home-based treatment (Borrione et al., 2020; Charvet et al., 2020). Moreover, tDCS is not associated with many of the most common adverse events observed with antidepressant medications (i.e., appetite, libido, and sleep alterations), nor with the adverse cognitive effects of ECT (Brunoni et al., 2019).

In the treatment of MDD in general, and more specifically in the case of TRD, tDCS displays an augmentative and synergistic potential, both to an existing regimen of antidepressants, and/or to cognitive behavioral therapy (CBT) (Sathappan et al., 2019).

In this chapter, we provide a detailed overview of tDCS as a potentializing intervention in TRD, encompassing recent advancements and interesting lines for future research.

Managing Treatment-Resistant Depression. https://doi.org/10.1016/B978-0-12-824067-0.00018-9

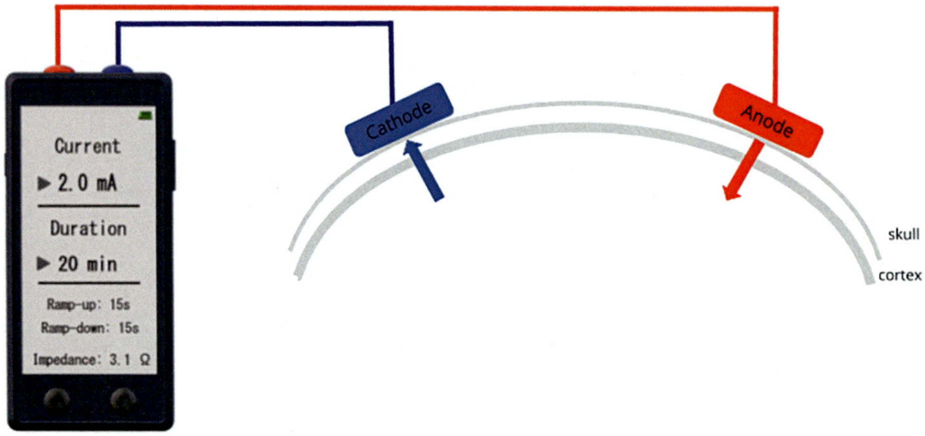

FIG. 18.1 Conventional tDCS montage. A conventional tDCS montage showing a battery-operated machine, conductive cables, an anode (positive electrode) and a cathode (negative electrode) placed over the scalp. The current flows from the anode to the cathode, is nonfocal, and does not elicit action potentials.

History and development

Animal studies involving basic aspects of direct electrical current stimulation in the brain began in the 1950s, with observations that brief periods of intracerebral or epidural stimulation could produce long-lasting effects on neuronal membrane resting potential, depending on the polarity of the electrodes: the anode was found to increase the rate of neuronal firing, while the cathode led to hyperpolarization of neurons and to the decrease of the neuronal firing rate (Bindman et al., 1964; Creutzfeldt et al., 1962). However, since the 1960s, it has been known that opposite polarization effects might also be observed in individual cortical layers, suggesting that effects of direct current stimulation might depend on other factors besides polarity, such as current direction and neuronal geometry (Purpura and Mcmurtry, 1965).

Perhaps the first placebo-controlled clinical trial to use noninvasive electrical brain polarization in the treatment of depressed patients was published in 1964, with the observation of clinical improvement in depressive patients receiving active treatment, especially regarding anxiety, agitation, and somatic symptom domains (Costain et al., 1964). In this cross-over trial involving 24 inpatients, the electrical current was passed through 2 electrodes (anodes) placed over each eyebrow, while the cathode was placed on one leg, using a protocol of 0.25 mA during 8 h per day, for 12 consecutive days (Costain et al., 1964). The protocol used in this trial was based on previous observations of optimal antidepressant response with current amplitudes at a maximum of 0.3 mA; however, antidepressant effects were observed to change in other amplitudes, depending on current density parameters (Lippold and Redfearn, 1964). At the time of the first brain polarization trials for depression, limitations in methodology and incipient neurobiological knowledge related to the disorder became obstacles in the development of the technique, which went gradually into oblivion over the next 40 years (Fregni et al., 2006).

Beginning around the year 2000, renewed observations involving the application of tDCS in humans began to shed deeper light on the technique's longer-lasting behavioral and therapeutic effects. In a seminal study by Nitsche and Paulus (2000), noninvasive electrical stimulation of the human motor cortex was shown to alter motor-evoked potentials as elicited by single-TMS pulses (Nitsche and Paulus, 2000). In this study, anodal stimulation led to an increases in cortical excitability, while cathodal stimulation led to subsequent decreases in this parameter; the authors also noted that strength and duration of after-effects could be controlled by modifications of current intensity and duration of stimulation (Nitsche and Paulus, 2000).

Later, it was observed that patients with depression commonly display an asymmetrical pattern of functional interhemispheric activity, with a reduction in prefrontal activity of the left hemisphere, and an increase in the right in comparison to heathy individuals (Eschweiler et al., 2000; Mayberg et al., 2000). This neurobiological model offered a rationale for the most commonly used tDCS montages in the treatment of MDD, whereby the "excitatory" anode is usually placed over the left hemisphere, while the cathode is placed over the right hemisphere (Borrione et al., 2020).

The first modern RCT involving the use of tDCS in MDD was published in 2006 (Fregni et al., 2006). In this trial, 10 patients with MDD received 5 days of anodal stimulation (1 mA for 20 min/day) over the left dorsolateral prefrontal cortex (DLPFC) or sham. Results showed four treatment responders in the active group,no responders in the sham group, and no reports of major adverse effects (Fregni et al., 2006).

These positive initial results gradually led to dozens of trials involving the use of tDCS for MDD, with varying individual degrees of efficacy, but with overall moderate effect sizes versus placebo. Additionally, consistent patient tolerability and general safety allowed for the continued research of this NIBS modality, as shall be described below.

Mechanisms of action

The exact mechanism of action of tDCS is still to be fully elucidated. At the cellular level, tDCS is believed to induce a change in neuronal membrane potentials, affecting the probability and timing of spikes (neural firing rates) (Chase et al., 2020; Purpura and Mcmurtry, 1965). In vitro experiments using rat motor cortex brain slices indicated that anodal stimulation produces partial depolarization of the neuronal soma, with the cathode producing somatic hyperpolarization, respectively increasing or decreasing the probability of triggering action potentials (Radman et al., 2009).

Initial neurophysiological studies with tDCS applied to the motor area also pointed in the same direction. When accessing motor-evoked potentials (MEPs) elicited by simple TMS pulses, researchers showed an increase in amplitude after anodal and a decrease after cathodal stimulation (Nitsche and Paulus, 2000). The latest research, however, suggests that this relationship is more complex and that parameters such as current intensity and duration of stimulation may have a nonlinear relationship with the effects of tDCS on cortical excitability. For instance, a study has shown that the application of both anodal and cathodal tDCS to the motor cortex, with an intensity of 2 mA for 20 min, led to an increase in the amplitude of MEPs, changing the direction of cathodal effects to excitability enhancement (Batsikadze et al., 2013).

More recent evidence also expands on classic findings and suggests that tDCS impacts neural functioning also at a network level, with its local effects spreading to other regions anatomically and functionally connected to the target area (Stagg et al., 2018). The current intensity commonly used in clinical studies in humans (typically 2 mA) is not sufficient to elicit action potentials. Therefore, tDCS would act influencing the ongoing oscillatory dynamics of nodes of the network and in that sense, selectively affecting the "active" neurons, which are receiving synaptic inputs (in contrast to the neurons "at rest" of the inactive networks at the time of stimulation) (Reato et al., 2010).

Therefore, for depression, most of the studies that place the anode over the left DLPFC use this region as a "window" to enter large-scale brain networks like the default mode network (DMN), for which the DLPFC is a key hub. The DMN is believed to be involved in self-referential processing, affective cognition, and emotion regulation (Andrews-Hanna et al., 2010). Facilitating its endogenous activity and normalizing functional connectivity with key regions within the network (like the subgenual anterior cingulate cortex, sACC) could potentially lead to an improvement in depressive symptoms, as shown after a course of TMS treatment (Andrews-Hanna et al., 2010; Liston et al., 2014).

Moreover, due to the diffuse nature of tDCS, it can stimulate other regions and potentially, deeper structures that are in the path of the current between the electrodes (Huang et al., 2017). Although evidence suggests that tDCS may functionally target neuronal dynamics (Rahman et al., 2017), the arrangement of the electrodes (montage) also plays an important role in the final effect of the stimulation (Truong and Bikson, 2018).

The long-term effects of tDCS are believed to be due to synaptic plasticity modulated by LTP- (long-term potentiation) and LTD- (long-term depression)-like mechanisms (Kronberg et al., 2017). Pharmacological studies in humans (Nitsche et al., 2003a) and animal electrophysiology experiments (Marquez-Ruiz et al., 2012; Podda et al., 2016) corroborate this hypothesis, although the exact biophysical mechanism involved is still unknown (Kronberg et al., 2020).

Clinical efficacy of tDCS for TRD patients

The clinical efficacy of tDCS for the treatment of depression has been extensively investigated in the past two decades. The first RCT of tDCS for depression was conducted by Fregni and colleagues in 2006 (Fregni et al., 2006). In this pilot trial, the authors evaluated the effectiveness of tDCS (1 mA, 20 min, 5 consecutive sessions) compared to a sham intervention in 10 adult patients with a diagnosis of depression, and found a significant improvement in depressive symptoms in the active group (Fregni et al., 2006). Afterwards, a larger RCT ($n = 40$) with antidepressant-free patients was conducted also showing clinical improvement in the active tDCS arm (2 mA, 20 min, 10 sessions) (Boggio et al., 2008).

However, a few subsequent RCTs were conducted with medicated and treatment-resistant patient samples, failing to show any significant difference between active and sham tDCS for depression improvement. For instance, Blumberger et al. recruited 24 adult patients with treatment-resistant depression (defined as failure to achieve remission or inability to tolerate two trials of an antidepressant medication from separate classes) and applied 15 tDCS sessions (one treatment per weekday), delivered at 2 mA for 20 min (Blumberger et al., 2012). Both the continuous outcome (depression scale improvement) and remission rates did not differ between active and sham tDCS after the 3 weeks of treatment (Blumberger et al., 2012). Palm et al. also conducted a trial with 22 treatment-resistant subjects (TRD defined with the same criteria used by

Blumberger et al., 2012) that were taking antidepressants in a stable regimen (Palm et al., 2012). The authors applied a cross-over design (10 active and 10 sham sessions) with tDCS intensity of 1 mA or 2 mA for 20 min (Palm et al., 2012). The first 10 patients were treated with 1 mA and the other 12 with 2 mA. Again, the authors failed to show any meaningful therapeutic effects of tDCS compared to sham for TRD patients, even applying both stimulation intensities (Palm et al., 2012).

Later, Brunoni and colleagues performed the *SELECT-TDCS* study, a factorial clinical trial evaluating the combination of tDCS and sertraline (50 mg/day) compared to tDCS, sertraline, or sham alone (Brunoni et al., 2013b). Overall, the authors found that the efficacy of tDCS as monotherapy was similar to the antidepressant drug, while the combined group (tDCS + sertraline) presented superior depression improvement compared to the other arms (Brunoni et al., 2013b). Interestingly, secondary analyses suggested that treatment resistance (more than one failed antidepressant trial) was a predictor variable for a lower treatment response (Brunoni et al., 2013b). TRD was also associated with lower tDCS efficacy in all predictor models conducted by a subsequent individual patient data meta-analysis of six randomized controlled trials ($n=289$ patients), in which tDCS effectiveness for depression was assessed (Brunoni et al., 2016). Moreover, based on the previous findings of Blumberger et al. and Palm et al., a guideline on the therapeutic use of tDCS considered the technique "probably ineffective" for TRD patients (Lefaucheur et al., 2017).

More recently, Brunoni and colleagues conducted the *ELECT-tDCS*, a noninferiority trial comparing the effects of tDCS and escitalopram for depression and showed that tDCS treatment was inferior to the antidepressant drug (20 mg/day), although superior to sham (Brunoni et al., 2017). This was the largest tDCS trial for depression to date, with 245 participants, of which 30% of the included patients in the tDCS group were resistant to at least 1 antidepressant treatment in the current episode, or with > 4 treatment failures over the patient's lifetime (Brunoni et al., 2017). Secondary analyses of this study also showed that TRD patients presented a negative association with depression improvement, when comparing active tDCS to sham intervention (Brunoni et al., 2017).

Results from an individual patient data (IPD) meta-analysis also confirmed the tDCS effects presenting a negative association with TRD patients (Brunoni et al., 2016). Nonetheless, two recent meta-analyses that showed tDCS having a moderate effect for depression (n = 25, Hedges' g = 0.46, 95% confidence interval (CI): 0.22–0.70 and n = 9, $\beta=0.31$, 95% CI: 0.15– 0.47, respectively) (Moffa et al., 2020; Razza et al., 2020) were not able to identify this negative association. Some hypotheses for these new findings suggest that the results of the first IPD meta-analysis were probably driven by the large study of Brunoni et al. conducted in 2013 (*SELECT-TDCS*), with a low-treatment–resistant sample showing clinical efficacy of tDCS (Brunoni et al., 2013b), whereas other studies with smaller sample sizes and higher treatment resistance showed no effects of tDCS (Bennabi et al., 2015; Blumberger et al., 2012; Palm et al., 2012).

Finally, a very recent guideline focusing on the evaluations of the effects of tDCS for neurological and psychiatric disorders considered anodal tDCS on the left DLPFC definitely effective (Level A) in improving depression in major depressive disorder, even for treatment-resistant patients (Fregni et al., 2021). Thus, the findings of the recent studies in the tDCS literature have been supporting the evidence that tDCS effects are not significantly discrepant between resistant and not treatment-resistant depression subjects. Therefore, even though the intervention is not yet approved by the U.S. FDA for depression, and its acute antidepressant effects are currently considered inferior when compared to TMS and ECT (Milev et al., 2016), tDCS has been presenting considerable advantages to other NIBS techniques such as portability, easier application and lower cost.

Adverse effects

Several studies have already systematically assessed the safety, tolerability and acceptability of tDCS application for both healthy subjects and patients with neuropsychiatric conditions, enabling more reliable findings regarding these outcomes for tDCS treatment. In general, safety, acceptability, and tolerability are known, respectively, as the potential damage caused by a protocol, the rate of drop-outs for any reasons, and the impact of an adverse effect (AE) that does not cause functional and structural damage (Brunoni et al., 2016). AEs are defined as any undesirable effect associated with the use of a medical intervention and can be classified as mild, moderate, serious or severe. The latter correspond to outcomes such as death, hospitalization, life-threatening events, disability or permanent damage, congenital anomaly/birth defect, or other serious important medical events (i.e., seizures/convulsions that do not result in hospitalization) (Brunoni et al., 2016).

The first study to evaluate AEs related to tDCS application in humans was conducted by Nitsche and colleagues (Nitsche et al., 2003b). They reported the findings of approximately 500 participants receiving a tDCS current of 1 mA ($35\,cm^2$ electrodes) for 20 min over the motor cortex, and only mild transient tingling sensations under the electrodes and/or light flashes when the stimulation was switched on or off abruptly were reported (Nitsche et al., 2003b). Posteriorly, a similar study by Poreisz et al. (2007) was conducted with data from previous studies of their group to evaluate safety aspects of tDCS over six different cortical areas (Poreisz et al., 2007). Observations from a total of 567 tDCS sessions on

102 participants were summarized. None reported serious AEs, however, 70.6% of the participants reported a mild tingling sensation, 35.3% presented moderate fatigue, 30.4% felt a light itching sensation under the electrodes during the tDCS session, 21.6% reported a slight burning sensation, 15.7% reported pain sensation, and 10.8% presented difficulties in concentrating. Headaches and nausea were less reported (4.9% and 2.9%, respectively) (Poreisz et al., 2007).

Two systematic reviews (SR) were conducted aiming to evaluate the safety, tolerability and acceptability of tDCS in several conditions. The first SR, conducted by Brunoni et al., 2011, summarized data of 209 studies, of which 56% (117 studies) reported any AE. The most common AEs reported were itching (active vs. sham group: 39.3% vs. 32.9%), tingling (22.2% vs. 18.3%), headache (14.8% vs. 16.2%), burning sensation (8.7% vs. 8.7%), and discomfort (10.4% vs. 10%) (Brunoni et al., 2011). According to the findings, all described AEs were considered mild, and no significant differences were found between active and sham groups. More recently, a study conducted a SR and meta-analysis to investigate acceptability and tolerability of tDCS treatment for several conditions (Aparício et al., 2016). Regarding acceptability, the authors showed that 6% and 7.2% of participants dropped-out from active and sham groups, respectively. However, almost half of the studies did not report dropouts and only 23.4% of the included experiments presented the reasons for discontinuations. Among the reasons, the most frequent were protocol violation and the presence of AEs. Regarding tolerability, the authors found tingling being the most common AE, followed by skin redness, itching, headache, burning sensation and sleepiness. Nevertheless, several studies did not report AEs and, when described, no systematic method was applied among the studies (Aparício et al., 2016).

Adverse effects in depression trials

Although the above findings show only mild adverse effects after tDCS application, some points should be underscored for depression trials. Loo et al. reported a case of suicide of a patient with depression the day after receiving an active tDCS session. The authors did not find any clinical changes, associated with the stimulation session, thus the event was not attributed to the tDCS application (Loo et al., 2010).

Episodes of mania and hypomania have also been reported in studies investigating the effects of tDCS for unipolar and bipolar depression (Berlow et al., 2019). For instance, the largest study of tDCS for unipolar depression ($n=245$) reported two patients presenting episodes of new-onset mania after the beginning of the tDCS protocol (Brunoni et al., 2017). Moreover, a recent meta-analysis showed that patients with unipolar depression present an increased risk of treatment-emergent mania when compared to the sham group. However, they could not predict whether these symptoms were caused by the tDCS application or due to a natural history of the condition (Berlow et al., 2019).

In turn, most antidepressant treatments carry a risk of inducing treatment-emergent mania, and it is still unclear whether tDCS treatment presents a higher risk when compared to other treatments. Other AEs, such as behavioral or cognitive effects were never reported in the literature of tDCS for depression (Martin et al., 2014; Moreno et al., 2020). Although some clinical trials of tDCS for unipolar depression patients have already reported (hypo)mania switch episodes, they were not yet directly associated with the tDCS application (Brunoni et al., 2017). Thus, the low rate of reported adverse effects during depression treatment and other conditions makes tDCS a safe and well-tolerable intervention. However, before tDCS application, a few relative contraindications must be evaluated, such as skin problems (dermatitis, psoriasis or eczema), metallic implants in the head (under or in between the electrodes), subjects with neurological history (neurosurgery, epilepsy, malformation), and special populations (children and elderly) must be carefully considered (Antal et al., 2017).

Summary and future directions

Contrary to other NIBS modalities (e.g., rTMS, theta-burst stimulation-TBS, deep TMS), tDCS has not yet received FDA clearance for the treatment of MDD. Nevertheless, tDCS has been extensively studied in all three important regulatory aspects (efficacy, safety and tolerability), and besides being deemed very safe and tolerable, its effectiveness as a treatment for MDD has been recently considered "definitely effective" in improving depression in MDD (anodal tDCS to the left DLPFC, level A) (Fregni et al., 2021). Furthermore, it should be noted that in a previous meta-analysis, even though absolute effect sizes for TMS modalities were superior to those of tDCS, their confidence intervals overlapped (Mutz et al., 2018), raising questions as to which of these two NIBS interventions is superior to the other, and also, to what extent (Brunoni, 2019). This is also complemented by the results of a recent umbrella review showing tDCS with "high" level of evidence for depression, together with high frequency rTMS (Razza et al., 2020).

On the other hand, regardless of the question of superiority among different NIBS modalities, tDCS, unlike TMS and ECT, is portable and may be safely and effectively performed at home (Alonzo et al., 2019; Charvet et al., 2020). Home-based tDCS, with remote monitoring by mental health professionals, may also allow for longer continuation and maintenance phases in the treatment of MDD, as transportation and logistical burdens would be diminished (Charvet et al., 2015). This approach has already been performed for the treatment of MDD in a pilot trial (n = 34), presenting significant improvement in mood scores from baseline to the end of the acute phase and a low rate of discontinuation (Alonzo et al., 2019) and in a case series (n = 5), with promising results (Borrione et al., 2021). However, despite these encouraging findings, it should be noted that these trial and case series lacked a control group, and only patients with adequate computer literacy were selected, which limits its generalizability. Therefore, although home-based tDCS seems a promising approach, trials with larger and broader MDD samples are needed in order to reach more definite conclusions.

Besides the investigation of home-based alternatives, the ongoing development of other strategies to improve tDCS delivery and antidepressant efficacy is deemed extremely necessary. In this regard, three approaches are worth mentioning: (1) the individualization of current dosing parameters, (2) the combination of tDCS with cognitive-behavioral interventions, and (3) the study of biomarkers and clinical predictors associated with tDCS response.

Currently, tDCS is generally delivered within fixed dosing parameters, which might help explain both within- and between-study variabilities regarding stimulation effects (López-Alonso et al., 2014; Moffa et al., 2020). The electric field that effectively reaches the brain can be substantially different across subjects due to individual characteristics, such as size and shape of the head, skull thickness and other parameters of brain anatomy (Opitz et al., 2015). However, recent developments based on individuals' functional or structural scans and computational modeling may allow additional investigations clarifying the correlations between electric field strength delivered by tDCS to specific brain regions of interest (ROIs), the interaction mechanism between this field with the neurons and clinical response. For instance, the simulated electrical field strength in the anterior cingulate cortex (ACC) and left DLPFC has been found to be inversely associated with negative affect and depression scores, unveiling a potential mechanism of action related to tDCS (Suen et al., 2020). Despite the limitations and complexity involved in computational modeling (Salvador et al., 2019), this methodology might ultimately provide grounds for individualized current dosing and electrode montages in future tDCS protocols (Lisanby, 2017).

Secondly, the combination of tDCS with cognitive and/or behavioral interventions is also very appealing, as tDCS targets brain areas that are also activated by the latter. This can potentially enhance clinical results through a process known as "target engagement," in which it is speculated that the effects of tDCS is greatly influenced by the functional state of the targeted cortical region (Dedoncker et al., 2016; Sathappan et al., 2019). For instance, both tDCS and CBT have been observed to engage the prefrontal cortex and ACC (Sathappan et al., 2019), brain areas which have been previously associated with maladaptive rumination and brooding in depressive states (Williams, 2016). In a recent pilot study, patients with TRD received active tDCS (20 min, 2 mA applied to the left DLPFC) on eight consecutive days, and were randomly assigned to a mindfulness-based cognitive therapy (MBCT) or relaxation, immediately after the tDCS sessions (Monnart et al., 2019). In this study, patients who received the combination of tDCS and MBCT displayed better maintenance of clinical improvement (Monnart et al., 2019). Other previous studies involving the combination of tDCS and cognitive control training in a small sample of patients with MDD have shown mixed results (Brunoni et al., 2014; Segrave et al., 2014). An ongoing multicenter trial evaluating the efficacy of tDCS associated with group CBT (Bajbouj et al., 2018) should help shed more light on the probable synergistic effects of NIBS and psychotherapy, in the context of "target engagement" and "dynamic duos" (Bajbouj et al., 2018; Sathappan et al., 2019).

Finally, the growing use in psychiatric trials of techniques like functional magnetic resonance image (fMRI), single-photon emission computed tomography (SPECT), positron emission tomography (PET), electroencephalography (EEG) and serum levels of several hormones to measure biological markers (Fidalgo et al., 2014) may help to clarify the mechanisms of action of tDCS, identify "optimal" parameters of stimulation and eventually guide patient selection more prone to respond to treatment (Borrione et al., 2020). Among the potential biomarkers that have already been explored, we can highlight baseline resting-state EEG oscillatory activity (Al-Kaysi et al., 2016), polymorphisms in brain-derived neurotrophic factor (BDNF) (Brunoni et al., 2013a) and the 5-HTTLPR serotonin transporters (Brunoni et al., 2013a), and brain volumes in specific ROIs (for instance, larger gray matter volumes in the left DLPFC at baseline have been associated with better tDCS antidepressant response) (Bulubas et al., 2019). Furthermore, tDCS responders have been found to display greater improvements in dysphoria and retardation symptoms, as compared to nonresponders, offering a more nuanced understanding of what could be expected from tDCS in treatment-resistant samples (Alonzo et al., 2013).

Moreover, the accumulation of big data and the use of a machine learning (ML) approach can also assist in clarifying the gaps in knowledge about the predictive role of biological and clinical characteristics that are associated with better tDCS response in TRD, potentially allowing the identification of preclinical changes which could guide adjustments in stimulation parameters or treatment regimen (Nitsche et al., 2019). To the best of our knowledge, there is no published research using ML to predict tDCS response based on clinical samples (Borrione et al., 2020).

Conclusions

TDCS is a NIBS intervention that delivers subthreshold electrical currents to neuronal populations, usually via two electrodes, inducing a shift in resting membrane potential, either toward depolarization or hyperpolarization, depending on stimulation parameters.

TDCS has been increasingly investigated in diverse neuropsychiatric disorders, presenting itself as a safe and feasible intervention. The results of recent studies have shown tDCS montages consisting of the positive electrode (anode) over the left dorsolateral prefrontal cortex (DLPFC) and the negative electrode (cathode) over the right DLPFC to be moderately and definitely effective for MDD (level A).

Interestingly, the first trials and reviews with TRD samples showed a negative association with tDCS therapeutic effects. However, recent meta-analyses did not present TRD as a possible predictor of negative response for tDCS. Moreover, a recent guideline considered tDCS as an effective treatment for depression, even for treatment-resistant patients.

Recent studies showed home-use tDCS, individualized dosing, tDCS associated with cognitive and behavioral interventions, the discovery of biomarkers and use of machine learning and big data, as being future directions to help diminish existing knowledge gaps in this most interesting of NIBS modalities.

References

Al-Kaysi, A.M., Al-Ani, A., Loo, C.K., Breakspear, M., Boonstra, T.W., 2016. Predicting brain stimulation treatment outcomes of depressed patients. Conf. Proc. IEEE Eng. Med. Biol. Soc. 2016, 5266–5269. https://doi.org/10.1109/EMBC.2016.7591915.

Alonzo, A., Chan, G., Martin, D., Mitchell, P.B., Loo, C., 2013. Transcranial direct current stimulation (tDCS) for depression: analysis of. J. Affect. Disord. 150 (1), 91–95. https://doi.org/10.1016/j.jad.2013.02.027.

Alonzo, A., Fong, J., Ball, N., Martin, D., Chand, N., Loo, C., 2019. Pilot trial of home-administered transcranial direct current stimulation. J. Affect. Disord. 252, 475–483. https://doi.org/10.1016/j.jad.2019.04.041.

Andrews-Hanna, J.R., Reidler, J.S., Sepulcre, J., Poulin, R., Buckner, R.L., 2010. Functional-anatomic fractionation of the brain's default network. Neuron 65 (4), 550–562. https://doi.org/10.1016/j.neuron.2010.02.005.

Antal, A., Alekseichuk, I., Bikson, M., Brockmöller, J., Brunoni, A.R., Chen, R., Cohen, L.G., Dowthwaite, G., Ellrich, J., Flöel, A., Fregni, F., George, M.S., Hamilton, R., Haueisen, J., Herrmann, C.S., Hummel, F.C., Lefaucheur, J.P., Liebetanz, D., Loo, C.K., Paulus, W., 2017. Low intensity transcranial electric stimulation: safety, ethical, legal. Clin. Neurophysiol. 128 (9), 1774–1809. https://doi.org/10.1016/j.clinph.2017.06.001.

Aparício, L.V.M., Guarienti, F., Razza, L.B., Carvalho, A.F., Fregni, F., Brunoni, A.R., 2016. A systematic review on the acceptability and tolerability of transcranial. Brain Stimul. 9 (5), 671–681. https://doi.org/10.1016/j.brs.2016.05.004.

Bajbouj, M., Aust, S., Spies, J., Herrera-Melendez, A.-L., Mayer, S.V., Peters, M., Plewnia, C., Fallgatter, A.J., Frase, L., Normann, C., et al., 2018. PsychotherapyPlus: augmentation of cognitive behavioral therapy (CBT) with. Eur. Arch. Psychiatry Clin. Neurosci. 268 (8), 797–808. https://idp.springer.com/authorize/casa?redirect_uri=https://link.springer.com/article/10.1007/s00406-017-0859-x&casa_token=LR2o4nXNpYQAAAAA:Vq8BVUT4kfpvGkgx6CZzE4-8ymyAvq-LYWNgoC4BegPeOskaOmQOVBnMq9wGTXSvH-H36pZyLKi7Cq7-s3M.

Batsikadze, G., Moliadze, V., Paulus, W., Kuo, M.-F., Nitsche, M.A., 2013. Partially non-linear stimulation intensity-dependent effects of direct. J. Physiol. 591 (7), 1987–2000. https://doi.org/10.1113/jphysiol.2012.249730.

Bennabi, D., Nicolier, M., Monnin, J., Tio, G., Pazart, L., Vandel, P., Haffen, E., 2015. Pilot study of feasibility of the effect of treatment with tDCS in. Clin. Neurophysiol. 126 (6), 1185–1189. https://doi.org/10.1016/j.clinph.2014.09.026.

Berlow, Y.A., Zandvakili, A., Carpenter, L.L., Philip, N.S., 2019. Transcranial direct current stimulation for unipolar depression and risk. Brain Stimul. https://doi.org/10.1016/j.brs.2019.03.025. 2019/03/31.

Bindman, L.J., Lippold, O.C.J., Redfearn, J.W.T., 1964. The action of brief polarizing currents on the cerebral cortex of the rat. J. Physiol. 172 (3), 369–382. https://physoc.onlinelibrary.wiley.com/doi/pdf/10.1113/jphysiol.1964.sp007425.

Blumberger, D.M., Tran, L.C., Fitzgerald, P.B., Hoy, K.E., Daskalakis, Z.J., 2012. A randomized double-blind sham-controlled study of transcranial direct. Front. Psychiatry 3, 74. https://doi.org/10.3389/fpsyt.2012.00074.

Boggio, P.S., Rigonatti, S.P., Ribeiro, R.B., Myczkowski, M.L., Nitsche, M.A., Pascual-Leone, A., Fregni, F., 2008. A randomized, double-blind clinical trial on the efficacy of cortical. Int. J. Neuropsychopharmacol. 11 (2), 249–254. https://doi.org/10.1017/S1461145707007833.

Borrione, L., Bellini, H., Razza, L.B., Avila, A.G., Baeken, C., Brem, A.-K., Busatto, G., Carvalho, A.F., Chekroud, A., Daskalakis, Z.J., Deng, Z.-D., Downar, J., Gattaz, W., Loo, C., Lotufo, P.A., da Martin, M.G.M., McClintock, S.M., O'Shea, J., Padberg, F., Passos, I.C., Salum, G.A., Vanderhasselt, M.-A., Fraguas, R., Benseñor, I., Valiengo, L., Brunoni, A.R., 2020. Precision non-implantable neuromodulation therapies: a perspective for the depressed brain. Braz. J. Psychiatry 42 (4), 403–419. https://doi.org/10.1590/1516-4446-2019-0741. 32187319.

Borrione, L., Klein, I., Razza, L.B., Suen, P., Brunoni, A.R., 2021. Use of app-based psychological interventions in combination with home-use transcranial direct current stimulation for the treatment of major depressive disorder: a case series. J. Affect. Disord. 288, 189–190. https://doi.org/10.1016/j.jad.2021.04.013.

Brunoni, A.R., 2019. tDCS in depression: quo usque tandem? J. Affect. Disord. 256, 431–432. https://doi.org/10.1016/j.jad.2019.06.020.

Brunoni, A.R., Amadera, J., Berbel, B., Volz, M.S., Rizzerio, B.G., Fregni, F., 2011. A systematic review on reporting and assessment of adverse effects. Int. J. Neuropsychopharmacol. 14 (8), 1133–1145. https://doi.org/10.1017/S1461145710001690.

Brunoni, A.R., Kemp, A.H., Shiozawa, P., Cordeiro, Q., Valiengo, L.C.L., Goulart, A.C., Coprerski, B., Lotufo, P.A., Brunoni, D., Perez, A.B.A., Fregni, F., Benseñor, I.M., 2013a. Impact of 5-HTTLPR and BDNF polymorphisms on response to sertraline versus. Eur. Neuropsychopharmacol. 23 (11), 1530–1540. https://doi.org/10.1016/j.euroneuro.2013.03.009.

Brunoni, A.R., Valiengo, L., Baccaro, A., Zanão, T.A., de Oliveira, J.F., Goulart, A., Fregni, F., 2013b. The Sertraline Vs Electrical Current Therapy for Treating Depression. JAMA Psychiatry. 23389323.

Brunoni, A.R., Boggio, P.S., De Raedt, R., Benseñor, I.M., Lotufo, P.A., Namur, V., Valiengo, L.C.L., Vanderhasselt, M.A., 2014. Cognitive control therapy and transcranial direct current stimulation for. J. Affect. Disord. 162, 43–49. https://doi.org/10.1016/j.jad.2014.03.026.

Brunoni, A.R., Moffa, A.H., Fregni, F., Palm, U., Padberg, F., Blumberger, D.M., Daskalakis, Z.J., Bennabi, D., Haffen, E., Alonzo, A., Loo, C.K., 2016. Transcranial direct current stimulation for acute major depressive. Br. J. Psychiatry 208 (6), 522–531. https://doi.org/10.1192/bjp.bp.115.164715.

Brunoni, A.R., Moffa, A.H., Sampaio-Junior, B., Borrione, L., Moreno, M.L., Fernandes, R.A., Veronezi, B.P., Nogueira, B.S., Aparicio, L.V.M., Razza, L.B., Chamorro, R., Tort, L.C., Fraguas, R., Lotufo, P.A., Gattaz, W.F., Fregni, F., Bensenor, I.M., Elect-Tdcs Investigators, 2017. Trial of electrical direct-current therapy versus escitalopram for. N. Engl. J. Med. 376 (26), 2523–2533. https://doi.org/10.1056/NEJMoa1612999.

Brunoni, A.R., Sampaio-Junior, B., Moffa, A.H., Aparicio, L.V., Gordon, P., Klein, I., Rios, R.M., Razza, L.B., Loo, C., Padberg, F., Valiengo, L., 2019. Noninvasive brain stimulation in psychiatric disorders: a primer. Braz. J. Psychiatry 41 (1), 70–81. https://doi.org/10.1590/1516-4446-2017-0018.

Bulubas, L., Padberg, F., Bueno, P.V., Duran, F., Busatto, G., Amaro Jr., E., Benseñor, I.M., Lotufo, P.A., Goerigk, S., Gattaz, W., Keeser, D., Brunoni, A.R., 2019. Antidepressant effects of tDCS are associated with prefrontal gray matter. Brain Stimul. 12 (5), 1197–1204. https://doi.org/10.1016/j.brs.2019.05.006.

Charvet, L.E., Kasschau, M., Datta, A., Knotkova, H., Stevens, M.C., Alonzo, A., Loo, C., Krull, K.R., Bikson, M., 2015. Remotely-supervised transcranial direct current stimulation (tDCS) for. Front. Syst. Neurosci. 9, 26. https://doi.org/10.3389/fnsys.2015.00026.

Charvet, L.E., Shaw, M.T., Bikson, M., Woods, A.J., Knotkova, H., 2020. Supervised transcranial direct current stimulation (tDCS) at home: a guide for clinical research and practice. Brain. Stimul. 13 (3), 686–693. https://doi.org/10.1016/j.brs.2020.02.011.

Chase, H.W., Boudewyn, M.A., Carter, C.S., Phillips, M.L., 2020. Transcranial direct current stimulation: a roadmap for research, from. Mol. Psychiatry 25 (2), 397–407. https://doi.org/10.1038/s41380-019-0499-9.

Costain, R., Redfearn, J.W.T., Lippold, O.C.J., 1964. A controlled trial of the therapeutic effects of polarization of the brain. Br. J. Psychiatry 110 (469), 786–799. https://pdfs.semanticscholar.org/35ea/fefeb54ac3f2e497b374dd522e87bdc911e9.pdf.

Creutzfeldt, O.D., Fromm, G.H., Kapp, H., 1962. Influence of transcortical d-c currents on cortical neuronal activity. Exp. Neurol. 5, 436–452. https://doi.org/10.1016/0014-4886(62)90056-0.

Dedoncker, J., Brunoni, A.R., Baeken, C., Vanderhasselt, M.-A., 2016. A systematic review and meta-analysis of the effects of transcranial. Brain Stimul. 9 (4), 501–517. https://doi.org/10.1016/j.brs.2016.04.006.

Eschweiler, G.W., Wegerer, C., Schlotter, W., Spandl, C., Stevens, A., Bartels, M., Buchkremer, G., 2000. Left prefrontal activation predicts therapeutic effects of repetitive. Psychiatry Res. 99 (3), 161–172. https://doi.org/10.1016/s0925-4927(00)00062-7.

Fidalgo, T.M., Leon Morales-Quezada, J., Muzy, G.S.C., Chiavetta, N.M., Mendonca, M.E., Santana, M.V.B., Goncalves, O.F., Brunoni, A.R., Fregni, F., 2014. Biological markers in noninvasive brain stimulation trials in major. J. ECT 30 (1), 47–61. https://doi.org/10.1097/yct.0b013e31828b34d8.

Fregni, F., Boggio, P.S., Nitsche, M.A., Marcolin, M.A., Rigonatti, S.P., Pascual-Leone, A., 2006. Treatment of major depression with transcranial direct current stimulation. Bipolar Disord. 8 (2), 203–204. https://doi.org/10.1111/j.1399-5618.2006.00291.x.

Fregni, F., El-Hagrassy, M.M, Pacheco-Barrios, K., Carvalho, S., Leite, J., Simis, M., Brunelin, J., Nakamura-Palacios, E.M., Marangolo, P., Venkatasubramanian, G., San-Juan, D., Caumo, W., Bikson, M., Brunoni, A.R., Neuromodulation Center Working Group, 2021. Evidence-based guidelines and secondary meta-analysis for the use of transcranial direct current stimulation in neurological and psychiatric disorders. Int. J. Neuropsychopharmacol. 24 (4), 256–313. https://doi.org/10.1093/ijnp/pyaa051.

Huang, Y., Liu, A.A., Lafon, B., Friedman, D., Dayan, M., Wang, X., Bikson, M., Doyle, W.K., Devinsky, O., Parra, L.C., 2017. Measurements and models of electric fields in the human brain during. elife 6. https://doi.org/10.7554/eLife.18834.

Kronberg, G., Bridi, M., Abel, T., Bikson, M., Parra, L.C., 2017. Direct current stimulation modulates LTP and LTD: activity dependence and. Brain Stimul. 10 (1), 51–58. https://doi.org/10.1016/j.brs.2016.10.001.

Kronberg, G., Rahman, A., Sharma, M., Bikson, M., Parra, L.C., 2020. Direct current stimulation boosts hebbian plasticity in vitro. Brain Stimul. 13 (2), 287–301. https://doi.org/10.1016/j.brs.2019.10.014.

Lefaucheur, J.-P., Antal, A., Ayache, S.S., Benninger, D.H., Brunelin, J., Cogiamanian, F., Cotelli, M., De Ridder, D., Ferrucci, R., Langguth, B., Marangolo, P., Mylius, V., Nitsche, M.A., Padberg, F., Palm, U., Poulet, E., Priori, A., Rossi, S., Schecklmann, M., Paulus, W., 2017. Evidence-based guidelines on the therapeutic use of transcranial direct. Clin. Neurophysiol. 128 (1), 56–92. https://doi.org/10.1016/j.clinph.2016.10.087.

Lefaucheur, J.-P., Wendling, F., 2019. Mechanisms of action of tDCS: a brief and practical overview. Neurophysiol. Clin. 49 (4), 269–275. https://doi.org/10.1016/j.neucli.2019.07.013.

Lippold, O.C., Redfearn, J.W., 1964. Mental changes resulting from the passage of small direct currents through. Br. J. Psychiatry 110, 768–772. https://doi.org/10.1192/bjp.110.469.768.

Lisanby, S.H., 2017. Noninvasive brain stimulation for depression - the devil is in the dosing. N. Engl. J. Med. 376 (26), 2593–2594. https://doi.org/10.1056/NEJMe1702492.

Liston, C., Chen, A.C., Zebley, B.D., Drysdale, A.T., Gordon, R., Leuchter, B., Voss, H.U., Casey, B.J., Etkin, A., Dubin, M.J., 2014. Default mode network mechanisms of transcranial magnetic stimulation in. Biol. Psychiatry 76 (7), 517–526. https://doi.org/10.1016/j.biopsych.2014.01.023.

Loo, C.K., Sachdev, P., Martin, D., Pigot, M., Alonzo, A., Malhi, G.S., Lagopoulos, J., Mitchell, P., 2010. A double-blind, sham-controlled trial of transcranial direct current. Int. J. Neuropsychopharmacol. 13 (1), 61–69. https://doi.org/10.1017/S1461145709990411.

López-Alonso, V., Cheeran, B., Río-Rodríguez, D., Fernández-del-Olmo, M., 2014. Inter-individual variability in response to non-invasive brain stimulation. Brain Stimul. 7 (3), 372–380. https://doi.org/10.1016/j.brs.2014.02.004.

Marquez-Ruiz, J., Leal-Campanario, R., Sanchez-Campusano, R., Molaee-Ardekani, B., Wendling, F., Miranda, P.C., Ruffini, G., Gruart, A., Delgado-Garcia, J.M., 2012. Transcranial direct-current stimulation modulates synaptic mechanisms. Proc. Natl. Acad. Sci. U. S. A. 109 (17), 6710–6715. https://doi.org/10.1073/pnas.1121147109.

Martin, D.M., Liu, R., Alonzo, A., Green, M., Loo, C.K., 2014. Use of transcranial direct current stimulation (tDCS) to enhance cognitive. Exp. Brain Res. 232 (10), 3345–3351. https://doi.org/10.1007/s00221-014-4022-x.

Mayberg, H.S., Brannan, S.K., Tekell, J.L., Silva, J.A., Mahurin, R.K., McGinnis, S., Jerabek, P.A., 2000. Regional metabolic effects of fluoxetine in major depression: serial. Biol. Psychiatry 48 (8), 830–843. https://doi.org/10.1016/s0006-3223(00)01036-2.

Milev, R.V., Giacobbe, P., Kennedy, S.H., Blumberger, D.M., Daskalakis, Z.J., Downar, J., Modirrousta, M., Patry, S., Vila-Rodriguez, F., Lam, R.W., MacQueen, G.M., Parikh, S.V., Ravindran, A.V., Canmat Depression Work Group, 2016. Canadian network for mood and anxiety treatments (CANMAT) 2016 clinical. Can. J. Psychiatr. 61 (9), 561–575. https://doi.org/10.1177/0706743716660033.

Moffa, A.H., Brunoni, A.R., Nikolin, S., Loo, C.K., 2018. Transcranial direct current stimulation in psychiatric disorders: A. Psychiatr. Clin. North Am. 41 (3), 447–463. https://doi.org/10.1016/j.psc.2018.05.002.

Moffa, A.H., Martin, D., Alonzo, A., Bennabi, D., Blumberger, D.M., Benseñor, I.M., Daskalakis, Z., Fregni, F., Haffen, E., Lisanby, S.H., Padberg, F., Palm, U., Razza, L.B., Sampaio-Jr, B., Loo, C., Brunoni, A.R., 2020. Efficacy and acceptability of transcranial direct current stimulation (tDCS) for major depressive disorder: an individual patient data meta-analysis. Prog. Neuro-Psychopharmacol. Biol. Psychiatry 99, 109836. https://doi.org/10.1016/j.pnpbp.2019.109836. 109836.

Monnart, A., Vanderhasselt, M.-A., Schroder, E., Campanella, S., Fontaine, P., Kornreich, C., 2019. Treatment of resistant depression: a pilot study assessing the efficacy of. Front. Psychiatry 10, 730. https://doi.org/10.3389/fpsyt.2019.00730.

Moreno, M.L., Goerigk, S.A., Bertola, L., Suemoto, C.K., Razza, L.B., Moffa, A.H., Veronezi, B.P., Tort, L., Nogueira, B.S., Gattaz, W.F., Fraguas, R., Padberg, F., Lotufo, P.A., Benseñor, I.M., Brunoni, A.R., 2020. Cognitive changes after tDCS and escitalopram treatment in major. J. Affect. Disord. 263, 344–352. https://doi.org/10.1016/j.jad.2019.12.009.

Mutz, J., Edgcumbe, D.R., Brunoni, A.R., Fu, C.H.Y., 2018. Efficacy and acceptability of non-invasive brain stimulation for the. Neurosci. Biobehav. Rev. 92, 291–303. https://doi.org/10.1016/j.neubiorev.2018.05.015.

Mutz, J., Vipulananthan, V., Carter, B., Hurlemann, R., Fu, C.H.Y., Young, A.H., 2019. Comparative efficacy and acceptability of non-surgical brain stimulation. BMJ 364, l1079. https://doi.org/10.1136/bmj.l1079.

Nitsche, M.A., Paulus, W., 2000. Excitability changes induced in the human motor cortex by weak. J. Physiol. 527 (Pt. 3), 633. https://www.ncbi.nlm.nih.gov/pmc/articles/pmc2270099.

Nitsche, M.A., Fricke, K., Henschke, U., Schlitterlau, A., Liebetanz, D., Lang, N., Henning, S., Tergau, F., Paulus, W., 2003a. Pharmacological modulation of cortical excitability shifts induced by. J. Physiol. 553 (Pt. 1), 293–301. https://doi.org/10.1113/jphysiol.2003.049916.

Nitsche, M.A., Liebetanz, D., Lang, N., Antal, A., Tergau, F., Paulus, W., 2003b. Safety criteria for transcranial direct current stimulation (tDCS) in. Clin. Neurophysiol. 114 (11), 2220–2222. author reply 2222-3 https://doi.org/10.1016/s1388-2457(03)00235-9.

Nitsche, M.A., Knotkova, H., Woods, A.J., Bikson, M., Knotkova, H., Nitsche, M.A., Bikson, M., Woods, A.J., 2019. Challenges, open questions and future direction in transcranial direct. In: Practical Guide to Transcranial Direct Current Stimulation: Principles. Springer International Publishing, pp. 627–639, https://doi.org/10.1007/978-3-319-95948-1_21.

Opitz, A., Paulus, W., Will, S., Antunes, A., Thielscher, A., 2015. Determinants of the electric field during transcranial direct current. NeuroImage 109, 140–150. https://doi.org/10.1016/j.neuroimage.2015.01.033.

Palm, U., Schiller, C., Fintescu, Z., Obermeier, M., Keeser, D., Reisinger, E., Pogarell, O., Nitsche, M.A., Möller, H.-J., Padberg, F., 2012. Transcranial direct current stimulation in treatment resistant depression. Brain Stimul. 5 (3), 242–251. https://doi.org/10.1016/j.brs.2011.08.005.

Podda, M.V., Cocco, S., Mastrodonato, A., Fusco, S., Leone, L., Barbati, S.A., Colussi, C., Ripoli, C., Grassi, C., 2016. Anodal transcranial direct current stimulation boosts synaptic plasticity. Sci. Rep. 6, 22180. https://doi.org/10.1038/srep22180.

Poreisz, C., Boros, K., Antal, A., Paulus, W., 2007. Safety aspects of transcranial direct current stimulation concerning. Brain Res. Bull. 72 (4–6), 208–214. https://doi.org/10.1016/j.brainresbull.2007.01.004.

Purpura, D.P., Mcmurtry, J.G., 1965. Intracellular activities and evoked potential changes during polarization. J. Neurophysiol. 28, 166–185. https://doi.org/10.1152/jn.1965.28.1.166.

Radman, T., Ramos, R.L., Brumberg, J.C., Bikson, M., 2009. Role of cortical cell type and morphology in subthreshold and. Brain Stimul. 2 (4), 215–228. 228.e1-3 https://doi.org/10.1016/j.brs.2009.03.007.

Rahman, A., Lafon, B., Parra, L.C., Bikson, M., 2017. Direct current stimulation boosts synaptic gain and cooperativity in vitro. J. Physiol. 595 (11), 3535–3547. https://doi.org/10.1113/JP273005.

Razza, L.B., Palumbo, P., Moffa, A.H., Carvalho, A.F., Solmi, M., Loo, C.K., Brunoni, A.R., 2020. A systematic review and meta-analysis on the effects of transcranial. Depress. Anxiety. https://doi.org/10.1002/da.23004.

Reato, D., Rahman, A., Bikson, M., Parra, L.C., 2010. Low-intensity electrical stimulation affects network dynamics by. J. Neurosci. 30 (45), 15067–15079. https://doi.org/10.1523/JNEUROSCI.2059-10.2010.

Salvador, R., Truong, D.Q., Bikson, M., Opitz, A., Dmochowski, J., Miranda, P.C., Knotkova, H., Nitsche, M.A., Bikson, M., Woods, A.J., 2019. Role of computational modeling for dose determination. In: Practical Guide to Transcranial Direct Current Stimulation: Principles. Springer International Publishing, pp. 233–262, https://doi.org/10.1007/978-3-319-95948-1_9.

Sathappan, A.V., Luber, B.M., Lisanby, S.H., 2019. The dynamic duo: combining noninvasive brain stimulation with cognitive. Prog. Neuro-Psychopharmacol. Biol. Psychiatry 89, 347–360. https://doi.org/10.1016/j.pnpbp.2018.10.006.

Segrave, R.A., Arnold, S., Hoy, K., Fitzgerald, P.B., 2014. Concurrent cognitive control training augments the antidepressant efficacy. Brain Stimul. 7 (2), 325–331. https://doi.org/10.1016/j.brs.2013.12.008.

Stagg, C.J., Antal, A., Nitsche, M.A., 2018. Physiology of transcranial direct current stimulation. J. ECT 34 (3), 144–152. https://doi.org/10.1097/YCT.0000000000000510.

Suen, P.J.C., Doll, S., Batistuzzo, M.C., Busatto, G., Razza, L.B., Padberg, F., Mezger, E., Bulubas, L., Keeser, D., Deng, Z.-D., Brunoni, A.R., 2020. Association between tDCS computational modeling and clinical outcomes in. Eur. Arch. Psychiatry Clin. Neurosci. https://doi.org/10.1007/s00406-020-01127-w.

Truong, D.Q., Bikson, M., 2018. Physics of transcranial direct current stimulation devices and their. J. ECT 34 (3), 137–143. https://doi.org/10.1097/YCT.0000000000000531.

Williams, L.M., 2016. Precision psychiatry: a neural circuit taxonomy for depression and anxiety. Lancet Psychiatry 3 (5), 472–480. https://doi.org/10.1016/S2215-0366(15)00579-9.

Further reading

Razza, L.B., Afonso Dos Santos, L., Borrione, L., Bellini, H., Branco, L.C., Cretaz, E., Duarte, D., Ferrão, Y., Galhardoni, R., Quevedo, J., Simis, M., Fregni, F., Correll, C.U., Padberg, F., Trevizol, A., Daskalakis, Z.J., Carvalho, A.F., Solmi, M., Brunoni, A.R., 2021. Appraising the effectiveness of electrical and magnetic brain stimulation techniques in acute major depressive episodes: an umbrella review of meta-analyses of randomized controlled trials. Braz. J. Psychiatry 43 (5), 514–524. https://doi.org/10.1590/1516-4446-2020-1169.

Chapter 19

ECT for treatment-resistant depression

Karina Drake[a] and Daniel F. Maixner[b]

[a]Department of Psychiatry, University of Colorado School of Medicine, University of Colorado, Boulder, CO, United States, [b]Department of Psychiatry, University of Michigan Medical School, University of Michigan, Ann Arbor, MI, United States

Introduction

Electroconvulsive therapy (ECT) is the oldest biologic treatment in psychiatry still in use and continues to play a crucial role in the management of treatment-resistant depression (TRD). ECT has been shown to be significantly more effective than pharmacotherapy in the management of TRD, with many studies citing remission rates of 50%–60% compared to 10%–40% with further medication adjustments (Dierckx et al., 2012; Husain et al., 2004; Rush et al., 2006). Rates of remission are even higher, nearing 80%–90%, in depression with certain features such as catatonia or psychosis (Petrides et al., 2001).

Beyond immediate symptom reduction, ECT has also been shown to reduce the rate of psychiatric hospitalizations, long-term risk of suicide, and all-cause mortality in the TRD population (Fink et al., 2014; Olfson et al., 1998; Slade et al., 2017). In addition, a study using a cost-effectiveness model projected that ECT decreases the time an individual experiences uncontrolled depression from 50% of life years to 33%–37% over a 4-year period. This study also found ECT to be a cost-effective option after two failed antidepressant trials, making it a reasonable strategy earlier in treatment resistance due to its excellent health-economic advantages (Ross et al., 2018).

Despite the known benefits of ECT, the rate of utilization in the TRD population remains low. In a 2017 analysis, only 1.5% of psychiatric inpatients with a severe affective disorder received ECT while hospitalized (Slade et al., 2017). In a review of a large insurance claims database of individuals with depression, fewer than 1% ever received ECT (Gibson et al., 2010). The reasons for the under use of ECT are likely multifactorial including stigma, concerns over adverse side effects, and lack of access. Despite often negative public opinion of ECT, studies have consistently demonstrated its safety and efficacy, even in special populations including children, pregnant women, and geriatric patients. Furthermore, the practice of ECT has evolved since its inception with several advances reducing side effect burden while maintaining efficacy.

ECT basics

ECT involves passing a carefully controlled electrical stimulus through the scalp to induce a generalized seizure. The stimulus is delivered by two electrodes creating an electrical current which causes rapid discharge of nerve impulses and subsequent seizure activity. Two common electrode placements are used. Right unilateral (RUL) involves one electrode placed at the vertex of the head and the other electrode on the right temple (Fig. 19.1). Both electrodes are set on the temples for the bitemporal (BT) placement (Fig. 19.2). RUL ECT is generally associated with less cognitive side effects, while BT ECT appears to have a faster speed of response (Kellner et al., 2010). The seizure is self-limited and usually lasts between 20 and 60 s. ECT is done under general anesthesia including the use of muscle relaxants such as succinylcholine that prevents any physical convulsion from occurring. Seizure duration and morphology is monitored by electroencephalography (EEG). An acute ECT course is typically 9–15 treatments occurring 2–3 times per week with significant variability depending on symptom response and underlying diagnoses. Response is typically not seen until treatments 6–8 but can also vary widely. Once a patient reaches remission or a response plateau, he or she may be transitioned to a taper or maintenance ECT if relapse risk is considered high for an individual.

History and development

In the early 1900s, there was increasing interest in the use of "physical therapies" for the treatment of neuropsychiatric illness after malarial fever was shown to successfully treat neurosyphilis (Shorter, 2004). This interest spread to psychiatric

Managing Treatment-Resistant Depression. https://doi.org/10.1016/B978-0-12-824067-0.00019-0

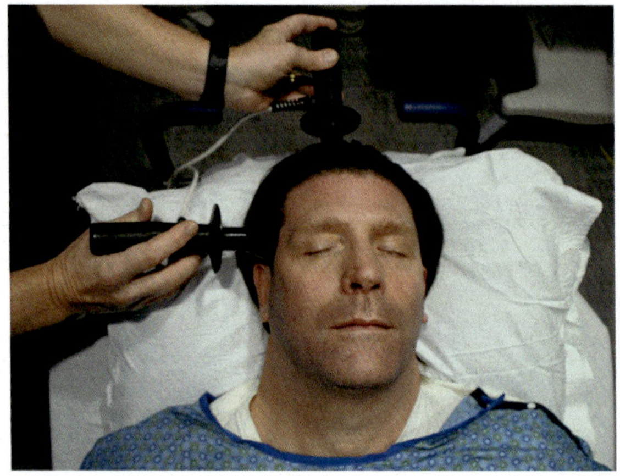

FIG. 19.1 Right unilateral electrode placement.

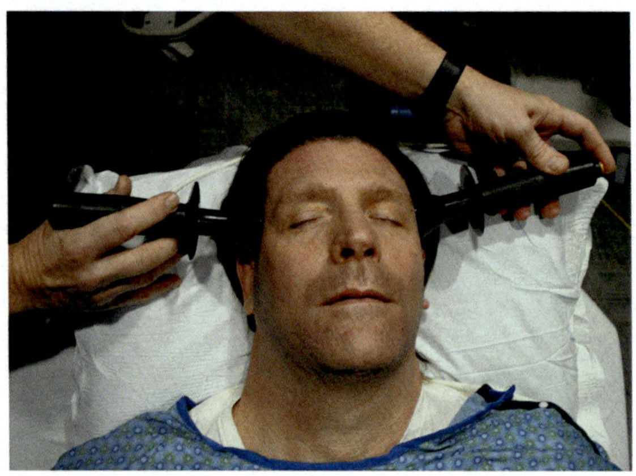

FIG. 19.2 Bitemporal electrode placement.

disease where physicians explored strategies to directly treat the brain rather than relying on psychotherapy only. This prompted the development of convulsive therapies by Dr. Ladislas von Meduna in 1934 in Budapest, Hungary. Dr. Meduna hypothesized that there was an "antagonism" between epilepsy and schizophrenia and therefore started to use convulsive therapy to treat psychosis (Shorter, 2004). He initially used camphor and then later pentylenetetrazole to chemically induce convulsions with noticeable reduction in psychotic symptoms (Gazdag and Ungvari, 2019). As a result of his success, convulsive treatment units sprang up all over Europe and the United States (Shorter, 2004).

Chemically induced convulsive therapy was quickly taken over by ECT after Dr. Ugo Cerletti showed greater success and consistency inducing convulsions by applying electricity directly to the brain (Suleman, 2020). Since ECT showed good efficacy in reducing psychotic symptoms in schizophrenia patients, its use quickly spread to the treatment of other affective disorders including depression (Shorter, 2004). Due to positive results, ECT became the standard treatment for individuals hospitalized with depression by the 1950s (Shorter, 2004).

As ECT became more widely used, a number of important advances were made. "Modified ECT" was developed where general anesthetics and then muscle relaxants were used (initially curare in 1940 and then succinylcholine in 1952) to sedate patients and minimize complications of physical seizures (Gazdag and Ungvari, 2019). Unilateral ECT use began in 1949 to reduce cognitive side effects since it avoids directly stimulating the dominant hemisphere in the majority of individuals therefore reducing the impact on verbal memory (Gazdag and Ungvari, 2019). Brief pulse ECT use began shortly thereafter (Gazdag and Ungvari, 2019). Brief pulse ECT is a pulsatile wave-form and is more efficient in stimulating neurons and inducing seizures than sinusoidal stimulation of older machines which is associated with more prominent memory side effects. Brief pulse devices are the standard of care in modern ECT practice.

Despite its efficacy and widespread use, ECT almost vanished from psychiatric practice in the 1960s. The rapid shift in reduced ECT use arose from a change in public opinion about the treatment due to the rise of the antipsychiatry movement (Shorter, 2004). Much of the public's outrage regarding ECT came from the impact of Ken Kesey's book "One Flew Over the Cuckoo's Nest" which was published in 1962 and later turned into a widely popular film in 1975 (Shorter, 2004).

ECT use remained lower until the 1980s when a gradual resurgence began. One of the biggest turning points was the summary of the NIH Consensus Conference on Electroconvulsive Therapy published in JAMA in October of 1985 that noted, "Not a single controlled study has shown another form of treatment to be superior to ECT in the short-term management of severe depression." Since then, ECT has been accepted in psychiatry as the gold standard for treatment of refractory mood disorders and its use throughout the United States has been slowly increasing (Shorter, 2004).

Mechanisms of action

The mechanisms of action of ECT are not fully understood and there is no unifying theory on how ECT exerts its powerful antidepressant effects. However, over the years a number of theories have been investigated. These have included ECT's effects on monoamines, seizure threshold/anticonvulsant action, the neuroendocrine system, and molecular pathways associated with neurogenesis such as increased brain-derived neurotrophic factor (BDNF). Other recent functional and structural neuroimaging studies have also produced interesting findings.

The monoamine neurotransmitter system has long been a focus of neurobiology studies in ECT, and many studies have sought to discover changes in norepinephrine, serotonin, and dopamine metabolites in bodily fluids including cerebrospinal fluid, urine, or blood as a result of treatment (Mann, 1998). As an example of this work, Nikisch and Mathé (2008) showed that the CSF monoamine metabolites of 5-hydroxyindoleacetic acid (5-HIAA) and homovanillic acid were both significantly elevated following ECT.

During a course of ECT, it is well known that seizure thresholds often rise and seizure durations shorten. This led to the anticonvulsant theory of ECT and the search for biologic correlates such as changes in gamma-aminobutyric acid (GABA) in animal models and ictal and postictal EEG findings which appear to indicate brain inhibitory processes are in play (Sackeim, 1999). Despite the prevalent anticonvulsant effects of ECT, there is no consistent data to suggest a clear association between clinical outcomes in depression and increases in seizure threshold (Duthie et al., 2015; Seymour, 2020).

ECT induces an array of acute neuroendocrine effects including release of adrenocorticotropic hormone, prolactin, and cortisol. Hypothalamic-pituitary-adrenal (HPA) dysregulation became a primary focus as a possible marker for melancholic or psychotic depression. Past studies show that dexamethasone suppression test (DST) abnormalities are corrected with ECT, but are not strongly connected to depression improvement (McKay and Zakzanis, 2010; Haskett, 2014).

Modern concepts in ECT mechanisms have also focused on neuroplasticity and neurogenesis. Brain-derived neurotrophic factor (BDNF), a widely distributed nervous system neurotrophic protein, has roles in neurotransmitter modulation, neural genesis, and neuronal plasticity. Animal and human ECT studies have been conducted for over 20 years investigating BDNF and its effects. In animal models, it has been shown that neurogenesis, synaptogenesis, and glial proliferation in the hippocampus occur with electroconvulsive seizures (ECS) (Bouckaert et al., 2014). In humans, a metaanalysis of nine ECT studies in depression found that BDNF serum levels were increased after ECT. These findings suggest that BDNF may be a potential biomarker of response in patients receiving ECT for depression.

Since it is not possible to readily examine the microscopic, molecular, or neuroplastic changes ECT has on the human brain, tools such as high-resolution magnetic resonance imaging (MRI) or functional MRI have helped bridge the knowledge gap by exploring possible *structural* and *connectivity* changes. In recent years, *structural* studies have focused on areas of the medial temporal lobe such as the hippocampus and amygdala which generally show volume increases with ECT. However, pooled data from the Global ECT-Magnetic Resonance Imaging Research Collaboration (GEMRIC) have not correlated these changes to positive ECT outcomes (Oltedal et al., 2018). However, Mulders et al. (2020) found that ECT response may be associated to structural changes in other areas in cortical, striatal, and lateral prefrontal areas also implicated in depression. Functional *connectivity* techniques and analyses are complex. In depression, there is a "hyperconnectivity" concept where mood networks exhibit increased resting state connectivity. This hyperconnectivity may be a biomarker for depression and may also be reduced with treatments. An early small study significantly showed such a reduction in frontal connectivity after ECT (Perrin et al., 2012). Other recent studies have examined hippocampal *connectivity* in ECT showing varied changes associated with ECT response (Leaver et al., 2020; Takamiya et al., 2020). Despite the complexities of ECT's neurophysiologic effects, advancements continue to be made on many fronts to foster our understanding of how ECT may work, but from past and recent research, ECT appears to likely have a multitude of potential mechanisms.

Clinical studies

General Efficacy Profile: ECT is not only one of the oldest biologic treatment in psychiatry, but it is also the most effective treatment for severe mood disorders, treatment refractory schizophrenia, catatonia, and neuroleptic malignant syndrome (NMS). Due to the speed of response with ECT which is typically 2–4 weeks compared to 4–6 weeks at the earliest with medications, ECT is the treatment of choice when rapid improvement is needed due to life threatening psychiatric symptoms (Weiner and Reti, 2017). ECT has also been deemed a safe treatment as reflected by the recent Federal Drug Administration's (FDA) reclassification ECT devices from Class III to the lower risk category Class II (Federal Drug Administration, 2018).

Details of ECT's effectiveness in treating unipolar and bipolar depression will be discussed at length below, but ECT has shown similar, if not higher, response rates, in treating mania, catatonia, and psychosis. ECT is an important treatment strategy listed in the national and international guidelines for the treatment of mania, especially in life threatening or treatment refractory cases (American Psychiatric Association, 2002; Grunze et al., 2009). ECT has similarly demonstrated effectiveness at rapidly reducing catatonic symptoms regardless of underlying etiology and is the treatment of choice in patients whose catatonia does not fully respond to benzodiazepines (Luchini et al., 2015; Medda et al., 2015; Raveendranathan et al., 2012). While ECT is less commonly used in the United States for schizophrenia and schizoaffective disorders, it has been shown to be effective at reducing psychotic symptoms that do not respond to antipsychotics (Tharyan and Adams, 2005). Recent studies suggest synergistic, antipsychotic effects when ECT is combined with Clozaril, producing 50% response rates in patients who did not respond to clozapine alone (Petrides et al., 2015).

General Efficacy in Treatment-Resistant Depression: Clinical studies have consistently shown the efficacy of ECT in TRD. While few true recent randomized control trials exist due to excellent efficacy of ECT in older trials and the inherent ethical issues of administering general anesthesia with sham ECT, data from several large observational studies for the treatment of depression highlight extremely positive results. One such large study led by the Columbia University Consortium was published in 2001. The goal of the study was to examine relapse rates among ECT remitters maintained on different psychotropic medication regimens versus placebo following an acute treatment course. Two hundred and ninety participants with severe depression completed an acute bilateral or right unilateral ECT course. At the end of the treatment course, 55% of subjects had reached remission defined as a 60% reduction in Hamilton Depression (HAM-D) scale scores as well as a total HAM-D of less than or equal to 10 (Sackeim et al., 2001).

Another large ECT study conducted by the Consortium for Research on ECT (CORE) assessed acute depression outcomes in a first phase followed by a second phase of continued medications vs continued ECT. During the first phase, the study included 253 participants with an HAMD score of ≥21. The participants received bitemporal ECT three times weekly with repeated HAMD ratings. By the end of the study, remission (defined by achieving HAMD scores less than or equal to 10 on 2 consecutive assessments) was achieved in 75% of the subjects. Of note, remission was achieved relatively early in the treatment course by 34% of subjects before their 6th treatment, and 65% reached remission at or before their 10th ECT (Husain et al., 2004).

In recent decades, ECT studies have centered around determining the efficacy and side effect profiles of various electrode placements. Many studies comparing right unilateral (RUL) vs bitemporal (BT) generally show that RUL is associated with less anterograde and retrograde amnesia compared to BT placement (Squire and Slater, 1983; Rosenberg and Pettinati, 1984; Steif et al., 1986; Sobin et al., 1995; Sackeim et al., 2000; Lisanby et al., 2000). However, in order to maintain similar efficacy to BT ECT, the electrical dose delivered for RUL ECT needs to be significantly above an individual's seizure threshold. BT ECT is very effective at doses approximately two times above seizure threshold while RUL ECT is most effective at five to six times above seizure threshold. This was elucidated by Sackeim et al. (2000) comparing the efficacy of RUL ECT at three dosage levels above threshold (1.5 times threshold-*low-dosage RUL*, 2.5 times threshold-*moderate-dosage RUL*, and 6 times threshold-*high dosage RUL*) versus BT ECT at 1.5 times threshold. Ultimately, *high dose RUL* and BT ECT were found to have similar remission rates (65%), but the remission rates for low and moderate dosage RUL ECT were only 30% and 35%, respectively. Of note, the study did find that BT ECT resulted in greater acute cognitive side effects than any dosage of unilateral ECT and that retrograde memory difficulties persisted even 2 months after ECT with BT ECT.

One of the other important ECT advancements has been modifications to stimulus parameters, specifically pulse width. The electrical stimulus delivered during an ECT treatment is composed of brief rectangular square-wave pulses which historically ranged from 0.5 to 2 ms in width. As the optimal pulse width to induce neuronal depolarization is estimated to be 0.1–0.2 ms, the use of narrower pulse widths was proposed as a more efficient way of inducing a seizure. Since less electricity is delivered to the neuron when depolarized, it was hypothesized that a smaller pulse width would cause less neuronal irritation and therefore fewer cognitive side effects (Sackeim et al., 2008). There has now been a shift from brief pulse

(0.5–2.0 ms) to ultrabrief pulse ECT (<0.5 ms) especially with RUL ECT. While studies comparing the effectiveness of the two vary, ultrabrief pulse ECT has been shown to have fewer cognitive side effects in several domains (Tor et al., 2015).

Direct comparison to other treatments

ECT versus antidepressants

ECT has been compared to pharmacotherapy in TRD patients with ECT consistently showing superior efficacy. In a naturalistic study of 1495 patients admitted over a 12-year period to an inpatient unit, Black et al. found that a significantly greater number of patients receiving ECT were considered to have "marked improvement" (70%) by provider report at the time of discharge compared to those that received an adequate antidepressant trial (48%) (Black et al., 1987). Several meta-analyses have shown similar results. A metaanalysis by Kho et al. (2003) included 16 randomized control trials comparing ECT to other forms of treatment for depression including various antidepressants (specifically SSRIs, SNRIs, TCAs). When response rates from with antidepressants were compared to ECT, ECT showed a nearly one standard deviation superiority in efficacy to other treatments (Kho et al., 2003). Similar results were found in a 2003 metaanalysis conducted by the UK ECT Group in 2003 which found that ECT was significantly more effective than pharmacotherapy (18 trials, 1144 participants, SES −0·80, 95% CI, −1·29 to −0·29) (UK ECT Review Group, 2003).

ECT versus transcranial magnetic stimulation (TMS)

TMS is a focal, noninvasive brain stimulation procedure. TMS uses magnetic fields to create pulses that enter the cortex and either excite or inhibit specific regions of the brain. In the 1990s, it was found that repeat stimulation of the left prefrontal cortex produces antidepressant effects (Perera et al., 2016). The FDA subsequently approved TMS in 2008 for the treatment of adult patients with major depressive disorder who had failed previous antidepressant medication at or above minimal effective dose. Several large, randomized control trials on the effectiveness of TMS exist with response rates ranging from 40% to 60% and remission rates ranging from 30% to 40% (Connolly et al., 2012; George, 2010; McDonald et al., 2011; Mantovani et al., 2012).

Based on previously observed response and remission rates alone, ECT demonstrates greater effectiveness in treatment-resistant depression. Studies have also been done directly comparing response rates of ECT versus TMS with similar results (Xie et al., 2013; Magnezi et al., 2016). Xie et al. (2013) published a metaanalysis in 2013 which included nine high quality randomized control trials that compared TMS and ECT in 395 patients with treatment-resistant depression. While ECT (*df*: 1; F: 32.72; $P<.0001$) and TMS (*df*: 1; F: 6.85; $P<.012$) were found to be effective in reducing depressive symptoms, ECT was superior to TMS (*df*: 1; F: 6.81; $P<.012$). Pretreatment mean HAM-D scores for the ECT and TMS groups were 27 and 26 respectively. With ECT, mean posttreatment HAM-D scores fell to 11 while the mean score with TMS fell only to 18. It is important to note though, that while ECT has shown superior efficacy, studies do suggest that TMS is better tolerated with fewer side effects (Xie et al., 2013; Magnezi et al., 2016; Cusin and Dougherty, 2012).

ECT versus ketamine

Ketamine is primarily a glutamatergic *n*-methyl-D-aspartate receptor antagonist that has an increasing role in the management of TRD after it was found to have rapid antidepressant effects when administered intravenously at subanesthetic doses (Phillips et al., 2020). Several randomized, placebo controlled trials have since been done showing the antidepressant effects in unipolar and bipolar depressed patients (Newport et al., 2015; Kishimoto et al., 2016; Phillips et al., 2020). However, results have consistently shown that while ketamine produces an antidepressant effect quickly in hours after a single infusion, the effect is time limited and dissipates after several days (Newport et al., 2015; Kishimoto et al., 2016; Phillips et al., 2020).

A large, multisite randomized control trial comparing the effectiveness of ketamine versus ECT called the "electroconvulsive therapy (ECT) vs. Ketamine in patients with Treatment-resistant Depression: The ELEKT-D study protocol" is currently underway but some preliminary data has been released. The primary outcomes of the study are HAM-D scores as well as suicidality. Twenty-six adult subjects with depression received 0.5 mg/kg ketamine infusion once, 36 subjects received 4 ketamine infusions over 2 weeks, and 47 subjects received ECT 3 times per week over 3 weeks. Suicidality and HAM-D scores in all groups decreased but were lower in both ketamine groups. However, 5 weeks after the last ketamine infusion, suicidality and HAM-D scores in the ketamine groups began to increase while scores remained stable in the ECT group even 3 months after the last treatment. This suggests that ketamine may have a more immediate antidepressant effects compared to ECT but the durability of ECT response may be more sustained. It is important to note though that subjects in the ECT arm only received nine treatments which is less than a typical index course, potentially limiting effectiveness. Further results are therefore needed to more definitively understand the different effectiveness of ECT versus ketamine (Mathew et al., 2019).

Bipolar depression

ECT is effective in treating bipolar disorder across the spectrum of symptoms including mania, mixed state episodes, depression, and bipolar illness with catatonia features. The study of bipolar depression is of particular interest since differences in response rates to antidepressant medications for bipolar depression and unipolar depression exists. Studies have historically shown limited benefit of antidepressants medications for bipolar depression, thus potentially inducing a more treatment-resistant form of depression. However, a 2012 metaanalysis of 6 prospective and retrospective studies of ECT in unipolar and bipolar depression with a strict definition of response (a final HAMD score of < 10 which is consistent with mission level) response noted similar rates for unipolar (51.5%) and bipolar depression (53.2%) (Dierckx et al., 2012). In a large naturalistic/retrospective study of over 500 bipolar ECT patients treated between 2006 and 2011, Perugi et al. (2017) found response rates of 75% for mania, 73% for mixed state, and 81% for catatonia features. In bipolar depression specifically, the response rate was 68%. Another retrospective study of a geriatric group of 34 bipolar depressed subjects, Morcos et al. (2021) found that the majority of subjects experienced remission or response (85%) on the Clinical Global Impression (CGI) scale with good cognitive tolerability of both right unilateral and bilateral ECT. Finally, a recent metaanalysis by Bahji et al. (2019) identified 19 unipolar and bipolar depression ECT studies that included 1803 unipolar and 553 bipolar depression subjects. Important findings of this study include a slightly higher statistically significant rate of response for bipolar depression (77%) vs unipolar depression (74%), similar remission rates of 52% for each group, and statistically significantly fewer treatments were required for bipolar depression subjects to achieve response. Overall, the data in treating bipolar depression with ECT suggests that outcomes are at least on par with ECT outcomes in unipolar depression with substantial response and remission rates and the possibility that ECT may induce a somewhat speedier recovery for bipolar depression patients.

Depression with psychotic features

Depressed individuals with cooccurring psychotic or delusional symptoms have been shown to respond more rapidly to ECT with higher remission rates compared to their nonpsychotically depressed counterparts. This has been demonstrated in numerous studies including a subsequent analysis of data from the CORE study. Seventy-seven of the 253 study participants met criteria for MDD with psychotic features and remission rate was 96% in this cohort with faster response times (Petrides et al., 2001). Similar outcomes were documented in a 2019 cohort study conducted in a psychiatric hospital in Belgium seeking ECT response predictors. Seventy-three subjects with depression were enrolled. Thirty-three had psychotic depression as diagnosed by the Psychotic Depression Assessment Score (PDAS). Clinical outcomes were tracked while the subjects received ultra-brief pulse right unilateral, bifrontal, or bitemporal ECT two times per week. At the conclusion of the study, 81% of the cohort had a significant response with 63.9% in remission defined by Montgomery Asberg Depression Rating Scale (MADRS) ≤ 10. The presence of psychotic symptoms as indicated by a higher PDAS score consistently predicted remission (van Diermen et al., 2019).

Overall, the degree and speed of improvement in psychotic depression is prominently higher and faster with ECT compared to psychotropic medications. Studies cite responses rates (defined as a 50% reduction in symptoms on HAM-D) with antidepressant monotherapy around 20%–40% after 5–6 weeks of treatment. Combination antidepressant and antipsychotic therapy is higher, with rates ranging from 60% to 80%, also after several weeks of treatment. In comparison, ECT results in 60%–80% of individuals achieving full remission with early response seen as early as 3–5 treatments (Petrides et al., 2001; van Diermen et al., 2019). Since individuals with psychotic depression tend to be severely ill with significant mood disturbance, suicidal thinking, neurovegetative symptoms, weight loss, and failure to thrive, the rapid response provided by ECT can be of significant benefit over medications especially when symptoms are life threatening.

Suicidality

Suicide rates in patients with mood disorders are estimated at 4%–20%, and 50%–70% of persons who complete suicide do so when depressed (Conwell et al., 2002; Bostwick and Pankratz, 2000). ECT has been shown to rapidly reduce those rates in several studies and should therefore be considered as an effective treatment for acute suicidality in depressed patients.

Brådvik and Berglund (2006) studied factors related to suicide attempts in 1206 severely depressed patients. They found that ECT significantly reduced the frequency of suicide attempts compared to subjects treated with psychotropic medications. Twenty percent (23/113) of subjects who received antidepressant medications attempted suicide versus 8% (11/133) of those receiving ECT ($\chi^2=7.49$, $P<0.01$). The incidence dropped even further to 2% (1/62) in subjects who received ECT as well as pharmacotherapy ($\chi^2=11.04$, $P<0.001$) (Brådvik and Berglund, 2006).

In a multicenter National Institute of Mental Health funded study, Kellner et al. (2005) looked at the effect three times per week bitemporal ECT had on suicidal thoughts and acts in unipolar depressed subjects. Suicidal intent was assessed at baseline and before each ECT on item 3 of the HAMD. Of the 444 subjects in the study, 29.5% reported suicidal thoughts and acts at baseline. After 1 week of ECT, 38.2% of these subjects reported no suicidal intent which further increased to 61.1% of subjects after six ECT. By the end of the ECT series, 80.9% had no suicidal intent (Kellner et al., 2005).

However, despite data suggesting ECT reduces suicidal ideation and suicide attempts, death by suicide can occur in the time frame around an ECT course. Østergaard et al. (2014) described data from the Danish Health and Medicines Authority on all-cause mortality 30 days after ECT. Between 2000 and 2007, The Danish National Patient Registry tallied a total of 99,728 ECT with 78 deaths occurring in the studied period. While this report was primarily focused on determining the procedural mortality of ECT, 20 of the 78 deaths were by suicide second to cardiopulmonary disease. Since the population receiving ECT is typically the most severely depressed and may be at higher risk for suicide, assessments for suicidal ideation should be done before, during, and after ECT (Østergaard et al., 2014).

Special populations

Children and adolescents

ECT remains an infrequently used treatment for children and adolescents. Little is known about the incidence of ECT use in children and adolescents within the United States, but studies in other countries suggest <1% of the patients receiving ECT are under the age of 18 (Wang et al., 2018). Given its limited use, data on the effectiveness of ECT for treatment refractory depression in children and adolescents is similarly limited, and there are no current clinical trials of moderate to large sample sizes. Much of the understanding and practice of ECT for adolescents is therefore based on studies in adults (Ghaziuddin et al., 2020). However, smaller studies and metaanalyses do exist.

While not limited to mood disorders, Bertagnoli and Borchardt (1990) reviewed all publications from 1947 to 1990 where patients between the ages of 5 and 18 received ECT. They found a reported positive response in all publications except for one which was a small case series involving four patients (Bertagnoli and Borchardt, 1990). Another review, conducted by Walter and Rey (2003), included 396 individuals all under the age of 18 who received ECT for a treatment-resistant mood disorder. Since few of the included studies used standardized outcome measures, the authors assessed marked improvement by clinical descriptions indicating subjects were functioning near or at their premorbid baseline following ECT. By the end of ECT, 53% of subjects demonstrated marked improvement of full remission of symptoms (Walter and Rey, 2003). Ghaziuddin et al. (2020) studied 54 adolescents with refractory mood disorders who received their first course of ECT before the age of 18. The mean number of treatments received during the index course was 13.7 ± 6.3 with a 52.8% response rate (defined by a Clinical Global Impression Score of ≤ 2) with 15.1% achieving remission (defined by a CGI score = 1). The study also found a reduction in suicidal ideation and self-injurious behaviors following completion of the treatment course (Ghaziuddin et al., 2020).

In the available literature, practice guidelines for the use of ECT in adolescents have been published (Ghaziuddin et al., 2004). This paper recommends that ECT be considered in adolescents when the following three criteria have been met: (1) A diagnosis of a severe and persistent mood disorder has been made such as major depression with or without psychotic features or mania; (2) The patient's symptoms include severe suicidality, refusal to eat or drink, or are severe or life threatening; and (3) There has been a lack of treatment response to at least two adequate trials of appropriate psychopharmacologic agents accompanied by any other indicated treatment modalities. However, ECT can be tried earlier in the treatment course if adequate medication trials are not possible because of a patient's inability to tolerate psychopharmacologic agents or waiting for a response would be life threatening.

Pregnant women

Depression is a common disorder during pregnancy with studies citing rates from 15% to 29% (Vesga-López et al., 2008). In cases where symptoms are severe or life threatening such as high suicide risk, catatonia, medication-resistant illness, psychotic agitation, or severe physical decline, ECT should be considered a safe and effective treatment option in pregnancy. In fact, both the American Psychiatric Association and the American College of Obstetricians and Gynecologists have made statements endorsing ECT for the treatment of refractory or life threatening depression during pregnancy.

The efficacy and safety of ECT in pregnancy has been well documented over the past 50 years. Four systematic reviews of the use of ECT in pregnancy exist (Leiknes et al., 2015; Anderson and Reti, 2009; Miller, 1994; Pompili et al., 2014) as well as two metaanalyses (Sinha et al., 2017; Rose et al., 2020). Results from these studies suggest that the speed of response and response rate for pregnant women is similar to any other individual receiving ECT (Leiknes et al., 2015). While the risk of adverse events is also similar, there are additional concerns that need to be considered when ECT is performed during pregnancy.

Due to physiologic changes, pregnant patients are at increased risk of aspiration and aortocaval compression which can be addressed by pretreatment with sodium citrate and elevation of the right hip. Anesthesiologists may also prefer to intubate patients during the brief ECT procedure to prevent aspiration especially in mid or late pregnancy. There are also potential risks to the pregnancy itself. The most commonly cited issue is premature contractions and/or preterm labor which are more frequently seen in the second and third trimesters (Leiknes et al., 2015; Anderson and Reti, 2009). However, the frequency was still relatively low (3.5%) (Anderson and Reti, 2009; Ray-Griffith et al., 2016) and not clearly related to ECT (Ward et al., 2018). Other adverse events such as vaginal bleeding were reported at low rates (Anderson and Reti, 2009; Leiknes et al., 2015), typically in the first trimester, and the vast majority of cases spontaneously resolved without any interventions. No significant difference was noted in the miscarriage rate in women receiving ECT compared to the general population.

In terms of risk to the fetus, the most commonly noted adverse event during ECT was fetal heart rate variability or arrhythmia (Miller, 1994). In a review of 339 cases of pregnant women receiving ECT, Anderson and Reti found that there were only 11 fetal complications including 2 deaths, neither of which was directly attributed to ECT (Anderson and Reti, 2009). Furthermore, ECT during pregnancy has not been associated with congenital anomalies, either morphological or behavioral (Ward et al., 2018).

Geriatric

ECT is often an important part of resistant depression treatment in geriatric patients. Similar to other age groups, ECT treats severe depression, mania, psychosis and catatonia in older adults, and studies have consistently demonstrated its efficacy and safety. In fact, older age has repeatedly been shown to be a predictor of increased likelihood of response as well as fast speed of response (Rhebergen et al., 2015; Rosen et al., 2016; Kellner et al., 2016).

Numerous studies have examined the effectiveness of ECT in the treatment of geriatric depression. Spaans et al. (2015) found that geriatric patients receiving ECT had remission rates of 63.8% compared to 33.3% with an antidepressant trial (Spaans et al., 2015). As mentioned above, rates of response and remission in geriatric patients are higher compared to mixed age groups with studies citing 60%–80% of patients reaching full remission at the end of a treatment course (Spaans et al., 2015; Kellner et al., 2016). One of the most recent and comprehensive studies of ECT response in geriatric depressed patients, the Prolonging Remission In Depressed Elderly (PRIDE) study, was a multiphase protocol with an acute ECT course followed by a maintenance phase to monitor for continued response. The acute phase of the study included 240 patients above 60 years of age diagnosed with unipolar depression. All patients received right unilateral, ultrabrief pulse ECT three times per week in combination with venlafaxine. Sixty-two percent of the patients reached remission (defined as a HAM-D score ≤ 10 on two consecutive ratings that did not increase >3 points on the second consecutive rating or it remained ≤ 6), while 70% responded (defined by having at least a 50% reduction in score from baseline to last assessment). The mean number of treatments it took to achieve remission was only 7.3 (Kellner et al., 2016). Despite the excellent rates of remission for ECT, providers are often hesitant to refer geriatric patients due to concerns about cognitive side effects with treatment. However, study results vary on the severity and duration of these effects with some studies showing minimal to no effect at all. In fact, global cognitive function remained relatively stable during the ECT treatment course in the PRIDE study, with no significant difference between baseline and post-ECT mean Mini-Mental Status Exam (MMSE) scores (27.5 [SD = 2.4] and 27.6 [SD = 2.6], respectively). Since risk factors for increased anterograde and retrograde amnesia with ECT include increased age, underlying brain disease, and limited cognition at baseline, geriatric patients have been shown to experience more memory side effects with ECT (Meyer et al., 2018). However, these effects resolve in the vast majority of elderly patients, and ECT has not been shown to cause long term changes in cognition.

Maintenance ECT (M-ECT)

ECT is a treatment that is usually discontinued after a successful acute course, and then maintenance medications are initiated. However, it has been known for years that a substantial risk of relapse with rates of 50% or higher within 6–12 months after ECT exists (Sackeim et al., 2001). In the ECT literature, there are two terms applied to ECT after an initial or index course. *Continuation ECT* refers to less frequent ECT (every week to 4 weeks) tapered within 6 months postacute ECT and is used to prevent a rapid relapse. *Maintenance ECT* is defined as ECT administered at intervals beyond 6 months from the acute ECT course and is used to maintain a response and prevent recurrence of a new episode. Most practitioners tend to use *maintenance ECT (M-ECT)* to generally mean any ECT administered postacute ECT, and this term will be used for the purpose of this overview.

Descriptions of M-ECT have been detailed in the literature for a number of psychiatric disorders since the late 1940s (Geoghegan and Stevenson, 1949; Bourne, 1954; Karliner and Wehrheim, 1965). The bulk of past research of M-ECT for depression comes from naturalistic, case controlled matched, and retrospective studies. Outcomes typically have focused on a duration to relapse (6 months to 2 years) and rehospitalization prevention. A number of studies have also compared these outcomes between M-ECT and maintenance medication groups after ECT. The most common findings include reduced time spent in the hospital, decreased rehospitalizations, and reduced relapses when compared to maintenance medications (Clarke et al., 1989; Vanelle et al., 1994; Schwarz et al., 1995; Petrides et al., 1994; Gagné et al., 2000).

With the need for more systematic study of M-ECT, the Consortium for Research in ECT (CORE) in the early 2000s embarked on a large four site study of M-ECT (n=89) or maintenance medication treatment (n=95) with nortriptyline and lithium augmentation after all subjects achieved remission with acute ECT. Subjects were monitored for relapse over 6 months. Both M-ECT and maintenance medications had similar relapse rates of approximately 33% (Kellner et al., 2006). Although no statistical difference was found, both treatment groups resulted in reduced relapse rates that were lower than the commonly seen 50% or higher relapse rates in other studies. Other limitations of this study that may have led to M-ECT being no better than maintenance medications included a rigid quick taper of ECT and M-ECT subjects were medication free during the study period which is unlike routine clinical practice. There have been only a few other prospective studies of M-ECT with concomitant medications vs maintenance medications alone in the past decade, but a metaanalysis of the available data from 5 studies with 436 subjects confirms that M-ECT plus medication therapy has superior benefit in reducing risk of relapse for 1 year after remission from acute ECT. However, with only a small number of randomized trials, there remains a need for more rigorous prospective study of M-ECT (Elias et al., 2018).

In real-world clinical practice, M-ECT is commonly employed while patients are also on antidepressant medications. The goal is to induce remission or a plateaued response acutely followed by a taper of ECT. Patients should be monitored closely by tracking depression symptoms with a thorough assessment including the use of well validated depression scales, when possible. Frequency of ECT can be adjusted based on improvement or relapsing symptoms. Patients who are prone to rapid relapse during M-ECT can often be "recaptured" with increasing ECT frequency temporarily to 2–3 times per week while medications and other treatments are maximized. For some patients with extremely resistant depression, they may respond well to ECT acutely, but relapse when M-ECT is tapered or discontinued. In these situations, M-ECT can be implemented for years at frequencies ranging from every 2 to 12 weeks without causing cumulative cognitive deficits (Gill and Kellner, 2019). When used in conjunction with medications and other treatments, M-ECT is an important treatment strategy in managing resistant depression and can be individualized with flexible schedules ranging from short tapers to even longer-term less frequent ECT when needed to prevent relapse and maintain wellness.

Safety and adverse effects of ECT

Mortality and safety

As with any procedure or surgery under general anesthesia, ECT carries some risk of medical morbidity and mortality. Although each ECT is a very brief procedure (10–20 min), physiologically a seizure is generated, and transient effects on the cardiovascular system can occur including elevated blood pressure and tachycardia. ECT patients typically undergo a thorough medical history and physical exam to assess medical comorbidities and risks. Overall, for all procedures under anesthesia, the commonly understood mortality rate is 3.3–3.7 deaths per 100,000 inductions (Roy and Overdyk, 1997; Bainbridge et al., 2012). Many studies over 35 years show similar or more favorable mortality rates with ECT. Kramer (1985) described a mortality rate of 2 deaths per 100,000 from data in California from 1977 to 1983. Since 1993, the state of Texas requires quarterly reports on deaths within 14 days of ECT whether mortality is directly associated with ECT or not. Recently, a 16-year review of the Texas data encompassing 166,711 ECT from 1998 to 2013 found a death rate of 2.4

deaths per 100,000 ECT at 1 day after ECT (Dennis et al., 2017). In a similarly large study of ECT data in Ontario, Canada, death rate on the day of ECT was 4 deaths per 100,000 ECT (Blumberger et al., 2017). Finally, a large analysis of 15 studies with data from 32 countries including 766,180 ECT sessions generated a calculated death rate of 2.1 deaths per 100,000 ECT (Tørring et al., 2017). Some conditions may elevate risk, and these include intracranial space-occupying lesions or hemorrhage, recent myocardial infraction, or high anesthesia risk score per American Society of Anesthesiologists class 4 or 5. However, with proper medical care and the overall low mortality rate data, the risk of mortality with ECT remains low, and the American Psychiatric Association (2001) indicates that ECT has *no absolute* contraindications.

Cognition

The study of memory side effects in ECT is complex because of the varied assessment instruments and study designs over the years. In addition, modern ECT employs lower energy brief electrical pulses and is different from the high energy sinusoidal waveform machines from decades past. It is also generally agreed that right unilateral electrode placement is associated with less cognitive side effects (Abrams, 2002). Described here are the three major forms of memory issues of posttreatment/postictal confusion, retrograde amnesia, and anterograde amnesia along with a discussion of the importance of monitoring memory during ECT and strategies to minimize cognitive issues.

Postictal confusion

This occurs in the immediate postoperative period during recovery from anesthesia. Patients may have transient disorientation that commonly resolves within a few minutes to a few hours. A review of the literature suggests that postictal confusion is associated with catatonic features, cerebrovascular disease, Parkinson's disease, dementia, bitemporal electrode placement, high stimulus intensity, or longer seizures (Tsujii et al., 2019). Occasionally, patients may experience more severe delirium with restlessness or agitation requiring an intervention such intravenous benzodiazepine or other medication to calm the patient and provide a smoother recovery from ECT and anesthesia medications.

Retrograde amnesia

Retrograde amnesia is the difficulty recalling information acquired before beginning ECT. This is generally described as spotty and tends to be more prominent in the weeks to the few months prior to ECT. Memories for public events appear more sensitive than personal information (Lisanby et al., 2000). There is the risk of persistence of retrograde amnesia and may occur more with older sine waveform stimulation or bitemporal ECT (Semkovska and McLoughlin, 2013).

Anterograde amnesia

Anterograde amnesia is the difficulty retaining newly learned information. This occurs during the ECT course, but typically clears after ECT is discontinued in a matter of days to a few weeks. Unlike the risk of some persistent retrograde amnesia complaints, anterograde amnesia has not been shown to be permanent in many studies (Semkovska and McLoughlin, 2010). In fact, a number of studies have shown cognitive improvements, compared to baseline measures pre-ECT, weeks to months after a successful ECT course once depression has resolved and anterograde amnesia side effects have dissipated (Weiner and Reti, 2017).

Monitoring cognition and strategies to minimize side effects

Similar to tracking depression symptoms during treatment over time, clinicians should monitor cognitive side effects during ECT to detect changes early and make clinical adjustments to minimize side effects (Hermida et al., 2018; Weiner and Reti, 2017). Brief assessment tools are often used prior to ECT and at regular intervals during the ECT course. The most common instruments are the Mini-Mental Status Exam (MMSE) or the MoCA Test (Folstein et al., 1975; MoCA TEST, n.d.). However, these are screening tests for cognitive problems that do not capture all memory domains affected by ECT. Recently, the Electroconvulsive therapy Cognitive Assessment (ECCA) tool is a newer brief scale that is more sensitive to ECT-induced global cognitive function difficulties and includes measures of retrograde memory/autobiographical memory and subjective assessments of both patient and a family member (Hermida et al., 2020). Strategies to minimize memory side effects at the start of ECT or if memory side effects develop include a number of ECT techniques. Using or switching to right unilateral ECT from bitemporal ECT reduces risk of memory side effects. Electrical dosing is now often titrated with a low level of energy to generate the initial seizure. Over the past two decades, the duration of the brief pulse waveform has shortened to more efficiently excite neurons to produce a seizure. Finally, ECT frequency in the United States usually begins three times per week, but reducing ECT frequency to twice a week can reduce memory side effects without compromising the antidepressant benefits (Prudic, 2008).

Other common side effects

One of the most common side effects of ECT is the immediate postoperative headache. It is often described as throbbing or migraine-like. The etiology is not well understood but may be due to direct muscle stimulation of the head during the brief ECT stimulation or may have a vascular origin. The incidence of headache is estimated to be in the range of 33% or higher. Treatment is symptomatic or prophylactic with nonsteroidal antiinflammatory drugs (NSAIDs) or migraine medications such as triptans (Mulder and Grootens, 2020). Nausea following ECT can be managed symptomatically as well with medications before or after ECT. Muscle aches following treatment can arise from sensitivity to the fasciculations during depolarizing muscle relaxation from succinylcholine and is also managed with pain medications commonly.

Future directions

Future directions for ECT and related seizure therapies include a focus on ways to induce seizures that maintain effectiveness while minimizing side effects. One emerging technique is Magnetic Seizure Therapy (MST) where seizures are induced through magnetic pulses covered elsewhere in this book. Only the superficial cortex is stimulated during seizure induction but a generalized seizure does then occur (Tang et al., 2020). Studies on another emerging technique is focal electrically administered seizure therapy (FEAST) which uses unidirectional rather than bidirectional current as well as a novel, asymmetric node placement to improve the efficiency of seizure induction (Spellman et al., 2009). Studies on the effectiveness of FEAST are also limited with only a few open label studies. However, results thus far show similar effectiveness to ECT with limited to no cognitive side effects (Spellman et al., 2009).

The pursuit of biomarkers such as the genetics of depression may also shed light on the neurobiology of depression. One such study underway is the Genomics of ECT International Consortium (GenECT-ic). This project is not only established to investigate the genomic underpinnings of resistant depression with an aim of assessing 30,000 patients referred for ECT from around the world, but also to study the genetic contribution of ECT treatment response. This study is underway and will be the largest clinical study of ECT known (Baune et al., 2019; Soda et al., 2020).

Summary

ECT is a treatment developed over 80 years ago, and through the decades, scientific studies hold that ECT is a safe and effective strategy for depression and many other neuropsychiatric disorders. Depression knows no age limit, and ECT is exceptionally effective across the life span from the young to very old. When severe and resistant depression strikes, ECT can be life-saving and is an extremely fast treatment with improvement occurring within a few weeks versus many weeks to months with pharmacotherapy or psychotherapy.

With past clinical depression studies of various treatment modalities, the focus has generally been on *response* which is typically defined as a 50% reduction of depression symptoms on well-validated rating scales. More recently in the past 15–20 years, studies such as STAR*D focus on achieving *remission*, which refers to very low or single-digit scores on depression scales at the end of treatment. Historically, remission rates in medication trials range from 22% to 40% (Mendlewicz, 2008), but ECT is the most effective treatment and has a remission rate between 50% and 80% across many retrospective and prospective studies. No other novel depression strategy can boast of such consistently positive remission rates. Maintenance ECT is also another strategy that is often employed to prevent relapses and maintain wellness usually in combination with pharmacotherapy.

Despite ECT's excellent outcomes for treatment-resistant depression, ECT is a treatment that is used sparingly in the United States for only a small fraction of the TRD population. Prior estimates from the 1990s indicated only 100,000 patients per year received ECT (Hermann et al., 1995) and at the current US population, this number may now only be 150–200,000 ECT patients per year. For this smaller patient population, there are also concerns that access to ECT could be negatively affected by a reduction in ECT use and availability in hospitals over the past 20 years (Case et al., 2013; Wilkinson et al., 2018).

ECT is not an antiquated somatic therapy that should be limited or fade away, but is actually an evolving treatment with ongoing explorations to reduce side effects, find predictors of response, and understand better the biologic underpinnings of ECT's effects. For the foreseeable future, ECT will remain a vital tool in combating treatment-resistant depression.

References

Abrams, R., 2002. Electroconvulsive Therapy, fourth ed. Oxford University Press.

American Psychiatric Association, 2001. The Practice of Electroconvulsive Therapy: Recommendations for Treatment, Training, and Privileging (A Task Force Report of the American Psychiatric Association), second ed. American Psychiatric Publishing.

American Psychiatric Association, 2002. Practice guideline for the treatment of patients with bipolar disorder (revision). Am. J. Psychiatry 159 (4 Suppl), 1–50.

Anderson, E.L., Reti, I.M., 2009. ECT in pregnancy: a review of the literature from 1941 to 2007. Psychosom. Med. 71 (2), 235–242. https://doi.org/10.1097/PSY.0b013e318190d7ca.

Bahji, A., Hawken, E.R., Sepehry, A.A., Cabrera, C.A., Vazquez, G., 2019. ECT beyond unipolar major depression: systematic review and meta-analysis of electroconvulsive therapy in bipolar depression. Acta Psychiatr. Scand. 139 (3), 214–226. https://doi.org/10.1111/acps.12994.

Bainbridge, D., Martin, J., Arango, M., Cheng, D., 2012. Perioperative and anaesthetic-related mortality in developed and developing countries: a systematic review and meta-analysis. Lancet 380 (9847), 1075–1081. https://doi.org/10.1016/S0140-6736(12)60990-8.

Baune, B.T., Soda, T., Sullivan, P.F., Zandi, P., 2019. The genomics of electroconvulsive therapy international consortium (GenECT-ic). Lancet Psychiatry 6 (10). https://doi.org/10.1016/S2215-0366(19)30343-8, e23.

Bertagnoli, M.W., Borchardt, C.M., 1990. A review of ECT for children and adolescents. J. Am. Acad. Child Adolesc. Psychiatry 29 (2), 302–307.

Black, D.W., Winokur, G., Nasrallah, A., 1987. The treatment of depression: electroconvulsive therapy v antidepressants: a naturalistic evaluation of 1,495 patients. Compr. Psychiatry 28 (2), 169–182.

Blumberger, D.M., Seitz, D.P., Herrmann, N., Kirkham, J.G., Ng, R., Reimer, C., Kurdyak, P., Gruneir, A., Rapoport, M.J., Daskalakis, Z.J., Mulsant, B.H., Vigod, S.N., 2017. Low medical morbidity and mortality after acute courses of electroconvulsive therapy in a population-based sample. Acta Psychiatr. Scand. 136 (6), 583–593. https://doi.org/10.1111/acps.12815.

Bostwick, J.M., Pankratz, V.S., 2000. Affective disorders and suicide risk: a reexamination. Am. J. Psychiatry 157 (12), 1925–1932.

Bouckaert, F., Sienaert, P., Obbels, J., Dols, A., Vandenbulcke, M., Stek, M., Bolwig, T., 2014. ECT: its brain enabling effects: a review of electroconvulsive therapy-induced structural brain plasticity. J. ECT 30 (2), 143–151. https://doi.org/10.1097/YCT.0000000000000129.

Bourne, H., 1954. Convulsion dependence. Lancet 267 (6850), 1193–1196.

Brådvik, L., Berglund, M., 2006. Long-term treatment and suicidal behavior in severe depression: ECT and antidepressant pharmacotherapy may have different effects on the occurrence and seriousness of suicide attempts. Depress. Anxiety 23 (1), 34–41.

Case, B.G., Bertollo, D.N., Laska, E.M., Price, L.H., Siegel, C.E., Olfson, M., Marcus, S.C., 2013. Declining use of electroconvulsive therapy in U.S. general hospitals is not restricted to unipolar depression. Biol. Psychiatry 74 (10), e19–e20. https://doi.org/10.1016/j.biopsych.2013.01.039.

Clarke, T.B., Coffey, C.E., Hoffman, G.W., Weiner, R.D., 1989. Continuation therapy for depression using outpatient electroconvulsive therapy. Convuls. Ther. 5 (4), 330–337.

Connolly, K.R., Helmer, A., Cristancho, M.A., Cristancho, P., O'Reardon, J.P., 2012. Effectiveness of transcranial magnetic stimulation in clinical practice post-FDA approval in the United States: results observed with the first 100 consecutive cases of depression at an academic medical center. J. Clin. Psychiatry 73 (4), e567–e573. https://doi.org/10.4088/JCP.11m07413.

Conwell, Y., Duberstein, P.R., Caine, E.D., 2002. Risk factors for suicide in later life. Biol. Psychiatry 52 (3), 193–204.

Cusin, C., Dougherty, D.D., 2012. Somatic therapies for treatment-resistant depression: ECT, TMS, VNS, DBS. Biol Mood Anxiety Disord. 2, 14. https://doi.org/10.1186/2045-5380-2-14.

Dennis, N.M., Dennis, P.A., Shafer, A., Weiner, R.D., Husain, M.M., 2017. Electroconvulsive therapy and all-cause mortality in Texas, 1998-2013. J. ECT 33 (1), 22–25. https://doi.org/10.1097/YCT.0000000000000340.

Dierckx, B., Heijnen, W.T., van den Broek, W.W., Birkenhäger, T.K., 2012. Efficacy of electroconvulsive therapy in bipolar versus unipolar major depression: a meta-analysis. Bipolar Disord. 14 (2), 146–150. https://doi.org/10.1111/j.1399-5618.2012.00997.x.

Duthie, A.C., Perrin, J.S., Bennett, D.M., Currie, J., Reid, I.C., 2015. Anticonvulsant mechanisms of electroconvulsive therapy and relation to therapeutic efficacy. J. ECT 31 (3), 173–178. https://doi.org/10.1097/YCT.0000000000000210.

Elias, A., Phutane, V.H., Clarke, S., Prudic, J., 2018. Electroconvulsive therapy in the continuation and maintenance treatment of depression: systematic review and meta-analyses. Aust. N. Z. J. Psychiatry 52 (5), 415–424. https://doi.org/10.1177/0004867417743343.

Federal Drug Administration (FDA), 2018. Neurological Devices; Reclassification of Electroconvulsive Therapy Devices; Effective Date of Requirement for Premarket Approval for Electroconvulsive Therapy Devices for Certain Specified Intended Uses. https://www.federalregister.gov/documents/2018/12/26/2018-27809/neurological-devices-reclassification-of-electroconvulsive-therapy-devices-effective-date-of#:~:text=The%20Food%20and%20Drug%20Administration%20(FDA)%20is%20issuing%20a%20final,and%20older%20who%20are%20treatment.

Fink, M., Kellner, C.H., McCall, W.V., 2014. The role of ECT in suicide prevention. J. ECT 30 (1), 5–9. https://doi.org/10.1097/YCT.0b013e3182a6ad0d.

Folstein, M.F., Folstein, S.E., McHugh, P.R., 1975. "Mini-mental state". A practical method for grading the cognitive state of patients for the clinician. J. Psychiatr. Res. 12 (3), 189–198.

Gagné, G.G., Furman, M.J., Carpenter, L.L., Price, L.H., 2000. Efficacy of continuation ECT and antidepressant drugs compared to long-term antidepressants alone in depressed patients. Am. J. Psychiatry 157 (12), 1960–1965.

Gazdag, G., Ungvari, G.S., 2019. Electroconvulsive therapy: 80 years old and still going strong. World J. Psychiatry 9 (1), 1–6. https://doi.org/10.5498/wjp.v9.i1.1.

Geoghegan, J.J., Stevenson, G.H., 1949. Prophylactic electroshock. Am. J. Psychiatry 105 (7), 494–496.

George, M.S., 2010. Transcranial magnetic stimulation for the treatment of depression. Expert Rev. Neurother. 10 (11), 1761–1772. https://doi.org/10.1586/ern.10.95.

Ghaziuddin, N., Kutcher, S.P., Knapp, P., Bernet, W., Arnold, V., Beitchman, J., Benson, R.S., Bukstein, O., Kinlan, J., McClellan, J., Rue, D., Shaw, J.A., Stock, S., Kroeger Ptakowski, K., 2004. Practice parameter for use of electroconvulsive therapy with adolescents. J. Am. Acad. Child Adolesc. Psychiatry 43 (12), 1521–1539.

Ghaziuddin, N., Shamseddeen, W., Gettys, G., Ghaziuddin, M., 2020. Electroconvulsive therapy for the treatment of severe mood disorders during adolescence: a retrospective chart review. J. Child Adolesc. Psychopharmacol. 30 (4), 235–243. https://doi.org/10.1089/cap.2019.0054.

Gibson, T.B., Jing, Y., Smith Carls, G., Kim, E., Bagalman, J.E., Burton, W.N., Tran, Q.-V., Pikalov, A., Goetzel, R.Z., 2010. Cost burden of treatment resistance in patients with depression. Am. J. Manag. Care 16 (5), 370–377.

Gill, S.P., Kellner, C.H., 2019. Clinical practice recommendations for continuation and maintenance electroconvulsive therapy for depression: outcomes from a review of the evidence and a consensus workshop held in Australia in May 2017. J. ECT 35 (1), 14–20. https://doi.org/10.1097/YCT.0000000000000484.

Grunze, H., Vieta, E., Goodwin, G.M., Bowden, C., Licht, R.W., Moller, H.-J., Kasper, S., 2009. The world Federation of Societies of biological psychiatry (WFSBP) guidelines for the biological treatment of bipolar disorders: update 2009 on the treatment of acute mania. World J. Biol. Psychiatry 10 (2), 85–116. https://doi.org/10.1080/15622970902823202.

Haskett, R.F., 2014. Electroconvulsive therapy's mechanism of action: neuroendocrine hypotheses. J. ECT 30 (2), 107–110. https://doi.org/10.1097/YCT.0000000000000143.

Hermann, R.C., Dorwart, R.A., Hoover, C.W., Brody, J., 1995. Variation in ECT use in the United States. Am. J. Psychiatry 152 (6), 869–875. https://doi.org/10.1176/ajp.152.6.869.

Hermida, A.P., Glass, O.M., Shafi, H., McDonald, W.M., 2018. Electroconvulsive therapy in depression: current practice and future direction. Psychiatr. Clin. North Am. 41 (3), 341–353. https://doi.org/10.1016/j.psc.2018.04.001.

Hermida, A.P., Goldstein, F.C., Loring, D.W., McClintock, S.M., Weiner, R.D., Reti, I.M., Janjua, A.U., Ye, Z., Peng, L., Tang, Y.-L., Galendez, G.C., Husain, M.M., Maixner, D.F., Riva-Posse, P., McDonald, W.M., 2020. ElectroConvulsive therapy cognitive assessment (ECCA) tool: a new instrument to monitor cognitive function in patients undergoing ECT. J. Affect. Disord. 269, 36–42. https://doi.org/10.1016/j.jad.2020.03.010.

Husain, M.M., Rush, A.J., Fink, M., Knapp, R., Petrides, G., Rummans, T., Biggs, M.M., O'Connor, K., Rasmussen, K., Litle, M., Zhao, W., Bernstein, H.J., Smith, G., Mueller, M., McClintock, S.M., Bailine, S.H., Kellner, C.H., 2004. Speed of response and remission in major depressive disorder with acute electroconvulsive therapy (ECT): a consortium for research in ECT (CORE) report. J. Clin. Psychiatry 65 (4), 485–491.

Karliner, W., Wehrheim, H.K., 1965. Maintenance convulsive treatments. Am. J. Psychiatry 121, 1113–1115.

Kellner, C.H., Fink, M., Knapp, R., Petrides, G., Husain, M., Rummans, T., Mueller, M., Bernstein, H., Rasmussen, K., O'connor, K., Smith, G., Rush, A.J., Biggs, M., McClintock, S., Bailine, S., Malur, C., 2005. Relief of expressed suicidal intent by ECT: a consortium for research in ECT study. Am. J. Psychiatry 162 (5), 977–982.

Kellner, C.H., Knapp, R.G., Petrides, G., Rummans, T.A., Husain, M.M., Rasmussen, K., Mueller, M., Bernstein, H.J., O'Connor, K., Smith, G., Biggs, M., Bailine, S.H., Malur, C., Yim, E., McClintock, S., Sampson, S., Fink, M., 2006. Continuation electroconvulsive therapy vs pharmacotherapy for relapse prevention in major depression: a multisite study from the consortium for research in electroconvulsive therapy (CORE). Arch. Gen. Psychiatry 63 (12), 1337–1344.

Kellner, C.H., Knapp, R., Husain, M.M., Rasmussen, K., Sampson, S., Cullum, M., McClintock, S.M., Tobias, K.G., Martino, C., Mueller, M., Bailine, S.H., Fink, M., Petrides, G., 2010. Bifrontal, bitemporal and right unilateral electrode placement in ECT: randomised trial. Br. J. Psychiatry J. Ment. Sci. 196 (3), 226–234. https://doi.org/10.1192/bjp.bp.109.066183.

Kellner, C.H., Husain, M.M., Knapp, R.G., McCall, W.V., Petrides, G., Rudorfer, M.V., Young, R.C., Sampson, S., McClintock, S.M., Mueller, M., Prudic, J., Greenberg, R.M., Weiner, R.D., Bailine, S.H., Rosenquist, P.B., Raza, A., Kaliora, S., Latoussakis, V., Tobias, K.G., Lisanby, S.H., 2016. Right unilateral ultrabrief pulse ECT in geriatric depression: phase 1 of the PRIDE study. Am. J. Psychiatry 173 (11), 1101–1109.

Kho, K.H., van Vreeswijk, M.F., Simpson, S., Zwinderman, A.H., 2003. A meta-analysis of electroconvulsive therapy efficacy in depression. J. ECT 19 (3), 139–147.

Kishimoto, T., Chawla, J.M., Hagi, K., Zarate, C.A., Kane, J.M., Bauer, M., Correll, C.U., 2016. Single-dose infusion ketamine and non-ketamine N-methyl-d-aspartate receptor antagonists for unipolar and bipolar depression: a meta-analysis of efficacy, safety and time trajectories. Psychol. Med. 46 (7), 1459–1472. https://doi.org/10.1017/S0033291716000064.

Kramer, B.A., 1985. Use of ECT in California, 1977-1983. Am. J. Psychiatry 142 (10), 1190–1192.

Leaver, A.M., Vasavada, M., Kubicki, A., Wade, B., Loureiro, J., Hellemann, G., Joshi, S.H., Woods, R.P., Espinoza, R., Narr, K.L., 2020. Hippocampal subregions and networks linked with antidepressant response to electroconvulsive therapy. Mol. Psychiatry. https://doi.org/10.1038/s41380-020-0666-z.

Leiknes, K.A., Cooke, M.J., Jarosch-von Schweder, L., Harboe, I., Høie, B., 2015. Electroconvulsive therapy during pregnancy: a systematic review of case studies. Arch. Womens Ment. Health 18 (1), 1–39. https://doi.org/10.1007/s00737-013-0389-0.

Lisanby, S.H., Maddox, J.H., Prudic, J., Devanand, D.P., Sackeim, H.A., 2000. The effects of electroconvulsive therapy on memory of autobiographical and public events. Arch. Gen. Psychiatry 57 (6), 581–590.

Luchini, F., Medda, P., Mariani, M.G., Mauri, M., Toni, C., Perugi, G., 2015. Electroconvulsive therapy in catatonic patients: efficacy and predictors of response. World J. Psychiatry 5 (2), 182–192. https://doi.org/10.5498/wjp.v5.i2.182.

Magnezi, R., Aminov, E., Shmuel, D., Dreifuss, M., Dannon, P., 2016. Comparison between neurostimulation techniques repetitive transcranial magnetic stimulation vs electroconvulsive therapy for the treatment of resistant depression: patient preference and cost-effectiveness. Patient Prefer. Adherence 10, 1481–1487. https://doi.org/10.2147/PPA.S105654.

Mann, J.J., 1998. Neurobiological correlates of the antidepressant action of electroconvulsive therapy. J. ECT 14 (3), 172–180.

Mantovani, A., Pavlicova, M., Avery, D., Nahas, Z., McDonald, W.M., Wajdik, C.D., Holtzheimer, P.E., George, M.S., Sackeim, H.A., Lisanby, S.H., 2012. Long-term efficacy of repeated daily prefrontal transcranial magnetic stimulation (TMS) in treatment-resistant depression. Depress. Anxiety 29 (10), 883–890. https://doi.org/10.1002/da.21967.

Mathew, S.J., Wilkinson, S.T., Altinay, M., Asghar-Ali, A., Chang, L.C., Collins, K.A., Dale, R.M., Hu, B., Krishnan, K., Kellner, C.H., Malone, D.A., Murrough, J.W., Ostroff, R.B., Sanacora, G., Shao, M., Anand, A., 2019. ELEctroconvulsive therapy (ECT) vs. ketamine in patients with treatment-resistant depression: the ELEKT-D study protocol. Contemp. Clin. Trials 77, 19–26. https://doi.org/10.1016/j.cct.2018.12.009.

McDonald, W.M., Durkalski, V., Ball, E.R., Holtzheimer, P.E., Pavlicova, M., Lisanby, S.H., Avery, D., Anderson, B.S., Nahas, Z., Zarkowski, P., Sackeim, H.A., George, M.S., 2011. Improving the antidepressant efficacy of transcranial magnetic stimulation: maximizing the number of stimulations and treatment location in treatment-resistant depression. Depress. Anxiety 28 (11), 973–980. https://doi.org/10.1002/da.20885.

McKay, M.S., Zakzanis, K.K., 2010. The impact of treatment on HPA axis activity in unipolar major depression. J. Psychiatr. Res. 44 (3), 183–192. https://doi.org/10.1016/j.jpsychires.2009.07.012.

Medda, P., Toni, C., Mariani, M.G., De Simone, L., Mauri, M., Perugi, G., 2015. Electroconvulsive therapy in 197 patients with a severe, drug-resistant bipolar mixed state: treatment outcome and predictors of response. J. Clin. Psychiatry 76 (9), 1168–1173. https://doi.org/10.4088/JCP.14m09181.

Mendlewicz, J., 2008. Towards achieving remission in the treatment of depression. Dialogues Clin. Neurosci. 10 (4), 371–375.

Meyer, J.P., Swetter, S.K., Kellner, C.H., 2018. Electroconvulsive therapy in geriatric psychiatry: a selective review. Psychiatr. Clin. North Am. 41 (1), 79–93. https://doi.org/10.1016/j.psc.2017.10.007.

Miller, L.J., 1994. Use of electroconvulsive therapy during pregnancy. Hosp. Community Psychiatry 45 (5), 444–450.

MoCA TEST. (n.d.). https://www.mocatest.org/.

Morcos, N., Maixner, S., Maixner, D.F., 2021. Electroconvulsive therapy for bipolar depression in older adults. J. ECT 37 (3), 182–188. https://doi.org/10.1097/YCT.0000000000000755.

Mulder, L.A.C., Grootens, K.P., 2020. The incidence of post-electroconvulsive therapy headache: a systematic review. J. ECT 36 (3), e22–e28. https://doi.org/10.1097/YCT.0000000000000677.

Mulders, P.C.R., Llera, A., Beckmann, C.F., Vandenbulcke, M., Stek, M., Sienaert, P., Redlich, R., Petrides, G., Oudega, M.L., Oltedal, L., Oedegaard, K.J., Narr, K.L., Magnusson, P.O., Kessler, U., Jorgensen, A., Espinoza, R., Enneking, V., Emsell, L., Dols, A., Tendolkar, I., 2020. Structural changes induced by electroconvulsive therapy are associated with clinical outcome. Brain Stimul. 13 (3), 696–704. https://doi.org/10.1016/j.brs.2020.02.020.

Newport, D.J., Carpenter, L.L., McDonald, W.M., Potash, J.B., Tohen, M., Nemeroff, C.B., 2015. Ketamine and other NMDA antagonists: early clinical trials and possible mechanisms in depression. Am. J. Psychiatry 172 (10), 950–966. https://doi.org/10.1176/appi.ajp.2015.15040465.

Nikisch, G., Mathé, A.A., 2008. CSF monoamine metabolites and neuropeptides in depressed patients before and after electroconvulsive therapy. Eur. Psychiatry 23 (5), 356–359. https://doi.org/10.1016/j.eurpsy.2008.03.003.

Olfson, M., Marcus, S., Sackeim, H.A., Thompson, J., Pincus, H.A., 1998. Use of ECT for the inpatient treatment of recurrent major depression. Am. J. Psychiatry 155 (1), 22–29.

Oltedal, L., Narr, K.L., Abbott, C., Anand, A., Argyelan, M., Bartsch, H., Dannlowski, U., Dols, A., van Eijndhoven, P., Emsell, L., Erchinger, V.J., Espinoza, R., Hahn, T., Hanson, L.G., Hellemann, G., Jorgensen, M.B., Kessler, U., Oudega, M.L., Paulson, O.B., Dale, A.M., 2018. Volume of the human hippocampus and clinical response following electroconvulsive therapy. Biol. Psychiatry 84 (8), 574–581. https://doi.org/10.1016/j.biopsych.2018.05.017.

Østergaard, S.D., Bolwig, T.G., Petrides, G., 2014. No causal association between electroconvulsive therapy and death: a summary of a report from the Danish health and medicines authority covering 99,728 treatments. J. ECT 30 (4), 263–264. https://doi.org/10.1097/YCT.0000000000000169.

Perera, T., George, M.S., Grammer, G., Janicak, P.G., Pascual-Leone, A., Wirecki, T.S., 2016. The clinical TMS society consensus review and treatment recommendations for TMS therapy for major depressive disorder. Brain Stimul. 9 (3), 336–346. https://doi.org/10.1016/j.brs.2016.03.010.

Perrin, J.S., Merz, S., Bennett, D.M., Currie, J., Steele, D.J., Reid, I.C., Schwarzbauer, C., 2012. Electroconvulsive therapy reduces frontal cortical connectivity in severe depressive disorder. Proc. Natl. Acad. Sci. U. S. A. 109 (14), 5464–5468. https://doi.org/10.1073/pnas.1117206109.

Perugi, G., Medda, P., Toni, C., Mariani, M.G., Socci, C., Mauri, M., 2017. The role of electroconvulsive therapy (ECT) in bipolar disorder: effectiveness in 522 patients with bipolar depression, mixed-state, mania and catatonic features. Curr. Neuropharmacol. 15 (3), 359–371. https://doi.org/10.2174/1570159X14666161017233642.

Petrides, G., Dhossche, D., Fink, M., Francis, A., 1994. Continuation ECT: relapse prevention in affective disorders. Convuls. Ther. 10 (3), 189–194.

Petrides, G., Fink, M., Husain, M.M., Knapp, R.G., Rush, A.J., Mueller, M., Rummans, T.A., O'Connor, K.M., Rasmussen, K.G., Bernstein, H.J., Biggs, M., Bailine, S.H., Kellner, C.H., 2001. ECT remission rates in psychotic versus nonpsychotic depressed patients: a report from CORE. J. ECT 17 (4), 244–253.

Petrides, G., Malur, C., Braga, R.J., Bailine, S.H., Schooler, N.R., Malhotra, A.K., Kane, J.M., Sanghani, S., Goldberg, T.E., John, M., Mendelowitz, A., 2015. Electroconvulsive therapy augmentation in clozapine-resistant schizophrenia: a prospective, randomized study. Am. J. Psychiatry 172 (1), 52–58. https://doi.org/10.1176/appi.ajp.2014.13060787.

Phillips, J.L., Norris, S., Talbot, J., Hatchard, T., Ortiz, A., Birmingham, M., Owoeye, O., Batten, L.A., Blier, P., 2020. Single and repeated ketamine infusions for reduction of suicidal ideation in treatment-resistant depression. Neuropsychopharmacology 45 (4), 606–612. https://doi.org/10.1038/s41386-019-0570-x.

Pompili, M., Dominici, G., Giordano, G., Longo, L., Serafini, G., Lester, D., Amore, M., Girardi, P., 2014. Electroconvulsive treatment during pregnancy: a systematic review. Expert Rev. Neurother. 14 (12), 1377–1390. https://doi.org/10.1586/14737175.2014.972373.

Prudic, J., 2008. Strategies to minimize cognitive side effects with ECT: aspects of ECT technique. J. ECT 24 (1), 46–51. https://doi.org/10.1097/YCT.0b013e31815ef238.

Raveendranathan, D., Narayanaswamy, J.C., Reddi, S.V., 2012. Response rate of catatonia to electroconvulsive therapy and its clinical correlates. Eur. Arch. Psychiatry Clin. Neurosci. 262 (5), 425–430. https://doi.org/10.1007/s00406-011-0285-4.

Ray-Griffith, S.L., Coker, J.L., Rabie, N., Eads, L.A., Golden, K.J., Stowe, Z.N., 2016. Pregnancy and electroconvulsive therapy: a multidisciplinary approach. J. ECT 32 (2), 104–112. https://doi.org/10.1097/YCT.0000000000000297.

Rhebergen, D., Huisman, A., Bouckaert, F., Kho, K., Kok, R., Sienaert, P., Spaans, H.-P., Stek, M., 2015. Older age is associated with rapid remission of depression after electroconvulsive therapy: a latent class growth analysis. Am. J. Geriatr. Psychiatry 23 (3), 274–282. https://doi.org/10.1016/j.jagp.2014.05.002.

Rose, S., Dotters-Katz, S.K., Kuller, J.A., 2020. Electroconvulsive therapy in pregnancy: safety, best practices, and barriers to care. Obstet. Gynecol. Surv. 75 (3), 199–203. https://doi.org/10.1097/OGX.0000000000000763.

Rosen, B.H., Kung, S., Lapid, M.I., 2016. Effect of age on psychiatric rehospitalization rates after electroconvulsive therapy for patients with depression. J. ECT 32 (2), 93–98. https://doi.org/10.1097/YCT.0000000000000271.

Rosenberg, J., Pettinati, H.M., 1984. Differential memory complaints after bilateral and unilateral ECT. Am. J. Psychiatry 141 (9), 1071–1074.

Ross, E.L., Zivin, K., Maixner, D.F., 2018. Cost-effectiveness of electroconvulsive therapy vs pharmacotherapy/psychotherapy for treatment-resistant depression in the United States. JAMA Psychiat. 75 (7), 713–722. https://doi.org/10.1001/jamapsychiatry.2018.0768.

Roy, C.R., Overdyk, F.J., 1997. The adult patient. In: Morell, R.C., Eichorn, J.H. (Eds.), Patient Safety in Anesthetic Practice. Churchill Livingstone.

Rush, A.J., Trivedi, M.H., Wisniewski, S.R., Nierenberg, A.A., Stewart, J.W., Warden, D., Niederehe, G., Thase, M.E., Lavori, P.W., Lebowitz, B.D., McGrath, P.J., Rosenbaum, J.F., Sackeim, H.A., Kupfer, D.J., Luther, J., Fava, M., 2006. Acute and longer-term outcomes in depressed outpatients requiring one or several treatment steps: a STAR*D report. Am. J. Psychiatry 163 (11), 1905–1917.

Sackeim, H.A., 1999. The anticonvulsant hypothesis of the mechanisms of action of ECT: current status. J. ECT 15 (1), 5–26.

Sackeim, H.A., Prudic, J., Devanand, D.P., Nobler, M.S., Lisanby, S.H., Peyser, S., Fitzsimons, L., Moody, B.J., Clark, J., 2000. A prospective, randomized, double-blind comparison of bilateral and right unilateral electroconvulsive therapy at different stimulus intensities. Arch. Gen. Psychiatry 57 (5), 425–434.

Sackeim, H.A., Haskett, R.F., Mulsant, B.H., Thase, M.E., Mann, J.J., Pettinati, H.M., Greenberg, R.M., Crowe, R.R., Cooper, T.B., Prudic, J., 2001. Continuation pharmacotherapy in the prevention of relapse following electroconvulsive therapy: a randomized controlled trial. JAMA 285 (10), 1299–1307.

Sackeim, H.A., Prudic, J., Nobler, M.S., Fitzsimons, L., Lisanby, S.H., Payne, N., Berman, R.M., Brakemeier, E.-L., Perera, T., Devanand, D.P., 2008. Effects of pulse width and electrode placement on the efficacy and cognitive effects of electroconvulsive therapy. Brain Stimul. 1 (2), 71–83. https://doi.org/10.1016/j.brs.2008.03.001.

Schwarz, T., Loewenstein, J., Isenberg, K.E., 1995. Maintenance ECT: indications and outcome. Convuls. Ther. 11 (1), 14–23.

Semkovska, M., McLoughlin, D.M., 2010. Objective cognitive performance associated with electroconvulsive therapy for depression: a systematic review and meta-analysis. Biol. Psychiatry 68 (6), 568–577. https://doi.org/10.1016/j.biopsych.2010.06.009.

Semkovska, M., McLoughlin, D.M., 2013. Measuring retrograde autobiographical amnesia following electroconvulsive therapy: historical perspective and current issues. J. ECT 29 (2), 127–133. https://doi.org/10.1097/YCT.0b013e318279c2c9.

Seymour, J., 2020. Commentary and update on the contribution of the GABA hypothesis to understanding the mechanism of action of electroconvulsive therapy. J. ECT. https://doi.org/10.1097/YCT.0000000000000711.

Shorter, W., 2004. The history of ECT: unsolved mysteries. Psychiatr. Times 21 (2). https://www.psychiatrictimes.com/view/history-ect-unsolved-mysteries.

Sinha, P., Goyal, P., Andrade, C., 2017. A meta-review of the safety of electroconvulsive therapy in pregnancy. J. ECT 33 (2), 81–88. https://doi.org/10.1097/YCT.0000000000000362.

Slade, E.P., Jahn, D.R., Regenold, W.T., Case, B.G., 2017. Association of electroconvulsive therapy with psychiatric readmissions in US hospitals. JAMA Psychiat. 74 (8), 798–804. https://doi.org/10.1001/jamapsychiatry.2017.1378.

Sobin, C., Sackeim, H.A., Prudic, J., Devanand, D.P., Moody, B.J., McElhiney, M.C., 1995. Predictors of retrograde amnesia following ECT. Am. J. Psychiatry 152 (7), 995–1001.

Soda, T., McLoughlin, D.M., Clark, S.R., Oltedal, L., Kessler, U., Haavik, J., Bousman, C., Smith, D.J., Bioque, M., Clements, C.C., Loo, C., Vila-Rodriguez, F., Minelli, A., Mickey, B.J., Milev, R., Docherty, A.R., Langan Martin, J., Achtyes, E.D., Arolt, V., Baune, B.T., 2020. International consortium on the genetics of electroconvulsive therapy and severe depressive disorders (Gen-ECT-ic). Eur. Arch. Psychiatry Clin. Neurosci. 270 (7), 921–932. https://doi.org/10.1007/s00406-019-01087-w.

Spaans, H.-P., Sienaert, P., Bouckaert, F., van den Berg, J.F., Verwijk, E., Kho, K.H., Stek, M.L., Kok, R.M., 2015. Speed of remission in elderly patients with depression: electroconvulsive therapy v. medication. Br. J. Psychiatry J. Ment. Sci. 206 (1), 67–71. https://doi.org/10.1192/bjp.bp.114.148213.

Spellman, T., Peterchev, A.V., Lisanby, S.H., 2009. Focal electrically administered seizure therapy: a novel form of ECT illustrates the roles of current directionality, polarity, and electrode configuration in seizure induction. Neuropsychopharmacology 34 (8), 2002–2010. https://doi.org/10.1038/npp.2009.12.

Squire, L.R., Slater, P.C., 1983. Electroconvulsive therapy and complaints of memory dysfunction: a prospective three-year follow-up study. Br. J. Psychiatry J. Ment. Sci. 142, 1–8.

Steif, B.L., Sackeim, H.A., Portnoy, S., Decina, P., Malitz, S., 1986. Effects of depression and ECT on anterograde memory. Biol. Psychiatry 21 (10), 921–930.

Suleman, R., 2020. A brief history of electroconvulsive therapy. Am. J. Psychiatry 16 (1), 6.

Takamiya, A., Kishimoto, T., Hirano, J., Nishikata, S., Sawada, K., Kurokawa, S., Yamagata, B., Kikuchi, T., Mimura, M., 2020. Neuronal network mechanisms associated with depressive symptom improvement following electroconvulsive therapy. Psychol. Med., 1–8. https://doi.org/10.1017/S0033291720001518.

Tang, V.M., Blumberger, D.M., Dimitrova, J., Throop, A., McClintock, S.M., Voineskos, D., Downar, J., Knyahnytska, Y., Mulsant, B.H., Fitzgerald, P.B., Daskalakis, Z.J., 2020. Magnetic seizure therapy is efficacious and well tolerated for treatment-resistant bipolar depression: an open-label clinical trial. J. Psychiatry Neurosci. 45 (5), 313–321. https://doi.org/10.1503/jpn.190098.

Tharyan, P., Adams, C.E., 2005. Electroconvulsive therapy for schizophrenia. Cochrane Database Syst. Rev. 2, CD000076.

Tor, P.-C., Bautovich, A., Wang, M.-J., Martin, D., Harvey, S.B., Loo, C., 2015. A systematic review and meta-analysis of brief versus ultrabrief right unilateral electroconvulsive therapy for depression. J. Clin. Psychiatry 76 (9), e1092–e1098.

Tørring, N., Sanghani, S.N., Petrides, G., Kellner, C.H., Østergaard, S.D., 2017. The mortality rate of electroconvulsive therapy: a systematic review and pooled analysis. Acta Psychiatr. Scand. 135 (5), 388–397. https://doi.org/10.1111/acps.12721.

Tsujii, T., Uchida, T., Suzuki, T., Mimura, M., Hirano, J., Uchida, H., 2019. Factors associated with delirium following electroconvulsive therapy: a systematic review. J. ECT 35 (4), 279–287. https://doi.org/10.1097/YCT.0000000000000606.

UK ECT Review Group, 2003. Efficacy and safety of electroconvulsive therapy in depressive disorders: a systematic review and meta-analysis. Lancet 361 (9360), 799–808.

van Diermen, L., Versyck, P., van den Ameele, S., Madani, Y., Vermeulen, T., Fransen, E., Sabbe, B.G.C., van der Mast, R.C., Birkenhäger, T.K., Schrijvers, D., 2019. Performance of the psychotic depression assessment scale as a predictor of ECT outcome. J. ECT 35 (4), 238–244. https://doi.org/10.1097/YCT.0000000000000610.

Vanelle, J.M., Loo, H., Galinowski, A., de Carvalho, W., Bourdel, M.C., Brochier, P., Bouvet, O., Brochier, T., Olie, J.P., 1994. Maintenance ECT in intractable manic-depressive disorders. Convuls. Ther. 10 (3), 195–205.

Vesga-López, O., Blanco, C., Keyes, K., Olfson, M., Grant, B.F., Hasin, D.S., 2008. Psychiatric disorders in pregnant and postpartum women in the United States. Arch. Gen. Psychiatry 65 (7), 805–815. https://doi.org/10.1001/archpsyc.65.7.805.

Walter, G., Rey, J.M., 2003. Has the practice and outcome of ECT in adolescents changed? Findings from a whole-population study. J. ECT 19 (2), 84–87.

Wang, S., Yang, C., Jia, J., Zhou, Y., Zheng, Y., 2018. Use of electroconvulsive therapy in adolescents with schizophrenia in China. Child Adolesc. Psychiatry Ment. Health 12, 49. https://doi.org/10.1186/s13034-018-0254-z.

Ward, H.B., Fromson, J.A., Cooper, J.J., De Oliveira, G., Almeida, M., 2018. Recommendations for the use of ECT in pregnancy: literature review and proposed clinical protocol. Arch. Womens Ment. Health 21 (6), 715–722. https://doi.org/10.1007/s00737-018-0851-0.

Weiner, R.D., Reti, I.M., 2017. Key updates in the clinical application of electroconvulsive therapy. Int. Rev. Psychiatry 29 (2), 54–62. https://doi.org/10.1080/09540261.2017.1309362.

Wilkinson, S.T., Agbese, E., Leslie, D.L., Rosenheck, R.A., 2018. Identifying recipients of electroconvulsive therapy: data from privately insured Americans. Psychiatr. Serv. 69 (5), 542–548. https://doi.org/10.1176/appi.ps.201700364.

Xie, J., Chen, J., Wei, Q., 2013. Repetitive transcranial magnetic stimulation versus electroconvulsive therapy for major depression: a meta-analysis of stimulus parameter effects. Neurol. Res. 35 (10), 1084–1091. https://doi.org/10.1179/1743132813Y.0000000245.

Chapter 20

MST for treatment-resistant depression

Cory Weissman (R.)[a], Daniel Blumberger (M.)[b,c,d], and Zafiris Daskalakis (J.)[a]

[a]*Department of Psychiatry, School of Medicine UC San Diego Health, University of California San Diego, San Diego, CA, United States,* [b]*General Adult Psychiatry and Health Systems Division, CAMH, Toronto, ON, Canada,* [c]*Temerty Centre for Therapeutic Brain Intervention, CAMH, Toronto, ON, Canada,* [d]*Department of Psychiatry, University of Toronto, Toronto, ON, Canada*

Introduction

Magnetic seizure therapy (MST), as the name implies, is a treatment that involves the induction of therapeutic seizures generated through magnetic fields. The impetus for developing this treatment was the goal of achieving comparable therapeutic effects to ECT, while minimizing its cognitive side effect profile. ECT is known to be a highly effective treatment for treatment-resistant depression (TRD) (UK ECT Review Group, 2003), and this is reviewed in detail in this book in Chapter 19. As highlighted in this chapter, ECT is the most effective treatment for depression with a remission rate of up to 75% in TRD (Kellner et al., 2010). Although it has impressive treatment effects, a significant proportion of patients do not tolerate the treatment due to troublesome side effects that include postictal confusion (Leechuy et al., 1988), anterograde and retrograde memory loss (Sobin et al., 1995), and other cognitive sequelae (Sackeim et al., 2008). In fact, difficulties forming new memories can last up to several months after completion of a treatment course of ECT (Calev et al., 1989, 1991; Ng et al., 2000; Shapira et al., 1998), and although rare, the loss of past memories, formed prior to a treatment course, remains as a serious adverse effect (Fraser et al., 2008).

ECT has been optimized and fine-tuned to improve its tolerability. For example, the development of right unilateral ultrabrief pulse width ECT demonstrated similar efficacy in treating depression, with fewer cognitive side effects in comparison to standard ECT techniques (Sackeim et al., 2008; McClintock et al., 2014). Evidence suggests that higher electrical field charge can predict cognitive outcomes for ECT, and that RUL ECT tends to have far fewer adverse cognitive effects than bilateral ECT (McClintock et al., 2014). This advancement in ECT delivery was made possible with knowledge of the neurophysiology of neuronal depolarization (Sackeim et al., 2008; Peterchev et al., 2010). This supports the notion that rational approaches to improvements in convulsive therapy are warranted and may lead to additional advances in the future. Even with these advances in ECT practice, cognitive side effects and negative stigma still greatly limit its use and accessibility, as less than 1% of patients with TRD end up receiving this treatment (Wilkinson et al., 2018). Therefore, the development of MST brings the promise of a fresh approach with the hope for little to no cognitive side effects, presumably as a result of more focal stimulation and an altogether different method of stimulus (McClintock et al., 2011). Based on the concept of electrical shunting from impedance of the skull and cerebrospinal fluid, electrical fields produced through electromagnetic induction could in theory greatly improve targeting of brain structures and eliminate cognitive side effects. The hope for MST is that it will achieve similar efficacy to ECT, but with a superior cognitive side effect profile to reduce the morbidity for patients with severe psychiatric and treatment-resistant illness.

In this chapter, we first review the origins of MST and then explore recent advances in the use of MST in TRD and related conditions. We then focus on research exploring potential mechanisms of action of MST. This chapter includes a review of the safety, efficacy, and adverse effects of MST.

History and development

The initial development of MST emerged out of work in the mid-1980s on the magnetic stimulation of nerve and muscle tissue in healthy individuals and patients with neurological conditions. Barker was the first to report on subconvulsive magnetic stimulation in the human brain, and described its relative ease of use, limited adverse effects, and the great clinical potential of this technology, in comparison to pure electric stimulation (Barker et al., 1987). Barker found significant ease in dealing with the high resistance structures of the skull in order to produce muscle contractions in the contralateral limb, which was previously seen with electric impulses (Barker et al., 1985). This early work was with magnetic stimulators

Managing Treatment-Resistant Depression. https://doi.org/10.1016/B978-0-12-824067-0.00020-7

capable of frequencies only as high as 0.33 Hz, and so development of magnetic stimulators capable of much higher frequencies of stimulation would later be essential for the advancement of this field in psychiatry, and medicine in general (Pascual-Leone et al., 1993).

Soon after Barker's work researchers using magnetic stimulators capable of repetitive transcranial magnetic stimulation (rTMS) showed that, while overall safe, rTMS had the capability to induce generalized tonic-clonic seizures in healthy individuals without epilepsy with frequencies as low as 10 Hz stimulation (Pascual-Leone et al., 1993). While unwanted in this particular study, this finding would later be the basis for purposefully induced therapeutic seizures with magnetic stimulation. In 1994, Sackeim commented on the potential of this technology and described how MST could drastically increase the specificity of brain areas targeted in convulsive type treatment, which could theoretically reduce or eliminate side effects seen with ECT treatment; brought about by discharge of deeper brain structures (Sackeim, 1994). This then spurred the first investigations of magnetic seizure therapy in mammals and primates, and later in human subjects.

Lisanby et al. performed the seminal translational research and early troubleshooting that brought MST from the bench to the bedside, with the development of optimal treatment parameters and targeting of specific brain regions (Lisanby and Deng, 2015). The first MST device could produce frequencies of 25 Hz, and would not induce seizures reliably in macaques (Lisanby and Deng, 2015). Yet, they were able to demonstrate consistent seizures with a device capable of a treatment frequency of 40 Hz with a 6.3 s train duration and machine output of 90% intensity. With this device though the authors describe struggles with overheating coils, which would be a major limitation in later clinical trials (Lisanby et al., 2001a). A subsequent report demonstrated the first successful use of MST in a human, a case report of a young female patient diagnosed with TRD who received four MST treatments during a course of ECT (Lisanby et al., 2001b). In this case, 40 Hz stimulation was administered over the right prefrontal cortex for 4 s; the patient did experience modest clinical improvement (Lisanby et al., 2001b).

MST devices capable of 50 Hz frequency for 8 s train durations were developed next (Lisanby and Deng, 2015), and in 2003, Kosel et al. published a case report on the first full MST treatment course for a patient with TRD with these treatment parameters (Kosel et al., 2003). This patient was in remission after 12 treatments, and importantly, there were minimal side effects, no memory issues, and a fast reorientation time after treatments, with overall improvement in verbal and nonverbal learning tasks (Kosel et al., 2003). Concurrent with the above human studies, a histological analysis in 12 macaques randomized to full courses of ECT, MST and sham demonstrated a clear absence of structural brain damage with MST treatment (Dwork et al., 2004). These early reports suggested both safety and efficacy of MST, and they set the stage for several case series and small pilot studies on the safety and feasibility of MST as a treatment for TRD.

In the first clinical trial to prospectively investigate the effects of MST in patients with TRD, Lisanby et al. studied the effect of MST in 10 inpatients in a within-subject comparison to ECT during an acute treatment course (Lisanby et al., 2003). In this study, the authors treated these inpatients with two MST treatments, with a maximum output of 60 Hz treatment at 100% output for 6.6 s with successful tonic-clonic seizures in all 10 patients (Lisanby et al., 2003). The authors experimented with three different coil arrangements and with different coil types as well in this study (Lisanby et al., 2003). They found an overall more favorable cognitive side effect profile posttreatment with MST treatments when compared to ECT treatments on the same patients during the same treatment course (Lisanby et al., 2003). This study, however, was not designed to compare the efficacy of MST to that of ECT. It is also important to note that the authors reasoned that the maximum treatment output was limited with the MST technology of that time, and they suggested that a direct comparison of efficacy and tolerability for equivalently dosed MST and ECT would be needed in the future. We review studies that address these specific issues below.

By this point, the High Dose MST device was developed to answer the issue of equivalent dosing for ECT and MST, as it was able to provide stimulation at 6 × the seizure threshold, identical to that used in right unilateral ECT. This MST device also provides 100 Hz frequency for a duration of 10 s (Lisanby and Deng, 2015) and is currently used in major clinical trials of MST. Kirov et al. were the first to test this machine in a pilot, feasibility study in 11 patients with MDD or schizoaffective disorder (Kirov et al., 2008). They tested both frontal and vertex stimulation targets with more consistent seizures found with the vertex stimulation approach. No adverse effects were seen, and similar to that seen in the above studies, reorientation time after stimulation was much quicker with MST compared to ECT in these patients. Further technical advances in the delivery of MST would allow for more reliable seizure induction in the frontal cortex target, the main target in recent, large clinical trials described below.

Clinical studies

Recent MST studies generally incorporate larger patient samples and use more rigorous clinical trial designs than those above, which is now culminating in the large, prospective RCTs currently underway in the field. Below we explore these recent clinical studies being completed by a variety of research groups in several countries.

White et al., initially reported on the essential components of anesthesia for patients undergoing MST versus ECT. This was a prospective, nonrandomized controlled trial in which 10 patients underwent MST with 10 matched controls. One key finding from this study confirmed previous reports that MST produced a marked reduction in reorientation times compared with ECT. Another finding was that patients required less succinylcholine in MST treatment, and that this dose reduction is needed to reduce the risk of waking paralysis (White et al., 2006). Authors also found that the patients who underwent MST required less antihypertensive medication, had less postictal suppression after seizures, and they also found, while still significant, there was less improvement in depression scores as indexed by the Hamilton Rating Scale for Depression. This may have been due to underdosing of patients receiving MST (1.3 × seizure threshold with MST versus 2.5 × with ECT), which was a common problem in early MST trials, as described above. There was an equivalent hemodynamic response from the two treatments. Of note, the more rapid reorientation time seen with MST here is a consistent finding evident in larger clinical trials that will be elaborated on in sections below.

In one of the earlier studies aiming to directly compare the efficacy and safety profile of MST with ECT in patients with unipolar and bipolar depression, Kayser et al. did not find any difference in clinical improvement between the groups ($N = 20$, total) (Kayser et al., 2011). MST treatment was delivered with 100 Hz frequency over the vertex for up to 6 s duration, compared to RUL ECT with a pulse width of 0.5 ms. There were less cognitive side effects seen with MST versus ECT on a test of neglect, and orientation time was significantly better with MST. In this study, there were no issues with eliciting reliable seizures with MST, which had been raised as a concern in the past. Nonetheless, this was an open label study of small sample size. This group has also reported on a mixed sample of 26 patients with MDD and BD treated with 100 Hz MST up to 10 s of stimulation with 100% machine output, with the ability to treat at 6 × seizure threshold. In this study, treatment was for up to 12 sessions (Kayser et al., 2015). Authors found a 69% response rate and 46.2% remission rate in this study, with half of responders relapsing several months after treatment. They did not find differences in seizure characteristics between responders and nonresponders. Cognitive outcomes were mostly positive, with improvements in visuospatial learning and memory, visual perception, executive functioning, and worsening in verbal fluency. Kayser et al. followed this study with an analysis of predictors of response to MST with 12 additional patients added to the above group ($N = 38$) (Kayser et al., 2019). They found a significant higher frequency of family history of mood disorders, fewer past MDEs, lower baseline HRSD scores, and potentially few symptoms of anhedonia and somatic anxiety in MST responders. Remitters did tend to have less anxiety symptoms as well (Kayser et al., 2019).

In their most recent report, Kayser et al. focused on the neuropsychological outcomes of MST in a group of 10 patients diagnosed with MDD or BDII (Kayser et al., 2020). Patients again were treated with 100 Hz stimulation over the vertex. Six patients responded and four remitted, and importantly, other than decreases in the German test RWT formal-lexical-verbal fluency, there were no significant decreases in other cognitive domains, including retrograde amnesia on the Autobiographical Memory Interview-Short Form (AMI-SF). There have been difficulties in properly assessing autobiographical memory in MST studies, because of inherent issues in the AMI-SF; this issue and a further exploration of the cognitive sequelae of MST in TRD is expanded on below.

Fitzgerald et al. published several studies important for the development of MST. In 2013, they published the first study in which MST was delivered three times per week, similar to common practices for acute ECT courses; MST was delivered at 100 Hz frequency over 10 s with the vertex as the brain target (Fitzgerald et al., 2013). Thirteen patients with TRD were given up to 18 MST treatments, and while response and remission rates were lower than most other studies reported here (38.5% and 15%, respectively), improvements in depression were significant, and there were no significant adverse cognitive effects, including with memory and orientation time. As described previously by White et al. (2006), authors here encountered the issue of early wakening paralysis due to early reorientation with MST. They concluded that alternative brain targets, including frontal and unilateral approaches should be explored, which would become the focus of later studies . In a follow-up study, Fitzgerald et al. completed the largest study to date comparing MST to ECT in TRD, with RUL ECT of 1.0 ms pulse width at three times seizure threshold, in a double-blind RCT. While underpowered ($N = 37$), they did not find a significant difference in the primary depression outcomes between the two groups. They did find relatively low efficacy in both groups actually (Fitzgerald et al., 2018), possibly due to the chronicity of illness of this group and use of propofol as the anesthetic agent, which is known to decrease the quality and duration of seizures (Hoyer et al., 2014). For MST treatment, the vertex was targeted at 100 Hz frequency at 3 times per week up to 15 treatments. The authors did find that the cognitive measure of digit symbol coding improved more so with MST over ECT.

In a recent rater-blinded nonrandomized comparator trial, Zhang et al. compared treatments of 100 Hz MST over the vertex to bifrontal ECT in patients with major depressive episodes, although not clearly with TRD, for up to six treatments (Zhang et al., 2020). While the duration of this treatment trial appears to be short in comparison to trials above, authors found no significant differences in depression outcomes between groups, and did replicate findings of faster recovery of consciousness, orientation and breathing after treatment with MST compared to ECT; this extended to improvements in formal cognitive measures of memory and attention (Zhang et al., 2020).

Recently, Daskalakis et al. published the largest open-label trial of MST in TRD to date ($N = 86$). Patients in this study were treated, importantly, with prefrontal MST two-to-three times per week in patient blocks divided by treatment frequency (25, 50–60, and 100 Hz treatment) up to a maximum of 24 treatments, or until remission of their depression (Daskalakis et al., 2019). Authors here reported a finding of potentially higher clinical efficacy on overall depression for patients receiving high frequency (100 Hz) versus low frequency (25 Hz) MST, with a response and remission rate of 41.7% and 33.5%, respectively, in this 100 Hz treatment group. High frequency MST was associated with shorter reorientation time after treatment as well. As alluded to above, while there was decline in the specific cognitive test AMI-SF in this study, there are methodological issues with this scale, as declining scores equivalent to those in this MST treatment group have been found with pharmacotherapy alone (Daskalakis et al., 2019; Kessler et al., 2014). Authors also found improvements in visuospatial memory and delayed recall in this study. Adverse events included induction of mania in one patient, one fall and shoulder dislocation, and one case of superficial burn from MST coil. The much anticipated follow-up study, comparing MST to ECT directly in a randomized controlled trial, is now currently underway (ClinicalTrials.gov Identifier: NCT03191058).

Researchers are now starting to capitalize on these early, promising results to explore novel treatment delivery schedules for MST in TRD patients. Wang et al. posited that since MST has fewer cognitive side effects than ECT, it could potentially be tolerable for patients to receive MST treatments every day for 6 continuous working days. In their small case series using this treatment schedule, these authors provided 100 Hz MST over the vertex and demonstrated early efficacy in two of three patients with no cognitive sequelae, and some actual improvement in cognition during this interval. They did not have any extended follow-up period to assess for relapse though (Wang et al., 2019). In a similar, larger follow-up study, 15 patients were treated with 100 Hz vertex stimulation for 10 s on 6 consecutive weekdays. There was a 60% response and 47% remission rate, with some improvement in measures of memory and other neurocognitive profiles. There were no serious adverse events and no dropouts. However, there was heterogeneity in the severity of the patients' depressive illness, some experiencing apparent mild-to-moderate episodes with limited treatment refractoriness, and many patients were started on new antidepressants the day before MST treatment began (Wang et al., 2020). Nonetheless, the main takeaway from these recent studies is that with the milder adverse effect profile and novel mechanism of action, there is room for flexible experimentation with the MST delivery schedule that ECT cannot allow for given its cognitive side effect profile.

Adverse effects and interactions

Study of the adverse effects of MST has expectedly focused on cognitive sequelae, as described above. In general, MST has proven to be cognitively safe, and potentially pro-cognitive, in the treatment of TRD. Further evidence for this is described below. We also explore in this section other adverse effects of MST. It is important to note that the number of patients who have received MST to date is still relatively small, so there is a possibility of unforeseen, rare effects emerging in future studies.

One of the main motivating factors for the development of MST, as described above, is to create a treatment without the debilitating cognitive side effects of ECT and with similar efficacy. McClintock et al. performed a systematic review of the neurocognitive effects of MST in computer, animal and human models. They highlight the shorter reorientation time, limited retrograde amnesia, and complete absence of anterograde amnesia with MST compared to ECT, and deem MST cognitively safe (McClintock et al., 2011). Polster et al. performed an open-label study comparing specifically the effects of 100 Hz MST over the vertex and RUL ECT on memory retrieval, including cued memory, in patients diagnosed with TRD (Polster et al., 2015). They used a within subject open-label design that lacked blinding, and treated patients with 10–12 treatments. Patients receiving MST in this study performed equally well to healthy controls on tests of delayed recall, while those receiving ECT performed worse (Polster et al., 2015). There were no differences between groups on cued recall (Polster et al., 2015). As described above, the evidence to date suggests that the cognitive sequelae of MST in patients with TRD is minimal, and drastically less than for those receiving ECT. The definitive head-to-head studies investigating these effects and treatment mechanisms are underway now.

In their early work, Lisanby et al. described that at the neuropathological level there have been no neuropathological changes found with MST treatment, and limited hippocampal remodeling, when compared to ECT, in animal models (Lisanby and Deng, 2015). In terms of physiological response, in comparison to ECT in human patients, MST appears to deliver less of a sympathetic surge postseizure, and less parasympathetic mediated bradycardia (Lisanby and Deng, 2015); this is in addition to less physical and cognitive side effects, corroborated in later studies and discussed thoroughly above.

Also as described above, similar to ECT, in MST treatment there is the risk of wakening paralysis, especially during dose titrations in the initial treatment phase. While not associated with any lethality or long-term physical effects, this potential adverse effect is unpleasant for patients.

There are other potential adverse effects not related to cognition in MST treatment. As described above, due to the risk of magnetic coils overheating, there is a risk of superficial burns with this treatment (Daskalakis et al., 2019). Noda et al. (2015) also describe a potential adverse effect of MST in bipolar depression, which is also seen with ECT treatment: they describe two patients that achieved remission of their depressive episodes with subsequent emergence of mania or hypomania requiring pharmacotherapy to resolve this mood switch (Noda et al., 2015). It was not entirely clear in these cases though that MST caused the switch in mood, as it may have been due to lithium taper in one patient, and due to natural evolution of disease in the other case. Nonetheless, manic switch from MST treatment appears to be a potential adverse effect shared in common with ECT.

Mechanism of action

There is limited understanding of the mechanisms of action of MST for relieving depression in patients diagnosed with TRD. It is assumed that the convulsion induced by MST is a necessary component, but this has not been confirmed in sham-controlled trials, as previously done with ECT (UK ECT Review Group, 2003). Nonetheless, as MST is a relatively novel treatment, our understanding of its mechanism benefits from concurrent advances in neuroscience occurring over the past few decades. Many of the above described clinical studies were completed with parallel, add-on investigations into the neurobiological mechanisms involved. Below we summarize this intriguing work into the mechanism of action of MST in TRD, with a focus on biomarkers and predictors of response with electroencephalogram (EEG), transcranial magnetic stimulation-EEG (TMS-EEG) and neuroimaging technologies.

One consistent finding in studies of MST compared to ECT is less postictal suppression after MST. Researchers have also described shorter seizure durations, lower ictal EEG amplitude, as well as a significant effect of head diameter on the variability of MST effects (Lisanby and Deng, 2015). The significance of these differences are explored below. However, before further exploring results from human-based studies, it is important to review the evidence from computational models of MST. Computational models are needed for estimations of electric fields produced by MST in humans, so that we can create better hypotheses of the mechanisms of MST in the brain. In vivo estimations of electric fields in live humans are not possible, and animal models are limited in themselves both by measurement technique and marked differences in neurophysiology compared to humans. To address this issue, Deng et al. completed a thorough and revealing study, which was the first computational model to compare electric field strength between MST and ECT (Deng et al., 2011). Authors used a simplified spherical head model and compared the treatment parameters of multiple MST and ECT configurations, including double-cone MST and bilateral and right unilateral ECT. They found that MST consists of a higher activation threshold, yet similar electric field strength to ECT, and importantly, that MST electrical fields attenuate markedly more rapidly as a function of brain depth, with a field strength approaching zero at the center of the brain when compared to ECT. ECT tends to produce a strong field throughout the entire brain, especially in the case of bilateral ECT. In fact, when compared to ECT with standard parameters of 800 mA current and 0.3 ms pulse width, MST has 10–60 times more focal stimulation (Deng et al., 2011). This finding is essential to understand the current, major hypothesis as to why MST does not produce the cognitive side effects that ECT does: the area of focus with MST does not include deeper brain structures involved in memory like the hippocampus. This notion is elaborated on above in the clinical studies section of this chapter.

Neuroimaging studies investigating the effects of MST with pre-MST and post-MST imaging have been helpful in elucidating the mechanism of action of MST. In the first study published in this area, Kosel et al. reported the first case of a full MST treatment course in a patient with TRD (Kosel et al., 2003). Along with clinical improvement, the authors observed increased frontoparietal and basal ganglia perfusion posttreatment on single photon emission computed tomography (SPECT) imaging (Kosel et al., 2003). In a study involving 10 patients diagnosed with TRD, Hoy et al. used fluorodeoxyglucose-positron emission tomography (FDG-PET) pre- and post- at least 6 completed MST treatments, and they found increased metabolism in several brain areas posttreatment, especially left-sided brain increases; specifically increases were observed in the basal ganglia (globus pallidus, substantia nigra, and putamen), as well as the orbitofrontal cortex (OFC), medial-prefrontal cortex (MPFC), and dorsolateral prefrontal cortex (DLPFC). The authors suggest that they found a "rebalancing" of signals in limbic-cortical regions (Hoy et al., 2013). This was one of the first studies to clearly demonstrate potential differences in mechanism between ECT and MST in a human subject, as ECT treatment is associated with decreases in brain activity, but here authors found clear, consistent and robust increases in brain activity, more in line with the mechanism of action of rTMS (Hoy et al., 2013). Lastly, in one study completed in parallel with a clinical trial described above, Kayser et al. treated 26 patients with MDD and BD with 100 Hz MST up to 12 sessions, and used FDG-PET imaging as an investigational biomarker (Kayser et al., 2015). They found increased brain metabolism in the frontal cortex and decreased activity in the left striatum in the 12 patients included in this analysis.

Several researchers have used simple analyses of EEG data to investigate the mechanism of action of MST. In conjunction with one of their clinical studies, Kayser et al. analyzed bifrontal EEG output to characterize the seizure quality in MST versus ECT (Kayser et al., 2013). They measured seizures in 7 patients diagnosed with TRD, and 1 with BD, whom received 12 MST treatments over the vertex at 100 Hz for up to 10 s. This was compared in open-label fashion and intraindividually to an ECT course of brief pulse width. They found no visible differences in both motor and EEG seizures between ECT and MST, although seizures were nonsignificantly longer with ECT, especially the polyspike phase. Similar to other studies, they did also find a significantly quicker reorientation time with MST versus ECT, approximately 2 versus 8 min (Kayser et al., 2013). Also, it is important to note that there was no significant clinical improvement in either group, suggesting a possible failed study (Kayser et al., 2013). Soehle et al. completed a similar analysis on separate groups of patients diagnosed with TRD receiving either MST or ECT, and added an analysis of the bispectral index (BIS), a measure of depth of anesthesia (Soehle et al., 2014). They found significantly longer seizures with ECT compared to MST, and also found faster recovery time and eye opening with MST. Interestingly, BIS was significantly lower in the ECT group at all time points compared to MST, demonstrating a lower level of wakefulness associated with this treatment, even when anesthesia doses were controlled for (Soehle et al., 2014). Last, Backhouse et al. analyzed seizure quality in patients receiving MST at low, moderate and high frequencies (25, 50–60, and 100 Hz) with fronto-mastoid EEG, and found better quality seizures in moderate and low frequency MST compared to high frequency based on a scoring system derived from the ECT literature (Backhouse et al., 2018). Seizure quality correlated with improvement in suicidality (Weissman et al., 2020), but not with overall depression. Authors here hypothesized that shorter seizures, and presumably lower seizure quality in high frequency MST, which actually correlated with better overall depression improvement in MST, may be due to early transition in seizures to the slow wave phase, which may be reflective of better pyramidal cell entrainment. Overall, it is still unclear what constitutes a good quality seizure in MST treatment, and whether this is an essential component of its mechanism.

Building on this early work in EEG, several groups have attempted to further explore the complexities and richness of EEG data in order to better understand MST mechanism of action. Using resting state EEG (RS-EEG) in 21 patients with MDD, Hill et al. found that MST led to upregulation of synchronization in the theta band subnetworks of the bilateral frontal and parieto-occipital areas, possibly suggesting an overlapping mechanism of action with rTMS (Hill et al., 2020). These authors also found decreased beta spectrum connectivity, and clinically, they demonstrated that change in theta connectivity predicted HDRS-24 improvement with a sensitivity of 89% and specificity of 67%. In a similar analysis of 34 patients with TRD treated with MST and ECT, Farzan et al. analyzed RS-EEG data and showed greater reductions in gamma frequency and in the complexity of fine timescale signal in the fronto-central and parieto-occipital regions; these changes were associated with improvement in depression for both ECT and MST. Also, greater increases in course time scales was linked to worsening cognition in ECT, which was not seen in MST (Farzan et al., 2017). One theory authors discussed here is that MST may mirror the positive changes in brain oscillations with ECT, by decoupling aberrant frontoparietal and default mode network (DMN) circuit connections in the brain, without the negative changes of increasing slow oscillations, which are associated with cognitive impairment. Atluri et al. also used RS-EEG in a comparison of TRD patients receiving ECT ($N = 22$) and MST ($N = 24$), as well as healthy controls. They analyzed brain microstates, which correspond to networks in the brain, and found that an increase in State A duration and a decrease in State B, C and D frequency associated with depression improvement with ECT. They also found that for MST, change in suicidality correlated with shorter duration of states A, B, C and a higher frequency of state D, possibly through alterations in the DMN, and that a decrease in the frequency of state B correlated with improved cognition for MST (Atluri et al., 2018). Similar to above, this group found that increases in slow cortical oscillations correlated with ECT and not with MST (Atluri et al., 2018). Overall, while complex and exploratory, these studies suggest a different mechanism of action on brain circuitry between MST and ECT, with evidence suggesting that the lack of cognitive adverse effects with MST may be correlated with a lack of increase in slow wave oscillations that occurs consistently with ECT. It is still unclear how these brain dynamics equate to improvement of depression in MST treatment, and how varying MST treatment frequencies affect this. Further work in these areas is needed.

A relatively novel neurobiological technology that can probe brain function combines transcranial magnetic stimulation and EEG (TMS-EEG). TMS-EEG has recently been applied as a potential biomarker for MST with the hopes of providing nuanced neurophysiologic information that can guide MST treatment selection for patients. Only one study to date has investigated this technique as a marker specifically for depression response in MST. In this study, Hadas et al. looked at TMS-EEG as an index of connectivity between the subgenual cingulate cortex (SGC) and the DLPFC. They assessed this in 31 patients receiving MST, with TMS-EEG measures pretreatment and posttreatment (Hadas et al., 2020). Reduced connectivity between the SGC and DLPFC correlated with depression response on the HRSD-24, and reductions in hippocampal activation were associated with a decline in cognition, as indexed by the MoCA scale, generally agreeing with results from the above studies in other technological modalities (Hadas et al., 2020). Sun et al. published two studies investigating TMS-EEG as a biomarker to predict response of suicidality in patients with TRD treated with MST. They found that improvement in suicidal ideation (indexed by the Beck Scale for Suicidal Ideation) was correlated with changes in the EEG measures N100 and long-interval cortical inhibition (LICI) over the DLPFC, and they suggest that these measures are possible correlates of GABA interneuron function, potentially recruited by MST through transsynaptic depolarization (Sun et al., 2016). In their follow-up study, Sun et al. found significant correlation between change in LICI over frontal and central electrodes and change in suicidal ideation in responders; the reduction in LICI identified those achieving resolution of suicidal ideation with 90% sensitivity and 88% specificity (Sun et al., 2018). It is hypothesized that the decreases in GABA interneuron function account for increased neuroplasticity, indexed by changes in LICI, correlating with improvements in suicidality.

Researchers have explored alternative biomarkers in recent MST studies. One study explored heart rate variability (HRV) as a biomarker for treatment response in MST. HRV is known to be decreased in patients diagnosed with MDD (Bassett, 2016). Noda et al. measured HRV pre- and post-MST treatment (Noda et al., 2020). Authors here found an association between baseline HRV (i.e., root mean square of successive r-r interval [RMSSD]) and improvement in executive function, but not depressive symptomatology. This improvement in cognitive functioning was hypothesized to be mediated through normalization of DMPFC and dorsal anterior cingulate functioning, known to be important in the regulation of cardiovascular physiology.

In sum, exciting research happening now is beginning to explore and explain the neurobiological underpinnings of the mechanisms of action of MST in TRD. There is evidence suggesting dual, parallel antidepressive and antisuicidal effects in TRD patients, with cognitive sparing effects.

Other indications

The majority of studies on MST in humans have focused on MDD, and specifically TRD. Yet, there have been a number of more recent, exploratory studies on additional potential psychiatric indications for MST. There are several studies focused solely on bipolar depression. Kayser et al. was first to publish on this in a case report of a patient with BD treated with 100 Hz MST for nine treatments with good effect, and a partial relapse after 6 months (Kayser et al., 2009). Noda et al. demonstrated the safe and efficacious use of frontal MST in an adolescent who was diagnosed with bipolar disorder II depression, , treated with 18 MST sessions (Noda et al., 2014). Tang et al. most recently published the largest trial of MST for patients diagnosed with BD in 26 patients completing an adequate trial of MST (vertex 100 Hz, or frontal 25, 50, 60, or 100 Hz) (Tang et al., 2020a). There was a 38.5% response rate and 23% remission rate with limited cognitive sequelae in this study; one patient in the study did have emergence of hypomania that was responsive to medication treatment. In parallel with the prospective, randomized trial underway in unipolar depression, a large randomized trial of ECT versus MST is currently underway for bipolar depression (ClinicalTrials.gov Identifier: NCT03641300).

Several secondary analyses have explored other indications for MST. Two studies report on suicidality itself as a treatment target for MST in both unipolar (Weissman et al., 2020) and bipolar depression (Tang et al., 2020b), and authors suggest that this may be an independent treatment target separate from core unipolar and bipolar depressive illness. There has been one negative trial on the use of MST in obsessive compulsive disorder (Tang et al., 2020c), and one small trial in schizophrenia suggesting feasibility and potential efficacy for MST (Tang et al., 2017). Other exploratory investigations are underway, including for borderline personality disorder (ClinicalTrials.gov Identifier: NCT03361826).

Conclusion

MST is an emerging interventional psychiatric treatment with growing evidence to support its efficacy and safety in the treatment of TRD. Research is demonstrating that the mechanisms of action of MST on the brain differ from those of ECT, with apparently far fewer cognitive side effects. This is indeed an exciting time for research in MST, as years of translational work are now being followed by large, impactful clinical trials. MST is poised to become an essential, first line treatment for patients struggling with TRD.

References

Atluri, S., et al., 2018. Selective modulation of brain network dynamics by seizure therapy in treatment-resistant depression. Neuroimage Clin. 20, 1176–1190. https://doi.org/10.1016/j.nicl.2018.10.015.

Backhouse, F.A., et al., 2018. Characteristics of ictal EEG in magnetic seizure therapy at various stimulation frequencies. Clin. Neurophysiol. 129, 1770–1779. https://doi.org/10.1016/j.clinph.2018.03.025.

Barker, A.T., Jalinous, R., Freeston, I.L., 1985. Non-invasive magnetic stimulation of human motor cortex. Lancet 1, 1106–1107. https://doi.org/10.1016/s0140-6736(85)92413-4.

Barker, A.T., Freeston, I.L., Jalinous, R., Jarratt, J.A., 1987. Magnetic stimulation of the human brain and peripheral nervous system: an introduction and the results of an initial clinical evaluation. Neurosurgery 20, 100–109. https://doi.org/10.1097/00006123-198701000-00024.

Bassett, D., 2016. A literature review of heart rate variability in depressive and bipolar disorders. Aust. N. Z. J. Psychiatry 50, 511–519. https://doi.org/10.1177/0004867415622689.

Calev, A., et al., 1989. Distinct memory impairments following electroconvulsive therapy and imipramine. Psychol. Med. 19, 111–119.

Calev, A., et al., 1991. Disorientation and bilateral moderately suprathreshold titrated ECT. Convuls. Ther. 7, 99–110.

Daskalakis, Z.J., et al., 2019. Magnetic seizure therapy (MST) for major depressive disorder. Neuropsychopharmacology. https://doi.org/10.1038/s41386-019-0515-4.

Deng, Z.D., Lisanby, S.H., Peterchev, A.V., 2011. Electric field strength and focality in electroconvulsive therapy and magnetic seizure therapy: a finite element simulation study. J. Neural Eng. 8. https://doi.org/10.1088/1741-2560/8/1/016007, 016007.

Dwork, A.J., et al., 2004. Absence of histological lesions in primate models of ECT and magnetic seizure therapy. Am. J. Psychiatry 161, 576–578. https://doi.org/10.1176/appi.ajp.161.3.576.

Farzan, F., et al., 2017. Brain temporal complexity in explaining the therapeutic and cognitive effects of seizure therapy. Brain 140, 1011–1025. https://doi.org/10.1093/brain/awx030.

Fitzgerald, P.B., et al., 2013. Pilot study of the clinical and cognitive effects of high-frequency magnetic seizure therapy in major depressive disorder. Depress. Anxiety 30, 129–136. https://doi.org/10.1002/da.22005.

Fitzgerald, P.B., et al., 2018. A pilot study of the comparative efficacy of 100 Hz magnetic seizure therapy and electroconvulsive therapy in persistent depression. Depress. Anxiety 35, 393–401. https://doi.org/10.1002/da.22715.

Fraser, L.M., O'Carroll, R.E., Ebmeier, K.P., 2008. The effect of electroconvulsive therapy on autobiographical memory: a systematic review. J. ECT 24, 10–17. https://doi.org/10.1097/YCT.0b013e3181616c26. 00124509-200803000-00003 [pii].

Hadas, I., et al., 2020. Subgenual cingulate connectivity and hippocampal activation are related to MST therapeutic and adverse effects. Transl. Psychiatry 10, 392. https://doi.org/10.1038/s41398-020-01042-7.

Hill, A.T., et al., 2020. Resting-state electroencephalographic functional network alterations in major depressive disorder following magnetic seizure therapy. Prog. Neuropsychopharmacol. Biol. Psychiatry. https://doi.org/10.1016/j.pnpbp.2020.110082, 110082.

Hoy, K.E., et al., 2013. Effect of magnetic seizure therapy on regional brain glucose metabolism in major depression. Psychiatry Res. 211, 169–175. https://doi.org/10.1016/j.pscychresns.2012.08.003.

Hoyer, C., Kranaster, L., Janke, C., Sartorius, A., 2014. Impact of the anesthetic agents ketamine, etomidate, thiopental, and propofol on seizure parameters and seizure quality in electroconvulsive therapy: a retrospective study. Eur. Arch. Psychiatry Clin. Neurosci. 264, 255–261. https://doi.org/10.1007/s00406-013-0420-5.

Kayser, S., Bewernick, B., Axmacher, N., Schlaepfer, T.E., 2009. Magnetic seizure therapy of treatment-resistant depression in a patient with bipolar disorder. J. ECT 25, 137–140. https://doi.org/10.1097/YCT.0b013e31817dc45a.

Kayser, S., et al., 2011. Antidepressant effects, of magnetic seizure therapy and electroconvulsive therapy, in treatment-resistant depression. J. Psychiatr. Res. 45, 569–576. https://doi.org/10.1016/j.jpsychires.2010.09.008.

Kayser, S., Bewernick, B.H., Hurlemann, R., Soehle, M., Schlaepfer, T.E., 2013. Comparable seizure characteristics in magnetic seizure therapy and electroconvulsive therapy for major depression. Eur. Neuropsychopharmacol. 23, 1541–1550. https://doi.org/10.1016/j.euroneuro.2013.04.011.

Kayser, S., et al., 2015. Magnetic seizure therapy in treatment-resistant depression: clinical, neuropsychological and metabolic effects. Psychol. Med. 45, 1073–1092. https://doi.org/10.1017/S0033291714002244.

Kayser, S., Bewernick, B.H., Wagner, S., Schlaepfer, T.E., 2019. Clinical predictors of response to magnetic seizure therapy in depression: a preliminary report. J. ECT 35, 48–52. https://doi.org/10.1097/YCT.0000000000000495.

Kayser, S., Bewernick, B.H., Wagner, S., Schlaepfer, T.E., 2020. Effects of magnetic seizure therapy on anterograde and retrograde amnesia in treatment-resistant depression. Depress. Anxiety 37, 125–133. https://doi.org/10.1002/da.22958.

Kellner, C.H., et al., 2010. Bifrontal, bitemporal and right unilateral electrode placement in ECT: randomised trial. Br. J. Psychiatry 196, 226–234. https://doi.org/10.1192/bjp.bp.109.066183. 196/3/226 [pii].

Kessler, U., et al., 2014. The effect of electroconvulsive therapy on neurocognitive function in treatment-resistant bipolar disorder depression. J. Clin. Psychiatry 75, e1306–e1313. https://doi.org/10.4088/JCP.13m08960.

Kirov, G., et al., 2008. Quick recovery of orientation after magnetic seizure therapy for major depressive disorder. Br. J. Psychiatry 193, 152–155. https://doi.org/10.1192/bjp.bp.107.044362.

Kosel, M., Frick, C., Lisanby, S.H., Fisch, H.U., Schlaepfer, T.E., 2003. Magnetic seizure therapy improves mood in refractory major depression. Neuropsychopharmacology 28, 2045–2048. https://doi.org/10.1038/sj.npp.1300293.

Leechuy, I., Abrams, R., Kohlhaas, J., 1988. ECT-induced postictal delirium and electrode placement. Am. J. Psychiatry 145, 880–881.

Lisanby, S.H., Deng, Z.D., 2015. In: Reti, I.M. (Ed.), Brain Stimulation: Methodologies and Interventions. Wiley-Blackwell, pp. 123–144.

Lisanby, S.H., Luber, B., Finck, A.D., Schroeder, C., Sackeim, H.A., 2001a. Deliberate seizure induction with repetitive transcranial magnetic stimulation in nonhuman primates. Arch. Gen. Psychiatry 58, 199–200. https://doi.org/10.1001/archpsyc.58.2.199.

Lisanby, S.H., Schlaepfer, T.E., Fisch, H.U., Sackeim, H.A., 2001b. Magnetic seizure therapy of major depression. Arch. Gen. Psychiatry 58, 303–305. https://doi.org/10.1001/archpsyc.58.3.303.

Lisanby, S.H., Luber, B., Schlaepfer, T.E., Sackeim, H.A., 2003. Safety and feasibility of magnetic seizure therapy (MST) in major depression: randomized within-subject comparison with electroconvulsive therapy. Neuropsychopharmacology 28, 1852–1865. https://doi.org/10.1038/sj.npp.1300229.

McClintock, S.M., Tirmizi, O., Chansard, M., Husain, M.M., 2011. A systematic review of the neurocognitive effects of magnetic seizure therapy. Int. Rev. Psychiatry 23, 413–423. https://doi.org/10.3109/09540261.2011.623687.

McClintock, S.M., et al., 2014. Multifactorial determinants of the neurocognitive effects of electroconvulsive therapy. J. ECT 30, 165–176. https://doi.org/10.1097/YCT.0000000000000137.

Ng, C., et al., 2000. Efficacy and cognitive effects of right unilateral electroconvulsive therapy. J. ECT 16, 370–379.

Noda, Y., et al., 2014. Magnetic seizure therapy in an adolescent with refractory bipolar depression: a case report. Neuropsychiatr. Dis. Treat. 10, 2049–2055. https://doi.org/10.2147/NDT.S71056.

Noda, Y., et al., 2015. Magnetic seizure therapy-induced mania: a report of 2 cases. J. ECT 31, e4–e6. https://doi.org/10.1097/YCT.0000000000000145.

Noda, Y., et al., 2020. Vagally mediated heart rate variability is associated with executive function changes in patients with treatment-resistant depression following magnetic seizure therapy. Neuromodulation. https://doi.org/10.1111/ner.13262.

Pascual-Leone, A., et al., 1993. Safety of rapid-rate transcranial magnetic stimulation in normal volunteers. Electroencephalogr. Clin. Neurophysiol. 89, 120–130. https://doi.org/10.1016/0168-5597(93)90094-6.

Peterchev, A.V., Rosa, M.A., Deng, Z.D., Prudic, J., Lisanby, S.H., 2010. Electroconvulsive therapy stimulus parameters: rethinking dosage. J. ECT 26, 159–174. https://doi.org/10.1097/YCT.0b013e3181e48165.

Polster, J.D., Kayser, S., Bewernick, B.H., Hurlemann, R., Schlaepfer, T.E., 2015. Effects of electroconvulsive therapy and magnetic seizure therapy on acute memory retrieval. J. ECT 31, 13–19. https://doi.org/10.1097/YCT.0000000000000130.

Sackeim, H., 1994. Magnetic stimulation therapy and ECT. Convuls. Ther. 10, 255–258.

Sackeim, H.A., et al., 2008. Effects of pulse width and electrode placement on the efficacy and cognitive effects of electroconvulsive therapy. Brain Stimul. 1, 71–83. https://doi.org/10.1016/j.brs.2008.03.001.

Shapira, B., et al., 1998. Cost and benefit in the choice of ECT schedule. Twice versus three times weekly ECT. Br. J. Psychiatry 172, 44–48.

Sobin, C., et al., 1995. Predictors of retrograde amnesia following ECT. Am. J. Psychiatry 152, 995–1001.

Soehle, M., Kayser, S., Ellerkmann, R.K., Schlaepfer, T.E., 2014. Bilateral bispectral index monitoring during and after electroconvulsive therapy compared with magnetic seizure therapy for treatment-resistant depression. Br. J. Anaesth. 112, 695–702. https://doi.org/10.1093/bja/aet410.

Sun, Y., et al., 2016. Indicators for remission of suicidal ideation following magnetic seizure therapy in patients with treatment-resistant depression. JAMA Psychiat. 73, 337–345. https://doi.org/10.1001/jamapsychiatry.2015.3097.

Sun, Y., et al., 2018. Magnetic seizure therapy reduces suicidal ideation and produces neuroplasticity in treatment-resistant depression. Transl. Psychiatry 8, 253. https://doi.org/10.1038/s41398-018-0302-8.

Tang, V.M., et al., 2017. Magnetic seizure therapy in treatment-resistant schizophrenia: a pilot study. Front. Psych. 8, 310. https://doi.org/10.3389/fpsyt.2017.00310.

Tang, V.M., et al., 2020a. Magnetic seizure therapy is efficacious and well tolerated for treatment-resistant bipolar depression: an open-label clinical trial. J. Psychiatry Neurosci. 45, 313–321. https://doi.org/10.1503/jpn.190098.

Tang, V.M., et al., 2020b. Magnetic seizure therapy for the treatment of suicidality in bipolar depression. Biol. Psychiatry. https://doi.org/10.1016/j.biopsych.2020.09.020.

Tang, V.M., et al., 2020c. A pilot study of magnetic seizure therapy for treatment-resistant obsessive-compulsive disorder. Depress. Anxiety. https://doi.org/10.1002/da.23097.

UK ECT Review Group, 2003. Efficacy and safety of electroconvulsive therapy in depressive disorders: a systematic review and meta-analysis. Lancet 361, 799–808. https://doi.org/10.1016/S0140-6736(03)12705-5.

Wang, J., et al., 2019. Accelerated magnetic seizure therapy for treatment of major depressive disorder: a report of 3 cases. J. ECT 35, 135–138. https://doi.org/10.1097/YCT.0000000000000521.

Wang, J., et al., 2020. Accelerated magnetic seizure therapy (aMST) for treatment of major depressive disorder: a pilot study. J. Affect. Disord. 264, 215–220. https://doi.org/10.1016/j.jad.2019.12.022.

Weissman, C.R., et al., 2020. Magnetic seizure therapy for suicidality in treatment-resistant depression. JAMA Netw. Open 3. https://doi.org/10.1001/jamanetworkopen.2020.7434, e207434.

White, P.F., et al., 2006. Anesthetic considerations for magnetic seizure therapy: a novel therapy for severe depression. Anesth Analg 103, 76–80. table of contents https://doi.org/10.1213/01.ane.0000221182.71648.a3.

Wilkinson, S.T., Agbese, E., Leslie, D.L., Rosenheck, R.A., 2018. Identifying recipients of electroconvulsive therapy: data from privately insured Americans. Psychiatr. Serv. 69, 542–548. https://doi.org/10.1176/appi.ps.201700364.

Zhang, J., et al., 2020. Shorter recovery times and better cognitive function—a comparative pilot study of magnetic seizure therapy and electroconvulsive therapy in patients with depressive episodes. Brain Behav. https://doi.org/10.1002/brb3.1900, e01900.

Chapter 21

VNS for treatment-resistant depression

Scott T. Aaronson[a,b] and Gregory Samuel Barber[b]

[a]*Institute for Advanced Diagnostics and Therapeutics, Sheppard Pratt, Baltimore, MD, United States,* [b]*Department of Psychiatry, University of Maryland School of Medicine, Baltimore, MD, United States*

History

Interest in the effects of vagus nerve stimulation (VNS) on the central nervous system began in the early 20th century, with Bailey and Bremer demonstrating in 1938 that electrical stimulation of the vagus nerve of cats resulted in increased electrographical activity of cortical structures (Bailey and Bremer, 1938). Dell and Olson discovered a suggestive link between the vagus nerve and affective disorders in 1951 in their own study of cats, demonstrating an evoked response in the amygdala in response to vagal stimulation (Dell and Olson, 1951). As a clinical intervention, VNS was first investigated for its potential role in treating epilepsy. Beginning in the 1980s, Zabara (1992) began animal experiments that demonstrated VNS's ability to both terminate acute seizure activity and prevent seizures chronically with continuous stimulation. The first human implanted VNS device occurred in 1988 by Penry and Dean in four patients with intractable partial seizures, of whom two obtained complete seizure control (Penry and Dean, 1990). VNS for medication refractory epilepsy received approval in Europe in 1994 and in the United States in 1997 (Schachter, 2002).

These epilepsy trials suggested a potential role for VNS in the treatment of depression. Reports of improved mood and cognition in patients receiving VNS for epilepsy came out of the trials used to gain VNS's initial United States Food and Drug Administration (FDA) approval (Handforth et al., 1998). The role of anticonvulsant medications, including carbamazepine, lamotrigine, and valproate, in the treatment of mood disorders further hinted at a common mechanism underlying treatments for epilepsy and depression. Perhaps, researchers suspected, VNS would provide the same shared benefit.

Initially, this hypothesis gained strength through undercontrolled clinical observations in patients whose mood improved after receiving VNS for epilepsy (Ben-Menachem et al., 1994; Handforth et al., 1998). These observations developed into more rigorously designed interventions exploring mood in epilepsy patients who had received VNS—several of which supported VNS's mood-improving effects (Aldenkamp et al., 2001; Elger et al., 2000; Harden et al., 2000).

The first VNS device was implanted specifically for depression in 1998 at the Medical University of South Carolina as part of a multicenter pilot trial which further established VNS's antidepressant effects (Rush et al., 2000). On the basis of four clinical trials in patients who had failed two to six treatments for their unipolar or bipolar depression (considered to be treatment-resistant depression (TRD)) which included a feasibility trial (Sackeim et al., 2001); a randomized, sham-controlled 3 month-clinical trial (Rush et al., 2005a); a long-term open-label extension of the 3-month randomized trial (Rush et al., 2005a,b); and a long-term observational study comparing subjects receiving standard-of-care treatments with subjects receiving VNS (Nahas et al., 2005), the United States Federal Drug Administration (FDA) granted approval for the use VNS in treatment-resistant depression in 2005 even though the 3-month randomized controlled trial did not show significance in the primary depression outcome measure. This unusual determination was largely due to the absence of proven effective alternatives for this treatment-resistant population.

Subsequently, in 2007, the United States Center for Medicare and Medicaid Services (CMS) decided against coverage of VNS for TRD. This decision was based on the failure of the initial randomized controlled trial (RCT) to demonstrate separation in response between the active VNS group and the sham VNS group during the first 12 weeks of treatment. Subsequent years of research would reveal that the length of stimulation time required for VNS to achieve clinical efficacy is far longer than this 12-week endpoint—closer, in fact, to 6 months to 1 year (Rush et al., 2005b; Aaronson and Conway, 2018). Nevertheless, the adoption of vagus nerve stimulation since its initial approval for depression has been impeded by this early trial as well as by a later dose-finding trial in which subjects were randomized to low-, medium-, or high-dose stimulation (Aaronson et al., 2013). While this study built on the results of the earlier RCT and made the primary end point 6 months after implantation, the efficacy of even modest, low-dose stimulation hampered demonstration of

Managing Treatment-Resistant Depression. https://doi.org/10.1016/B978-0-12-824067-0.00021-9

significantly improved efficacy of higher doses, though each group showed remarkable improvement. Most private insurers have followed the CMS decision and also do not cover the significant costs associated with the VNS device and surgical implantation. Costs of the device and implantation have been estimated at about $28,000; however, further economic analysis has suggested that, 8 years after device implantation, savings associated with VNS for TRD may be in the range of $23,000–$41,000 (Cohen and Allen, 2008). Despite these potential savings, the noncoverage determination placed VNS largely out of reach of the typical patient suffering from treatment-resistant depression during the first decade of its FDA approval (Rush et al., 2005b).

Researchers have continued investigating VNS's efficacy for TRD in intervening years. To date, some 5000 patients have been implanted with VNS devices for treatment of depression in various clinical trials (de Leon et al., 2019). Notable among this body of literature is the VNS for TRD registry, which was established by LivaNova, the manufacturer of the VNS device, as part of the FDA's initial approval in 2005 (Aaronson et al., 2017). This registry followed 500 patients who had received VNS for TRD and 300 who received treatment as usual for over 5 years. Aaronson et al.'s (2017) report on these patients suggested improvements in depression in those with VNS compared to those receiving treatment as usual. As a result of these findings, CMS reopened its noncoverage decision in 2018. Currently, VNS for TRD falls under the "coverage with evidence development" category, indicating CMS's participation in funding a new randomized controlled trial to further development of the VNS for TRD's evidence base (Centers for Medicare and Medicaid Services, 2019).

The VNS device and its mechanism of action

The vagus nerve comprises a complex web of efferent and afferent fibers involved in diverse bodily functions, including autonomic control, motor action, and various sensory modalities. VNS targets the afferent fibers of the left cervical vagus nerve, which carry visceral, somatic, and gustatory sensation to the brainstem (Conway and Xiong, 2018). While the precise mechanism of action of VNS in depression is not entirely understood, there are several anatomical and neurobiological theories that help elucidate its antidepressant effects.

The VNS device

VNS entails surgical placement of electrode coils around the left vagus nerve. The device itself includes the stimulating electrodes, a pulse generator which serves as a battery and provides current to the electrodes, and electrode extenders which connect the electrodes to the pulse generator (Edwards et al., 2017; de Leon et al., 2019; Carreno and Frazer, 2017) (Fig. 21.1).

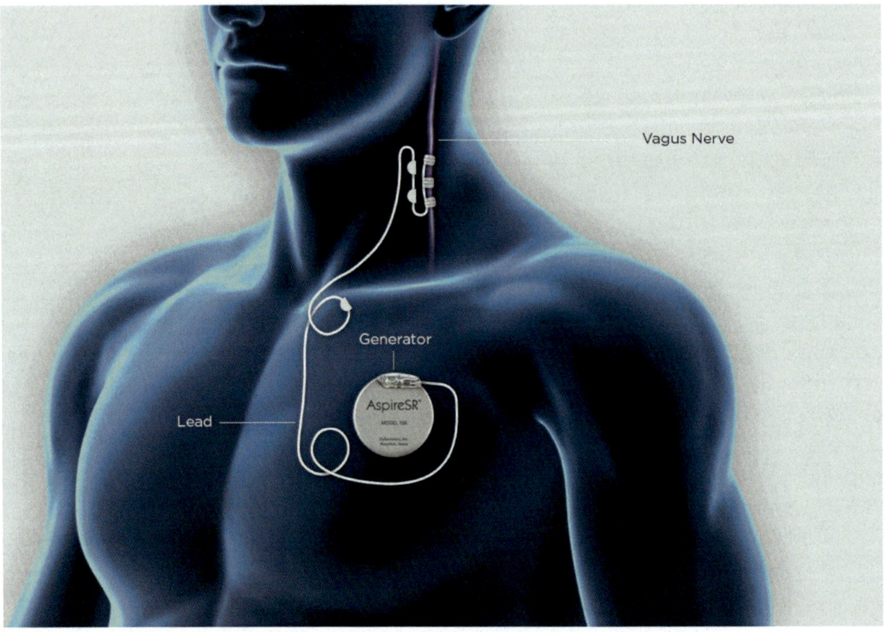

FIG. 21.1 The vagus nerve stimulation device. From (N.d.) LivaNova USA, Inc. The VNS device comprises a pulse generator, which sits in a surgically-created subcutaneous pocket in the left chest, stimulating electrodes, which coil around the left vagus nerve via a separate surgical incision, and a lead (electrode extenders), which is tunneled under the skin to connect the generator and the electrodes.

The vagus nerve

The anatomy of the vagus nerve itself also facilitates central propagation of the stimulation signal. The vagus nerve primarily comprises afferent fibers with low stimulation threshold (Foley and DuBois, 1937). What efferent fibers are present (approximately 20% of the total nerve fibers) have a higher stimulation threshold than the afferents. The preponderance of afferent fibers combined with their lower stimulation threshold helps direct stimulated impulses primarily toward the brain and away from visceral targets. Nevertheless, VNS does have effects on downstream organs. For example, VNS can cause bradycardia—an effect that is mitigated by stimulating the left vagus nerve instead of the right (Edwards et al., 2017). The left vagus nerve supplies parasympathetic input to the sinoatrial (SA) node while the right vagus nerve supplies the atrioventricular (AV) node. Since the AV node sits within the cardiac conduction pathway, right vagus nerve stimulation would potentially cause intracardiac conduction abnormalities. Left vagus stimulation would avoid such disruption, since the SA node lies at the very beginning of the pathway (Mauro et al., 2009). VNS is also known to cause GI side effects including dyspepsia, putatively a result of parasympathetic overstimulation of the enteric system.

Stimulation parameters

VNS combines four distinct stimulation parameters—including current (mA), frequency (Hz), pulse width (s), and duty cycle (the repeating period (in seconds) of active stimulation followed by a period of no stimulation)—which are independently manipulated to obtain a target "dose." After 2 weeks of postoperative healing, the device is activated and the dose adjusted, typically over a 2-week titration period. Starting at 0.25–0.75 mA, the current in gradually increased over this time, most often to between 1.0 and 2.0 mA, to ensure tolerability and to monitor for side effects. The typical frequency ranges between 20 and 30 Hz, with frequencies above 50 Hz potentially causing permanent damage of the vagus nerve (Agnew and McCreery, 1990). A pulse width of 250 s is common, and the duty cycle starts at 30 s on and 300 s off, so 10% of the time it is sending current.

Neurobiological effects of VNS

The essential brainstem pathway involved with VNS's antidepressant effect includes the tractus solitarius, which links vagal afferents from their CNS input in the medulla to the nucleus tractus solitarius. These fibers travel upstream to the pontine parabrachial nucleus, the raphe nuclei, and the locus coeruleus (Carreno and Frazer, 2017). Stimulation of the raphe nucleus, which contains serotonergic cell bodies, and the locus coeruleus, which contains noradrenergic cell bodies, is thought to play an essential role in VNS's antidepressant effects. Decussating afferent fibers ensure bilateral central nervous system effects despite unilateral peripheral vagal stimulation.

Higher level brain structures, including the hypothalamus, thalamus, nucleus accumbens, amygdala, and stria terminalis, are also implicated in VNS's mechanism of antidepressant action (Conway and Xiong, 2018). These regions each receive fibers directly from the nucleus tractus solitarius, and are known to mediate mood and depression. Cortical regions of affective regulation, including the anterior insula, lateral prefrontal cortex, and infralimbic cortex, also receive vagal afferent projections (Henry, 2002). Furthermore, the insula communicates with other cortical structures, including the ventrolateral and orbital prefrontal cortex, which are also considered important targets of downstream VNS effects (Saleem et al., 2008) (Fig. 21.2).

Brain imaging studies such as Bohning et al. (2001), have used fMRI to identify multiple regions, including the bilateral orbitofrontal cortex, parietooccipital cortex, hypothalamus, the left temporal cortex, and amygdala, that were activated while patients received VNS. Importantly, these brain regions are all known to be associated with major depression. Other studies have investigated the impact of various stimulation parameters, including stimulation frequency and pulse width, on regional brain activation in the subacute (Zobel et al., 2005) and chronic (Pardo et al., 2008) stages of treatment.

The timecourse of VNS's antidepressant effects have also been elucidated in brain imaging studies. Conway et al. (2013) and Nahas et al. (2007), for example, found alterations in brain response to VNS after 30 weeks of stimulation compared with the period immediately after initiation of stimulation (Conway et al., 2013; Nahas et al., 2007). These changes, including decreased cerebral metabolic glucose (CMRGlu) in the right dorsolateral prefrontal cortex, correlate with the timing of VNS's antidepressant effect. Conway et al. (2013) also noted that increased CMRGlu in the left substantia nigra and ventral tegmental area at 12 months of stimulation correlated with VNS response, while those who did not respond to VNS had decreased CMRGlu in this region. As the primary dopaminergic brainstem region, the VTA's activation here supports the hypothesis that VNS's antidepressant effects are also mediated in part by dopamine.

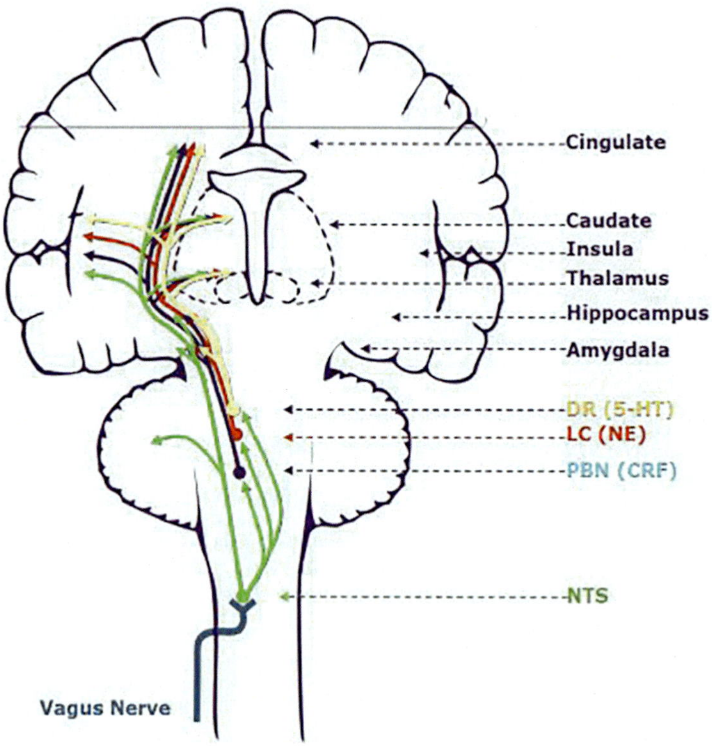

FIG. 21.2 Afferent projections of the vagus nerve. The vagus nerve's afferent projections synapse in the medulla of the brainstem at the nucleus tractus solitarius (NTS, in *green*). Fibers travel upstream to the parabrachial nucleus (PBN), locus coeruleus (LC, *red*), and dorsal raphe nucleus (DR, *yellow*). The LC projects noradrenergic synapses and the DR projects serotonergic synapses centrally to various brain regions essential to depression, including the hypothalamus, thalamus, amygdala, and various cortical regions, including the anterior insular cortex, lateral prefrontal cortex, and infralimbic cortex. From the insula, other targets of downstream VNS effects, including the ventrolateral and orbital prefrontal cortex, are linked.

In an effort to establish potential biological predictors of VNS response, Conway et al. (2012) performed prestimulation FDG-PET scans on patients who went on to receive a course of VNS. They found that the combination of low anterior insular cortex (AIC) CMRGlu and high orbitofrontal cortex (OFC) CMRGlu predicted VNS response after 12 months of treatment.

Animal models of VNS support a unique antidepressant mechanism of action for VNS compared to antidepressant medications. In several studies, sustained VNS has been shown to increase the base firing rate of noradrenergic neurons in the locus coeruleus while suppressing inhibitory GABAergic interneurons (Aston-Jones et al., 2004; Dorr and Debonnel, 2006; Manta et al., 2012). Serotonergic firing in the raphe nucleus is also increased by sustained VNS after 2 weeks of treatment (Dorr and Debonnel, 2006; Manta et al., 2009). Antidepressant medications, on the other hand, primarily act via effects on neurotransmitter release and inhibitory autoreceptor desensitization (Blier, 2003). VNS has not been found to cause desensitization of serotonergic and noradrenergic receptors, as is the case in antidepressant medications (Dorr and Debonnel, 2006).

Antiinflammatory effects of VNS

Other studies have considered an antiinflammatory mechanism of VNS's role in TRD (Carreno and Frazer, 2017). The vagus nerve is known to modulate inflammatory pathways (Pavlov and Tracey, 2012), while inflammation is increasingly acknowledged in the pathophysiology of depression (Suarez et al., 2003). Proinflammatory cytokines are known to be elevated levels in depressed patients, and are implicated in behavioral changes associated with depression (Felger and Lotrich, 2013). One small clinical trial of VNS in TRD demonstrated increased circulating levels of antiinflammatory cytokines in patients after 3 months of VNS (Corcoran et al., 2005). VNS's potential antiinflammatory mechanism is further supported by its efficacy in various inflammatory diseases, including rheumatoid arthritis and Crohn's disease (Sinniger et al., 2020). A clinical trial in patients receiving VNS for rheumatoid arthritis demonstrated that VNS inhibited proinflammatory cytokine production (Koopman et al., 2016).

Combined, these studies paint VNS's complex role in mediating brain areas crucial to affect, starting in the peripheral nervous system, passing through the brainstem and midbrain, and on to higher cortical structures directly involved in mood. VNS's relative lack of invasiveness (compared with therapeutics which directly involve the central nervous system, like deep brain stimulation) is striking in that it demonstrates how a peripheral intervention can broadly modulate upstream targets. More research is required to clarify the relative importance of the various pathways thought to play a role in VNS's antidepressant effects. Nevertheless, VNS's effect on several targets known to play a role in depression, as well as the correlation of onset of metabolic changes in these targets with onset of VNS's antidepressant effects, suggest that researchers are on the right path.

Indications

VNS is considered in patients suffering from treatment-resistant depression, which the United States FDA defines as four antidepressant treatment failures. The VNS registry, created after VNS for TRD gained FDA approval in 2005, included patients who met this criterion (Aaronson et al., 2017). The registry included patients with both unipolar and bipolar depression, patients who had failed electroconvulsive therapy in the current episode, patients who had been depressed for more than 3 years, and patients who suffered with co-morbid anxiety—features which typically exclude such patients from other clinical trials (Centers for Medicare and Medicaid Services, 2019).

Patients in the VNS registry had 20 years of depressive illness on average. They had an average of 8.2 prior treatment failures, and MADRS scores averaging 33.1 on average—a number consistent with marked-to-severe depression (Aaronson et al., 2017). These metrics indicate that VNS should be considered for those patients with the very most severe burden of depressive illness.

As discussed in de Leon et al. (2019), VNS may have a unique role in treatment of bipolar depression, where FDA approved treatments are scarce. Large trials of VNS in depression have included patients with both unipolar and bipolar depression, and results demonstrate similar efficacy between these groups (Aaronson et al., 2013, 2017; Rush et al., 2000, 2005a,b).

de Leon et al. (2019) also discuss the relationship between personality disorders and TRD, noting that VNS should perhaps be avoided in patients with a strong history of personality pathology, including those with a history of severe self-injurious and parasuicidal behaviors. Patients with unrealistic expectations and poor frustration tolerance may not adapt well to VNS, particularly given the invasiveness of this treatment and the possible need for explanation should treatment not be tolerated. Because VNS requires months to demonstrate clinical efficacy, implantation should also be avoided in acutely suicidal patients. The process of pursuing implantation should not delay addressing the acute safety needs of the patient.

Clinical studies

Early studies of VNS for depression

Early studies of VNS for TRD were of mixed quality and demonstrated mixed efficacy. Of the 33 studies included in one systematic analysis, just six were conducted with double-blind, randomized, sham-controlled designs (Cimpianu et al., 2017). Follow up periods, primary outcomes, and stimulation parameters were heterogeneous across studies, making it more difficult to reach broad conclusions from VNS's early trials.

The first study of VNS in patients with depression was an open label, nonrandomized, single-arm study conducted by Rush et al. (2000). Thirty patients with both unipolar and bipolar depression received 10 weeks of VNS, resulting in a 40% response rate based on both HAM-D and CGI-I scores and 50% response rate based on MADRS scores.

The first randomized controlled trial of VNS for TRD was performed in 235 patients with nonpsychotic depression by Rush et al. (2005a). Of these patients, all of whom underwent surgical implantation of the VNS device, 112 were randomized to active treatment while 110 received sham treatment. The primary outcome measure—change in Hamilton Depression Rating Scale after 10 weeks of treatment—demonstrated no statistically significant difference between active and control groups. Of the three secondary outcome assessments (MADRS, CGI-I, and IDS-SR), only the IDS-SR demonstrated a statistically significant difference between active and sham treatment. Subsequent studies would reveal that the short follow-up period likely accounted for the largely negative results of this trial (Aaronson et al., 2017). Moreover, the first 2 weeks of the active stimulation period were spent on dose titration, meaning patients only received full treatment for 8 weeks. Despite the limitations of this study's design, CMS made their noncoverage decision for VNS in TRD in 2007 despite the negative results from this trial.

Patients from this study were subsequently included in a long-term, noncontrolled efficacy trial, during which patients from both the active and sham group of the acute phase could receive active VNS for 9 months (Rush et al., 2005b). This study, which used the change in Hamilton Depression Rating Scale over the 9-month treatment period as its primary outcome, demonstrated a 27.2% response rate and a 15.8% remission rate. Similar results were achieved in the secondary outcomes, including MADRS scales. Response rates doubled from the beginning of the long-term phase of trial to its end—a result that contributes to the current understanding that VNS requires between 6 and 12 months of active treatment to achieve clinical efficacy. As the authors acknowledged, the lack of a control group in this chronic study prevents definitive assertion that the improvements are attributable to VNS.

VNS dose-finding study

In the only "dose-finding" trial for VNS to date, Aaronson et al. (2013) compared the effects of low-, medium-, and high-intensity stimulation by varying output current and pulse width. Duty cycle and pulse frequency were uniform across the groups. This multicenter trial assessed 316 patients during a 22-week acute phase, followed by a 28-week long-term phase to assess durability of response. A total of 102 patients received low-dose stimulation (0.25 mA, 130 μs), 101 medium-dose (0.5–1.0 mA, 250 μs), and 107 high-dose (1.25–1.5 mA, 250 μs). Results from both the acute and chronic phases did not demonstrate statistically significant separation between the low, medium, and high-dose groups in the primary or secondary outcome assessments. Nevertheless, all three groups did demonstrate overall significant improvements in depression ratings compared to baseline scores. This outcome suggests that even low-dose stimulation is effective in treating depression, and is considered the main reason for the lack of separation. More patients in the medium- and high-dose groups (9%–11%) did experience remission by the end of the acute phase compared with the low-dose group (5%–6%). Also notable was the finding that patients in the low-dose group were more likely to have a depressive relapse at 50 weeks compared with those in the medium- and high-dose groups. Thus, while low-dose stimulation was effective in treating depression, there were other important clinical differences between the groups.

The VNS registry study

In the largest study of VNS in TRD to date, Aaronson et al. (2017), compared 489 VNS patients with 276 patients receiving treatment as usual. This was a 5-year, prospective, open-label, nonrandomized, and observational registry study established by the manufacturer of the VNS device after it gained FDA approval in 2005. Patients in the VNS arm were among the most depressed and treatment resistant, with 8.2 failed antidepressant treatments on average, 1.8 lifetime suicide attempts on average, mean baseline MADRS scores indicating moderate to severe depression, and 57% of patients with prior ECT exposure. The results of this study demonstrated VNS's striking potential to help such patients.

The primary outcome looked at antidepressant response based on MADRS scores at 5 years of follow-up. On this measure, 67.6% of VNS patients achieved antidepressant response, while 40.9% of TAU patients responded, as demonstrated in Fig. 21.3 (Fig. 21.4).

Adverse effects

VNS is generally well-tolerated (Daban et al., 2008; Rush et al., 2005a). Aaronson et al.'s (2017) study found no additional side effect burden of VNS compared with treatment as usual. Nevertheless, VNS does entail certain adverse effects related to its mechanism of action. These effects should be considered and discussed with patients prior to implantation. Surgical placement of the VNS device carries the risks associated with any surgical intervention, including infection, pain, and skin irritation at the surgical site (Aaronson and Conway, 2018). While some patients may have the device implanted under local anesthesia, others will require general anesthesia. VNS may therefore be inappropriate in patients whose medical co-morbidities place them at high risk of complications from general anesthesia. Preoperative clearance by a primary care provider should be sought before implantation is pursued.

The VNS implantation procedure may cause damage to the vagus nerve, potentially resulting in impaired cardiac and digestive function. The surgery may cause hematoma, fluid accumulation, and histotoxic reactions against the implanted device. Other adverse surgical outcomes include migration of the pulse generator, and slippage, disconnection, corrosion, or breakage of the leads. These complications may require additional surgical intervention and device replacement to repair (United States Food and Drug Administration, 2005).

Adverse effects may result from the physiology of vagus nerve stimulation itself. A common side effect is voice alteration and hoarseness (Sackeim et al., 2001). The incidence of hoarseness varies by study from 25% to 68%, with the effect becoming less prominent at 12–24 months after implantation. This effect results from stimulation of the left recurrent laryngeal nerve, a branch of the vagus nerve that innervates structures of larynx and controls the vocal folds. Down-titration of

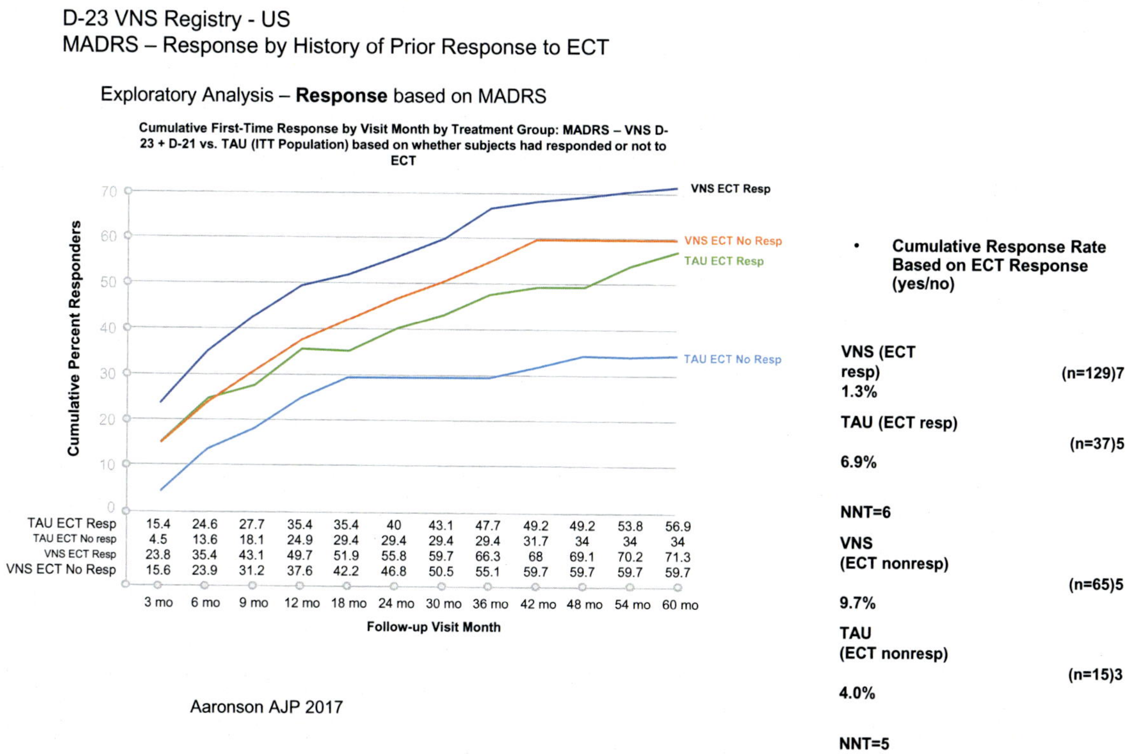

FIG. 21.3 VNS response by history of prior response to ECT. VNS is effective for patients both with and without a history of ECT response, relative to patients who did not receive VNS. Patients with a history of ECT response who received active VNS had the greatest response rate, based on change in MADRS score, at 71.3%. Patients with a history of ECT nonresponse also benefitted from VNS, with a response rate of 59.7—a rate comparable to patients with a history of ECT response who did not receive VNS (56.9%). Those with a history of ECT nonresponse who did not receive VNS had the lowest response rate at 34%.

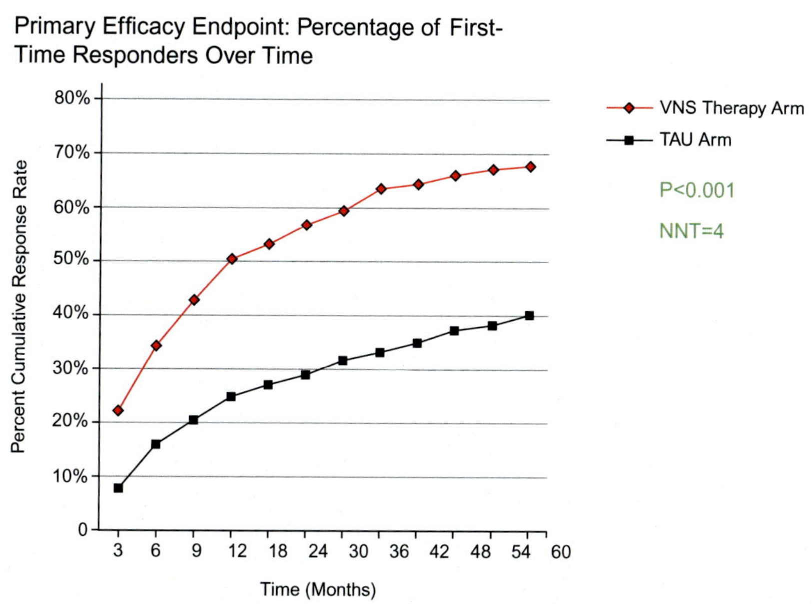

FIG. 21.4 VNS response rates. At 5 years (60 months) of follow up, patients receiving active VNS achieved a 67.6% response rate as measured by the change MADRS score, compared with a 40.9% response rate for those not receiving VNS.

stimulation intensity can mitigate this effect. Up to 25% of patients may experience cough, again the result of stimulation of the laryngeal branches of the vagus nerve (Schlaepfer et al., 2008). Dyspepsia, neck pain, dysphagia, apnea, and dyspnea are other side effects attributed to VNS.

Certain reports suggest that VNS places patients at slight risk of developing of mania, including in patients with no prior history of mania. One case report documents two such patients who developed manic symptoms 8 and 9 months after placement of the device (Salloum et al., 2017). After monitoring VNS patients naturalistically for 1 year, Rush et al. (2005b) documented three patients who developed manic symptoms. One of these patients had diagnosed bipolar disorder, another had a history of treatment-induced mania, and the third had no history of mania. Whether these patients would have gone on to develop mania even in the absence of VNS is unknown. Nevertheless, these cases suggest that while rare, the risk of VNS-induced mania exists, and clinicians should routinely monitor patients for the development of such symptoms.

Interactions

VNS is contraindicated in patients with a history of left or bilateral cervical vagotomy, as this precludes placement of the VNS device. It should be used with caution in patients with a history of cardiac conduction abnormalities. The risk of acute cardiac events, including heart block, is not well established in patients receiving VNS who have a history of cardiac conduction abnormalities, including reentry pathways. Consultation with a cardiologist is recommended prior to the decision to implant such patients, who may be evaluated for bradyarrhythmias with postimplant electrocardiograms and Holter monitoring. Patients with VNS implants require special precautions when obtaining MRI procedures; the manufacturer gives specific instructions for managing the device before, during, and after MRI.

There are no known interactions between VNS and antidepressant medications. During various VNS trials, patients have been maintained on their antidepressant medication regimens with no reported drug-VNS interactions (Aaronson et al., 2017). Studies have shown no risk with concurrent electroconvulsive therapy (ECT) and VNS, so long as the VNS device is properly shut off during ECT (Nahas et al., 2005). ECT has no negative effects on the VNS device, and the presence of the VNS device does not lead to unexpected or additional complications from ECT (Burke and Husain, 2006). Repetitive transcranial magnetic stimulation therapy (rTMS) can be pursued safely, but with caution, in patients who have received VNS implants (Philip et al., 2014). The VNS device contains ferromagnetic components which could theoretically be induced by rTMS pulses. Nevertheless, TMS is safe in VNS patients if precautions are taken to not activate the TMS coil near the components of the VNS device in the chest or neck. When the TMS coil is maintained at an appropriate distance from the VNS device (i.e., only near the scalp where TMS occurs), there is minimal interaction between TMS's magnetic fields and the VNS device's metal.

Transcutaneous vagus nerve stimulation

Research efforts are also underway to attempt vagus nerve stimulation by noninvasive means. Transcutaneous vagus nerve stimulation (tcVNS) exploits the anatomy of the vagus nerve and its wandering branches to avoid surgical implantation of an internal VNS device. A transcutaneous electrical nerve stimulator attached to a transmitting ear clip stimulates the auricular branch of the vagus nerve, which courses superficially through the tragus of the outer ear and carries afferent fibers to the vagus nerve and then the CNS. Researchers hypothesize that stimulating this branch of the vagus nerve could lead to many of the same antidepressant effects as direct stimulation of the vagus nerve itself, while obviating the risks and expenses of surgery (Jian et al., 2018). Several small trials have begun to investigate tcVNS for depression and anxiety, but the study populations generally consist of nonsevere or nontreatment-resistant depression patients. One study using tcVNS on patients with mild or moderate depression found significant decreases in HAM-D scores after 4 weeks compared to sham tcVNS (Rong et al., 2016). Another study also demonstrated changes to function amygdala connectivity in patients receiving tcVNS, offering evidence that tcVNS may indeed result in similar brain changes to implanted VNS (Liu et al., 2016). While this technique is theoretically promising, research in the future should seek further proof of concept that tcVNS is a viable and efficacious treatment strategy for severe, treatment-resistant depression.

Conclusion

Vagus nerve stimulation is an effective intervention for patients with the most severe and difficult to treat depression, and should be considered in patients with treatment-resistant depression who are able to undergo surgical implantation of the device. After the implantation, the side effect burden is generally mild and tolerable, though some patience is required before the clinical efficacy of VNS becomes apparent. Unfortunately, access to VNS remains limited. However, reassessment of CMS's noncoverage decision will hopefully expand its availability and affordability in coming years. While VNS's effects on brain structures implicated in depression are increasingly understood, a full accounting of its mechanism of action requires further research. More research is also required to establish VNS's place in the treatment algorithm for treatment-resistant depression. While VNS is safe in patients receiving concurrent antidepressant medication, ECT, and rTMS, there is no current guidance from the literature on which patients should receive which intervention at a particular time in the course of their illness. All told, VNS will likely play an important role in treatment-resistant depression in the coming years.

References

Aaronson, S.T., Conway, C.R., 2018. Vagus nerve stimulation: changing the paradigm for chronic severe depression? Psychiatr. Clin. N. Am. 41 (3), 409–418. https://doi.org/10.1016/j.psc.2018.05.001.

Aaronson, S.T., Carpenter, L.L., Conway, C.R., Reimherr, F.W., Lisanby, S.H., Schwartz, T.L., Moreno, F.A., Dunner, D.L., Lesem, M.D., Thompson, P.M., Husain, M., Vine, C.J., Banov, M.D., Bernstein, L.P., Lehman, R.B., Brannon, G.E., Keepers, G.A., O'Reardon, J.P., Rudolph, R.L., Bunker, M., 2013. Vagus nerve stimulation therapy randomized to different amounts of electrical charge for treatment-resistant depression: acute and chronic effects. Brain Stimul. 6 (4), 631–640. https://doi.org/10.1016/j.brs.2012.09.013.

Aaronson, S.T., Sears, P., Ruvuna, F., Bunker, M., Conway, C.R., Dougherty, D.D., Reimherr, F.W., Schwartz, T.L., Zajecka, J.M., 2017. A 5-year observational study of patients with treatment-resistant depression treated with vagus nerve stimulation or treatment as usual: comparison of response, remission, and suicidality. Am. J. Psychiatr. 174 (7), 640–648. https://doi.org/10.1176/appi.ajp.2017.16010034.

Agnew, W.F., McCreery, D.B., 1990. Considerations for safety with chronically implanted nerve electrodes. Epilepsia 31, S27–S32. https://doi.org/10.1111/j.1528-1157.1990.tb05845.x.

Aldenkamp, A.P., Van De Veerdonk, S.H.A., Majoie, H.J.M., Berfelo, M.W., Evers, S.M.A.A., Kessels, A.G.H., Renier, W.O., Wilmink, J., 2001. Effects of 6 months of treatment with vagus nerve stimulation on behavior in children with Lennox-Gastaut syndrome in an open clinical and nonrandomized study. Epilepsy Behav. 2 (4), 343–350. https://doi.org/10.1006/ebeh.2001.0218.

Aston-Jones, G., Zhu, Y., Card, J.P., 2004. Numerous GABAergic afferents to locus ceruleus in the pericerulear dendritic zone: possible interneuronal pool. J. Neurosci. 24 (9), 2313–2321. https://doi.org/10.1523/JNEUROSCI.5339-03.2004.

Bailey, P., Bremer, F., 1938. A sensory cortical representation of the vagus nerve: with a note on the effects of low blood pressure on the cortical electrogram. J. Neurophysiol. 1 (5), 405–412. https://doi.org/10.1152/jn.1938.1.5.405.

Ben-Menachem, E., Mañon-Espaillat, R., Ristanovic, R., Wilder, B.J., Stefan, H., Mirza, W., Tarver, W.B., Wernicke, J.F., 1994. Vagus nerve stimulation for treatment of partial seizures: 1. A controlled study of effect on seizures. Epilepsia 35 (3), 616–626. https://doi.org/10.1111/j.1528-1157.1994.tb02482.x.

Blier, P., 2003. The pharmacology of putative early-onset antidepressant strategies. Eur. Neuropsychopharmacol. 13 (2), 173–176. https://doi.org/10.1016/S0924-977X(02.

Bohning, D.E., Lomarev, M.P., Denslow, S., Nahas, Z., Shastri, A., George, M.S., 2001. Feasibility of vagus nerve stimulation-synchronized blood oxygenation level-dependent functional MRI. Investig. Radiol. 36 (8), 470–479. https://doi.org/10.1097/00004424-200108000-00006.

Burke, M.J., Husain, M.M., 2006. Concomitant use of vagus nerve stimulation and electroconvulsive therapy for treatment-resistant depression. J. ECT 22 (3), 218–222. https://doi.org/10.1097/01.yct.0000230364.04240.52.

Carreno, F.R., Frazer, A., 2017. Vagal nerve stimulation for treatment-resistant depression. Neurotherapeutics 14 (3), 716–727. https://doi.org/10.1007/s13311-017-0537-8.

Centers for Medicare & Medicaid Services, 2019. Decision Memo for Vagus Nerve Stimulation (VNS) for Treatment Resistant Depression (TRD) (Administrative file CAG-00313R2).

Cimpianu, C.L., Strube, W., Falkai, P., Palm, U., Hasan, A., 2017. Vagus nerve stimulation in psychiatry: a systematic review of the available evidence. J. Neural Transm. 124 (1), 145–158. https://doi.org/10.1007/s00702-016-1642-2.

Cohen, L.J., Allen, J.C., 2008. Estimating the potential savings with vagus nerve stimulation for treatment-resistant depression: a payer perspective. Curr. Med. Res. Opin. 24 (8), 2203–2217. https://doi.org/10.1185/03007990802229050.

Conway, C.R., Xiong, W., 2018. The mechanism of action of vagus nerve stimulation in treatment-resistant depression: current conceptualizations. Psychiatr. Clin. N. Am. 41 (3), 395–407. https://doi.org/10.1016/j.psc.2018.04.005.

Conway, C.R., Chibnall, J.T., Gangwani, S., Mintun, M.A., Price, J.L., Hershey, T., Giuffra, L.A., Bucholz, R.D., Christensen, J.J., Sheline, Y.I., 2012. Pretreatment cerebral metabolic activity correlates with antidepressant efficacy of vagus nerve stimulation in treatment-resistant major depression: a potential marker for response? J. Affect. Disord. 139 (3), 283–290. https://doi.org/10.1016/j.jad.2012.02.007.

Conway, C.R., Chibnall, J.T., Gebara, M.A., Price, J.L., Snyder, A.Z., Mintun, M.A., Craig, A.D., Cornell, M.E., Perantie, D.C., Giuffra, L.A., Bucholz, R.D., Sheline, Y.I., 2013. Association of cerebral metabolic activity changes with vagus nerve stimulation antidepressant response in treatment-resistant depression. Brain Stimul. 6 (5), 788–797. https://doi.org/10.1016/j.brs.2012.11.006.

Corcoran, C., Connor, T.J., O'Keane, V., Garland, M.R., 2005. The effects of vagus nerve stimulation on pro- and anti-inflammatory cytokines in humans: a preliminary report. Neuroimmunomodulation 12 (5), 307–309. https://doi.org/10.1159/000087109.

Daban, C., Martinez-Aran, A., Cruz, N., Vieta, E., 2008. Safety and efficacy of vagus nerve stimulation in treatment-resistant depression. A systematic review. J. Affect. Disord. 110 (1–2), 1–15. https://doi.org/10.1016/j.jad.2008.02.012.

de Leon, V.C., Drysdale, A.T., Conway, C.R., Aaronson, S.T., 2019. Predictors of response for vagus nerve stimulation in treatment-resistant depression. Personalized Med. Psychiatry, 32–36. https://doi.org/10.1016/j.pmip.2019.05.001.

Dell, P., Olson, R., 1951. Thalamic, cortical and cerebelar projections of vagal visceral afferences. C. R. Seances Soc. Biol. Fil. 145, 1084–1088.

Dorr, A.E., Debonnel, G., 2006. Effect of vagus nerve stimulation on serotonergic and noradrenergic transmission. J. Pharmacol. Exp. Ther. 318 (2), 890–898. https://doi.org/10.1124/jpet.106.104166.

Edwards, C.A., Kouzani, A., Lee, K.H., Ross, E.K., 2017. Neurostimulation devices for the treatment of neurologic disorders. Mayo Clin. Proc. 92 (9), 1427–1444. https://doi.org/10.1016/j.mayocp.2017.05.005.

Elger, G., Hoppe, C., Falkai, P., Rush, A.J., Elger, C.E., 2000. Vagus nerve stimulation is associated with mood improvements in epilepsy patients. Epilepsy Res. 42 (2–3), 203–210. https://doi.org/10.1016/S0920-1211(00)00181-9.

Felger, J.C., Lotrich, F.E., 2013. Inflammatory cytokines in depression: neurobiological mechanisms and therapeutic implications. Neuroscience 246, 199–229. https://doi.org/10.1016/j.neuroscience.2013.04.060.

Foley, J.O., DuBois, F.S., 1937. Quantitative studies of the vagus nerve in the cat. J. Nerv. Ment. Dis. 86 (5), 587.

Handforth, A., DeGiorgio, C.M., Schachter, S.C., Uthman, B.M., Naritoku, D.K., Tecoma, E.S., Henry, T.R., Collins, S.D., Vaughn, B.V., Gilmartin, R.C., Labar, D.R., Morris, G.L., Salinsky, M.C., Osorio, I., Ristanovic, R.K., Labiner, D.M., Jones, J.C., Murphy, J.V., Ney, G.C., Wheless, J.W., 1998. Vagus nerve stimulation therapy for partial-onset seizures: a randomized active-control trial. Neurology 51 (1), 48–55. https://doi.org/10.1212/WNL.51.1.48.

Harden, C.L., Pulver, M.C., Ravdin, L.D., Nikolov, B., Halper, J.P., Labar, D.R., 2000. A pilot study of mood in epilepsy patients treated with vagus nerve stimulation. Epilepsy Behav. 1 (2), 93–99. https://doi.org/10.1006/ebeh.2000.0046.

Henry, T.R., 2002. Therapeutic mechanisms of vagus nerve stimulation. Neurology 59 (6), S3–S14. https://doi.org/10.1212/wnl.59.6_suppl_4.s3.

Jian, K., Jiliang, F., Joel, P., Shaoyuan, L., Peijing, R., 2018. Treating depression with transcutaneous auricular vagus nerve stimulation: state of the art and future perspectives. Front. Psychiatry. https://doi.org/10.3389/fpsyt.2018.00020.

Koopman, F.A., Chavan, S.S., Miljko, S., Grazio, S., Sokolovic, S., Schuurman, P.R., Mehta, A.D., Levine, Y.A., Faltys, M., Zitnik, R., Tracey, K.J., Tak, P.P., 2016. Vagus nerve stimulation inhibits cytokine production and attenuates disease severity in rheumatoid arthritis. Proc. Natl. Acad. Sci. U. S. A. 113 (29), 8284–8289. https://doi.org/10.1073/pnas.1605635113.

Liu, J., Fang, J., Wang, Z., Rong, P., Hong, Y., Fan, Y., Wang, X., Park, J., Jin, Y., Liu, C., Zhu, B., Kong, J., 2016. Transcutaneous vagus nerve stimulation modulates amygdala functional connectivity in patients with depression. J. Affect. Disord. 205, 319–326. https://doi.org/10.1016/j.jad.2016.08.003.

Manta, S., Dong, J., Debonnel, G., Blier, P., 2009. Enhancement of the function of rat serotonin and norepinephrine neurons by sustained vagus nerve stimulation. J. Psychiatry Neurosci. 34 (4), 272–280. http://www.cma.ca/multimedia/staticContent/HTML/N0/l2/jpn/vol-34/issue-4/pdf/pg272.pdf.

Manta, S., El Mansari, M., Blier, P., 2012. Novel attempts to optimize vagus nerve stimulation parameters on serotonin neuronal firing activity in the rat brain. Brain Stimul. 5 (3), 422–429. https://doi.org/10.1016/j.brs.2011.04.005.

Mauro, M.P.S., Patronelli, F., Spinelli, E., Cordero, A., Covello, D., Gorostiaga., 2009. Nerves of the heart: a comprehensive review with a clinical point of view. Neuroanatomy 8, 26–31.

Nahas, Z., Marangell, L.B., Husain, M.M., Rush, A.J., Sackeim, H.A., Lisanby, S.H., Martinez, J.M., George, M.S., 2005. Two-year outcome of vagus nerve stimulation (VNS) for treatment of major depressive episodes. J. Clin. Psychiatry 66 (9), 1097–1104. https://doi.org/10.4088/JCP.v66n0902.

Nahas, Z., Teneback, C., Chae, J.H., Mu, Q., Molnar, C., Kozel, F.A., Walker, J., Anderson, B., Koola, J., Kose, S., Lomarev, M., Bohning, D.E., George, M.S., 2007. Serial vagus nerve stimulation functional MRI in treatment-resistant depression. Neuropsychopharmacology 32 (8), 1649–1660. https://doi.org/10.1038/sj.npp.1301288.

Pardo, J.V., Sheikh, S.A., Schwindt, G.C., Lee, J.T., Kuskowski, M.A., Surerus, C., Lewis, S.M., Abuzzahab, F.S., Adson, D.E., Rittberg, B.R., 2008. Chronic vagus nerve stimulation for treatment-resistant depression decreases resting ventromedial prefrontal glucose metabolism. NeuroImage 42 (2), 879–889. https://doi.org/10.1016/j.neuroimage.2008.04.267.

Pavlov, V.A., Tracey, K.J., 2012. The vagus nerve and the inflammatory reflex – linking immunity and metabolism. Nat. Rev. Endocrinol. 8 (12), 743–754. https://doi.org/10.1038/nrendo.2012.189.

Penry, J.K., Dean, J.C., 1990. Prevention of intractable partial seizures by intermittent vagal stimulation in humans: preliminary results. Epilepsia 31, S40–S43. https://doi.org/10.1111/j.1528-1157.1990.tb05848.x.

Philip, N.S., Carpenter, S.L., Carpenter, L.L., 2014. Safe use of repetitive transcranial magnetic stimulation in patients with implanted vagus nerve stimulators. Brain Stimul. 7 (4), 608–612. https://doi.org/10.1016/j.brs.2014.04.001.

Rong, P., Liu, J., Wang, L., Liu, R., Fang, J., Zhao, J., Zhao, Y., Wang, H., Vangel, M., Sun, S., Ben, H., Park, J., Li, S., Meng, H., Zhu, B., Kong, J., 2016. Effect of transcutaneous auricular vagus nerve stimulation on major depressive disorder: a nonrandomized controlled pilot study. J. Affect. Disord. 195, 172–179. https://doi.org/10.1016/j.jad.2016.02.031.

Rush, A.J., George, M.S., Sackeim, H.A., Marangell, L.B., Husain, M.M., Giller, C., Nahas, Z., Haines, S., Simpson, R.K., Goodman, R., 2000. Vagus nerve stimulation (VNS) for treatment-resistant depressions: a multicenter study. Biol. Psychiatry 47 (4), 276–286. https://doi.org/10.1016/S0006-3223(99)00304-2.

Rush, A.J., Marangell, L.B., Sackeim, H.A., George, M.S., Brannan, S.K., Davis, S.M., Howland, R., Kling, M.A., Rittberg, B.R., Burke, W.J., Rapaport, M.H., Zajecka, J., Nierenberg, A.A., Husain, M.M., Ginsberg, D., Cooke, R.G., 2005a. Vagus nerve stimulation for treatment-resistant depression: a randomized, controlled acute phase trial. Biol. Psychiatry 58 (5), 347–354. https://doi.org/10.1016/j.biopsych.2005.05.025.

Rush, A.J., Sackeim, H.A., Marangell, L.B., George, M.S., Brannan, S.K., Davis, S.M., Lavori, P., Howland, R., Kling, M.A., Rittberg, B., Carpenter, L., Ninan, P., Moreno, F., Schwartz, T., Conway, C., Burke, M., Barry, J.J., 2005b. Effects of 12 months of vagus nerve stimulation in treatment-resistant depression: a naturalistic study. Biol. Psychiatry 58 (5), 355–363. https://doi.org/10.1016/j.biopsych.2005.05.024.

Sackeim, H.A., Rush, A.J., George, M.S., Marangell, L.B., Husain, M.M., Nahas, Z., Johnson, C.R., Seidman, S., Giller, C., Haines, S., Simpson, R.K., Goodman, R.R., 2001. Vagus nerve stimulation (VNS™) for treatment-resistant depression: efficacy, side effects, and predictors of outcome. Neuropsychopharmacology 25 (5), 713–728. https://doi.org/10.1016/S0893-133X(01)00271-8.

Saleem, K.S., Kondo, H., Price, J.L., 2008. Complementary circuits connecting the orbital and medial prefrontal networks with the temporal, insular, and opercular cortex in the macaque monkey. J. Comp. Neurol. 506 (4), 659–693. https://doi.org/10.1002/cne.21577.

Salloum, N.C., Gott, B.M., Conway, C.R., 2017. Sustained remission in patients with treatment-resistant depression receiving vagal nerve stimulation: a case series. Brain Stimul. 10 (5), 997–1000. https://doi.org/10.1016/j.brs.2017.06.001.

Schachter, S.C., 2002. Vagus nerve stimulation therapy summary: five years after FDA approval. Neurology 59 (6), S15–S20. https://doi.org/10.1212/wnl.59.6_suppl_4.s15.

Schlaepfer, T.E., Frick, C., Zobel, A., Maier, W., Heuser, I., Bajbouj, M., O'Keane, V., Corcoran, C., Adolfsson, R., Trimble, M., Rau, H., Hoff, H.J., Padberg, F., Müller-Siecheneder, F., Audenaert, K., Van Den Abbeele, D., Matthews, K., Christmas, D., Stanga, Z., Hasdemir, M., 2008. Vagus nerve stimulation for depression: efficacy and safety in a European study. Psychol. Med. 38 (5), 651–661. https://doi.org/10.1017/S0033291707001924.

Sinniger, V., Pellissier, S., Fauvelle, F., Trocmé, C., Hoffmann, D., Vercueil, L., Cracowski, J.L., David, O., Bonaz, B., 2020. A 12-month pilot study outcomes of vagus nerve stimulation in Crohn's disease. Neurogastroenterol. Motil. 32 (10). https://doi.org/10.1111/nmo.13911.

Suarez, E.C., Krishnan, R.R., Lewis, J.G., 2003. The relation of severity of depressive symptoms to monocyte-associated proinflammatory cytokines and chemokines in apparently healthy men. Psychosom. Med. 65 (3), 362–368. https://doi.org/10.1097/01.PSY.0000035719.79068.2B.

United States Food and Drug Administration, 2005. Summary of Safety and Effectiveness Data (PMA P970003/S207).

Zabara, J., 1992. Inhibition of experimental seizures in canines by repetitive vagal stimulation. Epilepsia 33 (6), 1005–1012. https://doi.org/10.1111/j.1528-1157.1992.tb01751.x.

Zobel, A., Joe, A., Freymann, N., Clusmann, H., Schramm, J., Reinhardt, M., Biersack, H.J., Maier, W., Broich, K., 2005. Changes in regional cerebral blood flow by therapeutic vagus nerve stimulation in depression: an exploratory approach. Psychiatry Res. Neuroimaging 139 (3), 165–179. https://doi.org/10.1016/j.pscychresns.2005.02.010.

Chapter 22

Deep brain stimulation for treatment-resistant depression

Patricio Riva-Posse
Department of Psychiatry and Behavioral Sciences, Emory University School of Medicine, Atlanta, GA, United States

Introduction

Neurosurgical interventions for the treatment of psychiatric illnesses have been used since the dawn of the somatic treatments in neuropsychiatry (Abosch and Cosgrove, 2008). The use of surgery in psychiatry existed before there were effective pharmacological strategies, such as antidepressants or antipsychotics (Moniz, 1994). Egas Moniz, a Portuguese neurologist, partnered with neurosurgeon Almeida Lima to perform a frontal leucotomy in 1935 in a 63-year-old woman suffering from depression, anxiety, paranoia, hallucinations, and anxiety. It was for this work, described as "one of the most important discoveries ever made in psychiatric medicine," that he received the 1949 Nobel Prize for Physiology or Medicine (Haas, 2003). American neurologist Walter Freeman, with neurosurgeon James Watts, categorized the different types of operations, delineating specific landmarks and selecting the techniques for the lesions. A "minimal approach" involving more anterior lesions was performed in patients with affective disorders, while patients with schizophrenia and more severe symptomatology were given a "radical lobotomy," usually more posterior and extensive (Mahoney and Green, 2020). Case reports were inconsistent in their methodological reporting, but would estimate "positive responses" in 30%–70%, and techniques were varied: patients with for melancholic features would receive anterior cingulotomy, stereotactic subcaudate tractotomy, limbic leucotomy, and anterior capsulotomy (Abosch and Cosgrove, 2008; Volpini et al., 2017). From today's perspective, it may appear that these interventions were unsafe, but they were in many cases interventions of last resort to bring relief to patients when there were no other available options, as antipsychotics only became available in the 1950s, and antidepressants after that. Since early stages, psychosurgery discussions have invoked controversy, as the widespread and indiscriminate use of the transorbital lobotomy in the mid-20th century resulted in profound ethical ramifications that persist to this day (Staudt et al., 2019). The emergence of psychopharmacology contributed to a less invasive approach to the treatment of psychiatric disorders, and for a number of decades, neurosurgical interventions fell out of favor, with both societal and professional criticism regarding the significant and largely underreported adverse events, and the lack of objectivity and scientific rigor (Hoffman, 1949).

However, there are patients for whom standard treatments with pharmacology, psychotherapy, and noninvasive neuromodulation are not effective. A new paradigm in the conceptualization of neuropsychiatric disorders as dysfunctional brain networks, advances in functional stereotactic neurosurgery, and successes in treatment of neurological disorders such as Parkinson's disease made the exploration of neurosurgical interventions for severe depression possible again. Neuromodulation techniques, as opposed to lesions, have the possibility of reversibility of side effects. New neuroimaging methods, identifying structural and functional interconnectivity of relevant brain areas allowed the progress beyond a static lesion model (Williams, 2016). Recognition of these cortical and subcortical nodes, the pathways between them, functional changes observed in pathological states, and mood changes when these areas were targets of surgical ablation for other neuropsychiatric disorders opened the tangible possibility of implementing invasive neurostimulation in the form of deep brain stimulation (DBS). Brain regions for invasive and noninvasive stimulation are located in the same networks across psychiatric and neurological diseases. From there, the concept framing for understanding brain stimulation as a network phenomenon has generated specific hypotheses regarding optimization of brain stimulation therapies that are being tested (Fox et al., 2014).

In DBS, electrodes are implanted surgically through small holes in the skull in specific areas of the brain (usually in both hemispheres), depending on the desired symptoms targeted. In Parkinson's disease, for example, the most common sites for implantation are the subthalamic nucleus and the globus pallidus. Stereotaxy is used to determine the final location of the electrode, as well as the trajectory of the lead, and the surgery is planned with the aid of neuroimaging that is acquired in

Managing Treatment-Resistant Depression. https://doi.org/10.1016/B978-0-12-824067-0.00022-0

advance. Once implanted, the electrodes are connected, via extension cables that travel under the skin, to a programmable pacemaker device. Once the stimulation is turned on, this battery sends electrical pulses that modulate the local region as well as the neural circuits that communicate that area to the rest of the regions involved in the functioning of that circuit. The programming allows for multiple parameter settings, as each of the leads has multiple contacts (usually separated by a few millimeters), as well as voltage or current, frequency, and pulse width (Okun, 2012).

DBS acts on the cells and fibers located closest to the implanted electrode, in most cases inhibiting cells and exciting fibers (McIntyre and Hahn, 2010; Vitek, 2002). DBS has electrical, chemical, and other neural-network influences on brain tissue. Once modulation of pathologically functioning neurocircuits was introduced and the physiological basis of depression started to be understood, treatment-resistant depression emerged as a candidate to develop treatments. Mood neurocircuitry is located throughout cortical and subcortical areas, with different nodes and hubs. Therefore, it is not surprising that DBS for depression has been explored in multiple different targets. While the exact mechanism of action of DBS is still unclear, it has become clear that it involves modulation of pathologically functioning circuits. And despite the "connectomic" framework of the development of DBS for mood disorders, targets are still described as anatomically located. These include the subcallosal cingulate white matter (SCC), ventral capsule/ventral striatum or anterior limb of the internal capsule (VC/VS or ALIC), nucleus accumbens (NAcc), the supero-lateral medial forebrain bundle (MFB), the inferior thalamic peduncle (ITP), and the lateral habenula (LHb).

Worldwide, several groups have explored DBS for depression in these targets. The initial publication documenting DBS in depression came out in 2005. More than 15 years later, there have been open-label series and clinical trials with varying results, but advances in the mechanisms of antidepressant effects, newer technologies to deliver and sense stimulation as well as imaging have moved the field forward although it is still considered experimental (Riva-Posse, 2020). Considering the prevalence of depression, the levels of treatment resistance, and the continuing high rates of suicide, the total number of patients who have received DBS for TRD remains in the low hundreds, while numbers in several orders of magnitude could potentially benefit from this intervention (Dandekar et al., 2018). The pattern that happened in many of these targets started with initial promising results of open-label trials that were not successfully replicated in randomized, sham-controlled, double-blind studies. There are several reasons that can explain this discordance. Importantly, there may be a number of explanations that are related to study design, patient selection, or lack of clear biomarkers that would define this clinically diverse syndrome. The enthusiasm to replicate initial open-label findings pushed the scientific community to implement clinical trials that did not account for many variables that were (and still are) unknown, for example in patient, target and parameter selection, as well as trying to imitate outcome measures that are better suited for pharmacological trials. Something similar happened in vagus nerve stimulation, where in spite of having shown treatment efficacy and received Food and Drug Administration approval, has not been covered by healthcare companies and agencies (Rush and George, 2006). The cumulative evidence that has been generated in the last decade will hopefully give the clinical trials that may be conducted in the future a better chance of success, as several aspects involving neurosurgical refinement of the techniques have advanced, trial design has been critically discussed, and the promise of biomarkers appears closer to be fulfilled (Scangos et al., 2021) (Table 22.1).

Targets for DBS in treatment-resistant depression

Fig. 22.1.

Subcallosal cingulate white matter

The anterior cingulate cortex plays a crucial role in the pathophysiology of depression. This region's involvement has been verified through physiology, and functional and structural imaging data (Dunlop et al., 2017). The subcallosal aspect of the anterior cingulate is a critical node in mood regulation networks involved in negative mood and antidepressant treatment response (Fox et al., 2014). Neuroimaging studies demonstrated that the SCC is activated during sadness induction in healthy volunteers, the SCC is hyperactive in patients with depression when compared with healthy volunteers, and successful treatment of depression is associated with normalization of SCC function.

The metabolism in the SCC (studied mostly through positron emission tomography) is positively correlated with depression and anxiety severity (Drevets et al., 2002; Kennedy et al., 2001; Osuch et al., 2000). When normal controls underwent a sadness induction experiment, increases in blood flow in the area were documented (Liotti et al., 2000). Patients who responded to antidepressant therapy had a decline in metabolism from an abnormal elevation toward normality (Mayberg et al., 2000). These changes observed in the SCC were not isolated, and other brain region changes in treatment supported the concept of neurocircuitry based changes in mood states. Resting state functional connectivity between SCC and

TABLE 22.1 Relevant deep brain stimulation trials for major depressive disorder.

Study	Target	Subjects	Design	Follow-up period	Results	Impact
Mayberg et al. (2005)	SCC	6	OLS	6 months	66% response, 50% remission	First DBS for depression study. Positive clinical results, PET changes in local and remote areas
Lozano et al. (2008)	SCC	20	OLS	1 year	55% response, 33% remission	Showed sustained mood improvement at 1 year
Kennedy et al. (2011)	SCC	20	OLS	3 years	64% response, 43%–50% remission	Long-term stimulation is safe and provides sustained efficacy
Puigdemont et al. (2012)	SCC	8	OLS	1 year	62.5% response, 50% remission	Second independent study to show efficacy of SCC-DBS
Holtzheimer et al. (2012)	SCC	17 (seven bipolar 2 patients)	OLS	2 years	41% response at 6 months, 92% response at 2 years (11/12 patients)	Showed long-term efficacy and safety of SCC-DBS
Ramasubbu et al. (2013)	SCC	4	OLS	6 months	50% response	Longer pulse widths may have a role in antidepressant effects
Merkl et al. (2013)	SCC	6	OLS	24–36 weeks	33% response, 33% remission	SCC-DBS causes acute and chronic antidepressant effects
Puigdemont et al. (2015)	SCC	5	COS	6 months	Active phase: 4/5 sustained response Sham phase: 2/5 relapsed	Continuous electrical stimulation is necessary to avoid relapse. Slow return of symptoms
Accolla et al. (2016)	SCC	5	OLS	24 months	20% response	Posterior gyrus rectus is a viable target for DBS in MDD (more ventral than SCC)
Holtzheimer et al. (2017)	SCC	90	RCT	6 months double-blind, then open-label follow-up	Primary endpoint: Response: 20% (active) vs 17% (sham); Remission: 5% (active) vs 7% (sham). 30 month response rate: 48%	Not clinically significant at primary endpoint. Increased response rates over long-term open-label phase
Merkl et al. (2018)	SCC	8	RCT then OLS	24–48 months	37% response at 6 months, 51% response (average), 33% remission in long-term follow-up	No clinical significance during double-blind phase (8 weeks); long-term treatment with significant response
Eitan et al. (2018)	SCC	9	RCT-COS	13 months	44% response	Long-term high-frequency stimulation appears to be better than low frequency
Riva-Posse et al., 2018	SCC	11	OLS	1 year	81.8% response, 54% remission	Tractography-based target selection DBS improves outcomes
Crowell et al. (2019)	SCC	28	OLS	8 years	Response > 50%, remission > 30% between years 2–8. 75% were responders for more than half of their participation	Treatment response to SCC DBS is sustained over time. Longest follow-up period published

Continued

TABLE 22.1 Relevant deep brain stimulation trials for major depressive disorder—cont'd

Study	Target	Subjects	Design	Follow-up period	Results	Impact
Malone et al. (2009)	VC/VS	15	OLS	1 year	53% response; 40% remission	Significant improvement in depressive symptoms
Malone (2010)	VC/VS	17	OLS	14–67 months	71% response at last follow-up, 35% remission	Long-term sustained improvement in depressive symptoms
Dougherty et al. (2015)	VC/VS	30	RCT, then OLS	16-week RCT, then 24-month follow-up	Primary endpoint: 20% (active) vs 14.3% (sham); long-term follow-up: 20%–26.7%	First RCT for DBS in depression
Bergfeld et al. (2016)	vALIC (VC/VS)	25 (OLS); 16 (COS)	OLS then RCT-COS	52 week optimization phase (OLS), then 2 week COS	40% response (OLS phase); 16 patients on the COS had lower HAM-D (*P*<.001) during active DBS compared to sham	vALIC DBS showed significant decrease of depressive symptoms. Discontinuation of stimulation was followed by rapid return of symptoms
Bewernick et al. (2012)	NAcc	11	OLS	48 months	45% response	Response found in patients with long-term follow-up
Millet et al. (2014)	NAcc	4	OLS	6 months	50% response during extended stimulation	NAcc appears more promising target than caudate nucleus
Schlaepfer et al. (2013)	MFB	7	OLS	12–33 weeks	86% response, 57.1% remission	Rapid reduction of symptoms within 2 days, response with long-term treatment
Fenoy et al. (2016)	MFB	4	OLS	26 weeks	75% response	Rapid reduction of symptoms within 7 days and response with long-term treatment
Fenoy et al. (2018)	MFB	6	OLS	1 year	80% response (4/5, one withdrawal, one lost to follow-up)	Expansion from previous study with longer follow-up describing durable response
Coenen et al. (2019)	MFB	16	2 month RCT, then OLS	1 year	100% response, 50% remission at 1 year, 10 achieved response at 1 week	2-Month blinded phase did not show difference in sham vs active stimulation. Sustained responses starting after surgery
Jimenez et al. (2005)	ITP	1	CR	18 weeks	HDRS-17 scale decrease from 42% to 6. 100% remission	ITP DBS with promising antidepressant effects
Raymaekers et al. (2017)	ITP vs vALIC (VC/VS)	7	COS	3–8 years	Response in both targets	6/7 Participants preferred vALIC stimulation over ITP. 2 suicides
Sartorius et al. (2010)	LHb	1	CR	4 months	Remission	Highlighted antidepressive effects of LHb DBS; cessation of DBS current resulted in return of symptoms

OLS, open-label study; *RCT*, randomized control trial; *COS*, crossover study; *CR*, case report; *SCC*, subcallosal cingulate; *VC/VS*, ventral capsule/ventral striatum; *vALIC*, ventral anterior limb of the internal capsule; *NAcc*, nucleus accumbens; *MFB*, medial forebrain bundle; *ITP*, inferior thalamic peduncle; *LHb*, lateral habenula; *HDRS*, Hamilton Depression Rating Scale.

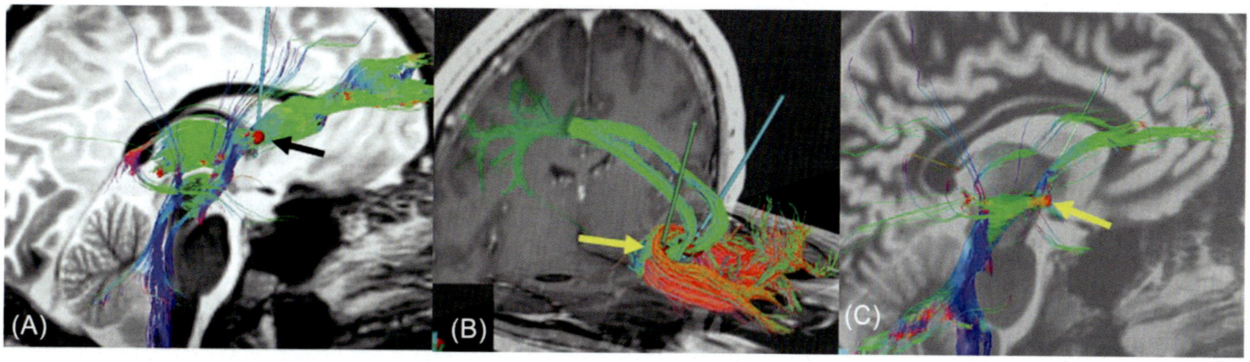

FIG. 22.1 Deterministic tractography imaging of three DBS targets. (A) ventral capsule/ventral striatum target, (B) subcallosal cingulate target, and (C) medial forebrain bundle target. *(Credit: KiSueng Choi, PhD.)*

thalamus within the default mode network (DMN) is significantly greater in depressed subjects. The length of the depressive episode is positively correlated with functional connectivity in the SCC in depressed subjects (Greicius et al., 2007). Meta-analytic findings show reliably increased functional connectivity between the DMN and SCC predicting levels of depressive rumination (Hamilton et al., 2015; Dutta et al., 2014). Other targets for DBS were originally designed imitating the lesion model (i.e., pallidotomy and globus pallidus stimulation in movement disorders, or capsulotomy and ventral capsule DBS in psychiatric illnesses). The SCC, on the other hand, was not considered a target for ablation previously (it is more medial than the subcaudate tractotomy lesion target). Functional imaging highlights this region as a primary dynamic modulator within a larger, multicomponent mood regulation system (Fox et al., 2012).

Clinical studies

The first publication of DBS in depression was published in 2005. Six severely treatment-resistant depressed patients were implanted in the SCC. After 6 months of stimulation four of the six patients responded (50% decline in depression severity from baseline), with three of them achieving symptomatic remission. The Hamilton Depression Rating Scale (HDRS-17) was reduced in 71% (Mayberg et al., 2005). That original cohort of six patients was expanded to 20 patients. One month after surgery, 35% of patients met criteria for response with 10% of patients in remission. After 6 months, 60% of patients were responders and 35% met criteria for remission. Notably, the response was largely maintained at 12 months and beyond (Lozano et al., 2008). After 1 year of stimulation, 62.5% of patients were responders, and the response rate after 2 years was 46%, with 75% response rate after 3 years of stimulation. The baseline severity of patients included in these studies was very high and with long duration in the index episode, approximating seven years. Only 10% were employed and had previously tried and relapsed after ECT. SCC DBS did not only generate a sustained symptomatic improvement, but patients regained overall function in physical health and social functioning, although two patients died of suicide during depressive relapses (Kennedy et al., 2011). After these initial promising reports, several single and multicenter studies were published, demonstrating comparable degrees of efficacy in open-label designs (Table 22.1). Seventeen patients (seven of them with chronic depression in bipolar 2 disorder) were implanted at Emory University. After 6 months of stimulation, the response rate was 41% (7/17), and 92% after 2 years of stimulation (12 of the 17 patients having reached the latter time point by the time of publication) (Holtzheimer et al., 2012). Both patients with unipolar and bipolar 2 depression responded similarly. Again, no side effects were attributable to stimulation, such as mania or cognitive changes (Moreines et al., 2014). Other centers have described similar outcomes, with many case series (Lozano et al., 2012; ; Merkl et al., 2013; Ramasubbu et al., 2013). Puigdemont et al. reported outcomes in eight patients with severe TRD with 6 months response and remission rates of 87.5% and 37.5%, respectively (Puigdemont et al., 2012). These dramatic improvements were sustained after 12 months, with 62.5% response and 50% remission rates. Another group in Germany implanted six subjects and explored acute effects of stimulation, but also described long-term antidepressant effects, with two of the six patients in remission of depression after 6 months of stimulation. As described in the other case series in the SCC, high voltage stimulation did not cause side effects (Merkl et al., 2013). An early metaanalysis of four studies by Berlim et al. confirmed the sustained positive results up to 1 year, with response and remission rates of 36.6% and 16.7%, 53.9% and 24.1%, and 39.9% and 26.3% at follow-up endpoints of 3, 6, and 12 months respectively (Berlim et al., 2014). There was an additional multicenter study conducted in three different sites in Canada that implanted 21 patients (Lozano et al., 2012). Forty-eight percent of patients were responders at 6 months.

An industry-sponsored, double-blind, randomized controlled multicenter study was conducted, with the initial intention of recruiting 200 patients in North America. The study was halted after a futility analysis determined that the likelihood of it achieving its primary outcome was low (Holtzheimer et al., 2017). Ninety patients had been implanted by the time the study was stopped. Its primary outcome was response (40% reduction in Montgomery–Åsberg Depression Rating Scale (MADRS) from baseline) averaged over months 4–6 of a double-blind phase (24 weeks). Participants were 2:1 randomized to either active ($n=60$) or sham ($n=30$) stimulation for the initial 6 months, then all participants received active simulation. There was no statistically significant difference in response during the blinded phase (twelve (20%) patients in the stimulation group vs five (17%) patients in the control group). These results were lower than the reports in the prior studies, but interestingly the long-term (open-label) follow-up phase described a gradual increase in response to treatment with half of the patients endorsing positive responses of chronic DBS.

Long-term sustained antidepressant effectiveness has been demonstrated in other groups as well (Merkl et al., 2018). A case series long-term report with up to 8 years of follow-up described that in 28 patients implanted at a single center, response was ≥50% and remission rates were ≥30%, after the second year since surgery. Crowell et al. reported that three-quarters of all participants met the treatment-response criterion for more than half of their duration of participation in the study, with 21% of all patients demonstrating continuous response to treatment from the first year onward. There were three patients who dropped out, and there were no suicides during the follow-up period (Crowell et al., 2019). Stimulation appears to be most effective with high frequency parameters, and delivery of stimulation needs to be uninterrupted, with return of symptoms over periods of weeks if it were to be stopped (Eitan et al., 2018).

The realization that variability in the surgical implantation was crucial generated a change from a standard anatomical coordinate-based to a connectomic approach. The use of tractography-based target selection was derived from the initial analysis of white matter structural connectivity in responders to SCC DBS, and then was implemented prospectively (Riva-Posse et al., 2014, 2018). Up until then, the SCC target had chosen the surgical region for implantation of the DBS leads extrapolating the functional imaging findings that implicated the subcallosal region and Brodmann Area 25 in depression, but lacked precision, especially in an area with high anatomical variability. Knowledge of the white matter fibers and the network that were stimulated in the SCC became more apparent when newer methods combining tractography imaging and engineering methods that estimated the volume of activated tissue were combined (Lujan et al., 2013). Small differences in electrode location caused substantial differences in the activated pathways and confirmed widespread network changes associated with DBS induced antidepressant effects (Riva-Posse et al., 2014). Using models that calculated volume of stimulation with individualized stimulation parameters, each contact had a unique set of white matter fibers. All the DBS responders shared a combination of white matter tracts connecting the SCC to the rest of the cingulate cortex (via the cingulum bundle), bilateral medial frontal cortices (through the forceps minor), subcortical nuclei and the thalamus (via uncinate fasciculus and frontostriatal fibers). With this response fingerprint, it became possible to test the hypothesis prospectively. Eleven subjects were implanted using target selection for the DBS that was based on the connectivity map that was present in responders (Riva-Posse et al., 2018). The response rate increased, with 8/11 (72.7%) of patients improving by more than 50% from baseline after 6 months, with an additional subject becoming a responder at the 12-month time point (9/11, 81.8% response rate). Prospective targeting allowed for personalized, patient-specific, target selection. This targeting method confirmed and validated the conceptualization of a network model with the cingulate as a hub, where engagement of remote areas of the depression network is needed for the adequate antidepressant effect. Furthermore, there have been additional reports that confirmed the network model rationale behind the SCC DBS. Distinct patterns of white matter activation have been found to be related to the intraoperative responses when there are changes in autonomic behavior (heart rate elevation) as well as positive antidepressant responses (Choi et al., 2015; Riva-Posse et al., 2019). Tractography-based targeting in the SCC not only resulted in better clinical outcomes in the months after implantation, but has also resulted in faster and more robust acute behavioral responses in the operating room (Riva-Posse et al., 2020). It has also allowed collecting more reliable electrophysiological signals, derived from a certainty of the ideal location of the stimulating contacts (Veerakumar et al., 2019).

Ventral capsule/ventral striatum (ventral anterior limb of internal capsule)

The ventral capsule/ventral striatum (VC/VS), also referred to as the ventral anterior limb of the internal capsule (vALIC), has been identified as another target for both depression and anxiety. For many years and still in use, ablative surgery in the form of anterior capsulotomy has been an effective treatment for treatment refractory psychiatric disorders since 1949 (Ballantine et al., 1967; Miguel et al., 2019). Recognizing the marked improvement in depressive symptoms in open-label trials of VC/VS DBS for OCD, patients with TRD were enrolled in studies. Over the years, the VC/VS target has migrated posteriorly, almost to the level of the anterior commissure, based solely on clinical experience of better outcomes associated with more posterior stimulation (Davidson et al., 2020; Greenberg et al., 2006). While initially replicating the old ablative approach, newer studies have demonstrated that stimulation in different contacts of the DBS leads along the span of the capsule generated distinct patterns of activation along the cortico-striatal-thalamic-cortical network (Dougherty et al., 2016).

Clinical studies

The surgical technique for VC/VS DBS involves stereotactic anatomic targeting, using high-resolution volumetric magnetic resonance imaging for planning. The anterior limb of the internal capsule and the VS can be easily visualized on T2 and inversion recovery images for direct targeting. Typical coordinates are 4–10 mm lateral to the midline, 3–5 mm ventral to the anterior commissure, and 1–3 mm anterior to the posterior border of the anterior commissure. DBS electrodes are implanted bilaterally through the shaft of the internal capsule with the most ventral aspect in the NAcc (Greenberg et al., 2010; Malone et al., 2009). In recent years, the introduction of tractography in this target for OCD, and in other targets for depression, has changed the older standard anatomical approach (Liebrand et al., 2019). The first open-label study commented on VC/VS-DBS of 17 TRD patients and found response rates of 53% at 12-month follow-up and 71% at the last follow-up after 14–67 months (and remission in 40%) (Malone et al., 2009; Malone, 2010). Such open-label results were promising, and a multicenter randomized clinical trial was designed with the goal of providing a higher level of evidence supporting this target for treatment in TRD. Importantly, advanced targeting technologies that would be considered essential for the surgery were not applied. DBS for OCD had already received a humanitarian device exemption in 2009 (Dougherty et al., 2015; Richardson et al., 2015). A 16-week, randomized, sham-controlled trial in the United States did not find clinical significance in treatment response rates. Only 3 of the 15 patients randomized to active stimulation (20%) had achieved response, while 2 of the 14 (14.3%) who received sham therapy responded. The response rates in the open-label phase of 24 months of follow-up were also low, with 20%–26.7% achieving response at any time during that period (Dougherty et al., 2015).

A different study, designed and conducted in the Netherlands, had better results although they were not directly comparable to the initial clinical trial. Patients had an initial open-label optimization phase that lasted 52 weeks. This phase described a 40% overall response rate in 25 patients. Mean depression scores decreased 28.3% (HDRS-17) and 30% (MADRS). Of these 25 patients, 16 (9 responders and 7 nonresponders) were entered in a randomized crossover period (active-sham vs sham-active). All responders had a return of symptoms within less than 2 weeks when stimulation was discontinued (Bergfeld et al., 2016). There are reasons to think that different targets may have variable times to have relapse in symptoms when stimulation is stopped (Kilian et al., 2019). Of the original 25 patients who had entered the clinical trial, 21 entered and 18 completed the maintenance phase for the following 2 years. 44% of patients were classified as responder. Patients who had not responded during the initial optimization phase did not become responders afterwards (van der Wal et al., 2020).

Studies of DBS for OCD in this same target have started to identify neuroimaging techniques such as tractography, identifying certain fibers that are responsible for response in patients (Liebrand et al., 2019; Fridgeirsson et al., 2020). In OCD, active stimulation of the capsule fibers closer to the MFB than the anterior thalamic radiations was associated with better treatment outcome ($P=.04$; $r^2=0.34$). Confirming the differences in tractography imaging and standard anatomical imaging (like in SCC), stimulation sites were largely overlapping and could not differentiate responder status, suggesting response is independent of the anatomically defined electrode position. As it was discussed earlier, the concept of the optimal target has evolved, from thinking of the target as a coordinate-determined gray or white matter structure, to conceiving of the target as "tapping into" a circuit connects to several regions of the symptomatic network (Karas et al., 2018). Innovative trial design, introduction of newer imaging methods, and engagement of specific biomarkers, whether physiologic, imaging, or neuropsychological, will continue to elucidate the mechanisms of action of DBS and making this target a viable therapy for patients with TRD (Drysdale et al., 2017).

Nucleus accumbens

The nucleus accumbens (NAcc) has been implicated as a neuromodulation target since the earlier days of this technology (Bishop et al., 1963). It is known to play key roles in the cognitive processing of motivation, reward (i.e., incentive salience, pleasure, and positive reinforcement), and reinforcement learning (Floresco, 2015). Preclinical animal studies have long described the antianhedonic effects of electrical stimulation of the accumbens (Lim et al., 2015). In animal models of depression, the reward system is dysfunctional, and is reset by chronic administration of an antidepressant (Berton et al., 2006).

Clinical studies

While septal stimulation had been conducted many decades ago, the "modern" beginnings of DBS in NAcc started when a group in Germany implanted three patients with severe refractory depression (Schlaepfer et al., 2008). The three participants improved "immediately" in their depression when the stimulator was on, and worsened in all three patients when the stimulator was turned off. Using FDG-PET, significant changes in brain metabolism as a function of the stimulation in fronto-striatal networks were observed. The cohort was extended to 10 participants. After 1 year of stimulation, half (50%) were responders and three patients were in remission (Bewernick et al., 2010). Similar PET results were seen in this larger cohort, with decreased metabolism in the SCC and in prefrontal regions including orbital prefrontal cortex. NAcc DBS also has sustained and durable antidepressant effects, with the same open-label cohort describing response rates of 45% (5 of 11) at 48 months (Bewernick et al., 2012). A different group in France implanted four subjects with the intention of doing a pilot multicenter prospective, noncomparative, and open trial. Patients initially received caudate stimulation, and after lack of antidepressant response, accumbens stimulation. This change in the DBS stimulation target caused that three of the four subjects had positive changes in mood (Millet et al., 2014). No cognitive or other major adverse effects were reported in either trial (Grubert et al., 2011). The literature provides promising evidence of NAcc DBS being a treatment for TRD; however, small samples have prevented definitive conclusions regarding its antidepressant efficacy.

Medial forebrain bundle

The MFB is a part of the mesolimbic-mesocortical dopamine reward system (Gálvez et al., 2015). Researchers have hypothesized the role of MFB stimulation as either a pathway to regulate descending glutaminergic fibers from the mPFC to the VTA and modulate dopamine and stimulate cortical brain regions. The MFB has been considered for a long time as an essential structure related to psycho-behavioral functioning in both animals and human subjects (Coenen et al., 2009, 2012a). There have been descriptions of its function involving affect regulation, as well as the reward-seeking system (Panksepp, 2005, 2010). The MFB, with its fibers traveling between the ventral tegmental area and the NAcc and structures beyond the reward-seeking system, appears to be involved in the pathophysiology of MDD. Interestingly, while other targets within the reward system (ventral striatum and nucleus accumbens) had already been tried, the difficulties identifying it in standard imaging, had delayed its potential target for DBS in depression (Coenen et al., 2012b).

Clinical studies

The surgical technique of DBS in MFB invariably requires of deterministic tractography to individually map the target area. The MFB is not visible in standard neuroimaging, instead requiring the acquisition of special tractography sequences. This apparent disadvantage in the target selection resulted, in the end, in avoiding the pitfalls that both SCC and VC/VS had, when the initial clinical trials were guided by less precise implantations.

The original pilot study in Germany assessed safety and efficacy of DBS in MFB in seven patients with TRD (Schlaepfer et al., 2013). Strikingly, the antidepressant effects were very rapid in their onset, even faster than NAcc previously described. The authors described instant intraoperative antidepressant-like effects such as appetitive motivation. In the acute intraoperative reports, MFB stimulation usually causes increased vigilance, eye contact, and engagement in conversation, as well as self-reported feelings of energy and motivation concordant with mood improvement (Fenoy et al., 2016). Within the first week of stimulation, patients already had significant responses, and response and remission rates of 86% and 57.1%, respectively. Not only the antidepressant effects were evident early on, but they were sustained in time. Half of those patients continued to have sustained remission at 1-year which continued until the 4-year follow-up (Fenoy et al., 2018). A separate group in the United States replicated the early effects. Three of six patients had > 50% decrease in depression scores after a week poststimulation initiation relative to baseline (Fenoy et al., 2018). An interesting fact in these two case-series is that the response to DBS relies on the preservation of the anatomical target. In the original cohort, the patient who did not respond to DBS had a minor intracranial hemorrhage in the trajectory of the lead, therefore irreversibly lesioning the intended target of stimulation. In the second cohort, the patient that did not respond to DBS had reduced frontal connectivity to the MFB when the DTI was analyzed. Both centers have published their long-term results, and the antidepressant effectiveness is sustained, and the neurocognitive tests did not find changes in cognition or impulsivity (Fenoy et al., 2018; Bewernick et al., 2017, 2018). Due to its proximity to oculomotor control areas in the midbrain, strabismus and double vision are the most significant stimulation related adverse effects of DBS in this target were blurred vision and strabismus. Similarly to the rapid onset of antidepressant effects, there seems to be an abrupt return of depressive symptomatology if stimulation is discontinued (Kilian et al., 2019).

More recently, a clinical study with a randomized-controlled onset of stimulation was reported (Coenen et al., 2019). Sixteen patients were implanted in a double-blind randomized-control (DBS active vs sham) condition for the initial 8 weeks after surgery, at which point blinding was lifted and all patients received active stimulation. Similar to the first study, all 16 patients reached the response criterion, and most patients ($n = 10$) responded within a week; 50% of patients were classified as remitters after 1 year of stimulation (MADRS < 10). Both groups (active/sham) demonstrated an antidepressant effect. Thus, it appeared that the sham-stimulation phase of 2 months was not long enough to separate treatment from other effects. The study reached its primary outcome at 12 months of stimulation, with all participants responding to DBS and a 56% mean decline in MADRS (50% of patients in remission). This lack of difference at this endpoint highlights the difficulties in trial design for DBS. It would be hard to justify that a trial with the sustained antidepressant results as these have reported would be due to placebo effects. Patients with TRD usually have little response to antidepressant interventions, and these are usually short lived. But any regulatory agency who would have asked for a sham versus active trial as the primary efficacy outcome would have deemed this study inconclusive. The authors of this trial rightfully comment that slow, careful, and adaptive study development in DBS for depression is appropriate.

Inferior thalamic peduncle

The ITP is part of the interplay between the dorsomedial thalamus and the orbitofrontal cortex (Kopell and Greenberg, 2008). Depression models have shown a dysregulation in this connection (Drevets, 2000). A case report regarding ITP DBS showed a decrease in depression scores early after surgery, and sustained after 8 months of stimulation using high frequency DBS and very large pulse widths (3.5 V, 450 μs pulse width) (Jimenez et al., 2005). There was a return of depressive symptoms when stimulation was interrupted, and the antidepressant effects were recaptured after stimulation was resumed with sustained effects (Jimenez et al., 2013). One trial attempted to evaluate the difference between stimulation in the internal capsule and the inferior thalamic peduncle. The group consisted of seven patients with TRD who were implanted and underwent blinded cross-over periods of 2 months per target. Due to low number of participants, it was not possible to determine superiority of one target over the other. All patients ($n = 7$) were followed up for at least 3 years (3–8 years) after implantation. Six patients completed the first crossover and five patients completed the second. Two participants died of suicide. Numerically, it appeared as if depression scores were lower for capsule stimulation. Three out of five subjects had response to ITP stimulation. Patients preferred the ALIC/BNST target with no major adverse effects (Raymaekers et al., 2017). Further clinical studies are needed to further validate the ITP as a DBS target for depression.

Lateral habenula

The LHb is an epithalamic structure that regulates serotonergic raphe nucleus activity and modulates dopaminergic midbrain functions (Boulos et al., 2017). Increased activation of the lateral habenular nucleus leads to the down regulation of the serotonergic, noradrenergic, dopaminergic systems and stimulation of the hypothalamic-pituitary-adrenal (HPA) axis (Sartorius and Henn, 2007). This overactivity in the habenula in depressive states may be counteracted by functional inhibition of the lateral habenula with DBS. The hypothesis is based on the findings of a clinical imaging study examining the habenula after tryptophan depletion and several animal models of depression (Lim et al., 2015). The acute antidepressant properties of ketamine (a rapid acting antidepressant) may be mediated through the habenula (Gass et al., 2019).

Clinical studies

Despite its promising neurobiological support, there have only been two case reports in the literature (Geisler and Trimble, 2008). In one case, the authors reported remission of depressive symptoms after 4 months of habenular stimulation. Interestingly, cessation of DBS current resulted in the return of depressive symptoms (Sartorius et al., 2010). Further clinical trials, case series, and studies of comparison between targets are under way to further examine the feasibility as a DBS target.

Conclusions

Psychiatric disorders are complex, with a constellation of symptoms and signs that are manifestations of multiple interconnected relevant circuits (mood, reward, anxiety/fear, homeostasis, cognition, etc.). Functional neurosurgery has been at the forefront of the advances in clinical neurosciences in the last couple of decades. Newer surgical techniques allow for more precise implantation, minimizing complications, and directly monitoring the physiologic changes caused by stimulation in the affected neurocircuitry (Lozano and Lipsman, 2013). The progress in DBS for depression will come not from a single discovery or breakthrough, but with a concerted effort by all the parties involved (basic scientists, neurosurgeons, psychiatrists, computational scientists, and biomedical engineering, to name some of them). Identification of biomarkers of illness, response, and change will play a role in the design of more precise targets for engagement (Dougherty, 2019; Yuen et al., 2021). Likewise, this will impact patient selection and trial design. The new concept of "connectomics" aims to map the structure of the brain at several levels from individual neurons to nodes and networks, and this framework is well suited to investigate key questions on the complex functions of the brain (Hagmann et al., 2010). The refinement of patient selection, and biomarker engagement with therapy, has evolved with incremental success in the field of Parkinson's disease, even aiding in the determination of target selection depending on clinical characteristics (Paschen and Deuschl, 2018). The field of neuromodulation in psychiatry is ripe for interventions that bring hope for many patients with severe, treatment-resistant disorders. The initial enthusiasm generated by small, open-label and mostly single-center studies was followed by a "loss of innocence," doubt and pessimism when the results of multicenter, randomized, double-blind trials did not replicate the positive results (Mayberg et al., 2016; Schlaepfer, 2015). In the end, strong clinical evidence will be needed to justify the invasiveness of these interventions for a large portion of patients with TRD who have not responded to other treatments. While initially disappointing in their primary outcomes, several cohorts of patients who received DBS in different targets have shown that when patients respond, they have sustained antidepressant effects. An exploratory metaanalysis published in 2014, gathering data from the DBS case series published to that date, estimated that DBS was 71% more likely than sham treatment to be of benefit for patients with TRD (Smith, 2014). However, it only included a total of 125 patients from eight different studies (only one randomized double blind control study in VC/VS). Another study that selected 10 clinical trials (with only two RCTs) combined a total of 190 patients receiving DBS in different targets. Patients on active treatment had a significantly higher response rates than sham stimulation (OR = 5.50; $P < .0001$) and reductions in mean depression score (SMD = −0.42; $P = .006$).

For a large number of patients with chronic depression, the endless cycle of response and relapse is disheartening and contributes to lesser engagement in functional rehabilitation. The sustained responses in DBS and other surgical interventions such as VNS are one of the justifications for the higher initial surgical risk, and possibly the financial costs of the interventions (Conway et al., 2018).

Advanced neuroimaging techniques are changing the clinical practice of psychiatric neurosurgery. DTI targeting methods are being used routinely in SCC, VC/VS, and MFB DBS not only in MDD but in a majority of conditions explored. A connectomic approach to DBS surgery opens up the reality of personalized treatment, where a particular target is decided on the involvement of certain white matter structures and symptoms that respond to specific circuit modulation. A notable example happened with medial prefrontal cortex/uncinate fasciculus DBS. Researchers intentionally steered the current in the DBS lead to stimulate the uncinate fasciculus, resulting in a dramatic reduction of posttraumatic stress disorder symptoms. The rationale for selecting this target and electrode contact configuration had a strong translational component, as this white matter bundle has an important role of modulating the basolateral amygdala.

The keys for success in the near future of DBS for depression will rely on the integration of advances in imaging, neurophysiology, and clinical expertise to plan new multicenter trials that will replicate, on a larger scale, the observations of different research groups, thus ensuring a safe and long-lasting treatment option for the TRD population.

References

Abosch, A., Cosgrove, G.R., 2008. Biological basis for the surgical treatment of depression. Neurosurg. Focus. 25 (1), E2.

Accolla, E.A., Aust, S., Merkl, A., Schneider, G.H., Kühn, A., Bajbouj, M., Draganski, B., 2016. Deep brain stimulation of the posterior gyrus rectus region for treatment resistant depression. J. Affect. Disord. 194, 33–37. https://doi.org/10.1016/j.jad.2016.01.022 (Epub 2016 Jan 13; PMID: 26802505).

Ballantine Jr., H.T., Cassidy, W.L., Flanagan, N.B., Marino Jr., R., 1967. Stereotaxic anterior cingulotomy for neuropsychiatric illness and intractable pain. J. Neurosurg. 26 (5), 488–495.

Bergfeld, I.O., Mantione, M., Hoogendoorn, M.L., Ruhe, H.G., Notten, P., van Laarhoven, J., et al., 2016. Deep brain stimulation of the ventral anterior limb of the internal capsule for treatment-resistant depression: a randomized clinical trial. JAMA Psychiatry 73 (5), 456–464.

Berlim, M.T., McGirr, A., Van den Eynde, F., Fleck, M.P., Giacobbe, P., 2014. Effectiveness and acceptability of deep brain stimulation (DBS) of the subgenual cingulate cortex for treatment-resistant depression: a systematic review and exploratory meta-analysis. J. Affect. Disord. 159, 31–38.

Berton, O., McClung, C.A., Dileone, R.J., Krishnan, V., Renthal, W., Russo, S.J., et al., 2006. Essential role of BDNF in the mesolimbic dopamine pathway in social defeat stress. Science 311 (5762), 864–868.

Bewernick, B.H., Hurlemann, R., Matusch, A., Kayser, S., Grubert, C., Hadrysiewicz, B., et al., 2010. Nucleus accumbens deep brain stimulation decreases ratings of depression and anxiety in treatment-resistant depression. Biol. Psychiatry 67 (2), 110–116.

Bewernick, B.H., Kayser, S., Sturm, V., Schlaepfer, T.E., 2012. Long-term effects of nucleus accumbens deep brain stimulation in treatment-resistant depression: evidence for sustained efficacy. Neuropsychopharmacology 37 (9), 1975–1985.

Bewernick, B.H., Kayser, S., Gippert, S.M., Switala, C., Coenen, V.A., Schlaepfer, T.E., 2017. Deep brain stimulation to the medial forebrain bundle for depression- long-term outcomes and a novel data analysis strategy. Brain Stimul. 10 (3), 664–671.

Bewernick, B.H., Kilian, H.M., Schmidt, K., Reinfeldt, R.E., Kayser, S., Coenen, V.A., et al., 2018. Deep brain stimulation of the supero-lateral branch of the medial forebrain bundle does not lead to changes in personality in patients suffering from severe depression. Psychol. Med. 48 (16), 2684–2692.

Bishop, M.P., Elder, S.T., Heath, R.G., 1963. Intracranial self-stimulation in man. Science 140 (3565), 394–396.

Boulos, L.J., Darcq, E., Kieffer, B.L., 2017. Translating the habenula-from rodents to humans. Biol. Psychiatry 81 (4), 296–305.

Choi, K.S., Riva-Posse, P., Gross, R.E., Mayberg, H.S., 2015. Mapping the "depression switch" during intraoperative testing of subcallosal cingulate deep brain stimulation. JAMA Neurol. 72 (11), 1252–1260.

Coenen, V.A., Honey, C.R., Hurwitz, T., Rahman, A.A., McMaster, J., Burgel, U., et al., 2009. Medial forebrain bundle stimulation as a pathophysiological mechanism for hypomania in subthalamic nucleus deep brain stimulation for Parkinson's disease. Neurosurgery 64 (6), 1106–1114. discussion 14-5.

Coenen, V.A., Panksepp, J., Hurwitz, T.A., Urbach, H., Madler, B., 2012a. Human medial forebrain bundle (MFB) and anterior thalamic radiation (ATR): imaging of two major subcortical pathways and the dynamic balance of opposite affects in understanding depression. J. Neuropsychiatry Clin. Neurosci. 24 (2), 223–236.

Coenen, V.A., Schlaepfer, T.E., Allert, N., Madler, B., 2012b. Diffusion tensor imaging and neuromodulation: DTI as key technology for deep brain stimulation. Int. Rev. Neurobiol. 107, 207–234.

Coenen, V.A., Bewernick, B.H., Kayser, S., Kilian, H., Bostrom, J., Greschus, S., et al., 2019. Superolateral medial forebrain bundle deep brain stimulation in major depression: a gateway trial. Neuropsychopharmacology 44 (7), 1224–1232. https://doi.org/10.1038/s41386-019-0369-9. [Epub 2019 Mar 13; PMID: 30867553; PMCID: PMC6785007.

Conway, C.R., Kumar, A., Xiong, W., Bunker, M., Aaronson, S.T., Rush, A.J., 2018. Chronic vagus nerve stimulation significantly improves quality of life in treatment-resistant major depression. J. Clin. Psychiatry 79 (5), 52–59.

Crowell, A.L., Riva-Posse, P., Holtzheimer, P.E., Garlow, S.J., Kelley, M.E., Gross, R.E., et al., 2019. Long-term outcomes of subcallosal cingulate deep brain stimulation for treatment-resistant depression. Am. J. Psychiatry 176 (11), 949–956.

Dandekar, M.P., Fenoy, A.J., Carvalho, A.F., Soares, J.C., Quevedo, J., 2018. Deep brain stimulation for treatment-resistant depression: an integrative review of preclinical and clinical findings and translational implications. Mol. Psychiatry 23 (5), 1094–1112.

Davidson, B., Lipsman, N., Meng, Y., Rabin, J.S., Giacobbe, P., Hamani, C., 2020. The use of tractography-based targeting in deep brain stimulation for psychiatric indications. Front. Hum. Neurosci. 14 (503), 1–8.

Dougherty, D.D., 2019. Will deep brain stimulation help move precision medicine to the clinic in psychiatry? Biol. Psychiatry 85 (9), 706–707.

Dougherty, D.D., Rezai, A.R., Carpenter, L.L., Howland, R.H., Bhati, M.T., O'Reardon, J.P., et al., 2015. A randomized sham-controlled trial of deep brain stimulation of the ventral capsule/ventral striatum for chronic treatment-resistant depression. Biol. Psychiatry 78 (4), 240–248.

Dougherty, D.D., Chou, T., Corse, A.K., Arulpragasam, A.R., Widge, A.S., Cusin, C., et al., 2016. Acute deep brain stimulation changes in regional cerebral blood flow in obsessive-compulsive disorder. J. Neurosurg. 125 (5), 1087–1093.

Drevets, W.C., 2000. Functional anatomical abnormalities in limbic and prefrontal cortical structures in major depression. Prog. Brain Res. 126, 413–431.

Drevets, W.C., Bogers, W., Raichle, M.E., 2002. Functional anatomical correlates of antidepressant drug treatment assessed using PET measures of regional glucose metabolism. Eur. Neuropsychopharmacol. 12 (6), 527–544.

Drysdale, A.T., Grosenick, L., Downar, J., Dunlop, K., Mansouri, F., Meng, Y., et al., 2017. Resting-state connectivity biomarkers define neurophysiological subtypes of depression. Nat. Med. 23 (1), 28–38.

Dunlop, B.W., Rajendra, J.K., Craighead, W.E., Kelley, M.E., McGrath, C.L., Choi, K.S., et al., 2017. Functional connectivity of the subcallosal cingulate cortex and differential outcomes to treatment with cognitive-behavioral therapy or antidepressant medication for major depressive disorder. Am. J. Psychiatry 174 (6), 533–545.

Dutta, A., McKie, S., Deakin, J.F., 2014. Resting state networks in major depressive disorder. Psychiatry Res. 224 (3), 139–151.

Eitan, R., Fontaine, D., Benoit, M., Giordana, C., Darmon, N., Israel, Z., et al., 2018. One year double blind study of high vs low frequency subcallosal cingulate stimulation for depression. J. Psychiatr. Res. 96, 124–134. https://doi.org/10.1016/j.jpsychires.2017.09.026 (PMID: 29032294).

Fenoy, A.J., Schulz, P., Selvaraj, S., Burrows, C., Spiker, D., Cao, B., et al., 2016. Deep brain stimulation of the medial forebrain bundle: distinctive responses in resistant depression. J. Affect. Disord. 203, 143–151.

Fenoy, A.J., Schulz, P.E., Selvaraj, S., Burrows, C.L., Zunta-Soares, G., Durkin, K., et al., 2018. A longitudinal study on deep brain stimulation of the medial forebrain bundle for treatment-resistant depression. Transl. Psychiatry 8 (1), 111.

Floresco, S.B., 2015. The nucleus accumbens: an interface between cognition, emotion, and action. Annu. Rev. Psychol. 66, 25–52.

Fox, M.D., Buckner, R.L., White, M.P., Greicius, M.D., Pascual-Leone, A., 2012. Efficacy of transcranial magnetic stimulation targets for depression is related to intrinsic functional connectivity with the subgenual cingulate. Biol. Psychiatry 72 (7), 595–603.

Fox, M.D., Buckner, R.L., Liu, H., Chakravarty, M.M., Lozano, A.M., Pascual-Leone, A., 2014. Resting-state networks link invasive and noninvasive brain stimulation across diverse psychiatric and neurological diseases. Proc. Natl. Acad. Sci. 111 (41), E4367.

Fridgeirsson, E.A., Figee, M., Luigjes, J., van den Munckhof, P., Schuurman, P.R., van Wingen, G., et al., 2020. Deep brain stimulation modulates directional limbic connectivity in obsessive-compulsive disorder. Brain 143 (5), 1603–1612.

Gálvez, J.F., Keser, Z., Mwangi, B., Ghouse, A.A., Fenoy, A.J., Schulz, P.E., et al., 2015. The medial forebrain bundle as a deep brain stimulation target for treatment resistant depression: a review of published data. Prog. Neuro-Psychopharmacol. Biol. Psychiatry 58, 59–70.

Gass, N., Becker, R., Reinwald, J., Cosa-Linan, A., Sack, M., Weber-Fahr, W., et al., 2019. Differences between ketamine's short-term and long-term effects on brain circuitry in depression. Transl. Psychiatry 9 (1), 172.

Geisler, S., Trimble, M., 2008. The lateral habenula: no longer neglected. CNS Spectr. 13 (6), 484–489.

Greenberg, B.D., Malone, D.A., Friehs, G.M., Rezai, A.R., Kubu, C.S., Malloy, P.F., et al., 2006. Three-year outcomes in deep brain stimulation for highly resistant obsessive-compulsive disorder. Neuropsychopharmacology 31 (11), 2384–2393.

Greenberg, B.D., Gabriels, L.A., Malone Jr., D.A., Rezai, A.R., Friehs, G.M., Okun, M.S., et al., 2010. Deep brain stimulation of the ventral internal capsule/ventral striatum for obsessive-compulsive disorder: worldwide experience. Mol. Psychiatry 15 (1), 64–79.

Greicius, M.D., Flores, B.H., Menon, V., Glover, G.H., Solvason, H.B., Kenna, H., et al., 2007. Resting-state functional connectivity in major depression: abnormally increased contributions from subgenual cingulate cortex and thalamus. Biol. Psychiatry 62 (5), 429–437.

Grubert, C., Hurlemann, R., Bewernick, B.H., Kayser, S., Hadrysiewicz, B., Axmacher, N., et al., 2011. Neuropsychological safety of nucleus accumbens deep brain stimulation for major depression: effects of 12-month stimulation. World J. Biol. Psychiatry 12 (7), 516–527.

Haas, L.F., 2003. Neurological stamp. Egas Moniz (1874–1955). J. Neurol. Neurosurg. Psychiatry 74 (5), 653.

Hagmann, P., Cammoun, L., Gigandet, X., Gerhard, S., Grant, P.E., Wedeen, V., et al., 2010. MR connectomics: principles and challenges. J. Neurosci. Methods 194 (1), 34–45.

Hamilton, J.P., Farmer, M., Fogelman, P., Gotlib, I.H., 2015. Depressive rumination, the default-mode network, and the dark matter of clinical neuroscience. Biol. Psychiatry 78 (4), 224–230.

Hoffman, J.L., 1949. Clinical observations concerning schizophrenic patients treated by prefrontal leukotomy. N. Engl. J. Med. 241 (6), 233–236.

Holtzheimer, P.E., Husain, M.M., Lisanby, S.H., Taylor, S.F., Whitworth, L.A., McClintock, S., et al., 2017. Subcallosal cingulate deep brain stimulation for treatment-resistant depression: a multisite, randomised, sham-controlled trial. Lancet Psychiatry 4 (11), 839–849.

Holtzheimer, P.E., Kelley, M.E., Gross, R.E., Filkowski, M.M., Garlow, S.J., Barrocas, A., Wint, D., Craighead, M.C., Kozarsky, J., Chismar, R., Moreines, J.L., Mewes, K., Posse, P.R., Gutman, D.A., Mayberg, H.S., 2012. Subcallosal cingulate deep brain stimulation for treatment-resistant unipolar and bipolar depression. Arch. Gen. Psychiatry 69 (2), 150–158. https://doi.org/10.1001/archgenpsychiatry.2011.1456 (Epub 2012 Jan 2; PMID: 22213770; PMCID: PMC4423545).

Holtzheimer et al 2012: https://pubmed.ncbi.nlm.nih.gov/22213770/.

Jimenez, F., Velasco, F., Salin-Pascual, R., Hernandez, J.A., Velasco, M., Criales, J.L., et al., 2005. A patient with a resistant major depression disorder treated with deep brain stimulation in the inferior thalamic peduncle. Neurosurgery 57 (3), 585–593. discussion -93.

Jimenez, F., Nicolini, H., Lozano, A.M., Piedimonte, F., Salin, R., Velasco, F., 2013. Electrical stimulation of the inferior thalamic peduncle in the treatment of major depression and obsessive compulsive disorders. World Neurosurg. 80 (3–4). S30.e17-25.

Karas, P.J., Lee, S., Jimenez-Shahed, J., Goodman, W.K., Viswanathan, A., Sheth, S.A., 2018. Deep brain stimulation for obsessive compulsive disorder: evolution of surgical stimulation target parallels changing model of dysfunctional brain circuits. Front. Neurosci. 12, 998.

Kennedy, S.H., Evans, K.R., Kruger, S., Mayberg, H.S., Meyer, J.H., McCann, S., et al., 2001. Changes in regional brain glucose metabolism measured with positron emission tomography after paroxetine treatment of major depression. Am. J. Psychiatry 158 (6), 899–905.

Kennedy, S.H., Giacobbe, P., Rizvi, S.J., Placenza, F.M., Nishikawa, Y., Mayberg, H.S., et al., 2011. Deep brain stimulation for treatment-resistant depression: follow-up after 3 to 6 years. Am. J. Psychiatry 168 (5), 502–510.

Kilian, H.M., Meyer, D.M., Bewernick, B.H., Spanier, S., Coenen, V.A., Schlaepfer, T.E., 2019. Discontinuation of superolateral medial forebrain bundle deep brain stimulation for treatment-resistant depression leads to critical relapse. Biol. Psychiatry 85 (6), e23–e24.

Kopell, B.H., Greenberg, B.D., 2008. Anatomy and physiology of the basal ganglia: implications for DBS in psychiatry. Neurosci. Biobehav. Rev. 32 (3), 408–422.

Liebrand, L.C., Caan, M.W.A., Schuurman, P.R., van den Munckhof, P., Figee, M., Denys, D., et al., 2019. Individual white matter bundle trajectories are associated with deep brain stimulation response in obsessive-compulsive disorder. Brain Stimul. 12 (2), 353–360.

Lim, L.W., Prickaerts, J., Huguet, G., Kadar, E., Hartung, H., Sharp, T., et al., 2015. Electrical stimulation alleviates depressive-like behaviors of rats: investigation of brain targets and potential mechanisms. Transl. Psychiatry 5, e535.

Liotti, M., Mayberg, H.S., Brannan, S.K., McGinnis, S., Jerabek, P., Fox, P.T., 2000. Differential limbic—cortical correlates of sadness and anxiety in healthy subjects: implications for affective disorders. Biol. Psychiatry 48 (1), 30–42.

Lozano, A.M., Lipsman, N., 2013. Probing and regulating dysfunctional circuits using deep brain stimulation. Neuron 77 (3), 406–424.

Lozano, A.M., Mayberg, H.S., Giacobbe, P., Hamani, C., Craddock, R.C., Kennedy, S.H., 2008. Subcallosal cingulate gyrus deep brain stimulation for treatment-resistant depression. Biol. Psychiatry 64 (6), 461–467.

Lozano, A.M., Giacobbe, P., Hamani, C., Rizvi, S.J., Kennedy, S.H., Kolivakis, T.T., et al., 2012. A multicenter pilot study of subcallosal cingulate area deep brain stimulation for treatment-resistant depression. J. Neurosurg. 116 (2), 315–322.

Lujan, J.L., Chaturvedi, A., Choi, K.S., Holtzheimer, P.E., Gross, R.E., Mayberg, H.S., et al., 2013. Tractography-activation models applied to subcallosal cingulate deep brain stimulation. Brain Stimul. 6 (5), 737–739.

Mahoney, D.E., Green, A.L., 2020. Psychosurgery: history of the neurosurgical management of psychiatric disorders. World Neurosurg. 137, 327–334.

Malone Jr., D.A., 2010. Use of deep brain stimulation in treatment-resistant depression. Cleve. Clin. J. Med. 77 (Suppl. 3), S77–S80. https://doi.org/10.3949/ccjm.77.s3.14 (PMID: 20622083).

Malone Jr., D.A., Dougherty, D.D., Rezai, A.R., Carpenter, L.L., Friehs, G.M., Eskandar, E.N., et al., 2009. Deep brain stimulation of the ventral capsule/ventral striatum for treatment-resistant depression. Biol. Psychiatry 65 (4), 267–275.

Mayberg, H.S., Brannan, S.K., Tekell, J.L., Silva, J.A., Mahurin, R.K., McGinnis, S., et al., 2000. Regional metabolic effects of fluoxetine in major depression: serial changes and relationship to clinical response. Biol. Psychiatry 48 (8), 830–843.

Mayberg, H.S., Lozano, A.M., Voon, V., McNeely, H.E., Seminowicz, D., Hamani, C., et al., 2005. Deep brain stimulation for treatment-resistant depression. Neuron 45 (5), 651–660.

Mayberg, H.S., Riva-Posse, P., Crowell, A.L., 2016. Deep brain stimulation for depression: keeping an eye on a moving target. JAMA Psychiatry 73 (5), 439–440.

McIntyre, C.C., Hahn, P.J., 2010. Network perspectives on the mechanisms of deep brain stimulation. Neurobiol. Dis. 38 (3), 329–337.

Merkl, A., Schneider, G.H., Schonecker, T., Aust, S., Kuhl, K.P., Kupsch, A., et al., 2013. Antidepressant effects after short-term and chronic stimulation of the subgenual cingulate gyrus in treatment-resistant depression. Exp. Neurol. 249, 160–168.

Merkl, A., Aust, S., Schneider, G.H., Visser-Vandewalle, V., Horn, A., Kuhn, A.A., et al., 2018. Deep brain stimulation of the subcallosal cingulate gyrus in patients with treatment-resistant depression: a double-blinded randomized controlled study and long-term follow-up in eight patients. J. Affect. Disord. 227, 521–529. https://doi.org/10.1016/j.jad.2017.11.024 (PMID: 29161674).

Miguel, E.C., Lopes, A.C., McLaughlin, N.C.R., Noren, G., Gentil, A.F., Hamani, C., et al., 2019. Evolution of gamma knife capsulotomy for intractable obsessive-compulsive disorder. Mol. Psychiatry 24 (2), 218–240.

Millet, B., Jaafari, N., Polosan, M., Baup, N., Giordana, B., Haegelen, C., et al., 2014. Limbic versus cognitive target for deep brain stimulation in treatment-resistant depression: accumbens more promising than caudate. Eur. Neuropsychopharmacol. 24 (8), 1229–1239.

Moniz, E., 1994. Prefrontal leucotomy in the treatment of mental disorders. 1937. Am. J. Psychiatry 151 (6 Suppl), 236–239.

Moreines, J.L., McClintock, S.M., Kelley, M.E., Holtzheimer, P.E., Mayberg, H.S., 2014. Neuropsychological function before and after subcallosal cingulate deep brain stimulation in patients with treatment-resistant depression. Depress. Anxiety 31 (8), 690–698.

Okun, M.S., 2012. Deep-brain stimulation for Parkinson's disease. N. Engl. J. Med. 367 (16), 1529–1538.

Osuch, E.A., Ketter, T.A., Kimbrell, T.A., George, M.S., Benson, B.E., Willis, M.W., et al., 2000. Regional cerebral metabolism associated with anxiety symptoms in affective disorder patients. Biol. Psychiatry 48 (10), 1020–1023.

Panksepp, J., 2005. Affective consciousness: core emotional feelings in animals and humans. Conscious. Cogn. 14 (1), 30–80.

Panksepp, J., 2010. Affective neuroscience of the emotional BrainMind: evolutionary perspectives and implications for understanding depression. Dialogues Clin. Neurosci. 12 (4), 533–545.

Paschen, S., Deuschl, G., 2018. Patient evaluation and selection for movement disorders surgery: the changing spectrum of indications. Prog. Neurol. Surg. 33, 80–93.

Puigdemont, D., Pérez-Egea, R., Portella, M., Molet, J., de Diego-Adeliño, J., Gironell, A., et al., 2012. Deep brain stimulation of the subcallosal cingulate gyrus: further evidence in treatment-resistant major depression. Int. J. Neuropsychopharmacol. 15 (1), 121–133. https://doi.org/10.1017/S1461145711001088 (PMID: 21777510).

Puigdemont, D., Portella, M., Pérez-Egea, R., Molet, J., Gironel, A., de Diego-Adeliño, J., Martín, A., Rodríguez, R., Àlvarez, E., Artigas, F., Pérez, V., 2015. A randomized double-blind crossover trial of deep brain stimulation of the subcallosal cingulate gyrus in patients with treatment-resistant depression: a pilot study of relapse prevention. J. Psychiatry Neurosci. 40 (4), 224–231. https://doi.org/10.1503/jpn.130295 (PMID: 25652752; PMCID: PMC4478055).

Ramasubbu, R., Anderson, S., Haffenden, A., Chavda, S., Kiss, Z.H., 2013. Double-blind optimization of subcallosal cingulate deep brain stimulation for treatment-resistant depression: a pilot study. J. Psychiatry Neurosci. 38 (3), 120–160.

Raymaekers, S., Luyten, L., Bervoets, C., Gabriels, L., Nuttin, B., 2017. Deep brain stimulation for treatment-resistant major depressive disorder: a comparison of two targets and long-term follow-up. Transl. Psychiatry 7 (10), e1251.

Richardson, R.M., Ghuman, A.S., Karp, J.F., 2015. Results of the first randomized controlled trial of deep brain stimulation in treatment-resistant depression. Neurosurgery 77 (2), N23–N24.

Riva-Posse, P., 2020. Why is deep brain stimulation for treatment-resistant depression a needed treatment option? Braz. J. Psychiatry 42 (4), 344–346.

Riva-Posse, P., Choi, K.S., Holtzheimer, P.E., Crowell, A.L., Garlow, S.J., Rajendra, J.K., et al., 2018. A connectomic approach for subcallosal cingulate deep brain stimulation surgery: prospective targeting in treatment-resistant depression. Mol. Psychiatry 23, 843–849. https://doi.org/10.1038/mp.2017.59.

Riva-Posse, P., Choi, K.S., Holtzheimer, P.E., McIntyre, C.C., Gross, R.E., Chaturvedi, A., et al., 2014. Defining critical white matter pathways mediating successful subcallosal cingulate deep brain stimulation for treatment-resistant depression. Biol. Psychiatry 76 (12), 963–969.

Riva-Posse, P., Inman, C.S., Choi, K.S., Crowell, A.L., Gross, R.E., Hamann, S., et al., 2019. Autonomic arousal elicited by subcallosal cingulate stimulation is explained by white matter connectivity. Brain Stimul. 12 (3), 743–751.

Riva-Posse, P., Crowell, A.L., Wright, K., Waters, A.C., Choi, K., Garlow, S.J., et al., 2020. Rapid antidepressant effects of deep brain stimulation and their relation to surgical protocol. Biol. Psychiatry 88 (8), e37–e39.

Rush, A.J., George, M.S., 2006. Reply. Biol. Psychiatry 60 (12), 1382–1383.

Sartorius, A., Henn, F.A., 2007. Deep brain stimulation of the lateral habenula in treatment resistant major depression. Med. Hypotheses 69 (6), 1305–1308.

Sartorius, A., Kiening, K.L., Kirsch, P., von Gall, C.C., Haberkorn, U., Unterberg, A.W., et al., 2010. Remission of major depression under deep brain stimulation of the lateral habenula in a therapy-refractory patient. Biol. Psychiatry 67 (2), e9–e11.

Scangos, K.W., Makhoul, G.S., Sugrue, L.P., Chang, E.F., Krystal, A.D., 2021. State-dependent responses to intracranial brain stimulation in a patient with depression. Nat. Med. 27 (2), 229–231.

Schlaepfer, T.E., 2015. Deep brain stimulation for major depression-steps on a long and winding road. Biol. Psychiatry 78 (4), 218–219.

Schlaepfer, T.E., Cohen, M.X., Frick, C., Kosel, M., Brodesser, D., Axmacher, N., et al., 2008. Deep brain stimulation to reward circuitry alleviates anhedonia in refractory major depression. Neuropsychopharmacology 33 (2), 368–377.

Schlaepfer, T.E., Bewernick, B.H., Kayser, S., Madler, B., Coenen, V.A., 2013. Rapid effects of deep brain stimulation for treatment-resistant major depression. Biol. Psychiatry 73 (12), 1204–1212.

Smith, D.F., 2014. Exploratory meta-analysis on deep brain stimulation in treatment-resistant depression. Acta Neuropsychiatr. 26 (6), 382–384.

Staudt, M.D., Herring, E.Z., Gao, K., Miller, J.P., Sweet, J.A., 2019. Evolution in the treatment of psychiatric disorders: from psychosurgery to psychopharmacology to neuromodulation. Front. Neurosci. 13, 108.

van der Wal, J.M., Bergfeld, I.O., Lok, A., Mantione, M., Figee, M., Notten, P., et al., 2020. Long-term deep brain stimulation of the ventral anterior limb of the internal capsule for treatment-resistant depression. J. Neurol. Neurosurg. Psychiatry 91 (2), 189–195.

Veerakumar, A., Tiruvadi, V., Howell, B., Waters, A.C., Crowell, A.L., Voytek, B., et al., 2019. Field potential 1/f activity in the subcallosal cingulate region as a candidate signal for monitoring deep brain stimulation for treatment-resistant depression. J. Neurophysiol. 122 (3), 1023–1035.

Vitek, J.L., 2002. Mechanisms of deep brain stimulation: excitation or inhibition. Mov. Disord. 17 (Suppl. 3), S69–S72.

Volpini, M., Giacobbe, P., Cosgrove, G.R., Levitt, A., Lozano, A.M., Lipsman, N., 2017. The history and future of ablative neurosurgery for major depressive disorder. Stereotact. Funct. Neurosurg. 95 (4), 216–228.

Williams, L.M., 2016. Precision psychiatry: a neural circuit taxonomy for depression and anxiety. Lancet Psychiatry 3 (5), 472–480.

Yuen, J., Rusheen, A.E., Price, J.B., Barath, A.S., Shin, H., Kouzani, A.Z., et al., 2021. Biomarkers for deep brain stimulation in animal models of depression. Neuromodulation. https://doi.org/10.1111/ner.13483 (PMID: 34105847; Online ahead of print).

Chapter 23

Other neurosurgical interventions for treatment-resistant depression

Ziad Nahas

Department of Psychiatry and Behavioral Sciences, University of Minnesota, Medical School, Minneapolis, MN, United States

Introduction

The use of somatic interventions to control or treat mental symptoms dates back to ancient times (Clower and Finger, 2001; Kalinowsky, 1986; Tourney, 1967). Evidence for burr holes drilled into the skull to "cure the demons" goes back to the Neolithic age. The notions that convulsions and fever may help mental disorders have been known since Hippocrates, while in medieval times, make-believe surgeries were performed to extract the "stone of madness."

In the 17th century, Descartes hypothesized that the ventricles were the reservoir of vital fluids and basis for the rational mind. This deemphasized the brain's role. Conversely, modern somatic treatments for mental illness are rooted in the conceptualization that neural tissue is responsible for behavior. This started in 1796, when Gall proposed that different brain regions were responsible for different functions, a system he called phrenology. Although this notion was revolutionary and would prove essential to our current understanding of brain function, he and his followers were later involved in pseudoscience and contributed little to the functional neuroanatomy of the mind (Critchley, 1965).

Almost a decade later, modern scientific conceptualization of brain functions and localizations began to emerge from animal experiments, cadaver dissections, and clinical observations conducted by Broca (1865), Jackson (1873), and others. Also, the notion of neuronal transmission based on electrochemical signals replaced the 19th century hydraulic neuronal transmission model (Tourney, 1967). These ideas of regional brain functional localization and electrochemical neuronal transmission would later evolve into the contemporary neuronal network models that guide and inform much of the current applications of somatic treatments in disorders such as obsessive compulsive disorder and depression. As the biological underpinnings of the complexity of the mind are still being worked out, current somatic interventions are being used both for understanding the neurobiology of mental illness as well as for treating disease (George et al., 2007).

Treatment-resistant depression (TRD) patients have long lasting depressive episodes, are prone to frequent relapses and have limited treatment options (Rush et al., 2006). They are unable to effectively situate themselves in relation to others, to control negative emotions and divert attention from negative experiences to cope with daily life. They present us with major clinical and socioeconomic challenges with cost estimates exceeding \$44 billion (Greenberg et al., 2003). A better understanding of the functional neuroanatomy of mood regulation and socialization will lead to optimal and effective treatments. The success rate with pharmacological treatments for TRD is low (Trivedi et al., 2004). Even after achieving relief from depressive symptoms, 35%–70% will relapse often precipitated by medication poor compliance or discontinuation of therapy. These patients become more vulnerable to stress (Kendler et al., 2000) and the course of their illness worsens leading to longer episode durations and shorter interepisodes periods in between. Brain stimulation therapies (BSTs) directly modulate brain function and regulate mood. Each presents with unique characteristics that define its role in the TRD therapeutic landscape. These various therapies have been discussed in more details in previous chapters (ADD number for chapters on ECT, MST, rTMS, tDCS, VNS, and DBS respectively). This chapter will discuss the current applications of other neurosurgical procedures not otherwise discussed, focusing on prefrontal cortical stimulation (PCS) and neurosurgery for psychiatric conditions.

PCS: The over simplification of the functional neuroanatomy of mood regulation may lead to missed opportunities in treatment development

Epidural cortical stimulation

Epidural cortical stimulation involves the placement of multicontact stimulating paddles over specific cortical regions and connected to a pacemaker-like generator. Chronic epidural cortical stimulation of the motor (EmCS) or sensory areas has

Managing Treatment-Resistant Depression. https://doi.org/10.1016/B978-0-12-824067-0.00023-2

been used over a decade to treat intractable pain syndromes, enhance recovery from stroke and for motor disorders like Parkinson's disease (Priori and Lefaucheur, 2007). EmCS induces its therapeutic effects at stimulation intensities below the threshold for overt motor responses. These results rule out the activation of large corticospinal cells, with possible candidates being nonpyramidal intracortical axons (Priori and Lefaucheur, 2007). The prefrontal cortex similar to the motor cortex is a six-layered structure that contains cell bodies and axons that run tangentially or perpendicular to its surface. Axons are better than neurons as the targets of cortical stimulation because membrane properties allow fibers to be more excitable than cell bodies (McIntyre and Grill, 2002). These axons could be short fibers of small inhibitory interneurons within the frontal cortex, as well as afferent or efferent fibers connected with distant cortical or subcortical structures (Garcia-Larrea et al., 2000). Various cortical inhibitory neurons are good for local activation owing to their low threshold for electrical stimulation (Wang et al., 2004). These cells exert strong inhibition on the underlying cellular network in the prefrontal cortex itself, locally modulating its output. Reciprocal connections between cortical and limbic regions could increase the responsiveness of neural circuits by generating a positive feedback loop. This echoes prefrontal TMS functional neuroimaging dosing studies, where local and distal effects in subcortical brain regions appear to be both intensity (Nahas et al., 2001a; Li et al., 2004; Kose et al., 2008) and frequency dependent (Nahas et al., 2001b).

Epidural prefrontal cortical stimulation (EpCS): Mayberg proposed that illness remission occurs when there is appropriate modulation of dysfunctional limbic-cortical interactions, an effect facilitated by various forms of treatment (Mayberg et al., 1999). Antidepressant drugs or DBS may initially regulate subcortical areas and later lead to a modulation of prefrontal regions (down-top model). The reverse may be observed with cognitive psychotherapy (top-down model), TMS, ECT, or EpCS. EpCS is an innovative new treatment through adaptive prefrontal governance of limbic structures. Two regions of the prefrontal cortex are prime candidates for cortical brain stimulation to treat depression through modulation of local and subcortical activity.

1. The frontopolar region [Brodmann Area (BA) 10] has distinctly higher number of dendritic spines per cell and lower density of cell bodies than any other prefrontal region (Petrides and Pandya, 2007). Its rich connections are directed to the cingulate cortex (including subgenual cortex or BA25 and precuneus/posterior cingulate), the orbitofrontal and the dorsolateral prefrontal cortex. It integrates higher level cognitive and emotional processing. It controls emotions, memory and motivation. It is also involved in self-reference, reflective self-awareness (Raichle and Snyder, 2007) and in attributing mental states to others ("the theory of mind") (Vogeley et al., 2001). High activity of BA10 is associated with depression.
2. The midlateral frontal regions [BA9 and 46] maintain preferential bidirectional connections with multimodal temporal areas on the one hand, and with paralimbic areas, such as the cingulate, the retrosplenial and the rostral temporal cortex, on the other hand (Petrides and Pandya, 2007). They play a critical role in organizing, monitoring, verification of information, attending to emotion stimuli and in reappraisal. The dorsal regions are involved in the monitoring of information in working memory and the ventral regions are involved in active judgments on information held in posterior cortical association regions that are necessary for active retrieval and encoding of information. Depression is associated with gray matter thinning in nearby BA45.

The experience with EpCS for treatment of depression remains very limited. Aside from our work targeting bilateral prefrontal stimulation with four leads, an industry sponsored small trial with unilateral left DLPFC was also carried out. When we began the EpCS work, our primary objective was to pilot its potential safety and therapeutic benefits in TRD patients. We also set out to support our choices of stimulation sites (BA10 and 46) and better characterize their specific roles in mood regulation and socialization. We also were interested in exploring the bilateral hemispheres lead placements, the frequency and most critically, the intermittent on/off duty cycle. The bilateral four paddles EpCS approach we pioneered in 2008 showed significant improvements in depressive symptoms and a number of behavioral measures associated with in self-awareness, internal monitoring and regulatory executive functions. At 7-month follow-up, the average improvement from preimplant baseline on the Hamilton Rating Scale for Depression ($HRSD_{24}$) (Nahas et al., 2010) was 54.9% (±37.7). Three implanted subjects reached full remission and maintained their clinical status at 5 years follow up (this study will be discussed in more details below). Separately from our work, an industry-sponsored study (North Star) reported on the feasibility of *unilateral* left dorsolateral prefrontal EpCS (Dougherty et al., 2008). Twelve MDD TRD patients were randomized to active or sham single blind for 8 weeks treatment with an adaptive open design follow up. No significant difference across conditions was noted after 2 months. Active left DLPFC EpCS proceeded to have a gradual improvement from 8 to 16 weeks with 21% ± 23 and 26% ± 29 changes from baseline. The results also showed that the placement of their single cortical lead paddle over left DLPFC was critical for their response rate. The more anterior the paddle was, the better the response. Interestingly, we had shown a similar relationship in a larger cohort of 59 patients enrolled in an RCT with noninvasive left DLPFC rTMS (Herbsman et al., 2009).

As mentioned earlier, the bilateral medial and lateral EpCS we undertook was an open study and included five participants. All implants took place within a period of 9 months. The mean age was 44.2 (±9.4). Four were women and three were diagnosed with recurrent Major Depressive Disorder whereas two others had Bipolar Affective Disorder I, depressed type. All were unemployed and three were receiving disability. The average length of depressive illness was 25.6 (±8.3) years. The average length of the current depressive episode was 3 years, 7 months (±38 months). It is noteworthy that participants had received an average of 9.8 (±5.3) unsuccessful clinical treatments during the current MDE. Of these CNS-active compounds, an average of 6.2 (±2.1) were classic antidepressant treatments of which 5.8 (±2.05) met ATHF criteria for trial adequacy. Two had failed a trial of a monoamine oxidase inhibitor or buprenorphine augmentation, one failed intravenous ketamine infusion, a trial of clozapine and another trial of mifepristone for nonpsychotic depressive symptoms. Four of the patients received prior treatments with ECT, TMS, or VNS. They enrolled the study taking on average 6 (±2.3) psychotropic drugs which were held constant. An expert psychiatrist, independent of the study team, evaluated all patients to insure they were well-informed of their clinical options. Consents were obtained in the presence of a patient advocate also independent of the study team. All subjects underwent comprehensive repeated assessments. A postoperative spiral CT scan confirmed accuracy of lead placements. We allowed 2–3 weeks for surgical recovery before initiating therapy.

(a) *EpCS intraoperative testing under controlled conditions:* Once all four leads were stereotactically in place over bilateral anterior prefrontal pole (BA10) and midlateral prefrontal cortex (BA46), they were connected to external stimulators. Discontinuing the propofol infusion then lifted the moderate sedation. With the patient fully alert, we conducted *a single-blind sham-controlled parametric testing* of bilateral anterior or midlateral leads at 0, 2, and 4–5 V and 60 Hz. Order was randomized. Each train of stimulation lasted about 2–3 min and patients were instructed to record their subjective experience by responding to 13 questions in the form of a visual analog scale. Questions were presented on a laptop computer placed on the patient's abdomen and were typically phrased "I feel SAD," "I feel ANXIOUS," "I feel SETTLED emotionally," "I feel ATTENTIVE to my surroundings," etc. (Fig. 23.2). Active EpCS was associated with significant decreases compared to baseline in patients' subjective sadness (mean change = 34.65%; $P = .05$) and anxiety levels (mean change = −45.81%; $P = .034$) despite this small sample size. Active stimulation was also associated with over 100% improvement in self-ratings of "improved pleasure," "sense of involvement," and "attention," but these effects were specifically pronounced in two subjects which led to a large standard deviation and were not statistically significant. None of the sham conditions led to significant changes from baseline. Two patients distinctly noted improved attention, increased brightness and alertness with lateral stimulation. They commented: "I feel attentive," "feel better and I can talk now," "I can think clearer." A patient noted during anterior frontal pole stimulation as if a "weight is lifting off my shoulder," "I feel calm," and another stated "and although I am worried, I feel dissociated from it. I can think back at my worry." Subject 2 failed to notice any positive changes across all tested parameters, became very discouraged and dysphoric. His testing was interrupted early. This TRD patient is now in full remission with chronic EpCS (Fig. 23.1).

(b) Clinical outcomes: Patients did not improve clinically 2–3 weeks postoperatively with the EpCS leads turned off as often seen in DBS studies due to local inflammation around implanted electrodes in subcortical regions, but rather showed a 5% worsening of symptoms associated with postoperative recovery. However, once the leads were activated, patients showed a significant improvement over time ($df = 4$, $F = 4.867$, $P = .009$). After 4 months of active stimulation the group showed a mean $HDRS_{24}$ improvement of 36% (±39) which increased to 55% (±38) at 7 months. Two patients achieved remission at 4 months and three out of five were remitted at 7 months. Mean scores of the MADRS at baseline, 4 and 7 months were 32 ± 7.6, 22 ± 6.6, and 14.6 ± 6.1 respectively. MADRS percent change from baseline significantly improved over time (ANOVA: $df = 4$, $F = 3.886$, $P = .022$). The IDS-SR scores at baseline, 4 and 7 months were 45.8 ± 16.2, 31.4 ± 25.3, and 19 ± 18.48. They also were significantly improved (ANOVA: $df = 4$, $F = 4.419$, $P = .014$). One patient reached remission after 2 weeks of stimulation based on IDS-SR ≤ 15 and remained well until her 7 month follow up. Two additional patients reached remission (HRSD < 10 for 2-week period); one at 4 months and the second at 7 months follow up. Last observation carried forward data at 5 years remains encouraging with three patients still in remission (Williams et al., 2016).

Disrupted emotion regulation in depression and EpCS

Emotion regulation involves the ability to modulate the intensity and quality of responses to emotional stimuli, involving both automatic and controlled regulatory processes. It involves an executive control of the limbic structures by the prefrontal cortex. Disruptions of emotional control are associated with poor inhibitory feedback between medial and dorsal frontal cortex, and limbic subcortical and medial temporal structures (Ochsner et al., 2002). They play a central role in

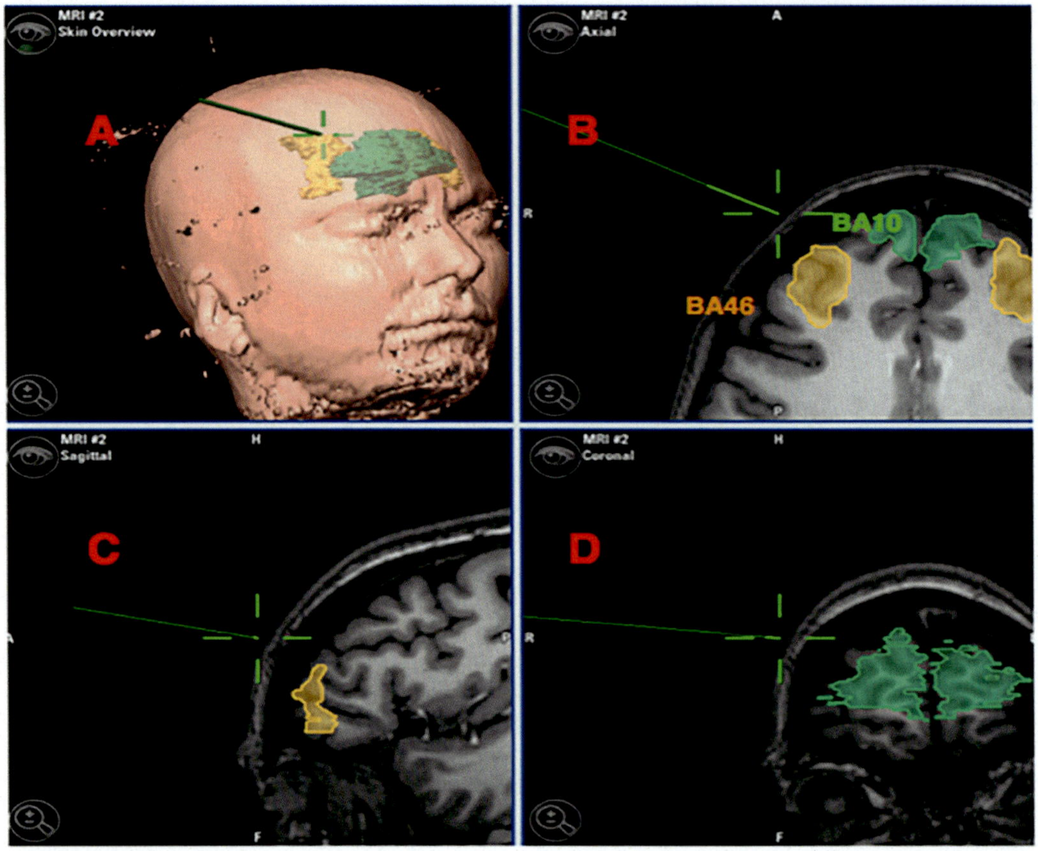

FIG. 23.1 Stereotactic identification of right Brodmann Area 46: BrainLab pointer guides the lead placement with visualization of cortical region of interest in 3D surface render mode (A), in transverse (B), sagittal (C), and coronal (D) views. Brodmann Area 46 is in *Yellow* and Brodmann Area 10 in *Green*.

mood and anxiety disorders. In fact, a feature of depression is the inability to disengage from negative memories, feelings and thoughts and engage the outside world with flexibility. Negative emotions can also distract away from other cognitive demands and make patients not able to cope with daily life needs (Drevets et al., 2002). The late positive potential (LPP) reflects an early electrocortical response to emotional stimuli that indexes sustained attention. It is sensitive to both relatively automatic (Hajcak et al., 2006) and conscious emotion regulation strategies. To examine the influence of EpCS on emotional appraisal, we recorded the ERPs caused by viewing emotional taken from the International Affective Picture System (IAPS) during stimulation under single-blind cross-randomized across five conditions (bilateral 60Hz BA10 or BA46 @ 2 or 4V and no stimulation). In a series of studies (Fig. 23.2), we demonstrated that the amplitude of the LPP is sensitive to changes in meaning and to emotion regulation instructions. Most importantly, we have also demonstrated that direct bilateral (and not bimedial) EpCS acts similarly to cognitive reappraisal and significantly reduces LPP (Hajcak et al., 2010b) making the lateral prefrontal cortex a good anatomical target for future EpCS studies.

Deficits in mentalization in depression and EpCS

Mentalization allows individuals to estimate others' states of mind. It is critical for social bonds, adaptation and richer relationships. Pathological states such as depression degrade mentalization (Uekermann et al., 2008) by limiting emotional involvement and leading to disconnected experiences from others. Lower social functions are themselves associated with higher risks for symptom recurrence 1 year after recovering from a major depressive episode (Uekermann et al., 2008). Mentalization recruits several complementary limbic, attention and high-level executive networks including the frontopolar cortex. Reading the Mind in the Eyes Task (Baron-Cohen et al., 2001) (Fig. 23.3) consists of presentations of a set of eyes expression for which the subject is asked to determine the person's emotional state. It tests how well the participant "tune in" to other mental state. It is also limited to items with complex mental states. The control task is to identify the gender in the photograph. We have shown that unmedicated depressed patient show a distinct increase in ventrolimbic activity

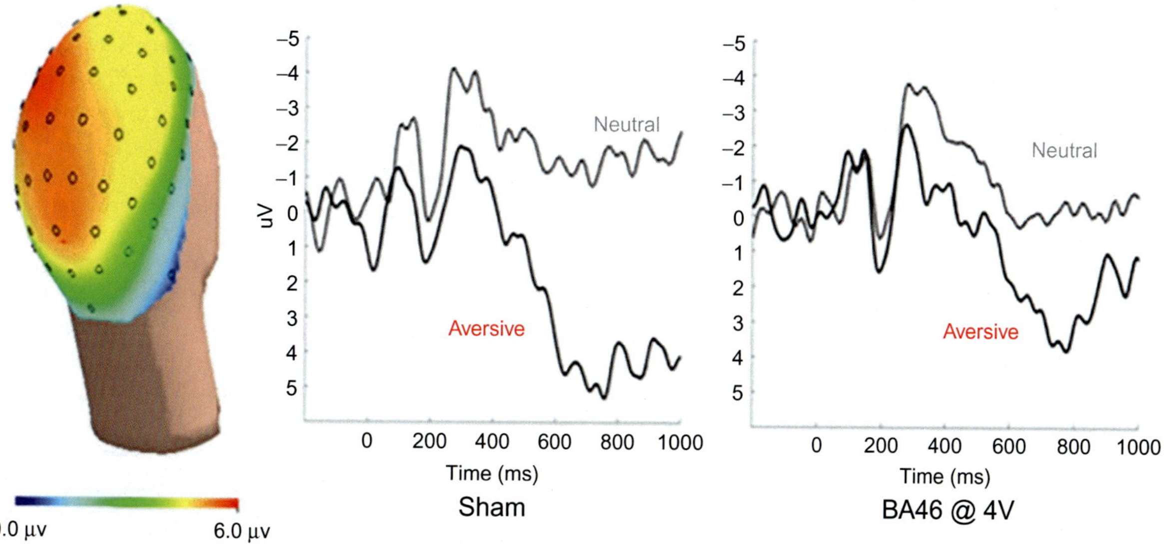

FIG. 23.2 Scalp distribution of the evoked response potentials (ERP) difference between aversive and neutral pictures from 400 to 1000 ms following picture presentation (left); ERPs at centro-parietal recording sites time-locked to picture onset (middle, right) for aversive *(dark)* and neutral *(light)* trials during the sham condition (middle) and when BA46 was stimulated at 4 V (right). Picture onset occurred at 0 ms and negative is plotted up.

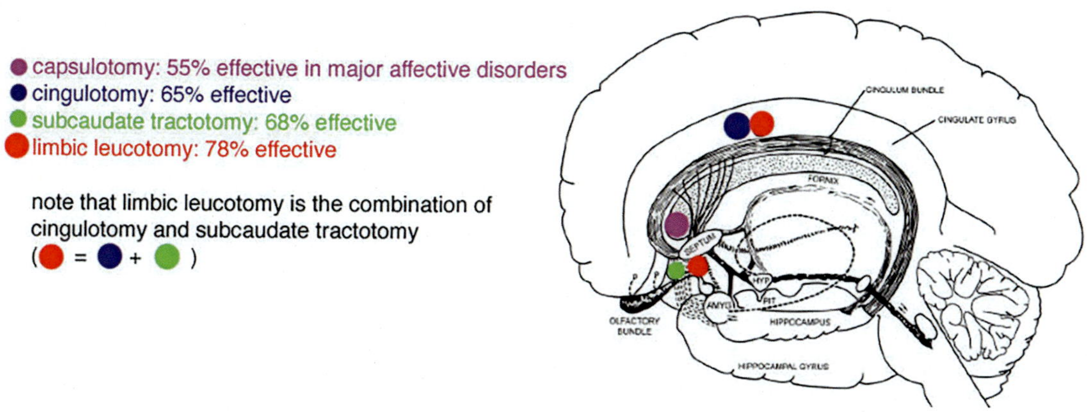

FIG. 23.3 The four lesions in psychiatric neurosurgeries.

while engaged in a mental attribution task compared to healthy controls. Conversely, healthy controls show more activity in frontopolar regions (Pincus et al., 2010). Direct stimulation with EpCS to those same frontopolar regions, in TRD patients engaged in a similar task, increases their reaction time compared to no stimulation (Hajcak et al., 2010a), supporting the choice of frontopolar regions as another anatomic target for future EpCS studies.

EpCS also appears to modulate alpha frontal activity and can be used as a biomarker for optimizing treatment outcomes. These EEG results are remarkably similar to a long standing literature spanning decades on frontal alpha activity and ECT: Postictal inhibition of frontal lobes is the most replicated positive predictor of outcomes and depressive symptoms (Sackeim et al., 1996). These results also are reminiscent of recent findings in which increased alpha activity at baseline predicted responders to 12 week treatment with SSRIs (Bruder et al., 2008).

Comparisons across various BSTs

Given the available BST options available, the big challenge is to provide a balance between the safety and tolerability of the method, its practicality, its associated costs and obviously its clinical efficacy across the full range of treatment-resistance spectrum. ECT is associated with cognitive impairments and high relapse rates. Twenty percent of responsive patients relapse in the first month following the last ECT treatment and up to 60% after 6 months. TMS is an acute treatment for mild to moderate TRD with only limited data on maintenance regimen and relapse preventions (Janicak et al., 2010). VNS is a surgical procedure that delivers chronic treatment but is not supported by Class 1 evidence (Nahas et al., 2003) although an large randomized controlled trial is currently underway. And finally, the open reports of continuous high frequency DBS in depression remain encouraging despite two industry sponsored RCTs were stopped after futility analyses. It remains to be seen whether intermittent EpCS will show significant results in a RCT, currently under design, and without the risks associated with inserting the leads into brain tissue? Separately, we used Bayesian optimization (BayesOpt) to program deep brain stimulation (DBS) in Parkinson's disease patients. BayesOpt allowed a more efficient exploration of a larger number of settings than otherwise possible using classic parametric permutations and consequently improved rigidity. We intend to apply these same analytic methods and optimize EpCS' continuous modulation of the attentional and default networks and the treatment of depressive symptoms over time.

Psychiatric neurosurgery

Of all the somatic therapies in psychiatric practice today, psychiatric neurosurgery requires the most knowledge about functional neuroanatomy since it is the most radical and irreversible of all interventions. It was only in the last years of the 19th century that rational approaches to psychosurgery were first tried, pursuant to well-publicized clinical cases like Phineas Gage whose frontal lobe lesion in a mining accident showed that frontal lesions could alter a person's personality. In 1894, a Swiss surgeon named Burkhardt performed an operation to selectively destroy the frontal lobes of several psychotic patients in an effort to control their symptoms. Subsequently, Fulton and Jackobsen reported that frontal ablation lessened anxiety in chimpanzees. From these observations, Moniz (1994) argued that by severing the connections between different brain regions, one could force impulses and thoughts to "re-channel" and in effect, coax the brain to "reorganize." This provisional conclusion about brain organization in the late 1930s launched frontal lobotomy as a treatment for psychiatric disorders. Overcrowded mental asylums and the use of highly morbid convulsive therapy at the time also aided the popularity of lobotomy. Moniz and Lima's first lesion technique was to inject alcohol in the bilateral frontal lobes of asylum patients. They reported "improvement" in 14 out 20 subjects. They later developed the leucotomy (a tool to interrupt white matter tracts) and described a rather large surgical target in the frontal lobes. Moniz was the first psychiatrist to receive the Nobel prize in 1949 for his contributions in treating psychosis with leucotomy.

Walter Freeman and James Watts introduced the procedure to the United States. They modified it so the disconnection was carried out through bilateral burr holes placed in the inferior frontal region at the level of the coronal suture. The leucotomy spatula was introduced blindly and swept back and forth. Freeman later devised the so-called "ice-pick transorbital leucotomy," performed typically on patients who were postictal from ECT. He inserted a sharp blade under the eyelid through the thin bone into the orbitofrontal lobe. Many of the patients were reported to be improved although a great number of them were observed to have become amotivational and to have lost their capacity to be emotional (Mclardy and Meyer, 1949). This crude intervention was zealously promoted by Freeman himself and is likely the source of many of the controversies and stigma surrounding neurosurgery for psychiatric conditions.

Currently, the practice of psychiatric neurosurgery is much more restricted and regulated. Candidates are evaluated by multidisciplinary teams and must meet stringent criteria for severe resistance to conventional multimodal therapies (Read and Greenberg, 2009). The majority has either refractory OCD, anxiety or mood disorders. Treatment of schizophrenia with this modality has fallen out of favor. Current surgical procedures are much more refined and specific in their targets. There are four distinct stereotactic approaches that continue to be used: cingulotomy, subcaudate tracheotomy, limbic leucotomy, and anterior capsulotomy. They are generally performed bilaterally and share similar complication and risks profiles. These include minor symptoms such as headaches, low-grade fever, confusion, and isolated seizures. The most serious complication (intracranial hemorrhages) is rare (0.4% in some case series). Mortality directly related to any of the procedures is very rare. Lethargy, personality, cognitive and behavioral changes are more specific to each intervention (Volpini et al., 2017).

The aim of a cingulotomy is to lesion the cingulate fasciculus with thermocoagulation approximately 2–2.5 cm posterior to the tip of frontal horns, 7 mm lateral to the midline and 1 mm above the ventricular roof. It is the most reported neurosurgical procedure for psychiatric disease in North America and likely the safest. Studies show a range of 30%–60% of significant improvement with affective disorders demonstrating the highest rate of response and OCD the lowest (Tippin and Henn, 1982). The more recent and largest prospective studies where in OCD and reported 32%–48% improvement (Dougherty et al., 2002; Jung et al., 2006).

The goal of a subcaudate tractotomy is to interrupt the fibers from the orbitofrontal cortex to the thalamus. This intervention was designed to minimize cognitive and personality impairments. The target is the substantia innominata (white matter beneath the head of the caudate). The lesion is created with radioactive rods with a half-life of 68 h. The target area lies at the antero-posterior level of the planum sphenoidale, extending from 6 to 18 mm from the midline, being 20 mm long in an antero-posterior direction. Studies have shown significant relief in up to 45% of severely ill subjects. Affective disorders are more likely to respond, although OCD symptoms also improve (Hodgkiss et al., 1995).

Limbic leucotomy is a combination of the two previous procedures (Montoya et al., 2002). Three 6 mm lesions are placed in the posterior inferior medial quadrant of each frontal lobe, along with two lesions in each cingulated gyrus. Results in OCD have shown up to 89% improvement up to 16 months postoperatively with improvement in cognitive functions in some patients.

Finally, the aim of anterior capsulotomy is to disconnect fronto-thalamic fibers as they pass through the anterior limb of the internal capsule, between the head of the caudate and putamen. One of the earlier indications for this procedure was schizophrenia, but this has fallen out of favor. Reports indicate greatest responses in OCD and depression. Interestingly, a mean 10% weight gain is common (Lippitz et al., 1999). Over the past decade, significant challenges have been overcome to allow transcranial focused ultrasound (TFU) to deliver spatially restricted energy into the nervous system at a wide range of intensities. Magnetic resonance guided high-intensity focused ultrasound produces reliable permanent lesions within the brain, and recently a Phase 1 clinical trial is underway to treat depression.

Clinical improvement with all four interventions is progressive over several months. Unfortunately, given the limited use of standardized rating scales across different sites and the open nature of these reports, it is hard to compare results. A review (Cosgrove and Rauch, 1995) used the "much improved" clinical outcome measure and found that in OCD capsulotomy was 67% effective followed by limbic leucotomy (61%), cingulotomy (56%), and subcaudate tractotomy (50%). In mood disorders, limbic leucotomy was 78% effective, subcaudate tractotomy was 68% effective, cingulotomy was 65% effective, and finally anterior capsulotomy was 55% effective. Leiphart's review attempted to match procedures and disorders for optimal results (Leiphart and Valone, 2010). They found OCD patients responded the best to anterior capsulotomy, whereas bipolar disorder and depression had the greatest reported improvements from anterior cingulotomy. Significant technical challenges have been overcome to allow transcranial focused ultrasound to deliver spatially restricted energy into the nervous system at a wide range of intensities. MRI guided high-intensity focused ultrasound (MRgFUS) produces reliable permanent lesions within the brain and would be preferable to gamma knife radiation neurosurgery (Jolesz, 2009). Proof of concepts have been reported in OCD (Kim et al., 2018; Jung et al., 2015) and depression (Davidson et al., 2020; Kinfe et al., 2020). With the advent of deep brain stimulation, all these areas are potential candidates for neuromodulation with "reversible lesions." Theoretically, one could expect double-blind controlled studies with lead-in sham arms where the implanted wires are not activated. Please refer to Chapter 22 for a current and more detailed discussion of DBS in TRD.

Future directions

Treating neuropsychiatric conditions with somatic nonpharmacological approaches is rooted in early 20th century exploration of the mind as a complex interplay of neuronal networks. Recent advancements in nanotechnologies and computational sciences, let alone neuroscience, neuroimaging and better understanding of brain functions allow for more empirically-based and precisely targeted neuromodulations (Elder et al., 2008). The 21st century is likely to witness a refinement of the interventions highlighted in this chapter, as well as many new therapies. Whereas MRgFSU is ablative, low-intensity focused ultrasound (LIFU) has been reported to both excite and inhibit neural activity reversibly. To this date, LIFU has not been applied to treat TRD (Kinfe et al., 2020; Luo et al., 2002) yet it has been attempted to modulate hippocampal blood-brain barrier to treat Alzheimer's disease (Rezai et al., 2020). The introduction of genetic information or genetically modified cells for functional augmentation, restoration or ablation is feasible (Sjulson et al., 2016). The discovery of CNS neurotrophic agents and the perfection of delivery systems could lead to implantable pumps and novel drug delivery devices (Kaplitt and Lozano, 2001). Even neuroprostheses are now conceivable (Kilgore et al., 2020). We may gradually be entering a new land of psychiatric interventions where the mind and brain are treated as a unified but multifaceted entity.

Acknowledgments

Interventional Psychiatry Program in the Department of Psychiatry and Behavioral Science at the University of Minnesota.

References

Baron-Cohen, S., et al., 2001. The "Reading the Mind in the Eyes" Test revised version: a study with normal adults, and adults with Asperger syndrome or high-functioning autism. J. Child Psychol. Psychiatry 42 (2), 241–251.

Broca, P., 1865. Sur le siege de la faculte du language articule. Bull. Mém. Soc. Anthropol. Paris 6, 377.

Bruder, G.E., et al., 2008. Electroencephalographic alpha measures predict therapeutic response to a selective serotonin reuptake inhibitor antidepressant: pre- and post-treatment findings. Biol. Psychiatry 63 (12), 1171–1177.

Clower, W.T., Finger, S., 2001. Discovering trepanation: the contribution of Paul Broca. Neurosurgery 49 (6), 1417–1425. discussion. 1425–6, 2002.

Cosgrove, G.R., Rauch, S.L., 1995. Psychosurgery. Neurosurg. Clin. N. Am. 6, 167–176.

Critchley, M., 1965. Neurology's debt to F. J. Gall (1758–1828). Br. Med. J. 5465, 775–781.

Davidson, B., et al., 2020. Magnetic resonance-guided focused ultrasound capsulotomy for treatment-resistant psychiatric disorders. Oper. Neurosurg. (Hagerstown) 19 (6), 741–749.

Dougherty, D.D., et al., 2002. Prospective long-term follow-up of 44 patients who received cingulotomy for treatment-refractory obsessive-compulsive disorder. Am. J. Psychiatry 159 (2), 269–275.

Dougherty, D.D., et al., 2008. Feasibility study of an implantable cortical stimulation system for patients with major depressive disorder. In: Krystal, J.H. (Ed.), Society of Biological Psychiatry 63rd Annual Meeting. Biol Psychiatry: Washington, DC, p. 930s.

Drevets, W.C., Bogers, W., Raichle, M.E., 2002. Functional anatomical correlates of antidepressant drug treatment assessed using PET measures of regional glucose metabolism. Eur. Neuropsychopharmacol. 12 (6), 527–544.

Elder, J.B., Liu, C.Y., Apuzzo, M.L., 2008. Neurosurgery in the realm of 10(-9), Part 2: applications of nanotechnology to neurosurgery—present and future. Neurosurgery 62 (2), 269–284. discussion 284–285.

Garcia-Larrea, L., et al., 2000. Functional imaging and neurophysiological assessment of spinal and brain therapeutic modulation in humans. Arch. Med. Res. 31 (3), 248–257.

George, M.S., et al., 2007. Brain stimulation for the treatment of psychiatric disorders. Curr. Opin. Psychiatry 20 (3), 250–254.

Greenberg, P.E., et al., 2003. The economic burden of depression in the United States: how did it change between 1990 and 2000? J. Clin. Psychiatry 64 (12), 1465–1475.

Hajcak, G., Moser, J.S., Simons, R.F., 2006. Attending to affect: appraisal strategies modulate the electrocortical response to arousing pictures. Emotion 6 (3), 517–522.

Hajcak, G., Anderson, B.S., Arana, A., Borckardt, J., Takacs, I., George, M.S., Nahas, Z., 2010a. Dorsolateral prefrontal cortex stimulation modulates electrocortical measures of visual attention: evidence from direct bilateral epidural cortical stimulation in treatment-resistant mood disorder. Neuroscience 170 (1), 281–288. https://doi.org/10.1016/j.neuroscience.2010.04.069 (Epub 2010 May 6. PMID: 20451585).

Hajcak, G., MacNamara, A., Olvet, D.M., 2010b. Event-related potentials, emotion, and emotion regulation: an integrative review. Dev. Neuropsychol. 35 (2), 129–155.

Herbsman, T., et al., 2009. More lateral and anterior prefrontal coil location is associated with better repetitive transcranial magnetic stimulation antidepressant response. Biol. Psychiatry 66 (5), 509–515.

Hodgkiss, A.D., et al., 1995. Outcome after the psychosurgical operation of stereotactic subcaudate tractotomy, 1979–1991. J. Neuropsychiatry Clin. Neurosci. 7 (2), 230–234.

Jackson, J.H., 1873. Observations on the localisation of movements in the cerebral hemispheres. In: West Riding Lunatic Asylum Medical Reports. vol. 3. Smith, Elder, and Co., London, pp. 175–190.

Janicak, P., et al., 2010. Durability of acute response to TMS in the treatment of major depression: relapse during a continuation pharmacotherapy extension study. Brain Stimul. 3 (4), 187–199. https://doi.org/10.1016/j.brs.2010.07.003.

Jolesz, F.A., 2009. MRI-guided focused ultrasound surgery. Annu. Rev. Med. 60, 417–430. https://doi.org/10.1146/annurev.med.60.041707.170303.

Jung, H.H., et al., 2006. Bilateral anterior cingulotomy for refractory obsessive-compulsive disorder: long-term follow-up results. Stereotact. Funct. Neurosurg. 84 (4), 184–189.

Jung, H.H., et al., 2015. Bilateral thermal capsulotomy with MR-guided focused ultrasound for patients with treatment-refractory obsessive-compulsive disorder: a proof-of-concept study. Mol. Psychiatry 20 (10), 1205–1211.

Kalinowsky, L.B., 1986. History of convulsive therapy. Ann. N. Y. Acad. Sci. 462, 1–4.

Kaplitt, M.G., Lozano, A.M., 2001. Surgical drug delivery for neurodegenerative diseases. Clin. Neurosurg. 48, 127–144.

Kendler, K.S., Thornton, L.M., Gardner, C.O., 2000. Stressful life events and previous episodes in the etiology of major depression in women: an evaluation of the "kindling" hypothesis. Am. J. Psychiatry 157 (8), 1243–1251.

Kilgore, K.L., et al., 2020. Powering strategies for implanted multi-function neuroprostheses for spinal cord injury. Healthc. Technol. Lett. 7 (3), 81–86.

Kim, S.J., et al., 2018. A study of novel bilateral thermal capsulotomy with focused ultrasound for treatment-refractory obsessive-compulsive disorder: 2-year follow-up. J. Psychiatry Neurosci. 43 (5), 327–337.

Kinfe, T., et al., 2020. Incisionless MR-guided focused ultrasound: technical considerations and current therapeutic approaches in psychiatric disorders. Expert Rev. Neurother. 20 (7), 687–696.

Kose, S., et al., 2008. Does the effect of prefrontal rTMS on cortico-limbic circuits vary as a function of mood state? Interleaved TMS-fMRI scans in treatment resistant depression patients in the NIMH OPT-TMS Study. In: Annual Meeting of the American College of Neuropsychopharmacology. Nature: Scottsdale, AZ.

Leiphart, J.W., Valone 3rd, F.H., 2010. Stereotactic lesions for the treatment of psychiatric disorders. J. Neurosurg. 113 (6), 1204–1211.

Li, X., et al., 2004. Interleaved transcranial magnetic stimulation/functional MRI confirms that lamotrigine inhibits cortical excitability in healthy young men. Neuropsychopharmacology 29 (7), 1395–1407.

Lippitz, B.E., et al., 1999. Lesion topography and outcome after thermocapsulotomy or gamma knife capsulotomy for obsessive-compulsive disorder: relevance of the right hemisphere. Neurosurgery 44 (3), 452–458. discussion 458–460.

Luo, J., et al., 2002. Subthalamic GAD gene therapy in a Parkinson's disease rat model. Science 298 (5592), 425–429.

Mayberg, H.S., et al., 1999. Reciprocal limbic-cortical function and negative mood: converging PET findings in depression and normal sadness. Am. J. Psychiatry 156 (5), 675–682.

McIntyre, C.C., Grill, W.M., 2002. Extracellular stimulation of central neurons: influence of stimulus waveform and frequency on neuronal output. J. Neurophysiol. 88 (4), 1592–1604.

Mclardy, T., Meyer, A., 1949. Anatomical correlates of improvement after leucotomy. J. Ment. Sci. 95, 182–192.

Moniz, E., 1994. Prefrontal leucotomy in the treatment of mental disorders. 1937. Am. J. Psychiatry 151, 236–239.

Montoya, A., et al., 2002. Magnetic resonance imaging-guided stereotactic limbic leukotomy for treatment of intractable psychiatric disease. Neurosurgery 50 (5), 1043–1049. discussion 1049–1052.

Nahas, Z., et al., 2001a. Unilateral left prefrontal transcranial magnetic stimulation (TMS) produces intensity-dependent bilateral effects as measured by interleaved BOLD fMRI. Biol. Psychiatry 50 (9), 712–720.

Nahas, Z., et al., 2001b. Brain effects of TMS delivered over prefrontal cortex in depressed adults: role of stimulation frequency and coil-cortex distance. J. Neuropsychiatry Clin. Neurosci. 13 (4), 459–470.

Nahas, Z., et al., 2003. Serial vagus nerve stimulation functional MRI (VNS/fMRI) in treatment-resistant depression. Biol. Psychiatry 53 (1S–217S), 600.

Nahas, Z., et al., 2010. Bilateral epidural prefrontal cortical stimulation for treatment-resistant depression. Biol. Psychiatry 67 (2), 101–109.

Ochsner, K.N., et al., 2002. Rethinking feelings: an FMRI study of the cognitive regulation of emotion. J. Cogn. Neurosci. 14 (8), 1215–1229.

Petrides, M., Pandya, D., 2007. Efferent association pathways from the rostral prefrontal cortex in the macaque monkey. J. Neurosci. 27 (43), 11573–11586.

Pincus, D., et al., 2010. Inverse effects of oxytocin on attributing mental activity to others in depressed and healthy subjects: a double-blind placebo controlled fMRI study. Front. Psychiatry 1 (134), 1–10.

Priori, A., Lefaucheur, J., 2007. Chronic epidural motor cortical stimulation for movement disorders. Lancet Neurol. 6 (3), 279–286.

Raichle, M.E., Snyder, A.Z., 2007. A default mode of brain function: a brief history of an evolving idea. Neuroimage 37 (4), 1083–1089.

Read, C.N., Greenberg, B.D., 2009. Psychiatric neurosurgery 2009: review and perspective. Semin. Neurol. 29 (3), 256–265.

Rezai, A.R., et al., 2020. Noninvasive hippocampal blood-brain barrier opening in Alzheimer's disease with focused ultrasound. Proc. Natl. Acad. Sci. U. S. A. 117 (17), 9180–9182.

Rush, A.J., et al., 2006. Report by the ACNP Task Force on response and remission in major depressive disorder. Neuropsychopharmacology 31 (9), 1841–1853.

Sackeim, H.A., et al., 1996. The effects of electroconvulsive therapy on quantitative electroencephalograms. Relationship to clinical outcome [see comments]. Arch. Gen. Psychiatry 53 (9), 814–824.

Sjulson, L., et al., 2016. Cell-specific targeting of genetically encoded tools for neuroscience. Annu. Rev. Genet. 50, 571–594.

Tippin, J., Henn, F.A., 1982. Modified leukotomy in the treatment of intractable obsessional neurosis. Am. J. Psychiatry 139, 1601–1603.

Tourney, G., 1967. A history of therapeutic fashions in psychiatry, 1800–1966. Am. J. Psychiatry 124, 784–796.

Trivedi, M.H., et al., 2004. Clinical results for patients with major depressive disorder in the Texas Medication Algorithm Project. Arch. Gen. Psychiatry 61 (7), 669–680.

Uekermann, J., et al., 2008. Executive function, mentalizing and humor in major depression. J. Int. Neuropsychol. Soc. 14 (1), 55–62.

Vogeley, K., et al., 2001. Mind reading: neural mechanisms of theory of mind and self-perspective. Neuroimage 14 (1 Pt 1), 170–181.

Volpini, M., et al., 2017. The history and future of ablative neurosurgery for major depressive disorder. Stereotact. Funct. Neurosurg. 95 (4), 216–228.

Wang, Y., et al., 2004. Anatomical, physiological and molecular properties of Martinotti cells in the somatosensory cortex of the juvenile rat. J. Physiol. 561 (Pt 1), 65–90.

Williams, N.R., et al., 2016. Five-year follow-up of bilateral epidural prefrontal cortical stimulation for treatment-resistant depression. Brain Stimul. 9 (6), 897–904.

Chapter 24

Treatment-resistant psychotic depression

Julia A. Koretski[a] and Anthony J. Rothschild[b]

[a]*Department of Psychiatry, University of Massachusetts Chan Medical School, Worcester, MA, United States,* [b]*Department of Psychiatry, University of Massachusetts Chan Medical School and Umass Memorial Healthcare, Worcester, MA, United States*

Introduction

Major depressive disorder with psychotic features, or psychotic depression, is a serious psychiatric illness that presents with a combination of depressed mood and psychosis. The psychosis commonly manifests through nihilistic-type delusions with overly self-critical beliefs, severe guilt, paranoia, and the belief that bad things are about to happen (Rothschild, 2009; Schatzberg and Rothschild, 1992; Thakur et al., 1999). Psychotic depression is considered a more severe form of depression with increased symptomatology, greater cognitive impairment, more brain atrophy, more pronounced psychomotor disturbance, increased hospitalizations, and greater morbidity and mortality (Jääskeläinen et al., 2018; Lykouras and Gournellis, 2009; Zaninotto et al., 2015). The evidence in the current literature also suggests that the presence of psychosis in major depression presents a higher lifetime risk for completed suicide and doubles the risk of a suicide attempt in the acute phase of the disorder (Gournellis et al., 2018; Zalpuri and Rothschild, 2016), while baseline suicidality predicts poorer response to treatment in adults with psychotic depression (Bingham et al., 2017).

The lifetime prevalence of psychotic depression in the general population varies between 0.35% and 1% (Jääskeläinen et al., 2018), although the community rates in older adults have been reported to be between 1.4% and 3% (Baldwin and Jolley, 1986; Blazer, 1994). However, the prevalence of psychotic depression is considerably higher among depressed individuals in the community or in hospital settings. A general population study found that 18.5% of respondents with major depression had psychotic symptoms (Ohayon and Schatzberg, 2002), and studies in inpatient settings have estimated that up to 53% of depressed adults are psychotic (Gournellis and Lykouras, 2006). Despite the relatively high prevalence, the misconception that psychotic depression is not a common illness persists in part due to misdiagnosis (Rothschild et al., 2006). Rothschild et al. (2008) reported in 2008 that 27% of patients with a diagnosis of psychotic depression were initially incorrectly diagnosed at four academic medical centers, with the rate thought to be a conservative estimate of the rate in the general clinical population as patients with comorbid conditions were excluded. Their findings suggest that clinicians are frequently missing the psychosis rather than the mood disorder, and in many cases they note the symptom (i.e., guilt, persecution, poverty) but do not recognize that the symptom is delusional.

In addition to frequent misdiagnosis or underdiagnosis, psychotic depression is subsequently often undertreated. Studies have shown a persisting low rate of adequate dose and duration of pharmacologic treatment of psychotic depression (Andreescu et al., 2007; Mulsant et al., 1997). Andreescu and colleagues reported that only 5% of patients with psychotic depression received an adequate combination of an antidepressant and antipsychotic medication. Unfortunately, therapies are further limited by a lack of interest of the pharmaceutical industry in developing medications for the treatment of psychotic depression (Rothschild, 2021). At the time of this writing, there are no Food and Drug Administration (FDA)–approved treatments for psychotic depression.

Evidence-based treatment of psychotic depression

The American Psychiatric Association (APA) Practice Guidelines recommend either electroconvulsive therapy (ECT) or combination pharmacotherapy with antidepressant and antipsychotic medications for the acute treatment of psychotic depression (American Psychiatric Association, 2010). Meta-analytic evidence supports this recommendation (Farahani and Correll, 2012; Wijkstra et al., 2006). In 2012, Farahani and Correll identified eight randomized, placebo-controlled acute-phase studies in adults with psychotic depression conducted before 2011 and yielded six antidepressant-antipsychotic combination treatment comparisons with antidepressant monotherapy and four comparisons with antipsychotic monotherapy. Their metaanalysis concluded that the combined use of antidepressants and antipsychotics was superior to either monotherapy strategy in the acute treatment of psychotic depression.

Managing Treatment-Resistant Depression. https://doi.org/10.1016/B978-0-12-824067-0.00024-4

Based on these recommendations, we define treatment resistance in psychotic depression as failure to respond to an adequate trial of an antidepressant-antipsychotic combination *and* a trial of ECT. Of note, some groups have argued that patients with psychotic depression should not be labeled as treatment-resistant until they have had a full course of bilateral ECT (Petrides et al., 2001).

Combination pharmacotherapy with antidepressant and antipsychotic medications

In the paragraphs below, we discuss the three combinations of newer antidepressants and antipsychotic medications that have been studied in double-blind, randomized, controlled trials (see Table 24.1).

Fluoxetine plus olanzapine

In 2004, Rothschild and colleagues reported on two multicenter randomized, double-blind, placebo-controlled trials that compared the combination of the selective serotonin reuptake inhibitor, fluoxetine, and the second-generation antipsychotic olanzapine with olanzapine monotherapy or placebo in 249 hospitalized patients with psychotic depression (Rothschild et al., 2004). In both trials, the subjects were randomized to an olanzapine monotherapy group ($n=43$ and 47), a placebo group ($n=50$ and 44), or an olanzapine-fluoxetine combination therapy group ($n=22$ and 23). The olanzapine monotherapy group initially received an evening dose of 10 mg of olanzapine followed by a morning dose of placebo; olanzapine could be titrated up to 20 mg daily maximum. The mean daily doses of olanzapine were 11.9 and 14.0 mg/day in the olanzapine monotherapy group of each study. The olanzapine-fluoxetine combination therapy group initially received an evening dose of 10 mg of olanzapine followed by a morning dose of 20 mg of fluoxetine; olanzapine could be titrated at each visit up to 20 mg daily maximum while the fluoxetine dose remained fixed until week 3, when it could be increased by 20 mg at each visit if the subject had reached the daily maximum olanzapine dose. In the olanzapine-fluoxetine combination therapy group of each study, the mean daily doses of olanzapine were 12.4 and 13.9 mg/day, and the mean daily doses of fluoxetine were 23.5 and 22.6 mg/day. Treatment duration was 8 weeks. The primary outcome measure was the 24-item Hamilton Depression Rating Scale (HAM-D) total score.

The results of the first trial showed significantly greater endpoint improvement in depressive symptoms (−20.9) among patients in the olanzapine-fluoxetine combination therapy group compared with the placebo group (−10.4, $P=.001$). This significant difference was present within 7 days of treatment, and the effect was maintained at every subsequent visit for the duration of the 8-week trial. The olanzapine-fluoxetine combination group also had significantly higher response rate (63.6%) than the placebo (28.0%, $P=.004$) or olanzapine monotherapy (34.9%, $P=.027$) groups. Olanzapine monotherapy did not demonstrate statistically significant effectiveness on overall symptom improvement in psychotic depression.

The second trial did not reveal any statistically significant differences between the three treatment groups, although the numeric pattern was similar to the first trial, with the combination pharmacotherapy group having the highest response rate and the placebo group having the lowest. This finding may be in part due to the study design which limited fluoxetine dosing according to olanzapine dosing, such that most subjects received only a starting dose of fluoxetine (20 mg daily). Higher doses of fluoxetine might have produced greater reductions in depressive symptoms or higher remission rates. Additionally, the study was initially powered to show a difference between olanzapine monotherapy and placebo rather than olanzapine-fluoxetine combination therapy and placebo. This resulted in a small sample size of the combination pharmacotherapy group, which limited statistical power.

TABLE 24.1 Combinations of newer antidepressant and antipsychotic medications with demonstrated efficacy in randomized controlled trials on treatment of psychotic depression.

Study	Sample size	Combination	Statistically significant results
Rothschild et al. (2004)	249	Fluoxetine + olanzapine	*Reduction in HAM-D score*: Trial 1: Fluoxetine + olanzapine > olanzapine and placebo Trial 2: Fluoxetine + olanzapine = olanzapine = placebo
Meyers et al. (2009)	259	Sertraline + olanzapine	*Remission*: Sertraline + olanzapine > olanzapine
Wijkstra et al. (2010)	122	Venlafaxine + quetiapine	*Response*: Venlafaxine + quetiapine > venlafaxine Venlafaxine + quetiapine = imipramine *Remission*: Venlafaxine + quetiapine > imipramine Venlafaxine + quetiapine = venlafaxine

Sertraline plus olanzapine

The Study of the Pharmacotherapy of Psychotic Depression (STOP-PD) (Meyers et al., 2009) examined the tolerability and efficacy of combination pharmacotherapy using a serotonergic antidepressant (sertraline) and an atypical antipsychotic (olanzapine) in the acute treatment of psychotic depression. This NIMH-funded randomized controlled trial included 259 subjects with psychotic depression across a broad spectrum of illness severity; of those, 117 were younger than 60 and 142 were age 60 years or older. Participants were randomly assigned to combination treatment of sertraline plus olanzapine (129 subjects) or olanzapine plus placebo (130 subjects). Olanzapine was administered openly, while sertraline and placebo were administered under double-blind conditions. The daily dosages of medications were increased gradually as tolerated. Initially, subjects received 5 mg of olanzapine and 50 mg of sertraline/placebo per day (frail elderly subjects were started on 2.5 mg of olanzapine and 25 mg of sertraline/placebo per day); daily dosages were increased to 10 mg of olanzapine and 100 mg of sertraline/placebo by the end of week 1, followed by 15 mg of olanzapine and 150 mg of sertraline/placebo by the end of week 2, with maximum daily doses of 20 mg of olanzapine and 200 mg of sertraline/placebo allowed for residual symptoms. Treatment duration was 12 weeks. The primary outcome measure was a response on the HAM-D. Remission was defined as a HAM-D score of 10 or lower at two consecutive assessments and required the absence of delusions, as classified by the Schedule of Affective Disorders and Schizophrenia (SADS) delusion severity score of 1 at the second remission of depression assessment. Subjects who achieved a HAM-D score of 10 or lower for the first time at week 12 were assessed again at week 13 to determine whether the 2-week duration of remission criterion was met.

The STOP-PD study results showed that the combination pharmacotherapy of sertraline and olanzapine was associated with higher remission rates during the trial than olanzapine monotherapy (OR, 1.28; 95% CI, 1.12–1.47; $P<.001$) among patients with psychotic depression. The benefits of the combination pharmacotherapy became more apparent as the clinical trial progressed, particularly from week 8 to the end of the trial. 41.9% of subjects who underwent combination pharmacotherapy were in remission at their last assessment compared with 23.9% of subjects treated with olanzapine monotherapy (Chi square = 9.53, $P=.002$). Furthermore, combination pharmacotherapy was comparably superior in both younger (OR, 1.25; 95% CI, 1.05–1.50; $P=.02$) and older (OR, 1.34; 95% CI, 1.09–1.66; $P=.01$) participants. The two treatments had comparable tolerability across age groups. The older subjects were more likely to experience pedal edema than the younger subjects, but the older subjects were no more likely to experience sedation, somnolence, falls, or greater extrapyramidal symptoms. The younger subjects gained significantly more weight than the older subjects (mean (SD), 6.5 [6.6] kg vs 3.3 [4.9] kg, $P=.001$). Both age groups had significant increases in cholesterol and triglyceride concentrations, but statistically significant increases in glucose occurred only in the younger subjects. There were no differences in incident akathisia or tardive dyskinesia by age group.

Venlafaxine plus quetiapine

In a double-blind, multicenter randomized controlled trial in the Netherlands, Wijkstra et al. (2010) compared the efficacy and safety of imipramine monotherapy, venlafaxine monotherapy, and venlafaxine plus quetiapine combination therapy among 122 hospitalized patients with psychotic depression. Participants were 18–65 years old. 42 subjects received imipramine monotherapy, 39 subjects received venlafaxine monotherapy, and 41 subjects received venlafaxine-quetiapine combination therapy. There was no placebo control group given the severity of patients' symptoms. Daily dosages used were as follows: imipramine was initiated at 75 mg/day and adjusted to a plasma level of 200–300 μg/mL with a maximum daily dose of 450 mg; venlafaxine was initiated at 75 mg/day and gradually tapered up to 375 mg/day on day 17; quetiapine was initiated at 100 mg/day and gradually increased to 600 mg/day on day 10. Treatment duration was 7 weeks. The primary outcome measure was a response on HAM-D that showed a 50% or greater decrease in score from baseline to study end-point, and a final HAM-D score of 14 or less. Remission was defined as a final HAM-D score of 7 or less.

Results showed that, for the primary outcome measure of response, the venlafaxine-quetiapine combination therapy was significantly more effective than venlafaxine monotherapy (OR, 3.96; 95% CI, 1.53–9.75), while there were no significant differences between venlafaxine-quetiapine and imipramine (OR, 1.75; 95% CI, 0.72–4.25), or between imipramine and venlafaxine (OR, 2.20; 95% CI, 0.89–5.41). In contrast to treatment response, venlafaxine-quetiapine combination therapy was not significantly more effective than venlafaxine monotherapy in improving remission rates; however, the combination pharmacotherapy was significantly more effective than imipramine alone. There was also no significant difference between imipramine and venlafaxine. Based on these findings, Wijkstra and colleagues concluded that venlafaxine plus quetiapine combination therapy was more effective than venlafaxine monotherapy on the primary outcome measure of treatment response.

Electroconvulsive therapy

Electroconvulsive therapy is widely considered to be the most effective treatment in severe major depression (Pagnin et al., 2004). It is particularly effective in patients with depression with psychotic features (Birkenhäger et al., 2003; Petrides et al., 2001; Van Diermen et al., 2018). In their 2018 metaanalysis looking at predictors of ECT response and remission rates in major depressive disorder, van Diermen and colleagues showed that presence of psychotic symptoms had an OR of 1.69 ($P<.001$) for response and 1.47 ($P=.001$) for remission. Expert guidelines recommend ECT as a first-line treatment of acute psychotic depression (American Psychiatric Association, 2010; Bauer et al., 2013; Milev et al., 2016). Unfortunately, the relative efficacy of ECT compared with pharmacotherapies is not known due to a lack of prospective controlled trials.

In 2001, Petrides and colleagues provided data that quantified the efficacy of bilateral ECT in psychotic depression (Petrides et al., 2001) and confirmed that the presence of psychosis in patients with major depression is a predictor of a favorable outcome with bilateral ECT. These data came from the first phase of a collaborative study of continuation treatments after successful ECT courses by the Consortium for Research in ECT (CORE). The authors reported the results for 253 patients ($n=77$ psychotic, $n=176$ nonpsychotic) who received bitemporal ECT three times a week until remission, defined as a 24-item HAM-D score of 10 or lower after two consecutive treatments, and a decrease of at least 60% from baseline. For the completers of the acute ECT course ($n=217$), the overall remission rate was 87% (189/217). The remission rate for the psychotic depressed patients was 95%, while for the nonpsychotic depressed patients it was 83% ($P<.01$). The odds of becoming a remitter for all patients was two times higher among the psychotic depressed than among the nonpsychotic depressed patients (OR, 2.0; 95% CI, 1.02–3.96). Additionally, symptomatic improvement, as measured by the HAM-D, was more robust and appeared sooner in the psychotic patients compared with the nonpsychotic patients.

Despite the demonstrated efficacy of ECT for the acute treatment of psychotic depression, early relapse rates after successful ECT are high (Sackeim et al., 2001), although not all studies are in agreement (Birkenhäger et al., 2005). The remission rate with ECT in community settings has also been shown to be substantially less than that in controlled clinical trials (Prudic et al., 2004). Prudic and colleagues showed that the intent-to-treat remission rates from a large cohort of adults treated with ECT in various community facilities were in the range of 30%–47%, a marked difference from the 70% to 90% remission rate typically reported in clinical trials (American Psychiatric Association, 2010; Petrides et al., 2001). This disparity between clinical outcomes in research and community settings was more marked for remission than for response rates. The low remission rates in routine practice might be explained by greater rates of comorbid psychiatric and medical conditions that are associated with poorer ECT outcomes in the community than in the patient populations studied in the clinical trials.

Continuation and maintenance pharmacotherapy

Despite the demonstrated efficacy of ECT in the acute episode of psychotic depression, studies show high rate of relapse within days to weeks after the completion of ECT course, and patients with psychotic depression are more likely to relapse following ECT compared to patients with nonpsychotic depression (Prudic et al., 2004; Sackeim et al., 2001). Intensive pharmacotherapy has been shown to reduce post-ECT relapse rates (Sackeim et al., 2001). The relapse rate in psychotic depression after ECT has been reported to be approximately 50% in studies where lithium was used after ECT (see later). Sackeim et al. (2001) reported on a randomized, double-blind study of maintenance pharmacotherapy of unipolar depression following a successful ECT course with nortriptyline monotherapy, nortriptyline plus lithium, or placebo. Over the 24-week trial, the relapse rate for placebo was 84% (95% CI, 70%–99%); the relapse rate for nortriptyline monotherapy was 60% (95% CI, 41%–79%); and for nortriptyline-lithium combination therapy, the relapse rate was 39% (95% CI, 19%–59%). The parametric survival analysis indicated that across the treatment conditions, medication-resistant nonpsychotic patients had a higher relapse rate than patients with psychotic depression. The relapse rates were 50.0% for psychotic patients ($n=28$), 55.6% for nonpsychotic patients without medication resistance ($n=9$), and 72.2% for nonpsychotic medication-resistant patients ($n=36$). Prudic et al. (2013) reported on a sample of 319 patients with an episode of major depression (either unipolar or bipolar) who were treated with moderate dosage bilateral ECT or high dosage right unilateral ECT; of those who met post-ECT remission criteria, 122 patients (including 25.41% = 31 with psychotic depression) were randomized to post-ECT continuation treatment with nortriptyline plus lithium or venlafaxine plus lithium. There was no difference in the relapse rates of those with psychotic depression or nonpsychotic depression with a 50% relapse in both groups within 6 months, regardless of what antidepressant lithium was combined with.

Few studies on continuation and maintenance pharmacotherapy for psychotic depression have been published to date. In 2001, Meyers and colleagues reported on a randomized controlled trial that compared the benefits and risks of antidepressant-antipsychotic combination pharmacotherapy (nortriptyline or sertraline plus perphenazine) with those of antidepressant monotherapy (nortriptyline or sertraline) during a 26-week period in 28 older patients with psychotic depression who had achieved symptom remission after completion of an ECT course (Meyers et al., 2001). Results showed that the relapse frequency was nonsignificantly greater in combination therapy than in monotherapy participants. However, participants in the combination therapy group were more likely to experience adverse effects of treatment, including 43% incidence of tardive dyskinesia after 6 months of perphenazine treatment.

The STOP-PD II multicenter randomized double-blind placebo-controlled trial ("Sustaining Remission of Psychotic Depression") assessed the benefits and risks of continuing antipsychotic medication in patients with psychotic depression once the episode of psychotic depression has responded to treatment with sertraline and olanzapine (Flint et al., 2013, 2019). The study included 126 patients between the ages of 18 and 85 years who met criteria for current major depressive episode with at least one associated delusion (with or without hallucinations) and scored 21 or higher on the 17-item HAM-D. The study had three phases: acute, stabilization, and randomization. In the open-label acute phase, participants received a combination of sertraline (median dosage was 150 mg/day, interquartile range was 150–200 mg/day) plus olanzapine (median dosage was 15 mg/day, interquartile range was 15–20 mg/day). As soon as the patients with psychotic depression met criteria for remission, defined as the absence of delusions and hallucinations and a 17-item HAM-D score of 10 or less for 2 consecutive weeks, participants entered the open-label sertraline and olanzapine 8-week stabilization phase. Participants who met criteria for near-remission, defined as the absence of delusions and hallucinations, a HAM-D score of 11–15 with 50% or more reduction in baseline HAM-D score, and being rated as "very much improved" or "much improved" on the Clinical Global Impression scale, were also eligible to enter the stabilization phase. Participants who still met full-remission or near-remission criteria at the end of the stabilization phase and who had a Mini-Mental State Examination (MMSE) score of 24 or higher were eligible for the third phase, a 36-week randomized controlled trial (RCT). All participants continued to take open-label sertraline for the duration of the trial. During the RCT, they were randomized under double-blind conditions to either continue olanzapine or switch from olanzapine to placebo over a 4-week taper of olanzapine. Participants were assessed weekly for the first 8 weeks and once every 4 weeks thereafter until study completion at week 36, relapse, or early termination. Risk of relapse was the primary outcome and required one of the following: (1) meeting criteria for major depressive episode; (2) 17-item HAM-D score of 18 or higher; (3) symptoms of psychosis (delusions or hallucinations); or (4) other significant clinical worsening, including having a suicide plan or attempting suicide, developing symptoms of mania or hypomania, or being hospitalized in a psychiatric unit. Secondary outcomes included anthropometric and metabolic measures such as weight, waist circumference, fasting cholesterol and triglyceride levels, fasting glucose, and hemoglobin A1c.

Of the 126 randomized participants, 13 of 64 (20.3%) randomized to olanzapine and 34 of 62 (54.8%) to placebo experienced a relapse (hazard ratio, 0.25; 95% CI, 0.13–0.48; $P<.001$). The effect of olanzapine on the daily rate of anthropometric and metabolic measures significantly differed from placebo for weight (0.13 lb.; 95% CI, 0.11–0.15; adjusted $P<.001$), waist circumference (0.009 in.; 95% CI, 0.004–0.014; adjusted $P=.002$), and total cholesterol (0.29 mg/dL; 95% CI, 0.13–0.45; adjusted $P=.003$) but was not significantly different for low-density lipoprotein cholesterol (0.04 mg/dL; 95% CI, −0.01 to 0.10; adjusted $P=.57$), high-density lipoprotein cholesterol (−0.01 mg/dL; 95% CI, −0.03 to 0.01; adjusted $P=.99$), triglyceride (−0.153 mg/dL; 95% CI, −0.306 to 0.004; adjusted $P=.25$), glucose (−0.02 mg/dL; 95% CI, −0.12 to 0.08; adjusted $P=.99$), or hemoglobin A1c levels (−0.0002 mg/dL; 95% CI, −0.0021 to 0.0016; adjusted $P=.99$). The authors concluded that continuing the combined treatment with sertraline plus olanzapine compared with sertraline plus placebo reduced the risk of relapse over 36 weeks, with a number needed to treat of 2.8. However, continuation of olanzapine was associated with weight gain. The study did not provide information on the optimal duration of treatment with olanzapine following remission of psychotic depression.

Options to consider for treatment-resistant psychotic depression

If *both* adequate trials of antidepressant and antipsychotic combination and ECT have been tried with limited efficacy, several other treatment options have been reported in the literature that may be worth considering (see Table 24.2).

Older medications (tricyclic antidepressants combined with typical antipsychotics)

In 1985, Spiker and colleagues reported on a double-blind randomized clinical trial that compared the amitriptyline-perphenazine combination with amitriptyline alone and perphenazine alone in the treatment of 51 hospitalized patients with psychotic depression (Spiker et al., 1985). Participants were 18–65 years old with a HAM-D score of 15 or higher at the time of study entry. 17 patients received amitriptyline monotherapy (mean dose 217.6 ± 46.7 mg/day), 16 patients received perphenazine monotherapy (mean dose 49.8 ± 15.4 mg/day), and 18 patients were treated with the

TABLE 24.2 Treatments that have been suggested to have efficacy in psychotic depression.[a]

Treatment	References
Amitriptyline plus perphenazine	Spiker et al. (1985)
Lithium	Birkenhäger et al. (2009), Ebert (1997), Rothschild et al. (1993), Nelson and Mazure (1986), Price et al. (1983)
Clozapine	Jeyapaul and Vieweg (2006), Hrdlicka (2002), Ranjan and Meltzer (1996)
Ketamine/esketamine	Zarrinnegar et al. (2019), Ajub and Lacerda (2018), Bartova et al. (2018), da Frota Ribeiro et al. (2016)
Acetylcholinesterase inhibitors	Smart et al. (2018)
Transcranial magnetic stimulation (TMS)	Freitas et al. (2012)
Psychotherapy	Gaudiano et al. (2013), Ružić et al. (2010), Gaudiano et al. (2007)
Benzodiazepines	Shiwaku et al. (2020)
Glucocorticoid antagonists	Block et al. (2018), Belanoff et al. (2001)

[a] *Mainly anecdotal case reports and case series except for the Spiker et al. (1985) study which was a small RCT.*

amitriptyline-perphenazine combination therapy (mean dose of 170.0 ± 45.5 mg/day of amitriptyline and 54.2 ± 16.8 mg/day of perphenazine). Treatment duration was 5 weeks. The treatment response criteria were defined as a 50% reduction in HAM-D and Brief Psychiatric Rating Scale (BPRS) and a final HAM-D score of less than 12. The overall response rate of patients treated with the amitriptyline-perphenazine combination therapy was 78% (14 of 18 patients), in contrast to 41% (14 of 18 patients) treated with amitriptyline alone, and 19% (3 of 16 patients) treated with perphenazine alone. The group of patients treated with amitriptyline-perphenazine combination therapy had a significantly higher response rate ($P < .05$) even after controlling for the plasma levels of amitriptyline and nortriptyline (Spiker et al., 1986).

Given these findings, Anton and Burch (1990) explored whether the efficacy of amitriptyline-perphenazine combination could be matched by monotherapy with amoxapine, an antidepressant with dopamine antagonist activity pharmacologically similar to antipsychotics. They conducted a randomized, double-blind study of 38 patients with psychotic depression which lasted 4 weeks, and defined the treatment response criteria as a 50% reduction in HAM-D score. Patients in each group showed similar improvement in depression and psychosis: the response rate was 71% in the amoxapine monotherapy group, and 81% in the amitriptyline-perphenazine combination therapy group. Notably, extrapyramidal symptoms were significantly more frequent in the patients treated with amitriptyline-perphenazine combination therapy than in those treated with amoxapine alone.

Lithium

Augmentation with lithium is recommended by some expert guidelines (American Psychiatric Association, 2010; Hamoda and Osser, 2008; Suehs et al., 2008; Royal Australian and New Zealand College of Psychiatrists (RANZCP), 2004) when the combination of an antidepressant and antipsychotic fails to achieve full remission of psychotic depression. However, the literature on its efficacy is limited by a lack of prospective, controlled trials. Several small uncontrolled studies (Birkenhäger et al., 2009; Ebert, 1997; Nelson and Mazure, 1986; Price et al., 1983; Rothschild et al., 1993) have suggested that lithium augmentation of the antidepressant-antipsychotic combination may be effective in management of treatment-resistant psychotic depression. Price et al. (1983) described the results of lithium augmentation in six patients with psychotic depression who were refractory to combined TCA-antipsychotic treatment. Clinical response was noted in five of the six patients. Three of the responders improved dramatically over the course of 48 h to 2 weeks, and two others had a more gradual response. In an open comparison of a lithium-amitriptyline combined therapy with a haloperidol-amitriptyline combined therapy in 40 patients with psychotic depression, Ebert (1997) described greater improvement of some depressive symptoms with fewer side effects in patients taking the lithium combination. Of note, 100% (7 of 7) of patients with bipolar psychotic depression, but only 61.5% (8 of 13) of patients with unipolar psychotic depression in the lithium-amitriptyline combination group ($n = 20$), were significantly improved by lithium. These results are in accordance with findings of other small uncontrolled studies that have looked at lithium augmentation for psychotic depression. Nelson and Mazure (1986) conducted a retrospective chart review of 21 patients with psychotic depression who had not responded to treatment with haloperidol or a desipramine-perphenazine combination. Their study showed that 89% (8 of 9) of patients with bipolar psychotic depression, but only 25% (3 of 12) of patients with unipolar psychotic depression, recovered when lithium was added to their treatment regimen. In another study of eight patients with psychotic depression refractory to a fluoxetine-perphenazine combination therapy (Rothschild et al., 1993), 100% (3 of 3) of patients with bipolar psychotic depression responded to lithium augmentation, in contrast to none of the five patients with unipolar psychotic depression ($P<.01$).

Clozapine

Clozapine is an atypical antipsychotic with greater serotonergic, adrenergic, and histaminergic blocking effects relative to its dopaminergic blocking activity (Lieberman et al., 1989). It has consistently been shown to be superior to other antipsychotics for the treatment of schizophrenia (Lewis et al., 2006; McEvoy et al., 2006; Nucifora et al., 2017; Siskind et al., 2016) and is considered the only effective treatment option for neuroleptic-resistant schizophrenia (Khokhar et al., 2018; Remington et al., 2016). Ranjan and Meltzer (1996) reported on three cases of treatment-refractory psychotic depression in which clozapine improved both psychotic and depressive symptomatology, and was used successfully as a maintenance therapy for up to 5 years. The patients in each case ranged in age from 32 to 50 years old, and had been unable to maintain sustained symptomatic improvement with both numerous psychotropic medication trials and ECT. Clozapine dosages ranged from 75 to 650 mg/day. One case relapsed when clozapine was stopped and recovered shortly after it was reinstituted. In another case, improvement in depression occurred several months before improvement in psychosis, suggesting an independent effect of clozapine on mood.

Jeyapaul and Vieweg (2006) reported on another case of a hospitalized 39-year-old female with psychotic depression who had been resistant to treatment with a combination of antidepressant, antipsychotic, mood stabilizer, and concomitant ECT therapy. Four months following initiation of clozapine, which was titrated up to 350 mg/day, the patient showed significant improvement in mood, psychosis, and functional abilities and a reduction in her BPRS score from 63 to 21. The improvement was sustained for 2 years on a combination of clozapine and an antidepressant. Similarly, Hrdlicka (2002) described another case of treatment-refractory psychotic depression in a hospitalized 48-year-old female who was able to achieve a marked improvement in symptoms with the combination treatment of clozapine and the tetracyclic antidepressant maprotiline. Remission was sustained for 2.5 years following discharge on maintenance therapy of clozapine and maprotiline.

Ketamine/esketamine

There is mounting evidence for robust and rapid-onset effectiveness of ketamine and its enantiomer, *S*-ketamine (esketamine), in nonpsychotic treatment-resistant major depressive disorder (Berman et al., 2000; Daly et al., 2018; Fedgchin et al., 2019; McGirr et al., 2015; Molero et al., 2018). In March 2019, intranasal esketamine was approved by the Food and Drug Administration for the treatment of adults with nonpsychotic treatment-resistant depression in the United States. Evidence also suggests that ketamine has rapid effects on suicidal ideation (Price et al., 2014; Witt et al., 2020), even with as little as a single intravenous dose (Bartoli et al., 2017; Grunebaum et al., 2018; Wilkinson et al., 2018).

Studies of ketamine use in treatment of patients with psychotic disorders are sparse, likely due to its potential to exacerbate psychotic symptoms. However, there is emerging evidence that the concerns about possible exacerbation of psychotic symptoms with ketamine treatment may be unwarranted. One case study described robust antisuicidal and antidepressant effects of repeat intravenous *S*-ketamine infusions, administered three times weekly for 3 weeks in total, with no psychotic or dissociative side effects in a young female with schizophrenia and severe postpsychotic depression (Bartova et al., 2018). In 2016, da Frota Ribeiro and colleagues reported on a small case series in which two patients with severe depression with psychotic features showed dramatic mood improvement and resolution of psychotic symptoms through the course of treatment with ketamine (da Frota Ribeiro et al., 2016). The first patient was a 52-year-old female with treatment-resistant psychotic depression who presented with depressive symptoms, auditory hallucinations, paranoid delusions, and active suicidal ideation. She reported mild dissociative symptoms during the ketamine infusion (0.5 mg/kg continuous intravenous infusion over 40 min); within hours, resolution of auditory hallucinations and paranoia as well as dramatic improvement in mood was noted. She was able to be discharged home on adjusted medications after three ketamine infusions. The second patient was a 55-year-old female with refractory schizoaffective disorder who presented with severe depressive symptoms, psychosis, catatonia, and active suicidal ideation. She also exhibited dramatic improvement in mood and resolution of psychotic symptoms and suicidal ideation following ketamine infusion. She received a total of five ketamine infusions prior to discharge on oral pharmacotherapy. Similarly, Ajub and Lacerda (2018) reported on a case study of four patients with treatment-resistant depression with psychotic features who were treated with either intravenous or subcutaneous administration of esketamine. Three out of four patients demonstrated either significant improvement or complete resolution of both mood and psychotic symptoms, and the effect was maintained for 2–4 weeks. Zarrinnegar et al. (2019) recently reported on a safe and successful use of intravenous ketamine in an adolescent patient with treatment-resistant psychotic depression. The 15-year-old patient received a course of six ketamine infusions dosed at 0.5 mg/kg per infusion over a period of 3 weeks, starting with three infusions a week and gradually tapering down to once a week, with gradual resolution of depressive, psychotic, and suicidal symptoms which was maintained for several months. In all of the above described cases, the patients experienced mild-to-moderate short-lasting increases in dissociative or psychotic symptoms during treatment, but they had no impact on the long-term outcome of these patients with psychosis.

Cholinergic system agents

Hofstatter and Girgis (1980) suggested that cholinergic prevalence together with a monoaminergic deficit may be the underlying neurotransmitter imbalance in psychotic depression, and that correction of this neurotransmitter imbalance by reducing the acetylcholine predominance or increasing the level of monoaminergic functioning—or a combination of both approaches—should guide all treatment methods for the relief of psychotic depression. As described above, the antidepressant tricyclic medications which combine a central anticholinergic action with an aminergic potentiating function do have some evidence for efficacy in treatment of psychotic depression. Conversely, there is emerging evidence for procholinergic effects of acetylcholinesterase inhibitors in management of psychotic symptoms (Blom et al., 2015; Patel et al., 2010; Singh et al., 2012). Smart et al. (2018) recently reported on a case of a 53-year-old male with psychotic depression with visual, auditory, and olfactory hallucinations that remained refractory despite treatments with multiple antidepressants, alone and in combination, lithium and carbamazepine augmentation, various antipsychotics including clozapine, ECT, bilateral anterior capsulotomy, and anterior cingulotomy. The patient's hallucinations resolved fully within a month of initiation of rivastigmine (4.2 mg/day transdermal patch) with concurrent improvement in mood. Unfortunately, the patient suffered intolerable side effects from rivastigmine but was able to largely remain symptom-free for 18 months on galantamine maintenance (up to 8 mg/day). He has required intermittent rivastigmine therapy during brief relapses with resolution of symptoms, and has since remained in partial remission.

Transcranial magnetic stimulation

Comparisons of transcranial magnetic stimulation (TMS) and ECT generally show similar rates of response in patients with nonpsychotic depression (Fitzgerald, 2004; Grunhaus et al., 2000), but evidence for TMS efficacy in the treatment of psychotic depression is limited. The evidence in the current literature demonstrates that patients with psychotic depression respond significantly better to ECT than TMS, whereas those without psychosis have a comparative response to both neurostimulation modalities (Grunhaus et al., 2000; Martin et al., 2002). Hebel et al. (2020) recently suggested that antipsychotic medications themselves may attenuate antidepressive effects of repetitive TMS (rTMS) independently of confounding factors. While a Cochrane review of TMS effectiveness in treatment of psychotic symptoms in patients with schizophrenia (Dougall et al., 2015) was inconclusive, one group described a case of effective treatment of auditory hallucinations with TMS in a patient with psychotic depression (Freitas et al., 2012). The patient was a 48-year-old male with refractory psychotic depression who was a nonresponder to ECT. Following an initial course of 15 consecutive daily sessions of high-frequency (20 Hz) rTMS to the left dorsolateral prefrontal cortex (DLPFC) over the course of 3 weeks, his depressive symptoms were greatly improved but auditory hallucinations persisted. Two months later (patient received once weekly maintenance rTMS sessions during this time), he completed a second rTMS course for the treatment of auditory hallucinations, which consisted of 10 consecutive daily sessions of low-frequency (1 Hz) rTMS to the left temporo-parietal cortex over the course of 2 weeks. A significant reduction of the patient's hallucinatory experience was achieved, as assessed by clinician-rated scoring on the 11-item Auditory Hallucinations subscale of the Psychotic Symptoms Rating Scales (PSYRATS-AH), and was sustained for 1-year of follow-up, with rTMS maintenance sessions only targeting the DLPFC for depression. These results suggest that different neural networks or different dysfunctions of overlapping networks sustain mood and psychotic features of psychotic depression. Other evidence supportive of these findings is currently lacking. To date, there have been no randomized controlled studies examining the effectiveness of TMS for the treatment of psychotic depression.

Psychotherapy

Combining pharmacotherapy and psychotherapy has been shown to produce improvements in efficacy when treating nonpsychotic depressed patients (Friedman et al., 2004). However, data on either psychotherapy or combined treatment among patients with psychotic depression are lacking because patients exhibiting psychotic features have historically been excluded from trials on psychotherapy for depression. Gaudiano et al. (2005) were among the first groups to compare treatment response in psychotic and nonpsychotic depressed patients who received combined pharmacotherapy plus an efficacious psychotherapy for depression, such as behavior therapy, cognitive therapy, or family therapy. In that study, patients with psychotic depression showed a poorer response in terms of depression severity to combined pharmacotherapy and psychotherapy compared with those with nonpsychotic depression. It should be noted that the psychotherapies provided were not adapted specifically for patients with psychotic depression.

In a later study, Gaudiano et al. (2007) examined the efficacy of acceptance and commitment therapy (ACT) among 18 hospitalized patients with psychotic depression as part of a larger clinical trial. Nine participants were randomly assigned to enhanced treatment as usual (ETAU) and the other nine underwent ETAU plus ACT sessions. The number of ACT sessions varied based on the patients' length of stay, with the average of three sessions. The results showed that 44% of the patients with psychotic depression in the ACT group showed clinically significant improvement by discharge on BPRS total scores compared with 0% of the ETAU only group (≥2 SD change from before to after treatment, $\chi^2=5.14$, $P<.05$). The rates of clinically significant change on BPRS mood symptoms subscale were 70% in the ACT group compared with 30% in the ETAU only group ($\chi^2=3.60$, $P=.058$). No significant differences were found between the rates of clinically significant change on BPRS positive symptoms subscale. In 2013, Gaudiano and colleagues conducted an open trial of a new behavioral intervention that combined elements of behavioral activation and ACT for depression and psychosis, referred to as acceptance-based depression and psychosis therapy (ADAPT), which was developed specifically for patients with psychotic depression (Gaudiano et al., 2013). Fourteen patients with psychotic depression were provided with up to 6 months of ADAPT in combination with pharmacotherapy. Results showed that patients achieved clinically significant and sustained improvements in both depressive and psychotic symptoms. Patients also demonstrated significant improvements in psychosocial functioning over time. These results suggest that patients with psychotic depression might benefit from psychotherapy, especially if it is individualized. A Croatian case report on integrative approach to treatment-resistant psychotic depression in a 34-year-old woman (Ružić et al., 2010) showed that integrative treatment including personalized psychotherapy resulted in a relatively quick recovery, which provides additional anecdotal evidence to support the above findings.

Benzodiazepines

Shiwaku et al. (2020) reported in a recent study that adjunctive use of benzodiazepines may reduce relapse/recurrence rates in patients with psychotic depression. Their investigation aimed to identify whether the use of benzodiazepines is associated with a lower risk of relapse or recurrence of major depression in some patients, and the characteristics of these patients. They identified 108 patients who were hospitalized with major depressive disorder, including 26 patients with psychotic depression, and followed them for 12 months. 83 patients received benzodiazepines (14 of them were patients with psychotic depression), and 25 patients did not (12 of them were patients with psychotic depression). There was no significant difference between groups of patients with psychotic depression with respect to age, severity assessed by HAM-D, time from onset to hospitalization, use of antidepressants and antipsychotics, modified ECT, and the length of hospitalization. Remission was defined as a HAM-D score of <8 for 2 consecutive weeks. The results of the study showed that benzodiazepine use did not have a significant effect on relapse/recurrence when all participants were analyzed in a single group. However, 75% of psychotic depressed patients who did not receive benzodiazepines had relapsed at the end of the 12-month follow-up compared with 21.2% of psychotic depressed patients who received benzodiazepines during the study period (hazard ratio (95% CI): 5.141 (1.591–16.61), $P=.0040$). The effect of benzodiazepines on time to depression relapse/recurrence in patients with psychotic depression was dose-dependent. No adverse effects of benzodiazepines during the study period were reported. However, these results must be interpreted with caution as the study was not prospective and participants were not randomized. Data from a randomized controlled trial is needed to confirm the effect of benzodiazepines on psychotic depression.

Glucocorticoid antagonists

Another possible option for patients with treatment-resistant psychotic depression is mifepristone, a glucocorticoid antagonist. In 1985, Schatzberg and colleagues proposed that reducing the effects of cortisol by lowering cortisol levels or antagonizing cortisol's effects at glucocorticoid receptors in the brain could decrease psychotic symptoms in patients with psychotic depression (Schatzberg et al., 1985). Initial reports of treatment of psychotic depression with mifepristone were encouraging (Belanoff et al., 2001). However, in larger double-blind studies, statistical significance was less consistently observed, although mifepristone produced response rates numerically superior to placebo (Block et al., 2018).

Block et al. (2018) combined the results of several RCTs sponsored by Corcept Therapeutics that did not achieve statistical significance and showed that mifepristone plasma levels ≥ 1637 ng/mL predicted clinical response to mifepristone in patients with psychotic depression. Mifepristone was never submitted to the FDA for approval for the treatment of psychotic depression, and given that its use is restricted, it may be difficult to obtain for patients suffering from psychotic depression.

Summary

Major depression with psychotic features is a significant public health problem that is associated with considerable morbidity and mortality, but is unfortunately often misdiagnosed or underdiagnosed. Accurate diagnosis is critical because failing to recognize psychotic features of a major depressive episode or improperly diagnosing a patient with another psychotic illness defers the use of effective treatment modalities. There are two forms of accepted treatment of psychotic depression that are currently recommended: electroconvulsive therapy or a combination of antipsychotic and antidepressant medications. Both have been well-studied and their efficacy has been documented in the literature. Failure to respond to both of these treatments given in adequate dose and duration should be considered treatment resistance in psychotic depression. The literature on alternative treatment options for treatment-resistant psychotic depression is primarily based on anecdotal evidence. Based on our review of the literature, either lithium augmentation of the antidepressant/antipsychotic medications or the combination of amitriptyline and perphenazine should be considered first for patients with treatment-resistant psychotic depression.

References

Ajub, E., Lacerda, A.L.T., 2018. Efficacy of esketamine in the treatment of depression with psychotic features: a case series. Biol. Psychiatry 83 (1), e15–e16. https://doi.org/10.1016/j.biopsych.2017.06.011.

American Psychiatric Association, 2010. Practice Guideline for the Treatment of Patients With Major Depressive Disorder, third ed. American Psychiatric Association. https://psychiatryonline.org/pb/assets/raw/sitewide/practice_guidelines/guidelines/mdd.pdf.

Andreescu, C., Mulsant, B.H., Peasley-Miklus, C., Rothschild, A.J., Flint, A.J., Heo, M., Caswell, M., Whyte, E.M., Meyers, B.S., 2007. Persisting low use of antipsychotics in the treatment of major depressive disorder with psychotic features. J. Clin. Psychiatry 68 (2), 194–200. https://doi.org/10.4088/JCP.v68n0203.

Anton, R.F., Burch, E.A., 1990. Amoxapine versus amitriptyline combined with perphenazine in the treatment of psychotic depression. Am. J. Psychiatr. 147 (9), 1203–1208. https://doi.org/10.1176/ajp.147.9.1203.

Baldwin, R.C., Jolley, D.J., 1986. The prognosis of depression in old age. Br. J. Psychiatry 149, 574–583. https://doi.org/10.1192/bjp.149.5.574.

Bartoli, F., Riboldi, I., Crocamo, C., Di Brita, C., Clerici, M., Carrà, G., 2017. Ketamine as a rapid-acting agent for suicidal ideation: a meta-analysis. Neurosci. Biobehav. Rev. 77, 232–236. https://doi.org/10.1016/j.neubiorev.2017.03.010.

Bartova, L., Papageorgiou, K., Milenkovic, I., Dold, M., Weidenauer, A., Willeit, M., Winkler, D., Kasper, S., 2018. Rapid antidepressant effect of S-ketamine in schizophrenia. Eur. Neuropsychopharmacol. 28 (8), 980–982. https://doi.org/10.1016/j.euroneuro.2018.05.007.

Bauer, M., Pfennig, A., Severus, E., Whybrow, P.C., Angst, J., Möller, H.J., Adli, M., Benkert, O., Bschor, T., Tadić, A., Holsboer-Trachsler, E., Anderson, I., Baldwin, D., Cookson, J.C., Grunze, H., Katona, C., Paykel, E.S., Tylee, A., Ayuso-Gutierrez, J.L., et al., 2013. World Federation of Societies of Biological Psychiatry (WFSBP) guidelines for biological treatment of unipolar depressive disorders, part 1: update 2013 on the acute and continuation treatment of unipolar depressive disorders. World J. Biol. Psychiatry 14 (5), 334–385. https://doi.org/10.3109/15622975.2013.804195.

Belanoff, J.K., Flores, B.H., Kalezhan, M., Sund, B., Schatzberg, A.F., 2001. Rapid reversal of psychotic depression using mifepristone. J. Clin. Psychopharmacol. 21 (5), 516–521. https://doi.org/10.1097/00004714-200110000-00009.

Berman, R.M., Cappiello, A., Anand, A., Oren, D.A., Heninger, G.R., Charney, D.S., Krystal, J.H., 2000. Antidepressant effects of ketamine in depressed patients. Biol. Psychiatry 47 (4), 351–354. https://doi.org/10.1016/S0006-3223(99)00230-9.

Bingham, K.S., Rothschild, A.J., Mulsant, B.H., Whyte, E.M., Meyers, B.S., Banerjee, S., Szanto, K., Flint, A.J., 2017. The association of baseline suicidality with treatment outcome in psychotic depression. J. Clin. Psychiatry 78 (8), 1149–1154. Physicians Postgraduate Press Inc. https://doi.org/10.4088/JCP.16m10881.

Birkenhäger, T.K., Pluijms, E.M., Lucius, S.A.P., 2003. ECT response in delusional versus non-delusional depressed inpatients. J. Affect. Disord. 74 (2), 191–195. https://doi.org/10.1016/S0165-0327(02)00005-8.

Birkenhäger, T.K., Van Den Broek, W.W., Mulder, P.G.H., De Lely, A., 2005. One-year outcome of psychotic depression after successful electroconvulsive therapy. J. ECT 21 (4), 221–226. https://doi.org/10.1097/01.yct.0000183269.62735.89.

Birkenhäger, T.K., Van Den Broek, W.W., Wijkstra, J., Bruijn, J.A., Van Os, E., Boks, M., Verkes, R.J., Janzing, J.G.E., Nolen, W.A., 2009. Treatment of unipolar psychotic depression: an open study of lithium addition in refractory psychotic depression. J. Clin. Psychopharmacol. 29 (5), 513–515. https://doi.org/10.1097/JCP.0b013e3181b6744e.

Blazer, D., 1994. Epidemiology of late-life depression. In: Schneider, L., Reynolds, C., Lebowitz, B., Friedhoff, A. (Eds.), Diagnosis and Treatment of Depression in Late Life. American Psychiatric Press, pp. 9–20.

Block, T.S., Kushner, H., Kalin, N., Nelson, C., Belanoff, J., Schatzberg, A., 2018. Combined analysis of mifepristone for psychotic depression: plasma levels associated with clinical response. Biol. Psychiatry 84 (1), 46–54. https://doi.org/10.1016/j.biopsych.2018.01.008.

Blom, J.D., Coebergh, J.A.F., Lauw, R., Sommer, I.E.C., 2015. Musical hallucinations treated with acetylcholinesterase inhibitors. Front. Psychiatry 6, 46. https://doi.org/10.3389/fpsyt.2015.00046.

da Frota Ribeiro, C., Sanacora, G., Hoffman, R., Ostroff, R., 2016. The use of ketamine for the treatment of depression in the context of psychotic symptoms. Biol. Psychiatry 79 (9), e65–e66.

Daly, E.J., Singh, J.B., Fedgchin, M., Cooper, K., Lim, P., Shelton, R.C., Thase, M.E., Winokur, A., Luc Van, N., Manji, H., Drevets, W.C., 2018. Efficacy and safety of intranasal esketamine adjunctive to oral antidepressant therapy in treatment-resistant depression. JAMA Psychiatry 75 (2), 139. https://doi.org/10.1001/jamapsychiatry.2017.3739.

Dougall, N., Maayan, N., Soares-Weiser, K., McDermott, L.M., McIntosh, A., 2015. Transcranial magnetic stimulation for schizophrenia. Cochrane Database Syst. Rev. https://doi.org/10.1093/schbul/sbv121.

Ebert, D., 1997. Lithium-TCA combination treatment of psychotic depression: comparison with TCA-neuroleptic treatment. J. Clin. Psychopharmacol. 17 (2), 129–130. https://doi.org/10.1097/00004714-199704000-00018.

Farahani, A., Correll, C.U., 2012. Are antipsychotics or antidepressants needed for psychotic depression? A systematic review and meta-analysis of trials comparing antidepressant or antipsychotic monotherapy with combination treatment. J. Clin. Psychiatry 73 (4), 486–496. https://doi.org/10.4088/JCP.11r07324.

Fedgchin, M., Trivedi, M., Daly, E.J., Melkote, R., Lane, R., Lim, P., Vitagliano, D., Blier, P., Fava, M., Liebowitz, M., Ravindran, A., Gaillard, R., Ameele, H.V.D., Preskorn, S., Manji, H., Hough, D., Drevets, W.C., Singh, J.B., 2019. Efficacy and safety of fixed-dose esketamine nasal spray combined with a new oral antidepressant in treatment-resistant depression: results of a randomized, double-blind, active-controlled study (TRANSFORM-1). Int. J. Neuropsychopharmacol. 22 (10), 616–630. https://doi.org/10.1093/ijnp/pyz039.

Fitzgerald, P.B., 2004. Repetitive transcranial magnetic stimulation and electroconvulsive therapy: complementary or competitive therapeutic options in depression? Australas. Psychiatry 12 (3), 234–238. https://doi.org/10.1111/j.1039-8562.2004.02113.x.

Flint, A.J., Meyers, B.S., Rothschild, A.J., Whyte, E.M., Mulsant, B.H., Rudorfer, M.V., Marino, P., 2013. Sustaining remission of psychotic depression: rationale, design and methodology of STOP-PD II. BMC Psychiatry 13, 38. https://doi.org/10.1186/1471-244X-13-38.

Flint, A.J., Meyers, B.S., Rothschild, A.J., Whyte, E.M., Alexopoulos, G.S., Rudorfer, M.V., Marino, P., Banerjee, S., Pollari, C.D., Wu, Y., Voineskos, A.N., Mulsant, B.H., 2019. Effect of continuing olanzapine vs placebo on relapse among patients with psychotic depression in remission: the STOP-PD II randomized clinical trial. JAMA 322 (7), 622–631. https://doi.org/10.1001/jama.2019.10517.

Freitas, C., Pearlman, C., Pascual-Leone, A., 2012. Treatment of auditory verbal hallucinations with transcranial magnetic stimulation in a patient with psychotic major depression: one-year follow-up. Neurocase 18 (1), 57–65. https://doi.org/10.1080/13554794.2010.547864.

Friedman, M.A., Detweiler-Bedell, J.B., Leventhal, H.E., Horne, R., Keitner, G.I., Miller, I.W., 2004. Combined psychotherapy and pharmacotherapy for the treatment of major depressive disorder. Clin. Psychol. Sci. Pract. 11 (1), 47–68. https://doi.org/10.1093/clipsy/bph052.

Gaudiano, B.A., Beevers, C.G., Miller, I.W., 2005. Differential response to combined treatment in patients with psychotic versus nonpsychotic major depression. J. Nerv. Ment. Dis. 193 (9), 625–628. https://doi.org/10.1097/01.nmd.0000177791.33649.69.

Gaudiano, B.A., Miller, I.W., Herbert, J.D., 2007. The treatment of psychotic major depression: is there a role for adjunctive psychotherapy? Psychother. Psychosom. 76 (5), 271–277. https://doi.org/10.1159/000104703.

Gaudiano, B.A., Nowlan, K., Brown, L.A., Epstein-Lubow, G., Miller, I.W., 2013. An open trial of a new acceptance-based behavioral treatment for major depression with psychotic features. Behav. Modif. 37 (3), 324–355. https://doi.org/10.1177/0145445512465173.

Gournellis, R., Lykouras, L., 2006. Psychotic (delusional) major depression in the elderly: a review. Curr. Psychiatr. Rev. 2 (2), 235–244. https://doi.org/10.2174/157340006776876003.

Gournellis, R., Tournikioti, K., Touloumi, G., Thomadakis, C., Michalopoulou, P.G., Christodoulou, C., Papadopoulou, A., Douzenis, A., 2018. Psychotic (delusional) depression and suicidal attempts: a systematic review and meta-analysis. Acta Psychiatr. Scand. 137 (1), 18–29. https://doi.org/10.1111/acps.12826.

Grunebaum, M.F., Galfalvy, H.C., Choo, T.H., Keilp, J.G., Moitra, V.K., Parris, M.S., Marver, J.E., Burke, A.K., Milak, M.S., Sublette, M.E., Oquendo, M.A., Mann, J.J., 2018. Ketamine for rapid reduction of suicidal thoughts in major depression: a midazolam-controlled randomized clinical trial. Am. J. Psychiatr. 175 (4), 327–335. https://doi.org/10.1176/appi.ajp.2017.17060647.

Grunhaus, L., Dannon, P.N., Schreiber, S., Dolberg, O.H., Amiaz, R., Ziv, R., Lefkifker, E., 2000. Repetitive transcranial magnetic stimulation is as effective as electroconvulsive therapy in the treatment of nondelusional major depressive disorder: an open study. Biol. Psychiatry 47 (4), 314–324. https://doi.org/10.1016/S0006-3223(99)00254-1.

Hamoda, H.M., Osser, D.N., 2008. The psychopharmacology algorithm project at the Harvard south shore program: an update on psychotic depression. Harv. Rev. Psychiatry 16 (4), 235–247. https://doi.org/10.1080/10673220802277904.

Hebel, T., Abdelnaim, M., Deppe, M., Langguth, B., Schecklmann, M., 2020. Attenuation of antidepressive effects of transcranial magnetic stimulation in patients whose medication includes drugs for psychosis. J. Psychopharmacol. 34 (10), 1119–1124. https://doi.org/10.1177/0269881120922965.

Hofstatter, L., Girgis, M., 1980. Psychotic depression: the common therapeutic principle. South. Med. J. 73 (7), 870–872. https://doi.org/10.1097/00007611-198007000-00011.

Hrdlicka, M., 2002. Combination of clozapine and maprotiline in refractory psychotic depression. Eur. Psychiatry 17 (8), 484. https://doi.org/10.1016/S0924-9338(02)00711-3.

Jääskeläinen, E., Juola, T., Korpela, H., Lehtiniemi, H., Nietola, M., Korkeila, J., Miettunen, J., 2018. Epidemiology of psychotic depression – systematic review and meta-analysis. Psychol. Med. 48 (6), 905–918. https://doi.org/10.1017/S0033291717002501.

Jeyapaul, P., Vieweg, R., 2006. A case study evaluating the use of clozapine in depression with psychotic features. Ann. General Psychiatry 5, 20.

Khokhar, J.Y., Henricks, A.M., Sullivan, E.D.K., Green, A.I., 2018. Unique effects of clozapine: a pharmacological perspective. In: Advances in Pharmacology. vol. 82. Academic Press Inc, pp. 137–162, https://doi.org/10.1016/bs.apha.2017.09.009.

Lewis, S.W., Barnes, T.R.E., Davies, L., Murray, R.M., Dunn, G., Hayhurst, K.P., Markwick, A., Lloyd, H., Jones, P.B., 2006. Randomized controlled trial of effect of prescription of clozapine versus other second-generation antipsychotic drugs in resistant schizophrenia. Schizophr. Bull. 32 (4), 715–723. https://doi.org/10.1093/schbul/sbj067.

Lieberman, J.A., Kane, J.M., Johns, C.A., 1989. Clozapine: guidelines for clinical management. J. Clin. Psychiatry 50 (9), 329–338.

Lykouras, L., Gournellis, R., 2009. Psychotic (delusional) major depression: new vistas. Curr. Psychiatr. Rev. 5 (1), 1–28. https://doi.org/10.2174/157340009787315271.

Martin, J.L., Barbanoj, M.J., Schlaepfer, T.E., Clos, S., Perez, V., Kulisevsky, J., Gironell, A., 2002. Transcranial magnetic stimulation for treating depression. Cochrane Database Syst. Rev. 2, CD003493.

McEvoy, J.P., Lieberman, J.A., Stroup, T.S., Davis, S.M., Meltzer, H.Y., Rosenheck, R.A., Swartz, M.S., Perkins, D.O., Keefe, R.S.E., Davis, C.E., Severe, J., Hsiao, J.K., 2006. Effectiveness of clozapine versus olanzapine, quetiapine, and risperidone in patients with chronic schizophrenia who did not respond to prior atypical antipsychotic treatment. Am. J. Psychiatr. 163 (4), 600–610. https://doi.org/10.1176/ajp.2006.163.4.600.

McGirr, A., Berlim, M.T., Bond, D.J., Fleck, M.P., Yatham, L.N., Lam, R.W., 2015. A systematic review and meta-analysis of randomized, double-blind, placebo-controlled trials of ketamine in the rapid treatment of major depressive episodes. Psychol. Med. 45 (4), 693–704. https://doi.org/10.1017/S0033291714001603.

Meyers, B.S., Klimstra, S.A., Gabriele, M., Hamilton, M., Kakuma, T., Tirumalasetti, F., Alexopoulos, G.S., 2001. Continuation treatment of delusional depression in older adults. Am. J. Geriatr. Psychiatr. 9 (4), 415–422. https://doi.org/10.1097/00019442-200111000-00010.

Meyers, B.S., Flint, A.J., Rothschild, A.J., Mulsant, B.H., Whyte, E.M., Peasley-Miklus, C., Papademetriou, E., Leon, A.C., Heo, M., Alexopoulos, G.S., Bruce, M.L., Klimstra, S., English, J., Gabriele, M., Santana, F., Addonizio, G., De Asis, J., Baran, X.Y., Kalyam, B., et al., 2009. A double-blind randomized controlled trial of olanzapine plus sertraline vs olanzapine plus placebo for psychotic depression: the study of pharmacotherapy of psychotic depression (STOP-PD). Arch. Gen. Psychiatry 66 (8), 838–847. https://doi.org/10.1001/archgenpsychiatry.2009.79.

Milev, R.V., Giacobbe, P., Kennedy, S.H., Blumberger, D.M., Daskalakis, Z.J., Downar, J., Modirrousta, M., Patry, S., Vila-Rodriguez, F., Lam, R.W., MacQueen, G.M., Parikh, S.V., Ravindran, A.V., 2016. Canadian Network for Mood and Anxiety Treatments (CANMAT) 2016 clinical guidelines for the management of adults with major depressive disorder: Section 4. Neurostimulation treatments. Can. J. Psychiatr. 61 (9), 561–575. https://doi.org/10.1177/0706743716660033.

Molero, P., Ramos-Quiroga, J.A., Martin-Santos, R., Calvo-Sánchez, E., Gutiérrez-Rojas, L., Meana, J.J., 2018. Antidepressant efficacy and tolerability of ketamine and esketamine: a critical review. CNS Drugs 32 (5), 411–420. https://doi.org/10.1007/s40263-018-0519-3.

Mulsant, B.H., Haskett, R.F., Prudic, J., Thase, M.E., Malone, K.M., Mann, J.J., Pettinati, H.M., Sackeim, H.A., 1997. Low use of neuroleptic drugs in the treatment of psychotic major depression. Am. J. Psychiatr. 154 (4), 559–561. https://doi.org/10.1176/ajp.154.4.559.

Nelson, J.C., Mazure, C.M., 1986. Lithium augmentation in psychotic depression refractory to combined drug treatment. Am. J. Psychiatr. 143 (3), 363–366. https://doi.org/10.1176/ajp.143.3.363.

Nucifora, F.C., Mihaljevic, M., Lee, B.J., Sawa, A., 2017. Clozapine as a model for antipsychotic development. Neurotherapeutics 14 (3), 750–761. https://doi.org/10.1007/s13311-017-0552-9.

Ohayon, M.M., Schatzberg, A.F., 2002. Prevalence of depressive episodes with psychotic features in the general population. Am. J. Psychiatr. 159 (11), 1855–1861. https://doi.org/10.1176/appi.ajp.159.11.1855.

Pagnin, D., De Queiroz, V., Pini, S., Cassano, G.B., 2004. Efficacy of ECT in depression: a meta-analytic review. J. ECT 20 (1), 13–20. https://doi.org/10.1097/00124509-200403000-00004.

Patel, S., Attard, A., Jacobsen, P., 2010. Acetylcholinesterase inhibitors (AChEi's) for the treatment of visual hallucinations in schizophrenia: a review of the literature. BMC Psychiatry 10, 69.

Petrides, G., Fink, M., Husain, M.M., Knapp, R.G., Rush, A.J., Mueller, M., Rummans, T.A., O'Connor, K.M., Rasmussen, K.G., Bernstein, H.J., Biggs, M., Bailine, S.H., Kellner, C.H., 2001. ECT remission rates in psychotic versus nonpsychotic depressed patients: a report from CORE. J. ECT 17 (4), 244–253. https://doi.org/10.1097/00124509-200112000-00003.

Price, L.H., Conwell, Y., Nelson, J.C., 1983. Lithium augmentation of combined neuroleptic-tricyclic treatment in delusional depression. Am. J. Psychiatr. 140 (3), 318–322. https://doi.org/10.1176/ajp.140.3.318.

Prudic, J., Olfson, M., Marcus, S.C., Fuller, R.B., Sackeim, H.A., 2004. Effectiveness of electroconvulsive therapy in community settings. Biol. Psychiatry 55 (3), 301–312. https://doi.org/10.1016/j.biopsych.2003.09.015.

Price, R.B., Iosifescu, D.V., Murrough, J.W., Chang, L.C., Al Jurdi, R.K., Iqbal, S.Z., Soleimani, L., Charney, D.S., Foulkes, A.L., Mathew, S.J., 2014. Effects of ketamine on explicit and implicit suicidal cognition: a randomized controlled trial in treatment-resistant depression. Depress. Anxiety 31 (4), 335–343. https://doi.org/10.1002/da.22253.

Prudic, J., Haskett, R.F., McCall, W.V., Isenberg, K., Cooper, T., Rosenquist, P.B., Mulsant, B.H., Sackeim, H.A., 2013. Pharmacological strategies in the prevention of relapse after electroconvulsive therapy. J. ECT 29 (1), 3–12. https://doi.org/10.1097/YCT.0b013e31826ea8c4.

Ranjan, R., Meltzer, H.Y., 1996. Acute and long-term effectiveness of clozapine in treatment-resistant psychotic depression. Biol. Psychiatry 40 (4), 253–258. https://doi.org/10.1016/0006-3223(95)00305-3.

Remington, G., Lee, J., Agid, O., Takeuchi, H., Foussias, G., Hahn, M., Fervaha, G., Burton, L., Powell, V., 2016. Clozapine's critical role in treatment resistant schizophrenia: ensuring both safety and use. Expert Opin. Drug Saf. 15 (9), 1193–1203. https://doi.org/10.1080/14740338.2016.1191468.

Rothschild, A.J., 2009. Clinical Manual for the Diagnosis and Treatment of Psychotic Depression. American Psychiatric Press.

Rothschild, A.J., Samson, J.A., Bessette, M.P., Carter-Campbell, J.T., 1993. Efficacy of the combination of fluoxetine and perphenazine in the treatment of psychotic depression. J. Clin. Psychiatry 54 (9), 338–342.

Rothschild, A.J., Williamson, D.J., Tohen, M.F., Schatzberg, A., Andersen, S.W., Van Campen, L.E., Sanger, T.M., Tollefson, G.D., 2004. A double-blind, randomized study of olanzapine and olanzapine/fluoxetine combination for major depression with psychotic features. J. Clin. Psychopharmacol. 24 (4), 365–373. https://doi.org/10.1097/01.jcp.0000130557.08996.7a.

Rothschild, AJ, 2021. Why Is There No Food and Drug Administration-Approved Medication for Major Depression With Psychotic Features? J. Clin. Psychopharmacol. 41 (4), 359–361. https://doi.org/10.1097/JCP.0000000000001433.

Rothschild, A.J., Mulsant, B.H., Meyers, B.S., Flint, A.J., 2006. Challenges in differentiating and diagnosing psychotic depression. Psychiatr. Ann. 36 (1), 40–46. https://doi.org/10.3928/00485713-20060101-09.

Rothschild, A.J., Winer, J., Flint, A.J., Mulsant, B.H., Whyte, E.M., Heo, M., Fratoni, S., Gabriele, M., Kasapinovic, S., Meyers, B.S., 2008. Missed diagnosis of psychotic depression at 4 academic medical centers. J. Clin. Psychiatry 69 (8), 1293–1296. https://doi.org/10.4088/JCP.v69n0813.

Royal Australian and New Zealand College of Psychiatrists (RANZCP), 2004. Australian and New Zealand clinical practice guidelines for the treatment of depression. Aust. N. Z. J. Psychiatry 38, 389–407.

Ružić, K., Knez, R., Grahovac, T., Dadić-Hero, E., Graovac, M., 2010. Integrative approach to treatment of a patient with psychotic depression. Psychiatr. Danub. 22 (2), 370–372.

Sackeim, H.A., Haskett, R.F., Mulsant, B.H., Thase, M.E., Mann, J.J., Pettinati, H.M., Greenberg, R.M., Crowe, R.R., Cooper, T.B., Prudic, J., 2001. Continuation pharmacotherapy in the prevention of relapse following electroconvulsive therapy: a randomized controlled trial. J. Am. Med. Assoc. 285 (10), 1299–1307. https://doi.org/10.1001/jama.285.10.1299.

Schatzberg, A.F., Rothschild, A.J., 1992. Psychotic (delusional) major depression: should it be included as a distinct syndrome in DSM-IV? Am. J. Psychiatr. 149 (6), 733–745. https://doi.org/10.1176/ajp.149.6.733.

Schatzberg, A.F., Rothschild, A.J., Langlais, P.J., Bird, E.D., Cole, J.O., 1985. A corticosteroid/dopamine hypothesis for psychotic depression and related states. J. Psychiatr. Res. 19 (1), 57–64. https://doi.org/10.1016/0022-3956(85)90068-8.

Shiwaku, H., Fujita, M., Takahashi, H., 2020. Benzodiazepines reduce relapse and recurrence rates in patients with psychotic depression. J. Clin. Med. 9 (6), 1938. https://doi.org/10.3390/jcm9061938.

Singh, J., Kour, K., Jayaram, M.B., 2012. Acetylcholinesterase inhibitors for schizophrenia. Cochrane Database Syst. Rev. https://doi.org/10.1002/14651858.cd007967.pub2.

Siskind, D., McCartney, L., Goldschlager, R., Kisely, S., 2016. Clozapine v. first- and second-generation antipsychotics in treatment-refractory schizophrenia: systematic review and meta-analysis. Br. J. Psychiatry 209 (5), 385–392. https://doi.org/10.1192/bjp.bp.115177261.

Smart, C., McAllister-Williams, H., Cousins, D.A., 2018. Acetylcholinesterase inhibitors in treatment-resistant psychotic depression. Ther. Adv. Psychopharmacol. 8 (1), 59–61. https://doi.org/10.1177/2045125317718810.

Spiker, D.G., Weiss, J.C., Dealy, R.S., Griffin, S.J., Hanin, I., Neil, J.F., Perel, J.M., Rossi, A.J., Soloff, P.H., 1985. The pharmacological treatment of delusional depression. Am. J. Psychiatr. 142 (4), 430–436. https://doi.org/10.1176/ajp.142.4.430.

Spiker, D.G., Perel, J.M., Hanin, I., Dealy, R.S., Griffin, S.J., Soloff, P.H., Cofsky Weiss, J., 1986. The pharmacological treatment of delusional depression: part II. J. Clin. Psychopharmacol. 6 (6), 339–342.

Suehs, B., Argo, T.R., Bendele, S.D., Crismon, M.L., Trivedi, M.H., Kurian, B., 2008. Texas Medication Algorithm Project Procedural Manual: Major Depressive Disorder Algorithms. Texas Department of State Health Services.

Thakur, M., Hays, J., Krishnan, K.R.R., 1999. Clinical, demographic and social characteristics of psychotic depression. Psychiatry Res. 86 (2), 99–106. https://doi.org/10.1016/S0165-1781(99)00030-X.

Van Diermen, L., Van Den Ameele, S., Kamperman, A.M., Sabbe, B.C.G., Vermeulen, T., Schrijvers, D., Birkenhäger, T.K., 2018. Prediction of electroconvulsive therapy response and remission in major depression: meta-analysis. Br. J. Psychiatry 212 (2), 71–80. https://doi.org/10.1192/bjp.2017.28.

Wijkstra, J., Lijmer, J., Balk, F.J., Geddes, J.R., Nolen, W.A., 2006. Pharmacological treatment for unipolar psychotic depression: systematic review and meta-analysis. Br. J. Psychiatry 188 (MAY), 410–415. https://doi.org/10.1192/bjp.bp.105.010470.

Wijkstra, J., Burger, H., Van Den Broek, W.W., Birkenhäger, T.K., Janzing, J.G.E., Boks, M.P.M., Bruijn, J.A., Van Der Loos, M.L.M., Breteler, L.M.T., Ramaekers, G.M.G.I., Verkes, R.J., Nolen, W.A., 2010. Treatment of unipolar psychotic depression: a randomized, double-blind study comparing imipramine, venlafaxine, and venlafaxine plus quetiapine. Acta Psychiatr. Scand. 121 (3), 190–200. https://doi.org/10.1111/j.1600-0447.2009.01464.x.

Wilkinson, S.T., Ballard, E.D., Bloch, M.H., Mathew, S.J., Murrough, J.W., Feder, A., Sos, P., Wang, G., Zarate Jr., C.A., Sanacora, G., 2018. The effect of a single dose of intravenous ketamine on suicidal ideation: a systematic review and individual participant data meta-analysis. Am. J. Psychiatr. 175 (2), 150–158. https://doi.org/10.1176/appi.ajp.2017.17040472.

Witt, K., Potts, J., Hubers, A., Grunebaum, M.F., Murrough, J.W., Loo, C., Cipriani, A., Hawton, K., 2020. Ketamine for suicidal ideation in adults with psychiatric disorders: a systematic review and meta-analysis of treatment trials. Aust. N. Z. J. Psychiatry 54 (1), 29–45. https://doi.org/10.1177/0004867419883341.

Zalpuri, I., Rothschild, A.J., 2016. Does psychosis increase the risk of suicide in patients with major depression? A systematic review. J. Affect. Disord. 198, 23–31. https://doi.org/10.1016/j.jad.2016.03.035.

Zaninotto, L., Guglielmo, R., Calati, R., Ioime, L., Camardese, G., Janiri, L., Bria, P., Serretti, A., 2015. Cognitive markers of psychotic unipolar depression: a meta-analytic study. J. Affect. Disord. 174, 580–588. https://doi.org/10.1016/j.jad.2014.11.027.

Zarrinnegar, P., Kothari, J., Cheng, K., 2019. Successful use of ketamine for the treatment of psychotic depression in a teenager. J. Child Adolesc. Psychopharmacol. 29 (6), 472–473. https://doi.org/10.1089/cap.2019.0028.

Chapter 25

Evidence-based psychotherapy for treatment-resistant depression

Jesse H. Wright[a], John C. Markowitz[b], Jennifer Wood[a], Jessica Reis[a], and Tracy D. Eells[a]

[a]*Department of Psychiatry and Behavioral Sciences, University of Louisville, Louisville, Kentucky,* [b]*Department of Psychiatry at Columbia University, New York City, NY, United States*

Introduction

There is strong evidence from a large number of studies that several psychotherapies are as effective for depression as pharmacotherapy (Cuijpers et al., 2011a, 2020). Moreover, a combined approach employing both psychotherapy and pharmacotherapy typically provides better outcomes than either treatment alone (Cuijpers et al., 2011a, 2020). Yet there has been limited research on psychotherapies for treatment-resistant depression (TRD), and categorization systems for TRD rarely have included psychotherapy, despite its potential value in reducing symptoms and in directly addressing many of the common problems in TRD. As this chapter illustrates, psychotherapies can treat inherent TRD difficulties such as demoralization, hopelessness, complicated grief, interpersonal conflicts and deficits, behavioral deactivation and inertia, and entrenched low self-esteem. The lack of attention to psychotherapy for TRD may be due, in part, to the concept of treatment resistance having developed within psychopharmacology, which narrowly defined it as failure to respond to antidepressant medications.

We begin this chapter with a brief review of treatment guidelines and metaanalyses that support use of psychotherapy for depression in general before turning to evidence for the effectiveness of psychotherapy in TRD. After capsule summaries of several commonly used psychotherapies and their application for depression that has not responded to previous treatments, we propose possible criteria for the adequacy of a psychotherapy trial for TRD.

Effectiveness of psychotherapy for depression

Treatment guidelines of the American Psychiatric Association (2010) and the American Psychological Association (2018) recommend several evidence-based psychotherapies for depression (Table 25.1).

Effectiveness of psychotherapy for TRD

The overall findings of randomized, controlled trials of psychotherapy for TRD have been examined in systematic reviews and metaanalyses (van Bronswijk et al., 2019; Li et al., 2018; Ijaz et al., 2018). Drawing from these analyses and from our literature search, we summarize the current evidence base for psychotherapy for TRD in Table 25.2.

In a metaanalysis of multiple psychotherapies for TRD, van Bronswijk et al. (2019) examined seven different approaches in 21 trials. Because they included 10 studies of chronic depression which may have included patients who did not meet stated criteria for TRD, the results are not specific to depression refractory to antidepressant medication. Also, definitions of TRD, and of comparators such as treatment as usual (TAU), varied from trial to trial. Nevertheless, this metaanalysis reported a moderate overall effect size of 0.42 for addition of psychotherapy to treatment as usual (TAU) compared to TAU alone. The most frequently studied psychotherapy in this metaanalysis was CBT ($N = 7$). All studies of CBT were performed with patients who met criteria for TRD. The relative effects of the different therapies were not reported.

A separate metaanalysis of CBT for TRD included 6 randomized controlled trials with a total of 847 participants (Li et al., 2018). Four of the studies used standard CBT or a variant tailored to treatment-resistant depression, while two investigations utilized mindfulness-based CBT (MCBT). The blending of approaches makes this analysis difficult to interpret. However, the results favored CBT compared to a control group (most commonly TAU). The pooled standard mean difference was -0.42, indicating that CBT was superior in reducing depression symptoms immediately after intervention.

Managing Treatment-Resistant Depression. https://doi.org/10.1016/B978-0-12-824067-0.00025-6

TABLE 25.1 Effectiveness of psychotherapy for depression.

Psychotherapy	American Psychiatric Association Treatment Guidelines 2010	American Psychological Association Clinical Practice Guideline 2019	Cuijpers et al. Metaanalysis 2011
Cognitive-behavioral therapy (CBT)	Recommended	Recommended for adolescent and adult patients (individual format) Conditionally recommended for older adult patients (individual format) Recommended for older adult patients (group format)	*N*[a] = 91 ES = 0.67
Interpersonal psychotherapy (IPT/IPT-A)	Recommended	Recommended for adult and adolescent patients	*N* = 16 ES = 0.63
Short-term psychodynamic therapy	Recommended	Recommended for adult patients	*N* = 5 ES = 0.69
Problem-solving therapy (PST)	Recommended	Recommended for adult patients	*N* = 13 ES = 0.83
Supportive psychotherapy	Not included	Recommended for adult patients	*N* = 14 ES = 0.57
Mindfulness-based cognitive therapy (MBCT)	Not included	Recommended for adult patients	Not included
Behavioral activation (BA)	Not included	Not included	*N* = 10 ES = 0.87
Self-control therapy	Not included	Not included	*N* = 6 ES = 0.45
Acceptance and commitment therapy (ACT)	Not included	Not included	Not included
Cognitive-behavioral analysis system of psychotherapy (CBASP)	Not included	Insufficient evidence for adult patients	Not included
Emotion focused therapy for depression	Not included	Not included	Not included
Reminiscence/life review therapy	Not included	Recommended for use with older adult patients (group format)	Not included

[a] *N, number of studies included in this metaanalysis; ES, mean effect size.*

TABLE 25.2 Evidence for psychotherapy for TRD.

Psychotherapy	Evidence base for effectiveness for TRD
Cognitive-behavior therapy	+++
Interpersonal psychotherapy	0/+
Brief psychodynamic psychotherapy	+
Problem-solving therapy	0
Supportive psychotherapy	0
Mindfulness-based cognitive therapy	+++

Key: 0, no randomized controlled trials (RCTs) for TRD; +, 1 RCT for TRD showing beneficial effect; +++, three or more RCTs for TRD showing beneficial effect.

Where remission data were reported, CBT had a significantly higher remission rate (25.8%) than the control condition (14.2%).

Among studies of standard CBT or CBT tailored for TRD (Wiles et al., 2013; Nakagawa et al., 2017; Watkins et al., 2011; Rush et al., 2006; Thase et al., 2007; Hauksson et al., 2017; Minelli et al., 2019), the largest investigation was the CoBalT trial (Wiles et al., 2013) (N=469)—a comparison of CBT plus TAU versus TAU alone in primary care that found a robust advantage for CBT. CBT response rates more than doubled (46.1%) those of TAU (21.6%). A smaller trial of rumination-focused CBT (Watkins et al., 2011) also found significantly better outcome in CBT than TAU. One study of different treatment formats found that individual CBT, but not group CBT, led to significantly better improvement in TRD than TAU (Hauksson et al., 2017). Taken together, studies of CBT for TRD indicate that this widely used psychotherapy is an effective intervention for depression resistant to antidepressant medication.

Although interpersonal psychotherapy (IPT) has been studied extensively for MDD (Cuijpers et al., 2011b, 2020), only one RCT has tested IPT specifically for TRD. Souza et al. (2016) recruited 40 adults with TRD in a tertiary care setting in Brazil and randomly assigned them to adjunctive IPT (n=17) in addition to pharmacotherapy (TAU), or to TAU alone (n=23). The definition of TRD was modest: "failure to respond to one antidepressant medication in adequate dose [equivalent of amitriptyline ≥75 mg] and duration [≥4 weeks]." In a study the authors acknowledge was underpowered, both groups significantly improved over time (IPT+TAU, d=0.93; TAU, d=0.73), without significant between-group differences in ratings on the Hamilton Rating Scale for Depression. Response (IPT+TAU=35.5%; TAU=22.2%) and remission (IPT+TAU=28.6%; TAU=16.7%) rates also did not significantly differ.

Another approach to assessing the value of IPT for TRD is to consider the related entity of persistent depressive disorder. Patients with this condition report deeply engrained symptoms and outlook (Angst et al., 2009), and many, if not most, prove hard to treat and thus may qualify for TRD (Eisendrath et al., 2016). Because IPT was originally designed to address a current life crisis in acute major depression, Markowitz and associates developed an adaptation (IPT-D) to treat what DSM-IV then termed dysthymic disorder (Markowitz, 1998). IPT-D lowered symptoms in a randomized trial of patients with "pure" DSM-IV dysthymic disorder (i.e., with chronic mild but not "double" depression) (Markowitz et al., 2005), but fared less well than sertraline or combined IPT+sertraline, and roughly as well as an active brief supportive therapy comparator. This outcome supports a conclusion that many TRD patients are most likely to benefit from the combination of a vigorous pharmacotherapy trial in conjunction with an evidence-based psychotherapy (van Bronswijk et al., 2019).

Brief psychodynamic psychotherapy has been studied in one trial for TRD. Town et al. (2017) compared the effectiveness of Intensive Short-Term Dynamic Psychotherapy (ISTDP) (Davanloo, 1987) with TAU in 60 patients whose depression did not remit after at least one antidepressant course. Among the participants who enrolled in the study, at least 34% had failed two or more pharmacotherapy trials from two distinct antidepressant medication classes for the current depressive episode. After a maximum 20-session course of treatment, analyses showed that change in depressive symptoms over time was significantly greater in the ISTDP group than in TAU. Relative to TAU, patients who received ISTDP were significantly more likely at the 6-month follow-up to have remitted (36.0% vs 3.7%) or partially remitted (48.0% vs 18.5%).

Long-term psychoanalytic treatment also has been investigated for TRD. Fonagy et al. (2015) randomly assigned 129 primary care patients either to long-term psychoanalytic psychotherapy plus TAU or to TAU alone. Treatment lasted up to 18 months. Treatment resistance was defined as having two failed treatment attempts, one of which must have been an antidepressant medication and the other was either an antidepressant medication or a psychological intervention. Remission rates were low in both conditions at the end of treatment (9.4% for those receiving psychodynamic treatment vs 6.5% for those in TAU only), as well as at a 42-month follow-up (14.9% vs 4.4%). Partial remission rates did not significantly differ at the end of treatment (32.1% vs 23.9%) but did favor psychoanalytic treatment at 24-month (38.8% vs 19.2%) and 42-month (30.0% vs 4.4%) follow-up assessments. These authors suggest that long-term psychoanalytic psychotherapy may be useful in improving the long-term outcome of treatment-resistant depression and that end-of-treatment evaluations or short follow-ups may miss the emergence of delayed therapeutic benefit. One additional study has been conducted on long-term psychoanalytic therapy for long-standing depression (Leuzinger-Bohleber et al., 2019), but did not specifically define TRD or require it for inclusion.

We found no randomized, controlled trials of problem-solving therapy (PST) or supportive therapy (SP) for TRD. Although SP was included in the only metaanalysis (van Bronswijk et al., 2019) on psychotherapy effectiveness for TRD, the results should be interpreted with caution because study inclusion criteria required chronic depression, not TRD. In the REVAMP trial (Kocsis et al., 2009), patients with chronic depression (defined as having MDD plus persistent depressive symptoms for more than 2 years) were randomized to either receive 12 weeks of continued pharmacotherapy plus CBASP, continued pharmacotherapy plus SP, or continued pharmacotherapy alone. Outcomes did not differ between the two psychotherapy options or medication management alone. Another RCT (Schramm et al., 2017) compared CBASP and SP in patients with early-onset (before age 21) chronic depression who were not taking antidepressant medication. Of note, most participants expressed a strong preference for psychotherapy over medication. Patients receiving CBASP reported significantly less depressive symptoms than those in SP at the end of the acute phase; however, the actual differences were quite small and barely reached statistical significance. Both CBASP and SP led to significant improvements in depressive symptoms and quality of life.

There have been three RCTs of mindfulness-based cognitive therapy (MBCT) for TRD (Chiesa et al., 2015; Eisendrath et al., 2016; Cladder-Micus et al., 2018). A study of 43 patients nonresponsive to antidepressant medication found that depressive improvement was significantly greater for MBCT than a psychoeducational control group (Chiesa et al., 2015). In a much larger study ($N=173$), Eisendrath et al. (2016) reported that MBCT yielded better outcomes than a health enhancement program (physical fitness, music therapy, and nutritional counseling) in reducing depressive symptoms and improving treatment response rates, but not remission. In contrast, another trial (Cladder-Micus et al., 2018) found that MBCT plus TAU did not outperform TAU on the primary outcome measure, depressive symptoms; yet produced significantly higher remission rates than TAU alone. MBCT also performed better on measures of quality of life, mindfulness skills, self-compassion, and rumination. The overall results of these three RCTS of MBCT indicate a positive benefit for this form of psychotherapy for TRD.

Psychotherapy methods for TRD

Cognitive-behavior therapy

The most studied psychotherapy for depression (Cuijpers et al., 2011a, 2020) and TRD (van Bronswijk et al., 2019; Li et al., 2018; Wiles et al., 2013; Nakagawa et al., 2017; Watkins et al., 2011; Rush et al., 2006; Thase et al., 2007; Hauksson et al., 2017; Minelli et al., 2019), cognitive-behavior therapy (CBT) has become a standard, widely available treatment. CBT methods are anchored in a collaborative-empirical therapeutic relationship in which therapists actively engage patients in a process of detecting and modifying maladaptive cognitions and reversing dysfunctional behavioral patterns (Wright et al., 2017). The treatment is usually short-term (12–20 sessions), but briefer and longer courses of therapy have been studied and found to be useful (Wright et al., 2017).

Because patients with TRD may have often tried many previous treatment options and feel demoralized about the chances of response to yet another treatment, the CBT approach to TRD typically focuses directly on hopelessness and treatment burnout (Wright, 2003). With in-session demonstrations of the value of spotting and changing depressive thoughts and behaviors, and use of action plans for between sessions, therapists try to stimulate hope that progress is possible. Standard CBT methods such as identifying and changing negative automatic thoughts, behavioral activation, and problem solving are used, as in the treatment of acute depression. However, TRD is conceptualized as a deeply ingrained condition that may require higher levels of activity from the therapist, especially early in treatment. There is an emphasis on rehabilitation of patients who have lost much of their "conditioning" to work effectively, engage in interpersonal relationships, and enjoy pleasurable activities. Thus, behavioral methods such as activity scheduling and step-by-step plans to manage effortful tasks often have a central role in treatment of patients with TRD (Wright, 2003). By gradually increasing activity level, social engagement, and involvement in pleasurable activities, the patient with TRD is participating in a process similar to physical therapy. CBT for long-standing depression also emphasizes a search for the patient's strengths and positive features that have been obscured by chronic negativism but offer potential in a path toward recovery.

Another feature of CBT for TRD is the use of a comprehensive, multidimensional treatment plan. If both CBT and pharmacotherapy are delivered by a psychiatrist, the integration of both treatments is a fundamental principle of treatment (Wright et al., 2010). When split treatment is delivered by a physician and a nonmedical therapist, effective communication and coordination is recommended (Wright et al., 2010). The CBT elements of therapy can facilitate pharmacotherapy and other methods, such as engaging in physical exercise and getting full medical evaluations and treatment, by addressing hopelessness and the inertia of TRD that has prevented the patient from taking actions that could help reduce depression. Nonadherence to pharmacotherapy is a particularly important target of CBT for treatment-resistant depression. The cognitive-behavioral therapist can elicit automatic thoughts and core beliefs that may be interfering with adherence and can devise behavioral interventions to increase the likelihood that medications and other treatments will be used to their full potential (Wright et al., 2010).

Interpersonal psychotherapy

Interpersonal psychotherapy (IPT) is a manualized, time-limited treatment focusing on life events, affects, and their connection (Weissman et al., 2018). This treatment method offers the patient two general principles: (1) depression is a debilitating but treatable medical illness that is not the patient's fault; and (2) depression does not arise in a vacuum but has a social context: upsetting life events evoke negative emotions and can precipitate mood episodes in vulnerable individuals; and depressed mood and functioning induce negative life events (Weissman et al., 2018). The treatment defines a life crisis—*complicated bereavement* following the death of a significant other; a *role dispute* with a significant other; a major life event *role transition*; or interpersonal *isolation*—for the patient to address within a brief time frame (generally 12–16 weeks), but assigns no formal homework. This time-limited approach helps patients to understand their depression in a social environmental context and, rather than avoiding painful negative effects (Markowitz and Milrod, 2011), to recognize them as useful social signals in order to resolve the painful life situation. In doing so, patients mobilize protective social supports and learn new social skills.

Although there has been no specific adaptation of IPT for TRD, IPT has been adapted for chronic depression (IPT-D) (Markowitz, 1998; Markowitz et al., 2005). Beyond the standard IPT approach, some features of IPT-D may benefit patients with TRD. IPT-D focuses on the patient's inevitable difficulty in distinguishing between chronic, often lifelong depression and his or her personality. These patients may see themselves as defective, depressive, hopeless, even toxic people whom no one can really like, leading them to maintain distance from others. The goal of therapy is, in part, an "iatrogenic role transition," engineered by the therapy, to help the patient distinguish a healthy self beneath the mood syndrome and to recognize pessimism, resignation, and a sense of defectiveness as longstanding depressive symptoms, not who the patient actually is (Kocsis et al., 2009).

Because chronically depressed patients tend to be passive, conflict-avoidant, and uncomfortable with strong affects, IPT-D focuses on eliciting and validating negative effects such as anger, helping the patient to understand them as appropriate signals in reaction to mistreatment or unfairness in interpersonal encounters, rather than as "bad" emotions. Patients with persistent depression need encouragement and repeated role play to practice confronting others, something they may have avoided throughout life even when not clinically depressed. Discovering that one can own and use one's feelings to set boundaries with others can be a liberating and mood-enhancing experience, providing mastery over one's environment and a sense of personal control.

Another crucial aspect of IPT for patients with TRD is emphasis on the medical model. They are not defective themselves, but individuals with a tough illness that has not responded to prior treatments. They have not "failed the treatments," the treatments have failed them. It's important for the therapist to generate realistic therapeutic optimism that this next course, taking a very different, interpersonally focused approach from past pharmacotherapy or psychotherapy trials, may be the one that relieves them of TRD.

Psychodynamic psychotherapy

Although the theoretical underpinnings of contemporary psychodynamic treatment for depression vary widely (Busch et al., 2016), most treatments focus on how past experiences and conflictual relationships affect current functioning, including how the relationship between the patient and the therapist reflects these conflicts. Psychodynamic treatments also aim to facilitate insight and address how patients avoid difficult topics. Themes that tend to emerge include the impact of perceived or actual losses, conflict about the expression of anger, and avoidance of shame (Busch et al., 2016).

Short-term psychodynamic treatment for depression may include approximately 16–20 sessions, whereas long-term treatment may include 50 or more sessions. Regardless of the specific variant of psychodynamic treatment employed, shorter treatments tend to focus on symptom relief, whereas long-term treatment emphasizes changing core personality features and maladaptive interpersonal patterns.

The single completed trial of brief psychodynamic therapy for TRD used a manualized form of this treatment that emphasizes emotional factors that lead to, exacerbate, and perpetuate depression, as well as increasing the patient's tolerance for anxiety (Davanloo, 1987). The trial report did not discuss adaptations for persons who had not responded to other interventions for depression (Town et al., 2017). Although we found no work focused on specific adaptations of psychodynamic treatment to TRD, the guiding assumption may be that exploration of internal conflicts and unconscious issues are not addressed in other psychosocial treatment modalities or in pharmacotherapy, and thus may be used to facilitate change.

Problem-solving therapy

This evidence-based, CBT-related psychosocial intervention focuses on training in adaptive problem-solving attitudes and skills (Bell and D'Zurilla, 2009). Sometimes referred to as structured problem solving, PST is a prescriptive model for solving and effectively coping with major and minor life stressors. Its two essential components are: (1) general orientation, and (2) problem-solving skills. Developing a problem-solving orientation is a meta-cognitive process that teaches individuals to see problems as challenges to be solved and thus serves a motivational function. Problem-solving skills are developed by working through real-life stressors with a structured four-pronged approach: (1) problem definition and formulation, (2) generation of alternative solutions, (3) decision-making strategies, and (4) solution implementation and verification (Nezu et al., 2013).

Although no randomized controlled studies have evaluated the effectiveness of PST in treatment-resistant depression, PST experts believe this approach can be adapted for this population. A positive problem-solving orientation can be used to build hope and view treatment resistance and core symptomatology as workable problems. Skill building in problem solving may thus focus on dealing with particular difficulties associated with not getting better. For example, the therapist and patient might work on identifying strategies to better handle disappointment, decreased motivation, and associated feelings of distress. Other common challenges in TRD such as interpersonal conflict, job loss, and financial stress can also be targeted in problem-solving therapy.

Supportive psychotherapy

Supportive psychotherapy (SP) is often thought of as an ambiguous therapeutic entity in which the therapist provides empathy, kindness and encouragement with the aim of supporting patients in an atmosphere of safety through the symptoms of psychiatric illness or life struggles. It is typically poorly defined, not implemented methodically, and is rarely investigated as the primary treatment intervention for any psychiatric disorder (Markowitz, 2014). Although many clinicians say they provide supportive psychotherapy, they may be offering an eclectic, potentially confusing amalgam of various therapeutic techniques. In its best form, SP is a specific therapeutic approach based on a solid theoretical framework (Markowitz, 2014; Grover et al., 2020; Harari, 2014).

Supportive psychotherapy can be a manualized, affect-focused approach that typically involves identifying and reinforcing mature defense mechanisms, strengthening reality testing, encouraging use of adaptive coping mechanisms, and bolstering self-esteem, all within a holding environment that fosters trust, structure, and physical and emotional safety (Markowitz, 2014; Grover et al., 2020). Specific techniques may include reflecting the patient's concerns, helping the patient name conscious emotions, highlighting the patient's strengths, providing reassurance, and gentle reframing.

Several manualized versions of SP have been developed including Brief Supportive Psychotherapy, a time-limited affect-focused treatment focusing on the "common factors" inherent to most forms of psychotherapy including empathy, therapeutic alliance, and optimism for improvement; and Short Psychodynamic Supportive Psychotherapy, a time limited approach rooted in psychoanalytic theory and technique but focusing on affective and relational content from the supportive side of the supportive-exploratory continuum. These versions offer uniform and replicable forms of SP for both clinical and research use (Markowitz, 2014; de Maat et al., 2008).

Despite the lack of empirical research for supportive psychotherapy in TRD, other considerations may lend credence to the use of this approach. It is well known that comorbid diagnoses, of which personality disorders, substance use, and medical illness are common in the TRD population (Bennabi et al., 2019; Holtzheimer, 2010), may influence outcome. Though not studied directly with TRD, evidence supports the efficacy of SP in depressive disorders with comorbid personality disorders (Kool et al., 2003; Koppers et al., 2019), alcohol use (Markowitz et al., 2008), and medical/neurologic illness including breast cancer (Blanco et al., 2019), diabetic foot syndrome (Simson et al., 2008), and TBI (Ashman et al., 2014). In addition, SP may be more acceptable for patients with severe symptoms or marked functional impairment. Harari (2014) postulates that SP is specifically helpful for patients with chronic or recurring disability because the goals of minimizing deterioration and maximizing competence can be performed within the limitations of the illness, almost irrespective of severity. Supportive psychotherapy does not require the insight, curiosity or motivation necessary in other modalities and may be more tolerable than other forms of therapy which focus on exploration of the unconscious, challenging firmly held beliefs, addressing avoidance behaviors, or completing homework (Grover et al., 2020).

Finally, there is the benefit of accessibility. In a survey of psychiatrists, SP was the most commonly used psychotherapy with 36% saying they provided it in their practice (Tanielian et al., 2001). Though the definition of SP was vague, usage patterns have likely changed since this survey was conducted, and therapists' subjective self-report of their treatment have limited reliability, SP still may be more likely to be available in clinical practice than other forms of psychotherapies for TRD.

Mindfulness-based cognitive-behavior therapy

When first developed for mood disorders, mindfulness-based cognitive therapy (MBCT) was applied only for relapse prevention in patients who had little or no active depression. The rationale was that persons with symptoms such as psychomotor agitation, poor concentration, and low energy would have difficulty participating in mindful meditation (Segal et al., 2002; Sipe and Eisendrath, 2014). Subsequently, MCBT was adapted for the treatment of acute depression and found to be effective for short-term symptom relief (Wang et al., 2018). The key differences between standard CBT and MBCT are the use of mindfulness practices in the latter to increase awareness and attentional control. Instead of attempting to question ruminative, self-condemning thoughts and modify them as in standard CBT, patients are taught to disengage from such thoughts as a way of interrupting ruminative, depression-inducing thinking (Sipe and Eisendrath, 2014).

Modifications of MCBT for treating persons with TRD have been described by Sipe and Eisendrath (2014) who recommend that mindfulness be used to help patients focus on "noticing and accepting that which is in the present moment." Patients learn that their thoughts and feelings are "mental events rather than aspects of self or direct reflections of truth." Because patients with TRD have often withdrawn from activities that can offer positive reinforcement, MBCT also teaches skills to help patients engage in constructive, positive behaviors (Sipe and Eisendrath, 2014).

Although MBCT for TRD is based on a standard manual for depression (Segal et al., 2013), several modifications are recommended (Sipe and Eisendrath, 2014). Psychoeducation is provided on the natural course of depression with an emphasis on avoiding self-blame (Sipe and Eisendrath, 2014). Another key modification is reduced reliance on sitting meditation that patients with TRD may have difficulty in sustaining. Mindful movement such as walking meditation, yoga, and brief "breathing spaces" can take the place of longer sitting meditations when needed. Metaphors and exercises drawn from acceptance and commitment therapy are used to help patients effectively use mindfulness meditation and to distinguish between the useful process of acceptance and the negative effects of resignation (Sipe and Eisendrath, 2014).

Definition of adequate treatment with psychotherapy

Clinical researchers assessing treatment resistance and defining TRD often attempt to define an adequate course of pharmacotherapy and the needed number of failed medication trials (Segal et al., 2013; Sackeim, 2001; Thase and Rush, 1997; Souery et al., 1999; Fava, 2003; Fekadu et al., 2009). Most such systems have not considered psychotherapy in categorizing the level of treatment resistance (Sackeim, 2001; Thase and Rush, 1997; Souery et al., 1999; Fava, 2003; Fekadu et al., 2009; Conway et al., 2017). Nor has defining the adequacy of a course of psychotherapy received sufficient attention. Such a definition would be useful in both TRD research and in treatment planning with patients with depression unresponsive to previous pharmacotherapy trials. Although constructing such a definition is more complex than for psychopharmacotherapy, we outline an approach here.

We cannot absolutely define an adequate psychotherapy course because numerous factors such as psychiatric comorbidity, interpersonal problems, medical illnesses, and presence or absence of personal resources may influence the need for intensity and duration of treatment. Different therapies have also been tested at different durations and doses. Limited evidence hints that patients may require varied amounts of psychotherapy to achieve remission (Shapiro et al., 1994; Molenaar et al., 2011; Dekker et al., 2005; Cuijpers et al., 2013). However, studies that compared varying lengths of CBT or short-term psychodynamic therapy (8 vs 16 sessions) for depression reported no advantage for increased duration of treatment (Shapiro et al., 1994; Molenaar et al., 2011; Dekker et al., 2005; Cuijpers et al., 2013). Perhaps the most telling report on the influence of length of therapy was a meta-regression analysis of 70 studies which discovered no significant relationships between number of psychotherapy sessions and treatment outcome (Cuijpers et al., 2013).

The mean number of sessions actually completed in psychotherapy RCTs may yield the best available information on how many sessions are needed for a reasonable trial. Because a Cochrane review of psychotherapy for TRD (Ijaz et al., 2018) reported the mean number of sessions completed were 9–15 for CBT, 11 for IPT, and 16 for short-term dynamic therapy, we conclude that a range of 9–16 sessions may constitute an adequate course of psychotherapy. Longer or shorter treatment courses may be appropriate depending on treatment type, complexity of comorbid illnesses, and other potential influences on treatment outcome.

A definition of adequacy of a psychotherapy trial could include whether or not the therapist has been trained in the approach, follows the basic methods for the respective treatment, and is competently delivering therapy. For some therapies such as CBT and IPT, certification is available, and measures of competence have been developed (Wright et al., 2017). Yet such standards may set an impractically high bar for delivering evidence-based psychotherapy to large numbers of patients. Lacking measures of training adequacy and therapist competence, clinicians assessing patients' prior treatment experience can ask targeted questions to determine if the patient actually did participate in a therapy such as CBT, IPT, or MBCT. Questions could include: "What did you learn in the therapy?" "What methods did the therapist use?" "Did you do 'homework' between sessions?" "Can you describe the key elements of the therapy?" "What principles of the therapy are you using now?"

Determination of psychotherapy adequacy also may be influenced by the large number of computer-assisted CBT (CCBT) studies, which greatly reduced the amount of clinician time without impairing treatment effectiveness (Richards and Richardson, 2012; Wright et al., 2019; Thase et al., 2018; Carlbring et al., 2017). Investigations comparing CCBT with standard face-to-face CBT for depression have found no differences in outcome (Thase et al., 2018; Carlbring et al., 2017). In addition CCBT has performed as well as standard CBT in patients with chronic or severe depression (Thase et al., 2018). Engagement in CCBT provides patients a degree of standardization and "quality control" of therapy that may be absent from traditional psychotherapy. Patients participate in online exercises and skill-building practices that provide core CBT learning experiences in a reliable and measurable way (Wright and Mishkind, 2020).

Conclusion

Psychotherapy has been omitted from most categorization systems for TRD, yet a strong argument can be made for considering psychotherapy for persons who haven't responded to other treatments and withholding the concept of treatment-resistance until they have undergone a course of evidence-based antidepressant psychotherapy. Principal reasons to consider psychotherapy are: (1) solid evidence for effectiveness of many psychotherapies for depression; (2) demonstrated effectiveness of several psychotherapies for TRD; and (3) the possibility that psychotherapy may have selective benefits for certain patients.

This chapter detailed some of the core methods of various psychotherapies (CBT, interpersonal psychotherapy, brief psychodynamic therapy, problem-solving therapy, supportive therapy, and mindfulness-based cognitive therapy) that can address common problems in TRD. Examples of problems that may not respond fully to yet another trial of pharmacotherapy include hopelessness, treatment burnout, behavioral deconditioning, chronic low self-esteem, interpersonal conflicts, losses, and deficits, and lack of mindfulness skills. We conclude from review of available evidence that an adequate trial of one of the short-term psychotherapies for depression should range from 9 to 16 sessions and be conducted by a clinician well-trained in the approach who can deliver the core elements of this form of psychotherapy. Computer-assisted psychotherapy can provide standardization of treatment delivery along with improved efficiency of therapy. Future clinical and research work on TRD should include psychotherapy as an important therapeutic option.

Disclosures

Dr. Jesse Wright is an author of the *Good Days Ahead* (GDA) computer program used in an investigation cited in this article, has an equity interest, and serves as a consultant to Mindstreet, Inc. developer and distributors of GDA. His conflict of interest is managed with an agreement with the University of Louisville. He receives book royalties from American Psychiatric Publishing, Inc., Guilford Press, and Simon and Schuster, and has received grant support from the Agency for Healthcare Research and Quality and the Oticon Foundation.

Dr. John Markowitz receives salary support from New York State Psychiatric Institute and book royalties from American Psychiatric Publishing, Basic Books, and Oxford University Press.

Dr. Tracy D. Eells receives book royalties from Guilford Press and the American Psychological Association.

Drs. Jessica Reis and Jennifer Wood have no disclosures to report.

References

American Psychiatric Association, 2010. Practice Guideline for the Treatment of Patients with Major Depressive Disorder, third ed. Available from: http://www.psychiatryonline.com/pracGuideTopic_7.aspx.

American Psychological Association, 2018. Clinical Practice Guideline for the Treatment of Depression in Children, Adolescents, and Young, Middle-Aged, and Older Adults, Draft. August 21. Available from: http://www.apa.org/about/offices/directorates/guidelines/clinical-practice.aspx.

Angst, J., Gamma, A., Rossler, W., Vladeta, A., Klein, D.N., 2009. Long-term depression versus episodic major depression: results from the prospective Zurich study of a community sample. J. Affect. Disord. 115 (1–2), 112–121. https://doi.org/10.1016/j.jad.2008.09.023.

Ashman, T., Cantor, J.B., Tsaousides, T., Spielman, L., Gordon, W., 2014. Comparison of cognitive behavioral therapy and supportive psychotherapy for the treatment of depression following traumatic brain injury: a randomized controlled trial. J. Head Trauma Rehabil. 29 (6), 467–478. https://doi.org/10.1097/HTR.0000000000000098.

Bell, A.C., D'Zurilla, T.J., 2009. Problem-solving therapy for depression: a meta-analysis. Clin. Psychol. Rev. 29 (4), 348–353. https://doi.org/10.1016/j.cpr.2009.02.003. Epub 2009 Feb 26 19299058.

Bennabi, D., Charpeaud, T., Yrondi, A., Genty, J.-B., Destouches, S., et al., 2019. Clinical guidelines for the management of treatment-resistant depression: French recommendations from experts, the French Association for Biological Psychiatry and Neuropsychopharmacology and the fondation FondaMental. BMC Psychiatry 19, 262. https://doi.org/10.1186/s12888-019-2237-x.

Blanco, C., Markowitz, J.C., Hellerstein, D.J., Nezu, A.M., Wall, M., et al., 2019. A randomized trial of interpersonal psychotherapy, problem solving therapy, and supportive psychotherapy for major depressive disorder in women with breast cancer. Breast Cancer Res. Treat. 173 (2), 353–364. https://doi.org/10.1007/s10549-018-4994-5.

Busch, F., Rudden, M., Shapiro, T., 2016. Psychodynamic Treatment of Depression, second ed. American Psychiatric Association Publishing, Arlington, VA.

Carlbring, P., Andersson, G., Cuijpers, P., Riper, H., Hedman-Lagerlof, E., 2017. Internet-based vs. face-to-face cognitive behavior therapy for psychiatric and somatic disorders: an updated systematic review and meta-analysis. Cogn. Behav. Ther. 47 (1), 1–18. https://doi.org/10.1080/16506073.2017.1401115.

Chiesa, A., Castagner, V., Andrisano, C., Serretti, A., Mandelli, L., et al., 2015. Mindfulness-based cognitive therapy vs. psycho-education for patients with major depression who did not achieve remission following antidepressant treatment. Psychiatry Res. 226 (2–3), 474–483. https://doi.org/10.1016/j.psychres.2015.02.003.

Cladder-Micus, M.B., Speckens, A.E.M., Vrijsen, J.N., Donders, A.R.T., Becker, E.S., et al., 2018. Mindfulness-based cognitive therapy for patients with chronic, treatment-resistant depression: a pragmatic randomized controlled trial. Depress. Anxiety, 1–11. https://doi.org/10.1002/da.22788.

Conway, C.R., George, M.S., Sackeim, H.A., 2017. Toward an evidence-based, operational definition of treatment-resistant depression: when enough is enough. JAMA Psychiatry 74 (1), 9–10. https://doi.org/10.1001/jamapsychiatry.2016.2586.

Cuijpers, P., Andersson, G., Donker, T., van Straten, A., 2011a. Psychological treatment of depression: results of a series of meta-analyses. Nord J. Psychiatry 65 (6), 354–364. https://doi.org/10.3109/08039488.2011.596570.

Cuijpers, P., Geraedts, A.S., van Oppen, P., Andersson, G., Markowitz, J.C., et al., 2011b. Interpersonal psychotherapy of depression: a meta-analysis. Am. J. Psychiatry 168 (6), 581–592. https://doi.org/10.1176/appi.ajp.2010.10101411.

Cuijpers, P., Huibers, M., Ebert, D.D., Koole, S.L., Andersson, G., 2013. How much psychotherapy is needed to treat depression? A metaregression analysis. J. Affect. Disord. 149 (1–3), 1–13. https://doi.org/10.1016/j.jad.2013.02.030.

Cuijpers, P., Noma, H., Karyotaki, E., Vinkers, C.H., Cipriani, A., et al., 2020. A network meta-analysis of the effects of psychotherapies, pharmacotherapies and their combination in the treatment of adult depression. World Psychiatry 19 (1), 92–107. https://doi.org/10.1002/wps.20701.

Davanloo, H., 1987. Intensive short-term dynamic psychotherapy with highly resistant depressed patients: part I. Restructuring ego's regressive defenses. Int. J. Short-Term Psychother. 2, 99–132.

de Maat, S., Dekker, J., Schoevers, R., van Aalst, G., Gijsbers-van Wijk, C., et al., 2008. Short psychodynamic supportive psychotherapy, antidepressants, and their combination in the treatment of major depression: a mega-analysis based on three randomized clinical trials. Depress. Anxiety 25 (7), 565–574. https://doi.org/10.1002/da.20305.

Dekker, J., Molenaar, P.J., Kool, S., Van Aalst, G., Peen, J., et al., 2005. Dose-effect relations in time-limited combined psycho-pharmacological treatment for depression. Psychol. Med. 35 (1), 47–58. https://doi.org/10.1017/s0033291704002685.

Eisendrath, S.J., Gillung, E., Delucchi, K.L., Segal, Z.V., Nelson, J.C., et al., 2016. A randomized controlled trial of mindfulness-based cognitive therapy for treatment-resistant depression. Psychother. Psychosom. 85 (2), 99–110. https://doi.org/10.1159/000442260.

Fava, M., 2003. Diagnosis and definition of treatment-resistant depression. Biol. Psychiatry 53 (8), 649–659. https://doi.org/10.1016/s0006-3223(03)00231-2.

Fekadu, A., Wooderson, S., Donaldson, C., Markopoulou, K., Masterson, B., et al., 2009. A multidimensional tool to quantify treatment resistance in depression: the Maudsley staging method. J. Clin. Psychiatry 70 (2), 177–184. https://doi.org/10.4088/jcp.08m04309.

Fonagy, P., Rost, F., Carlyle, J.A., McPherson, S., Thomas, R., et al., 2015. Pragmatic randomized controlled trial of long-term psychoanalytic psychotherapy for treatment-resistant depression: the Tavistock Adult Depression Study (TADS). World Psychiatry 14 (3), 312–321. https://doi.org/10.1002/wps.20267.

Grover, S., Avasthi, A., Jagiwala, M., 2020. Clinical practice guidelines for practice of supportive psychotherapy. Indian J. Psychiatry. 62 (Suppl. S2), S173–S182. Available from: https://www.indianjpsychiatry.org/text.asp?2020/62/8/173/276095.

Harari, E., 2014. Supportive psychotherapy. Australas. Psychiatry 22 (5), 440–442. https://doi.org/10.1177/1039856214546674.

Hauksson, P., Ingibergsdottir, S., Gunnarsdottir, T., Jonsdottir, I.H., 2017. Effectiveness of cognitive behaviour therapy for treatment-resistant depression with psychiatric comorbidity: comparison of individual versus group CBT in an interdisciplinary rehabilitation setting. Nord. J. Psychiatry 71 (6), 465–472. https://doi.org/10.1080/08039488.2017.1331263.

Holtzheimer, P.E., 2010. Advances in the management of treatment-resistant depression. Focus 8 (4), 488–500. https://doi.org/10.1176/foc.8.4.foc488.

Ijaz, S., Davies, P., Williams, C.J., Kessler, D., Lewis, G., et al., 2018. Psychological therapies for treatment-resistant depression in adults (review). Cochrane Database Syst. Rev. (5). https://doi.org/10.1002/14651858.CD010558.pub2, CD010558. www.cochranelibrary.com.

Kocsis, J.H., Gelenberg, A.J., Rothbaum, B.O., Klein, D.N., Trivedi, M.H., et al., 2009. Cognitive behavioral analysis system of psychotherapy (CBASP) and brief supportive psychotherapy for augmentation of antidepressant nonresponse in chronic depression: the REVAMP Trial. Arch. Gen. Psychiatry 66 (11), 1178–1188. https://doi.org/10.1001/archgenpsychiatry.2009.144.

Kool, S., Dekker, J., Duijsens, I.J., de Jonghe, F., Puite, B., 2003. Efficacy of combined therapy and pharmacotherapy for depressed patients with or without personality disorders. Harv. Rev. Psychiatry 11 (3), 133–141. https://doi.org/10.1080/10673220303950.

Koppers, D., Kool, M., Van, H., Driessen, E., Peen, J., Dekker, J., 2019. The effect of comorbid personality disorder on depression outcome after short-term psychotherapy in a randomized clinical trial. BJPsych. Open 5 (4), 1–9. https://doi.org/10.1192/bjo.2019.47.

Leuzinger-Bohleber, M., Hautzinger, M., Fiedler, G., Keller, W., Bahrke, U., et al., 2019. Outcome of psychoanalytic and cognitive-behavioural long-term therapy with chronically depressed patients: a controlled trial with preferential and randomized allocation. Can. J. Psychiatry 64 (1), 47–58. https://doi.org/10.1177/0706743718780340.

Li, J.-M., Zhang, Y., Su, W.-J., Liu, L.-L., Gong, H., et al., 2018. Cognitive behavioral therapy for treatment-resistant depression: a systematic review and meta-analysis. Psychiatry Res. 268, 243–250. https://doi.org/10.1016/j.psychres.2018.07.020.

Markowitz, J.C., 1998. Interpersonal Psychotherapy for Dysthymic Disorder. American Psychiatric Press, Washington, DC.

Markowitz, J.C., 2014. What is supportive psychotherapy? Focus 12 (3), 285–289. https://doi.org/10.1176/appi.focus.12.3.285.

Markowitz, J.C., Milrod, B., 2011. The importance of responding to negative affect in psychotherapies. Am. J. Psychiatry 168 (2), 124–128. https://doi.org/10.1176/appi.ajp.2010.10040636.

Markowitz, J.C., Kocsis, J.H., Bleiberg, K.L., Christos, P.J., Sacks, M.H., 2005. A comparative trial of psychotherapy and pharmacotherapy for "pure" dysthymic patients. J. Affect. Disord. 89 (1–3), 167–175. https://doi.org/10.1016/j.jad.2005.10.001.

Markowitz, J.C., Kocsis, J.H., Christos, P., Bleiberg, K., Carlin, A., 2008. Pilot study of interpersonal psychotherapy versus supportive psychotherapy for dysthymic patients with secondary alcohol abuse or dependence. J. Nerv. Ment. Dis. 196 (6), 468–474. https://doi.org/10.1097/NMD.0b013e31817738f1.

Minelli, A., Zampieri, E., Sacco, C., Bazzanella, R., Mezzetti, N., et al., 2019. Clinical efficacy of trauma-focused psychotherapies in treatment-resistant depression (TRD) in-patients: a randomized, controlled pilot-study. Psychiatry Res. 273, 567–574. https://doi.org/10.1016/j.psychres.2019.01.070.

Molenaar, P.J., Boom, Y., Peen, J., Schoevers, R.A., Van, R., et al., 2011. Is there a dose-effect relationship between the number of psychotherapy sessions and improvement of social functioning? Br. J. Clin. Psychol. 50 (3), 268–282. https://doi.org/10.1348/014466510X516975.

Nakagawa, A., Mitsuda, D., Sado, M., Takayuki, A., Daisuke, F., et al., 2017. Effectiveness of supplementary cognitive-behavioral therapy for pharmacotherapy-resistant depression: a randomized controlled trial. J. Clin. Psychiatry 78 (8), 1126–1135. https://doi.org/10.4088/JCP.15m10511.

Nezu, A.M., Nezu, C.M., D'Zurilla, T.J., 2013. Problem-Solving Therapy: A Treatment Manual. Springer Publishing, New York.

Richards, D., Richardson, T., 2012. Computer-based psychological treatments for depression: a systematic review and meta-analysis. Clin. Psychol. Rev. 32 (4), 329–342. https://doi.org/10.1016/j.cpr.2012.02.004.

Rush, A.J., Trivedi, M.H., Wisniewski, S.R., Nierenberg, A.A., Stewart, J.W., et al., 2006. Acute and longer-term outcomes in depressed outpatients requiring one or several treatment steps: a STAR*D report. Am. J. Psychiatry 163 (11), 1905–1917. https://doi.org/10.1176/ajp.2006.163.11.1905.

Sackeim, H.A., 2001. The definition and meaning of treatment-resistant depression. J. Clin. Psychiatry 62 (Suppl. 16), 10–17. 11480879.

Schramm, E., Kriston, L., Zobel, I., Bailer, J., Wambach, K., et al., 2017. Effect of disorder-specific vs nonspecific psychotherapy for chronic depression: a randomized clinical trial. JAMA Psychiat. 74 (3), 233–242. https://doi.org/10.1001/jamapsychiatry.2016.3880. 28146251.

Segal, Z.V., Williams, J.M.G., Teasdale, J.D., 2002. Mindfulness-Based Cognitive Therapy for Depression: A New Approach to Preventing Relapse. Guilford Publications, New York.

Segal, Z.V., Williams, J.M.G., Teasdale, J.D., 2013. Mindfulness-Based Cognitive Therapy for Depression, second ed. The Guilford Press, New York.

Shapiro, D.A., Barkham, M., Rees, A., Hardy, G.E., Reynolds, S., et al., 1994. Effects of treatment duration and severity of depression on the effectiveness of cognitive-behavioral and psychodynamic-interpersonal psychotherapy. J. Consult. Clin. Psychol. 62 (3), 522–534. https://doi.org/10.1037/0022-006x.62.3.522.

Simson, U., Nawarotzky, U., Friese, G., Porck, W., Schottenfeld-Naor, Y., et al., 2008. Psychotherapy intervention to reduce depressive symptoms in patients with diabetic foot syndrome. Diabet. Med. 25 (2), 206–212. https://doi.org/10.1111/j.1464-5491.2007.02370.x.

Sipe, W.E.B., Eisendrath, S.J., 2014. Mindfulness-Based Cognitive Therapy for Treatment-Resistant Depression. Mindfulness-Based Treatment Approaches. Elsevier Publishing, Amsterdam, https://doi.org/10.1016/B978-0-12-416031-6.00003-7.

Souery, D., Amsterdam, J., de Montigny, C., Lecrubier, Y., Montgomery, S., et al., 1999. Treatment resistant depression: methodological overview and operational criteria. Eur. Neuropsychopharmacol. 9 (1–2), 83–91. https://doi.org/10.1016/s0924-977x(98)00004-2.

Souza, L.H., Salum, G.A., Mosqueiro, B.P., Caldieraro, M.A., Guerra, T.A., et al., 2016. Interpersonal psychotherapy as add-on for treatment-resistant depression: a pragmatic randomized controlled trial. J. Affect. Disord. 193, 373–380. https://doi.org/10.1016/j.jad.2016.01.004.

Tanielian, T., Marcus, S., Suarez, A., Pincus, H., 2001. Datapoints: trends in psychiatric practice, 1988–1998: II. Caseload and treatment characteristics. Psychiatr. Serv. 52 (7), 880. https://doi.org/10.1176/appi.ps.52.7.880.

Thase, M.E., Rush, A.J., 1997. When at first you don't succeed: sequential strategies for antidepressant nonresponders. J. Clin. Psychiatry 58 (Suppl. 13), 23–29.

Thase, M.E., Friedman, E.S., Biggs, M.M., Wisniewski, S.R., Trivedi, M.H., et al., 2007. Cognitive therapy versus medication in augmentation and switch strategies as second-step treatments: a STAR*D report. Am. J. Psychiatry 164 (5), 739–752. https://doi.org/10.1176/ajp.2007.164.5.739.

Thase, M.E., Wright, J.H., Eells, T.D., Barrett, M.S., Wisniewski, S.R., et al., 2018. Improving the efficiency of psychotherapy for depression: computer-assisted versus standard CBT. Am. J. Psychiatry 175 (3), 242–250. https://doi.org/10.1176/appi.ajp.2017.17010089.

Town, J.M., Abbass, A., Stride, C., Bernier, D., 2017. A randomised controlled trial of Intensive Short-Term Dynamic Psychotherapy for treatment resistant depression: the Halifax Depression Study. J. Affect. Disord. 214, 15–25. https://doi.org/10.1016/j.jad.2017.02.035.

van Bronswijk, S., Moopen, N., Beijers, L., Ruhe, H.G., Peeters, F., 2019. Effectiveness of psychotherapy for treatment-resistant depression: a meta-analysis and meta-regression. Psychol. Med. 49 (3), 366–379. https://doi.org/10.1017/S003329171800199X. Epub 2018 Aug 24.

Wang, Y.-Y., Li, X.-H., Zheng, W., Xu, Z.-Y., Ng, C.H., et al., 2018. Mindfulness-based interventions for major depressive disorder: a comprehensive meta-analysis of randomized controlled trials. J. Affect. Disord. 229, 429–436. https://doi.org/10.1016/j.jad.2017.12.093.

Watkins, E.R., Mullan, E., Wingrove, J., Rimes, K., Steiner, H., et al., 2011. Rumination-focused cognitive-behavioural therapy for residual depression: phase II randomized controlled trial. Br. J. Psychiatry 199 (4), 317–322. https://doi.org/10.1192/bjp.bp.110.090282.

Weissman, M.M., Markowitz, J.C., Klerman, G.L., 2018. The Guide to Interpersonal Psychotherapy. Oxford University Press, New York.

Wiles, N., Thomas, L., Abel, A., Ridgway, N., Turner, N., et al., 2013. Cognitive behavioural therapy as an adjunct to pharmacotherapy for primary care-based patients with treatment resistant depression: results of the CoBalT randomized controlled trial. Lancet 381 (9864), 375–384. https://doi.org/10.1016/S0140-6736(12)61552-9.

Wright, J.H., 2003. Cognitive-behavior therapy for chronic depression. Psychiatr. Ann. 33, 777–784.

Wright, J.H., Mishkind, M., 2020. Computer-Assisted CBT and Mobile Apps for Depression: Assessment and Integration into Clinical Care. pp. 1–7, https://doi.org/10.1176/appi.focus.20190044. Focus.psychiatryonline.org.

Wright, J.H., Sudak, D., Turkington, D., Thase, M.E., 2010. High-Yield Cognitive-Behavior Therapy for Brief Sessions: An Illustrated Guide. American Psychiatric Press, Inc., Arlington, VA.

Wright, J.H., Brown, G.K., Thase, M.E., Basco, M.R., 2017. Learning Cognitive-Behavior Therapy: An Illustrated Guide, second ed. American Psychiatric Publishing, Inc, Arlington, VA.

Wright, J.H., Owen, J.J., Eells, T.D., Richards, D., Richardson, T., et al., 2019. Computer-assisted cognitive-behavior therapy for depression: a systematic review and meta-analysis. J. Clin. Psychiatry 80 (2). https://doi.org/10.4088/JCP.18r12188, 18r12188.

Chapter 26

Treatment-resistant bipolar depression

Alexandre Paim Diaz[a,b], Brisa S. Fernandes[a,b], Joao Quevedo[a,b,c,d], Marsal Sanches[a], and Jair C. Soares[a,b]

[a]*Center of Excellence on Mood Disorders, Faillace Department of Psychiatry and Behavioral Sciences, McGovern Medical School, The University of Texas Health Science Center at Houston (UTHealth), Houston, TX, United States,* [b]*Translational Psychiatry Program, Faillace Department of Psychiatry and Behavioral Sciences, McGovern Medical School, The University of Texas Health Science Center at Houston (UTHealth), Houston, TX, United States,* [c]*Neuroscience Graduate Program, The University of Texas MD Anderson Cancer Center UTHealth Graduate School of Biomedical Sciences, Houston, TX, United States,* [d]*Translational Psychiatry Laboratory, Graduate Program in Health Sciences, University of Southern Santa Catarina (UNESC), Criciúma, SC, Brazil*

Introduction

Depressive symptoms are present during almost 70% of the symptomatic period in bipolar disorder patients (BD) (Kupka et al., 2007). Overall, patients with BD spent more than one-third of the time with subsyndromal or syndromal depression (17.6% and 18.4%, respectively, for BD type I; 17.8% and 19.1%, respectively, for BD type II) (Kupka et al., 2007). Depression in BD is also highly associated with suicidality. Plans et al. (2019) did a systematic review of the relationship between suicide and BD diagnosis and reported rates 20–30 higher than for the general population. Genetic factors, psychiatric comorbidity, and predominant depressive polarity were among the associated factors (Plans et al., 2019). These clinical characteristics may contribute to the substantial economic burden associated with BD. Bessonova et al. estimated the economic burden associated with BD in the United States. The results pointed to more than $195 billion per year, with the indirect costs (lost work productivity for both parents and caregivers, and unemployment) accounting for 72%–80% of the total (Bessonova et al., 2020). For many patients, the burden of bipolar depression is compounded by well-known delays in accurately diagnosing bipolar disorders more broadly. For instance, Buoli et al. (2020) found an association between a longer duration of untreated illness and a depressive polarity of the first episode and more lifetime depressive episodes.

Different treatments are indicated for the management of bipolar depression (Yatham et al., 2018). However, there are far fewer evidence-supported therapeutic options for treating bipolar depression than for treating manic episodes. Moreover, about one-quarter to one-third of patients with bipolar depression lack response to two or more treatments, a crucial, unmet, clinical challenge in psychiatry (Mendlewicz et al., 2010; Fornaro et al., 2020; Parker and Graham, 2017). In this book chapter, we will discuss the concept of treatment-resistant bipolar depression (TRBD), its epidemiology and neurobiology, in addition to promising interventions, pharmacological and nonpharmacological, for its treatment.

Concept

The concept of treatment-resistant depression (TRD) has traditionally been grounded on unipolar depression. In a study published in the 60s, Haskovec and Rysanek (1967) investigated reserpine's role in treating "imipramine-resistant" patients. The concept of treatment-resistance considered in this study was "endogenous depression" with no improvement after at least 3 weeks of administration of imipramine, with daily doses up to 300 mg. The diagnosis of endogenous depression included: "prolonged pathological depressive emotional disposition, loss of interest and energy, psychomotor inhibition, sleep disturbances, anorexia with loss of weight, feeling of hopelessness, self-reproach, morning depression and an indication of suicidal tendencies." The outcome was defined according to a clinical scale that included seven symptoms of depression, taking into account how long the improvement was observed, and it was categorized as follows: "good and stable improvement"; "good improvement lasting several days"; "transient of no improvement" (Haskovec and Rysanek, 1967). Davidson et al. (1978) included patients with depression, according to the Feighner criteria, who "had failed to respond to conventional treatment" in a trial with a pharmacological combination compared with electroconvulsive therapy (ECT) (Davidson et al., 1978; Feighner et al., 1972). Despite the mentioned remission and response and the evaluation through psychometric scales, there is no objective definition of how the outcome was measured (Davidson et al., 1978). More straightforward inclusion criteria can be found in the study of De Montigny in which lithium was used for patients

Managing Treatment-Resistant Depression. https://doi.org/10.1016/B978-0-12-824067-0.00026-8

with depression who did not respond to tricyclic antidepressants after 3 weeks of treatment, meaning a decreased total scored in the Hamilton Depression Rating Scale (HDRS) less than 40% (De Montigny et al., 1981; Hamilton, 1960). MacEwan and Remick (1988) reported patients treated and followed in a Mood Disorder Clinic. The concepts the authors used are more closely aligned with what current studies usually consider, although their remission criteria, if used today, would be classified instead as response (50% decrease in the depression scores from the previous evaluation) (MacEwan and Remick, 1988) (Table 26.1).

More recently, the concept of resistance as nonresponse to at least two trials of antidepressants has become more frequent, and also it has been almost the rule for TRBD trials (Table 26.2).

Table 26.2 provides a summary of TRBD definition according to different randomized controlled trials. This definition was refined by a consensus of internationally recognized experts in the field, resulting in a proposal for a multitherapy-resistant bipolar depression (MTRBD) concept (Hidalgo-Mazzei et al., 2019). These authors pointed out that a clear definition of TRBD could help guide when clinicians could consider novel and nonstandard interventions to treat patients. In summary, TRBD was defined as nonremission for 8 consecutive weeks after two different adequate trials of "at least two recommended monotherapy treatments or at least one monotherapy treatment and another combination treatment" for a patient with a current moderate or severe (bipolar) major depressive episode. MTRBD also requires "the addition of failure of at least one trial with an antidepressant, a psychological treatment and a course of electroconvulsive therapy." The authors further include a list of the required medications, their adequate therapeutic dose ranges, minimum lithium plasma levels, and the characteristics of therapeutic trials of cognitive-behavioral therapy and the ECT (at least 12 sessions, bilateral) needed to satisfy MTRBD criteria (Hidalgo-Mazzei et al., 2019).

TABLE 26.1 Evolution of the treatment-resistant depression concept in the last decades.[a]

Authors and year	Definition of depression	Definition of TRD	Outcome	Outcome measure
Haskovec and Rysanek (1967)	"Endogenous depression"	No improvement after at least 3 weeks of administration of imipramine	"Good and stable improvement"	According to a clinical scale that included seven symptoms of depression
Davidson et al. (1978)	Feighner criteria	Patients who "had failed to respond to conventional treatment"	Reduction in depression scores	Psychometric scales, including HDRS and BDI
De Montigny et al. (1981)	RDC	Decreased total scored in the HDRS less than 40% after treatment with a TCA	Reduction in depression scores	HDRS
MacEwan and Remick (1988)	DSM-III and RDC criteria	Unsuccessful treatment with at least two different antidepressants or an antidepressant and/or a course of ECT	Full or complete remission	HDRS decreased more than 50% and a CGI score of "very much improved" or "much improved"
Shelton et al. (2005)	DSM-IV	Failure to respond to a SSRI and subsequently failed to respond to nortriptyline	Mean change in the MADRS scores	MADRS
Husain et al. (2019)	DSM-5	Failed to respond to at least two adequate antidepressant trials	Change in the MADRS scores from baseline to week 12	MADRS

[a] *Concepts related to unipolar depression; BDI, Beck Depression Inventory (Beck et al., 1961); CGI, Clinical Global Impressions Scale (Department of Health EaW, 1976); ECT, electroconvulsive therapy; HDRS, Hamilton Depression Rating Scale (Hamilton, 1960); MADRS, Montgomery-Asberg Depression Rating Scale; RDC, Research Diagnostic Criteria (Spitzer et al., 1978); SSRI, selective serotonin reuptake inhibitors; TCA, tricyclic antidepressants; TRD, treatment-resistant depression; DSM-III, Diagnostic and Statistical Manual of Mental Disorders, 3rd edition (American Psychiatric Association, 1980).*

TABLE 26.2 Characteristics of treatment-resistant bipolar depression definitions according to randomized controlled trial studies.

Authors and year	Definition of depression	Definition of TRBD	Outcome	Outcome measure
Tavares et al. (2017)	DSM-IV criteria and confirmed by the MINI	Failure to achieve remission with ≥ two interventions approved as first, second, or third-line therapies for BD according to CANMAT guidelines	Change in HDRS-17 from baseline to week 4	HDRS-17
Schoeyen et al. (2015)	DSM-IV-TR, BD I or II	Indication for ECT and lack of response to two trials for at least 6 weeks	Longitudinal profile of weekly MADRS scores during the intervention. Remission (MADRS scores ≤12) as a secondary outcome	MADRS
Xu et al. (2015)	DSM-IV	Nonresponsive to at least one AD confirmed by the ATHF	Percent change in MADRS score from baseline	MADRS
Lally et al. (2014)	SCID-I for DSM-IV, BD I or BD II	Failed to respond to at least one adequate AD	Changes in SHAPS scores from the baseline	SHAPS
Diazgranados et al. (2010)	SCID-I for DSM-IV, BD I or BD II	Failed of at least one adequate AD trial (assessed by the ATHF), and to have failed a prospective open trial of a mood stabilizer (lithium or valproate)	Changes in MADRS scores	MADRS
Nierenberg et al. (2006)	M.I.N·I, BD I or BD II	A well-documented failure (e.g., a medical chart was available) to respond to at least two trials of AD or an AD and mood stabilizer regimen or not responded to treatment in first 12 weeks of standard or randomized care pathways for bipolar depression	Recovery as defined by no more than two symptoms meeting DSM-IV threshold criteria for a mood episode and no significant symptoms present for 8 weeks	DSM-IV
Rush et al. (2005)	DSM-IV-TR, BD I or II	Unsatisfactory response to at least two adequate trials of different classes of AD medication, but not more than six, regardless of AD category	Response, defined as a 50% reduction relative to the mean HRSD score at the baseline	HDRS
Goldberg et al. (2004)	SCID-I for DSM-IV	Not responded to at least two adequate trials of standard AD with concomitant mood stabilizers during the current episode	Response, defined as improvement in HDRS score of 50% or more over the baseline score	HDRS

AD, antidepressant; *ATHF*, Antidepressant Treatment History Form (Sackeim, 2001); *BD*, bipolar disorders; *BD I*, bipolar disorders type I; *BD II*, bipolar disorders type II; *BDI*, Beck Depression Inventory; *CANMAT*, The Canadian Network for Mood and Anxiety Treatments (Yatham et al., 2013); *DSM-IV-TR*, Diagnostic and Statistical Manual of Mental Disorders, 4th edition, revised (American Psychiatric Association, 2000); *ECT*, electroconvulsive therapy; *HDRS*, Hamilton Depression Rating Scale (Hamilton, 1960); *MADRS*, Montgomery-Asberg Depression Rating Scale (Montgomery and Asberg, 1979); *MDD*, major depressive episode; *MINI*, Mini-International Neuropsychiatric Interview (Sheehan et al., 1998); *SCID-I*, Structured Clinical Interview for DSM Disorders (First et al., 1996); *SHAPS*, Snaith-Hamilton Pleasure Scale (Ameli et al., 2014); *TRBD*, treatment-resistant bipolar depression.

Frequency, adherence and guidelines compliance

Mendlewicz et al., for the Group for the Study of Resistant Depression (GSRD), investigated clinical factors associated with TRBD. Of 375 patients with BD type I or II who had received at least 4 weeks of adequate treatment with mood stabilizers and an antidepressant, 99 (26.4%) were classified as treatment-resistant. The criteria for TRBD considered for this study was a failure to reach a score of less than 17 in the Hamilton Depression Rating Scale (HAM-D 17) after at least two adequate consecutive antidepressant trials during the current episode (Mendlewicz et al., 2010). A similar frequency was found in the study of Fornaro et al. (2020). The authors investigated clinical predictors of TRBD, defined as lack of response to at least two established treatments for bipolar depression, with an exploratory-hypothesis generating approach. Fifteen of the 53 participants with BD type I or II filled the criteria per TRBD (28.3%) (Fornaro et al., 2020). Parker and Graham (2017) compared three groups of patients categorized according to the number of medications received regarding several demographic and clinical characteristics. Of the 1034 participants (128 with BD type I and 906 with BD type II), close to one-third of them had received five or more medications, the authors' criteria as TRBD (Parker and Graham, 2017). Despite no studies have addressed the prevalence of patients with BD and treatment-resistant depression; these findings suggest that about one-quarter to one-third of the patients would lack response to two or more pharmacological treatments.

Variables associated with treatment resistance in bipolar depression

Socio-demographic and clinical variables

Fornaro et al. (2020) investigated predictors of treatment-resistant in patients with bipolar depression. From the 53 individuals with bipolar depression, 15 (28.3%) had a history of lack of response to at least two adequate treatments for depression. The presence of mixed features, including symptoms such as increased energy, psychomotor activity, and irritability, correctly classified more than 70% of patients with TRBD (Fornaro et al., 2020). Parker et al. reported a higher frequency of females among patients with TRBD (64%), early age of depression onset (17.3 years), and a considerable number of medical conditions considering the mean age of the sample (about three medical conditions in a sample with 38.6 years old on average). Psychiatric comorbidity burden was also sizeable: almost half presented generalized anxiety disorder, panic disorder, and obsessive-compulsive disorder: Patients with TRBD also reported a substantial family history of psychiatric disorder (83.1% of depression and more than half of alcohol problems). Functional impairment was also highly-evident, as only 23% of the sample reported being engaged in full-time work (Parker and Graham, 2017). Mendlewicz et al. (2010) found that patients with TRBD presented an odds ratio around twofold for the severe intensity of the last depressive episode, melancholic features, social phobia, and suicide risk.

Cognitive impairment has been recognized as a hallmark of BD and seems to be present regardless of clinical mood states (Cullen et al., 2016). Bo et al. reviewed cognitive performance in patients with BD compared to healthy controls (HC) using The Measurement and Treatment Research to Improve Cognition Schizophrenia Consensus Cognitive Battery (MCCB) (Bo et al., 2017). Seven studies, including 467 patients with BD and 570 HC, were included in their meta-analyses, with results showing overall lower performance in all seven domains evaluated for the former group. The larger effect size was for processing speed (>0.8), followed by visual and verbal learning (0.5), and social cognition, reasoning, and problem-solving (0.2) (Bo et al., 2017). Kessler et al. evaluated neurocognitive function in patients with TRBD and investigated whether patients with types I and II would present distinct cognitive profiles. The authors found that both groups presented global impairment of clinically significant severity, which was more evident for BD type I, especially processing speed (Kessler et al., 2013).

Neurobiological characteristics

The neurobiological characterization of BD could help identify novel targets for intervention, optimizing the efficacy, safety, and broadening the array of options for its treatment. Moreover, it would facilitate the characterization of illness biotypes contributing to precise and personalized therapeutic strategies (Fernandes et al., 2017). However, its etiology and pathophysiology are complex, involving several biological mechanisms, including dysfunctions in serotoninergic, noradrenergic, and glutamatergic systems, neurotrophic factors, inflammation, and external stressors, which challenges identifying critical pathways for intervention. Scaini et al. reviewed biological findings associated with BD, including genetic, biochemical, and neuroimaging (Scaini et al., 2020). According to this study, BD is highly heritable (70%–80%), and patients exhibit DNA methylation, alterations in the levels of micro-RNAs, and long noncoding RNAs, all potentially contributing to the phenotypic clinical presentation of the disease. Some studies also showed changes in brain-derived neurotrophic factor (BDNF), nerve growth factor (NGF), neurotrophin-3 (NT-3), and NT-4, which are associated with synaptic formation, neuronal growth, and regulation of neuronal survival processes (Scaini et al., 2020). Dysfunctions in the immune system and alterations on the inflammatory marker are also extensively found among patients with BD. Fernandes et al. reported increased C-reactive protein levels (CPR) in individuals with BD during all mood states, which were not associated with symptoms severity (Fernandes et al., 2016). Postmortem studies showed inflammation in several brain regions, like the dorsolateral prefrontal cortex, anterior cingulate cortex, amygdala, and hippocampus (Giridharan et al., 2020). Among the alterations, the authors reported those related to glial density and cytoarchitecture, and inflammatory gene and protein expression (Giridharan et al., 2020). Excessive oxidative stress and mitochondrial dysfunction have also been found in patients with BD, as changes in mitochondrial morphology, high-energy phosphates and pH, and downregulation of proteins related to mitochondrial respiration (Scaini et al., 2016). Finally, neuroimaging studies have shown several structural and connectivity alterations in patients with BD compared with HC, as increased putamen volume, decreased volume of the amygdala, gray matter reductions in the insula, dorsolateral and orbitofrontal regions, and decreased connectivity in fronto-limbic tracts, cingulum, corpus callosum, and uncinate fasciculus (Diaz et al., 2020).

Although numerous studies have examined BD's neurobiology, the majority of this work focused on patients with treatment-resistant depression has done so in the context of unipolar depression—not bipolar depression (Schosser et al., 2012; Bartova et al., 2019; Yang et al., 2019; Fabbri et al., 2021). That being said, some of the studies that investigated interventions for TRBD (discussed in detail below) targeted a number of the biological mechanisms discussed above. Next, we will discuss important clinical issues concerning the management of TRBD more broadly. We will then summarize intervention studies for patients with TRBD, specifically, focusing on the results of randomized controlled trials that enrolled patients with rigorously-defined treatment-resistant illness.

Management of treatment-resistant bipolar depression

Initial assessment

When a clinician is treating a patient with bipolar depression who has not had an adequate response to sequential trials, it is imperative to carefully reevaluate the diagnosis to, first, confirm the existing diagnosis of BD and, second, to investigate missed or inadequately-treated psychiatric and nonpsychiatric comorbidities, assess treatment adherence, investigate whether past treatments for BD have been delivered at proper doses and for sufficient periods of time, and to evaluate the quality of (and adherence to) psychosocial interventions (Bobo and Shelton, 2021).

In a survey of individuals with BD, 48% of the respondents did not receive the diagnosis of BD until they had consulted with three or more professionals, and 10% received the diagnosis only after visiting with seven or more professionals (Lish et al., 1994). These findings highlight the challenging nature of diagnosing BD, even in mental health specialty settings; therefore, the chance of misdiagnosis should always be considered when dealing with treatment-resistant patients. Efforts at clarifying the BP diagnosis may be enhanced by obtaining additional histories from family members and other collateral informants and by using structured or semistructured interviews (Singh and Rajput, 2006). BD is often confused with borderline personality disorder, given a number of shared clinical features such as impulsivity and emotional dysregulation (Bayes et al., 2021). In the study of Ruggero et al. (2010), among 145 patients who had reported a diagnosis of BD, the diagnosis could not be confirmed in 56.5% of the sample after using a gold-standard structured interview. Both conditions are also highly comorbid, conferring an additional challenge in the management of the patients. A systematic review found that 21.6% of people with BD may also meet diagnostic criteria for borderline personality disorder and 18.5% of individuals with borderline personality disorder may beet diagnostic criteria for BD (Fornaro et al., 2016). A high frequency of cooccurring substance abuse or dependence and anxiety disorders has also been reported in BD patients (Goldberg, 2001). Regarding nonpsychiatric comorbidities in patients with BD, Forty et al. found a self-reported frequency of asthma, diabetes, dyslipidemia, and hypertension in patients with BD of 19.2%, 6%, 19.2%, and 15%, respectively. The degree of burden from comorbid general medical conditions have been associated with longer illness duration, a greater number of psychiatric admissions, worse functioning, and ECT treatment in patients with BD (Forty et al., 2014).

BD is a chronic illness that usually requires a complex treatment approach, including distinct pharmacological interventions—each with different dosage, adverse effect, and efficacy profiles—and adjunctive psychological interventions. About one-third of BD patients in the United States are receiving polytherapy, with almost three psychotropic medications per person on average (Baldessarini et al., 2008). Thus, patients who present a lack of response to several treatments for bipolar depression need careful evaluation of current and past treatments, with an eye toward nonadherence, inadequately conducted therapeutic trials (due to inadequate treatment duration and suboptimal dose/blood levels where applicable), and the use of interventions not supported by the best scientific evidence available before classifying patients as having TRBD. Greene et al. (2018) reviewed adherence and pharmacological treatment profiles among patients with BD in the United States. The authors reported that only 30% of the patients remained on the prescribed mood stabilizer after 1 year of treatment. The results also showed an association between common comorbidities, such as a history of alcohol or drug abuse, and higher illness complexity (including moderate to severe BD, an indication of self-harm, or comorbidity other than substance abuse) with lower treatment adherence (Greene et al., 2018). Low conformity with guidelines for treatment could also prevent patients from responding to evidence-based interventions, although some studies have shown no significant association between good compliance with guideline-based practices and clinical outcomes in young patients with BD (Sunder et al., 2020; Goldberg et al., 2015). On the other hand, Bessonova et al. (2020) showed that both suboptimal medication adherence and clinical management were among the main drivers of the direct economic costs of BD, which were estimated at around $50 billion in the United States.

Psychosocial interventions

Psychosocial interventions have a crucial role in the management of patients with BD. In a systematic review, Miklowitz et al. found that cognitive behavioral therapy (CBT), family or conjoint therapy, and interpersonal therapy were associated with stabilizing depressive symptoms. Also, the findings showed the benefits of psychoeducation with guided practice of illness management skills in a family or group format in preventing recurrences and a better retention profile of brief psychoeducation and family therapy (Miklowitz et al., 2021). Other advantages of psychosocial interventions to treat patients with BD include educating patients about their illness, addressing comorbid psychiatric disorders, promoting self-monitoring, managing stress, addressing problems with relationships and work, and developing a plan to cope with an acute crisis (Bobo and Shelton, 2021). Swartz and Swanson (2014) reviewed published studies that focused on the clinical effects of evidence-based, bipolar-specific psychotherapies including psychoeducation, CBT, functional remediation, family-focused psychotherapy, interpersonal and social rhythm therapy, and integrated care management. However, as the authors point out, most randomized controlled trials that compared two psychosocial interventions did not find considerable differences. This suggests that the psychosocial intervention of choice may depend on treatment availability and a continuous reassessment of psychosocial needs in consultation with the patient and their family members (if appropriate) (Swartz and Swanson, 2014). Guideline-based recommendations for various psychosocial interventions as adjuncts to pharmacotherapy for acute bipolar depression and bipolar maintenance treatment were summarized by Murray (2018). The information provided highlighted the role of psychoeducation as a first line recommendation (Canadian Network for Mood and Anxiety Treatments (CANMAT)/International Society for Bipolar Disorders (ISBD) and British Association for Psychopharmacology) and level 1 evidence (Royal Australian and New Zealand College of Psychiatrists (RANZCP)) for bipolar maintenance and CBT as a second line (CANMAT/ISBD) and level 1 evidence (RANZCP) for acute bipolar depression (Murray, 2018).

Pharmacological treatment

Rigorous pharmacological treatment is fundamental to the management of psychiatric symptoms in patients with BD. Bahji et al. reviewed randomized controlled trials for the treatment of bipolar depression and performed a network meta-analysis of 50 randomized trials (totaling 11,448 participants) to investigate the comparative efficacy and tolerability of pharmacological treatments. The network meta-analysis supported the efficacy of divalproex, olanzapine/fluoxetine, olanzapine, quetiapine, cariprazine, and lamotrigine for bipolar depression, compared to placebo. Sertraline, paroxetine, and lithium in monotherapy were not effective compared to placebo (Bahji et al., 2020). The negative result with lithium contrasts with a number of reports supporting lithium for treating bipolar depression and for reducing the risk of suicidal behavior. For instance, from a nationwide cohort study, Antolin-Concha et al. (2020) found that lithium was associated with a significant decrease risk of suicide in patients with BD who had been hospitalized. Lithium discontinuation has also been associated with high risk of illness recurrence and risk of suicide (Baldessarini et al., 1999). These findings make the well-documented declines in the use of lithium for bipolar disorder management all-the-more troubling. Indeed, a 20-year observational study documenting the trends in the pharmacological treatment in patients with diagnosed BD showed a decreasing percentage of visits for BD that included prescription of lithium—from 30.4% to 17.6%. In comparison, there was an increase in second-generation antipsychotics prescriptions from 12.4% to 51.4% (Rhee et al., 2020).

The sequencing of medication therapy for bipolar depression is addressed by several practice guidelines (Yatham et al., 2018; Malhi et al., 2021; Goodwin et al., 2016). Integration of pharmacotherapy with psychosocial treatment is usually considered optimal. First-line medication considerations often prioritize monotherapy with treatments supported by randomized trials (e.g., quetiapine, lurasidone), most of which enrolled people with bipolar I depression. If monotherapy is ineffective, augmentation with mood stabilizers such as lithium, divalproex, or lamotrigine is usually recommended. If monotherapy with a particular agent is intolerable or if the addition of mood stabilizers as augmenting agents is ineffective, switching is usually indicated. Next step treatment options supported by controlled evidence include switching to the alternative evidence-based monotherapy, olanzapine plus fluoxetine, divalproex (if not trialed already), or mood stabilizer combination therapy (e.g., lithium plus divalproex or lamotrigine). If these therapeutic options are ineffective or poorly tolerated, additional treatment options include alternative monotherapies (e.g., lamotrigine, lithium, carbamazepine, olanzapine, or cariprazine), other mood stabilizer combinations (e.g., lithium plus carbamazepine or lamotrigine if not already trialed), and combination therapy with olanzapine plus lithium or divalproex. As noted earlier, some patients may respond to combination therapy with lithium or divalproex plus an antidepressant such as a selective serotonin reuptake inhibitor (e.g., fluoxetine) or bupropion; or combination therapy with a second-generation antipsychotic (usually quetiapine, lurasidone, or olanzapine) plus an antidepressant. Therapeutic options for multidrug-resistant bipolar depression include ECT, as reviewed further below.

To address partial responses, once misdiagnosis, treatment nonadherence, drug-drug interactions, medication side effects, psychosocial stressors and/or substance abuse destabilizing patient's clinical condition are ruled out, drug dosage optimization and/or add-on pharmacotherapy may be preferable before considering switch medication (Bobo, 2017). For patients on second-generation antipsychotics, augmentation with lamotrigine, lithium, or valproate are options unless the patient has a history of nonresponse with these medications (Bobo, 2017). The efficacy of lamotrigine as an add-on treatment on lithium was investigated in a multicenter, double-blind, placebo-controlled trial. The findings showed a benefit for lamotrigine compared to placebo (van der Loos et al., 2009). A systematic review and meta-analysis of six randomized controlled trials of add-on modafinil, armodafinil, and lisdexamfetamine compared to placebo reported a significant association of these medications with remission. The psychostimulants and stimulant-like drugs were not associated with a significant increase in adverse effects or dropouts than placebo (Tsapakis et al., 2020).

There are surprisingly few evidence-based therapeutic options for depressed patients with bipolar II disorder despite how common bipolar II disorder is encountered in clinical settings and the predominance of depressive episodes or symptoms. Of the available options, quetiapine, lithium, and lamotrigine may be the best-supported (Swartz and Thase, 2011). The CANMAT and the International Society for Bipolar Disorders (ISBD) 2018 guidelines for the management of patients with bipolar disorders include as adjunctive options for the treatment of bipolar II depression the following medications in combination with mood stabilizing drugs: bupropion (second-line recommendation, level 2 of evidence), agomelatine (third-line, level 4), eicosapentaenoic acid (third-line, level 4), ketamine (third-line, level 3), *N*-acetylcysteine (third-line, level 4), pramipexole (third-line, level 3), and T3/T4 thyroid hormones (third-line, level 4) (Yatham et al., 2018).

For breakthrough depressive episodes in patients who responded initially to treatment, a similar approach to partial responses is suggested, including the investigation of an inadequate dose of medication, poor adherence to pharmacotherapy and psychosocial therapies, drug-drug interactions that may contribute to lower medication effectiveness, the use of concomitant medications (e.g., antidepressants) that may destabilize mood, and potential changes in thyroid functioning induced or not by lithium therapy (Bobo, 2017).

The use of antidepressants to treat bipolar depression in its acute phase is controversial owing to limited efficacy in large clinical trials, risk of mood destabilization (particularly in people with bipolar I disorder), and the risk of polar mood switching (Ghaemi, 2012). Much ambiguity in the literature remains about the role of antidepressants in people with bipolar depression. Most experienced clinicians have encountered bipolar patients who require adjunctive antidepressants to achieve remission from depressive symptoms and prevent depressive relapses once well. Moreover, the risk of polar mood switching with antidepressants in people with bipolar disorders appears to be lower than what was once believed, especially when antidepressants are combined with mood stabilizing medications. Therefore, we avoid the use of antidepressants in BD patients in absence of concomitant mood stabilizing medications with antimanic effects such as conventional mood stabilizers (e.g., lithium, divalproex, carbamazepine) and thymoleptic atypical antipsychotic drugs.

Next, we will discuss interventional studies including investigational/nonstandard options for patients with TRBD, focusing on randomized controlled trials.

Interventions for treatment-resistant bipolar depression

Pharmacological

Ketamine

Ketamine is an *N*-methyl-D-aspartate receptor (NMDA-R) antagonist, which single administration has been associated with depressive symptoms improvement (Abdallah et al., 2016). Zarate et al. investigated ketamine for the treatment of depression in 15 participants with a history of failure to at least one adequate antidepressant trial and a lack of response to a prospective open trial with a mood stabilizer while at the National Institute of Mental Health (NIMH) in a double-blind, randomized, crossover, placebo-controlled study (Zarate et al., 2012). Participants of this study had 9.7 mean lifetime antidepressant trials, and the length of the current depressive episode of 20.9 months. Ketamine was significantly more efficacious than placebo in decreasing depressive symptoms and suicidal ideation at 40 min postinfusion, and almost 80% of subjects presented a treatment response to ketamine at some point during the study and none to placebo. The main side effects observed were dissociative symptoms that happened only at the 40-min evaluation (Zarate et al., 2012).

Armodafinil

Armodafinil is an isomer of modafinil, a dopamine and norepinephrine transporter inhibitor (Calabrese et al., 2010; Minzenberg et al., 2018). Calabrese et al. evaluated the efficacy of adjunctive armodafinil 150 mg/day compared to adjunctive placebo to treat depressive symptoms in patients with BD and no response to lithium, olanzapine, or valproic acid.

Adjunctive armodafinil was associated with a significant reduction of depressive symptoms compared to placebo ($P=.04$). However, the intervention was also associated with more adverse events as diarrhea, insomnia, and headaches (Calabrese et al., 2010).

N-Acetyl cysteine

N-Acetyl cysteine (NAC) is a precursor of l-cysteine and reduced glutathione (GSH), which has a role in antioxidant defense, oxidation-reduction reaction-regulated signal transduction, and regulation of immune responses (Bavarsad Shahripour et al., 2014). In a double-blind, randomized placebo-controlled trial, Berk et al. investigated NAC 1 g twice daily compared to placebo to treat depression in participants with BD and moderate severity of the symptomatology. The results showed that at week 24, NAC was significantly associated with reducing depressive symptoms, evident from the 20-week visit. NAC group also presented a significant improvement in measures of quality of life. There was no difference between the groups regarding the frequency of adverse effects (Berk et al., 2008).

Celecoxib

Several studies have shown the association between depression and dysfunctions in the immune system (Fernandes et al., 2016; Brunoni et al., 2020; Hunt et al., 2020). Celecoxib is a cyclooxygenase-2 inhibitor that has been investigated to treat mood disorders because of its antiinflammatory properties. Halaris et al. investigated the potential of celecoxib in reducing depressive symptoms among patients with TRBD in a 10-week randomized, double-blind, placebo-controlled study. Participants were randomized to receive either escitalopram and celecoxib, or escitalopram and placebo. TRBD was defined as no response to at least two adequate trials of antidepressant medications or an atypical antipsychotic or mood stabilizer. Results showed that the association between escitalopram and celecoxib was significantly associated with response and remission rates, with almost 80% of the group presenting response and 63% remission (vs 45% and 10% for the placebo group, respectively) (Halaris et al., 2020).

Nonpharmacological

Noninvasive

Bright light therapy

Bright light therapy (BLT) was compared to a sham negative ion generator in a randomized controlled trial for nonseasonal depression in patients with BD ($n=50$) and unipolar depression ($n=40$) (Chojnacka et al., 2016). The protocol included 30 min of the intervention or placebo (sham negative ion generator) after they awoke in the morning for 14 days. Results showed that among the subset of patients with treatment-resistant depression, considered as a lack of response despite two adequate antidepressant trials, the intervention was more efficacious than placebo. The authors also reported that the intervention was associated with only mild side effects (most commonly slight agitation and headaches) that disappeared during the study (Chojnacka et al., 2016).

Intermittent theta burst stimulation

Intermittent theta burst stimulation (iTBS) is a repetitive transcranial magnetic stimulation (rTMS) protocol that uses short trains of stimuli at high frequency (Oberman et al., 2011). Bulteau et al. investigated twice-daily neuronavigated iTBS for 26 participants with TRBD, according to the study's criteria, a MADRS score of >20 despite adequate treatment for at least 1 month (Bulteau et al., 2019). The target was the left dorsolateral prefrontal cortex. Despite no significant differences compared with sham, participants from the active iTBS group presented response and remission rates of 72% and 42%, respectively. Also, the intervention was well tolerated, and there was no mood switch reported during the 6-week follow-up (Bulteau et al., 2019).

Deep transcranial magnetic stimulation

Deep transcranial magnetic stimulation (dTMS) uses an H1 coil for transcranial magnetic stimulation (TMS) that generates electrical fields with greater penetration compared with figure-of-eight coil. Tavares et al. applied dTMS or sham to participants with TRBD in a randomized, parallel-group clinical trial. TRBD was defined as lack of remission after two or more interventions according with international guidelines (Tavares et al., 2017). Participants of this study had a median of two depressive episodes, the current duration of the depressive episode was 6 months, and almost 60% presented severe depressive symptoms. dTMS was superior to sham in decreasing depressive symptoms at week 4 ($P=.03$) and there was a trend for association regarding response rates (48% vs 24%, respectively, $P=.08$) (Tavares et al., 2017). The intervention was not associated with treatment-emergent mania or hypomania.

Invasive

Electroconvulsive therapy

Schoeyen et al. compared ECT with algorithm-based pharmacological treatment for patients with TRBD. In this study, TRBD criteria were nonresponse to at least two trials with mood stabilizers or antidepressants in adequate doses during 6 weeks or more (Schoeyen et al., 2015). ECT protocol was three ECT sessions a week for up to 6 weeks (up to 18 ECT sessions). Stimulation electrodes were placed at the right unilateral. Results showed that the intervention was associated with a significant decrease in depressive symptoms than the algorithm-based pharmacological treatment group ($P=.002$). The response rates were also significantly higher in the ECT group (73.9% vs 35% in the comparison group; $P=.01$), despite no differences regarding remission rates. Memory problems were reported equally between the groups, but probable relation to ECT in the intervention group (Schoeyen et al., 2015).

Vagus nerve stimulation

Vagus nerve stimulation (VNS) electrodes deliver a low-frequency and chronic electrical signal to the left cervical vagus nerve from an implantable generator. VNS is approved by the FDA for the treatment of recurrent unipolar and bipolar depression. The mechanism of action associated with the improvement of depressive symptoms includes the direct stimulation of brain stem structures and modulation of limbic and cortical regions' activity (Nemeroff et al., 2006). Rush et al. compared active VNS versus sham to treat depression in participants with unipolar depression or BD and a current episode for 2 years or a minimum of four lifetime major depressive episodes. For patients with BD, it was also required contraindication, intolerance, or resistance to lithium. More than half of the participants had a history of ECT treatment and psychiatric hospitalization (Rush et al., 2005). The intervention was well tolerated, but significant improvement of depressive symptoms was found only for the secondary outcome measure (Rush et al., 2005). Recently, results from a 5-year prospective open-label registry showed that VNS was associated with a higher likelihood of achieving response compared to treatment as usual alone, as well as a reduction in suicidality (McAllister-Williams et al., 2020).

Deep brain stimulation

Ramasubbu et al. investigated the use of short pulse width (SPW) and long pulse width (LPW) subcallosal cingulate deep brain stimulation (DBS) to treat patients with resistant depression. This double-blind, randomized controlled parallel, and crossover trial included patients diagnosed with major depressive disorder (MDD) or BD (Ramasubbu et al., 2020). The criteria for TRD comprised a current depressive episode of more than 1 year of duration and failure to respond to four different classes of antidepressants, with various augmentation or combination strategies (atypical antipsychotics, anticonvulsants, psychotherapy, lithium, and ECT). Of the five (23%) participants included with BD (all BD type II), three presented more than 50% reduction on the Hamilton depression rating scale (HDRS) scores at 6 months of follow-up. Results showed a significant decrease in HDRS scores at 6 months of the follow-up with no significant differences between SPW and LPW regarding safety and efficacy, considering all participants' findings (Ramasubbu et al., 2020).

Potential biomarkers of treatment-response

Some studies have investigated treatment response biomarkers in patients with BD and treatment-resistant depression, which could help provide more personalized treatments. Edberg et al. found lower levels of monocyte chemoattractant protein-1 (MCP-1) among nonresponders to a combination between celecoxib and escitalopram, suggesting a better response to this treatment for patients with higher levels of MCP-1 (Edberg et al., 2020). Response to repeated ketamine administration (six intravenous infusions over 12 days) was associated with kynurenine pathway metabolites in the study of Zhou et al. that included patients with unipolar and bipolar depression with suicidality or history of failure to at least two antidepressant treatments (Zhou et al., 2018). The authors found higher serum concentrations of kynurenic acid and kynurenic acid/kynurenine ratio at 24 h associated with lower depressive symptoms at days 13 and 26 following the sixth infusion (Zhou et al., 2018). Response to ketamine treatment was also associated with markers of circadian timekeeping in patients with unipolar and bipolar treatment-resistant depression, providing clinical evidence for the association between ketamine's efficacy and the circadian system (Duncan et al., 2017). Villaseñor et al. found differences in the mitochondrial β-oxidation of fatty acids between responders and nonresponders to ketamine in patients with TRBD (Villasenor et al., 2014). Clark et al. evaluated the association between rostral anterior cingulate (rACC) glutamate/glutamine concentration and treatment response to DBS of the subcallosal cingulate in 16 patients with TRD, two of them with BD. The authors reported that lower baseline glutamate could predict both reduction in depressive symptoms and response to treatment at 6 months (Clark et al., 2020). The investigation of biomarkers for treatment-response could guide personalized strategies, and the characterization of the neurobiology associated with treatment-resistant in patients with bipolar depression could help identify more homogeneous groups of patients; both approaches could optimize the identification of novel therapeutic targets and the efficacy of existing treatments.

Conclusions

Depression in BD accounts for most of the disease's symptomatic period and a significant part of the patient's life. Also, it contributes to functional impairment, suicidality, and the economic burden associated with BD. Despite safe and effective treatments available to treat symptoms in patients with BD, about one-quarter of the patients with depression lack response to two or more trials. Several pharmacological and nonpharmacological interventions have been investigated in randomized clinical trials for these clinical situations showing promising results, adding to a lower rate of adverse events. Nonetheless, most of the studies have evaluated symptoms improvement in the acute phase of the treatment, so the long-lasting effect of these interventions in protecting against recurrence is unclear. The neurobiological characterization of psychiatric disorders could help in identifying novel targets for treatments. However, the high heterogeneity associated with depression in BD, both clinically and biologically, may challenge the identification of more precise and personalized therapeutic interventions, as well as biomarkers for treatment response. Thus, studies addressing the biology of BD according to distinct units of analysis, from genetics to neuroimaging, in the context of different environmental stressors and resilient factors, could help identify safe, efficacious, and individualized treatments, and deserve further investigation.

References

Abdallah, C.G., Adams, T.G., Kelmendi, B., Esterlis, I., Sanacora, G., Krystal, J.H., 2016. Ketamine's mechanism of action: a path to rapid-acting antidepressants. Depress. Anxiety 33 (8), 689–697.

Ameli, R., Luckenbaugh, D.A., Gould, N.F., Holmes, M.K., Lally, N., Ballard, E.D., et al., 2014. SHAPS-C: the Snaith-Hamilton pleasure scale modified for clinician administration. PeerJ 2, e429.

American Psychiatric Association, 1980. Diagnostic and Statistical Manual of Mental Disorders, third ed. American Psychiatric Association, Washington, DC.

American Psychiatric Association, 2000. Diagnostic and Statistical Manual of Mental Disorders, fourth ed. Text Revision. American Psychiatric Association, Washington, DC.

Antolin-Concha, D., Lahteenvuo, M., Vattulainen, P., Tanskanen, A., Taipale, H., Vieta, E., et al., 2020. Suicide mortality and use of psychotropic drugs in patients hospitalized due to bipolar disorder: a Finnish nationwide cohort study. J. Affect. Disord. 277, 885–892.

Bahji, A., Ermacora, D., Stephenson, C., Hawken, E.R., Vazquez, G., 2020. Comparative efficacy and tolerability of pharmacological treatments for the treatment of acute bipolar depression: a systematic review and network meta-analysis. J. Affect. Disord. 269, 154–184.

Baldessarini, R.J., Tondo, L., Hennen, J., 1999. Effects of lithium treatment and its discontinuation on suicidal behavior in bipolar manic-depressive disorders. J. Clin. Psychiatry 60 (Suppl. 2), 77–84. discussion 111-6.

Baldessarini, R., Henk, H., Sklar, A., Chang, J., Leahy, L., 2008. Psychotropic medications for patients with bipolar disorder in the United States: polytherapy and adherence. Psychiatr. Serv. 59 (10), 1175–1183.

Bartova, L., Dold, M., Kautzky, A., Fabbri, C., Spies, M., Serretti, A., et al., 2019. Results of the European Group for the Study of Resistant Depression (GSRD)—basis for further research and clinical practice. World J. Biol. Psychiatry 20 (6), 427–448.

Bavarsad Shahripour, R., Harrigan, M.R., Alexandrov, A.V., 2014. N-acetylcysteine (NAC) in neurological disorders: mechanisms of action and therapeutic opportunities. Brain Behav. 4 (2), 108–122.

Bayes, A., Spoelma, M.J., Hadzi-Pavlovic, D., Parker, G., 2021. Differentiation of bipolar disorder versus borderline personality disorder: a machine learning approach. J. Affect. Disord. 288, 68–73.

Beck, A.T., Ward, C.H., Mendelson, M., Mock, J., Erbaugh, J., 1961. An inventory for measuring depression. Arch. Gen. Psychiatry 4, 561–571.

Berk, M., Copolov, D.L., Dean, O., Lu, K., Jeavons, S., Schapkaitz, I., et al., 2008. N-acetyl cysteine for depressive symptoms in bipolar disorder—a double-blind randomized placebo-controlled trial. Biol. Psychiatry 64 (6), 468–475.

Bessonova, L., Ogden, K., Doane, M.J., O'Sullivan, A.K., Tohen, M., 2020. The economic burden of bipolar disorder in the United States: a systematic literature review. Clinicoecon. Outcomes Res. 12, 481–497.

Bo, Q., Mao, Z., Li, X., Wang, Z., Wang, C., Ma, X., 2017. Use of the MATRICS consensus cognitive battery (MCCB) to evaluate cognitive deficits in bipolar disorder: a systematic review and meta-analysis. PLoS One 12 (4), e0176212.

Bobo, W.V., 2017. The diagnosis and management of bipolar I and II disorders: clinical practice update. Mayo Clin. Proc. 92 (10), 1532–1551.

Bobo, W.V., Shelton, R.C., 2021. Bipolar major depression in adults: general principles of treatment. In: Solomon, D., Keck, P. (Eds.), UpToDate Topic Reviews.

Brunoni, A.R., Supasitthumrong, T., Teixeira, A.L., Vieira, E.L., Gattaz, W.F., Bensenor, I.M., et al., 2020. Differences in the immune-inflammatory profiles of unipolar and bipolar depression. J. Affect. Disord. 262, 8–15.

Bulteau, S., Beynel, L., Marendaz, C., Dall'Igna, G., Pere, M., Harquel, S., et al., 2019. Twice-daily neuronavigated intermittent theta burst stimulation for bipolar depression: a randomized sham-controlled pilot study. Neurophysiol. Clin. 49 (5), 371–375.

Buoli, M., Cesana, B.M., Fagiolini, A., Albert, U., Maina, G., de Bartolomeis, A., et al., 2020. Which factors delay treatment in bipolar disorder? A nationwide study focussed on duration of untreated illness. Early Interv. Psychiatry 15 (5), 1136–1145. https://doi.org/10.1111/eip.13051 (Epub 2020 Oct 15).

Calabrese, J.R., Ketter, T.A., Youakim, J.M., Tiller, J.M., Yang, R., Frye, M.A., 2010. Adjunctive armodafinil for major depressive episodes associated with bipolar I disorder: a randomized, multicenter, double-blind, placebo-controlled, proof-of-concept study. J. Clin. Psychiatry 71 (10), 1363–1370.

Chojnacka, M., Antosik-Wojcinska, A.Z., Dominiak, M., Bzinkowska, D., Borzym, A., Sokol-Szawlowska, M., et al., 2016. A sham-controlled randomized trial of adjunctive light therapy for non-seasonal depression. J. Affect. Disord. 203, 1–8.

Clark, D.L., MacMaster, F.P., Brown, E.C., Kiss, Z.H.T., Ramasubbu, R., 2020. Rostral anterior cingulate glutamate predicts response to subcallosal deep brain stimulation for resistant depression. J. Affect. Disord. 266, 90–94.

Cullen, B., Ward, J., Graham, N.A., Deary, I.J., Pell, J.P., Smith, D.J., et al., 2016. Prevalence and correlates of cognitive impairment in euthymic adults with bipolar disorder: a systematic review. J. Affect. Disord. 205, 165–181.

Davidson, J., McLeod, M., Law-Yone, B., Linnoila, M., 1978. A comparison of electroconvulsive therapy and combined phenelzine-amitriptyline in refractory depression. Arch. Gen. Psychiatry 35 (5), 639–642.

De Montigny, C., Grunberg, F., Mayer, A., Deschenes, J.P., 1981. Lithium induces rapid relief of depression in tricyclic antidepressant drug nonresponders. Br. J. Psychiatry 138, 252–256.

Department of Health EaW, 1976. Clinical Global Impressions. ECDEU Assessment Manual (ADM). Department of Health EaW, Washington, DC, pp. 218–222.

Diaz, A.P., Bauer, I.E., Sanches, M., Soares, J.C., 2020. Neuroanatomic and functional neuroimaging findings. In: Young, A.H., Juruena, M.F. (Eds.), Current Topics in Behavioral Neurosciences Bipolar Disorder: From Neuroscience to Treatment. Springer Berlin Heidelberg, Berlin, Heidelberg, pp. 1–24.

Diazgranados, N., Ibrahim, L., Brutsche, N.E., Newberg, A., Kronstein, P., Khalife, S., et al., 2010. A randomized add-on trial of an N-methyl-D-aspartate antagonist in treatment-resistant bipolar depression. Arch. Gen. Psychiatry 67 (8), 793–802.

Duncan Jr., W.C., Slonena, E., Hejazi, N.S., Brutsche, N., Yu, K.C., Park, L., et al., 2017. Motor-activity markers of circadian timekeeping are related to ketamine's rapid antidepressant properties. Biol. Psychiatry 82 (5), 361–369.

Edberg, D., Hoppensteadt, D., Walborn, A., Fareed, J., Sinacore, J., Halaris, A., 2020. Plasma MCP-1 levels in bipolar depression during cyclooxygenase-2 inhibitor combination treatment. J. Psychiatr. Res. 129, 189–197.

Fabbri, C., Kasper, S., Zohar, J., Souery, D., Montgomery, S., Albani, D., et al., 2021. Drug repositioning for treatment-resistant depression: hypotheses from a pharmacogenomic study. Prog. Neuro-Psychopharmacol. Biol. Psychiatry 104, 110050.

Feighner, J.P., Robins, E., Guze, S.B., Woodruff Jr., R.A., Winokur, G., Munoz, R., 1972. Diagnostic criteria for use in psychiatric research. Arch. Gen. Psychiatry 26 (1), 57–63.

Fernandes, B.S., Steiner, J., Molendijk, M.L., Dodd, S., Nardin, P., Goncalves, C.A., et al., 2016. C-reactive protein concentrations across the mood spectrum in bipolar disorder: a systematic review and meta-analysis. Lancet Psychiatry 3 (12), 1147–1156.

Fornaro, M., Orsolini, L., Marini, S., De Berardis, D., Perna, G., Valchera, A., et al., 2016. The prevalence and predictors of bipolar and borderline personality disorders comorbidity: systematic review and meta-analysis. J. Affect. Disord. 195, 105–118.

Fernandes, B.S., Williams, L.M., Steiner, J., Leboyer, M., Carvalho, A.F., Berk, M., 2017. The new field of 'precision psychiatry. BMC Med 15 (1), 80.

First, M., Spitzer, R., Gibbon, M., Williams, J., 1996. Structured Clinical Interview for DSM-IV Axis I Disorders, Clinical Version (SCID-CV). Americam Psychiatric Press, Inc, Washington, DC.

Fornaro, M., Fusco, A., Novello, S., Mosca, P., Anastasia, A., De Blasio, A., et al., 2020. Predictors of treatment resistance across different clinical subtypes of depression: comparison of unipolar vs. bipolar cases. Front. Psychiatry 11, 438.

Forty, L., Ulanova, A., Jones, L., Jones, I., Gordon-Smith, K., Fraser, C., et al., 2014. Comorbid medical illness in bipolar disorder. Br. J. Psychiatry 205 (6), 465–472.

Ghaemi, S.N., 2012. Antidepressants in bipolar depression: the clinical debate. Aust. N. Z. J. Psychiatry 46 (4), 298–301.

Giridharan, V.V., Sayana, P., Pinjari, O.F., Ahmad, N., da Rosa, M.I., Quevedo, J., et al., 2020. Postmortem evidence of brain inflammatory markers in bipolar disorder: a systematic review. Mol. Psychiatry 25 (1), 94–113.

Goldberg, J.F., 2001. Bipolar disorder with comorbid substance abuse: diagnosis, prognosis, and treatment. J. Psychiatr. Pract. 7 (2), 109–122.

Goldberg, J.F., Burdick, K.E., Endick, C.J., 2004. Preliminary randomized, double-blind, placebo-controlled trial of pramipexole added to mood stabilizers for treatment-resistant bipolar depression. Am. J. Psychiatry 161 (3), 564–566.

Goldberg, J.F., Freeman, M.P., Balon, R., Citrome, L., Thase, M.E., Kane, J.M., et al., 2015. The American society of clinical psychopharmacology survey of psychopharmacologists' practice patterns for the treatment of mood disorders. Depress. Anxiety 32 (8), 605–613.

Goodwin, G.M., Haddad, P.M., Ferrier, I.N., Aronson, J.K., Barnes, T., Cipriani, A., et al., 2016. Evidence-based guidelines for treating bipolar disorder: revised third edition recommendations from the British Association for Psychopharmacology. J. Psychopharmacol. 30 (6), 495–553.

Greene, M., Paladini, L., Lemmer, T., Piedade, A., Touya, M., Clark, O., 2018. Systematic literature review on patterns of pharmacological treatment and adherence among patients with bipolar disorder type I in the USA. Neuropsychiatr. Dis. Treat. 14, 1545–1559.

Halaris, A., Cantos, A., Johnson, K., Hakimi, M., Sinacore, J., 2020. Modulation of the inflammatory response benefits treatment-resistant bipolar depression: a randomized clinical trial. J. Affect. Disord. 261, 145–152.

Hamilton, M., 1960. A rating scale for depression. J. Neurol. Neurosurg. Psychiatry 23, 56–62.

Haskovec, L., Rysanek, K., 1967. The action of reserpine in imipramine-resistant depressive patients. Clinical and biochemical study. Psychopharmacologia 11 (1), 18–30.

Hidalgo-Mazzei, D., Berk, M., Cipriani, A., Cleare, A.J., Florio, A.D., Dietch, D., et al., 2019. Treatment-resistant and multi-therapy-resistant criteria for bipolar depression: consensus definition. Br. J. Psychiatry 214 (1), 27–35.

Hunt, C., Macedo, E.C.T., Suchting, R., de Dios, C., Cuellar Leal, V.A., Soares, J.C., et al., 2020. Effect of immune activation on the kynurenine pathway and depression symptoms—a systematic review and meta-analysis. Neurosci. Biobehav. Rev. 118, 514–523.

Husain, M.I., Chaudhry, I.B., Khoso, A.B., Husain, M.O., Rahman, R.R., Hamirani, M.M., et al., 2019. Adjunctive simvastatin for treatment-resistant depression: study protocol of a 12-week randomised controlled trial. BJPsych Open 5 (1), e13.

Kessler, U., Schoeyen, H.K., Andreassen, O.A., Eide, G.E., Hammar, A., Malt, U.F., et al., 2013. Neurocognitive profiles in treatment-resistant bipolar I and bipolar II disorder depression. BMC Psychiatry 13, 105.

Kupka, R.W., Altshuler, L.L., Nolen, W.A., Suppes, T., Luckenbaugh, D.A., Leverich, G.S., et al., 2007. Three times more days depressed than manic or hypomanic in both bipolar I and bipolar II disorder. Bipolar Disord. 9 (5), 531–535.

Lally, N., Nugent, A.C., Luckenbaugh, D.A., Ameli, R., Roiser, J.P., Zarate, C.A., 2014. Anti-anhedonic effect of ketamine and its neural correlates in treatment-resistant bipolar depression. Transl. Psychiatry 4, e469.

Lish, J.D., Dime-Meenan, S., Whybrow, P.C., Price, R.A., Hirschfeld, R.M., 1994. The national depressive and manic-depressive association (DMDA) survey of bipolar members. J. Affect. Disord. 31 (4), 281–294.

MacEwan, G.W., Remick, R.A., 1988. Treatment resistant depression: a clinical perspective. Can. J. Psychiatr. 33 (9), 788–792.

Malhi, G.S., Bell, E., Bassett, D., Boyce, P., Bryant, R., Hazell, P., et al., 2021. The 2020 Royal Australian and New Zealand College of Psychiatrists clinical practice guidelines for mood disorders. Aust. N. Z. J. Psychiatry 55 (1), 7–117.

McAllister-Williams, R.H., Sousa, S., Kumar, A., Greco, T., Bunker, M.T., Aaronson, S.T., et al., 2020. The effects of vagus nerve stimulation on the course and outcomes of patients with bipolar disorder in a treatment-resistant depressive episode: a 5-year prospective registry. Int. J. Bipolar Disord. 8 (1), 13.

Mendlewicz, J., Massat, I., Linotte, S., Kasper, S., Konstantinidis, A., Lecrubier, Y., et al., 2010. Identification of clinical factors associated with resistance to antidepressants in bipolar depression: results from an European Multicentre Study. Int. Clin. Psychopharmacol. 25 (5), 297–301.

Miklowitz, D.J., Efthimiou, O., Furukawa, T.A., Scott, J., McLaren, R., Geddes, J.R., et al., 2021. Adjunctive psychotherapy for bipolar disorder: a systematic review and component network meta-analysis. JAMA Psychiatry 78 (2), 141–150.

Minzenberg, M.J., Yoon, J.H., Soosman, S.K., Carter, C.S., 2018. Altered brainstem responses to modafinil in schizophrenia: implications for adjunctive treatment of cognition. Transl. Psychiatry 8 (1), 58.

Montgomery, S.A., Asberg, M., 1979. A new depression scale designed to be sensitive to change. Br. J. Psychiatry 134, 382–389.

Murray, G., 2018. Adjunctive psychosocial interventions for bipolar disorder: some psychotherapeutic context for the Canadian network for mood and anxiety treatments (CANMAT) & international society for bipolar disorders (ISBD) guidelines. Bipolar Disord. https://doi.org/10.1111/bdi.12655 (2018 May 8. Online ahead of print).

Nemeroff, C.B., Mayberg, H.S., Krahl, S.E., McNamara, J., Frazer, A., Henry, T.R., et al., 2006. VNS therapy in treatment-resistant depression: clinical evidence and putative neurobiological mechanisms. Neuropsychopharmacology 31 (7), 1345–1355.

Nierenberg, A.A., Ostacher, M.J., Calabrese, J.R., Ketter, T.A., Marangell, L.B., Miklowitz, D.J., et al., 2006. Treatment-resistant bipolar depression: a STEP-BD equipoise randomized effectiveness trial of antidepressant augmentation with lamotrigine, inositol, or risperidone. Am. J. Psychiatry 163 (2), 210–216.

Oberman, L., Edwards, D., Eldaief, M., Pascual-Leone, A., 2011. Safety of theta burst transcranial magnetic stimulation: a systematic review of the literature. J. Clin. Neurophysiol. 28 (1), 67–74.

Parker, G.B., Graham, R.K., 2017. Clinical characteristics associated with treatment-resistant bipolar disorder. J. Nerv. Ment. Dis. 205 (3), 188–191.

Plans, L., Barrot, C., Nieto, E., Rios, J., Schulze, T.G., Papiol, S., et al., 2019. Association between completed suicide and bipolar disorder: a systematic review of the literature. J. Affect. Disord. 242, 111–122.

Ramasubbu, R., Clark, D.L., Golding, S., Dobson, K.S., Mackie, A., Haffenden, A., et al., 2020. Long versus short pulse width subcallosal cingulate stimulation for treatment-resistant depression: a randomised, double-blind, crossover trial. Lancet Psychiatry 7 (1), 29–40.

Rhee, T.G., Olfson, M., Nierenberg, A.A., Wilkinson, S.T., 2020. 20-year trends in the pharmacologic treatment of bipolar disorder by psychiatrists in outpatient care settings. Am. J. Psychiatry 177 (8), 706–715.

Ruggero, C.J., Zimmerman, M., Chelminski, I., Young, D., 2010. Borderline personality disorder and the misdiagnosis of bipolar disorder. J. Psychiatr. Res. 44 (6), 405–408.

Rush, A.J., Marangell, L.B., Sackeim, H.A., George, M.S., Brannan, S.K., Davis, S.M., et al., 2005. Vagus nerve stimulation for treatment-resistant depression: a randomized, controlled acute phase trial. Biol. Psychiatry 58 (5), 347–354.

Sackeim, H.A., 2001. The definition and meaning of treatment-resistant depression. J. Clin. Psychiatry 62 (Suppl. 16), 10–17.

Scaini, G., Rezin, G.T., Carvalho, A.F., Streck, E.L., Berk, M., Quevedo, J., 2016. Mitochondrial dysfunction in bipolar disorder: evidence, pathophysiology and translational implications. Neurosci. Biobehav. Rev. 68, 694–713.

Scaini, G., Valvassori, S.S., Diaz, A.P., Lima, C.N., Benevenuto, D., Fries, G.R., et al., 2020. Neurobiology of bipolar disorders: a review of genetic components, signaling pathways, biochemical changes, and neuroimaging findings. Braz. J. Psychiatry 42 (5), 536–551.

Schoeyen, H.K., Kessler, U., Andreassen, O.A., Auestad, B.H., Bergsholm, P., Malt, U.F., et al., 2015. Treatment-resistant bipolar depression: a randomized controlled trial of electroconvulsive therapy versus algorithm-based pharmacological treatment. Am. J. Psychiatry 172 (1), 41–51.

Schosser, A., Serretti, A., Souery, D., Mendlewicz, J., Zohar, J., Montgomery, S., et al., 2012. European Group for the Study of Resistant Depression (GSRD)—where have we gone so far: review of clinical and genetic findings. Eur. Neuropsychopharmacol. 22 (7), 453–468.

Sheehan, D.V., Lecrubier, Y., Sheehan, K.H., Amorim, P., Janavs, J., Weiller, E., et al., 1998. The mini-international neuropsychiatric interview (M.I.N.I.): the development and validation of a structured diagnostic psychiatric interview for DSM-IV and ICD-10. J. Clin. Psychiatry 59 (Suppl. 20), 22–33. quiz 4-57.

Shelton, R.C., Williamson, D.J., Corya, S.A., Sanger, T.M., Van Campen, L.E., Case, M., et al., 2005. Olanzapine/fluoxetine combination for treatment-resistant depression: a controlled study of SSRI and nortriptyline resistance. J. Clin. Psychiatry 66 (10), 1289–1297.

Singh, T., Rajput, M., 2006. Misdiagnosis of bipolar disorder. Psychiatry (Edgmont) 3 (10), 57–63.

Spitzer, R.L., Endicott, J., Robins, E., 1978. Research diagnostic criteria: rationale and reliability. Arch. Gen. Psychiatry 35 (6), 773–782.

Sunder, P., Chia, M.F., Filia, K., Macneil, C., Hasty, M., Davey, C., et al., 2020. Does guideline-concordant care predict naturalistic outcomes in youth with early stage bipolar I disorder? J. Affect. Disord. 278, 23–32.

Swartz, H.A., Swanson, J., 2014. Psychotherapy for bipolar disorder in adults: a review of the evidence. Focus (Am. Psychiatr. Publ.) 12 (3), 251–266.

Swartz, H.A., Thase, M.E., 2011. Pharmacotherapy for the treatment of acute bipolar II depression: current evidence. J. Clin. Psychiatry 72 (3), 356–366.

Tavares, D.F., Myczkowski, M.L., Alberto, R.L., Valiengo, L., Rios, R.M., Gordon, P., et al., 2017. Treatment of bipolar depression with deep TMS: results from a double-blind, randomized, parallel group, sham-controlled clinical trial. Neuropsychopharmacology 42 (13), 2593–2601.

Tsapakis, E.M., Preti, A., Mintzas, M.D., Fountoulakis, K.N., 2020. Adjunctive treatment with psychostimulants and stimulant-like drugs for resistant bipolar depression: a systematic review and meta-analysis. CNS Spectr., 1–12.

van der Loos, M.L., Mulder, P.G., Hartong, E.G., Blom, M.B., Vergouwen, A.C., de Keyzer, H.J., et al., 2009. Efficacy and safety of lamotrigine as add-on treatment to lithium in bipolar depression: a multicenter, double-blind, placebo-controlled trial. J. Clin. Psychiatry 70 (2), 223–231.

Villasenor, A., Ramamoorthy, A., Silva dos Santos, M., Lorenzo, M.P., Laje, G., Zarate Jr., C., et al., 2014. A pilot study of plasma metabolomic patterns from patients treated with ketamine for bipolar depression: evidence for a response-related difference in mitochondrial networks. Br. J. Pharmacol. 171 (8), 2230–2242.

Xu, A.J., Niciu, M.J., Lundin, N.B., Luckenbaugh, D.A., Ionescu, D.F., Richards, E.M., et al., 2015. Lithium and valproate levels do not correlate with ketamine's antidepressant efficacy in treatment-resistant bipolar depression. Neural Plast. 2015, 858251.

Yang, C., Wardenaar, K.J., Bosker, F.J., Li, J., Schoevers, R.A., 2019. Inflammatory markers and treatment outcome in treatment resistant depression: a systematic review. J. Affect. Disord. 257, 640–649.

Yatham, L.N., Kennedy, S.H., Parikh, S.V., Schaffer, A., Beaulieu, S., Alda, M., et al., 2013. Canadian Network for Mood and Anxiety Treatments (CANMAT) and International Society for Bipolar Disorders (ISBD) collaborative update of CANMAT guidelines for the management of patients with bipolar disorder: update 2013. Bipolar Disord 15 (1), 1–44.

Yatham, L.N., Kennedy, S.H., Parikh, S.V., Schaffer, A., Bond, D.J., Frey, B.N., et al., 2018. Canadian Network for Mood and Anxiety Treatments (CANMAT) and International Society for Bipolar Disorders (ISBD) 2018 guidelines for the management of patients with bipolar disorder. Bipolar Disord. 20 (2), 97–170.

Zarate Jr., C.A., Brutsche, N.E., Ibrahim, L., Franco-Chaves, J., Diazgranados, N., Cravchik, A., et al., 2012. Replication of ketamine's antidepressant efficacy in bipolar depression: a randomized controlled add-on trial. Biol. Psychiatry 71 (11), 939–946.

Zhou, Y., Zheng, W., Liu, W., Wang, C., Zhan, Y., Li, H., et al., 2018. Antidepressant effect of repeated ketamine administration on kynurenine pathway metabolites in patients with unipolar and bipolar depression. Brain Behav. Immun. 74, 205–212.

Chapter 27

Treatment-resistant depression in children and adolescents

Melissa DeFilippis and Karen Dineen Wagner
Department of Psychiatry and Behavioral Sciences, University of Texas Medical Branch, Galveston, TX, United States

Introduction

Treatment-resistant depression in children and adolescents is defined as clinically significant depression that has not responded to treatment with a selective serotonin reuptake (SSRI) for at least 8 weeks by the TORDIA (Treatment of SSRI-Resistant Depression in Adolescents) trial (Brent et al., 2009). Dwyer et al. (2020) define treatment-resistant depression in youth as clinically significant depression that fails to respond to an adequate treatment trial of one evidence-based antidepressant or one adequate psychotherapy trial, either in two separate treatment trials or used in combination. They further define treatment-refractory depression as two failed trials of evidence-based antidepressant treatment and a failed adequate psychotherapy trial. A ketamine study in adolescents defined treatment-resistant depression as a failure to respond to at least two antidepressant medications (Cullen et al., 2018). Based on a systematic review of interventions for treatment-resistant depression in adolescents, Boylan et al. (2020) proposed that treatment-resistant depression be defined as failure to respond to two evidence based trials of SSRI antidepressants of adequate dose and duration in combination with psychotherapy. The lack of a consistent definition for treatment-resistant depression in children and adolescents is problematic and can be confusing for clinicians trying to critically appraise the evidence and find appropriate treatments for their patients.

Depression in youth

A diagnosis of major depressive disorder in children and adolescents is made using the same Diagnostic and Statistical Manual, 5th edition (DSM 5) diagnostic criteria for adults, with the exception of irritability as a possible mood state in children and teens, even in the absence of depressed mood (APA, 2013). The lifetime prevalence of depression in adolescents has been reported to be approximately 12% (Merikangas et al., 2010). National Survey on Drug Use and Health data has shown an increase in the prevalence of major depressive disorder from 8.1% to 13.2% among adolescents aged 12–17 years, from the year 2005 to the year 2017 (Twenge et al., 2019). This represents an increase of 52% in the prevalence of major depressive disorder in adolescents over the course of 12 years. Of the children and adolescents diagnosed with major depressive disorder, about 40% will not respond to initial treatment (Bridge et al., 2007).

Even in youth who do respond to the first treatments used, recurrence of depressive symptoms is also a significant problem. In a study examining long term outcomes of 140 adolescents with depression receiving treatment with therapy and/or medication, 93% of youth experienced full remission from their initial depressive episode, but most went on to experience recurrence of depression (53%) and development of a nonmood disorder (79%). Only 15% were reported as having no subsequent depressive episodes or nonmood disorders at follow-up (mean 6 years) (Melvin et al., 2013).

Depression in adolescence has been shown to be associated with significant psychosocial morbidity, including failure to finish school, unemployment, and pregnancy/parenthood in a metaanalysis of long-term outcome studies (Clayborne et al., 2019). Suicide is a greater risk in patients who experience symptoms of depression prior to adulthood, and about 50% of depressed adults experience symptom onset before the age of 18 (Williams et al., 2012). Suicide is now the second leading cause of death in individuals aged 10–34 years (CDC, 2017). Research examining the association of suicidality in youth with treatment-resistant depression shows that youth with reduced right superior temporal gyrus volume may have a greater risk of suicide attempts, suggesting a possible biological marker to help assess risk in this population (McLellan et al., 2018).

More severe illness and the presence of suicidality are associated with poor response to treatment in children and adolescents with depression. Children with comorbid psychiatric disorders, increased levels of hopelessness, and increased family

Managing Treatment-Resistant Depression. https://doi.org/10.1016/B978-0-12-824067-0.00027-X

stress also have a higher risk of poor treatment response. Children with better global functioning prior to treatment may respond better to treatment, as well as children with depressive episodes that are shorter in duration (Emslie et al., 2011). Early treatment response has been shown to be a predictor of positive outcomes in pediatric depression and may also predict long-term response and recovery. In a metaanalysis of randomized placebo-controlled studies in children and adolescents, the greatest benefits of SSRIs occurred early in treatment and were minimal after 4 weeks (Varigonda et al., 2015). The Treatment of Adolescents with Depression (TADS) study showed that early responders at 12 weeks of treatment were more likely to achieve full recovery at 3-year follow-up (Curry et al., 2011).

Treatment-resistant depression studies

The TORDIA trial examined the efficacy of SSRIs, serotonin-norepinephrine reuptake inhibitors (SNRIs), and cognitive behavior therapy (CBT) in the treatment of pediatric SSRI treatment-resistant depression (Brent et al., 2009). The study included 334 patients aged 12–18 years with treatment-resistant depression, defined as no response to at least 8 weeks of treatment with an SSRI at a dosage of at least 40 mg of fluoxetine or its equivalent. The participants were randomized to receive one of four different treatments for the course of the 12-week study. The four treatment groups included switching to a different SSRI, switching to a different SSRI plus CBT, switching to venlafaxine, or switching to venlafaxine plus CBT. Response was defined as a Clinical Global Impression-Improvement score (CGI-I) ≥ 2 (Guy, 1976) and a reduction in the Children's Depression Rating Scale-Revised (CDRS-R) (Poznanski and Mokros, 1996) score of at least 50% from baseline. Switching to either medication and adding CBT produced the highest response rate (54.8%). The response rate in the SSRI group (47%) did not differ significantly from the venlafaxine group (48.2%), but more adverse effects were observed in the venlafaxine group. Increased diastolic blood pressure and pulse and skin problems were observed at a significantly higher rate when compared to the SSRI group. Switching to a second SSRI and adding CBT is recommended in youth who have failed their first antidepressant trial, as SSRIs may have fewer adverse effects.

The TORDIA trial showed that remission at week 24 was associated with early response (Emslie et al., 2010). When comparing the participants who were in remission at 24 weeks with those who were not, the remitted group had significantly less severe symptoms by week 6 of the study. Early response to treatment may be predictive of remission at a much earlier stage than the recommended acute treatment duration of 12 weeks. An analysis of the TORDIA data using a Bayesian hierarchical model showed that switching to an SSRI vs venlafaxine produced a greater response and quicker time to response (Suresh et al., 2020). When examining these results in youth with significant comorbid anxiety, however, venlafaxine performed as well as the SSRIs, suggesting clinicians may consider switching to an SNRI earlier in treatment after a failed response to an initial SSRI in children with treatment-resistant depression and comorbid anxiety.

Long-term follow-up in the TORDIA trial showed that by 72 weeks, about 61% of the participants had achieved remission, with no difference between treatment groups in remission rate or time to remission (Vitiello et al., 2011). Certain factors were associated with a lower chance of remitting, including substance use, severity of depression, and more impairment in functioning at baseline. Anhedonia was shown to be a predictor of poor recovery with SSRIs, with longer time to remission and fewer depression-free days when compared to other depressive symptoms (McMakin et al., 2012). Treatment nonadherence was found to be a common cause of treatment nonresponse (Woldu et al., 2011). A history of nonsuicidal self-injury and hopelessness were significant predictors for suicide attempts throughout the study (Asarnow et al., 2011). Venlafaxine was associated with more treatment-emergent suicidality when compared to the SSRI group, a finding which further strengthened the recommendation to try a second SSRI prior to switching to an SNRI (Brent et al., 2009).

Substance use was fairly common in the TORDIA trial, with about 28% of participants reporting repeated substance use (between three and nine times in the previous month), and youth who had lower substance use at baseline or who decreased their substance use during the study showed better response to treatment during the study (Goldstein et al., 2009). Parent conflict was an important factor in response to treatment in the TORDIA trial (Rengasamy et al., 2013). Higher parent-reported conflict predicted lower rates of remission, though there was no effect seen on treatment response. Youth who received more than nine CBT sessions had a higher rate of response than those who had nine or less sessions, emphasizing the importance of an adequate number of sessions for treatment-resistant depression in youth (Kennard et al., 2009a). Overall, the TORDIA trial showed a response rate of about 50% when switching to a second SSRI. With approximately 40% of depressed youth not responding to initial treatment (from acute depression treatment studies), this means that approximately 20% of depressed youth (half of initial nonresponders) will not respond to the first two treatments tried. There is inadequate evidence guiding clinicians in treatment decisions for these remaining youth.

Augmentation of antidepressants with antipsychotic medications has shown some benefit in adult depression studies, but there is currently very little evidence examining their efficacy in children and adolescents with depression, and no randomized controlled trials in this population. A case series examining the efficacy of quetiapine (median dose 200 mg/day) augmentation in 10 adolescents with treatment-resistant depression showed a 70% response rate (CGI-I ≥ 2) (Pathak et al., 2005). A case report described improvement in depressive symptoms with adjunctive clozapine for an adolescent with treatment-resistant depression (Whitlock et al., 2020). Large, randomized controlled treatment studies are needed before recommending this as an evidence-based augmentation strategy in children and adolescents with depression. The potential side effects with antipsychotics are concerning, and children and adolescents seem to be at an increased risk of experiencing cardiometabolic side effects with the second-generation antipsychotics (Correll et al., 2009). Other augmentation strategies using medications may include lithium, bupropion, and mirtazapine, but are based on adult studies, with no randomized controlled trials showing efficacy or safety of these strategies in the pediatric population (Hughes et al., 2007).

The FDA has approved repetitive transcranial magnetic stimulation (rTMS) as a treatment for treatment-resistant depression in adults, but it is not approved for use in the pediatric population. A randomized controlled trial examining the efficacy of TMS for treatment-resistant depression in adolescents included a double-blind design using a sham TMS treatment for the control group (Croarkin et al., 2020). The trial included 103 adolescents, aged 12–21 years, with treatment-resistant depression. This was a 6-week study, with a total of 30 treatments given as monotherapy. Participants were randomized to receive left prefrontal 10 Hz TMS or sham TMS treatment (including a sham coil that appeared identical to the active coil, an acoustically matched profile, and a mild percussive sensation to mimic active treatment). The primary outcome measure was change in the Hamilton Depression Rating Scale (HAM-D-24) (Hamilton, 1967) from baseline to endpoint. There was no significant difference between the active TMS treatment group and sham TMS treatment group on the primary outcome measure, with change in scores being − 11.1 and − 10.6, respectively. There were also no significant differences between TMS and sham TMS in response (41.7% and 36.4%, respectively) or remission rates (29.2% and 29.0%, respectively). Participants in the active treatment group reported side effects, including headaches, eye pain, nausea, and facial twitching, which were observed significantly more than in the sham treatment group.

A recent systematic review showed 14 studies examining the efficacy of TMS in adolescents with depression, eight of which were open-label, multisubject trials (Hett et al., 2021). The other six studies were post hoc analyses of the open label studies and one 3-year follow-up of an open trial. The eight open-label studies included a total of 142 adolescents with depression, all of which showed TMS having some effect on reducing symptoms of depression. The review concluded the studies were of either poor or fair quality, with a high risk of bias. A recent randomized sham-controlled trial compared active TMS and sham TMS in 103 adolescents aged 12–21 with treatment-resistant depression. Each group received 30 daily treatments over 6 weeks. There was no significant difference between the two groups on the primary outcome measure, a change in the Hamilton Depression Rating Scale from baseline to endpoint (Croarkin et al., 2021). More large randomized controlled trials are needed before routinely recommending TMS for adolescents with treatment-resistant depression.

There are no randomized controlled trials examining the efficacy of electroconvulsive therapy (ECT) in children and adolescents with depression. A retrospective chart review examined the efficacy of ECT in 54 adolescents with treatment-resistant mood disorders, including unipolar and bipolar depression (Ghaziuddin et al., 2020). The youth received their first ECT treatment course prior to the age of 18. Response was defined as a Clinical Global Impressions (CGI) score ≤ 2 and was assessed after initial treatment and at 1-year follow-up. The mean number of treatments was 13.7 for initial treatment and the response rate was 52.8%. The study showed that 82% of the youth were considered responders at 1-year follow-up, with a reduction in suicidal ideation and self-injurious behaviors. Prolonged seizures were common during ECT (74%) and long-term side effects were memory loss limited to the time surrounding ECT treatment. A retrospective observational study including 190 adolescents and young adults, aged 16–25 years, examined the efficacy of ECT for multiple psychiatric diagnoses, including 128 subjects with treatment-resistant depression and showed significant improvement on self-reported depressive symptoms as a group (the study did not report on percent of responders separately) (Benson et al., 2019). A case series of 13 adolescents aged 15–18 years with treatment-resistant depression showed a mean of 14 ECT sessions to be effective for reducing depressive symptoms in 10 of the 13 (77%) participants (Zhand et al., 2015).

A retrospective study of ECT in adolescents with treatment-resistant depression showed an average improvement of 46% in the participants' symptoms, with a third of participants experiencing at least 60% improvement in symptoms (Hegeman et al., 2008a,b). A retrospective case report showed significant improvement in 83% of adolescents (aged 14–17 years) with treatment-resistant depression receiving ECT. The study included six subjects, and ECT was well-tolerated by the adolescents in the study (Ghaziuddin et al., 2011). Side effects of ECT can include transient cognitive impairment, headache, prolonged seizures, and nausea. The American Academy of Child and Adolescent Psychiatry developed practice parameters for the use of ECT in adolescents (Neera et al., 2004). The recommendation is that ECT should be considered for severe, persistent, and significantly disabling depression in adolescents who have failed at least two adequate trials of appropriate psychopharmacological agents along with other appropriate treatments.

Ketamine is another treatment being used in adults with treatment-resistant depression, but there are no randomized controlled studies in children and adolescents. An open-label study examined the efficacy of intravenous ketamine in 13 adolescents aged 12–18 years with treatment-resistant depression (Cullen et al., 2018). The participants were administered six ketamine (0.5 mg/kg) infusions over 2 weeks. Response (≥50% reduction in CDRS-R score from baseline to endpoint) was seen in five (38%) of the adolescents, with relapse occurring within 2 weeks in two of these responders. Adverse effects reported included transient blood pressure changes, transient (within 1 h) dissociative symptoms, dysphoria, and nausea. A systematic review on ketamine treatments for use in adolescents included the above study and two case reports in adolescents with treatment-resistant depression (Kim et al., 2020). Ketamine was shown to improve depressive symptoms in some patients. The authors concluded that ketamine may have potential for treatment-resistant depression in youth but considering ketamine in treatment guidelines for adolescents with treatment-resistant depression is premature, given the limited evidence of efficacy and safety in this patient population. A commentary on esketamine use in adolescents raised concerns about repetitive and chronic exposures to esketamine in this patient population (Zimmermann et al., 2020). Animal studies of esketamine on developing brains show permanent negative effects on the neurodevelopmental process in the prefrontal cortex, which is a risk that could be heightened in children and adolescents with genetic risk factors for schizophrenia. Ketamine and esketamine lack sufficient efficacy and safety data to recommend for use in children and adolescents with treatment-resistant depression.

Adjunctive L-methylfolate was examined as a potential treatment for treatment-resistant depression in 10 adolescents (mean age 14) in a case series (Dartois et al., 2019). Eighty percent of patients had a single mutation among two methylene tetrahydrofolate reductase (MTHFR) gene variants, indicative of reduced MTHFR activity. The dose of L-methylfolate ranged from 2 to 15 mg/day. In the study, 80% of participants showed improvement in symptoms of depression, irritability, and anxiety. L-Methylfolate was well-tolerated. More rigorous studies are required in this patient population before L-methylfolate augmentation can be recommended for adolescent treatment-resistant depression.

A medication consensus treatment guideline was developed for major depressive disorder in youth (Hughes et al., 2007) but has not been updated. The recommendation was initial treatment with an SSRI, and then switching to a second SSRI if no response. This recommendation is now consistent with the findings of the TORDIA study. If there is a partial response to treatment with an SSRI, augmentation with a second antidepressant may be considered, although this has not been studied in children and adolescents. If there is no response to the second SSRI tried, an alternate antidepressant from a different class (venlafaxine, bupropion, mirtazapine, duloxetine) is recommended. Desvenlafaxine and vilazodone have now been studied in youth with depression and may be considered in the list of alternate antidepressants. If there is still no response, the clinician should reassess symptoms. Newer antidepressants should be used with caution due to limited safety and efficacy data in children and adolescents.

Antidepressant studies for acute treatment

It is important to use evidence-based treatments for acute depressive episodes in children and adolescents in order to decrease the risk of treatment-resistant cases. SSRIs are the medications with the most evidence supporting their use in the treatment of depression in children and adolescents. Studies examining the efficacy of SSRIs in acute pediatric depressive episodes led to Food and Drug Administration (FDA) approval for two SSRIs in this population. Fluoxetine is approved for children with major depression aged 8 years and older, and escitalopram is approved for adolescents with major depression aged 12 years and older. Three randomized controlled trials examining the efficacy of fluoxetine in the acute treatment of major depressive disorder showed it to be superior to placebo in children and adolescents (Emslie et al., 1997, 2002; March et al., 2004). Doses used in these studies ranged from 10 to 40 mg/day, with two studies using fixed dosing of 20 mg/day. Escitalopram has one positive randomized, placebo-controlled trial in adolescents, aged 12–17 (Emslie et al., 2009), which contributed to its FDA approval for adolescent depression. Dosing ranged from 10 to 20 mg/day. Sertraline and citalopram are not FDA approved for the treatment of pediatric depression, but they have shown benefit in randomized trials and may be considered in this patient population (Wagner et al., 2003, 2004).

Other antidepressants have been studied in children and adolescents with major depressive disorder, with negative results. Escitalopram was not found to be superior to placebo in a randomized controlled trial that included younger children, aged 6–17 years (Wagner et al., 2006). Paroxetine was not found to be superior to placebo on most outcome measures in the treatment of major depressive disorder in children and adolescents in three randomized controlled trials (Berard et al., 2006; Emslie et al., 2006; Keller et al., 2001). Citalopram did not separate from placebo in a randomized controlled trial in adolescent inpatients with major depressive disorder (Von Knorring et al., 2006).

Venlafaxine ER was not superior to placebo in the treatment of pediatric major depressive disorder in two placebo-controlled trials (Emslie et al., 2007). In two 10-week randomized controlled comparison trials of duloxetine in the treatment of pediatric major depressive disorder, duloxetine and fluoxetine were not superior to placebo (Atkinson et al., 2014; Emslie et al., 2014). The efficacy of desvenlafaxine versus fluoxetine for major depressive disorder in children and adolescents aged 7–18 was examined in an 8-week randomized controlled trial (Weihs et al., 2018). Neither the desvenlafaxine nor the fluoxetine treatment groups were shown to be superior to placebo. A metaanalysis of five trials, including 973 patients, found that SNRIs (duloxetine and venlafaxine) were not superior to placebo in the acute treatment of major depressive disorder in children and adolescents, aged 7–18 years (Xu et al., 2016). Adverse effects reported in these studies included abdominal pain and dizziness in the venlafaxine studies and nausea, headache, and nasopharyngitis in the duloxetine studies, but showed the treatments to be generally well-tolerated.

Mirtazapine has had negative unpublished randomized controlled trials in children and adolescents with major depressive disorder (Cheung et al., 2005). A randomized controlled trial examined the efficacy of selegiline transdermal system in the treatment of major depressive disorder in 308 adolescents and did not show a difference between active treatment and placebo (DelBello et al., 2014). Vilazodone was examined as a potential treatment for 473 children and adolescents (aged 7–17 years) with major depressive disorder in an 8-week randomized, double-blind, placebo-controlled trial (Findling et al., 2020). Treatment groups included vilazodone (15 or 30 mg/day), fluoxetine (20 mg/day), or placebo. Neither vilazodone nor fluoxetine were superior to placebo, but vilazodone was well-tolerated. Vilazodone was similarly not superior to placebo for the acute treatment of depression in another randomized controlled trial including adolescents with major depressive disorder (Durgam et al., 2018). A small, open-label trial of bupropion showed positive results in children with comorbid ADHD, but no large, controlled studies have been conducted in children or adolescents with major depressive disorder (Daviss et al., 2001). Vortioxetine has shown promise as a possible treatment for pediatric depression in an open-label study including youth with depressive or anxiety disorders, but no randomized-controlled trials have been completed with this medication (Findling et al., 2018).

A recent metaanalysis of antidepressants in children and adolescents showed that fluoxetine compared to placebo is more efficacious than other antidepressants for major depressive disorder (Boaden et al., 2020). It also showed venlafaxine to be associated with an increased risk of treatment-emergent suicidality in depression studies. A metaanalysis of antidepressants in the treatment of pediatric depression showed a placebo response rate of 50% in treatment studies, which is high, especially when considering the antidepressant treatment response rates of 60% (Bridge et al., 2007). This makes it very difficult for active treatments to show efficacy in treatment studies.

Despite the number of negative trials of antidepressants, these studies provide important safety data which clinicians may find helpful when considering treatment with antidepressants in children and adolescents that are off label. When treating with medication, clinicians should ensure patients have an adequate trial of medication with adequate dosing. Doses of medications should be safely increased to maximum recommended doses in youth depending upon clinical response. In the Treatment for Adolescents with Depression (TADS) study, response and remission rates improved significantly between 12 weeks (acute treatment) and 36 weeks with fluoxetine treatment (Kennard et al., 2009a). A discontinuation study including 102 youth with acute response to fluoxetine at 12 weeks, showed continuation treatment with fluoxetine to be superior to placebo in the 6-month follow-up study. Relapse rates were significantly higher in the placebo group (69.2%) compared to the fluoxetine continuation group (42%), with a shorter time to relapse in the placebo group (Emslie et al., 2008). A meta-analysis including 882 depressed youth showed that antidepressants significantly reduced the risk of relapse and recurrence of depressive episodes in children and adolescents, with a relapse/recurrence rate of 40.9% in the youth on antidepressants versus 66% in the youth on placebo (Cox et al., 2012). Recommendations include keeping patients on treatments for at least a year after response is seen in order to avoid relapse (Wagner and Pliszka, 2013). Youth with higher number of recurrences and longer recovery periods should have a longer maintenance period of treatment, as should those with chronic depressive episodes, with recommendations to continue treatment longer than a year (Hughes et al., 2007).

Common side effects associated with SSRIs include gastrointestinal symptoms, headache, insomnia, somnolence, and dry mouth (Emslie et al., 2002; Keller et al., 2001; Wagner et al., 2004, 2006). The risk of QT interval prolongation on EKG is potentially increased with some antidepressants. A metaanalysis examining this association found a modest but significant effect of SSRI use on QT interval prolongation, with citalopram causing more prolongation than other SSRIs (Beach et al., 2014). The FDA recommends citalopram dosing of 40 mg/day or less to mitigate this risk (FDA, 2012). The risk of QT interval prolongation was not observed in other studies examining this side effect with SSRIs, including citalopram (Van Haelst et al., 2014; Zivin et al., 2013). A study examining the connection between QT interval prolongation and psychotropic medications found minimal evidence connecting clinically meaningful QT prolongation with most classes of psychotropics and recommended clinicians focus on other risk factors for QT prolongation when prescribing psychiatric medications (Beach et al., 2018). Less common but possible side effects with SSRIs include, but are not limited to, decreased bone density, increased risk of bleeding, metabolic effects, liver toxicity, extrapyramidal symptoms, serotonin syndrome, tics, and myoclonus (Hammerness et al., 2006; Jerrell, 2010; Rizzoli et al., 2012; Voican et al., 2014).

The black box warning on all antidepressants regarding the risk of suicidality applies to children, adolescents, and young adults (FDA, 2004). After the FDA warning was posted, a decrease in antidepressant prescriptions was seen among children and adolescents without the expected increase in psychotherapy treatments, and suicide rates increased (Bridge et al., 2008; Kurian et al., 2007; Libby et al., 2009). A study using validated death data in the United States from 1990 to 2017 showed a downward trend of suicide deaths in adolescents the decade preceding the FDA warning (Lu et al., 2020). After the FDA warning, declines were seen in both diagnosis of depression and in depression treatment. The suicide trend reversed from 4.4 suicide deaths per 100,000 adolescents in 2005 to 7.2 in 2017, a change that was seen almost immediately after the warning. The study estimates that this may have resulted in 2365 excess suicide deaths in adolescents in the United States from 2005 to 2010. While these effects cannot be directly attributed to the FDA warning, they are concerning. A meta-analysis of studies concluded that the benefits of antidepressants in the pediatric population outweigh the risks. This study showed a number needed to treat of 10 and a number needed to harm of 112 (Bridge et al., 2007). Though the TORDIA trial found a possible increased risk of treatment emergent side effects with venlafaxine compared to SSRIs, another study found no difference for risk of suicide attempts between the different antidepressants in children and adolescents (Cooper et al., 2014). Clinicians should follow dosing guidelines when starting youth on antidepressants and titrate carefully, as evidence shows there may be a connection between self-injurious behaviors in youth starting at a higher than modal dose of these medications (Miller et al., 2014). Any child or adolescent started on an antidepressant should be followed closely by their treating physician, especially early in treatment, for response and possible adverse effects.

Psychotherapy studies for acute treatment

CBT is a treatment option clinicians should consider at any point in treatment with depressed youth. The TORDIA trial showed that combination treatment with medication and CBT was most effective for treatment-resistant depression (Brent et al., 2008). There are not any other studies examining the efficacy of CBT in youth with treatment-resistant depression, but there is ample evidence supporting its use in acute depressive episodes. In a randomized trial examining the efficacy of CBT in 175 adolescents with major depressive disorder, CBT was shown to be superior to systematic behavior family therapy and individual nondirective supportive therapy in both response and remission rates (Brent et al., 1997). There have been multiple studies showing similar benefits of CBT in the treatment of child and adolescent depression (Clarke et al., 1999; Lewinsohn et al., 1990; Rosselló and Bernal, 1999; Wood et al., 1996). The TADS study showed no difference between the CBT group and the placebo group in acute treatment; however, long-term follow-up results showed response and remission rates both increased significantly over time in the CBT group (Kennard et al., 2009b; March et al., 2007). Response rates went from 43% to 65% to 81% at 12, 18, and 36 weeks, respectively. Remission rates also increased with time, going from 19% (12 weeks) to 27% (18 weeks) to 64% (36 weeks). This suggests CBT monotherapy may require more time to see improvement in depressive symptoms.

Interpersonal therapy has evidence supporting its use in children and adolescents with depressive disorders. A 16-week randomized clinical trial of 63 adolescents with depression examined the efficacy of IPT modified for depressed adolescents (IPT-A) and was found to be superior to treatment as usual (Mufson et al., 2004). Several other studies have shown similar efficacy of IPT-A in the treatment of adolescent depression (Mufson et al., 1999; Tze-Chun et al., 2009; Young et al., 2010). IPT is a good option to consider at any point in the treatment of depressed youth.

A multicenter, observer-blinded, randomized controlled comparison trial (IMPACT) examined the efficacy of CBT, short-term psychoanalytical therapy, and brief psychological intervention in 470 adolescents (aged 11–17 years) with major depressive disorder (Goodyer et al., 2017). All three treatment groups showed improvement, with no significant difference on outcome measures, which included self-reported symptoms on the Mood and Feelings Questionnaire (MFQ) (Daviss et al., 2006). These results suggest even brief psychological interventions may have some benefit in depressed adolescents. Family-based interventions have been shown to be effective for depressive symptoms in adolescents. Attachment-based family therapy (ABFT) includes weekly therapy sessions focusing on the parent and child relationship. It was shown to be helpful for decreasing depressive symptoms and suicidal ideation in a randomized controlled trial of 66 adolescents (Diamond et al., 2010).

A recent metaanalysis of 55 randomized psychotherapy trials for youth aged 4–18 years with depression showed an overall modest effect size of 0.36 at posttreatment and 0.21 at follow-up (average 42 weeks after posttreatment) with psychotherapy (Eckshtain et al., 2020). Effects were significantly larger for IPT-A than for CBT and for youth-reported outcomes than parent-reported. A metaanalysis examining treatment effects with psychotherapy in depression across the age span showed significantly lower effect sizes in children and adolescents when compared to adults, with effect sizes of 0.35 in children, 0.55 in adolescents, and 0.98 in young adults (Cuijpers et al., 2020). A literature review of psychotherapy studies showed similar findings, with CBT being classified as *possibly efficacious* in children with depression and CBT and IPT-A being classified as *well-established, effective treatments* in adolescents with depression (Weersing et al., 2017).

Complementary alternative medicine

Complementary alternative treatments have been examined as possible alternatives to traditional antidepressants in children and adolescents. Studies examining the efficacy of omega-3 fatty acids in the treatment of acute pediatric depression have mixed results. A 10-week randomized controlled trial in adolescents aged 12–19 years with major depressive disorder did not find any significant difference between omega-3 fatty acids (eicosatetraenoic acid (EPA) and docosahexaenoic acid (DHA)) and placebo on any of the outcome measures (Gabbay et al., 2018). A 16-week, double-blind randomized controlled study included 28 children with major depressive disorder, aged 6–12 years, randomized to either omega-3 (EPA and DHA) or placebo (Nemets et al., 2006). Response and remission rates were found to be significantly higher in the omega-3 treatment group as compared to placebo. Omega-3 fatty acids appeared to be well-tolerated in both studies and may be considered as adjunctive treatment in children and adolescents with depression, but more evidence examining their efficacy in this population is needed. There are no randomized controlled trials examining the efficacy of augmentation of antidepressants with Omega-3 fatty acids in pediatric depression or in treatment-resistant pediatric depression.

Vitamin C was studied as an adjunctive treatment to antidepressants for major depressive disorder in 24 youth aged 7–14 years in a 6-month, double-blind, placebo-controlled trial (Amr et al., 2013). The treatment groups included fluoxetine (10–20 mg/day) plus placebo and fluoxetine (10–20 mg/day) plus vitamin C (1000 mg/day). The group receiving combined treatment with fluoxetine and vitamin C showed significant improvement when compared to the group receiving fluoxetine and placebo. It was well-tolerated with no major adverse effects reported in the study. Larger studies are needed to examine the efficacy of adjunctive vitamin C in the treatment of pediatric depression. The efficacy of vitamin D supplementation on depressive symptoms was evaluated in a randomized controlled trial of 113 youth with vitamin D deficiency and depressive symptoms (Libuda et al., 2020). No significant difference was found between the vitamin D and placebo groups on depressive symptoms at endpoint. The efficacy of zinc supplementation (10 mg of zinc oxide) for acute pediatric depression was examined in a 6-month, double-blind, placebo-controlled trial in 674 school-aged children, and was not found to be superior to placebo on any of the outcome measures (DiGirolamo et al., 2010). Other complementary alternative treatments have been studied in small, open-label studies or in case reports but lack randomized controlled trials examining their efficacy in pediatric depression.

Combination treatment

In cases where depressed patients may not be fully responding to medication or psychotherapy alone, combination treatment with medication and psychotherapy may be considered. Studies have shown benefit with combination treatment in children and adolescents with depression. The TADS study showed combination treatment (fluoxetine plus CBT) to be most effective compared to fluoxetine alone, CBT alone, and placebo in a 12-week study (March et al., 2004). A greater decrease in suicidal ideation, acceleration in treatment response, and a 1.5-fold increased probability of sustained early response to treatment was also found in the combination treatment group (Kratochvil et al., 2006). A recent metaanalysis of 71 trials showed that fluoxetine alone or in combination with CBT was superior to CBT alone and psychotherapy for moderate to severe depressive disorder in children and adolescents (Zhou et al., 2020). The addition of CBT may be effective in preventing relapse when added to antidepressant medication. Two sequential treatment studies included youth who responded to acute treatment with fluoxetine who were then randomized to receive continuation treatment for 6 months with fluoxetine alone or fluoxetine plus CBT. The youth who had the addition of CBT to their treatment had a significantly lower risk of relapse by the end of the studies (Kennard et al., 2008, 2014). These findings further support the recommendation to add CBT at any point in treatment to improve outcomes in depressed youth.

The ADAPT (Adolescent Depression and Psychotherapy Trial) examined the efficacy of CBT alone compared with CBT plus SSRI in adolescents who had failed to respond to brief psychotherapy (Goodyer et al., 2008). Youth in both treatment groups showed response (much or very much improved on CGI-I), but the response rates did not differ significantly between the groups at weeks 12 or 28. Suicidality decreased in both treatment groups, with no significant difference between the groups, suggesting CBT may have a protective effect for depressed youth with suicidality.

A systematic review showed CBT in combination with antidepressants to be effective for depressive symptoms and associated with lower rates of relapse in children and adolescents (Viswanathan et al., 2020). The addition of CBT to treatment with an SSRI may increase response rates (Brent et al., 2008; March et al., 2004). The benefit of CBT in improving response rates as well as preventing relapse supports the use of CBT at any phase of treatment in youth with treatment-resistant depression.

Factors associated with treatment nonresponse

If an adequate evidence-based treatment trial is not effective in treating an acute depressive episode in a child or adolescent patient, the clinician should consider possible diagnostic, medical, or psychosocial contributing factors. Revisiting diagnostic criteria may be necessary to ensure a different diagnosis, such as bipolar disorder, has not been missed is an important step (Maalouf et al., 2011). Clinicians should also discuss with parents the importance of monitoring their children to ensure they are compliant with the prescribed treatments, as noncompliance can be a contributor to treatment failure. The possible effects of comorbid disorders on the depression diagnosis and treatment may need to be considered. Medical comorbidities can also affect treatment response or contribute to presenting depressive symptoms (Kovacs et al., 1996; Lewinsohn et al., 1996). Psychosocial factors, such as peer relationships, bullying, abuse, or parental mental illness can affect response and should always be assessed as part of the complete evaluation when a child is in treatment. A population-based study including 1443 children followed from 5 months to 15 years showed that maternal depression in the early childhood years (prior to age 5) was associated with higher levels of adolescent depression, generalized anxiety, and social phobia symptoms (Côté et al., 2018). Bullying from peers during childhood (ages 6–12) was a significant mediator in this study, accounting for a 35.9% association with major depression.

A study examining the connection between peer victimization and mental health disorders also examined a school-based treatment program targeting bullying (Tiiri et al., 2020). This study found that youth experiencing traditional or cyberbullying victimization had higher levels of mental health problems. A nationwide school-based antibullying program was associated with a 30% decrease in traditional victimization and an increase in child-reported bullying to adults. This emphasizes the importance of school and parental involvement in decreasing peer victimization and improving mental health outcomes in youth. In another study, peer and parental support was associated with maintained improvement in depressive symptoms in children who received targeted cognitive-behavioral family interventions (Healy and Sanders, 2017). The study found that children with depressive symptoms 6 months after the intervention were more likely to have lower levels of peer support and facilitative parenting immediately following the intervention and increased social victimization as reported by teachers. Certain stressors in early childhood seem to increase the risk for depressive symptoms. A meta-analysis of 62 articles on early life stressors found that sexual abuse, physical abuse, death of a family member, domestic violence, and emotional abuse were associated with a significantly increased risk of a major depression diagnosis before age 18 (LeMoult et al., 2020).

Genetic differences in metabolism of medications (P450 polymorphisms) may be considered when children do not respond to initial evidence-based treatments (Schosser and Kasper, 2009). With the advances in genetic testing, pharmacogenetic testing is a topic parents may bring up with the clinician in appointments, especially in treatment-resistant cases. While these tests are becoming easier to access, it is important to discuss with parents that the tests do not predict response to treatment, and caution should be taken when interpreting these test results. The FDA recently released a statement about these pharmacogenetic tests which stated that claims of genetic tests to predict response to medications may not have evidence to support this use for most medications, including antidepressants (FDA, 2018). The FDA cautioned that the relationship between genetic variations and effectiveness of antidepressant medications has not been established, and that clinicians should not make changes to a patient's medication regimen based on the results of these tests. The American Academy of Child and Adolescent Psychiatry (AACAP) policy statement on pharmacogenetic testing recommends clinicians avoid using testing results to select treatments for children and adolescents due to lack of evidence supporting its use in this population (AACAP, 2020). The statement notes that many factors affect treatment response in children, and genetic variations are best addressed by the clinician with careful medication management and monitoring.

Conclusion

Depression is a common psychiatric diagnosis in youth, with lifetime prevalence in adolescence similar to that seen in adults. It can lead to a wide array of impairment in social, academic, and family functioning. A diagnosis of depression increases the risk of suicide, the second leading cause of death in patients aged 10–34, highlighting the importance of finding effective treatments for youth with depression. Using evidence-based treatment in the acute management of depression in children and adolescents may lower the risk of the development of treatment-resistant depression. Based on consensus guidelines and the evidence-base, SSRIs are recommended as initial medication treatment for children and adolescents with major depressive disorder. If there is no treatment response after an adequate dose and duration of an SSRI, switch to another evidence-based SSRI is indicated. If response is still inadequate, switching to a non-SSRI (SNRIs, for example) is recommended. Augmentation with a second medication or combination treatment with two antidepressant medications may be needed if there is a partial response to the antidepressant, though there is limited evidence in pediatric depression.

Psychotherapy, particularly CBT or IPT-A, may be considered at any time in treatment to improve response, and there is evidence suggesting that CBT prevents relapse in youth with acute response to treatment. Other somatic treatments, such as rTMS and ECT have been used for treatment of youth with major depression who have failed to respond to medications and psychotherapy. Ketamine is currently being studied for treatment-resistant depression in youth. Though the body of research in pediatric depression treatment is growing, the evidence examining effective treatments for treatment-resistant depression in youth remains limited, and more studies are needed to help guide clinicians in managing these challenging cases.

References

AACAP, 2020. Policy statement: clinical Use of Pharmacogenetic Tests in Prescribing Psychotropic Medications for Children and Adolescents. AACAP.

Amr, M., El-Mogy, A., Shams, T., Vieira, K., Lakhan, S.E., 2013. Efficacy of vitamin C as an adjunct to fluoxetine therapy in pediatric major depressive disorder: a randomized, double-blind, placebo-controlled pilot study. Nutr. J. 12 (1). https://doi.org/10.1186/1475-2891-12-31.

APA, 2013. Diagnostic and Statistical Manual of Mental Disorders. APA.

Asarnow, J.R., Porta, G., Spirito, A., Emslie, G., Clarke, G., Wagner, K.D., Vitiello, B., Keller, M., Birmaher, B., McCracken, J., Mayes, T., Berk, M., Brent, D.A., 2011. Suicide attempts and nonsuicidal self-injury in the treatment of resistant depression in adolescents: findings from the TORDIA study. J. Am. Acad. Child Adolesc. Psychiatry 50 (8), 772–781. https://doi.org/10.1016/j.jaac.2011.04.003.

Atkinson, S.D., Prakash, A., Zhang, Q., Pangallo, B.A., Bangs, M.E., Emslie, G.J., March, J.S., 2014. A double-blind efficacy and safety study of duloxetine flexible dosing in children and adolescents with major depressive disorder. J. Child Adolesc. Psychopharmacol. 24 (4), 180–189. https://doi.org/10.1089/cap.2013.0146.

Beach, S.R., Kostis, W.J., Celano, C.M., Januzzi, J.L., Ruskin, J.N., Noseworthy, P.A., Huffman, J.C., 2014. Meta-analysis of selective serotonin reuptake inhibitor-associated QTc prolongation. J. Clin. Psychiatry 75 (5), e441–e449. https://doi.org/10.4088/JCP.13r08672.

Beach, S.R., Celano, C.M., Sugrue, A.M., Adams, C., Ackerman, M.J., Noseworthy, P.A., Huffman, J.C., 2018. QT prolongation, torsades de pointes, and psychotropic medications: a 5-year update. Psychosomatics 59 (2), 105–122. https://doi.org/10.1016/j.psym.2017.10.009.

Benson, N.M., Seiner, S.J., Bolton, P., Fitzmaurice, G., Meisner, R.C., Pierce, C., Busch, A.B., 2019. Acute phase treatment outcomes of electroconvulsive therapy in adolescents and young adults. J. ECT 35 (3), 178–183. https://doi.org/10.1097/YCT.0000000000000562.

Berard, R., Fong, R., Carpenter, D.J., Thomason, C., Wilkinson, C., 2006. An international, multicenter, placebo-controlled trial of paroxetine in adolescents with major depressive disorder. J. Child Adolesc. Psychopharmacol. 16 (1–2), 59–75. https://doi.org/10.1089/cap.2006.16.59.

Boaden, K., Tomlinson, A., Cortese, S., Cipriani, A., 2020. Antidepressants in children and adolescents: meta-review of efficacy, tolerability and suicidality in acute treatment. Front. Psychiatry 11. https://doi.org/10.3389/fpsyt.2020.00717.

Boylan, K., MacQueen, G., Kirkpatrick, R., Lee, J., Santaguida, P.L., 2020. A systematic review of interventions for treatment resistant major depressive disorder in adolescents. Eur. Child Adolesc. Psychiatry 29 (4), 433–443. https://doi.org/10.1007/s00787-019-01341-5.

Brent, D.A., Holder, D., Kolko, D., Birmaher, B., Baugher, M., Roth, C., Iyengar, S., Johnson, B.A., 1997. A clinical psychotherapy trial for adolescent depression comparing cognitive, family, and supportive therapy. Arch. Gen. Psychiatry 54 (9), 877–885. https://doi.org/10.1001/archpsyc.1997.01830210125017.

Brent, D., Emslie, G., Clarke, G., Wagner, K.D., Asarnow, J.R., Keller, M., Vitiello, B., Ritz, L., Iyengar, S., Abebe, K., Birmaher, B., Ryan, N., Kennard, B., Hughes, C., DeBar, L., McCracken, J., Strober, M., Suddath, R., Spirito, A., et al., 2008. Switching to another SSRI or to venlafaxine with or without cognitive behavioral therapy for adolescents with SSRI-resistant depression: the TORDIA randomized controlled trial. JAMA 299 (8), 901–913. https://doi.org/10.1001/jama.299.8.901.

Brent, D.A., Emslie, G.J., Clarke, G.N., Asarnow, J., Spirito, A., Ritz, L., et al., 2009. Predictors of spontaneous and systematically assessed suicidal adverse events in the treatment of SSRI-resistant depression in adolescents (TORDIA) study. Am. J. Psychiatry 166 (4), 418–426. https://doi.org/10.1176/appi.ajp.2008.08070976.

Bridge, J.A., Iyengar, S., Salary, C.B., Barbe, R.P., Birmaher, B., Pincus, H.A., Ren, L., Brent, D.A., 2007. Clinical response and risk for reported suicidal ideation and suicide attempts in pediatric antidepressant treatment: a meta-analysis of randomized controlled trials. J. Am. Med. Assoc. 297 (15), 1683–1696. https://doi.org/10.1001/jama.297.15.1683.

Bridge, J.A., Greenhouse, J.B., Weldon, A.H., Campo, J.V., Kelleher, K.J., 2008. Suicide trends among youths aged 10 to 19 years in the United States, 1996–2005. JAMA 300 (9), 1025–1026. https://doi.org/10.1001/jama.300.9.1025.

CDC, 2017. United States mortality tables by 5y age, race, sex (1999–2014). In: National Vital Statistics System. CDC.

Cheung, A.H., Emslie, G.J., Mayes, T.L., 2005. Review of the efficacy and safety of antidepressants in youth depression. J. Child Psychol. Psychiatry 46 (7), 735–754. https://doi.org/10.1111/j.1469-7610.2005.01467.x.

Clarke, G.N., Rohde, P., Lewinsohn, P.M., Hops, H., Seeley, J.R., 1999. Cognitive-behavioral treatment of adolescent depression: efficacy of acute group treatment and booster sessions. J. Am. Acad. Child Adolesc. Psychiatry 38 (3), 272–279. https://doi.org/10.1097/00004583-199903000-00014.

Clayborne, Z.M., Varin, M., Colman, I., 2019. Systematic review and meta-analysis: adolescent depression and long-term psychosocial outcomes. J. Am. Acad. Child Adolesc. Psychiatry 58 (1), 72–79. https://doi.org/10.1016/j.jaac.2018.07.896.

Cooper, W.O., Callahan, T., Shintani, A., Fuchs, D.C., Shelton, R.C., Dudley, J.A., Graves, A.J., Ray, W.A., 2014. Antidepressants and suicide attempts in children. Pediatrics 133 (2), 204–210. https://doi.org/10.1542/peds.2013-0923.

Correll, C.U., Manu, P., Olshanskiy, V., Napolitano, B., Kane, J.M., Malhotra, A.K., 2009. Cardiometabolic risk of second-generation antipsychotic medications during first-time use in children and adolescents. JAMA 302 (16), 1765–1773. https://doi.org/10.1001/jama.2009.1549.

Côté, S.M., Ahun, M.N., Herba, C.M., Brendgen, M., Geoffroy, M.C., Orri, M., Liu, X., Vitaro, F., Melchior, M., Boivin, M., Tremblay, R.E., 2018. Why is maternal depression related to adolescent internalizing problems? A 15-year population-based study. J. Am. Acad. Child Adolesc. Psychiatry 57 (12), 916–924. https://doi.org/10.1016/j.jaac.2018.04.024.

Cox, G.R., Fisher, C.A., De Silva, S., Phelan, M., Akinwale, O.P., Simmons, M.B., Hetrick, S.E., 2012. Interventions for preventing relapse and recurrence of a depressive disorder in children and adolescents. Cochrane Database Syst. Rev. 2012 (11). https://doi.org/10.1002/14651858.CD007504.pub2.

Croarkin, P.E., Elmaadawi, A.Z., Aaronson, S.T., Schrodt, G.R., Holbert, R.C., Verdoliva, S., Heart, K.L., Demitrack, M.A., Strawn, J.R., 2020. Left prefrontal transcranial magnetic stimulation for treatment-resistant depression in adolescents: a double-blind, randomized, sham-controlled trial. Neuropsychopharmacology. https://doi.org/10.1038/s41386-020-00829-y.

Croarkin, P.E., Elmaadawi, A.Z., Aaronson, S.T., et al., 2021. Left prefrontal transcranial magnetic stimulation for treatment-resistant depression in adolescents: A double-blind, randomized, sham-controlled trial. Neuropsychopharmacology 46 (2), 462–469. https://doi.org/10.1038/s41386-020-00829-y.

Cuijpers, P., Karyotaki, E., Eckshtain, D., Ng, M.Y., Corteselli, K.A., Noma, H., Quero, S., Weisz, J.R., 2020. Psychotherapy for depression across different age groups: a systematic review and meta-analysis. JAMA Psychiatry 77 (7), 694–702. https://doi.org/10.1001/jamapsychiatry.2020.0164.

Cullen, K.R., Amatya, P., Roback, M.G., Albott, C.S., Westlund Schreiner, M., Ren, Y., Eberly, L.E., Carstedt, P., Samikoglu, A., Gunlicks-Stoessel, M., Reigstad, K., Horek, N., Tye, S., Lim, K.O., Klimes-Dougan, B., 2018. Intravenous ketamine for adolescents with treatment-resistant depression: an open-label study. J. Child Adolesc. Psychopharmacol. 28 (7), 437–444. https://doi.org/10.1089/cap.2018.0030.

Curry, J., Silva, S., Rohde, P., Ginsburg, G., Kratochvil, C., Simons, A., Kirchner, J., May, D., Kennard, B., Mayes, T., Feeny, N., Albano, A.M., Lavanier, S., Reinecke, M., Jacobs, R., Becker-Weidman, E., Weller, E., Emslie, G., Walkup, J., et al., 2011. Recovery and recurrence following treatment for adolescent major depression. Arch. Gen. Psychiatry 68 (3), 263–270. https://doi.org/10.1001/archgenpsychiatry.2010.150.

Dartois, L.L., Stutzman, D.L., Morrow, M., 2019. L-Methylfolate augmentation to antidepressants for adolescents with treatment-resistant depression: a case series. J. Child Adolesc. Psychopharmacol. 29 (5), 386–391. https://doi.org/10.1089/cap.2019.0006.

Daviss, W.B., Bentivoglio, P., Racusin, R., Brown, K.M., Bostic, J.Q., Wiley, L., 2001. Bupropion sustained release in adolescents with comorbid attention-deficit/hyperactivity disorder and depression. J. Am. Acad. Child Adolesc. Psychiatry 40 (3), 307–314. https://doi.org/10.1097/00004583-200103000-00010.

Daviss, W.B., Birmaher, B., Melhem, N.A., Axelson, D.A., Michaels, S.M., Brent, D.A., 2006. Criterion validity of the Mood and Feelings Questionnaire for depressive episodes in clinic and non-clinic subjects. J. Child Psychol. Psychiatry Allied Discip. 47 (9), 927–934. https://doi.org/10.1111/j.1469-7610.2006.01646.x.

DelBello, M.P., Hochadel, T.J., Portland, K.B., Azzaro, A.J., Katic, A., Khan, A., Emslie, G., 2014. A double-blind, placebo-controlled study of selegiline transdermal system in depressed adolescents. J. Child Adolesc. Psychopharmacol. 24 (6), 311–317. https://doi.org/10.1089/cap.2013.0138.

Diamond, G.S., Wintersteen, M.B., Brown, G.K., Diamond, G.M., Gallop, R., Shelef, K., Levy, S., 2010. Attachment-based family therapy for adolescents with suicidal ideation: a randomized controlled trial. J. Am. Acad. Child Adolesc. Psychiatry 49 (2), 122–131. https://doi.org/10.1097/00004583-201002000-00006.

DiGirolamo, A.M., Ramirez-Zea, M., Wang, M., Flores-Ayala, R., Martorell, R., Neufeld, L.M., Ramakrishnan, U., Sellen, D., Black, M.M., Stein, A.D., 2010. Randomized trial of the effect of zinc supplementation on the mental health of school-age children in Guatemala. Am. J. Clin. Nutr. 92 (5), 1241–1250. https://doi.org/10.3945/ajcn.2010.29686.

Durgam, S., Chen, C., Migliore, R., Prakash, C., Edwards, J., Findling, R.L., 2018. A phase 3, double-blind, randomized, placebo-controlled study of vilazodone in adolescents with major depressive disorder. Pediatr. Drugs 20 (4), 353–363. https://doi.org/10.1007/s40272-018-0290-4.

Dwyer, J.B., Stringaris, A., Brent, D.A., Bloch, M.H., 2020. Annual research review: defining and treating pediatric treatment-resistant depression. J. Child Psychol. Psychiatry Allied Discip. 61 (3), 312–332. https://doi.org/10.1111/jcpp.13202.

Eckshtain, D., Kuppens, S., Ugueto, A., Ng, M.Y., Vaughn-Coaxum, R., Corteselli, K., Weisz, J.R., 2020. Meta-analysis: 13-year follow-up of psychotherapy effects on youth depression. J. Am. Acad. Child Adolesc. Psychiatry 59 (1), 45–63. https://doi.org/10.1016/j.jaac.2019.04.002.

Emslie, G.J., John Rush, A., Weinberg, W.A., Kowatch, R.A., Hughes, C.W., Carmody, T., Rintelmann, J., 1997. A double-blind, randomized, placebo-controlled trial of fluoxetine in children and adolescents with depression. Arch. Gen. Psychiatry 54 (11), 1031–1037. https://doi.org/10.1001/archpsyc.1997.01830230069010.

Emslie, G.J., Heiligenstein, J.H., Wagner, K.D., Hoog, S.L., Ernest, D.E., Brown, E., Nilsson, M., Jacobson, J.G., 2002. Fluoxetine for acute treatment of depression in children and adolescents: a placebo-controlled, randomized clinical trial. J. Am. Acad. Child Adolesc. Psychiatry 41 (10), 1205–1215. https://doi.org/10.1097/00004583-200210000-00010.

Emslie, G.J., Wagner, K.D., Kutcher, S., Krulewicz, S., Fong, R., Carpenter, D.J., Lipschitz, A., Machin, A., Wilkinson, C., 2006. Paroxetine treatment in children and adolescents with major depressive disorder: a randomized, multicenter, double-blind, placebo-controlled trial. J. Am. Acad. Child Adolesc. Psychiatry 45 (6), 709–719. https://doi.org/10.1097/01.chi.0000214189.73240.63.

Emslie, G.J., Findling, R.L., Yeung, P.P., Kunz, N.R., Li, Y., 2007. Venlafaxine ER for the treatment of pediatric subjects with depression: results of two placebo-controlled trials. J. Am. Acad. Child Adolesc. Psychiatry 46 (4), 479–488. https://doi.org/10.1097/chi.0b013e31802f5f03.

Emslie, G.J., Kennard, B.D., Mayes, T.L., Nightingale-Teresi, J., Carmody, T., Hughes, C.W., et al., 2008. Fluoxetine versus placebo in preventing relapse of major depression in children and adolescents. Am. J. Psychiatry 165 (4), 459–467. https://doi.org/10.1176/appi.ajp.2007.07091453.

Emslie, G.J., Ventura, D., Korotzer, A., Tourkodimitris, S., 2009. Escitalopram in the treatment of adolescent depression: a randomized placebo-controlled multisite trial. J. Am. Acad. Child Adolesc. Psychiatry 48 (7), 721–729. https://doi.org/10.1097/CHI.0b013e3181a2b304.

Emslie, G.J., Mayes, T., Porta, G., Vitiello, B., Clarke, G., Wagner, K.D., Asarnow, J.R., Spirito, A., Birmaher, B., Ryan, N., Kennard, B., DeBar, L., McCracken, J., Strober, M., Onorato, M., Zelazny, J., Keller, M., Iyengar, S., Brent, D., 2010. Treatment of resistant depression in adolescents (TORDIA): week 24 outcomes. Am. J. Psychiatr. 167 (7), 782–791. https://doi.org/10.1176/appi.ajp.2010.09040552.

Emslie, G.J., Kennard, B.D., Mayes, T.L., 2011. Predictors of treatment response in adolescent depression. Pediatr. Ann. 40 (6), 300–306. https://doi.org/10.3928/00904481-20110512-05.

Emslie, G.J., Prakash, A., Zhang, Q., Pangallo, B.A., Bangs, M.E., March, J.S., 2014. A double-blind efficacy and safety study of duloxetine fixed doses in children and adolescents with major depressive disorder. J. Child Adolesc. Psychopharmacol. 24 (4), 170–179. https://doi.org/10.1089/cap.2013.0096.

FDA, 2004. Postmarket Drug Safety Information for Patients and Providers: Suicidality in Children and Adolescents Being Treated With Antidepressant Medications. FDA.

FDA, 2012. Drug Safety Communication: Abnormal Heart Rhythms Associated With High Doses of Celexa (Citalopram Hydrobromide). FDA.

FDA, 2018. United States Food & Drug Association Statement. FDA.

Findling, R.L., Robb, A.S., DelBello, M.P., Huss, M., McNamara, N.K., Sarkis, E.H., Scheffer, R.E., Poulsen, L.H., Chen, G., Lemming, O.M., Auby, P., 2018. A 6-month open-label extension study of vortioxetine in pediatric patients with depressive or anxiety disorders. J. Child Adolesc. Psychopharmacol. 28 (1), 47–54. https://doi.org/10.1089/cap.2017.0047.

Findling, R.L., McCusker, E., Strawn, J.R., 2020. A randomized, double-blind, placebo-controlled trial of vilazodone in children and adolescents with major depressive disorder with twenty-six-week open-label follow-up. J. Child Adolesc. Psychopharmacol. 30 (6), 355–365. https://doi.org/10.1089/cap.2019.0176.

Gabbay, V., Freed, R.D., Alonso, C.M., Senger, S., Stadterman, J., Davison, B.A., Klein, R.G., 2018. A double-blind placebo-controlled trial of omega-3 fatty acids as a monotherapy for adolescent depression. J. Clin. Psychiatry 79 (4). https://doi.org/10.4088/JCP.17m11596.

Ghaziuddin, N., Dumas, S., Hodges, E., 2011. Use of continuation or maintenance electroconvulsive therapy in adolescents with severe treatment-resistant depression. J. ECT 27 (2), 168–174. https://doi.org/10.1097/YCT.0b013e3181f665e4.

Ghaziuddin, N., Shamseddeen, W., Gettys, G., Ghaziuddin, M., 2020. Electroconvulsive therapy for the treatment of severe mood disorders during adolescence: a retrospective chart review. J. Child Adolesc. Psychopharmacol. 30 (4), 235–243. https://doi.org/10.1089/cap.2019.0054.

Goldstein, B.I., Shamseddeen, W., Spirito, A., Emslie, G., Clarke, G., Wagner, K.D., Asarnow, J.R., Vitiello, B., Ryan, N., Birmaher, B., Mayes, T., Onorato, M., Zelazny, J., Brent, D.A., 2009. Substance use and the treatment of resistant depression in adolescents. J. Am. Acad. Child Adolesc. Psychiatry 48 (12), 1182–1192. https://doi.org/10.1097/CHI.0b013e3181bef6e8.

Goodyer, I.M., Dubicka, B., Wilkinson, P., Kelvin, R., Roberts, C., Byford, S., Breen, S., Ford, C., Barrett, B., Leech, A., Rothwell, J., White, L., Harrington, R., 2008. A randomised controlled trial of cognitive behaviour therapy in adolescents with major depression treated by selective serotonin reuptake inhibitors. The ADAPT trial. Health Technol. Assess. 12 (14). https://doi.org/10.3310/hta12140.

Goodyer, I.M., Reynolds, S., Barrett, B., Byford, S., Dubicka, B., Hill, J., Holland, F., Kelvin, R., Midgley, N., Roberts, C., Senior, R., Target, M., Widmer, B., Wilkinson, P., Fonagy, P., 2017. Cognitive behavioural therapy and short-term psychoanalytical psychotherapy versus a brief psychosocial intervention in adolescents with unipolar major depressive disorder (IMPACT): a multicentre, pragmatic, observer-blind, randomised controlled superiority trial. Lancet Psychiatry 4 (2), 109–119. https://doi.org/10.1016/S2215-0366(16)30378-9.

Guy, W., 1976. ECDEU Assessment Manual for Psychopharmacology-Revised. Department of Health, Education, and Welfare, Public Health Service, Alcohol, Drug Abuse, and Mental Health Administration, NIMH Psychopharmacology Research Branch, Division of Extramural Research Programs, pp. 218–222.

Hamilton, M., 1967. Development of a rating scale for primary depressive illness. Br. J. Soc. Clin. Psychol. 6 (4), 278–296. https://doi.org/10.1111/j.2044-8260.1967.tb00530.x.

Hammerness, P.G., Vivas, F.M., Geller, D.A., 2006. Selective serotonin reuptake inhibitors in pediatric psychopharmacology: a review of the evidence. J. Pediatr. 148 (2), 158–165. https://doi.org/10.1016/j.jpeds.2005.09.028.

Healy, K., Sanders, M., 2017. Antecedents of treatment resistant depression in children victimized by peers. Child Psychiatry Hum. Dev. 48 (1), 107–119. https://doi.org/10.1007/s10578-016-0658-z.

Hegeman, J.M., Doesborgh, S.J., van Niel, M.C., van Megen, H.J., 2008a. The efficacy of electroconvulsive therapy in adolescents. A restrospective study. Tijdschr. Psychiatr. 50 (1), 23–31.

Hegeman, J.M., Doesborgh, S.J.C., Van Niel, M.C., Van Megen, H.J.G.M., 2008b. De effectiviteit van elektroconvulsietherapie bij adolescenten: Een retrospectief vergelijkend onderzoek. Tijdschr. Psychiatr. 50 (1), 23–31.

Hett, D., Rogers, J., Humpston, C., Marwaha, S., 2021. Repetitive transcranial magnetic stimulation (rTMS) for the treatment of depression in adolescence: a systematic review. J. Affect. Disord. 278, 460–469. https://doi.org/10.1016/j.jad.2020.09.058.

Hughes, C.W., Emslie, G.J., Crismon, M.L., Posner, K., Birmaher, B., Ryan, N., Jensen, P., Curry, J., Vitiello, B., Lopez, M., Shon, S.P., Pliszka, S.R., Trivedi, M.H., Barcelona, J., Hopkins, C., Jensen, P.S., Mallett, R., Musquiz, S., Robinson, V., et al., 2007. Texas children's medication algorithm project: update from Texas consensus conference panel on medication treatment of childhood major depressive disorder. J. Am. Acad. Child Adolesc. Psychiatry 46 (6), 667–686. https://doi.org/10.1097/chi.0b013e31804a859b.

Jerrell, J.M., 2010. Neuroendocrine-related adverse events associated with antidepressant treatment in children and adolescents. CNS Neurosci. Ther. 16 (2), 83–90. https://doi.org/10.1111/j.1755-5949.2009.00106.x.

Keller, M.B., Ryan, N.D., Strober, M., Klein, R.G., Kutcher, S.P., Birmaher, B., Hagino, O.R., Koplewicz, H., Carlson, G.A., Clarke, G.N., Emslie, G.J., Feinberg, D., Geller, B., Kusumakar, V., Papatheodorou, G., Sack, W.H., Sweeney, M., Wagner, K.D., Weller, E.B., et al., 2001. Efficacy of paroxetine in the treatment of adolescent major depression: a randomized, controlled trial. J. Am. Acad. Child Adolesc. Psychiatry 40 (7), 762–772. https://doi.org/10.1097/00004583-200107000-00010.

Kennard, B.D., Emslie, G.J., Mayes, T.L., Nightingale-Teresi, J., Nakonezny, P.A., Hughes, J.L., Jones, J.M., Tao, R., Stewart, S.M., Jarrett, R.B., 2008. Cognitive-behavioral therapy to prevent relapse in pediatric responders to pharmacotherapy for major depressive disorder. J. Am. Acad. Child Adolesc. Psychiatry 47 (12), 1395–1404. https://doi.org/10.1097/CHI.0b013e31818914a1.

Kennard, B.D., Clarke, G.N., Weersing, V.R., Asarnow, J.R., Shamseddeen, W., Porta, G., Berk, M., Hughes, J.L., Spirito, A., Emslie, G.J., Keller, M.B., Wagner, K.D., Brent, D.A., 2009a. Effective components of TORDIA cognitive-behavioral therapy for adolescent depression: preliminary findings. J. Consult. Clin. Psychol. 77 (6), 1033–1041. https://doi.org/10.1037/a0017411.

Kennard, B.D., Silva, S.G., Tonev, S., Rohde, P., Hughes, J.L., Vitiello, B., Kratochvil, C.J., Curry, J.F., Emslie, G.J., Reinecke, M., March, J., 2009b. Remission and recovery in the treatment for adolescents with depression study (TADS): acute and long-term outcomes. J. Am. Acad. Child Adolesc. Psychiatry 48 (2), 186–195. https://doi.org/10.1097/CHI.0b013e31819176f9.

Kennard, B.D., Emslie, G.J., Mayes, T.L., Nakonezny, P.A., Jones, J.M., Foxwell, A.A., King, J., 2014. Sequential treatment with fluoxetine and relapse-prevention CBT to improve outcomes in pediatric depression. Am. J. Psychiatr. 171 (10), 1083–1090. https://doi.org/10.1176/appi.ajp.2014.13111460.

Kim, S., Rush, B.S., Rice, T.R., 2020. A systematic review of therapeutic ketamine use in children and adolescents with treatment-resistant mood disorders. Eur. Child Adolesc. Psychiatry. https://doi.org/10.1007/s00787-020-01542-3.

Kovacs, M., Mukerji, P., Iyengar, S., Drash, A., 1996. Psychiatric disorder and metabolic control among youths with IDDM: a longitudinal study. Diabetes Care 19 (4), 318–323. https://doi.org/10.2337/diacare.19.4.318.

Kratochvil, C., Emslie, G., Silva, S., McNulty, S., Walkup, J., Curry, J., Reinecke, M., Vitiello, B., Rohde, P., Feeny, N., Casat, C., Pathak, S., Weller, E., May, D., Mayes, T., Robins, M., March, J., 2006. Acute time to response in the Treatment for Adolescents with Depression Study (TADS). J. Am. Acad. Child Adolesc. Psychiatry 45 (12), 1412–1418. https://doi.org/10.1097/01.chi.0000237710.73755.14.

Kurian, B.T., Ray, W.A., Arbogast, P.G., Fuchs, D.C., Dudley, J.A., Cooper, W.O., 2007. Effect of regulatory warnings on antidepressant prescribing for children and adolescents. Arch. Pediatr. Adolesc. Med. 161 (7), 690–696. https://doi.org/10.1001/archpedi.161.7.690.

LeMoult, J., Humphreys, K.L., Tracy, A., Hoffmeister, J.A., Ip, E., Gotlib, I.H., 2020. Meta-analysis: exposure to early life stress and risk for depression in childhood and adolescence. J. Am. Acad. Child Adolesc. Psychiatry 59 (7), 842–855. https://doi.org/10.1016/j.jaac.2019.10.011.

Lewinsohn, P.M., Clarke, G.N., Hops, H., Andrews, J., 1990. Cognitive-behavioral treatment for depressed adolescents. Behav. Ther. 21 (4), 385–401. https://doi.org/10.1016/S0005-7894(05)80353-3.

Lewinsohn, P.M., Seeley, J.R., Hibbard, J., Rohde, P., Sack, W.H., 1996. Cross-sectional and prospective relationships between physical morbidity and depression in older adolescents. J. Am. Acad. Child Adolesc. Psychiatry 35 (9), 1120–1129. https://doi.org/10.1097/00004583-199609000-00009.

Libby, A.M., Orton, H.D., Valuck, R.J., 2009. Persisting decline in depression treatment after FDA warnings. Arch. Gen. Psychiatry 66 (6), 633–639. https://doi.org/10.1001/archgenpsychiatry.2009.46.

Libuda, L., Timmesfeld, N., Antel, J., Hirtz, R., Bauer, J., Führer, D., Zwanziger, D., Öztürk, D., Langenbach, G., Hahn, D., Ring, S., Peters, T., Hinney, A., Bühlmeier, J., Hebebrand, J., Grasemann, C., Föcker, M., 2020. Effect of vitamin D deficiency on depressive symptoms in child and adolescent psychiatric patients: results of a randomized controlled trial. Eur. J. Nutr. 59 (8), 3415–3424. https://doi.org/10.1007/s00394-020-02176-6.

Lu, C.Y., Penfold, R.B., Wallace, J., et al., 2020. Increases in suicide deaths among adolescents and young adults following US Food and Drug Administration antidepressant boxed warnings and declines in depression care. Psychiatric Research and Clinical Practice 2 (2), 43–52. https://doi.org/10.1176/appi.prcp.20200012.

Maalouf, F.T., Atwi, M., Brent, D.A., 2011. Treatment-resistant depression in adolescents: review and updates on clinical management. Depress. Anxiety 28 (11), 946–954. https://doi.org/10.1002/da.20884.

March, J.S., Silva, S., Petrycki, S., Curry, J., Wells, K., Fairbank, J., et al., 2004. Fluoxetine, cognitive-behavioral therapy, and their combination for adolescents with depression: Treatment for Adolescents with Depression Study (TADS) randomized controlled trial. J. Am. Med. Assoc. 292 (7), 807–820. https://doi.org/10.1001/jama.292.7.807.

March, J.S., Silva, S., Petrycki, S., Curry, J., Wells, K., Fairbank, J., Burns, B., Domino, M., McNulty, S., Vitiello, B., Severe, J., Casat, C., Kolker, J., Riedal, K., Feeny, N., Findling, R., Stull, S., Baab, S., Weller, E.B., et al., 2007. The Treatment for Adolescents with Depression Study (TADS): long-term effectiveness and safety outcomes. Arch. Gen. Psychiatry 64 (10), 1132–1144. https://doi.org/10.1001/archpsyc.64.10.1132.

McLellan, Q., Wilkes, T.C., Swansburg, R., Jaworska, N., Langevin, L.M., MacMaster, F.P., 2018. History of suicide attempt and right superior temporal gyrus volume in youth with treatment-resistant major depressive disorder. J. Affect. Disord. 239, 291–294. https://doi.org/10.1016/j.jad.2018.07.030.

McMakin, D.L., Olino, T.M., Porta, G., Dietz, L.J., Emslie, G., Clarke, G., Wagner, K.D., Asarnow, J.R., Ryan, N.D., Birmaher, B., Shamseddeen, W., Mayes, T., Kennard, B., Spirito, A., Keller, M., Lynch, F.L., Dickerson, J.F., Brent, D.A., 2012. Anhedonia predicts poorer recovery among youth with selective serotonin reuptake inhibitor treatment-resistant depression. J. Am. Acad. Child Adolesc. Psychiatry 51 (4), 404–411. https://doi.org/10.1016/j.jaac.2012.01.011.

Melvin, G.A., Dudley, A.L., Gordon, M.S., Ford, S., Taffe, J., Tonge, B.J., 2013. What happens to depressed adolescents? A follow-up study into early adulthood. J. Affect. Disord. 151 (1), 298–305. https://doi.org/10.1016/j.jad.2013.06.012.

Merikangas, K.R., He, J.P., Burstein, M., Swanson, S.A., Avenevoli, S., Cui, L., Benjet, C., Georgiades, K., Swendsen, J., 2010. Lifetime prevalence of mental disorders in U.S. adolescents: results from the National Comorbidity Survey Replication-Adolescent Supplement (NCS-A). J. Am. Acad. Child Adolesc. Psychiatry 49 (10), 980–989. https://doi.org/10.1016/j.jaac.2010.05.017.

Miller, M., Swanson, S.A., Azrael, D., Pate, V., Stürmer, T., 2014. Antidepressant dose, age, and the risk of deliberate self-harm. JAMA Intern. Med. 174 (6), 899–909. https://doi.org/10.1001/jamainternmed.2014.1053.

Mufson, L., Weissman, M.M., Moreau, D., Garfinkel, R., 1999. Efficacy of interpersonal psychotherapy for depressed adolescents. Arch. Gen. Psychiatry 56 (6), 573–579. https://doi.org/10.1001/archpsyc.56.6.573.

Mufson, L., Dorta, K.P., Wickramaratne, P., Nomura, Y., Olfson, M., Weissman, M.M., 2004. A randomized effectiveness trial of interpersonal psychotherapy for depressed adolescents. Arch. Gen. Psychiatry 61 (6), 577–584. https://doi.org/10.1001/archpsyc.61.6.577.

Neera, G., Kutcher, S.P., Penelope, K., 2004. Practice parameter for use of electroconvulsive therapy with adolescents. J. Am. Acad. Child Adolesc. Psychiatry, 1521–1539. https://doi.org/10.1097/01.chi.0000142280.87429.68.

Nemets, H., Nemets, B., Apter, A., Bracha, Z., Belmaker, R.H., 2006. Omega-3 treatment of childhood depression: a controlled, double-blind pilot study. Am. J. Psychiatr. 163 (6), 1098–1100. https://doi.org/10.1176/ajp.2006.163.6.1098.

Pathak, S., Johns, E.S., Kowatch, R.A., 2005. Adjunctive quetiapine for treatment-resistant adolescent major depressive disorder: a case series. J. Child Adolesc. Psychopharmacol. 15 (4), 696–702. https://doi.org/10.1089/cap.2005.15.696.

Poznanski, E., Mokros, H., 1996. Children's Depression Rating Scale Revised (CDRS-R). Western Psychological Services.

Rengasamy, M., Mansoor, B.M., Hilton, R., Porta, G., He, J., Emslie, G.J., Mayes, T., Clarke, G.N., Wagner, K.D., Keller, M.B., Ryan, N.D., Birmaher, B., Shamseddeen, W., Asarnow, J.R., Brent, D.A., 2013. The bi-directional relationship between parent-child conflict and treatment outcome in treatment-resistant adolescent depression. J. Am. Acad. Child Adolesc. Psychiatry 52 (4), 370–377. https://doi.org/10.1016/j.jaac.2013.01.012.

Rizzoli, R., Cooper, C., Reginster, J.Y., Abrahamsen, B., Adachi, J.D., Brandi, M.L., Bruyère, O., Compston, J., Ducy, P., Ferrari, S., Harvey, N.C., Kanis, J.A., Karsenty, G., Laslop, A., Rabenda, V., Vestergaard, P., 2012. Antidepressant medications and osteoporosis. Bone 51 (3), 606–613. https://doi.org/10.1016/j.bone.2012.05.018.

Rosselló, J., Bernal, G., 1999. The efficacy of cognitive-behavioral and interpersonal treatments for depression in Puerto Rican adolescents. J. Consult. Clin. Psychol. 67 (5), 734–745. https://doi.org/10.1037/0022-006X.67.5.734.

Schosser, A., Kasper, S., 2009. The role of pharmacogenetics in the treatment of depression and anxiety disorders. Int. Clin. Psychopharmacol. 24 (6), 277–288. https://doi.org/10.1097/YIC.0b013e3283306a2f.

Suresh, V., Mills, J.A., Croarkin, P.E., Strawn, J.R., 2020. What next? A Bayesian hierarchical modeling re-examination of treatments for adolescents with selective serotonin reuptake inhibitor-resistant depression. Depress. Anxiety 37 (9), 926–934. https://doi.org/10.1002/da.23064.

Tang, T.C., Jou, S.H., Ko, C.H., Huang, S.Y., Yen, C.F., 2009. Randomized study of school-based intensive interpersonal psychotherapy for depressed adolescents with suicidal risk and parasuicide behaviors. Psychiatry Clin. Neurosci., 463–470. https://doi.org/10.1111/j.1440-1819.2009.01991.x.

Tiiri, E., Luntamo, T., Mishina, K., Sillanmäki, L., Brunstein Klomek, A., Sourander, A., 2020. Did bullying victimization decrease after nation-wide school-based antibullying program? A time-trend study. J. Am. Acad. Child Adolesc. Psychiatry 59 (4), 531–540. https://doi.org/10.1016/j.jaac.2019.03.023.

Twenge, J.M., Cooper, A.B., Joiner, T.E., Duffy, M.E., Binau, S.G., 2019. Age, period, and cohort trends in mood disorder indicators and suicide-related outcomes in a nationally representative dataset, 2005–2017. J. Abnorm. Psychol. 128 (3), 185–199. https://doi.org/10.1037/abn0000410.

Van Haelst, I.M.M., Van Klei, W.A., Doodeman, H.J., Warnier, M.J., De Bruin, M.L., Kalkman, C.J., Egberts, T.C.G., 2014. QT interval prolongation in users of selective serotonin reuptake inhibitors in an elderly surgical population: a cross-sectional study. J. Clin. Psychiatry 75 (1), 15–21. https://doi.org/10.4088/JCP.13m08397.

Varigonda, A.L., Jakubovski, E., Taylor, M.J., Freemantle, N., Coughlin, C., Bloch, M.H., 2015. Systematic review and meta-analysis: early treatment responses of selective serotonin reuptake inhibitors in pediatric major depressive disorder. J. Am. Acad. Child Adolesc. Psychiatry 54 (7), 557–564. https://doi.org/10.1016/j.jaac.2015.05.004.

Viswanathan, M., Kennedy, S.M., McKeeman, J., Christian, R., Coker-Schwimmer, M., Cook Middleton, J., et al., 2020. Treatment of Depression in Children and Adolescents: A Systematic Review. Agency for Healthcare Research and Quality.

Vitiello, B., Emslie, G., Clarke, G., Wagner, K.D., Asarnow, J.R., Keller, M.B., Birmaher, B., Ryan, N.D., Kennard, B., Mayes, T.L., DeBar, L., Lynch, F., Dickerson, J., Strober, M., Suddath, R., McCracken, J.T., Spirito, A., Onorato, M., Zelazny, J., et al., 2011. Long-term outcome of adolescent depression initially resistant to selective serotonin reuptake inhibitor treatment: a follow-up study of the TORDIA sample. J. Clin. Psychiatry 72 (3), 388–396. https://doi.org/10.4088/JCP.09m05885blu.

Voican, C.S., Corruble, E., Naveau, S., Perlemuter, G., 2014. Antidepressant-induced liver injury: a review for clinicians. Am. J. Psychiatr. 171 (4), 404–415. https://doi.org/10.1176/appi.ajp.2013.13050709.

Von Knorring, A.L., Olsson, G.I., Thomsen, P.H., Lemming, O.M., Hultén, A., 2006. A randomized, double-blind, placebo-controlled study of citalopram in adolescents with major depressive disorder. J. Clin. Psychopharmacol. 26 (3), 311–315. https://doi.org/10.1097/01.jcp.0000219051.40632.d5.

Wagner, K.D., Ambrosini, P., Rynn, M., Wohlberg, C., Yang, R., Greenbaum, M.S., Childress, A., Donnelly, C., Deas, D., 2003. Efficacy of sertraline in the treatment of children and adolescents with major depressive disorder: two randomized controlled trials. J. Am. Med. Assoc. 290 (8), 1033–1041. https://doi.org/10.1001/jama.290.8.1033.

Wagner, K., Pliszka, S., 2013. In: Schatzberg, A, Nemeroff, C (Eds.), Essentials of Clinical Psychopharmacology, third. American Psychiatric Publishing, Arlington, VA, pp. 693–750.

Wagner, K.D., Robb, A.S., Findling, R.L., Jin, J., Gutierrez, M.M., Heydorn, W.E., 2004. A randomized, placebo-controlled trial of citalopram for the treatment of major depression in children and adolescents. Am. J. Psychiatr. 161 (6), 1079–1083. https://doi.org/10.1176/appi.ajp.161.6.1079.

Wagner, K.D., Jonas, J., Findling, R.L., Ventura, D., Saikali, K., 2006. A double-blind, randomized, placebo-controlled trial of escitalopram in the treatment of pediatric depression. J. Am. Acad. Child Adolesc. Psychiatry 45 (3), 280–288. https://doi.org/10.1097/01.chi.0000192250.38400.9e.

Weersing, V.R., Jeffreys, M., Do, M.C.T., Schwartz, K.T.G., Bolano, C., 2017. Evidence base update of psychosocial treatments for child and adolescent depression. J. Clin. Child Adolesc. Psychol. 46 (1), 11–43. https://doi.org/10.1080/15374416.2016.1220310.

Weihs, K.L., Murphy, W., Abbas, R., Chiles, D., England, R.D., Ramaker, S., Wajsbrot, D.B., 2018. Desvenlafaxine versus placebo in a fluoxetine-referenced study of children and adolescents with major depressive disorder. J. Child Adolesc. Psychopharmacol. 28 (1), 36–46. https://doi.org/10.1089/cap.2017.0100.

Whitlock, K., Smith, N., Poulsen, R.J., Coffey, B.J., 2020. Treatment-resistant depression in an adolescent treated with clozapine: weighing the options in a young suicidal patient. J. Child Adolesc. Psychopharmacol. 30 (5), 342–344. https://doi.org/10.1089/cap.2020.29184.bjc.

Williams, J.M.G., Barnhofer, T., Crane, C., Duggan, D.S., Shah, D., Brennan, K., Krusche, A., Crane, R., Eames, C., Jones, M., Radford, S., Russell, I.T., 2012. Pre-adult onset and patterns of suicidality in patients with a history of recurrent depression. J. Affect. Disord. 138 (1–2), 173–179. https://doi.org/10.1016/j.jad.2011.12.011.

Woldu, H., Porta, G., Goldstein, T., Sakolsky, D., Perel, J., Emslie, G., Mayes, T., Clarke, G., Ryan, N.D., Birmaher, B., Wagner, K.D., Asarnow, J.R., Keller, M.B., Brent, D., 2011. Pharmacokinetically and clinician-determined adherence to an antidepressant regimen and clinical outcome in the TORDIA trial. J. Am. Acad. Child Adolesc. Psychiatry 50 (5), 490–498. https://doi.org/10.1016/j.jaac.2011.01.018.

Wood, A., Harrington, R., Moore, A., 1996. Controlled trial of a brief cognitive-behavioural intervention in adolescent patients with depressive disorders. J. Child Psychol. Psychiatry Allied Discip. 37 (6), 737–746. https://doi.org/10.1111/j.1469-7610.1996.tb01466.x.

Xu, Y., Bai, S.J., Lan, X.H., Qin, B., Huang, T., Xie, P., 2016. Randomized controlled trials of serotonin-norepinephrine reuptake inhibitor in treating major depressive disorder in children and adolescents: a meta-analysis of efficacy and acceptability. Braz. J. Med. Biol. Res. 49 (6). https://doi.org/10.1590/1414-431X20164806.

Young, J.F., Mufson, L., Gallop, R., 2010. Preventing depression: a randomized trial of interpersonal psychotherapy-adolescent skills training. Depress. Anxiety 27 (5), 426–433. https://doi.org/10.1002/da.20664.

Zhand, N., Courtney, D.B., Flament, M.F., 2015. Use of electroconvulsive therapy in adolescents with treatment-resistant depressive disorders a case series. J. ECT 31 (4), 238–245. https://doi.org/10.1097/YCT.0000000000000236.

Zhou, X., Teng, T., Zhang, Y., Del Giovane, C., Furukawa, T.A., Weisz, J.R., Li, X., Cuijpers, P., Coghill, D., Xiang, Y., Hetrick, S.E., Leucht, S., Qin, M., Barth, J., Ravindran, A.V., Yang, L., Curry, J., Fan, L., Silva, S.G., et al., 2020. Comparative efficacy and acceptability of antidepressants, psychotherapies, and their combination for acute treatment of children and adolescents with depressive disorder: a systematic review and network meta-analysis. Lancet Psychiatry 7 (7), 581–601. https://doi.org/10.1016/S2215-0366(20)30137-1.

Zimmermann, K.S., Richardson, R., Baker, K.D., 2020. Esketamine as a treatment for paediatric depression: questions of safety and efficacy. Lancet Psychiatry 7 (10), 827–829. https://doi.org/10.1016/S2215-0366(19)30521-8.

Zivin, K., Pfeiffer, P.N., Bohnert, A.S.B., Ganoczy, D., Blow, F.C., Nallamothu, B.K., Kales, H.C., 2013. Evaluation of the FDA warning against prescribing citalopram at doses exceeding 40 mg. Am. J. Psychiatr. 170 (6), 642–650. https://doi.org/10.1176/appi.ajp.2013.12030408.

Chapter 28

Treatment-resistant depression in geriatrics

Olusola Ajilore, Alexandra Dyer Merg, and Victor Alejandro Valencia
Department of Psychiatry, University of Illinois College of Medicine, Chicago, IL, United States

Introduction

There has been a considerable increase in the number of older individuals in the United States, with those over the age of 85 representing the fastest growing segment of the population (Sadock and Dan Ruiz, 2017). The global population has also seen an increase in the proportion of older individuals, especially among developing countries (Sadock and Dan Ruiz, 2017). We must appreciate this demographic change by recognizing the many biopsychosocial elements unique to this diverse population. As clinicians and researchers, we must always remain vigilant to the ever-changing and specific needs of our patients. This will require particular focus on the complex network of associations and relationships that govern the everyday lives of people. This is crucial when discussing mental health, especially in the older patient population. The mental health of older individuals has profound consequences for not only quality of life but also morbidity and mortality. This is deeply concerning given that many psychiatric disorders are often underdiagnosed and therefore untreated in geriatric patients (Sadock and Dan Ruiz, 2017; Park and Unützer, 2011; Skoog et al., 1993; Birrer and Vemuri, 2004; Park et al., 2017; Himelhoch et al., 2004; Lökk and Delbari, 2010; Brodaty et al., 2001). Even when recognized, some psychiatric disorders have prolonged and tortuous treatment courses. This is often the case for late-life depression (LLD). In general, it has been determined that depression is among one of the leading causes of disability-adjusted life years (DALYs) in the world (Park and Unützer, 2011). It is well known that depression has devastating health consequences in older adults that include but are not limited to social deprivation, decreased quality of life, cognitive decline, and suicide (Palsson and Skoog, 1997).

The development of psychiatric disorders later in life has significant health consequences for these individuals. There are complex bidirectional interactions that occur between medical and psychiatric diseases that often leads to worse outcomes overall (Himelhoch et al., 2004; Aziz and Steffens, 2013; Goldman and Schafer, 2020). As an example, there is significant worsening of cardiovascular outcomes in patients with coronary heart disease and comorbid depression (van Melle et al., 2004). It has also been determined that patients with depression are at increased risk of diabetic complications due to worse glycemic control (Gehlawat et al., 2013). This remains relevant when discussing healthcare outcomes for geriatric patients given their increased burden of medical disease in combination with the historical under recognition of psychiatric disorders (Sadock and Dan Ruiz, 2017; Skoog et al., 1993; Duan et al., 2017). In addition to the unique health burden assumed by aging populations, there are many psychosocial factors that influence the overall wellbeing of geriatric patients (Gerst-Emerson and Jayawardhana, 2015). It has been determined that psychological and social variables such as environmental mastery, sense of purpose, and autonomy were crucial elements related to the diagnosis of major depression in older individuals (Aziz and Steffens, 2013; Davison et al., 2012). In addition to unique psychosocial stressors experienced later in life, older individuals are also at increased risk for depression due to age specific biologic changes (Palsson and Skoog, 1997; Smith et al., 2015). There is mounting evidence that supports unique pathophysiology for LLD with wide-ranging medical etiologies (Sadock and Dan Ruiz, 2017). This has profound consequences for disease presentation, clinical course, symptom management, treatment response, and prognosis (Sadock and Dan Ruiz, 2017; Alexopoulos et al., 1999). In this chapter, we will focus primarily on LLD with particular emphasis on treatment resistance.

Epidemiology

There are many epidemiological studies that exclude older adults, focusing more on younger populations when investigating psychiatric disorders (Sadock and Dan Ruiz, 2017). This is due to the misconception that psychiatric disorders are rare in older individuals (Sadock and Dan Ruiz, 2017). Although there is epidemiological evidence suggesting an overall lower rate of depression among this population, we must consider how underdiagnosis impacts the validity of these studies. It has been well documented that depression in geriatric patients is often not recognized, underdiagnosed, and inadequately

Managing Treatment-Resistant Depression. https://doi.org/10.1016/B978-0-12-824067-0.00028-1

treated (Sadock and Dan Ruiz, 2017; Park and Unützer, 2011; Skoog et al., 1993; Birrer and Vemuri, 2004; Park et al., 2017; Himelhoch et al., 2004; Lökk and Delbari, 2010; Brodaty et al., 2001). There is data to suggest that depression in older adults is not fully assessed due to differences in clinical presentation when compared to younger individuals (Fiske et al., 2009; Rodda et al., 2011; Rapp et al., 2005). There is a paucity of training leading to poor clinical competency in the area of geriatric depression which has further exacerbated the level of underdiagnosis (Sadock and Dan Ruiz, 2017). In addition, we must also be cognizant of changes in diagnostic criteria or algorithms over time and how this can impact prevalence data (Busfield, 2012; Karlsson et al., 2010). Therefore, late-life depression has the potential to be higher than traditionally reported in the geriatric population.

In general, there is a 20% lifetime prevalence for major depressive disorder (Sadock and Dan Ruiz, 2017). In terms of psychiatric disorders presenting later in life, depression remains one of the most frequent diagnoses with the majority of individuals having no previous psychiatric history (Smith et al., 2015; Lebowitz, 1997; Blazer and Williams, 1980). It has been determined that over 50% of elderly patients with depression are experiencing their first episode (Brodaty et al., 2001; Meyers and Alexopoulos, 1988). The overall prevalence of major depression in adults over the age of 65 was roughly 1% with rates of 1.5% in women and 0.5% in men (Sadock and Dan Ruiz, 2017). Depressive symptoms do not preclude the diagnosis of other psychiatric diseases and are not mutually exclusive. There is evidence to suggest that depression occurring later in life increases the risk of developing other psychiatric disorders. This is particularly true when considering alcohol use disorder, which is three to four times more likely to occur in the setting of geriatric depression (Devanand, 2002). These psychiatric comorbidities may contribute to increased treatment resistance in geriatric depression. For example, comorbid anxiety is associated with poorer treatment response (Driscoll et al., 2007). Treatment nonresponse in geriatric depression has also been associated with clinical factors such as higher baseline depressive severity, longer duration, more guilt, and greater functional impairment (Smagula et al., 2015).

There are many sociodemographic factors to consider when discussing depression (Williams et al., 2007; Cornwell and Waite, 2009; Lasser et al., 2002). Some of these risk factors specific to LLD include but are not limited to social isolation, bereavement, low income or educational level, divorced or separated marital status, sleep problems, and prior history of depression (Cornwell and Waite, 2009; Koenig and Blazer, 1992). Moreover, low level of perceived social support has been shown to be one of the risk factors for poor response to psychotherapy in LLD (Solomonov, 2021).

There are studies revealing 50%–60% of patients overall do not achieve adequate response following initial treatment for depression (Fava, 2003). Other studies revealed that between 29% and 46% of patients with depression did not achieve therapeutic effect from antidepressant medications even after adequate dose and duration (Fava and Davidson, 1996). One study had determined that only 19% of elderly individuals with depressive disorders were on antidepressant medications (Palsson and Skoog, 1997). It was found that treatment resistance was present in roughly 18%–40% of the elderly population (Bonner and Howard, 1995). There are some survey data that report even higher rates with nearly 55%–88% of elderly patients with major depression not responding to selective serotonin reuptake inhibitor (SSRI) or SNRI medication (Arandjelovic et al., 2016).

Pathophysiology

There is growing evidence to suggest that late life depression (LLD) has multiple potential etiologies which contributes to its unique presentation when compared to depression occurring earlier in life. It has been well documented that comorbid medical disease can impact the accurate diagnosis and treatment of depression due to confounding symptomatology (Lökk and Delbari, 2010; Burke et al., 2019). There are cerebrovascular, metabolic, inflammatory, immune, neurocognitive, and genetic changes that can influence both the clinical presentation and treatment response of LLD (Sadock and Dan Ruiz, 2017; Burke et al., 2019). There are also specific neural structures and cerebral networks that have been implicated in the pathophysiology of LLD (Tadayonnejad and Ajilore, 2014). The conglomeration of these elements provides insight into mechanisms that potentially govern treatment resistance of depression within the geriatric population.

This hypothesis proposes that cerebrovascular disease plays a critical role in the development of depression in older individuals. The development of LLD secondary to these changes has been given the term "vascular depression" (Alexopoulos et al., 1999). This was originally suggested due to the increased prevalence of depression among individuals with stroke and other cerebrovascular risk factors (CVRFs) (Lökk and Delbari, 2010; Mast et al., 2004). This is supported by neuroimaging studies. There is notable gray matter atrophy in the frontal and hippocampal regions of the brain along with subcortical lacunes, micro infarcts, and micro hemorrhages seen in patients with LLD (Rapp et al., 2005). Of note, the rates of clinically defined vascular depression (35.2%) and poststroke depression (36.4%) are similar among geriatric patients (Mast et al., 2004). It is believed that the underlying pathophysiology of vascular depression involves hypoperfusion and inflammation that leads to the development of white matter lesions (WMLs) that disrupt neural systems such as the cortico-striato-pallido-thalamo-cortical (CSPTC) pathway among others (Alexopoulos et al., 1999). As an example, consider the "depression-executive dysfunction (DED)" hypothesis which results from abnormal communication between limbic and dorsal cortical structures resulting from lesions in white matter networks (Taylor et al., 2013). There is evidence to suggest that WML burden is related to treatment resistance in LLD and is associated with poor response to antidepressant medications with nonremission, early relapse, and recurrence (Taylor et al., 2003). This is supported by imaging studies demonstrating that white matter integrity in particular networks is associated with poor response to antidepressant medications, specifically those involved in serotonin reuptake inhibition (Sheline et al., 2010).

There are also investigations focusing on the relationship between neurocognitive disorders and LLD (Burke et al., 2019). It has been suggested that frontal-subcortical abnormality leads to depressive mood symptoms in patients with dementia and that 10%–30% of elderly individuals with Alzheimer's disease experienced depressive mood symptoms (Nowrangi et al., 2015). It has been found that the default mode, executive control, and salience networks are involved in LLD. Of these networks, abnormal functioning of the default mode network (DMN) appears to be involved with Alzheimer's disease pathology. It was found that moderate-severe treatment resistance in patients with LLD had amyloid PET findings similar to patients with confirmed Alzheimer's disease (Burke et al., 2019). The amyloid burden in these patients was located in the precuneus, parietal, temporal, and occipital areas. This has led researchers to believe that treatment resistance in LLD might represent a prodrome to Alzheimer's disease and reflect its early changes (Burke et al., 2019).

There are other various age-related inflammatory, metabolic, immune and genetic processes that influence neural networks impacting the development of treatment resistance in LLD. There was an association between the presence of a single nucleotide polymorphism (SNP) which encoded for IL-1β and treatment response to antidepressant medication (Sadock and Dan Ruiz, 2017). There have been genetic studies that have demonstrated an association between specific serotonin receptors and response to antidepressant medications. The study revealed that homozygous gene expression of the serotonin 2A-receptor led to an increased treatment response to citalopram.[1]

Additionally, patients with elevated levels of TNF-α did not respond well to escitalopram. Interestingly, patients also responded poorly to escitalopram if they had elevated baseline CRP levels, with greater response found in patients with low baseline levels (CRP < 1 mg/L).[1] The contrary was true when considering nortriptyline. It was found that patients had a greater response to the medications with elevated levels of CRP and responded poorly with lower levels.[1] There have also been studies focused on insula metabolism and how this impacts treatment response to escitalopram in patients with LLD. A positron emission-based study found that insula hypometabolism was associated with a poor response to escitalopram with hypermetabolism leading to more a favorable response.[1] These studies provide further insight into the challenges faced with developing an effective treatment plans for patients with LLD.

Clinical presentation

The clinical diagnosis of geriatric depression, or late-life depression (LLD), is based on the same criteria as depression in younger individuals. The Diagnostic and Statistical Manual for Mental Disorders Fifth Addition (DSM 5) indicates that an individual must display five of nine criteria, including depressed affect, for a 2-week period in order to be diagnosed with a major depressive episode. These criteria include anhedonia, change in weight or appetite, insomnia or hypersomnia, psychomotor agitation or retardation, fatigue or loss of energy, inappropriate guilt or worthlessness, decreased concentration, recurrent thoughts of death or passive or active suicidal ideation (DSM 5).

Symptoms of depression in late life are more similar to early life than they are dissimilar (Nelson et al., 2005). However, clinical presentation of depression can be different in patients of advanced age, and symptom frequency and severity may vary. Quality and duration of sleep changes later in life regardless of depression, making insomnia much more likely in LLD than hypersomnia, which can be a symptom of depression earlier in life (Nelson, 2018). A significant decrease in appetite and resultant weight loss are also very common in late life depression, more so than earlier in life. Individuals over age 65 with depression often have multiple medical comorbidities that can contribute to worsening mood. Additionally, they may also present with physical symptoms that make underlying depression more likely to remain undiagnosed (Kok and Reynolds, 2017). Examples of these presenting complaints include cognitive or memory issues, fatigue, weight loss, pain, social withdrawal, increased substance or medication intake.

Cognitive impairment is a presenting symptom that is associated with LLD. In the past, the cognitive decline noted in depressed patients was termed *pseudodementia*, indicating that the cognitive impairment would fully resolve with treatment of the depression. However, this is no longer an accurate term since it has been determined that the cognitive impairment is as real and potentially irreversible as that caused by any other etiology (Morimoto et al., 2015). Additionally, these patients are at higher risk of developing dementia in the future (Nelson, 2018). Cognitive impairment may be a symptom of depression, or, alternatively, patients with a higher predisposition for, or in the early stages of, Alzheimer's dementia, vascular dementia and other major neurocognitive disorders may be more likely to develop depression. The directionality of this association remains unclear.

When evaluating geriatric patients for depression, cognitive impairment may make it more difficult for them to report symptoms themselves. It is often helpful to obtain history from caretakers or family members during an assessment. In addition to history-taking, a comprehensive evaluation for late like depression includes administering symptom rating scales and completing a cognitive examination, a medication review, and laboratory testing (Kok and Reynolds, 2017). Rating scales that are commonly used to assess symptoms as quantitatively as possible are the Geriatric Depression Scale (GDS) and the Personality Health Questionnaire 9 (PHQ-9) (Kok and Reynolds, 2017). A cognitive examination includes administering a screening tool like the Mini Mental Status Examination (MMSE). It is also important to review medications since polypharmacy and specific medications used to treat other illnesses could be contributing to low mood. Laboratory testing for nutrient deficiencies, anemia, or changes in thyroid hormone are recommended in order to assess for easily reversible medical conditions that could be contributing to depressive symptoms (Taylor, 2014).

The course of geriatric depression is often chronic and resistant to treatment. In a study published in 2015, 61% of geriatric participants had chronic depressive symptoms over a 2-year period, even with treatment throughout the entire study (Comijs et al., 2015). In a study done by Kok et al. (2012), remission was achieved in one third of all elderly patients treated with an antidepressant. This demonstrates the high rate of treatment-resistance in this population. Additionally, in a study done by Deng et al. in 2018, over half of the depressed elderly patients who achieved symptom remittance experienced recurrence (Deng et al., 2018). Due to the difficulty in treating late life depression a wide variety of treatment options must be considered.

Current treatment options

Pharmacotherapy for the treatment of LLD includes antidepressants that are used for depression in all ages. SSRIs are first line, as they are in younger patients with depression. In a review done in 2018, specific SSRIs sertraline and paroxetine and an SNRI, duloxetine, were effective for symptom improvement in LLD. Vortioxetine, a newer antidepressant, also demonstrated efficacy (Beyer and Johnson, 2018). Escitalopram and sertraline have been recommended over other SSRIs due to less risk of adverse effects or interactions with other medications (Lenze and Ajam Oughli, 2019).

However, the results demonstrating the efficacy of antidepressant monotherapy is mixed. In a metaanalysis including 15 randomized controlled trials, antidepressants were found to be ineffective in patients over 65 years of age (Tedeschini et al., 2011). Antidepressants have also been shown to decrease in efficacy as age increases (Calati et al., 2013). The reason for this difference in efficacy of antidepressants based on age remains unclear. It could be due to the increased number of medical comorbidities that contribute to depressive symptomatology, or physicians' concerns about increasing doses of psychotropic medication in patients of advanced age on multiple other medications, making them less likely to increase a medication to the dose necessary for symptom remission (Barry et al., 2012; Kok and Reynolds, 2017).

When first line medications prove to be ineffective other treatments are considered. These include adjunct treatment with lithium or atypical antipsychotics, electroconvulsive therapy, transcranial magnetic stimulation, and ketamine. Lithium augmentation has been investigated and in one study had an overall response rate of 42% (Cooper et al., 2011). In another study, lithium augmentation was more effective for geriatric patients with depression than younger adults (Buspavanich et al., 2019). Atypical antipsychotics are also used as augmentation agents. Some patients with late-life depression may benefit from the addition of SGAs. Aripiprazole has been shown to be efficacious, and adverse events of SGAs were not serious in elderly patients (Cakir and Senkal, 2016).

Another newer treatment for depression is intravenous ketamine, which has also been considered for use in LLD that does not respond to other medications. A few case reports have looked at ketamine for treatment-resistant depression in patients of advanced age. In a 2017 review of these case reports by da Frota Ribeiro, two showed evidence of leading to improvement without significant side effects. The relative safety of intravenous ketamine in the elderly was demonstrated in further studies by the mild, transient adverse effects seen by this patient group (da Frota Ribeiro and Riva-Posse, 2017). However, in another study by Bryant et al., the geriatric population was unable to maintain an antidepressant response to intravenous ketamine over time, signifying that ketamine may have diminished efficacy for the elderly(Bryant et al., 2019). Whether this is a viable treatment option for the geriatric population remains unclear.

When choosing a treatment regimen for patients with LLD, it is important to weigh the benefits of starting medication with the risks of adverse effects. SSRIs are first-line for depression for all age groups and are generally thought to have a more favorable side effect profile than the older classes of antidepressants: tricyclics and monoamine oxidase inhibitors. SSRIs do not have the cardiac and anticholinergic adverse effects of tricyclic antidepressants, which makes them safer for elderly patients. However, there is still concern that paroxetine could have adverse outcomes in the geriatric population due to anticholinergic properties; however, studies show no increases in mortality, dementia risk, or cognitive measures (Beyer and Johnson, 2018). Also, SSRIs have been associated with increased risk of falls and osteoporosis (Kok and

Reynolds, 2017). Duloxetine, an SNRI, has also been associated with increased risk of falls due to its possible drug-drug interactions (Lenze and Ajam Oughli, 2019). Aripiprazole, which has been shown to be effective for augmentation for treatment-resistant LLD, comes with the risk of akathisia and parkinsonism (Lenze et al., 2015). It is crucial that the risk of side effects be carefully weighed with the benefits of treatment, especially in a population with higher rates of medical comorbidities and polypharmacy that can worsen outcomes.

Electroconvulsive shock therapy is treatment option when pharmacologic agents lack sufficient efficacy or are not well tolerated. New data add to the evidence demonstrating that ECT is a highly effective, safe, and well-tolerated antidepressant treatment option for geriatric patients (Geduldig and Kellner, 2016). In a study published in 2016 investing right unilateral ultra-brief ECT for elderly patients, there was a 61% remission rate (Kellner et al., 2016). However, the study did not include specifically treatment-resistant participants, and the efficacy for this group is likely to be less. Although concerns have been raised for the adverse effects associated with ECT in the geriatric population, the use of anesthesia poses the most risk, as opposed to the ECT itself (Nelson, 2018).

Transcranial magnetic stimulation is another nonpharmacological option for treatment-resistant depression that can be considered for LLD. In a study in 2019 looking at the use of this procedure on patients of advanced age, participants receiving bilateral rTMS experienced greater remission rates (40%) compared to unilateral (0%) or sham (0%) groups. Response to rTMS in the Hamilton Depression Rating Scale scores similarly favored the efficacy of bilateral rTMS (Trevizol et al., 2019).

Placebo-controlled RCTs

Multiple randomized placebo-controlled trials (RCTs) have been done to determine the efficacy of monotherapy for the treatment of late life depression. One trial comparing treatment of geriatric patients with depression with fluoxetine versus placebo found fluoxetine to be more likely to lead to symptom improvement (Evans et al., 1997). Another study was done to investigate the efficacy of paroxetine for the treatment of LLD, showing that 12.5 and 25 mg of paroxetine were effective compared to placebo for the treatment of LLD (Rapaport et al., 2008).

Given the high rate of treatment resistance when monotherapy is used, other studies have looked at the efficacy of psychotropic medications used as adjunct treatment. A randomized double-blind placebo-controlled trial looked at citalopram and methylphenidate as a combination regimen for geriatric patients with depression. The results of this study showed that the combination of these medications lead to greater improvement in mood and at a faster rate than either drug with placebo (Lavretsky et al., 2020). In a randomized, double-blind, placebo-controlled trial in 2015 studying aripiprazole used as an adjunct treatment, participants in the aripiprazole group were more likely to achieve remission than those in the placebo group (Lenze et al., 2015). In this study, aripiprazole was used with and SSRI or SNRI that had not been effective alone.

In 2020, another randomized double-blind placebo-controlled trial looked at the combination of escitalopram and memantine for the treatment of elderly patients with late life depression and cognitive impairment. The combination was as efficacious as escitalopram and placebo for depression, but much more efficacious than escitalopram and placebo for improving cognitive outcomes (Lavretsky et al., 2020).

It is important to note that most RCTs have exclusion criteria preventing the participation of patients with severe medical comorbidities, which limits the generalizability to the geriatric population since so many of these patients have other illnesses (Kok and Reynolds, 2017).

Future directions

In a review of refractory LLD, the overall response rate for all active pharmacological treatments investigated was 52% (Cooper et al., 2011). This leaves much room for improvement in our treatment of depression in this population. Current research must be evaluated in order to identify shortcomings and to design studies aimed at filling the gaps. Three future priority areas identified in a review in 2016 were focusing on patient-centered and culturally sensitive care, involving caregivers outside of the traditional healthcare team, and choosing different care settings (Hoeft et al., 2016). Paying close attention to ethnic and cultural differences when research is conducted is especially important, since presentation of depression in the elderly can vary widely, and therefore, treatment must be personalized. For example, suicidal ideation and hopelessness are symptoms more commonly expressed by White British than Black Africans and South Asians (Mansour et al., 2020).

In addition to improved cultural awareness when determining future treatment options, the advances being developed in neuroscience and neuroimaging techniques should be utilized. Functional magnetic resonance imaging (fMRI) has recently been used more to compare individuals with late life depression to healthy controls (Respino et al., 2020). Using fMRI techniques to identify areas of the brain that are functioning abnormally in late life depression will help develop treatment options to specifically target these specific areas.

In addition to the use of neuroimaging to create a better diagnostic understanding of late life depression, the role of biomarkers, genetics and hormone levels in disease pathophysiology should also be further investigated. In a recent 2020 study, single nucleotide polymorphisms in CYP19A1, an enzyme implicated in the conversion of androgen to estrogen, were associated with increased risk of late life depression (Ancelin et al., 2020). Elevations in the biomarker C-reactive protein (CRP) have also been shown to be associated with late life depression, and variants in the CRP gene can alter susceptibility to the development of the disorder (Ancelin et al., 2015). Further research to better understand the clinical and treatment implications of these findings is imperative in order to create new and more specific treatments that will have better efficacy than our current management for patients of advanced age.

References

Alexopoulos, G.S., et al., 1999. Vascular depression: a new view of late-onset depression. Dialogues Clin. Neurosci. 1 (2), 68–80.

Ancelin, M.L., Farré, A., Carrière, I., Ritchie, K., Chaudieu, I., Ryan, J., 2015. C-reactive protein gene variants: independent association with late-life depression and circulating protein levels. Transl. Psychiatry 5 (1), e499.

Ancelin, M.L., Norton, J., Canonico, M., Scarabin, P.Y., Ritchie, K., Ryan, J., 2020. Aromatase (CYP19A1) gene variants, sex steroid levels, and late-life depression. Depress. Anxiety 37 (2), 146–155.

Arandjelovic, K., Eyre, H.A., Lavretsky, H., 2016. Clinicians' views on treatment-resistant depression: 2016 survey reports. Am. J. Geriatr. Psychiatry 24 (10), 913–917.

Aziz, R., Steffens, D.C., 2013. What are the causes of late-life depression? Psychiatr. Clin. North Am. 36 (4), 497–516.

Barry, L.C., Abou, J.J., Simen, A.A., Gill, T.M., 2012. Under-treatment of depression in older persons. J. Affect. Disord. 136 (3), 789–796.

Beyer, J.L., Johnson, K.G., 2018. Advances in pharmacotherapy of late-life depression. Curr. Psychiatry Rep. 20 (5), 34.

Birrer, R.B., Vemuri, S.P., 2004. Depression in later life: a diagnostic and therapeutic challenge. Am. Fam. Physician 69 (10), 2375–2382.

Blazer, D., Williams, C.D., 1980. Epidemiology of dysphoria and depression in an elderly population. Am. J. Psychiatry 137 (4), 439–444.

Bonner, D., Howard, R., 1995. Treatment-resistant depression in the elderly. Int. Psychogeriatr. 7 (S1), 83–94.

Brodaty, H., et al., 2001. Early and late onset depression in old age: different aetiologies, same phenomenology. J. Affect. Disord. 66 (2–3), 225–236.

Bryant, K.A., Altinay, M., Finnegan, N., Cromer, K., Dale, R.M., 2019. Effects of repeated intravenous ketamine in treatment-resistant geriatric epression: a case series. J. Clin. Psychopharmacol. 39 (2), 158–161.

Burke, A.D., et al., 2019. Diagnosing and treating depression in patients with Alzheimer's disease. Neurol. Ther. 8 (2), 325–350.

Busfield, J., 2012. Challenging claims that mental illness has been increasing and mental well-being declining. Soc. Sci. Med. 75 (3), 581–588.

Buspavanich, P., Behr, J., Stamm, T., Schlattmann, P., Bschor, T., Richter, C., Hellweg, R., Heinz, A., Berger, M., Hindinger, C., Rentzsch, J., 2019. Treatment response of lithium augmentation in geriatric compared to non-geriatric patients with treatment-resistant depression. J. Affect. Disord. 251, 136–140.

Cakir, S., Senkal, Z., 2016. Atypical antipsychotics as add-on treatment in late-life depression. Clin. Interv. Aging 11, 1193.

Calati, R., Signorelli, M.S., Balestri, M., Marsano, A., De Ronchi, D., Aguglia, E., Serretti, A., 2013. Antidepressants in elderly: metaregression of double-blind, randomized clinical trials. J. Affect. Disord. 147 (1–3), 1–8.

Comijs, H.C., Nieuwesteeg, J., Kok, R., van Marwijk, H.W., van der Mast, R.C., Naarding, P., Voshaar, R.C., Verhaak, P., de Waal, M.W., Stek, M.L., 2015. The two-year course of late-life depression; results from the Netherlands study of depression in older persons. BMC Psychiatry 15 (1), 20.

Cooper, C., Katona, C., Lyketsos, K., Blazer, D., Brodaty, H., Rabins, P., de Mendonça Lima, C.A., Livingston, G., 2011. A systematic review of treatments for refractory depression in older people. Am. J. Psychiatry 168 (7), 681–688.

Cornwell, E.Y., Waite, L.J., 2009. Social disconnectedness, perceived isolation, and health among older adults. J. Health Soc. Behav. 50 (1), 31–48.

da Frota Ribeiro, C.M., Riva-Posse, P., 2017. Use of ketamine in elderly patients with treatment-resistant depression. Curr. Psychiatry Rep. 19 (12), 107.

Davison, T.E., et al., 2012. Biopsychosocial factors related to depression in aged care residents. J. Affect. Disord. 142 (1–3), 290–296.

Deng, Y., McQuoid, D.R., Potter, G.G., Steffens, D.C., Albert, K., Riddle, M., Beyer, J.L., Taylor, W.D., 2018. Predictors of recurrence in remitted late-life depression. Depress. Anxiety 35 (7), 658–667.

Devanand, D.P., 2002. Comorbid psychiatric disorders in late life depression. Biol. Psychiatry 52 (3), 236–242.

Driscoll, H.C., et al., 2007. Getting better, getting well: understanding and managing partial and non-response to pharmacological treatment of non-psychotic major depression in old age. Drugs Aging 24 (10), 801–814.

Duan, W., et al., 2017. How great is the medical burden of disease on the aged? Research based on "System of Health Account 2011". Health Qual. Life Outcomes 15 (1), 134.

Evans, M., Hammond, M., Wilson, K., Lye, M., Copeland, J., 1997. Placebo-controlled treatment trial of depression in elderly physically ill patients. Int. J. Geriatr. Psychiatry. 12 (8), 817–824.

Fava, M., 2003. Diagnosis and definition of treatment-resistant depression. Biol. Psychiatry 53 (8), 649–659.

Fava, M., Davidson, K.G., 1996. Definition and epidemiology of treatment-resistant depression. Psychiatr. Clin. N. Am. 19 (2), 179–200.

Fiske, A., Wetherell, J.L., Gatz, M., 2009. Depression in older adults. Annu. Rev. Clin. Psychol. 5, 363–389.

Geduldig, E.T., Kellner, C.H., 2016. Electroconvulsive therapy in the elderly: new findings in geriatric depression. Curr. Psychiatry Rep. 18 (4), 40.

Gehlawat, P., et al., 2013. Diabetes with comorbid depression: role of SSRI in better glycemic control. Asian J. Psychiatr. 6 (5), 364–368.

Gerst-Emerson, K., Jayawardhana, J., 2015. Loneliness as a public health issue: the impact of loneliness on health care utilization among older adults. Am. J. Public Health 105 (5), 1013–1019.

Goldman, L., Schafer, A.I., 2020. Goldman-Cecil Medicine. Elsevier.

Himelhoch, S., et al., 2004. Chronic medical illness, depression, and use of acute medical services among medicare beneficiaries. Med. Care 42 (6), 512–521.

Hoeft, T.J., Hinton, L., Liu, J., Unützer, J., 2016. Directions for effectiveness research to improve health services for late-life depression in the United States. Am. J. Geriatr. Psychiatry. 24 (1), 18–30.

Karlsson, L., et al., 2010. Minor change in the diagnostic threshold leads into major alteration in the prevalence estimate of depression. J. Affect. Disord. 122 (1–2), 96–101.

Kellner, C.H., Husain, M.M., Knapp, R.G., McCall, W.V., Petrides, G., Rudorfer, M.V., Young, R.C., Sampson, S., McClintock, S.M., Mueller, M., Prudic, J., Greenberg, R.M., Weiner, R.D., Bailine, S.H., Rosenquist, P.B., Raza, A., Kaliora, S., Latoussakis, V., Tobias, K.G., Briggs, M.C., Liebman, L.S., Geduldig, E.T., Teklehaimanot, A.A., Lisanby, S.H., CORE/PRIDE Work Group, 2016. Right unilateral ultrabrief pulse ECT in geriatric depression: phase 1 of the PRIDE study. Am. J. Psychiatry 173 (11), 1101–1109.

Koenig, H.G., Blazer, D.G., 1992. Epidemiology of geriatric affective disorders. Clin. Geriatr. Med. 8 (2), 235–251.

Kok, R.M., Nolen, W.A., Heeren, T.J., 2012. Efficacy of treatment in older depressed patients: a systematic review and meta-analysis of double-blind randomized controlled trials with antidepressants. J Affect. Disord. 141 (2–3), 103–115.

Kok, R., Reynolds, C., 2017. Management of depression in older adults: a review. JAMA 317 (20), 2114–2122.

Lasser, K.E., et al., 2002. Do minorities in the United States receive fewer mental health services than whites? Int. J. Health Serv. 32 (3), 567–578.

Lavretsky, H., Laird, K.T., Krause-Sorio, B., Heimberg, B.F., Yeargin, J., Grzenda, A., Wu, P., Thana-Udom, K., Ercoli, L.M., Siddarth, P., 2020. A randomized double-blind placebo-controlled trial of combined escitalopram and memantine for older adults with major depression and subjective memory complaints. Am. J. Geriatr. Psychiatry 28 (2), 178–190.

Lebowitz, B.D., 1997. Diagnosis and treatment of depression in late life. Consensus statement update. JAMA 278 (14), 1186–1190.

Lenze, E.J., Ajam Oughli, H., 2019. Antidepressant Treatment for Late-Life Depression: Considering Risks and Benefits. J. Am. Geriatr. Soc. 67 (8), 1555–1556.

Lenze, E.J., Mulsant, B.H., Blumberger, D.M., Karp, J.F., Newcomer, J.W., Anderson, S.J., Dew, M.A., Butters, M.A., Stack, J.A., Begley, A.E., Reynolds III, C.F., 2015. Efficacy, safety, and tolerability of augmentation pharmacotherapy with aripiprazole for treatment-resistant depression in late life: a randomised, double-blind, placebo-controlled trial. Lancet 386 (10011), 2404–2412.

Lökk, J., Delbari, A., 2010. Management of depression in elderly stroke patients. Neuropsychiatr. Dis. Treat. 6, 539–549.

Mansour, R., Tsamakis, K., Rizos, E., Perera, G., Das-Munshi, J., Stewart, R., Mueller, C., 2020. Late-life depression in people from ethnic minority backgrounds: differences in presentation and management. J. Affect. Disord. 264, 340–347.

Mast, B.T., MacNeill, S.E., Lichtenberg, P.A., 2004. Post-stroke and clinically-defined vascular depression in geriatric rehabilitation patients. Am. J. Geriatr. Psychiatry 12 (1), 84–92.

Meyers, B.S., Alexopoulos, G.S., 1988. Geriatric depression. Med. Clin. N. Am. 72 (4), 847–866.

Morimoto, S.S, Kanellopoulos, D., Manning, K.J., Alexopoulos, G.S., 2015. Diagnosis and treatment of depression and cognitive impairment in late-life. Ann. N Y Acad. Sci. 1345 (1).

Nelson, J.C., 2018. Management of late-life depression. Handb. Exp. Pharmacol. 250, 389–413.

Nelson, J.C., Clary, C.M., Leon, A.C., Schneider, L.S., 2005. Symptoms of late-life depression: frequency and change during treatment. Am. J. Geriatr. Psychiatry 13 (6), 520–526.

Nowrangi, M.A., Lyketsos, C.G., Rosenberg, P.B., 2015. Principles and management of neuropsychiatric symptoms in Alzheimer's dementia. Alzheimers Res. Ther. 7 (1), 12.

Palsson, S., Skoog, I., 1997. The epidemiology of affective disorders in the elderly. Int. Clin. Psychopharmacol. 12, S3–S14.

Park, M., Unützer, J., 2011. Geriatric depression in primary care. Psychiatr. Clin. North Am. 34 (2), 469–487. ix–x.

Park, S.-C., et al., 2017. Screening for depressive disorder in elderly patients with chronic physical diseases using the Patient Health Questionnaire-9. Psychiatry Investig. 14 (3), 306–313.

Rapaport, M.H., Lydiard, R.B., Edward, C.D., Carfagno, I.B., Lipschitz, A., 2008. Low doses of controlled-release paroxetine in the treatment of late-life depression: a randomized, placebo-controlled trial. J. Clin. Psychiatry 70 (1), 46–57.

Rapp, M.A., et al., 2005. Neuropsychological differences between late-onset and recurrent geriatric major depression. Am. J. Psychiatry 162 (4), 691–698.

Respino, M., Hoptman, M.J., Victoria, L.W., Alexopoulos, G.S., Solomonov, N., Stein, A.T., Coluccio, M., Morimoto, S.S., Blau, C.J., Abreu, L., Burdick, K.E., Liston, C., Gunning, F.M., 2020. Cognitive control network homogeneity and executive functions in late-life depression. Biol. Psychiatry Cogn. Neurosci. Neuroimaging 5 (2), 213–221. https://doi.org/10.1016/j.bpsc.2019.10.013.

Rodda, J., Walker, Z., Carter, J., 2011. Depression in older adults. BMJ 343, d5219.

Sadock, B., Dan Ruiz, P., 2017. Kaplan and Sadock's Comprehensive Textbook of Psychiatry. Wolters Kluwer, Philadelphia.

Sheline, Y.I., et al., 2010. Support for the vascular depression hypothesis in late-life depression: results of a 2-site, prospective, antidepressant treatment trial. Arch. Gen. Psychiatry 67 (3), 277–285.

Skoog, I., et al., 1993. Mental disorders and the use of psychotropic drugs in an 85-year-old urban population. Int. Psychogeriatr. 5 (1), 33–48.

Smagula, S.F., et al., 2015. Antidepressant response trajectories and associated clinical prognostic factors among older adults. JAMA Psychiat. 72 (10), 1021–1028.

Smith, M., Haedtke, C., Shibley, D., 2015. Evidence-based practice guideline: late-life depression detection. J. Gerontol. Nurs. 41 (2), 18–25.

Solomonov, N., et al., 2021. Modifiable predictors of nonresponse to psychotherapies for late-life depression with executive dysfunction: a machine learning approach. Mol. Psychiatry 26 (9), 5190–5198.

Tadayonnejad, R., Ajilore, O., 2014. Brain network dysfunction in late-life depression: a literature review. J. Geriatr. Psychiatry Neurol. 27 (1), 5–12.

Taylor, W.D., et al., 2003. White matter hyperintensity progression and late-life depression outcomes. Arch. Gen. Psychiatry 60, 1090–1096.

Taylor, W.D., 2014. Depression in the elderly. N. Engl. J. Med. 371 (13), 1228–1236.

Taylor, W.D., Aizenstein, H.J., Alexopoulos, G.S., 2013. The vascular depression hypothesis: mechanisms linking vascular disease with depression. Mol. Psychiatry 18, 963–974.

Tedeschini, E., Levkovitz, Y., Iovieno, N., Ameral, V.E., Nelson, J.C., Papakostas, G.I., 2011. Efficacy of antidepressants for late-life depression: a meta-analysis and meta-regression of placebo-controlled randomized trials. In: Database of Abstracts of Reviews of Effects (DARE): Quality-assessed Reviews [Internet]. Centre for Reviews and Dissemination (UK).

Trevizol, A.P., Goldberger, K.W., Mulsant, B.H., Rajji, T.K., Downar, J., Daskalakis, Z.J., Blumberger, D.M., 2019. Unilateral and bilateral repetitive transcranial magnetic stimulation for treatment-resistant late-life depression. Int. J. Geriatr. Psychiatry 34 (6), 822–827.

van Melle, J.P., et al., 2004. Prognostic association of depression following myocardial infarction with mortality and cardiovascular events: a meta-analysis. Psychosom. Med. 66 (6), 814–822.

Williams, D.R., et al., 2007. Prevalence and distribution of major depressive disorder in African Americans, Caribbean blacks, and non-Hispanic whites. Arch. Gen. Psychiatry 64 (3), 305.

Chapter 29

Identification and management of anxious depression in patients with treatment-resistant depression

Rachel Hershenberg[a], Mark Rapaport[b], and Kamile Weischedel[b]

[a]*Department of Psychiatry & Behavioral Sciences, Emory University, Atlanta, GA, United States,* [b]*Department of Psychiatry, University of Utah, Salt Lake City, UT, United States*

Part I: Identification

Depression is a debilitating illness. The World Health Organization reports that globally 264 million people of all ages suffer from depression. The National Institute of Health reports that about 7% of all adults in the United States experience at least one episode of depression in their lives. Although depression is a treatable illness, access to treatment can be limited due to several factors including financial, cultural, and logistical limitations. Depression is one of the main causes of disability in the United States and all around the world, and suicide is the second leading cause of death in 15- to 29-year-olds.

Certain characteristics of the disorder can complicate and impede treatment response. For example, history of trauma, periods of medication noncompliance, and other health problems can impair recovery. In addition to external stressors, depression itself can have features that result in the illness to become more resistant to treatment. Anxiety is one of these features, and its prevalence is noted to be significantly high. The National Institute of Health reports an estimated 31% of US adults experience any anxiety disorder at some point in their lives and an estimated 19% of US adults had any anxiety disorder in the past year. The same source reports that, among adults with any anxiety disorder within the past year, an estimated 22.8% experience serious impairment.

In this chapter, we will review depression with anxious distress, also referred to as "anxious depression," and focus on its prevalence, clinical features, and treatment options including pharmacotherapies and neuromodulation techniques such as electroconvulsive therapy (ECT), ketamine, and transcranial magnetic stimulation (TMS).

Defining anxious depression

The Diagnostic and Statistical Manual-5 (DSM-5; American Psychiatric Association, 2013) includes 7 specifiers: with mixed features, melancholic features, psychotic features, catatonia, peripartum onset, seasonal pattern, and anxious distress. These specifiers can guide the clinician in their treatment plan. DSM-5 requires at least 2 of the following 5 symptoms to meet criteria for anxious distress: (1) feeling keyed up or tense; (2) feeling unusually restless; (3) difficulty concentrating because of worry; (4) fear that something awful might happen; and/or (5) feeling that the individual might lose control.

Aligning with the DSM-5 criteria, Zimmerman and colleagues developed a brief scale to standardize the measurement of depression with anxious distress. A total of 773 psychiatric outpatients completed the Clinically Useful Depression Outcome Scale (CUDOS) with the DSM-5 anxious distress specifier (A). The study found CUDOS-A to be a valid and reliable measure of the DSM-5 anxious distress specifier for major depressive disorder (Zimmerman et al., 2014).

In research trials, a common operationalization of anxious depression (which is frequently referenced to in his chapter), consists of symptoms of major depressive disorder and high levels of anxiety, defined as a baseline anxiety/somatization factor score of ≥ 7 from the 17-item or 21-item Hamilton Depression Rating Scale (Hamilton, 1960). This score includes a summary of the following items: hypochondriasis, insight, general and gastrointestinal somatic symptoms, and psychic and somatic anxiety (Ionescu et al., 2013a,b).

About half of outpatients with major depressive disorder have clinically meaningful levels of anxiety (Fava et al., 2008a,b,c). The Sequenced Treatment Alternatives to Relieve Depression (STAR*D) study, a clinical trial funded by the

Managing Treatment-Resistant Depression. https://doi.org/10.1016/B978-0-12-824067-0.00029-3

National Institute of Health, assessed the effectiveness of depression treatments in patients diagnosed with major depressive disorder, in both primary and specialty care settings (NIH). It is the largest and longest study ever conducted to evaluate depression treatment (NIH). In level 1 of Star*D, 53.2% of patients were found to have anxious depression. Patients with anxious depression had lower remission rates, higher number of visits, and less time in treatment compared to those without anxious depression (Fava et al., 2008a,b,c). The same study also found that side effect frequency, intensity, and burden were significantly greater in the group with anxious depression.

Once clinicians identify anxious depression in a patient who has already demonstrated a degree of treatment resistance, what should be the next step? The treatment strategy working with depressed patients with cooccurring anxiety tends to be complicated. Next, we review the evidence-base behind pharmacological strategies, as well as neuromodulation, psychotherapy, and emerging treatment approaches when working with TRD patients who present with cooccurring anxiety.

Part II: Management

Pharmacological strategies

Patients with anxious depression overall have a poorer outcome than depressed patients without a significant anxiety component. Although they respond to SSRIs, SNRIs, TCAs and MAOIs, this group of patients usually cannot maintain response or remission, compared to patients with pure depression (Ionescu et al., 2014a,b). In this section, we will review data on monotherapy and augmentation strategies across classes of medications for patients with anxious depression (see Table 29.1). For additional reading, we also refer the reader to a comprehensive metaanalysis conducted by Ionescu and colleagues on the pharmacologic treatment of anxious depression (Ionescu et al., 2014a,b). As emphasized in this review, patients with anxious depression are less likely to experience remission compared to patients with pure depression. Moreover, despite comparable effectiveness with SSRIs and SNRIs, given their side effect profile, TCAs and MAOIs are not deemed first-line treatment for this population. Both SSRIs and SNRIs are considered safe, effective, and tolerable first-line treatments for patients with anxious depression.

Monotherapy with SSRIs

About two decades ago, Fava and colleagues examined tolerability and efficacy of different selective serotonin reuptake inhibitors (Fava et al., 2000). The group conducted a multisite, randomized, double-blind study using head-to-head comparison of three SSRIs, fluoxetine, sertraline and paroxetine, in 108 patients with anxious depression, defined by a score of 16 or greater on the HAMD-17 and 7 or greater on the HAM-D-Anxiety/Somatization Factor. After placebo responders were eliminated, patients were randomized to 4 weeks of double-blind treatment with fluoxetine (20 mg/day), sertraline (50 mg/day), or paroxetine (20 mg/day). Patients who achieved a CGI-Severity score of 1 or 2 at week three and week four of active treatment continued therapy at their original dose for an additional 6 weeks. Patients not achieving the required CGI-Severity scores entered an optional 1- to 6-week dose titration phase. During this phase, the clinician was allowed to titrate dosages (20–60 mg/day for fluoxetine, 50–200 mg for sertraline, and 20–60 mg for paroxetine). The dose that was found effective (a CGI-Severity score of 1 or 2 for two consecutive weeks) was continued for an additional 6 weeks. Patients received 10–16 weeks of active drug therapy. Baseline-to-endpoint improvement scores were similar in fluoxetine, sertraline, and paroxetine groups. Change-over-time improvement was similar among the antidepressants, except that fluoxetine and sertraline had greater improvement at week one compared to paroxetine. Overall, all treatments were well tolerated.

Feiger and colleagues conducted a post hoc pooled combined analysis of 5 double blind studies to examine response and remission rates in outpatients treated with sertraline or fluoxetine who were suffering from anxious-depression and severe depression (Feiger et al., 2003). Data were pooled from randomized, double-blind studies comparing fluoxetine versus sertraline for the treatment of major depression. Clinical outcome was assessed via the Hamilton Depression Rating Scale (HAM-D) and the Clinical Global Impression-Improvement scale (CGI-I). One thousand and eighty-eight patients were randomized, with 654 (60%) meeting criteria for anxious depression and 212 (19%) meeting criteria for high severity depression. A total of 334 patients were on sertraline and 320 patients were on fluoxetine. Treatment response in anxious depression was similar for both sertraline and fluoxetine, with $\geq$70% response rate and 47% remission rate for both treatment groups (Ionescu et al., 2014a,b).

SSRIs with anxiolytics

It is not uncommon to combine a first-line antidepressant with an anxiolytic for a short period of time, especially at the beginning of a medication trial or transitioning from one agent to another or when anxiety is difficult to control and debilitating. In a post hoc analysis of a 3-week, randomized, double-blind study, cotherapy with fluoxetine and clonazepam was

TABLE 29.1 Psychopharmacology for anxious depression.

Study	Participants	Design	Outcome
Fava et al. (2000)	MDD plus HDRS A/S score ≥ 7 at baseline	Randomized, double-blind, w/fluoxetine, sertraline, or paroxetine. 10–16 week study	All treatments were well tolerated. Similar response and remission rates
Feiger et al. (2003)	MDD plus HDRS A/S score ≥ 7 at baseline	Post hoc, 5 randomized, double blind studies comparing fluoxetine versus sertraline	Treatment response in anxious depression was found similar for both sertraline and fluoxetine, with ≥ 70% response rate; 47% remission rate for both treatment groups
Papakostas et al. (2008)	MDD plus HDRS A/S score ≥ 7 at baseline	Post hoc, randomized, double-blind, placebo-controlled. Fluoxetine and placebo versus fluoxetine and clonazepam	Difference in remission rates between the cotherapy and monotherapy groups with anxious depression were numerically (not statistically) higher than remission rates for nonanxious depression (Ionescu et al., 2014a,b)
Kim et al. (2011)	MDD and HARS ≥ 18	Randomized, open-label, controlled study with mirtazapine versus paroxetine. 60 participants	Mirtazapine and paroxetine were found equally effective and well tolerated for the depressive symptoms in MDD patients with the high level of anxiety symptoms. Mirtazapine was, however, more effective in reducing the anxiety symptoms than paroxetine in the early weeks of treatment
Papakostas et al. (2008)	*MDD and HDRS A/S score ≥ 7*	Post hoc, 10 double-blind, randomized studies, bupropion compared with SSRIs	Response rate greater in SSRIs. No difference in remission rates
Fava et al. (2007)	MDD plus HDRS A/S score ≥ 7 at baseline	Open label study with duloxetine. Total of 249 patients, 109 with anxious depression	Anxious depressed on duloxetine had greater improvements in total HAMD17 scores and HAMD17 A/S Factor scores compared to nonanxious depressed. Similar remission and response rates between groups, but anxious depressed's median time to respond was shorter than the nonanxious depressed
Howland et al. (2008)	Same group as above	Post hoc, open label, duloxetine	Higher scores on HDRS A/S predicted poorer response and remission
Nelson (2010)	First analysis: MDD plus HDRS A/S score ≥ 7 at baseline Second analysis: MDD plus scores greater than the median for the sum of HDRS psychic and somatic anxiety item	Post hoc, 11 double-blind, placebo-controlled trials with duloxetine	First analysis: Higher response rate with anxious depression; however, anxious depression had a lower chance for remission due to greater severity of depression Second analysis: No difference in response rates but lower remission rates in anxious depression
Versiani et al. (1999)	*MDD and HDRS*	Randomized, double-blind, amitriptyline versus fluoxetine	Similar response rates, however, fluoxetine is better tolerated
Tollefson et al. (1994)	MDD (DSM or NIMH Research Domain Criteria), 1 study used bipolar, plus HDRS A/S ≥ 7	Post hoc metaanalysis, 19 double-blind, fluoxetine versus placebo and fluoxetine versus TCAs	No difference in response between TCA and fluoxetine groups

Continued

TABLE 29.1 Psychopharmacology for anxious depression—cont'd

Study	Participants	Design	Outcome
Thase et al. (2012)	MDD plus HDRS A/S score ≥ 7 at baseline	Post hoc, 2 randomized, double-blind, placebo controlled studies of quetiapine extended release with 788 patients with anxious depression and 180 with nonanxious depression	No significant difference in response to quetiapine extended release between patients with anxious depression and without anxious depression, however, those with anxious depression with significantly more adverse events than those with nonanxious depression
Trivedi et al. (2008)	MDD plus HDRS A/S score ≥ 7 at baseline	Post hoc pooled data from 2 randomized, double-blind, placebo-controlled trials; 384 patients with anxious depression (aripiprazole: $n = 183$, placebo: $n = 201$) were compared to 259 patients with nonanxious depression (aripiprazole: $n = 138$, placebo: $n = 121$)	Adjunctive aripiprazole was found to be an effective and well-tolerated treatment for patients with major depression presenting with either anxious or atypical features
Papakostas et al. (2020a,b)	MDD and HDRS A/S	Secondary analysis on multicenter, randomized, double-blind, placebo-controlled study	Adjunctive pimavanserin was associated with significant improvement in response and remission
Fava et al. (2008a,b,c) Farabaugh et al. (2012)	MDD and HDRS A/S	STAR*D Study Level 1: post hoc, open label to citalopram Level 2: post hoc open label randomized to switch to bupropion sustained release, sertraline, venlafaxine extended release, or cognitive therapy or to augment citalopram with bupropion, buspirone, or cognitive therapy	In Level 1 remission was significantly less likely and took longer to occur in the participants with anxious depression. The same group also experienced greater side effects In Level 2, patients with anxious depression were found significantly worse in both the psychopharmacology and cognitive therapy switching and augmentation options
Delini-Stula et al. (1995)	MDD; one study with dysthymia with or without double depression	Post hoc metaanalysis of 40 studies (38 double-blind and 2 single-blind); total of 2416 patients comparing moclobemide, imipramine, maprotiline, amitriptyline, mianserin, and placebo	Sedative and nonsedative antidepressants had comparable efficacy for treating anxious depression and previous treatment failures lowered the response rate to trial drugs

compared with fluoxetine monotherapy (Papakostas et al., 2010). Of a total of 80 patients with MDD, 46 were identified as having anxious depression. The difference in remission rates between the cotherapy and monotherapy groups for patients with anxious depression did not reach statistical significance, though at trend level suggested higher rates of remission for the cotherapy.

SSRIs versus mirtazapine

In a randomized, open-label, controlled study, mirtazapine was compared with paroxetine in terms of efficacy and tolerability in treating patients with MDD with high anxiety (Kim et al., 2011). A total of 60 patients with MDD and a score above 18 on the Hamilton Anxiety Rating Scale (HARS) were included in the trial. Patients were randomly assigned to mirtazapine (15–30 mg/day) or paroxetine (10–20 mg/day) for 8 weeks. Patients were assessed with the HARS and the 17-item Hamilton Depression Rating Scale (HDRS) at weeks 1, 2, 4, and 8 after treatment. Mirtazapine and paroxetine were found equally effective and well tolerated for the depressive symptoms in MDD patients with a high level of anxiety symptoms. Mirtazapine was, however, more effective in reducing anxiety symptoms than paroxetine in the early weeks of treatment.

SSRIs versus bupropion

Efficacy of bupropion was compared to SSRIs (escitalopram, fluoxetine, sertraline, and paroxetine) in anxious depression (Papakostas et al., 2008). Data from 10 double-blind, randomized studies with 2122 participants were pooled. Anxious depression was defined as a 17-item Hamilton Rating Scale for Depression (HAM-D-17) anxiety-somatization factor score ≥ 7. Among patients with anxious depression ($N = 1275$), response rates were greater with the SSRI group than the bupropion group based on the HAM-D-17 and the Hamilton Rating Scale for Anxiety; however, remission rates did not differ.

Monotherapy with SNRIs

Clinically, many practitioners hesitate to use SNRIs to treat anxiety symptoms, as this group of medications can cause a transient increase in anxiety symptoms. We next review data on the effectiveness of SNRIs in this group of patients.

Fava et al. examined the efficacy and tolerability of duloxetine in anxious versus nonanxious depression in a 12 week open-label study (2007). Anxious depression was defined by a HAMD17 Anxiety/Somatization Factor score greater than or equal to 7, of which 109 of 249 (43.8%) patients met criteria. Participants in a major depressive episode ($N = 249$) were followed on duloxetine treatment for up to 12 weeks. Anxious depressed patients on duloxetine showed greater reductions in total HAMD17 scores and HAMD17 Anxiety/Somatization Factor scores compared to nonanxious depression. Remission and response rates at endpoint were similar between groups, but patients with anxious depression evidenced a significantly shorter median time to respond than the group with nonanxious depression. Moreover, an additional study utilizing the same dataset examined factors predicting reduced response to duloxetine (Howland et al., 2008) and found that higher HDRS A/S scores were associated with lower rates of response and remission when continuous scores were examined (Ionescu et al., 2014a,b).

A pooled analysis from 11 double-blind, placebo-controlled trials of duloxetine examined whether specific antidepressant effects are reduced in anxious depression during treatment of MDD with duloxetine (Nelson, 2010). Eleven trials included 2841 patients of whom 1326 were classified as anxious. Anxious depression was defined by ≥ 7 on the anxiety/somatization factor of the HAMD (HDRS A/S). Change on the HAMD was greater with duloxetine compared to placebo, in both anxious and nonanxious groups. Patients with anxious depression had a higher rate of response; however, their chance of remission was lower due to greater severity of depression. A second analysis was conducted with anxious depressed patients who were defined as MDD plus scores greater than the median for the sum of HDRS psychic and somatic anxiety items. While patients with anxious depression evidenced similar response rates to treatment compared to the nonanxious depressed group, they were noted to have significantly lower remission rates compared to nonanxious depressed patients.

Taken together, selective norepinephrine reuptake inhibitors should be considered safe and effective medications for patients with anxious depression. That said, it is of note that while response can happen relatively quickly, remission rates tend to be lower in anxious depression compared with nonanxious depressed patients.

Tricyclic monotherapy

As reviewed next, the efficacy between SSRIs and tricyclic antidepressants in the treatment of depression are not significantly different; importantly, however, more patients on a tricyclic antidepressant withdraw from treatment because of side effects (MacGillivray et al., 2003). Because of their advantages with regard to safety, dosing, and tolerability, SSRIs are usually chosen over TCAs initially, though are appropriate to prescribe if SSRIs (or SNRIs) fail to achieve full efficacy or lose their efficacy, and the patient can tolerate the TCA.

Versiani et al. (1999) compared the effects of fluoxetine versus amitriptyline in patients with anxious depression. In this double-blind comparison, 157 patients were randomized to either fluoxetine or amitriptyline for 8 weeks. Assessment instruments included: 21-item Hamilton Rating Scale for Depression, Hamilton Rating Scale for Anxiety, Clinical Global Impressions, Raskin Depression Scale, and Covi Anxiety Scale. Fluoxetine was given at a fixed dose of 20 mg/day and amitriptyline was given in a range of 50 to 250 mg/day (mean of 138.1 mg/day). Fluoxetine was comparable to amitriptyline in all efficacy measures except the HAMD sleep factor. As expected, side effects were more frequent and more severe in the amitriptyline group.

Tollefson et al. (1994) conducted a post hoc metaanalysis of 19 randomized, double-blind clinical trials comparing fluoxetine versus placebo, and fluoxetine versus TCAs. A total of 3183 patients were diagnosed with major depression; a total of 631 with anxious depression were in the fluoxetine versus placebo group and a total of 511 patients with anxious depression were in the fluoxetine versus TCA analysis. The efficacy of fluoxetine and TCAs was comparable across all measures for all groups. Discontinuation due to adverse events was significantly higher with fluoxetine than placebo and with TCAs than fluoxetine.

In summary, even though tricyclic antidepressants have been available since the early 1950s and they have been proven as effective as first-line treatment options such as SSRIs and SNRIs, given their relatively high side effect profile and withdrawal rates, clinicians are advised to use this group of medications only after first-line options fail. Tricyclics can be added to the existing dose of the SSRI/SNRI or started after a wash out period.

Antipsychotics are another group of psychotropics clinicians use for the treatment of anxious depression. Importantly, side effects from antipsychotics range from mild (e.g., sedation diabetes) to major (e.g., tremor, tardive dyskinesia) and can be lethal. Therefore, physicians should adhere carefully to monitoring guidelines should they chose to utilize antipsychotics as monotherapy or adjunctive treatment, both options which we briefly review below.

Antipsychotic monotherapy

A post hoc study that pooled data from 2 randomized, double-blind, placebo controlled studies of quetiapine extended release (50–300 mg/day) was conducted to evaluate its efficacy in anxious depression (Thase et al., 2012). A total of 788 of 968 patients included in the analysis were categorized with anxious depression (defined as HAM-D anxiety/somatization factor score ≥ 7) and 180 (18.6%) were categorized as nonanxious. Outcomes included change from randomization at Week 6 in Montgomery Åsberg Depression Rating Scale (MADRS) total score for patients with anxious and nonanxious depression. There was no significant difference in response to quetiapine extended release between patients with and without anxious depression. Adverse events were seen more often in the anxious depression group.

Antipsychotic augmentation

Trivedi et al. (2008) assessed the efficacy of adjunctive aripiprazole to standard antidepressant therapy (ADT) for patients with major depressive disorder. In this post hoc study, data was obtained from 2 randomized, placebo-controlled trials. 384 patients with anxious depression (aripiprazole: $n=183$, placebo: $n=201$) were compared to 259 patients with nonanxious depression (aripiprazole: $n=138$, placebo: $n=121$). Adjunctive aripiprazole was found to be an effective and well-tolerated treatment for patients with major depression, regardless of anxiety. Remission rates were significantly higher with adjunctive aripiprazole versus adjunctive placebo.

Pimavanserin is an atypical antipsychotic approved for the treatment of Parkinson's disease psychosis and is under research for schizophrenia and Alzheimer's disease psychosis. The original multicenter, randomized, double-blind, placebo-controlled CLARITY trial conducted by Fava et al. (2019) tested the efficacy and safety of adjunctive pimavanserin on depressed patients who did not respond to SSRIs or SNRIs. The study found pimavanserin to be effective, as assessed with changes on the HDRS-17. A secondary study conducted by Papakostas et al. (2020a,b) examined the efficacy of Pimavanserin for patients with anxious depression; compared to placebo, both response and remission rates were significant, based on changes in HAMD-A/S factor scores.

Inter-class comparisons

Fava et al. (2008a,b,c) and Farabaugh et al. (2012) conducted post hoc analyses on the STAR*D study to compare antidepressant treatment as well as cognitive therapy outcomes for patients with anxious and nonanxious major depression. In Level 1 of STAR*D, a total of 2876 adult outpatients with major depressive disorder received citalopram. Nonremitters and patients who did not tolerate citalopram then entered the Level 2 Phase of the study where they were randomly assigned either to switch to sustained-release bupropion ($N=239$), sertraline ($N=238$), extended-release venlafaxine ($N=250$), or cognitive therapy, or to continue taking citalopram and receive augmentation with sustained-release bupropion ($N=279$) or buspirone (N=286). Treatment lasted up to 14 weeks in each level. Patients were designated as having anxious depression if their anxiety/somatization factor score from the 17-item Hamilton Depression Rating Scale (HAM-D) was 7 or higher at baseline. In Level 1 of STAR*D, 53.2% of patients had anxious depression. Remission was significantly less likely and took longer to occur in the participants with anxious depression. The same group also experienced greater side effect frequency, intensity, and burden, as well as the number of serious adverse events. Similarly, in Level 2, patients with anxious depression fared significantly worse in both the psychopharmacology and cognitive therapy switching and augmentation options.

Although not commonly used since the advent of the SSRIs, monoamine oxidase inhibitors are effective medications in the treatment of depression. A post hoc metaanalysis of 40 studies (38 double-blind and 2 single-blind) were conducted to compare the efficacy of nonsedative, reversible MAO-I moclobemide, with imipramine, and sedative antidepressants (amitriptyline, mianserin, maprotiline), and placebo (Delini-Stula et al., 1995). A total of 2416 patients were included in the analysis (moclobemide n=1215; imipramine n=619; sedative antidepressants n=340). Sedative antidepressants included cases on maprotiline (n=122), amitriptyline (n=80), and mianserin (n=40). Findings demonstrated that the sedative and nonsedative antidepressants had comparable efficacy for treating anxious depression, and previous treatment failures lowered the response rate to trial drugs.

In sum, various medications from multiple classes of psychotropics can be utilized in the treatment of anxious depression. Given the safety, efficacy, and tolerability profile, SSRIs and SNRIs are deemed the first-line treatment for this group of patients. When patients demonstrate nonresponse or limited response to maximal dose of SSRIs, a next step is usually switching or adding an SNRI, and if still not effective, then considering augmentation strategies such as adding an anxiolytic for a limited period of time, or using an antipsychotic while monitoring the side effect profile. Depending on the patient's side effect tolerability, TCAs and MAOIs can also be effective treatment options. Tricyclics may be associated with difficult side effects but, if tolerated, are likely to help a patient with anxious depression achieve a clinically adequate response. Data on the efficacy of moclobemide (a nonsedative MAOI) suggest that this medication is as effective as imipramine and sedative antidepressants in treating anxious depression, however, just like TCAs, it is not considered a first-line treatment given the side effect profile and tolerability. TCAs can be added to the SSRI/SNRI while an MAOI must be initiated following a complete washout of the initial antidepressant(s).

Given the resistant nature of anxious depression, research focused on interventional treatments, including ketamine, electroconvulsive therapy, and transcranial magnetic stimulation, has surged in recent years, which we turn to next.

Interventional psychiatric approaches: Ketamine

A pioneering study examining baseline anxious depression as a predictor of treatment outcome was conducted by Ionescu and colleagues as secondary analyses of a larger clinical trial (Ionescu et al., 2014a,b). In this study, patients received a single, open-label intravenous infusion (0.5 mg/kg) over 40 min and were assigned to the placebo (rather than riluzole) condition. Patients with anxious depression were defined categorically by a Hamilton Depression Rating Scale Anxiety Somatization (HAMD-AS) score $\geq$7, which comprised 58% (n=15) of their sample. Given poorer response to traditional monoaminergic antidepressants, authors hypothesized that anxious depressed patients would evidence poorer outcomes, and analyses compared groups following a single infusion of ketamine over a course of 28 days. Contrary to expectations, the anxious group showed comparable or greater improvements in symptoms of depression as measured with the MADRS and the HDRS; the largest effect was seen on day 2. Moreover, anxious depressed patients had slower time to relapse as measured on the MADRS. Notably, with regard to change in anxious symptoms, improvements in anxiety as measured with the Hamilton Anxiety Rating Scale were greater for the anxious depressed group for all but two time points. Dissociative and psychotic symptoms during the infusion were not different between groups. Taken together, this was one of the first pieces of evidence that anxious depressed patients may respond appropriately to ketamine—but it was far from conclusive, particularly given the small sample size. That said, it was consistent with an earlier metaanalysis showing that comorbid anxiety disorders at baseline do not contribute to significant heterogeneity in outcomes up to 7 days post infusion among TRD patients receiving IV ketamine (Romeo et al., 2015).

Follow-up studies have been largely consistent. For example, among patients with treatment-resistant bipolar depression (Ionescu et al., 2015), secondary analyses were completed on patients in two separate but identical randomized, placebo-controlled crossover trials of ketamine combined with lithium or valproate monotherapy. All analyzed patients received a single, open-label intravenous infusion of ketamine (0.5 mg/kg) over 40 min. Consistent with Ionescu et al. (2014b), anxious depressed patients (n=21, 58%), categorized based on HDRS-AS, did not evidence worse outcomes on depression (MADRS, HDRS) or anxiety (Hamilton Anxiety Rating Scale). Findings suggested that bipolar TRD patients with anxious depression may not respond worse to a single infusion of ketamine, despite their being more severe at baseline.

Consistently, in another multisite, double-blind, placebo-controlled trial (Salloum et al., 2019), 99 unipolar TRD participants received a single infusion of IV ketamine or midazolam and patients with and without anxious depression (similarly defined by HAMD-AS scores) were compared at 1 and 3 days postinfusion. At days 1 and 3, collapsing across ketamine dose, there were no significant interactions, indicating the treatment group did not impact score change. When stratifying based on infusion dosage, anxious TRD patients who received 0.1 mg/kg did worse as measured by the MADRS and HAMD-6 than TRD patients without anxious depression (at the same 0.1 mg/kg dose). This effect was not significant at the other three doses of ketamine (0.2, 0.5, 1.0 mg/kg). The anxious TRD group had lower levels of dissociation reported—one post hoc explanation was that this group may have had higher proportion of benzodiazepine use, which could blunt dissociative symptoms. This study, along with Ionescu et al. (2014a,b, 2015), have been cited in support of ketamine's appropriateness in patients with anxious depression, though these studies have not been powered to detect this exploratory aim, and comparisons have focused on changes in mood (and, to some extent, anxiety), following a single infusion of ketamine.

Of note, a recent study contradicts this growing body of research. Specifically, Chen et al. (2020) conducted a randomized, placebo-controlled trial of 71 TRD patients, randomizing patients to one of three groups—a single 40-min intravenous saline infusion mixed with 0.0, 0.2, or 0.5 mg/kg of ketamine and similarly comparing patients with or without anxious depression based on HAMD-AS scores. Patients *without* anxious depression did significantly better at 0.5 and 0.2 mg/kg compared to placebo; in contrast, patient with anxious depression did not do significantly better with 0.5 or 0.2 mg/kg compared to placebo. Comparing anxious to nonanxious depression, the depressive symptom reduction between groups was specifically detected at 0.5 mg/kg and was not significantly different between anxious and nonanxious depressed patients at 0.2 mg/kg or placebo. In reconciling these findings, the authors wondered if anxious TRD patients might benefit from a higher dose, such as 1.0 mg/kg.

Importantly, to increase ecological validity of these findings, recent work has examined anxiety as a predictor of treatment response following an acute course of IV ketamine. McIntyre et al. (2020) analyzed 209 adults with unipolar or bipolar depression and categorized anxious depression based on item responses to the QIDS-SR16 and the GAD7. Following patients naturalistically over a course of four infusions, QIDS-SR16, GAD7, and suicidal ideation improved among all patients, and change in QIDS and GAD7 was more pronounced among patients with anxious-distress. Consistently, in a sample of 97 TRD patients with unipolar or bipolar depression (Wang et al., 2019a), patients were classified as anxious TRD (consistent with the HAMD-AS scores described previously), as melancholic (defined based on the SCID-5 criteria), as melancholic-anxious (if they met criteria for both), or as no subtype. Following patients over six ketamine infusions, patients with melancholic and melancholic-anxious features showed a worse trajectory, including in changes on the MADRS and Hamilton Anxiety Rating Scale; in contrast, patients in the anxious group had greater MADRS reductions compared to those in the melancholic and melancholic anxious groups, and patients in the anxious group had faster score reductions on the HAMA compared to the melancholic group and those without a subtype group.

Taken together, recent research crossing from randomized clinical trials to open label effectiveness trials have generally supported the appropriateness of IV ketamine for patients presenting with anxious distress at baseline (though see Chen et al., 2020 for an exception), and it is notable that this finding has been replicated in studies with minimal exclusion criteria (McIntyre et al., 2020; Wang et al., 2019a).

Extrapolating from related areas of research

Additional ways to consider if ketamine may be appropriate for patients with cooccurring anxiety is to examine the impact and tolerability of ketamine on patients with primary anxiety disorders, as in the preclinical literature there is evidence that ketamine may have an anxiolytic effect (Fortress et al., 2018; Parise et al., 2013). Though a comprehensive clinical review is beyond the scope of this chapter, we highlight seminal work from which to extrapolate below and secondly direct the reader to a recent review by Banov et al. (2020) for a more thorough discussion.

One relevant study was conducted by Glue and colleagues in an ascending single-dose, uncontrolled, open label study in 12 patients with refractory DSM-IV GAD and/or SAD (Glue et al., 2017). To analyze response in treatment-resistant anxiety and reduce potential confounds due to depression, patients with MADRS scores of 20 or greater were excluded. The study found support for rapid anxiolytic effects, which wore off over 3 to 7 days. Moreover, examining dose response relationships, the maximum reduction in anxiety was greatest for 0.5 to 1 mg/kg, and the greatest duration of anxiolytic effects was for 1 mg/kg. Authors summarized their data as the first evidence that ketamine improves symptoms of anxiety in nondepressed patients with treatment-resistant GAD and SAD and was also safe and well-tolerated. Follow-up work conducted by Taylor et al. (2018) was largely consistent, demonstrating dose-response relationships for anxiolytic effects as well as dissociative side effects and changes in blood pressure and heart rate, in a more rigorous experimental design among nondepressed patients with treatment-resistant GAD and/or SAD.

Though no longer listed as anxiety disorders in the DSM, to increase comparison of ketamine data with the ECT data (described next), we also briefly review the literature on the use of ketamine in patients with OCD and PTSD.

Obsessive compulsive disorder

In the first randomized, placebo-controlled crossover study of OCD patients (Rodriguez et al., 2013) receiving a single dose of IV ketamine (0.5 mg/kg), patients receiving ketamine ($n=8$) showed a significant improvement in obsessional thoughts compared to patients receiving placebo ($n=7$). One-week postinfusion, half of the patients who received ketamine showed more than a 35% reduction on the Y-BOCS, indicating that treatment response may persist for at least a week. Additional research has tested if ketamine may work synergistically with cognitive behavioral therapy (CBT), as it may enhance plasticity and extinction learning. In 10 unmedicated OCD outpatients with intrusive obsessions (Rodriguez et al., 2013), patients received a single 0.5 mg/kg dose of IV ketamine followed by 10 sixty-minute exposure sessions over 2 weeks. Eight of the 10 patients reported a rapid reduction in obsessive severity (measured the OCD Visual Analog Scale) up to 230 min postinfusion (for 7 patients). OCD severity, measured with the YBOCS, significantly decreased over time including at week 2 and week 4, suggesting the potential for ketamine and CBT to be used together to maintain and extend gains, a finding consistent with a case report of treatment-resistant OCD in which the patient received intensive CBT with concurrent racemic ketamine hydrochloride (50 mg) twice weekly, delivered in five 10 mg doses over a 20-min period (Adams et al., 2017).

Posttraumatic stress disorder

IV ketamine has also been examined for patients with treatment-resistant posttraumatic stress disorder (PTSD), with or without cooccurring TRD. In one of the major proof-of-concept RCTs ($n=41$), Feder et al. (2014) showed that a single 0.5 mg/kg infusion (compared to midazolam) led to significant reductions in PTSD severity 24 h after infusion, and this effect remained significant after adjusting for baseline and 24-h depressive severity. Importantly, as there are concerns that ketamine can exacerbate PTSD symptoms including dissociation, this trial found that dissociative symptoms after treatment had resolved by the next assessment 120 min from the start of the infusion, only 1 patient dropped out due to dissociative effects, and there was no emergence of psychotic or manic symptoms. Expanding this finding, Zeng et al. (2013) conducted post hoc analyses of three RCTs of TRD patients (unipolar or bipolar) with trauma history or comorbid PTSD receiving a single dose of 0.5 mg/kg of ketamine. Results demonstrated no significant increase in psychosis, dissociation, or anxiety in the trauma group up to 7 days postinfusion. Moreover, in an open-label trial, Albott et al. (2018) treated 15 Veterans with comorbid PTSD and TRD with 6 doses of 0.5 mg/kg of ketamine over 12 days. Consistent with previous studies, there were significant improvements in PTSD and depression and minimal problems with dissociative side effects. None of the patients experienced a worsening of PTSD during the trial.

Taken together, the preponderance of data suggests that IV ketamine is appropriate for patients with cooccurring symptoms of anxiety. More work is needed to understand the trajectory of change in symptoms of anxiety following treatment, but core symptoms of depression seem likely to improve. Importantly, there is evidence that patients with primary anxiety disorders can tolerate and respond to the treatment, including treatment-resistant OCD and PTSD, providing tangential support that TRD patients with anxious depression may similarly experience anxiolytic as well as antidepressant benefits. Thus, it may be reasonable to infer that IV ketamine is not contraindicated and may be beneficial for TRD patients who present with cooccurring anxiety, though more research is needed.

Practical considerations and contraindications

Benzodiazepine and naltrexone use in anxiety patients

Data suggests that benzodiazepine and naltrexone use may interfere with the therapeutic effects of ketamine (Albott et al., 2017; Frye et al., 2015; Andrashko et al., 2019; Williams et al., 2018). Best practices include withholding benzodiazepines at least 24 h prior to an infusion (Bobo et al., 2020) and, if feasible, to withhold naltrexone for the duration of the treatment course.

Anxiety about the treatment

Anxiety is the most common acute psychiatric side-effect of IV Ketamine (followed by agitation or irritability) and is the second most common psychiatric reason for withdrawal from a ketamine trial, following worsening mood (Short et al., 2018). Little empirical research examines if baseline anxiety is a predictor of an anxious response to ketamine. Of note, initial evidence suggests that *anxiety about the treatment* may predict poorer response—and that low levels of openness to experience, rather than state or trait anxiety, may be a predictor of that treatment anxiety. For example, after a series of

6 infusions, 17 responders and 14 nonresponders did not differ in baseline anxiety but did differ in "anxious ego disintegration," more specifically driven by the following subscales of the 5-Dimensional Altered States of Consciousness Rating Scale (5D-ASC; Dittrich, 1998): experiences of negative derealization, reduced self-control, and reduced body control. These constructs were inversely related to openness to experience—but were not related to state or trait anxiety at baseline (Aust et al., 2019). Though more research is warranted, this finding suggests that baseline anxiety is not synonymous with anxiety about the treatment, and expectations about the treatment itself could potentially be a point of intervention prior to or after the first infusion.

Thus, anxiety alone may not be a contraindication for ketamine. The most caution should be taken in high-risk patients exhibiting nonmodifiable risk factors such as those with a psychotic illness, active substance use, or any condition that can result in dire medical consequences as a result of elevation in blood pressure (aneurysm, uncontrolled hypertension) or intracranial pressure secondary to ketamine use (Banov et al., 2020). Many clinicians would like their patients to be off any recreational substances for at least 3 to 6 months prior to starting ketamine. Modifiable factors such as benzodiazepine use and expectancies and anxiety following the treatment can and should be addressed on an ongoing basis to increase the likelihood of success.

Non-IV ketamine

To our knowledge, no studies have examined anxious depression as a predictor of outcome among patients receiving intranasal ketamine. In a randomized, double-blind, crossover study examining the efficacy of esketamine in TRD patients (Lapidus et al., 2014), 18 MDD patients completed 2 treatment days with intranasal ketamine hydrochloride (50 mg) or saline solution. Ketamine was superior to placebo in improving anxiety symptoms at 24 h as measured by HAM-A scores. In another seminal esketamine trial (Singh et al., 2016), patients with primary anxiety disorders were excluded from the trial, and no anxiety outcomes were reported. In a recent metaanalysis of esketamine augmentation in TRD (Papakostas et al., 2020a,b), anxiety was not included as a predictor of heterogeneity or as an outcome. Thus, it remains inconclusive to what extent intranasal ketamine may improve symptoms of anxiety.

In a recent investigation, oral ketamine was investigated in a multiple dose, open-label, uncontrolled study of patients with treatment-resistant DSM-5 major depressive episode (MDE) and/or SAD and/or GAD who had previous positive responses to IV ketamine (Glue et al., 2020). Seven patients were administered flexible doses of a 60 to 240 mg tablet over 4 days. Patients significantly improved on measures of anxiety (as measured by the HAM-A and Fear Questionnaire) and depression (as measured by the MADRS) and reported only mild adverse events including headache, dizziness, and dissociation.

Thus, more research will be critical to understanding how efficacious alternative forms of administration may be among TRD patients with significant cooccurring anxiety.

Summary and recommendations: Ketamine

IV ketamine appears to be appropriate for TRD patients with cooccurring anxiety (see Table 29.2). The majority of this research has defined anxious depressed patients based on scores to six items from a larger structured clinical interview for depression (HDRS). Ketamine's appropriateness for patients with cooccurring anxiety can be inferred from secondary analyses of clinical trials that examine anxious depression as a predictor of response following a single infusion of IV ketamine, from initial effectiveness work following patients during an acute course of ketamine, as well as from extrapolating from research on ketamine's effectiveness in patients with primary anxiety disorders.

Opportunities for future research will be to continue to expand outcomes to more measures of anxiety and quality of life, in addition to symptoms of depression, over an acute course of treatment. Doing so will make results more robust and generalizable. Moreover, reporting on treatment retention as well as ratings of quality of life, and more consistent measures of anxiety as an outcome, will help to contextualize how patients with anxious depression respond to an acute course of treatment. Future research could benefit from examining comorbid diagnoses more specifically—that said, a strength of the research base to date is the standard operationalization of anxious depression, making comparison across trials possible. With regard to dose optimization, future research will benefit from testing dose response relationships for patients with anxious depression, as a higher dose (1.0 mg/kg) may lead to improved outcomes in patients with anxiety, compared to the standard 0.5 mg/kg; and future research can examine if prolonged maintenance treatment may help to sustain gains. Finally, it will be important to examine expectancies about and anxiety from the treatment itself and if these psychological factors can be modified prior to or subsequent to the start of treatment.

TABLE 29.2 Ketamine and anxious depression.

Study	Diagnosis	Comparison	Dose, # of treatments	Primary outcome	Anxiety outcome	Notes
Ionescu et al. (2014b)	26 unipolar TRD patients	Anxious depression (n=15, 58%) versus no anxious depression	0.5 mg/kg, single infusion	Anxious depressed patients slower time to relapse on MADRS	Anxious depressed patients have greater improvements on the Hamilton Anxiety Rating Scale	Secondary analysis of a larger clinical trial
Ionescu et al. (2015)	36 TRD patients, bipolar depression	Anxious depression (n=21, 58%) versus no anxious depression	0.5 mg/kg, single infusion	Anxious depression not associated with worse outcomes on MADRS or HDRS	Anxious depression not associated with worse outcomes on the Hamilton Anxiety Rating Scale	Secondary analyses, 2 separate RCTs, ketamine + lithium or valproate monotherapy
Salloum et al. (2019)	99 unipolar TRD patients	Anxious depression (n=45, 45%) versus no anxious depression	0.1, 0.2, 0.5, or 1.0 mg/kg, single infusion	Anxious TRD patients did worse as measured by MADRS and HAMD6 only at 0.1 mg/kg	–	Secondary analyses, RCT comparing ketamine to midazolam
Chen et al. (2020)	71 Taiwanese unipolar TRD patients	Anxious depression (n=42, 59%) versus no anxious depression	0.0, 0.2, or 0.5 mg/kg, single infusion	At 0.5 mg/kg, nonanxious patients do significantly better than placebo on the HDRS; anxious patients do not	–	Secondary analysis or RCT
McIntyre et al. (2020)	209 patients with unipolar (n=117) or bipolar (n=26) TRD	Anxious distress (n=94, 45%)	2 infusions at 0.5 mg/kg, dose optimized up to 0.75 mg/kg if needed for infusions 3 and 4	Anxious distress patients have greater reductions on QIDS-SR16	Anxious distress patients have greater reductions on GAD7	Naturalistic study, open label
Wang et al. (2019a)	97 TRD patients unipolar (n=77) or bipolar (n=20) TRD	Anxious, melancholic, melancholic-anxious, or no subtype	6 infusions of 0.5 mg/kg delivered over 2 weeks	Anxious patients have greater MADRS reductions compared to melancholic and melancholic anxious patients	Anxious patients have faster score reductions on the HAMA compared to melancholic and no subtype groups	Single-arm, open-label study
Intranasal ketamine						
Lapidus et al. (2014)	18 unipolar TRD patients	Full sample	Crossover design, 2 treatment days, Intranasal ketamine hydrochloride (50 mg) or saline solution	Significant improvement on the MADRS at 24 h after ketamine compared to placebo	Significant improvement on Hamilton Anxiety Rating Scale scores compared to placebo	
Oral ketamine						
Glue et al. (2020)	7 patients with TRD and/or SAD and/or GAD with previous positive response to IV ketamine	Full sample	Multiple dose open-label flexible dose, range 60 mg–240 mg, over 4 days	All seven patients had 50% or greater reductions in MADRS compared with baseline	All seven patients had 50% or greater reductions in Hamilton Anxiety Rating Scale compared with baseline. Six of seven patients had 50% or greater reductions on the fear questionnaire compared with baseline	Open-label, uncontrolled study

Interventional psychiatric approaches: ECT

Considering the size and the strength of the evidence-base for ECT in TRD, there has been a paucity of studies examining the role of ECT in patients with concurrent anxiety, particularly in recent years. Initial ECT research conceptualized concurrent anxiety within the purview of neurotic depression, in contrast to melancholic depression, and as characterized predominantly by somatic preoccupation. Summarizing the literature up until 1982, Fink (1982) stated, "the presence of anxiety, hysteria, phobia, and hypochondriasis are the principle contraindications to ECT. While the literature lacks well-designed studies to support this conclusion, the burden of the experience does not suggest that controlled comparisons would be helpful."

Reviewing the literature, there was a scant number of TRD studies that examine concurrent anxiety as a predictor of treatment outcomes. Even more broadly, there is little research demonstrating the extent to which ECT leads to changes in anxiety symptoms in TRD patients more generally. Indeed, surveying the major ECT publications, there is little information on the relationship between concurrent anxiety and outcomes across the major ECT multisite randomized clinical trials (e.g., Kellner et al., 2017; Fink, 2014; Østergaard et al., 2020), in ECT outcomes in naturalistic settings (e.g., Dunner et al., 2006), in metaanalytic reviews (e.g., Van Diermen et al., 2018), in qualitative reviews (e.g., Meyer et al., 2020), or in clinical practice recommendations (e.g., Gill and Kellner, 2019). Relevant literature is summarized below.

Anxiety as a predictor of ECT outcomes

As part of the CORE multisite trial, one study examined a specific facet of anxiety, *somatization*, as a predictor of remission with BT ECT (Rasmussen et al., 2004). Somatization was defined as a sum of item 11 (somatic anxiety) and item 15 (hypochondriasis) from the HAMD-24. Results showed that high somatization predicted lower sustained remission rates, specifically among patients with less severe depression (defined as HAMD-24 scores of 21–32). Among patients with more severe depression (defined as HAMD-24 scores above 32), there was no association between somatization and remission. Item 10 (psychic anxiety) was not associated with remission for either depression groups.

Since 2004, one naturalistic, prospective follow-up study analyzed 70 bipolar patients in a depressed (51%) or mixed (49%) episode who received a bilateral course of ECT twice weekly (Medda et al., 2020). At baseline, 47% (n=33) of patients met criteria for lifetime anxiety disorders, which included Panic Disorder, Social Phobia, and GAD. Patients were followed naturalistically, with a mean duration of follow-up at 57 weeks. Across the whole sample, 93% of patients maintained a partial or full therapeutic response for more than 90% of the follow-up period. Analyzing baseline factors associated with remission and time to relapse, depressed (compared to mixed) patients had higher rates of remission. Importantly, patients with anxiety disorders relapsed in a shorter time; length of the index episode and the presence of psychotic symptoms were not associated with time to relapse.

Two studies have examined concurrent anxiety as a predictor of outcomes among unipolar geriatric TRD patients. In the first (Flint and Rifat, 1997), a staging method of treatment for depressed elderly patients was used, beginning with 6 weeks of nortriptyline, followed by adjunctive lithium, followed by phenelzine (with or without lithium augmentation) and, finally, a course of ECT or 6 weeks of fluoxetine (with or without lithium augmentation). Unilateral ECT was given three times per week. Hospital Anxiety and Depression Scale (HAD-A) scores were used to dichotomize patients at baseline as anxious and nonanxious. In the analysis of the whole sample, 83.8% of patients responded to one of the lines of treatment, and by the end of the protocol, anxious depressed patients had a lower rate of response in intent to treat analyses and a higher rate of attrition. Specifically, with regard to ECT, six anxious and two nonanxious patients entered the third-line treatment; four received ECT. Only one patient, who was nonanxious, responded to ECT.

In another naturalistic follow-up of elderly, unipolar depressed TRD patients, specifically using the Duke Staged Approach for depression (Steffens and McQuoid, 2005), patients started with an SSRI; if needed, received augmentation with bupropion or switched to venlafaxine; if needed, received augmentation with a tricyclic or lithium; and last, or as needed throughout, received ECT. Use of benzodiazepines was allowed. Patients were evaluated at least every 3 months, and patients with 3 months to 2 years of follow-up data were analyzed. The final sample was 204 older depressed patients, average age 70, 30.9% of whom had symptoms of GAD. Remitters (as defined on the MADRS) were less likely to have GAD symptoms. Moreover, greater GAD symptoms were associated with greater time to achieve remission, even when controlling for functional impairment, stress and social support, and use of benzodiazepines (which was also an independent negative predictor of outcome). Notably, these analyses do not include remission specifically from ECT.

Considered more broadly as a treatment strategy within a continuum of care, findings suggest that comorbid anxiety may be a negative prognostic factor in depressed elderly patients (Steffens and McQuoid, 2005), a finding consistent with Flint and Rifat (1997).

Anxiety as an outcome in ECT

Two studies examined the extent to which an acute course of ECT broadly impacts symptoms of anxiety. In the most recent effectiveness study (Huang et al., 2019), 116 Taiwanese inpatients with unipolar MDD requiring ECT were recruited. Patients were medication free starting 3 days before treatment and during the ECT course, except for anxiolytic or sedative-hypnotic medications as needed. Standard bitemporal ECT was given either two or three times per week, with a maximum of 12 treatments. Analyses focused on two subscales of the HRSD17, anxiety/somatization and "core depression" symptoms, comprised of items for depressed mood, guilt, suicide, work and activities, loss of interest or pleasures, and psychomotor retardation. HRSD scores were collected across the baseline and four additional assessment points during the acute course of treatment. Findings showed that core depression and anxiety symptoms significantly improved, demonstrating large effect sizes. Notably, the reduction of core symptoms was significantly greater than the anxiety reduction. Moreover, in cross-lagged structural equation models, only one temporal relationship emerged as significant—specifically, core symptom changes at assessment 2 were significantly associated with anxiety/somatization subscale score changes at assessment 3—the opposite direction did not emerge as significant. Findings suggest that ECT does improve symptoms of anxiety, but that these changes are less dramatic than depression symptom changes, and early reduction in depression symptoms do not consistently predict subsequent improvement in anxiety symptoms (Huang et al., 2019).

One earlier investigation similarly examined change in anxiety symptoms following a naturalistic course of ECT (Langius-Eklöf and Samuelsson, 2009). The study was comprised of 15 depressed inpatients receiving ECT, ranging in age from 34 to 76. The total number of ECT sessions was 3 to 20 (mean 9.2, SD 4.1). Little detail is provided about the participants or the course of ECT, likely because the primary purpose of the study was to validate a self-report instrument of a different construct (sense of cohesion), hypothesized to be trait-like and nonresponsive to treatment. With regard to change in anxiety, there was a significant decrease in anxiety symptoms from baseline to a week after the final ECT treatment. Unlike the Huang et al. (2019) study, the timing or degree of change in anxiety was not compared with change in depression.

In summary, a small number of articles examine how anxiety changes from treatment or if anxiety impacts outcomes in TRD patients. Operational definitions of anxiety are varying. Most are naturalistic studies or small samples from which to draw conclusions. The most consistent finding suggests that anxiety is a negative prognostic factor, particularly with regard to the timing of therapeutic gains, either defined as length of time to achieve remission or time to relapse.

Extrapolating from related areas of research

Additional evidence that helps to consider if ECT may be appropriate for patients with cooccurring anxiety is to examine the impact and tolerability of ECT on anxious patients (with or without TRD). Though neither are grouped with the anxiety disorders in the current DSM-5, the majority of relevant research has been conducted on patients with OCD and PTSD, briefly reviewed next.

Obsessive compulsive disorder

Summaries of the literature on OCD have been mixed. In a chart review of 420 patients treated in an OCD clinic between 1998 and 2013 (Lins-Martins et al., 2015), ECT was prescribed in only five patients (1.2%), all of whom had cooccurring mood disorders. Based on Clinical Global Impression (CGI) scores recorded in the medical record, four of five patients improved in their mood disorder symptoms; no patients reportedly improved in their OCD symptoms, and two reported transient worsening of their OCD symptoms. In a qualitative literature review conducted by Bais et al. (2014) summarizing the case studies to date, 10 ECT sessions in 9 treatment-resistant patients with OCD led to a 30% reduction in obsessive-compulsive symptoms; however, authors noted that for the preponderance of patients, maintenance ECT sessions were required for 6 to 24 months to retain treatment effects. A more recent review (Dos Santos-Ribeiro et al., 2019) included 52 case studies ($n=69$) that described individual responses of obsessive-compulsive-related disorders to ECT, including tic disorders and self-injurious behaviors. Again, no RCTs were located during the literature search. Out of 69 patients, 73.4% demonstrated a "positive response" (operational definitions varying across studies) and patients with positive responses to ECT were likely to report previous nonresponse to antipsychotics and antidepressants. This recent work has been interpreted as encouraging evidence that ECT can be considered as a treatment option for patients with obsessive-compulsive-related disorders (Kellner et al., 2019), though controlled clinical trials are lacking and the evidence-base is far from clear-cut. That said, inferring from the OCD literature for patients with TRD, OCD does not seem to be a contraindication for the treatment itself—though it would be hypothesized based on the available literature that mood symptoms may be more responsive to ECT than the OC symptoms themselves.

Post traumatic stress disorder

Clinical summaries of the literature on ECT as an indication for PTSD converge on the recommendation that PTSD is not a contraindication for ECT (Youssef et al., 2017; Beynel et al., 2020; Koek et al., 2019). For example, ECT outperforms medication in patients with PTSD and MDD as measured with the CGI and, over an 8-year follow-up period, the ECT group showed significant reductions in suicide mortality, cardiovascular mortality, and all-cause mortality rates relative to the group that did not receive ECT. Though some data suggests that symptoms of PTSD do not improve as much as MDD symptoms (see, e.g., Novakovic et al., 2011), other research suggests that PTSD symptoms can improve as early as the third ECT session and that PTSD symptom improvements are independent from depression improvements (Margoob et al., 2010). After reviewing the literature of 3 empirical studies and 5 case reports published between 1958 and 2016, Youssef et al. (2017) concluded that while it is not clear if PTSD symptoms improve only because of improvement in depression or because there is a "true anti-PTSD effect," there are no indications thus far that ECT causes harm among patients with PTSD.

Taken together across PTSD and OCD, neither comorbid conditions would appear to be a contraindication for the treatment itself.

Practical considerations and contraindications

Benzodiazepine use in anxiety patients

Similar to ketamine, data suggests that benzodiazepine use may interfere with the therapeutic effects of ECT. Importantly, concomitant benzodiazepine use may affect unilateral and bilateral ECT differently, as benzodiazepine use is contraindicated specifically for RUL but not BL or BT ECT (Janjua et al., 2020; Tang et al., 2017). In contrast, for example, patients on benzodiazepines receiving a course of BT ECT demonstrated comparable or stronger outcomes than the nonbenzodiazepine group (Delamarre et al., 2019). Thus, recommendations would be a gradual reduction or discontinuation of benzodiazepines before the first RUL ECT; alternatively, a course of bilateral ECT may be the preference among patients taking benzodiazepines and not responding to RUL.

Anxiety about the treatment

ECT-related anxiety has been studied most extensively by Obbels et al. (2017, 2020a,b). In a 2017 review of the literature, summarizing studies published between 1958 and 2013, Obbels et al. (2017) found that anywhere from 14% to 75% of patients endorsed ECT-related anxiety. Reviewing data across 9 studies of interventions designed to improve ECT-related anxiety, initial support has been found for chlorpromazine, meprobamate, psychoeducation, and animal-assisted therapy (Obbels et al., 2017). Notably, however, ECT-related anxiety up until the 2017 review lacked a consistent operational definition and assessments of the anxiety were taken at various time points, often subsequent to the acute course. To improve the measurement of this construct, Obbels and colleagues have recently validated a 17-item "ECT-Related Anxiety Questionnaire" (ERAQ; Obbels et al., 2020a), which has been supported as a one-factor model of ECT-related anxiety, yielding a global score reflecting the severity of a patient's ECT-related anxiety, which can be completed prior to each session. Sample items include "I am anxious about temporary memory loss due to ECT," "I am anxious about permanent memory loss due to ECT," "I am anxious about the convulsion during the ECT procedure," and "I am anxious about what others would think of my treatment." In the only prospective study that has been conducted to-date, Obbels et al. (2020b) used the ERAQ prior to each session of ECT in a cohort of unipolar and bipolar depressed patients who received twice weekly RUL (or BT as needed). Findings showed that females and patients with psychotic depression had higher ERAQ scores at baseline. Patients with psychotic depression were more likely to decrease in their ERAQ scores whereas ERAQ scores of patients with nonpsychotic depression were likely to remain stable during the ECT course. Last, decreases in ECT-related anxiety were associated with decreases in depression severity.

The extent to which anxiety disorders are correlated with ECT-related anxiety, to our knowledge, has not been addressed. Future prospective research can incorporate the ERAQ to improve consistency of measurement and give insight into the relationship between anxious depression, comorbid anxieties, and anxiety specific to the ECT treatment itself.

Summary and recommendations: ECT

Overall, there is a noticeable lack of data examining the efficacy of ECT for TRD patients with cooccurring anxiety (see Table 29.3). The limited data examining anxiety as a predictor of treatment outcome suggests that comorbid anxiety or somatization (a variant of anxious depression) predicts a poorer response, particularly lower sustained remission and time

TABLE 29.3 ECT and anxious depression.

Study	Diagnosis	Comparison, sample size	ECT parameters	Depression outcome	Anxiety outcome	Notes
Rasmussen et al. (2004)	TRD, unipolar MDD	253 patients, comparing somatization v no somatization at baseline	BT ECT	Remission as defined by HAMD24 For patients with less severe depression, high somatization predicts lower sustained remission	Not reported	Part of CORE randomized multisite trial Sample size in each group not presented
Medda et al. (2020)	TRD, bipolar disorder, current episode mixed (49%) or depressed (51%)	Lifetime anxiety disorders (n=33, 47.1%) v. no lifetime anxiety disorders	Bilateral ECT, twice weekly	Patients with anxiety disorders relapsed in a shorter time	–	Naturalistic follow-up
Flint and Rifat (1997)	TRD geriatric patients, unipolar	Hospital Anxiety Depression Scale-Anxiety scores, anxious (n=53, 53.5%) v. non anxious (n=46, 46.5%)	Staging method, unilateral ECT 3×/week if needed	Anxious patients had lower rates of response as measured with HAM-D and higher rate of attrition. 1 out of 4 were ECT responders, responder was nonanxious	–	Naturalistic follow-up
Steffens and McQuoid (2005)	TRD geriatric patients, unipolar	204 geriatric patients, n=63 (30.9%) with comorbid GAD v no GAD	Duke staging method, ECT parameters not included	Remitters as defined by MADRS were less likely to have GAD symptoms, and GAD associated with longer time to achieve remission	–	Analyses do not analyze response to ECT specifically but across all stages of treatment
Anxiety as an outcome						
Huang et al. (2019)	TRD, unipolar MDD	n=116	Bitemporal ECT, 2 or 3 times/week, maximum 12 treatments		Anxiety symptoms improve, but less so than core depression symptoms	Naturalistic outcomes, Taiwanese inpatients
Langius-Eklöf and Samuelsson (2009)	TRD	15 inpatients	Total number of ECT sessions ranged from 3 to 20, no details provided on ECT parameters		Significant decrease in anxiety symptoms	Purpose of the study was a validation of a self-report instruction assessing "sense of cohesion"

to relapse. When measuring anxiety as an outcome, across TRD patients more broadly, initial data suggests that symptoms of anxiety do improve, but not as significantly as symptoms of depression.

Inferring from related disorders, there is an increasing body of work showing that patients with significant comorbid PTSD or OCD can tolerate and respond to ECT, though data is mixed on how much the treatment improves these secondary symptoms rather than exclusively improving the mood disorder symptoms themselves. Nevertheless, this body of work suggests that comorbid conditions that are associated with anxiety features such as OCD and PTSD do not necessarily indicate intolerance of the treatment. Taken together, it may be that comorbid anxiety is not a contraindication for ECT but could instead signal that more careful maintenance planning may be in order, given the initial evidence for shorter time to relapse.

Additionally, for anxious patients who take concomitant benzodiazepines, cessation may be optimal prior to starting a course of RUL; if that is not feasible, BT or BL may be considered.

Which patients are likely to experience anxiety about the treatment itself is an important area for future research. Initial data suggest that women and patients with psychotic disorders may be more likely to experience this anxiety at the beginning of ECT treatment, and much more work is needed to understand how to improve this anxiety over time.

Overall much more research is needed to examine anxiety as a predictor and an outcome among TRD patients receiving a course of ET. Defining patients broadly with the HRSD-AS would facilitate ease of comparison with the ketamine literature, which is important given trials that are underway and plans to conduct head to head comparisons of ECT and ketamine (e.g., Mathew et al., 2019) and opportunities to test the combined effectiveness of ketamine with ECT (e.g., Kellner and Iosifescu, 2017).

Interventional psychiatric approaches: Transcranial magnetic stimulation

Repetitive transcranial magnetic stimulation (rTMS) is a neurostimulation technique where neuromagnetic pulses are used to stimulate certain areas of the brain to treat specific neurologic and psychiatric conditions. Several trials (O'Reardon et al., 2007; George et al., 2010; Ray et al., 2011; Kozel and George, 2002) and at least one metaanalysis (Gaynes et al., 2014) have demonstrated the efficacy of rTMS in depression. Repetitive TMS was approved for the treatment of depression by the FDA in 2008. Although ECT is still considered the most effective neurostimulation technique for the treatment of depression, its invasive nature and complications make patients and physicians first consider TMS.

There are several parameters rTMS rely on for its effects: position of the coil, frequency, intensity, duration of stimulation, waveform, and number of stimuli per session. Frequency can vary between treatments for different conditions. ≥5 Hz frequency is considered high-frequency (HF) and ≤1 Hz is considered low-frequency (LF). Evidence has suggested that LF rTMS is "inhibitory" while HF rTMS is "excitatory" (Guo et al., 2017). Intermittent Theta Burst Stimulation (iTBS) provides 3 min treatment to patients who are given a train of bursts at high frequencies. Treatment can be modified, using some of these parameters, based on psychiatric condition. Depression can be treated either via LF rTMS to the right DLPFC or via HF rTMS to the left DLPFC. Certain anxiety disorders with hyperexcitatory states in brain circuits, such as OCD, tend to respond to LF rTMS (Mantovani et al., 2006; Gomes et al., 2012; Hawken et al., 2016).

Evidence-based guidelines on the therapeutic use of repetitive transcranial magnetic stimulation (Lefaucheur et al., 2020) summarize recommendations for rTMS efficacy based on clinical indications. rTMS is effective in the treatment of depression and to a lower degree of evidence, it is also effective in some anxiety disorders. The data on the efficacy of rTMS for comorbid conditions is still limited, and more randomized, controlled studies with larger sample sizes are needed. Below we review the literature on rTMS for anxious depression and for depression with comorbid anxiety disorders.

Anxious depression

Diefenbach et al. (2013) examined the effectiveness of rTMS in a small group of TRD patients with ($n=8$) and without ($n=24$) anxiety. As indicated by response on the HAMD, both depression and anxiety symptoms improved with TMS. Improvements in depression were similar in anxious and nonanxious groups, and the anxious group demonstrated a more significant improvement in anxiety symptoms.

Trevizol et al. (2021) examined data obtained from the multisite, randomized, noninferiority THREE-D Trial to assess the effect of rTMS (high-frequency left sided; HFL) and iTBS on anxiety symptoms of patients with MDD. HAM-D, HAMD A/S factor, and the Brief Symptom Inventory (BSI-A) were utilized to assess improvement. All participants had failed 1–3 adequate antidepressant trials or had been unable to tolerate at least two separate trials of antidepressants and were randomly assigned in a 1:1 ratio to receive iTBS ($n=199$) or HFL ($n=189$). Findings demonstrated that both HFL-TMS and iTBS significantly improved symptoms of both depression and anxiety.

In a post hoc analysis by Chen et al. (2019), data were obtained from 3 randomized, double-blind trials that used a combination of RLF/BL rTMS, RLF/BL rTMS, and LHF/RLF rTMS protocols to assess the effectiveness of rTMS in treating anxiety symptoms in depression. A total of 697 patients were pooled for analysis from three trials, and only patients with depression severity in the moderate-severe range (scoring > 13 on the HAMD) were included. All patients had failed to respond to at least two antidepressant medications for at least 6 weeks in the current illness episode, qualifying their illness as stage II treatment-resistant depression (Thase and Rush, 1997). The 17-item Hamilton Depression Rating Scale (HAMD) and the 21-item Beck Anxiety Inventory (BAI) were used to assess change in symptoms. Across the three pooled studies, 150 patients received LHF rTMS, 312 received RLF rTMS, and 235 received BL rTMS. Consistent with Trevizol et al. (2021), all three rTMS protocols resulted in significant antidepressant and anxiolytic effects, equivalent between treatment conditions.

Comorbid conditions

Clarke et al. (2019) analyzed a sample of 248 treatment-resistant depressed patients, of whom 172 had one or more comorbid anxiety disorders including GAD, OCD, Panic Disorder with Agoraphobia, PTSD, Social Anxiety, and Agoraphobia. HAMA, HAMD, MADRS, and ZUNG scores were used as assessment tools. Patients, regardless of anxiety status, demonstrated significant improvements in ratings of depression and anxiety.

In a small pilot study (White and Tavakoli, 2015) of 13 patients with MDD and comorbid GAD who received between 24 and 36 treatments of bilateral rTMS, 11 out of 13 (84.6%) patients' anxiety symptoms achieved remission, achieving a score of < 5 on the GAD-7, and 10 out of 13 patients (76.9%) achieved a HAM-D-21 score of < 8 for depressive symptoms.

A randomized double-blind study with 25 patients (Mantovani et al., 2013) assessed the effectiveness of rTMS on patients with panic disorder with comorbid MDD. Twenty-five patients were assigned 4 weeks of active or sham rTMS to the right dorsolateral prefrontal cortex (DLPFC). Response was defined as a ≥ 40% decrease on the panic disorder severity scale and a ≥ 50% decrease on the Hamilton depression rating scale. At the end of the randomized phase, patients were offered the option of receiving open-label rTMS for an additional 4 weeks. At the end of week 4, results revealed significantly greater improvement in panic symptoms with active (50%) compared with sham rTMS (8%), but no significant difference in symptoms of depression. After 8 weeks of active rTMS, response rates were improved for both panic (67%) and depression symptoms (50%). Clinical improvement was sustained at 6-month follow-up.

Kumar et al. (2018) conducted a retrospective analysis of 13 treatment-resistant patients diagnosed with comorbid panic disorder and depression to examine the effectiveness of rTMS in patients with MDD and panic disorder. Participants received 20 sessions of HF rTMS over a 4 week period. The Hamilton Depression Rating Scale (HDRS) and Panic Disorder Severity Scale (PDSS) were completed at baseline (prior to starting rTMS) and after treatment #20. There was a significant reduction in the mean PDSS and HDRS scores after completion of 20 sessions of rTMS from baseline; the change in severity of panic disorder was not correlated with change in depression severity, possibly indicating an independent effect of rTMS on panic symptoms.

In a prospective, open-label study with 40 patients, Carpenter et al. (2018) examined the efficacy of rTMS on PTSD comorbid with MDD. Symptoms were assessed using the PTSD Checklist (PCL-5) and Inventory of Depressive Symptomatology, Self-Report (IDS-SR). Patients with comorbid MDD and PTSD experienced significant improvement in symptoms of both conditions when treated with 5-Hz TMS. Of note, there was no comparison group with MDD only.

Philip et al. (2019) investigated the efficacy of synchronized TMS (sTMS) on comorbid PTSD and MDD in 22 Veterans. The principle inclusion criteria were comorbid PTSD and MDD, defined as score of > 33 on the PTSD Checklist for DSM-5 (PCL) and score of > 10 on the Quick Inventory of Depressive Symptoms, Self-Report (QIDS). Ten participants were randomized to active, and 13 to sham sTMS. All participants received 4 weeks of active or sham sTMS, 5 treatments per week. Clinical assessments included the PCL and QIDS. All participants demonstrated significant reductions in PTSD and MDD symptoms. The active stimulation group demonstrated greater reductions in a count of PTSD moderate-to-severe symptoms, PTSD symptom severity, and depression severity.

Contrary to these positive findings, a study conducted by Lisanby et al. (2009) demonstrated different results in distinct phases of the trial. In the randomized, sham-controlled phase, a comorbid anxiety disorder did not negatively predict the treatment outcome; however, in the open-label extension phase, anxiety negatively impacted treatment response to TMS. Similarly, Brakemeier et al. (2008) investigated response rates in depressed patients treated with rTMS and found that responders had less anxiety at baseline.

In sum, the growing body of evidence regarding the effect of anxiety on response to TMS has been somewhat mixed (e.g., Lisanby et al., 2009; Brakemeier et al., 2008) though generally in support that rTMS can be effective in reducing symptoms of both depression and anxiety, with seemingly equivocal findings across modalities of administration (see Table 29.4).

TABLE 29.4 TMS and anxious depression.

Study	Participants	Design	Outcome
Diefenbach et al. (2013)	MDD with anxiety	32 TRD patients; 8 with and 24 without anxious depression All patients received rTMS	Both depression and anxiety symptoms improved with rTMS
Trevizol et al. (2021)	THREE-D trial participants with MDD and anxiety	Multisite, randomized, noninferiority trial. rTMS and iTBS. A total of 388 patients	Both iTBS and rTMS improve not only depression but also anxiety symptoms
Chen et al. (2019)	MDD and comorbid anxiety disorder	Post hoc analysis of 3 randomized, double-blind studies (697 patients) using different rTMS protocols	All three rTMS protocols resulted in antidepressant and anxiolytic effects
Clarke et al. (2019)	MDD and comorbid anxiety disorder	Naturalistic dataset from an Australian rTMS service. A total of 248 patients; 172 with a comorbid anxiety disorder All patients received rTMS	Significant improvement in depression regardless of anxiety status. All anxiety disorders had significant reductions in depression and anxiety scores
White and Tavakoli (2015)	MDD comorbid with GAD	Open label trial. 13 patients. Patients received 24–36 treatments	Significant effects in treating both depression and anxiety
Mantovani et al. (2013)	MDD comorbid with panic disorder	25 patients, 4 weeks of active versus sham TMS to R DLPFC. Additional 4 week open phase with TMS available after the first 4 weeks	By the end of 4 weeks, significant improvement only in panic disorder symptoms. At the end of the 8 week, response rates were significantly better for both depression and panic disorder
Kumar et al. (2018)	MDD comorbid with panic disorder	Retrospective analysis of 13 TR patients. All receive 20 treatments of rTMS	Significant reductions in depression and panic disorder symptoms
Carpenter et al. (2018)	PTSD comorbid with MDD	Open label study with 40 patients	5 Hz rTMS effective for both depression and PTSD
Philip et al. (2019)	MDD with comorbid PTSD	22 veterans, 4 weeks of active versus sham TMS. Efficacy of synchronized TMS	Significant reductions in depression and PTSD symptoms

Emerging treatments

Psychedelics

Psilocybin is the psychedelic mostly widely studied in the context of TRD. More broadly, both psilocybin and LSD have been examined for patients with depression and anxiety from life threatening illness (e.g., Gasser et al., 2014; Griffiths et al., 2016) and ayahuasca has been examined for recurrent MDD (Osório et al., 2015). Across these emerging studies, there is evidence of sustained antidepressant and anxiolytic effects as well as evidence that the treatments are well tolerated and without persisting adverse effects (Muttoni et al., 2019). Carhart-Harris and colleagues with the Imperial College group have led seminal work examining psilocybin-assisted psychotherapy (see Kelly et al., 2019 for a brief review), and a major phase 2b dose finding double-blind clinical trial of psilocybin with psychological support for patients with TRD is underway (ClinicalTrials.gov Identifier: NCT03775200). As reviewed in Wheeler and Dyer (2020), mechanistic research thus far suggests that psilocybin-assisted psychotherapy leads to: reduced cerebral blood flow in the temporal lobes and amygdala (Carhart-Harris et al., 2017); to increased amygdala response to fearful face stimuli—with stronger reactions associated with stronger outcomes (Roseman et al., 2018); and to faster emotional face processing (Stroud et al., 2018). Symptom-wise improvements on anxiety and depression scales are seen at three and six months, and one study found changes in personality marked by increased extraversion and openness to experience and decreased neuroticism (see Wheeler and Dyer, 2020). Last, in qualitative research, initial data suggests psilocybin may work by moving patients from disconnection to connection and from emotional avoidance to acceptance (Watts et al., 2017).

Though we could not locate any data about anxious depression as a specific predictor of treatment response, as above, there is data that psilocybin can reduce symptoms of anxiety for patients with life threatening illness (see Muttoni et al., 2019; Wheeler and Dyer, 2020), which could suggest that it may also be helpful for TRD patients with cooccurring anxiety. That said, much of this research has been open label and more work is underway.

Vague nerve stimulation

Vagus Nerve Stimulation (VNS) was approved by the US FDA in 2005 as an adjunctive treatment for TRD patients with both unipolar and bipolar depression, though in 2007 the Center for Medicare and Medicaid Services (CMS) issued a "noncoverage" decision. In 2019, based on accumulating evidence, CMS began coverage again, specifically for patients participating in a "coverage with evidence" RCT. A comprehensive, systematic review and metaanalysis by Bottomley et al. (2020) reports on 22 VNS for TRD studies; of the 22, only one included a measure of anxiety as a secondary outcome. Specifically, in a single arm VNS + TAU Study (Trottier-Duclos et al., 2018), 7 unipolar and 3 bipolar patients were followed over 6 years postimplant and showed clinically and statistically significant improvements on depression, quality of life, and symptoms of anxiety as measured with the Hamilton Anxiety Rating Scale. Notably, of two current VNS trials, RECOVER (ClinicalTrials.gov Identifier: NCT03887715) and RESTORE-LIFE (ClinicalTrials.gov Identifier: NCT03320304), only one measure of anxiety, the GAD7, is included in one of the trials (RESTORE-LIFE). Thus, how VNS may impact anxiety outcomes, particularly for anxious TRD patients, is far from understood. That said, VNS is nonsystemic and can be combined with other existing treatments. Thus, for anxious TRD patients whose depression may be likely to respond to a neuromodulation approach but whose anxiety may persist, VNS may be an interesting consideration, as it can be safely combined with other medications and could even be shut off to permit ECT if needed acutely.

One area from which to extrapolate is the burgeoning literature on transcutaneous vagus nerve stimulation (tVNS), which allows noninvasive stimulation of the auricular branch of the vagus nerve via the concha of the outer ear. This body of research shows that tVNS promotes neural activation in areas that reduce worry such as the prefrontal cortex and anterior cingulate (Yakunina et al., 2017), that it may reduce the frequency of worry thoughts in participants with high levels of worry (Burger et al., 2019), and that it may reduce symptoms of both depression and anxiety in MDD patients (Rong et al., 2016). Thus, considering tVNS as a proof of concept for traditional VNS could suggest that secondary anxiety symptoms could be likely to improve, though again much more research is needed.

Deep brain stimulation

Nucleus Accumbens deep brain stimulation (DBS) for TRD has been shown to improve symptoms of anxiety. For example, in small groups of unipolar TRD patients, symptoms of anxiety as measured with the Hamilton Anxiety Rating Scale significantly reduced from baseline to a year (Bewernick et al., 2010, 2012), reductions that were maintained at 2-year follow-up (Bewernick et al., 2012). Depression responders were also shown to have more pronounced reduction in anxiety than nonresponders (Bewernick et al., 2012).

To infer how anxious depressed TRD patients might respond to DBS, it is reasonable to extrapolate from the literature on DBS for other anxiety-based disorders, for which the largest evidence-base has accumulated for OCD. For example, DBS in the anterior limb of the internal capsule and nucleus accumbens region (ALIC-NAcc) has been shown to improve symptoms of OCD but to leave symptoms of both depression and anxiety (measured with the State-Trait Anxiety Inventory) unchanged (Huys et al., 2019). An open label case series of DBS with placements for both the nucleus accumbens and bed nucleus stria terminalis demonstrated that OCD symptoms as well as depression subscales on the DASS improved, whereas anxiety subscale scores on the DASS deteriorated (Farrand et al., 2018). With unilateral DBS in the right nucleus accumbens, OCD and depression improved significantly but anxiety as measured with the State-Trait Anxiety Inventory and the Hamilton Anxiety Rating Scale did not (Huff et al., 2010). In contrast, DBS with quadripolar stimulating leads implanted bilaterally in the ventral capsule/ventral striatum showed evidence of improvements in OCD as well as both depression and anxiety, with improvement on the Hamilton Anxiety Rating Scale lasting at 36 months (Greenberg et al., 2006). Across these small studies, though there is some evidence that broad-based symptoms of anxiety may improve (Greenberg et al., 2006), OCD and depressive symptoms seem to more consistently and appreciably improve across studies and areas of stimulation.

Last, one case report using bilateral DBS to the basolateral amygdala for combat-related PTSD showed meaningful improvements in overall PTSD symptoms using the CAPS as well as qualitative improvements in overall mood, including anger and irritability after 8 months of stimulation (Langevin et al., 2016). Broad-based symptoms of anxiety were not reported.

Taken together, there is some evidence from Nucleus Accumbens DBS for TRD that comorbid symptoms of anxiety will improve. However, for patients with primary OCD, symptoms of anxiety may respond significantly less so than the specific target of OCD symptomatology. More work is needed to understand how anxious TRD patients may respond to DBS, with the most likely of targets the Nucleus Accumbens.

Psychotherapy

As described throughout, anxious depression predicts worse clinical outcomes including poor response to first-line pharmacological strategies. Psychotherapy can provide a critical opportunity to address residual symptoms, particularly if psychiatric treatments improve the cardinal symptoms of depression much more so than the symptoms of anxiety.

For example, patients with anxious depression and residual symptoms after 6 weeks of escitalopram had longer time to remission and lower rates of remission; however, when sequentially referred to a course of interpersonal psychotherapy (IPT), anxious patients did not ultimately have a poorer response (Greenlee et al., 2010). This finding is consistent with support for the sequential model in TRD, showing that staged administration of psychotherapy after response to acute-phase pharmacotherapy, either alone or in combination with antidepressant drugs, may play a role in reducing MDD relapse and recurrence (Guidi et al., 2016; Ijaz et al., 2018).

When making a referral to psychotherapy, the most important considerations include the timing of the psychotherapy as well as the type of the psychotherapy.

Timing of psychotherapy

In the chronic depression literature, more broadly there is support for the sequential model, offering psychotherapy after a course of pharmacotherapy, either alone or in combination with antidepressant drugs; doing so may be particularly important for patients with anxiety, which is unlikely to be addressed by one course of a monotherapy (Cosci et al., 2020; Guidi et al., 2016). More specifically with TRD patients, there is emerging data showing the benefits of pairing psychotherapy with Ketamine, ECT, and TMS, with the shared hypothesis that the somatic intervention can create synaptic plasticity and thereby create new opportunities for learning of coping skills that can improve residual symptoms and decrease relapse likelihood (Wilkinson et al., 2019). For example, patients who responded to a course of IV ketamine went on to receive 12 sessions of CBT, which significantly extended time to relapse (Wilkinson et al., 2017). Similar findings were demonstrated with RUL ECT responders who then engaged in a course of CBT (compared to maintenance ECT or medication only), evidencing significantly higher rates of sustained response at both 6 and 12 months compared to the other two conditions (Brakemeier et al., 2014). TMS has been examined concurrent with psychotherapy, showing that CBT during TMS sessions (Donse et al., 2018), and brief behavioral activation prior to the day's TMS session (Russo et al., 2018), can improve response and remission rates compared to TMS alone.

Though most models explore psychotherapy as concurrent to or subsequent to the somatic intervention, if anxiety is a barrier to initiating the somatic treatment itself, then psychotherapy could be a critical precursor. As reviewed previously, there is not compelling data that anxious depression is synonymous with anxiety about the treatment; that said, clinically this may still be the case, and assessing the patient's fears about the recommended treatment may facilitate joint collaboration and decision making about staging of interventions.

Which psychotherapy

There is no data, to our knowledge, that examines treatment matching for anxious TRD patients. Metaanalyses of psychotherapy with TRD patients more broadly provides evidence that adjunctive psychotherapy is cost effective, improves response and remission rates over the short, medium and long-term (though data is more robust for short-term), and is as acceptable as antidepressant care with regard to patient retention (v dropout) (Ijaz et al., 2018). In this same Cochrane review (Ijaz et al., 2018), no differences were found between the treatments included in the RCTs, including CBT, group dialectical behavior therapy, intensive short-term dynamic therapy, and interpersonal psychotherapy. Another metaanalysis similarly did not favor one treatment over another, which included CBT, Cognitive Behavioral Analysis Systems of Psychotherapy, Mindfulness Based Cognitive Therapy, and Interpersonal Psychotherapy, though did find a larger effect size in studies that utilized a group rather than individual format (van Bronswijk et al., 2019). Thus, it may be reasonable to recommend a specific psychotherapy based on other aspects of evidence-based care including case conceptualization (e.g., additional perpetuating factors that contribute to treatment resistance), patient preferences, and logistics (e.g., available providers' respective expertise, cost, location).

Summary and recommendations: Psychotherapy

The TRD literature more broadly suggests augmenting psychiatric care with psychotherapy. Doing so may be particularly important in the care of anxious TRD patients who may be more likely to have residual symptoms of anxiety following antidepressant care and whose residual symptoms put them at continual risk of relapse and recurrence. Most models have examined psychotherapy as subsequent to a somatic intervention. This model may reduce patient burden as well as help patients to better engage with psychotherapy once they get initial benefit, thereby increasing hope and capitalizing on new windows for learning. The TMS literature, perhaps because of the frequency of visits for a 6-week course, has examined psychotherapy concurrently as a way to reduce patient burden. Optimal timing of psychotherapy is an open research question, and so good case formulation and patient preferences may most realistically dictate these decisions.

Part III: Clinical summary

To identify a patient with anxious depression, options include assessing categorical, comorbid diagnoses or dimensional assessment. Diagnostic comorbidities can be identified using structured or semistructured interview, such as the Mini International Neuropsychiatric Interview (MINI; Lecrubier et al., 1997), Structured Clinical Interview for DSM-5 (SCID; First et al., 2016), or the Anxiety Disorder Interview Schedule for DSM-5 (ADIS-5; Brown and Barlow, 2014). A growing number of research groups are defining the presence of anxious depression as MDD with high levels of anxiety operationalized dimensionally, such as total scores on the 6 items from the HDRS-AS subscale (Ionescu et al., 2013a,b), or considered using the DSM specifier "with anxious distress" defined by the presence of two or more of five symptoms on most days during a MDE, including *feeling keyed up or tense*; *feeling unusually restless*; *difficulty concentrating because of worry*; *fear that something awful may happen*; or *feeling that the individual might lose control* (Uher et al., 2014; Zimmerman et al., 2014).

Patients with anxious depression may represent a more severe and difficult to treat subgroup, even among TRD patients. Thus, appropriate psychoeducation with patients about expectations for the pace of change may help to offset future frustration and improve the therapeutic alliance. While patients may request to initiate or maintain on benzodiazepines, risk of developing tolerance, addiction, cognitive impairments, falls, and disinhibition should be remembered before clinicians prescribe these medications. This group of medications may also interfere with efficacy of interventional psychiatric approaches including ECT and ketamine. Reasonable pharmacological strategies include SSRIs, SNRIs, TCAs, MAOIs and antipsychotics. Considering the tolerability, efficacy, and side effect profiles, SSRIs and SNRIs should be the first-line treatment options employed.

If a patient's symptoms remain problematic, it may be helpful to consider psychotherapy as well as initiation of another biological treatment; psychotherapy following or concurrent with an interventional psychiatric approach may help to capitalize on neurogenesis and be an optimal time to acquire new coping skills. Moreover, psychotherapy may help to address any anxiety about additional treatments as well as help to target residual anxiety symptoms that may be likely to remain, even if core depressive symptoms improve.

Core symptoms are likely to improve among anxious TRD patients who go on to receive ketamine, ECT, or TMS. More research is needed to better understand relative advantages and disadvantages of each of these treatments for patients with anxious TRD, and it is not unreasonable at this stage to consider patient preferences, logistics, and other aspects of their presentation, for example, cooccurring melancholic features may be an indicator for ECT rather than ketamine (Wang et al., 2019b). Anxious TRD patients may require a longer maintenance course to slow down risk of relapse, particularly with regard to ECT. Emerging treatments to consider are psilocybin and, should severe symptoms persist, VNS or DBS.

References

Part I references

American Psychiatric Association, 2013. Diagnostic and Statistical Manual of Mental Disorders, fifth ed. American Psychiatric Association, Arlington, VA.

Zimmerman, M., Chelminski, I., Young, D., Dalrymple, K., Walsh, E., Rosenstein, L., 2014. A clinically useful self-report measure of the DSM-5 anxious distress specifier for major depressive disorder. J. Clin. Psychiatry 75 (6), 601–607.

Ionescu, D.F., Niciu, M.J., Henter, I.D., Zarate Jr., C.A., 2013a. Defining anxious depression: a review of the literature. CNS Spectr. 18 (5), 252.

Fava, M., Rush, A.J., Alpert, J.E., Balasubramani, G.K., Wisniewski, S.R., Carmin, C.N., Trivedi, M.H., 2008a. Difference in treatment outcome in outpatients with anxious versus nonanxious depression: a STAR* D report. Am. J. Psychiatry 165 (3), 342–351.

Psychopharmacology references

Ionescu, D.F., Niciu, M.J., Richards, E.M., Zarate Jr., C.A., 2014a. Pharmacologic treatment of dimensional anxious depression: a review. Prim. Care Companion CNS Disord. 16 (3). PCC.13r01621.

Fava, M., Rosenbaum, J.F., Hoog, S.L., Tepner, R.G., Kopp, J.B., Nilsson, M.E., 2000. Fluoxetine versus sertraline and paroxetine in major depression: tolerability and efficacy in anxious depression. J. Affect. Disord. 59 (2), 119–126.

Feiger, A.D., Flament, M.F., Boyer, P., Gillespie, J.A., 2003. Sertraline versus fluoxetine in the treatment of major depression: a combined analysis of five double-blind comparator studies. Int. Clin. Psychopharmacol. 18 (4), 203–210.

Papakostas, G.I., Clain, A., Ameral, V.E., Baer, L., Brintz, C., Smith, W.T., Londborg, P.D., Glaudin, V., Painter, J.R., Fava, M., 2010. Fluoxetine-clonazepam cotherapy for anxious depression: an exploratory, post-hoc analysis of a randomized, double blind study. Int. Clin. Psychopharmacol. 25 (1), 17–21.

Kim, J.E., Yoon, S.J., Kim, J., Jung, J.Y., Jeong, H.S., Cho, H.B., Shin, E., Lyoo, I.K., Kim, T.S., 2011. Efficacy and tolerability of mirtazapine in treating major depressive disorder with anxiety symptoms: an 8-week open-label randomised paroxetine-controlled trial. Int. J. Clin. Pract. 65 (3), 323–329.

Papakostas, G.I., Stahl, S.M., Krishen, A., et al., 2008. Efficacy of bupropion and the selective serotonin reuptake inhibitors in the treatment of major depressive disorder with high levels of anxiety (anxious depression): a pooled analysis of 10 studies. J. Clin. Psychiatry 69 (8), 1287–1292.

Fava, M., Martinez, J.M., Greist, J., Marangell, L.B., Brown, E., Chen, L., Wohlreich, M.M., 2007. The efficacy and tolerability of duloxetine in the treatment of anxious versus non-anxious depression: a post-hoc analysis of an open-label outpatient study. Ann. Clin. Psychiatry 19 (3), 187–195.

Howland, R.H., Wilson, M.G., Kornstein, S.G., Clayton, A.H., Trivedi, M.H., Wohlreich, M.M., Fava, M., 2008. Factors predicting reduced antidepressant response: experience with the SNRI duloxetine in patients with major depression. Ann. Clin. Psychiatry 20 (4), 209–218.

Nelson, J.C., 2010. Anxiety does not predict response to duloxetine in major depression: results of a pooled analysis of individual patient data from 11 placebo-controlled trials. Depress. Anxiety 27 (1), 12–18.

MacGillivray, S., Arroll, B., Hatcher, S., Ogston, S., Reid, I., Sullivan, F., Williams, B., Crombie, I., 2003. Efficacy and tolerability of selective serotonin reuptake inhibitors compared with tricyclic antidepressants in depression treated in primary care: systematic review and meta-analysis. BMJ 326 (7397), 1014.

Versiani, M., Ontiveros, A., Mazzotti, G., Ospina, J., Dávila, J., Mata, S., Pacheco, A., Plewes, J., Tamura, R., Palacios, M., 1999. Fluoxetine versus amitriptyline in the treatment of major depression with associated anxiety (anxious depression): a double-blind comparison. Int. Clin. Psychopharmacol. 14 (6), 321–327.

Tollefson, G.D., Holman, S.L., Sayler, M.E., Potvin, J.H., 1994. Fluoxetine, placebo, and tricyclic antidepressants in major depression with and without anxious features. J. Clin. Psychiatry 55 (2), 50–59.

Thase, M.E., Demyttenaere, K., Earley, W.R., Gustafsson, U., Udd, M., Eriksson, H., 2012. Extended release quetiapine fumarate in major depressive disorder: analysis in patients with anxious depression. Depress. Anxiety 29 (7), 574–586.

Trivedi, M.H., Thase, M.E., Fava, M., Nelson, C.J., Yang, H., Qi, Y., Tran, Q.V., Pikalov, A., Carlson, B.X., Marcus, R.N., Berman, R.M., 2008. Adjunctive aripiprazole in major depressive disorder: analysis of efficacy and safety in patients with anxious and atypical features. J. Clin. Psychiatry 69 (12), 1928–1936.

Fava, M., Dirks, B., Freeman, M.P., Papakostas, G.I., Shelton, R.C., Thase, M.E., Trivedi, M.H., Liu, K., Stankovic, S., 2019. A phase 2, randomized, double-blind, placebo-controlled study of adjunctive pimavanserin in patients with major depressive disorder and an inadequate response to therapy (CLARITY). J. Clin. Psychiatry 80 (6). 19m12928.

Papakostas, G.I., Fava, M., Freeman, M.P., Shelton, R.C., Thase, M.E., Jha, M.K., Trivedi, M.H., Dirks, B., Liu, K., Stankovic, S., 2020a. Effect of pimavanserin on anxious depression in patients with major depression and an inadequate response to previous therapy: secondary analysis of the clarity study. Int. Clin. Psychopharmacol. 35 (6), 313–321.

Fava, M., Rush, A.J., Alpert, J.E., Balasubramani, G.K., Wisniewski, S.R., Carmin, C.N., Biggs, M.M., Zisook, S., Leuchter, A., Howland, R., Warden, D., Trivedi, M.H., 2008b. Difference in treatment outcome in outpatients with anxious versus nonanxious depression: a STAR*D report. Am. J. Psychiatry 165 (3), 342–351.

Farabaugh, A., Alpert, J., Wisniewski, S.R., Otto, M.W., Fava, M., Baer, L., Perlis, R., Friedman, E., Nyer, M., Bitran, S., Balasubramani, G.K., Inamori, A., Trivedi, M., Thase, M.E., 2012. Cognitive therapy for anxious depression in STAR(*) D: what have we learned? J. Affect. Disord. 142 (1–3), 213–218.

Delini-Stula, A., Mikkelsen, H., Angst, J., 1995. Therapeutic efficacy of antidepressants in agitated anxious depression—a meta-analysis of moclobemide studies. J. Affect. Disord. 35 (1–2), 21–30.

Hamilton, M., 1960. A rating scale for depression. J. Neurol. Neurosurg. Psychiatry 23 (1), 56–62.

Ketamine references

Ionescu, D.F., Luckenbaugh, D.A., Niciu, M.J., Richards, E.M., Slonena, E.E., Voort, J.L.V., Zarate Jr., C.A., 2014b. Effect of baseline anxious depression on initial and sustained antidepressant response to ketamine. J. Clin. Psychiatry 75 (9), 932–938.

Romeo, B., Choucha, W., Fossati, P., Rotge, J.Y., 2015. Meta-analysis of short-and mid-term efficacy of ketamine in unipolar and bipolar depression. Psychiatry Res. 230 (2), 682–688.

Ionescu, D.F., Luckenbaugh, D.A., Niciu, M.J., Richards, E.M., Zarate Jr., C.A., 2015. A single infusion of ketamine improves depression scores in -patients with anxious bipolar depression. Bipolar Disord. 17 (4), 438–443.

Salloum, N.C., Fava, M., Freeman, M.P., Flynn, M., Hoeppner, B., Hock, R.S., Mathew, S.J., 2019. Efficacy of intravenous ketamine treatment in anxious versus nonanxious unipolar treatment-resistant depression. Depress. Anxiety 36 (3), 235–243.

Chen, M.H., Lin, W.C., Wu, H.J., Bai, Y.M., Li, C.T., Tsai, S.J., Su, T.P., 2020. Efficacy of low-dose ketamine infusion in anxious vs nonanxious depression: revisiting the adjunctive ketamine study of Taiwanese patients with treatment-resistant depression. CNS Spectr. 26 (4), 362–367.

McIntyre, R.S., Rodrigues, N.B., Lipsitz, O., Nasri, F., Gill, H., Lui, L.M., Lee, Y., 2020. The effectiveness of intravenous ketamine in adults with treatment-resistant major depressive disorder and bipolar disorder presenting with prominent anxiety: results from the Canadian Rapid Treatment Center of Excellence. J. Psychopharmacol. 35 (2), 128–136. https://doi.org/10.1177/0269881120954048. PMID: 33040665 [Epub 2020 Oct 11].

Wang, C., Zhou, Y., Zheng, W., Liu, W., Zhan, Y., Li, H., Li, M.D., 2019a. Association between depression subtypes and response to repeated-dose intravenous ketamine. Acta Psychiatr. Scand. 140 (5), 446–457.

Fortress, A.M., Smith, I.M., Pang, K.C., 2018. Ketamine facilitates extinction of avoidance behavior and enhances synaptic plasticity in a rat model of anxiety vulnerability: implications for the pathophysiology and treatment of anxiety disorders. Neuropharmacology 137, 372–381.

Parise, E.M., Alcantara, L.F., Warren, B.L., Wright, K.N., Hadad, R., Sial, O.K., Bolaños-Guzmán, C.A., 2013. Repeated ketamine exposure induces an enduring resilient phenotype in adolescent and adult rats. Biol. Psychiatry 74 (10), 750–759.

Banov, M.D., Young, J.R., Dunn, T., Szabo, S.T., 2020. Efficacy and safety of ketamine in the management of anxiety and anxiety spectrum disorders: a review of the literature. CNS Spectr. 25 (3), 331–342.

Glue, P., Medlicott, N.J., Harland, S., Neehoff, S., Anderson-Fahey, B., Le Nedelec, M., McNaughton, N., 2017. Ketamine's dose-related effects on anxiety symptoms in patients with treatment refractory anxiety disorders. J. Psychopharmacol. 31 (10), 1302–1305.

Taylor, J.H., Landeros-Weisenberger, A., Coughlin, C., Mulqueen, J., Johnson, J.A., Gabriel, D., Bloch, M.H., 2018. Ketamine for social anxiety disorder: a randomized, placebo-controlled crossover trial. Neuropsychopharmacology 43 (2), 325–333.

Rodriguez, C.I., Kegeles, L.S., Levinson, A., Feng, T., Marcus, S.M., Vermes, D., Simpson, H.B., 2013. Randomized controlled crossover trial of ketamine in obsessive-compulsive disorder: proof-of-concept. Neuropsychopharmacology 38 (12), 2475–2483.

Adams, T., Bloch, M., Pittenger, C., 2017. Intranasal ketamine and cognitive-behavioral therapy for treatment refractory obsessive-compulsive disorder. J. Clin. Psychopharmacol. 37 (2), 269.

Feder, A., Parides, M.K., Murrough, J.W., Perez, A.M., Morgan, J.E., Saxena, S., Iosifescu, D., 2014. Efficacy of intravenous ketamine for treatment of chronic posttraumatic stress disorder: a randomized clinical trial. JAMA Psychiat. 71 (6), 681–688.

Zeng, M.C., Niciu, M.J., Luckenbaugh, D.A., Ionescu, D.F., Mathews, D.C., Richards, E.M., Zarate, C.A., 2013. Acute stress symptoms do not worsen in posttraumatic stress disorder and abuse with a single subanesthetic dose of ketamine. Biol. Psychiatry 73 (12), e37–e38.

Albott, C.S., Lim, K.O., Forbes, M.K., Erbes, C., Tye, S.J., Grabowski, J.G., Shiroma, P.R., 2018. Efficacy, safety, and durability of repeated ketamine infusions for comorbid posttraumatic stress disorder and treatment-resistant depression. J. Clin. Psychiatry 79 (3), e1–e8.

Albott, C.S., Shiroma, P.R., Cullen, K.R., Johns, B., Thuras, P., Wels, J., Lim, K.O., 2017. The antidepressant effect of repeat dose intravenous ketamine is delayed by concurrent benzodiazepine use. J. Clin. Psychiatry 78 (3), 308–309.

Frye, M.A., Blier, P., Tye, S.J., 2015. Concomitant benzodiazepine use attenuates ketamine response: implications for large scale study design and clinical development. J. Clin. Psychopharmacol. 35 (3), 334–336.

Andrashko, V., Novak, T., Horacek, J., Klirova, M., Brunovsky, M., 2019. Concurrent benzodiazepines undermine the antidepressant effect of ketamine. Eur. Neuropsychopharmacol. 29, S389–S390.

Williams, N.R., Heifets, B.D., Blasey, C., Sudheimer, K., Pannu, J., Pankow, H., Schatzberg, A.F., 2018. Attenuation of antidepressant effects of ketamine by opioid receptor antagonism. Am. J. Psychiatry 175 (12), 1205–1215.

Bobo, W.V., Riva-Posse, P., Goes, F.S., Parikh, S.V., 2020. Next-step treatment considerations for patients with treatment-resistant depression that responds to low-dose intravenous ketamine. Focus 18 (2), 181–192.

Short, B., Fong, J., Galvez, V., Shelker, W., Loo, C.K., 2018. Side-effects associated with ketamine use in depression: a systematic review. Lancet Psychiatry 5 (1), 65–78.

Dittrich, A., 1998. The standardized psychometric assessment of altered states of consciousness (ASCs) in humans. Pharmacopsychiatry 31 (S 2), 80–84.

Aust, S., Gärtner, M., Basso, L., Otte, C., Wingenfeld, K., Chae, W.R., Grimm, S., 2019. Anxiety during ketamine infusions is associated with negative treatment responses in major depressive disorder. Eur. Neuropsychopharmacol. 29 (4), 529–538.

Lapidus, K.A., Levitch, C.F., Perez, A.M., Brallier, J.W., Parides, M.K., Soleimani, L., Murrough, J.W., 2014. A randomized controlled trial of intranasal ketamine in major depressive disorder. Biol. Psychiatry 76 (12), 970–976.

Singh, J.B., Fedgchin, M., Daly, E., Xi, L., Melman, C., De Bruecker, G., Drevets, W.C., 2016. Intravenous esketamine in adult treatment-resistant depression: a double-blind, double-randomization, placebo-controlled study. Biol. Psychiatry 80 (6), 424–431.

Papakostas, G.I., Salloum, N.C., Hock, R.S., Jha, M.K., Murrough, J.W., Mathew, S.J., Fava, M., 2020b. Efficacy of esketamine augmentation in major depressive disorder: a meta-analysis. J. Clin. Psychiatry 81 (4), e1–e8.

Glue, P., Medlicott, N.J., Neehoff, S., Surman, P., Lam, F., Hung, N., Hung, C.T., 2020. Safety and efficacy of extended release ketamine tablets in patients with treatment-resistant depression and anxiety: open label pilot study. Ther. Adv. Psychopharmacol. 10, 2045125320922474.

ECT references

Fink, M., 1982. ECT in anxiety: an appraisal. Am. J. Psychother. 36 (3), 371–378.

Kellner, C., Husain, M., Knapp, R., McCall, W., Petrides, G., Rudorfer, M., Prudic, J., 2017. The prolonging remission in depressed elderly (PRIDE) trial (NCT01028508): results from the randomised phase. Eur. Neuropsychopharmacol. 27, S543.

Fink, M., 2014. What was learned: studies by the consortium for research in ECT (CORE) 1997–2011. Acta Psychiatr. Scand. 129 (6), 417–426.

Østergaard, S.D., Speed, M.S., Kellner, C.H., Mueller, M., McClintock, S.M., Husain, M.M., Lisanby, S.H., 2020. Electroconvulsive therapy (ECT) for moderate-severity major depression among the elderly: data from the PRIDE study. J. Affect. Disord. 274, 1134–1141.

Dunner, D.L., Rush, A.J., Russell, J.M., Burke, M., Woodard, S., Wingard, P., Allen, J., 2006. Prospective, long-term, multicenter study of the naturalistic outcomes of patients with treatment-resistant depression. J. Clin. Psychiatry 67 (5), 688–695.

Van Diermen, L., Van Den Ameele, S., Kamperman, A.M., Sabbe, B.C., Vermeulen, T., Schrijvers, D., Birkenhäger, T.K., 2018. Prediction of electroconvulsive therapy response and remission in major depression: meta-analysis. Br. J. Psychiatry 212 (2), 71–80.

Meyer, J.P., Swetter, S.K., Kellner, C.H., 2020. Electroconvulsive therapy in geriatric psychiatry: a selective review. Clin. Geriatr. Med. 36 (2), 265–279.

Gill, S.P., Kellner, C.H., 2019. Clinical practice recommendations for continuation and maintenance electroconvulsive therapy for depression: outcomes from a review of the evidence and a consensus workshop held in Australia in May 2017. J. ECT 35 (1), 14–20.

Rasmussen, K.G., Snyder, K.A., Knapp, R.G., Mueller, M., Yim, E., Husain, M.M., Kellner, C.H., 2004. Relationship between somatization and remission with ECT. Psychiatry Res. 129 (3), 293–295.

Medda, P., Barbuti, M., Novi, M., Boccolini, A., Tripodi, B., De Simone, L., Perugi, G., 2020. Naturalistic follow-up in bipolar patients after successful electroconvulsive therapy. J. Affect. Disord. 271, 152–159.

Flint, A.J., Rifat, S.L., 1997. Anxious depression in elderly patients: response to antidepressant treatment. Am. J. Geriatr. Psychiatry 5 (2), 107–115.

Steffens, D.C., McQuoid, D.R., 2005. Impact of symptoms of generalized anxiety disorder on the course of late-life depression. Am. J. Geriatr. Psychiatry 13 (1), 40–47.

Huang, C.J., Lin, C.H., Wu, J.I., Yang, W.C., 2019. The relationship between depression symptoms and anxiety symptoms during acute ECT for patients with major depressive disorder. Int. J. Neuropsychopharmacol. 22 (10), 609–615.

Langius-Eklöf, A., Samuelsson, M., 2009. Sense of coherence and psychiatric morbidity in terms of anxiety and depression in patients with major depression before and after electric convulsive treatment. Scand. J. Caring Sci. 23 (2), 375–379.

Lins-Martins, N.M., Yücel, M., Tovar-Moll, F., Rodrigues, E.C., Fontenelle, L.F., 2015. Electroconvulsive therapy in obsessive-compulsive disorder: a chart review and evaluation of its potential therapeutic effects. J. Neuropsychiatr. Clin. Neurosci. 27 (1), 65–68.

Bais, M., Figee, M., Denys, D., 2014. Neuromodulation in obsessive-compulsive disorder. Psychiatr. Clin. N. Am. 37 (3), 393–413. https://doi.org/10.1016/j.psc.2014.06.003.

Dos Santos-Ribeiro, S., Andrade, J.S., Quintas, J.N., Batista, K.B., Yücel, M., Fontenelle, L.F., 2019. A systematic review of the utility of electroconvulsive therapy in broadly defined obsessive-compulsive related disorders. Eur. Neuropsychopharmacol. 29, S507–S508.

Kellner, C.H., Wachtel, L.E., Dhossche, D., 2019. Electroconvulsive therapy is helpful for patients with obsessive-compulsive disorder-related disorders: a response to dos Santos-Ribeiro et al. Prim. Care Companion CNS Disord. 21 (3), 18l02414.

Youssef, N.A., McCall, W.V., Andrade, C., 2017. The role of ECT in posttraumatic stress disorder: a systematic review. Ann. Clin. Psychiatry 29 (1), 62–70.

Beynel, L., Appelbaum, L.G., Kimbrel, N.A., 2020. Neurobiology and neuromodulation of emotion in PTSD. In: Emotion in Posttraumatic Stress Disorder. Academic Press, pp. 175–210.

Koek, R.J., Roach, J., Athanasiou, N., van't Wout-Frank, M., Philip, N.S., 2019. Neuromodulatory treatments for post-traumatic stress disorder (PTSD). Prog. Neuro-Psychopharmacol. Biol. Psychiatry 92, 148–160.

Novakovic, V., Sher, L., Lapidus, K.B., Mindes, J., Golier J, A., Yehuda, R., 2011. Brain stimulation in posttraumatic stress disorder. Eur. J. Psychotraumatol. 2 (1), 5609.

Margoob, M.A., Ali, Z., Andrade, C., 2010. Efficacy of ECT in chronic, severe, antidepressant-and CBT-refractory PTSD: an open, prospective study. Brain Stimul. 3 (1), 28–35.

Janjua, A.U., Dhingra, A.L., Greenberg, R., McDonald, W.M., 2020. The efficacy and safety of concomitant psychotropic medication and electroconvulsive therapy (ECT). CNS Drugs 34 (5), 509–520.

Tang, V.M., Pasricha, A.N., Blumberger, D.M., Voineskos, D., Pasricha, S., Mulsant, B.H., Daskalakis, Z.J., 2017. Should benzodiazepines and anticonvulsants be used during electroconvulsive therapy?: a case study and literature review. J. ECT 33 (4), 237–242.

Delamarre, L., Galvao, F., Gohier, B., Poulet, E., Brunelin, J., 2019. How much do benzodiazepines matter for electroconvulsive therapy in patients with major depression? J. ECT 35 (3), 184–188.

Obbels, J., Verwijk, E., Bouckaert, F., Sienaert, P., 2017. ECT-related anxiety: a systematic review. J. ECT 33 (4), 229–236.

Obbels, J., Vanbrabant, K., Verwijk, E., Bouckaert, F., Vansteelandt, K., Sienaert, P., 2020a. Monitoring electroconvulsive therapy–related anxiety: the ECT-related anxiety questionnaire. J. ECT 36 (3), 180–186.

Obbels, J., Vansteelandt, K., Verwijk, E., Lambrichts, S., Bouckaert, F., Sienaert, P., 2020b. Understanding ECT-related anxiety: a prospective study. Acta Psychiatr. Scand. 142 (2), 132–140.

Mathew, S.J., Wilkinson, S.T., Altinay, M., Asghar-Ali, A., Chang, L.C., Collins, K.A., Malone, D.A., 2019. ELEctroconvulsive therapy (ECT) vs. ketamine in patients with treatment-resistant depression: the ELEKT-D study protocol. Contemp. Clin. Trials 77, 19–26.

Kellner, C.H., Iosifescu, D.V., 2017. Ketamine and ECT: better alone than together? Lancet Psychiatry 4 (5), 348–349.

TMS references

Diefenbach, G.J., Bragdon, L., Goethe, J.W., 2013. Treating anxious depression using repetitive transcranial magnetic stimulation. J. Affect. Disord. 151 (1), 365–368.

O'Reardon, J.P., Solvason, H.B., Janicak, P.G., Sampson, S., Isenberg, K.E., Nahas, Z., McDonald, W.M., Avery, D., Fitzgerald, P.B., Loo, C., Demitrack, M.A., George, M.S., Sackeim, H.A., 2007. Efficacy and safety of transcranial magnetic stimulation in the acute treatment of major depression: a multisite randomized controlled trial. Biol. Psychiatry 62 (11), 1208–1216.

George, M.S., Lisanby, S.H., Avery, D., McDonald, W.M., Durkalski, V., Pavlicova, M., Anderson, B., Nahas, Z., Bulow, P., Zarkowski, P., Holtzheimer 3rd, P.E., Schwartz, T., Sackeim, H.A., 2010. Daily left prefrontal transcranial magnetic stimulation therapy for major depressive disorder: a sham-controlled randomized trial. Arch. Gen. Psychiatry 67 (5), 507–516.

Ray, S., Nizamie, S.H., Akhtar, S., Praharaj, S.K., Mishra, B.R., Zia-ul-Haq, M., 2011. Efficacy of adjunctive high frequency repetitive transcranial magnetic stimulation of left prefrontal cortex in depression: a randomized sham controlled study. J. Affect. Disord. 128 (1–2), 153–159.

Gaynes, B.N., Lloyd, S.W., Lux, L., Gartlehner, G., Hansen, R.A., Brode, S., Jonas, D.E., Swinson Evans, T., Viswanathan, M., Lohr, K.N., 2014. Repetitive transcranial magnetic stimulation for treatment-resistant depression: a systematic review and meta-analysis. J. Clin. Psychiatry 75 (5), 477–489.

Kozel, F.A., George, M.S., 2002. Meta-analysis of left prefrontal repetitive transcranial magnetic stimulation (rTMS) to treat depression. J. Psychiatr. Pract. 8 (5), 270–275.

Clarke, E., Clarke, P., Gill, S., Paterson, T., Hahn, L., Galletly, C., 2019. Efficacy of repetitive transcranial magnetic stimulation in the treatment of depression with comorbid anxiety disorders. J. Affect. Disord. 252, 435–439.

White, D., Tavakoli, S., 2015. Repetitive transcranial magnetic stimulation for treatment of major depressive disorder with comorbid generalized anxiety disorder. Ann. Clin. Psychiatry 27 (3), 192–196.

Thase, M.E., Rush, A.J., 1997. When at first you don't succeed: sequential strategies for antidepressant nonresponders. J. Clin. Psychiatry 58 (Suppl. 13), 23–29.

Mantovani, A., Aly, M., Dagan, Y., Allart, A., Lisanby, S.H., 2013. Randomized sham controlled trial of repetitive transcranial magnetic stimulation to the dorsolateral prefrontal cortex for the treatment of panic disorder with comorbid major depression. J. Affect. Disord. 144 (1–2), 153–159.

Fava, M., Rush, A.J., Alpert, J.E., Balasubramani, G.K., Wisniewski, S.R., Carmin, C.N., Biggs, M.M., Zisook, S., Leuchter, A., Howland, R., Warden, D., Trivedi, M.H., 2008c. Difference in treatment outcome in outpatients with anxious versus nonanxious depression: a STAR*D report. Am. J. Psychiatry 165 (3), 342–351.

Kumar, S., Singh, S., Parmar, A., Verma, R., Kumar, N., 2018. Effect of high-frequency repetitive transcranial magnetic stimulation (rTMS) in patients with comorbid panic disorder and major depression. Australas. Psychiatry 26 (4), 398–400.

Philip, N.S., Aiken, E.E., Kelley, M.E., Burch, W., Waterman, L., Holtzheimer, P.E., 2019. Synchronized transcranial magnetic stimulation for posttraumatic stress disorder and comorbid major depression. Brain Stimul. 12 (5), 1335–1337.

Guo, Q., Li, C., Wang, J., 2017. Updated review on the clinical use of repetitive transcranial magnetic stimulation in psychiatric disorders. Neurosci. Bull. 33 (6), 747–756.

Trevizol, A.P., Downar, J., Vila-Rodriguez, F., Konstantinou, G., Daskalakis, Z.J., Blumberger, D.M., 2021. Effect of repetitive transcranial magnetic stimulation on anxiety symptoms in patients with major depression: an analysis from the THREE-D trial. Depress. Anxiety 38 (3), 262–271.

Chen, L., Hudaib, A.R., Hoy, K.E., Fitzgerald, P.B., 2019. Is rTMS effective for anxiety symptoms in major depressive disorder? An efficacy analysis comparing left-sided high-frequency, right-sided low-frequency, and sequential bilateral rTMS protocols. Depress. Anxiety 36 (8), 723–731.

Mantovani, A., Lisanby, S.H., Pieraccini, F., Ulivelli, M., Castrogiovanni, P., Rossi, S., 2006. Repetitive transcranial magnetic stimulation (rTMS) in the treatment of obsessive-compulsive disorder (OCD) and Tourette's syndrome (TS). Int. J. Neuropsychopharmacol. 9 (1), 95–100.

Gomes, P.V., Brasil-Neto, J.P., Allam, N., Rodrigues de Souza, E., 2012. A randomized, double-blind trial of repetitive transcranial magnetic stimulation in obsessive-compulsive disorder with three-month follow-up. J. Neuropsychiatr. Clin. Neurosci. 24 (4), 437–443.

Hawken, E.R., Dilkov, D., Kaludiev, E., Simek, S., Zhang, F., Milev, R., 2016. Transcranial magnetic stimulation of the supplementary motor area in the treatment of obsessive-compulsive disorder: a multi-site study. Int. J. Mol. Sci. 17 (3), 420.

Lisanby, S.H., Husain, M.M., Rosenquist, P.B., Maixner, D., Gutierrez, R., Krystal, A., Gilmer, W., Marangell, L.B., Aaronson, S., Daskalakis, Z.J., Canterbury, R., Richelson, E., Sackeim, H.A., George, M.S., 2009. Daily left prefrontal repetitive transcranial magnetic stimulation in the acute treatment of major depression: clinical predictors of outcome in a multisite, randomized controlled clinical trial. Neuropsychopharmacology 34 (2), 522–534.

Brakemeier, E.L., Wilbertz, G., Rodax, S., Danker-Hopfe, H., Zinka, B., Zwanzger, P., Grossheinrich, N., Várkuti, B., Rupprecht, R., Bajbouj, M., Padberg, F., 2008. Patterns of response to repetitive transcranial magnetic stimulation (rTMS) in major depression: replication study in drug-free patients. J. Affect. Disord. 108 (1–2), 59–70.

Lefaucheur, J.P., Aleman, A., Baeken, C., Benninger, D.H., Brunelin, J., Di Lazzaro, V., Filipović, S.R., Grefkes, C., Hasan, A., Hummel, F.C., Jääskeläinen, S.K., Langguth, B., Leocani, L., Londero, A., Nardone, R., Nguyen, J.P., Nyffeler, T., Oliveira-Maia, A.J., Oliviero, A., Padberg, F., Ziemann, U., 2020. Evidence-based guidelines on the therapeutic use of repetitive transcranial magnetic stimulation (rTMS): an update (2014-2018). Clin. Neurophysiol. 131 (2), 474–528.

Carpenter, L.L., Conelea, C., Tyrka, A.R., Welch, E.S., Greenberg, B.D., Price, L.H., Niedzwiecki, M., Yip, A.G., Barnes, J., Philip, N.S., 2018. 5 Hz repetitive transcranial magnetic stimulation for posttraumatic stress disorder comorbid with major depressive disorder. J. Affect. Disord. 235, 414–420.

Psychotherapy references

Greenlee, A., Karp, J.F., Dew, M.A., Houck, P., Andreescu, C., Reynolds III, C.F., 2010. Anxiety impairs depression remission in partial responders during extended treatment in late-life. Depress. Anxiety 27 (5), 451–456.

Guidi, J., Tomba, E., Fava, G.A., 2016. The sequential integration of pharmacotherapy and psychotherapy in the treatment of major depressive disorder: a meta-analysis of the sequential model and a critical review of the literature. Am. J. Psychiatry 173 (2), 128–137.

Ijaz, S., Davies, P., Williams, C.J., Kessler, D., Lewis, G., Wiles, N., 2018. Psychological therapies for treatment-resistant depression in adults. Cochrane Database Syst. Rev. https://doi.org/10.1002/14651858.CD010558.pub2. Issue 5. Art. No.: CD010558.

Cosci, F., Guidi, J., Mansueto, G., Fava, G.A., 2020. Psychotherapy in recurrent depression: efficacy, pitfalls, and recommendations. Expert. Rev. Neurother. https://doi.org/10.1080/14737175.2020.1804870.

Wilkinson, S.T., Holtzheimer, P.E., Gao, S., Kirwin, D.S., Price, R.B., 2019. Leveraging neuroplasticity to enhance adaptive learning: the potential for synergistic somatic-behavioral treatment combinations to improve clinical outcomes in depression. Biol. Psychiatry 85 (6), 454–465.

Wilkinson, S.T., Wright, D., Fasula, M.K., Fenton, L., Griepp, M., Ostroff, R.B., Sanacora, G., 2017. Cognitive behavior therapy may sustain antidepressant effects of intravenous ketamine in treatment-resistant depression. Psychother. Psychosom. 86 (3), 162–167.

Brakemeier, E.L., Merkl, A., Wilbertz, G., Quante, A., Regen, F., Bührsch, N., Heuser, I., 2014. Cognitive-behavioral therapy as continuation treatment to sustain response after electroconvulsive therapy in depression: a randomized controlled trial. Biol. Psychiatry 76 (3), 194–202.

Donse, L., Padberg, F., Sack, A.T., Rush, A.J., Arns, M., 2018. Simultaneous rTMS and psychotherapy in major depressive disorder: clinical outcomes and predictors from a large naturalistic study. Brain Stimul. 11 (2), 337–345.

Russo, G.B., Tirrell, E., Busch, A., Carpenter, L.L., 2018. Behavioral activation therapy during transcranial magnetic stimulation for major depressive disorder. J. Affect. Disord. 236, 101–104.

van Bronswijk, S., Moopen, N., Beijers, L., Ruhe, H.G., Peeters, F., 2019. Effectiveness of psychotherapy for treatment-resistant depression: a meta-analysis and meta-regression. Psychol. Med. 49 (3), 366–379.

Psychedelics references

Osório, F.L., Sanches, R.F., Macedo, L.R., Santos, R.G., Maia-de-Oliveira, J.P., WichertAna, L., et al., 2015. Antidepressant effects of a single dose of ayahuasca in patients with recurrent depression: a preliminary report. Rev. Bras. Psiquiatr. 37 (1), 13–20. https://doi.org/10.1590/1516-4446-2014-1496.

Gasser, P., Holstein, D., Michel, Y., Doblin, R., Yazar-Klosinski, B., Passie, T., 2014. Safety and efficacy of lysergic acid diethylamide-assisted psychotherapy for anxiety associated with life-threatening diseases. J. Nerv. Ment. Dis. 202 (7), 513–520. https://doi.org/10.1097/NMD.0000000000000113.

Griffiths, R.R., Johnson, M.W., Carducci, M.A., Umbricht, A., Richards, W.A., Richards, B.D., et al., 2016. Psilocybin produces substantial and sustained decreases in depression and anxiety in patients with life-threatening cancer: a randomized doubleblind trial. J. Psychopharmacol. 30 (12), 1181–1197. https://doi.org/10.1177/0269881116675513.

Muttoni, S., Ardissino, M., John, C., 2019. Classical psychedelics for the treatment of depression and anxiety: a systematic review. J. Affect. Disord. 258, 11–24.

Kelly, J.R., Baker, A., Babiker, M., Burke, L., Brennan, C., O'Keane, V., 2019. The psychedelic renaissance: the next trip for psychiatry? Ir. J. Psychol. Med., 1–5.

Wheeler, S.W., Dyer, N.L., 2020. A systematic review of psychedelic-assisted psychotherapy for mental health: an evaluation of the current wave of research and suggestions for the future. Psychol. Conscious. Theory Res. Pract. 7 (3), 279.

Carhart-Harris, R.L., Roseman, L., Bolstridge, M., Demetriou, L., Pannekoek, J.N., Wall, M.B., Nutt, D.J., 2017. Psilocybin for treatment resistant depression: FMRI-measured brain mechanisms. Sci. Rep. 7, 13187. https://doi.org/10.1038/s41598-017-13282-7.

Roseman, L., Demetriou, L., Wall, M.B., Nutt, D.J., Carhart-Harris, R.L., 2018. Increased amygdala responses to emotional faces after psilocybin for treatment-resistant depression. Neuropharmacology 142, 263–269. https://doi.org/10.1016/j.neuropharm.2017.12.041.

Stroud, J.B., Freeman, T.P., Leech, R., Hindocha, C., Lawn, W., Nutt, D.J., Carhart-Harris, R.L., 2018. Psilocybin with psychological support improves emotional face recognition in treatment resistant depression. Psychopharmacology 235, 459–466. https://doi.org/10.1007/s00213-017-4754-y.

Watts, R., Day, C., Krzanowski, J., Nutt, D., Carhart-Harris, R., 2017. Patients' accounts of increased "connectedness" and "acceptance" after psilocybin for treatment-resistant depression. J. Humanist. Psychol. 57, 520–564. https://doi.org/10.1177/0022167817709585.

VNS references

Trottier-Duclos, F., Desbeaumes Jodoin, V., Fournier-Gosselin, M.-P., Richer, F., Desjardins, N., Tieu, S., et al., 2018. A 6-year follow-up study of vagus nerve stimulation effect on quality of life in treatment-resistant depression: a pilot study. J. ECT 34, E58–E60.

Bottomley, J.M., LeReun, C., Diamantopoulos, A., Mitchell, S., Gaynes, B.N., 2020. Vagus nerve stimulation (VNS) therapy in patients with treatment resistant depression: a systematic review and meta-analysis. Compr. Psychiatry 98, 152156.

Burger, A.M., Van der Does, W., Thayer, J.F., Brosschot, J.F., Verkuil, B., 2019. Transcutaneous vagus nerve stimulation reduces spontaneous but not induced negative thought intrusions in high worriers. Biol. Psychol. 142, 80–89.

Yakunina, N., Kim, S.S., Nam, E.-C., 2017. Optimization of transcutaneous vagus nerve stimulation using functional MRI. Neuromodulation 20 (3), 290–300. https://doi.org/10.1111/ner.12541.

Rong, P., Liu, J., Wang, L., Liu, R., Fang, J., Zhao, J., Kong, J., 2016. Effect of transcutaneous auricular vagus nerve stimulation on major depressive disorder: a nonrandomized controlled pilot study. J. Affect. Disord. 195, 172–179. https://doi.org/10.1016/j.jad.2016.02.031.

DBS references

Bewernick, B.H., Kayser, S., Sturm, V., Schlaepfer, T.E., 2012. Long-term effects of nucleus accumbens deep brain stimulation in treatment-resistant depression: evidence for sustained efficacy. Neuropsychopharmacology 37 (9), 1975–1985.

Bewernick, B.H., Hurlemann, R., Matusch, A., Kayser, S., Grubert, C., Hadrysiewicz, B., Brockmann, H., 2010. Nucleus accumbens deep brain stimulation decreases ratings of depression and anxiety in treatment-resistant depression. Biol. Psychiatry 67 (2), 110–116.

Huys, D., Kohl, S., Baldermann, J.C., Timmermann, L., Sturm, V., Visser-Vandewalle, V., Kuhn, J., 2019. Open-label trial of anterior limb of internal capsule–nucleus accumbens deep brain stimulation for obsessive-compulsive disorder: insights gained. J. Neurol. Neurosurg. Psychiatry 90 (7), 805–812.

Farrand, S., Evans, A.H., Mangelsdorf, S., Loi, S.M., Mocellin, R., Borham, A., Velakoulis, D., 2018. Deep brain stimulation for severe treatment-resistant obsessive-compulsive disorder: an open-label case series. Aust. N. Z. J. Psychiatry 52 (7), 699–708.

Huff, W., Lenartz, D., Schormann, M., Lee, S.H., Kuhn, J., Koulousakis, A., Sturm, V., 2010. Unilateral deep brain stimulation of the nucleus accumbens in patients with treatment-resistant obsessive-compulsive disorder: outcomes after one year. Clin. Neurol. Neurosurg. 112 (2), 137–143.

Greenberg, B.D., Malone, D.A., Friehs, G.M., Rezai, A.R., Kubu, C.S., Malloy, P.F., Rasmussen, S.A., 2006. Three-year outcomes in deep brain stimulation for highly resistant obsessive–compulsive disorder. Neuropsychopharmacology 31 (11), 2384–2393.

Langevin, J.P., Koek, R.J., Schwartz, H.N., Chen, J.W., Sultzer, D.L., Mandelkern, M.A., Krahl, S.E., 2016. Deep brain stimulation of the basolateral amygdala for treatment-refractory posttraumatic stress disorder. Biol. Psychiatry 79 (10), e82–e84.

Clinical summary references

Brown, T.A., Barlow, D.H., 2014. Anxiety and Related Disorders Interview Schedule for DSM-5 (ADIS-5L). Client Interview Schedule.

Lecrubier, Y., Sheehan, D.V., Weiller, E., et al., 1997. The MINI international neuropsychiatric interview (MINI). A short diagnostic structured interview: reliability and validity according to the CIDI. Eur. Psychiatry 12, 224–231.

First, M.B., Williams, J.B., Karg, R.S., Spitzer, R.L., 2016. User's Guide for the SCID-5-CV Structured Clinical Interview for DSM-5® Disorders: Clinical Version. American Psychiatric Publishing.

Ionescu, D.F., Niciu, M.J., Henter, I.D., Zarate Jr., C.A., 2013b. Defining anxious depression: a review of the literature. CNS Spectr. 18 (5), 252.

Uher, R., Payne, J.L., Pavlova, B., Perlis, R.H., 2014. Major depressive disorder in DSM-5: implications for clinical practice and research of changes from DSM-IV. Depress. Anxiety 31 (6), 459–471.

Wang, C., Zhou, Y., Zheng, W., Liu, W., Zhan, Y., Li, H., Li, M.D., 2019b. Association between depression subtypes and response to repeated-dose intravenous ketamine. Acta Psychiatr. Scand. 140 (5), 446–457.

Chapter 30

Managing treatment-resistant depression with comorbid personality disorders

Vitaliy L. Voytenko, PsyD[a,b] and Steven K. Huprich, PhD[a,c]

[a]*Division of Psychiatry and Behavioral Medicine, Michigan State University College of Human Medicine, Grand Rapids, MI, United States,* [b]*Alliance Clinical Associates, Wheaton, IL, United States,* [c]*Department of Psychology, University of Detroit Mercy, Detroit, MI, United States*

Personality disorders are some of the most common comorbid psychiatric conditions in major depressive disorder (MDD). The combination of MDD and personality disorders provides a real challenge to clinicians, as treatment of patients with these comorbid conditions is hampered by significant deficits in psychosocial and interpersonal functioning, often leading to poorer outcomes and treatment resistance. This chapter explores the relationship between personality disorders and TRD by synthesizing relevant empirical data and their implications for clinical practice and discusses alternative, personality-derived theoretical models of TRD. We begin by reviewing the definition and criteria for personality disorders according to the Diagnostic and Statistical Manual of Mental Disorders, 5th Edition (*DSM-5*; American Psychiatric Association, 2013), along with the Alternative DSM-5 Model for Personality Disorders included in Section III: Emerging Measures and Models of the manual. Then, we discuss the relationship between depression and personality dysfunction, including the challenges in assessing each of the conditions in comorbid presentations. We then turn our attention to reviewing the evidence of the influence of comorbid personality disorders on the functioning of patients with MDD and treatment outcome for MDD. The two sections that follow discuss the relationship between depression and specific personality disorders. In order to allow for sufficient depth of discussion, we chose to focus on two types of personality pathology, Borderline Personality Disorder (BPD) and Narcissistic Personality Disorder (NPD), because of their frequent comorbidity with TRD in clinical practice. An additional factor in our selection was the predominance of the diagnosis of BPD in female and NPD in male patients. In each of the two sections, we review the literature on effective biological, psychological, and combined treatments for the corresponding personality disorders and the influence of the personality pathology on the treatment outcome in cooccurring MDD. We provide case studies to illustrate each of the comorbid clinical presentations. We then discuss alternative, personality-derived models of TRD through the framework of the Section III of the DSM-5, the historical concept of Depressive Personality Disorder, and a novel empirically derived construct of *malignant self-regard.* Finally, we distill the implications for clinical practice in managing TRD with comorbid personality disorders in the "Clinical summary" section of this chapter.

Personality disorders in DSM-5

The DSM-5 describes personality disorders in both Section II (Diagnostic Criteria and Codes) and Section III (Emerging Measures and Models). Section II describes personality disorders and their criteria in a manner that is nearly identical to the DSM-IV-TR (American Psychiatric Association, 2001), which describes 10 discrete personality disorders with specific criteria and instructions on how combinations of these criteria lead one to be diagnosed with the disorder. A personality disorder is "an enduring pattern of inner experience and behavior that deviates markedly from the expectations of the individual's culture, is pervasive and inflexible, has an onset in adolescence or early adulthood, is stable over time, and leads to distress or impairment" (p. 645). The Section III description of personality disorders, known as the Alternative DSM-5 Model for Personality Disorders (AMPD), is a product of the DSM-5 Personality and Personality Disorders Work Group and represents a radical departure from the current diagnostic approach. This model was included because of considerable problems that have been leveled against the current categorical approach, including the "failure to 'carve nature at its joints', the lack of meaningful or well-validated boundary between normal and disordered personality, excessive heterogeneity within diagnostic categories, excessive diagnostic cooccurrence across categories, inadequate coverage of the full range of personality difficulties seen in clinical practice, questionable temporal stability of diagnoses, dissatisfaction among clinicians, and inadequate scientific foundation" (p. 45, Samuel and Griffin, 2015).

Managing Treatment-Resistant Depression. https://doi.org/10.1016/B978-0-12-824067-0.00030-X

Because of these concerns, the AMPD was included in Section III of DSM-5, with the encouragement for the model to be further tested and validated. Here, "personality disorders are characterized by impairments in personality functioning and pathological personality traits" (p. 761). Personality functioning refers to the individual's representations of both self and others, also known as interpersonal functioning. Within the *self* domain, individuals are assessed for the quality of their *identity* (i.e., experience of oneself as unique, separateness from others, self-esteem stability, and the sense of who one is as a person) and *self-directedness* (i.e., a sense of one's ability to plan and act on personally meaningful short-term and long-term goals). Within the interpersonal domain, individuals are assessed for *empathy* (i.e., the ability to understand, appreciate, and tolerate other's points of view and needs, as well as knowing how one's actions affects others) and *intimacy* (i.e., the ability to have meaningful, in-depth relationships with others). Additionally, pathological personality traits are organized into five broad trait domains, each of which is composed of several facets. The domains and facets are as follows:

Negative affectivity (emotional lability, anxiousness, separation insecurity, submissiveness, hostility, perseveration, depressivity, suspiciousness, and restricted affectivity)
Detachment (withdrawal, intimacy avoidance, anhedonia, depressivity, restricted affectivity, suspiciousness)
Antagonism (manipulativeness, deceitfulness, grandiosity, attention-seeking, callousness, hostility)
Disinhibition (irresponsibility, impulsivity, distractibility, risk taking, rigid perfectionism, or lack thereof)
Psychoticism (unusual beliefs and experiences, eccentricity, cognitive and perceptual dysregulation)

It should be noted that certain facets can be part of differing trait domains (e.g., depressivity and restricted affectivity) and that the facets per domain vary in number. These trait domains were derived from extensive empirical investigation and are strongly grounded in empirical investigations of trait theory in personality psychology (Krueger et al., 2012). To be diagnosed with a personality disorder by way of the AMPD, an individual must have "moderate or greater impairment in personality (self/interpersonal) functioning" and "one or pathological personality traits"; p. 761). How the AMPD is related to TRD will be discussed below. However, suffice it to say that the AMPD offers a much wider range of personality pathology than the typical taxonomy and incorporates chronic affective states (e.g., depressivity, emotional lability, anxiousness, and hostility) into its assessment framework.

The relationship between personality disorders and depression

The combination of personality disorders and MDD is one of the most common comorbid psychiatric presentations (Hirschfeld, 1999; Oldham et al., 1995). Prevalence rates in community and clinical samples span a wide range from 20% to 85% (Corruble et al., 1996; Grant et al., 2005; Hasin et al., 2005; Lenzenweger et al., 2007; Skodol et al., 1999), with the average being approximately 50%. Prevalence data on comorbid personality disorders in TRD are virtually nonexistent, although one small study comparing rigorously-defined treatment-resistant depressed outpatients with their non-TRD counterparts did not find treatment resistance to be associated with increased rates of personality disturbance (Petersen et al., 2002).

From a conceptual standpoint, the development and nature of the association between depression and personality pathology remain poorly understood. These constructs are interrelated, but the relationship between them is likely complex and multifactorial (Klein et al., 2014; Thase, 1996). Thase (1996) summarizes relevant theoretical work by identifying three possible operative models in the relationship between depression and personality disorders: predisposition, complication, and coaggregation. *Predisposition* proposes that patients with certain personality traits or trait constellations, including neuroticism and low extraversion/detachment, are more susceptible to developing depressive disorders. *Complication* refers to pathoplastic effects of personality pathology on the expression or course of depression, such as increased severity or chronicity and poorer response to standard treatments. Conversely, the complication model can account for the development or exacerbation of personality pathology as a result of recurrent or persistent depressive episodes. This conceptual model is most applicable to elucidating the association between personality disorders and TRD and most relevant to the topic of this book chapter. Finally, *coaggregation* suggests that when a patient's coexisting vulnerabilities for depression and personality pathology are simultaneously expressed, the personality disorder becomes a precursor to or an attenuated manifestation of depression.

Klein et al. (2014) review additional conceptual work to expand the discussion of the interrelatedness of depression and maladaptive personality traits/personality disorders. They highlight several additional explanations, including (a) depression and personality disorders have common causes, (b) the content and diagnostic criteria of the two conditions overlap, and (c) personality traits/disorders represent consequences or "scars" of depressive illness. The role of life stress and its shared association with depression and personality pathology deserves special mention here. Thase (1996) observes that the relationship between personality traits/disorder and depression is intensified by adverse life events, which often serve as triggers or perpetuating factors in major depressive episodes. In addition, there is an interactive and reciprocal relationship between life stress and personality pathology, whereby individuals with maladaptive personality traits (e.g., antisocial, narcissistic, borderline) have more unstable life circumstances (e.g., work, financial, and legal problems) and interpersonal relations (e.g., marital discord, estrangement from family), which, in turn, increase their risk of being exposed to various stressors (Thase, 1996).

Assessment of comorbid personality and depressive disorders

The complexity of interrelatedness between depression and personality pathology makes the task of disentangling depressive symptoms from those of personality disorders exceptionally difficult in clinical settings. One challenge lies in the fact that the presence of depressive phenomenology may color the expression of personality and therefore complicate its assessment and vice versa. Furthermore, there is evidence to suggest that patients with MDD tend to overreport symptoms of personality pathology (De Fruyt et al., 2006; Farabaugh et al., 2004), which makes the assessment of personality while the patient is acutely depressed problematic. While clinicians can reliably evaluate current depressive symptoms using patient self-report, the depressed patient's description of their long-standing, habitual interpersonal patterns and self-perceptions are likely to be negatively influenced state effects of the illness, yielding a potentially biased assessment of personality functioning (Ganellen, 2007; Huprich et al., 2011). Conversely, personality pathology may color the quality of the patient's experience of MDD and lead to overreporting of depressive symptoms. For example, one literature review found that patients with BPD tend to score higher on self-report measures of depressive symptoms, compared to clinician-administered scales (Silk, 2010). In fact, patients with BPD without MDD may score as highly on self-rated symptom inventories as BPD patients with MDD and as highly as patients with MDD without BPD (Levy et al., 2007). Finally, an initial, cross-sectional evaluation of an adult patient based primarily on self-report may not allow for accurate assessment of their personality traits/disorders, which are, by definition, longitudinal constructs with a developmental origin that typically manifest themselves by late adolescence (Thase, 1996).

There are several possible strategies for addressing the challenge of assessment of comorbid depressive and personality disorders. One such strategy involves integration of collateral information in the evaluation process. Reports from collateral sources, such as long-term romantic partners or spouses, parents, and adult children can improve the accuracy of the assessment by providing additional data points to patient self-report, including longitudinal data (Zimmerman, 1994). However, this approach is not without its limitations, as it potentially introduces informant bias to the assessment process. In particular, familial risk of depression increases the probability that the informants themselves may be suffering from a mood disorder, which, in turn, can color their reports of the identified patient's behavior or interpersonal patterns. In addition, the negative impact of the identified patient's debilitating depressive illness or problematic personality traits on the informants' quality of life and relationship with the proband (Markowitz et al., 2007), may negatively bias their descriptions (Barnett and Gotlib, 1988; Thase, 1996).

Another strategy for navigating the challenge of accurately assessing personality disorders and depression in comorbid presentations involves incorporating clinician-administered semistructured interviews in the assessment process. Although clinicians typically rely on unstructured interviews in diagnosing personality disorders (Klein et al., 2014), structured and semistructured interviews have shown greater interrater reliability (Farmer, 2000; Widiger and Samuel, 2005). Examples of commonly used measures for personality disorders include Structured Interview for DSM-IV Personality (SIDP-IV; Pfohl et al., 1997), Structured Clinical Interview for DSM-5 Personality Disorders (SCID-5-PD; First et al., 2016), Personality Inventory for DSM-5 (PID-5; Al-Dajani et al., 2016), and the DSM-5 Level of Personality Functioning Scale (LPFS; Morey, 2017). The assessment of depression can also be aided by incorporating structured interviews and self-administered inventories. Joiner et al. (2005) proposed an evidence-based toolkit of measures with adequate psychometric properties for optimal depression assessment in clinical settings: (a) the Structural Clinical Interview for DSM (SCID), with most recent clinician version being SCID-5-CV (Osório et al., 2019) to establish a formal mood disorder diagnosis; (b) the Mini-International Neuropsychiatric Interview (MINI; Sheehan et al., 1998) module on melancholic features to supplement the SCID; (c) the Seasonal Pattern Assessment Questionnaire (SPAQ; Raheja et al., 1996) to assess for a seasonal component; (d) the Beck Depression Inventory-II (BDI-II; Wang and Gorenstein, 2013) to assess depressive symptom severity and short-term change in symptoms; and (e) the Longitudinal Interval Follow-Up Evaluation (LIFE; Keller et al., 1987) to assess remission of MDD.

The impact of personality disorders on functioning and treatment outcome in patients with MDD

Very few studies have examined the role of personality disorders on psychosocial functioning of patients with MDD. The best available evidence comes from the Collaborative Longitudinal Personality Disorders Study (CLPS; Gunderson et al., 2000), a large prospective multisite study with over 600 participants recruited at Brown, Columbia, Harvard, and Yale University medical schools. In a cross-section analysis of data obtained at 36-month follow-up, Skodol et al. (2005) compared the psychosocial functioning and well-being of patients with comorbid MDD and schizotypal, borderline, avoidant, or obsessive-compulsive personality disorder to those with MDD. They utilized the Medical Outcomes Study Short-Form Health Survey (SF-36; Hays et al., 1995), a widely used self-report inventory that assesses health-related quality of life. They found that personality disorder significantly contributed to functional impairment and diminished sense of well-being in at least three areas measured by the SF-36: *emotional role limitations* (i.e., problems with daily activities or work), *social functioning* (i.e., interpersonal relationships with relatives, friends, or neighbors), and *general health perceptions* (i.e., attitudes toward one's health, both present and anticipated, and one's appraisal of their tendency to get sick). These results were consistent with the findings from the baseline data from CLPS (Corruble et al., 1996; Grant et al., 2005; Hasin et al., 2005; Lenzenweger et al., 2007; Skodol et al., 1999), which showed that patients with comorbid MDD and personality disorders experienced greater impairment in employment and interpersonal relationships, compared to those suffering from MDD alone.

In a longitudinal, naturalistic follow-along study using data from the first 2 years of the CLPS cohort, Markowitz et al. (2007) compared three distinct groups of patients: (a) those with baseline MDD and personality disorder diagnoses, whose personality pathology remained stable and did not remit during the study period; (b) those with the same baseline comorbid presentation, but whose personality disorders remitted over time; and (c) those with MDD who did not meet the criteria for a personality disorder diagnosis at baseline. Using the LIFE psychosocial scales, which assess employment, interactions with friends, spouse/partner, and parents, recreation, global social adjustment, and the DSM-IV Axis V Global Assessment Functioning (GAF), they found that the presence of a persistent personality disorder predicted impairment in psychosocial functioning over and above the debilitating effects of comorbid MDD. They also found that the patients whose personality disorders remitted were more likely to experience a remission of MDD and improvement in social and occupational functioning. Markowitz et al. (2007) concluded that the burden of personality disorders in depressed patients extended beyond suffering from symptoms to psychosocial functioning and that the clinical course of comorbid personality disorders directly influenced outcome of MDD.

Numerous studies have examined the pathoplastic effects of personality pathology on the course and treatment outcome of depression. There is considerable evidence that patents with MDD suffering from a comorbid personality disorder have an earlier age of illness onset, experience more severe symptoms and longer duration of depressive episodes and are at higher risk of relapse after remission (Grilo et al., 2005, 2010; Skodol et al., 2011; Tyrer et al., 1990, 1997). The literature on the impact of personality pathology on depression treatment outcome has been more mixed and has been the subject of a long-standing debate in psychiatry going back nearly 70 years (Hobson, 1953). In order to synthesize the findings from scores of relevant original studies, several narrative and systematic literature reviews and metaanalyses have been published in the last two decades. In a narrative review of over 50 studies, Mulder (2002) found that the prognostic significance of personality pathology for treatment outcome in patients with MDD depended on study design, with the best-designed studies reporting the least amount of impact. A systematic literature review and metaanalysis of six randomized controlled trials (RCT) of pharmacotherapy for depressed patients with or without personality disorders also found no difference between the two groups (Kool et al., 2005). On the other hand, a more inclusive systematic review and metaanalysis of 34 studies, incorporating cohort, case-control, and RCT methodologies reported poorer outcomes for the comorbid group (Newton-Howes et al., 2006). This finding supported the conclusions of several earlier narrative reviews that found comorbid personality disorders to predict poorer outcome of depression treatment, including pharmacotherapy and electroconvulsive therapy (ECT; DeBattista and Mueller, 2001; Ilardi and Edward Head, 1994; Reich and Vasile, 1993; Reich and Green, 1991).

In attempt to bring clarity to this debate, the authors of several previous reviews came together as the Personality Disorder and Depression Outcome Group (PDDOG) and published the most comprehensive systematic literature review and metaanalysis on this topic to date, encompassing a total of 58 studies (Newton-Howes et al., 2014). Of the included studies, 35 were prospective case series and 17 RCTs. In the metaanalysis, Newton-Howes et al. (2014) showed that, irrespective of treatment modality, a diagnosis of personality disorder in patients with MDD was associated with a poorer treatment outcome, with an overall odds ratio of 2.16 (1.83–2.56). The authors also noted that, in the included studies, the treatment was limited to depressive disorders and did not address coexisting personality pathology, highlighting a potential reason for treatment resistance in MDD. They concluded that assessing personality status and treating cooccurring personality disorders were essential for optimizing treatment response in depressed patients. Furthermore, in the most up-to-date narrative review of recent literature on this topic, Young (2020) highlights additional evidence to support the conclusion that untreated personality pathology can negatively affect the clinical course and treatment outcome in MDD. These findings are consistent with the clinical experience of the authors of this book chapter and many other clinicians working with patients with chronic and treatment-resistant depression.

Borderline personality disorder and depression

Clinical presentation

Borderline personality disorder (BPD) is a serious mental illness characterized by "a pervasive pattern of instability of interpersonal relationships, self-image, and affects, and marked impulsivity that begins by early adulthood and is present in a variety of contexts" (American Psychiatric Association, 2013, p. 663). Individuals with BPD frequently experience extreme fear of abandonment and make frantic and often self-sabotaging efforts to avoid real or perceived rejection, such as responding with inappropriate anger, suicidal threats, or parasuicidal gestures. This dynamic is produced by an intense fear of being alone and an overreliance on others for a sense of personal identity and worth. BPD patients struggle with individuation and often experience their identity as "fused" with that of other people. At the same time, their interpersonal relationships are marked by instability and intensity, often resulting in significant fluctuations from idealization to debasement or derision, depending on their perception of the other individual's readiness to meet their needs or prove their loyalty. Their descriptions of other people may lack complexity and are often extreme: for example, "the best human being ever" or "a total loser." These individuals' self-image and sense of self are also highly unstable, and they are prone to intense shame, guilt and self-condemnation. They experience these feelings as so intolerable, that they often resort to radical and ultimately counterproductive mitigating measures, such as self-mutilation and suicidal behaviors. Individuals with BPD also display significant impulsivity, including behaviors that prove to be self-damaging (e.g., unprotected sex, excessive spending, substance abuse, unsafe driving, etc.). Finally, BPD is characterized by affective volatility, frequently resulting in depressive states marked by intense albeit transitory dysphoria, several hours to a few days in duration, anger and hostility. These fluctuations stem from BPD-associated mood reactivity, which, in turn, is triggered by a real or imagined abandonment or anticipated rejection. In some cases, the affective dysregulation can be accompanied by transient paranoid ideation or dissociative experiences.

Due to its characteristic instability of emotions and predominance of negative affect, BPD can often mimic a depressive disorder. Both BPD and MDD are marked by dysphoric affect and increased risk for suicidal behavior, which makes the task of determining whether the depressive symptoms should be attributed to the cooccurring depression or to the personality disorder itself, more complicated. Furthermore, individuals with this type of personality pathology appear to have a heightened subjective experience of depression, so much so that patients with BPD without MDD often score as highly on standardized self-rated depression measures as BPD patients with MDD and as highly as patients with MDD without BPD (Silk, 2010). In addition, patients with comorbid MDD and BPD tend to endorse greater severity of symptoms on self-report measures as compared to clinician-rated severity (Stanley and Wilson, 2006). Finally, the experience of depression in individuals with MDD is often qualitatively different from that in BPD patients, with the former being marked by melancholy, anhedonia, and apathy, accompanied by neurovegetative symptoms, and the latter by a sense of boredom, alienation, emptiness and inner "badness," as well as anger and hostility (Beatson and Rao, 2012; Ceresa et al., 2020; Klein et al., 2014).

BPD comorbidity and its impact on the course and treatment outcome in MDD

BPD is a relatively common mental illness, with prevalence rates of approximately 1.7% in the general population and 15%–28% in patients receiving psychiatric care in inpatient and outpatient settings (Gunderson et al., 2000). BPD is also by far the most-studied personality disorder, and its interface with depression has received a great deal of attention in professional literature. The lifetime prevalence of cooccurrence of depression with this type of personality pathology is extremely high, with recent studies suggesting that 61%–85% of individuals with BPD eventually meet the criteria for MDD (Gunderson et al., 2008; Zanarini et al., 2004b).

In light of the frequent comorbidity encountered in both research and clinical practice, a number of studies have explored the possibility that the two conditions have shared causes, with some authors going as far as conceptualizing BPD as a variant of a mood disorder (Akiskal et al., 1985; Akiskal, 2002). A recent review by Goodman et al. (2010) has in fact revealed an overlap in neurobiology of BPD and MDD, including reduced serotonin neurotransmission, amygdala hyperreactivity to emotional stimuli, and volume abnormalities in the anterior cingulate cortex and hippocampus. At the same time, the review also identified significant differences in brain region involvement, neurohormonal activation, heritability, symptomatology, prognosis, and pharmacotherapy response. The above similarities notwithstanding, the consensus of expert opinion remains that BPD and MDD are separate and distinct disorders (Beatson and Rao, 2012; Galione and Zimmerman, 2010; Gunderson et al., 2004; Klein et al., 2014; Silk, 2010; Stanley and Wilson, 2006; Young, 2020) that may share certain biological, temperamental, and early environmental risk factors (Klein et al., 2014; Thase, 1996).

There has been a long-standing consensus of expert opinion that antidepressant medication is not as effective in ameliorating the symptoms of major depression cooccurring with BPD as in MDD without personality pathology (Beatson and Rao, 2012; Gunderson, 2008). In a recent systematic literature review synthesizing the findings of three longitudinal studies (Gunderson et al., 2004; Riihimäki et al., 2014; Skodol et al., 2011) with a total sample of nearly 2400 patients, Ceresa et al. (2020) reported that the presence of comorbid BPD in fact slightly worsens the outcome of MDD. Cooccurring BPD is also associated with earlier onset of depression, longer duration of Major Depressive Episodes (MDE), increased rate of persistence of depressive symptoms, and lower rate of remission (Ceresa et al., 2020; Sullivan et al., 1994). These findings are consistent with the results of an earlier review of the impact of personality pathology on treatment outcome in major depression (Mulder, 2002) that reported that high neuroticism scores, which are characteristic of BPD (Widiger, 2005), were associated with poor prognosis, particularly in long-term outcomes. Furthermore, in a 6-year prospective study, Grilo et al. (2010) found that high neuroticism scores also predicted earlier relapse and shorter remission of major depression, suggesting that that depressed patients with cooccurring BPD respond more poorly to pharmacological treatment. In addition, in a prospective epidemiological study using a large nationally-representative sample from the National Epidemiologic Survey on Alcoholism and Related Conditions, Skodol et al. (2011) observed that BPD accounted for over half of cases of persistent depression at the 3-year follow-up. BPD robustly predicted persistence of MDD, even after controlling for age at onset of depression, the number of previous episodes, duration of the current episode, family history, treatment, and other personality disorders. Finally, in another longitudinal study from CLPS that followed a sample of BPD patients with and without cooccurring MDD for 3 years, Gunderson et al. (2004) found that the rate of remission of BPD was not affected by the presence of comorbid MDD; however, the remission rate of MDD was significantly reduced by the presence of coexisting BPD. When considered together, the above evidence identifies BPD as a key risk factor for TRD.

It should be noted here that the prognostic significance of comorbid BPD extends to somatic treatments for major depression, although the available evidence is more limited. A narrative review by DeBattista and Mueller (2001) suggested that depressed patients receiving ECT had poorer outcomes on some measures in the presence of coexisting BPD. This finding was supported by more recent original studies (Feske, 2004; Ward et al., 2020). For example, in a prospective longitudinal study, Feske et al. (2004) compared the outcome of ECT in three different groups of depressed patients: (a) those with BPD, (b) those with PDs other than BPD, and (c) those without a PD and found that the BPD group showed the least amount of symptomatic improvement following treatment completion. In addition, in a recent retrospective chart review of patients with MDD receiving ECT and transcranial magnetic stimulation (TMS), Ward et al. (2020) found that the presence of coexisting borderline personality traits modestly predicted poorer response to ECT. Interestingly, the authors also found that borderline personality traits had no impact on the outcome of TMS. The available data does indicate that MDD could be successfully treated with ECT in patients with cooccurring BPD, leading some researchers to speculate that "the depressed patient appears to have two distinct disorders, one with is responsive to ECT and the other which is not" (DeBattista and Mueller, 2001, p. 91).

Treatment

In light of the negative prognostic impact of BPD on cooccurring major depression, a proper initial treatment of BPD can improve the chances of achieving remission in MDD. In fact, Gunderson et al. (2004) found that improvements in BPD were followed by improvements in MDD, but not vice versa. This section will provide an overview of treatments for BPD and review the available limited data on treatments for the combination of MDD and BPD.

Pharmacotherapy for BPD

There are no medications approved by the US Food and Drug Administration (FDA) for the treatment of BPD to date. Nevertheless, pharmacotherapy for BPD is common in clinical practice, with approximately 80% of patients with this personality disorder regularly taking medications and over 40% taking three or more medications daily (Gunderson et al., 2008; Zanarini et al., 2004a). A 2010 Cochrane systematic review and metaanalysis by Lieb et al. (2010) analyzed 27 RCTs, including drug vs placebo, drug vs drug, and single drug vs combined treatment that tested first- and second-generation antipsychotics, mood stabilizers, antidepressants, and omega-3 fatty acids in adult patients with BPD. The review found the mood stabilizers topiramate, lamotrigine, and valproate semisodium, and the second-generation antipsychotics aripiprazole and olanzapine to have most beneficial effects in treating core facets of the psychopathology of BPD. However, the available evidence did not support effectiveness of pharmacotherapy in reducing the overall severity of BPD. In addition, the authors noted that the robustness of the findings was limited, due to the low number of studies per drug tested and small samples in included trials. Interestingly, the effectiveness of SSRIs, endorsed by the American Psychiatric Association's practice guidelines (American Psychiatric Association Practice Guidelines, 2001) as a first-line treatment for BPD, was not supported by the corresponding evidence. In another systematic literature review, Stoffers and Lieb (2014) came to similar conclusions. The authors found some evidence for beneficial effects of second-generation antipsychotics, mood stabilizers, and omega-3 fatty acids, but not SSRIs, while noting that the overall evidence base was "unsatisfying" (Stoffers and Lieb, 2014).

Ingenhoven et al. (2010) took a slightly different approach in their metaanalysis of 21 placebo-controlled RCTs of psychotropic medications in "severe" personality disorders (i.e., BPD and/or schizotypal personality disorder) by focusing on the drugs' effect on three specific symptom domains: cognitive-perceptual symptoms, impulsive-behavioral dyscontrol, and affective dysregulation. The third domain, in turn, was comprised of four subdomains: depressed mood, anxiety, anger, and mood lability. Study drugs were grouped into three categories: antipsychotics (which included both atypical and typical antipsychotic medications), antidepressants, and mood stabilizers. The data synthesis revealed that antipsychotics had a moderate effect on cognitive-perceptual disturbances and a moderate-to-large effect on anger; antidepressants were associated with small effects for anxiety and anger, but were ineffective for depressed mood or impulsive-behavioral dyscontrol; mood stabilizers, on the other hand, were associated with very large effects for impulsive-behavioral dyscontrol, as well as anger, and had a large effect on anxiety and a moderate effect on depressed mood (Ingenhoven et al., 2010). The metaanalysis also found that mood stabilizers were superior to antipsychotics and antidepressants in improving global functioning. In contrast, a recent update of published, unpublished, and ongoing studies by Stoffers-Winterling et al. (2020) suggested that there was no evidence of beneficial effects of mood stabilizer lamotrigine in routine care for BPD. In addition, the newest studies did not support the use of fluoxetine for treating self-harm and suicidal behaviors associated with BPD (Stoffers-Winterling et al., 2020).

Stoffers-Winterling et al. (2020) concluded that, despite its prevalence in everyday clinical practice, the use of psychoactive drugs for the treatment BPD was not supported by current evidence. It should be noted here that this conclusion is reflected in the U.K. National Institute for Health and Clinical Excellence (NICE) guideline on treatment and management of BPD, which recommends that pharmacotherapy should not be used specifically for BPD. Furthermore, based on the available evidence, Francois et al. (2015) suggested that medication should only be used as an adjunctive treatment aimed at crisis stabilization and maximizing the patient's use of psychotherapy.

Psychotherapy for BPD

Psychotherapy is widely considered to be the primary treatment for BPD. In the last two decades, there has been a proliferation of studies evaluating the effectiveness of psychotherapeutic approaches developed specifically for this personality disorder, as well as adaptations of existing treatments. In fact, a 2020 Cochrane systematic review (Storebø et al., 2020) included 75 RCTs with over 4500 participants, reflecting an exponential increase in the amount of available evidence since the publication of the previous Cochrane review in 2012 (Stoffers et al., 2012), which included 28 trials involving a total of 1804 subjects. In the updated review, Storebø et al. (2020) covered over 16 different kinds of psychotherapy of one to 36 months in duration conducted mostly in outpatient settings with patients who were mostly female and whose mean age ranged from 15 to 46 years; the comparator interventions included treatment-as-usual (TAU), waitlist, and other active treatments. The authors reported that, compared to TAU, psychotherapy resulted in a clinically relevant reduction in BPD symptom severity (moderate-quality evidence) and suicidality (low-quality evidence). In addition, they found that it could potentially reduce self-harm (low-quality evidence) and depression scores (very low-quality evidence), while improving psychosocial functioning (low-quality evidence). When compared to waiting list or no treatment, psychotherapy was more effective at improving BPD symptom severity and depression. While not all included studies addressed side effects of treatments, those that did found no adverse reactions to psychotherapy. The most studied treatment approaches among those included in the review were dialectical behavior therapy (DBT) and mentalization-based treatment (MBT).

Another influential review deserves a mention here. Cristea et al. (2017) conducted a systematic literature review and metaanalysis of 33 clinical trials with a total 2256 participants randomized to psychotherapy alone or to a control intervention. The authors found that DBT and psychodynamic treatments, such as MBT and transference-focused psychotherapy

(TFP), significantly improved BPD-relevant outcomes, such as severity of symptoms, self-harm, and suicidal behavior, compared to control interventions. There were no differences between DBT and psychodynamic approaches, as both types of treatment produced significant albeit small effect sizes. Interestingly, despite its prominence in research and clinical practice, cognitive-behavioral therapy (CBT) for BPD did not differentiate from control interventions.

TFP and MBT are described in some detail in the section on Narcissistic Personality Disorder and depression below. DBT is by far the most studied of the effective psychotherapies for BPD, as well as most widely used, particularly in the United States (Gunderson et al., 2008; Zanarini et al., 2004a). It was originally developed by Marsha Linehan as a form of cognitive-behavioral therapy for chronic suicidal and parasuicidal behavior (Linehan et al., 1991) and eventually developed into a comprehensive, manualized treatment for BPD. Linehan theorized that the self-destructive behavior characteristic of BPD stemmed from a lack of important interpersonal, self-regulation, and distress tolerance skills. Within the DBT theoretical framework, the treatment is organized around a hierarchy of behavioral targets: (1) minimize suicidal behaviors and parasuicidal gestures (e.g., suicide attempts, self-injury), (2) decrease behaviors undermining treatment (e.g., no-shows, excessive between-session phone calls), (3) minimize quality of life-interfering behaviors (e.g., alcohol misuse, binge-eating), and (4) increase self-regulation skills (e.g., distress tolerance, mindfulness) (Panos et al., 2014). These therapeutic goals are achieved through a multipronged treatment approach encompassing skills groups, individual therapy, between-session phone coaching, and care coordination within the patient's treatment team. In summary, DBT is an intensive manualized treatment that requires specialized training and a significant investment of time and effort by the clinician.

Treatment for comorbid depression and BPD

There are no published treatment trials for comorbid TRD and BPD and only a handful of studies evaluating effectiveness of treatments for cooccurring MDD and BPD. Two separate trials of SSRIs demonstrated some effectiveness. In particular, Fava et al. (1994) found that an 8-week trial of fluoxetine 20 mg/day led to an improvement of both depressive and personality pathology symptoms, as measured by the 17-item Hamilton Rating Scale for Depression (HAM-D-17), Personality Diagnostic Questionnaire-Revised (PDQ-R), and the frequency of the diagnosis of BPD before and after treatment. In addition, a double-blind trial of sertraline 50–150 mg/day or citalopram 20–60 mg/day (Ekselius and von Knorring, 1998) resulted in significant reductions in the frequency of BPD diagnosis and depression scores. And in a small 6-week trial of tricyclic antidepressants (i.e., clomipramine or desipramine) (Sullivan et al., 1994), an MDD + BPD subsample showed similar significant improvement of depressive symptoms to patients with MDD with another personality disorder and MDD without a personality disorder diagnosis. More clinical trials of pharmacotherapy for the comorbid MDD + BPD and TRD + BPD presentations are desperately needed.

Clinical illustration[a]

Sharon is a 29-year-old recently divorced heterosexual Caucasian female who was referred to a TRD specialty clinic by her outpatient treatment team for a one-time consultation due to chronic depression with suicidal ideation, extreme anxiety, relationship problems, and limited response to treatment. She has been psychiatrically hospitalized five times in the past 3 years as a result of suicide attempts. She acknowledged having attempted suicide at least 10 other times. All of the suicide attempts involved intentional overdosing on prescription medications. She also acknowledged some past nonsuicidal self-injurious behavior. She indicated that common triggers for these behaviors were relationship difficulties, including marital separation and divorce.

Sharon has been under the care of a psychiatrist and a Masters-level therapist for the last 3 years, with a working diagnosis of Major Depressive Disorder, Recurrent, Severe and Panic Disorder. Within this time period, she completed several unsuccessful antidepressant medication trials of adequate duration and intensity. There are some indications that she did not always remember to take her medications regularly and at times refused to take medication stating that they did not work. She is currently prescribed bupropion extended-release 300 mg qam and fluvoxamine 200 mg qam. She also completed a series of 12 acute and 6 maintenance ECT treatments in the past year. She indicates that these treatments were ineffective in addressing her depression and believes that they have had a lasting negative effect on her memory. Sharon has also been in weekly psychotherapy for the last 3 years. The progress notes indicate the use of cognitive-behavioral, interpersonal, and acceptance and commitment therapy modalities. In regards to her psychotherapy, Sharon states that she is "sick of talking about herself." Despite the ongoing treatment, she reports experiencing unremitting severe depressive symptoms that significantly interfere with her daily functioning.

a. A composite case.

Sharon's parents divorced when she was 10 years old. She has an older brother to whom she never felt close and from whom she is currently estranged. She was very close with her mother who died of ALS 3 years ago. Outside of her parents' divorce, she describes growing up in a "typical American household" and denies experiencing significant trauma, abuse, or neglect. Socially, she reports that she always had friends growing up, but also that she could be "hot or cold" with friends. She lost a group of close friends following a break-up in a romantic relationship in high school, which precipitated her first depressive episode. Sharon acknowledges that she currently has very few friends and is largely isolated socially. Since separating from her husband, she has had a string of unsuccessful romantic relationships and has been uncharacteristically promiscuous. She feels that she needs "constant attention" from men. Currently, her social network is limited to her father and two friends, both of whom live out of state. She lives in a rented townhouse and sublets one room to a graduate student. She shares custody of her 2-year-old daughter with her ex-husband, from whom she has been divorced for 1 year. Sharon works part-time as a cashier at a local hardware store, and, despite having a college degree, believes that she will never make more than the minimum wage. She reports having no hobbies or interests and states that she "doesn't do anything." When she is not taking care of her daughter, she spends much of her free time sleeping and browsing social media on her phone. She used to enjoy literature, art, and music, but says that she "cannot remember the last time she read a book."

Sharon's comprehensive consultation at the specialty clinic included a thorough review of history, an interview with a clinical psychologist, who also administered a battery of psychological tests geared for TRD, and a reevaluation by a psychiatrist. The results of the diagnostic interviews and psychological testing indicated the presence of major depression, anxiety, and personality pathology marked by borderline and depressive traits. She appears to experience intense irritable and depressed affects with few outlets for them. Her daily experience is marked by guilt and shame. Her relational style is marked by an approach-avoidance conflict and she vacillates between an intense desire for closeness and withdrawal. She exhibits extreme fear of rejection and reacts strongly to real or perceived abandonment. She often finds herself acting in ways that ultimately bring about the very thing she fears. She has a generally pessimistic, negative outlook on life. Cognitively, she is prone to all-or-nothing thinking, which appears to drive some of her distress. She has strong negative affectivity punctuated by lability, disinhibited behaviors, and significant instability in her sense of self and interpersonal functioning. These traits have become more pronounced in recent years, although they have been present throughout her life to a greater or lesser extent. This constellation of personality and emotional features is best captured by the diagnosis of Borderline Personality Disorder.

Narcissistic personality disorder and depression

Clinical presentation

Narcissistic Personality Disorder (NPD) is characterized by an exaggerated sense of self-importance and expectations that others appreciate and recognize one's uniqueness and specialness (American Psychiatric Association, 2013). Individuals with NPD often present as entitled and struggle to see the perspectives of others, often using others mainly to meet their own needs without appreciating the needs of others. On the surface, NPD individuals may not seem likely to present for treatment for depression; however, they actually do make their way to psychotherapy and/or pharmacotherapy quite often, due to the high level of fluctuation in their self-esteem (Dawood and Pincus, 2018; Ronningstam, 2005). Narcissistic patients often find themselves negatively impacted by the criticism or disappointment of others. And while they are likely to externalize these problems to others, NPD patients may become depressed and suicidal (Fjermestad-Noll et al., 2019, 2020; Kernberg, 2004), with NPD being uniquely associated with an increasing risk of suicide attempts over a 10-year period (Ansell et al., 2015) and an increased risk for suicide completion (Giner et al., 2013). Depression in NPD patients often presents with fluctuations in the depressive state, ruminations about the hostility of others, fluctuation in suicidal ideation triggered by external events, which is accompanied by parasuicidal behavior, and a lack of neurovegetative symptoms (Fjermestad-Noll et al., 2019). Their chronic depressive episodes subsequently make them a particularly challenging group of patients to treat.

Individuals with NPD are prone to prolonged depressive episodes due to repeated work and relational failures, perceived social humiliation, and defeat. Chronic experiences like these predispose the patient toward feelings of shame and aggression, with acting out one's aggressive feelings on others as a consequence. For instance, a patient might be passed up for a job promotion, invoking feelings of depression, humiliation, and shame. However, in order to avoid feeling this level of rejection, the patient verbally lashes out at colleagues, disparaging the individual who was selected for the promotion and invoking concerns about the unfairness and poor judgment of those making the decisions. While such reactions might be short-lived, the NPD patient might continue to act out in some aggressive ways but may also become self-destructive, for example, by drinking too much alcohol, or even driving an automobile while intoxicated. In fact, high rates of substance abuse disorders are commonly reported in NPD patients (Simonsen and Simonsen, 2011). Given their proclivity to deny their vulnerability (Kernberg, 1984), NPD patients may appear angry or frustrated, yet their underlying depression can often be detected within the first few sessions of psychotherapy.

Prevalence of comorbidity of NPD and depression

Clemence et al. (2009) found that in a sample of patients diagnosed with narcissistic PD, 47% met criteria for Major Depressive Disorder, and 42% for Dysthymia. Similar comorbidity rates were reported by Fava et al. (1996) and Tedlow et al. (1999). In addition, in a study of Narcissistic PD subtypes, Russ et al. (2008) found that the vulnerable subtype showed particular proneness to depression. However, more recently, Ronningstam (2005) suggested that the rates of Dysthymia were much higher in NPD than the rates of MDD. There is a dearth of research on the effects of Narcissistic PD on the treatment of depression. However, Orth et al. (2016) reviewed six longitudinal studies with a combined total of 2717 subjects and found that individuals with high self-esteem have a lower risk for developing depression, regardless of whether or not they were narcissistic.

Treatment

It is not uncommon for NPD patients who present with depression to be treated with an SSRI, as it is not uncommon for most patients with personality disorders to be provided with a medication upon consultation with a primary care physician or psychiatrist. In fact, it has been observed that both pharmacotherapy and psychotherapy may be utilized to treat depressed patients with personality pathology (Kernberg, 2004; Koeningsberg et al., 2000). However, recent reviews of literature have failed to provide any empirical support for the efficacy of pharmacotherapy for NPD (Silk, 2010). There are likely many reasons for these failures, such as the difficulty in detecting and subsequently treating NPD, the poor adherence that many NPD patients have in pharmacotherapy or psychotherapy, and the possible premature discontinuation of medication in patients with NPD who often resist acknowledging their need for medication. Furthermore, many NPD patients are unlikely to seek psychiatric treatment for depression. Consequently, pharmacotherapy for NPD currently lacks empirical support.

A frequently discussed treatment for NPD in the literature is a psychoanalytically-informed treatment known as transference-focused psychotherapy (TFP; Yeomans et al., 2015). First applied to the treatment of Borderline Personality Disorder (BPD), TFP is designed to elicit the maladaptive representations of self and other that exist in the therapeutic relationship, and elucidate the process by which these representations lead to affective and behavioral dysregulation (Diamond and Hersh, 2020). Past research on TFP with BPD patients has demonstrated that these patients recover from their chronic depressivity (Clarkin et al., 2007) and develop more adaptive and realistic representations of self and other. Diamond and Hersh (2020) report that TFP is associated with a lower dropout rate in NPD patients than other types of psychotherapy, and that its efficacy may be attributed to challenging and changing the grandiose and vulnerable aspects of the self-representations. For example, NPD patients may feel the therapist is being overly critical of judgmental toward them and begin to feel depressed or ashamed. In TFP, the therapist identifies these depressive and shameful feelings and thus interprets how and why they developed and how they are being applied inappropriately to the current situation between the patient and therapist.

Another treatment recently described in the literature is mentalization-based treatment (Drozek and Unruh, 2020). In this theoretical framework, NPD is understood to develop from noncontingent mirroring of the child by the parents, such that the mirroring reflects the parents' experiences of the child and not their experiences directly. For instance, the parent might be overly effusive about the child's accomplishments while not recognizing or attending to his or her vulnerable affective states, such as depression. Consequently, the child (and adult) adapts a more grandiose representation known as the *narcissistic alien self*, which overvalues the parent's effusive praise but pays little to no attention to the more vulnerable states. This leads the child to a disconnection from his or her primary affective experience. MBT seeks to bring these experiences into conscious awareness, so that the NPD patient can learn to evaluate his or her experiences more accurately and within the context in which they occur. Thus, depressive or shameful feelings can be experienced more immediately and with appropriate context to make sense of why they are there and how they might be managed. While there is no empirical evidence yet for MBT applied to NPD patients, treatment successes have been reported in case studies and clinical experience (Drozek and Unruh, 2020).

Many authors have articulated that NPD is marked by both feelings of grandiosity and vulnerability, with some suggesting that there are two distinct phenotypes of NPD (Pincus and Lukowitsky, 2010). Those who are more grandiose are typically socially charming and outgoing, desire admiration, and have fantasies about success, yet are often seen as self-absorbed, struggle to recognize others' feelings and motives, and tend to have shallow relationships. Those who are more vulnerable tend to struggle with self-doubt and feelings of inferiority, have difficulty depending on and trusting others, often feel depressed and easily hurt, and regularly experience shame and embarrassment. While these phenotypes, or subtypes, have been described in the literature, there is a growing consensus that patients with NPD tend to have both grandiose and vulnerable aspects to their personalities, with one being more dominant than the other. Indeed, what has been described above are the reactions of NPD patients who experience both grandiose and vulnerable self-representations. However, a sizable body of literature describes the very different expressions of these subtypes based upon the type of traits they exhibit (Miller and Maples, 2011), response to treatment (Ellison et al., 2013), and response to ego threat (Besser and Zeigler-Hill, 2010). To account for these differences, Pincus and colleagues (e.g., Pincus et al., 2013; Pincus and Roche, 2011) recommend the use of the term *pathological narcissism* to encompass the wide range of emotional, social, and behavior expressions that are part and parcel of NPD. While this concept has been a useful addition to the literature, this conceptualization of narcissism has some challenges, particularly as it is related to the chronic nature of depressivity that is experienced by patients with NPD.

Clinical illustration

Ethan is a 26-year-old Caucasian male who is seeking treatment over difficulty with a relationship break-up 3 years ago. He reported prior treatment with both an SSRI medication and psychotherapy but discontinued the medication after 4 months when he began using marijuana periodically. Although he had previously been in psychotherapy for about 2 months, he discontinued after the therapist reportedly "shamed" him for how he treated his previous girlfriend. Ethan also devalued the therapist because she was about his age and, in his perception, did not have enough experience to help him. He sought out the current therapist for his expertise and maturity.

Ethan indicated that he has been depressed "on and off for years," thinking he would never get better. He reported some suicidal ideation, but no plan or intent. He believed his former girlfriend was "perfect" for him, and, when she broke up with him (although later he indicated that the break-up was mutual), he became very depressed. Since then, he dated intermittently, getting sexually involved with a woman early in their relationship, only later to decide she was not good enough for him or did not measure up to his previous girlfriend. The therapist began to identify the patient's relational pattern as that marked by an idealized sense of self and devaluation of women, which helped relieve any depressive affect he might have about the relationship ending badly. When the therapist would offer an observation on this pattern, Ethan would say, "that is really insightful" or "that's a really good comment." Yet, as the therapy relationship unfolded, Ethan wondered if there might be anything he would learn about psychotherapy that might allow him to undermine the work he was doing with the therapist.

Though the therapy utilized a transference-focused approach (discussed earlier) toward integrating the disparate and split-off aspects of the self and others, it also addressed depressive episodes as they arose. For instance, one time, after a brief romantic relationship did not end well, the patient smoked marijuana and contacted his former, idealized girlfriend. This led him to become very depressed, feeling suicidal, and then smoking too much marijuana and drinking too much alcohol. He missed work, and the therapist explored the extent to which his actions could be detrimental to both his physical health and work vitality. Ethan avoided discussion of his depressive affect; yet, the therapist persisted and addressed his resistance. This led the patient to eventually talk about his hopelessness in ever being able to get over the breakup, which evoked sadness and depression in session. Subsequent sessions found Ethan returning to consider these feelings, though often another problem would arise centering around a relational template of grandiosity and vulnerability, which would be explored and interpreted. The identification and explanation of these templates offered a new perspective to the patient about how his mood state was linked to his ideas about himself and others. He began to find himself feeling more stable and resilient to events that would otherwise be distressing. Over time, he moved more toward a stance of greater curiosity about himself and began feeling less depressed.

Alternative, personality-derived models of TRD

The DSM-5 alternative model of personality disorder diagnosis and assessment (AMPD)

As noted earlier, the AMPD was included in the DSM-5 for ongoing empirical and clinical evaluation. This model makes explicit that *depressivity* and *anhedonia* are personality trait facets and within the Negative Affectivity and Detachment trait domains. Individuals with high levels of depressivity are described as regularly feeling down, miserable, and/or hopeless. They have difficulty recovering from these moods and are often pessimistic about the future. They also have frequent feelings of shame and/or guilt, low self-worth, and thoughts of suicide or suicidal behavior. Individuals with high levels of anhedonia lack enjoyment in life's experiences and avoid engaging in them. Thus, the AMPD explicitly emphasizes that these problems in mood are actually problems in personality. These traits are described as being relatively stable across time, having their onset in adolescence or early adulthood, and persisting across different situations. Furthermore, the AMPD discusses problems with levels of personality functioning. Patients with TRD would be anticipated to have problems in their personality functioning. Specifically, their sense of self is likely to be negatively affected by low self-esteem or self-esteem instability, as well as difficulties with engaging in personally meaningful goals. Their interpersonal functioning also is likely to be negatively affected, as they struggle to relate to the experiences of others when their own experience is so pervasively negative, and their ability to develop deep relationships might be severely hampered by their limited emotional "bandwidth" and psychological "unattractiveness" to others. An expanded discussion of TRD features through the lens of the AMPD could be found in Huprich (2013).

It is critical to note that the concept of personality is central to the understanding of TRD in modern psychiatry. This idea is not new, having been discussed for decades and reviewed by Huprich (1998). Unfortunately, personality has received little attention in treatment outcome studies (e.g., STAR*D) seeking to understand why some depressed patients do not recover from brief psychotherapies and pharmacotherapy. As discussed later, insights from personality theory might prove to be very useful in the conceptualization and treatment of some patients with TRD.

Depressive personality disorder

In DSM-IV (American Psychiatric Association, 1994), the diagnosis of Depressive Personality Disorder (DPD) was introduced. Drawing upon a long history in the clinical literature (see Huprich, 1998), DPD was meant to fill the gap in the literature by describing those patients whose chronic depressive affect was embedded more within their personality structure and less in symptoms. Features of DPD included depressed affect, pessimism, self-criticism, negative feelings toward others, proneness to feeling guilty, and feelings of inadequacy. In fact, DPD was differentiated from DSM-IV Dysthymic Disorder, whose symptoms were associated with somatic abnormalities, such as low energy and motivation, difficulty concentrating, and disruptions in eating or sleeping patterns. DPD was proposed for inclusion in future editions of the DSM given the wealth of clinical experience that documents that certain patients' distress seems rooted in their personality organization and developmental history, making them poor candidates for pharmacotherapy and potentially good candidates for psychotherapy.

Research on DPD led to many interesting findings. First, studies indicated that the overlap between DPD and Dysthymia in clinical and nonclinical samples ranged between 18% and 95% (reviewed in Huprich, 2001), with an average rate of comorbidity being approximately 50%. Second, despite efforts to develop psychometrically sound measures for assessing and differentiating DPD and Dysthymia, most instruments failed to effectively differentiate patients with these conditions, thus limiting their clinical utility. Third, no randomized control trial of treatment for DPD was ever conducted, though a few studies evaluated how DPD traits affected outcome of treatment of Major Depressive Disorder. One study found DPD to interfere with successful treatment (Shahar et al., 2003), while others found no adverse effects of DPD on treatment outcome (Ryder et al., 2010; Saulsman et al., 2006). However, in one study, DPD traits had a negative effect on outcome of interpersonal psychotherapy of 16–20 weeks' duration (Ryder et al., 2010). In that same study, antidepressant medication from multiple pharmacological classes was found to be effective in treating MDD without a negative effect of DPD traits.

Other studies have documented that DPD patients are commonly seen in outpatient clinical settings (Ganellen, 2007; Huprich et al., 2011), are more prone to major depressive and dysthymic disorders and suicide attempts (e.g., Klein and Shih, 1998; McDermut et al., 2003), and spend more time in psychotherapy than those with other mood disorders, often with less improvement (Markowitz et al., 2007). In short, DPD patients may be the prototype of what it means to have treatment-resistant depression. However, studies on treatment-resistant depression often rule out personality disorders and have not included assessment for DPD. In fact, it could be because of the underlying personality structure that a number of treatments for MDD fail, resulting in TRD.

Unfortunately, the Personality and Personality Disorders Workgroup of DSM-5 failed to recommend DPD as a diagnostic category in DSM-5. It might have been that the issues of comorbidity with other mood disorders led to its exclusion from the manual (Ryder et al., 2006); however, no formal justification was provided for this decision (Huprich, 2012). In fact, the workgroup appeared to violate its own rules for the inclusion of DPD as reliable and valid diagnostic category (Huprich, 2012). Nonetheless, ideas were proposed for what might become of DPD (Ganellen, 2007; Huprich et al., 2011), including the need to consider underlying psychological and personality processes when reformulating the construct, which has demonstrated decades of interest clinically. Out of this concern, a new construct was derived.

Malignant self-regard

In a comprehensive review of the DPD literature, Huprich (2014) identified a number of personality patterns that were common in several personality disorders and related conditions, which included not only DPD, but also masochistic or self-defeating personalities, and vulnerable expressions of narcissistic personality. Among the shared qualities are depression proneness; feelings of guilt, shame, and inadequacy; self-criticism; hypersensitive self-focus; pessimism; perfectionism in the context of "grandiose" or idealized fantasies; desire for approval from others; masochism; and poor anger management. Huprich (2014) described this constellation of features as *malignant self-regard* (MSR), which identifies a condition of the underlying self-structure. Based in personality theory, MSR captures many features that are commonly shared among patients who have, among other things, persistent depression that is often resistant to treatment. Several studies have validated the MSR construct in university students, nonclinical adult population, psychiatric outpatients, and medical outpatients in different parts of the world. Similarly, factor analytic studies have confirmed that the MSR construct is composed of a single factor of features of DPD, Self-Defeating PD, vulnerable narcissism, and depressive symptoms (Huprich, 2018, 2019; Huprich et al., 2016, 2018; Huprich and Nelson, 2014; Lengu et al., 2015; Pedone et al., 2018). Even when controlling for depressive symptoms, the same personality variables continue to load on a single factor, thus supporting the coherence of these personality patterns independent of depressive symptoms. Also of note, MSR and vulnerable narcissism have been further considered in light of their connection with chronic depression. Guidelines for the differentiation of the two constructs, particularly with regard to the manifestation of and clinical attention needed for chronic depression, are provided in the Appendix for the interested reader.

Clinical illustration

A detailed discussion of someone with high levels of MSR (and also meeting criteria for DSM-IV Depressive PD) has been reported by Huprich (2019). This patient, Mark, possessed all the features of MSR and came into treatment for chronic depression which seemed not to get better with time. He had been in brief, cognitive-behavioral psychotherapy while studying at the university, which appeared to have minimal effectiveness. He decided to enter psychotherapy again at age 27, as he struggled with depression and suicidal ideation, as well as having ambivalent feelings about his girlfriend and his work. Though he had not tried pharmacotherapy, he was resistant to considering it, as he feared potential unintended effects of medication.

Treatment began from a psychodynamic perspective, which initially yielded some positive effects when Mark began to understand how limited his agency was, and that there were some choices he could make that would remove him from unpleasant situations. However, as the treatment continued, his masochistic attitudes increased, and he regularly believed that his problems were entirely of his making. His suicidal ideation increased, at which time he agreed to consult a psychiatrist for medication. Mark tried three SSRI medications, all of which failed. The first medication produced intolerable side effects and had to be discontinued, while others had trials of adequate duration and intensity. Even with increases in dosage, the medications did not result in a clinically significant response. Eventually, his psychiatrist recommended ketamine treatment. He opted to try it, but discontinued after a few sessions when he had a very strong experience of dissociation and derealization which was terrifying to him.

As psychotherapy unfolded, Mark changed jobs after putting up with some very damaging conditions. However, a very clear pattern of self and interpersonal relatedness was observed, in which he saw himself as inadequate to manage the stressors that came his way. He struggled to be perfect or to function at such a high level that he could feel more powerful and agentic, yet could not internalize the adversity of the conditions in which he found himself. This pattern had clear associations to many events in his childhood and adolescence, in which he tolerated loss after loss and several family moves with little attention to or awareness of how painful such transitions were for him. Even though he believed that he would find happiness after graduating from the university, his low self-esteem and perceived problems in self-efficacy continued to dominate his thinking.

Mark seemed to find greater resiliency and an investment in his own needs, however, as the relationship with his girlfriend was coming to an end. In this case, he became angry over her unfounded criticism of his behavior. From the patient's description of the relationship, it became clear for the first time that she was regularly devaluing him and putting him down, while he masochistically tolerated her demands and expectations all the while being criticized for being so depressed. In this case, Mark fought for himself, his need for fair treatment and attention, and took a more proactive stance in pointing out the ways his girlfriend was being emotionally insensitive. This led to a reduction in his depressive symptoms and a more engaged attitude toward himself and his future.

Clinical summary

Comorbid personality disorders in depressed patients present a real challenge for clinicians. Cooccurring personality pathology negatively impacts the course and treatment outcome in major depression and is associated with poorer overall functioning. Many patients with these coexisting conditions are thus identified as having treatment-resistant depression. Chronic and severe depressive symptoms can sometimes overshadow underlying personality pathology in the assessment process and dominate treatment focus. It is possible that failure to diagnose and treat the underlying personality disorder represents one reason why otherwise effective antidepressant treatments fail. Consequently, clinicians should carefully assess for personality pathology in all TRD patients and implement a comprehensive treatment plan addressing each of the comorbid conditions. Research evidence indicates that treating the coexisting personality disorder can lead to an improvement in both personality and depressive symptoms. Alternative, personality-derived models of TRD show much promise in being able to successfully integrate personality and depressive symptoms into a single construct. Utilization of these models, in some cases, may offer a more robust conceptualization of TRD, thus allowing for the development of treatments strategies that more adequately address the full range of problems associated with this presentation.

Appendix

As noted earlier, narcissistic personalities are marked by both grandiosity and vulnerability in the same individual. It has been speculated that a purely vulnerable subtype of narcissism exists, separate from the grandiose subtype (see Pincus and Lukowitsky, 2010). However, Huprich et al. (2018) evaluated individuals for pathological narcissism (both the grandiose and vulnerable manifestations) and MSR. They found that MSR and vulnerable narcissism were both associated with higher levels of neuroticism, lower levels of extraversion, and lower levels of self-esteem.[b] High levels of both MSR and vulnerable narcissism also predicted trait levels of anger and anxiety, as well as levels of state anger that is held internally (i.e., not expressed outwardly), when individuals received negative feedback on their performance. This led Huprich et al. (2018) to consider whether MSR and vulnerable narcissism were the same construct.

However, it was argued by Huprich (2019, 2020) and Huprich et al. (2018) that while these two constructs might be the same, it would be an error to suggest that these two constructs fall within the umbrella of pathological narcissism, as described by Pincus and colleagues (Pincus and Lukowitsky, 2010; Pincus et al., 2013). Instead, Huprich et al. (2018) noted that those who are labeled as vulnerably narcissistic are at risk of having their chronic depressivity overlooked for its debilitating effects on their quality of life and potential for suicide. He also suggested that describing those with MSR (or depressive personality as a historical precursor) as vulnerably narcissistic inappropriately labels these patients' perfectionistic strivings as "grandiose." Such labels could hurt patients whose perfectionistic (or "grandiose") aspirations are driven by a strong unconscious need to be cared for and accepted, which is different than the motivations of grandiosity in a narcissistic patient, who is motivated to maintain an insecure self-esteem. Furthermore, the depressive patient has limited or minimal conscious awareness for recognition or praise and often balks at the idea that the recognition is something wanted or sought after (Huprich, 2014, 2019). Given such strong empirical overlap between the constructs, but yet important clinical differences in their experiences, Huprich (2019, 2020) suggested that vulnerable narcissism should not be used to describe a type of narcissistic personality, since narcissistic patients have both grandiosity and vulnerability. In other words, the subtype idea should be abandoned. He went on to argue that the concept of MSR should be used to describe what has been called the "vulnerable narcissist," which is most like what Kernberg (1984) has described as the depressive-masochistic personality.

In order to differentiate the depressive experiences of those with high levels of MSR from those with narcissistic personality pathology, Huprich (2020) offers three guiding principles for assessment and treatment. First, the narcissistic patient will always have manifestations of unmet grandiosity needs that present themselves in some way, even if presentation is one that is depressive. Clinicians might observe the depressive symptoms being reported in light of the patient's sense of entitlement or unmet recognition as being unique or distinct. In contrast, MSR/depressive personality patients do not view others' failings as being the source of their depressive symptoms; rather, it is the person themselves who believes they are inadequate or not good enough.

Second, the depressive personality/MSR patient has high standards and expectations for themselves, and others are often seen as functioning even better. Unlike the narcissistic patient, however, the depressive/MSR patient has a greater capacity for moral functioning, such that they treat others with more kindness than themselves. Even when slighted, the depressive patient would be likely to keep their anger inside and not express it, unlike the narcissistic patient who would be more verbally expressive with their anger.

Third, while narcissistic and depressive/MSR patients have an ego ideal that is perfectionistic, the narcissistic patient has some degree of confidence in their grandiose or perfectionistic strivings, while the depressive patient sees this quality as

b. An influential review by Mulder (2002) found that high levels of neuroticism in a comorbid personality pathology predicted poor prognosis in MDD, particularly when long-term outcome was taken into account.

never fulfilled, despite having achieved a number of successes. Stated differently, both patients have troubles with their aspirations—the narcissistic patient overvalues or overstates their accomplishments, while the depressive patient undervalues them. Successful treatment of the depressive/MSR patient will require the identification and internalization of the desire for a receptive or caring other, who does not require perfection, while successful treatment of the narcissistic patient requires the internalization of the loss of what they cannot be or the limits that they have reached.

References

Akiskal, H.S., 2002. The bipolar spectrum—the shaping of a new paradigm in psychiatry. Curr. Psychiatry Rep. 4 (1), 1–3. https://doi.org/10.1007/s11920-002-0001-1.

Akiskal, H.S., Chen, S.E., Davis, G.C., Puzantian, V.R., Kashgarian, M., Bolinger, J.M., 1985. Borderline: an adjective in search of a noun. J. Clin. Psychiatry 46 (2), 41–48.

Al-Dajani, N., Gralnick, T.M., Bagby, R.M., 2016. A psychometric review of the personality inventory for DSM-5 (PID-5): current status and future directions. J. Pers. Assess. 98 (1), 62–81. https://doi.org/10.1080/00223891.2015.1107572.

American Psychiatric Association, 1994. Diagnostic and Statistical Manual of Mental Disorders, fourth ed. American Psychiatric Publishing.

American Psychiatric Association, 2001. Diagnostic and Statistical Manual of Mental Disorders (4th ed., Text Revision). American Psychiatric Publishing.

American Psychiatric Association, 2013. Diagnostic and Statistical Manual of Mental Disorders (DSM-5), fifth ed. American Psychiatric Publishing, https://doi.org/10.1176/appi.books.9780890425596.

American Psychiatric Association Practice Guidelines, 2001. Practice guideline for the treatment of patients with borderline personality disorder. Am. J. Psychiatry 158 (10 Suppl), 1–52.

Ansell, E.B., Wright, A.G.C., Markowitz, J.C., Sanislow, C.A., Hopwood, C.J., Zanarini, M.C., Yen, S., Pinto, A., McGlashan, T.H., Grilo, C.M., 2015. Personality disorder risk factors for suicide attempts over 10 years of follow-up. Personal. Disord. Theory Res. Treat. 6 (2), 161–167. https://doi.org/10.1037/per0000089.

Barnett, P.A., Gotlib, I.H., 1988. Psychosocial functioning and depression: distinguishing among antecedents, concomitants, and consequences. Psychol. Bull. 104 (1), 97–126. https://doi.org/10.1037/0033-2909.104.1.97.

Beatson, J.A., Rao, S., 2012. Depression and borderline personality disorder. Med. J. Aust. 199 (S6), S24–S27. https://doi.org/10.5694/mja12.10474.

Besser, A., Zeigler-Hill, V., 2010. The influence of pathological narcissism on emotional and motivational responses to negative events: the roles of visibility and concern about humiliation. J. Res. Pers. 44 (4), 520–534. https://doi.org/10.1016/j.jrp.2010.06.006.

Ceresa, A., Esposito, C.M., Buoli, M., 2020. How does borderline personality disorder affect management and treatment response of patients with major depressive disorder? A comprehensive review. J. Affect. Disord. https://doi.org/10.1016/j.jad.2020.11.111.

Clarkin, J.F., Levy, K.N., Lenzenweger, M.F., Kernberg, O.F., 2007. Evaluating three treatments for borderline personality disorder: a multiwave study. Am. J. Psychiatry 164 (6), 922–928. https://doi.org/10.1176/ajp.2007.164.6.922.

Clemence, A.J., Perry, J.C., Plakun, E.M., 2009. Narcissistic and borderline personality disorders in a sample of treatment refractory patients. Psychiatr. Ann. 39 (4), 175–184. https://doi.org/10.3928/00485713-20090401-05.

Corruble, E., Ginestet, D., Guelfi, J.D., 1996. Comorbidity of personality disorders and unipolar major depression: a review. J. Affect. Disord. 37 (2–3), 157–170.

Cristea, I.A., Gentili, C., Cotet, C.D., Palomba, D., Barbui, C., Cuijpers, P., 2017. Efficacy of psychotherapies for borderline personality disorder: a systematic review and meta-analysis. JAMA Psychiat. 74 (4), 319. https://doi.org/10.1001/jamapsychiatry.2016.4287.

Dawood, S., Pincus, A.L., 2018. Pathological narcissism and the severity, variability, and instability of depressive symptoms. Personal. Disord. Theory Res. Treat. 9 (2), 144–154. https://doi.org/10.1037/per0000239.

De Fruyt, F., Van Leeuwen, K., Bagby, R.M., Rolland, J.-P., Rouillon, F., 2006. Assessing and interpreting personality change and continuity in patients treated for major depression. Psychol. Assess. 18 (1), 71–80. https://doi.org/10.1037/1040-3590.18.1.71.

DeBattista, C., Mueller, K., 2001. Is electroconvulsive therapy effective for the depressed patient with comorbid borderline personality disorder? J. ECT 17 (2), 91–98. https://doi.org/10.1097/00124509-200106000-00002.

Diamond, D., Hersh, R.G., 2020. Transference-focused psychotherapy for narcissistic personality disorder: an object relations approach. J. Pers. Disord. 34, 159–176. https://doi.org/10.1521/pedi.2020.34.supp.159.

Drozek, R.P., Unruh, B.T., 2020. Mentalization-based treatment for pathological narcissism. J. Pers. Disord. 34, 177–203. https://doi.org/10.1521/pedi.2020.34.supp.177.

Ekselius, L., von Knorring, L., 1998. Personality disorder comorbidity with major depression and response to treatment with sertraline or citalopram. Int. Clin. Psychopharmacol. 13 (5), 205–211. https://doi.org/10.1097/00004850-199809000-00003.

Ellison, W.D., Levy, K.N., Cain, N.M., Ansell, E.B., Pincus, A.L., 2013. The impact of pathological narcissism on psychotherapy utilization, initial symptom severity, and early-treatment symptom change: a naturalistic investigation. J. Pers. Assess. 95 (3), 291–300. https://doi.org/10.1080/00223891.2012.742904.

Farabaugh, A., Mischoulon, D., Fava, M., Guyker, W., Alpert, J., 2004. The overlap between personality disorders and major depressive disorder (MDD). Ann. Clin. Psychiatry 16 (4), 217–224. https://doi.org/10.1080/10401230490522043.

Farmer, R.F., 2000. Issues in the assessment and conceptualization of personality disorders. Clin. Psychol. Rev. 20 (7), 823–851. https://doi.org/10.1016/s0272-7358(99)00014-8.

Fava, M., Bouffides, E., Pava, J.A., McCarthy, M.K., Steingard, R.J., Rosenbaum, J.F., 1994. Personality disorder comorbidity with major depression and response to fluoxetine treatment. Psychother. Psychosom. 62 (3–4), 160–167. https://doi.org/10.1159/000288918.

Fava, M., Alpert, J.E., Borus, J.S., Nierenberg, A.A., Pava, J.A., Rosenbaum, J.F., 1996. Patterns of personality disorder comorbidity in early-onset versus late- onset major depression. Am. J. Psychiatry 153 (10), 1308–1312. https://doi.org/10.1176/ajp.153.10.1308.

Feske, U., 2004. Clinical outcome of ECT in patients with major depression and comorbid borderline personality disorder. Am. J. Psychiatry 161 (11), 2073–2080. https://doi.org/10.1176/appi.ajp.161.11.2073.

Feske, U., Mulsant, B.H., Pilkonis, P.A., Soloff, P., Dolata, D., Sackeim, H.A., Haskett, R.F., 2004. Clinical outcome of ECT in patients with major depression and comorbid borderline personality disorder. Am. J. Psychiatry 161 (11). https://doi.org/10.1176/appi.ajp.161.11.2073. 2073–80.

First, M.B., Williams, J.B., Benjamin, L.S., Spitzer, R.L., 2016. SCID-5-PD: Structured Clinical Interview for DSM-5 Personality Disorders. American Psychiatric Publishing.

Fjermestad-Noll, J., Ronningstam, E., Bach, B., Rosenbaum, B., Simonsen, E., 2019. Characterological depression in patients with narcissistic personality disorder. Nord. J. Psychiatry 73 (8), 539–545. https://doi.org/10.1080/08039488.2019.1664630.

Fjermestad-Noll, J., Ronningstam, E., Bach, B.S., Rosenbaum, B., Simonsen, E., 2020. Perfectionism, shame, and aggression in depressive patients with narcissistic personality disorder. J. Pers. Disord. 34, 25–41. https://doi.org/10.1521/pedi.2020.34.supp.25.

Francois, D., Roth, S.D., Klingman, D., 2015. The efficacy of pharmacotherapy for borderline personality disorder: a review of the available randomized controlled trials. Psychiatr. Ann. 45 (8), 431–437. https://doi.org/10.3928/00485713-20150803-09.

Galione, J., Zimmerman, M., 2010. A comparison of depressed patients with and without borderline personality disorder: implications for interpreting studies of the validity of the bipolar spectrum. J. Pers. Disord. 24 (6), 763–772. https://doi.org/10.1521/pedi.2010.24.6.763.

Ganellen, R.J., 2007. Assessing normal and abnormal personality functioning: strengths and weaknesses of self-report, observer, and performance-based methods. J. Pers. Assess. 89 (1), 30–40. https://doi.org/10.1080/00223890701356987.

Giner, L., Blasco-Fontecilla, H., Mercedes Perez-Rodriguez, M., Garcia-Nieto, R., Giner, J., Guija, J.A., Rico, A., Barrero, E., Luna, M.A., De Leon, J., Oquendo, M.A., Baca-Garcia, E., 2013. Personality disorders and health problems distinguish suicide attempters from completers in a direct comparison. J. Affect. Disord. 151 (2), 474–483. https://doi.org/10.1016/j.jad.2013.06.029.

Goodman, M., New, A.S., Triebwasser, J., Collins, K.A., Siever, L., 2010. Phenotype, endophenotype, and genotype comparisons between borderline personality disorder and major depressive disorder. J. Pers. Disord. 24 (1), 38–59. https://doi.org/10.1521/pedi.2010.24.1.38.

Grant, B.F., Hasin, D.S., Stinson, F.S., Dawson, D.A., Patricia Chou, S., June Ruan, W., Huang, B., 2005. Co-occurrence of 12-month mood and anxiety disorders and personality disorders in the US: results from the national epidemiologic survey on alcohol and related conditions. J. Psychiatr. Res. 39 (1), 1–9. https://doi.org/10.1016/j.jpsychires.2004.05.004.

Grilo, C.M., Sanislow, C.A., Shea, M.T., Skodol, A.E., Stout, R.L., Gunderson, J.G., Yen, S., Bender, D.S., Pagano, M.E., Zanarini, M.C., Morey, L.C., McGlashan, T.H., 2005. Two-year prospective naturalistic study of remission from major depressive disorder as a function of personality disorder comorbidity. J. Consult. Clin. Psychol. 73 (1), 78–85. https://doi.org/10.1037/0022-006X.73.1.78.

Grilo, C.M., Stout, R.L., Markowitz, J.C., Sanislow, C.A., Ansell, E.B., Skodol, A.E., Bender, D.S., Pinto, A., Shea, M.T., Yen, S., Gunderson, J.G., Morey, L.C., Hopwood, C.J., McGlashan, T.H., 2010. Personality disorders predict relapse after remission from an episode of major depressive disorder: a 6-year prospective study. J. Clin. Psychiatry 71 (12), 1629–1635. https://doi.org/10.4088/JCP.08m04200gre.

Gunderson, J.G., 2008. Borderline Personality Disorder: A Clinical Guide. American Psychiatric Publishing.

Gunderson, J.G., Shea, M.T., Skodol, A.E., McGlashan, T.H., Morey, L.C., Stout, R.L., Zanarini, M.C., Grilo, C.M., Oldham, J.M., Keller, M.B., 2000. The Collaborative Longitudinal Personality Disorders Study: development, aims, design, and sample characteristics. J. Pers. Disord. 14 (4), 300–315. https://doi.org/10.1521/pedi.2000.14.4.300.

Gunderson, J.G., Morey, L.C., Stout, R.L., Skodol, A.E., Shea, M.T., McGlashan, T.H., Zanarini, M.C., Grilo, C.M., Sanislow, C.A., Yen, S., 2004. Major depressive disorder and borderline personality disorder revisited: longitudinal interactions. J. Clin. Psychiatry 65 (8), 1049–1056. http://www.psychiatrist.com/JCP/article/Pages/major-depressive-disorder-borderline-personality-disorder.aspx?sclick=1.

Gunderson, J.G., Stout, R.L., Sanislow, C.A., Shea, M.T., McGlashan, T.H., Zanarini, M.C., Daversa, M.T., Grilo, C.M., Yen, S., Skodol, A.E., 2008. New episodes and new onsets of major depression in borderline and other personality disorders. J. Affect. Disord. 111 (1), 40–45. https://doi.org/10.1016/j.jad.2008.01.026.

Hasin, D.S., Goodwin, R.D., Stinson, F.S., Grant, B.F., 2005. Epidemiology of major depressive disorder: results from the National Epidemiologic Survey on alcoholism and related conditions. Arch. Gen. Psychiatry 62 (10), 1097–1106. https://doi.org/10.1001/archpsyc.62.10.1097.

Hays, R.D., Wells, K.B., Sherbourne, C.D., Rogers, W., Spritzer, K., 1995. Functioning and well-being outcomes of patients with depression compared with chronic general medical illnesses. Arch. Gen. Psychiatry 52 (1), 11–19. https://doi.org/10.1001/archpsyc.1995.03950130011002.

Hirschfeld, R.M., 1999. Personality disorders and depression: comorbidity. Depress. Anxiety 10 (4), 142–146. https://doi.org/10.1002/(sici)1520-6394(1999)10:4<142::aid-da2>3.0.co;2-q.

Hobson, R.F., 1953. Prognostic factors in electric convulsive therapy. J. Neurol. Neurosurg. Psychiatry 16 (4), 275–281. https://doi.org/10.1136/jnnp.16.4.275.

Huprich, S.K., 1998. Depressive personality disorder: theoretical issues, clinical findings, and future research questions. Clin. Psychol. Rev. 18 (5), 477–500. https://doi.org/10.1016/S0272-7358(97)00104-9.

Huprich, S.K., 2001. The overlap of depressive personality disorder and dysthymia, reconsidered. Harv. Rev. Psychiatry 9 (4), 158–168. https://doi.org/10.1080/10673220127895.

Huprich, S.K., 2012. Considering the evidence and making the most empirically informed decision about depressive personality disorder in DSM-5. Personal. Disord. Theory Res. Treat. 3 (4), 470–482. https://doi.org/10.1037/a0027765.

Huprich, S.K., 2013. New directions for an old construct: depressive personality research in the DSM-5 era. Personal. Ment. Health 7 (3), 213–222. https://doi.org/10.1002/pmh.1217.

Huprich, S.K., 2014. Malignant self-regard: a self-structure enhancing the understanding of masochistic, depressive, and vulnerably narcissistic personalities. Harv. Rev. Psychiatry 22 (5), 295–305. https://doi.org/10.1097/HRP.0000000000000019.

Huprich, S.K., 2018. Malignant self-regard: finding a place for depressive, masochistic, and vulnerably narcissistic personalities in modern day classification. Persönlichkeitsstörungen: Theorie Therapie 22, 113–132.

Huprich, S.K., 2019. Personality-driven depression: the case for malignant self-regard (and depressive personalities). J. Clin. Psychol. 75 (5), 834–845. https://doi.org/10.1002/jclp.22760.

Huprich, S.K., Nelson, S.M., 2014. Malignant self-regard: accounting for commonalities in vulnerably narcissistic, depressive, self-defeating, and masochistic personality disorders. Compr. Psychiatry 55 (4), 989–998. https://doi.org/10.1016/j.comppsych.2013.09.017.

Huprich, S.K., 2020. Commentary on the special issue: critical distinctions between vulnerable narcissism and depressive personalities. J. Pers. Disord. 34 (Suppl), 207–209. https://doi.org/10.1521/pedi.2020.34.supp.207.

Huprich, S.K., Bornstein, R.F., Schmitt, T.A., 2011. Self-report methodology is insufficient for improving the assessment and classification of Axis II personality disorders. J. Pers. Disord. 25 (5), 557–570. https://doi.org/10.1521/pedi.2011.25.5.557.

Huprich, S.K., Lengu, K., Evich, C., 2016. Interpersonal problems and their relationship to depression, self-esteem, and malignant self-regard. J. Pers. Disord. 30 (6), 742–761. https://doi.org/10.1521/pedi_2015_29_227.

Huprich, S.K., Macaluso, M., Baade, L., Zackula, R., Jackson, J., Kitchens, R., 2018. Malignant self-regard in clinical outpatient samples. Psychiatry Res. 266, 253–261. https://doi.org/10.1016/j.psychres.2018.03.005.

Ilardi, S.S., Edward Head, W.C., 1994. Personality pathology and response to somatic treatments for major depression: a critical review. Depression 2 (4), 200–217. https://doi.org/10.1002/depr.3050020404.

Ingenhoven, T., Lafay, P., Rinne, T., Passchier, J., Duivenvoorden, H., 2010. Effectiveness of pharmacotherapy for severe personality disorders: meta-analyses of randomized controlled trials. J. Clin. Psychiatry 71 (1), 14–25.

Joiner Jr., T.E., Walker, R.L., Pettit, J.W., Perez, M., Cukrowicz, K.C., 2005. Evidence-based assessment of depression in adults. Psychol. Assess. 17 (3). https://doi.org/10.1037/1040-3590.17.3.267. 267–77.

Keller, M.B., Lavori, P.W., Friedman, B., Nielsen, E., Endicott, J., McDonald-Scott, P., Andreasen, N.C., 1987. The Longitudinal Interval Follow-up Evaluation. A comprehensive method for assessing outcome in prospective longitudinal studies. Arch. Gen. Psychiatry 44 (6), 540–548. https://doi.org/10.1001/archpsyc.1987.01800180050009.

Kernberg, O.F., 1984. Severe Personality Disorders. Yale University Press.

Kernberg, O.F., 2004. Aggressivity, narcissism, and self-destructiveness in the psychotherapeutic relationship: new developments in the psychopathology and psychotherapy of severe personality disorders. In: Aggressivity, Narcissism, and Self-Destructiveness in the Psychotherapeutic Relationship: New Developments in the Psychopathology and Psychotherapy of Severe Personality Disorders. Yale University Press, pp. 1–271. http://yalepress.yale.edu/yupbooks/book.asp?isbn=9780300101805.

Klein, D.N., Shih, J.H., 1998. Depressive personality: associations with DSM-III-R mood and personality disorders and negative and positive affectivity, 30-month stability, and prediction of course of Axis I depressive disorders. J. Abnorm. Psychol., 319–327. https://doi.org/10.1037//0021-843x.107.2.319.

Klein, D., Bufferd, S., Ro, E., Clark, L.A., 2014. Depression and comorbidity: personality disorders. In: Richards, S., O'Hara, M. (Eds.), The Oxford Handbook of Depression and Comorbidity. Oxford University Press.

Koeningsberg, H.W., Kernberg, O.F., Stone, M.H., Applebaum, A., Yeomans, F.E., Diamond, D., 2000. Borderline Patients: Extending the Limits of Treatability. Basic Books.

Kool, S., Schoevers, R., de Maat, S., Van, R., Molenaar, P., Vink, A., Dekker, J., 2005. Efficacy of pharmacotherapy in depressed patients with and without personality disorders: a systematic review and meta-analysis. J. Affect. Disord. 88 (3), 269–278. https://doi.org/10.1016/j.jad.2005.05.017.

Krueger, R.F., Derringer, J., Markon, K.E., Watson, D., Skodol, A.E., 2012. Initial construction of a maladaptive personality trait model and inventory for DSM-5. Psychol. Med. 42 (9), 1879–1890. https://doi.org/10.1017/S0033291711002674.

Lengu, K.J., Evich, C.D., Nelson, S.M., Huprich, S.K., 2015. Expanding the utility of the malignant self-regard construct. Psychiatry Res. 229 (3), 801–808. https://doi.org/10.1016/j.psychres.2015.07.087.

Lenzenweger, M.F., Lane, M.C., Loranger, A.W., Kessler, R.C., 2007. DSM-IV personality disorders in the National Comorbidity Survey Replication. Biol. Psychiatry 62 (6), 553–564. https://doi.org/10.1016/j.biopsych.2006.09.019.

Levy, K.N., Edell, W.S., McGlashan, T.H., 2007. Depressive experiences in inpatients with borderline personality disorder. Psychiatry Q. 78 (2), 129–143.

Lieb, K., Birgit, V., Gerta, R., Antje, T., Stoffers, J.M., 2010. Pharmacotherapy for borderline personality disorder: Cochrane systematic review of randomised trials. Br. J. Psychiatry 196 (1), 4–12. http://dx.doi.org.proxy1.cl.msu.edu/10.1192/bjp.bp.108.062984.

Linehan, M.M., Armstrong, H.E., Suarez, A., Allmon, D., Heard, H.L., 1991. Cognitive-behavioral treatment of chronically parasuicidal borderline patients. Arch. Gen. Psychiatry 48 (12), 1060–1064.

Markowitz, J.C., Skodol, A.E., Petkova, E., Cheng, J., Sanislow, C.A., Grilo, C.M., Gunderson, J.G., McGlashan, T.H., 2007. Longitudinal effects of personality disorders on psychosocial functioning of patients with major depressive disorder. J. Clin. Psychiatry 68 (02), 186–193. https://doi.org/10.4088/JCP.v68n0202.

McDermut, W., Zimmerman, M., Chelminski, I., 2003. The construct validity of depressive personality disorder. J. Abnorm. Psychol. 109, 419–427.

Miller, J.D., Maples, J., 2011. Trait personality models of narcissistic personality disorder, grandiose narcissism, and vulnerable narcissism. In: Campbell, W.K., Miller, J.D. (Eds.), Handbook of Narcissism and Narcissistic Personality Disorder: Theoretical Approaches, Empirical Findings, and Treatments. Wiley, pp. 71–88.

Morey, L.C., 2017. Development and initial evaluation of a self-report form of the DSM-5 Level of Personality Functioning Scale. Psychol. Assess. 29 (10), 1302–1308. https://doi.org/10.1037/pas0000450.

Mulder, R.T., 2002. Personality pathology and treatment outcome in major depression: a review. Am. J. Psychiatry 159 (3), 359–371. https://doi.org/10.1176/appi.ajp.159.3.359.

Newton-Howes, G., Tyrer, P., Johnson, T., 2006. Personality disorder and the outcome of depression: meta-analysis of published studies. Br. J. Psychiatry 188 (1), 13–20. http://dx.doi.org.proxy1.cl.msu.edu/10.1192/bjp.188.1.13.

Newton-Howes, G., Tyrer, P., Johnson, T., Mulder, R., Kool, S., Dekker, J., Schoevers, R., 2014. Influence of personality on the outcome of treatment in depression: systematic review and meta-analysis. J. Pers. Disord. 28 (4), 577–593. https://doi.org/10.1521/pedi_2013_27_070.

Oldham, J.M., Skodol, A.E., Kellman, H.D., Hyler, S.E., Doidge, N., Rosnick, L., Gallaher, P.E., 1995. Comorbidity of axis I and axis II disorders. Am. J. Psychiatry 152 (4), 571–578. https://doi.org/10.1176/ajp.152.4.571.

Orth, U., Robins, R.W., Meier, L.L., Conger, R.D., 2016. Refining the vulnerability model of low self-esteem and depression: disentangling the effects of genuine self-esteem and narcissism. J. Pers. Soc. Psychol. 110 (1), 133–149. https://doi.org/10.1037/pspp0000038.

Osório, F.L., Loureiro, S.R., Hallak, J.E.C., Machado-de-Sousa, J.P., Ushirohira, J.M., Baes, C.V.W., Apolinario, T.D., Donadon, M.F., Bolsoni, L.M., Guimarães, T., Fracon, V.S., Silva-Rodrigues, A.P.C., Pizeta, F.A., Souza, R.M., Sanches, R.F., Dos Santos, R.G., Martin-Santos, R., Crippa, J.A.S., 2019. Clinical validity and intrarater and test-retest reliability of the Structured Clinical Interview for DSM-5—Clinician Version (SCID-5-CV). Psychiatry Clin. Neurosci. 73 (12), 754–760. https://doi.org/10.1111/pcn.12931.

Panos, P.T., Jackson, J.W., Hasan, O., Panos, A., 2014. Meta-analysis and systematic review assessing the efficacy of dialectical behavior therapy (DBT). Res. Soc. Work. Pract. 24 (2), 213–223. https://doi.org/10.1177/1049731513503047.

Pedone, R., Huprich, S.K., Nelson, S.M., Cosenza, M., Carcione, A., Nicolò, G., Semerari, A., Colle, L., 2018. Expanding the validity of the malignant self-regard construct in an Italian general population sample. Psychiatry Res. 270, 688–697. https://doi.org/10.1016/j.psychres.2018.10.059.

Petersen, T., Hughes, M., Papakostas, G.I., Kant, A., Fava, M., Rosenbaum, J.F., Nierenberg, A.A., 2002. Treatment-resistant depression and Axis II co-morbidity. Psychother. Psychosom. 71 (5), 269–274. https://doi.org/10.1159/000064808.

Pfohl, B., Blum, N., Zimmerman, M., 1997. Structured Interview for DSM-IV Personality: Sidp-IV. American Psychiatric Publishing.

Pincus, A.L., Lukowitsky, M.R., 2010. Pathological narcissism and narcissistic personality disorder. Annu. Rev. Clin. Psychol. 6, 421–446. https://doi.org/10.1146/annurev.clinpsy.121208.131215.

Pincus, A.L., Roche, M.J., 2011. Narcissistic grandiosity and narcissistic vulnerability. In: Campbell, W.K., Miller, J.D. (Eds.), Handbook of Narcissism and Narcissistic Personality Disorder: Theoretical Approaches, Empirical Findings, and Treatments. Wiley, pp. 31–40.

Pincus, A.L., Roche, M.J., Good, E.W., 2013. Narcissistic personality disorder and pathological narcissism. In: Blaney, P.H., Krueger, R.F., Millon, T. (Eds.), Oxford Textbook of Psychopathology. Oxford Press.

Raheja, S.K., King, E.A., Thompson, C., 1996. The Seasonal Pattern Assessment Questionnaire for identifying seasonal affective disorders. J. Affect. Disord. 41 (3), 193–199. https://doi.org/10.1016/s0165-0327(96)00087-0.

Reich, J.H., Green, A.I., 1991. Effect of personality disorders on outcome of treatment. J. Nerv. Ment. Dis. 179 (2), 74–82. https://doi.org/10.1097/00005053-199102000-00003.

Reich, J.H., Vasile, R.G., 1993. Effect of personality disorders on the treatment outcome of axis I conditions: an update. J. Nerv. Ment. Dis. 181 (8), 475–484. https://doi.org/10.1097/00005053-199308000-00002.

Riihimäki, K., Vuorilehto, M., Isometsä, E., 2014. Borderline personality disorder among primary care depressive patients: a five-year study. J. Affect. Disord. 155, 303–306. https://doi.org/10.1016/j.jad.2013.10.050.

Ronningstam, E., 2005. Identifying and Understanding the Narcissistic Personality. Oxford Press.

Ryder, A.G., Schuller, D.R., Bagby, R.M., 2006. Depressive personality and dysthymia: evaluating symptom and syndrome overlap. J. Affect. Disord. 91 (2–3), 217–227. https://doi.org/10.1016/j.jad.2006.01.008.

Russ, E., Shedler, J., Bradley, R., Westen, D., 2008. Refining the construct of narcissistic personality disorder: diagnostic criteria and subtypes. Am. J. Psychiatry 165 (11). https://doi.org/10.1176/appi.ajp.2008.07030376. 1473–81.

Ryder, A.G., Quilty, L.C., Vachon, D.D., Bagby, R.M., 2010. Depressive personality and treatment outcome in major depressive disorder. J. Pers. Disord. 24 (3), 392–404. https://doi.org/10.1521/pedi.2010.24.3.392.

Samuel, D.B., Griffin, S.A., 2015. A critical evaluation of retaining personality categories and types. In: Huprich, S.K. (Ed.), Personality Disorders: Toward Theoretical and Empirical Integration in Diagnosis and Assessment. American Psychological Association, pp. 43–62.

Saulsman, L.M., Coall, D.A., Nathan, P.R., 2006. The association between depressive personality and treatment outcome for depression following a group cognitive-behavioral intervention. J. Clin. Psychol. 62 (9), 1181–1196. https://doi.org/10.1002/jclp.20278.

Shahar, G., Zuroff, D.C., Blatt, S.J., Pilkonis, P.A., 2003. Role of perfectionism and personality disorder features in response to brief treatment for depression. J. Consult. Clin. Psychol. 71 (3), 629–633. https://doi.org/10.1037/0022-006X.71.3.629.

Sheehan, D.V., Lecrubier, Y., Sheehan, K.H., Amorim, P., Janavs, J., Weiller, E., Hergueta, T., Baker, R., Dunbar, G.C., 1998. The Mini-International Neuropsychiatric Interview (M.I.N.I.): the development and validation of a structured diagnostic psychiatric interview for DSM-IV and ICD-10. J. Clin. Psychiatry 59 (Suppl. 20), 22–33.

Silk, K.R., 2010. The quality of depression in borderline personality disorder and the diagnostic process. J. Pers. Disord. 24 (1), 25–37. https://doi.org/10.1521/pedi.2010.24.1.25.

Simonsen, S., Simonsen, E., 2011. Comorbidity between narcissistic personality disorder and Axis I diagnoses. In: Handbook of Narcissism and Narcissistic Personality Disorder: Theoretical Approaches, Empirical Findings, and Treatments. John Wiley & Sons, pp. 239–247.

Skodol, A.E., Stout, R.L., McGlashan, T.H., Grilo, C.M., Gunderson, J.G., Shea, M.T., Morey, L.C., Zanarini, M.C., Dyck, I.R., Oldham, J.M., 1999. Co-occurrence of mood and personality disorders: a report from the Collaborative Longitudinal Personality Disorders Study (CLPS). Depress. Anxiety 10 (4), 175–182.

Skodol, A.E., Grilo, C.M., Pagano, M.E., Bender, D.S., Gunderson, J.G., Shea, M.T., Yen, S., Zanarini, M.C., McGlashan, T.H., 2005. Effects of personality disorders on functioning and well-being in major depressive disorder. J. Psychiatr. Pract. 11 (6). https://doi.org/10.1097/00131746-200511000-00002. 363–8.

Skodol, A.E., Grilo, C.M., Keyes, K.M., Geier, T., Grant, B.F., Hasin, D.S., 2011. Relationship of personality disorders to the course of major depressive disorder in a nationally representative sample. Am. J. Psychiatry 168 (3), 257–264. https://doi.org/10.1176/appi.ajp.2010.10050695.

Stanley, B., Wilson, S.T., 2006. Heightened subjective experience of depression in borderline personality disorder. J. Pers. Disord. 20 (4), 307–318. https://doi.org/10.1521/pedi.2006.20.4.307.

Stoffers, J.M., Lieb, K., 2014. Pharmacotherapy for borderline personality disorder—current evidence and recent trends. Curr. Psychiatry Rep. 17 (1), 534. https://doi.org/10.1007/s11920-014-0534-0.

Stoffers, J.M., Völlm, B.A., Rücker, G., Timmer, A., Huband, N., Lieb, K., 2012. Psychological therapies for people with borderline personality disorder. Cochrane Database Syst. Rev. 8, CD005652. https://doi.org/10.1002/14651858.CD005652.pub2.

Stoffers-Winterling, J., Storebø, O.J., Lieb, K., 2020. Pharmacotherapy for borderline personality disorder: an update of published, unpublished and ongoing studies. Curr. Psychiatry Rep. 22 (8), 37. https://doi.org/10.1007/s11920-020-01164-1.

Storebø, O.J., Stoffers-Winterling, J.M., Völlm, B.A., Kongerslev, M.T., Mattivi, J.T., Jørgensen, M.S., Faltinsen, E., Todorovac, A., Sales, C.P., Callesen, H.E., Lieb, K., Simonsen, E., 2020. Psychological therapies for people with borderline personality disorder. Cochrane Database Syst. Rev. 5, CD012955. https://doi.org/10.1002/14651858.CD012955.pub2.

Sullivan, P.F., Joyce, P.R., Mulder, R.T., 1994. Borderline personality disorder in major depression. J. Nerv. Ment. Dis. 182 (9), 508–516. https://doi.org/10.1097/00005053-199409000-00006.

Tedlow, J., Leslie, V., Keefe, B.R., Alpert, J., Nierenberg, A.A., Rosenbaum, J.F., Fava, M., 1999. Axis I and Axis II disorder comorbidity in unipolar depression with anger attacks. J. Affect. Disord. 52 (1–3), 217–223. https://doi.org/10.1016/S0165-0327(98)00074-3.

Thase, M.E., 1996. The role of Axis II comorbidity in the management of patients with treatment-resistant depression. Psychiatr. Clin. N. Am. 19 (2), 287–309. https://doi.org/10.1016/S0193-953X(05)70289-6.

Tyrer, P., Seivewright, N., Ferguson, B., Murphy, S., Darling, C., Brothwell, J., Kingdon, D., Johnson, A.L., 1990. The Nottingham Study of Neurotic Disorder: relationship between personality status and symptoms. Psychol. Med. 20 (2), 423–431. https://doi.org/10.1017/s0033291700017736.

Tyrer, P., Gunderson, J., Lyons, M., Tohen, M., 1997. Extent of comorbidity between mental state and personality disorders. J. Pers. Disord. 11 (3), 242–259. https://doi.org/10.1521/pedi.1997.11.3.242.

Wang, Y.-P., Gorenstein, C., 2013. Psychometric properties of the Beck Depression Inventory-II: a comprehensive review. Rev. Bras. Psiquiatr. 35 (4), 416–431. https://doi.org/10.1590/1516-4446-2012-1048.

Ward, H.B., Yip, A., Siddiqui, R., Morales, O.G., Seiner, S.J., Shan, H., 2020. Borderline personality traits do not influence response to TMS. J. Affect. Disord. https://doi.org/10.1016/j.jad.2020.11.054.

Widiger, T.A., 2005. A temperament model a borderline personality disorder. In: Zanarini, M.C. (Ed.), Borderline Personality Disorder. Taylor and Francis, pp. 63–81.

Widiger, T.A., Samuel, D.B., 2005. Evidence-based assessment of personality disorders. Psychol. Assess. 17 (3), 278–287. https://doi.org/10.1037/1040-3590.17.3.278.

Yeomans, F.E., Clarkin, J.F., Kernberg, O.F., 2015. Transference-Focused Psychotherapy for Borderline Personality Disorder: A Clinical Guide. American Psychiatric Publishing.

Young, M., 2020. Prognostic significance of personality disorders in patients with major depressive disorder. Curr. Treat. Options Psychiatry. https://doi.org/10.1007/s40501-020-00227-7.

Zanarini, M.C., Frankenburg, F.R., Hennen, J., Reich, D.B., Silk, K.R., 2004a. Axis I comorbidity in patients with borderline personality disorder: 6-year follow-up and prediction of time to remission. Am. J. Psychiatry 161 (11), 2108–2114. https://doi.org/10.1176/appi.ajp.161.11.2108.

Zanarini, M.C., Frankenburg, F.R., Hennen, J., Silk, K.R., 2004b. Mental health service utilization by borderline personality disorder patients and Axis II comparison subjects followed prospectively for 6 years. J. Clin. Psychiatry 65 (1), 28–36. https://doi.org/10.4088/jcp.v65n0105.

Zimmerman, M., 1994. Diagnosing personality disorders. A review of issues and research methods. Arch. Gen. Psychiatry 51 (3), 225–245. https://doi.org/10.1001/archpsyc.1994.03950030061006.

Chapter 31

Managing treatment-resistant depression with comorbid substance use disorders

Kimberly R. Stubbs[a], Jack Van Bezooyen[b], and Yi-lang Tang[b,c]

[a]Department of Psychiatry, Wright State University, Dayton, OH, United States, [b]Department of Psychiatry and Behavioral Sciences, Emory University, Atlanta, GA, United States, [c]Substance Abuse Treatment Program, Atlanta VA Medical Center, Decatur, GA, United States

Introduction

Overview: Double-trouble

Substance use disorders (SUDs) and depressive disorders, including major depressive disorder (MDD), dysthymia, and substance induced mood (depressive) disorder (SIMD) are among the most prevalent psychiatric disorders in the world (Kessler et al., 1994; Quello et al., 2005). Not only are both independently common in the general population, they frequently cooccur in patients leading to challenges in assessment, differential diagnosis, treatment, and clinical management. Therefore, this chapter will focus on this area of clinical importance with a specific focus on the comorbidity between depression and alcohol use disorder (AUD), opioid use disorder (OUD), and stimulant (cocaine and methamphetamine) use disorder.

Throughout this chapter, the terms of drug dependence, drug addiction, alcohol and drug abuse will be used as most research conducted in the field used these terms, based on previous versions of the DSM systems (such as DSM-IV). However, it should be noted that these terms are no longer used in DSM-5 terminology and criteria, and a distinction will be made where relevant.

SUDs are common in the general population with the most recent data showing the prevalence of 12-month and lifetime AUD (DSM-IV) were 13.9% and 29.1%, respectively (Grant et al., 2015). The same survey also revealed prevalence rates of 12-month and lifetime substance use disorders (amphetamine, cannabis, club drug, cocaine, hallucinogen, heroin, nonheroin opioid, sedative/tranquilizer, and/or solvent/inhalant use disorders) were 3.9% and 9.9%, respectively.

SUDs are common among individuals with depressive disorders. One early study, the Epidemiologic Catchment Area Study (Regier et al., 1990) found that, among individuals with mood disorders, nearly one-third (32%) had a cooccurring SUD. Among those with lifetime major depression, 16.5% met the diagnosis of an AUD and 18% met the diagnosis of a substance use disorder. The National Comorbidity Survey found that, compared with individuals with no mood disorders, those with depression were approximately twice as likely to have a SUD (Kessler et al., 1994).

On the other hand, depressive disorders are also common among individuals with SUDs. One study (Grant et al., 2004) found that people with alcohol dependence (DSM-IV) are 3.7 times more likely to also have MDD, and 2.8 times more likely to have dysthymia, in the previous year. Among people in treatment for DSM-IV AUD, almost 33% met criteria for MDD in the past year, and 11% met criteria for dysthymia. In another analysis of 6050 former drinkers (Hasin and Grant, 2002), the authors found that the risk of major depression during the past year was 4.2 times greater among respondents with a history of alcohol dependence than among those without such a history. Importantly, this relationship was not attenuated after controlling confounding factors. Additionally, current major depression has been associated with worse clinical outcome among substance-dependent patients (Nunes and Levin, 2008).

SUDs and depressive disorders tend to impact each other in several ways. Individuals with both SUDs and depression often present with a more severe clinical course and worse outcomes than individuals who have only one or the other (Quello et al., 2005; Hasin et al., 2002). They tend to be more difficult to manage and are more likely to be treatment-resistant (Kelly and Daley, 2013; Kelly et al., 2012). Outcomes and prognosis of patients with both SUDs and depression tend to be worse than either disorder alone (Hasin et al., 2002; Greenfield et al., 1998). Some found that patients with comorbid depression and SUD had a higher risk of suicidal behavior (Young et al., 1994; Conner et al., 2019, 2014).

Managing Treatment-Resistant Depression. https://doi.org/10.1016/B978-0-12-824067-0.00031-1

Hypotheses and mechanisms of comorbidity

When considering the temporal relationship between SUDs and depressive disorders, there are the following possibilities: (1) the depressive disorder preceded the SUDs, (2) the SUDs preceded the depressive disorder, and (3) the depressive disorder and SUDs occurred nearly in the same time. After treatment, there are the following possibilities: (1) the depressive disorder symptoms improve or remit before that of the SUDs, (2) the SUDs improve or remit before the symptoms of the depressive disorder, and (3) both disorders improve or remit in the same time frame.

The mere correlation or coexistence between SUDs and depressive disorders does not imply a casual relationship in either direction, although some have proposed that it is possible that the pathological effects of a depressive disorder or SUD may actually increase risk for the other (disorder fostering disorder hypothesis) (Quello et al., 2005). With regard the possible mechanisms underlying their cooccurrence, several possibilities have been suggested: (1) Common risk factors contribute to both depressive disorders and SUDs, such as genetic, neurobiological and environmental risk factors (Cerda et al., 2010). Environmental factors, such as childhood adverse experience and chronic stress may cause alterations in the hypothalamic–pituitary–adrenal (HPA) axis and those changes may affect limbic brain circuits that are involved in important functions such as motivation, learning, and adaptation. These functions are known to be impaired in individuals with both SUDs and depressive disorders (Kelly and Daley, 2013; Villanueva, 2013). Additionally, the association between childhood adversities and mental disorders (including SUDs) is often nonspecific (Kessler et al., 2010). (2) Depressive disorders may contribute to drug use and substance use disorders (Quello et al., 2005; Baigent, 2012). Some mental health conditions have been identified as risk factors for developing a substance use disorder (Baigent, 2012), and some even propose that people with mental illnesses may use drugs or alcohol as a form of self-medication (Santucci, 2012; Khantzian, 1997). (3) Substance use and SUDs contribute to the development of depressive disorder. Substance use may change the brain in ways that make a person more likely to develop a mental illness (Brook et al., 2002). Additionally, substance use often cause or worsen psychosocial stressors, including financial difficulty, relationship problems, unemployment, unstable housing and legal issues, all of which may increase the risk of depression (Cohen et al., 2007).

Challenges in diagnosis and assessments

The diagnosis and assessment of patients with both depressive disorders and substance use disorders often pose clinical challenges (Morojele et al., 2012; Ross and Peselow, 2012). Several factors may contribute to the complexity and difficult in assessing and diagnosing these patients: (1) Patients often have multiple comorbid diagnoses, including other substance use disorder, anxiety disorders, personality disorders, and depressive disorders. Polysubstance use often makes it hard to establish a clear timeline of different symptoms and describe their symptoms clearly; (2) Cooccurring disorders often obscure the presentations of one another, making it difficult to make appropriate diagnoses (Iqbal et al., 2019). Chronic use of central nervous system depressants, such as alcohol, benzodiazepines, and opioids, may lead to anhedonia, depressed mood, and sleep disorders (such as insomnia) (Quello et al., 2005); cocaine and other stimulants may also cause presentations similar to mania or hypomania or psychotic symptoms. (3) Patients may be poor historians or no family members are available to provide collateral information; (4) Some patients may have cognitive impairments related to their substance use that may or may not improve once abstinent for a sufficient period of time (Maillard et al., 2020).

To clarify whether a patient's presentation is part of a primary psychiatric disorder or a psychiatric disorder that is a consequence of substance use (SUDs), a detailed history is essential, including the timeline of the onset and changes of different symptoms and substance use. If possible, clinicians should obtain the history of symptoms and function during past periods of sobriety. Collateral information from family members and other significant others can also be very helpful. Even with a careful history, it may be necessary to employ the use of structured interview instruments such as the Structured Clinical Interview for DSM-5 (SCID-5) or the Diagnostic Interview Schedule (DIS). When there is no clear separation in the timeline or onset of the two disorders, sometimes the only way to clarify the diagnosis is to reassess the symptoms after patients are abstinent for a suggested period of 4 weeks according to the DSM-5. However, some patients may need earlier assessment and intervention sooner due to the severity of the symptoms and other factors. Clinicians need to use their own clinical judgment.

Treatment considerations and sequence of treatment implementations

Several significant barriers exist for patients with comorbid disorders. Overall, they have poor treatment adherence (Magura et al., 2014; DeMarce et al., 2008) and higher rates of treatment dropout (Kelly and Daley, 2013; Torrens et al., 2012), which negatively affects outcomes, including treatment response and functions (Hasin et al., 2002; Kelly and Daley, 2013; Kelly et al., 2012).

Despite these challenges, research in the past few decades shows significant progress on treating patients with comorbid depression and SUDs (Kelly et al., 2012; Torrens et al., 2012; Murthy et al., 2019; Murthy and Chand, 2012).

Evidence-based treatments exist for both depression and SUDs. However, when treating patients with dual diagnosis, it is important to apply these methods in a way that produces the best outcomes for both disorders.

With respect to the implementation of treatments for the comorbid disorders, three common models have been adopted: the sequential, parallel, and integrated models.

The sequential approach: This approach suggests treating one disorder prior to addressing the other disorder. The sequential approach often prioritizes treating active substance use first, as it is believed that patients with active substance use are less likely to benefit from treatment of depression. It may also help with differential diagnosis by reassessing depressive symptoms and their severity after sobriety is achieved. Sometimes, treating depression first may also reduce patient's substance use and make them more motivated to change their behavior. It may also be a pragmatic treatment strategy, at least in the early stage, for patients with dual diagnosis who may only have interest or expectation in treating one of the disorders.

The parallel approach: This approach requires that specific treatments for both disorders are provided simultaneously, although not necessarily by the same provider or even in the same facility. This approach may be sensible and realistic from a practical standpoint, given that in the current treatment culture, addiction and mental health settings generally are separated and efforts to unify and integrate treatment services have been slow. However, several limitations need to be mentioned. One common challenge is coordination and communication among providers. Second, as the patients need to be engaged in treatments for both disorders, they can easily feel overburdened and the approach generally may be inefficient.

The Integrated Approach: The integrated approaches are similar to the parallel approach, but with two additional features: both disorders are treated by a single provider (or a team) and treatment addresses the functional interrelationship of the comorbid disorders. The integrated approach has been found to be consistently superior compared with separate treatment of each diagnosis (Kelly and Daley, 2013; Torrens et al., 2012; Mueser and Gingerich, 2013; Wusthoff et al., 2014; Lydecker et al., 2010). The integrated approach often involves using cognitive behavioral therapy strategies to boost interpersonal and coping skills and other approaches that support motivation and functional recovery (Mueser and Gingerich, 2013). Based on the range of potential advantages associated with integrated therapy, expert opinion strongly suggests adopting this approach to treating depression and SUDs. However, many changes need to be made, including changing the current organization of care, updating guidelines, revising insurance reimbursement policies, and training current providers and change their practice models.

Alcohol use disorder and major depressive disorder

Epidemiology

A recent comprehensive literature review on the relationship between alcohol use disorder (AUD) and MDD confirmed a previously reported association between AUD and MDD. That is, that the presence of either disorder doubled the risks of the second disorder. Further data analysis suggests that the association between AUD and MDD cannot be fully accounted for by common factors, and there seems to be a causal relationship. Specifically, the data suggest that AUD contributes to the risk of MDD, rather than vice versa (Boden and Fergusson, 2011).

MDD and alcohol dependence (DSM-IV) are among the most prevalent mental disorders worldwide and commonly cooccurring (Grant et al., 2015; Pettinati et al., 2013). The 12-month prevalence of depression in adults is 5.3% and lifetime prevalence is 13.3% (Stinson et al., 2005) and the 12-month prevalence of alcohol dependence is 13.9% (Grant et al., 2015). The prevalence of depression is higher in people with alcohol dependence and each of these disorders is associated with a significant risk of developing the other (Agabio et al., 2018). Treatment of both cooccurring disorders is imperative as past year AUD remission has been associated with a threefold-reduced risk of depression (Agosti and Levin, 2006).

Clinical presentation

Differentiating SIMD from MDD involves examining the timing of onset of AUD and depressive symptoms to determine if depressive episodes were established during a period of sustained abstinence (Schuckit et al., 2007).

Diagnosis is particularly challenging because of overlapping symptoms such as insomnia and psychomotor agitation/depression, poor concentration, and feelings of guilt/worthlessness (McHugh and Weiss, 2019; von Knorring et al., 1983).

Studies showed that patients with SIMD have better short-term reduction in symptoms after treatment, as compared to patients with major depression and cooccurring AUD (Foulds et al., 2015). However, SIMD may not be a stable diagnosis as evidenced by the finding that a high percentage of patients with SIMD also have major depression after sustained sobriety is achieved (Brown et al., 2011; Schwartz et al., 1990; Nunes and Levin, 2004).

Alcohol-induced depression disorder is defined by remission of depression during periods of abstinence; however, studies show that alcohol-induced depression is associated with risk for onset of a later MDD. A minimum of 4 weeks of abstinence may be necessary to distinguish alcohol-induced depression from primary depressive disorder in the absence of reliable symptom history (Brown et al., 1995).

Treatment

Clinicians treating patients with AUD may be reluctant to initiate antidepressant medications for depressive symptoms for a number of reasons, including concerns about distinguishing from SIMD versus MDD (Schwartz et al., 1990; Nunes and Levin, 2004). The chronology of symptom onset, symptom severity, and duration should be considered when determining potential treatment interventions (Schuckit et al., 2006). Temporary alcohol-induced depressive symptoms may result from the effects of intoxication or withdrawal, and tends to resolve within 1 month of abstinence from alcohol (Pettinati et al., 2013; Schuckit et al., 2006). As such, antidepressant treatment may not be required in this scenario, and it may be prudent to focus on treatment of AUD in cases of mild depression. However, when the cooccurring disorders are of significant duration or severity pharmacologic treatment of both disorders may be required, particularly patients with significant risk of suicidality (Schuckit et al., 2006). Studies showed that antidepressants may reduce the severity of depression in patients with cooccurring AUD, and increase the rate of abstinence and decrease severity of alcohol consumption (the number of drinks per drinking days) (Agabio et al., 2018). Additionally, the magnitude of benefit of antidepressants for patients with AUD and depression is similar to that of those with depression alone (McHugh and Weiss, 2019; von Knorring et al., 1983).

Concurrent MDD and AUD present specific safety concerns that should be assessed prior to determining appropriate treatment setting and modality. Cooccurring alcohol and depression increase the risk of self-harm, suicide, and unintended harm to self or others such as workplace injuries and as such, hospitalization should be considered given this elevated risk. For nonsuicidal patients with cooccurring depression and AUD, outpatient management may be appropriate and motivational interviewing techniques may be useful in assessing the patient's readiness for change.

When deciding on specific medications for the treatment of depression in the context of AUD, bupropion should be used with caution in patients with severe AUD, and avoided in those with a history of alcohol-withdrawal seizure or other complicated alcohol withdrawal. Sedating antidepressants, such as sedating TCAs and mirtazapine should be used cautiously due to the risk of increased sedating effects with alcohol (McIntyre et al., 2012). Earlier literature suggested that less sedating tricyclic antidepressants, such as nortriptyline may be more effective in reducing depressive symptoms in people with cooccurring depression and AUD as compared to selective serotonin reuptake inhibitors (SSRIs) (McHugh and Weiss, 2019).

Recently, combinations of medications for both depression and AUD, such as sertraline with naltrexone, and acamprosate with escitalopram have shown promise for reducing depressive symptoms and maintaining abstinence in patients with cooccurring disorders (Pettinati et al., 2010; Witte et al., 2012). A study aimed at evaluating efficacy of combining approved medications for depression and alcohol dependence, found that depressed, alcohol-dependent patients treated with the combination of sertraline and naltrexone had delayed relapse to heavy drinking and fewer adverse events as compared to a group taking single-medication treatment (Pettinati et al., 2010). One randomized control trial evaluated a group of adults with cooccurring major depression and alcohol dependence who were randomized to citalopram or placebo plus naltrexone. Both the citalopram and placebo groups showed substantial improvements in mood and drinking outcomes in this 12-week study. The combination of naltrexone and citalopram was well tolerated overall as evidenced by the similar reporting of adverse events between the two groups in this study (Adamson et al., 2015). Animal studies have looked at the combined effects of acamprosate and escitalopram on ethanol consumption in mice, and one human study found this combination of medications to be well tolerated (Witte et al., 2012). These findings support the use of combination pharmacotherapies to address cooccurring major depression and alcohol dependence.

Utilization of psychotherapy is often understudied in patients with cooccurring depression and AUD despite evidence to support its effectiveness in both AUD and depression (McHugh and Weiss, 2019; von Knorring et al., 1983). Additionally, recent evidence suggests that behavioral activation therapy may also be useful for treating cooccurring depression and AUD (Ross et al., 2016).

Integral to the treatment of comorbid AUD and MDD is directing efforts specifically toward the AUD. Positive trial data exist for multiple medications as well as psychotherapies.

Among all medications for AUD, naltrexone or acamprosate should be offered to patients with moderate-to-severe AUD who intend to reduce alcohol consumption or achieve abstinence, especially those who prefer pharmacotherapy or have failed nonpharmacological treatments alone. Disulfiram, an old medication that is still in use, is often recommended to patients who have a goal of achieving abstinence. For disulfiram, supervision is important in maximizing the benefits.

Additionally, topiramate or gabapentin can also be offered; however, clinicians need to know that they are used as off-label (American Psychiatric Association, 2018; Department of Veterans Affairs, Department of Defense, 2015). Motivational interview, contingency management, cognitive behavioral therapy, mindfulness-based interventions, and family/couples therapy for AUD all have positive trial data for achieving abstinence, reducing number of drinking days, and reducing harms associated with drinking alcohol (Knox et al., 2019).

Conclusion

Distinguishing mood symptoms that are substance-related from independent mood disorders is a pivotal part of the clinical assessment and may impact treatment decisions. Evaluating mood symptoms during periods of sustained abstinence, 4 weeks per DMS5 guidelines, is ideal. Seeking collateral information from family, friends, and other healthcare providers regarding the timeline of symptoms may also be useful. Toxicology screens and laboratory tests such as CDT, GGT, or blood alcohol concentration or alcohol metabolites may also be helpful in clarifying the history of substance use. During the diagnostic assessment, clinicians should also keep in mind alcohol intoxication symptoms, such as mood lability and agitation, that may mimic major depression or mania. Alcohol withdrawal symptoms, such as sleep disturbance and dysphoria may mimic major depression as well. If an AUD is established, then a goal of abstinence or reduction in alcohol use should be implemented prior to diagnosis and treatment of depression.

In people with cooccurring AUD and depression, antidepressants may be useful for the treatment of depression, AUD, or both (Agabio et al., 2018). Early intervention is supported by the finding that patients receiving comprehensive integrated treatment for AUD and depression have the most significant reduction in symptoms within the first 3 weeks of beginning treatment (Foulds et al., 2015). Treatment with antidepressants should be considered for severe or persistent depressive symptoms. SSRIs should be considered as first line due to their good tolerability and safety characteristics in the setting of AUD. Clinicians may consider switching to a non-SSRI antidepressant if SSRI trials fail. Antidepressant side effect profiles should be considered when selecting a medication, for example, avoiding antidepressants that are more sedating, have high overdose risk or lower the seizure threshold. Substance use treatment should be initiated concurrently with addressing depressive symptoms (Nunes and Weiss, 2014).

Limitations in the available evidence include the large number of studies showing high or unclear risk of bias (Agabio et al., 2018). Other limiting factors include the uncertainty of diagnosis of depression in people affected by AUD, inadequacy of reporting psychosocial interventions, and timeframe concerns related to the antidepressants reaching full effect. Additionally, there are limited studies that compare one antidepressant to another antidepressant or other medication. Two metaanalyses concluded that antidepressant medication is more effective than placebo in improving outcomes among patients with cooccurring alcohol and depressive disorders (Schuckit et al., 2007; Nunes and Levin, 2004).

Opioid use disorder and major depressive disorder

Epidemiology

Of the estimated 44,965 suicides in the United States in 2016, about 15% were related to drug overdose according to the CDC (Oquendo and Volkow, 2018). According to the same survey, the proportion of suicides that were opioid overdoses rose from 2.2% to 4.3% between 1999 and 2001. Evidence from clinical studies support the association between prescription opioid analgesic use and risk of new-depression episodes (Scherrer et al., 2017). One study suggests that new depressive episodes related to chronic opioid dependence are at least as severe, if not more severe, than new depressive episodes unrelated to chronic opioid use (Scherrer et al., 2017). In comparison to individuals with one disorder alone, individuals with comorbid disorders often demonstrate a more severe course of illness, including increased social and occupational impairment, disability, and economic burden (Scherrer et al., 2017). Comorbidity has also been shown to be associated with poorer treatment outcome, higher rates of relapse, and rehospitalization (Gros et al., 2013). Of note, both MDD and SIMD are risk factors for future depressive episodes (Nunes et al., 2006).

Clinical presentation

The following symptoms of opioid intoxication resemble major depression or dysthymia: apathy, dysphoria, and psychomotor agitation. Opioid withdrawal symptoms that resemble depressive disorders include irritability, dysphoria, insomnia, and fatigue. Clinicians should also keep in mind the time of last opioid use and typical time course of intoxication and withdrawal syndromes, which is of variable length of time depending on the half-life of the opioid. For patients with chronic opioid exposure, their acute withdrawal symptoms may be followed by more chronic signs of a protracted abstinence syndrome, which may last for weeks even months (Nunes and Weiss, 2014).

Chronic depression, even if more mild symptoms, should be taken seriously among patients with opioid and other substance use disorders due to the association with worse clinical outcome in patients with dual diagnoses (Nunes and Levin, 2008).

Assessment of depression in patients with cooccurring opioid dependence is difficult by confounding issues such as state at the time of evaluation (Rounsaville et al., 1982). Patients who are acutely intoxicated may not accurately relay the extent of depressive symptoms, while those who are experiencing withdrawal symptoms may exaggerate the depressive symptoms over the preceding weeks (Rounsaville et al., 1982).

Studies of patients with OUD and cooccurring depression show that elevated scores on depression symptoms scales improve substantially after initiation of abstinence (Strain et al., 1991). Efforts to address withdrawal symptoms and begin treatment to achieve abstinence should always be a first step in the treatment of patients with both OUD and depression. Clinicians should also keep in mind that some cases of depression may persist despite abstinence of reductions in opioid use. A careful clinical history as well as a period of abstinence may be helpful in distinguishing which patients may have a primary depressive disorder that should be addressed while patient is receiving substance use treatment (Nunes and Weiss, 2014). The DSM-5 suggests an abstinence period of 4 weeks; however, clinical judgment should be used.

Treatment

For treatment of severe and/or obviously independent depressive symptoms in patients with active opioid use disorder (OUD), antidepressants may be considered. SSRIs are considered first line of treatment in these patients in part due to their low level of sedation as compared to other antidepressants and overall lower risk of adverse interactions with substances of abuse (Nunes and Weiss, 2014). Other antidepressants can be considered as well if patients do not respond to SSRIs. In clinical studies, use of TCAs in patients with a cooccurring SUDs shows evidence of substantial improvement of depressive symptoms; however, TCAs are associated with more side effects and toxicity than other new generation antidepressants (Nunes and Levin, 2008). Several recent studies demonstrated that, at low doses, buprenorphine is an efficacious, well-tolerated, and safe option in reducing depressive symptoms, serious suicidal ideation, and nonsuicidal self-injury, even in patients with treatment-resistant depression (Serafini et al., 2018). Placebo-controlled trials that offered a manual-guided psychosocial intervention for all participants tend to have high placebo depression response rates (Nunes and Levin, 2004). This finding emphasizes the importance of considering behavioral interventions as part of the treatment plan for cooccurring disorders. Cognitive behavioral interventions and 12-step facilitation were utilized in the aforementioned placebo-controlled trials. These substance use disorder interventions have components that focus on managing mood symptoms in addition to helping reduce substance use (Nunes and Weiss, 2014).

Integral to the treatment of comorbid OUD and MDD is directing efforts specifically toward the OUD as pursuing abstinence without treatment has relapse rates above 90%. The following medications have positive trial data: methadone, buprenorphine/naloxone, sublingual buprenorphine, buprenorphine implants, intramuscular long-acting buprenorphine, and intramuscular long-acting naltrexone. Methadone maintenance requires referral to a federally certified opioid use treatment center, whereas buprenorphine requires special certification in the treatment of opioid disorder and X-waiver from on the prescriber's DEA license. Intramuscular naltrexone is an effective medication; however, drop off rates tend to be high. Behavioral therapy may be effective for relapse prevention in those who decline medication; however, medication-assisted treatment (MAT) is noted to be twice as effective (Coffa and Snyder, 2019).

Conclusion

Comprehensive treatment is recommended for patient with both OUD and MDD, and it includes evidence-based behavioral interventions (Nunes and Weiss, 2014) and medications. Antidepressants used concurrently with other treatment modalities for OUD can be helpful for managing cooccurring mood symptoms and support recovery in some patients (Nunes and Weiss, 2014).

When considering antidepressant medications for patients with both OUD and depression, clinicians should consider factors such as past history of independent depression, family history, or past positive response to antidepressant medication. Current evidence supports the use of SSRIs as first line for patients with dual diagnosis due to their safety profiles and overall tolerability. Additionally, evidence-based psychosocial interventions have also been shown to improve mood symptoms (Nunes and Levin, 2008).

Stimulant (cocaine and methamphetamine) use disorder and major depressive disorder

Epidemiology

According to the 2018 National Survey on Drug Use and Health, in the demographic over 18 years of age, cocaine use was estimated at 6.25 million uses in the last year with 0.97 million meeting criteria for cocaine use disorder (0.4% of the population) (Substance Abuse and Mental Health Services Administration, 2019). Methamphetamine (MA) use was estimated at 1.87 million users in the last year with 1.03 million meeting criteria for methamphetamine use disorder (0.4% of the population) (Substance Abuse and Mental Health Services Administration, 2019).

Cocaine use disorder is highly associated with SIMD as opposed to independent MDD. However, female sex, age of onset < 18, and comorbid psychiatric diagnoses are associated with independent depression (Dakwar et al., 2011).

Up to 74% of active methamphetamine users have experienced independent MDD or dysthymia with disproportionate levels of females meeting criteria for independent MDD (Conway et al., 2006). Lifetime diagnosis of MDD, a Beck's Depression Inventory (BDI) score above the moderate range at time of active use, having received treatment for AUD, intravenous methamphetamine use, female sex, and history of suicide attempt are associated with the persistence of MDD at longitudinal follow-up, despite abstinence (Glasner-Edwards et al., 2009).

Clinical presentation

Accurate diagnosis is challenging given the overlap in symptomatology and time consideration of cocaine and methamphetamine-induced depressive disorder (encompassing cocaine and methamphetamine use and withdrawal) and MDD. Cocaine and methamphetamine use and withdrawal can be associated with severe dysphoria, irritability, anxiety, fatigue, poor concentration, hypersomnia, hyperphagia, and suicidal ideation (Hart et al., 2008). DSM-5 criteria suggest the cutoff of 4 weeks postabstinence to diagnose independent MDD.

In addition to common presentations described above, cocaine use has been found to be associated with several other medical conditions, such as cardiovascular disease, kidney disease, movement disorders, and dystonic reactions with antipsychotic medications (Lange and Hillis, 2001; Pendergraft 3rd and Niles, 2014; van Harten et al., 1998). In addition to neurovegetative symptoms, MA use is associated with psychosis, violence, anxiety, suicidality, and various health consequences including higher all-cause mortality (Kuo et al., 2011), cardiovascular disease (Sliman et al., 2016), stroke (Swor et al., 2019), and high-risk sexual behavior (Hittner, 2016).

A detailed history of presubstance use depressive episodes as well as obtaining collateral information may reveal that depression preceded the onset of cocaine use, in which event, it is essential to direct treatment toward both the substance use disorder as well as the likely MDD, as untreated MDD is associated with higher cocaine use and risk of relapse (Nunes et al., 2006). Of note, the presence and severity of comorbid depressive symptoms does not affect ability of those with cocaine use disorder to engage in substance use disorder treatment or achieve remission (Milby et al., 2015). Regular urine drug screening is also recommended with the detection of cocaine typically being positive up to 3 days from last use as concurrent cocaine use complicates the treatment of depression. When assessing individuals with cocaine use disorder, severity of the illness may additionally direct treatment recommendations. Specifically when assessing individuals with methamphetamine use disorder, severity of the illness and route of administration may direct treatment and are associated with different prognoses, with IV/IM use being associated with the worst outcomes (Zweben et al., 2004).

Treatment

Cocaine

The following antidepressants given as monotherapy or in combination with a psychological intervention have been studied for reducing depressive symptoms in the context of cocaine use disorder. Imipramine, while it demonstrated a reduction in depressive symptoms, drop out was 85% at 12 weeks (Nunes et al., 1995). Desipramine with relapse prevention therapy was associated with an improvement in depression, but not cocaine use (McDowell et al., 2005). Nefazodone resulted in decreased cocaine use; however, the groups were not appropriately matched in terms of baseline use (Ciraulo et al., 2005). Sertraline with weekly CBT was demonstrated to increase length of time until relapse on cocaine in patients with comorbid depressive symptoms (Oliveto et al., 2012), while the addition of gabapentin to sertraline does not change outcomes (Mancino et al., 2014). Venlafaxine does not show a statistically significant change in cocaine use or depression (Raby et al., 2014).

In terms of psychological interventions, weekly CBT was associated with an increased length of time until relapse and increased engagement but not significant change in depressive symptoms over the control condition (Carroll et al., 1995). Motivational interviewing has been investigated in comorbid cocaine use disorder and MDD and demonstrated increased engagement and fewer hospitalizations (Daley et al., 1998). Contingency management was found to be effective at reducing depressive symptoms in patients with cocaine use disorder with depressive symptoms (Miguel et al., 2017; Petry et al., 2013).

Methamphetamine

All psychological interventions evaluated for addressing comorbid depressive symptoms and methamphetamine use disorder are with cognitive behavioral therapy (CBT) and all studies demonstrated a reduction in methamphetamine use. However, no statistically significant changes were noted with respect to depressive symptoms with the exception of a mild reduction using the stepped care model (Baker et al., 2005; Murphy et al., 2007; Peck et al., 2005).

Sertraline, bupropion, and mirtazapine have been found to be ineffective at reducing depressive symptoms (Cruickshank et al., 2008; Elkashef et al., 2008; Hellem, 2016; Newton et al., 2006; Shoptaw et al., 2008; Zorick et al., 2011); however, bupropion + behavioral therapy was found to reduce amount of methamphetamine use in mild methamphetamine users (Newton et al., 2006). The opposite was demonstrated in a trial of sertraline + CBT, with increased incidence of cravings and methamphetamine use (Zorick et al., 2011). Aripiprazole, while studied in the context of methamphetamine induced psychosis, should be mentioned as it is sometimes used as an adjunct in the treatment of depression. Aripiprazole was found to increase methamphetamine cravings and methamphetamine use and may potentiate the effects of methamphetamine (Newton et al., 2008); therefore, it should be avoided. One trial of methylphenidate demonstrated decreased methamphetamine cravings, withdrawal symptoms, and depressive symptoms (Rezaei et al., 2015). There is one trial in HIV positive men where Modafinil + CBT was found to reduce depressive symptoms in those with high but not low BDI scores at baseline (McElhiney et al., 2009). One trial of the nutritional supplement citicoline demonstrated a reduction in depressive symptoms, but no significant changes in methamphetamine use, in those diagnosed with unipolar or bipolar disorder with depressed episode (Brown and Gabrielson, 2012).

In a study evaluating aerobic exercise and strength training, three times a week, in an inpatient substance use treatment facility, participants were found to experience statistically significant improvement in their Beck's Anxiety Inventory and Beck's Depression Inventory scores relative to controls in methamphetamine abstinent patients (Rawson et al., 2015a,b).

Conclusion

There are no FDA-approved medications or psychological interventions effective for the treatment of comorbid independent depression and cocaine or methamphetamine use disorder, though various approaches have demonstrated reduction in substance use, which has demonstrated a reduction in depressive symptoms via abstinence.

Where positive trial data exist for the treatment of depression in the context of ongoing cocaine use, results are limited by small sample size, high dropout rate, and/or lack of generalizability; therefore, no strong recommendations can be currently made.

It is in this context that no medication or psychotherapy can be strongly recommended at this time for the treatment of depression with concomitant cocaine or methamphetamine use. Of note, sertraline and aripiprazole should be avoided when the patient is using methamphetamine as they may worsen cravings and/or potentiate the effect of methamphetamine use. However, it is recommended that patients be referred for cocaine use disorder—specific psychotherapy and/or inpatient substance use rehabilitation based on severity of cocaine use disorder.

Summary of the chapter: Practical considerations

SUDs and depressive disorders are both common and they also tend to cooccur, although the mechanisms are unclear. Patients with SUDs should be evaluated for common psychiatric disorders including MDD, and patients with primary psychiatric illness should be evaluated for comorbid substance use. A thorough assessment for cooccurring disorders is critical when directing patients to the appropriate treatment modality. Substance-related depressive symptoms may self-resolve after a period of abstinence in some patients. For diagnostic clarification, it is generally recommended to monitor for up to 4 weeks after abstinence is achieved.

The initial patient assessment should determine stability and assess for safety given the elevated risk for self-harming and suicidal ideation in these patients. Hospitalizations should be considered for the patient with acute risk of self-harm or acute exacerbation of other psychiatric symptoms. For patients appropriate for outpatient treatment, an integrative care model is recommended and treatments often include therapy focusing on substance use recovery and depressive symptoms. In the meantime, medications should be considered when indicated, including medications for depressive symptoms (antidepressants) and specific medications for certain types of SUDs (such as naltrexone for AUD, buprenorphine for OUD).

References

Adamson, S.J., Sellman, J.D., Foulds, J.A., et al., 2015. A randomized trial of combined citalopram and naltrexone for nonabstinent outpatients with co-occurring alcohol dependence and major depression. J. Clin. Psychopharmacol. 35 (2), 143–149.

Agabio, R., Trogu, E., Pani, P.P., 2018. Antidepressants for the treatment of people with co-occurring depression and alcohol dependence. Cochrane Database Syst. Rev. 4, CD008581.

Agosti, V., Levin, F.R., 2006. The effects of alcohol and drug dependence on the course of depression. Am. J. Addict. 15 (1), 71–75.

American Psychiatric Association, 2018. The American Psychiatric Association practice guideline for the pharmacological treatment of patients with alcohol use disorder. In: APA Practice Guidelines, first ed. American Psychiatric Association Publishing, Washington, DC, p. 1. online resource.

Baigent, M., 2012. Managing patients with dual diagnosis in psychiatric practice. Curr. Opin. Psychiatry 25 (3), 201–205.

Baker, A., Lee, N.K., Claire, M., et al., 2005. Brief cognitive behavioural interventions for regular amphetamine users: a step in the right direction. Addiction 100 (3), 367–378.

Boden, J.M., Fergusson, D.M., 2011. Alcohol and depression. Addiction 106 (5), 906–914.

Brook, D.W., Brook, J.S., Zhang, C., Cohen, P., Whiteman, M., 2002. Drug use and the risk of major depressive disorder, alcohol dependence, and substance use disorders. Arch. Gen. Psychiatry 59 (11), 1039–1044.

Brown, E.S., Gabrielson, B., 2012. A randomized, double-blind, placebo-controlled trial of citicoline for bipolar and unipolar depression and methamphetamine dependence. J. Affect. Disord. 143 (1–3), 257–260.

Brown, D.R., Ahmed, F., Gary, L.E., Milburn, N.G., 1995. Major depression in a community sample of African Americans. Am. J. Psychiatry 152 (3), 373–378.

Brown, R.A., Ramsey, S.E., Kahler, C.W., et al., 2011. A randomized controlled trial of cognitive-behavioral treatment for depression versus relaxation training for alcohol-dependent individuals with elevated depressive symptoms. J. Stud. Alcohol Drugs 72 (2), 286–296.

Carroll, K.M., Nich, C., Rounsaville, B.J., 1995. Differential symptom reduction in depressed cocaine abusers treated with psychotherapy and pharmacotherapy. J. Nerv. Ment. Dis. 183 (4), 251–259.

Cerda, M., Sagdeo, A., Johnson, J., Galea, S., 2010. Genetic and environmental influences on psychiatric comorbidity: a systematic review. J. Affect. Disord. 126 (1–2), 14–38.

Ciraulo, D.A., Knapp, C., Rotrosen, J., Sarid-Segal, O., Ciraulo, A.M., LoCastro, J., Greenblatt, D.J., Leiderman, D., 2005. Nefazodone treatment of cocaine dependence with comorbid depressive symptoms. Addiction 100 (Suppl 1), 23–31.

Coffa, D., Snyder, H., 2019. Opioid use disorder: medical treatment options. Am. Fam. Physician 100 (7), 416–425.

Cohen, S., Janicki-Deverts, D., Miller, G.E., 2007. Psychological stress and disease. JAMA 298 (14), 1685–1687.

Conner, K.R., Gamble, S.A., Bagge, C.L., He, H., Swogger, M.T., Watts, A., Houston, R.J., 2014. Substance-induced depression and independent depression in proximal risk for suicidal behavior. J. Stud. Alcohol Drugs 75 (4), 567–572.

Conner, K.R., Bridge, J.A., Davidson, D.J., Pilcher, C., Brent, D.A., 2019. Metaanalysis of mood and substance use disorders in proximal risk for suicide deaths. Suicide Life Threat. Behav. 49 (1), 278–292.

Conway, K.P., Compton, W., Stinson, F.S., Grant, B.F., 2006. Lifetime comorbidity of DSM-IV mood and anxiety disorders and specific drug use disorders: results from the National Epidemiologic Survey on alcohol and related conditions. J. Clin. Psychiatry 67 (2), 247–257.

Cruickshank, C.C., Montebello, M.E., Dyer, K.R., Quigley, A., Blaszczyk, J., Tomkins, S., Shand, D., 2008. A placebo-controlled trial of mirtazapine for the management of methamphetamine withdrawal. Drug Alcohol Rev. 27 (3), 326–333.

Dakwar, E., Nunes, E.V., Bisaga, A., Carpenter, K.C., Mariani, J.P., Sullivan, M.A., Raby, W.N., Levin, F.R., 2011. A comparison of independent depression and substance-induced depression in cannabis-, cocaine-, and opioid-dependent treatment seekers. Am. J. Addict. 20 (5), 441–446.

Daley, D.C., Salloum, I.M., Zuckoff, A., Kirisci, L., Thase, M.E., 1998. Increasing treatment adherence among outpatients with depression and cocaine dependence: results of a pilot study. Am. J. Psychiatry 155 (11), 1611–1613.

DeMarce, J.M., Lash, S.J., Stephens, R.S., Grambow, S.C., Burden, J.L., 2008. Promoting continuing care adherence among substance abusers with co-occurring psychiatric disorders following residential treatment. Addict. Behav. 33 (9), 1104–1112.

Department of Veterans Affairs, Department of Defense, Defense DoVAaDo (Eds.), 2015. Management of Substance Use Disorder (SUD). Department of Veterans Affairs, Department of Defense.

Elkashef, A.M., Rawson, R.A., Anderson, A.L., et al., 2008. Bupropion for the treatment of methamphetamine dependence. Neuropsychopharmacology 33 (5), 1162–1170.

Foulds, J.A., Douglas Sellman, J., Adamson, S.J., Boden, J.M., Mulder, R.T., Joyce, P.R., 2015. Depression outcome in alcohol dependent patients: an evaluation of the role of independent and substance-induced depression and other predictors. J. Affect. Disord. 174, 503–510.

Glasner-Edwards, S., Marinelli-Casey, P., Hillhouse, M., Ang, A., Mooney, L.J., Rawson, R., 2009. Methamphetamine treatment project corporate a. depression among methamphetamine users: association with outcomes from the Methamphetamine treatment project at 3-year follow-up. J. Nerv. Ment. Dis. 197 (4), 225–231.

Grant, B.F., Stinson, F.S., Dawson, D.A., Chou, S.P., Dufour, M.C., Compton, W., Pickering, R.P., Kaplan, K., 2004. Prevalence and co-occurrence of substance use disorders and independent mood and anxiety disorders: results from the National Epidemiologic Survey on alcohol and related conditions. Arch. Gen. Psychiatry 61 (8), 807–816.

Grant, B.F., Goldstein, R.B., Saha, T.D., et al., 2015. Epidemiology of DSM-5 alcohol use disorder: results from the National Epidemiologic Survey on alcohol and related conditions III. JAMA Psychiatry 72 (8), 757–766.

Greenfield, S.F., Weiss, R.D., Muenz, L.R., Vagge, L.M., Kelly, J.F., Bello, L.R., Michael, J., 1998. The effect of depression on return to drinking: a prospective study. Arch. Gen. Psychiatry 55 (3), 259–265.

Gros, D.F., Milanak, M.E., Brady, K.T., Back, S.E., 2013. Frequency and severity of comorbid mood and anxiety disorders in prescription opioid dependence. Am. J. Addict. 22 (3), 261–265.

Hart, C.L., Gunderson, E.W., Perez, A., Kirkpatrick, M.G., Thurmond, A., Comer, S.D., Foltin, R.W., 2008. Acute physiological and behavioral effects of intranasal methamphetamine in humans. Neuropsychopharmacology 33 (8), 1847–1855.

Hasin, D.S., Grant, B.F., 2002. Major depression in 6050 former drinkers: association with past alcohol dependence. Arch. Gen. Psychiatry 59 (9), 794–800.

Hasin, D., Liu, X., Nunes, E., McCloud, S., Samet, S., Endicott, J., 2002. Effects of major depression on remission and relapse of substance dependence. Arch. Gen. Psychiatry 59 (4), 375–380.

Hellem, T.L., 2016. A review of Methamphetamine dependence and withdrawal treatment: a focus on anxiety outcomes. J. Subst. Abus. Treat. 71, 16–22.

Hittner, J.B., 2016. Meta-analysis of the association between methamphetamine use and high-risk sexual behavior among heterosexuals. Psychol. Addict. Behav. 30 (2), 147–157.

Iqbal, M.N., Levin, C.J., Levin, F.R., 2019. Treatment for substance use disorder with co-occurring mental illness. Focus (Am. Psychiatr. Publ.) 17 (2), 88–97.

Kelly, T.M., Daley, D.C., 2013. Integrated treatment of substance use and psychiatric disorders. Soc. Work Public Health 28 (3–4), 388–406.

Kelly, T.M., Daley, D.C., Douaihy, A.B., 2012. Treatment of substance abusing patients with comorbid psychiatric disorders. Addict. Behav. 37 (1), 11–24.

Kessler, R.C., McGonagle, K.A., Zhao, S., Nelson, C.B., Hughes, M., Eshleman, S., Wittchen, H.U., Kendler, K.S., 1994. Lifetime and 12-month prevalence of DSM-III-R psychiatric disorders in the United States. Results from the National Comorbidity Survey. Arch. Gen. Psychiatry 51 (1), 8–19.

Kessler, R.C., McLaughlin, K.A., Green, J.G., et al., 2010. Childhood adversities and adult psychopathology in the WHO world mental health surveys. Br. J. Psychiatry 197 (5), 378–385.

Khantzian, E.J., 1997. The self-medication hypothesis of substance use disorders: a reconsideration and recent applications. Harv. Rev. Psychiatry 4 (5), 231–244.

Knox, J., Hasin, D.S., Larson, F.R.R., Kranzler, H.R., 2019. Prevention, screening, and treatment for heavy drinking and alcohol use disorder. Lancet Psychiatry 6 (12), 1054–1067.

Kuo, C.J., Liao, Y.T., Chen, W.J., Tsai, S.Y., Lin, S.K., Chen, C.C., 2011. Causes of death of patients with methamphetamine dependence: a record-linkage study. Drug Alcohol Rev. 30 (6), 621–628.

Lange, R.A., Hillis, L.D., 2001. Cardiovascular complications of cocaine use. N. Engl. J. Med. 345 (5), 351–358.

Lydecker, K.P., Tate, S.R., Cummins, K.M., McQuaid, J., Granholm, E., Brown, S.A., 2010. Clinical outcomes of an integrated treatment for depression and substance use disorders. Psychol. Addict. Behav. 24 (3), 453–465.

Magura, S., Mateu, P.F., Rosenblum, A., Matusow, H., Fong, C., 2014. Risk factors for medication non-adherence among psychiatric patients with substance misuse histories. Ment. Health Subst. Use 7 (4), 381–390.

Maillard, A., Poussier, H., Boudehent, C., Lannuzel, C., Vicente, A., Vabret, F., Cabe, N., Pitel, A.L., 2020. Short-term neuropsychological recovery in alcohol use disorder: a retrospective clinical study. Addict. Behav. 105, 106350.

Mancino, M.J., McGaugh, J., Chopra, M.P., et al., 2014. Clinical efficacy of sertraline alone and augmented with gabapentin in recently abstinent cocaine-dependent patients with depressive symptoms. J. Clin. Psychopharmacol. 34 (2), 234–239.

McDowell, D., Nunes, E.V., Seracini, A.M., Rothenberg, J., Vosburg, S.K., Ma, G.J., Petkova, E., 2005. Desipramine treatment of cocaine-dependent patients with depression: a placebo-controlled trial. Drug Alcohol Depend. 80 (2), 209–221.

McElhiney, M.C., Rabkin, J.G., Rabkin, R., Nunes, E.V., 2009. Provigil (modafinil) plus cognitive behavioral therapy for methamphetamine use in HIV+ gay men: a pilot study. Am. J. Drug Alcohol Abuse 35 (1), 34–37.

McHugh, R.K., Weiss, R.D., 2019. Alcohol use disorder and depressive disorders. Alcohol Res. 40 (1), arcr.v40.1.01.

McIntyre, R.S., Alsuwaidan, M., Goldstein, B.I., et al., 2012. The Canadian network for mood and anxiety treatments (CANMAT) task force recommendations for the management of patients with mood disorders and comorbid metabolic disorders. Ann. Clin. Psychiatry 24 (1), 69–81.

Miguel, A.Q.C., Madruga, C.S., Cogo-Moreira, H., et al., 2017. Contingency management targeting abstinence is effective in reducing depressive and anxiety symptoms among crack cocaine-dependent individuals. Exp. Clin. Psychopharmacol. 25 (6), 466–472.

Milby, J.B., Conti, K., Wallace, D., Mennemeyer, S., Mrug, S., Schumacher, J.E., 2015. Comorbidity effects on cocaine dependence treatment and examination of reciprocal relationships between abstinence and depression. J. Consult. Clin. Psychol. 83 (1), 45–55.

Morojele, N.K., Saban, A., Seedat, S., 2012. Clinical presentations and diagnostic issues in dual diagnosis disorders. Curr. Opin. Psychiatry 25 (3), 181–186.

Mueser, K.T., Gingerich, S., 2013. Treatment of co-occurring psychotic and substance use disorders. Soc. Work Public Health 28 (3–4), 424–439.

Murphy, S.A., Lynch, K.G., Oslin, D., McKay, J.R., TenHave, T., 2007. Developing adaptive treatment strategies in substance abuse research. Drug Alcohol Depend. 88 (Suppl 2), S24–S30.

Murthy, P., Chand, P., 2012. Treatment of dual diagnosis disorders. Curr. Opin. Psychiatry 25 (3), 194–200.

Murthy, P., Mahadevan, J., Chand, P.K., 2019. Treatment of substance use disorders with co-occurring severe mental health disorders. Curr. Opin. Psychiatry 32 (4), 293–299.

Newton, T.F., Roache, J.D., De La Garza, R., 2nd, et al., 2006. Bupropion reduces methamphetamine-induced subjective effects and cue-induced craving. Neuropsychopharmacology 31 (7), 1537–1544.

Newton, T.F., Reid, M.S., De La Garza, R., et al., 2008. Evaluation of subjective effects of aripiprazole and methamphetamine in methamphetamine-dependent volunteers. Int. J. Neuropsychopharmacol. 11 (8), 1037–1045.

Nunes, E.V., Levin, F.R., 2004. Treatment of depression in patients with alcohol or other drug dependence: a meta-analysis. JAMA 291 (15), 1887–1896.

Nunes, E.V., Levin, F.R., 2008. Treatment of co-occurring depression and substance dependence: using meta-analysis to guide clinical recommendations. Psychiatr. Ann. 38 (11), nihpa128505.

Nunes, E.V., Weiss, R.D., 2014. Co-occuring addictive and mood disorders. In: Ries, R., Miller, S.C., Saitz, R., Fiellin, D.A., American Society of Addiction Medicine (Eds.), The ASAM Principles of Addiction Medicine, fifth ed. Wolters Kluwer Health/Lippincott Williams & Wilkins, Philadelphia, pp. 1300–1332 (Chapter 86).

Nunes, E.V., McGrath, P.J., Quitkin, F.M., Ocepek-Welikson, K., Stewart, J.W., Koenig, T., Wager, S., Klein, D.F., 1995. Imipramine treatment of cocaine abuse: possible boundaries of efficacy. Drug Alcohol Depend. 39 (3), 185–195.

Nunes, E.V., Liu, X., Samet, S., Matseoane, K., Hasin, D., 2006. Independent versus substance-induced major depressive disorder in substance-dependent patients: observational study of course during follow-up. J. Clin. Psychiatry 67 (10), 1561–1567.

Oliveto, A., Poling, J., Mancino, M.J., et al., 2012. Sertraline delays relapse in recently abstinent cocaine-dependent patients with depressive symptoms. Addiction 107 (1), 131–141.

Oquendo, M.A., Volkow, N.D., 2018. Suicide: a silent contributor to opioid-overdose deaths. N. Engl. J. Med. 378 (17), 1567–1569.

Peck, J.A., Reback, C.J., Yang, X., Rotheram-Fuller, E., Shoptaw, S., 2005. Sustained reductions in drug use and depression symptoms from treatment for drug abuse in methamphetamine-dependent gay and bisexual men. J. Urban Health 82 (1 Suppl 1), i100–i108.

Pendergraft 3rd, W.F., Niles, J.L., 2014. Trojan horses: drug culprits associated with antineutrophil cytoplasmic autoantibody (ANCA) vasculitis. Curr. Opin. Rheumatol. 26 (1), 42–49.

Petry, N.M., Alessi, S.M., Rash, C.J., 2013. Contingency management treatments decrease psychiatric symptoms. J. Consult. Clin. Psychol. 81 (5), 926–931.

Pettinati, H.M., Oslin, D.W., Kampman, K.M., Dundon, W.D., Xie, H., Gallis, T.L., Dackis, C.A., O'Brien, C.P., 2010. A double-blind, placebo-controlled trial combining sertraline and naltrexone for treating co-occurring depression and alcohol dependence. Am. J. Psychiatry 167 (6), 668–675.

Pettinati, H.M., O'Brien, C.P., Dundon, W.D., 2013. Current status of co-occurring mood and substance use disorders: a new therapeutic target. Am. J. Psychiatry 170 (1), 23–30.

Quello, S.B., Brady, K.T., Sonne, S.C., 2005. Mood disorders and substance use disorder: a complex comorbidity. Sci. Pract. Perspect. 3 (1), 13–21.

Raby, W.N., Rubin, E.A., Garawi, F., et al., 2014. A randomized, double-blind, placebo-controlled trial of venlafaxine for the treatment of depressed cocaine-dependent patients. Am. J. Addict. 23 (1), 68–75.

Rawson, R.A., Chudzynski, J., Gonzales, R., Mooney, L., Dickerson, D., Ang, A., Dolezal, B., Cooper, C.B., 2015a. The impact of exercise on depression and anxiety symptoms among abstinent methamphetamine-dependent individuals in a residential treatment setting. J. Subst. Abus. Treat. 57, 36–40.

Rawson, R.A., Chudzynski, J., Mooney, L., et al., 2015b. Impact of an exercise intervention on methamphetamine use outcomes post-residential treatment care. Drug Alcohol Depend. 156, 21–28.

Regier, D.A., Farmer, M.E., Rae, D.S., Locke, B.Z., Keith, S.J., Judd, L.L., Goodwin, F.K., 1990. Comorbidity of mental disorders with alcohol and other drug abuse. Results from the Epidemiologic Catchment Area (ECA) Study. JAMA 264 (19), 2511–2518.

Rezaei, F., Emami, M., Zahed, S., Morabbi, M.J., Farahzadi, M., Akhondzadeh, S., 2015. Sustained-release methylphenidate in methamphetamine dependence treatment: a double-blind and placebo-controlled trial. Daru 23, 2.

Ross, S., Peselow, E., 2012. Co-occurring psychotic and addictive disorders: neurobiology and diagnosis. Clin. Neuropharmacol. 35 (5), 235–243.

Ross, J., Teesson, M., Lejuez, C., et al., 2016. The efficacy of behavioural activation treatment for co-occurring depression and substance use disorder (the activate study): a randomized controlled trial. BMC Psychiatry 16, 221.

Rounsaville, B.J., Weissman, M.M., Crits-Christoph, K., Wilber, C., Kleber, H., 1982. Diagnosis and symptoms of depression in opiate addicts. Course and relationship to treatment outcome. Arch. Gen. Psychiatry 39 (2), 151–156.

Santucci, K., 2012. Psychiatric disease and drug abuse. Curr. Opin. Pediatr. 24 (2), 233–237.

Scherrer, J.F., Salas, J., Schneider, F.D., et al., 2017. Characteristics of new depression diagnoses in patients with and without prior chronic opioid use. J. Affect. Disord. 210, 125–129.

Schuckit, M.A., Smith, T.L., Chacko, Y., 2006. Evaluation of a depression-related model of alcohol problems in 430 probands from the San Diego prospective study. Drug Alcohol Depend. 82 (3), 194–203.

Schuckit, M.A., Smith, T.L., Danko, G.P., et al., 2007. A comparison of factors associated with substance-induced versus independent depressions. J. Stud. Alcohol Drugs 68 (6), 805–812.

Schwartz, C.E., Dorer, D.J., Beardslee, W.R., Lavori, P.W., Keller, M.B., 1990. Maternal expressed emotion and parental affective disorder: risk for childhood depressive disorder, substance abuse, or conduct disorder. J. Psychiatr. Res. 24 (3), 231–250.

Serafini, G., Adavastro, G., Canepa, G., De Berardis, D., Valchera, A., Pompili, M., Nasrallah, H., Amore, M., 2018. The efficacy of buprenorphine in major depression, treatment-resistant depression and suicidal behavior: a systematic review. Int. J. Mol. Sci. 19 (8), 2410.

Shoptaw, S., Heinzerling, K.G., Rotheram-Fuller, E., et al., 2008. Randomized, placebo-controlled trial of bupropion for the treatment of methamphetamine dependence. Drug Alcohol Depend. 96 (3), 222–232.

Sliman, S., Waalen, J., Shaw, D., 2016. Methamphetamine-associated congestive heart failure: increasing prevalence and relationship of clinical outcomes to continued use or abstinence. Cardiovasc. Toxicol. 16 (4), 381–389.

Stinson, F.S., Grant, B.F., Dawson, D.A., Ruan, W.J., Huang, B., Saha, T., 2005. Comorbidity between DSM-IV alcohol and specific drug use disorders in the United States: results from the National Epidemiologic Survey on alcohol and related conditions. Drug Alcohol Depend. 80 (1), 105–116.

Strain, E.C., Stitzer, M.L., Bigelow, G.E., 1991. Early treatment time course of depressive symptoms in opiate addicts. J. Nerv. Ment. Dis. 179 (4), 215–221.

Substance Abuse and Mental Health Services Administration, 2019. Key Substance use and Mental Health Indicators in the United States: Results from the 2018 National Survey on Drug Use and Health HHS Publication No. PEP19-5068, NSDUH Series H-54. Center for Behavioral Health Statistics and Quality, Substance Abuse and Mental Health Services Administration, Rockville, MD.

Swor, D.E., Maas, M.B., Walia, S.S., Bissig, D.P., Liotta, E.M., Naidech, A.M., Ng, K.L., 2019. Clinical characteristics and outcomes of methamphetamine-associated intracerebral hemorrhage. Neurology 93 (1), e1–e7.

Torrens, M., Rossi, P.C., Martinez-Riera, R., Martinez-Sanvisens, D., Bulbena, A., 2012. Psychiatric co-morbidity and substance use disorders: treatment in parallel systems or in one integrated system? Subst. Use Misuse 47 (8–9), 1005–1014.

van Harten, P.N., van Trier, J.C., Horwitz, E.H., Matroos, G.E., Hoek, H.W., 1998. Cocaine as a risk factor for neuroleptic-induced acute dystonia. J. Clin. Psychiatry 59 (3), 128–130.

Villanueva, R., 2013. Neurobiology of major depressive disorder. Neural Plast. 2013, 873278.

von Knorring, A.L., Cloninger, C.R., Bohman, M., Sigvardsson, S., 1983. An adoption study of depressive disorders and substance abuse. Arch. Gen. Psychiatry 40 (9), 943–950.

Witte, J., Bentley, K., Evins, A.E., Clain, A.J., Baer, L., Pedrelli, P., Fava, M., Mischoulon, D., 2012. A randomized, controlled, pilot study of acamprosate added to escitalopram in adults with major depressive disorder and alcohol use disorder. J. Clin. Psychopharmacol. 32 (6), 787–796.

Wusthoff, L.E., Waal, H., Grawe, R.W., 2014. The effectiveness of integrated treatment in patients with substance use disorders co-occurring with anxiety and/or depression—a group randomized trial. BMC Psychiatry 14, 67.

Young, M.A., Fogg, L.F., Scheftner, W.A., Fawcett, J.A., 1994. Interactions of risk factors in predicting suicide. Am. J. Psychiatry 151 (3), 434–435.

Zorick, T., Sugar, C.A., Hellemann, G., Shoptaw, S., London, E.D., 2011. Poor response to sertraline in methamphetamine dependence is associated with sustained craving for methamphetamine. Drug Alcohol Depend. 118 (2–3), 500–503.

Zweben, J.E., Cohen, J.B., Christian, D., Galloway, G.P., Salinardi, M., Parent, D., Iguchi, M., Methamphetamine Treatment Project, 2004. Psychiatric symptoms in methamphetamine users. Am. J. Addict. 13 (2), 181–190.

Chapter 32

Managing treatment-resistant depression in the setting of chronic pain

Ahmad Subhi Abu Mohammad and Sagar V. Parikh
Department of Psychiatry, University of Michigan, Ann Arbor, MI, United States

Among the many challenges of treating depression are the complications that arise when the depression is concurrent with other medical conditions. Not surprisingly, when a patient has two conditions, insufficient treatment of one condition may worsen the presentation or treatment outcome of the second condition. This is particularly true when dealing with concurrent mood problems and pain problems. Such complications arise from the bidirectional relationship between depression and pain (Gambassi, 2009). Individuals with serious acute pain that progresses to chronic pain are at higher risk for developing various psychiatric disorders, particularly depression. Patients with depression experience various physical symptoms, and various attributes of depression, including reduced energy and reduced ability to exercise, may, in turn, predispose to exacerbation of pain conditions. Chronic pain often generates pessimistic thoughts and feelings of frustration and helplessness; in turn, this cognitive frame allows for the evolution of depression as a clinical problem. At an even more fundamental level, the neural pathways for pain and the neural pathways for mood disorders share some common brain regions and certainly share many common neurotransmitters, particularly serotonin as a major signaling molecule (Goesling et al., 2013; Han and Pae, 2015). As the body tries to compensate for one condition by altering some attribute of the neurotransmitter in the initially affected system, this may in turn cause disruption of pathways that use the same neurotransmitter in the other system. Furthermore, once the individual has progressed to have depression and chronic pain, treatment for each condition independently may either help or hurt the other condition. Finally, when two disorders are present, it is common for both the patient and the clinician to focus treatment efforts on one disorder with the hope that any other sequelae will fade with the treatment of what is perceived to be the primary disorder. Treat the pain, and the depression might disappear, or treat the depression and the pain may lessen—so goes the mantra. Unfortunately, the reality is that by focusing on only one of two concurrent disorders, both conditions ultimately get treated more poorly.

In this chapter on the intersection between refractory depression and chronic pain disorders, we will begin by defining what constitutes chronic pain disorders since the definition of chronic or refractory depression is to be found elsewhere in this volume. We will look at the prevalence of pain disorders and in particular look at a couple of practical strategies to monitor the pain in the same spirit of measurement-based care that we typically apply to the management of mood disorders. Once we have established these common definitions in matrix, we systematically explore several key questions around the interface of depression and pain. From an epidemiologic perspective, we clarify how common pain is in depressed patients and conversely how common depression is in pain patients. Next, we moved to treatment by evaluating whether the presence of depression affects clinical outcomes in chronic pain treatment and the parallel question of whether pain affects depression outcomes. Finally, we look at the various treatments that have been explored for the interface of treating depression in the context of a chronic pain disorder.

Definition and measurement of chronic pain disorders

The International Association for the Study of Pain (IASP) defines chronic pain as pain that lasts beyond the expected time required for normal tissue healing and/or is intractable to treatment. From purely the pain perspective, it is estimated that just over 20% of the US adult population had a chronic pain disorder in 2016, representing approximately 50,000,000 adults. The impact of chronic pain is enormous in terms of suffering for the individual and families but also for society in terms of direct and indirect economic costs, which have been estimated to be over half a trillion dollars in 2010. Proper assessment and management of chronic pain is itself a major medical subspecialty and is certainly beyond the scope of this chapter. However, it is reasonable that the clinician treating depression and particularly refractory depression will need to be

Managing Treatment-Resistant Depression. https://doi.org/10.1016/B978-0-12-824067-0.00032-3

able to ask basic questions about the presence or absence of pain disorders and be able to measure whether the pain disorder is getting better or worse while depression is being treated.

For any patient, assessment of the various types of symptoms of pain along with their impact on social and occupational functioning is essential. Measuring that is exactly the same as measuring the social and occupational impact of a mood disorder, so a clinician skilled in treating depression can simply apply the same approach in measuring impairment related to pain. But getting a precise metric for the pain is also necessary. Much like depression, where many clinician and patient-rated scales exist, the pain literature is also full of many patient and clinician measures of pain. We have recently published a practical guide to measurement based care for depression, highlighting how to use commonly available rating scales and how frequently to implement them in daily practice (Hong et al., 2021). In a parallel way, two pain questionnaires, namely the McGill Pain Questionnaire (MPQ) and the Brief Pain Inventory (BPI), have been validated and are widely used and quite practical even for the clinician who does not specializes in pain. Both of these scales were originally validated for one type of pain and in the English language; subsequently, both scales have been validated in multiple languages and across diverse types of pain. Both scales are self-administered and take less than 5 min to complete. We review both questionnaires below and suggest that the clinician pick one scale and use this periodically—perhaps monthly in tandem with a mood rating scale—in the assessment of a patient who has a mood disorder and a pain disorder.

The McGill Pain Questionnaire and the Brief Pain Inventory

The McGill Pain Questionnaire (Melzack, 1975) and its more practical short form, the SF-MPQ (Melzack, 1987), evaluate sensory, affective-emotional, evaluative, and temporal aspects of the pain experience of the patient. The SF-MPQ consists of 15 questions which are divided into 11 sensory questions which identified the nature of the pain such as "sharp or shooting, etc." and for affective verbal descriptions such as "sickening, fearful, etc." The patient is asked to rate the intensity of each descriptor on a scale from 0 (none) to 3 (severe). Three pain scores are calculated: the sensory, the affective, and the total pain index; these are totaled to get an overall score, much like the PHQ-9. Typically, less than 5 min is needed for the patient to fill out the scale and to calculate a score. It is free to use, and widely available on the internet in multiple languages (Melzack, 1975).

The Brief Pain Inventory is a numeric rating scale from 0 to 10 that assesses the severity of pain and the degree of interference with functioning (Cleeland et al., 1994). It measures both pain intensity and pain interference with different life domains. It includes six questions that ask for the intensity and temporal nature of the pain, including a diagram of the body on which the patient marks where the pain is felt. One question includes seven specific items to rate the impact of the pain, during the past 24 h, on functioning, mood, and enjoyment of life, again scored 0–10. Two other questions summarize pain treatment and asked the patient to rate relief of pain from the treatment. Since these questions cover a range of topics, there is no simple algorithm to generate an overall score; instead, specific answers to each question need to be compared from visit to visit. The scale is under copyright but is free for clinical practice.

Summary of recommendations for assessment practices in comorbid pain and depression scenarios

Measurement-based care is enormously helpful for the successful treatment of most medical conditions, including either depression or pain disorders. Given the additional complexity when there are two disorders, it is even more important to ensure that each condition is properly measured and monitored over time. We, therefore, make the following suggestions:

1. As part of an assessment for a mood disorder, at a minimum ask the patient to complete the patient self-report form for depression, the Patient Health Questionnaire (PHQ-9), and the patient self-report for General Anxiety Disorder (GAD-7). Ideally, this would be complemented by a clinician-assessed scale; we recommend the Montgomery-Asberg Depression Rating Scale.
2. As part of an assessment for a mood disorder, ask if there is a significant problem with pain, and document its location and intensity. If the patient indicates that the pain is significantly impairing, ask the patient to complete either the SF-MPQ or the BPI.
3. In ongoing care, use a mood rating scale at regular intervals, such as once per month during active treatment without remission. During the same appointment, a single question should be asked about any chronic pain problems; if present, again administer the same pain questionnaire previously administered.
4. In addition to reviewing depression treatment, ask about patient preferences and current practices of pain management. If the patient needs treatment of pain, explore with the patient and consider whether a referral to an outside resource for pain management is necessary.

How common is depression in pain patients?

Two recent reviews (DeJean et al., 2013; IsHak et al., 2018) have clarified the prevalence of major depression across chronic medical illnesses more broadly and also specifically in a variety of chronic pain disorders. The findings vary enormously in terms of reported rates of prevalence since the studies that are summarized in these reviews span many different medical and psychiatric settings and use a variety of measurement instruments. While some of the underlying studies have reported low rates of prevalence, most studies across multiple settings have reported that a substantial percentage, usually below 50%, of patients with chronic pain disorders, have major depression. Other studies have clarified that those with multiple pain symptoms are three to five times as likely to have major depression as those without pain symptoms. Furthermore, there appears to be a "dose response" relationship where those with more pain are more likely to experience clinical depression. The highest rates of clinical depression were described in pain clinics (52%), rheumatology or orthopedic clinics (56%), and in dental clinics addressing facial pain (85%) as reported by IsHak and colleagues.

How common is pain in depression patients?

Painful symptoms are very common in patients with major depression. However, given the wide variety of pain symptoms and the wide variety of tools to measure both pain and depression, it is difficult to accurately report the prevalence of a chronic pain disorder in a patient with major depression. As reported by Lepine and Briley, an analysis by the World Health Organization from primary care centers around the world suggested that 22% of all primary care patients suffer from debilitating pain and that these patients with a chronic pain disorder are four times as likely to have a serious problem with anxiety or depression than primary care patients without pain. Furthermore, Lepine and Briley noted that 75% of patients with major depression being treated in a primary care setting had a significant painful physical symptom, most commonly headaches, stomach aches, and neck and back pain. They concluded that chronic pain should be seen as a harbinger of depression and not an explanatory factor. Furthermore, they felt that chronic pain should be seen as a cardinal symptom of depression requiring elimination in order to achieve good treatment outcomes for depression (Lépine and Briley, 2004).

IsHak and colleagues, in their 2018 review, explored this question with a bidirectional approach, starting with 14 studies where individuals were first identified as depressed and then assessed for pain. Next, they examined 42 articles looking at patients with a painful condition who were then subsequently assessed for depression. They found multiple difficulties in evaluating this evidence given that studies varied greatly in sample size, thoroughness of the pain assessment, and variability in the assessment of depression. Nevertheless, they did suggest that the prevalence of significant pain in depressed patients ranged from 15% to 100% with a mean prevalence of 65%. Together, the two review papers strongly emphasize high rates of pain symptoms if not full-fledged chronic pain disorder in patients with depression. They also emphasize that the presence of such pain complicates and potentially prolongs the treatment of depression, underscoring the need for mental health professionals to assess and help manage pain even if their primary goal is to treat a mood disorder. In essence, effective management of major depression requires attention to the management of the pain condition, whether it is a severe and chronic pain disorder or a more isolated significant pain symptom. Thus, there is the need for routine assessment of pain symptoms in patients with depression—ideally with the McGill Pain Questionnaire or the Brief Pain Inventory!

What is the impact of pain and depression comorbidity, in terms of outcomes either for depression or for pain?

As noted earlier, chronic pain conditions have a substantial impact on suffering, overall functioning, and financial burden for both the individual and for society. Similarly, major depressive disorders are well known to be impairing across a variety of functional domains and have recently emerged to be the major cause of disability worldwide, according to the World Health Organization (Depression and Other Common Mental Disorders: Global Health Estimates. Geneva: World Health Organization; 2017). When an individual has the misfortune to experience both a pain disorder and mood disorder, it can only be expected that suffering increases and functioning is even further impaired. More specifically, it would be reasonable to predict that pain symptoms would worsen depression outcomes, and depression symptoms would worsen pain outcomes. Indeed, extensive studies by Von Korff et al. (1988) identified a number of pain complaints that are associated with increased symptoms of depression. These researchers also found that the worse the pain at baseline, the worse the depression outcome both in terms of symptoms and functioning. As reviewed by (IsHak and colleagues (2018)), a number of studies have substantiated this point, demonstrating that inadequate pain control clearly worsens depressive symptoms over time and conversely that improvement in pain symptoms was associated with a decrease in depressive symptoms.

Approaches to treatment of comorbid pain and depression

Both pain and depression share etiologies that are a blend of neurobiological, psychological, and environmental factors. Since the environmental or social determinants of health are beyond the scope of this chapter, medications remain a cardinal intervention that is widely applicable and feasible. Psychological factors and psychological interventions are common to both pain and disorder management and depression management, but access and implementation of psychological treatments for either disorder remains problematic. In particular, while cognitive behavior therapy for depression is widely available in a variety of formats such as individual therapy, group interventions, and online intervention via apps and websites, cognitive behavior therapy for pain disorders or other kinds of psychosocial treatments for pain disorders is much less available. Psychosocial interventions for pain disorders have a substantial evidence base and are widely used in specialized pain clinics, but the interventions are often specific to the type of pain and are not readily accessible to the average practitioner who wishes to prescribe it. Two useful articles include a comprehensive summary of psychosocial processes in pain (Edwards et al., 2016) and a brief review of psychosocial treatments (Sturgeon, 2014). In addition, a key resource in this area is the 2019 book, Handbook of Psychosocial Interventions for Chronic Pain: An Evidence-Based Guide, edited by Andrea Kohn Maikovich-Fong; Routledge, 2019. Additionally, some other "mind-body" interventions may be useful for comorbid pain and depression, such as yoga. A recent review summarizes this evidence (Sud et al., 2021).

Medication treatment of comorbid pain and depression

Evidence-based medicine is predicated on the use of well-designed and well-conducted research studies, preferably randomized controlled trials. Such well-designed studies would include rigorously established diagnoses, carefully defined treatments, and standardized outcome scales that are statistically analyzed according to widely accepted methods. Furthermore, when there are multiple studies for the treatment of a particular condition, a properly designed metaanalysis should be conducted to synthesize all high-quality evidence into a more definitive answer.

As we have established to our earlier review, almost none of these conditions are met when examining the issue of comorbid depression and pain disorders. While there are a significant number of studies of treatments, they suffer from the same kinds of problems identified in studies of prevalence and outcome—great variability in how either pain or depression is diagnosed or measured, how the patient population is defined, and what outcome is used. It is also unclear how a metaanalysis of say multiple trials for one type of pain such as back pain and a particular medication would be applicable to another type of pain concomitant with depression. In order to deal with this heterogeneity of research, we will summarize key medication treatments and the relevant evidence. Most of the evidence reviewed will be individual studies, although there are occasional review articles and very rarely a med analysis that is available for interpretation. Therefore, clinical practice will depend on the clinician interpreting the evidence and identifying those parts of the evidence which may be relevant for the patient that they are trying to treat. We begin our medication treatment summary by exploring key review articles. Subsequently, we report evidence for specific medications, including whether this is supported by reviews or high-quality evidence, or other types of studies.

Overall results from antidepressants

Since tricyclic antidepressants have been marketed for over 50 years, they have been extensively used both for depression and for pain disorders. Few placebo-controlled studies, however, have been conducted, and so no metaanalysis is available. In a 2003 review by Bair and colleagues (Bair et al., 2003), specific positive findings in pain relief was summarized by conditions such as pain syndromes like fibromyalgia, irritable bowel syndrome, and migraines; for these conditions, a variety of antidepressants provided pain relief. In fact, a broader review of antidepressants for conditions featuring "somatic symptoms" (pain was only one type of somatic disorder included), found TCAs more useful than SSRIs in relieving somatic symptoms. Bair et al. also found 22 studies where individuals with comorbid pain and depression were treated, with the pain outcome as the primary outcome, and depression as the main secondary outcome. The review noted that 18 studies involved tricyclics, while four used SSRIs. Most of the tricyclic studies were uncontrolled, but demonstrated improvement in both pain and depression, while the results from SSRIs were less clear. A more recent review by Gebhardt et al. (2016) used a more rigorous study selection process, insisting on only placebo-controlled studies, and found only 19 studies, with 11 testing only SNRIs (principally venlafaxine and duloxetine), 3 testing both SNRIs and SSRIs, 3 only testing SSRIs, and just 2 studies employing tricyclics. Based on comparison to placebo, TCAs surprisingly did not beat placebo for analgesic relief, while both SSRIs and SNRIs beat placebo with a small effect size. The authors surmised that on balance, in view of the many positive uncontrolled studies of TCAs, the expectation is that TCAs should also have a small but positive benefit in reducing pain in individuals with comorbid depression. In summary, SNRIs should be the first choice in treating pain and depression, while SSRIs and TCAs have some use. For more specific conditions such as neuropathic pain or musculoskeletal pain, no generalizations about can be made, but there are positive studies with a variety of compounds (Bair et al., 2003; Kroenke et al., 2019).

Key antidepressants for special consideration

Amitriptyline is the most widely studied and used TCA for a variety of somatic conditions, with most studies reporting a primary outcome in the somatic condition and occasionally reporting depressive symptom change. A key limitation of this literature is the choice of dosing; for most somatic conditions, doses employed are usually well below the doses shown to be effective for depression. Such studies often find 25–50 mg of amitriptyline to be helpful for the pain or other somatic disorder—for instance, Uçeyler et al. (2008) summarized that 10 of 11 RCTs found amitriptyline reduced the pain of fibromyalgia. The majority of patients in such studies of somatic conditions may have depressive symptoms but not major depression, so only a few studies report improvement in depression. Since amitriptyline is considered a second-line antidepressant due to its high side effect burden and potential toxicity in overdose, it should not be used for comorbid major depression and pain unless SNRIs and SSRIs have both been tried first.

Venlafaxine is widely used in comorbid pain and depression, but clinical studies are few, resulting in a lack of review articles. Bradley et al., 2003 conducted the only significant study of chronic pain and depression treated with venlafaxine and reported improvement in both pain and depression scores after 1 year in this uncontrolled trial. Another trial of venlafaxine versus placebo in depressed spinal cord injury patients with pain showed benefit only on secondary measures of efficacy (Richards et al., 2015). As such, while compelling evidence is lacking, the previous metaanalysis of SNRIs had shown potential value, and venlafaxine's potent efficacy in moderate and severe depression makes it worth considering here. A derivative compound, **desvenlafaxine**, unfortunately, did not show efficacy compared to placebo for the treatment of fibromyalgia pain (Welsch et al., 2018), while depression was not assessed. While desvenlafaxine is a safe and effective treatment for major depression, no evidence exists to support the use of desvenlafaxine in pain and depression, or in refractory depression.

Duloxetine is the only modern antidepressant with an extensive research portfolio in various pain or somatic conditions. In fact, it is FDA-approved for many conditions including GAD, MDD, fibromyalgia, diabetic peripheral neuropathy, and musculoskeletal pain. The evidence for direct effect on pain, largely independent of the impact of the medication on depression, is substantial for fibromyalgia (Marangell et al., 2011), neuropathy (Wong et al., 2007), and musculoskeletal pain, both arthritic and post orthopedic surgery (Gao et al., 2019; Koh et al., 2019). In the situation of a specific study of comorbid depression and pain, no specific trial of MDD patients with pain was done with depression outcomes as a primary outcome. Nevertheless, improvement in pain and depression symptoms (via secondary outcomes in RCTs) is demonstrated across a variety of patient populations and types of pain. Hence, duloxetine should be considered as a first-line treatment for those individuals with pain and depression comorbidity.

Isolated trials suggest the utility of several other antidepressants. **Bupropion** has a single trial of individuals with NO depression but with neuropathic pain showed improvement in pain (Semenchuk et al., 2001). A large observational study of patients with chronic pain and depression showed no clear efficacy of **mirtazapine** (Freynhagen et al., 2006). One recent study of **vortioxetine** in a specific condition—burning mouth syndrome—with modest benefit (Adamo et al., 2021). No studies of vortioxetine's effect on comorbid depression and pain were found. One placebo-controlled trial has demonstrated benefit in both pain and depression in fibromyalgia from **quetiapine**, which is approved as a monotherapy treatment for depression despite its original development as an antipsychotic (McIntyre et al., 2014). No relevant studies were found involving **vilazodone or levomilnacipran** (Semenchuk et al., 2001).

Ketamine has emerged as a major new treatment for refractory depression but also has a long history in the management of chronic pain. Ketamine has multiple levels of complexity; it has two enantiomers, multiple metabolites some of which may be active, and affects multiple brain systems, including particularly the glutamate system, but also the opioid, monoamine, and cholinergic systems. Furthermore, ketamine also has multiple widely used methods of administration—most often IV infusion of racemic ketamine, but also intranasal, sublingual, intramuscular, and oral routes. For TRD, IV administration of racemic ketamine has established value in treating depression and also independently in reducing suicidality (McIntyre et al., 2021; Wilkinson et al., 2018). The two enantiomers, (R) and (S), each may have different efficacy and side effect profiles (Jelen et al., 2021). One of these, esketamine, has been approved as an add-on treatment for TRD via intranasal administration (Papakostas et al., 2020). Preliminary evidence also indicates the efficacy of oral and sublingual ketamine for TRD (Rosenblat et al., 2019).

Predating ketamine's use in TRD was off-label use for pain of many different types, sometimes refractory chronic pain, sometimes for very specific types of pain, and often as part of palliative care in cancer-related pain (Culp et al., 2020). While such applications are often based on uncontrolled studies, the evidence is sufficiently compelling that many health care entities provide guidelines for its routine administration (Tran and McCormack, 2020). Its use in pain, however, differs in one dramatic way from psychiatric use of ketamine—much like other analgesics, it is often administered several times a day.

Differing frequencies and dosing of ketamine, compounded by the fact that ketamine for pain is more commonly administer orally while ketamine for depression is commonly by IV, complicate recommendations about the use of ketamine in comorbid TRD and pain. In the absence of any specific studies, only expert opinion can provide recommendations that must remain tentative. Firstly, treat TRD with ketamine in a fashion that is recommended and effective for that person's TRD. While it is common to administer either IV ketamine or intranasal esketamine twice per week initially, patients may use ketamine less frequently after the depression improves. However, the pain benefits of ketamine do not last beyond 24 h. As such, other analgesics will particularly be needed for pain management beyond the 24-h period. Alternatively, it may be possible to administer lower doses of ketamine, particularly orally, on a daily basis to manage the pain and derive some benefit for the treatment of depression. Here, data from a metaanalysis of oral ketamine (Rosenblat et al., 2019) may provide support for the administration of 100–200 mg daily for TRD; this dosage is at the low end of dosages for oral ketamine in pain conditions but may be useful for the less severe pain situations. Careful evaluation of the side effects of such frequent and potentially higher-than-usual dosing of ketamine will be essential, as will management of any addiction potential and careful monitoring of other analgesics.

Conclusion

Management of TRD with comorbid pain is difficult for the clinician. It is even more difficult for the patient. Pain complicates depression, and depression complicates pain. There is little value in arbitrarily deciding which comes first and aggravates the other (the chicken or egg first dilemma). Instead, humane management compels careful measurement of both pain and depression, and consistent monitoring of both outcomes when pursuing the improvement of TRD. We have identified easy-to-use pain measures that are self-reports (such as the short form of the McGill Pain Questionnaire) that may be used in conjunction with traditional psychiatric scales like the PHQ-9, and recommend this strategy to evaluate and monitor progress in patients with these two conditions. But measurement of outcomes by even simpler methods can be useful, such as asking the patient to rate the pain in the preceding week on a 1 (no pain) to 10 (maximum pain). Once both TRD and pain are measured, both should be monitored and specifically targeted in order to improve both conditions. Improvement of pain management will often improve depression symptoms; improvement of depression will often reduce pain. Since it is often unwieldy to have the patient have a separate "pain physician," it is useful for the psychiatrist to measure and manage both conditions. Fortunately, a number of antidepressants—particularly duloxetine and venlafaxine—have special value in TRD and in pain disorders, and should be considered first. Other antidepressants may also have a role, but ketamine will likely emerge in the future as a major advance not only to TRD but also to TRD with comorbid pain.

References

Adamo, D., Pecoraro, G., Coppola, N., Calabria, E., Aria, M., Mignogna, M., 2021. Vortioxetine versus other antidepressants in the treatment of burning mouth syndrome: an open-label randomized trial. Oral Dis. 27 (4), 1022–1041. https://doi.org/10.1111/odi.13602.

Bair, M.J., Robinson, R.L., Katon, W., Kroenke, K., 2003. Depression and pain comorbidity: a literature review. Arch. Intern. Med. 163 (20), 2433–2445.

Bradley, R.H., Barkin, R.L., Jerome, J., DeYoung, K., Dodge, C.W., 2003. Efficacy of venlafaxine for the long term treatment of chronic pain with associated major depressive disorder. Am. J. Ther. 10 (5), 318–323.

Cleeland, C.S., Gonin, R., Hatfield, A.K., Edmonson, J.H., Blum, R.H., Stewart, J.A., Pandya, K.J., 1994. Pain and its treatment in outpatients with metastatic cancer. N. Engl. J. Med. 330 (9), 592–596.

Culp, C., Kim, H.K., Abdi, S., 2020. Ketamine use for cancer and chronic pain management. Front. Pharmacol. 11. https://doi.org/10.3389/fphar.2020.599721, 599721.

DeJean, D., Giacomini, M., Vanstone, M., Brundisini, F., 2013. Patient experiences of depression and anxiety with chronic disease: a systematic review and qualitative meta-synthesis. Ont. Health Technol. Assess. Ser. 13 (16), 1–33.

Edwards, R.R., Dworkin, R.H., Sullivan, M.D., Turk, D.C., Wasan, A.D., 2016. The role of psychosocial processes in the development and maintenance of chronic pain. J. Pain 17 (9 Suppl), T70–T92. https://doi.org/10.1016/j.jpain.2016.01.001.

Freynhagen, R., Muth-Selbach, U., Lipfert, P., Stevens, M.F., Zacharowski, K., Tölle, T.R., von Giesen, H.-J., 2006. The effect of mirtazapine in patients with chronic pain and concomitant depression. Curr. Med. Res. Opin. 22 (2), 257–264.

Gambassi, G., 2009. Pain and depression: the egg and the chicken story revisited. Arch. Gerontol. Geriatr. 49 (Suppl. 1), 103–112. https://doi.org/10.1016/j.archger.2009.09.018.

Gao, S.-H., Huo, J.-B., Pan, Q.-M., Li, X.-W., Chen, H.-Y., Huang, J.-H., 2019. The short-term effect and safety of duloxetine in osteoarthritis: a systematic review and meta-analysis. Medicine 98 (44). https://doi.org/10.1097/MD.0000000000017541, e17541.

Gebhardt, S., Heinzel-Gutenbrunner, M., König, U., 2016. Pain relief in depressive disorders: a meta-analysis of the effects of antidepressants. J. Clin. Psychopharmacol. 36 (6), 658–668.

Goesling, J., Clauw, D.J., Hassett, A.L., 2013. Pain and depression: an integrative review of neurobiological and psychological factors. Curr. Psychiatry Rep. 15 (12), 421. https://doi.org/10.1007/s11920-013-0421-0.

Han, C., Pae, C.-U., 2015. Pain and depression: a neurobiological perspective of their relationship. Psychiatry Investig. 12 (1), 1–8. https://doi.org/10.4306/pi.2015.12.1.1.

Hong, R.H., Murphy, J.K., Michalak, E.E., Chakrabarty, T., Wang, Z., Parikh, S.V., Culpepper, L., Yatham, L.N., Lam, R.W., Chen, J., 2021. Implementing measurement-based care for depression: practical solutions for psychiatrists and primary care physicians. Neuropsychiatr. Dis. Treat. 17, 79–90. https://doi.org/10.2147/NDT.S283731.

IsHak, W.W., Wen, R.Y., Naghdechi, L., Vanle, B., Dang, J., Knosp, M., Dascal, J., Marcia, L., Gohar, Y., Eskander, L., Yadegar, J., Hanna, S., Sadek, A., Aguilar-Hernandez, L., Danovitch, I., Louy, C., 2018. Pain and depression: a systematic review. Harv. Rev. Psychiatry 26 (6), 352–363. https://doi.org/10.1097/HRP.0000000000000198.

Jelen, L.A., Young, A.H., Stone, J.M., 2021. Ketamine: a tale of two enantiomers. J. Psychopharmacol. (Oxford, England) 35 (2), 109–123. https://doi.org/10.1177/0269881120959644.

Koh, I.J., Kim, M.S., Sohn, S., Song, K.Y., Choi, N.Y., In, Y., 2019. Duloxetine reduces pain and improves quality of recovery following total knee arthroplasty in centrally sensitized patients: a prospective, randomized controlled study. J. Bone Joint Surg. Am. 101 (1), 64–73. https://doi.org/10.2106/JBJS.18.00347.

Kroenke, K., Baye, F., Lourens, S.G., Evans, E., Weitlauf, S., McCalley, S., Porter, B., Matthias, M.S., Bair, M.J., 2019. Automated self-management (ASM) vs. ASM-enhanced collaborative care for chronic pain and mood symptoms: the CAMMPS randomized clinical trial. J. Gen. Intern. Med. 34 (9), 1806–1814. https://doi.org/10.1007/s11606-019-05121-4.

Lépine, J.-P., Briley, M., 2004. The epidemiology of pain in depression. Hum. Psychopharmacol. 19 (Suppl. 1), S3–S7.

Marangell, L.B., Clauw, D.J., Choy, E., Wang, F., Shoemaker, S., Bradley, L., Mease, P., Wohlreich, M.M., 2011. Comparative pain and mood effects in patients with comorbid fibromyalgia and major depressive disorder: secondary analyses of four pooled randomized controlled trials of duloxetine. Pain 152 (1), 31–37. https://doi.org/10.1016/j.pain.2010.05.029.

McIntyre, A., Paisley, D., Kouassi, E., Gendron, A., 2014. Quetiapine fumarate extended-release for the treatment of major depression with comorbid fibromyalgia syndrome: a double-blind, randomized, placebo-controlled study. Arthritis Rheumatol. (Hoboken, NJ) 66 (2), 451–461. https://doi.org/10.1002/art.38228.

McIntyre, R.S., Rosenblat, J.D., Nemeroff, C.B., Sanacora, G., Murrough, J.W., Berk, M., Brietzke, E., Dodd, S., Gorwood, P., Ho, R., Iosifescu, D.V., Lopez Jaramillo, C., Kasper, S., Kratiuk, K., Lee, J.G., Lee, Y., Lui, L.M.W., Mansur, R.B., Papakostas, G.I., et al., 2021. Synthesizing the evidence for ketamine and esketamine in treatment-resistant depression: an international expert opinion on the available evidence and implementation. Am. J. Psychiatry 178 (5), 383–399. https://doi.org/10.1176/appi.ajp.2020.20081251.

Melzack, R., 1975. The McGill Pain Questionnaire: major properties and scoring methods. Pain 1 (3), 277–299. https://doi.org/10.1016/0304-3959(75)90044-5.

Melzack, R., 1987. The short-form McGill Pain Questionnaire. Pain 30 (2), 191–197. https://doi.org/10.1016/0304-3959(87)91074-8.

Papakostas, G.I., Salloum, N.C., Hock, R.S., Jha, M.K., Murrough, J.W., Mathew, S.J., Iosifescu, D.V., Fava, M., 2020. Efficacy of esketamine augmentation in major depressive disorder: a meta-analysis. J. Clin. Psychiatry 81 (4). https://doi.org/10.4088/JCP.19r12889.

Richards, J.S., Bombardier, C.H., Wilson, C.S., Chiodo, A.E., Brooks, L., Tate, D.G., Temkin, N.R., Barber, J.K., Heinemann, A.W., McCullumsmith, C., Fann, J.R., 2015. Efficacy of venlafaxine XR for the treatment of pain in patients with spinal cord injury and major depression: a randomized, controlled trial. Arch. Phys. Med. Rehabil. 96 (4), 680–689. https://doi.org/10.1016/j.apmr.2014.11.024.

Rosenblat, J.D., Carvalho, A.F., Li, M., Lee, Y., Subramanieapillai, M., McIntyre, R.S., 2019. Oral ketamine for depression: a systematic review. J. Clin. Psychiatry 80 (3). https://doi.org/10.4088/JCP.18r12475.

Semenchuk, M.R., Sherman, S., Davis, B., 2001. Double-blind, randomized trial of bupropion SR for the treatment of neuropathic pain. Neurology 57 (9), 1583–1588.

Sturgeon, J.A., 2014. Psychological therapies for the management of chronic pain. Psychol. Res. Behav. Manage. 7, 115–124. https://doi.org/10.2147/PRBM.S44762.

Sud, A., Lai, K.S.P., Cheng, D.K.-Y., Chung, C., Pico-Espinosa, O.J., Rice, D.B., 2021. Mind-body interventions for depressive symptoms in chronic pain: a systematic review of meta-analyses. Pain Physician 24 (1), 61–72.

Tran, K., McCormack, S., 2020. Ketamine for Chronic Non-Cancer Pain: A Review of Clinical Effectiveness, Cost-Effectiveness, and Guidelines (No. 33231962). Canadian Agency for Drugs and Technologies in Health.

Uçeyler, N., Häuser, W., Sommer, C., 2008. A systematic review on the effectiveness of treatment with antidepressants in fibromyalgia syndrome. Arthritis Rheum. 59 (9), 1279–1298. https://doi.org/10.1002/art.24000.

Von Korff, M., Dworkin, S.F., Le Resche, L., Kruger, A., 1988. An epidemiologic comparison of pain complaints. Pain 32 (2), 173–183. https://doi.org/10.1016/0304-3959(88)90066-8.

Welsch, P., Üçeyler, N., Klose, P., Walitt, B., Häuser, W., 2018. Serotonin and noradrenaline reuptake inhibitors (SNRIs) for fibromyalgia. Cochrane Database Syst. Rev. 2, CD010292. https://doi.org/10.1002/14651858.CD010292.pub2.

Wilkinson, S.T., Ballard, E.D., Bloch, M.H., Mathew, S.J., Murrough, J.W., Feder, A., Sos, P., Wang, G., Zarate, C.A., Sanacora, G., 2018. The effect of a single dose of intravenous ketamine on suicidal ideation: a systematic review and individual participant data meta-analysis. Am. J. Psychiatry 175 (2), 150–158. https://doi.org/10.1176/appi.ajp.2017.17040472.

Wong, M., Chung, J.W.Y., Wong, T.K.S., 2007. Effects of treatments for symptoms of painful diabetic neuropathy: systematic review. BMJ 335 (7610), 87.

Chapter 33

Inhaled gases for treatment-resistant major depression

Charles R. Conway[a], Brian J. Mickey[b,c,d], Ben J.A. Palanca[a,e], Scott C. Tadler[b,c], and Peter Nagele[f]

[a]*Department of Psychiatry, Washington University School of Medicine, St. Louis, MO, United States,* [b]*Department of Psychiatry, The University of Utah, Salt Lake City, UT, United States,* [c]*Department of Anesthesiology, The University of Utah, Salt Lake City, UT, United States,* [d]*Department of Bioengineering, The University of Utah, Salt Lake City, UT, United States,* [e]*Department of Anesthesiology, Washington University School of Medicine, St. Louis, MO, United States,* [f]*Department of Anesthesia and Critical Care, The University of Chicago, Chicago, IL, United States*

Introduction

Fortunately, most individuals who suffer from major depressive disorder (MDD) respond to first-line antidepressant treatments [pharmacotherapy and psychotherapy; (American Psychiatric Association, 2010; Rush et al., 2006)]. Over the course of the past 20 years, there has been increasing awareness that a significant subset of patients suffer from a more virulent form of MDD that does not respond to a series of standard antidepressant treatments. These patients are said to suffer from "treatment-resistant major depression" (TRMD). Though the field of psychiatry has yet to agree upon how many failed treatments constitute TRMD, there is emerging consensus that failure to respond to two first-line treatments from different antidepressant classes of adequate dose and duration is a clinically useful definition, as the likelihood of responding to a third treatment is substantially lower (Conway et al., 2017; Gaynes et al., 2020).

Though a relatively small subset (20%–30%) of MDD patients have TRMD, given the relatively high lifetime prevalence of MDD in the United States (10%–20%), approximately 7–20 million Americans suffer from TRMD, which represents a significant and increasingly prevalent public health concern (Hasin et al., 2018; Rush et al., 2012; Thomas et al., 2013). Evidence supports that TRMD patients are at significantly higher risk for suicide attempts, have greater rates of hospitalization and health care utilization, and per patient are much more costly to treat than non-TRMD MDD patients (Amos et al., 2018; Bergfeld et al., 2018; Corey-Lisle et al., 2002; Crown et al., 2002; Gibson et al., 2010; Greenberg et al., 2004; Lin et al., 2019).

Over the course of the past 15 years, there have been important discoveries allowing the investigation and clinical application of novel approaches to the treatment of TRMD. One of these novel treatment classes explored is inhaled (gaseous) agents. This chapter reviews three emerging inhaled treatments which target TRMD: nitrous oxide (laughing gas), isoflurane, and xenon. We also briefly discuss the use of aromatic compounds in the treatment of mood disorders.

Nitrous oxide

Colloquially referred to as "laughing gas," nitrous oxide is a colorless, odorless anesthetic gas with a long history of medical use dating back 150 years (Maze et al., 2008). The current clinical uses of nitrous oxide are primarily as an inhalational anesthetic agent or employed during short, painful procedures in dentistry, labor and delivery, and emergency medicine (Huang and Nathaniel, 2016; Likis et al., 2014; Tobias, 2013). Most recently, nitrous oxide has emerged as a potential treatment for depression, suicidal ideation, and posttraumatic stress disorder.

Nitrous oxide for major depressive disorder

Similar to ketamine, which has rapid efficacy for treatment-resistant depression, nitrous oxide is a highly potent *N*-methyl-D-aspartate receptor (NMDAR) antagonist. This knowledge, along with the well-established euphoric effects of nitrous oxide, prompted speculation that it may have clinical antidepressant effects.

Nagele et al. (2015) conducted a proof-of-concept double-blind, prospective, cross-over trial of 20 TRMD patients using inhaled nitrous oxide. TRMD patients underwent serial treatments with an hour-long inhalation with either 50% nitrous

Managing Treatment-Resistant Depression. https://doi.org/10.1016/B978-0-12-824067-0.00033-5

oxide/50% oxygen (active treatment); or 50% nitrogen/50% oxygen (placebo). These sessions were randomized in order and separated by 1 week. To be eligible for the trial, patients had to have failed at least 2 adequate dose-duration antidepressant trials (confirmed by medical record review), and could not have recent history of substance abuse (past 12 months, except nicotine), any lifetime history of psychotic or bipolar illness, or severe personality disorder. Patients were allowed to maintain their existing antidepressant treatments but were requested not to start any new antidepressant treatment or modify their treatment dosage in the 4 weeks prior to trial entry. To avoid nitrous oxide's potential euphoric effects being confounded with antidepressant effects, the group selected 24 h postinhalation as the primary endpoint for assessment of antidepressant effects. TRMD patients enrolled in the trial suffered from sustained and severe refractory MDD: patients had a mean of 19 years of lifetime MDD, history of 8 average failed antidepressant treatments, and 4 had a history of failing to respond to ECT.

The results of the pilot trial were promising. At 24 h postinhalation, there was a statistically significant reduction in depressive symptoms in those who had received nitrous oxide versus placebo. Those receiving nitrous oxide experienced a mean decrease of 5.5 points on the Hamilton Depression Rating Scale (HDRS; 95 CI −2.5 to −8.5 points) at 24 h. There were 4 TRMD patients achieving antidepressant response (≥50% reduction from baseline HDRS score) and one achieving remission at 24 h; whereas only one placebo patient achieved response, and no placebo patients achieved remission (Nagele et al., 2015).

Interestingly, not all antidepressant symptoms improved equally, i.e., certain depressive symptoms appeared to be more responsive than others to nitrous oxide, including anxiety, depressed mood, suicidal ideation, and guilt, which has directed future studies (see "Future studies of nitrous oxide" section below). Similar to the observations of ketamine's rapid antidepressant effects in TRMD, the effects of a single inhalation of nitrous oxide were frequently sustained; several patients who were randomized to receive nitrous oxide first showed sustained antidepressant effects on assessment at 1 week. Hence, this prompted future cross-over studies of nitrous oxide in TRMD having longer durations between exposures.

Also of importance, nitrous oxide was well-tolerated; the side effects were infrequent, minor, and short-lived. These included nausea and vomiting (15%), headache (10%), and paradoxical anxiety (10%) in a small subset of patients.

Most recently, Nagele et al. (2021) conducted a double-blind, randomized, cross-over dose finding study of nitrous oxide. In this trial, 24 patients with severe TRD were randomly assigned to receive a one-hour inhalation with either (1) 50% nitrous oxide, (2) 25% nitrous oxide, or (3) placebo (air/oxygen). The primary outcome was the Hamilton Depression Rating Scale (HDRS-21). Both nitrous groups combined demonstrated significant improvement versus placebo ($P = 0.01$); however, there was no acute difference in antidepressant effects between 25% and 50% nitrous oxide ($P = 0.58$). However, adverse events (nausea, headache, dissociative feeling) were statistically higher in the 50% group ($P < 0.001$). There was a suggestion that 50% nitrous may have had more sustained antidepressant benefit. Further study of optimal dose finding is warranted.

Though limited in scope, this study (and our group's continued studies in this area) support that nitrous oxide may have several advantages over ketamine as a rapid acting antidepressant. Unlike ketamine, existing data do not support emergence of psychosis as a common side effect, and nitrous oxide inhalation does not influence systemic blood pressure. Though nitrous oxide does have limited abuse potential (see "Abuse potential of nitrous oxide" section below), it has considerably less addictive potential than ketamine. Last, given that nitrous oxide is not hepatically or renally metabolized, it is rapidly cleared from the system, allowing patients to safely operate motor vehicles following treatments, whereas ketamine has a markedly longer recovery time and requires a driver postadministration.

Clinical administration of nitrous oxide for TRMD

From a regulatory standpoint, nitrous oxide is not regulated by the Drug Enforcement Agency. Occupational Safety and Health Administration (OSHA) does not have standards for nitrous exposure. The National Institute for Occupational Safety and Health, NIOSH, however, has a recommended exposure limit of 25 ppm in closed rooms.

Potential effects on psychomotor and cognitive function in health care workers leads to the need of minimizing leaks in gas delivery circuits and adequate exhaust through wall-suction waste gas scavenging systems. Closely fitting masks minimize leakage around the face. Even with a closed system, nitrous oxide is not effective as a single-agent general anesthetic.

Medical grade nitrous oxide can be acquired from gas distributors as *E*-cylinder tanks that can be dispensed through FDA-approved inhalation circuits or anesthesia ventilators. Nitrous oxide and oxygen/air flow meters allow accurate dispensing. Proportioning systems link the flows of oxygen and nitrous oxide to prevent the administration of mixtures that would induce hypoxia. Infrared spectroscopic gas analyzers and infrared pulse oximeters allow additional safety monitoring for detecting potential hypoxemia. Suction and airway equipment should be readily available. Staff trained in administering sedatives and airway support ensure safety, as per American Society of Anesthesia guidelines on sedation and analgesia by nonanesthesiologists.

Patient preparation and selection may minimize the risk of adverse events even for short inhalations. Fasting guidelines should be considered to minimize the possibility of aspiration events. A history of motion sickness or postoperative vomiting may portend susceptibility to the side effect of nausea. Respiratory depressant effects of narcotics and sedatives are augmented by nitrous oxide. As nitrous oxide accumulates in tissues with collected air, it is contraindicated in patients with pneumothorax, pneumomediastinum, or gastrointestinal distention. Patients with B12 deficiency or particular mutations in the methylenetetrahydrofolate reductase (*MTHFR*) gene should be reconsidered given the inhibition of methionine synthase by nitrous oxide.

With adherence to these considerations the clinical administration of nitrous oxide can be administered safely for TRMD.

Putative potential mechanism of action of nitrous oxide in TRMD

Despite over 150 years of clinical use for analgesia and anesthesia, how nitrous oxide brings about its effects is not fully understood. Below we discuss possible mechanisms of action.

Effects on the NMDAR

Extensive research over the past 20 years has conclusively demonstrated that nitrous oxide acts as a noncompetitive inhibitor of NMDARs by blocking NMDAR responses at approximately a half-maximal effective concentration of 30%–40%; however, nitrous oxide is only a partial NMDAR antagonist at concentrations up to 80% (Jevtović-Todorović et al., 1998; Maze et al., 2008; Mennerick et al., 1998; Zorumski et al., 2015). The effects of nitrous oxide do not alter the decay of NMDAR-mediated currents, and are weakly voltage dependent (Mennerick et al., 1998). Additionally, work in animal models provides supporting evidence that nitrous acts in vivo via NMDAR-antagonistic behavioral effects (Nagele et al., 2004; Sato et al., 2005). Further, nitrous oxide also acts as a weak inhibitor of AMPA/kainite-type glutamate receptors.

Other nitrous oxide receptor interactions

In addition to acting on the NMDAR, nitrous oxide is known to act on numerous other receptors, including GABA-A, GABA-C, serotonin-3, potassium and calcium channel receptors, nicotinic acetylcholine receptors, alpha-adrenergic receptors, and opioid receptors.

Nitrous oxide, unlike many other anesthetic agents, has weak potentiating effects at the GABA-A receptor (Mennerick et al., 1998), while weakly inhibiting the GABA-C receptors (Yamakura and Harris, 2000). Nitrous oxide inhibits low voltage activated (LVA, T-type) calcium channels, producing depression of LVA currents at 80% nitrous oxide (Todorovic et al., 2001), without having an effect on the high voltage activated (L-type) calcium currents (Mennerick et al., 1998). Further, nitrous oxide is a weak antagonist of serotonin-type 3 receptors, and partially inhibits a subtype of nicotinic acetylcholine receptors (Yamakura and Harris, 2000). Nitrous oxide is an agonist at the two-pore domain potassium channel (TREK-1), which is expressed throughout the central nervous system (Gruss et al., 2004), as well as activating alpha-adrenergic receptors associated with antinociceptive effects in the brainstem and spinal cord in rat models (Orii et al., 2002; Sawamura et al., 2000).

Finally, nitrous oxide appears to have modulatory effects on the endogenous opioid system, which could be linked to both analgesia and psychotropic effects (Daras et al., 1983; Fang et al., 1997; Ohashi et al., 2003; Ori et al., 1989). This is potentially highly relevant in TRMD, as evidence suggests that ketamine has diminished antidepressant effects when combined with an opioid antagonist (Williams et al., 2018). Existing evidence supports that nitrous oxide may preferentially act on kappa-type opiate receptors over mu-type receptors (Fukagawa et al., 2014; Koyama and Fukuda, 2010).

Nitrous oxide effects on brain circuits

There is currently limited understanding of the effects of nitrous oxide on brain pathways. Animal models suggest that nitrous oxide significantly impacts the hippocampus: nitrous oxide brings about disinhibition of network-driven hippocampal spike firing (Nagashima et al., 2005). Another study demonstrated that nitrous oxide can increase cellular proliferation in dentate gyrus, possibly via neurogenesis (Chamaa et al., 2018).

Limited electroencephalography studies in humans demonstrated that nitrous oxide decreases functional connectivity in superficial parietal networks (Kuhlmann et al., 2013) and leads to an increase in theta activity and a decrease in frontal slow wave activity (delta) with drug discontinuation (Foster and Liley, 2013). How these changes are associated with proposed behavioral changes is not yet understood.

Abuse potential of nitrous oxide

There is longstanding history of recreational use of nitrous oxide, with descriptions of "laughing gas parties" during the Victorian era (van Amsterdam et al., 2015), and recreational use which continues to the present. Currently, recreational nitrous oxide use typically involves inhalation with either bulbs or balloons and is primarily limited to teens and young adults, with an estimated prevalence of 20% in this age group (Nagele et al., 2018). Contributing to the sustained popularity of recreational nitrous oxide is its wide availability, low cost, and relative safety. Though there are reports of psychological dependence and tolerance (Gillman, 1986), such cases are relatively rare, with most recreational users not developing physical or psychological dependence: over 90% report using once per month or less frequently (van Amsterdam et al., 2015).

Acute effects include rapid onset of euphoria, excitement, and dissociation, typically resolving within minutes. Though relatively rare, long-term abuse of nitrous oxide can be associated with permanent neurological damage, including spinal cord degeneration (Lan et al., 2019), which is thought to be brought about in part by nitrous oxide-induced vitamin B12 inactivation (Oussalah et al., 2019). Other known safety concerns previously associated with nitrous oxide use include dyspnea, headache, impaired judgment, and even impaired fertility with repeated exposure (NIH U.S. National Library of Medicine, 2020). Further, high concentration inhaled nitrous oxide (> 70%) can lead to hypoxia and associated sequalae. As the use of nitrous oxide expands for the TRMD indication, greater awareness of this abuse potential and associated dangers will be necessary.

Future studies of nitrous oxide

Along with the exciting emergence of another potentially efficacious rapid-acting antidepressant, many questions remain. Similar to studies demonstrating reduced suicidal ideation using ketamine, Nagele et al. (2015) demonstrated that in addition to rapidly decreasing depressive symptoms, nitrous oxide also lowered suicidal ideation, prompting current investigation of antisuicidal properties. Further, consistent with animal studies demonstrating the NMDAR antagonism occurring at nitrous oxide concentrations far below the heretofore tested concentration of 50% (Jevtović-Todorović et al., 1998), dose-finding studies in patients with TRMD are currently underway. Additionally, more precise functional brain imaging studies to test the acute and chronic effects of nitrous oxide along with elucidating mechanism of action, on both nondepressed controls and TRMD patients should be undertaken.

Isoflurane

Introduction

Isoflurane can be dosed to induce cortical burst suppression on the electroencephalogram. The burst suppression state has been likened to the transient postictal generalized electroencephalographic suppression seen after electroconvulsive therapy (ECT). Initial clinical trials suggest that burst suppression anesthesia has antidepressant activity similar to ECT while avoiding associated adverse effects which limit its use. The use of anesthetic induced burst suppression with isoflurane (or propofol, see below) is an area of active interest and investigation.

Pharmacology

Isoflurane is a halogenated methyl ethyl ether compound and isomer of enflurane. It was approved for use in 1979. Its anesthetic effects are largely attributed to GABA-A agonism but other pharmacological effects have been reported including NMDA antagonism (Petrenko et al., 2013). Isoflurane is well-tolerated and, relative to previously developed inhaled anesthetics, it demonstrates cardiovascular safety and decreased renal and hepatic toxicity (Eger, 1981). Of particular interest, isoflurane produces predictable, well-characterized, dose-related changes on the electroencephalogram (EEG) (Eger et al., 1971). At high doses, isoflurane administration is associated with EEG burst suppression—a state during which bursts of activity alternate with periods of quiescence. For isoflurane, this burst suppression pattern initially occurs at approximately 1.5 MAC and progresses to form a near isoelectric pattern at approximately 2 MAC. As discussed below, this burst suppression activity may have therapeutic effects in TRMD.

Clinical studies

Anesthetic-induced cortical burst suppression was first suggested as a potential alternative to ECT in the 1980s. Langer et al. (1985) observed that after an ECT-induced seizure, patients experienced a transient period (i.e., minutes) of postictal suppression on EEG. They speculated that a seizure was not necessary for antidepressant efficacy, but rather the induction of EEG suppression was more important to ECT's therapeutic effects. In support of this idea, postictal suppression has been suggested to be at least as predictive of ECT efficacy as seizure duration (Kranaster et al., 2013; Mayur, 2006). Therefore, they hypothesized that isoflurane-induced burst suppression (described as "isoflurane narcotherapy") could potentially provide similar antidepressant efficacy to ECT while avoiding seizure-associated adverse effects, which include headache, body aches, and cognitive impairment.

Langer et al. (1985) published the first open-label trial of isoflurane-induced burst suppression (ISO) for TRMD in which significant reduction of depressive symptoms was experienced by 9 of 11 patients. Subsequent open-label work by others also suggested that ISO may have similar antidepressant therapeutic efficacy to ECT (Carl et al., 1988; Engelhardt et al., 1993). Langer et al. (1995) reported the results of a small double-blind randomized controlled trial comparing the antidepressant efficacy of six ISO treatments ($n = 9$) to six ECT treatments ($n = 10$) in women. The ISO group was treated to a "predominant burst suppression pattern" on EEG and maintained for 15 min. Patients were treated twice a week. Assessments were done prior to each treatment, at series end, and 2 weeks posttreatment series. ISO patients showed a significant and superior antidepressant response to ECT. At series end, six of nine showed a clinically significant (>50% improvement) response and three of nine achieved remission. Effects were durable, and at the 2-week follow-up, there were six of nine responders and five of nine remitters. ISO patients suffered no adverse cognitive effects, but rather demonstrated neuropsychological improvements across several domains. In contrast, the ECT cohort patients showed significant cognitive deficits.

Not all burst suppression anesthesia studies have reported similar impressive antidepressive outcomes. Greenberg et al. (1987) treated six elderly depressed patients (including five aged 74–82 years), in an open-label pilot study, with little reported improvement. Predictably, hypotension was an issue for these elderly ISO-treated patients and two patients (both with past medical histories of hypertension and one with coronary artery disease) were not able to complete the series. García-Toro et al. (2001) reported a small randomized trial of sevoflurane, a gas anesthetic related to isoflurane in which subjects underwent four 1-h sevoflurane-induced burst suppression treatments ($n = 9$) versus four sham treatments (induced using propofol but allowed to awaken) ($n = 7$). In this study, two of the nine sevoflurane patients showed more than 50% decrease in depression scores, and two more showed between 25% and 50% improvement. A 24% decrease in depression scores was observed, but given the small sample size, this result did not reach statistical significance. Of interest, however, for those responding to treatment, the effects lasted for at least for 1–3 months. There are several methodological considerations worth noting in this trial. First, sevoflurane was administered at lower relative concentration [average sevoflurane 2.5% (~1.25 MAC) versus isoflurane 2.5% (~2 MAC) maintenance]. Second, sevoflurane, even when administered at equipotent doses, is likely less effective than isoflurane at inducing burst suppression (Murrell et al., 2008). Third, patients received only four treatments versus six for Langer and 8–12 for a typical ECT series. While patients showed improvement in the sevoflurane study, it is not surprising that the antidepressant effects were smaller than those reported for ISO. As a result of these mixed and negatively interpreted data, there was a pause in related work for a number of years.

Recently, encouraging studies by a number of groups have renewed interest in burst suppression anesthesia for TRMD. Tadler et al. (2009) reported the first preclinical experiment of ISO's antidepressant-like effects. In that study, the forced swim test (FST) was used. In the FST, mice are placed in an inescapable cylinder of water, and efforts to escape are quantified over 6 min. Immobility in the FST is interpreted as a measure of behavioral despair. Conversely, increased efforts to escape are associated with antidepressant efficacy (Can et al., 2012). In Tadler's study at 24 h post-ISO treatment, mice showed increased mobility relative to controls and comparable to the behavioral effects of a conventional antidepressant, desipramine. Antila et al. (2017) subsequently showed antidepressant-like effects in rodents treated with isoflurane (2% for 30 min) assessed both acutely, using the FST, and 6 days postexposure, in a test of learned helplessness. Brown et al. (2018) reported that isoflurane dosed to achieve burst suppression (3% induction and 2% maintenance for 2 h) prevents the development of learned helplessness in rats and reverses it in mice. Interestingly, halothane, which fails to induce burst suppression even at high doses, shows no antidepressant effect with equivalent high-dose administration.

Given the initially encouraging results of burst suppression anesthesia, methodological differences between Langer's and later work, and supportive animal work, a small clinical trial was performed, similar to Langer's initial work (Weeks et al., 2013). In the study, eight patients were treated open-label with 10 sessions ISO and compared to 20 patients treated naturalistically with bifrontal ECT. ISO was dosed to produce a predominant burst suppression pattern on cortical EEG (burst suppression ratio > 80%) for 15 min per session. ISO showed an antidepressant response (>50% improvement in Hamilton depression score) in 75% of patients and an identical rate of full remission as ECT (50%). Improvement was durable and continued to the 4-week follow-up for most patients.

Adverse effects and toxicology

No serious adverse events have been reported in clinical trials of isoflurane-induced burst suppression. Weeks 3rd et al. (2013) reported that while ECT led to temporary declines in memory and cognitive function, ISO patients actually improved in many neuropsychological domains. Expected dose-dependent decreases in blood pressure were the most significant effect of isoflurane administration. Hypotension was managed with intravenous isotonic fluid coadministration, Trendelenburg positioning, and intermittent phenylephrine boluses as necessary. Treatments were well tolerated by patients with minimal associated nausea and vomiting facilitated by the use of a proprietary isoflurane adsorption device used at emergence.

Summary and future directions

In summary, for isoflurane, there is preliminary support, both preclinical and clinical, that ISO has rapid, robust and durable antidepressant effects and lacks ECT-associated adverse effects. Patients awaken alert, without confusion. It also lacks the psychotomimetic side effects associated with ketamine. Whether the induction of burst suppression using ISO or other anesthetics will ultimately provide a viable alternative to ECT with its undesirable seizure-associated adverse effects remains to be definitively shown. A well-powered blinded randomized placebo-controlled trial of isoflurane anesthesia for TRMD has yet to be completed.

While current evidence suggests that isoflurane doses adequate to induce burst suppression may be necessary to realize an antidepressant effect, the optimal dosing, duration, and number of treatments is not known. Post-ECT cortical suppression lasts only a few minutes. Langer et al. (1985, 1995) noted antidepressant effects with as little as 5 min of suppression in open-label studies, but 15 min was targeted in the RCT. Greenberg et al. (1987) aimed for a 1 h burst suppression treatment interval, administered fewer treatments, used a lower relative dose, and did not observe as robust effects as others. Weeks et al. (2013), using isoflurane, more closely replicated Langer's work with similar outcomes. Animal studies have shown effects with varied doses and durations of isoflurane. Clarifying the optimal dose, duration, and frequency of treatments will be important in future investigations, as will elucidating associated molecular mechanisms.

Subsequent to isoflurane studies, an open-label trial has provided the first evidence that the induction of burst suppression using another anesthetic agent, propofol, may have antidepressant effects similar to ECT (Mickey et al., 2018). As of 2020, ongoing work (led by SCT and BJM) is testing the antidepressant effects of deep propofol anesthesia in a double-blind, randomized, controlled trial.

If anesthetic drugs do indeed have clinical antidepressant effects, a better understanding of their mechanisms will be an important step in refining these interventions. Rantamaki et al. have recently suggested that ketamine, nitrous oxide, and ECT may act by similar mechanisms to initiate synaptic changes that result in antidepressant efficacy (Kohtala et al., 2018). EEG delta wave prominence in the early posttreatment period are associated with immediate early gene transcription and activation of BDNF-TrkB signaling. Identification of translational biomarkers like this, which can be investigated in humans and animal models, will be critical for understanding therapeutic mechanisms and developing optimized interventions.

Xenon

Xenon is a noble gas with anesthetic, analgesic, neuroprotective, and cardioprotective properties. Xenon has been proposed as a therapeutic agent for depression and other psychiatric conditions when administered at subanesthetic doses. As of late 2020, no definitive evidence of antidepressant effects in humans has yet been reported.

Chemistry

This unique gas is a mono-atomic gas element that exists in the atmosphere in trace concentrations. Xenon is manufactured through the distillation of liquefied air (Neice and Zornow, 2016). Like other noble gases, xenon does not react with other chemicals under ordinary conditions.

Pharmacology

Mechanisms of action

Despite the fact that xenon is chemically nonreactive, it is biologically active. Xenon inhibits NMDA receptor function by binding the glycine site, and this mechanism of action is thought to mediate its hypnotic and neuroprotective effects (Franks et al., 1998; Harris et al., 2013; Preckel et al., 2006). Direct actions at other glutamate receptors, potassium channels, and other targets have also been reported (Preckel et al., 2006). Along with nitrous oxide and ketamine, xenon is one of the few anesthetics that do not act directly at GABA receptors (Franks et al., 1998).

Pharmacokinetics

The minimum alveolar concentration (MAC) for surgical anesthesia has been estimated between 51% and 71%, with lower values in women, and xenon's MAC-awake is ~33% (Sanders et al., 2004). Both onset and offset of action are more rapid than other inhaled agents (Hou et al., 2016; Sanders et al., 2004). Because xenon is chemically nonreactive, it is excreted via the lungs without biotransformation by renal or hepatic systems (Maze and Laitio, 2020; Sanders et al., 2004).

In addition to the inhalation route, xenon has been administered intravenously and intraperitoneally to experimental animals. The gas can be dissolved in saline (Shao et al., 2020) or encapsulated in liposomes which can be subsequently uncaged using ultrasound (Peng et al., 2013). Oral formulations of xenon have been developed for administration to humans (Yin et al., 2019).

Indications

The anesthetic properties of xenon were discovered more than 70 years ago (Cullen and Gross, 1951) but only in the 2000s did xenon receive regulatory approval as an anesthetic agent in Europe and Russia (Tonner, 2006). Although not yet an approved anesthetic in the United States, certain radioisotopes of xenon (e.g., ^{133}Xe) are approved by the FDA as nuclear medicine contrast agents.

Clinical studies

General efficacy profile

Pivotal clinical trials have demonstrated the efficacy of xenon as a general anesthetic along with its potential cardioprotective and neuroprotective benefits (Sanders et al., 2004; Tonner, 2006).

Efficacy for treatment-resistant depression

Several recent investigations have examined the antidepressant-like and anxiolytic-like effects of xenon in rodents. An early behavioral study using a rat model of posttraumatic stress disorder found that administration of subanesthetic xenon gas interfered with reconsolidation of fear memories (Meloni et al., 2014). Dandekar et al. (2018) administered xenon-containing liposomes intravenously to male rats and applied ultrasound over the carotid artery to uncage the xenon gas. Antidepressant-like effects were found at 6 mg/kg but not 9 mg/kg. Dobrovolsky et al. (2019) administered xenon gas at a subanesthetic dose to rats and found improvements in some anxiety-like and social behaviors but no significant antidepressant-like effects. Finally, Shao et al. (2020) administered subanesthetic xenon gas via inhalation or xenon-rich saline intraperitoneally to male mice and found antidepressant-like and anxiolytic-like effects at 30 min.

Xenon is being studied as a potential treatment for major depression, panic disorder, and posttraumatic stress disorder, but few clinical studies have yet been published. Dobrovolsky et al. (2017) reported an open-label clinical trial of subanesthetic xenon for panic disorder. Eighty-one patients received a series of 6 treatments, each lasting up to 4 min. The inhaled xenon/oxygen mixture was titrated from 15%/85% to 30%/70% over 9 days. Scores on the anxiety subscale of the Hospital Anxiety and Depression Scale decreased by ~60% on average after 6 treatments, improvements on the depression subscale were reported, and treatments were generally well-tolerated (6% dropout). More recently, Aftanas et al. (2020) reported a blinded, randomized controlled, cross-over trial of xenon versus placebo in 14 healthy men. Participants received a single treatment with either xenon/nitrogen/oxygen (25%/45%/30%) or nitrogen/oxygen (70%/30%) for 15 min. Self-reported positive emotions increased with xenon in a subset of participants when assessed 12 min after drug discontinuation. Beneficial effects of xenon for major depressive disorder have been described in abstract form (Aftanas et al., 2017), but a full report has not been published as of late 2020.

Adverse effects and toxicology

Compared to volatile anesthetics and propofol, postoperative nausea and vomiting are more common with xenon (Law et al., 2016), but xenon provides better hemodynamic stability and faster recovery (Hou et al., 2016; Law et al., 2016). Studies to date indicate that xenon is nontoxic and, in fact, can be organoprotective (Maze and Laitio, 2020; Preckel et al., 2006).

Summary and future directions

Xenon is an NMDA receptor antagonist that holds potential as a treatment for depression and other neuropsychiatric disorders. Preclinical studies indicate this agent has antidepressant-like and anxiolytic-like effects in some assays. No controlled clinical trials of xenon's antidepressant or antianxiety effects have yet appeared in the literature.

As of late 2020, several active clinical trials of xenon are registered in the United States and European Union for indications that include major depressive disorder, bipolar disorder, posttraumatic stress disorder, and panic disorder.

The use of inhalants in complimentary medicine: Aromatherapy

Within the past 25 years, there has been a significant increase in the use of complimentary alternative medicine (CAM) techniques to attempt to treat depression. Broadly defined, CAM is a collection of healing resources (such as medical practices and products) for the prevention, diagnosis, and treatment of diseases, which are complimentary to mainstream medicine (Ernst et al., 1995). The use of CAM in the treatment of depression has become widespread: recent estimates support that CAM is employed by approximately 54% of depressed patients (Van Der Watt et al., 2008).

Emerging from the CAM field in the past 30 years is the use of aromatic compounds, most especially "essential oils" in the treatment of depression. Aromatherapy is the therapeutic use of plant-derived highly concentrated oils (Herz, 2009; Perry and Perry, 2006). Currently, there is no standardization of dosage or dilution used; hence, it is difficult to do comparative studies (Tisserand and Balacs, 1995).

Though to date these compounds have not been systematically targeted in TRMD, recent controlled trials have attempted to verify the antidepressant efficacy in different depressed populations, particularly in patient groups for whom standard antidepressant treatments are potentially unsafe or poorly tolerated (e.g., pregnancy, cancer patients). For a recent summary of these findings, see Sanchez-Vidana et al. (2017).

Currently, there are four controlled studies of aromatherapy in MDD (Conrad and Adams, 2012; Graham et al., 2003; Igarashi, 2013; Sehhatie et al., 2015), and of these four studies, only one reported antidepressant efficacy (Conrad and Adams, 2012; Matsumoto et al., 2014).

Summary and future directions

The field of CAM, and aromatherapy in particular, is only recently undergoing careful scientific inquiry. Given the widespread use of these treatments and the relative potential advantages (sensitive populations, safety in combination with other treatments), future carefully designed study of these treatments is critical. The current use aromatherapy for MDD currently suffers from a lack of standardization of dosing, delivery methods, and frequency of treatments; all areas that would benefit from additional research.

Acknowledgment

Special thanks to Britt Gott, MS, for her editorial input and assistance with the compilation of this chapter.

References

Aftanas, L., Akhmetova, O., Brack, I., Danilenko, K., Khabarov, A., Nikolenko, E., 2017. 787. Xenon in sub-anesthetic doses for treatment of major depression: a proof-of-concept placebo-controlled pilot study. Biol. Psychiatry 81 (10), S319–S320.

Aftanas, L.I., Bazanova, O.M., Khabarov, A.N., Pustovoyt, S.M., 2020. Psychophysiological analysis of thymoleptic effects of xenon in humans. Bull. Exp. Biol. Med. 169 (2), 183–186. https://doi.org/10.1007/s10517-020-04846-1.

American Psychiatric Association, 2010. Practice Guideline for the Treatment of Patients with Major Depressive Disorder. American Psychiatric Association.

Amos, T.B., Tandon, N., Lefebvre, P., Pilon, D., Kamstra, R.L., Pivneva, I., Greenberg, P.E., 2018. Direct and indirect cost burden and change of employment status in treatment-resistant depression: a matched-cohort study using a us commercial claims database. J. Clin. Psychiatry 79 (2). https://doi.org/10.4088/JCP.17m11725.

Antila, H., Ryazantseva, M., Popova, D., Sipilä, P., Guirado, R., Kohtala, S., Yalcin, I., Lindholm, J., Vesa, L., Sato, V., Cordeira, J., Autio, H., Kislin, M., Rios, M., Joca, S., Casarotto, P., Khiroug, L., Lauri, S., Taira, T., Rantamäki, T., 2017. Isoflurane produces antidepressant effects and induces TrkB signaling in rodents. Sci. Rep. 7 (1). https://doi.org/10.1038/s41598-017-08166-9.

Bergfeld, I.O., Mantione, M., Figee, M., Schuurman, P.R., Lok, A., Denys, D., 2018. Treatment-resistant depression and suicidality. J. Affect. Disord. 235, 362–367. https://doi.org/10.1016/j.jad.2018.04.016.

Brown, P., Zanos, P., Wang, L., Elmer, G., Gould, T., Shepard, P.D., 2018. Isoflurane but not halothane prevents and reverses helpless behavior: a role for EEG burst suppression? Int. J. Neuropsychopharmacol. 21 (8). https://doi.org/10.1093/ijnp/pyy029. 29554264.

Can, A., Dao, D., Arad, M., 2012. The mouse forced swim test. J. Vis. Exp. 59, 3638.

Carl, C., Engelhardt, W., Teichmann, G., Fuchs, G., 1988. Open comparative study with treatment-refractory depressed patients: electroconvulsive therapy–anesthetic therapy with isoflurane (preliminary report). Pharmacopsychiatry 21 (6), 432–433. https://doi.org/10.1055/s-2007-1017040.

Chamaa, F., Bahmad, H.F., Makkawi, A.K., Chalhoub, R.M., Al-Chaer, E.D., Bikhazi, G.B., Nahas, Z., Abou-Kheir, W., 2018. Nitrous oxide induces prominent cell proliferation in adult rat hippocampal dentate gyrus. Front. Cell. Neurosci. 12. https://doi.org/10.3389/fncel.2018.00135.

Conrad, P., Adams, C., 2012. The effects of clinical aromatherapy for anxiety and depression in the high risk postpartum woman – a pilot study. Complement. Ther. Clin. Pract. 18 (3), 164–168. https://doi.org/10.1016/j.ctcp.2012.05.002.

Conway, C.R., George, M.S., Sackeim, H.A., 2017. Toward an evidence-based, operational definition of treatment-resistant depression: when enough is enough. JAMA Psychiatry 74 (1), 9–10. https://doi.org/10.1001/jamapsychiatry.2016.2586.

Corey-Lisle, P.K., Birnbaum, H.G., Greenberg, P.E., Marynchenko, M.B., Claxton, A.J., 2002. Identification of a claims data "signature" and economic consequences for treatment-resistant depression. J. Clin. Psychiatry 63 (8), 717–726. https://doi.org/10.4088/JCP.v63n0810.

Crown, W.H., Finkelstein, S., Berndt, E.R., Ling, D., Poret, A.W., Rush, A.J., Russell, J.M., 2002. The impact of treatment-resistant depression on health care utilization and costs. J. Clin. Psychiatry 63 (11), 963–971. https://doi.org/10.4088/JCP.v63n1102.

Cullen, S.C., Gross, E.G., 1951. The anesthetic properties of xenon in animals and human beings, with additional observations on krypton. Science 113 (2942), 580–582. https://doi.org/10.1126/science.113.2942.580.

Dandekar, M.P., Peng, T., McPherson, D.D., Quevedo, J., Soares, J.C., Huang, S.L., 2018. Intravenous infusion of xenon-containing liposomes generates rapid antidepressant-like effects. Prog. Neuro-Psychopharmacol. Biol. Psychiatry 86, 140–149. https://doi.org/10.1016/j.pnpbp.2018.03.011.

Daras, C., Cantrill, R.C., Gillman, M.A., 1983. [3H]naloxone displacement: evidence for nitrous oxide as opioid receptor agonist. Eur. J. Pharmacol. 89 (1–2), 177–178. https://doi.org/10.1016/0014-2999(83)90626-X.

Dobrovolsky, A., Ichim, T.E., Ma, D., Kesari, S., Bogin, V., 2017. Xenon in the treatment of panic disorder: an open label study. J. Transl. Med. 15 (1). https://doi.org/10.1186/s12967-017-1237-1.

Dobrovolsky, A.P., Gedzun, V.R., Bogin, V.I., Ma, D., Ichim, T.E., Sukhanova, I.A., Malyshev, A.V., Dubynin, V.A., 2019. Beneficial effects of xenon inhalation on behavioral changes in a valproic acid-induced model of autism in rats. J. Transl. Med. 17 (1). https://doi.org/10.1186/s12967-019-02161-6.

Eger, E.I., 1981. Isoflurane. Anesthesiology, 559–576. https://doi.org/10.1097/00000542-198111000-00014.

Eger, E.I., Stevens, W.C., Cromwell, T.H., 1971. The electroencephalogram in man anesthetized with forane. Anesthesiology 35 (5), 504–508. https://doi.org/10.1097/00000542-197111000-00010.

Engelhardt, W., Carl, G., Hartung, E., 1993. Intra-individual open comparison of burst-suppression-isoflurane-anaesthesia versus electroconvulsive therapy in the treatment of severe depression. Eur. J. Anaesthesiol. 10 (2), 113–118.

Ernst, E., Resch, K.L., Mills, S., Hill, R., Mitchell, A., Willoughby, M., White, A., 1995. Complementary medicine – a definition. Br. J. Gen. Pract. 45 (398), 506.

Fang, F., Guo, T.Z., Davies, M.F., Maze, M., 1997. Opiate receptors in the periaqueductal gray mediate analgesic effect of nitrous oxide in rats. Eur. J. Pharmacol. 336 (2–3), 137–141. https://doi.org/10.1016/S0014-2999(97)01219-3.

Foster, B.L., Liley, D.T.J., 2013. Effects of nitrous oxide sedation on resting electroencephalogram topography. Clin. Neurophysiol. 124 (2), 417–423. https://doi.org/10.1016/j.clinph.2012.08.007.

Franks, N.P., Dickinson, R., Sousa, S.L., Hall, A.C., Lieb, W.R., 1998. How does xenon produce anaesthesia? Nature 396 (6709).

Fukagawa, H., Koyama, T., Fukuda, K., 2014. κ-Opioid receptor mediates the antinociceptive effect of nitrous oxide in mice. Br. J. Anaesth. 113 (6), 1032–1038. https://doi.org/10.1093/bja/aeu254.

García-Toro, M., Segura, C., González, A., Perelló, J., Valdivia, J., Salazar, R., Tarancón, G., Campoamor, F., Salva, J., De La Fuente, L., Romera, M., 2001. Inefficacy of burst-suppression anesthesia in medication-resistant major depression: a controlled trial. J. ECT 17 (4), 284–288. https://doi.org/10.1097/00124509-200112000-00009.

Gaynes, B.N., Lux, L., Gartlehner, G., Asher, G., Forman-Hoffman, V., Green, J., Boland, E., Weber, R.P., Randolph, C., Bann, C., Coker-Schwimmer, E., Viswanathan, M., Lohr, K.N., 2020. Defining treatment-resistant depression. Depress. Anxiety 37 (2), 134–145. https://doi.org/10.1002/da.22968.

Gibson, T.B., Jing, Y., Carls, G.S., Kim, E., Bagalman, J.E., Burton, W.N., Tran, Q.V., Pikalov, A., Goetzel, R.Z., 2010. Cost burden of treatment resistance in patients with depression. Am. J. Manag. Care 16 (5), 370–377.

Gillman, M.A., 1986. Nitrous oxide, an opioid addictive agent. Review of the evidence. Am. J. Med. 81 (1), 97–102. https://doi.org/10.1016/0002-9343(86)90189-0.

Graham, P.H., Browne, L., Cox, H., Graham, J., 2003. Inhalation aromatherapy during radiotherapy: results of a placebo-controlled double-blind randomized trial. J. Clin. Oncol. 21 (12), 2372–2376. https://doi.org/10.1200/JCO.2003.10.126.

Greenberg, L.B., Gage, J., Vitkun, S., Fink, M., 1987. Isoflurane anesthesia therapy: a replacement for ECT in depressive disorders? Convuls. Ther. 3 (4), 269–277.

Greenberg, P., Corey-Lisle, P.K., Birnbaum, H., Marynchenko, M., Claxton, A., 2004. Economic implications of treatment-resistant depression among employees. PharmacoEconomics 22 (6), 363–373. https://doi.org/10.2165/00019053-200422060-00003.

Gruss, M., Bushell, T.J., Bright, D.P., Lieb, W.R., Mathie, A., Franks, N.P., 2004. Two-pore-domain K+ channels are a novel target for the anesthetic gases xenon, nitrous oxide, and cyclopropane. Mol. Pharmacol. 65 (2), 443–452. https://doi.org/10.1124/mol.65.2.443.

Harris, K., Armstrong, S.P., Campos-Pires, R., Kiru, L., Franks, N.P., Dickinson, R., 2013. Neuroprotection against traumatic brain injury by xenon, but not argon, is mediated by inhibition at the n-methyl-D-aspartate receptor glycine site. Anesthesiology 119 (5), 1137–1148. https://doi.org/10.1097/ALN.0b013e3182a2a265.

Hasin, D.S., Sarvet, A.L., Meyers, J.L., Saha, T.D., Ruan, W.J., Stohl, M., Grant, B.F., 2018. Epidemiology of adult DSM-5 major depressive disorder and its specifiers in the United States. JAMA Psychiatry 75 (4), 336–346. https://doi.org/10.1001/jamapsychiatry.2017.4602.

Herz, R.S., 2009. Aromatherapy facts and fictions: a scientific analysis of olfactory effects on mood, physiology and behavior. Int. J. Neurosci. 119 (2), 263–290. https://doi.org/10.1080/00207450802333953.

Hou, B., Li, F., Ou, S., Yang, L., Zhou, S., 2016. Comparison of recovery parameters for xenon versus other inhalation anesthetics: systematic review and meta-analysis. J. Clin. Anesth. 29, 65–74. https://doi.org/10.1016/j.jclinane.2015.10.018.

Huang, C., Nathaniel, J., 2016. Nitrous oxide, from the operating room to the emergency department. Curr. Emerg. Hosp. Med. Rep., 11–18. https://doi.org/10.1007/s40138-016-0092-3.

Igarashi, T., 2013. Physical and psychologic effects of aromatherapy inhalation on pregnant women: a randomized controlled trial. J. Altern. Complement. Med. 19 (10), 805–810. https://doi.org/10.1089/acm.2012.0103.

Jevtović-Todorović, V., Todorovć, S.M., Mennerick, S., Powell, S., Dikranian, K., Benshoff, N., Zorumski, C.F., Olney, J.W., 1998. Nitrous oxide (laughing gas) is an NMDA antagonist, neuroprotectant and neurotoxin. Nat. Med., 460–463. https://doi.org/10.1038/nm0498-460.

Kohtala, S., Theilmann, W., Rosenholm, Penna, L., Karabulut, G., Uusitalo, S., Järventausta, K., Yli-Hankala, A., Yalcin, I., Matsui, N., Wigren, H., Rantamäki, T., 2018. Cortical excitability and activation of TrkB signaling during rebound slow oscillations are critical for rapid antidepressant responses. Mol. Neurobiol. 56 (6). https://doi.org/10.1007/s12035-018-1364-6. Epub.

Koyama, T., Fukuda, K., 2010. Involvement of the κ-opioid receptor in nitrous oxide-induced analgesia in mice. J. Anesth. 24 (2), 297–299. https://doi.org/10.1007/s00540-010-0886-5.

Kranaster, L., Plum, P., Hoyer, C., Sartorius, A., Ullrich, H., 2013. Burst suppression: a more valid marker of postictal central inhibition? J. ECT 29 (1), 25–28. https://doi.org/10.1097/YCT.0b013e3182622c0e.

Kuhlmann, L., Foster, B.L., Liley, D.T.J., 2013. Modulation of functional EEG networks by the NMDA antagonist nitrous oxide. PLoS One 8 (2). https://doi.org/10.1371/journal.pone.0056434.

Lan, S.Y., Kuo, C.Y., Chou, C.C., Kong, S.S., Hung, P.C., Tsai, H.Y., Chen, Y.C., Lin, J.J., Chou, I.J., Lin, K.L., 2019. Recreational nitrous oxide abuse related subacute combined degeneration of the spinal cord in adolescents – a case series and literature review. Brain Dev. 41 (5), 428–435. https://doi.org/10.1016/j.braindev.2018.12.003.

Langer, G., Neumark, J., Koinig, G., Graf, M., Schönbeck, G., 1985. Rapid psychotherapeutic effects of anesthesia with isoflurane (ES narcotherapy) in treatment-refractory depressed patients1. Neuropsychobiology 14 (3), 118–120. https://doi.org/10.1159/000118216.

Langer, G., Karazman, R., Neumark, J., Saletu, B., Schönbeck, G., Grünberger, J., Dittrich, R., Petricek, W., Hoffmann, P., Linzmayer, L., Anderer, P., Steinberger, K., 1995. Isoflurane narcotherapy in depressive patients refractory to conventional antidepressant drug treatment. Neuropsychobiology 31 (4), 182–194. https://doi.org/10.1159/000119190.

Law, L.S.C., Lo, E.A.G., Gan, T.J., 2016. Xenon anesthesia: a systematic review and meta-analysis of randomized controlled trials. Anesth. Analg. 122 (3), 678–697. https://doi.org/10.1213/ANE.0000000000000914.

Likis, F.E., Andrews, J.C., Collins, M.R., Lewis, R.M., Seroogy, J.J., Starr, S.A., Walden, R.R., McPheeters, M.L., 2014. Nitrous oxide for the management of labor pain: a systematic review. Anesth. Analg. 118 (1), 153–167. https://doi.org/10.1213/ANE.0b013e3182a7f73c.

Lin, J., Szukis, H., Sheehan, J.J., Alphs, L., Menges, B., Lingohr-Smith, M., Benson, C., 2019. Economic burden of treatment-resistant depression among patients hospitalized for major depressive disorder in the United States. Psychiatr. Res. Clin. Pract., 68–76. https://doi.org/10.1176/appi.prcp.20190001.

Matsumoto, T., Asakura, H., Hayashi, T., 2014. Effects of olfactory stimulation from the fragrance of the Japanese citrus fruit yuzu (Citrus junos Sieb. Ex Tanaka) on mood states and salivary chromogranin A as an endocrinologic stress marker. J. Altern. Complement. Med. 20 (6), 500–506. https://doi.org/10.1089/acm.2013.0425.

Mayur, P., 2006. Ictal electroencephalographic characteristics during electroconvulsive therapy: a review of determination and clinical relevance. J. ECT 22 (3), 213–217. https://doi.org/10.1097/01.yct.0000235922.14623.39.

Maze, M., Laitio, T., 2020. Neuroprotective properties of xenon. Mol. Neurobiol. 57 (1), 118–124. https://doi.org/10.1007/s12035-019-01761-z.

Maze, M., Sanders, R.D., Weimann, J., 2008. Biologic effects of nitrous oxide: a mechanistic and toxicologic review. Anesthesiology 109 (4), 707–722. https://doi.org/10.1097/ALN.0b013e3181870a17.

Meloni, E.G., Gillis, T.E., Manoukian, J., Kaufman, M.J., 2014. Xenon impairs reconsolidation of fear memories in a rat model of post-traumatic stress disorder (PTSD). PLoS One 9 (8). https://doi.org/10.1371/journal.pone.0106189.

Mennerick, S., Jevtovic-Todorovic, V., Todorovic, S.M., Shen, W., Olney, J.W., Zorumski, C.F., 1998. Effect of nitrous oxide on excitatory and inhibitory synaptic transmission in hippocampal cultures. J. Neurosci. 18 (23), 9716–9726. https://doi.org/10.1523/jneurosci.18-23-09716.1998.

Mickey, B.J., White, A.T., Arp, A.M., Leonardi, K., Torres, M.M., Larson, A.L., Odell, D.H., Whittingham, S.A., Beck, M.M., Jessop, J.E., Sakata, D.J., Bushnell, L.A., Pierson, M.D., Solzbacher, D., Jeremy Kendrick, E., Weeks, H.R., Light, A.R., Light, K.C., Tadler, S.C., 2018. Propofol for treatment-resistant depression: a pilot study. Int. J. Neuropsychopharmacol. 21 (12), 1079–1089. https://doi.org/10.1093/ijnp/pyy085.

Murrell, J.C., Waters, D., Johnson, C.B., 2008. Comparative effects of halothane, isoflurane, sevoflurane and desflurane on the electroencephalogram of the rat. Lab. Anim. 42 (2), 161–170. https://doi.org/10.1258/la.2007.06019e.

Nagashima, K., Zorumski, C.F., Izumi, Y., 2005. Nitrous oxide (laughing gas) facilitates excitability in rat hippocampal slices through γ-aminobutyric acid A receptor-mediated disinhibition. Anesthesiology 102 (1), 230–234. https://doi.org/10.1097/00000542-200501000-00034.

Nagele, P., Metz, L.B., Crowder, C.M., 2004. Nitrous oxide (N_2O) requires the N-methyl-D-aspartate receptor for its action in *Caenorhabditis elegans*. Proc. Natl. Acad. Sci. U. S. A. 101 (23), 8791–8796. https://doi.org/10.1073/pnas.0402825101.

Nagele, P., Duma, A., Kopec, M., Gebara, M.A., Parsoei, A., Walker, M., Janski, A., Panagopoulos, V.N., Cristancho, P., Miller, J.P., Zorumski, C.F., Conway, C.R., 2015. Nitrous oxide for treatment-resistant major depression: a proof-of-concept trial. Biol. Psychiatry 78 (1), 10–18. https://doi.org/10.1016/j.biopsych.2014.11.016.

Nagele, P., Zorumski, C.F., Conway, C., 2018. Exploring nitrous oxide as treatment of mood disorders: basic concepts. J. Clin. Psychopharmacol. 38 (2), 144–148. https://doi.org/10.1097/JCP.0000000000000837.

Nagele, P., Palanca, B.J., Gott, B., Brown, F., Barnes, L., Nguyen, T., Xiong, W., Salloum, N.C., Espejo, G.D., Lessov-Schlaggar, C.N., Jain, N., Cheng, W.W.L., Komen, H., Yee, B., Bolzenius, J.D., Janski, A., Gibbons, R., Zorumski, C.F., Conway, C.R., 2021. A phase 2 trial of inhaled nitrous oxide for treatment-resistant major depression. Sci. Transl. Med. 13 (597). PMID: 34108247.

Neice, A.E., Zornow, M.H., 2016. Xenon anesthesia for all, or only a select few? Anaesthesia 71 (11), 1267–1272. https://doi.org/10.1111/anae.13569.

NIH U.S. National Library of Medicine, 2020. PubChem Nitrous Oxide Compound Summary. PubChem Database.

Ohashi, Y., Guo, T., Orii, R., Maze, M., Fujinaga, M., 2003. Brain stem opioidergic and GABAergic neurons mediate the antinociceptive effect of nitrous oxide in fischer rats. Anesthesiology 99 (4), 947–954. https://doi.org/10.1097/00000542-200310000-00030.

Ori, C., Ford-Rice, F., London, E.D., 1989. Effects of nitrous oxide and halothane on μ and κ opioid receptors in guinea-pig brain. Anesthesiology 70 (3), 541–544. https://doi.org/10.1097/00000542-198903000-00027.

Orii, R., Ohashi, Y., Guo, T., Nelson, L.E., Hashimoto, T., Maze, M., Fujinaga, M., 2002. Evidence for the involvement of spinal cord α1 adrenoceptors in nitrous oxide-induced antinociceptive effects in Fischer rats. Anesthesiology 97 (6), 1458–1465. https://doi.org/10.1097/00000542-200212000-00018.

Oussalah, A., Julien, M., Levy, J., Hajjar, O., Franczak, C., Stephan, C., Laugel, E., Wandzel, M., Filhine-Tresarrieu, P., Green, R., Guéant, J.-L., 2019. Global burden related to nitrous oxide exposure in medical and recreational settings: a systematic review and individual patient data meta-analysis. J. Clin. Med. 551. https://doi.org/10.3390/jcm8040551.

Peng, T., Britton, G.L., Kim, H., Cattano, D., Aronowski, J., Grotta, J., Mcpherson, D.D., Huang, S.L., 2013. Therapeutic time window and dose dependence of xenon delivered via echogenic liposomes for neuroprotection in stroke. CNS Neurosci. Ther. 19 (10), 773–784. https://doi.org/10.1111/cns.12159.

Perry, N., Perry, E., 2006. Aromatherapy in the management of psychiatric disorders: clinical and neuropharmacological perspectives. CNS Drugs 20 (4), 257–280. https://doi.org/10.2165/00023210-200620040-00001.

Petrenko, A., Yamakura, T., Sakimura, K., Baba, H., 2013. Defining the role of NMDA receptors in anesthesia: are we there yet? Eur. J. Pharmacol. 723. https://doi.org/10.1016/j.ejphar.2013.11.039. Epub.

Preckel, B., Weber, N.C., Sanders, R.D., Maze, M., Schlack, W., 2006. Molecular mechanisms transducing the anesthetic, analgesic, and organ-protective actions of xenon. Anesthesiology 105 (1), 187–197. https://doi.org/10.1097/00000542-200607000-00029.

Rush, A.J., Trivedi, M.H., Wisniewski, S.R., Nierenberg, A.A., Stewart, J.W., Warden, D., Niederehe, G., Thase, M.E., Lavori, P.W., Lebowitz, B.D., McGrath, P.J., Rosenbaum, J.F., Sackeim, H.A., Kupfer, D.J., Luther, J., Fava, M., 2006. Acute and longer-term outcomes in depressed outpatients requiring one or several treatment steps: A STAR*D report. Am. J. Psychiatr. 163 (11), 1905–1917. https://doi.org/10.1176/ajp.2006.163.11.1905.

Rush, A.J., Wisniewski, S.R., Zisook, S., Fava, M., Sung, S.C., Haley, C.L., Chan, H.N., Gilmer, W.S., Warden, D., Nierenberg, A.A., Balasubramani, G.K., Gaynes, B.N., Trivedi, M.H., Hollon, S.D., 2012. Is prior course of illness relevant to acute or longer-term outcomes in depressed out-patients? A STAR*D report. Psychol. Med. 42 (6), 1131–1149. https://doi.org/10.1017/S0033291711002170.

Sanchez-Vidana, D.I., Ngai, S.P., He, W., Chow, J.K., Lau, B.W., Tsang, H.W., 2017. The effectiveness of aromatherapy for depressive symptoms: a systematic review. Evid. Based Complement. Alternat. Med. 2017, 5869315.

Sanders, R.D., Ma, D., Maze, M., 2004. Xenon: elemental anaesthesia in clinical practice. Br. Med. Bull. 71, 115–135. https://doi.org/10.1093/bmb/ldh034.

Sato, Y., Kobayashi, E., Murayama, T., Mishina, M., Seo, N., 2005. Effect of N-methyl-D-aspartate receptor ε1 subunit gene disruption of the action of general anesthetic drugs in mice. Anesthesiology 102 (3), 557–561. https://doi.org/10.1097/00000542-200503000-00013.

Sawamura, S., Kingery, W.S., Davies, M.F., Agashe, G.S., Clark, J.D., Kobilka, B.K., Hashimoto, T., Maze, M., 2000. Antinociceptive action of nitrous oxide is mediated by stimulation of noradrenergic neurons in the brainstem and activation of α2B adrenoceptors. J. Neurosci. 20 (24), 9242–9251. https://doi.org/10.1523/jneurosci.20-24-09242.2000.

Sehhatie, F., Mirgafourvand, M., Niri, Z.V., 2015. The effect of non-pharmaceutical methods of labor pain relief on mothers' postpartum depression: a randomized controlled trial. Int. J. Women's Health Reprod. Sci. 3 (1), 48–55. https://doi.org/10.15296/ijwhr.2015.08.

Shao, J., Meng, L., Yang, Z., Yu, P., Song, L., Gao, Y., Gong, M., Meng, C., Shi, H., 2020. Xenon produces rapid antidepressant- and anxiolytic-like effects in lipopolysaccharide-induced depression mice model. Neuroreport, 387–393. https://doi.org/10.1097/WNR.0000000000001415.

Tadler, S., Light, A., Hughen, R., 2009. Isoflurane demonstrates antidepressant-like activity in a mouse model of depression (abstract). Anesth. Analg. 108, 212.

Thomas, L., Kessler, D., Campbell, J., Morrison, J., Peters, T.J., Williams, C., Lewis, G., Wiles, N., 2013. Prevalence of treatment-resistant depression in primary care: cross-sectional data. Br. J. Gen. Pract. 63 (617), e852–e858. https://doi.org/10.3399/bjgp13X675430.

Tisserand, R., Balacs, T., 1995. Essential Oil Safety: A Guide for Health Care Professionals. Elsevier.

Tobias, J.D., 2013. Applications of nitrous oxide for procedural sedation in the pediatric population. Pediatr. Emerg. Care 29 (2), 245–265. https://doi.org/10.1097/PEC.0b013e318280d824.

Todorovic, S.M., Jevtovic-Todorovic, V., Mennerick, S., Perez-Reyes, E., Zorumski, C.F., 2001. Cav3.2 channel is a molecular substrate for inhibition of T-type calcium currents in rat sensory neurons by nitrous oxide. Mol. Pharmacol. 60 (3), 603–610.

Tonner, P.H., 2006. Xenon: one small step for anaesthesia…? Curr. Opin. Anaesthesiol. 19 (4), 382–384. https://doi.org/10.1097/01.aco.0000236136.85356.13.

van Amsterdam, J., Nabben, T., van den Brink, W., 2015. Recreational nitrous oxide use: prevalence and risks. Regul. Toxicol. Pharmacol. 73 (3), 790–796. https://doi.org/10.1016/j.yrtph.2015.10.017.

Van Der Watt, G., Laugharne, J., Janca, A., 2008. Complementary and alternative medicine in the treatment of anxiety and depression. Curr. Opin. Psychiatry 21 (1), 37–42. https://doi.org/10.1097/YCO.0b013e3282f2d814.

Weeks 3rd, H.R., Tadler, S.C., Smith, K.W., Iacob, E., Saccoman, M., White, A.T., Landvatter, J.D., Chelune, G.J., Suchy, Y., Clark, E., Cahalan, M.K., Bushnell, L., Sakata, D., Light, A.R., Light, K.C., 2013. Antidepressant and neurocognitive effects of isoflurane anesthesia versus electroconvulsive therapy in refractory depression. PLoS One 8 (7). https://doi.org/10.1371/journal.pone.0069809.

Williams, N.R., Heifets, B.D., Blasey, C., Sudheimer, K., Pannu, J., Pankow, H., Hawkins, J., Birnbaum, J., Lyons, D.M., Rodriguez, C.I., Schatzberg, A.F., 2018. Attenuation of antidepressant effects of ketamine by opioid receptor antagonism. Am. J. Psychiatr. 175 (12), 1205–1215. https://doi.org/10.1176/appi.ajp.2018.18020138.

Yamakura, T., Harris, R.A., 2000. Effects of gaseous anesthetics nitrous oxide and xenon on ligand-gated ion channels: comparison with isoflurane and ethanol. Anesthesiology 93 (4), 1095–1101. https://doi.org/10.1097/00000542-200010000-00034.

Yin, X., Moody, M.R., Hebert, V., Klegerman, M.E., Geng, Y.J., Dugas, T.R., McPherson, D.D., Kim, H., Huang, S.L., 2019. Oral delivery of xenon for cardiovascular protection. Sci. Rep. 9 (1). https://doi.org/10.1038/s41598-019-50515-3.

Zorumski, C.F., Nagele, P., Mennerick, S., Conway, C.R., 2015. Treatment-resistant major depression: rationale for NMDA receptors as targets and nitrous oxide as therapy. Front. Psychiatry 6. https://doi.org/10.3389/fpsyt.2015.00172.

Chapter 34

Drugs under investigation for treatment-resistant depression

Ana Paula Costa[a], Rodrigo Machado-Vieira[a], and Manuella P. Kaster[b]

[a]*Faillace Department of Psychiatry and Behavioral Sciences at McGovern Medical School, The University of Texas Health Science Center at Houston (UTHealth), Houston, TX, United States,* [b]*Department of Biochemistry, Federal University of Santa Catarina, Florianopolis, SC, Brazil*

Medications targeting the hypothalamic–pituitary–adrenal (HPA) axis

Dysfunction in the hypothalamic–pituitary–adrenal (HPA) axis is one of the most well-established biological alterations in psychiatry (Schule et al., 2009). The first observations of an elevation in basal cortisol levels in patients with depression were made almost half a century ago by Board and colleagues (Board et al., 1956) and multiple pieces of evidence contributed to this idea, including the role of stress in the precipitation of depressive episodes in susceptible individuals, the observation that depression is frequent in patients with abnormalities of cortisol production, such as Cushing's disease (Lin et al., 2020), and that exogenous corticosteroid administration is associated with increased prevalence of depression (Starkman, 2013). In fact, in patients with MDD, higher cortisol and adrenocorticotropic hormone (ACTH) levels are found in plasma, cerebrospinal fluid, and urine, suggesting that drug treatments that directly lower cortisol levels should reduce depressive symptomatology. However, HPA axis dysfunction differs by severity and subtype of depression and has been found to be linked to nonresponse to antidepressants and relapse following successful treatment (Schule et al., 2009). This idea opened the investigation of a broad range of new treatment options related to direct interventions in the HPA axis in patients with MDD and treatment-resistant depression.

Mifepristone

The synthetic steroid mifepristone, also known as RU-486, is an antagonist with a high affinity at progesterone and glucocorticoid receptors (GR) (Nieman et al., 1985). Following oral administration of a single dose, mifepristone is rapidly absorbed, with a peak plasma concentration occurring approximately 90 min after ingestion. Despite rapid degradation and poor brain penetration, mifepristone can reach the brain, leading to a decrease in GR-mediated negative feedback, which increases cortisol and ACTH secretion. However, the mechanism of action induced by mifepristone has not been fully elucidated, and it is speculated that it may act by a rebound increase in GR function. Other effects could be attributed to an indirect effect of higher circulating cortisol levels on mineralocorticoid receptors, which may lead to its downregulation and resetting of the HPA axis (Lupien and McEwen, 1997; Dalm et al., 2019).

Early work highlighted the potential for mifepristone to normalize endocrine parameters in patients with MDD (Kling et al., 1989; Krishnan et al., 1992). Murphy and colleagues (1993) conducted the first open-trial study, including four patients with chronic severe depression who were resistant to conventional therapies. Mifepristone (200 mg/day, up to 8 weeks) produced a 16% to 66% improvement in Hamilton Depression Rating Scale scores in three patients (Murphy et al., 1993). Recently, studies are focusing on mifepristone's potential to treat psychotic depression, a form of depression with pronounced HPA axis dysfunction and hypercortisolemia (Posener et al., 2000). Randomized controlled trials also suggested positive effects of mifepristone for psychotic depression (DeBattista et al., 2000; Block et al., 2018). However, data interpretation and the possible application in clinical settings were challenged by Carroll and Rubin (2006). The positive findings have been followed by a larger negative trial that found no statistically significant differences between mifepristone and placebo on reduction in psychotic symptoms (Blasey et al., 2011). However, a positive linear correlation between plasma concentration of mifepristone and efficacy was observed. The authors argued that mifepristone has good tolerability and might elicit more robust responses at higher doses.

One interesting characteristic of antiglucocorticoid strategies for the treatment of psychotic depression is that their effects appear to initiate a rapid, short-term clinical response. Thus, we can speculate that they might be helpful for TRD,

Managing Treatment-Resistant Depression. https://doi.org/10.1016/B978-0-12-824067-0.00034-7

used to increase the efficacy of treatment regimens in medicated patients, or initiate a response that can be maintained with conventional treatments (Gallagher and Young, 2006).

Metyrapone

Metyrapone inhibits the enzyme 11-β hydroxylase, which catalyzes the conversion of 11-deoxycortisol to cortisol. This effect reduces cortisol synthesis and attenuates the HPA axis negative feedback, resulting in increased ACTH, desidroepiandrosterona (DHEA), and 11-deoxycortisol levels after repeated administration (Jahn et al., 2004; Otte et al., 2007). Metyrapone is well tolerated and rapidly absorbed in humans following oral administration, and blood levels peak 1 h after ingestion.

Since the late 1970s, metyrapone has been suggested for the treatment of psychiatric complications of Cushing's disease (Jeffcoate et al., 1979). However, evidence supporting the role of metyrapone in treatment-resistant depression is limited and inconclusive. Most studies have utilized metyrapone as an augmenting agent for antidepressants, and no double-blind, randomized controlled trials of metyrapone monotherapy are available so far.

Murphy and colleagues (1991) described a patient with TRD in an early case report responding to metyrapone and aminoglutethimide, another inhibitor of cortisol synthesis, who remained in remission for more than 2 years (Murphy et al., 1991). Also, the authors conducted an open-label study using one or more steroid suppressive agents (aminoglutethimide, ketoconazole, and/or metyrapone) in eight patients with TRD for 2 months. The study shown that six patients achieved symptoms remission for 5 months after the end of the treatments, and two patients had partial remission (Murphy et al., 1991). Positive responses for patients with TRD treated with compounds that decrease corticosteroid biosynthesis (aminoglutethimide, ketoconazole) combined with metyrapone were also observed in 13 out of 20 patients (Ghadirian et al., 1995). Metyrapone was also combined with hydrocortisone in a single-blind crossover study to assess the therapeutic value of blocking and replacing glucocorticoids. Treatment with metyrapone for 2 weeks produced a significant reduction in symptoms in 8 patients with depression compared to placebo, even in the presence of normal replacement levels of hydrocortisone (O'Dwyer et al., 1995). Iizuka and colleagues treated six patients with TRD with Metyrapone (doses that varied up to a maximum of 2 g per day) for 4 weeks. Three patients achieved complete remission, and one went into partial remission.

Metyrapone was also used as an augmentation strategy of serotonergic antidepressants. In an open-trial study conducted by Rogoz and colleagues, metyrapone produced augmentation of imipramine's effects in patients with treatment-resistant depression (Rogoz et al., 2004). Additionally, in a placebo-controlled double-blind study including 63 patients treated with nefazodone or fluvoxamine, metyrapone (250 mg four times daily for 3 weeks) also showed improvement compared to placebo (Jahn et al., 2004). However, in recent studies, negative results were observed. In a double-blind, randomized, placebo-controlled trial including 83 treated with metyrapone and 82 with placebo, metyrapone, in addition to standard serotonergic antidepressants, did not produce augmentation effects when compared to placebo. Moreover, Strawbridge et al. (2021), in a double-blind, randomized controlled trial including patients with TRD, no benefits were observed with metyrapone augmentation to ongoing serotonergic antidepressants. These negative findings might be related to the relative absence of HPA axis dysfunction in this population (Strawbridge et al., 2021).

It is important to highlight that metyrapone might act through different mechanisms that do not seem to involve a simple reduction in plasma cortisol. It is speculated that some of the potential beneficial effects for TRD might be associated with GR upregulation, changes in serotonergic 5-HT1A receptors sensitivity, activation of the MR, or effects induced by other hormones in the steroid pathway, including 11-deoxycortisol, ACTH, and DHEA (Sigalas et al., 2012).

Serum and glucocorticoid-inducible kinase I (SGKI)

SGK1 is a member of the serum/glucocorticoid-regulated kinase (SGK) family, a ubiquitously expressed serine/threonine kinase regulated by a wide variety of factors, including glucocorticoids (Dattilo et al., 2020). In the brain, SGK1 expression is positively regulated by glucocorticoids (Sarabdjitsingh et al., 2010). Moreover, SGK1 is a downstream target of GR signaling and exerts a positive feedback loop, favoring nuclear translocation and activation of GR (Anacker et al., 2013). Besides this role in GR regulation, SGK1 has multiple effects relevant for brain function, including the regulation of ion channels, neuronal activity, BDNF, and VEGF signaling, cell proliferation and neurogenesis, and apoptosis (Dattilo et al., 2020; Numakawa et al., 2013).

SGK1 has been investigated as a potential target in psychiatric conditions, including depression (Dattilo et al., 2020). Most data associating SGK1 and depression come from preclinical studies. In preclinical models of stress or chronic corticosterone administration, different nutraceutical compounds with antidepressant-like effects counteract the increase in cytosolic GR and SGK1 levels in the hippocampus and prefrontal cortex (Dattilo et al., 2020). Moreover, SGK1 is upregulated in the striatum and hippocampus of mice treated with ketamine, a drug recently approved for treatment-resistant depression (Qiao et al., 2019). However, human studies are by far more complex. SGK1 polymorphisms were associated with susceptibility to depression in a Chinese cohort of coronary heart disease patients. Additionally, SGK1 mRNA increased in the blood of drug-free depressed patients (Cattaneo et al., 2020). On the other hand, Frodl et al. (2012) found that MDD patients with less peripheral expression of SGK1 had smaller hippocampal volumes, suggesting SGK1 as a marker of reduced activation of the glucocorticoid system. Collectively, the causative role of SGK1 in MDD is made more confusing in the presence of different splicing variants, altered distribution in distinct cell types, and different levels of expression. However, in light of the potential mechanisms associated with SGK1, compounds targeting SGK1 might represent potential treatment strategies that deserve future investigation, especially for their effect in counteracting stress-induced impairments in neuroplasticity and neurogenesis.

Fludrocortisone

Fludrocortisone is a synthetic adrenal steroid that acts as an MR agonist. In the brain, although both GRs and MRs inhibit cortisol secretion through negative feedback, MRs have 10 times the affinity of GRs to glucocorticoids and are vital to controlling the HPA responses (de Kloet, 2013).

Clinical studies examining MR function in MDD revealed mixed results. Polymorphisms of the MR gene were associated with MDD (Kuningas et al., 2007). Besides that, increased MR function was found in patients with MDD, and an upregulation in MR was found in postmortem hypothalamic tissue of patients with MDD. On the other hand, other studies found a decreased MR density in postmortem anterior hippocampus and prefrontal cortex samples of MDD subjects (Medina et al., 2013) and in suicide victims (Lopez et al., 1998), suggesting that stimulating MR might be beneficial in MDD. In fact, the previously described results with GR antagonists reinforce this notion by pointing that a possible shift in MR/GR balance toward predominant MR effects after GR antagonism is responsible for the beneficial effects of GR antagonists such as mifepristone (Lupien and McEwen, 1997; Dalm et al., 2019). MR dysfunction has been demonstrated in TRD (Juruena et al., 2013). Juruena et al. (2013) showed that fludrocortisone's ability to inhibit cortisol secretion was attenuated in patients with psychotic depression, suggesting impaired MR function. No studies assessed the effects of fludrocortisone as monotherapy in MDD (Juruena et al., 2013). Nevertheless, two studies failed to show beneficial effects of fludrocortisone treatment as monotherapy on depressive symptoms in patients with chronic fatigue syndrome (Rowe et al., 2001; Blockmans et al., 2003). However, the stronger evidence supporting the beneficial effects of fludrocortisone for MDD came from its use as an adjuvant treatment to classical antidepressants. In a randomized controlled clinical trial including 64 patients, fludrocortisone used as add-on therapy to standard antidepressants accelerated the antidepressant response (Otte et al., 2010). Thus, more studies are necessary, especially in patients with TRD, but it is possible that fludrocortisone acts as an accelerator at least in a subgroup of patients.

Medication targeting the melatonergic system

Agomelatine

Agomelatine is an agonist at melatonergic MT1/MT2 receptors and antagonist at serotonergic 5-HT2C receptors. One of the most important pharmacological properties of agomelatine is its pro-chronobiological effect. Melatonin is a hormone involved in the regulation of the sleep–wake cycle, and depressed patients have decreased concentrations of melatonin as well as dysfunctions in the circadian rhythms. Agomelatine was demonstrated to improve disrupted sleep–wake rhythms, enhance dopamine and noradrenaline levels in the frontal cortex, and produce neuroplasticity and neurogenesis effects (De Berardis et al., 2015). The recommended therapeutic dose of agomelatine is 25–50 mg/day (Gahr, 2014).

The antidepressant efficacy of agomelatine and better tolerability, compared to other antidepressants, was evaluated in several controlled clinical trials (Lanfumey et al., 2013). One randomized clinical trial, including 164 patients treated with agomelatine and 160 patients treated with escitalopram, showed that agomelatine was better tolerated than escitalopram and displayed long-term benefits on sleep–wake quality and emotional responses in depressive patients (Corruble et al., 2013). Agomelatine was also studied in combination with other antidepressants, especially selective serotonin reuptake inhibitors. However, only a few studies evaluated agomelatine for TRD, most of them with small samples and uncontrolled designs. Different case reports have looked at combinations of agomelatine with escitalopram, venlafaxine, and clomipramine, which resulted in improved outcomes in depression (Potmesil, 2019). Moreover, better remission rates and successful remission for TRD were achieved by combining agomelatine with moclobemide in a case report (Stuhec and Oravecz, 2016) and with bupropion in a sample of 15 patients (Suhs et al., 2015).

A larger 12-week open-label noninterventional study (VIVALDI study) evaluated the effects of agomelatine alone or in combination with other antidepressants in 3317 patients. According to the results, remission was achieved for 66.5% of the patients treated with agomelatine as monotherapy, 44.7% of the patients treated with agomelatine as add-on-therapy, and 50.9% of the patients who switched to agomelatine after another antidepressant. The results suggest that agomelatine improved depressive symptoms, daytime functioning, and sleep-wake rhythm (Laux et al., 2014). Agomelatine was also investigated in combination with electroconvulsive therapy (ECT) but without significant results in terms of baseline effect and prevention of relapse (Lin et al., 2020). Thus, considering the neurobiological targets of agomelatine, its safety profile, and its ability to improve the effectiveness of other antidepressants, larger studies with TRD might be indicated.

Medications targeting the glutamatergic system

Clinical and preclinical studies suggest that dysfunction of the glutamatergic system may play a role in the pathophysiology and treatment of MDD. Trullas and Skolnick (1990), substantiated by preclinical data from behavioral despair models, hypothesized that *N*-methyl-D-aspartate (NMDA) receptor antagonists would possess antidepressant properties. The clinical proof of concept came from the study of Berman et al. (2000) and was systematic and elegantly replicated by Zarate et al. (2006), leading to a revolution in the field of antidepressant drug discovery. Ketamine is, by far, the best studied of the glutamatergic modulators, as well as the most effective. In clinical studies of individuals with TRD, rapid reductions in depressive symptoms have been observed in response to subanesthetic intravenous (IV) doses of ketamine, an agent whose mechanism of action involves the modulation of glutamatergic signaling (Zarate et al., 2006, 2003).

Based on the discovery of ketamine, there has been intense interest in the repurposing and/or development of other glutamatergic modulators that produce rapid and efficacious antidepressants actions, both as monotherapy or adjunctive to other therapies. Some of the glutamatergic system modulators that are under investigation for treatment of TRD are discussed below.

Esketamine

Recently, the researchers found that ketamine's antidepressant effects appeared to be produced by one of its metabolites—(2*R*,6*R*)-hydroxynorketamine (HNK)—through an NMDA receptor-independent mechanism that seems to enhance AMPA receptor activation (Zanos et al., 2016). Esketamine is the *S*(+) enantiomer of ketamine and has a three- to fourfold higher affinity for NMDA receptors (Zhang and Hashimoto, 2019). Clinical trials are currently underway to study the efficacy of intranasal and intravenous esketamine in TRD. Singh et al. (2016) investigated the efficacy of intravenous esketamine in patients with TRD in a multicenter, randomized, placebo-controlled trial. Esketamine 0.20 mg/kg and 0.40 mg/kg had a robust and rapid (within 2 h) antidepressant effect (as assessed by the Montgomery–Asberg Depression Rating Scale - MADRS) compared to placebo (Singh et al., 2016). A phase III, double-blind, active-controlled, multicenter study conducted in 67 individuals with TRD evaluated the efficacy of each individual dose of esketamine nasal spray (28 mg, 56 mg, or 84 mg) compared to placebo. All three esketamine doses improved depressive symptoms in participants with TRD, as assessed by MADRS total score on day 8 (Daly et al., 2015). Another multicenter, placebo-controlled, double-blind, randomized, 12-week study investigated the efficacy of intranasal esketamine (84 mg) in 68 adults with MDD and active suicidal ideation. Intranasal esketamine significantly reduced depressive symptoms (MADRS) within 4 h, as well as suicidal thoughts, which were measured by Scale for Suicide Ideation (SSI) (Singh et al., 2015). Currently, clinical trials are in course to evaluate the efficacy and safety of switching elderly TRD participants from a prior unsuccessful antidepressant treatment to either intranasal esketamine plus a new oral antidepressant or to a new oral antidepressant plus intranasal placebo.

Rapastinel

Rapastinel (GLYX-13) is a positive allosteric modulator of NMDA receptors with glycine-like partial agonist properties (Zhang et al., 2008). Preclinical studies demonstrated that rapastinel has a promising behavioral profile, with ketamine-like effects. The results reported that GLYX-13 administration produces rapid antidepressant effects that last for 1 week. The mechanisms included an increase in synaptic number and function in the medial prefrontal cortex (mPFC), an effect mediated by the release of brain-derived neurotrophic factor (BDNF) (Kato et al., 2018).

A phase IIb safety and efficacy trial randomized 116 TRD unmedicated inpatients to receive either a single intravenous GLYX-13 or placebo administration over three to 15 min. Adjunctive GLYX-13 significantly improved depressive symptoms 1-week postadministration in patients who had received 5 or 10 mg/kg (IV) (Moskal et al., 2014). A recent pilot phase II randomized clinical trial (RCT) showed that a weekly intravenous dose of rapastinel (5 or 10 mg/kg, IV) reduced depressive symptoms as assessed by the HAMD 17 within 2 h, and this effect was maintained for around 7 days (Preskorn et al., 2015). Otherwise, a recently terminated phase III RCT failed to demonstrate a reduction in the depressive symptoms after adjunctive or monotherapy with rapastinel in TRD. However, another compound with a similar mechanism of rapastinel is being developed. NRX-1074 has similar effects to those of rapastinel, but with the advantage of being orally active, and might potentially be a novel antidepressant for TRD.

AVP-923 (Nuedexta)

AVP-923 is a dextromethorphan and quinidine fixed combination that is now being studied for TRD (FDA, 2010; EMA, 2013). Commonly known by its brand name, Nuedexta is the first FDA-approved treatment for pseudobulbar affect. Dextromethorphan is a nonselective and noncompetitive NMDA receptor antagonist and 1-receptor agonist. Additionally, quinidine acts as a cytochrome 2D6 enzyme inhibitor, which is usually used to increase dextromethorphan plasma concentrations and prolong its half-life (FDA, 2010; EMA, 2013). AVP-923 was tested in a small one-arm phase II study designed to evaluate its efficacy, safety, and tolerability in patients with TRD. The primary outcomes after 10 weeks demonstrated that in 20 patients with TDR who received oral doses of AVP-923 every 12 h, 45% of subjects remitted, and 35% responded (MADRS total score) (Murrough et al., 2017).

AVP-786

AVP-786 is a new investigational compound consisting of a combination of deuterated (d6)-dextromethorphan and an ultra-low dose of quinidine. This compound differs from Nuedexta because it contains deuterium, a heavier hydrogen isotope that reduces dextromethorphan first-pass liver metabolism and quinidine dosage. A 10-week, phase II randomized controlled trial is currently investigating the efficacy, safety, and tolerability of AVP 786 as adjunctive therapy in patients with MDD and inadequate response to antidepressant treatment. The study was recently completed, but the results have not been presented yet.

AXS-05

AXS-05 is an orally administrated, rapidly acting, investigational combination of dextromethorphan and bupropion. The purpose of adding norepinephrine-dopamine reuptake inhibitor bupropion to the formulation, in addition to its antidepressant effects, is to increase the bioavailability of dextromethorphan through inhibition of CYP2D6. AXS-05 is currently in a late-stage of clinical development for MDD, including TRD. AXS-05 is currently recruiting for an ongoing phase III clinical trial in patients with TRD (N ~ 350), where the active comparator is bupropion. With pharmacological properties similar to ketamine, AXS-05 represents a promising new investigational agent for TRD (Stahl, 2020).

Rislenemdaz (CERC-301/MK-0657)

Rislenemdaz/CERC-301 (MK-0657) is a highly selective, orally bioavailable NMDA receptor antagonist acting at the NMDA receptor 2B (NR2B) subunit. Preclinical studies demonstrated that CERC-301 has acute antidepressant-like effects in the forced swim test (FST), a predictive rodent model of antidepressant efficacy (Garner et al., 2015). A small RCT was completed with an oral formulation of the drug, which suggested antidepressant efficacy for Rislenemdaz/CERC-301 in TRD (Ibrahim et al., 2012). Another phase II RCT designed to evaluate the antidepressant effects of intermittent doses of CERC-301 in TRD is currently recruiting participants.

AV-101

AV-101 (L-4-chlorokyurenine or 4-CI-KYN) is a pro-drug that is rapidly converted in vivo into the active metabolite 7-chloro-kynurenic acid (7-Cl-KYNA), a potent and highly selective antagonist at the glycine-binding site of the NMDAR NR1 subunit. AV-101 has been designed as an oral formulation to be used once daily (Salituro et al., 1994). An animal study exploring potential ketamine-like antidepressant agents of AV-101 demonstrated rapid, dose-dependent, and persistent antidepressant effects following a single treatment, with no psychotomimetic effects (Zanos et al., 2015). However, there are no published results in humans so far. A phase II randomized controlled trial investigating the antidepressant efficacy of AV-101 in individuals with TRD is currently recruiting participants.

AZD6765 (Lanicemine)

AZD6765 (Lanicemine) is a low-affinity, nonselective NMDA receptor channel blocker. A recent phase II RCT designed to evaluate whether a single AZD6765 infusion (150 mg) could produce rapid antidepressant effects in TRD subjects showed some promising results (Zarate Jr. et al., 2013). The primary outcome measured was the MADRS, which was used to rate overall depressive symptoms at baseline and 60, 80, 110, and 230 min postinfusion and on days 1, 2, 3, and 7 postinfusion. AZD6765 infusion was associated with a rapid (within 80 min) response that was not as robust or sustained as the one observed after ketamine administration. Besides that, the effects of AZD6765 were short-lived and the response rate low, but no dissociative side effects were noted.

MK-1942

The structure and mechanism of action of MK-1942 were not yet disclosed. However, a phase II RCT is being conducted to assess the efficacy and safety of daily and intermittent dosing of MK-1942 compared to placebo among participants with TDR on a stable course of antidepressant therapy. The primary outcome of the trail is to see whether of MK-1942 treatment is superior to placebo in reducing depressive symptoms (MADRS). It is currently in clinical development (ClinicalTrials.gov Identifier: NCT04663321) with no results posted yet.

Nitrous oxide

Nitrous oxide (N_2O) is a gas used for inhalational general anesthesia and analgesia for short procedures. It possesses several biological properties, including the ability to act as a noncompetitive antagonist at NMDA receptors. N_2O has a broad mechanism of action similar to ketamine but is not believed to have strong addiction potential. A phase II, double-blind, placebo-controlled, crossover study has demonstrated that 2 h as well as at 24 h postinhalation, N_2O significantly improved antidepressant response rates (as measured by the HAM-D) compared to placebo; however, the size effect was not large. Moreover, no psychotomimetic effects were observed after the administration, but some adverse effects included anxiety, headache, and nausea/vomiting (Nagele et al., 2015). Further investigation is needed to evaluate the potential of N_2O to TRD.

Atypical antipsychotics

Combination antidepressants with other psychotropic medications remain a popular treatment strategy, especially for TRD. People with depression who also display psychotic symptoms are more likely to be treatment-resistant. Thus, the most robust evidence as augmentation of conventional antidepressant therapy is with several atypical antipsychotics, a class of medication primarily used to treat psychosis (McIntyre et al., 2014). Three atypical antipsychotic agents are FDA approved as adjunct (i.e., aripiprazole and quetiapine) or in combination (olanzapine-fluoxetine combination [OFC]) with antidepressant therapy (Spielmans et al., 2013). A variety of pharmacologically distinct agents has been evaluated in heterogeneous TRD samples. Currently, RCTs include identifying baseline determinants of treatment response and tolerability in order to reduce the proportion of individuals with insufficient outcomes.

Ziprasidone

Among the available drugs for the TRD treatment, second-generation antipsychotics (SGAs) have been reported as one possible adjunctive therapy for the management of TRD (Shelton and Papakostas, 2008). Ziprasidone is an antipsychotic approved by the FDA to treat schizophrenia and bipolar disorder, but it is used off-label for MDD. Preclinical studies have been shown in vitro inhibition of the neuronal serotonin and norepinephrine uptake, with a potency comparable to imipramine (Schmidt et al., 2001). A pilot trial with an open-label design demonstrated that 10 of 20 patients with TRD had increased antidepressant effects after 6 weeks of ziprasidone augmentation (Papakostas et al., 2004).

Brexpiprazole

Brexpiprazole is an atypical antipsychotic which acts as a partial agonist at dopamine D2 receptors and, possibly, as an antagonist at 5-HT2A receptors (Fornaro et al., 2019). Due to its mechanism of action, brexpiprazole has been discussed as a promising candidate for targeting both depressive and cognitive deficits observed in TRD. A recent phase III/IV trial was designed to investigate the maintenance of efficacy and safety profile during long-term treatment with brexpiprazole in adults with an inadequate response to antidepressant treatment. The study was recently completed, but results were not posted to ClinicalTrials.com.

Cariprazine

Cariprazine is a novel antipsychotic drug that exerts partial agonist properties at dopamine D2/D3 receptors with preferential affinity for dopamine D3 receptors. Besides, cariprazine also interacts with the serotonergic system as an antagonist of 5HT2B receptors and a partial agonist of 5HT1A. Animal studies revealed antianhedonic effects of cariprazine, suggesting a potential to treat TRD (Gyertyan et al., 2011). In patients, a phase II RCT aimed to evaluate the efficacy, safety, and tolerability of cariprazine as an adjunctive treatment in MDD patients who have had an inadequate response to standard antidepressant therapy. The results showed that compared with placebo, reduction in MADRS total score at week 8 was significantly greater with adjunctive cariprazine treatment (2–4.5 mg/d) (Durgam et al., 2016). However, in a phase III RCT which evaluated flexible doses of cariprazine as adjunctive therapy in TRD patients, no significant efficacy was observed compared with placebo.

Other drugs

Pilocybin

Psychedelics were first investigated as therapeutic agents in the 1960s, and although studies carried out then had several methodological shortcomings, they provided preliminary evidence that such compounds could be effective as a mood enhancer and potential antidepressant (Muttoni et al., 2019). Psilocybin is a naturally occurring plant alkaloid, which is rapidly converted by the body to psilocin, a partial agonist for several brain serotonergic receptors, particularly 5-HT2A (Halberstadt and Geyer, 2011). An open-label feasibility study assessed the efficacy and safety of adjunctive psilocybin in 12 patients with TRD (Carhart-Harris et al., 2016). The acute psychedelic effects peaked 2–3 h after dosing and declined to negligible levels at least 6 h after dosing. Depressive symptoms were markedly reduced (defined as $\geq$50% reduction in Beck Depression Inventory (BDI) score relative to baseline) 1 week and 3 months after high-dose (25 mg) treatment. A 6-month follow-up study, in which a further eight participants were enrolled, found that the antidepressant and anxiolytic effects of psilocybin were sustained and remained significant at 6 months posttreatment (Carhart-Harris et al., 2018). One RCT observed at pre- and posttreatment data with psilocybin using functional magnetic resonance imaging (fMRI) in 19 patients. All patients decreased depressive symptoms at 1-week posttreatment, and 47% met the criteria for responsiveness after 5 weeks (Carhart-Harris et al., 2017). Psilocybin is a promising paradigm for unresponsive depression that deserves further investigation in larger samples.

Cannabidiol

Cannabidiol (CBD), a nonpsychotomimetic cannabinoid present in the *Cannabis sativa* plant, seems to be a promising compound for MDD since it shows large-spectrum therapeutic potential in preclinical models and humans (Campos et al., 2017). However, its antidepressant properties have not been thoroughly investigated. The mechanisms involved in CBD-induced psychotropic effects involve the activation of serotonergic 5-HT1A receptors and peroxisome proliferator-activated (PPAR) receptors. Indeed, recent preclinical studies have been demonstrated that CBD effects may be related to rapid changes in synaptic plasticity in the mPFC through activation of the BDNF-TrkB signaling pathway (Sales et al., 2019). Currently, Epidiolex (CBD) is being evaluated in crossover, placebo-controlled, double-blind RCT including 10 depressed adults with TRD over 16 weeks. The antidepressant effects of CBD will be measured by standard rating scales. The results were not posted on ClinicalTrials.com so far.

Minocycline

Minocycline is a tetracycline antibiotic approved by the FDA for various bacterial infections (Food and Drug Administration, 2021). Preclinical studies have reported pleiotropic effects of minocycline in animal models and humans, including antiinflammatory and neuroprotective actions (Husain et al., 2015). Several clinical trials are underway to explore the efficacy of off-label minocycline treatment for augmentation in TRD. A small pilot double-blind placebo-controlled 12-week study demonstrated that minocycline treatment produced a large decrease in HAMD scores compared to the placebo

(Husain et al., 2015). These findings require replication in larger samples to confirm the efficacy and tolerability of this approach. Another RCT evaluating the antidepressant efficacy of adjunctive minocycline in TRD patients is currently recruiting participants.

Conclusions

Even after trying several different treatments, a substantial proportion of MDD patients do not respond. Besides that, the chance of successful response decreases after every new treatment is tried. The etiology of depression is multifactorial, and the pathophysiology itself remains unknown, especially for TRD. Additional research on the biological mechanisms can offer new treatment targets for specific groups of patients, acting in different mechanisms. Intense efforts are being made to understand the causes of TRD and develop new therapies. In the past several years, discoveries have launched a new era of antidepressant research, with the promise to contemplate different biological correlates and increase efficacy and response time. Many compounds are being evaluated for TRD alone or combined with other medications, with several promising new agents acting on the HPA axis, melatonergic system, glutamatergic system, and others. While these novel therapeutic targets have shown some good results, additional long-term and larger clinical trials are needed to evaluate the efficacy and monitor possible adverse effects. Additionally, relapse rates for responsive patients are very high after 2 years, and the studies should include long-term follow-ups. In conclusion, although preliminary, overall results reported by preclinical and clinical studies exploring the impact of drugs under investigation seem to indicate promising new options for TRD.

References

Anacker, C., Cattaneo, A., Musaelyan, K., Zunszain, P.A., Horowitz, M., Molteni, R., et al., 2013. Role for the kinase SGK1 in stress, depression, and glucocorticoid effects on hippocampal neurogenesis. Proc. Natl. Acad. Sci. U. S. A. 110 (21), 8708–8713.

Berman, R.M., Cappiello, A., Anand, A., Oren, D.A., Heninger, G.R., Charney, D.S., et al., 2000. Antidepressant effects of ketamine in depressed patients. Biol. Psychiatry 47 (4), 351–354.

Blasey, C.M., Block, T.S., Belanoff, J.K., Roe, R.L., 2011. Efficacy and safety of mifepristone for the treatment of psychotic depression. J. Clin. Psychopharmacol. 31 (4), 436–440.

Block, T.S., Kushner, H., Kalin, N., Nelson, C., Belanoff, J., Schatzberg, A., 2018. Combined analysis of mifepristone for psychotic depression: plasma levels associated with clinical response. Biol. Psychiatry 84 (1), 46–54.

Blockmans, D., Persoons, P., Van Houdenhove, B., Lejeune, M., Bobbaers, H., 2003. Combination therapy with hydrocortisone and fludrocortisone does not improve symptoms in chronic fatigue syndrome: a randomized, placebo-controlled, double-blind, crossover study. Am. J. Med. 114 (9), 736–741.

Board, F., Persky, H., Hamburg, D.A., 1956. Psychological stress and endocrine functions; blood levels of adrenocortical and thyroid hormones in acutely disturbed patients. Psychosom. Med. 18 (4), 324–333.

Campos, A.C., Fogaca, M.V., Scarante, F.F., Joca, S.R.L., Sales, A.J., Gomes, F.V., et al., 2017. Plastic and neuroprotective mechanisms involved in the therapeutic effects of Cannabidiol in psychiatric disorders. Front. Pharmacol. 8, 269.

Carhart-Harris, R.L., Bolstridge, M., Rucker, J., Day, C.M., Erritzoe, D., Kaelen, M., et al., 2016. Psilocybin with psychological support for treatment-resistant depression: an open-label feasibility study. Lancet Psychiatry 3 (7), 619–627.

Carhart-Harris, R.L., Roseman, L., Bolstridge, M., Demetriou, L., Pannekoek, J.N., Wall, M.B., et al., 2017. Psilocybin for treatment-resistant depression: fMRI-measured brain mechanisms. Sci. Rep. 7 (1), 13187.

Carhart-Harris, R.L., Bolstridge, M., Day, C.M.J., Rucker, J., Watts, R., Erritzoe, D.E., et al., 2018. Psilocybin with psychological support for treatment-resistant depression: six-month follow-up. Psychopharmacology 235 (2), 399–408.

Carroll, B.J., Rubin, R.T., 2006. Is mifepristone useful in psychotic depression? Neuropsychopharmacology 31 (12), 2793–2794 (author reply 5-7).

Cattaneo, A., Ferrari, C., Turner, L., Mariani, N., Enache, D., Hastings, C., et al., 2020. Whole-blood expression of inflammasome- and glucocorticoid-related mRNAs correctly separates treatment-resistant depressed patients from drug-free and responsive patients in the BIODEP study. Transl. Psychiatry 10 (1), 232.

Corruble, E., de Bodinat, C., Belaidi, C., Goodwin, G.M., 2013. Agomelatine study g. efficacy of agomelatine and escitalopram on depression, subjective sleep and emotional experiences in patients with major depressive disorder: a 24-wk randomized, controlled, double-blind trial. Int. J. Neuropsychopharmacol. 16 (10), 2219–2234.

Dalm, S., Karssen, A.M., Meijer, O.C., Belanoff, J.K., de Kloet, E.R., 2019. Resetting the stress system with a mifepristone challenge. Cell. Mol. Neurobiol. 39 (4), 503–522.

Daly, E.J., Singh, J.B., Fedgchin, M., et al., 2015. Intranasal esketamine in treatment-resistant depression, a dose response study - double-blind and open-label extension data. In: 54th Annual Meeting of the American College of Neuropsychopharmacology 2015 (ACNP), p. 74.

Dattilo, V., Amato, R., Perrotti, N., Gennarelli, M., 2020. The emerging role of SGK1 (serum- and glucocorticoid-regulated kinase 1) in major depressive disorder: hypothesis and mechanisms. Front. Genet. 11, 826.

De Berardis, D., Fornaro, M., Serroni, N., Campanella, D., Rapini, G., Olivieri, L., et al., 2015. Agomelatine beyond borders: current evidences of its efficacy in disorders other than major depression. Int. J. Mol. Sci. 16 (1), 1111–1130.

de Kloet, E.R., 2013. Functional profile of the binary brain corticosteroid receptor system: mediating, multitasking, coordinating, integrating. Eur. J. Pharmacol. 719 (1–3), 53–62.

DeBattista, C., Posener, J.A., Kalehzan, B.M., Schatzberg, A.F., 2000. Acute antidepressant effects of intravenous hydrocortisone and CRH in depressed patients: a double-blind, placebo-controlled study. Am. J. Psychiatry 157 (8), 1334–1337.

Durgam, S., Earley, W., Guo, H., Li, D., Nemeth, G., Laszlovszky, I., et al., 2016. Efficacy and safety of adjunctive cariprazine in inadequate responders to antidepressants: a randomized, double-blind, placebo-controlled study in adult patients with major depressive disorder. J. Clin. Psychiatry 77 (3), 371–378.

EMA, European Medicines Agency, 2013. Nuedexta. Product Information.

FDA, Food and Drug Administration, 2010. Drugs@FDA. Neudexta. Drug Approval Package.

Fornaro, M., Fusco, A., Anastasia, A., Cattaneo, C.I., De Berardis, D., 2019. Brexpiprazole for treatment-resistant major depressive disorder. Expert. Opin. Pharmacother. 20 (16), 1925–1933.

Frodl, T., Carballedo, A., Hughes, M.M., Saleh, K., Fagan, A., Skokauskas, N., et al., 2012. Reduced expression of glucocorticoid-inducible genes GILZ and SGK-1: high IL-6 levels are associated with reduced hippocampal volumes in major depressive disorder. Transl. Psychiatry 2, e88.

Gahr, M., 2014. Agomelatine in the treatment of major depressive disorder: an assessment of benefits and risks. Curr. Neuropharmacol. 12 (5), 287–398.

Gallagher, P., Young, A.H., 2006. Mifepristone (RU-486) treatment for depression and psychosis: a review of the therapeutic implications. Neuropsychiatr. Dis. Treat. 2 (1), 33–42.

Garner, R., Gopalakrishnan, S., McCauley, J.A., Bednar, R.A., Gaul, S.L., Mosser, S.D., et al., 2015. Preclinical pharmacology and pharmacokinetics of CERC-301, a GluN2B-selective N-methyl-D-aspartate receptor antagonist. Pharmacol. Res. Perspect. 3 (6), e00198.

Ghadirian, A.M., Engelsmann, F., Dhar, V., Filipini, D., Keller, R., Chouinard, G., et al., 1995. The psychotropic effects of inhibitors of steroid biosynthesis in depressed patients refractory to treatment. Biol. Psychiatry 37 (6), 369–375.

Gyertyan, I., Kiss, B., Saghy, K., Laszy, J., Szabo, G., Szabados, T., et al., 2011. Cariprazine (RGH-188), a potent D3/D2 dopamine receptor partial agonist, binds to dopamine D3 receptors in vivo and shows antipsychotic-like and procognitive effects in rodents. Neurochem. Int. 59 (6), 925–935.

Halberstadt, A.L., Geyer, M.A., 2011. Multiple receptors contribute to the behavioral effects of indoleamine hallucinogens. Neuropharmacology 61 (3), 364–381.

Husain, M.I., Chaudhry, I.B., Rahman, R.R., Hamirani, M.M., Qurashi, I., Khoso, A.B., et al., 2015. Minocycline as an adjunct for treatment-resistant depressive symptoms: study protocol for a pilot randomised controlled trial. Trials 16, 410.

Ibrahim, L., Diaz Granados, N., Jolkovsky, L., Brutsche, N., Luckenbaugh, D.A., Herring, W.J., et al., 2012. A randomized, placebo-controlled, crossover pilot trial of the oral selective NR2B antagonist MK-0657 in patients with treatment-resistant major depressive disorder. J. Clin. Psychopharmacol. 32 (4), 551–557.

Jahn, H., Schick, M., Kiefer, F., Kellner, M., Yassouridis, A., Wiedemann, K., 2004. Metyrapone as additive treatment in major depression: a double-blind and placebo-controlled trial. Arch. Gen. Psychiatry 61 (12), 1235–1244.

Jeffcoate, W.J., Silverstone, J.T., Edwards, C.R., Besser, G.M., 1979. Psychiatric manifestations of Cushing's syndrome: response to lowering of plasma cortisol. Q. J. Med. 48 (191), 465–472.

Juruena, M.F., Pariante, C.M., Papadopoulos, A.S., Poon, L., Lightman, S., Cleare, A.J., 2013. The role of mineralocorticoid receptor function in treatment-resistant depression. J. Psychopharmacol. 27 (12), 1169–1179.

Kato, T., Fogaca, M.V., Deyama, S., Li, X.Y., Fukumoto, K., Duman, R.S., 2018. BDNF release and signaling are required for the antidepressant actions of GLYX-13. Mol. Psychiatry 23 (10), 2007–2017.

Kling, M.A., Whitfield Jr., H.J., Brandt, H.A., Demitrack, M.A., Kalogeras, K., Geracioti, T.D., et al., 1989. Effects of glucocorticoid antagonism with RU 486 on pituitary-adrenal function in patients with major depression: time-dependent enhancement of plasma ACTH secretion. Psychopharmacol. Bull. 25 (3), 466–472.

Krishnan, K.R., Reed, D., Wilson, W.H., Saunders, W.B., Ritchie, J.C., Nemeroff, C.B., et al., 1992. RU486 in depression. Prog. Neuro-Psychopharmacol. Biol. Psychiatry 16 (6), 913–920.

Kuningas, M., de Rijk, R.H., Westendorp, R.G., Jolles, J., Slagboom, P.E., van Heemst, D., 2007. Mental performance in old age dependent on cortisol and genetic variance in the mineralocorticoid and glucocorticoid receptors. Neuropsychopharmacology 32 (6), 1295–1301.

Lanfumey, L., Mongeau, R., Hamon, M., 2013. Biological rhythms and melatonin in mood disorders and their treatments. Pharmacol. Ther. 138 (2), 176–184.

Laux, G., Huttner, N.A., group Vs., 2014. Subgroup analysis of the non-interventional study VIVALDI: agomelatine in treatment-naive patients, in combination therapy and after treatment switch. Int. J. Psychiatry Clin. Pract. 18 (2), 86–96.

Lin, T.Y., Hanna, J., Ishak, W.W., 2020. Psychiatric symptoms in Cushing's syndrome: a systematic review. Innov. Clin. Neurosci. 17 (1–3), 30–35.

Lopez, J.F., Chalmers, D.T., Little, K.Y., Watson, S.J., 1998. A.E. Bennett Research Award. Regulation of serotonin1A, glucocorticoid, and mineralocorticoid receptor in rat and human hippocampus: implications for the neurobiology of depression. Biol. Psychiatry 43 (8), 547–573.

Lupien, S.J., McEwen, B.S., 1997. The acute effects of corticosteroids on cognition: integration of animal and human model studies. Brain Res. Brain Res. Rev. 24 (1), 1–27.

McIntyre, R.S., Filteau, M.J., Martin, L., Patry, S., Carvalho, A., Cha, D.S., et al., 2014. Treatment-resistant depression: definitions, review of the evidence, and algorithmic approach. J. Affect. Disord. 156, 1–7.

Medina, A., Seasholtz, A.F., Sharma, V., Burke, S., Bunney Jr., W., Myers, R.M., et al., 2013. Glucocorticoid and mineralocorticoid receptor expression in the human hippocampus in major depressive disorder. J. Psychiatr. Res. 47 (3), 307–314.

Moskal, J.R., Burch, R., Burgdorf, J.S., Kroes, R.A., Stanton, P.K., Disterhoft, J.F., et al., 2014. GLYX-13, an NMDA receptor glycine site functional partial agonist enhances cognition and produces antidepressant effects without the psychotomimetic side effects of NMDA receptor antagonists. Expert Opin. Investig. Drugs 23 (2), 243–254.

Murphy, B.E., Dhar, V., Ghadirian, A.M., Chouinard, G., Keller, R., 1991. Response to steroid suppression in major depression resistant to antidepressant therapy. J. Clin. Psychopharmacol. 11 (2), 121–126.

Murphy, B.E., Filipini, D., Ghadirian, A.M., 1993. Possible use of glucocorticoid receptor antagonists in the treatment of major depression: preliminary results using RU 486. J. Psychiatry Neurosci. 18 (5), 209–213.

Murrough, J.W., Wade, E., Sayed, S., Ahle, G., Kiraly, D.D., Welch, A., et al., 2017. Dextromethorphan/quinidine pharmacotherapy in patients with treatment resistant depression: a proof of concept clinical trial. J. Affect. Disord. 218, 277–283.

Muttoni, S., Ardissino, M., John, C., 2019. Classical psychedelics for the treatment of depression and anxiety: a systematic review. J. Affect. Disord. 258, 11–24.

Nagele, P., Duma, A., Kopec, M., Gebara, M.A., Parsoei, A., Walker, M., et al., 2015. Nitrous oxide for treatment-resistant major depression: a proof-of-concept trial. Biol. Psychiatry 78 (1), 10–18.

Nieman, L.K., Chrousos, G.P., Kellner, C., Spitz, I.M., Nisula, B.C., Cutler, G.B., et al., 1985. Successful treatment of Cushing's syndrome with the glucocorticoid antagonist RU 486. J. Clin. Endocrinol. Metab. 61 (3), 536–540.

Numakawa, T., Adachi, N., Richards, M., Chiba, S., Kunugi, H., 2013. Brain-derived neurotrophic factor and glucocorticoids: reciprocal influence on the central nervous system. Neuroscience 239, 157–172.

O'Dwyer, A.M., Lightman, S.L., Marks, M.N., Checkley, S.A., 1995. Treatment of major depression with metyrapone and hydrocortisone. J. Affect. Disord. 33 (2), 123–128.

Otte, C., Lenoci, M., Metzler, T., Yehuda, R., Marmar, C.R., Neylan, T.C., 2007. Effects of metyrapone on hypothalamic-pituitary-adrenal axis and sleep in women with post-traumatic stress disorder. Biol. Psychiatry 61 (8), 952–956.

Otte, C., Hinkelmann, K., Moritz, S., Yassouridis, A., Jahn, H., Wiedemann, K., et al., 2010. Modulation of the mineralocorticoid receptor as add-on treatment in depression: a randomized, double-blind, placebo-controlled proof-of-concept study. J. Psychiatr. Res. 44 (6), 339–346.

Papakostas, G.I., Petersen, T.J., Nierenberg, A.A., Murakami, J.L., Alpert, J.E., Rosenbaum, J.F., et al., 2004. Ziprasidone augmentation of selective serotonin reuptake inhibitors (SSRIs) for SSRI-resistant major depressive disorder. J. Clin. Psychiatry 65 (2), 217–221.

Posener, J.A., DeBattista, C., Williams, G.H., Chmura Kraemer, H., Kalehzan, B.M., Schatzberg, A.F., 2000. 24-hour monitoring of cortisol and corticotropin secretion in psychotic and nonpsychotic major depression. Arch. Gen. Psychiatry 57 (8), 755–760.

Potmesil, P., 2019. What combinations of agomelatine with other antidepressants could be successful during the treatment of major depressive disorder or anxiety disorders in clinical practice? Ther. Adv. Psychopharmacol. 9, 2045125319855206.

Preskorn, S., Macaluso, M., Mehra, D.O., Zammit, G., Moskal, J.R., Burch, R.M., et al., 2015. Randomized proof of concept trial of GLYX-13, an N-methyl-D-aspartate receptor glycine site partial agonist, in major depressive disorder nonresponsive to a previous antidepressant agent. J. Psychiatr. Pract. 21 (2), 140–149.

Qiao, J., Sun, Y., Wu, J., Wang, L., 2019. Investigation of the underling mechanism of ketamine for antidepressant effects in treatment-refractory affective disorders via molecular profile analysis. Exp. Ther. Med. 18 (1), 580–588.

Rogoz, Z., Skuza, G., Wojcikowski, J., Daniel, W.A., Wrobel, A., Dudek, D., et al., 2004. Effect of metyrapone supplementation on imipramine therapy in patients with treatment-resistant unipolar depression. Pol. J. Pharmacol. 56 (6), 849–855.

Rowe, P.C., Calkins, H., DeBusk, K., McKenzie, R., Anand, R., Sharma, G., et al., 2001. Fludrocortisone acetate to treat neurally mediated hypotension in chronic fatigue syndrome: a randomized controlled trial. JAMA 285 (1), 52–59.

Sales, A.J., Fogaca, M.V., Sartim, A.G., Pereira, V.S., Wegener, G., Guimaraes, F.S., et al., 2019. Cannabidiol induces rapid and sustained antidepressant-like effects through increased BDNF signaling and synaptogenesis in the prefrontal cortex. Mol. Neurobiol. 56 (2), 1070–1081.

Salituro, F.G., Tomlinson, R.C., Baron, B.M., Palfreyman, M.G., McDonald, I.A., Schmidt, W., et al., 1994. Enzyme-activated antagonists of the strychnine-insensitive glycine/NMDA receptor. J. Med. Chem. 37 (3), 334–336.

Sarabdjitsingh, R.A., Isenia, S., Polman, A., Mijalkovic, J., Lachize, S., Datson, N., et al., 2010. Disrupted corticosterone pulsatile patterns attenuate responsiveness to glucocorticoid signaling in rat brain. Endocrinology 151 (3), 1177–1186.

Schmidt, A.W., Lebel, L.A., Howard Jr., H.R., Zorn, S.H., 2001. Ziprasidone: a novel antipsychotic agent with a unique human receptor binding profile. Eur. J. Pharmacol. 425 (3), 197–201.

Schule, C., Baghai, T.C., Eser, D., Rupprecht, R., 2009. Hypothalamic-pituitary-adrenocortical system dysregulation and new treatment strategies in depression. Expert. Rev. Neurother. 9 (7), 1005–1019.

Shelton, R.C., Papakostas, G.I., 2008. Augmentation of antidepressants with atypical antipsychotics for treatment-resistant major depressive disorder. Acta Psychiatr. Scand. 117 (4), 253–259.

Sigalas, P.D., Garg, H., Watson, S., McAllister-Williams, R.H., Ferrier, I.N., 2012. Metyrapone in treatment-resistant depression. Ther. Adv. Psychopharmacol. 2 (4), 139–149.

Singh, J.B., Fedgchin, M., Daly, E., et al., 2015. Intravenous esketamine in adult treatment-resistant depression: double-blind, double-randomization, placedo-controlled study. Biol. Psychiatry 173, 816–826.

Singh, J.B., Fedgchin, M., Daly, E., Xi, L., Melman, C., De Bruecker, G., et al., 2016. Intravenous esketamine in adult treatment-resistant depression: a double-blind, double-randomization, placebo-controlled study. Biol. Psychiatry 80 (6), 424–431.

Spielmans, G.I., Berman, M.I., Linardatos, E., Rosenlicht, N.Z., Perry, A., Tsai, A.C., 2013. Adjunctive atypical antipsychotic treatment for major depressive disorder: a meta-analysis of depression, quality of life, and safety outcomes. PLoS Med. 10 (3), e1001403.

Stahl, S.M., 2020. Dextromethorphan/bupropion: a novel oral NMDA (N-methyl-d-aspartate) receptor antagonist with multimodal activity-addendum. CNS Spectr. 25 (6), 803.

Starkman, M.N., 2013. Neuropsychiatric findings in Cushing syndrome and exogenous glucocorticoid administration. Endocrinol. Metab. Clin. N. Am. 42 (3), 477–488.

Strawbridge, R., Jamieson, A., Hodsoll, J., Ferrier, I.N., McAllister-Williams, R.H., Powell, T.R., et al., 2021. The role of inflammatory proteins in anti-glucocorticoid therapy for treatment-resistant depression. J. Clin. Med. 10 (4).

Stuhec, M., Oravecz, R., 2016. Moclobemide as add-on therapy to agomelatine in a patient with treatment-resistant major depressive disorder: a psychopharmacological case. Wien. Klin. Wochenschr. 128 (7–8), 295–298.

Suhs, K.W., Correll, C., Eberlein, C.K., Pul, R., Frieling, H., Bleich, S., et al., 2015. Combination of agomelatine and bupropion for treatment-resistant depression: results from a chart review study including a matched control group. Brain Behav. 5 (4), e00318.

Zanos, P., Piantadosi, S.C., Wu, H.Q., Pribut, H.J., Dell, M.J., Can, A., et al., 2015. The prodrug 4-chlorokynurenine causes ketamine-like antidepressant effects, but not side effects, by NMDA/GlycineB-site inhibition. J. Pharmacol. Exp. Ther. 355 (1), 76–85.

Zanos, P., Moaddel, R., Morris, P.J., Georgiou, P., Fischell, J., Elmer, G.I., et al., 2016. NMDAR inhibition-independent antidepressant actions of ketamine metabolites. Nature 533 (7604), 481–486.

Zarate Jr., C.A., Du, J., Quiroz, J., Gray, N.A., Denicoff, K.D., Singh, J., et al., 2003. Regulation of cellular plasticity cascades in the pathophysiology and treatment of mood disorders: role of the glutamatergic system. Ann. N. Y. Acad. Sci. 1003, 273–291.

Zarate Jr., C.A., Singh, J.B., Carlson, P.J., Brutsche, N.E., Ameli, R., Luckenbaugh, D.A., et al., 2006. A randomized trial of an N-methyl-D-aspartate antagonist in treatment-resistant major depression. Arch. Gen. Psychiatry 63 (8), 856–864.

Zarate Jr., C.A., Mathews, D., Ibrahim, L., Chaves, J.F., Marquardt, C., Ukoh, I., et al., 2013. A randomized trial of a low-trapping nonselective N-methyl-D-aspartate channel blocker in major depression. Biol. Psychiatry 74 (4), 257–264.

Zhang, K., Hashimoto, K., 2019. An update on ketamine and its two enantiomers as rapid-acting antidepressants. Expert. Rev. Neurother. 19 (1), 83–92.

Zhang, X.L., Sullivan, J.A., Moskal, J.R., Stanton, P.K., 2008. A NMDA receptor glycine site partial agonist, GLYX-13, simultaneously enhances LTP and reduces LTD at Schaffer collateral-CA1 synapses in hippocampus. Neuropharmacology 55 (7), 1238–1250.

Chapter 35

Chronotherapeutics for treatment-resistant depression

Sara Dallaspezia[a] **and Francesco Benedetti**[a,b]

[a]*Division of Neuroscience, Psychiatry and Psychobiology Unit, IRCCS San Raffaele Scientific Institute, Milan, Italy,* [b]*Division of Neuroscience, Psychiatry and Psychobiology Unit, Vita-Salute San Raffaele University, Milan, Italy*

Introduction

According to the World Health Organization (WHO), bipolar or unipolar depression is the leading cause of ill health and disability worldwide. Depression affects nearly 20% of the population with an increase of more than 18% having been observed between 2005 and 2015 (WHO: WHO I Depression [Internet]. Depression. 2019). Depression is associated with a reduced quality of life, with the WHO predicting several years ago that it would become the greatest cause of disability worldwide by 2030 (Wittchen et al., 2011). People affected by major depression show increased mortality, with an estimated 50% risk increase compared to nondepressed individuals (Osby et al., 2001), with suicide accounting for nearly 20% of the deaths (Wulsin et al., 1999).

Most antidepressant drugs need several weeks to achieve therapeutic effects and several trials show that no difference can be expected during the first 2 weeks between active and placebo treatments (Walsh et al., 2002). Moreover, long-term use of antidepressant drugs often generates significant side effects, leading to treatment discontinuation and consequent relapses. When facing bipolar depression, iatrogenic mania or rapid cycling are associated risks. Indeed, all effective standard antidepressant medications have been associated with treatment-emergent mania and the administration of chronic antidepressant drugs has been linked to an acceleration of the rate of cycling without an induction to actual mania (Sachs, 1996).

Furthermore, antidepressant drugs show low-to-moderate response in most patients, with at least 40% of patients treated for depression not responding to the first administered antidepressant medication, and several further treatment trials not showing an antidepressant effect in at least one-half of these patients (Bartova et al., 2019; Crown et al., 2002). This clinical condition characterized by an inadequate response to two or more consecutive antidepressant drugs administered at adequate doses for an adequate length is referred to as treatment-resistant depression (TRD), and it is a persistent source of morbidity and mortality for patients and contributes substantially to the burden of disease (McIntyre et al., 2014; Berlim and Turecki, 2007). Comparing unipolar with bipolar depressed patients, unresponsiveness to antidepressant drugs affects up to 30% of unipolar depressed patients and was found to be 1.6 times more common in people affected by bipolar disorder. Moreover, bipolar patients showed a loss of response to antidepressant drugs 3.4 times more frequently than unipolar ones (Ghaemi et al., 2004).

Resistance to antidepressant treatment has been associated with functional impairment and poorer quality of life, increased mortality and suicide risk, and higher relapse rates (Fekadu et al., 2009). Because of the morbidity linked to treatment-resistant depression, the development of novel, rapidly acting antidepressants is crucial for patients affected by this condition.

Many augmentation strategies have been developed to increase the efficacy of antidepressant drugs. However, for many patients, medication augmentation may be risky or demanding treatment strategies in terms of adverse effects and drug interactions. A growing evidence base supports the usefulness of chronotherapeutic interventions, which are nonpharmaceutical psychiatric treatments developed from the study of circadian rhythm organization and sleep physiology. These techniques exploit the control of the exposure to environmental stimuli that act on biological rhythms. They include sleep deprivation and sleep phase advancement, which manipulate sleep-wake cycle, and light therapy which acts through exposure to light.

Chronobiological interventions had been widely used as augmentation strategies for several years in clinical settings. Although relatively few studies have focused on patients with well-defined TRD, recent clinical studies have confirmed their efficacy in TRD and other forms of difficult to treat depression.

Managing Treatment-Resistant Depression. https://doi.org/10.1016/B978-0-12-824067-0.00035-9

Light therapy

The scientific approach to the treatment of depression with bright light was introduced by Rosenthal and his colleagues in the 1980s for winter seasonal affective disorder. Over the subsequent years, several trials extended the use of light therapy (LT) to nonseasonal unipolar depression (Al-Karawi and Jubair, 2016). More recently, the utility and safety of light therapy (LT) was confirmed in patients with bipolar depression (Dallaspezia and Benedetti, 2020).

The conventional LT system consists of a set of white fluorescent bulbs installed in a box with a diffusing screen that filters out ultraviolet (UV)-wavelength light. During LT sessions, patients sit in front of a light box. The therapeutic response to LT critically depends on the time of delivery, with best results having been found when the therapy is administered considering the personal circadian phase, influenced by the onset of melatonin secretion. Since this chronobiological phenomenon strongly correlates with Morningness-Eveningness Questionnaire (MEQ) scores, a predictive algorithm based on MEQ scores has been developed to define the individual optimal timing of LT administration, and it is now regularly used in clinical settings (Terman et al., 2001). The optimum dose of light is 10,000 lx for 30 min/day. Although lower intensities of light have antidepressant efficacy, substantially longer exposure durations are needed (Terman et al., 1990). Moreover, some observations have suggested that bipolar patients could be highly sensitive to the antidepressant properties of light, including at intensities as low as 500 lx (Lewy et al., 1985). When combined with standard antidepressant drugs, LT has been shown to hasten recovery, with patients perceiving benefits during the first week of treatment (Benedetti et al., 2003; Martiny et al., 2005). Moreover, after 1 month of treatment, depressed patients who were given LT augmentation therapy had a 30% greater reduction in depression severity than patients treated with drug plus placebo. These augmentation effects were similar between patients who received LT and selective serotonergic antidepressant drugs and those who received LT and tricyclic antidepressants (Kripke, 1998). The beneficial effects of LT are also clinically evident in patients with unipolar and bipolar major depression after LT is added to ongoing treatment with ineffective medications, highlighting its potential utility in people with TRD (Prasko et al., 2010; Camardese et al., 2015). The effectiveness of adjunctive LT in patients with TRD was demonstrated in both the acute phase of treatment and after LT discontinuation, with patients remaining euthymic for up to 8 weeks following the end of treatment (Camardese et al., 2015). In a recent study focusing on TRD, LT was found to accelerate antidepressant response to neuromodulation treatments such as high-frequency repetitive transcranial magnetic stimulation, showing a reduction of depressive symptomatology as early as during the first week of treatment (Barbini et al., 2021).

Sleep deprivation therapy and combined chronotherapies

The first clinical report about antidepressant effect of sleep deprivation was published in 1959 by Schulte, suggesting that sleep deprivation (SD) might benefit depressed patients. This opinion was based on anecdotal reports from three depressed patients who improved clinically after accidentally staying awake all night (Schulte, 1959). It was not until the 1970s that the first experimental trials were conducted focusing on this intervention's efficacy for depressed patients (Pflug and Tolle, 1971a,b).

Sleep deprivation therapy

In the early studies, SD was observed to produce rapid benefits in the broadly defined depressive syndrome, although better effects were shown in endogenous primary depression and in bipolar depression when compared to unipolar depressive disorder (Barbini et al., 1998). To achieve the best antidepressant effects, the wake period includes the extension of daytime wake into the night, and lasts about 36 h until the evening of the following day (referred to as total sleep deprivation, or TSD), but the period of sleep deprivation can also be limited to the second half of the night and the following day, thus allowing sleep during the first half of the night (Dallaspezia and Benedetti, 2014). SD shows very rapid antidepressant effects with clinical improvement being seen in the morning after a single night awake. The reported response rates to SD range from 50% to 80%, with a mean response rate of 60% across all diagnostic subgroups (Wirz-Justice and Van Den Hoofdakker, 1999) Unfortunately, the clinical efficacy of SD as a sole therapy seems to be limited by early depressive relapses (in up to 80% of SD responders) after a subsequent recovery sleep night. Indeed, during the days following a night of sleep deprivation, most patients experience a progressive worsening of depression, eventually returning to the same levels of depressive symptom severity that were observed at baseline (Giedke and Schwarzler, 2002). After only one night of SD, only 5% of early responders will maintain clinical remission and, after treatment discontinuation, the same proportion of relapse was found regardless of the number of sleep deprivation nights.

Sleep deprivation and light therapy

Methods for increasing and sustaining the efficacy of SD via combinatorial strategies have been reported in numerous studies, mainly involving the concomitant use of other chronotherapeutic interventions and mood stabilizers such as lithium salts. LT, used during and after SD, was found not only to maintain the clinical improvement generated by both partial (Neumeister et al., 1996) and repeated total SD (Colombo et al., 2000), but also to increase the overall antidepressant effect. Our group in Milan has developed a treatment protocol that includes repeated TSD and LT that results in a lengthening of the sleep-wake period from the usual duration of 24 to 48 h. According to this protocol, patients are administered three consecutive TSD cycles (days 1–6). On days 1, 3, and 5, patients are totally sleep deprived for 36 h and they are allowed to sleep during the night of days 2, 4, and 6. Patients are administered LT (exposure for 30 min to a 10,000 lx light) at 03:00 a.m. during the TSD night and in the morning after recovery sleep, half an hour after awakening, between 8 a.m. and 9 a.m. The addition of this combined chronobiological treatment (SD plus LT) to continued treatment with antidepressant medication and mood stabilizers was found to be useful in the treatment of drug-resistant bipolar patients, with an acute antidepressant response observed in 44% of patients who did not show a response to antidepressant drugs (Benedetti et al., 2005). More recently, the efficacy of TSD + LT was observed also in unipolar TRD patients (Sikkens et al., 2019). Moreover, starting lithium salts together with chronotherapeutics and prolonging LT for 1 week after the three TSD cycles increased the acute antidepressant response to up to 70% of bipolar TRD, with 53% of patients achieving a sustained euthymia (Benedetti et al., 2014). Moreover, this triple therapy was found to rapidly decrease suicidality regardless of clinical response. Indeed, it also resulted in decreased depressive symptoms even in nonresponder patients, who obtained an observable benefit from the treatment despite not achieving a categorical response (Benedetti et al., 2014).

Sleep deprivation and sleep phase advance

Sleep Phase Advance (SPA) has also been used to avoid relapses after a recovery night of sleep. SPA therapy consists of advancing the timing of the sleep-wake cycle, meaning that the time of sleep onset and awakening from sleep are moved forward, before the usual waking time. Early studies advanced sleep phases on a 1-week schedule; however, more recent studies used a 3-day sleep advancing schedule and demonstrated similar efficacy: after one night of TSD, bedtime started at 5 p.m. during the first recovery night, and then it was delayed daily by 2 h until the conventional bedtime (Voderholzer et al., 2003). This combined therapy leads to an improvement of depressive symptoms (greater than 30% from baseline) in 80% of treated patients, with 68% of patients maintaining their improvement beyond the end of treatment. When considering TRD patients, a positive antidepressant effect was found with pharmacotherapy augmented by TSD and SPA. Indeed, up to 50% of TRD patients showed a clinical response (Kurczewska et al., 2019).

Triple chronotherapy

Most recently, a triple chronotherapy protocol was developed that used a combination of LT, TSD, and SPA to treat both unipolar and bipolar depression (D'agostino et al., 2020). Response rates ranged from 50% to 84% across the individual studies, with up to 61.5% of responding patients maintaining clinical improvement 1 month after the end of treatment. Moreover, triple chronotherapy administered with concomitant pharmacotherapy produced a rapid improvement in depressive symptoms, which endured for as long as 9 weeks (Echizenya et al., 2013; Martiny et al., 2012; Wu et al., 2009). When triple chronotherapy was used to treat TRD patients, a clinically significant improvement of depressive symptoms in terms of both objective and subjective mood ratings was observed, and 61.5% of patients maintained the antidepressant effect until the end of the observation period (20 days after the end of treatment) (Echizenya et al., 2013).

Conclusions

For patients with major depression, chronobiological therapies have response effects sizes that are similar to the most effective antidepressant drugs, but with a much shorter latency of therapeutic effects. Chronobiological therapies appear to have only marginal or absent side effects, and can therefore be considered for the rapid treatment of depressed patients for whom drug therapy may be contraindicated or for whom medications may not be desired, such as pregnant women and people undergoing treatment with medications that should not be combined with antidepressants.

In the specific case of using chronobiological therapies in the management of TRD patients, promising results have also been documented, especially when such treatment is used adjunctively with ongoing pharmacotherapy. However, only a few studies have focused on the antidepressant efficacy of chronobiological therapies in patients with well-defined TRD.

In conclusion, chronobiological therapies are relatively safe, rapidly acting interventions that could fill therapeutic gaps—such as a long lag time to therapeutic benefit, treatment-limiting adverse effects, and potentially significant drug-drug interactions—that are left by traditional antidepressant drugs for treating mood disorders, even when TRD is taken into account. Further research is needed, however, to optimize chronotherapeutic protocols in order to both increase antidepressant response rates and extend the duration of acute antidepressant effects.

References

Al-Karawi, D., Jubair, L., 2016. Bright light therapy for nonseasonal depression: meta-analysis of clinical trials. J. Affect. Disord. 198, 64–71.

Barbini, B., Colombo, C., Benedetti, F., Campori, E., Bellodi, L., Smeraldi, E., 1998. The unipolar-bipolar dichotomy and the response to sleep deprivation. Psychiatry Res. 79, 43–50.

Barbini, B., Attanasio, F., Manfredi, E., Cavallini, M.C., Zanardi, R., Colombo, C., 2021. Bright light therapy accelerates the antidepressant effect of repetitive transcranial magnetic stimulation in treatment resistant depression: a pilot study. Int. J. Psychiatry Clin. Pract. 18, 1–3.

Bartova, L., Dold, M., Kautzky, A., et al., 2019. Results of the European Group for the Study of resistant depression (GSRD)—basis for further research and clinical practice. World J. Biol. Psychiatry 20, 427–448.

Benedetti, F., Colombo, C., Pontiggia, A., Bernasconi, A., Florita, M., Smeraldi, E., 2003. Morning light treatment hastens the antidepressant effect of citalopram: a placebo-controlled trial. J. Clin. Psychiatry 64, 648–653.

Benedetti, F., Barbini, B., Fulgosi, M.C., et al., 2005. Combined total sleep deprivation and light therapy in the treatment of drug-resistant bipolar depression: acute response and long-term remission rates. J. Clin. Psychiatry 66, 1535–1540.

Benedetti, F., Riccaboni, R., Locatelli, C., Poletti, S., Dallaspezia, S., Colombo, C., 2014. Rapid treatment response of suicidal symptoms to lithium, sleep deprivation, and light therapy (chronotherapeutics) in drug-resistant bipolar depression. J. Clin. Psychiatry 75, 133–140.

Berlim, M.T., Turecki, G., 2007. Definition, assessment, and staging of treatment-resistant refractory major depression: a review of current concepts and methods. Can. J. Psychiatry 52, 46–54.

Camardese, G., Leone, B., Serrani, R., et al., 2015. Augmentation of light therapy in difficult-to-treat depressed patients: an open-label trial in both unipolar and bipolar patients. Neuropsychiatr. Dis. Treat. 11, 2331–2338.

Colombo, C., Lucca, A., Benedetti, F., Barbini, B., Campori, E., Smeraldi, E., 2000. Total sleep deprivation combined with lithium and light therapy in the treatment of bipolar depression: replication of main effects and interaction. Psychiatry Res. 95, 43–53.

Crown, W.H., Finkelstein, S., Berndt, E.R., et al., 2002. The impact of treatment-resistant depression on health care utilization and costs. J. Clin. Psychiatry 63, 963–971.

D'agostino, A., Ferrara, P., Terzoni, S., et al., 2020. Efficacy of triple chronotherapy in unipolar and bipolar depression: a systematic review of the available evidence. J. Affect. Disord. 276, 297–304.

Dallaspezia, S., Benedetti, F., 2014. Sleep deprivation therapy for depression. Curr. Top. Behav. Neurosci. 25, 483–502.

Dallaspezia, S., Benedetti, F., 2020. Antidepressant light therapy for bipolar patients: a meta-analyses. J. Affect. Disord. 274, 943–948.

Echizenya, M., Suda, H., Takeshima, M., Inomata, Y., Shimizu, T., 2013. Total sleep deprivation followed by sleep phase advance and bright light therapy in drug-resistant mood disorders. J. Affect. Disord. 144, 28–33.

Fekadu, A., Wooderson, S.C., Markopoulo, K., Donaldson, C., Papadopoulos, A., Cleare, A.J., 2009. What happens to patients with treatment-resistant depression? A systematic review of medium to long term outcome studies. J. Affect. Disord. 116, 4–11.

Ghaemi, S.N., Rosenquist, K.J., Ko, J.Y., Baldassano, C.F., Kontos, N.J., Baldessarini, R.J., 2004. Antidepressant treatment in bipolar versus unipolar depression. Am. J. Psychiatry 161, 163–165.

Giedke, H., Schwarzler, F., 2002. Therapeutic use of sleep deprivation in depression. Sleep Med. Rev. 6, 361–377.

Kripke, D.F., 1998. Light treatment for nonseasonal depression: speed, efficacy, and combined treatment. J. Affect. Disord. 49, 109–117.

Kurczewska, E., Ferensztajn-Rochowiak, E., Jasinska-Mikolajczyk, A., Chlopocka-Wozniak, M., Rybakowski, J.K., 2019. Augmentation of pharmacotherapy by sleep deprivation with sleep phase advance in treatment-resistant depression. Pharmacopsychiatry 52, 186–192.

Lewy, A.J., Nurnberger, J.I.J.R., Wehr, T.A., et al., 1985. Supersensitivity to light: possible trait marker for manic-depressive illness. Am. J. Psychiatry 142, 725–727.

Martiny, K., Lunde, M., Unden, M., Dam, H., Bech, P., 2005. Adjunctive bright light in non-seasonal major depression: results from clinician-rated depression scales. Acta Psychiatr. Scand. 112, 117–125.

Martiny, K., Refsgaard, E., Lund, V., et al., 2012. A 9-week randomized trial comparing a chronotherapeutic intervention (wake and light therapy) to exercise in major depressive disorder patients treated with duloxetine. J. Clin. Psychiatry 73, 1234–1242.

McIntyre, R.S., Filteau, M.J., Martin, L., et al., 2014. Treatment-resistant depression: definitions, review of the evidence, and algorithmic approach. J. Affect. Disord. 156, 1–7.

Neumeister, A., Goessler, R., Lucht, M., Kapitany, T., Bamas, C., Kasper, S., 1996. Bright light therapy stabilizes the antidepressant effect of partial sleep deprivation. Biol. Psychiatry 39, 16–21.

Osby, U., Brandt, L., Correia, N., Ekbom, A., Sparen, P., 2001. Excess mortality in bipolar and unipolar disorder in Sweden. Arch. Gen. Psychiatry 58, 844–850.

Pflug, B., Tolle, R., 1971a. Therapy of endogenous depression using sleep deprivation. Practical and theoretical consequences. Nervenarzt 42, 117–124.

Pflug, B., Tolle, R., 1971b. Disturbance of the 24-hour rhythm in endogenous depression and the treatment of endogenous depression by sleep deprivation. Int. Pharmacopsychiatry 6, 187–196.

Prasko, J., Brunovsky, M., Latalova, K., et al., 2010. Augmentation of antidepressants with bright light therapy in patients with comorbid depression and borderline personality disorder. Biomed. Pap. Med. Fac. Univ. Palacky Olomouc Czech Repub. 154, 355–361.

Sachs, G.S., 1996. Treatment-resistant bipolar depression. Psychiatr. Clin. North Am. 19, 215–236.

Schulte, W., 1959. Sequelae of sleep deprivation. Med. Klin. (Munich) 54, 969–973.

Sikkens, D., Riemersma-Van Der Lek, R.F., Meesters, Y., RA, S., Haarman, B.C.M., 2019. Combined sleep deprivation and light therapy: clinical treatment outcomes in patients with complex unipolar and bipolar depression. J. Affect. Disord. 246, 727–730.

Terman, J.S., Terman, M., Schlager, D., et al., 1990. Efficacy of brief, intense light exposure for treatment of winter depression. Psychopharmacol. Bull. 26, 3–11.

Terman, J.S., Terman, M., Lo, E.S., Cooper, T.B., 2001. Circadian time of morning light administration and therapeutic response in winter depression. Arch. Gen. Psychiatry 58, 69–75.

Voderholzer, U., Valerius, G., Schaerer, L., et al., 2003. Is the antidepressive effect of sleep deprivation stabilized by a three day phase advance of the sleep period? A pilot study. Eur. Arch. Psychiatry Clin. Neurosci. 253, 68–72.

Walsh, B.T., Seidman, S.N., Sysko, R., Gould, M., 2002. Placebo response in studies of major depression: variable, substantial, and growing. JAMA 287, 1840–1847.

Wirz-Justice, A., Van Den Hoofdakker, R.H., 1999. Sleep deprivation in depression: what do we know, where do we go? Biol. Psychiatry 46, 445–453.

Wittchen, H.U., Jacobi, F., Rehm, J., et al., 2011. The size and burden of mental disorders and other disorders of the brain in Europe 2010. Eur. Neuropsychopharmacol. 21, 655–679.

Wu, J.C., Kelsoe, J.R., Schachat, C., et al., 2009. Rapid and sustained antidepressant response with sleep deprivation and chronotherapy in bipolar disorder. Biol. Psychiatry 66, 298–301.

Wulsin, L.R., GE, V., Wells, V.E., 1999. A systematic review of the mortality of depression. Psychosom. Med. 61, 6–17.

Chapter 36

Managing treatment resistant depression in primary care settings

Mohit Chauhan[a], Olivia E. Bogucki[b], William B. Leasure[b], Craig N. Sawchuk[b], Mark D. Williams[b], and David J. Katzelnick[b]

[a]*Division of Integrated Behavioral Health, Mayo Clinic, Jacksonville, FL, United States,* [b]*Division of Integrated Behavioral Health, Mayo Clinic, Rochester, MN, United States*

Scope of problem

Major Depressive Disorder (MDD) is a common, often recurrent or chronic illness associated with significant morbidity. According to the World Health Organization, an estimated 4.4% of the global population suffers from depression. MDD is among the five leading causes of years lived with disability (YLDs) in 2016 (Disease et al., 2017). Past-year depression prevalence increased significantly in the United States (U.S.) from 2005 to 2015 (Weinberger et al., 2018). Few studies have estimated the prevalence of treatment-resistant depression (TRD) in primary care. Information is further limited by lack of uniform criteria for defining TRD as detailed in Chapter 1. Using failure to respond to two antidepressants from different classes as definition for TRD, the InSight study conducted in Canada, found the prevalence of TRD in primary care to be about 22.0% (Rizvi et al., 2014).

Treatment for major depression is often initiated by primary care providers (PCPs) in the form of antidepressants and/or referral to psychotherapy (Maradiegue and Khan, 2013; Margaretten et al., 2013; The, 2016; Kato et al., 2018). Only about 20.0%–30.0% of patients in clinical practice respond to initial antidepressant treatment (Corey-Lisle et al., 2004). The large STAR*D trial, which recruited patients from primary care and psychiatric clinics, showed that even with close follow-up and structured treatment, close to 30.0% of patients have TRD (Warden et al., 2007).

Various clinical and psychosocial factors contributing to treatment resistance are relevant to primary care practices. Longer duration and severity of current depressive episode, previous failed trials of antidepressants and comorbid anxiety disorders have been associated with increased risk of treatment failure (The, 2016).

Depression comorbid with physical health problems contributes to poor response to depression treatment and poor outcome of physical health problems. The physiological impact of medical illnesses, the presence of other medications, and the burden of physical health problems contributes to lack of response to depression treatment. Also, this relationship is bidirectional with MDD being associated with poor compliance with treatment recommendations for the management of physical health problems, poor self-care, greater functional impairment, and poor quality of life.

Bidirectional relationship between diabetes and depression is well established. A meta-analysis of long-term longitudinal studies found significantly higher risk of incident diabetes in depressed patients (Rotella and Mannucci, 2013). Diabetes patients with depression have poor self-management leading to increased risk of complications and mortality (de Groot et al., 2001; Nouwen et al., 2019). Another meta-analysis found that close top 40% of individuals with TRD met criteria for metabolic syndrome, with men being at significantly higher risk. Of the individuals with TRD, less than a third of patients with hypertension and dyslipidemia were treated for these conditions (Godin et al., 2019). Similar to diabetes, prevalence of depression is higher in patients with cardiovascular disease (Rutledge et al., 2006) and is a risk factor for incident cardiovascular illness (Cao et al., 2020). Comorbid depression and cardiovascular disease is associated with poor outcomes for both (Fan et al., 2014; Gathright et al., 2017). Greater number of comorbid physical health problems has been associated with poor response to depression treatment (Warden et al., 2007).

TRD has been consistently associated with high economic burden in the form increased utilization of health services; increased cost of care, poor health-related quality of life, and lost productivity. Compared with non-TRD, patients with TRD had more emergency department visits, more and longer inpatient hospitalizations and higher per person per year health care costs. These patterns have been observed in both government and commercially insured patients (Li et al., 2020). Both general medical and depression care-related costs increase with an increase in the severity of TRD, accounting for close

Managing Treatment-Resistant Depression. https://doi.org/10.1016/B978-0-12-824067-0.00036-0

to 30% increase in medical expenditures (Olchanski et al., 2013; Sussman et al., 2019; Pilon et al., 2020; Zhdanava et al., 2020; Evans-Lacko and Knapp, 2016).

Challenges of managing TRD in primary care settings

A major challenge to adequately addressing TRD is the limited number of mental health providers trained to adequately treat the condition. There are too few psychiatrists, advanced practice providers, and therapists to address TRD alone while continuing to treat the myriad other mental health conditions, and primary care providers (PCP) struggle to connect patients with existing mental health providers (Cunningham, 2009; Thomas et al., 2009). Thus, supporting PCPs in delivering depression care is the key to expanding their ability to treat TRD. PCPs currently identify and treat the majority of patients with depression, which, combined with their generally long-term relationships with patients, creates natural opportunities for increasing their ability to treat TRD while improving their comfort level in doing so. It also provides opportunities to address comorbid medical conditions that are less likely to be addressed in mental health settings. A number of studies have demonstrated challenges PCPs identify in providing mental health care to patients, including limited time, inadequate knowledge, and lack of resources (Cooper et al., 2006; Loeb et al., 2012). Thus, it is important to incorporate strategies that help with efficiencies rather than simply adding more demands or requirements.

Collaborative care for management of depression in primary care (Fig. 36.1)

Integrating mental health services with general medical services to manage patients with chronic illnesses have been conclusively shown to improve outcomes for chronic physical and mental health problems. Collaborative care is a specific form of integration that provides team-based care including a behavioral health care manager, PCPs, and a psychiatric consultant. Other members, such as a pharmacist and social worker, can also be part of the team to address factors that may be contributing to the maintenance of TRD. Collaborative care teams conduct systematic case and population review using a registry facilitating timely, evidence- and measurement-based recommendations for patients' PCP (Gilbody et al., 2006; Unützer et al., 2002; Katon, 2009).

The U.S. Preventive Services Task Force (UPSTF) recommends screening for depression in all adults. UPSTF further recommends that positive screens should lead to further assessments, which is often difficult to accomplish in busy primary care practices (Zivin et al., 2009). A large number of patients referred to mental health services, especially the most

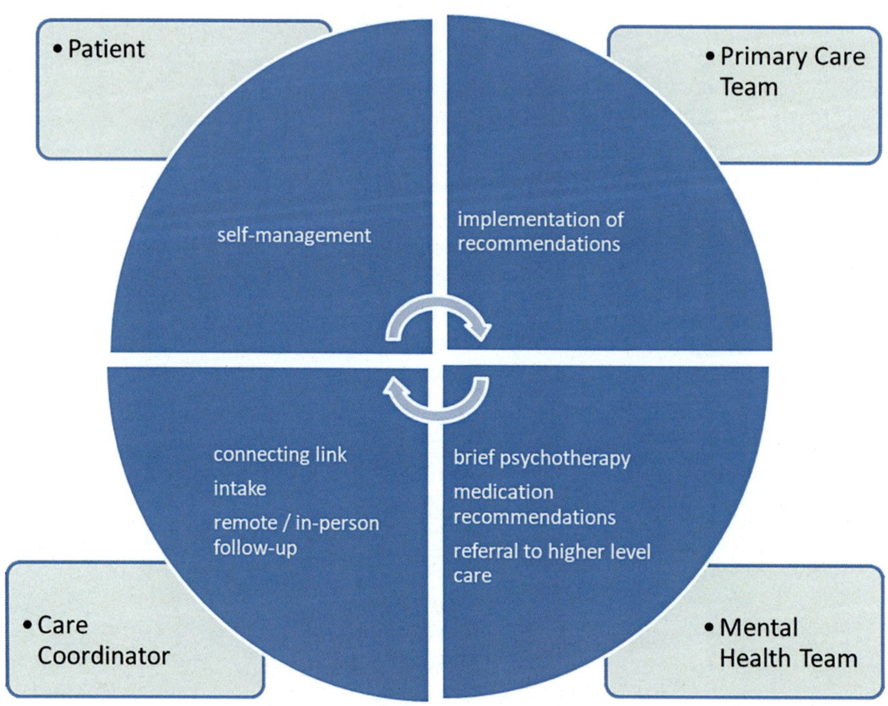

FIG. 36.1 Collaborative care model.

vulnerable, do not show up for initial appointment (Zivin et al., 2009; Ruggeri et al., 2020). Having a behavioral health care manager available for warm handoffs improves the chances of further evaluation being completed (Siu et al., 2016). In the collaborative care model, an initial assessment by the care manager includes the use of rating scales for measuring severity of depression; record of response to past treatment; and screening for comorbid conditions including bipolar disorder and substance use, which are all factors that can complicate treatment of depression. The collaborative care team then provides individualized treatment recommendations to the patient's PCP. Frequency of monitoring is based on severity of illness, adverse effects, and treatment adherence. The care manager also provides education and support to improve adherence, may use motivational interviewing, and incorporates psychotherapeutic techniques. Routine follow-up allows for early identification of failure or intolerance to pharmacological interventions and treatment is subsequently modified. Structured, manualized cognitive behavior therapy (CBT) can also be integrated in the treatment plan (Unützer et al., 2002; Katon and Unutzer, 2006).

More than 80 randomized controlled trials have found significantly better response and remission rates for depression with collaborative care compared to usual care. A systematic review of 79 RCTs which included 24,308 participants, showed significantly greater improvement in depression outcomes for adults with depression treated with the collaborative care model in the short-term and medium-term (Archer et al., 2012). A meta-analysis of 37 randomized studies that included 12,355 patients with depression in primary care settings showed that depression outcome benefits were maintained for up to 5 years (Gilbody et al., 2006).

In addition, collaborative care programs have been adapted to address both depression and chronic medical conditions. The TEAMCare and subsequent Care Of Mental, Physical, And Substance-use Syndromes (COMPASS) studies delivered collaborative care to patients with depression and comorbid diabetes and/or coronary heart disease. The initial trial found that patients engaged in collaborative care reported greater improvements in depression symptoms; blood sugar, cholesterol, and systolic blood pressure readings; quality of life ratings; and care satisfaction compared to those receiving usual care (Katon et al., 2010). Large-scale implementation reported similar, albeit somewhat less robust, results, indicating that such programs can be successfully translated into real-world practice (Rossom et al., 2017).

Components of collaborative care for depression

Pharmacotherapy

While pharmacotherapeutic strategies for TRD are covered in Chapters 7 through 16, there is evidence that the collaborative care model is associated with better implementation of available pharmacotherapy strategies with stepped intensification of care, resulting in a significantly greater number of antidepressant medication trials in patients enrolled in the intervention when compared to those under usual care (Dejesus et al., 2013). Several large-scale collaborative care programs for depression have been implemented in primary care. Compared to those receiving usual care, patients with MDD who engaged in the collaborative care program reported greater medication adherence, quality of care, and improvement depressive symptoms. Importantly, greater improvement in depressive symptoms was observed among patients who required a medication adjustment, suggesting that collaborative care may be most impactful for patients who do not initially respond to treatment (Katon et al., 1995). Large single-payer health care systems, such as the Veterans Administration (VA), have implemented collaborative care. The Translating Initiatives in Depression into Effective Solutions (TIDES) found high rates of medication adherence and treatment engagement among veterans enrolled in collaborative care. A significant decrease in depressive symptoms was reported over the course of treatment, which was similar in magnitude to veterans referred to mental health specialty (Rubenstein et al., 2010). The Telemedicine Enhanced Antidepressant Management (TEAM) study provided evidence for a telehealth-based collaborative care program delivered in rural VA clinics without on-site psychiatrists. Veterans participating in the program reported greater medication adherence, reductions in depression symptoms, and satisfaction of care compared to those receiving usual care (Fortney et al., 2007). This study provided preliminary evidence for telehealth-based collaborative care programs as a way to reach rural and underserved populations, which has since been expanded to other health care settings (Fortney et al., 2013, 2015; Hudson et al., 2015).

Psychotherapy

The presence of psychological services in primary care has steadily grown over the past two decades (Alexander et al., 2010). Behavioral activation (BA) and problem-solving therapy (PST) are the most common psychotherapy interventions employed across primary-care–based collaborative care programs for depression. BA is based on a behavioral model of depression. This model proposes that individuals who develop depression experience low-levels of positive reinforcement due to limited access to, availability of, or ability to engage in pleasant activities, which results in depressed mood and avoidance. The very nature of depressive symptoms, such as loss of interest, lack of motivation, poor energy, and social withdrawal, serve to maintain this negative cycle. Ultimately, this perpetuates the problem and leads to continued low levels of positive reinforcement (Dimidjian et al., 2011). BA is a brief, structured psychotherapy that aims to increase engagement in pleasant, masterful, and/or valued activities and decrease avoidance. This is accomplished through a number of different strategies including activity monitoring, activity scheduling, and problem solving (Kanter et al., 2010). PST is based on a cognitive-behavioral model of depression. This model postulates that individuals with depression exhibit deficits in problem-solving skills (Nezu, 1987) including a negative problem-solving orientation and maladaptive problem-solving style (D'Zurilla and Goldfried, 1971). PST is a brief psychotherapy approach that aims to teach effective problem-solving attitudes and skills. This is achieved through a step-by-step strategy in which the patient identifies the problem, generates potential solutions, selects and implements a solution, and evaluates the outcome with the option to trial another solution if necessary (Nezu et al., 2012).

The rationale behind selecting these two interventions include their strong evidence-base, short-term structure, and straight-forward theoretical framework and therapeutic techniques. The structure of these therapies aligns well with the population-based model of care practiced in primary care settings. While there are a number of different protocols available, both interventions typically consist of only 6–12 sessions (Dimidjian et al., 2011; Nezu, 2004). The fact that these therapies are relatively easy to understand conceptually and implement practically enables care managers, often without specialized training in mental health, to deliver these interventions with high treatment fidelity.

While, BA and PST are evidence-based treatments for depression, these have not been extensively studied in patients with TRD. Rather, TRD trials have investigated the use of CBT, which often includes the same strategies used in BA and PST (Li et al., 2018; van Bronswijk et al., 2019).

The efficacy of CBT for TRD has begun to be evaluated in primary care. Only one large-scale psychotherapy trial has been conducted (Bogucki et al., 2020). The Cognitive-Behavioral therapy as an Adjunct to pharmacotherapy for primary care patients with Treatment-resistant depression (CoBAulT) randomized clinical trial was conducted in primary care practices across the United Kingdom. Participants included primary care patients with TRD as operationalized as antidepressant medication adherent, but not responsive, for at least 6 weeks. Participants who engaged in individual CBT were significantly more likely to achieve response and remission at 6 (Ruggeri et al., 2020) and 46 months (Siu et al., 2016) compared to those receiving treatment as usual. Older participants were found to benefit more from CBT while participants with more severe depressive symptoms at baseline benefited less; however, these differences were relatively small and resulted in the authors ultimately recommending CBT for all patients with TRD (Button et al., 2015). While these results are promising, this trial has not yet been replicated in other primary care settings. Various psychotherapeutic interventions for TRD are further detailed in Chapter 25.

Collaborative model of care expands access to evidence-based psychotherapy to a larger proportion of the primary care population. Studies have employed a variety of providers as care managers including psychologists (Unützer et al., 2002; Solberg et al., 2013), nurses (Unützer et al., 2002; Rossom et al., 2017; Rubenstein et al., 2010; Fortney et al., 2007, 2013; Hudson et al., 2015; Solberg et al., 2013; Angstman et al., 2009), other medical professionals (Rossom et al., 2017; Solberg et al., 2013), and paraprofessionals (Richards et al., 2016). While outcomes differ slightly across studies, collaborative care can be competently delivered by care managers from diverse backgrounds. This model of care can also expand access to mental health specialists such as clinical psychologists, master's level clinicians, and clinical social workers for patients experiencing more severe or long-lasting depressive symptoms.

Advancing treatment in primary care settings

Improving capacity of primary care providers

Decision tools to assist PCPs, such as the Texas Medication Algorithm Project and the psychopharmacology algorithm project at the Harvard South Shore Program, have the potential to improve depression outcomes by facilitating systematic decision-making around evidenced-based treatments for TRD (Crismon et al., 1999; Giakoumatos and Osser, 2019). However, in studies of improving depression treatment in primary care by using algorithms, decision support, guideline implementation strategies, or education alone for PCPs the improvements have been limited (Kurian et al., 2009; Thompson et al., 2000; Kendrick et al., 2001; Worrall et al., 1999; Brown et al., 2000).

Project ECHO is a telementoring model of education in which participating clinicians remotely attend lecture and case-based learning sessions hosted by specialists at regular intervals to improve their capacity to manage complex medical conditions. It is intended to improve access to specialty care to communities that would otherwise have difficulty accessing specialty care, such as rural locations. As opposed to traditional telepsychiatry models in which one clinician interacts with one patient at a time, Project ECHO allows for dissemination of specialty expertise to many primary care clinicians simultaneously and incorporates time for practical discussion of cases submitted by participating clinicians. Project ECHO has not specifically been studied for TRD but has demonstrated positive outcomes for a number of behavioral health conditions (Brant Hager et al., 2018). Further research is needed to test this model of education and support for TRD.

Another approach to expand the capacity of PCPs to treat TRD is to make informal curbside consultation to psychiatric providers easily accessible (Raney et al., 2020). That approach is often a component of collaborative care models, but can also be leveraged in clinics or systems which have not implemented collaborative care. The Massachusetts Child Psychiatry Access Project (MCPAP) is an example of such a program that offers mental health consultation services to enrolled PCPs across Massachusetts, usually by telephone (Straus and Sarvet, 2014). It has been very well received by PCPs (Sheldrick et al., 2012). The model can be replicated on a smaller scale within medical systems or even by contracting with a consulting psychiatrist to provide education and assistance with coordination of care when needed. While not yet studied for TRD, it provides the opportunity to assist PCPs in taking additional steps to treat TRD than they might otherwise take with the guidance of a psychiatrist.

Integrating digital solutions

A variety of digital tools have become available in recent years to treat mental health conditions. Mental health applications have proliferated rapidly, and along with their availability has come claims of their efficacy for a variety of conditions (Torous and Firth, 2018). Although some have been tested in randomized controlled trials and proven efficacious (Seyffert et al., 2016; Firth et al., 2017), many do not incorporate evidence-based interventions, and even fewer have been studied to evaluate their outcomes (Psihogios et al., 2020; Powell et al., 2014). There is great promise that, over time, they will be studied in clinical trials and be evaluated specifically for TRD, and there have been some initial efforts to do so. A 2017 Japanese study compared the effectiveness of a smartphone CBT app as an adjunctive treatment for patients who had not responded to initial treatment with at least one antidepressant (Mantani et al., 2017). While the study population did not meet commonly used definitions for TRD, the intervention group had low dropout rates and those in the treatment arm scored lower on standard measures of depression severity and side effects than the control arm. The accessibility and affordability of smartphone apps make them ideal vehicles for distribution of psychotherapy for patients with TRD treated in primary care, but the ability for them to be produced and marketed without proper evaluation raises concerns about their quality. Further trials are needed to identify those most effective for TRD.

Telemental health, the remote delivery of mental health care through telecommunication technology, has an evidence base dating back several decades and is widely accepted by patients and providers. Studies comparing synchronous telemental health with in person care in primary and specialty care settings have shown similar treatment outcomes and levels of patient satisfaction, adherence, and health care cost (Bashshur et al., 2016; O'Reilly et al., 2007; Ruskin et al., 2004; Guaiana et al., 2021). The ubiquity of telemental health renders it able to be delivered to patients in a variety of settings (e.g., home, work, primary care clinic) by clinicians in a range of locations and adds flexibility to how mental health treatment can be incorporated into primary care settings. A growing body of literature is emerging on the value of being deliberate in developing and managing virtual teams to improve their efficacy in delivering telemental health (Giakoumatos and Osser, 2019). There is still a need to study telemental health specifically for TRD, but the studies that have shown it to be effective across a range of conditions including depression suggest it can be successful in treatment TRD as well.

Other digital solutions, such as asynchronous telemental health and e-consults, have been identified as ways to expand access to psychiatric care (Raney et al., 2020). Similar to other digital modalities they have not been rigorously studied specifically for TRD in primary care. Early literature on asynchronous telemental health is promising, suggesting potential for it to be implemented by the clinical team, maintain patient satisfaction, and improve access to care (O'Keefe et al., 2019). Likewise, e-consults have been shown to improve access to care, especially for non-Native English speakers and vulnerable populations, but improvements in other treatment outcomes have not yet been demonstrated (Lowenstein et al., 2017; Calderone et al., 2020).

Clinical summary (Table 36.1)

As is the case with most chronic illnesses, TRD requires the careful review of diagnosis, comorbidities, clinical, and psychosocial factors contributing to treatment failure. Long-term treatment with multiple medication trials, combination of antidepressants and augmenting agents along with psychotherapy is often required to manage TRD. If these measures fail, education and referral for use of novel agents such as ketamine or somatic therapies such as transcranial magnetic

TABLE 36.1 Advantages of collaborative care for management of treatment resistant depression in primary care.

Components of care	Collaborative care advantages
Screening	Quick initiation of care with integrated care coordinator and warm hand-off
Identification of treatment failure	Use of registry for routinely collecting patient reported outcome measures can facilitate early identification of treatment failure Further improvement in recognition of TRD through the use of algorithmic decision tools
Pharmacotherapy	Treatment recommendations for switching antidepressants, use of combination of antidepressants or augmenting agents recommended by psychiatrist during systematic case and population review Faster and more appropriate referral for somatic and novel therapies
Psychotherapy	Brief psychotherapy integrated in primary care practice making is more accessible and acceptable Care coordinators more familiar with area resources to make appropriate referrals for long-term psychotherapy
Follow-up	Proactive instead of reactive follow-up, reducing the chances of poorly compliant patients "falling through the cracks"
Managing medical and psychiatric comorbidities	Improved collaboration between primary care and mental health team leading to improved care for patients with multimorbidity

stimulation (TMS), electroconvulsive therapy (ECT), and deep brain stimulation (DBS) is required. This need for individualized treatment, as well as frequent medical and psychiatric comorbidity in patients with TRD, requires collaboration between mental health providers and PCPs. The collaborative care model allows PCPs to continue to manage depression even when it does not respond to initial treatment by providing them with additional treatment strategies that they might otherwise be less comfortable with or knowledgeable of, and offering back-up support and monitoring to assist patients proceeding with those treatments. Collaborative care is designed to focus intervention efforts toward those who are struggling or not responding to treatment and to help triage and refer such patients to a higher-step level of care when indicated, including to more specialized mental health services. Successful implementation of integrated care in primary care practices requires rigorous organizational change and involved collaborative care team, or intensive quality improvement programs (Archer et al., 2012; Rost et al., 2001; Sherbourne et al., 2001).

While collaborative care programs typically include those who have not initially responded to treatment, there has yet to be a study that has exclusively recruited patients with TRD. There has been a call to evaluate collaborative care for TRD (Unützer et al., 2002). Additional research is also needed to evaluate the efficacy and effectiveness of BA and PST for TRD. Given their potential to improve access to evidence based treatment strategies for TRD in primary care and to do so at a scale that might otherwise be difficult to achieve there is urgent need to better study the range of digital tools that are now available but are as of yet largely unproven for this indication.

References

Alexander, C.L., Arnkoff, D.B., Glass, C.R., 2010. Bringing psychotherapy to primary care: innovations and challenges. Clin. Psychol. Sci. Pract. 17 (3), 191–214.

Angstman, K.B., DeJesus, R.S., Williams, M.D., 2009. Initial implementation of a depression care manager model: an observational study of outpatient utilization in primary care clinics. Popul. Health Manag. 12 (5), 227–230.

Archer, J., et al., 2012. Collaborative care for depression and anxiety problems. Cochrane Database Syst. Rev. 10, CD006525.

Bashshur, R.L., et al., 2016. The empirical evidence for telemedicine interventions in mental disorders. Telemed. J. E Health 22 (2), 87–113.

Bogucki, O.E., et al., 2020. The role of the patient-centered medical home in treating depression. Curr. Psychiatry Rep. 22 (9), 47.

Brant Hager, M.D., et al., 2018. Leveraging behavioral health expertise: practices and potential of the project ECHO approach to virtually integrating care in underserved areas. Psychiatr. Serv. 69 (4), 366–369.

Brown, J.B., et al., 2000. Controlled trials of CQI and academic detailing to implement a clinical practice guideline for depression. Jt Comm. J. Qual. Improv. 26 (1), 39–54.

Button, K.S., et al., 2015. Moderators of response to cognitive behavioural therapy as an adjunct to pharmacotherapy for treatment-resistant depression in primary care. J. Affect. Disord. 174, 272–280.

Calderone, J., et al., 2020. Telepsychiatry and integrated primary care: setting expectations and creating an effective process for success. Mhealth 6, 29.

Cao, L., Sheng, C., Luo, G., Ou, J., 2020. Depression as a risk factor for developing heart failure: a meta-analysis of prospective cohort studies. J. Cardiovasc. Nurs. https://doi.org/10.1097/jcn.0000000000000760.

Cooper, S., et al., 2006. Running out of time: physician management of behavioral health concerns in rural pediatric primary care. Pediatrics 118 (1), e132–e138.

Corey-Lisle, P.K., et al., 2004. Response, partial response, and nonresponse in primary care treatment of depression. Arch. Intern. Med. 164 (11), 1197–1204.

Crismon, M.L., et al., 1999. The Texas medication algorithm project: report of the Texas consensus conference panel on medication treatment of major depressive disorder. J. Clin. Psychiatry 60 (3), 142–156.

Cunningham, P.J., 2009. Beyond parity: primary care physicians' perspectives on access to mental health care. Health Aff. 28 (3), w490–w501.

de Groot, M., et al., 2001. Association of depression and diabetes complications: a meta-analysis. Psychosom. Med. 63 (4), 619–630.

Dejesus, R.S., et al., 2013. Antidepressant medication use among patients with depression: comparison between usual care and collaborative care using care managers. Clin. Pract. Epidemiol. Ment. Health 9, 84–87.

Dimidjian, S., et al., 2011. The origins and current status of behavioral activation treatments for depression. Annu. Rev. Clin. Psychol. 7, 1–38.

Disease, G.B.D., Injury, I., Prevalence, C., 2017. Global, regional, and national incidence, prevalence, and years lived with disability for 328 diseases and injuries for 195 countries, 1990-2016: a systematic analysis for the global burden of Disease study 2016. Lancet 390 (10100), 1211–1259.

D'Zurilla, T.J., Goldfried, M.R., 1971. Problem solving and behavior modification. J. Abnorm. Psychol. 78 (1), 107–126.

Evans-Lacko, S., Knapp, M., 2016. Global patterns of workplace productivity for people with depression: absenteeism and presenteeism costs across eight diverse countries. Soc. Psychiatry Psychiatr. Epidemiol. 51 (11), 1525–1537.

Fan, H., et al., 2014. Depression after heart failure and risk of cardiovascular and all-cause mortality: a meta-analysis. Prev. Med. 63, 36–42.

Firth, J., et al., 2017. The efficacy of smartphone-based mental health interventions for depressive symptoms: a meta-analysis of randomized controlled trials. World Psychiatry 16 (3), 287–298.

Fortney, J.C., et al., 2007. A randomized trial of telemedicine-based collaborative care for depression. J. Gen. Intern. Med. 22 (8), 1086–1093.

Fortney, J.C., et al., 2013. Population level effectiveness of implementing collaborative care management for depression. Gen. Hosp. Psychiatry 35 (5), 455–460.

Fortney, J.C., et al., 2015. Telepsychiatry integration of mental health services into rural primary care settings. Int. Rev. Psychiatry 27 (6), 525–539.

Gathright, E.C., et al., 2017. Depression increases the risk of mortality in patients with heart failure: a meta-analysis. J. Psychosom. Res. 94, 82–89.

Giakoumatos, C.I., Osser, D., 2019. The psychopharmacology algorithm project at the harvard south shore program: an update on unipolar nonpsychotic depression. Harv. Rev. Psychiatry 27 (1), 33–52.

Gilbody, S., et al., 2006. Collaborative care for depression: a cumulative meta-analysis and review of longer-term outcomes. Arch. Intern. Med. 166 (21), 2314–2321.

Godin, O., et al., 2019. Prevalence of metabolic syndrome and associated factors in a cohort of individuals with treatment-resistant depression: results from the FACE-DR study. J. Clin. Psychiatry 80 (6).

Guaiana, G., Mastrangelo, J., Hendrikx, S., et al., 2021. A systematic review of the use of telepsychiatry in depression. Community Ment. Health J. 57, 93–100.

Hudson, T.J., et al., 2015. Reduction of patient-reported antidepressant side effects, by type of collaborative care. Psychiatr. Serv. 66 (3), 272–278.

Kanter, J.W., et al., 2010. What is behavioral activation? A review of the empirical literature. Clin. Psychol. Rev. 30 (6), 608–620.

Kato, E., et al., 2018. Missed opportunities for depression screening and treatment in the United States. J. Am. Board Fam. Med. 31 (3), 389–397.

Katon, W.J., 2009. Collaborative care: evidence-based models that improve primary care depressive outcomes. CNS Spectr. 14 (12 Suppl. 14), 10–13.

Katon, W., Unutzer, J., 2006. Collaborative care models for depression: time to move from evidence to practice. Arch. Intern. Med. 166 (21), 2304–2306.

Katon, W., et al., 1995. Collaborative management to achieve treatment guidelines. Impact on depression in primary care. JAMA 273 (13), 1026–1031.

Katon, W.J., et al., 2010. Collaborative care for patients with depression and chronic illnesses. N. Engl. J. Med. 363 (27), 2611–2620.

Kendrick, T., et al., 2001. Hampshire depression project: changes in the process of care and cost consequences. Br. J. Gen. Pract. 51 (472), 911–913.

Kurian, B.T., et al., 2009. A computerized decision support system for depression in primary care. Prim. Care Companion J. Clin. Psychiatry 11 (4), 140–146.

Li, J.M., et al., 2018. Cognitive behavioral therapy for treatment-resistant depression: a systematic review and meta-analysis. Psychiatry Res. 268, 243–250.

Li, G., et al., 2020. A retrospective analysis to estimate the healthcare resource utilization and cost associated with treatment-resistant depression in commercially insured US patients. PLoS One 15 (9), e0238843.

Loeb, D.F., et al., 2012. Primary care physician perceptions on caring for complex patients with medical and mental illness. J. Gen. Intern. Med. 27 (8), 945–952.

Lowenstein, M., et al., 2017. Psychiatric consultation at your fingertips: descriptive analysis of electronic consultation from primary care to psychiatry. J. Med. Internet Res. 19 (8), e279.

Mantani, A., et al., 2017. Smartphone cognitive behavioral therapy as an adjunct to pharmacotherapy for refractory depression: randomized controlled trial. J. Med. Internet Res. 19 (11), e373.

Maradiegue, A.H., Khan, F., 2013. Missed opportunities in primary care: the importance of identifying depression through screening, family history, and chronic disease management. J. Psychosoc. Nurs. Ment. Health Serv. 51 (2), 27–36.

Margaretten, M.E., et al., 2013. Missed opportunities for depression screening in patients with arthritis in the United States. J. Gen. Intern. Med. 28 (12), 1637–1642.

Nezu, A.M., 1987. A problem-solving formulation of depression: a literature review and proposal of a pluralistic model. Clin. Psychol. Rev. 7 (2), 121–144.

Nezu, A.M., 2004. Problem solving and behavior therapy revisited. Behav. Ther. 35 (1), 1–33.

Nezu, A.M., Nezu, C.M., D'Zurilla, T., 2012. Problem-Solving Therapy: A Treatment Manual. Springer Publishing Company.

Nouwen, A., et al., 2019. Longitudinal associations between depression and diabetes complications: a systematic review and meta-analysis. Diabet. Med. 36 (12), 1562–1572.

O'Keefe, M., White, K., Jennings, J.C., 2019. Asynchronous telepsychiatry: a systematic review. J. Telemed. Telecare, 1357633x19867189.

Olchanski, N., et al., 2013. The economic burden of treatment-resistant depression. Clin. Ther. 35 (4), 512–522.

O'Reilly, R., et al., 2007. Is telepsychiatry equivalent to face-to-face psychiatry? Results from a randomized controlled equivalence trial. Psychiatr. Serv. 58 (6), 836–843.

Pilon, D., et al., 2020. US integrated delivery networks perspective on economic burden of patients with treatment-resistant depression: a retrospective matched-cohort study. Pharmacoecon. Open 4 (1), 119–131.

Powell, A.C., Landman, A.B., Bates, D.W., 2014. In search of a few good apps. JAMA 311 (18), 1851–1852.

Psihogios, A.M., Stiles-Shields, C., Neary, M., 2020. The needle in the haystack: identifying credible mobile health apps for pediatric populations during a pandemic and beyond. J. Pediatr. Psychol. 45 (10), 1106–1113.

Raney, L., et al., 2020. Consultative approaches to leveraging the psychiatric workforce for larger populations in need of psychiatric expertise. Psychiatr. Serv. 71 (10), 1084–1087.

Richards, D.A., et al., 2016. Clinical effectiveness and cost-effectiveness of collaborative care for depression in UK primary care (CADET): a cluster randomised controlled trial. Health Technol. Assess. 20 (14), 1–192.

Rizvi, S.J., et al., 2014. Treatment-resistant depression in primary care across Canada. Can. J. Psychiatr. 59 (7), 349–357.

Rossom, R.C., et al., 2017. Impact of a national collaborative care initiative for patients with depression and diabetes or cardiovascular Disease. Focus 15 (3), 324–332.

Rost, K., et al., 2001. Improving depression outcomes in community primary care practice: a randomized trial of the quEST intervention. Quality enhancement by strategic teaming. J. Gen. Intern. Med. 16 (3), 143–149.

Rotella, F., Mannucci, E., 2013. Depression as a risk factor for diabetes: a meta-analysis of longitudinal studies. J. Clin. Psychiatry 74 (1), 31–37.

Rubenstein, L.V., et al., 2010. Using evidence-based quality improvement methods for translating depression collaborative care research into practice. Fam. Syst. Health 28 (2), 91–113.

Ruggeri, K., et al., 2020. Nudging New York: adaptive models and the limits of behavioral interventions to reduce no-shows and health inequalities. BMC Health Serv. Res. 20 (1), 363.

Ruskin, P.E., et al., 2004. Treatment outcomes in depression: comparison of remote treatment through telepsychiatry to in-person treatment. Am. J. Psychiatry 161 (8), 1471–1476.

Rutledge, T., et al., 2006. Depression in heart failure a meta-analytic review of prevalence, intervention effects, and associations with clinical outcomes. J. Am. Coll. Cardiol. 48 (8), 1527–1537.

Seyffert, M., et al., 2016. Internet-delivered cognitive behavioral therapy to treat insomnia: a systematic review and meta-analysis. PLoS One 11 (2), e0149139.

Sheldrick, R.C., Mattern, K., Perrin, E.C., 2012. Pediatricians' perceptions of an off-site collaboration with child psychiatry. Clin. Pediatr. (Phila) 51 (6), 546–550.

Sherbourne, C.D., et al., 2001. Long-term effectiveness of disseminating quality improvement for depression in primary care. Arch. Gen. Psychiatry 58 (7), 696–703.

Siu, A.L., et al., 2016. Screening for depression in adults: US preventive services task force recommendation statement. JAMA 315 (4), 380–387.

Solberg, L.I., et al., 2013. The DIAMOND initiative: implementing collaborative care for depression in 75 primary care clinics. Implement. Sci. 8, 135.

Straus, J.H., Sarvet, B., 2014. Behavioral health care for children: the Massachusetts child psychiatry access project. Health Aff. 33 (12), 2153–2161.

Sussman, M., et al., 2019. Economic burden of treatment-resistant depression on the U.S. health care system. J. Manag. Care Spec. Pharm. 25 (7), 823–835.

The, L., 2016. Screening for perinatal depression: a missed opportunity. Lancet 387 (10018), 505.

Thomas, K.C., et al., 2009. County-level estimates of mental health professional shortage in the United States. Psychiatr. Serv. 60 (10), 1323–1328.

Thompson, C., et al., 2000. Effects of a clinical-practice guideline and practice-based education on detection and outcome of depression in primary care: hampshire depression project randomised controlled trial. Lancet 355 (9199), 185–191.

Torous, J., Firth, J., 2018. Bridging the dichotomy of actual versus aspirational digital health. World Psychiatry 17 (1), 108–109.

Unützer, J., et al., 2002. Collaborative care management of late-life depression in the primary care setting: a randomized controlled trial. JAMA 288 (22), 2836–2845.

van Bronswijk, S., et al., 2019. Effectiveness of psychotherapy for treatment-resistant depression: a meta-analysis and meta-regression. Psychol. Med. 49 (3), 366–379.

Warden, D., et al., 2007. The STAR*D project results: a comprehensive review of findings. Curr. Psychiatry Rep. 9 (6), 449–459.

Weinberger, A.H., et al., 2018. Trends in depression prevalence in the USA from 2005 to 2015: widening disparities in vulnerable groups. Psychol. Med. 48 (8), 1308–1315.

Worrall, G., et al., 1999. Effectiveness of an educational strategy to improve family physicians' detection and management of depression: a randomized controlled trial. CMAJ 161 (1), 37–40.

Zhdanava, M., et al., 2020. Economic burden of treatment-resistant depression in privately insured U.S. patients with physical conditions. J. Manag. Care Spec. Pharm. 26 (8), 996–1007.

Zivin, K., et al., 2009. "No-shows": who fails to follow up with initial behavioral health treatment? Am. J. Manag. Care 15 (2), 105–112.

Chapter 37

Treatment-resistant depression in pregnancy, the postpartum period, and transition to menopause

Yardana Kaufman[a,*], Rebecca M. Eichenbaum[a,b,*], Sara V. Carlini[a], and Kristina M. Deligiannidis[a,c,d]
[a]The Zucker Hillside Hospital, Women's Behavioral Health Research, Northwell Health, Glen Oaks, NY, United States, [b]Montefiore Medical Center, Bronx, NY, United States, [c]Departments of Psychiatry and Obstetrics & Gynecology at the Zucker School of Medicine at Hofstra/Northwell, Hempstead, NY, United States, [d]Institute of Behavioral Science, The Feinstein Institutes for Medical Research, Manhasset, NY, United States

Introduction

The lifetime risk for depression is greater in women than men (Seedat et al., 2009) and is especially high during female reproductive events including the perimenstrual, peripartum, and perimenopausal phases. The increased depression prevalence in women during reproductive transitions is hypothesized to be due to psychological, social, and physiological changes including the unique temporal hormonal fluctuations associated with each transition. Treatments available for women suffering from depression during the reproductive life cycle are similar to those for the general population and include psychotherapy, pharmacotherapy, and neuromodulation. Limited data are available on treatment-resistant depression (TRD) during the peripartum and perimenopausal periods.

In this chapter, we will review the epidemiology, risk factors, symptomatology, and distinct treatment considerations for unipolar depression occurring during the peripartum and perimenopausal periods. We will also discuss the treatment of peripartum bipolar illness including data on the relative safety profiles of mood stabilizers and antipsychotics in pregnancy and lactation.

Unipolar peripartum depression

Diagnosis of unipolar peripartum depression

The Diagnostic and Statistical Manual Fifth Edition (DSM-5) defines peripartum depression as a major depressive episode with peripartum onset for patients whose symptoms arise during pregnancy or within 4 weeks of delivery (American Psychiatric Association (APA), 2013). Despite this narrow definition of the postpartum period, many clinicians and the World Health Organization (WHO) extend the definition of the postpartum period to up to 1 year after childbirth (Stewart et al., 2003). Peripartum depression is frequently distinguished by the presence of anxiety with excessive concern for the child's health and fear of harming the child. A 2017 study isolated peripartum depressive symptom patterns that differed in severity, time of onset, and type of symptoms, and categorized these distinct presentations into the following five subtypes: severe anxious depression, moderate anxious depression, anxious anhedonia, pure anhedonia, and resolved depression (Putnam et al., 2017).

Epidemiology of unipolar peripartum depression

Peripartum depression is one of the most common complications of childbirth and is a serious public health problem. In 2018, the Centers for Disease Control and Prevention (CDC) estimated the overall prevalence of postpartum depression in the United States (U.S.) to be 13%, with a range of 9.7%–23.5% by state (Bauman et al., 2020). According to a 2019 CDC analysis, the rate of depression diagnosis at delivery increased sevenfold from 2000 to 2015 (Haight et al., 2019). Among women with peripartum depression, 11.5% develop symptoms antenatally, 66.5% experience symptoms within 6 weeks of delivery, and 22% exhibit symptoms within 12 months of delivery (Stowe et al., 2005).

*Equal contribution.

Managing Treatment-Resistant Depression. https://doi.org/10.1016/B978-0-12-824067-0.00037-2

Despite the high prevalence of peripartum depression, it is frequently underdiagnosed and undertreated with resultant deleterious consequences to mothers and their offspring. Of women with antenatal depression, only 49.9% are identified in clinical settings, 13.6% receive treatment, 8.6% receive adequate treatment, and 4.8% achieve remission. Of women with postpartum depression, only 30.8% are identified in clinical settings, 15.8% receive treatment, 6.3% receive adequate treatment, and 3.2% achieve remission (Cox et al., 2016).

The U.S. Preventive Services Task Force and the American Psychiatric Association (APA) recommend screening for depression in pregnant and postpartum women (Byatt et al., 2018; Siu et al., 2016). The APA recommends screening with a validated tool twice during pregnancy, once in early pregnancy and again later in the pregnancy, as well as postpartum during pediatric visits during the first 6 months as recommended by the American Academy of Pediatrics (AAP) (Byatt et al., 2018; Earls et al., 2019). The Edinburgh Postnatal Depression Scale (EPDS) and the Patient Health Questionnaire 9 (PHQ-9) are two self-rating screening tools that are validated to identify peripartum depression (Flynn et al., 2011); however, overreliance on peripartum depression screening tools may lead to misdiagnosis if screening is not followed by thorough psychiatric diagnostic evaluation. In women with an EPDS score > 11/30, 31.4% had major depressive disorder (MDD), 13.1% had bipolar disorder, 60.8% had an anxiety disorder (Lydsdottir et al., 2014).

Untreated or undertreated antenatal depression can impair a mother's ability to function and care for herself resulting in inadequate weight gain and is associated with nicotine, alcohol, and substance use (Zuckerman et al., 1989). Antenatal depression is also associated with increased risk of miscarriage (Andersen et al., 2014), pregnancy-induced hypertension, preeclampsia, preterm or operative delivery, fetal growth restriction (Bansil et al., 2010), low birth weight (Grote et al., 2010), maternal–infant attachment difficulties (Perry et al., 2011), and impaired lactation (Dennis and McQueen, 2009). Additionally, offspring of depressed mothers are at increased risk of neurodevelopmental delay (Tuovinen et al., 2018) as well as emotional and behavioral problems, including depression (Pearson et al., 2013), hyperactivity, and conduct problems (Leis et al., 2014). Tragically, maternal suicide is the leading cause of direct maternal mortality in the first postpartum year (MBRRACE, 2018).

Risk factors associated with unipolar peripartum depression

Numerous risk factors are associated with peripartum depression. Biological risk factors include personal or family history of psychiatric illness and smoking, alcohol or substance use (Biaggi et al., 2016). Psychosocial risk factors include poor social support, stressful life events, adverse childhood experiences, domestic violence, and lifetime history of abuse (Biaggi et al., 2016). Demographic and economic factors include young or advanced maternal age, African American or Latina race, single status, unemployment, and low education level or socioeconomic status (Biaggi et al., 2016). Pregnancy-related factors include multiparity, unintended or unwanted pregnancy, present or past pregnancy complications, and pregnancy loss (Biaggi et al., 2016).

Overview of treatment of unipolar peripartum depression

The treatment of depression in pregnant and postpartum women involves use of similar evidence-based psychological, pharmacotherapeutic, and neuromodulation interventions as in nonperipartum women, with some special considerations. When developing an individualized treatment plan for a peripartum patient, the provider should consider factors including gestational age (i.e., early, middle, or late gestation), plans for lactation or lactation status, psychiatric diagnoses, the severity of current and past symptoms, a suicidal and homicidal risk assessment, past psychiatric history particularly related to past pregnancies, efficacy of prior and current psychiatric treatments to determine level of treatment-resistance, medical comorbidities, family psychiatric history, patient preferences for treatment, and ability to adhere to the recommended treatment plan.

Psychotherapeutic treatment of unipolar peripartum depression

An APA taskforce recommends either psychotherapy or antidepressants as first-line treatment for mild-to-moderate peripartum depression (Yonkers et al., 2009). Psychotherapy should also be considered as an adjunct to antidepressants in moderate-to-severe depression and in cases where women decline to take antidepressants. Both cognitive–behavioral therapy and interpersonal psychotherapy are evidence-based, effective forms of psychotherapy for peripartum depression (Claridge, 2014; Sockol, 2015). For a comprehensive review of the evidence base for psychological interventions in peripartum depression, we refer the reader to Stuart and Koleva (Stuart and Koleva, 2014).

Pharmacologic treatment of unipolar peripartum depression

Evaluating the safety and efficacy of antidepressant use for peripartum depression

Safety is an important consideration when choosing among psychopharmacotherapies for peripartum women. Past restrictions on randomized controlled trials (RCTs) of psychopharmacotherapies in pregnancy limited available data on safety and efficacy of even commonly used antidepressants. Most published pharmacotherapy studies are observational in design and therefore limited in accurate assessments of reproductive drug safety due to the potential for selection bias, recall bias, confounding, and exposure or outcome misclassification. While there are inconsistent data and quality of the literature, when evaluating the risk of antidepressant use during the peripartum period, it is crucial to ensure that the study adequately controlled for confounders, used a representative sample, and had results that have been replicated.

Antidepressants are indicated for moderate-to-severe peripartum depression, with selective serotonin reuptake inhibitors (SSRIs) typically considered first-line therapy. While there are no RCTs examining efficacy of antidepressants in antenatal depression, eight RCTs (Appleby et al., 1997; Bloch et al., 2012; Hantsoo et al., 2014; Milgrom et al., 2015; Misri et al., 2004; Sharp et al., 2010; Wisner et al., 2006; Yonkers et al., 2008) and four open-label studies (Misri et al., 2012; Roy et al., 1993; Stowe et al., 1995; Suri et al., 2001) have assessed use of SSRIs in treatment of postpartum depression. A meta-analysis of three studies comparing treatment with SSRIs with placebo for treatment of postpartum depression found that patients randomized to SSRIs were more likely to show response or remission of depressive symptoms at follow-up (Molyneaux et al., 2014). A systematic review reported that SSRIs, nortriptyline, and psychotherapy are effective for short-term treatment of postpartum depression, but there is not enough evidence to demonstrate clear superiority of either (De Crescenzo et al., 2014). Despite limited evidence, there is consensus supporting the use of antidepressants in both antenatal and postpartum depression.

There are no RCTs evaluating use of selective norepinephrine reuptake inhibitors (SNRIs), bupropion, mirtazapine, trazodone, nefazodone, tricyclic antidepressants (TCAs) (except nortriptyline), or monoamine oxidase inhibitors (MAOIs). Open-label studies support efficacy for venlafaxine (Cohen et al., 2001), desvenlafaxine (Misri et al., 2016), bupropion (Nonacs et al., 2005), and nefazodone (Suri et al., 2005) in treatment of postpartum depression.

General considerations for the antenatal use of antidepressants

When possible, any changes to a medication regimen should be made prior to conception to ensure symptom stability and to minimize exposure to the fetus. Abrupt discontinuation of antidepressants in pregnant women with a history of depression is associated with high risk of relapse without known benefit (Cohen et al., 2006a). Women with history of recurrent unipolar MDD or history of symptom recurrence associated with past reduction or discontinuation of pharmacotherapy should not discontinue their antidepressant during pregnancy. All psychotropic medications cross the placenta. When considering the use of antidepressants during pregnancy, it is preferable to prescribe those with a well-studied reproductive safety profile, often older antidepressants which have been studied over time, and to use a single antidepressant at the lowest effective dose rather than polypharmacy.

Treatment in pregnancy is complicated by physiologic and pharmacokinetic changes, including changes in gastrointestinal absorption, hepatic and renal blood flow, changes in glomerular filtration rate, and changes in both phase 1 (hepatic cytochrome P450) and phase 2 (uridine diphosphate glucuronosyltransferase) hepatic enzyme activities (Deligiannidis, 2010). Changes in maternal drug pharmacokinetics in conjunction with placental transfer and fetal drug metabolism affect fetal psychotropic drug exposure (Deligiannidis, 2010). Pregnancy-associated changes in absorption, distribution, metabolism, and elimination may lower psychotropic drug levels and possibly decrease treatment effects, particularly in late pregnancy. Furthermore, increased plasma volume and changes in protein binding may increase volume of distribution for lipophilic drugs (Yonkers et al., 1992). Dose increases are often clinically necessary for citalopram, escitalopram, clomipramine and imipramine after 20 weeks' gestation. In addition, antenatal dose requirements for fluoxetine, fluvoxamine, nortriptyline, paroxetine, and sertraline often increase in the third trimester (Deligiannidis et al., 2014). Patients should be educated on the likely need for dose increases as gestation progresses, and symptoms should be monitored closely for changes in severity that would indicate need such increases.

While there is inadequate evidence to recommend routine therapeutic drug monitoring (TDM) of antidepressants other than TCAs during the peripartum period, measurement of serum drug levels when using TCAs to monitor efficacy and prevent toxicity is clinically necessary (Deligiannidis et al., 2014). During pregnancy, we recommend monthly monitoring of TCA trough levels and frequent psychiatric symptom assessment. In the postpartum period, we recommend weekly monitoring of depressive symptoms and a review of systems to monitor for TCA side effects. The TCA dose often must slowly be decreased towards the preconception dose during the 2–6 weeks after delivery to reduce risk of TCA toxicity. We recommend checking a TCA blood level whenever adverse effects emerge in the early postpartum or at week six when nortriptyline level/dose ratios have been reported to peak.

When considering use of antidepressants during pregnancy, potential risk of under- or untreated depressive symptoms in both the mother and fetus must be weighed against the potential risk of antidepressant exposure for the following types of adverse events: (1) fetal organ malformation, (2) poor pregnancy and birth outcomes including spontaneous abortion, stillbirth, preterm birth, low birth weight and postpartum hemorrhage, (3) neonatal complications including poor neonatal adaptation syndrome and persistent pulmonary hypertension of the newborn, and (4) long-term neurobehavioral sequelae in the offspring. Regarding potential risks of antenatal antidepressant use, most studies have evaluated the safety of SSRIs and SNRIs, while less data are available on other antidepressants such as TCAs and MAOIs. For women with unipolar TRD who often do not achieve remission with nonpharmacotherapeutic interventions alone, the overall small absolute risks of antidepressant use during gestation often outweigh the risks of untreated depression. All women should be educated on the potential risks of pharmacotherapy during the peripartum period. The discussion of the risk/benefits/alternatives to treatment, the patient's capacity to make the treatment decision, and ideally the partner's agreement to treatment should be documented in the patient's chart.

Risk of teratogenesis with antenatal antidepressant use

Birth defects affect 3% of infants born in the U.S. each year (CDC, 2008). Any consideration of antidepressant effects on rate of teratogenesis must consider baseline rates. Most studies indicate that antidepressants are not associated with overall major congenital malformations (Grigoriadis et al., 2013b; Reis and Källén, 2013). While some studies have observed a correlation between antidepressant exposure and specific birth defects, (e.g., anencephaly, craniosynostosis, and omphalocele), only a small increase in the absolute risk (AR) has been observed (Alwan et al., 2007). Some studies have noted the potential for a small increase in AR for cardiovascular defects (Grigoriadis et al., 2013b), specifically septal defects, in antidepressant-exposed infants, particularly paroxetine; however, many other studies, after controlling for underlying maternal depressive illness and genetics, have found no association between antenatal antidepressant use and heart defects (Alwan et al., 2007; Huybrechts et al., 2014; Reis and Källén, 2013).

Pregnancy and birth complications associated with antenatal antidepressant use

Risk of pregnancy loss

Miscarriage is a relatively common outcome of pregnancy, with most studies reporting 12%–15% loss among recognized pregnancies by 20 weeks' gestation (Magnus et al., 2019). Across multiple studies, antenatal antidepressant use may contribute a small to nonexistent risk for spontaneous abortion after adjusting for potential confounders (Andersen et al., 2014; Ross et al., 2013). Antenatal antidepressant use does not appear to be associated with elevated risk of stillbirth, neonatal death or postneonatal death after controlling for potential confounders (Jimenez-Solem et al., 2013).

Risk of preterm birth

There are mixed data regarding association between antenatal antidepressant use and preterm birth (PTB), defined as delivery at <37 weeks of gestation. The PTB rate in U.S. infants is 10% (Martin et al., 2015). A large national registry study which adjusted for potential confounding factors found that risk of PTB was lower in neonates with antenatal exposure to SSRIs than in unexposed neonates (4.7% vs 5.4%) (Malm et al., 2015). A meta-analysis reported that depressed women with antenatal SSRI use had increased risk of PTB compared with controls (adjusted odds ratio (aOR) 1.24, 95% confidence interval (CI) 1.09–1.41) (Eke et al., 2016). Control groups included both women with depression but without SSRI exposure and women without depression or SSRI exposure. In a subgroup analysis, risk of PTB remained significant when comparing depressed women with SSRI use with depressed women treated with psychotherapy alone (6.8% vs 5.8%; OR 1.17, 95% CI 1.10–1.25). Other studies indicate that antenatal antidepressant exposure may be associated with PTB (Huang et al., 2014; Ross et al., 2013) but that the reduction in gestational age at birth (approximately 3 days shorter) may not be clinically significant.

Risk of low birth weight

Approximately 8.28% of neonates in the U.S. are born at a low birth weight (LBW), defined as <2500 g (Martin et al., 2019). There are mixed data regarding an association between antenatal SSRI exposure and LBW. A meta-analysis found that antenatal SSRI exposure was associated with LBW (relative risk (RR) 1.48, 95% CI 1.22–1.79) (Huang et al., 2014); however, there was significant heterogeneity across studies and few studies adequately controlled for confounding factors. A larger meta-analysis observed that neonates exposed to antenatal antidepressants (mostly SSRIs) had lower birth weights than unexposed newborns (Ross et al., 2013). However, the mean difference between exposed and unexposed babies was noted to be small (74 g) and unlikely to be clinically significant. Furthermore, when the control group was limited to neonates without antenatal antidepressant exposure born to depressed mothers, there was no longer a significant association between antidepressant exposure and lower birth weight. In summary, if there is an association, the AR increase for LBW appears to be small with antidepressant use, and studies have found that untreated antenatal depression is associated with LBW (Grote et al., 2010).

Risk for postpartum hemorrhage

Postpartum hemorrhage (PPH), classically defined as an estimated blood loss ≥500 mL after vaginal birth or ≥1000 mL after cesarean delivery, has been estimated to occur in 1%–3% of U.S. deliveries (Reale et al., 2020). Some studies have associated serotonergic antidepressant use with a small increased AR for PPH, hypothesized to be related to serotonin-induced inhibition of platelet aggregation (Palmsten et al., 2013). A large U.S. Medicaid database study found an increased risk for PPH in women who used serotonergic antidepressants close to delivery, after adjustment for maternal illness (adjusted relative risk (aRR) 1.42, 95% CI 1.27–1.57) (Palmsten et al., 2013). A significant study limitation was the inability to control for other potential confounders including alcohol, drug, and tobacco use or the use of over-the-counter medications. Other large studies have documented no increase in risk of PPH with antenatal serotonergic antidepressant use (Lupattelli et al., 2014).

Neonatal complications associated with antenatal antidepressant use

Poor neonatal adaptation syndrome

Antenatal exposure to antidepressants has been associated with poor neonatal adaptation syndrome (PNAS) in up to 30% of neonates (Grigoriadis et al., 2013a). PNAS is characterized by typically mild physical and neurobehavioral symptoms including abnormal irritability, sleep disturbances, lethargy, jitteriness, mild respiratory distress, hypothermia, feeding disturbances, altered muscle tone, tremor and rarely seizures (Grigoriadis et al., 2013a). PNAS is hypothesized to be due to a side effect of antidepressants persisting in the infant (antidepressant toxicity) or to antidepressant withdrawal in the neonate (discontinuation syndrome) (Howard et al., 2014). PNAS is a transient condition, usually developing hours after birth and self-resolving within days to 2 weeks (Grigoriadis et al., 2013a). One study observed persistent PNAS signs in newborns at 1 month of age, but they were subtle and not clinically significant (Salisbury et al., 2016). PNAS symptoms are typically managed with supportive care, rarely requiring neonatal intensive care unit (NICU) admission. One large national registry study found that 13.7% of infants with antenatal exposure to SSRIs were admitted to the NICU compared with 8.2% of unexposed babies in the general population (aOR 1.5, 95% CI 1.4–1.5) (Nörby et al., 2016); however, this study concluded that the AR for severe disease was low. It is not recommended to discontinue or lower the antidepressant dose before delivery because this approach does not reduce the risk of PNAS (Salisbury et al., 2016) but may increase the relapse risk of maternal depression.

Risk for persistent pulmonary hypertension of the newborn

Some studies have identified an association between late gestation antidepressant use with persistent pulmonary hypertension of the newborn (PPHN) (Chambers et al., 2006; Grigoriadis et al., 2014), a serious but rare condition, characterized by failure of the neonatal cardiopulmonary transition to ensue, that occurs in 1.9 per 1000 live births in the general population (Walsh-Sukys et al., 2000). Newborns with PPHN typically present within 24 h after birth with signs of respiratory distress and cyanosis that can progress to severe respiratory failure requiring intubation and mechanical ventilation (Walsh-Sukys et al., 2000). Risk factors include increased maternal age, smoking, group B streptococcus infection, diabetes mellitus, hypertension, substance abuse, meconium aspiration, cesarean delivery, and prematurity and postmaturity (Occhiogrosso et al., 2012). Among other known causes of PPHN, the pathophysiology of PPHN may involve serotonin-induced damage to the neonatal lungs from antenatal antidepressant exposure characterized by vasoconstriction of pulmonary vessels, leading to increased vascular resistance (Eddahibi et al., 2002).

A 2014 meta-analysis reported an associated between late gestation SSRI use and increased risk of PPHN (OR 2.50, 95% CI 1.32–4.73) (Grigoriadis et al., 2014); however, there was a moderate degree of heterogeneity across the studies. Furthermore, it is important to consider that since the frequency of PPHN in the general population is low, the AR increase related to late pregnancy antidepressant exposure must also be low. This meta-analysis estimated that the AR difference between late pregnancy SSRI exposure and no exposure in the development of PPHN was approximately 3 per 1000 infants and that approximately 350 women would have to be treated with SSRIs in late pregnancy to result in one additional case of PPHN. A study limitation included inadequate control for factors associated with PPHN that are also common in depressed women, including obesity, cesarean delivery, and PTB. Finally, a large observational study found that antenatal exposure to SSRIs was associated with a much lower risk of PPHN or no association at all (Huybrechts et al., 2015). In summary, there appears to be either a nonexistent or small increase in the AR for PPHN from 1.9 per 1000 in the general population to approximately 3 per 1000 with late gestation SSRI use.

Risk for psychopathology in offspring

Antenatal antidepressant exposure does not appear to be associated with increased risk for attention deficit hyperactivity disorder (ADHD) or autism in the offspring, based upon several studies which adjusted for potential confounders (Malm et al., 2016; Sujan et al., 2017). Antenatal antidepressant use may be associated with increased risk of depression in the offspring. A large national registry study which controlled for potential confounding factors found that the cumulative incidence of depression was greater in offspring (with antenatal SSRI exposure) by age 15, compared with offspring of women with psychiatric disorders without SSRI use (8% vs 2%) (Malm et al., 2016). Additionally, depression occurred more in offspring with antenatal SSRI exposure compared with offspring of women who used SSRIs prior but not during pregnancy (8% vs 3%). The strength of these conclusions, however, are limited due to potential residual confounding in the study by maternal postpartum depression, which has been shown to be linked with an increased risk of depression in offspring by 16 years of age (Pearson et al., 2013).

Antenatal use of atypical antidepressants in unipolar depression

Bupropion

There is inconsistent evidence regarding whether antenatal exposure to bupropion is associated with teratogenesis. While a few studies have reported an association between antenatal bupropion exposure and congenital heart defects, the type of defect differs across studies. A retrospective case-control study, which adjusted for potential confounding factors, concluded that there was a greater likelihood of first trimester bupropion exposure in neonates with left ventricular outflow tract obstruction (LVOTO) compared to newborns without malformations (OR 2.6, 95% CI 1.2–5.7) (Alwan et al., 2010); however, the AR was small (2.1 per 1000 births). Additionally, bupropion was not associated with other cardiac malformations, such as ventricular septal defects (VSDs). More recently, a large registry case-control study, which did not control for potential confounding from maternal mental illness, found that there was a greater likelihood of first trimester bupropion exposure in infants with VSDs than neonates without malformations (aOR 2.5, 95% CI 1.3–5.0); however, bupropion was not associated with other congenital heart defects, such as LVOTO defects (Louik et al., 2014). The studies are limited in that neither controlled for confounding by indication (using bupropion for depression or smoking), due to the small number of infants exposed to bupropion antenatally.

Several other studies have reported that bupropion is not a major teratogen. A large retrospective cohort study observed that first trimester bupropion exposure was not associated with an increased risk for overall congenital malformations or cardiac malformations (Cole et al., 2007). A separate large U.S. Medicaid database study evaluating first trimester bupropion exposure found no increased risk for any congenital cardiac malformations, VSDs, or right ventricular outflow tract obstruction (RVOTO) after adjustment for maternal illness (Huybrechts et al., 2014).

Bupropion is often a treatment choice for women with peripartum depression and comorbid ADHD or nicotine dependence. One multistate insurance claims database study found that fetal exposure to bupropion, particularly in the second trimester, was strongly associated with an increased risk for ADHD (Figueroa, 2010). While this study adjusted for some confounding factors (i.e., sex of the child, parental psychiatric diagnoses, and peripartum complications), it did not control adequately for tobacco, alcohol, or drug use, and there may have been confounding by undiagnosed parental ADHD. Future research is needed to evaluate the potential risk for ADHD in offspring exposed prenatally to bupropion.

Mirtazapine

Mirtazapine is not often used as a first-line therapy for antepartum depression as there is limited information available concerning its safety profile during pregnancy. The available data suggest that mirtazapine is not a major teratogen. In a systematic review of six observational studies, mirtazapine was not associated with increased incidence of major congenital malformations (Smit et al., 2016). Unlike other antidepressants, mirtazapine has antiemetic properties that have used in the treatment of hyperemesis gravidarum (Uguz et al., 2018), which is frequently associated with severe anxiety and depression. As such, mirtazapine may be a particularly favorable choice for pregnant women suffering from severe nausea and comorbid depression or anxiety. For women who have failed traditional SSRI or SNRI pharmacotherapies, mirtazapine may additionally be a beneficial choice.

Trazodone and nefazodone

Trazodone and nefazodone are not used as first-line therapy for antepartum depression or insomnia as they do not have well-studied reproductive safety profiles. Limited data on the antenatal use of trazodone and nefazodone are reassuring regarding the risk of teratogenesis and major pregnancy complications. One small prospective observational study of found that first trimester exposure to trazodone or nefazodone ($n=147$) was not associated with major malformations, miscarriages, or stillbirths (Einarson et al., 2003), but larger studies powered to find rare adverse effects are not available in the literature.

Antenatal use of TCAs in unipolar depression

TCAs are generally not used as first- or second-line therapy for antepartum depression, as they can be poorly tolerated due to their anticholinergic side effect profile. In addition, data on the risks of antenatal exposure to TCAs are limited and includes smaller sample sizes compared to those of SSRIs. TCAs should be reserved for use in women with TRD who have not responded to a SSRI or SNRI. Desipramine and nortriptyline are favored TCAs as they are less anticholinergic, and thereby less likely to worsen orthostatic hypotension and constipation during pregnancy. Most studies have found that exposure to antenatal TCA use is not associated with congenital malformations, including cardiac malformations (Ban et al., 2014; Huybrechts et al., 2014). Although TCAs as a class are not thought to be teratogenic, some data suggest that antenatal exposure to clomipramine may be associated with a modestly elevated risk of total malformations (OR 1.36, 95% CI 1.07–1.72), cardiovascular defects as a group (OR 1.63, 95% CI 1.12–2.36), and VSDs and/or atrial septal defects (aOR 1.84, 95% CI 1.13–2.97) (Reis and Kllén, 2010). A main limitation is that the analysis did not control for maternal depression, illicit drug use, or other nonantidepressant medication, and the results may have been due to confounding by indication.

Antenatal use of MAOIs in unipolar depression

MAOIs are not used as first- or second-line therapy for antepartum depression because of the potential for interactions with foods and other drugs as well as vasoconstrictive effects. In addition, data on the risks of antenatal exposure to MAOIs are primarily limited to case reports (Gracious and Wisner, 1997; Pavy et al., 1995). One study that evaluated fetal exposure to MAOIs with other antidepressants did not identify adverse infant outcomes (Ramos et al., 2008). MAOIs should generally be avoided during pregnancy due to insufficient data regarding their reproductive safety profile and restrictions associated with their use. MAOIs may be considered for use in women with TRD who have not responded to other agents.

Antidepressant use during lactation

Antidepressant use should not discourage women from breastfeeding, as the overwhelming benefits of breastfeeding usually outweigh the low risks of antidepressant use in women suffering from postpartum depression. Furthermore, postpartum depression is associated with decreased lactation duration and increased breastfeeding difficulties (Dennis and McQueen, 2009). Data about safety of infant exposure to antidepressants through breast milk are derived from small observational studies, as no RCTs have been performed. The relative infant dose (RID) is an estimate of infant drug exposure from breast milk derived from dividing the dose provided to the infant via breast milk (mg/kg/day) by the mother's weight-adjusted dose (mg/kg/day). For most drugs, a RID from breast milk of $<10\%$ of the weight-adjusted maternal dose is satisfactory while breastfeeding a healthy infant. Caution is required for drugs with a RID from breast milk of 10%–25% (Drugs and Lactation Database (LactMed), 2006a, b).

Most antidepressants are transferred to breast milk in low concentrations, with few reaching 10% of the maternal weight-adjusted dose (Chad et al., 2013). Paroxetine, sertraline, bupropion, duloxetine, fluvoxamine, and mirtazapine have RIDs $<3\%$, while citalopram, escitalopram, venlafaxine, and desvenlafaxine produce RIDs in breast milk closer to 10%. In contrast, fluoxetine has a RID slightly above the 10% guideline. Several TCAs may be used while breastfeeding because infant exposure has been found to be low and serious side effects in breastfed infants have not been observed, except for doxepin (Gentile, 2014). Nortriptyline is generally the preferred TCA for use with lactation due to its favorable safety profile. MAOIs are generally not used in breastfed infants as insufficient data are available about their use

(Newton and Hale, 2015). Limited data are available regarding the safety of serotonin modulators (trazodone, nefazodone, and vilazodone) in breastfed infants.

Nonspecific adverse signs to watch for in breastfed infants exposed to antidepressants include sedation, sleep disturbance, irritability, and poor feeding (Newton and Hale, 2015). Strategies to limit infant exposure such as advising mothers to take antidepressants immediately after breastfeeding or discarding breast milk produced at the time of highest drug concentration are not recommended, as they can potentially make breastfeeding more stressful and are not evidence-based.

Estradiol and progestin as treatments for unipolar postpartum depression

A small double-blind RCT found that women with severe postpartum depression treated with transdermal estradiol showed a greater and faster improvement in symptoms compared to those who received placebo patches (Gregoire et al., 1996). One large RCT was terminated prematurely after the investigators found that an 8-week course of treatment with transdermal estradiol for postpartum depression did not result in significantly increased serum estradiol concentrations (Wisner et al., 2015). Further evidence from large RCTs is needed before estradiol may be recommended as a therapeutic intervention for postpartum depression due to its associated risks for venous thromboembolism and impaired lactation (Moses-Kolko et al., 2009).

There is less evidence examining the effects of progesterone as treatment for postpartum depression. Some RCTs found that administration of synthetic progestogens, including intramuscular injections of norethisterone enanthate (Lawrie et al., 1998) and depot medroxyprogesterone acetate (Singata-Madliki et al., 2016), was associated with higher risk of developing postpartum depressive symptoms compared with those taking placebo (Lawrie et al., 1998) or with an intrauterine device (Singata-Madliki et al., 2016). In contrast, a retrospective review concluded that use of synthetic progestogens did not appear to predispose postpartum women to depressive symptoms (Tsai and Schaffir, 2010). A Cochrane review reported that synthetic progestogens should not be used to prevent postpartum depression and that they should be used cautiously in postpartum women (Dennis et al., 2008).

Neuroactive steroids for the treatment of postpartum depression

In 2019, the U.S. Food and Drug Administration (FDA) approved the first medication with an indication for unipolar postpartum depression in adult women. Brexanolone (ZULRESSO™) is a soluble, intravenous preparation of synthetic allopregnanolone, an endogenous neuroactive steroid and is administered as a postpartum 60-h peripheral IV infusion. Brexanolone is a potent, selective, positive allosteric modulator (PAM) of extrasynaptic and synaptic $GABA_A$ receptors. A recent series of open-label (Kanes et al., 2017b) and placebo-controlled RCTs (Kanes et al., 2017a; Meltzer-Brody et al., 2018) of brexanolone demonstrated rapid reduction of postpartum depression symptoms.

Three studies were completed in adult women, ≤6 months postpartum, with a diagnosis of peripartum depression with moderate-to-severe symptoms as rated by the 17-item Hamilton Rating Scale for Depression ($HAM\text{-}D_{17}$). Participants were randomized to receive placebo or brexanolone iv 90 μg/kg/h or 60 μg/kg/h for 60 h. The primary outcome measure was the least-squares mean change in $HAM\text{-}D_{17}$ total score from baseline to hour 60 and safety, tolerability and depression severity were assessed through day 30. Across a range of disease severities, at hour 60, there were significantly larger mean reductions from baseline in HAM-D total scores with brexanolone iv 90 μg/kg/h (−17.0; $P<0.001$) and brexanolone iv 60 μg/kg/h (−19.1; $P<0.001$) vs placebo (−12.8). Significant differences from placebo were observed at hour 24 (both dose groups $P=0.001$) and maintained through day 30 ($P\leq 0.021$) (Meltzer-Brody et al., 2018).

In clinical studies, brexanolone caused sedation and somnolence that required dose interruption or reduction in some patients during the infusion (5% of brexanolone-treated patients compared to 0% of placebo-treated patients). Some patients were also reported to have loss of consciousness or altered state of consciousness during the brexanolone injection infusion (4% of the brexanolone-treated patients compared with 0% of the placebo-treated patients). Time to full recovery from loss or altered state of consciousness, after dose interruption, ranged from 15 to 60 min (Meltzer-Brody et al., 2018). Brexanolone has a black box warning for excessive sedation and sudden loss of consciousness and is available only through a restricted program under a risk evaluation and mitigation strategy (REMS) to reduce the risk of serious adverse events. According to the package insert, brexanolone appears compatible with breastfeeding as the reported maximum RID during infusion is between 1% and 2% of the maternal weight adjusted dosage.

Zuranolone is an investigational product under development for the treatment of unipolar peripartum depression. Zuranolone is a potent and selective extrasynaptic and synaptic $GABA_A$-R PAM with a pharmacology distinct from benzodiazepines and has oral bioavailability (Althaus et al., 2020; Martinez Botella et al., 2017). Zuranolone was recently tested in an outpatient randomized, double-blind, placebo-controlled phase 3 study in severe unipolar peripartum depression

(NCT02978326) and results were presented at scientific conferences (Deligiannidis et al., 2021; Lasser et al., 2019). Participants included 151 adult women with severe peripartum depression who received 2 weeks of either zuranolone 30 mg or placebo and then were followed for 4 weeks. Zuranolone achieved the primary endpoint of a significant reduction in LS mean HAM-D_{17} total score vs placebo (−17.8 vs −13.6, $P=0.0029$). Significant differences favoring zuranolone vs placebo were observed at day 3 ($P=0.0255$) and were sustained through day 45 ($P=0.0027$). HAM-D response (72% vs 48%, $P=0.0050$) and remission rates (45% vs 23%, $P=0.0122$) were significantly greater in the zuranolone group compared to the placebo group at day 15, and these significant improvements were maintained through day 45 (response $P=0.0220$; remission $P=0.0102$). The most common (≥5%) adverse events in the zuranolone group were somnolence, headache, dizziness, upper respiratory tract infection, diarrhea, and sedation. There were no reports of loss of consciousness or syncope in either group. Oral zuranolone is currently being evaluated in an additional randomized, double-blind placebo-controlled trial in severe unipolar postpartum depression (NCT04442503).

Additionally, in development for the treatment of unipolar postpartum depression is the 3β-methylated synthetic analog of allopregnanolone, ganaxolone. Ganaxolone is an extrasynaptic and synaptic $GABA_A$-R PAM, similar to allopregnanolone, but unlike allopregnanolone, in binding studies, ganaxolone is reported to show no affinity for estrogen or progesterone receptors (Bialer et al., 2017). It has been tested in both IV and oral forms (NCT03460756; NCT03228394) but at the time of print, published results were not available.

Neuromodulation

Antenatal electroconvulsive therapy

There are no antenatal electroconvulsive therapy (ECT) clinical trials in the literature. Data are derived from case series and retrospective database reviews, mainly of unipolar peripartum depression with fewer reports for bipolar disorder. ECT has been approved for use in all three trimesters (Jaffe, 2002). Recent evidence suggests that ECT in pregnancy is low risk to both the mother and fetus and should be used when clinically warranted. Indications for ECT include severe depression, high suicide risk, catatonia, medication-resistant illness, psychotic agitation, and severe physical decline (Kim et al., 2015; Ward et al., 2018). The risks, adverse reactions, length of treatment, and response in pregnancy are like that observed in ECT in a nonpregnant population (Leiknes et al., 2015). Fetal death, major congenital malformations, and PTB appear to occur at rates like the psychiatric population not receiving ECT (Jablensky et al., 2005).

Older studies of ECT in pregnancy should be interpreted cautiously, as the techniques utilized were markedly different from the procedure today (Ward et al., 2018). Maternal and fetal adverse events have been variable, and it is difficult to establish a causal relationship. Some adverse events occurred weeks after ECT, and some complications occurred in the context of medical comorbidities (Leiknes et al., 2015). Two seminal reviews of ECT in pregnancy, which include a total of 339 cases, show a 10% rate of maternal and fetal complications, most of which are transient and nonlife threatening. The most frequent complications were vaginal bleeding, uterine contractions, abdominal pain, and fetal arrhythmias. There was one report of fetal death from status epilepticus indirectly attributed to ECT as the patient had comorbid medical conditions (Anderson and Reti, 2009).

In a recent review including 169 women with unipolar depression, bipolar depression, and psychotic depression throughout pregnancy, there was a 4.7% stillborn/neonatal mortality rate, and a 4.1% rate of major congenital malformations. The most common adverse obstetric outcome was preterm delivery (11%), while reduced fetal heart rate was the most common adverse event reported during ECT (8.8%) (Leiknes et al., 2015). A retrospective chart review of 33 antenatal women admitted inpatient and treated with ECT showed a complete response to treatment in 84.2% of patients with MDD, and a partial response to treatment in 15.8%. For women with bipolar disorder, 91.7% of the patients had a complete response and 8.3% had a partial response, rates similar to those in nonpregnant depressed women (Pagnin et al., 2004).

An obstetrician should be readily available in cases where there is an existing increased risk for preterm labor or vaginal bleeding. After week 24 of pregnancy, the woman should be positioned in pelvic tilt to decrease the risk of supine hypotension due to uterine pressure on the vena cava. Aspiration precautions should be taken, and succinylcholine can be used to reduce risk of decreased blood supply to the fetus in the case of a prolonged seizure (O'Reardon et al., 2011). The most common anesthetics cross the placenta, but are not known to be teratogenic (Anderson and Reti, 2009). There is no evidence to support a link between ECT and teratogenicity (Calaway et al., 2016).

Long-term effects of ECT in pregnancy are not well studied. In many cases the well-being of the mother in need of treatment will outweigh any potential risks (Kim et al., 2015). Recent case reports of pregnant women who received ECT for unipolar depression, bipolar depression, and primary psychotic disorder show normal postnatal development measured from 9 months of age to 3.5 years (Halmo et al., 2014; O'Reardon et al., 2011; Salzbrenner et al., 2011).

Postpartum ECT

There are no RCTs studying the use of ECT for postpartum women (Rabheru, 2001). ECT is considered the treatment of choice for severe and refractory postpartum psychosis and postpartum depression (Focht and Kellner, 2012). Case reports and case series have described positive treatment outcomes with women with treatment refractory postpartum illness who received ECT (Forray and Ostroff, 2007). ECT may be beneficial for women whose illness severity necessitates rapid resolution of symptoms and need not be reserved solely for treatment refractory cases. It should also be considered in cases of catatonia, or when women with postpartum psychosis or postpartum depression refuse to eat (Forray and Ostroff, 2007). ECT has been proposed as a first-line treatment for postpartum psychosis (Babu et al., 2013; Focht and Kellner, 2012). ECT should also be considered for treatment of postpartum depression with psychotic features, given the relatively longer median duration of illness compared with postpartum mania (Bergink et al., 2011).

Transcranial magnetic stimulation

Data on transcranial magnetic stimulation (TMS) for peripartum depression is minimal, with few adverse events reported in the existing literature. The first study of antenatal TMS was an open-label pilot study of low frequency, right-sided TMS of the dorsolateral prefrontal cortex (DLPFC) in 10 women with MDD. The mean HAM-D_{17} total score decreased by 60%, with 70% achieving response and 30% achieving remission. There were no reports of fetal arrhythmia, uterine contractions, PTB, NICU admissions, prolonged hospital stays or other adverse events (Kim et al., 2011). In an open-label trial on high-frequency left-sided TMS of the DLPFC in 30 antenatal depressed women, 41% had a $>50\%$ reduction in the HAM-D_{17} score and 21% achieved remission (HAM-D score <8). There were no reports of complications in the infants (Hizli Sayar et al., 2014). Case reports published to date have also been reassuring. Babies born to mothers who received TMS in the second or third trimester were born full-term and healthy (Nahas et al., 1999). There is one report of an infant born prematurely at 36 weeks to a mother who received TMS starting at 31 weeks' gestation as well as venlafaxine. The baby was irritable for 1 week but otherwise healthy (Zhang et al., 2010). Pelvic tilt should be employed in women at 24 weeks of pregnancy and beyond to prevent vena cava syndrome and decrease the risk of supine hypotension (Kim and Wang, 2014).

Vagus nerve stimulation

Vagus nerve stimulation (VNS) was approved for TRD in 2005 (Groves and Brown, 2005). There is one case report of VNS for antenatal depression. The woman with unipolar depression and medical comorbidities in the report had been receiving VNS for about 3 years prior to her pregnancy and continued to receive treatment at her regular parameters, and pharmacotherapy with citalopram and bupropion. No adverse peripartum events were reported (Husain et al., 2005).

Clinical summary for the treatment of unipolar peripartum depression

Peripartum depression is common, yet frequently underdiagnosed and undertreated. Untreated and undertreated unipolar peripartum depression is associated with adverse health outcomes for mothers and their offspring. The management of unipolar depression in the peripartum period follows the same treatment algorithm as that of the general population, along with unique considerations of dosing and the potential risk of medication use during pregnancy and lactation. Current best practice is to conduct an individualized risk–benefit discussion with the patient regarding medication use that considers her illness course and psychosocial factors, and weighs the benefit and potential risk of medication exposure vs the risk, to both mother and child, of exposure to maternal depression.

Current effective psychotherapies and pharmacotherapies for unipolar peripartum depression are available but underutilized. Psychotherapies are recommended as monotherapy for mild unipolar peripartum depression. Antidepressants are effective and indicated for moderate-to-severe or recurrent unipolar peripartum depression. When using pharmacotherapy, it is recommended to select medications based on individual and family history plus consideration of the reproductive safety data, and to use the lowest effective dose and fewest number of agents to achieve euthymia. Most evidence is reassuring regarding the safety of taking antidepressants (SSRIs, SNRIs, TCAs, bupropion, and mirtazapine) during pregnancy and lactation. While there are limited data for MAOIs, they may be considered in women with TRD who have not responded to other agents. Brexanolone is the only FDA-approved medication for postpartum depression and has rapid antidepressant effects. Neuromodulation may be useful in TRD, as well as for women who wish to avoid pharmacotherapy. ECT appears to be relatively safe and effective in pregnancy. Most complications observed are mild and limited, but more research is needed on this topic. There are little data safety and efficacy on TMS and VNS in pregnancy. These modalities may present a safe and efficacious treatment alternative and thus warrant future research.

Peripartum bipolar depression

Introduction

Women with bipolar disorder are at risk for symptom exacerbation across reproductive events including prior to menstruation (Teatero et al., 2014), during the peripartum period (Di Florio et al., 2018), and during perimenopause (Yatham et al., 2018). Antenatal bipolar illness is associated with a 66% increase in risk for postpartum episode (Viguera et al., 2007b). A personal history of postpartum depression, postpartum psychosis, or a family history of a first-degree relative with bipolar disorder increases the risk for peripartum illness onset. Women with bipolar disorder are 50% more likely to develop postpartum depression than women with MDD (Clark and Wisner, 2018). Women with bipolar I disorder report an approximately 50% risk of a peripartum major affective episode per pregnancy/postpartum period, while women with bipolar II disorder report an approximately 40% risk of an affective episode (Di Florio et al., 2013). Peripartum mood episodes in bipolar disorder tend to be depressive (Freeman et al., 2002; Viguera et al., 2000). Women with bipolar disorder are also at an increased risk for postpartum psychosis (Kendell et al., 1987).

Bipolar disorder during pregnancy is associated with adverse outcomes in both treated and untreated women (Bodén et al., 2012). There is an increased risk of preterm labor, cesarean section, infants who are small for gestational age, gestational hypertension, antepartum hemorrhage, and placenta previa. Infants of women with untreated bipolar disorder may also be at increased risk of microcephaly and neonatal hypoglycemia (Bodén et al., 2012). Additionally, untreated peripartum mood episodes are associated with intrauterine growth retardation, elevated levels of fetal stress hormones, and adverse neurodevelopmental outcomes (Jablensky et al., 2005). Critically, women with bipolar disorder are at an increased risk for suicide and infanticide (Kim et al., 2008). Despite the substantial risks of untreated symptoms, bipolar disorder in the peripartum period is underdiagnosed and undertreated (Freeman et al., 2002).

When determining appropriate treatment of bipolar disorder in pregnancy, it is important to keep in mind the limitations of the available evidence base. Most available data derive from observational cohort studies or medical record databases. RCTs examining mood stabilizers and second-generation antipsychotics (SGAs) in pregnancy and large prospective studies that follow women over time across the peripartum period are lacking.

Treatment overview of peripartum bipolar depression

Medication discontinuation has been associated with a twofold increase in the risk of recurrence (37% vs 85%), shorter time to first recurrence, and longer duration of recurrence. Symptoms occur most frequently in the first trimester of pregnancy, and depressive or mixed episodes are most likely. Rapid discontinuation further increases this risk. Gradual discontinuation of medication over a period > 14 days, was associated with a 20-week delay in reaching a 50% recurrence rate as compared to abrupt discontinuations of < 14 days (Viguera et al., 2007b).

The benefits of pharmacologic treatment of bipolar disorder in pregnancy and lactation often outweigh risks. Preconception counseling should be offered at least 3 months prior to considering pregnancy or immediately for women who have become pregnant (Yatham et al., 2018). The general guidelines used for the acute treatment of bipolar depression in the general population should be applied to pregnant women, with special attention to safety and dosing considerations (Yatham et al., 2018). The medication chosen for the acute treatment should ideally also be effective in the prevention of postpartum mood episodes of both polarities (Sharma et al., 2020). A single medication at a higher dose is preferable to polypharmacy. Medication switches should be minimized during pregnancy to limit exposure of the fetus to multiple medications. The medication selection process should be guided by a history of efficacy, prior exposure during pregnancy, and available reproductive safety. If a euthymic woman presents for preconception planning, a baseline therapeutic drug level should be obtained to use as a benchmark for TDM during pregnancy. The preconception period may also be used to establish the lowest effective dose of medication, in order to minimize potential exposure to the fetus while avoiding increasing the mother's risk of relapse. A comprehensive relapse prevention plan that includes prophylactic medication, minimization of sleep deprivation, and provision of psychosocial support should be established.

Pharmacotherapy for peripartum bipolar depression

Lithium

Lithium can be safely used in pregnancy with close TDM. Abrupt discontinuation of lithium in pregnancy has been associated with high relapse rates (Viguera et al., 2000). Women with mild and infrequent mood episodes may taper lithium prior to conception in lieu of an alternative agent. Women with a history of more severe episodes, who are not at high risk for relapse in the short term, should taper lithium prior to conception, and restart after organogenesis. Women with severe or frequent episodes of illness should continue lithium throughout pregnancy. Some women taking lithium and another agent may rely on monotherapy with the other agent for the first trimester, and restart lithium in the second trimester to minimize possible teratogenicity from lithium. When lithium is continued, it should be maintained at the lowest therapeutic range throughout pregnancy.

Dosing and therapeutic drug monitoring

The pharmacodynamics of lithium are influenced by weight, renal function, age, co-administered medications, pregnancy, and lactation (Grandjean and Aubry, 2009a). Renal lithium clearance increases by 30%–50% in pregnancy, causing a substantial dip in serum concentrations, and an increased risk of maternal relapse if the dose is not adjusted (Ward and Wisner, 2007). Lithium levels may drop 24% in the first trimester, 36% in the second trimester, and 21% in the third trimester (Wesseloo et al., 2017). Pregnancy-related complications can also alter maternal lithium concentration. Fluid loss due to hyperemesis gravidarum, or normal labor and delivery, can decrease intravascular volume and increase lithium levels. Preeclampsia may reduce renal clearance and increase lithium levels. Lithium dose must often therefore be increased to maintain effective concentrations. It is critical to establish the effective serum lithium concentration prior to pregnancy. When this is not available, the dose should be titrated to the lower end of the therapeutic range and monitored (Blake et al., 2008).

Lithium level should be monitored monthly or minimally at every trimester in a euthymic woman on a stable maintenance dose (Newport et al., 2005). Women who initiate or reinitiate lithium in pregnancy, or women with medical comorbidities affecting lithium absorption or clearance, may require weekly or bi-weekly lithium levels in the last month of pregnancy (Ward and Wisner, 2007). A sustained release formulation may be preferable as it may produce more stable lithium levels (Malhi et al., 2012). A level should be checked when the woman presents for delivery. Adequate hydration should be ensured, and nephrotoxins should be avoided (Blake et al., 2008; Galbally et al., 2010). Vascular volume rapidly decreases at delivery, and lithium clearance abruptly decreases to prepregnancy levels (Grandjean and Aubry, 2009b). A level should be drawn immediately after delivery and for any clinical worsening or change in psychiatric or medical status. Preconception dosing can be resumed immediately following delivery once the woman is medically stable (Ward and Wisner, 2007). If lithium was initiated in pregnancy, levels should be checked at the end of the first week postpartum, and repeated weekly to achieve a therapeutic level while preventing toxicity (Newport et al., 2005).

Teratogenicity

Data regarding the risk of Ebstein's anomaly and overall congenital heart defects are limited and conflicting (Patorno et al., 2017). Older retrospective studies and case reports suggested that first trimester lithium use was associated with a rate of Ebstein's anomaly of 1:1000, as compared to a background risk of 1:20,000 (Newport et al., 2005; Nora et al., 1974). The largest study of in utero lithium exposure to date concluded that the baseline risk of cardiac malformations without lithium exposure was 1.15%, derived from studying 1,322,955 nonexposed infants. Whereas the rate of cardiac malformations in infants exposed to lithium in utero was 2.41%, RR 1.65. The AR of cardiac malformations is 1.90% with lithium exposure. This effect is more modest than in older studies and was found to be dose related. The risk of RVOTO defects was 0.60 per 100 live births among infants exposed to lithium and 0.18 per 100 among unexposed infants (Patorno et al., 2017).

In a recent systematic review and meta-analysis analyzing over 1.3 million pregnancies, the risk of adverse outcomes associated with lithium exposure at any time in pregnancy was low but increased with first-trimester exposure or exposure to higher doses. Mothers with serum levels of 0.64 mEq/L and doses of 600 mg/day had reactive newborns without increased risk of cardiac malformations. Lithium exposure in pregnancy was associated with any congenital anomaly (OR = 1.81, NNH = 33) and cardiac anomalies (OR = 1.86, NNH = 71). The risk of cardiac anomalies was only significantly increased with first-trimester exposure (Fornaro et al., 2020). Women with first-trimester lithium exposure should receive fetal echocardiography and level II ultrasound at 16–18 weeks' gestation (Wesseloo et al., 2017).

Adverse obstetrical outcomes

An increased risk of PTB of infants has been reported among women treated with lithium in pregnancy (Troyer et al., 1993). However, this is confounded by the twofold increased risk of preterm delivery in women with bipolar disorder compared to women without a history of mental illness (Lee and Lin, 2010). A large cohort study of 10,575 women on lithium monotherapy or anticonvulsant monotherapy did not find an association with use of these drugs and PTB or placenta-mediated complications when adjusted for confounders. Moreover, women who remained on the medications throughout pregnancy were not at increased risk of ischemic placental disease or PTB compared to women who used mood stabilizers exclusively in the first half of pregnancy (Cohen et al., 2019).

Postpartum neonatal adverse effects

Fetal plasma lithium concentrations are almost equal to maternal plasma concentrations due to equilibration of ions across the placenta. Toxicity can occur in neonates exposed to high maternal blood levels with higher concentrations at delivery associated with more initial complications (Blake et al., 2008). Predominant symptoms include lethargy, hypotonicity, and poor suck reflexes, and can persist beyond 7 days (Woody et al., 1971). Neonatal nephrogenic diabetes insipidus (Mizrahi et al., 1979), cardiac arrhythmias (Wilson et al., 1983), floppy baby syndrome (Llewellyn et al., 1998), hypoglycemia, polyhydramnios, and reversible changes in thyroid function (Karlsson et al., 1975) have also been reported. Women treated with lithium in pregnancy should thus be considered obstetrical high-risk patients and should receive care from a maternal-fetal medicine specialist.

Neurodevelopmental outcomes

There have been no adverse neurodevelopment outcomes identified in the limited data available. Two small retrospective cohort studies showed no adverse effects on cognitive development at ages 5 and 15 after exposure to lithium in utero (Forsberg et al., 2012; van der Lugt et al., 2012).

Use of anticonvulsants in peripartum bipolar disorder

Most of the reproductive safety data on the use of anticonvulsants as mood stabilizers in pregnancy and postpartum is derived from epilepsy registry data (Morrow et al., 2006). This presents potential confounders, and more studies of women taking mood stabilizers for bipolar disorder are needed.

Valproic acid

Valproic acid (VPA) should be avoided in all women of childbearing age who are not surgically sterile due to valproate's association with serious neural tube defects (NTDs), reduced IQ, autism, and neurodevelopmental delays in children (Veroniki et al., 2017). When clinically appropriate, VPA should be discontinued at least 6 months prior to planning for conception to allow time for taper, initiation of an alternate agent, and adequate monitoring for euthymia. However, if the decision is made that VPA is the only appropriate medication for peripartum mood stabilization, the risks/benefits/alternatives to treatment should be discussed thoroughly with the mother and her partner and documented.

For the very rare instances when VPA is continued during pregnancy, a decrease in VPA level has been observed towards the end of pregnancy due to increased VPA clearance (Philbert et al., 1985). Several authors have noted that while total VPA level declines, free VPA level remains unchanged or increased. Free and total plasma levels of VPA should thus be measured (Koerner et al., 1989). Baseline VPA levels should be obtained preconception to identify optimal serum concentration for mood stabilization. Levels should be checked at least monthly to maintain this concentration (Krishnamurthy, 2012). VPA concentration decreases sharply in the immediate postpartum period. The dose should be tapered rapidly to avoid toxicity and to maintain preconception VPA levels (Philbert et al., 1985).

Teratogenicity

VPA has been shown consistently to have the highest rate of congenital malformations (8.2%–13.2%) among the antiepileptic drugs (AEDs) (Walker et al., 2009). Most malformations are structural abnormalities, particularly NTDs, with a 1%–3.8% risk with antenatal VPA use (Canger et al., 1999; Lindhout and Schmidt, 1986). Other abnormalities associated with antenatal VPA exposure include cardiovascular (Sodhi et al., 2001), craniofacial (Paulson and Paulson, 1981), and limb (Rodríguez-Pinilla et al., 2000) defects. A conglomerate of abnormalities known as "fetal valproate syndrome," characterized by distinct facial dysmorphism, congenital anomalies, and developmental delays, has been reported (Ardinger et al., 1988).

Adverse neonatal outcomes

Acute neonatal risks include hepatotoxicity (Kennedy and Koren, 1998), coagulopathies (Mountain et al., 1970), and neonatal hypoglycemia (Thisted and Ebbesen, 1993). Withdrawal symptoms have also been reported (Ebbesen et al., 2000).

Neurodevelopmental outcomes

Prenatal VPA exposure may also negatively impact neurocognitive development (Eriksson et al., 2005; Meador et al., 2009). In a network meta-analysis of 933 children exposed to AEDs, VPA was the only AED with a significant association with cognitive delay including an increased odds of language delay. It was also associated with greater odds of developing autism (Veroniki et al., 2017). These findings are consistent with a smaller meta-analysis in which in utero VPA exposure was associated with significant reductions in mean full scale IQ, verbal IQ and performance IQ (Banach et al., 2010).

Lamotrigine

Continuation of lamotrigine in pregnancy is associated with a decreased risk of both manic, hypomanic, and depressive episodes. A comparison among pregnant women with bipolar disorder who remained on lamotrigine vs women who discontinued showed a 30% relapse rate with continued lamotrigine use vs 100% relapse rate with discontinuation (Newport et al., 2008b).

Lamotrigine undergoes hepatic glucuronic acid conjugation to an inactive metabolite. Pregnancy-related sex steroid changes increase phase 2 glucuronidation, which increases lamotrigine clearance (Öhman et al., 2008). Pregnancy related changes in plasma protein binding, absorption, and transplacental transfer can also affect lamotrigine level. Lamotrigine clearance increases substantially at pregnancy onset and continues to increase progressively through the third trimester, potentially to greater than 330% from preconception (Öhman et al., 2008). Women with epilepsy, require an average dose increase of 250% to sustain therapeutic drug levels across pregnancy. A preconception level should serve as a guide for adjusting the dose throughout pregnancy when available (Krishnamurthy, 2012). Lamotrigine elimination rate rapidly declines within days of delivery, and plasma concentrations increase (Öhman et al., 2008). Women requiring a higher dose in pregnancy should taper their dose to the prepregnancy dose postpartum to avoid toxicity (Clark et al., 2013).

Teratogenicity

Lamotrigine has a favorable reproductive profile and has not been associated with an increased risk of congenital malformations. While early studies signaled a possible increased risk of cleft palate or cleft lip deformity with first trimester lamotrigine exposure (Holmes et al., 2008), more recent studies have not found this risk, or demonstrate only mild elevation above baseline risk (Cunnington et al., 2011; Dolk et al., 2016; Pariente et al., 2017). The risk of major congenital malformations (MCM) with lamotrigine monotherapy exposure in women with epilepsy (2.0%–2.9%) appears to be consistent with the rate major congenital malformations found in the general population (3%–5%) (Cunnington et al., 2011; Dolk et al., 2016; Hernández-Díaz et al., 2012). Similarly, a meta-analysis of 21 studies did not identify an association between lamotrigine exposure and increased risk of MCM (Pariente et al., 2017). There may be a potential dose-related difference in risk of cardiac malformations with utero lamotrigine exposure when dosed <300 mg, vs >300 mg (Campbell et al., 2014). However, this finding is inconsistent across studies (Cunnington et al., 2011).

Adverse obstetrical outcomes

Antenatal lamotrigine exposure has not been associated with pregnancy complications or adverse infant outcomes. A systematic review and meta-analysis of 21 studies found no increased risk of miscarriages, stillbirths, preterm deliveries, and small for gestational age neonates after prenatal lamotrigine exposure (Pariente et al., 2017).

Neurodevelopmental outcomes

Antenatal lamotrigine exposure has not been associated with reduced IQ, or verbal, nonverbal, or spatial abilities, or increased need for educational intervention (Cummings et al., 2011; Meador et al., 2013; Rihtman et al., 2013).

Carbamazepine

Carbamazepine (CBZ) is converted hepatically to an active metabolite, which then undergoes glucuronidation, conjugation, and hydroxylation to inactive compounds. Greater CYP3A4 activity in women due to progesterone effects may accelerate metabolism and reduce plasma levels of CBZ (Deligiannidis, 2010). However, the data on CBZ metabolism in pregnancy are conflicting. Some studies report declining total concentrations of CBZ during the second and third trimester (Bernus et al., 1995; De Santis et al., 2011), but others have not found a significant change in CBZ plasma clearance (Tomson et al., 1994; Yerby et al., 1990). A rapid taper to the prepregnancy CBZ dose following delivery is advised (Tomson et al., 1994).

Teratogenicity

Antenatal CBZ exposure may increase the rate of congenital anomalies. A meta-analysis of 1255 exposures found that CBZ during pregnancy increased the rate of congenital anomalies, most commonly NTDs, cleft palate, as well as cardiovascular and urinary tract anomalies. CBZ was more teratogenic in combination with other drugs than when used as monotherapy (Matalon et al., 2002). Antenatal CBZ exposure has also been associated with a fetal CBZ syndrome manifested by facial dysmorphia and fingernail hypoplasia (Moore et al., 2000).

Adverse neonatal outcomes

CBZ has been associated with neonatal transient hepatic toxicity and hyperbilirubinemia (Frey et al., 1990). Antenatal CBZ exposure has been linked to reduced fetal head growth, restricted uterine growth, and LBW in some studies (Almgren et al., 2009; Bertollini et al., 1987). Other studies have not found these outcomes to differ significantly from controls (Wide et al., 2000a).

Neurodevelopmental outcomes

It is unclear whether fetal CBZ exposure increases the risk of developmental delay (Moore et al., 2000; Wide et al., 2000b). In one study of children exposed to CBZ in utero, there was no effect on overall IQ or nonverbal and spatial abilities. However, IQ sub-score was lower in verbal ability compared to controls. This effect was not dose related. There was no association between CBZ exposure and increased educational intervention (Baker et al., 2015). A meta-analysis found that CBZ in pregnancy was not associated with reduced full-scale IQ and verbal IQ but was associated with lower performance IQ (Banach et al., 2010). Other studies report similar findings (Gaily et al., 2004; Meador et al., 2013).

Other anticonvulsants

There are limited data on the relative safety of other anticonvulsants used as mood stabilizers, such as oxcarbazepine and topiramate. Since they are not first line agents in bipolar disorder, they should be avoided in pregnancy.

Use of antipsychotics in peripartum bipolar disorder

Antipsychotics undergo hepatic metabolism via the CYP450 system. CYP3A4 and CYP2D6 are induced in pregnancy. Quetiapine is primarily metabolized by CYP3A4, while aripiprazole is primarily metabolized by CYP2D6 into an active metabolite, which is metabolized by CYP3A4 (Deligiannidis, 2010). A reduction in serum quetiapine and aripiprazole levels of > 50% may occur in pregnancy (Newport et al., 2002). Perphenazine (CYP2D6), ziprasidone (CYP3A4), haloperidol and risperidone (CYP2D6 and CY3A4) serum levels could also decline in pregnancy, even with intramuscular administration. In contrast, olanzapine and clozapine are primarily metabolized via CYP1A2, which has decreased activity during the second and third trimesters. Little or no change in levels of these medications has been observed. Serum antipsychotic drug concentrations return to prepregnancy levels within the first 4 weeks postpartum (Westin et al., 2018).

A study of placental passage of antipsychotics found that olanzapine, haloperidol, risperidone and quetiapine demonstrate incomplete placental passage. Of these antipsychotics, olanzapine had the highest placental passage (mean 72.1%), followed by haloperidol (mean 65.5%), risperidone (mean 49.2%), then quetiapine (mean 23.8%) (Newport et al., 2007).

Teratogenicity

There are considerable data on the reproductive safety of first-generation antipsychotics (FGAs). No clear increased risk of congenital malformations with the use of FGAs in pregnancy has been identified (Einarson and Boskovic, 2009; Huybrechts et al., 2016). High-potency FGAs are preferred over low-potency, as some data suggest a possible increased risk of congenital malformations with prenatal exposure to low-potency antipsychotics (Rumeau-Rouquette et al., 1977). Data support the use of haloperidol in pregnancy, including a prospective study of 188 women exposed to haloperidol who showed no increased rate of congenital anomalies compared to the control group (Diav-Citrin et al., 2005). Large studies have not detected a difference in the rate of malformations in infants exposed to FGAs vs SGAs (Coughlin et al., 2015; Reis and Källén, 2008).

Despite the widespread availability and use of SGAs, reliable reproductive safety data for these agents are sparse. Safety data are derived from case studies, manufacturer reports, and more recently a select number of larger cohort studies (Cohen et al., 2016). While studies are lacking, the available data largely support the use of SGAs and suggests that the teratogenic risk is not greatly raised above baseline (Cohen et al., 2016; Einarson and Boskovic, 2009; Petersen et al., 2016). More reproductive safety data are needed, especially for newer drugs like lurasidone, iloperidone, and brexpiprazole (Sharma et al., 2020). Risperidone may be a possible exception and may confer a small increased risk of congenital malformations (Ennis and Damkier, 2015; Huybrechts et al., 2016).

The National Pregnancy Registry for Atypical Antipsychotics, in an ongoing prospective cohort study, has not found a difference in rate of congenital malformations in infants exposed to SGAs compared to infants of women with psychiatric illness treated with other psychotropic medication (Cohen et al., 2016). However, some studies have suggested an increased risk of congenital malformations with SGA exposure. Among 130 infants exposed to FGAs and SGAs, there was a 6% risk of congenital malformations, particularly cardiac, compared to the 3.1% baseline rate in Australia (Kulkarni et al., 2014). A large prospective cohort study found that the odds of a major malformation, particularly cardiac, in women exposed to SGAs was 5.1%, nearly a twofold increase compared to unexposed controls (Habermann et al., 2013). However, given the varied methodology among the studies and the potential for detection bias when reporting cardiac anomalies, these results should be interpreted cautiously. Additionally, these studies used healthy comparison groups, introducing the significant confounding variable of psychiatric illness. These studies do not distinguish between agents, doses, or diagnosis of the pregnant women using the drugs.

Adverse obstetrical and neonatal outcomes

Both FGAs and SGAs in pregnancy have been associated with LBW (Coughlin et al., 2015; Reis and Källén, 2008). A study in Sweden found an excess of PTB and LBW in 570 infants exposed to SGAs and FGAs in the first trimester, but no certain effect on intrauterine growth (Reis and Källén, 2008). Underlying pathology or unidentified confounders may explain the excess risk. The authors concluded that in a risk–benefit analysis, these findings should not prevent necessary treatment to women (Reis and Källén, 2008).

A large-scale systematic review and meta-analysis found a significant association between antenatal antipsychotic exposure and increased risk of multiple adverse obstetric and neonatal outcomes, including neonates small for gestational age (OR 2.44) and preterm delivery (OR 1.86) (Coughlin et al., 2015). However, most studies included had limited adjustment for potential confounding variables, and no study adjusted for type or severity of underlying mental illness. The authors concluded that women who take antenatal antipsychotics are at higher risk for adverse obstetric and neonatal outcomes, but they could not conclude that antipsychotics cause this increased morbidity in pregnancy (Coughlin et al., 2015). Antenatal use of SGAs has also been associated with delivering large for gestational age infants (Newham et al., 2008). Pregnant women on SGAs should be closely monitored and screened for gestational diabetes mellitus (Raffi et al., 2019). It is not clear what other potential neonatal complications of SGAs may be if any. FGAs have been associated with transient extrapyramidal side effects in the neonate (Galbally et al., 2010).

Neurodevelopmental outcomes

There are no studies that examine neurodevelopmental effects of antipsychotic use in pregnant women with mood disorders. Available data come from observational studies of women with schizophrenia. One case-controlled prospective study suggests that FGAs may cause short-term developmental delays in cognitive, motor, social-emotional, and adaptive behavior, but not in language, bodyweight, or height. Differences were identified in offspring at 2 and 6 months of age, but not at 12 months. The authors note that the women in the exposed group were independently at greater risk for adverse outcomes (Peng et al., 2013). Case reports are consistent with the above study (Duran et al., 2008; Gupta and Grover, 2004; Lin et al., 2010).

Postpartum psychosis

Postpartum psychosis (PPP), also termed puerperal psychosis, is a psychiatric emergency that occurs in 1–2/1000 child-bearing women, usually within the first 2 weeks postpartum (Bergink et al., 2015). PPP is generally considered to be a bipolar spectrum illness, as clinical presentation, illness course, and family history tend to align with bipolar disorder (Chaudron and Pies, 2003; Sit et al., 2006). Initial symptoms are commonly insomnia, mood fluctuations, and obsessive concerns regarding the newborn. These may be followed by more severe symptoms including delusions, hallucinations, and disorganized behavior. Primary mood disturbance may be mania, depression, or a mixed state (Bergink et al., 2015). Untreated mothers with PPP have an estimated 4% rate of infanticide (D'Orban, 1979).

Lithium is effective in the prevention of PPP (Austin, 1992; Bergink et al., 2012). Based on a study of women with first-onset psychosis or mania in the postpartum period, Berjink et al. proposed a four-step treatment algorithm resulting in high remission rates in patients with first-episode PPP. The treatment involves the subsequent addition of benzodiazepines, antipsychotics, and lithium at set time points in the treatment course. Women who do not respond to pharmacotherapy proceed to ECT. The study included women with bipolar mania with psychotic features, bipolar mixed episodes, unipolar depression with psychotic features, and psychosis without prominent affective symptoms. Using this sequential treatment, 98.4% of patients achieved complete remission within the first three steps, and 79.9% of patients were in sustained remission at 9 months postpartum. Patients treated with lithium had a significantly lower rate of relapse compared with those treated with antipsychotic monotherapy (Bergink et al., 2015).

Pharmacotherapy use during lactation

Breastfeeding poses a challenge in the treatment of women with bipolar disorder. All psychotropic medications studied to date can enter human milk (Hostetter et al., 2000; Newport et al., 2002). When considering relative safety in lactation, the degree of exposure and the implications for infant health outcomes must be considered. The severity of the maternal disorder and functional impairment must be weighed against the therapeutic and side effects of medications, and the available scientific data regarding relative safety of psychotropic drugs in breastfeeding. Obtaining sufficient uninterrupted sleep to reduce a woman's risk of mania or depression should be factored into her decision regarding breastfeeding. Supplementation with formula or donor breastmilk is advisable in many cases.

Lithium

Breastfeeding with lithium use is controversial as safety data are limited (Newmark et al., 2019). Data suggest that lithium exposure through breast milk may confer less risk than previously considered (Moretti et al., 2003; Viguera et al., 2007a). In most cases, lithium transferred through breast milk is much less than the complete placental transfer of lithium (Newport et al., 2005). The concentration of lithium in the serum of breastfed infants ranges from 30% to 40% of maternal levels, with a RID range of 12%–30%, although findings vary greatly between studies (Galbally et al., 2010; Uguz and Sharma, 2016). Premature or LBW infants are more vulnerable to dehydration and therefore lithium toxicity. In a study of 10 mother–infant dyads, maternal serum, breast milk, and infant serum daily through concentrations of lithium averaged 0.76, 0.35, and 0.16 mEq/L, respectively, each lithium level approximately one-half the preceding level. No serious adverse events were observed. Four infants had minor and transient elevations of thyroid-stimulating hormone, blood urea nitrogen, and creatinine, which were not always associated with high infant serum lithium levels (Viguera et al., 2007a). Newmark et al. conducted a systematic review of lithium in breastfeeding, where adverse infant effects were reported in 9.4%. Abnormal lab values were reported in an additional 9.4% of the infants. Complications included cyanosis, lethargy, electrolyte abnormalities, and gross irregular twitching. Among the adverse effects reported, it is difficult to differentiate poor outcomes from other factors affecting infant health, concomitant medications, and gestational lithium exposure (Newmark et al., 2019). The longer-term significance of lithium exposure and its associated abnormalities is unclear.

The AAP and LactMed, an NIH supported database, state that lithium can be continued during lactation with careful monitoring of the infant, including careful monitoring of the maternal serum concentrations and possibly of infant serum concentrations (Drugs and Lactation Database (LactMed), 2006a, b; Ward et al., 2001). However, others consider lithium exposure through breast milk as possibly hazardous given the relatively high exposure and the theoretical risks, and recommend formula feeding (Galbally et al., 2018). Guidance on whether to breastfeed while taking lithium must be personalized to the mother–infant dyad.

Valproic acid

The infant/maternal ratio of serum drug concentration is lower in VPA compared to other mood stabilizers. The RID of VPA is low (0.68%) and is compatible with breastfeeding. Recent studies show serum levels in infants exposed through breastmilk in the range of 0.7%–6.6 μg/mL, corresponding to a mean of 2.1% of the maternal serum level. One adverse event, anemia and thrombocytopenia purpura, was reported in one of nine infants in recent studies (Uguz and Sharma, 2016).

Carbamazepine

The RID of CBZ (4.3%) is low and compatible with breastfeeding. The ranges for milk/plasma ratio, infant serum levels of CBZ, and infant/maternal ratio of serum CBZ concentrations were 0.0–1.4, 0.0–2.6, and 0.0–0.7, respectively (Uguz and Sharma, 2016). Due to confounders, it is unclear whether the cases of adverse events were due to CBZ exposure through breastmilk or another cause.

Oxcarbazepine

The literature is limited to two case reports on use of oxcarbazepine during lactation. The RID was 1.5%–1.7%, and infant/maternal ratio of serum oxcarbazepine concentration was <0.5–1.0. Oxcarbazepine use during lactation was not associated with adverse events in these studies (Gentile, 2003; Lutz et al., 2007).

Lamotrigine

Considerable amounts of lamotrigine are transmitted to the infant through breastmilk (Fotopoulou et al., 2009; Newport et al., 2008a). Lamotrigine RIDs range from 2.0% to 21.1%. However, lamotrigine use during lactation has rarely been associated with adverse effects and is thus considered low risk (Newport et al., 2008b; Uguz and Sharma, 2016). In a systematic review of 122 women taking lamotrigine during lactation, adverse events were reported in only 3.27% of infants. Of note, most women in the sample had epilepsy, and fewer women had bipolar disorder (Uguz and Sharma, 2016). Maternal dose has been shown to be a significant predictor of lamotrigine concentration in breastmilk, whereas time after maternal dose has not (Newport et al., 2008a).

Atypical antipsychotics

Data suggest that levels of SGAs transmitted in breast milk are low, with RIDs generally <3% of maternal dose (Galbally et al., 2010). Olanzapine, risperidone, and quetiapine have the most data available to inform safety in lactation with RIDs <10%, reflecting acceptable infant exposure (Uguz and Sharma, 2016). Data are limited for other SGAs such as aripiprazole and lurasidone. Adverse events are mild and infrequent (Clark and Wisner, 2018). Breastfeeding can be considered with concurrent infant monitoring. Lactation safety data for FGAs are minimal and derived mainly from case reports and case series. Data for haloperidol and chlorpromazine are largely reassuring, with one case report of drowsiness and lethargy with high-dose chlorpromazine found in breast milk (Gentile, 2015).

Clinical summary for the treatment of peripartum bipolar depression

The peripartum period is a time of increased risk for both first-onset and recurrent mood episodes in women with bipolar disorder. Untreated bipolar illness presents known risks to the fetus, so maintaining euthymia is paramount. Pharmacotherapy plays a key role in the treatment of acute antenatal symptoms, as well as in the prevention of relapse. Safety data on the use of psychotropics in pregnancy and lactation in women with bipolar disorder are sparse. However, the available data, derived from observational studies, are largely positive for lamotrigine and antipsychotics, and the teratogenicity associated with lithium in more recent studies is less than older data suggested. Medication choice should consider the available safety data as well as the mother's treatment history and desire to breastfeed. Given the limited data available, the known benefits of breastfeeding must be weighed against potential unknown effects of medication, and treatment recommendations may vary based on the unique needs of the mother–infant dyad.

Choosing an agent that is protective for both depression and mania is preferable in order to avoid exposure to polypharmacy. Ideally, the lowest effective dose should be determined during preconception. Pregnancy associated metabolic changes must also be considered, with careful TDM and dose adjustments when warranted. ECT also plays an important role the treatment of peripartum bipolar disorder. It should be used in treatment-resistant bipolar depression and considered for depression with psychotic features, catatonia, and cases which warrant rapid resolution of symptoms. It is also an effective treatment for PPP.

Depression during the transition to menopause

Introduction

The menopausal transition occurs at the end of the female reproductive life cycle: in the perimenopausal period, menstrual irregularity occurs as variability in primarily follicle stimulating hormone (FSH) increases, leading to concurrent variability in estradiol levels (Harlow et al., 2012). The postmenopausal period begins after 1 year of amenorrhea, which retroactively identifies the final menstrual period (FMP) (Harlow et al., 2012). FSH and estradiol continue to fluctuate for some time through the postmenopausal period; per the Stages of Reproductive Aging Workshop (STRAW)+10, an update to the landmark study seeking to establish scientific consensus on the menopausal transition, it may require up to 8 years after the FMP before hormones stabilize (Harlow et al., 2012; Soules et al., 2001). Despite a period of scientific debate on the question, longitudinal studies have established that the perimenopausal period is one of increased vulnerability to clinical depression in women (Bromberger et al., 2011; Cohen et al., 2006b; Freeman et al., 2004). Women without a prior episode of depression are also at increased risk (Cohen et al., 2006b; Freeman et al., 2006a), although prior episodes are both a risk factor for reoccurrence and higher symptom burden (Willi et al., 2020). Women who have demonstrated sensitivity to hormone fluctuations in the past through prior diagnoses of peripartum depression or premenstrual dysphoric disorder (PMDD) are also at increased risk of perimenopausal depression (Freeman et al., 2004, 2006a). This lends credence to the current theories regarding the pathophysiology of perimenopausal mood changes as relating to sensitivity in fluctuations of neuroactive hormones rather than their absolute values in the steady state (Gordon et al., 2015, 2019, 2020). Adverse effects on mood are only one of several symptom clusters observed in the perimenopausal period; others include vasomotor symptoms, cognitive symptoms, sleep disturbance, pain symptoms, tension, urinary incontinence, and changes in libido (Cray et al., 2012; Woods and Mitchell, 2016). Comprehensive treatment should target multiple symptom clusters whenever possible.

Treatment approaches to depression during the menopausal transition

Pharmacotherapy

Perimenopausal depression is commonly treated according to standard clinical practice of treating MDD: SSRIs, SNRIs, TCAs, and MAOIs are all used in clinical practice (Santoro et al., 2015). However, research on the specific efficacy of these agents in the perimenopausal period has been limited. Open-label trials have found favorable outcomes for mirtazapine, escitalopram, citalopram, duloxetine, venlafaxine, and vortioxetine, but there are methodological limitations (Freeman et al., 2006b, 2013; Joffe et al., 2001, 2007; Soares et al., 2003).

The most robust evidence available is for desvenlafaxine, and it has been shown to be significantly effective vs placebo on depressive symptoms during the peri- and postmenopausal periods per evidence from double-blind, randomized studies as well as large pooled analyses examining doses ranging from 50 mg per day to 200 mg per day (Kornstein et al., 2010, 2014a, 2015; Soares et al., 2010). However, a randomized, placebo-controlled comparative study failed to demonstrate superiority of desvenlafaxine over escitalopram (Soares et al., 2010). Similarly, venlafaxine extended release did not vary in efficacy from fluoxetine with respect to menopausal status or sex (Kornstein et al., 2014b), despite some earlier evidence suggesting diminished efficacy of SSRIs in older and postmenopausal women (Pinto-Meza et al., 2006; Thase et al., 2005).

A similar clinical controversy concerns demonstration of superior efficacy of SSRIs vs TCAs in premenopausal women (Kornstein et al., 2000; Vermeiden et al., 2010), findings that bring into focus broader questions about sex, hormones, and efficacy of antidepressants that warrant further study before clinical recommendations can be made.

With regards to intravenous ketamine for TRD, data available to date are analyses from one randomized, placebo-controlled study, which demonstrated no change in efficacy or tolerability related to sex or menopausal status (Freeman et al., 2019).

Concurrent treatment of vasomotor symptoms

Vasomotor symptoms, including acute onset of fever-like sensation of heat commonly referred to as "hot flashes," are experienced by most women in the perimenopausal period (Anon, 2014), and for some women, these symptoms are distressing enough to significantly interfere with quality of life (Avis et al., 2009) as well as exacerbate depressive symptoms (Natari et al., 2018; Stuenkel, 2018; Zeleke et al., 2017). Although there are many hormonal treatments for vasomotor symptoms, research has shown that the antidepressants venlafaxine, desvenlafaxine, escitalopram, citalopram, fluoxetine, and sertraline are also effective in managing these symptoms, although none are FDA approved for that indication (Guthrie et al., 2015; Handley and Williams, 2015). The only antidepressant currently FDA approved for treatment of vasomotor symptoms is paroxetine 7.5 mg per day, marketed as Brisdelle (Slaton et al., 2015).

Gabapentin, pregabalin, and clonidine have also been studied as potential treatments for vasomotor symptoms of perimenopause. A meta-analysis of four studies of clonidine vs placebo found decreases in frequency of vasomotor symptoms (Nelson et al., 2006). Gabapentin was efficacious compared to placebo in reducing severity and frequency of hot flashes according to a recent systematic review and meta-analysis; however, the same study did not demonstrate conclusive evidence for use of pregabalin largely due to methodological limitations (Shan et al., 2020). Neither treatment carries an FDA approval for the indication.

Concurrent treatment of concentration difficulties

In addition to the appropriate identification and treatment of underlying depression in perimenopause, there may be a role for use of psychostimulants in the treatment of concentration and cognition difficulties in perimenopausal women. One double-blind, placebo-controlled crossover trial of atomoxetine for nondepressed perimenopausal women with cognitive symptoms but without a history of ADHD found significant improvement compared to placebo in the total score of the Brown Attention Deficit Disorder Scale as well as multiple subscales (Neill Epperson et al., 2011).

Concurrent treatment of insomnia

The etiology of insomnia in depressed perimenopausal women is multidimensional, and a thorough review of pharmacotherapy for sleep is beyond the current scope of this chapter. However, sleep difficulties in perimenopausal women are sometimes related to vasomotor symptoms (Stuenkel, 2018), and treating those symptoms may lead to symptomatic improvement in insomnia.

Hormone replacement therapy

The evidence base for use of hormone replacement therapy (HRT) for perimenopausal depression is fraught with scientific controversy and complexity. Low-dose testosterone augmentation has been hypothesized to be a potential treatment for TRD in women, but a placebo-controlled study including both premenopausal and postmenopausal subjects failed to demonstrate efficacy (Dichtel et al., 2020). The hypothesis that prophylactic HRT could be used to prevent the onset of depression throughout menopause has largely not been supported by the evidence (Rubinow et al., 2015), however, one recent randomized, placebo-controlled trial of estrogen patch and oral progesterone found that this combination was significantly more effective than placebo in preventing development of clinically significant depressive symptoms during perimenopause in previously euthymic women (Gordon et al., 2018). The same study aligned with prior evidence that depression in the postmenopausal period has not been shown to be responsive to HRT (Gordon et al., 2018; Rubinow et al., 2015). While there is some contrary evidence that depression in the perimenopausal period may be successfully treated with HRT, there are significant risks associated with the initiation of HRT in older women, notably the increased risk of venous thromboembolism (Lee et al., 2020). Any trial of estrogen therapy for perimenopausal depression that has failed treatment with antidepressants should occur within a collaborative effort between a gynecologist and treating psychiatrist.

Clinical summary for the treatment of depression during the transition to menopause

The depression risk is increased in the menopausal transition, and its treatment is complicated by exacerbating factors such as vasomotor symptoms, concentration difficulties, and insomnia. Whenever clinically relevant, therapies that address multiple symptom domains should be utilized, such as the antidepressants with demonstrated efficacy for depression as well as vasomotor symptoms: desvenlafaxine, venlafaxine, and escitalopram. Psychostimulants should be considered as they may be effective even for women without a previous diagnosis of ADHD. Refractory cases warrant referral to an obstetrician-gynecologist for the initiation of collaborative care due to the controversial and complex evidence surrounding the use of estrogen replacement therapy in perimenopausal TRD. Finally, although a search yielded no studies of ECT within the specific population of women with perimenopausal depression, ECT has been demonstrated to be a safe and effective treatment option for TRD that should be considered in cases that fail to respond to pharmacotherapy (Geddes et al., 2003).

References

Almgren, M., Källén, B., Lavebratt, C., 2009. Population-based study of antiepileptic drug exposure in utero-influence on head circumference in newborns. Seizure. https://doi.org/10.1016/j.seizure.2009.09.002.

Althaus, A.L., Ackley, M.A., Belfort, G.M., Gee, S.M., Dai, J., Nguyen, D.P., Kazdoba, T.M., Modgil, A., Davies, P.A., Moss, S.J., Salituro, F.G., Hoffmann, E., Hammond, R.S., Robichaud, A.J., Quirk, M.C., Doherty, J.J., 2020. Preclinical characterization of zuranolone (SAGE-217), a selective neuroactive steroid GABAA receptor positive allosteric modulator. Neuropharmacology. https://doi.org/10.1016/j.neuropharm.2020.108333.

Alwan, S., Reefhuis, J., Rasmussen, S.A., Olney, R.S., Friedman, J.M., 2007. Use of selective serotonin-reuptake inhibitors in pregnancy and the risk of birth defects. N. Engl. J. Med. https://doi.org/10.1056/NEJMoa066584.

Alwan, S., Reefhuis, J., Botto, L.D., Rasmussen, S.A., Correa, A., Friedman, J.M., 2010. Maternal use of bupropion and risk for congenital heart defects. Am. J. Obstet. Gynecol. https://doi.org/10.1016/j.ajog.2010.02.015.

American Psychiatric Association (APA), 2013. Diagnostic and statistical manual of mental disorders (5th ed.). Am. J. Psychiatry. https://doi.org/10.1176/appi.books.9780890425596.744053.

Andersen, J.T., Andersen, N.L., Horwitz, H., Poulsen, H.E., Jimenez-Solem, E., 2014. Exposure to selective serotonin reuptake inhibitors in early pregnancy and the risk of miscarriage. Obstet. Gynecol. https://doi.org/10.1097/AOG.0000000000000447.

Anderson, E.L., Reti, I.M., 2009. ECT in pregnancy: a review of the literature from 1941 to 2007. Psychosom. Med. https://doi.org/10.1097/PSY.0b013e318190d7ca.

Anon, 2014. ACOG Practice Bulletin No. 141: management of menopausal symptoms. Obstet. Gynecol. 123 (1), 202–216. https://doi.org/10.1097/01.AOG.0000441353.20693.78.

Appleby, L., Warner, R., Whitton, A., Faragher, B., 1997. A controlled study of fluoxetine and cognitive-behavioural counselling in the treatment of postnatal depression. Br. Med. J. https://doi.org/10.1136/bmj.314.7085.932.

Ardinger, H.H., Atkin, J.F., Blackston, R.D., Elsas, L.J., Clarren, S.K., Livingstone, S., Flannery, D.B., Pellock, J.M., Harrod, M.J., Lammer, E.J., Majewski, F., Schinzel, A., Toriello, H.V., Hanson, J.W., 1988. Verification of the fetal valproate syndrome phenotype. Am. J. Med. Genet. https://doi.org/10.1002/ajmg.1320290123.

Austin, M.P.V., 1992. Puerperal affective psychosis: is there a case for lithium prophylaxis? Br. J. Psychiatry. https://doi.org/10.1192/bjp.161.5.692.

Avis, N.E., Colvin, A., Bromberger, J.T., Hess, R., Matthews, K.A., Ory, M., Schocken, M., 2009. Change in health-related quality of life over the menopausal transition in a multiethnic cohort of middle-aged women: study of Women's Health Across the Nation. Menopause 16 (5), 860–869. https://doi.org/10.1097/gme.0b013e3181a3cdaf.

Babu, G.N., Thippeswamy, H., Chandra, P.S., 2013. Use of electroconvulsive therapy (ECT) in postpartum psychosis—a naturalistic prospective study. Arch. Womens Ment. Health. https://doi.org/10.1007/s00737-013-0342-2.

Baker, G.A., Bromley, R.L., Briggs, M., Cheyne, C.P., Cohen, M.J., García-Fiñana, M., Gummery, A., Kneen, R., Loring, D.W., Mawer, G., Meador, K.J., Shallcross, R., Clayton-Smith, J., 2015. IQ at 6 years after in utero exposure to antiepileptic drugs: a controlled cohort study. Neurology. https://doi.org/10.1212/WNL.0000000000001182.

Ban, L., Gibson, J.E., West, J., Fiaschi, L., Sokal, R., Smeeth, L., Doyle, P., Hubbard, R.B., Tata, L.J., 2014. Maternal depression, antidepressant prescriptions, and congenital anomaly risk in offspring: a population-based cohort study. BJOG. https://doi.org/10.1111/1471-0528.12682.

Banach, R., Boskovic, R., Einarson, T., Koren, G., 2010. Long-term developmental outcome of Children of Women with epilepsy, unexposed or exposed prenatally to antiepileptic drugs: a meta-analysis of cohort studies. Drug Saf. https://doi.org/10.2165/11317640-000000000-00000.

Bansil, P., Kuklina, E.V., Meikle, S.F., Posner, S.F., Kourtis, A.P., Ellington, S.R., Jamieson, D.J., 2010. Maternal and fetal outcomes among women with depression. J. Womens Health 2002. https://doi.org/10.1089/jwh.2009.1387.

Bauman, B.L., Ko, J.Y., Cox, S., D'Angelo, D.V., Warner, L., Folger, S., Tevendale, H.D., Coy, K.C., Harrison, L., Barfield, W.D., 2020. Vital signs: postpartum depressive symptoms and provider discussions about perinatal depression—United States, 2018. MMWR Morb. Mortal. Wkly Rep. https://doi.org/10.15585/mmwr.mm6919a2.

Bergink, V., Lambregtse-van Den Berg, M.P., Koorengevel, K.M., Kupka, R., Kushner, S.A., 2011. First-onset psychosis occurring in the postpartum period: a prospective cohort study. J. Clin. Psychiatry. https://doi.org/10.4088/JCP.10m06648.

Bergink, V., Bouvy, P.F., Vervoort, J.S.P., Koorengevel, K.M., Steegers, E.A.P., Kushner, S.A., 2012. Prevention of postpartum psychosis and mania in women at high risk. Am. J. Psychiatry. https://doi.org/10.1176/appi.ajp.2012.11071047.

Bergink, V., Burgerhout, K.M., Koorengevel, K.M., Kamperman, A.M., Hoogendijk, W.J., Lambregtse-Van Den Berg, M.P., Kushner, S.A., 2015. Treatment of psychosis and mania in the postpartum period. Am. J. Psychiatry. https://doi.org/10.1176/appi.ajp.2014.13121652.

Bernus, I., Hooper, W.D., Dickinson, R.G., Eadie, M.J., 1995. Metabolism of carbamazepine and co-administered anticonvulsants during pregnancy. Epilepsy Res. https://doi.org/10.1016/0920-1211(95)00012-Y.

Bertollini, R., Källen, B., Mastroiacovo, P., Robert, E., 1987. Anticonvulsant drugs in monotherapy. Effect on the fetus. Eur. J. Epidemiol. https://doi.org/10.1007/BF00239754.

Biaggi, A., Conroy, S., Pawlby, S., Pariante, C.M., 2016. Identifying the women at risk of antenatal anxiety and depression: a systematic review. J. Affect. Disord. https://doi.org/10.1016/j.jad.2015.11.014.

Bialer, M., Johannessen, S.I., Levy, R.H., Perucca, E., Tomson, T., White, H.S., 2017. Progress report on new antiepileptic drugs: a summary of the Thirteenth Eilat Conference on New Antiepileptic Drugs and Devices (EILAT XIII). Epilepsia. https://doi.org/10.1111/epi.13634.

Blake, L.D., Lucas, D.N., Aziz, K., Castello-Cortes, A., Robinson, P.N., 2008. Lithium toxicity and the parturient: case report and literature review. Int. J. Obstet. Anesth. https://doi.org/10.1016/j.ijoa.2007.09.014.

Bloch, M., Meiboom, H., Lorberblatt, M., Bluvstein, I., Aharonov, I., Schreiber, S., 2012. The effect of sertraline add-on to brief dynamic psychotherapy for the treatment of postpartum depression: a randomized, double-blind, placebo-controlled study. J. Clin. Psychiatry. https://doi.org/10.4088/JCP.11m07117.

Bodén, R., Lundgren, M., Brandt, L., Reutfors, J., Andersen, M., Kieler, H., 2012. Risks of adverse pregnancy and birth outcomes in women treated or not treated with mood stabilisers for bipolar disorder: population based cohort study. BMJ. https://doi.org/10.1136/bmj.e7085.

Bromberger, J.T., Kravitz, H.M., Chang, Y.F., Cyranowski, J.M., Brown, C., Matthews, K.A., 2011. Major depression during and after the menopausal transition: study of Women's Health Across the Nation (SWAN). Psychol. Med. 41 (9), 1879–1888. https://doi.org/10.1017/S003329171100016X.

Byatt, N., Carter, D., Deligiannidis, K.M., Neill Epperson, C., Meltzer-Brody, S., Payne, J.L., Robinson, G., Silver, N.E., Stowe, Z., Sayres Van Niel, M., Wisner, K.L., Yonkers, K.N., 2018. Position Statement on Screening and Treatment of Mood and Anxiety Disorders During Pregnancy and Postpartum. https://www.google.com/search?q=APA+recommends+screening+pregnant+and+postpartum+women&rlz=1C1VFKB_enUS895US895&oq=APA+recommends+screening+pregnant+and+postpartum+women&aqs=chrome..69i57.9089j0j7&sourceid=chrome&ie=UTF-8.

Calaway, K., Coshal, S., Jones, K., Coverdale, J., Livingston, R., 2016. A systematic review of the safety of electroconvulsive therapy use during the first trimester of pregnancy. J. ECT. https://doi.org/10.1097/YCT.0000000000000330.

Campbell, E., Kennedy, F., Russell, A., Smithson, W.H., Parsons, L., Morrison, P.J., Liggan, B., Irwin, B., Delanty, N., Hunt, S.J., Craig, J., Morrow, J., 2014. Malformation risks of antiepileptic drug monotherapies in pregnancy: updated results from the UK and Ireland epilepsy and pregnancy registers. J. Neurol. Neurosurg. Psychiatry. https://doi.org/10.1136/jnnp-2013-306318.

Canger, R., Battino, D., Canevini, M.P., Fumarola, C., Guidolin, L., Vignoli, A., Mamoli, D., Palmieri, C., Molteni, F., Granata, T., Hassibi, P., Zamperini, P., Pardi, G., Avanzini, G., 1999. Malformations in offspring of women with epilepsy: a prospective study. Epilepsia. https://doi.org/10.1111/j.1528-1157.1999.tb00851.x.

CDC, 2008. Update on overall prevalence of major birth defects—Atlanta, Georgia, 1978–2005. MMWR Morb. Mortal. Wkly Rep. 57 (1), 1–5. Retrieved November 3, 2020 https://www.cdc.gov/mmwr/preview/mmwrhtml/mm5701a2.htm.

Chad, L., Pupco, A., Bozzo, P., Koren, G., 2013. Update on antidepressant use during breastfeeding. Can. Fam. Physician 59 (6), 633–634.

Chambers, C.D., Hernandez-Diaz, S., Van Marter, L.J., Werler, M.M., Louik, C., Jones, K.L., Mitchell, A.A., 2006. Selective serotonin-reuptake inhibitors and risk of persistent pulmonary hypertension of the newborn. N. Engl. J. Med. https://doi.org/10.1056/NEJMoa052744.

Chaudron, L.H., Pies, R.W., 2003. The relationship between postpartum psychosis and bipolar disorder: a review. J. Clin. Psychiatry. https://doi.org/10.4088/JCP.v64n1102.

Claridge, A.M., 2014. Efficacy of systemically oriented psychotherapies in the treatment of perinatal depression: a meta-analysis. Arch. Womens Ment. Health. https://doi.org/10.1007/s00737-013-0391-6.

Clark, C.T., Wisner, K.L., 2018. Treatment of peripartum bipolar disorder. Obstet. Gynecol. Clin. North Am. https://doi.org/10.1016/j.ogc.2018.05.002.

Clark, C.T., Klein, A.M., Perel, J.M., Helsel, J., Wisner, K.L., 2013. Lamotrigine dosing for pregnant patients with bipolar disorder. Am. J. Psychiatry. https://doi.org/10.1176/appi.ajp.2013.13010006.

Cohen, L.S., Viguera, A.C., Bouffard, S.M., Nonacs, R.M., Morabito, C., Collins, M.H., Ablon, J.S., 2001. Venlafaxine in the treatment of postpartum depression. J. Clin. Psychiatry. https://doi.org/10.4088/JCP.v62n0803.

Cohen, L.S., Altshuler, L.L., Harlow, B.L., Nonacs, R., Newport, D.J., Viguera, A.C., Suri, R., Burt, V.K., Hendrick, V., Reminick, A.M., Loughead, A., Vitonis, A.F., Stowe, Z.N., 2006a. Relapse of major depression during pregnancy in women who maintain or discontinue antidepressant treatment. JAMA. https://doi.org/10.1001/jama.295.5.499.

Cohen, L.S., Soares, C.N., Vitonis, A.F., Otto, M.W., Harlow, B.L., 2006b. Risk for new onset of depression during the menopausal transition: the Harvard study of moods and cycles. Arch. Gen. Psychiatry 63 (4), 385–390. https://doi.org/10.1001/archpsyc.63.4.385.

Cohen, L.S., Viguera, A.C., McInerney, K.A., Freeman, M.P., Sosinsky, A.Z., Moustafa, D., Marfurt, S.P., Kwiatkowski, M.A., Murphy, S.K., Farrell, A.M., Chitayat, D., Hernández-Díaz, S., 2016. Reproductive safety of second-generation antipsychotics: current data from the Massachusetts general hospital national pregnancy registry for atypical antipsychotics. Am. J. Psychiatry. https://doi.org/10.1176/appi.ajp.2015.15040506.

Cohen, J.M., Huybrechts, K.F., Patorno, E., Desai, R.J., Mogun, H., Bateman, B.T., Hernández-Díaz, S., 2019. Anticonvulsant mood stabilizer and lithium use and risk of adverse pregnancy outcomes. J. Clin. Psychiatry. https://doi.org/10.4088/JCP.18m12572.

Cole, J.A., Modell, J.G., Haight, B.R., Cosmatos, I.S., Stoler, J.M., Walker, A.M., 2007. Bupropion in pregnancy and the prevalence of congenital malformations. Pharmacoepidemiol. Drug Saf. https://doi.org/10.1002/pds.1296.

Coughlin, C.G., Blackwell, K.A., Bartley, C., Hay, M., Yonkers, K.A., Bloch, M.H., 2015. Obstetric and neonatal outcomes after antipsychotic medication exposure in pregnancy. Obstet. Gynecol. https://doi.org/10.1097/AOG.0000000000000759.

Cox, E.Q., Sowa, N.A., Meltzer-Brody, S.E., Gaynes, B.N., 2016. The perinatal depression treatment cascade: baby steps toward improving outcomes. J. Clin. Psychiatry. https://doi.org/10.4088/JCP.15r10174.

Cray, L.A., Woods, N.F., Herting, J.R., Mitchell, E.S., 2012. Symptom clusters during the late reproductive stage through the early postmenopause: observations from the Seattle Midlife Women's Health Study. Menopause 19 (8), 864–869. https://doi.org/10.1097/gme.0b013e31824790a6.

Cummings, C., Stewart, M., Stevenson, M., Morrow, J., Nelson, J., 2011. Neurodevelopment of children exposed in utero to lamotrigine, sodium valproate and carbamazepine. Arch. Dis. Child. https://doi.org/10.1136/adc.2009.176990.

Cunnington, M.C., Weil, J.G., Messenheimer, J.A., Ferber, S., Yerby, M., Tennis, P., 2011. Final results from 18 years of the International Lamotrigine Pregnancy Registry. Neurology. https://doi.org/10.1212/WNL.0b013e31821ccd18.

D'Orban, P.T., 1979. Women who kill their children. Br. J. Psychiatry. https://doi.org/10.1192/bjp.134.6.560.

De Crescenzo, F., Perelli, F., Armando, M., Vicari, S., 2014. Selective serotonin reuptake inhibitors (SSRIs) for post-partum depression (PPD): a systematic review of randomized clinical trials. J. Affect. Disord. https://doi.org/10.1016/j.jad.2013.09.019.

De Santis, M., De Luca, C., Mappa, I., Cesari, E., Quattrocchi, T., Visconti, D., Spagnuolo, T., Caruso, A., 2011. Antiepileptic drugs during pregnancy: pharmacokinetics and transplacental transfer. Curr. Pharm. Biotechnol. https://doi.org/10.2174/138921121342118201.0.

Deligiannidis, K.M., 2010. Therapeutic drug monitoring in pregnant and postpartum women: recommendations for SSRIs, lamotrigine, and lithium. J. Clin. Psychiatry 71 (5), 649–650. https://doi.org/10.4088/JCP.10ac06132gre.

Deligiannidis, K.M., Byatt, N., Freeman, M.P., 2014. Pharmacotherapy for mood disorders in pregnancy. J. Clin. Psychopharmacol. https://doi.org/10.1097/jcp.0000000000000087.

Deligiannidis, K.M., Meltzer-Brody, S., Gunduz-Bruce, H., Doherty, J., Jonas, J., Li, S., Sankoh, A.J., Silber, C., Campbell, A.D., Werneberg, B., Kanes, S.J., Lasser, R., 2021. Effect of zuranolone vs. placebo in postpartum depression: a randomized clinical trial. JAMA Psychiatry 78 (9), 951–959. https://doi.org/10.1001/jamapsychiatry.2021.1559.

Dennis, C.L., McQueen, K., 2009. The relationship between infant-feeding outcomes and postpartum depression: a qualitative systematic review. Pediatrics. https://doi.org/10.1542/peds.2008-1629.

Dennis, C.L., Ross, L.E., Herxheimer, A., 2008. Oestrogens and progestins for preventing and treating postpartum depression. Cochrane Database Syst. Rev. https://doi.org/10.1002/14651858.CD001690.pub2.

Di Florio, A., Forty, L., Gordon-Smith, K., Heron, J., Jones, L., Craddock, N., Jones, I., 2013. Perinatal episodes across the mood disorder spectrum. JAMA Psychiat. https://doi.org/10.1001/jamapsychiatry.2013.279.

Di Florio, A., Gordon-Smith, K., Forty, L., Kosorok, M.R., Fraser, C., Perry, A., Bethell, A., Craddock, N., Jones, L., Jones, I., 2018. Stratification of the risk of bipolar disorder recurrences in pregnancy and postpartum. Br. J. Psychiatry. https://doi.org/10.1192/bjp.2018.92.

Diav-Citrin, O., Shechtman, S., Ornoy, S., Arnon, J., Schaefer, C., Garbis, H., Clementi, M., Ornoy, A., 2005. Safety of haloperidol and penfluridol in pregnancy: a multicenter, prospective, controlled study. J. Clin. Psychiatry. https://doi.org/10.4088/JCP.v66n0307.

Dichtel, L.E., Carpenter, L.L., Nyer, M., Mischoulon, D., Kimball, A., Deckersbach, T., Dougherty, D.D., Schoenfeld, D.A., Fisher, L., Cusin, C., Dording, C., Trinh, N.H., Pedrelli, P., Yeung, A., Farabaugh, A., Papakostas, G.I., Chang, T., Shapero, B.G., Chen, J., Miller, K.K., 2020. Low-dose testosterone augmentation for antidepressant-resistant major depressive disorder in women: an 8-week randomized placebo-controlled study. Am. J. Psychiatry. https://doi.org/10.1176/appi.ajp.2020.19080844.

Dolk, H., Wang, H., Loane, M., Morris, J., Garne, E., Addor, M.C., Arriola, L., Bakker, M., Barisic, I., Doray, B., Gatt, M., Kallen, K., Khoshnood, B., Klungsoyr, K., Lahesmaa-Korpinen, A.M., Latos-Bielenska, A., Mejnartowicz, J.P., Nelen, V., Neville, A., De Jong-Van Den Berg, L.T.W., 2016. Lamotrigine use in pregnancy and risk of orofacial cleft and other congenital anomalies. Neurology. https://doi.org/10.1212/WNL.0000000000002540.

Drugs and Lactation Database (LactMed), 2006a. Internet, Lithium. National Library of Medicine, (US), Bethesda, MD.

Drugs and Lactation Database (LactMed), 2006b. Internet, Drugs and Lactation Database (LactMed)—Record Format. National Library of Medicine (US), Bethesda, MD. https://www.ncbi.nlm.nih.gov/books/NBK547442/.

Duran, A., Ugur, M.M., Turan, S., Emul, M., 2008. Clozapine use in two women with schizophrenia during pregnancy. J. Psychopharmacol. https://doi.org/10.1177/0269881107079171.

Earls, M.F., Yogman, M.W., Mattson, G., Rafferty, J., 2019. Incorporating recognition and management of perinatal depression into pediatric practice. Pediatrics. https://doi.org/10.1542/peds.2018-3259.

Ebbesen, F., Holsteen, V., Rix, M., Moeller, M., Joergensen, A., Hoseth, E., Kaad, P.H., 2000. Neonatal hypoglycaemia and withdrawal symptoms after exposure in utero to valproate. Arch. Dis. Child. Fetal Neonatal Ed. https://doi.org/10.1136/fn.83.2.f124.

Eddahibi, S., Raffestin, B., Hamon, M., Adnot, S., 2002. Is the serotonin transporter involved in the pathogenesis of pulmonary hypertension? J. Lab. Clin. Med. https://doi.org/10.1067/mlc.2002.122181.

Einarson, A., Boskovic, R., 2009. Use and safety of antipsychotic drugs during pregnancy. J. Psychiatr. Pract. https://doi.org/10.1097/01.pra.0000351878.45260.94.

Einarson, A., Bonari, L., Voyer-Lavigne, S., Addis, A., Matsui, D., Johnson, Y., Koren, G., 2003. A multicentre prospective controlled study to determine the safety of trazodone and nefazodone use during pregnancy. Can. J. Psychiatry. https://doi.org/10.1177/070674370304800207.

Eke, A.C., Saccone, G., Berghella, V., 2016. Selective serotonin reuptake inhibitor (SSRI) use during pregnancy and risk of preterm birth: a systematic review and meta-analysis. BJOG. https://doi.org/10.1111/1471-0528.14144.

Ennis, Z.N., Damkier, P., 2015. Pregnancy exposure to olanzapine, quetiapine, risperidone, aripiprazole and risk of congenital malformations. A systematic review. Basic Clin. Paharmacol. Toxicol. https://doi.org/10.1111/bcpt.12372.

Eriksson, K., Viinikainen, K., Mönkkönen, A., Äikiä, M., Nieminen, P., Heinonen, S., Kälviäinen, R., 2005. Children exposed to valproate in utero—population based evaluation of risks and confounding factors for long-term neurocognitive development. Epilepsy Res. https://doi.org/10.1016/j.eplepsyres.2005.06.001.

Figueroa, R., 2010. Use of antidepressants during pregnancy and risk of attention-deficit/hyperactivity disorder in the offspring. J. Dev. Behav. Pediatr. https://doi.org/10.1097/DBP.0b013e3181e5ac93.

Flynn, H.A., Sexton, M., Ratliff, S., Porter, K., Zivin, K., 2011. Comparative performance of the Edinburgh Postnatal Depression Scale and the Patient Health Questionnaire-9 in pregnant and postpartum women seeking psychiatric services. Psychiatry Res. https://doi.org/10.1016/j.psychres.2010.10.022.

Focht, A., Kellner, C.H., 2012. Electroconvulsive therapy (ECT) in the treatment of postpartum psychosis. J. ECT. https://doi.org/10.1097/YCT.0b013e3182315aa8.

Fornaro, M., Maritan, E., Ferranti, R., Zaninotto, L., Miola, A., Anastasia, A., Murru, A., Solé, E., Stubbs, B., Carvalho, A.F., Serretti, A., Vieta, E., Fusar-Poli, P., McGuire, P., Young, A.H., Dazzan, P., Vigod, S.N., Correll, C.U., Solmi, M., 2020. Lithium exposure during pregnancy and the postpartum period: a systematic review and meta-analysis of safety and efficacy outcomes. Am. J. Psychiatry. https://doi.org/10.1176/appi.ajp.2019.19030228.

Forray, A., Ostroff, R.B., 2007. The use of electroconvulsive therapy in postpartum affective disorders. J. ECT. https://doi.org/10.1097/yct.0b013e318074e4b1.

Forsberg, L., Berglund, G., Naver, L., Ljungdahl, M., Adler, M., Wide, K., 2012. 636 Bipolar disorder and pregnancy—neurodevelopmental outcome of children exposed to maternal illness with or without lithium during pregnancy. Arch. Dis. Child. https://doi.org/10.1136/archdischild-2012-302724.0636.

Fotopoulou, C., Kretz, R., Bauer, S., Schefold, J.C., Schmitz, B., Dudenhausen, J.W., Henrich, W., 2009. Prospectively assessed changes in lamotrigine-concentration in women with epilepsy during pregnancy, lactation and the neonatal period. Epilepsy Res. https://doi.org/10.1016/j.eplepsyres.2009.02.011.

Freeman, M.P., Wosnitzer Smith, K., Freeman, S.A., McElroy, S.L., Kmetz, G.F., Wright, R., Keck, P.E., 2002. The impact of reproductive events on the course of bipolar disorder in women. J. Clin. Psychiatry. https://doi.org/10.4088/JCP.v63n0403.

Freeman, E.W., Sammel, M.D., Liu, L., Gracia, C.R., Nelson, D.B., Hollander, L., 2004. Hormones and menopausal status as predictors of depression in women in transition to menopause. Arch. Gen. Psychiatry 61 (1), 62–70. https://doi.org/10.1001/archpsyc.61.1.62.

Freeman, E.W., Sammel, M.D., Lin, H., Nelson, D.B., 2006a. Associations of hormones and menopausal status with depressed mood in women with no history of depression. Arch. Gen. Psychiatry 63 (4), 375–382. https://doi.org/10.1001/archpsyc.63.4.375.

Freeman, M.P., Hill, R., Brumbach, B.H., 2006b. Escitalopram for perimenopausal depression: an open-label pilot study. J. Womens Health 15 (7), 857–861. https://doi.org/10.1089/jwh.2006.15.857.

Freeman, M.P., Hirschberg, A.M., Wang, B., Petrillo, L.F., Connors, S., Regan, S., Joffe, H., Cohen, L.S., 2013. Duloxetine for major depressive disorder and daytime and nighttime hot flashes associated with the menopausal transition. Maturitas 75 (2), 170–174. https://doi.org/10.1016/j.maturitas.2013.03.007.

Freeman, M.P., Papakostas, G.I., Hoeppner, B., Mazzone, E., Judge, H., Cusin, C., Mathew, S., Sanacora, G., Iosifescu, D., DeBattista, C., Trivedi, M.H., Fava, M., 2019. Sex differences in response to ketamine as a rapidly acting intervention for treatment resistant depression. J. Psychiatr. Res. 110, 166–171. https://doi.org/10.1016/j.jpsychires.2019.01.010.

Frey, B., Schubiger, G., Musy, J.P., 1990. Transient cholestatic hepatitis in a neonate associated with carbamazepine exposure during pregnancy and breast-feeding. Eur. J. Pediatr. https://doi.org/10.1007/BF02072057.

Gaily, E., Kantola-Sorsa, E., Hiilesmaa, V., Isoaho, M., Matila, R., Kotila, M., Nylund, T., Bardy, A., Kaaja, E., Granström, M.L., 2004. Normal intelligence in children with prenatal exposure to carbamazepine. Neurology. https://doi.org/10.1212/WNL.62.1.28.

Galbally, M., Snellen, M., Walker, S., Permezel, M., 2010. Management of antipsychotic and mood stabilizer medication in pregnancy: recommendations for antenatal care. Aust. N. Z. J. Psychiatr. https://doi.org/10.3109/00048670903487217.

Galbally, M., Bergink, V., Vigod, S.N., Buist, A., Boyce, P., Chandra, P., Kohan, R., Howard, L.M., 2018. Breastfeeding and lithium: is breast always best? Lancet Psychiatry. https://doi.org/10.1016/S2215-0366(18)30085-3.

Geddes, J., Carney, S., Cowen, P., Goodwin, G., Rogers, R., Dearness, K., Tomlin, A., Eastaugh, J., Freemantle, N., Lester, H., Harvey, A., Scott, A., 2003. Efficacy and safety of electroconvulsive therapy in depressive disorders: a systematic review and meta-analysis. Lancet 361 (9360), 799–808. https://doi.org/10.1016/S0140-6736(03)12705-5.

Gentile, S., 2003. Oxcarbazepine in pregnancy and lactation. Clin. Drug Investig. https://doi.org/10.2165/00044011-200323100-00008.

Gentile, S., 2014. Tricyclic antidepressants in pregnancy and puerperium. Expert Opin. Drug Saf. https://doi.org/10.1517/14740338.2014.869582.

Gentile, S., 2015. Managing antidepressant treatment in pregnancy and puerperium. Careful with that axe, Eugene. Expert Opin. Drug Saf. https://doi.org/10.1517/14740338.2015.1037273.

Gordon, J.L., Girdler, S.S., Meltzer-Brody, S.E., Stika, C.S., Thurston, R.C., Clark, C.T., Prairie, B.A., Moses-Kolko, E., Joffe, H., Wisner, K.L., 2015. Ovarian hormone fluctuation, neurosteroids, and HPA axis dysregulation in perimenopausal depression: a novel heuristic model. Am. J. Psychiatry 172 (3), 227–236. American Psychiatric Association https://doi.org/10.1176/appi.ajp.2014.14070918.

Gordon, J.L., Rubinow, D.R., Eisenlohr-Moul, T.A., Xia, K., Schmidt, P.J., Girdler, S.S., 2018. Efficacy of transdermal estradiol and micronized progesterone in the prevention of depressive symptoms in the menopause transition: a randomized clinical trial. JAMA Psychiat. 75 (2), 149–157. https://doi.org/10.1001/jamapsychiatry.2017.3998.

Gordon, J.L., Peltier, A., Grummisch, J.A., Sykes Tottenham, L., 2019. Estradiol fluctuation, sensitivity to stress, and depressive symptoms in the menopause transition: a pilot study. Front. Psychol. 10 (JUN). https://doi.org/10.3389/fpsyg.2019.01319.

Gordon, J.L., Sander, B., Eisenlohr-Moul, T.A., Sykes Tottenham, L., 2020. Mood sensitivity to estradiol predicts depressive symptoms in the menopause transition. Psychol. Med. https://doi.org/10.1017/S0033291720000483.

Gracious, B.L., Wisner, K.L., 1997. Phenelzine use throughout pregnancy and the puerperium: case report, review of the literature, and management recommendations. Depress. Anxiety 6 (3), 124–128. https://doi.org/10.1002/(SICI)1520-6394(1997)6:3<124::AID-DA6>3.0.CO;2-D.

Grandjean, E.M., Aubry, J.M., 2009a. Lithium: updated human knowledge using an evidence-based approach: part II: clinical pharmacology and therapeutic monitoring. CNS Drugs. https://doi.org/10.2165/00023210-200923040-00005.

Grandjean, E.M., Aubry, J.M., 2009b. Lithium: updated human knowledge using an evidence-based approach: part III: clinical safety. CNS Drugs. https://doi.org/10.2165/00023210-200923050-00004.

Gregoire, A.J.P., Kumar, R., Everitt, B., Henderson, A.F., Studd, J.W.W., 1996. Transdermal oestrogen for treatment of severe postnatal depression. Lancet. https://doi.org/10.1016/S0140-6736(96)91414-2.

Grigoriadis, S., VonderPorten, E.H., Mamisashvili, L., Eady, A., Tomlinson, G., Dennis, C.L., Koren, G., Steiner, M., Mousmanis, P., Cheung, A., Ross, L.E., 2013a. The effect of prenatal antidepressant exposure on neonatal adaptation: a systematic review and meta-analysis. J. Clin. Psychiatry. https://doi.org/10.4088/JCP.12r07967.

Grigoriadis, S., VonderPorten, E.H., Mamisashvili, L., Roerecke, M., Rehm, J., Dennis, C.L., Koren, G., Steiner, M., Mousmanis, P., Cheung, A., Ross, L.E., 2013b. Antidepressant exposure during pregnancy and congenital malformations: is there an association? A systematic review and meta-analysis of the best evidence. J. Clin. Psychiatry. https://doi.org/10.4088/JCP.12r07966.

Grigoriadis, S., VonderPorten, E.H., Mamisashvili, L., Tomlinson, G., Dennis, C.L., Koren, G., Steiner, M., Mousmanis, P., Cheung, A., Ross, L.E., 2014. Prenatal exposure to antidepressants and persistent pulmonary hypertension of the newborn: systematic review and meta-analysis. BMJ. https://doi.org/10.1136/bmj.f6932.

Grote, N.K., Bridge, J.A., Gavin, A.R., Melville, J.L., Iyengar, S., Katon, W.J., 2010. A meta-analysis of depression during pregnancy and the risk of preterm birth, low birth weight, and intrauterine growth restriction. Arch. Gen. Psychiatry. https://doi.org/10.1001/archgenpsychiatry.2010.111.

Groves, D.A., Brown, V.J., 2005. Vagal nerve stimulation: a review of its applications and potential mechanisms that mediate its clinical effects. Neurosci. Biobehav. Rev. https://doi.org/10.1016/j.neubiorev.2005.01.004.

Gupta, N., Grover, S., 2004. Safety of clozapine in 2 successive pregnancies. Can. J. Psychiatry. https://doi.org/10.1177/070674370404901211.

Guthrie, K.A., Lacroix, A.Z., Ensrud, K.E., Joffe, H., Newton, K.M., Reed, S.D., Caan, B., Carpenter, J.S., Cohen, L.S., Freeman, E.W., Larson, J.C., Manson, J.E., Rexrode, K., Skaar, T.C., Sternfeld, B., Anderson, G.L., 2015. Pooled analysis of six pharmacologic and nonpharmacologic interventions for vasomotor symptoms. Obstet. Gynecol. 126 (2), 413–422. https://doi.org/10.1097/AOG.0000000000000927.

Habermann, F., Fritzsche, J., Fuhlbrück, F., Wacker, E., Allignol, A., Weber-Schoendorfer, C., Meister, R., Schaefer, C., 2013. Atypical antipsychotic drugs and pregnancy outcome: a prospective, cohort study. J. Clin. Psychopharmacol. https://doi.org/10.1097/JCP.0b013e318295fe12.

Haight, S.C., Byatt, N., Moore Simas, T.A., Robbins, C.L., Ko, J.Y., 2019. Recorded diagnoses of depression during delivery hospitalizations in the United States, 2000-2015. Obstet. Gynecol. https://doi.org/10.1097/AOG.0000000000003291.

Halmo, M., Spodniaková, B., Nosá'Ová, P., 2014. Fetal spasms after the administration of electroconvulsive therapy in pregnancy: our experience. J. ECT. https://doi.org/10.1097/YCT.0000000000000101.

Handley, A.P., Williams, M., 2015. The efficacy and tolerability of SSRI/SNRIs in the treatment of vasomotor symptoms in menopausal women: a systematic review. J. Am. Assoc. Nurse Pract. 27 (1), 54–61. Blackwell Publishing Ltd https://doi.org/10.1002/2327-6924.12137.

Hantsoo, L., Ward-O'Brien, D., Czarkowski, K.A., Gueorguieva, R., Price, L.H., Epperson, C.N., 2014. A randomized, placebo-controlled, double-blind trial of sertraline for postpartum depression. Psychopharmacology (Berl). https://doi.org/10.1007/s00213-013-3316-1.

Harlow, S.D., Gass, M., Hall, J.E., Lobo, R., Maki, P., Rebar, R.W., Sherman, S., Sluss, P.M., De Villiers, T.J., 2012. Executive summary of the stages of reproductive aging workshop + 10: addressing the unfinished agenda of staging reproductive aging. J. Clin. Endocrinol. Metabol. 97 (4), 1159–1168. https://doi.org/10.1210/jc.2011-3362.

Hernández-Díaz, S., Smith, C.R., Shen, A., et al., 2012. Comparative safety of antiepileptic drugs during pregnancy. Neurology 78 (21), 1692–1699.

Hizli Sayar, G., Ozten, E., Tufan, E., Cerit, C., Kağan, G., Dilbaz, N., Tarhan, N., 2014. Transcranial magnetic stimulation during pregnancy. Arch. Womens Ment. Health. https://doi.org/10.1007/s00737-013-0397-0.

Holmes, L.B., Baldwin, E.J., Smith, C.R., Habecker, E., Glassman, L., Wong, S.L., Wyszynski, D.F., 2008. Increased frequency of isolated cleft palate in infants exposed to lamotrigine during pregnancy. Neurology. https://doi.org/10.1212/01.wnl.0000304343.45104.d6.

Hostetter, A., Ritchie, J.C., Stowe, Z.N., 2000. Amniotic fluid and umbilical cord blood concentrations of antidepressants in three women. Biol. Psychiatry. https://doi.org/10.1016/S0006-3223(00)00958-6.

Howard, L.M., Molyneaux, E., Dennis, C.L., Rochat, T., Stein, A., Milgrom, J., 2014. Non-psychotic mental disorders in the perinatal period. Lancet 384 (9956), 1775–1788.

Huang, H., Coleman, S., Bridge, J.A., Yonkers, K., Katon, W., 2014. A meta-analysis of the relationship between antidepressant use in pregnancy and the risk of preterm birth and low birth weight. Gen. Hosp. Psychiatry. https://doi.org/10.1016/j.genhosppsych.2013.08.002.

Husain, M.M., Stegman, D., Trevino, K., 2005. Pregnancy and delivery while receiving vagus nerve stimulation for the treatment of major depression: a case report. Ann. Gen. Psychiatry. https://doi.org/10.1186/1744-859X-4-16.

Huybrechts, K.F., Palmsten, K., Avorn, J., Cohen, L.S., Holmes, L.B., Franklin, J.M., Mogun, H., Levin, R., Kowal, M., Setoguchi, S., Hernández-Díaz, S., 2014. Antidepressant use in pregnancy and the risk of cardiac defects. N. Engl. J. Med. https://doi.org/10.1056/NEJMoa1312828.

Huybrechts, K.F., Bateman, B.T., Palmsten, K., Desai, R.J., Patorno, E., Gopalakrishnan, C., Levin, R., Mogun, H., Hernandez-Diaz, S., 2015. Antidepressant use late in pregnancy and risk of persistent pulmonary hypertension of the newborn. JAMA. https://doi.org/10.1001/jama.2015.5605.

Huybrechts, K.F., Hernández-Díaz, S., Patorno, E., Desai, R.J., Mogun, H., Dejene, S.Z., Cohen, J.M., Panchaud, A., Cohen, L., Bateman, B.T., 2016. Antipsychotic use in pregnancy and the risk for congenital malformations. JAMA Psychiat. https://doi.org/10.1001/jamapsychiatry.2016.1520.

Jablensky, A.V., Morgan, V., Zubrick, S.R., Bower, C., Yellachich, L.-A., 2005. Pregnancy, delivery, and neonatal complications in a population cohort of women with schizophrenia and major affective disorders. Am. J. Psychiatry, 79–91. https://doi.org/10.1176/appi.ajp.162.1.79.

Jaffe, R., 2002. The practice of electroconvulsive therapy: recommendations for treatment, training, and privileging: a Task Force Report of the American Psychiatric Association, 2nd ed. Am. J. Psychiatry. https://doi.org/10.1176/appi.ajp.159.2.331.

Jimenez-Solem, E., Andersen, J.T., Petersen, M., Broedbaek, K., Lander, A.R., Afzal, S., Torp-Pedersen, C., Poulsen, H.E., 2013. SSRI use during pregnancy and risk of stillbirth and neonatal mortality. Am. J. Psychiatry. https://doi.org/10.1176/appi.ajp.2012.11081251.

Joffe, H., Groninger, H., Soares, C., Nonacs, R., Cohen, L.S., 2001. An open trial of mirtazapine in menopausal women with depression unresponsive to estrogen replacement therapy. J. Womens Health Gend. Based Med. 10 (10), 999–1004. https://doi.org/10.1089/152460901317193576.

Joffe, H., Soares, C.N., Petrillo, L.F., Viguera, A.C., Somley, B.L., Koch, J.K., Cohen, L.S., 2007. Treatment of depression and menopause-related symptoms with the serotonin-norepinephrine reuptake inhibitor duloxetine. J. Clin. Psychiatry 68 (6), 943–950. https://doi.org/10.4088/JCP.v68n0619.

Kanes, S., Colquhoun, H., Gunduz-Bruce, H., Raines, S., Arnold, R., Schacterle, A., Doherty, J., Epperson, C.N., Deligiannidis, K.M., Riesenberg, R., Hoffmann, E., Rubinow, D., Jonas, J., Paul, S., Meltzer-Brody, S., 2017a. Brexanolone (SAGE-547 injection) in post-partum depression: a randomised controlled trial. Lancet. https://doi.org/10.1016/S0140-6736(17)31264-3.

Kanes, S.J., Colquhoun, H., Doherty, J., Raines, S., Hoffmann, E., Rubinow, D.R., Meltzer-Brody, S., 2017b. Open-label, proof-of-concept study of brexanolone in the treatment of severe postpartum depression. Hum. Psychopharmacol. https://doi.org/10.1002/hup.2576.

Karlsson, K., Lindstedt, G., Lundberg, P.A., Selstam, U., 1975. Transplacental lithium poisoning: reversible inhibition of fetal thyroid. Lancet. https://doi.org/10.1016/S0140-6736(75)92578-7.

Kendell, R., Chalmers, J., Platz, C., 1987. Epidemiology of the peurperal psychosis. Br. J. Psychiatry 150, 662–673.

Kennedy, D., Koren, G., 1998. Valproic acid use in psychiatry: issues in treating women of reproductive age. J. Psychiatry Neurosci. 23, 223–228.

Kim, D.R., Wang, E., 2014. Prevention of supine hypotensive syndrome in pregnant women treated with transcranial magnetic stimulation. Psychiatry Res. https://doi.org/10.1016/j.psychres.2014.04.001.

Kim, J.H., Choi, S.S., Ha, K., 2008. A closer look at depression in mothers who kill their children: is it unipolar or bipolar depression? J. Clin. Psychiatry. https://doi.org/10.4088/JCP.v69n1013.

Kim, D.R., Epperson, N., Paré, E., Gonzalez, J.M., Parry, S., Thase, M.E., Cristancho, P., Sammel, M.D., O'Reardon, J.P., 2011. An open label pilot study of transcranial magnetic stimulation for pregnant women with major depressive disorder. J. Womens Health. https://doi.org/10.1089/jwh.2010.2353.

Kim, D.R., Snell, J.L., Ewing, G.C., O'Reardon, J., 2015. Neuromodulation and antenatal depression: a review. Neuropsychiatr. Dis. Treat. https://doi.org/10.2147/NDT.S80480.

Koerner, M., Yerby, M., Friel, P., McCormick, K., 1989. Valproic acid disposition and protein binding in pregnancy. Ther. Drug Monit. https://doi.org/10.1097/00007691-198905000-00002.

Kornstein, S.G., Schatzberg, A.F., Thase, M.E., Yonkers, K.A., McCullough, J.P., Keitner, G.I., Gelenberg, A.J., Davis, S.M., Harrison, W.M., Keller, M.B., 2000. Gender differences in treatment response to sertraline versus imipramine in chronic depression. Am. J. Psychiatry 157 (9), 1445–1452. https://doi.org/10.1176/appi.ajp.157.9.1445.

Kornstein, S.G., Jiang, Q., Reddy, S., Musgnung, J.J., Guico-Pabia, C.J., 2010. Short-term efficacy and safety of desvenlafaxine in a randomized, placebo-controlled study of perimenopausal and postmenopausal women with major depressive disorder. J. Clin. Psychiatry 71 (8), 1088–1096. https://doi.org/10.4088/JCP.10m06018blu.

Kornstein, S.G., Clayton, A., Bao, W., Guico-Pabia, C.J., 2014a. Post hoc analysis of the efficacy and safety of desvenlafaxine 50 mg/day in a randomized, placebo-controlled study of perimenopausal and postmenopausal women with major depressive disorder. Menopause 21 (8), 799–806. https://doi.org/10.1097/GME.0000000000000178.

Kornstein, S.G., Pedersen, R.D., Holland, P.J., Nemeroff, C.B., Rothschild, A.J., Thase, M.E., Trivedi, M.H., Ninan, P.T., Keller, M.B., 2014b. Influence of sex and menopausal status on response, remission, and recurrence in patients with recurrent major depressive disorder treated with venlafaxine extended release or fluoxetine: analysis of data from the PREVENT study. J. Clin. Psychiatry 75 (1), 62–68. https://doi.org/10.4088/JCP.12m07841.

Kornstein, S.G., Clayton, A.H., Bao, W., Guico-Pabia, C.J., 2015. A pooled analysis of the efficacy of desvenlafaxine for the treatment of major depressive disorder in perimenopausal and postmenopausal women. J. Womens Health 24 (4), 281–290. https://doi.org/10.1089/jwh.2014.4900.

Krishnamurthy, K.B., 2012. Managing epilepsy during pregnancy: assessing risk and optimizing care. Curr. Treat. Options. Neurol. https://doi.org/10.1007/s11940-012-0184-7.

Kulkarni, J., Worsley, R., Gilbert, H., Gavrilidis, E., Van Rheenen, T.E., Wang, W., McCauley, K., Fitzgerald, P., 2014. A prospective cohort study of antipsychotic medications in pregnancy: the first 147 pregnancies and 100 one year old babies. PLoS One. https://doi.org/10.1371/journal.pone.0094788.

Lasser, R., Deligiannidis, K.M., Gunduz-Bruce, H., Silber, C., Sankoh, A., Li, S., Jonas, J., Doherty, J., Kanes, S., 2019. A phase 3, double-blind, placebo-controlled trial of SAGE-217 in postpartum depression: topline assessment of secondary efficacy measures of anxiety and depression. Am. Soc. Clin. Psychopharmacol. 116. https://pmg.joynadmin.org/documents/1005/5ce6b5d668ed3fae0195e2f3.pdf.

Lawrie, T.A., Hofmeyr, G.J., De Jager, M., Berk, M., Paiker, J., Viljoen, E., 1998. A double-blind randomised placebo controlled trial of postnatal norethisterone enanthate: the effect on postnatal depression and serum hormones. BJOG. https://doi.org/10.1111/j.1471-0528.1998.tb09940.x.

Lee, H.C., Lin, H.C., 2010. Maternal bipolar disorder increased low birthweight and preterm births: a nationwide population-based study. J. Affect. Disord. https://doi.org/10.1016/j.jad.2009.05.019.

Lee, S.R., Cho, M.K., Cho, Y.J., Chun, S., Hong, S.-H., Hwang, K.R., Jeon, G.-H., Joo, J.K., Kim, S.K., Lee, D.O., Lee, D.-Y., Lee, E.S., Song, J.Y., Yi, K.W., Yun, B.H., Shin, J.-H., Chae, H.D., Kim, T., 2020. The 2020 menopausal hormone therapy guidelines. J. Menopausal Med. 26 (2), 69. https://doi.org/10.6118/jmm.20000.

Leiknes, K.A., Cooke, M.J., Jarosch-von Schweder, L., Harboe, I., Høie, B., 2015. Electroconvulsive therapy during pregnancy: a systematic review of case studies. Arch. Womens Ment. Health. https://doi.org/10.1007/s00737-013-0389-0.

Leis, J.A., Heron, J., Stuart, E.A., Mendelson, T., 2014. Associations between maternal mental health and child emotional and behavioral problems: does prenatal mental health matter? J. Abnorm. Child Psychol. https://doi.org/10.1007/s10802-013-9766-4.

Lin, H.C., Chen, I.J., Chen, Y.H., Lee, H.C., Wu, F.J., 2010. Maternal schizophrenia and pregnancy outcome: does the use of antipsychotics make a difference? Schizophr. Res. https://doi.org/10.1016/j.schres.2009.10.011.

Lindhout, D., Schmidt, D., 1986. In-utero exposure to valproate and neural tube defects. Lancet. https://doi.org/10.1016/S0140-6736(86)91711-3.

Llewellyn, A., Stowe, Z.N., Strader, J.R., 1998. The use of lithium and management of women with bipolar disorder during pregnancy and lactation. J. Clin. Psychiatry.

Louik, C., Kerr, S., Mitchell, A.A., 2014. First-trimester exposure to bupropion and risk of cardiac malformations. Pharmacoepidemiol. Drug Saf. https://doi.org/10.1002/pds.3661.

Lupattelli, A., Spigset, O., Koren, G., Nordeng, H., 2014. Risk of vaginal bleeding and postpartum hemorrhage after use of antidepressants in pregnancy: a study from the Norwegian mother and child cohort study. J. Clin. Psychopharmacol. https://doi.org/10.1097/JCP.0000000000000036.

Lutz, U.C., Wiatr, G., Gaertner, H.J., Bartels, M., 2007. Oxcarbazepine treatment during breast-feeding: a case report. J. Clin. Psychopharmacol. https://doi.org/10.1097/JCP.0b013e31815a5819.

Lydsdottir, L.B., Howard, L.M., Olafsdottir, H., Thome, M., Tyrfingsson, P., Psych, C., Sigurdsson, J.F., 2014. The mental health characteristics of pregnant women with depressive symptoms identified by the Edinburgh postnatal depression scale. J. Clin. Psychiatry. https://doi.org/10.4088/JCP.13m08646.

Magnus, M.C., Wilcox, A.J., Morken, N.H., Weinberg, C.R., Håberg, S.E., 2019. Role of maternal age and pregnancy history in risk of miscarriage: prospective register based study. BMJ. https://doi.org/10.1136/bmj.l869.

Malhi, G.S., Tanious, M., Das, P., Berk, M., 2012. The science and practice of lithium therapy. Aust. N. Z. J. Psychiatr. https://doi.org/10.1177/0004867412437346.

Malm, H., Sourander, A., Gissler, M., Gyllenberg, D., Hinkka-Yli-Salomäki, S., McKeague, I.W., Artama, M., Brown, A.S., 2015. Pregnancy complications following prenatal exposure to SSRIs or maternal psychiatric disorders: results from population-based national register data. Am. J. Psychiatry. https://doi.org/10.1176/appi.ajp.2015.14121575.

Malm, H., Brown, A.S., Gissler, M., Gyllenberg, D., Hinkka-Yli-Salomäki, S., McKeague, I.W., Weissman, M., Wickramaratne, P., Artama, M., Gingrich, J.A., Sourander, A., 2016. Gestational exposure to selective serotonin reuptake inhibitors and offspring psychiatric disorders: a national register-based study. J. Am. Acad. Child Adolesc. Psychiatry. https://doi.org/10.1016/j.jaac.2016.02.013.

Martin, J.A., Brady, M.P.H., Hamilton, E., Osterman, M.J.K., Driscoll, A.K., Mathews, T.J., 2015. National vital statistics reports. In: National Vital Statistics Reports. vol. 66. Number 1, January 5, 2017. Vol. 66, Issue 1 http://www.cdc.gov/nchs/data_access/Vitalstatsonline.htm.

Martin, J.A., Hamilton, B.E., Osterman, M.J.K., Driscoll, A.K., 2019. Births: Final Data for 2018 Figure 1. Number and Rate of Triplet and Higher-Order Multiple Births: United States, 1980-2018. https://www.cdc.gov/nchs/products/index.htm.

Martinez Botella, G., Salituro, F.G., Harrison, B.L., Beresis, R.T., Bai, Z., Blanco, M.J., Belfort, G.M., Dai, J., Loya, C.M., Ackley, M.A., Althaus, A.L., Grossman, S.J., Hoffmann, E., Doherty, J.J., Robichaud, A.J., 2017. Neuroactive steroids. 2.3α-hydroxy-3β-methyl-21-(4-cyano-1H-pyrazol-1′-yl)-19-nor-5β-pregnan-20-one (SAGE-217): a clinical next generation neuroactive steroid positive allosteric modulator of the (γ-aminobutyric acid)A receptor. J. Med. Chem. https://doi.org/10.1021/acs.jmedchem.7b00846.

Matalon, S., Schechtman, S., Goldzweig, G., Ornoy, A., 2002. The teratogenic effect of carbamazepine: a meta-analysis of 1255 exposures. Reprod. Toxicol. https://doi.org/10.1016/S0890-6238(01)00199-X.

MBRRACE, 2018. Saving Lives, Improving Mothers' Care Lessons learned to inform maternity care from the UK and Ireland Confidential Enquiries into Maternal Deaths and Morbidity 2014–16. In: Maternal, Newborn and Infant Clinical Outcome Review Programme.

Meador, K.J., Baker, G.A., Browning, N., Clayton-Smith, J., Combs-Cantrell, D.T., Cohen, M., Kalayjian, L.A., Kanner, A., Liporace, J.D., Pennell, P.B., Privitera, M., Loring, D.W., 2009. Cognitive function at three years of age after fetal exposure to antiepileptic drugs. Obstet. Gynecol. Surv. https://doi.org/10.1097/01.ogx.0000356745.41545.57.

Meador, K.J., Baker, G.A., Browning, N., Cohen, M.J., Bromley, R.L., Clayton-Smith, J., Kalayjian, L.A., Kanner, A., Liporace, J.D., Pennell, P.B., Privitera, M., Loring, D.W., 2013. Fetal antiepileptic drug exposure and cognitive outcomes at age 6 years (NEAD study): a prospective observational study. Lancet Neurol. https://doi.org/10.1016/S1474-4422(12)70323-X.

Meltzer-Brody, S., Colquhoun, H., Riesenberg, R., Epperson, C.N., Deligiannidis, K.M., Rubinow, D.R., Li, H., Sankoh, A.J., Clemson, C., Schacterle, A., Jonas, J., Kanes, S., 2018. Brexanolone injection in post-partum depression: two multicentre, double-blind, randomised, placebo-controlled, phase 3 trials. Lancet. https://doi.org/10.1016/S0140-6736(18)31551-4.

Milgrom, J., Gemmill, A.W., Ericksen, J., Burrows, G., Buist, A., Reece, J., 2015. Treatment of postnatal depression with cognitive behavioural therapy, sertraline and combination therapy: a randomised controlled trial. Aust. N. Z. J. Psychiatr. https://doi.org/10.1177/0004867414565474.

Misri, S., Reebye, P., Corral, M., Milis, L., 2004. The use of paroxetine and cognitive-behavioral therapy in postpartum depression and anxiety: a randomized controlled trial. J. Clin. Psychiatry. https://doi.org/10.4088/JCP.v65n0913.

Misri, S., Abizadeh, J., Albert, G., Carter, D., Ryan, D., 2012. Restoration of functionality in postpartum depressed mothers: an open-label study with escitalopram. J. Clin. Psychopharmacol. https://doi.org/10.1097/JCP.0b013e31826867c9.

Misri, S., Swift, E., Abizadeh, J., Shankar, R., 2016. Overcoming functional impairment in postpartum depressed or anxious women: a pilot trial of desvenlafaxine with flexible dosing. Ther. Adv. Psychopharmacol. https://doi.org/10.1177/2045125316656297.

Mizrahi, E.M., Hobbs, J.F., Goldsmith, D.I., 1979. Nephrogenic diabetes insipidus in transplacental lithium intoxication. J. Pediatr. https://doi.org/10.1016/S0022-3476(79)80610-1.

Molyneaux, E., Howard, L.M., Mcgeown, H.R., Karia, A.M., Trevillion, K., 2014. Antidepressant treatment for postnatal depression. Cochrane Database Syst. Rev. https://doi.org/10.1002/14651858.CD002018.pub2.

Moore, S.J., Turnpenny, P., Quinn, A., Glover, S., Lloyd, D.J., Montgomery, T., Dean, J.C.S., 2000. A clinical study of 57 children with fetal anticonvulsant syndromes. J. Med. Genet. https://doi.org/10.1136/jmg.37.7.489.

Moretti, M.E., Koren, G., Verjee, Z., Ito, S., 2003. Monitoring lithium in breast milk: an individualized approach for breast-feeding mothers. Ther. Drug Monit. https://doi.org/10.1097/00007691-200306000-00017.

Morrow, J., Russell, A., Guthrie, E., Parsons, L., Robertson, I., Waddell, R., Irwin, B., McGivern, R.C., Morrison, P.J., Craig, J., 2006. Malformation risks of antiepileptic drugs in pregnancy: a prospective study from the UK Epilepsy and Pregnancy Register. J. Neurol. Neurosurg. Psychiatry. https://doi.org/10.1136/jnnp.2005.074203.

Moses-Kolko, E.L., Berga, S.L., Kalro, B., Sit, D.K.Y., Wisner, K.L., 2009. Transdermal estradiol for postpartum depression: a promising treatment option. Clin. Obstet. Gynecol. https://doi.org/10.1097/GRF.0b013e3181b5a395.

Mountain, K.R., Hirsh, J., Gallus, A.S., 1970. Neonatal coagulation defect due to anticonvulsant drug treatment in pregnancy. Lancet. https://doi.org/10.1016/S0140-6736(70)90636-7.

Nahas, Z., Bohning, D.E., Molloy, M.A., Oustz, J.A., Risch, S.C., George, M.S., 1999. Safety and feasibility of repetitive transcranial magnetic stimulation in the treatment of anxious depression in pregnancy: a case report. J. Clin. Psychiatry. https://doi.org/10.4088/JCP.v60n0111.

Natari, R.B., Clavarino, A.M., McGuire, T.M., Dingle, K.D., Hollingworth, S.A., 2018. The bidirectional relationship between vasomotor symptoms and depression across the menopausal transition: a systematic review of longitudinal studies. Menopause 25 (1), 109–120. Lippincott Williams and Wilkins https://doi.org/10.1097/GME.0000000000000949.

Neill Epperson, C., Pittman, B., Czarkowski, K.A., Bradley, J., Quinlan, D.M., Brown, T.E., 2011. Impact of atomoxetine on subjective attention and memory difficulties in perimenopausal and postmenopausal women. Menopause 18 (5), 542–548. https://doi.org/10.1097/gme.0b013e3181fcafd6.

Nelson, H.D., Vesco, K.K., Haney, E., Fu, R., Nedrow, A., Miller, J., Nicolaidis, C., Walker, M., Humphrey, L., 2006. Nonhormonal therapies for menopausal hot flashes: systematic review and meta-analysis. JAMA 295 (17), 2057–2071. American Medical Association https://doi.org/10.1001/jama.295.17.2057.

Newham, J.J., Thomas, S.H., MacRitchie, K., McElhatton, P.R., McAllister-Williams, R.H., 2008. Birth weight of infants after maternal exposure to typical and atypical antipsychotics: prospective comparison study. Br. J. Psychiatry. https://doi.org/10.1192/bjp.bp.107.041541.

Newmark, R.L., Bogen, D.L., Wisner, K.L., Isaac, M., Ciolino, J.D., Clark, C.T., 2019. Risk-Benefit assessment of infant exposure to lithium through breast milk: a systematic review of the literature. Int. Rev. Psychiatry. https://doi.org/10.1080/09540261.2019.1586657.

Newport, D.J., Hostetter, A., Arnold, A., Stowe, Z.N., 2002. The treatment of postpartum depression: minimizing infant exposures. J. Clin. Psychiatry.

Newport, D.J., Viguera, A.C., Beach, A.J., Ritchie, J.C., Cohen, L.S., Stowe, Z.N., 2005. Lithium placental passage and obstetrical outcome: implications for clinical management during late pregnancy. Am. J. Psychiatry. https://doi.org/10.1176/appi.ajp.162.11.2162.

Newport, D.J., Calamaras, M.R., DeVane, C.L., Donovan, J., Beach, A.J., Winn, S., Knight, B.T., Gibson, B.B., Viguera, A.C., Owens, M.J., Nemeroff, C.B., Stowe, Z.N., 2007. Atypical antipsychotic administration during late pregnancy: placental passage and obstetrical outcomes. Am. J. Psychiatry. https://doi.org/10.1176/appi.ajp.2007.06111886.

Newport, D.J., Pennell, P.B., Calamaras, M.R., Ritchie, J.C., Newman, M., Knight, B., Viguera, A.C., Liporace, J., Stowe, Z.N., 2008a. Lamotrigine in breast milk and nursing infants: determination of exposure. Pediatrics. https://doi.org/10.1542/peds.2007-3812.

Newport, D.J., Stowe, Z.N., Viguera, A.C., Calamaras, M.R., Juric, S., Knight, B., Pennell, P.B., Baldessarini, R.J., 2008b. Lamotrigine in bipolar disorder: efficacy during pregnancy. Bipolar Disord. https://doi.org/10.1111/j.1399-5618.2007.00565.x.

Newton, E.R., Hale, T.W., 2015. Drugs in breast milk. Clin. Obstet. Gynecol. https://doi.org/10.1097/GRF.0000000000000142.

Nonacs, R.M., Soares, C.N., Viguera, A.C., Pearson, K., Poitras, J.R., Cohen, L.S., 2005. Bupropion SR for the treatment of postpartum depression: a pilot study. Int. J. Neuropsychopharmacol. https://doi.org/10.1017/S1461145705005079.

Nora, J.J., Nora, A.H., Toews, W.H., 1974. Lithium, Ebstein's anomaly, and other congenital heart defects. Lancet. https://doi.org/10.1016/S0140-6736(74)91918-7.

Nörby, U., Forsberg, L., Wide, K., Sjörs, G., Winbladh, B., Källén, K., 2016. Neonatal morbidity after maternal use of antidepressant drugs during pregnancy. Pediatrics 138 (5). https://doi.org/10.1542/peds.2016-0181, e20160181.

O'Reardon, J.P., Cristancho, M.A., Von Andreae, C.V., Cristancho, P., Weiss, D., 2011. Acute and maintenance electroconvulsive therapy for treatment of severe major depression during the second and third trimesters of pregnancy with infant follow-up to 18 months: case report and review of the literature. J. ECT. https://doi.org/10.1097/YCT.0b013e3181e63160.

Occhiogrosso, M., Omran, S.S., Altemus, M., 2012. Persistent pulmonary hypertension of the newborn and selective serotonin reuptake inhibitors: lessons from clinical and translational studies. Am. J. Psychiatry. https://doi.org/10.1176/appi.ajp.2011.11040553.

Öhman, I., Beck, O., Vitols, S., Tomson, T., 2008. Plasma concentrations of lamotrigine and its 2-N-glucuronide metabolite during pregnancy in women with epilepsy. Epilepsia. https://doi.org/10.1111/j.1528-1167.2007.01471.x.

Pagnin, D., De Queiroz, V., Pini, S., Cassano, G.B., 2004. Efficacy of ECT in depression: a meta-analytic review. J. ECT. https://doi.org/10.1097/00124509-200403000-00004.

Palmsten, K., Hernández-Díaz, S., Huybrechts, K.F., Williams, P.L., Michels, K.B., Achtyes, E.D., Mogun, H., Setoguchi, S., 2013. Use of antidepressants near delivery and risk of postpartum hemorrhage: cohort study of low income women in the United States. BMJ. https://doi.org/10.1136/bmj.f4877.

Pariente, G., Leibson, T., Shulman, T., Adams-Webber, T., Barzilay, E., Nulman, I., 2017. Pregnancy outcomes following in utero exposure to lamotrigine: a systematic review and meta-analysis. CNS Drugs. https://doi.org/10.1007/s40263-017-0433-0.

Patorno, E., Huybrechts, K.F., Bateman, B.T., Cohen, J.M., Desai, R.J., Mogun, H., Cohen, L.S., Hernandez-Diaz, S., 2017. Lithium use in pregnancy and the risk of cardiac malformations. N. Engl. J. Med. https://doi.org/10.1056/NEJMoa1612222.

Paulson, G.W., Paulson, R.B., 1981. Teratogenic effects of anticonvulsants. Arch. Neurol. https://doi.org/10.1001/archneur.1981.00510030034003.

Pavy, T.J.G., Kliffer, A.P., Douglas, M.J., 1995. Anaesthetic management of labour and delivery in a woman taking long-term MAOI. Can. J. Anaesth. https://doi.org/10.1007/BF03011881.

Pearson, R.M., Evans, J., Kounali, D., Lewis, G., Heron, J., Ramchandani, P.G., O'Connor, T.G., Stein, A., 2013. Maternal depression during pregnancy and the postnatal period risks and possible mechanisms for offspring depression at age 18 years. JAMA Psychiat. https://doi.org/10.1001/jamapsychiatry.2013.2163.

Peng, M., Gao, K., Ding, Y., Ou, J., Calabrese, J.R., Wu, R., Zhao, J., 2013. Effects of prenatal exposure to atypical antipsychotics on postnatal development and growth of infants: a case-controlled, prospective study. Psychopharmacology (Berl). https://doi.org/10.1007/s00213-013-3060-6.

Perry, D.F., Ettinger, A.K., Mendelson, T., Le, H.N., 2011. Prenatal depression predicts postpartum maternal attachment in low-income Latina mothers with infants. Infant Behav. Dev. https://doi.org/10.1016/j.infbeh.2011.02.005.

Petersen, I., McCrea, R.L., Sammon, C.J., Osborn, D.P.J., Evans, S.J., Cowen, P.J., Freemantle, N., Nazareth, I., 2016. Risks and benefits of psychotropic medication in pregnancy: cohort studies based on UK electronic primary care health records. Health Technol. Assess. https://doi.org/10.3310/hta20230.

Philbert, A., Pedersen, B., Dam, M., 1985. Concentration of valproate during pregnancy, in the newborn and in breast milk. Acta Neurol. Scand. https://doi.org/10.1111/j.1600-0404.1985.tb00901.x.

Pinto-Meza, A., Usall, J., Serrano-Blanco, A., Suárez, D., Haro, J.M., 2006. Gender differences in response to antidepressant treatment prescribed in primary care. Does menopause make a difference? J. Affect. Disord. 93 (1–3), 53–60. https://doi.org/10.1016/j.jad.2006.02.010.

Putnam, K.T., Wilcox, M., Robertson-Blackmore, E., Sharkey, K., Bergink, V., Munk-Olsen, T., Deligiannidis, K.M., Payne, J., Altemus, M., Newport, J., Apter, G., Devouche, E., Viktorin, A., Magnusson, P., Penninx, B., Buist, A., Bilszta, J., O'Hara, M., Stuart, S., Meltzer-Brody, S., 2017. Clinical phenotypes of perinatal depression and time of symptom onset: analysis of data from an international consortium. Lancet Psychiatry. https://doi.org/10.1016/S2215-0366(17)30136-0.

Rabheru, K., 2001. The use of electroconvulsive therapy in special patient populations. Can. J. Psychiatry. https://doi.org/10.1177/070674370104600803.

Raffi, E.R., Nonacs, R., Cohen, L.S., 2019. Safety of psychotropic medications during pregnancy. Clin. Perinatol. https://doi.org/10.1016/j.clp.2019.02.004.

Ramos, É., St-André, M., Rey, É., Oraichi, D., Bérard, A., 2008. Duration of antidepressant use during pregnancy and risk of major congenital malformations. Br. J. Psychiatry. https://doi.org/10.1192/bjp.bp.107.042523.

Reale, S.C., Easter, S.R., Xu, X., Bateman, B.T., Farber, M.K., 2020. Trends in postpartum hemorrhage in the United States from 2010 to 2014. Anesth. Analg. https://doi.org/10.1213/ANE.0000000000004424.

Reis, M., Källén, B., 2008. Maternal use of antipsychotics in early pregnancy and delivery outcome. J. Clin. Psychopharmacol. https://doi.org/10.1097/JCP.0b013e318172b8d5.

Reis, M., Källén, B., 2013. Combined use of selective serotonin reuptake inhibitors and sedatives/hypnotics during pregnancy: risk of relatively severe congenital malformations or cardiac defects. A register study. BMJ Open. https://doi.org/10.1136/bmjopen-2012-002166.

Reis, M., Kllén, B., 2010. Delivery outcome after maternal use of antidepressant drugs in pregnancy: an update using Swedish data. Psychol. Med. https://doi.org/10.1017/S0033291709992194.

Rihtman, T., Parush, S., Ornoy, A., 2013. Developmental outcomes at preschool age after fetal exposure to valproic acid and lamotrigine: cognitive, motor, sensory and behavioral function. Reprod. Toxicol. https://doi.org/10.1016/j.reprotox.2013.06.001.

Rodríguez-Pinilla, E., Arroyo, I., Fondevilla, J., García, M.J., Martínez-Frías, M.L., 2000. Prenatal exposure to valproic acid during pregnancy and limb deficiencies: a case-control study. Am. J. Med. Genet. https://doi.org/10.1002/(SICI)1096-8628(20000228)90:5<376::AID-AJMG6>3.0.CO;2-V.

Ross, L.E., Grigoriadis, S., Mamisashvili, L., Vonderporten, E.H., Roerecke, M., Rehm, J., Dennis, C.L., Koren, G., Steiner, M., Mousmanis, P., Cheung, A., 2013. Selected pregnancy and delivery outcomes after exposure to antidepressant medication: a systematic review and meta-analysis. JAMA Psychiat. https://doi.org/10.1001/jamapsychiatry.2013.684.

Roy, A., Cole, K., Goldman, Z., Barris, M., 1993. Fluoxetine treatment of postpartum depression. Am. J. Psychiatry. https://doi.org/10.1176/ajp.150.8.1273a.

Rubinow, D., Johnson, S., Schmidt, P., Girdler, S., Gaynes, B., 2015. Efficacy of estradiol in perimenopausal depression: so much promise and so few answers. Depress. Anxiety 32 (8). https://doi.org/10.1002/DA.22391.

Rumeau-Rouquette, C., Goujard, J., Huel, G., 1977. Possible teratogenic effect of phenothiazines in human beings. Teratology. https://doi.org/10.1002/tera.1420150108.

Salisbury, A.L., O'Grady, K.E., Battle, C.L., Wisner, K.L., Anderson, G.M., Stroud, L.R., Miller-Loncar, C.L., Young, M.E., Lester, B.M., 2016. The roles of maternal depression, serotonin reuptake inhibitor treatment, and concomitant benzodiazepine use on infant neurobehavioral functioning over the first postnatal month. Am. J. Psychiatry. https://doi.org/10.1176/appi.ajp.2015.14080989.

Salzbrenner, S., Breeden, A., Jarvis, S., Rodriguez, W., 2011. A 48-year-old woman primigravid via in vitro fertilization with severe bipolar depression and preeclampsia treated successfully with electroconvulsive therapy. J. ECT. https://doi.org/10.1097/YCT.0b013e3181ca4d22.

Santoro, N., Epperson, C.N., Mathews, S.B., 2015. Menopausal symptoms and their management. Endocrinol. Metab. Clin. North Am. 44 (3), 497–515. W.B. Saunders https://doi.org/10.1016/j.ecl.2015.05.001.

Seedat, S., Scott, K.M., Angermeyer, M.C., Berglund, P., Bromet, E.J., Brugha, T.S., Demyttenaere, K., De Girolamo, G., Haro, J.M., Jin, R., Karam, E.G., Kovess-Masfety, V., Levinson, D., Medina Mora, M.E., Ono, Y., Ormel, J., Pennell, B.E., Posada-Villa, J., Sampson, N.A., Kessler, R.C., 2009. Cross-national associations between gender and mental disorders in the World Health Organization World Mental Health Surveys. Arch. Gen. Psychiatry. https://doi.org/10.1001/archgenpsychiatry.2009.36.

Shan, D., Zou, L., Liu, X., Shen, Y., Cai, Y., Zhang, J., 2020. Efficacy and safety of gabapentin and pregabalin in patients with vasomotor symptoms: a systematic review and meta-analysis. Am. J. Obstet. Gynecol. 222 (6), 564–579.e12. Mosby Inc https://doi.org/10.1016/j.ajog.2019.12.011.

Sharma, V., Sharma, P., Sharma, S., 2020. Managing bipolar disorder during pregnancy and the postpartum period: a critical review of current practice. Expert Rev. Neurother. https://doi.org/10.1080/14737175.2020.1743684.

Sharp, D.J., Chew-Graham, C.A., Tylee, A., Lewis, G., Howard, L., Anderson, I., Abel, K., Turner, K.M., Hollinghurst, S.P., Tallon, D., McCarthy, A., Peters, T.J., 2010. A pragmatic randomised controlled trial to compare antidepressants with a community-based psychosocial intervention for the treatment of women with postnatal depression: the RESPOND trial. Health Technol. Assess. https://doi.org/10.3310/hta14430.

Singata-Madliki, M., Hofmeyr, G.J., Lawrie, T.A., 2016. The effect of depot medroxyprogesterone acetate on postnatal depression: a randomised controlled trial. J. Fam. Plann. Reprod. Health Care. https://doi.org/10.1136/jfprhc-2015-101334.

Sit, D., Rothschild, A.J., Wisner, K.L., 2006. A review of postpartum psychosis. J. Womens Health. https://doi.org/10.1089/jwh.2006.15.352.

Siu, A.L., Bibbins-Domingo, K., Grossman, D.C., Baumann, L.C., Davidson, K.W., Ebell, M., García, F.A.R., Gillman, M., Herzstein, J., Kemper, A.R., Krist, A.H., Kurth, A.E., Owens, D.K., Phillips, W.R., Phipps, M.G., Pignone, M.P., 2016. Screening for depression in adults: US preventive services task force recommendation statement. JAMA. https://doi.org/10.1001/jama.2015.18392.

Slaton, R.M., Champion, M.N., Palmore, K.B., 2015. A review of paroxetine for the treatment of vasomotor symptoms. J. Pharm. Pract. 28 (3), 266–274. https://doi.org/10.1177/0897190014544785.

Smit, M., Dolman, K.M., Honig, A., 2016. Mirtazapine in pregnancy and lactation—a systematic review. Eur. Neuropsychopharmacol. https://doi.org/10.1016/j.euroneuro.2015.06.014.

Soares, C.N., Poitras, J.R., Prouty, J., Alexander, A.B., Shifren, J.L., Cohen, L.S., 2003. Efficacy of citalopram as a monotherapy or as an adjunctive treatment to estrogen therapy for perimenopausal and postmenopausal women with depression and vasomotor symptoms. J. Clin. Psychiatry 64 (4), 473–479. https://doi.org/10.4088/JCP.v64n0419.

Soares, C.N., Thase, M.E., Clayton, A., Guico-Pabia, C.J., Focht, K., Jiang, Q., Kornstein, S.G., Ninan, P., Kane, C.P., Cohen, L.S., 2010. Desvenlafaxine and escitalopram for the treatment of postmenopausal women with major depressive disorder. Menopause 17 (4), 700–711. https://doi.org/10.1097/gme.0b013e3181d88962.

Sockol, L.E., 2015. A systematic review of the efficacy of cognitive behavioral therapy for treating and preventing perinatal depression. J. Affect. Disord. https://doi.org/10.1016/j.jad.2015.01.052.

Sodhi, P., Poddar, B., Parmar, V., 2001. Fatal cardiac malformation in fetal valproate syndrome. Indian J. Pediatr. https://doi.org/10.1007/BF02722604.

Soules, M.R., Sherman, S., Parrott, E., Rebar, R., Santoro, N., Utian, W., Woods, N., 2001. Executive summary: stages of Reproductive Aging Workshop (STRAW) Park city, Utah, July, 2001. Menopause 8 (6), 402–407. https://doi.org/10.1097/00042192-200111000-00004.

Stewart, D.E., Robertson, E., Phil, M., Dennis, C., Grace, S.L., Wallington, T., 2003. Postpartum Depression: Literature Review of Risk Factors and Interventions. WHO Publication.

Stowe, Z.N., Casarella, J., Landry, J., Nemeroff, C.B., 1995. Sertraline in the treatment of women with postpartum major depression. Depression. https://doi.org/10.1002/depr.3050030109.

Stowe, Z.N., Hostetter, A.L., Newport, D.J., 2005. The onset of postpartum depression: implications for clinical screening in obstetrical and primary care. Am. J. Obstet. Gynecol. https://doi.org/10.1016/j.ajog.2004.07.054.

Stuart, S., Koleva, H., 2014. Psychological treatments for perinatal depression. Best Pract. Res. Clin. Obstet. Gynaecol. https://doi.org/10.1016/j.bpobgyn.2013.09.004.

Stuenkel, C.A., 2018. Vasomotor and related menopause symptoms. Clin. Obstet. Gynecol. 61 (3), 433–446. https://doi.org/10.1097/GRF.0000000000000385.

Sujan, A.C., Rickert, M.E., Öberg, A.S., Quinn, P.D., Hernández-Díaz, S., Almqvist, C., Lichtenstein, P., Larsson, H., D'Onofrio, B.M., 2017. Associations of maternal antidepressant use during the first trimester of pregnancy with preterm birth, small for gestational age, autism spectrum disorder, and attention-deficit/hyperactivity disorder in offspring. JAMA. https://doi.org/10.1001/jama.2017.3413.

Suri, R., Burt, V.K., Altshuler, L.L., Zuckerbrow-Miller, J., Fairbanks, L., 2001. Fluvoxamine for postpartum depression [7]. Am. J. Psychiatry. https://doi.org/10.1176/appi.ajp.158.10.1739.

Suri, R., Burt, V.K., Altshuler, L.L., 2005. Nefazodone for the treatment of postpartum depression [2]. Arch. Womens Ment. Health. https://doi.org/10.1007/s00737-005-0071-2.

Teatero, M.L., Mazmanian, D., Sharma, V., 2014. Effects of the menstrual cycle on bipolar disorder. Bipolar Disord. https://doi.org/10.1111/bdi.12138.

Thase, M.E., Entsuah, R., Cantillon, M., Kornstein, S.G., 2005. Relative antidepressant efficacy of venlafaxine and SSRIs: sex-age interactions. J. Womens Health 14 (7), 609–616. https://doi.org/10.1089/jwh.2005.14.609.

Thisted, E., Ebbesen, F., 1993. Malformations, withdrawal manifestations, and hypoglycaemia after exposure to valproate in utero. Arch. Dis. Child. https://doi.org/10.1136/adc.69.3_Spec_No.288.

Tomson, T., Lindbom, U., Ekqvist, B., Sundqvist, A., 1994. Epilepsy and pregnancy: a prospective study of seizure control in relation to free and total plasma concentrations of carbamazepine and phenytoin. Epilepsia. https://doi.org/10.1111/j.1528-1157.1994.tb02921.x.

Troyer, W.A., Pereira, G.R., Lannon, R.A., Belik, J., Yoder, M.C., 1993. Association of maternal lithium exposure and premature delivery. J. Perinatol. 13, 123–127.

Tsai, R., Schaffir, J., 2010. Effect of depot medroxyprogesterone acetate on postpartum depression. Contraception. https://doi.org/10.1016/j.contraception.2010.03.004.

Tuovinen, S., Lahti-Pulkkinen, M., Girchenko, P., Lipsanen, J., Lahti, J., Heinonen, K., Reynolds, R.M., Hämäläinen, E., Kajantie, E., Laivuori, H., Pesonen, A.K., Villa, P.M., Räikkönen, K., 2018. Maternal depressive symptoms during and after pregnancy and child developmental milestones. Depress. Anxiety. https://doi.org/10.1002/da.22756.

Uguz, F., Sharma, V., 2016. Mood stabilizers during breastfeeding: a systematic review of the recent literature. Bipolar Disord. https://doi.org/10.1111/bdi.12398.

Uguz, F., Turgut, K., Aydin, A., Ak, M., 2018. Low-dose mirtazapine in major depression developed after hyperemesis gravidarum: a case series. Am. J. Ther. https://doi.org/10.1097/MJT.0000000000000698.

van der Lugt, N.M., van de Maat, J.S., van Kamp, I.L., Knoppert-van der Klein, E.A.M., Hovens, J.G.F.M., Walther, F.J., 2012. Fetal, neonatal and developmental outcomes of lithium-exposed pregnancies. Early Hum. Dev. https://doi.org/10.1016/j.earlhumdev.2011.09.013.

Vermeiden, M., Van Den Broek, W.W., Mulder, P., Birkenhäger, T.K., 2010. Influence of gender and menopausal status on antidepressant treatment response in depressed inpatients. J. Psychopharmacol. 24 (4), 497–502. https://doi.org/10.1177/0269881109105137.

Veroniki, A.A., Rios, P., Cogo, E., Straus, S.E., Finkelstein, Y., Kealey, R., Reynen, E., Soobiah, C., Thavorn, K., Hutton, B., Hemmelgarn, B.R., Yazdi, F., D'Souza, J., Macdonald, H., Tricco, A.C., 2017. Comparative safety of antiepileptic drugs for neurological development in children exposed during pregnancy and breast feeding: a systematic review and network meta-analysis. BMJ Open. https://doi.org/10.1136/bmjopen-2017-017248.

Viguera, A.C., Nonacs, R., Cohen, L.S., Tondo, L., Murray, A., Baldessarini, R.J., 2000. Risk of recurrence of bipolar disorder in pregnant and nonpregnant women after discontinuing lithium maintenance. Am. J. Psychiatry. https://doi.org/10.1176/appi.ajp.157.2.179.

Viguera, A.C., Newport, D.J., Ritchie, J., Stowe, Z., Whitfield, T., Mogielnicki, J., Baldessarini, R.J., Zurick, A., Cohen, L.S., 2007a. Lithium in breast milk and nursing infants: clinical implications. Am. J. Psychiatry. https://doi.org/10.1176/ajp.2007.164.2.342.

Viguera, A.C., Whitfield, T., Baldessarini, R.J., Newport, D.J., Stowe, Z., Reminick, A., Zurick, A., Cohen, L.S., 2007b. Risk of recurrence in women with bipolar disorder during pregnancy: prospective study of mood stabilizer discontinuation. Am. J. Psychiatry. https://doi.org/10.1176/appi.ajp.2007.06101639.

Walker, S.P., Permezel, M., Berkovic, S.F., 2009. The management of epilepsy in pregnancy. BJOG. https://doi.org/10.1111/j.1471-0528.2009.02141.x.

Walsh-Sukys, M.C., Tyson, J.E., Wright, L.L., Bauer, C.R., Korones, S.B., Stevenson, D.K., Verter, J., Stoll, B.J., Lemons, J.A., Papile, L.A., Shankaran, S., Donovan, E.F., Oh, W., Ehrenkranz, R.A., Fanaroff, A.A., 2000. Persistent pulmonary hypertension of the newborn in the era before nitric oxide: practice variation and outcomes. Pediatrics. https://doi.org/10.1542/peds.105.1.14.

Ward, S., Wisner, K.L., 2007. Collaborative management of women with bipolar disorder during pregnancy and postpartum: pharmacologic considerations. J. Midwifery Womens Health. https://doi.org/10.1016/j.jmwh.2006.09.002.

Ward, R.M., Bates, B.A., Benitz, W.E., Burchfield, D.J., Ring, J.C., Walls, R.P., Walson, P.D., 2001. The transfer of drugs and other chemicals into human milk. Pediatrics. https://doi.org/10.1542/peds.108.3.776.

Ward, H.B., Fromson, J.A., Cooper, J.J., De Oliveira, G., Almeida, M., 2018. Recommendations for the use of ECT in pregnancy: literature review and proposed clinical protocol. Arch. Womens Ment. Health. https://doi.org/10.1007/s00737-018-0851-0.

Wesseloo, R., Wierdsma, A.I., Van Kamp, I.L., Munk-Olsen, T., Hoogendijk, W.J.G., Kushner, S.A., Bergink, V., 2017. Lithium dosing strategies during pregnancy and the postpartum period. Br. J. Psychiatry. https://doi.org/10.1192/bjp.bp.116.192799.

Westin, A.A., Brekke, M., Molden, E., Skogvoll, E., Castberg, I., Spigset, O., 2018. Treatment with antipsychotics in pregnancy: changes in drug disposition. Clin. Pharmacol. Ther. https://doi.org/10.1002/cpt.770.

Wide, K., Winbladh, B., Tomson, T., Källén, B., 2000a. Body dimensions of infants exposed to antiepileptic drugs in utero: observations spanning 25 years. Epilepsia. https://doi.org/10.1111/j.1528-1157.2000.tb00253.x.

Wide, K., Winbladh, B., Tomson, T., Sars-Zimmer, K., Berggren, E., 2000b. Psychomotor development and minor anomalies in children exposed to antiepileptic drugs in utero: a prospective population-based study. Dev. Med. Child Neurol. https://doi.org/10.1017/S0012162200000177.

Willi, J., Süss, H., Grub, J., Ehlert, U., 2020. Prior depression affects the experience of the perimenopause—findings from the Swiss Perimenopause Study. J. Affect. Disord. 277, 603–611. https://doi.org/10.1016/j.jad.2020.08.062.

Wilson, N., Forfar, J.C., Godman, M.J., 1983. Atrial flutter in the newborn resulting from maternal lithium ingestion. Arch. Dis. Child. https://doi.org/10.1136/adc.58.7.538.

Wisner, K.L., Hanusa, B.H., Perel, J.M., Peindl, K.S., Piontek, C.M., Sit, D.K.Y., Findling, R.L., Moses-Kolko, E.L., 2006. Postpartum depression: a randomized trial of sertraline versus nortriptyline. J. Clin. Psychopharmacol. https://doi.org/10.1097/01.jcp.0000227706.56870.dd.

Wisner, K.L., Sit, D.K.Y., Moses-Kolko, E.L., Driscoll, K.E., Prairie, B.A., Stika, C.S., Eng, H.F., Dills, J.L., Luther, J.F., Wisniewski, S.R., 2015. Transdermal estradiol treatment for postpartum depression: a pilot, randomized trial. J. Clin. Psychopharmacol. https://doi.org/10.1097/JCP.0000000000000351.

Woods, N.F., Mitchell, E.S., 2016. The Seattle Midlife Women's Health Study: a longitudinal prospective study of women during the menopausal transition and early postmenopause. Women's Midlife Health 2 (1). https://doi.org/10.1186/s40695-016-0019-x.

Woody, J.N., London, W.L., Wilbanks, G.D., 1971. Lithium toxicity in a newborn. Pediatrics 47, 94–96.

Yatham, L.N., Kennedy, S.H., Parikh, S.V., Schaffer, A., Bond, D.J., Frey, B.N., Sharma, V., Goldstein, B.I., Rej, S., Beaulieu, S., Alda, M., Mac Queen, G., Milev, R.V., Ravindran, A., O'Donovan, C., McIntosh, D., Lam, R.W., Vazquez, G., Kapczinski, F., Berk, M., 2018. Canadian Network for Mood and Anxiety Treatments (CANMAT) and International Society for Bipolar Disorders (ISBD) 2018 guidelines for the management of patients with bipolar disorder. Bipolar Disord. https://doi.org/10.1111/bdi.12609.

Yerby, M.S., Friel, P.N., McCormick, K., Koerner, M., Van Allen, M., Leavitt, A.M., Sells, C.J., Yerby, J.A., 1990. Pharmacokinetics of anticonvulsants in pregnancy: alterations in plasma protein binding. Epilepsy Res. https://doi.org/10.1016/0920-1211(90)90042-T.

Yonkers, K.A., Kando, J.C., Cole, J.O., Blumenthal, S., 1992. Gender differences in pharmacokinetics and pharmacodynamics of psychotropic medication. Am. J. Psychiatry. https://doi.org/10.1176/ajp.149.5.587.

Yonkers, K.A., Lin, H., Howell, H.B., Heath, A.C., Cohen, L.S., 2008. Pharmacologic treatment of postpartum women with new-onset major depressive disorder: a randomized controlled trial with paroxetine. J. Clin. Psychiatry. https://doi.org/10.4088/JCP.v69n0420.

Yonkers, K.A., Wisner, K.L., Stewart, D.E., Oberlander, T.F., Dell, D.L., Stotland, N., Ramin, S., Chaudron, L., Lockwood, C., 2009. The management of depression during pregnancy: a report from the American Psychiatric Association and the American College of Obstetricians and Gynecologists. Gen. Hosp. Psychiatry. https://doi.org/10.1016/j.genhosppsych.2009.04.003.

Zeleke, B.M., Bell, R.J., Billah, B., Davis, S.R., 2017. Vasomotor symptoms are associated with depressive symptoms in community-dwelling older women. Menopause 24 (12), 1365–1371. https://doi.org/10.1097/GME.0000000000000938.

Zhang, X., Liu, K., Sun, J., Zheng, Z., 2010. Safety and feasibility of repetitive transcranial magnetic stimulation (rTMS) as a treatment for major depression during pregnancy. Arch. Womens Ment. Health. https://doi.org/10.1007/s00737-010-0163-5.

Zuckerman, B., Amaro, H., Bauchner, H., Cabral, H., 1989. Depressive symptoms during pregnancy: relationship to poor health behaviors. Am. J. Obstet. Gynecol. https://doi.org/10.1016/0002-9378(89)90170-1.

Chapter 38

Lifestyle interventions for treatment-resistant depression

Marisa Toups
Singularity Psychiatry and Mental Health, Austin, TX, United States

Introduction

There are several common circumstances in which treatments outside the traditional Western Allopathic medical tradition can be useful for patients with treatment-resistant depression (TRD). While many patients benefit only partially from maximized medication and psychotherapy and desire additional evidence-based recommendations, a holistic approach has other advantages. First, many patients request alternative and complementary treatments with a goal of avoiding or reducing pharmacological therapies and their side effects. Second, behavioral changes that fall outside the commonly used psychotherapeutic approaches can be addressed via complementary approaches. Finally, patients may come from cultural or personal perspectives that value a mind-body approach which when used adjunctively with traditional treatment can improve engagement and adherence and strengthen the therapeutic alliance. Fortunately, many interventions have sufficient clinical evidence of efficacy and underlying scientific and mechanistic foundation to provide a strong menu of recommendations for mental health professionals. Here we will review diet, exercise, mindfulness, and closely related interventions, including some that cross domains.

Diet and nutrition

Although the idea the diet is an important determinant of health, including mental health, has persisted for thousands of years, there is little high-quality contemporary evidence defining "healthy diet" with respect to specific mental disorders. In general, diets are described by ratios of macronutrients and their subtypes (e.g., saturated vs unsaturated fats), by food type (e.g., legumes, dairy, vegetables), by micronutrient and/or fiber content (e.g., low sodium), or a combination of these (Ge et al., 2020). The concept of "health" in turn is defined based on long-term weight, cardiovascular, and metabolic outcomes (Cena and Calder, 2020). Thus, expert consensus defines a "healthy diet" as one with carbohydrate > protein > fat caloric macronutrient content, consisting largely of minimally processed plant-derived foods to ensure sufficient fiber, vitamins, and micronutrients (U.S.D.o.A.a.U.S.D.o.H.a.H. Services, 2020).

Most studies focused on diet with respect to major depressive disorder (MDD) are epidemiological, with some clinical trials of diet education interventions in clinical or population samples using self-reported assessments. Mediterranean diets and those conforming to Food and Drug Administration or European health agency guidelines are the most commonly studied. Overall, these data consistently support that healthier diets, grossly defined, are associated with better mood. At the population level, multiple metrics of diet health are associated with lower current and longitudinal depression symptoms (Lassale et al., 2019). Similarly, trials of diet education/planning interventions in various clinical populations find small decreases in depression symptoms (Firth et al., 2019). One small, low quality clinical trial has been specifically designed to reduce depression in a sample with investigator validated MDD; it found that a Mediterranean diet support program was superior to "social support" control (Jacka et al., 2017; Molendijk et al., 2018). However, three trials in obese depressed adults suggest that diet interventions designed to improve metabolic health and lower weight also improve depression symptoms (Linde et al., 2011; Ma et al., 2019; Pagoto et al., 2013). Unfortunately, there are no large, high-quality trials of direct diet interventions and none which recruited subjects who were identified as treatment resistant.

Despite the lack of TRD-specific data, the close relationship between mood and metabolic outcomes means that clinicians can confidently recommend the same diet recommended to the general population to patients. Additional recommendations may apply for those with TRD (McIntyre et al., 2012). Patients should be screened and treated appropriately for eating disorders, particularly binge eating disorder which has a high rate of comorbidity with depression (Baek et al., 2018;

Managing Treatment-Resistant Depression. https://doi.org/10.1016/B978-0-12-824067-0.00038-4

Woldeyohannes et al., 2016). The impact of commonly used adjunctive medications that can increase appetite and weight, as well as directly induce metabolic changes should be considered and minimized (Grajales et al., 2019). Good quality evidence supports the use of metformin or topiramate for metabolic risk reduction in patients on atypical antipsychotics regardless of indication (Zhuo et al., 2018).

Probiotics

Extensive evidence from animal models links the gut microbiome to MDD. Experimental inductions of depression-like behaviors induce changes in microbiome composition, and vice versa (Yang et al., 2020). Furthermore, comparisons of depressed human subjects to healthy controls consistently find differences in microbiome composition; although the nature of the differences is variable, a possible theme may be a reduction in taxa producing short chain fatty acids and other metabolically active or antiinflammatory compounds (Sanada et al., 2020). Lack of consensus on the nature of microbial changes in MDD has not impeded the investigation of probiotic supplements in clinical trials. Overall, there is a small reduction in depression symptoms across treatments (Huang et al., 2016; Liu et al., 2019), with those including more rigorously defined clinical samples showing a slightly larger effect (Sanada et al., 2020; Noonan et al., 2020). Only two nonrandomized studies recruited patients with rigorously defined TRD (Bambling et al., 2017; Miyaoka et al., 2018). This remains an area in which more work is needed to make clinical recommendations, particularly as the risks of probiotic supplements are poorly understood; cases of bacterial over-growth, including small bowel colonization, and infection in immunocompromised persons, have been reported (Didari et al., 2014).

Mindfulness

Broadly, mindfulness describes the practice of "bare attention" a common English translation of the Buddhist term *sati*, and in this sense is a subtype of traditional Buddhist meditation originating in China and best known in the western world via Japanese Zen Buddhism (Sharf, 2014). Modern mindfulness practice aims to foster mental calm, improve awareness of physical and mental states, and develop self-compassion (Crane et al., 2017). Clinically, mindfulness programs have been designed to improve tolerance of negative affect, reduce symptoms of depression, and facilitate healthy regulation of emotion and behavior (Bishop et al., 2004). Meditation can be practiced by patients independently, with support from a large variety of written, audio, and digital guides, or within a psychotherapeutic frame. Alternatively, manualized therapies incorporating meditation—"mindfulness-based interventions" (MBIs)—integrate secular mindfulness exercises (meditation) with group psychoeducation and psychotherapeutic techniques (Crane et al., 2017).

The two major MBIs are mindfulness-based stress reduction (MBSR) developed for patients with chronic, painful, or debilitating medical illness and comorbid psychiatric symptoms (Kabat-Zinn, 2013), and mindfulness-based cognitive therapy (MBCT), developed for relapse prevention in MDD (Segal et al., 2013). MBCT has been studied specifically for TRD in several trials, while MBSR has been applied more specifically to treatment resistance in those with chronic medical illness (Iosifescu et al., 2003). Other interventions may use concepts of mindfulness without meditation practice (e.g., Acceptance and Commitment Therapy); here we focus on therapies featuring the practice of regular mediation.

Overall, MBIs are superior to treatment-as-usual, and nonactive or active nonspecific controls and at least as effective as other evidence-based therapies; when used as augmentation MBIs have a larger effect size, suggesting that they may be more effective in TRD patients compared to the general MDD population (Goldberg et al., 2018; van Agteren et al., 2021; Wang et al., 2018). MBCT, unsurprisingly, may be more effective for relapse prevention than acute treatment (Parikh et al., 2016) but some studies in TRD have shown superiority to active control conditions (Chiesa et al., 2015; Eisendrath et al., 2016). MBSR appears to be effective for depression symptoms in a wide array of clinical conditions, with a small effect size, when compared to inactive control, and is more effective for anxiety and somatic symptoms (e.g., pain, fatigue) than for core symptoms of depression (Bohlmeijer et al., 2010; Haugmark et al., 2019). Limited data suggest similar efficacy for a primary diagnosis of MDD without comorbidity (McIndoo et al., 2016). Independent practice of mindfulness exercises (e.g., body scan, breathing or sitting meditation) using audio, or video guide materials is common, and more feasible than MBIs, though it appears to have less benefit (Blanck et al., 2018). Adherence may mitigate this gap in efficacy; however, there are little data from samples with well validated MDD or in which independent mindfulness is used as augmentation. Similarly, studies with active controls show little or no benefit to independent practice (Cavanagh et al., 2014).

Overall, it appears likely that patients with TRD can benefit from mindfulness interventions, but may require the higher level of structure and active support of an MBI to maintain engaged practice and obtain symptomatic improvement. MBCT has more applicability and evidence for use in TRD, but patients dealing with pain or loss of function from medical illness are good candidates for treatment augmentation with MBSR. While sparse cases of induction of psychotic or manic symptoms following intense meditation have been reported, mindfulness is considered low risk, and can be safely recommended for most patients.

Mindful movement

Mindful movement practices combine physical activity with mindful breathing and meditation. Although commonly used for wellness, these practices have been adapted less rigorously to treatment of mental illness, and have less data examining their use as treatment compared to MBIs (Larkey et al., 2009; Solloway et al., 2016). Yoga combines physical postures (asanas), breath control, and mindful attention, and has the largest evidence base for MDD. Modern practice may involve significant athletic, even aerobic, activity but this style of practice dates only to the mid-20th century, while older forms emphasize meditation (Singleton, 2010). Although there are methodological limitations to published studies (Elwy et al., 2014), yoga appears effective for depression symptoms compared to inactive control conditions and roughly equivalent to other exercise or mindfulness (Cramer et al., 2013; Ravindran et al., 2016). Data are decidedly more mixed for well-defined samples with MDD (Bieber et al., 2021; Cramer et al., 2017; Vollbehr et al., 2018). Two studies comparing samples with some degree of treatment resistance to active control were both negative (Butler et al., 2008; Uebelacker et al., 2017). Some data suggest that in clinical samples with more severe symptoms lower adherence may contribute to negative findings (Nauphal et al., 2019; Uebelacker et al., 2018). Additionally, the rate of yoga associated adverse events may be higher than for either aerobic exercise or meditation (Cramer et al., 2018, 2019).

With roots stretching back centuries in traditional Chinese Medicine, Tai Chi and Qi Gong share components of body posture, fluid movement, and synchronization with breathing (Yeung et al., 2018). Both practices show efficacy for symptoms of depression compared to inactive control conditions, and light activity (e.g., walking) though studies are of low quality (Liu et al., 2015; Wang et al., 2013). One study of moderate quality examined Tai Chi vs health education augmentation of SSRI in older depressed adults and found it superior for mood improvement (Lavretsky et al., 2011), but there are no data specifically pertaining to TRD. Tai Chi appears to be low risk (Wayne et al., 2014).

Both the quality of evidence and the support for efficacy of mindful movement are less strong than for MBIs, or for exercise (discussed later), in particular, there is a lack of support for use of mindful movement in patients with TRD. However, in patients who express a preference for mindful movement, they can be recommended as adjuncts, keeping in mind that patients who have low fitness/flexibility should use caution in undertaking a yoga practice due to its relatively high rate of minor musculoskeletal injuries. As with other mindful interventions, support structures which enhance adherence, such as supervision and group education may improve the benefits associated with mindful movement.

Exercise

There is significant evidence substantiating the efficacy of exercise as a treatment for MDD with moderate effect size, similar to other evidence supported treatments (Cooney et al., 2013; Krogh et al., 2017). Larger effects are found in studies including only inactive controls and smaller effects in studies with more rigorous trial designs (Cooney et al., 2013; Krogh et al., 2017; Schuch et al., 2016); however, the overall study quality is close to that of pharmacologic trials (Krogh et al., 2015). Generally, studies examining exercise as a combination or adjunctive treatment with antidepressant medication find similar results (Mura et al., 2014). There are lack of data specifically examining well-defined TRD, with only two very small trials (Mota-Pereira et al., 2011; Pilu et al., 2007).

The majority of trials have incorporated aerobic exercise, which is easier to standardize across subjects in units of kilocalories expended in a given study period, while a minority prescribe resistance training or a combination. It appears likely that these types of exercise are similar in efficacy for symptom reduction (Cooney et al., 2013; Gordon et al., 2018). Similarly, differences in dosing (time/week) and intensity (light, moderate, vigorous) seem to have small or no difference in symptom reduction, although this may be due to lower adherence to higher prescribed doses. Findings of no difference in trials with active controls such as walking or stretching (Trivedi et al., 2006) suggest that some of the efficacy of exercise comes from shared elements of behavioral activation. There are lack of data on moderators/mediators of outcome other than adherence, which is often low, and likely explains findings that supervision from a professional trainer, group or other structured programs may have superior efficacy (Krogh et al., 2017; Schuch et al., 2016). In general, enjoyment of exercise rather than self-efficacy (i.e., the feeling that one is engaging in a "should do" activity) is associated with long term engagement in exercise, and suggests the importance of patient preference in choosing among the various options for type, dosing, and intensity (Lewis et al., 2016). Surprisingly, few studies have rigorously collected and reported adverse event data, and risks vary substantially by type of exercise. All of these factors point toward recommendations of exercise with supervision from professionals on an individual or group basis in which patients can engage in activities that fit with their personal preferences and physical capabilities.

Acupuncture

Acupuncture is a centuries-old component of traditional Chinese medicine in which specified body points are stimulated, traditionally using fine needles, to manipulate a body force called Qi (Chon and Lee, 2013). It has been applied broadly to a variety of clinical conditions but the evidence base is limited by a failure to harmonize the concept of Qi with modern physiology and by low to very low quality of clinical trials (Li et al., 2020). Additionally, MDD is nosologically inconsistent with traditional Chinese medicine, so there is no consensus on which points should be targeted (Smith et al., 2018). Among many positive results, studies which use high quality design with adequate blinding of sham procedures find no effect (Smith et al., 2018). Of 13 studies acupuncture augmentation of antidepressant treatment, only two reported *any* blinding, and the overall effect size was negligible (Chan et al., 2015). Although most of the interventions included here share similar challenges in trial design and execution, high-level interpretation of acupuncture data suggests effects are likely due to placebo effect (Kong et al., 2013). Acupuncture cannot be recommended for depression or for TRD as a primary or secondary intervention. However, risks associated with acupuncture are low, and it may be helpful for patients who desire to seek it out for treatment.

Common mechanisms of action of lifestyle modifications

All of these lifestyle interventions share a foundational assumption that the health of the body and mind are linked, so it is not surprising that they share proposed and probable mechanisms of action based on the interaction of somatic systems with the nervous system. These interactions are mediated primarily by humoral signaling molecules and the autonomic nervous system (Critchley and Harrison, 2013; Dantzer et al., 2000; Moura-Assis et al., 2021; Pedersen, 2019); the brain processes somatic information via the insula-centered interoceptive system (Kleckner et al., 2017), and the hypothalamic system, important as a hub for metabolic control as well as the stress response (Waterson and Horvath, 2015). These systems are interlinked and modulate other brain circuits critical for depression, such as the limbic system, and have been implicated repeatedly in the neurobiology of MDD (Pariante and Lightman, 2008; Quadt et al., 2018).

Chronic increases in proinflammatory cytokines are now a well-established finding in MDD (Osimo et al., 2020), and likely form a central hub in the network of somatic mechanisms of action of diet, exercise, and mindfulness. Chronic inflammation is associated with illness (e.g., autoimmune disease), obesity, and glucocorticoid resistance, all of which increase the risk of depression (Anacker et al., 2011; Evans et al., 2005; Luppino et al., 2010). The rate of overweight and obesity in depression is about twice that of the general population, and adiposity appears to be the single largest predictor of inflammation in depressed adults (Ambrosio et al., 2018). Impairments in cortisol signaling and interoceptive autonomic nervous system efferent activity also increase peripheral inflammation by reducing feedback inhibition on the immune system, may contribute to further adiposity by impairing hypothalamic metabolic control and accurate sensation of hunger and satiety, respectively (Andersson and Tracey, 2012; Milaneschi et al., 2019; Quax et al., 2013). Inflammation alters functional connectivity to reduce hedonic drive and increase feels of social exclusion (Brydon et al., 2008; Harrison et al., 2009). Local CNS inflammation impairs neuroplasticity and neurogenesis by decreasing the production of neurotrophins and increasing that of neurotoxic kynurenine metabolites (Dantzer et al., 2008).

While data are not fully consistent, there is a probable clinical association between treatment resistance and chronic inflammation (Strawbridge et al., 2015; Woo et al., 2016), as well as some evidence that antiinflammatory therapies may successfully treat TRD (Yang et al., 2019). While it is probable that obesity associated inflammation largely explains the *consistency* of the long-term association between metabolically healthy diets and mood, measures of diet associated inflammation are often better predictors of depression symptoms (Lassale et al., 2019; Busch et al., 2013; Fabricatore et al., 2011; Jones et al., 2021; Chocano-Bedoya et al., 2013; Lucas et al., 2014). Exercise broadly decreases indices of inflammation, and some data support particular efficacy in those who have elevated cytokines at study entry (Eyre et al., 2013; Rethorst et al., 2013). Exercise also directly increases neurotrophin production and reduces the production of neurotoxic kynurenine metabolites, counteracting the impact of inflammation in the brain (Pedersen, 2019). Mindfulness and mindful movement decrease peripheral inflammation, though perhaps less robustly than exercise, and may also decrease cortisol resistance (Morgan et al., 2014; Pascoe et al., 2017a,b; So et al., 2019).

At the brain circuit level, lifestyle interventions may exert antidepressant effects by improving insula dependent interoception. Accurate interoception is related to higher resting functional connectivity between insular regions and between the insula and other parts of the salience network (Chong et al., 2017), and the ability to functionally disconnect and suppress insula activity during somatic attention tasks and those requiring attention to outside stimuli, respectively (Craig, 2009; Kuehn et al., 2016). Depressed subjects have lower interoceptive accuracy, increased resting activity and functional connectivity of the anterior insula compared to healthy subjects, and show impairments in their ability to modify insula activity and connectivity in tasks requiring shifting attention (Sliz and Hayley, 2012). Several studies have found that excess anterior insula activity correlates with treatment resistance to antidepressants or TMS (Sliz and Hayley, 2012; Geugies et al., 2019; Kelley et al., 2021). Although there are lack of data in depressed samples, mindfulness training improves interoceptive accuracy (Treves et al., 2019) and may normalize insula activity and integration into the salience network (Gotink et al., 2016; Young et al., 2018).

References

Ambrosio, G., et al., 2018. Depression and peripheral inflammatory profile of patients with obesity. Psychoneuroendocrinology 91, 132–141.

Anacker, C., et al., 2011. The glucocorticoid receptor: pivot of depression and of antidepressant treatment? Psychoneuroendocrinology 36 (3), 415–425.

Andersson, U., Tracey, K.J., 2012. Reflex principles of immunological homeostasis. Annu. Rev. Immunol. 30, 313–335.

Baek, J.H., et al., 2018. Binge eating, trauma, and suicide attempt in community adults with major depressive disorder. PLoS One 13 (6), e0198192.

Bambling, M., et al., 2017. A combination of probiotics and magnesium orotate attenuate depression in a small SSRI resistant cohort: an intestinal anti-inflammatory response is suggested. Inflammopharmacology 25 (2), 271–274.

Bieber, M., et al., 2021. Effects of body-oriented yoga: a RCT study for patients with major depressive disorder. Eur. Arch. Psychiatry Clin. Neurosci. 271, 1217–1229.

Bishop, S.R., et al., 2004. Mindfulness: a proposed operational definition. Clin. Psychol. Sci. Pract. 11 (3), 230–241.

Blanck, P., et al., 2018. Effects of mindfulness exercises as stand-alone intervention on symptoms of anxiety and depression: systematic review and meta-analysis. Behav. Res. Ther. 102, 25–35.

Bohlmeijer, E., et al., 2010. The effects of mindfulness-based stress reduction therapy on mental health of adults with a chronic medical disease: a meta-analysis. J. Psychosom. Res. 68 (6), 539–544.

Brydon, L., et al., 2008. Peripheral inflammation is associated with altered substantia nigra activity and psychomotor slowing in humans. Biol. Psychiatry 63 (11), 1022–1029.

Busch, A.M., et al., 2013. Reliable change in depression during behavioral weight loss treatment among women with major depression. Obesity (Silver Spring) 21 (3), E211–E218.

Butler, L.D., et al., 2008. Meditation with yoga, group therapy with hypnosis, and psychoeducation for long-term depressed mood: a randomized pilot trial. J. Clin. Psychol. 64 (7), 806–820.

Cavanagh, K., et al., 2014. Can mindfulness and acceptance be learnt by self-help?: a systematic review and meta-analysis of mindfulness and acceptance-based self-help interventions. Clin. Psychol. Rev. 34 (2), 118–129.

Cena, H., Calder, P.C., 2020. Defining a healthy diet: evidence for the role of contemporary dietary patterns in health and disease. Nutrients 12 (2), 334.

Chan, Y.Y., et al., 2015. The benefit of combined acupuncture and antidepressant medication for depression: a systematic review and meta-analysis. J. Affect. Disord. 176, 106–117.

Chiesa, A., et al., 2015. Mindfulness-based cognitive therapy vs. psycho-education for patients with major depression who did not achieve remission following antidepressant treatment. Psychiatry Res. 226 (2–3), 474–483.

Chocano-Bedoya, P.O., et al., 2013. Prospective study on long-term dietary patterns and incident depression in middle-aged and older women. Am. J. Clin. Nutr. 98 (3), 813–820.

Chon, T.Y., Lee, M.C., 2013. Acupuncture. Mayo Clin. Proc. 88 (10), 1141–1146.

Chong, J.S.X., et al., 2017. Salience network connectivity in the insula is associated with individual differences in interoceptive accuracy. Brain Struct. Funct. 222 (4), 1635–1644.

Cooney, G.M., et al., 2013. Exercise for depression. Cochrane Database Syst. Rev. 9, CD004366.

Craig, A.D., 2009. How do you feel-now? The anterior insula and human awareness. Nat. Rev. Neurosci. 10 (1), 59–70.

Cramer, H., et al., 2013. Yoga for depression: a systematic review and meta-analysis. Depress. Anxiety 30 (11), 1068–1083.

Cramer, H., et al., 2017. A systematic review of yoga for major depressive disorder. J. Affect. Disord. 213, 70–77.

Cramer, H., Ostermann, T., Dobos, G., 2018. Injuries and other adverse events associated with yoga practice: a systematic review of epidemiological studies. J. Sci. Med. Sport 21 (2), 147–154.

Cramer, H., et al., 2019. Adverse effects of yoga: a national cross-sectional survey. BMC Complement. Altern. Med. 19 (1), 190.

Crane, R.S., et al., 2017. What defines mindfulness-based programs? The warp and the weft. Psychol. Med. 47 (6), 990–999.

Critchley, H.D., Harrison, N.A., 2013. Visceral influences on brain and behavior. Neuron 77 (4), 624–638.

Dantzer, R., et al., 2000. Neural and humoral pathways of communication from the immune system to the brain: parallel or convergent? Auton. Neurosci. 85 (1–3), 60–65.

Dantzer, R., et al., 2008. From inflammation to sickness and depression: when the immune system subjugates the brain. Nat. Rev. Neurosci. 9 (1), 46–56.

Didari, T., et al., 2014. A systematic review of the safety of probiotics. Expert Opin. Drug Saf. 13 (2), 227–239.

Eisendrath, S.J., et al., 2016. A randomized controlled trial of mindfulness-based cognitive therapy for treatment-resistant depression. Psychother. Psychosom. 85 (2), 99–110.

Elwy, A.R., et al., 2014. A systematic scoping review of yoga intervention components and study quality. Am. J. Prev. Med. 47 (2), 220–232.

Evans, D.L., et al., 2005. Mood disorders in the medically ill: scientific review and recommendations. Biol. Psychiatry 58 (3), 175–189.

Eyre, H.A., Papps, E., Baune, B.T., 2013. Treating depression and depression-like behavior with physical activity: an immune perspective. Front. Psychiatry 4, 3.

Fabricatore, A.N., et al., 2011. Intentional weight loss and changes in symptoms of depression: a systematic review and meta-analysis. Int. J. Obes. 35 (11), 1363–1376.

Firth, J., et al., 2019. The effects of dietary improvement on symptoms of depression and anxiety: a meta-analysis of randomized controlled trials. Psychosom. Med. 81 (3), 265–280.

Ge, L., et al., 2020. Comparison of dietary macronutrient patterns of 14 popular named dietary programmes for weight and cardiovascular risk factor reduction in adults: systematic review and network meta-analysis of randomised trials. BMJ 369, m696.

Geugies, H., et al., 2019. Decreased functional connectivity of the insula within the salience network as an indicator for prospective insufficient response to antidepressants. NeuroImage Clin. 24, 102064.

Goldberg, S.B., et al., 2018. Mindfulness-based interventions for psychiatric disorders: a systematic review and meta-analysis. Clin. Psychol. Rev. 59, 52–60.

Gordon, B.R., et al., 2018. Association of efficacy of resistance exercise training with depressive symptoms: meta-analysis and meta-regression analysis of randomized clinical trials. JAMA Psychiatry 75 (6), 566–576.

Gotink, R.A., et al., 2016. 8-week Mindfulness Based Stress Reduction induces brain changes similar to traditional long-term meditation practice – a systematic review. Brain Cogn. 108, 32–41.

Grajales, D., Ferreira, V., Valverde, A.M., 2019. Second-generation antipsychotics and dysregulation of glucose metabolism: beyond weight gain. Cells 8 (11), 1336.

Harrison, N.A., et al., 2009. Inflammation causes mood changes through alterations in subgenual cingulate activity and mesolimbic connectivity. Biol. Psychiatry 66 (5), 407–414.

Haugmark, T., et al., 2019. Mindfulness- and acceptance-based interventions for patients with fibromyalgia – a systematic review and meta-analyses. PLoS One 14 (9), e0221897.

Huang, R., Wang, K., Hu, J., 2016. Effect of probiotics on depression: a systematic review and meta-analysis of randomized controlled trials. Nutrients 8 (8), 483.

Iosifescu, D.V., et al., 2003. The impact of medical comorbidity on acute treatment in major depressive disorder. Am. J. Psychiatry 160 (12), 2122–2127.

Jacka, F.N., et al., 2017. A randomised controlled trial of dietary improvement for adults with major depression (the 'SMILES' trial). BMC Med. 15 (1), 23.

Jones, R.A., et al., 2021. The impact of adult behavioural weight management interventions on mental health: a systematic review and meta-analysis. Obes. Rev. 22 (4), e13150.

Kabat-Zinn, J., 2013. Full Catastrophe Living: Using the Wisdom of your Body and Mind to Face Stress, Pain, and Illness, revised and updated ed. vol. xlv Bantam Books Trade Paperback, New York, p. 650.

Kelley, M.E., et al., 2021. Establishing evidence for clinical utility of a neuroimaging biomarker in major depressive disorder: prospective testing and implementation challenges. Biol. Psychiatry 90, 236–242.

Kleckner, I.R., et al., 2017. Evidence for a large-scale brain system supporting allostasis and interoception in humans. Nat. Hum. Behav. 1, 0069.

Kong, J., et al., 2013. Are all placebo effects equal? Placebo pills, sham acupuncture, cue conditioning and their association. PLoS One 8 (7), e67485.

Krogh, J., et al., 2015. DEPERROR: risks of systematic errors in drug and non-drug randomized clinical trials assessing intervention effects in patients with unipolar depression. J. Affect. Disord. 179, 121–127.

Krogh, J., et al., 2017. Exercise for patients with major depression: a systematic review with meta-analysis and trial sequential analysis. BMJ Open 7 (9), e014820.

Kuehn, E., et al., 2016. Interoceptive awareness changes the posterior insula functional connectivity profile. Brain Struct. Funct. 221 (3), 1555–1571.

Larkey, L., et al., 2009. Meditative movement as a category of exercise: implications for research. J. Phys. Act. Health 6 (2), 230–238.

Lassale, C., et al., 2019. Healthy dietary indices and risk of depressive outcomes: a systematic review and meta-analysis of observational studies. Mol. Psychiatry 24 (7), 965–986.

Lavretsky, H., et al., 2011. Complementary use of tai chi chih augments escitalopram treatment of geriatric depression: a randomized controlled trial. Am. J. Geriatr. Psychiatry 19 (10), 839–850.

Lewis, B.A., et al., 2016. Self-efficacy versus perceived enjoyment as predictors of physical activity behaviour. Psychol. Health 31 (4), 456–469.

Li, M., et al., 2020. The effectiveness and safety of acupuncture for depression: an overview of meta-analyses. Complement. Ther. Med. 50, 102202.

Linde, J.A., et al., 2011. A randomized controlled trial of behavioral weight loss treatment versus combined weight loss/depression treatment among women with comorbid obesity and depression. Ann. Behav. Med. 41 (1), 119–130.

Liu, X., et al., 2015. A systematic review and meta-analysis of the effects of Qigong and Tai Chi for depressive symptoms. Complement. Ther. Med. 23 (4), 516–534.

Liu, R.T., Walsh, R.F.L., Sheehan, A.E., 2019. Prebiotics and probiotics for depression and anxiety: a systematic review and meta-analysis of controlled clinical trials. Neurosci. Biobehav. Rev. 102, 13–23.

Lucas, M., et al., 2014. Inflammatory dietary pattern and risk of depression among women. Brain Behav. Immun. 36, 46–53.

Luppino, F.S., et al., 2010. Overweight, obesity, and depression: a systematic review and meta-analysis of longitudinal studies. Arch. Gen. Psychiatry 67 (3), 220–229.

Ma, J., et al., 2019. Effect of integrated behavioral weight loss treatment and problem-solving therapy on body mass index and depressive symptoms among patients with obesity and depression: the RAINBOW randomized clinical trial. JAMA 321 (9), 869–879.

McIndoo, C.C., et al., 2016. Mindfulness-based therapy and behavioral activation: a randomized controlled trial with depressed college students. Behav. Res. Ther. 77, 118–128.

McIntyre, R.S., et al., 2012. The Canadian Network for Mood and Anxiety Treatments (CANMAT) task force recommendations for the management of patients with mood disorders and comorbid metabolic disorders. Ann. Clin. Psychiatry 24 (1), 69–81.

Milaneschi, Y., et al., 2019. Depression and obesity: evidence of shared biological mechanisms. Mol. Psychiatry 24 (1), 18–33.

Miyaoka, T., et al., 2018. *Clostridium butyricum* MIYAIRI 588 as adjunctive therapy for treatment-resistant major depressive disorder: a prospective open-label trial. Clin. Neuropharmacol. 41 (5), 151–155.

Molendijk, M.L., Fried, E.I., Van der Does, W., 2018. The SMILES trial: do undisclosed recruitment practices explain the remarkably large effect? BMC Med. 16 (1), 243.

Morgan, N., et al., 2014. The effects of mind-body therapies on the immune system: meta-analysis. PLoS One 9 (7), e100903.

Mota-Pereira, J., et al., 2011. Moderate exercise improves depression parameters in treatment-resistant patients with major depressive disorder. J. Psychiatr. Res. 45 (8), 1005–1011.

Moura-Assis, A., Friedman, J.M., Velloso, L.A., 2021. Gut-to-brain signals in feeding control. Am. J. Physiol. Endocrinol. Metab. 320 (2), E326–E332.

Mura, G., et al., 2014. Exercise as an add-on strategy for the treatment of major depressive disorder: a systematic review. CNS Spectr. 19 (6), 496–508.

Nauphal, M., et al., 2019. Yoga for the treatment of depression: five questions to move the evidence-base forward. Complement. Ther. Med. 46, 153–157.

Noonan, S., et al., 2020. Food & mood: a review of supplementary prebiotic and probiotic interventions in the treatment of anxiety and depression in adults. BMJ Nutr. Prev. Health 3 (2), 351–362.

Osimo, E.F., et al., 2020. Inflammatory markers in depression: a meta-analysis of mean differences and variability in 5,166 patients and 5,083 controls. Brain Behav. Immun. 87, 901–909.

Pagoto, S., et al., 2013. Randomized controlled trial of behavioral treatment for comorbid obesity and depression in women: the Be Active Trial. Int. J. Obes. 37 (11), 1427–1434.

Pariante, C.M., Lightman, S.L., 2008. The HPA axis in major depression: classical theories and new developments. Trends Neurosci. 31 (9), 464–468.

Parikh, S.V., et al., 2016. Canadian Network for Mood and Anxiety Treatments (CANMAT) 2016 clinical guidelines for the management of adults with major depressive disorder: section 2. Psychological treatments. Can. J. Psychiatr. 61 (9), 524–539.

Pascoe, M.C., et al., 2017a. Mindfulness mediates the physiological markers of stress: systematic review and meta-analysis. J. Psychiatr. Res. 95, 156–178.

Pascoe, M.C., Thompson, D.R., Ski, C.F., 2017b. Yoga, mindfulness-based stress reduction and stress-related physiological measures: a meta-analysis. Psychoneuroendocrinology 86, 152–168.

Pedersen, B.K., 2019. Physical activity and muscle-brain crosstalk. Nat. Rev. Endocrinol. 15 (7), 383–392.

Pilu, A., et al., 2007. Efficacy of physical activity in the adjunctive treatment of major depressive disorders: preliminary results. Clin. Pract. Epidemiol. Ment. Health 3, 8.

Quadt, L., Critchley, H.D., Garfinkel, S.N., 2018. The neurobiology of interoception in health and disease. Ann. N. Y. Acad. Sci. 1428 (1), 112–128.

Quax, R.A., et al., 2013. Glucocorticoid sensitivity in health and disease. Nat. Rev. Endocrinol. 9 (11), 670–686.

Ravindran, A.V., et al., 2016. Canadian Network for Mood and Anxiety Treatments (CANMAT) 2016 clinical guidelines for the management of adults with major depressive disorder: section 5. Complementary and alternative medicine treatments. Can. J. Psychiatr. 61 (9), 576–587.

Rethorst, C.D., et al., 2013. Pro-inflammatory cytokines as predictors of antidepressant effects of exercise in major depressive disorder. Mol. Psychiatry 18 (10), 1119–1124.

Sanada, K., et al., 2020. Gut microbiota and major depressive disorder: a systematic review and meta-analysis. J. Affect. Disord. 266, 1–13.

Schuch, F.B., et al., 2016. Exercise as a treatment for depression: a meta-analysis adjusting for publication bias. J. Psychiatr. Res. 77, 42–51.

Segal, Z.V., Williams, J.M.G., Teasdale, J.D., 2013. Mindfulness-Based Cognitive Therapy for Depression, second ed. vol. xix Guilford Press, New York, p. 451.

Sharf, R., 2014. Mindfulness and mindlessness in early Chan. Philos. East West 64 (4), 933–964.

Singleton, M., 2010. Yoga Body: The Origins of Modern Posture Practice. Oxford University Press, Oxford, New York. viii 262 p.

Sliz, D., Hayley, S., 2012. Major depressive disorder and alterations in insular cortical activity: a review of current functional magnetic imaging research. Front. Hum. Neurosci. 6, 323.

Smith, C.A., et al., 2018. Acupuncture for depression. Cochrane Database Syst. Rev. 3, CD004046.

So, W.W.Y., et al., 2019. The neurophysiological and psychological mechanisms of qigong as a treatment for depression: a systematic review and meta-analysis. Front. Psychiatry 10, 820.

Solloway, M.R., et al., 2016. An evidence map of the effect of Tai Chi on health outcomes. Syst. Rev. 5 (1), 126.

Strawbridge, R., et al., 2015. Inflammation and clinical response to treatment in depression: a meta-analysis. Eur. Neuropsychopharmacol. 25 (10), 1532–1543.

Treves, I.N., et al., 2019. The relationship between mindfulness and objective measures of body awareness: a meta-analysis. Sci. Rep. 9 (1), 17386.

Trivedi, M.H., et al., 2006. TREAD: TReatment with exercise augmentation for depression: study rationale and design. Clin. Trials 3 (3), 291–305.

U.S.D.o.A.a.U.S.D.o.H.a.H. Services (Ed.), 2020. Dietary Guidelines for Americans, 2020–2025.

Uebelacker, L.A., et al., 2017. Adjunctive yoga v. health education for persistent major depression: a randomized controlled trial. Psychol. Med. 47 (12), 2130–2142.

Uebelacker, L.A., et al., 2018. Treatment credibility, expectancy, and preference: prediction of treatment engagement and outcome in a randomized clinical trial of hatha yoga vs. health education as adjunct treatments for depression. J. Affect. Disord. 238, 111–117.

van Agteren, J., et al., 2021. A systematic review and meta-analysis of psychological interventions to improve mental wellbeing. Nat. Hum. Behav. 5, 631–652.

Vollbehr, N.K., et al., 2018. Hatha yoga for acute, chronic and/or treatment-resistant mood and anxiety disorders: a systematic review and meta-analysis. PLoS One 13 (10), e0204925.

Wang, C.W., et al., 2013. The effect of qigong on depressive and anxiety symptoms: a systematic review and meta-analysis of randomized controlled trials. Evid. Based Complement. Alternat. Med. 2013, 716094.

Wang, Y.Y., et al., 2018. Mindfulness-based interventions for major depressive disorder: a comprehensive meta-analysis of randomized controlled trials. J. Affect. Disord. 229, 429–436.

Waterson, M.J., Horvath, T.L., 2015. Neuronal regulation of energy homeostasis: beyond the hypothalamus and feeding. Cell Metab. 22 (6), 962–970.

Wayne, P.M., et al., 2014. What do we really know about the safety of tai chi?: a systematic review of adverse event reports in randomized trials. Arch. Phys. Med. Rehabil. 95 (12), 2470–2483.

Woldeyohannes, H.O., et al., 2016. Binge eating in adults with mood disorders: results from the International Mood Disorders Collaborative Project. Obes. Res. Clin. Pract. 10 (5), 531–543.

Woo, Y.S., et al., 2016. Obesity and its potential effects on antidepressant treatment outcomes in patients with depressive disorders: a literature review. Int. J. Mol. Sci. 17 (1), 80.

Yang, C., et al., 2019. Inflammatory markers and treatment outcome in treatment resistant depression: a systematic review. J. Affect. Disord. 257, 640–649.

Yang, Z., et al., 2020. Updated review of research on the gut microbiota and their relation to depression in animals and human beings. Mol. Psychiatry 25 (11), 2759–2772.

Yeung, A., et al., 2018. Qigong and Tai-Chi for mood regulation. Focus (Am. Psychiatr. Publ.) 16 (1), 40–47.

Young, K.S., et al., 2018. The impact of mindfulness-based interventions on brain activity: a systematic review of functional magnetic resonance imaging studies. Neurosci. Biobehav. Rev. 84, 424–433.

Zhuo, C., et al., 2018. Topiramate and metformin are effective add-on treatments in controlling antipsychotic-induced weight gain: a systematic review and network meta-analysis. Front. Pharmacol. 9, 1393.

Chapter 39

Managing the risk of suicide in people with treatment-resistant depression

Peter J. Franz[a] and Jonathan E. Alpert[b]

[a]*Department of Psychology, Harvard University, Cambridge, MA, United States,* [b]*Department of Psychiatry and Behavioral Sciences, Albert Einstein College of Medicine, Montefiore Medical Center, Bronx, NY, United States*

Introduction

The management of suicide risk is often integral to the care of people with treatment-resistant depression (TRD). Suicide is a major public health concern worldwide, and yet it is a problem that remains poorly understood. Suicide occurs across a wide range of clinical presentations, including TRD, but the mechanisms accounting for suicide risk in TRD are not fully understood. Nevertheless, treatment guidelines for suicide risk in TRD are beginning to emerge, although more evidence is needed. In the sections that follow, we provide a brief overview of the phenomenology, risk, and treatment of suicidal thoughts and behaviors (STBs), and then we make recommendations for best practices for managing suicide risk for those with TRD.

Terms and definitions

Prior to considering management of suicide risk in TRD, it is worthwhile to define and clarify several key constructs related to suicide. Suicide, also referred to as suicide death, refers to the intentional ending of one's own life. A discussion of STBs can be structured according to a recently developed theoretical model describing the pathway to suicide. This model begins with thoughts about suicide and identifies the sequence of thoughts and behaviors that culminate in a suicide attempt (Millner et al., 2017). Suicidal ideation refers to thoughts about one's own death or suicide. Suicidal ideation is often considered either passive (i.e., the wish to no longer be alive) or active (i.e., the wish to take one's own life). Although active suicidal ideation is typically considered the more risky of the two, recent research suggests that passive and active ideation may be somewhat equivalent in terms of risk for suicide attempts, as well as other outcomes (Liu et al., 2020). A suicide plan refers to the selection of a method and place to attempt suicide (Millner et al., 2017). Suicide attempt refers to an intentional act undertaken with some intent to die as a result, regardless of whether death results. Nonsuicidal self-injury (NSSI; often referred to as "self-injury" or "self-harm") refers to self-directed, harmful acts without intent to die as a result, most often performed for the benefit of reducing distressing psychological states, such as negative emotions (Nock and Prinstein, 2004). NSSI is different than suicide, and is not the focus of this chapter, but NSSI is common among those with treatment-resistant depression (Asarnow et al., 2011), and is a top risk factor for a future suicide attempts (Franklin et al., 2017).

Suicide prevalence

Suicide is a public health concern of enormous magnitude. In the United States, suicide is the 2nd leading cause of death among those ages 10–35, and the 10th leading cause of death among all age groups (Centers for Disease Control and Prevention: National Center for Injury Prevention and Control, 2019). In 2018, over 48,000 Americans died by suicide. Around the world, approximately 800,000 people die by suicide each year, according to the World Health Organization (2018). Suicide accounts for 1.4% of all deaths worldwide, making it the 18th leading cause of death per year globally (WHO, 2018). To put these numbers into perspective, suicide accounts for more global deaths each year than war, automobile accidents, AIDS, and homicide combined (Centers for Disease Control and Prevention: National Center for Injury Prevention and Control, 2019; World Health Organization (WHO), 2018). In addition to suicide death, there are also an estimated 25 million suicide attempts, and 140 million people engaging in suicidal ideation annualy around the world (Borges et al., 2008).

Managing Treatment-Resistant Depression. https://doi.org/10.1016/B978-0-12-824067-0.00039-6

Suicide prevalence rates vary between countries, and some data have suggested that suicide is more common in developed versus developing countries (Borges et al., 2010). However, attempts to better understand this variation have not provided additional insights. Perhaps more illuminating is that several demographic characteristics seem to be commonly associated with suicide rates internationally. For example, males are more likely to die by suicide, whereas females are more likely to attempt and think about suicide (Borges et al., 2010; Kessler et al., 2005; World Health Organization, 2018). In addition, younger individuals engage in STBs at a higher rate than older individuals (Borges et al., 2010).

Some recent epidemiological data suggest that suicide may be on the rise, with rates in the United States increasing for both males and females between 2000 and 2016 (Hedegaard et al., 2018). This alarming statistic has drawn warranted attention from the media. However, considering a broader span of time adds nuance to the conclusion that suicide rates are increasing. During the 20 years prior to 2000, suicide rates in the US actually decreased (Fig. 39.1).

Unlike many of the other leading causes of death worldwide, including cancer, heart disease, Alzheimer's disease, stroke, and violence, we have not been able to effect a meaningful reduction in suicide rates. Our ability to generate knowledge about suicide and suicide prevention has been hampered in at least three ways. First, suicide is an extremely low base-rate phenomenon, with fewer than 11 out of every 100,000 people dying by suicide annually (WHO, 2018). From a statistical perspective, as discussed in greater detail below, predicting the occurrence of such a rare phenomenon requires very large samples of research participants. Second, those who die by suicide cannot be directly studied using the tools most common in mental health research. Although we can examine morphological or biochemical abnormalities in the brains of suicide victims, these approaches have not revealed major insights about the phenomenology of suicide, or how we may prevent it. Third, suicide has not been studied systematically for very long, and research efforts have largely focused on a small number of constructs theorized to cause suicide (Franklin et al., 2017). A primary goal of this chapter is to highlight what we do know about suicide generally and among those with treatment-resistant depression, and to discuss future research priorities to better understand and prevent this enigmatic tragedy.

Suicide risk factors

Identifying risk factors for suicide is one approach toward better understanding the phenomenology of suicide, as well as more effectively predicting suicide and intervening before it occurs. A risk factor refers to a prospective predictor of an outcome of interest. Risk factors precede the outcome in time, and they predict whether the outcome is likely to occur by dividing people into high- and low-risk categories (Kraemer, 1997). This is different than a correlate, which is associated with the outcome of interest, but the nature of that association is unclear. For example, if a cross-sectional study identified sleep loss as a correlate of suicidal ideation, it would be possible that suicidal ideation caused sleep loss—or vice-versa. Longitudinal data collection methods are therefore necessary to identify risk factors by establishing temporal precedence. It is important to note that identifying a risk factor does not necessarily indicate that the risk factor is causal. Causal risk factors are a specific type of risk factor in which *manipulation* of the risk factor creates an observable change in the

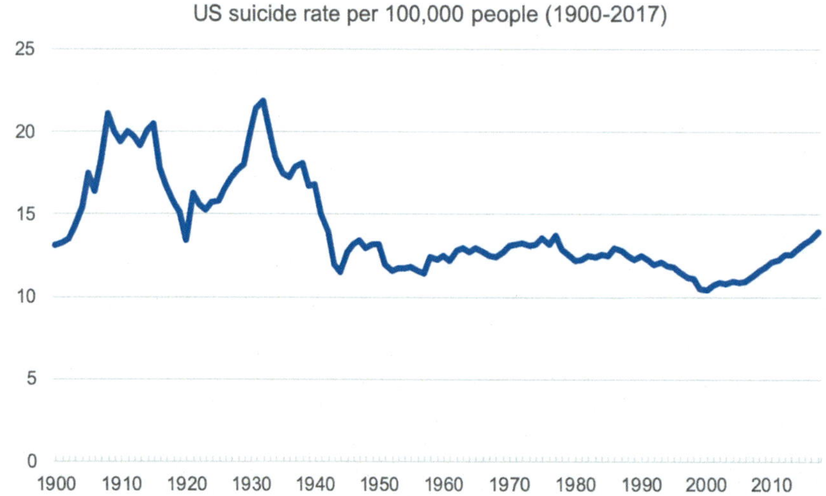

FIG. 39.1 United States suicide rate per capita from 1900 to 2017. *(Data are from CDC's publicly available National Vital Statistics System (https://www.cdc.gov/nchs/nvss/mortality_historical _data.htm).)*

likelihood of the outcome (Kraemer, 1997). Very few studies have attempted to influence risk factors to determine causal relationships with suicide, and thus our current knowledge of suicide risk is largely informed by risk factors not yet known to be causal.

A recent meta-analysis of the past five decades of suicide research, which included data only from longitudinal studies of suicide risk factors, has provided some of the most compelling evidence to date about risk for suicide (Franklin et al., 2017). This meta-analysis examined the prospective relationship between risk factors and STBs (suicidal ideation, suicide attempt, and suicide) across 365 separate studies. The relationship between each risk factor and STB outcome was quantified in this study with a weighted odds ratio. This weighted odds ratio provided an estimate of how much more likely the outcome is to occur for those who are high in the risk factor, while taking into account the size of the sample used for the original analysis. Odds ratios above one indicate that outcome is more likely to occur for those higher in the risk factor, whereas the opposite is true for odds ratios less than one. An odds ratio of one indicates that the outcome is just as likely to occur for those who are higher vs lower in the risk factor (i.e., the outcome is predicted at chance level by the risk factor).

Depression has long been considered a risk factor for suicide, and indeed, countless studies have demonstrated an association between depression and STBs. However, Franklin et al. (2017) found that depression was one of the five strongest risk factors only for suicidal ideation. For suicide attempt and suicide death, depression (categorized together with other forms of internalizing psychopathology) was rated sixth and ninth among the strongest risk factors, respectively. Data from other studies (some of which were included in the Franklin et al. (2017) meta-analysis) have suggested that the effect of depression on increased suicide risk may not result from depression itself, but rather from the conditions that are commonly comorbid with depression. For example, two studies that included large, representative samples (one from the United States, one international) found that depression was one of the strongest risk factors for suicidal ideation. However, among those with suicidal ideation, depression predicted future suicide attempt and suicide death no better than chance, whereas other common mental disorders, especially those characterized by anxiety and agitation (panic disorder, posttraumatic stress disorder) or by poor impulse control (substance use disorders), were far better predictors than depression of whether ideators would go on to make a suicide attempt (Nock et al., 2009a, 2010a). Moreover, the likelihood of making a suicide attempt was shown in each of these studies to be higher for those with greater numbers of comorbid mental disorders (Nock et al., 2009a, 2010a). In other words, individuals with one mental disorder were less likely to attempt suicide than those with two mental disorders, who were less likely than those with three, and so on. Together, this body of evidence suggests that depression alone often leads to thoughts about death and suicide, but other conditions lead depressed individuals to act on these thoughts. This may be particularly important for TRD, wherein comorbid mental disorders are particularly common (Albott et al., 2018).

According to the Franklin et al. (2017) meta-analysis, prior self-injurious thoughts and behaviors (including suicidal ideation, suicide attempt, and NSSI) were among the strongest risk factors for all STBs outcomes, as shown in Table 39.1.

Prior psychiatric treatment, including inpatient hospitalization, is another top risk factor for suicide (Franklin et al., 2017). The week following inpatient hospitalization is one of the highest risk periods for suicide attempts (Qin and Nordentoft, 2005). Although the factors that contribute to the significantly elevated risk of suicide in those recently discharged have not been adequately delineated, the nature of the reasons for inpatient hospitalization (e.g., STBs, emotional distress), and other challenges posthospitalization (e.g., persistent symptoms, disappointment with lack of greater improvement during hospitalization) are plausible contributors to risk.

Additional categories of risk factors, including physical illness and demographic characteristics were shown by Franklin et al. (2017) to be poor predictors of suicide. In fact, their effects on suicidal ideation were close to chance. In practical terms, this means that by flipping a coin, one could predict a person's suicidal ideation almost as accurately as one could by knowing that person's ethnicity or their history of asthma.

Alternative methods for suicide prediction

A fundamental goal of understating suicide risk is to be able to predict suicide so that we can ultimately prevent it. This is indeed a tall order. To predict the real-life occurrence of suicide requires identifying a person who is at risk and/or a context that puts an at-risk person in imminent danger of attempting suicide. Until recently, predicting suicide largely relied on self- or clinician-reported risk factors, like those discussed above. For example, in hospital emergency departments across the globe, clinical staff assess suicide risk by asking patients to self-report about risk factors like STB history, current suicide plans, and intent to die by suicide. Relying on self-report in this way is a method of suicide prediction with some limitations. First, those who are at risk are often motivated to conceal or deny their true intentions to act on suicidal thoughts. Revealing suicidal intentions can be viewed as shameful, and for those with strong intentions, communicating these can lead to unwanted intervention. One study found that 78% of people who died by suicide denied suicidal ideation

TABLE 39.1 Top 5 predictors across STB outcomes.

Rank	Risk factor	wOR	CI
	Top 5 risk factors for suicidal ideation		
1	Prior suicide ideation	3.5	(2.6, 4.8)
2	Hopelessness	3.3	(1.5, 7.2)
3	Depression	2.5	(1.4, 4.3)
4	Abuse history	1.9	(1.6, 2.3)
5	Anxiety	1.8	(1.3, 2.4)
	Top 5 risk factors for suicide attempt		
1	Prior NSSI	4.2	(2.9, 6.9)
2	Prior suicide attempt	3.4	(2.7, 4.3)
3	Suicide risk screening	2.5	(1.8, 4.4)
4	Axis II diagnosis	2.4	(1.9, 2.9)
5	Prior psychiatric hospitalization	2.3	(1.6, 3.4)
	Top 5 risk factors for suicide death		
1	Prior psychiatric hospitalization	3.6	(2.8, 4.5)
2	Prior suicide attempt	2.2	(1.7, 3.0)
3	Prior suicide ideation	2.2	(1.5, 3.4)
4	Low socioeconomic status	2.2	(1.3, 3.7)
5	Stressful life events	2.2	(1.6, 2.9)

wOR, weighted odds ratio (weighted by number of studies); *CI*, 95% confidence interval.

immediately prior to their death (Busch et al., 2003). Second, suicidal thoughts are now known to be transient, varying in presence and intensity throughout a single day (Coppersmith et al., 2019; Kleiman et al., 2017; Nock et al., 2009b), and so self-reported current suicide risk may not be an accurate indicator of risk a week or even a day later. Third, assessment of suicide risk through self-report often relies on patients' STB histories (i.e., assessing past STBs, which are a top risk factor, as discussed above), but these self-reports are based on retrospective recall, which can be biased and inaccurate (Schacter et al., 1998). Nonetheless, among suicide attempters presenting at a psychiatric emergency department, self-reported prediction of future suicide attempts (i.e., the patient responded to the question "what is the likelihood that you will make a suicide attempt in the future?) is a significant predictor of future suicide attempt within a 6-month follow-up period, even when controlling for key risk factors, such as depression and history of prior suicide attempts (Nock et al., 2010b). As such, self-reported suicide risk is likely the most useful and easily available indicator of suicide risk within a 6-month time frame.

Clinical care decisions for suicidal patients also typically rely on the subjective judgment of a clinician. This judgment is often based on intuition, but also the experience of seasoned clinicians. However, clinician prediction of future suicide attempt does not outperform actuarial prediction of suicide based on known risk factors, and neither of these predict suicide better than chance (Nock et al., 2010b). Although this finding casts doubt on the ability of clinicians to accurately predict suicide risk using their own subjective clinical judgment, it accords with prior research showing that clinical judgment often has limited utility for predicting important mental health outcomes (Dawes et al., 1989).

A more recent method for assessing suicide risk that has gained attention is the assessment of implicit cognitions about death and suicide. Implicit cognition refers to the influence of memories or perceptions on thoughts and behaviors that is outside of conscious awareness. In the same study by Nock and colleagues (2010b) referenced above, the research team compared a measure of implicit suicide-related cognition against self-reported, clinician-reported, and actuarial prediction of suicide attempts. They found that, even after controlling for the predictive effects of each of these alternative methods, a measure of implicit suicide-related cognition identified those who would go on to make suicide attempts 6 months into the future. Subsequent studies have supported the predictive validity of suicide-related implicit cognitions (Glenn et al. 2016,

2017a,b, 2019; Millner et al. 2019). Implicit cognition has two major benefits over more traditional suicide risk prediction methods. First, implicit cognitions, which are measured using responses to stimuli calculated down to the millisecond, are considered a more objective measure of suicide risk since they are resistant to explicit attempts at concealing stigmatized thoughts (Nock and Benaji, 2007). Second, implicit cognitions are shown to have exceptional specificity, in contrast to the vast majority of other prospective risk factors, which have low specificity for suicide. For example, two well-known risk factors for suicide, depression and sleep disturbance, each respectively predict anxiety and accidental falls better than they predict suicide (Kessler et al., 2008; Stone et al., 2014). However, optimizing the specificity of methods for suicide prediction is only half the battle.

Specificity, as exemplified above, refers to the ability to correctly identify the absence of the outcome of interest (i.e., true negatives; correctly classified nonattempters), but highly specific tests are prone to failing to identify those with the outcome of interest (i.e., false negatives; misclassified nonattempters). Sensitivity refers to the ability to correctly identify those with the outcome of interest (i.e., true positives, correctly-classified attempters), but highly sensitive tests are prone to misclassifying the absence of the outcome of interest (i.e., false positives, misclassified attempters). A major limitation of these previously-discussed approaches is finding the "optimal" balance between sensitivity and specificity. Ideally, we want to identify and intervene for a small number of people (high sensitivity) who are almost definitely going to attempt suicide (high specificity). Failing to intervene for people who actually attempt suicide because of low sensitivity is obviously problematic, but intervening for lots of people who aren't truly at risk because of low specificity unduly consumes valuable clinical resources and can lead to unnecessary involuntary hospitalizations. Self-report and behavioral measures of risk tend to be highly sensitive, but not highly specific. This is because suicide is an extremely low base rate behavior, and so the odds ratios often used to indicate risk for suicide can be misleading, increasing the difficulty of clinical decision-making. Artificial intelligence, otherwise known as machine learning (ML), approaches are beginning to show promise by combining a broad array of risk predictor types while offering the ability to determine the tradeoff between sensitivity and specificity.

Although ML approaches can enable researchers to measure a broad range of risk factors simultaneously (from electronic medical records and other rich sources of data) from extremely large and representative samples, ML has found only limited practical benefit beyond that of the more commonly used suicide risk prediction approaches already discussed. For example, Simon et al. (2018) surveyed outpatient mental health clinics' data and used an ML approach to classify individuals' level of risk based on key information obtained from their health records. Their approach revealed extremely high odds ratios for the risk metrics included, but misclassified many patients as high risk at the highest levels of sensitivity (only 5%–10% of those classified as highest risk by the algorithm actually attempted suicide). Barak-Corren et al. (2020) used a similar ML approach to classify individuals' risk for a first suicide attempt based on five health care systems' electronic medical records. This study also found high odds ratios for the risk metrics they surveyed, but their approach misclassified even more patients as high risk at the highest levels of sensitivity (only 1%–6% of those classified as highest risk by the algorithm actually attempted suicide). These results indicate that ML approaches can be highly sensitive, and so they are capable of correctly identifying nearly all truly high-risk people, but at the cost of erroneously identifying truly low-risk people as high risk. The clinical utility of ML approaches for suicide risk prediction essentially depends on the intervention that is offered. If, for example, a hospital system using ML to detect suicide risk among its patients planned to utilize a low-cost intervention (i.e., interventions requiring little human labor and low risk to the rights of patients), such as a weekly phone check-in for all high-risk people, then the ML approach can provide practical benefits. However, if the hospital system planned to use a higher-cost intervention, such as involuntarily hospitalization, for patients classified as high risk, then ML approaches to suicide risk prediction would be unsuitable in their current state. However, researchers are working to improve ML methods, and so this promising area of research may soon hold tremendous practical value to clinicians and hospital administrators.

Suicide risk in TRD

Although it may seem intuitive that TRD is associated with increased risk for STBs, very few studies have examined the association between suicide risk and TRD. In part, the dearth of studies in this area may stem from the lack of definitional consistency of TRD, or the possibility that TRD, though a notable clinical challenge, may not be a categorically distinct syndrome from other forms of depression (e.g., Major Depressive Disorder in an individual who responds to the first several treatments or persistent depressive symptoms that fail to meet diagnostic criteria for a major mood disorder and do not respond to multiple treatments). Nonetheless, in this section, we summarize what is currently known about suicide risk in TRD, and then provide theory- and evidence-based speculation about why suicide risk might be particularly high in TRD.

Approximately 30% of patients with TRD attempt suicide at least once in their lives (Hantouche et al., 2010). This figure is staggering in its own right, but when considered in relation to nonresistant depression, the problem of suicide in TRD is even more concerning. Compared with depressed patients whose symptoms either responded to intervention or had not yet failed to respond to two or more interventions, those with TRD experience a twofold increased rate of suicide attempts, and a 10-fold increase rate of suicide death (Bergfeld et al., 2018). There are several possible reasons that STBs occur in such high numbers among those with TRD, and these can be organized under the theoretical framework of one the most well-cited theories in suicide research: the Interpersonal Theory of Suicide (Joiner, 2005; Van Orden et al., 2011). According to the Interpersonal Theory, three interconnected factors cause STBs: thwarted belongingness, perceived burdensomeness, and hopelessness. Meta-analytic results support first-order and interacting associations between each of the theory's proposed risk factors and suicide (Chu et al., 2017). Below, we argue that each of these risk factors are likely to be high among those with TRD, and may be at least partially responsible for the high rates of STBs observed in this population.

The need to belong is a fundamental part of human nature. Thwarted belongingness refers to the unmet need to belong, resulting in loneliness. Social isolation, conflict with others, having a small social network, and social withdrawal are each components of thwarted belongness, and are common in depression (Beutel et al., 2017; Derntl et al., 2011; Erzen and Çikrikci, 2018; Ge et al., 2017; Wilkinson et al., 2009). Some of these aspects of thwarted belongingness are also elevated in TRD (Brent et al., 2009). A considerable body of evidence indicates that thwarted belongingness predicts STBs, and interacts with both perceived burdensomeness and hopelessness in its association with STBs (Chu et al., 2017). Additionally, the association between thwarted belongingness and STBs is especially strong in late adulthood (Chu et al., 2017), when risk for suicide in TRD is particularly high (Conwell et al., 2011). Taken together, these findings indicate that thwarted belongingness may be one mechanism linking TRD with STBs.

Suicide risk among those with TRD may also increase because these individuals construe themselves to be a burden on others (i.e., perceived burdensomeness), such as significant others and treaters. The construct of perceived burdensomeness encompasses the perception that one's life is less valuable to others than one's death. Perceived burdensomeness is thought to be driven by the eusocial nature of human beings, who, like other eusocial animals, depend on finite, communal resources and cooperation among individuals to ensure a group's survival (Joiner et al., 2016). Tenets of inclusive fitness suggest that under certain environmental scenarios, such as threat to the health or resources of a group, self-sacrifice can aid in the group's survival by reducing the threat to the group as a whole. This form of self-sacrifice has been observed in a wide array of nonhuman animals, such as bees, naked mole rats, and shrimp (Joiner et al., 2016). Human suicide can be considered a self-sacrifice similarly intended for social benefit when an individual perceives that their continued life prevents others from consuming needed resources, as mentioned above. A large number of studies have shown that self-reported perceived burdensomeness predicts suicide (Chu et al., 2017), and is associated with depression (Jahn et al., 2011). Because impairment in functioning among those with TRD often necessitates the provision of resources from others, perceived burdensomeness in TRD may increase suicide risk, though no studies to date have investigated this hypothesis.

The final and perhaps most obvious reason that STBs are so common in TRD is that TRD and STBs are each linked with hopelessness. Hopelessness, broadly defined as negative expectations about the future and often encompasses the belief that problems are inescapable, is a central component in multiple suicide theories (Beck et al., 1975; Van Orden et al., 2011). Hopelessness has been shown by some studies to be a relatively strong predictor of future STBs (Beck et al., 1985, 1990; Franklin et al., 2017). With such a history of failed attempts at symptom remediation, it is intuitive that those with TRD are prone to experience hopelessness (Papakostas et al., 2003). It is likely that those with TRD begin to feel hopeless because their attempts to find symptom remediation have failed to yield sustained positive changes and, instead, have been met with repeated disappointment. They may also feel hopeless because of their belief that concerns related to thwarted belongingness and perceived burdensomeness, as described above, are intractable (Chu et al., 2017). These individuals may see the future as increasingly bleak, and may estimate their chances of recovery to be very low. It is also possible that the relationship between hopelessness and TRD may be bidirectional, such that hopelessness results from repeated treatment nonresponse, but also predicts persistence of depression symptoms after treatment (Brent et al., 1988; Emslie et al., 2010). If so, this deleterious cycle may erode the motivation to seek treatment, and increase the appeal of suicide as a means of escape.

Treatment of suicide in TRD

To date, no treatments have been developed to prevent STBs specifically in TRD. However, several therapies, ranging from pharmacological, to neurostimulation, to behavioral, have shown some promise for reducing STBs in depressed patients, and in some cases, TRD. Below we outline many of these promising approaches, and provide speculative commentary on their utility for treating STBs in those with TRD.

Pharmacological treatments

Intravenous administration of ketamine has recently received considerable scientific and clinical attention as a treatment for depression, culminating in the approval by the Food and Drug Administration (FDA) of its isomer, esketamine, administered intranasally for resistant depression. Multiple studies have shown that a single dose of ketamine reduces depressive symptoms in as many as 70% of depressed patients, and that these antidepressant effects are apparent in as little as 24 h after administration (Berman et al., 2000; Zarate et al., 2006). Accordingly, the rapid pace and potential efficacy of ketamine has become an exciting area of investigation for STB remediation, as well. In a sample of TRD patients with histories of STBs, a single-dose administration of ketamine was associated with a significant decrease in the intensity of suicidal ideation 24 h after administration, though this difference may have been accounted for by reductions in depressive symptoms, which were also clinically significant (Price et al., 2009). In addition, this study found that ketamine administration was associated with a reduction in implicit escape-related cognitions. A similar follow-up study found that implicit suicide-related cognitions, as well as explicit suicidal ideation also decreased following administration of ketamine, relative to a psychoactive placebo (Price et al., 2014). It is important to note that one study found no association between ketamine and implicit death-related cognitions, though this study did find that repeated ketamine administration was associated with reductions in explicit suicidal ideation (Ionescu et al., 2016). Nonetheless, these findings are encouraging for two reasons. First, ketamine and esketamine may be effective treatments for those facing acute suicidal emergencies, such as those presenting to an emergency department following a suicide attempt, because ketamine and esketamine work rapidly, relative to more traditional antidepressant medications. Second, findings that ketamine may impact implicit cognition related to suicide provides an indication as to one mechanism by which this medication may exert its antisuicidal effect—namely, by reducing automatic thoughts about escape and death. Although more research is needed to determine whether ketamine or esketamine should be administered as a first-line treatment for those with TRD and STBs, the results reviewed above suggest that these approaches may be promising.

Lithium is another pharmacotherapy shown to be effective for reducing STBs in depressed patients, and may be a fruitful approach for this purpose in TRD. In fact, one recent meta-analysis found that lithium had a larger effect for STB reduction than for mood symptom reduction, suggesting potential specificity of lithium for STBs (Cipriani et al., 2013). Lithium is thought to exert antisuicidal effects through two possible mechanisms. First, and most relevant to TRD, lithium may help prevent relapse of depressive symptoms, thereby limiting ongoing psychological distress, such as hopelessness. Second, lithium is shown to reduce agitation and impulsivity (Kovacsics et al., 2009; O'Donnell and Gould, 2007), which are associated with STBs, as mentioned above (Nock et al., 2009a, 2010a). It is noteworthy, however, that the efficacy of lithium for reducing STBs has primarily been documented relative to placebo, and lithium's possible antisuicide efficacy relative to other pharmacotherapies in individuals with unipolar and bipolar mood disorders is an area of continued investigation.

Some research indicates that clozapine, an atypical antipsychotic usually prescribed for symptoms of schizophrenia or other psychotic disorders, may be effective for preventing suicide in high risk patients. In a study comparing clozapine with olanzapine, researchers recruiting psychotic patients (schizophrenia or schizoaffective disorder diagnoses) from 11 different countries found that those treated with clozapine had a lower rate of suicide attempts over a 2-year follow-up period (Meltzer et al., 2003). The patients treated with clozapine also required fewer hospitalizations following suicide attempts, as well as fewer acute rescue interventions, suggesting that clozapine may have implications for reducing the lethality of suicide attempts. Of note, over one quarter of the patients in both treatment groups in this study were considered treatment-resistant at the time of enrollment (Meltzer et al., 2003). One study has since corroborated the efficacy of clozapine for the treatment of STBs in schizophrenia, finding that clozapine was associated with reductions in both suicidal ideation and suicide attempts, based on retrospective medical record reviews (Modestin et al., 2005). Although intriguing, it is unclear whether these results would generalize to patients with TRD, since neither study examined the efficacy of clozapine for reducing STBs in depressed patients without a psychotic disorder, and the only treatment compared to clozapine was olanzapine, which is now infrequently prescribed in unipolar depression. Regardless, the putative mechanisms by which clozapine protects against STBs may be relevant to TRD. Clozapine is shown to reduce substance use in schizophrenia patients, likely by altering dopamine-mediated reward processing (Green et al., 1999). This impact on substance use is particularly noteworthy, given that depressed patients with comorbid substance abuse are at elevated risk for suicide (Nock et al., 2009a, 2010a).

Although the use of opioid medications to treat depressive symptoms fell out of common clinical practice in the 1950s (due to increasing concerns about abuse and overdose, as well as the discovery of tricyclic antidepressants and monoamine oxidase inhibitors), one opioid medication, buprenorphine, a mu opioid agonist and kappa opioid antagonist, could help treat STBs in patients with TRD. A study showed that patients treated with buprenorphine experienced a significantly larger reduction in suicidal ideation than patients who received placebo (Yovell et al., 2016). Reductions in suicidal ideation among the group treated with buprenorphine were apparent within a week of beginning treatment, and these reductions persisted through the conclusion of the 4-week trial. Critically, the sublingual dose of buprenorphine used in this study was very low (mean daily dose of 0.44 mg) and posed virtually no risk for overdose, and so outpatients were able to fill prescriptions of buprenorphine outside of the hospital study sites. This presents a major advantage over many of the other pharmacotherapies described above, such as ketamine, which must be administered repeatedly under direct medical supervision in order to achieve antisuicidal effects. Given that a recent meta-analysis indicated that buprenorphine is effective for reducing depressive symptoms in TRD patients (Serafini et al., 2018), buprenorphine may be a promising treatment for STBs in this population as well, although more research is needed.

Neurostimulation treatments

In addition to the pharmacological approaches discussed above, three neurostimulation approaches have also been evaluated as suicide prevention measures in those with depression: electroconvulsive therapy, deep brain stimulation, and vagus nerve stimulation. Electroconvulsive therapy is the most well-studied of these three neurostimulation approaches. Electroconvulsive therapy is shown to be highly effective for remediating depressive symptoms in TRD (Voineskos et al., 2020), but historically was considered a last-line suicide prevention among chronically suicidal depressed patients, due to potentially harmful side effects, including memory loss and seizures. However, published guidelines by the American Psychiatric Association in 2001 promoted the use (American Psychiatric Association, 2001) and continued refinement of ECT, leading to widespread utilization of electroconvulsive therapy as a treatment for TRD (McDonald et al., 2013). Since then, several studies have demonstrated that electroconvulsive therapy may be effective for reducing in STBs among depressed patients (Ahmadi et al., 2016; Kellner et al., 2005; Liang et al., 2018).

Though less well-established as a treatment for TRD than electroconvulsive therapy, studies suggest that vagus nerve stimulation is associated with reduced incidence of STBs in depressed patients. For example, one study of TRD patients showed that the group treated with vagus nerve stimulation experienced significantly fewer STBs than patients in the "treatment as usual" group (Aaronson et al., 2017). Additional studies have also reported suicide attempt rates in samples of TRD patients treated with vagus nerve stimulation (Bajbouj et al., 2010; Christmas et al., 2013; Cristancho et al., 2011). Although the number of STBs in these samples is promisingly low, it is impossible to conclude from these studies whether vagus nerve stimulation had a discernible impact on STBs in TRD, and so more work is needed in this area.

Similar to electroconvulsive therapy, little empirical research has investigated deep brain stimulation for reducing STBs. Two anecdotal reports of deep brain stimulation in depressed patients suggest the possibility that deep brain stimulation may be associated with increased suicide risk (Foncke et al., 2006; Mahgoub and Kotbi, 2009). However, other studies have concluded that the small number of STBs in TRD patients treated with deep brain stimulation during clinical trials was unrelated to this procedure, though no statistical inferences were provided to support this this lack of association (Bergfeld et al., 2016; Bewernick et al., 2012). Thus, the efficacy of deep brain stimulation for reducing STBs in TRD remains an open question.

Finally, repetitive transcranial magnetic stimulation is another neurostimulation approach that may help reduce STBs among depressed patients. Repetitive Transcranial Magnetic Stimulation was approved by the Food and Drug Administration in 2008 as a treatment for TRD. Since then, many studies have demonstrated its efficacy in remediating depressive symptoms (George et al., 2013). A recent retrospective study of patients who had been treated with repetitive transcranial magnetic stimulation showed that patients' self-reported suicidal ideation reduced significantly during the course of their treatment (Abdelnaim et al., 2020). Because this study was retrospective in nature, and suicide attempts were not measured, these results should be considered preliminary evidence for the effectiveness of repetitive transcranial magnetic stimulation for treating STBs.

Behavioral treatments

Cognitive Behavioral Therapy is one of the most commonly utilized forms of psychotherapy worldwide. Cognitive Behavioral Therapy aims to provide relief from the challenges of psychopathology and emotional suffering through a combination of strategies for emotion identification/regulation and in vivo exposure to distressing emotional states. A large body of evidence indicates that Cognitive Behavioral Therapy is effective in reducing STBs (D'Anci et al., 2019; Gøtzsche and Gøtzsche, 2017; Leavey and Hawkins, 2017). Some evidence also suggests that Cognitive Behavioral Therapy is effective for remediating depressive symptoms in TRD (Ijaz et al., 2018); however, no studies have evaluated changes in STBs after Cognitive Behavioral Therapy for TRD patients.

Dialectical Behavior Therapy combines many aspects of Cognitive Behavioral Therapy with mindfulness- and acceptance-based skills. Dialectical Behavior Therapy was created for self-injurious (those with STBs as well as NSSI) Borderline personality disorder patients who struggled with the putatively less flexible aspects of traditional cognitive behavioral therapy (Linehan, 1993). Dialectical Behavior Therapy has been evaluated as a treatment for STBs against treatment as usual (Goodman et al., 2016; McMain et al., 2017), waitlist (Wilks et al., 2018), community-based psychotherapy (Linehan et al., 2006), and more direct suicide risk management approaches (Andreasson et al., 2016). Although some of these studies have provided evidence for the efficacy of Dialectical Behavior Therapy in reducing STBs (Linehan et al., 2006; McMain et al., 2017; Wilks et al., 2018), results are mixed (Andreasson et al., 2016; Goodman et al., 2016). Though no studies to date have examined whether Dialectical Behavior Therapy is shown to reduce STBs in TRD, Dialectical Behavior Therapy is shown to reduce depressive symptoms in TRD (Ijaz et al., 2018).

Additional psychotherapeutic interventions, such as Interpersonal Psychotherapy, have shown some promise for reducing STBs (Heisel et al., 2009), and may be effective for TRD (Ijaz et al., 2018). However, more research regarding other therapy interventions are needed in order to identify them as candidates for reducing STBs in TRD.

Additional best practices for risk management

Patients experiencing a suicide crisis (i.e., an increase in the intensity of suicidal ideation that may include a plan and/or intent to act on these thoughts) require immediate intervention. Pharmacological, neurostimulation, and behavioral therapies are often insufficient for managing acute suicide risk without additional targeted interventions. Several clinical best practices, as well as novel crisis interventions for managing suicide risk are important to discuss in the context of this chapter on managing suicide risk among those with TRD.

Regular assessment of suicide risk is of paramount importance for clinicians treating any population at elevated risk for suicide, such as TRD. Because suicidal thoughts and intentions can vary from moment to moment (Coppersmith et al., 2019; Kleiman et al., 2017; Nock et al., 2009b), clinicians cannot rely on a patient's recent history as a determinant of current risk, and should seek to evaluate current suicidal ideation, plans, intentions, and access to methods, as well as any recent preparatory behaviors for suicide (e.g., stockpiling medications, purchasing weapons, etc.), recent nonsuicidal self-injury, and recent suicide attempts. These thoughts and behaviors, which represent the most concerning group of suicide risk factors (Franklin et al., 2017), should be assessed very regularly with a patient with TRD. Contrary to the somewhat popular misconception, asking questions about STBs is not associated with an increase in suicidal thoughts for the patient (Dazzi et al., 2014), and so clinicians should not feel deterred from asking about suicide for fear of negatively impacting the patient's risk. On the contrary, direct risk assessment using available information (e.g., patient interview) is perhaps the most accurate and readily available tool to identify the need for a clinician's support or a higher level of care. Similarly, thorough documentation of regular risk assessment, and of crisis interventions administered, is important for clinicians. By documenting the information obtained through risk assessment interviews, as well as all interventions offered, clinicians can create a record of a patient's changes in risk throughout therapy, and protect themselves from potential liability.

For those deemed to be at suicide risk through self-report of STBs or other risk factors, clinicians and patients should collaboratively agree upon a safety plan. A safety plan is a list of strategies an at-risk patient can take to keep themselves safe from suicide, NSSI, substance abuse, or other harmful behaviors. The steps of a safety plan are sequential, such that patients should start the first strategy listed, and move on to subsequent strategies if the last strategy attempted was ineffective. A template of a safety plan is shown in Fig. 39.2.

Safety plans differ from and are more effective for suicide prevention than safety contracts, which are written agreements between a clinician and a patient indicating the patient's commitment to remaining alive (Bryan et al., 2017). Safety contracts have been widely used by clinicians hoping to help prevent their patients from acting on suicidal thoughts. Safety contracts may have some benefits, including demonstration of the therapist's investment in the patient's safety, clarifying limits to the benefits of psychotherapy, and even as an assessment tool to identify a patient's suicidal intent or barriers to remaining safe (Kelly and Knudson, 2000). However, no empirical research indicates that safety contracts actually help prevent suicide. Furthermore, a patient's willingness to sign a contract does not indicate their ability to take the necessary steps to stop themselves form acting on suicidal thoughts (Kelly and Knudson, 2000), and patients who are experiencing an acute emotional crisis may forget about or undervalue a safety contract made with a therapist (Reid, 1999). By comparison, a formal safety plan is a far more realistic strategy to help patients because it provides step-by-step actions a patient can take to identify a suicidal crisis and keep themselves safe. Moreover, safety plans were demonstrated to be more effective for preventing suicides than safety contracts in a head-to-head randomized clinical trial. This study found that patients who completed safety plans with a clinician were 75% less likely to attempt suicide during a 6-month follow-up period than patients who completed a safety contract (Bryan et al., 2017).

Safety Plan

A list of steps you can take to make yourself safe

Step 1: Warning signs (situation, thoughts, emotions, behaviors)

1. ______________________________
2. ______________________________
3. ______________________________

Step 2: Things I can do to relax or distract myself

1. ______________________________
2. ______________________________
3. ______________________________

Step 3: Get support from people you know

1. Name________________ Phone________________
2. Name________________ Phone________________
3. Name________________ Phone________________

Step 4: Professionals I can ask for help

1. Clinician__________ Phone____________ Emergency # __________
2. Local Crisis Support________________ Phone______________
3. Suicide Prevention Lifeline Phone: 1-800-273-TALK (8255)

Step 5: Urgent Care

1. Address______________________________ Phone__________
2. Address______________________________ Phone__________
3. 9-1-1

Reasons to stay alive

1. ______________________________
2. ______________________________
3. ______________________________

FIG. 39.2 Safety Plan Template. Safety plans provide patients a hierarchical list of strategies to help themselves in the event of a crisis.

An ongoing, and perhaps growing challenge is providing suicide crisis support to patients virtually. In an age when patients often seek help or communicate their own risk using a cell phone or other digital device, alternative strategies for the ambulatory management of suicide risk have become necessary. The utilization of phone and text-based suicide crisis support services is now commonplace in the United States and around the world. A recent systematic review indicated that these types of services are most likely effective for mitigating suicide crises, though more high-quality evidence is needed (Hoffberg et al., 2020).

Additionally, researchers are increasingly collecting suicide-related data from patients in situ using cell phones and other digital devices. This approach, often termed ecological momentary assessment (EMA), allows researchers to capture aspects of STBs and their risk factors as they occur in daily life. Such approaches have granted unprecedented insights into the temporal dynamics of suicide risk (Kleiman et al., 2017) and protective factors (Coppersmith et al., 2019). However, the emergence of EMA as an approach to study STBs has also created a need to establish guidelines for managing emergent risk remotely. A group of suicide researchers and clinicians who treat suicidal patients have agreed upon a list of best practices for managing suicide risk during EMA studies (Nock et al., 2021). Several of these guidelines are important to mention. First, data should be reviewed as often as possible, and at minimum once per weekday. Second, patients' self-reported suicidal ideation and intent should be considered the primary means of risk estimation, and based on this estimation, participants should be categorized by risk level (e.g., low, moderate, high, imminent). Finally, the research/clinical team should respond to a participant who has been identified as imminent risk as quickly as possible, but ideally within 12 h of being alerted of the participant's level of risk.

During the time of the COVID-19 pandemic, ambulatory suicide crisis interventions have become increasingly important. Although no data have been published indicating whether rates of utilization of these services has increased over the course of COVID-19, telehealth for psychiatric patients has certainly become more commonplace in the past decade (Notredame et al., 2018). With psychiatrists, psychologists, social workers, and mental health counselors conducting visits virtually, new challenges for managing acute suicide crises have arisen. Without the ability to administer in-person risk assessments, which often take place within close proximity to emergency departments, clinicians may feel that their options for determining suicide risk and administering acute support are limited. By following many of the above-listed guidelines, these clinicians can feel more confident that they are providing evidence-based care to help mitigate suicide risk through telehealth.

Conclusions and future directions

Suicide is a devastating, complex, and enduring public health crisis. The risk of suicide among patients with TRD is relatively high. Although the exact mechanisms through which risk for suicide increases in TRD are unknown, there are many treatment options available to these patients that may help protect against STBs. To help mitigate risk for suicide, clinicians should seek to follow established guidelines for evidence-based care of TRD patients, much of which is listed in this chapter. Managing risk of suicide among individuals at high risk is often best accomplished, where possible, in treatment settings in which care can be shared among an interdisciplinary mental health team. Care by more than a single clinician may help address potential clinical blind spots as well as reduce the risk of clinician burnout and perceived burdensomeness on the part of the patient. In settings of protracted depression with STBs, clinicians should maintain a low threshold for seeking consultation. Refreshing the range of plausible treatment options often helps reduce the development of nihilism, frustration or disengagement by the clinician who may feel increasingly at a loss for ideas. Clinicians should endeavor to communicate realistic hope in the long-term while not overpromising relief from any single next intervention. It is usually far more meaningful to patients to know that their clinician(s) will stick with them with continued interest, empathy, expertise and expectation of eventual improvement than to receive assurances of short-term benefit which may not materialize. Additional research is necessary to better understand suicide risk in TRD, as well as ways to prevent suicide in this population. Below, we highlight key limitations and suggest promising future directions for research in this area.

Clinical trials testing the efficacy of pharmacotherapies for suicide in TRD often exclude acutely suicidal individuals (Simon et al., 2006). Although there can be tension between the goal of monitoring suicidal outcomes and the ethical imperative to prevent suicide in research participants, excluding these individual limits the ability to draw conclusions about the antisuicide properties of these medications. Future research should seek to include those at risk for suicide and utilize evidence-based risk management approaches, such as those described in the section on best clinical practices above.

Deep brain stimulation, vagus nerve stimulation, and transcranial magnetic stimulation are three promising, but understudied approaches to treating suicide risk in TRD. Extant studies of deep brain stimulation (Foncke et al., 2006; Mahgoub and Kotbi, 2009), vagus nerve stimulation (Bajbouj et al., 2010; Christmas et al., 2013; Cristancho et al., 2011) and transcranial magnetic stimulation (George et al., 2013) have not explicitly examined suicidal outcomes or have only examined suicidal ideation, and in some cases have collapsed suicidal and other outcomes into single "adverse events" categories. Although this method was likely intended to enhance the number of observations of adverse events, it obscures the ability to better understand associations between STBs and these novel approaches to treating TRD. Researchers can overcome this limitation by including larger samples to capture a greater number of STBs, and direct test the association between treatments and STB outcomes.

Finally, more research is needed to identify the mechanisms through which TRD is associated with STBs. In this chapter, we have provided one theoretical perspective that elucidates mechanisms potentially responsible for this association. However, many other perspectives exist that could help guide future investigation of the psychological and biological factors that link TRD with suicide. Through further research, the field will be able to improve existing therapies and develop novel approaches to prevent suicide in TRD.

References

Aaronson, S.T., Sears, P., Ruvuna, F., Bunker, M., Conway, C.R., Dougherty, D.D., Reimherr, F.W., Schwartz, T.L., Zajecka, J.M., 2017. A 5-year observational study of patients with treatment-resistant depression treated with vagus nerve stimulation or treatment as usual: comparison of response, remission, and suicidality. Am. J. Psychiatry 174 (7), 640–648. https://doi.org/10.1176/appi.ajp.2017.16010034.

Abdelnaim, M.A., Langguth, B., Deppe, M., Mohonko, A., Kreuzer, P.M., Poeppl, T.B., Hebel, T., Schecklmann, M., 2020. Anti-suicidal efficacy of repetitive transcranial magnetic stimulation in depressive patients: a retrospective analysis of a large sample. Front. Psychiatry 10. https://doi.org/10.3389/fpsyt.2019.00929.

Ahmadi, N., Moss, L., Simon, E., Nemeroff, C.B., Atre-Vaidya, N., 2016. Efficacy and long-term clinical outcome of comorbid posttraumatic stress disorder and major depressive disorder after electroconvulsive therapy. Depress. Anxiety 33 (7), 640–647. https://doi.org/10.1002/da.22451.

Albott, C.S., Lim, K.O., Forbes, M.K., Erbes, C., Tye, S.J., Grabowski, J.G., Thuras, P., Batres-Y-Carr, T.M., Wels, J., Shiroma, P.R., 2018. Efficacy, safety, and durability of repeated ketamine infusions for comorbid posttraumatic stress disorder and treatment-resistant depression. J. Clin. Psychiatry 79 (3). https://doi.org/10.4088/JCP.17m11634.

American Psychiatric Association, 2001. The Practice of Electroconvulsive Therapy: Recommendations for Treatment, Training and Privileging: A Task Force Report of the American Psychiatric Association.

Andreasson, K., Krogh, J., Wenneberg, C., Jessen, H.K.L., Krakauer, K., Gluud, C., Thomsen, R.R., Randers, L., Nordentoft, M., 2016. Effectiveness of dialectical behavior therapy versus collaborative assessment and management of suicidality treatment for reduction of self-harm in adults with borderline personality traits and disorder—a randomized observer-blinded clinical trial. Depress. Anxiety 33 (6), 520–530. https://doi.org/10.1002/da.22472.

Asarnow, J.R., Porta, G., Spirito, A., Emslie, G., Clarke, G., Wagner, K.D., Vitiello, B., Keller, M., Birmaher, B., McCracken, J., Mayes, T., Berk, M., Brent, D.A., 2011. Suicide attempts and nonsuicidal self-injury in the treatment of resistant depression in adolescents: findings from the TORDIA study. J. Am. Acad. Child Adolesc. Psychiatry 50 (8), 772–781. https://doi.org/10.1016/j.jaac.2011.04.003.

Bajbouj, M., Merkl, A., Schlaepfer, T.E., Frick, C., Zobel, A., Maier, W., O'Keane, V., Corcoran, C., Adolfsson, R., Trimble, M., Rau, H., Hoff, H.J., Padberg, F., Müller-Siecheneder, F., Audenaert, K., Van Den Abbeele, D., Matthews, K., Christmas, D., Eljamel, S., Heuser, I., 2010. Two-year outcome of vagus nerve stimulation in treatment-resistant depression. J. Clin. Psychopharmacol. 30 (3), 273–281. https://doi.org/10.1097/JCP.0b013e3181db8831.

Beck, A.T., Kovacs, M., Weissman, A., 1975. Hopelessness and suicidal behavior. JAMA 234, 1146–1149. https://doi.org/10.1001/jama.1975.03260240050026.

Beck, A.T., Steer, R.A., Kovacs, M., Garrison, B., 1985. Hopelessness and eventual suicide: a 10-year prospective study of patients hospitalized with suicidal ideation. Am. J. Psychiatry 142 (5), 559–563. https://doi.org/10.1176/ajp.142.5.559.

Barak-Corren, Y., Castro, V., Nock, M., et al., 2020. Validation of an electronic health record-based suicide risk prediction modeling approach across multiple health care systems. JAMA Netw. Open 3 (3), 1–13. https://doi.org/10.1001/jamanetworkopen.2020.1262. https://jamanetwork.com/journals/jamanetworkopen/fullarticle/2763237.

Beck, A.T., Brown, G., Berchick, R.J., Stewart, B.L., Steer, R.A., 1990. Relationship between hopelessness and ultimate suicide: a replication with psychiatric outpatients. Am. J. Psychiatry 147 (2), 190–195. https://doi.org/10.1176/ajp.147.2.190.

Bergfeld, I.O., Mantione, M., Hoogendoorn, M.L.C., Ruhé, H.G., Notten, P., Van Laarhoven, J., Visser, I., Figee, M., De Kwaasteniet, B.P., Horst, F., Schene, A.H., Van Den Munckhof, P., Beute, G., Schuurman, R., Denys, D., 2016. Deep brain stimulation of the ventral anterior limb of the internal capsule for treatment-resistant depression. JAMA Psychiatry 73 (5), 456–464. https://doi.org/10.1001/jamapsychiatry.2016.0152.

Bergfeld, I.O., Mantione, M., Figee, M., Schuurman, P.R., Lok, A., Denys, D., 2018. Treatment-resistant depression and suicidality. J. Affect. Disord. 235, 362–367. https://doi.org/10.1016/j.jad.2018.04.016.

Berman, R.M., Cappiello, A., Anand, A., Oren, D.A., Heninger, G.R., Charney, D.S., Krystal, J.H., 2000. Antidepressant effects of ketamine in depressed patients. Biol. Psychiatry 47 (4), 351–354. https://doi.org/10.1016/S0006-3223(99)00230-9.

Beutel, M.E., Klein, E.M., Brähler, E., Reiner, I., Jünger, C., Michal, M., Wiltink, J., Wild, P.S., Münzel, T., Lackner, K.J., Tibubos, A.N., 2017. Loneliness in the general population: prevalence, determinants and relations to mental health. BMC Psychiatry 17 (1). https://doi.org/10.1186/s12888-017-1262-x.

Bewernick, B.H., Kayser, S., Sturm, V., Schlaepfer, T.E., 2012. Long-term effects of nucleus accumbens deep brain stimulation in treatment-resistant depression: evidence for sustained efficacy. Neuropsychopharmacology 37 (9), 1975–1985. https://doi.org/10.1038/npp.2012.44.

Borges, G., Angst, J., Nock, M.K., Ruscio, A.M., Kessler, R.C., 2008. Risk factors for the incidence and persistence of suicide-related outcomes: a 10-year follow-up study using the National Comorbidity Surveys. J. Affect. Disord. 105 (1–3), 25–33. https://doi.org/10.1016/j.jad.2007.01.036.

Borges, G., Nock, M.K., Abad, J.M.H., Hwang, I., Sampson, N.A., Alonso, J., Andrade, L.H., Angermeyer, M.C., Beautrais, A., Bromet, E., Bruffaerts, R., De Girolamo, G., Florescu, S., Gureje, O., Hu, C., Karam, E.G., Kovess-Masfety, V., Lee, S., Levinson, D., et al., 2010. Twelve-month prevalence of and risk factors for suicide attempts in the world health organization world mental health surveys. J. Clin. Psychiatry 71 (12), 1617–1628. https://doi.org/10.4088/JCP.08m04967blu.

Brent, D., Perper, J., Goldstein, C., Kolko, D., Allan, M., Allman, C., Zelenak, J., 1988. Risk factors for adolescent suicide: a comparison of adolescent suicide victims with suicidal inpatients. Arch. Gen. psychiatry 45 (6), 581–588. https://doi.org/10.1001/archpsyc.1988.01800300079011.

Brent, D.A., Emslie, G.J., Clarke, G.N., Asarnow, J., Spirito, A., Ritz, L., Vitiello, B., Iyengar, S., Birmaher, B., Ryan, N.D., Zelazny, J., Onorato, M., Kennard, B., Mayes, T.L., DeBar, L.L., McCracken, J.T., Strober, M., Suddath, R., Leonard, H., et al., 2009. Predictors of spontaneous and systematically assessed suicidal adverse events in the treatment of SSRI-resistant depression in adolescents (TORDIA) study. Am. J. Psychiatry 166 (4), 418–426. https://doi.org/10.1176/appi.ajp.2008.08070976.

Bryan, C.J., Mintz, J., Clemans, T.A., Leeson, B., Burch, T.S., Williams, S.R., Maney, E., Rudd, M.D., 2017. Effect of crisis response planning vs. contracts for safety on suicide risk in U.S. Army Soldiers: a randomized clinical trial. J. Affect. Disord. 212, 64–72. https://doi.org/10.1016/j.jad.2017.01.028.

Busch, K.A., Fawcett, J., Jacobs, D.G., 2003. Clinical correlates of inpatient suicide. J. Clin. Psychiatry 64 (1), 14–19. https://doi.org/10.4088/JCP.v64n0105.

Centers for Disease Control and Prevention: National Center for Injury Prevention and Control, 2019. Web-based Injury Statistics Query and Reporting System (WISQARS). Source of data is the National Vital Statistics System from the National Center for Health Statistics. https://www.cdc.gov/injury/wisqars/LeadingCauses.html. (Accessed 5 October 2020).

Christmas, D., Steele, J.D., Tolomeo, S., Eljamel, M.S., Matthews, K., 2013. Vagus nerve stimulation for chronic major depressive disorder: 12-month outcomes in highly treatment-refractory patients. J. Affect. Disord. 150 (3), 1221–1225. https://doi.org/10.1016/j.jad.2013.05.080.

Chu, C., Buchman-Schmitt, J.M., Stanley, I.H., Hom, M.A., Tucker, R.P., Hagan, C.R., Rogers, M.L., Podlogar, M.C., Chiurliza, B., Ringer, F.B., Michaels, M.S., Patros, C.H.G., Joiner, T.E., 2017. The interpersonal theory of suicide: a systematic review and meta-analysis of a decade of cross-national research. Psychol. Bull. 143 (12), 1313–1345. https://doi.org/10.1037/bul0000123.

Cipriani, A., Hawton, K., Stockton, S., Geddes, J.R., 2013. Lithium in the prevention of suicide in mood disorders: updated systematic review and meta-analysis. BMJ 347 (7916). https://doi.org/10.1136/bmj.f3646.

Conwell, Y., Van Orden, K., Caine, E.D., 2011. Suicide in older adults. Psychiatr. Clin. N. Am. 34 (2), 451–468. https://doi.org/10.1016/j.psc.2011.02.002.

Coppersmith, D.D.L., Kleiman, E.M., Glenn, C.R., Millner, A.J., Nock, M.K., 2019. The dynamics of social support among suicide attempters: a smartphone-based daily diary study. Behav. Res. Ther. 120. https://doi.org/10.1016/j.brat.2018.11.016.

Cristancho, P., Cristancho, M.A., Baltuch, G.H., Thase, M.E., O'Reardon, J.P., 2011. Effectiveness and safety of vagus nerve stimulation for severe treatment-resistant major depression in clinical practice after FDA approval: outcomes at 1 year. J. Clin. Psychiatry 72 (10), 1376–1382. https://doi.org/10.4088/JCP.09m05888blu.

D'Anci, K.E., Uhl, S., Giradi, G., Martin, C., 2019. Treatments for the prevention and management of suicide. Ann. Intern. Med. 171 (5), 334–342. https://doi.org/10.7326/M19-0869.

Dawes, R., Faust, D., Meehl, P., 1989. Clinical versus actuarial judgment. Science 243 (4899), 1668–1674. https://doi.org/10.1126/science.2648573.

Dazzi, T., Gribble, R., Wessely, S., Fear, N.T., 2014. Does asking about suicide and related behaviours induce suicidal ideation? What is the evidence? Psychol. Med. 44 (16), 3361–3363. https://doi.org/10.1017/S0033291714001299.

Derntl, B., Seidel, E.M., Eickhoff, S.B., Kellermann, T., Gur, R.C., Schneider, F., Habel, U., 2011. Neural correlates of social approach and withdrawal in patients with major depression. Soc. Neurosci. 6 (5–6), 482–501. https://doi.org/10.1080/17470919.2011.579800.

Emslie, G.J., Mayes, T., Porta, G., Vitiello, B., Clarke, G., Wagner, K.D., Asarnow, J.R., Spirito, A., Birmaher, B., Ryan, N., Kennard, B., DeBar, L., McCracken, J., Strober, M., Onorato, M., Zelazny, J., Keller, M., Iyengar, S., Brent, D., 2010. Treatment of resistant depression in adolescents (TORDIA): week 24 outcomes. Am. J. Psychiatry 167 (7), 782–791. https://doi.org/10.1176/appi.ajp.2010.09040552.

Erzen, E., Çikrikci, Ö., 2018. The effect of loneliness on depression: a meta-analysis. Int. J. Soc. Psychiatry 64 (5), 427–435. https://doi.org/10.1177/0020764018776349.

Foncke, E.M.J., Schuurman, P.R., Speelman, J.D., 2006. Suicide after deep brain stimulation of the internal globus pallidus for dystonia. Neurology 66 (1), 142–143. https://doi.org/10.1212/01.wnl.0000191328.05752.e2.

Franklin, J.C., Ribeiro, J.D., Fox, K.R., Bentley, K.H., Kleiman, E.M., Huang, X., Musacchio, K.M., Jaroszewski, A.C., Chang, B.P., Nock, M.K., 2017. Risk factors for suicidal thoughts and behaviors: a meta-analysis of 50 years of research. Psychol. Bull. 143 (2), 187–232. https://doi.org/10.1037/bul0000084.

Ge, L., Yap, C.W., Ong, R., Heng, B.H., 2017. Social isolation, loneliness and their relationships with depressive symptoms: a population-based study. PLoS One 12 (8). https://doi.org/10.1371/journal.pone.0182145.

George, M.S., Taylor, J.J., Short, E.B., 2013. The expanding evidence base for rTMS treatment of depression. Curr. Opin. Psychiatry 26 (1), 13–18. https://doi.org/10.1097/YCO.0b013e32835ab46d.

Glenn, C.R., Kleiman, E.M., Cha, C.B., Nock, M.K., Prinstein, M.J., 2016. Implicit cognition about self-injury predicts actual self-injurious behavior: results from a longitudinal study of adolescents. J. Child Psychol. Psychiatry Allied Discip. 57 (7), 805–813. https://doi.org/10.1111/jcpp.12500.

Glenn, C.R., Kleiman, E.M., Coppersmith, D.D.L., Santee, A.C., Esposito, E.C., Cha, C.B., Nock, M.K., Auerbach, R.P., 2017a. Implicit identification with death predicts change in suicide ideation during psychiatric treatment in adolescents. J. Child Psychol. Psychiatry Allied Discip. 58 (12), 1319–1329. https://doi.org/10.1111/jcpp.12769.

Glenn, J.J., Werntz, A.J., Slama, S.J.K., Steinman, S.A., Teachman, B.A., Nock, M.K., 2017b. Suicide and self-injury-related implicit cognition: a large-scale examination and replication. J. Abnorm. Psychol. 126 (2), 199–211. https://doi.org/10.1037/abn0000230.

Glenn, C.R., Millner, A.J., Esposito, E.C., Porter, A.C., Nock, M.K., 2019. Implicit identification with death predicts suicidal thoughts and behaviors in adolescents. J. Clin. Child Adolesc. Psychol. 48 (2), 263–272. https://doi.org/10.1080/15374416.2018.1528548.

Goodman, M., Banthin, D., Blair, N.J., Mascitelli, K.A., Wilsnack, J., Chen, J., Messenger, J.W., Mercedes Perez-Rodriguez, M., Triebwasser, J., Koenigsberg, H.W., Goetz, R.R., Hazlett, E.A., New, A.S., 2016. A randomized trial of dialectical behavior therapy in high-risk suicidal veterans. J. Clin. Psychiatry 77 (12), e1591–e1600. https://doi.org/10.4088/JCP.15m10235.

Gøtzsche, P.C., Gøtzsche, P.K., 2017. Cognitive behavioural therapy halves the risk of repeated suicide attempts: systematic review. J. R. Soc. Med. 110 (10), 404–410. https://doi.org/10.1177/0141076817731904.

Green, A.I., Zimmet, S.V., Strous, R.D., Schildkraut, J.J., 1999. Clozapine for comorbid substance use disorder and schizophrenia: do patients with schizophrenia have a reward-deficiency syndrome that can be ameliorated by clozapine? Harv. Rev. Psychiatry 6 (6), 287–296. https://doi.org/10.3109/10673229909017206.

Hantouche, E., Angst, J., Azorin, J.M., 2010. Explained factors of suicide attempts in major depression. J. Affect. Disord. 127 (1–3), 305–308. https://doi.org/10.1016/j.jad.2010.04.032.

Hedegaard, H., Curtin, S., Warner, M., 2018. Suicide rates in the United States continue to increase. NCHS Data Brief (309), 1–8. https://www.cdc.gov/nchs/data/databriefs/db309.pdf. (Accessed 12 October 2020).

Heisel, M.J., Duberstein, P.R., Talbot, N.L., King, D.A., Tu, X.M., 2009. Adapting interpersonal psychotherapy for older adults at risk for suicide: preliminary findings. Prof. Psychol. Res. Pract. 40 (2), 156–164. https://doi.org/10.1037/a0014731.

Hoffberg, A.S., Stearns-Yoder, K.A., Brenner, L.A., 2020. The effectiveness of crisis line services: a systematic review. Front. Public Health 7. https://doi.org/10.3389/fpubh.2019.00399.

Ijaz, S., Davies, P., Williams, C.J., Kessler, D., Lewis, G., Wiles, N., 2018. Psychological therapies for treatment-resistant depression in adults. Cochrane Database Syst. Rev. 2018 (5). https://doi.org/10.1002/14651858.CD010558.pub2.

Ionescu, D.F., Swee, M.B., Pavone, K.J., Taylor, N., Akeju, O., Baer, L., Nyer, M., Cassano, P., Mischoulon, D., Alpert, J.E., Brown, E.N., Nock, M.K., Fava, M., Cusin, C., 2016. Rapid and sustained reductions in current suicidal ideation following repeated doses of intravenous ketamine. J. Clin. Psychiatry, e719–e725. https://doi.org/10.4088/jcp.15m10056.

Jahn, D.R., Cukrowicz, K.C., Linton, K., Prabhu, F., 2011. The mediating effect of perceived burdensomeness on the relation between depressive symptoms and suicide ideation in a community sample of older adults. Aging Ment. Health 15 (2), 214–220. https://doi.org/10.1080/13607863.2010.501064.

Joiner, T., 2005. Why People Die by Suicide. Why People Die by Suicide. Harvard University Press, Cambridge, MA.

Joiner, T.E., Hom, M.A., Hagan, C.R., Silva, C., 2016. Suicide as a derangement of the self-sacrificial aspect of eusociality. Psychol. Rev. 123 (3), 235–254. https://doi.org/10.1037/rev0000020.

Kellner, C.H., Fink, M., Knapp, R., Petrides, G., Husain, M., Rummans, T., Mueller, M., Bernstein, H., Rasmussen, K., O'Connor, K., Smith, G., Rush, A.J., Biggs, M., McClintock, S., Bailine, S., Malur, C., 2005. Relief of expressed suicidal intent by ECT: a consortium for research in ECT study. Am. J. Psychiatry 162 (5), 977–982. https://doi.org/10.1176/appi.ajp.162.5.977.

Kelly, K.T., Knudson, M.P., 2000. Are no-suicide contracts effective in preventing suicide in suicidal patients seen by primary care physicians? Arch. Fam. Med. 9 (10), 1119–1121. https://doi.org/10.1001/archfami.9.10.1119.

Kessler, R.C., Berglund, P., Demler, O., Jin, R., Merikangas, K.R., Walters, E.E., 2005. Lifetime prevalence and age-of-onset distributions of DSM-IV disorders in the national comorbidity survey replication. Arch. Gen. Psychiatry 62 (6), 593–602. https://doi.org/10.1001/archpsyc.62.6.593.

Kessler, R.C., Gruber, M., Hettema, J.M., Hwang, I., Sampson, N., Yonkers, K.A., 2008. Co-morbid major depression and generalized anxiety disorders in the National Comorbidity Survey follow-up. Psychol. Med. 38 (3), 365–374. https://doi.org/10.1017/S0033291707002012.

Kleiman, E.M., Turner, B.J., Fedor, S., Beale, E.E., Huffman, J.C., Nock, M.K., 2017. Examination of real-time fluctuations in suicidal ideation and its risk factors: results from two ecological momentary assessment studies. J. Abnorm. Psychol. 126 (6), 726–738. https://doi.org/10.1037/abn0000273.

Kovacsics, C.E., Gottesman, I.I., Gould, T.D., 2009. Lithium's antisuicidal efficacy: elucidation of neurobiological targets using endophenotype strategies. Annu. Rev. Pharmacol. Toxicol. 49, 175–198. https://doi.org/10.1146/annurev.pharmtox.011008.145557.

Kraemer, H., 1997. Coming to terms with the terms of risk. Arch. Gen. Psychiatry 54 (4), 337–343. https://doi.org/10.1001/archpsyc.1997.01830160065009.

Leavey, K., Hawkins, R., 2017. Is cognitive behavioural therapy effective in reducing suicidal ideation and behaviour when delivered face-to-face or via e-health? A systematic review and meta-analysis. Cogn. Behav. Ther. 46 (5), 353–374. https://doi.org/10.1080/16506073.2017.1332095.

Liang, C.S., Chung, C.H., Ho, P.S., Tsai, C.K., Chien, W.C., 2018. Superior anti-suicidal effects of electroconvulsive therapy in unipolar disorder and bipolar depression. Bipolar Disord. 20 (6), 539–546. https://doi.org/10.1111/bdi.12589.

Linehan, M.M., 1993. Dialectical behavior therapy for treatment of borderline personality disorder: implications for the treatment of substance abuse. In: NIDA Research Monograph Series, pp. 201–216. Issue 137.

Linehan, M.M., Comtois, K.A., Murray, A.M., Brown, M.Z., Gallop, R.J., Heard, H.L., Korslund, K.E., Tutek, D.A., Reynolds, S.K., Lindenboim, N., 2006. Two-year randomized controlled trial and follow-up of dialectical behavior therapy vs therapy by experts for suicidal behaviors and borderline personality disorder. Arch. Gen. Psychiatry 63 (7), 757–766. https://doi.org/10.1001/archpsyc.63.7.757.

Liu, R.T., Bettis, A.H., Burke, T.A., 2020. Characterizing the phenomenology of passive suicidal ideation: a systematic review and meta-analysis of its prevalence, psychiatric comorbidity, correlates, and comparisons with active suicidal ideation. Psychol. Med. 50 (3), 367–383. https://doi.org/10.1017/S003329171900391X.

Mahgoub, N.A., Kotbi, N., 2009. Acute depression and suicidal attempt following lowering the frequency of deep brain stimulation. J. Neuropsychiatr., 468. https://doi.org/10.1176/appi.neuropsych.21.4.468.

McDonald, W.M., Prudic, J., McCall, W.V., Husain, M.M., 2013. Update from the American Psychiatry Association's Task Force on Electroconvulsive Therapy. Am. J. Geriatr. Psychiatry, S39–S40. https://doi.org/10.1016/j.jagp.2012.12.078.

McMain, S.F., Guimond, T., Barnhart, R., Habinski, L., Streiner, D.L., 2017. A randomized trial of brief dialectical behaviour therapy skills training in suicidal patients suffering from borderline disorder. Acta Psychiatr. Scand. 135 (2), 138–148. https://doi.org/10.1111/acps.12664.

Meltzer, H., Alphs, L., Green, A., Altamura, C., Anand, R., Bertoldi, A., Bourgeois, M., 2003. Clozapine treatment for suicidality in schizophrenia: international suicide prevention trial (InterSePT). Arch. Gen. Psychiatry 60 (1), 82–91. https://doi.org/10.1001/archpsyc.60.7.735.

Millner, A.J., Lee, M.D., Nock, M.K., 2017. Describing and measuring the pathway to suicide attempts: a preliminary study. Suicide Life Threat. Behav. 47 (3), 353–369. https://doi.org/10.1111/sltb.12284.

Millner, A.J., Augenstein, T.M., Visser, K.H., Gallagher, K., Vergara, G.A., D'Angelo, E.J., Nock, M.K., 2019. Implicit cognitions as a behavioral marker of suicide attempts in adolescents. Arch. Suicide Res. 23 (1), 47–63. https://doi.org/10.1080/13811118.2017.1421488.

Modestin, J., Dal Pian, D., Agarwalla, P., 2005. Clozapine diminishes suicidal behavior: a retrospective evaluation of clinical records. J. Clin. Psychiatry 66 (4), 534–538. https://doi.org/10.4088/JCP.v66n0418.

Nock, M., Prinstein, M., 2004. A functional approach to the assessment of self-mutilative behavior. J. Consult. Clin. Psychol. 72 (5), 885. https://doi.org/10.1037/0022-006X.72.5.885. In this issue.

Nock, M.K., Benaji, M.R., 2007. Prevention of suicide ideation and attempts among adolescents using a breif perfomance-based test. J. Consult. Clin. Psychol. 22 (5), 629.

Nock, M.K., Hwang, I., Sampson, N., Kessler, R.C., Angermeyer, M., Beautrais, A., Borges, G., Bromet, E., Bruffaerts, R., De Girolamo, G., De Graaf, R., Florescu, S., Gureje, O., Haro, J.M., Hu, C., Huang, Y., Karam, E.G., Kawakami, N., Kovess, V., et al., 2009a. Cross-national analysis of the associations among mental disorders and suicidal behavior: findings from the WHO World Mental Health Surveys. PLoS Med. 6 (8). https://doi.org/10.1371/journal.pmed.1000123.

Nock, M.K., Prinstein, M.J., Sterba, S.K., 2009b. Revealing the form and function of self-injurious thoughts and behaviors: a real-time ecological assessment study among adolescents and young adults. J. Abnorm. Psychol. 118 (4), 816–827. https://doi.org/10.1037/a0016948.

Nock, M.K., Hwang, I., Sampson, N.A., Kessler, R.C., 2010a. Mental disorders, comorbidity and suicidal behavior: results from the National Comorbidity Survey Replication. Mol. Psychiatry, 868–876. https://doi.org/10.1038/mp.2009.29.

Nock, M.K., Park, J.M., Finn, C.T., Deliberto, T.L., Dour, H.J., Banaji, M.R., 2010b. Measuring the suicidal mind: implicit cognition predicts suicidal behavior. Psychol. Sci. 21 (4), 511–517. https://doi.org/10.1177/0956797610364762.

Nock, M., Kleiman, E., Abraham, M., Bentley, K., Brent, D., Buonopane, R., Castro-Ramirez, F., Cha, C., Dempsey, W., et al., 2021. Consensus statement on ethical & safety practices for conducting digital monitoring studies with people at risk of suicide and related behaviors. Psychiatr. Res. Clin. Pract. 3 (2), 57–66. https://doi.org/10.1176/appi.prcp.20200029.

Notredame, C.E., Grandgenèvre, P., Pauwels, N., Morgiève, M., Wathelet, M., Vaiva, G., Séguin, M., 2018. Leveraging the web and social media to promote access to care among suicidal individuals. Front. Psychol. 9. https://doi.org/10.3389/fpsyg.2018.01338.

O'Donnell, K.C., Gould, T.D., 2007. The behavioral actions of lithium in rodent models: leads to develop novel therapeutics. Neurosci. Biobehav. Rev. 31 (6), 932–962. https://doi.org/10.1016/j.neubiorev.2007.04.002.

Papakostas, G.I., Petersen, T., Pava, J., Masson, E., Worthington, J.J., Alpert, J.E., Fava, M., Nierenberg, A.A., 2003. Hopelessness and suicidal ideation in outpatients with treatment-resistant depression: prevalence and impact on treatment outcome. J. Nerv. Ment. Dis. 191 (7), 444–449. https://doi.org/10.1097/01.NMD.0000081591.46444.97.

Price, R.B., Nock, M.K., Charney, D.S., Mathew, S.J., 2009. Effects of intravenous ketamine on explicit and implicit measures of suicidality in treatment-resistant depression. Biol. Psychiatry 66 (5), 522–526. https://doi.org/10.1016/j.biopsych.2009.04.029.

Price, R.B., Iosifescu, D.V., Murrough, J.W., Chang, L.C., Al Jurdi, R.K., Iqbal, S.Z., Soleimani, L., Charney, D.S., Foulkes, A.L., Mathew, S.J., 2014. Effects of ketamine on explicit and implicit suicidal cognition: a randomized controlled trial in treatment-resistant depression. Depress. Anxiety 31 (4), 335–343. https://doi.org/10.1002/da.22253.

Qin, P., Nordentoft, M., 2005. Suicide risk in relation to psychiatric hospitalization. Arch. Gen. Psychiatry 62 (4), 427–432. https://doi.org/10.1001/archpsyc.62.4.427.

Reid, S., 1999. Perception and communication of risk, and the importance of dependability. Struct. Saf. 21 (4), 373–384. https://doi.org/10.1016/S0167-4730(99)00031-4.

Schacter, D.L., Norman, K.A., Koutstaal, W., 1998. The cognitive neuroscience of constructive memory. Annu. Rev. Psychol. 49, 289–318. https://doi.org/10.1146/annurev.psych.49.1.289.

Serafini, G., Adavastro, G., Canepa, G., De Berardis, D., Valchera, A., Pompili, M., Nasrallah, H., Amore, M., 2018. The efficacy of buprenorphine in major depression, treatment-resistant depression and suicidal behavior: a systematic review. Int. J. Mol. Sci. 19 (8). https://doi.org/10.3390/ijms19082410.

Simon, G.E., Savarino, J., Operskalski, B., Wang, P.S., 2006. Suicide risk during antidepressent treatment. Am. J. Psychiatry 163 (1), 41–47. https://doi.org/10.1176/appi.ajp.163.1.41.

Simon, G.E., Johnson, E., Lawrence, J.M., Rossom, R.C., Ahmedani, B., Lynch, F.L., Beck, A., Waitzfelder, B., Ziebell, R., Penfold, R.B., Shortreed, S.M., 2018. Predicting suicide attempts and suicide deaths following outpatient visits using electronic health records. Am. J. Psychiatry 175 (10), 951–960. https://doi.org/10.1176/appi.ajp.2018.17101167.

Stone, K.L., Blackwell, T.L., Ancoli-Israel, S., Cauley, J.A., Redline, S., Marshall, L.M., Ensrud, K.E., 2014. Sleep disturbances and risk of falls in older community-dwelling men: the outcomes of sleep disorders in older men (MrOS sleep) study. J. Am. Geriatr. Soc. 62 (2), 299–305. https://doi.org/10.1111/jgs.12649.

Van Orden, K.A., Witte, T.K., Cukrowicz, K.C., Braithwaite, S.R., Selby, E.A., Joiner Jr., T.E., 2011. The interpersonal theory of suicide. Psychol. Rev. 117 (2), 575–600. https://doi.org/10.1037/a0018697.

Voineskos, D., Daskalakis, Z.J., Blumberger, D.M., 2020. Management of treatment-resistant depression: challenges and strategies. Neuropsychiatr. Dis. Treat. 16, 221–234. https://doi.org/10.2147/NDT.S198774.

Wilkinson, P., Dubicka, B., Kelvin, R., Roberts, C., Goodyer, I., 2009. Treated depression in adolescents: predictors of outcome at 28 weeks. Br. J. Psychiatry 194 (4), 334–341. https://doi.org/10.1192/bjp.bp.108.052381.

Wilks, C.R., Lungu, A., Ang, S.Y., Matsumiya, B., Yin, Q., Linehan, M.M., 2018. A randomized controlled trial of an Internet delivered dialectical behavior therapy skills training for suicidal and heavy episodic drinkers. J. Affect. Disord. 232, 219–228. https://doi.org/10.1016/j.jad.2018.02.053.

World Health Organization (WHO), 2018. Crude Suicide Rates Per 100,000 Population. World Health Organization.

Yovell, Y., Bar, G., Mashiah, M., Baruch, Y., Briskman, I., Asherov, J., Lotan, A., Rigbi, A., Panksepp, J., 2016. Ultra-low-dose buprenorphine as a time-limited treatment for severe suicidal ideation: a randomized controlled trial. Am. J. Psychiatry 173 (5), 491–498. https://doi.org/10.1176/appi.ajp.2015.15040535.

Zarate, C.A., Singh, J.B., Carlson, P.J., Brutsche, N.E., Ameli, R., Luckenbaugh, D.A., Charney, D.S., Manji, H.K., 2006. A randomized trial of an N-methyl-D-aspartate antagonist in treatment-resistant major depression. Arch. Gen. Psychiatry 63 (8), 856–864. https://doi.org/10.1001/archpsyc.63.8.856.

Chapter 40

Maintaining motivation and preserving the therapeutic alliance as tools to overcome treatment-resistant depression

James C. West, Mary C. Vance, and Robert J. Ursano
Department of Psychiatry, Uniformed Services University of the Health Sciences, Bethesda, MD, United States

Thus far, this text has focused on the understanding and application of a broad range of interventions to improve outcomes in treatment-resistant depression. Understanding the breadth of psychopharmacologic and psychotherapeutic interventions permits tailoring treatment to specific patient characteristics to ensure the greatest response. However, there is another dimension to treatment-resistant depression that must be considered. In this age of biological psychiatry the rapid expansion of understanding of depressive illness and treatment options tends to capture increasingly more of both therapists' and patients' attention. It is important not to lose sight of the most commonplace, yet highly effective, mechanism of behavioral change—the therapeutic alliance. This chapter will develop the concept of therapeutic alliance and how it relates to treatment resistance in depression. We will consider how the treatment relationship can deviate from one of working alliance into unhealthy alliance and how this can contribute to treatment resistance. We will consider interventions to enhance therapeutic alliance and motivation to achieve the best outcome in treatment-resistant depression. Finally, we will consider the effect of telehealth and other technology-based interventions on therapeutic alliance.

We will use the term therapist to describe the physician or other provider offering the full range of psychotherapy, pharmacotherapy, and other interventions throughout this chapter. This is in deference to the fact that most of the literature around therapeutic alliance grows from psychotherapy theory and research. Therapy for the purposes of this chapter will include the full range of somatic interventions in addition to psychotherapy and other psychosocial interventions. Treatment resistance can broadly be thought of as biologically caused or behavioral or interpersonal. Biological treatment resistance would be manifested as poor response to adequate trials of medications at correct doses and treatment durations. Behavioral or interpersonal treatment resistance would be manifest as missed appointments, dropping out of treatment, not taking medications as prescribed, or not accurately reporting symptoms to the therapist. What we will show is that the concepts drawn from the psychotherapeutic relationships are broadly applicable regardless of the treatment modality.

Understanding the therapeutic Alliance

> *... therapeutic alliance is not only a prerequisite for therapeutic work, but often may be the main vehicle of change.*
>
> Horwitz (1974)

Therapeutic alliance, or working alliance, describes the reality-based nature of the relationship between therapist and patient. The primary elements of the therapeutic alliance are commonly identified as goal, task, and bond (Bordin, 1979). A therapeutic alliance is present with a patient whether the treatment is medication or psychotherapy. It is the working together and the interpersonal and behavioral elements of that task. *Goal* refers to the shared goals of the patient and the therapist. Goals in the therapeutic alliance can include specific symptoms from which the patient seeks relief. They can extend to understanding of thought distortions and interpersonal dynamics, or they can explore challenges with closeness or warmth (Summers and Barber, 2010). *Task* refers to those actions taken by both the patient and therapist to make progress toward therapeutic goals. For the patient, this may include taking medication in right dose and time and free and honest reflection in psychotherapy. Therapist tasks include attentive listening, putting aside biases to develop an understanding of the patient, accurate inquiry for information in assessment and appropriate prescribing. *Bond* refers to trust and attachment

Managing Treatment-Resistant Depression. https://doi.org/10.1016/B978-0-12-824067-0.00040-2

required of both patient and therapist to maintain an interpersonal relationship and continue to carry out the agreed upon tasks (Hilsenroth et al., 2012). Bond is the emotional attachment and interpersonal behaviors that allow the patient to feel safe in the relationship, trust the therapist, and experience empathy from the therapist. The therapist should have a feeling of working together jointly, commitment to the patient's well-being (even when the patient may not be able to sustain that hope) and a caring emotional attachment that maintains the separation and objectivity to provide effective therapy (Box 40.1; Summers and Barber, 2010).

Measures of therapeutic alliance examine both patient and therapist perceptions of working together. The Working Alliance Inventory (WAI) has been adapted over the years into several different versions, but all have in common measurement of both the therapist and patient assessment of their alliance (Hatcher and Gillaspy, 2006). Other measures of therapeutic alliance include the California Pharmacotherapy Alliance Scale (CALPAS) (Gaston, 1991), Helping Alliance Questionnaires (HAq-II) (Luborsky et al., 1996), and the Vanderbilt Psychotherapy Process Scale (Smith et al., 2003). Although designed for research these scales may be used in clinical settings to inform the therapist on the quality of the therapeutic alliance. The WAI includes items from Greenson's original conception of the therapeutic alliance, while the CALPAS incorporates additional features such as patient commitment and idealization of the therapist, and therapist understanding (Barber et al., 2014). Almost all of these scales incorporate subscales that assess dimensions of goal, task, and bond.

Transference plays a significant role in treatment outcomes. Transference represents the patient's perceptions of the patient-therapist relationship that are an echo from the past, expectations from the past that mold the present, and are derived from past relationships. Such expectations typically involve repetition of past relationships in ways that do not match the actual present situation (Greenson, 1965). The recognition and use of transference is also important in pharmacological based treatments. Patients' transference, such as being reminded of an untrustworthy parent, can contribute to ruptures in the therapeutic alliance which at their worst lead to termination of treatment. Therapeutic ruptures can manifest as withdrawal, in which the patient moves away from the work of treatment by missing appointments, or confrontation, in which the patient expresses anger or dissatisfaction with the therapist or attempts to pressure or control the therapist (Eubanks et al., 2018). Recognizing risk for a rupture in the relationship and knowing how to repair ruptures are important skills for therapists to learn and can be highly important to successful outcomes of treatment (Parth et al., 2017).

Box 40.1 Clinical example: Goal, task, and bond

Therapist: So, reviewing the symptoms you have described, and your history of medication trials, it seems you have had this depression to some extent for much of the last 18 months.

Patient: Yes, I am really hoping you have some better ideas on medications. I don't feel like they do anything.

Therapist: Let's talk a little bit about what you think better is going to feel like. Are there symptoms or other parts of your life as it is right now that you think need to get better?

Patient: This lack of sleep is really overwhelming me. No matter how much I try to get sleep, I am always up and down all night, and then I just can't function because I'm so tired.

Therapist: Anything else?

Patient: I don't know. I just feel so overwhelmed with work and the kids. I mean, they spend all day at daycare, and then I am so exhausted when I get home. I just want to lie down for a while, but I can't. If I do, then I just feel guilty when they wake me up because their video ended. Seriously, what kind of mom is that?

Therapist: That suggests several goals we might focus on in the near term. One is we need to ensure you are getting restful sleep. The other seems like we need to work on some of your thoughts and feelings about balancing work and home. Do those sound like reasonable targets to you?

Patient: That sounds pretty good.

Therapist: Getting there is going to take some effort on my part and also on your part. I can offer a couple of suggestions for medications that might work best for you. I would also recommend we start a course of cognitive-behavioral therapy to get at some of those thoughts and feelings. That's going to require some commitment on your part. You'll need to take the medications as prescribed and keep me informed of how they are working. Also, it may seem like a lot, but I would like you to come to weekly psychotherapy sessions at first so we can get to work.

Patient: Weekly, that seems like a lot.

Therapist: It's a commitment for both of us, but when you are here you will have my full attention. I want you to understand that this relationship is about me using my skills to help, and you being present to allow this process to work. We can space things out once we start to see some improvement.

Patient: OK.

Relationship between therapeutic alliance and treatment outcome

The quality of therapeutic alliance is related to outcome in depression treatment, whether that treatment is medication, psychotherapy, or the combination. Krupnick and colleagues demonstrated that the therapeutic alliance had a significant effect on treatment outcome for treatment with CBT, interpersonal therapy, imipramine, and even placebo (Krupnick et al., 1996). In both cognitive therapy and medication management, a stronger therapeutic alliance is associated with lower treatment dropout rates (Cooper et al., 2016). Comparing medication and placebo in clinical trials for depression, a positive therapeutic alliance is associated with greater symptomatic improvement. This effect is greater for placebo than active medication (Zilcha-Mano et al., 2015). Interestingly, there appears to be an interaction between the perceived therapeutic alliance, the act of taking a pill (either drug or placebo), and improvement in depressive symptoms (Leuchter et al., 2014). Meta-analysis has shown that the therapeutic alliance has a small to moderate effect size on psychotherapy outcome regardless of when in therapy it is measured or whether perceived by therapist or patient (Horvath et al., 2011). The alliance is also significantly associated with the time to relapse after a successful cognitive-behavioral therapy (Weck et al., 2013). One can imagine a good therapeutic alliance will encourage a patient to return to treatment earlier and to report early symptoms to the therapist. The benefit of the positive therapeutic alliance is not limited to psychiatric diagnoses. The positive therapeutic alliance between patients with chronic low back pain and their physical therapist is associated with greater improvement in pain and reduced disability over 8 weeks of treatment (Ferreira et al., 2013). It may be that in such treatments, the patient is more likely to accurately report symptom changes and to carry out prescribed exercises. This is a fascinating effect to consider as it is not difficult to draw parallels between the hopelessness patients with chronic low back pain experience with that experienced by patients with treatment-resistant depression. Strong therapeutic alliance appears to be associated with fewer suicidal thoughts and less self-harm behaviors (Dunster-Page et al., 2017).

Certain patient characteristics seem to contribute to the effect of the therapeutic alliance on treatment outcome. Patients who demonstrate a higher degree of submissiveness or low interpersonal agency appear to benefit from a more positive therapeutic alliance in treatment of depression (Gómez Penedo et al., 2020a). This may be due to a tendency to withhold constructively disclosing negative feelings toward a therapist and instead engaging in behaviors leading to therapeutic rupture. In addition to patients perceiving a more positive therapeutic alliance, it has also been shown that greater agreement between therapist and patient rating of the therapeutic alliance is associated with greater improvement in depressive symptoms in psychotherapy (Laws et al., 2017). The therapist and the patient have a shared view in such relationships rather than a discrepant one. Therapeutic alliance has been shown to mediate the effect of the personality traits of agreeableness, extroversion, and openness on treatment outcome in major depressive disorder (Kushner et al., 2016). Patients with depressive symptoms as a result of a personality disorder may respond better to psychodynamic psychotherapy, a psychotherapy in which the therapeutic alliance is a key focus of the work, equal to that of transference (Gabbard, 2014).

In contrast, a negative therapeutic alliance may contribute to treatment resistance in depression. An unhealthy alliance can emerge as the result of patient or therapist factors. For the patient, a negative alliance may be part of the patient maintaining the sick role as part of their identity, as part of oppositional stance toward the therapist's efforts to help them get well, and inability to develop trust in the therapist. The therapist may fail to appreciate and address psychosocial stressors if only focused on medication management and physical symptoms. The therapist may also introduce distance into the therapeutic relationship with a treatment-resistant patient, with feelings of pessimism and failure to help, leading to a weakened therapeutic alliance.

Techniques to improve therapeutic alliance and enhance motivation

The establishment and maintenance of the therapeutic alliance is an ongoing, iterative process that requires focused work on the part of both the therapist and the patient. A "set it and forget it" mindset is unlikely to be successful, as each interaction between the patient and therapist has the potential to change the direction and tone of the alliance. At all phases of treatment (assessment, initiation, continuation, and termination), the therapist should evaluate the strength and nature of the alliance, and adjust treatment techniques to enhance and sustain the alliance. These efforts take place both in the foreground (in the content of sessions) and in the background (in the therapist's way of relating to the patient and responding to the patients requests).

There are several frameworks and techniques that therapists may help foster a productive therapeutic alliance with a patient with treatment-resistant depression. There are some common therapist-level factors to keep in mind, which can strengthen or impede the therapeutic alliance. The concepts presented here are by no means a comprehensive summary of effective strategies for building the therapeutic alliance. Rather, they are intended as foundational guideposts to help

TABLE 40.1 Techniques to improve therapeutic alliance.

Focus	Tasks
Patient-focused tasks	Put the patient first: Listen to the patient's complaints, requests, and stressors Unconditional positive regard fostering realistic hope and optimism Work together toward understanding and healing
Relationship-focused tasks	Avoid ruptures of the relationship Recognize and repair alliance ruptures when they occur
Therapist-focused tasks	Clarify the meaning of events within the patient's life and thought Manage countertransference Maintain humility and awareness of the limitations of treatment Hold hope when the patient cannot

therapists establish both a frame of mind and a way of working with patients that enables the development of a trusting, supportive therapeutic alliance that motivates change (Table 40.1).

Patient-focused tasks

The therapist's perspective is constructed around accomplishing specific tasks. The therapist must ask specific questions to establish a psychiatric history and diagnosis, make specific treatment recommendations based on that information, and work toward specific goals in order to reduce symptoms and return the patient to a state of mental health. The patient, on the other hand, does not know or necessarily understand these specifics and may not share this agenda. Rather, they come to the therapist with a different set of questions and goals in mind, many of which have little or nothing to do with medical diagnosis and treatment. They may be seeking advice about personal matters; they may wish to make an administrative request, such as a letter for school or work; or they may just want to "vent" (Lazare, 1989). Given this discrepancy that often exists between the patient and the therapist perspectives, it is not difficult to imagine that the therapeutic alliance may start off on the wrong foot from the very beginning of treatment unless the therapist takes steps to recognize the patient's perspective and goals and prevent or resolve these discrepant agendas. This can require education on what is expectable and necessary to move forward and then to work toward common goals.

To that end, Lazare and Eisenthal's guidance is insightful to ensure the formation of a productive therapeutic alliance: "Take patients where they are. They may well know what they need. Even if you disagree, negotiations must start from the patient's perspective" (Lazare, 1989). Starting from the patient's perspective and negotiating the treatment are valuable stances to take particularly during the beginning phase of treatment, as the therapist-patient dynamic is being established. Elevating the patient's needs and preferences signals a patient-centered approach and respects their lived experience and autonomy. These attitudes are generally preferred by patients today and perhaps particularly by patients with treatment-resistant depression, who may have had previous experiences of losing control over their treatment course once they are labeled as "treatment-resistant" or "difficult." Acknowledging both real and perceived conflicts in treatment goals, and working through these conflicts in a caring, respectful process of negotiation to arrive at shared goals, further validates the patient's perspective and helps them understand the therapist's perspective as well (Lazare, 1989).

Unconditional positive regard which fosters realistic hope and optimism is another way of putting the patient first. Unconditional positive regard refers to the radical acceptance and support of a person as they are, without judgment, on the belief that this kind of facilitative nurturance enables the individual to tap into their existing, abundant resources and drive toward personal growth (Rogers, 1951; Myers, 2007). In this way, the stance of unconditional positive regard can be a powerful motivating factor for change. However, it requires the therapist to work toward a genuine belief in the patient's basic goodness and ability to recover from their illness, which may be effortful especially in cases of treatment-resistant mental illness characterized by the erosion of hope in the therapist-patient dynamic.

In working together toward understanding and healing, therapists should not assume that patients, even those with chronic depression who have been in treatment for a long time, understand the particulars of their illness from a medical perspective. In today's fast-paced clinical environment, the time allotted for counseling patients on medical decision-making grows increasingly short. The ability of patients to access large amounts of web-based information has led to the democratization of medical information also has the effect of spreading medical misinformation. Given these factors, a good place to start when establishing common goals for treatment is to ask the patient first for their understanding of their current problems and then to provide psychoeducation so that the patient has an accurate conception of their diagnosis, treatment options, and prognosis based on the medical model of mental illness. Although patients may choose not to believe this information or to align with it, it is useful for them to understand.

Relationship-focused tasks

Once the foundational task of putting the patient first is accomplished, attending to the therapeutic aspects of the therapist-patient relationship can go a long way toward strengthening the alliance. Even if the treatment being provided is primarily pharmacotherapy rather than psychotherapy, therapists can recognize that a treatment environment properly framed and attuned to the patient's needs can provide an emotional experience that does not replicate past stressful or traumatic life experiences by offering the patient to a safer and more empowering relationship (Alexander and French, 1946). Recognizing and understanding interventions from the perspective of the patient's transference can pave the way for a real relationship and avoid a ruptured relationship (Greenson and Wexler, 1969). Such a real, authentic relationship further enhances the therapeutic alliance, and may elicit a new learning of possible behaviors and feelings by the patient.

When a problem, strain, or impasse arises in the therapeutic alliance it is imperative that the therapist recognize such a therapeutic rupture in a timely manner and make attempts to repair it. Although this task may seem confrontational, in many cases it is easier to avoid points of disagreement than to talk about them. The process of openly addressing alliance ruptures in a warm, humble, and nonjudgmental way both assures the continuation of a productive therapeutic alliance and models for the patient a healthy way of coping with conflict.

Therapist-focused tasks

Beyond the alleviation of psychiatric symptoms and return to normal functioning, patients often wish to "understand" the challenges, including mental health challenges, that they have encountered in their lives. They seek to construct a coherent, integrated narrative of their emotions, experiences, and relationships that they can use to understand themselves and to help others understand them. This process of understanding, sometimes referred to as "meaning-making," can result in an empowering life story that brings closure or inspiration, or understanding the limits that one had wished were not present. Without intervention, patients with depression have a tendency to construct the future as pessimistic, negative failure, which obviously is a detriment to their mental health and can also be a detriment to the therapeutic alliance (e.g., a patient who rejects help because they do not believe they are worthy of help). Therefore, techniques that offer the patient an alternative, positive narrative or that help the patient understand in an accepting and "valuing of self" manner can be more constructive way and an indispensable tool for treatment. These techniques include narrative interviewing; constructing and sharing a psychodynamic formulation (Cabaniss et al., 2013); and using the Core Conflictual Relationship Theme (CCRT) method to identify a focal point for change (Book, 1998).

Just as patients may bring previous experiences, relationships, and personality structures to the table when interacting with therapists in their transference, therapists may do the same to patients in their countertransference. Countertransference reactions such as feeling helpless or incapable, or wondering if the patient elicits similar feelings from others in their life are normal and can provide useful data points for treatment. However, therapists who are not aware of their countertransference, or who have countertransference reactions that are difficult to control or that are outside the range of normal, may find that their responses disrupt the therapeutic alliance and impede the patient's treatment. Staying attuned to countertransference and seeking additional guidance or supervision as needed are important aspects of clinical practice generally.

Therapists should maintain humility and an awareness of the limitations of treatment. Although there are many advances to celebrate in the age of modern medicine and biological psychiatry, there is also still much that we do not know. In line with the recommendation to put the patient first and that eliciting the patient's perspective is paramount, maintaining humility about the extent of our medical knowledge is another way for therapists to signal respect for the patient's viewpoint and lived experience. When working with patients from different cultural backgrounds, showing cultural humility by acknowledging the impact of cultural factors on perspectives and demonstrating interest in and knowledge of the patient's culture goes a long way in strengthening the therapeutic alliance (Tasman et al., 2013).

A key role of the herapist is to hold hope when the patient cannot. In cases of severe and chronic mental illness such as treatment-resistant depression, hope for recovery may erode over time. This process can occur insidiously as treatment progresses and treatment after treatment fails to improve the patient's suffering. In some cases it can be ingrained into the therapist-patient relationship from the beginning when a patient is passed off to a new therapist in a way that excessively conveys lack of hope or frustration with the patient. Both patients and therapists can lose hope, and indeed many patients either enter or return to treatment after they have already lost hope. Holding hope for the patient in an absolute and consistent way that is expressed to the patient is a powerful tool in the treatment of mental illness. The therapist sustains a picture of who the patient was and can again be, even when the patients loses that picture. Hope cements the therapeutic alliance. In the case of a patient contemplating suicide, it may mean the difference between attempting to end their life and deciding to live another day. The therapist, by continuing to believe in the potential for a positive outcome, becomes a pivotal protective factor (Box 40.2).

Therapeutic alliance in telehealth and other technology-based interventions

Telehealth promises greater access to care and convenience for both the therapist and the patient. Its implementation will bring new dimensions and challenges to the therapeutic alliance. Telehealth includes a range of technology-assisted encounters from those conducted over synchronous, real-time video link to those conducted over telephone. The impact of telehealth on the therapeutic alliance is just beginning to be understood, but the data appear promising. One study examining the strength of the alliance and depression outcome comparing face-to-face cognitive-behavioral therapy with a blended model showed no difference in measured strength of the alliance, and depression improvement was comparable between groups (Kooistra et al., 2020). The techniques discussed in this chapter are likely to be equally effective in telehealth interventions. Research is needed to evaluate these techniques in telehealth.

Other technology-based interventions are available to patients and therapists for the treatment of depression. These include smartphone applications (apps) and other computer-based therapy interventions. These treatments offer advantages of greater accessibility for patients, who can participate at convenient time, and without requiring travel to a therapists' office. Various studies have looked at the impact of incorporating apps and other technology on the therapeutic alliance. Addition of internet- or app-based interventions to face-to-face treatment does not diminish the quality of therapeutic alliance and highlights the importance of the therapist incorporating education and follow up on the use and utility of these tools into the discussions with the patient. Reducing the number of face-to-face sessions and adding technology-based interventions does not reduce the quality of the therapeutic alliance (Wehmann et al., 2020). In guided internet-based intervention combined with email support, task and goal perceived by the patient at the start of therapy are associated with positive treatment outcomes (Gómez Penedo et al., 2020b), meaning the therapeutic alliance is a factor even in interventions in which there is no face-to-face component. Guided internet based cognitive-behavioral therapy has also been shown to be an effective preliminary treatment for patients with severe depression waiting for clinic services, and therapeutic alliance was a positive factor in predicting symptom improvement (Duffy et al., 2019).

Taken together, these findings on telehealth and other technology-based interventions seem very promising for their use in treatment-resistant depression. Combined thoughtfully, and with appropriate attention paid to ensuring the telehealth and

Box 40.2 Clinical example: Techniques to improve alliance

Therapist: Now that you have been in treatment for four months, I thought it might be a good time to talk about how you feel we are doing.

Patient: Well, I've been taking the second medication now for the last three months. It is not giving me the same stomach problems the first did, so I don't mind taking it. I don't know. I think it is helping.

Therapist: I remember at the beginning you told me that you were concerned about your sleep and balancing demands of work and being a parent. How are we doing with that?

Patient: You only gave me that sleep medication for a few weeks. I remember being really anxious when you said I should not need it anymore, and that you wanted to let the antidepressant work.

Therapist: I imagine that felt like I was not listening to you.

Patient: Maybe, but you let me talk about how worried I was, and you told me we could change back if we needed to. That helped me settle down a little. It turns out I sleep okay without medication now.

Therapist: How about work-life balance?

Patient: That CBT is helping some. Like, I recognize that I tend to catastrophize little parts of the day. It is easier for me to let myself off the hook if I am not a perfect parent all the time. So that's kind of better. But why is it I still feel tired most of the time? And, frankly, my mood is still pretty lousy.

Therapist: This is where our relationship is so important. I know a lot about treating depression, but as your case shows, there is something for me to learn from every patient's unique situation. It's good to keep in mind that we have a number of other options to consider as we need them. In the big scheme of things we are still fairly early in your treatment.

Patient: That's good news? (awkward laugh) It seems like we've been at this a long time.

Therapist: I hear that. I am very grateful that you have been so engaged in treatment so far. My hope, and my expectation, is that these small gains will continue. We have many more therapeutic tools to turn to.

other technology-based interventions are not meant to be a substitute for a real relationship, these interventions offer the promise of making treatment more accessible and tailored to the needs of each patient. Therapists employing such interventions should attend to any perception of using technology as a means to place distance between patient and therapist as this has the potential to lead to rupture in the alliance.

Clinical summary

In this chapter, we have discussed the role of therapeutic alliance in managing treatment-resistant depression. Therapists employing pharmacotherapy and other somatic therapies, as well as psychotherapy, should attend to the principles of goal, task and bond in developing the therapeutic alliance with the patient with treatment-resistant depression. A positive therapeutic alliance has been shown to lead to better treatment outcomes for a range of treatments. A negative alliance can contribute to the patient maintaining the sick role or reinforcing patient behaviors in opposition to treatment. There are techniques therapists can apply to enhance therapeutic alliance and are believe to lead to better treatment outcomes. Looking forward as telehealth and other technology-based interventions become more common, the therapeutic alliance will continue to be an important element to sustain positive outcomes from treatment.

Disclaimers

The opinions and assertions expressed herein are those of the author(s) and do not necessarily reflect the official policy or position of the Uniformed Services University or the Department of Defense.

References

Alexander, F., French, T., 1946. Psychoanalytic Therapy, Principles and Application. Ronald Press, New York.

Barber, J.P., Zilcha-Mano, S., Gallop, R., Barrett, M., McCarthy, K.S., Dinger, U., 2014. The associations among improvement and alliance expectations, alliance during treatment, and treatment outcome for major depressive disorder. Psychother. Res. 24 (3), 257–268.

Book, H., 1998. How to Practice Brief Psychodynamic Psychotherapy: The Core Conflictual Relationship Theme Method. American Psychological Association, Washington, DC.

Bordin, E.S., 1979. The generalizability of the psychoanalytic concept of the working alliance. Psychotherapy (Chic.) 16 (3), 252–260.

Cabaniss, D.C.S., Douglas, C.J., Graver, R.L., Schwartz, A.R., 2013. Psychodynamic Formulation. Wiley-Blackwell, Chichester.

Cooper, A.A., Strunk, D.R., Ryan, E.T., DeRubeis, R.J., Hollon, S.D., Gallop, R., 2016. The therapeutic alliance and therapist adherence as predictors of dropout from cognitive therapy for depression when combined with antidepressant medication. J. Behav. Ther. Exp. Psychiatry 50, 113–119.

Duffy, D., Enrique, A., Connell, S., Connolly, C., Richards, D., 2019. Internet-delivered cognitive behavior therapy as a prequel to face-to-face therapy for depression and anxiety: a naturalistic observation. Front. Psych. 10, 902.

Dunster-Page, C., Haddock, G., Wainwright, L., Berry, K., 2017. The relationship between therapeutic alliance and patient's suicidal thoughts, self-harming behaviours and suicide attempts: a systematic review. J. Affect. Disord. 223, 165–174.

Eubanks, C.F., Muran, J.C., Safran, J.D., 2018. Alliance rupture repair: a meta-analysis. Psychotherapy (Chic.) 55 (4), 508–519.

Ferreira, P.H., Ferreira, M.L., Maher, C.G., Refshauge, K.M., Latimer, J., Adams, R.D., 2013. The therapeutic alliance between clinicians and patients predicts outcome in chronic low back pain. Phys. Ther. 93 (4), 470–478.

Gabbard, G.O., 2014. Affective disorders. In: Psychodynamic Psychotherapy in Clinical Practice. American Psychiatric Publishing, Inc, Arlington, VA, pp. 219–255.

Gaston, L., 1991. Reliability and criterion-related validity of the California psychotherapy alliance scales—patient version. Psychol. Assess. 3 (1), 68–74.

Gómez Penedo, J.M., Babl, A., Krieger, T., Heinonen, E., Flückiger, C., Grosse, H.M., 2020a. Interpersonal agency as predictor of the within-patient alliance effects on depression severity. J. Consult. Clin. Psychol. 88 (4), 338–349.

Gómez Penedo, J.M., Babl, A.M., Grosse Holtforth, M., et al., 2020b. The association of therapeutic alliance with long-term outcome in a guided internet intervention for depression: secondary analysis from a randomized control trial. J. Med. Internet Res. 22 (3), e15824.

Greenson, R.R., 1965. The working alliance and the transference neurosis. Psychoanal. Q. 34, 155–181.

Greenson, R., Wexler, M., 1969. The non-transference relationship in the psycho-analytic situation. Int. J. Psychoanal. 50, 27–39.

Hatcher, R.L., Gillaspy, J.A., 2006. Development and validation of a revised short version of the working alliance inventory. Psychother. Res. 16 (1), 12–25.

Hilsenroth, M.J., Cromer, T.D., Ackerman, S.J., 2012. How to make practical use of therapeutic alliance research in your clinical work. In: Levy, R.A., Ablon, J.S., Kachele, H. (Eds.), Psychodynamic Psychotherapy Research: Evidence-Based Practice and Practice-Based Evidence. Springer Science, Humana Press, New York, NY, pp. 361–380.

Horvath, A.O., Del Re, A.C., Fluckiger, C., Symonds, D., 2011. Alliance in individual psychotherapy. Psychotherapy (Chic.) 48 (1), 9–16.

Horwitz, L., 1974. Clinical Prediction in Psychotherapy. Jason Aronson, New York, p. 157.

Kooistra, L., Ruwaard, J., Wiersma, J., van Oppen, P., Riper, H., 2020. Working alliance in blended versus face-to-face cognitive behavioral treatment for patients with depression in specialized mental health care. J. Clin. Med. 9 (2), 347.

Krupnick, J.L., Sotsky, S.M., Simmens, S., et al., 1996. The role of the therapeutic alliance in psychotherapy and pharmacotherapy outcome: findings in the National Institute of Mental Health Treatment of Depression Collaborative Research Program. J. Consult. Clin. Psychol. 64 (3), 532–539.

Kushner, S.C., Quilty, L.C., Uliaszek, A.A., McBride, C., Bagby, R.M., 2016. Therapeutic alliance mediates the association between personality and treatment outcome in patients with major depressive disorder. J. Affect. Disord. 201, 137–144.

Laws, H.B., Constantino, M.J., Sayer, A.G., et al., 2017. Convergence in patient-therapist therapeutic alliance ratings and its relation to outcome in chronic depression treatment. Psychother. Res. 27 (4), 410–424.

Lazare, A., 1989. Clinician/patient relations I: attending to the patient's perspective. In: Lazare, A. (Ed.), Outpatient Psychiatry. Williams and Wilkins, Baltimore.

Leuchter, A.F., Hunter, A.M., Tartter, M., Cook, I.A., 2014. Role of pill-taking, expectation and therapeutic alliance in the placebo response in clinical trials for major depression. Br. J. Psychiatry 205 (6), 443–449.

Luborsky, L., Barber, J.P., Siqueland, L., et al., 1996. The revised Helping Alliance Questionnaire (HAq-II): psychometric properties. J. Psychother. Pract. Res. 5 (3), 260–271.

Myers, D.G., 2007. Psychology, eight ed. Worth, New York.

Parth, K., Datz, F., Seidman, C., Löffler-Stastka, H., 2017. Transference and countertransference: a review. Bull. Menninger Clin. 81 (2), 167–211.

Rogers, C.R., 1951. Client-Centered Therapy: Its Current Practice, Implications and Theory. Houghton Mifflin, Boston.

Smith, S.R., Hilsenroth, M.J., Baity, M.R., Knowles, E.S., 2003. Assessment of patient and therapist perspectives of process: a revision of the vanderbilt Psychotherapy Process Scale. Am. J. Psychother. 57 (2), 195–205.

Summers, R.F., Barber, J.P., 2010. The therapeutic alliance: goal, task, and bond. In: Pschodynamic Psychotherapy: A Guide to Evidence-Based Practice. Guilford Press, New York, NY, pp. 71–89.

Tasman, A., Kay, J., Ursano, R., 2013. The Psychiatric Interview: Evaluation and Diagnosis. Wiley-Blackwell, Chichester.

Weck, F., Rudari, V., Hilling, C., et al., 2013. Relapses in recurrent depression 1 year after maintenance cognitive-behavioral therapy: the role of therapist adherence, competence, and the therapeutic alliance. Psychiatry Res. 210 (1), 140–145.

Wehmann, E., Köhnen, M., Härter, M., Liebherz, S., 2020. Therapeutic alliance in technology-based interventions for the treatment of depression: systematic review. J. Med. Internet Res. 22 (6), e17195.

Zilcha-Mano, S., Roose, S.P., Barber, J.P., Rutherford, B.R., 2015. Therapeutic alliance in antidepressant treatment: cause or effect of symptomatic levels? Psychother. Psychosom. 84 (3), 177–182.

Index

Note: Page numbers followed by *f* indicate figures and *t* indicate tables.

N

O

P

Q

R

S

U

V

W

X

Printed in the United States
by Baker & Taylor Publisher Services